Asphalt Mixtures and Materials 2018

Transportation Research Record: Journal of the Transportation Research Board

Volume 2672, Issue 28

ISBN: 978-1-5108-7832-7

Printed from e-media with permission by:

Curran Associates, Inc.
57 Morehouse Lane
Red Hook, NY 12571

CURRAN ASSOCIATES INC. proceedings.com

Some format issues inherent in the e-media version may also appear in this print version.

Published by Sage Publications USA

ISBN (Print) 978-1-5108-7832-7
ISBN 2018 Printed Set (All Issues) 978-1-5108-7735-1

Printed by Curran Associates, Inc. (2019)

Additional copies of this publication are available from:

Curran Associates, Inc.
57 Morehouse Lane
Red Hook, NY 12571 USA
Phone: 845-758-0400
Fax: 845-758-2633
Email: curran@proceedings.com
Web: www.proceedings.com

Volume 2672 Number 28 December 2018

TRR
JOURNAL OF THE TRANSPORTATION RESEARCH BOARD

Contents

Asphalt Mixtures and Materials

Articles

Performance Assessment of Interlayers with Different Tack Coats by Considering Loading Types and Failure Modes 1
Soroosh Amelian and Yong-Rak Kim

Conditioning and Testing Protocol Combinations to Detect Asphalt Mixture Damage 10
Rabeea W. Bazuhair, Carl V. Pittman, Isaac L. Howard, Walter S. Jordan III, James Michael Hemsley Jr, and Gaylon L. Baumgardner

Development and Validation of a Model to Predict Interface Bonding Between Pavement Layers 22
Ramendra Das, Louay N. Mohammad, Mostafa Elseifi, Wei Cao, and Samuel B. Cooper, Jr.

Effects of Regular and Nano Sized Hydrated Lime Fillers on Fatigue and Bond Strength Behavior of Asphalt Mastic 31
Aditya Kumar Das and Dharamveer Singh

Effect of Synthetic Fiber State on Mechanical Performance of Fiber Reinforced Asphalt Concrete 42
Hossein Noorvand, Ramadan Salim, Jose Medina, Jeffrey Stempihar, and B. Shane Underwood

Utilizing Network and Project-Level Data to Evaluate Impact of Reclaimed Asphalt Pavement and Recycled Asphalt Shingles on Laboratory and Field Performance in Texas 52
Berenice Salaices, David Teutli, Imad Abdallah, Anjan Kumar Siddagangaiah, and Soheil Nazarian

Laboratory Evaluation of Mixture Type on Highly Modified Asphalt Mixtures in Virginia 59
Benjamin F. Bowers

Laboratory Observation and Evaluation of Asphalt Blends of Reclaimed Asphalt Pavement Binder with Virgin Binder using SEM/EDS 69
Yanxu Jiang, Xingyu Gu, Zhou Zhou, Fujian Ni, and Qiao Dong

Effect of Partial Blending on High Content Reclaimed Asphalt Pavement (RAP) Mix Design and Mixture Properties 79
Shuai Yu, Shihui Shen, Xuli Zhou, and Xinghai Li

Laboratory Performance of Asphalt Mixtures Containing Re-Refined Engine Oil Bottoms Modified Asphalt Binders 88
Taesun You, Yucheng Shi, Louay N. Mohammad, and Samuel B. Cooper, III

Wearing Course Mixtures Prepared with High Reclaimed Asphalt Pavement Content Modified by Rejuvenators 96
Stephan Büchler, Augusto Cannone Falchetto, Axel Walther, Chiara Riccardi, Di Wang, and Michael P. Wistuba

A Mechanical Approach to Quantify Blending of Aged Binder from Recycled Materials in New Hot Mix Asphalt Mixtures 107
Milad Zokaei Ashtiani, Walaa S. Mogawer, and Alexander J. Austerman

Effect of Binder Modification and Recycled Asphalt Pavement on the Performance of Permeable Friction Course 119
Mark J. Brum, Walaa S. Mogawer, Kevin D. Stuart, and Alexander J. Austerman

Investigating the Performances of Plant-Produced High-Reclaimed Asphalt Pavement Content Warm Mix Asphalts 130
Walaa S. Mogawer, Kevin Stuart, Alexander J. Austerman, and Ahmed A. Soliman

Impact of Gyration Reduction and Design Specification Changes on Volumetric Properties of Virginia Dense-Graded Asphalt Mixtures 143
Stacey D. Diefenderfer, Benjamin F. Bowers, and Kevin K. McGhee

Influence of Recycled Asphalt Pavement on Interfacial Energy and Bond Strength of Asphalt Binder for Different Types of Aggregates 154
Ayyanna Habal and Dharamveer Singh

Effect of Mixing Time and Temperature on the Homogeneity of Asphalt Mixtures Containing Reclaimed Asphalt Pavement Material 167
Jiantao Wu, Quan Liu, Yu Wang, Jun Chen, Dawei Wang, Linlin Xie, and Cadnel Ago

Performance of Virginia's Early Foamed Warm Mix Asphalt Mixtures 178
Stacey D. Diefenderfer

Chemo-Rheological Study of Hardening of Epoxy Modified Bituminous Binders with the Finite Element Method 190
P. Apostolidis, X. Liu, C. Kasbergen, M.F.C. van de Ven, G. Pipintakos, and A. Scarpas

Comparison of the Relative Long-Term Field Performance among Various Warm Mix Asphalt (WMA) Pavements 200
Weiguang Zhang, Shihui Shen, and Shenghua Wu

Evaluation of Marshall Stability Design Principles: Applied to Cold In-Place Recycling 211
Ben C. Cox and Isaac L. Howard

Comparing Pressure Aging Vessel Time to Field Aging of Binder as a Function of Pavement Depth and Time 223
Braden T. Smith, Isaac L. Howard, Walter S. Jordan III, Codrin Daranga, and Gaylon L. Baumgardner

Modeling Shear Viscosity of Asphalt through Nonequilibrium Molecular Dynamics Simulation 235
Yongjie Ding, Baoshan Huang, and Xiang Shu

Analysis of Asphalt Oxidation by Means of Accelerated Testing and Environmental Conditions 244
Rafael E. Villegas-Villegas, Alejandra Baldi-Sevilla, José Pablo Aguiar-Moya, and Luis Loria-Salazar

Low-Temperature Vacuum Drying Procedure for Rapid Asphalt Emulsion Residue Recovery 256
Haritha Malladi, Meron Asnake, Andrew LaCroix, and Cassie Castorena

Low-Temperature Emulsion Performance-Graded Specification for Chip Seals 266
Cassie Castorena, Mohammad Ilias, Javon Adams, and Y. Richard Kim

Impact of Recycled Materials and Recycling Agents on Asphalt Binder Oxidative Aging Predictions 277
Sara Pournoman, Elie Y. Hajj, Nathan Morian, and Amy Epps Martin

Study of Asphalt Binder Fatigue with a New Dynamic Shear Rheometer Geometry 290
Panos Apostolidis, Cor Kasbergen, Amit Bhasin, Athanassios Scarpas, and Sandra Erkens

Effects of Light-Activated Self-Healing Polymers on the Rheological Behaviors of Asphalt Binder Containing Recycled Asphalt Shingles 301
Sharareh Shirzad, Marwa M. Hassan, Max A. Aguirre, Samuel Cooper, Jr., and Ioan I. Negulescu

Developing Simple Binder Indices for Cracking Resistance of Asphalt Binders at Intermediate and Low Temperatures 311
Raquel Moraes and Hussain Bahia

Nonrecoverable Behavior of Polymer Modified and Reclaimed Asphalt Pavement Modified Binder under Different Multiple Stress Creep Recovery Tests 324
Zhou Zhou, Xingyu Gu, Jiwang Jiang, Fujian Ni, and Yanxu Jiang

Rheological Behavior and Its Chemical Interpretation of Crumb Rubber Modified Asphalt Containing Warm-Mix Additives 337
Haopeng Wang, Xueyan Liu, Panos Apostolidis, and Tom Scarpas

Evaluation of Chemical and Rheological Aging Indices to Track Oxidative Aging of Asphalt Mixtures 349
Farhad Yousefi Rad, Michael D. Elwardany, Cassie Castorena, and Y. Richard Kim

Selecting a Laboratory Loose Mix Aging Protocol for the NCAT Top-Down Cracking Experiment 359
Chen Chen, Fan Yin, Pamela Turner, Randy C. West, and Nam Tran

Sine versus Haversine Displacement Waveform Comparison for Hot Mix Asphalt Four-Point Bending Fatigue Testing 372
Angel Mateos, Rongzong Wu, Erik Denneman, and John Harvey

Impact of Specimen Configuration and Characteristics on Illinois Flexibility Index 383
Jose Rivera-Perez, Hasan Ozer, and Imad L. Al-Qadi

Indirect Tensile Test (IDT) to Determine Asphalt Mixture Performance Indicators during Quality Control Testing in New Jersey 394
Thomas Bennert, Edwin Haas, and Edward Wass

Comparison of Different Micromechanical Models for Predicting the Effective Properties of Open Graded Mixes 404
Hong Zhang, Kumar Anupam, Athanasios Scarpas, and Cor Kasbergen

Comparison of Asphalt Mixtures Crack Resistance at Intermediate Temperatures using Advanced Test Methods and Theories 416
Wei Cao, Louay N. Mohammad, Peyman Barghabany, Samuel B. Cooper, III, and Saman Salari

Influence of Production Considerations on Balanced Mixture Designs 426
Alexander J. Austerman, Walaa S. Mogawer, and Kevin D. Stuart

Optimization of the Laboratory Fabrication of Small Specimens for Asphalt Mixture Performance Testing 438
Sonja Pape, Kangjin Lee, Cassie Castorena, and Y. Richard Kim

Effect of Loading Waveform Pattern and Rest Period on Fatigue Life of Asphalt Concrete Using Viscoelastic Continuum Damage Model 451
Waleed Abdelaziz Zeiada, Padmini P. Gudipudi, B. Shane Underwood, and Mena I. Souliman

Practical Method to Determine the Effect of Air Voids on the Dynamic Modulus of Asphalt Mixture 462
Xinjun Li and Jack Youtcheff

Effect of Mix Design Variables on Thermal Cracking Performance Parameters of Asphalt Mixtures 471
Mirkat Oshone, Debaroti Ghosh, Eshan V. Dave, Jo Sias Daniel, Joseph M. Voels, and Shongtao Dai

A Straightforward Procedure to Characterize Nonlinear Viscoelastic Response of Asphalt Concrete at High Temperatures 481
Mohammad Bazzaz, Masoud K. Darabi, Dallas N. Little, and Navneet Garg

Improving the Reliability of Damage Characteristic Curves in the Simplified Viscoelastic Continuum Damage Model 493
Kangjin Lee, Cassie Castorena, and Y. Richard Kim

Evaluation of Viscoelastic and Fracture Properties of Asphalt Mixtures with Long-Term Laboratory Conditioning 503
Reyhaneh Rahbar-Rastegar, Jo Sias Daniel, and Eshan V. Dave

TRB

TRANSPORTATION RESEARCH BOARD

CORE SPONSORS

Transportation Departments of the 50 States and the District of Columbia

Federal Government

U.S. Department of Transportation
 Federal Aviation Administration (FAA)
 Federal Highway Administration
 Federal Motor Carrier Safety Administration (FMCSA)
 Federal Railroad Administration (FRA)
 Federal Transit Administration
 National Highway Traffic Safety Administration (NHTSA)
 Office of the Assistant Secretary of Transportation for Research and Technology (OST-R)

U.S. Air Force Civil Engineer Center (AFCEC)
U.S. Coast Guard (USCG)
U.S. Department of Energy
U.S. Department of the Interior (DOI)
U.S. Environmental Protection Agency (EPA)

Nongovernmental Organizations

American Public Transportation Association (APTA)
Association of American Railroads (AAR)
California Air Resources Board

The National Academies of

SCIENCES • ENGINEERING • MEDICINE

The **National Academy of Sciences** was established in 1863 by an Act of Congress, signed by President Lincoln, as a private, nongovernmental institution to advise the nation on issues related to science and technology. Members are elected by their peers for outstanding contributions to research. Dr. Marcia McNutt is president.

The **National Academy of Engineering** was established in 1964 under the charter of the National Academy of Sciences to bring the practices of engineering to advising the nation. Members are elected by their peers for extraordinary contributions to engineering. Dr. C. D. Mote, Jr., is president.

The **National Academy of Medicine** (formerly the Institute of Medicine) was established in 1970 under the charter of the National Academy of Sciences to advise the nation on medical and health issues. Members are elected by their peers for distinguished contributions to medicine and health. Dr. Victor J. Dzau is president.

The three Academies work together as the National Academies of Sciences, Engineering, and Medicine to provide independent, objective analysis and advice to the nation and conduct other activities to solve complex problems and inform public policy decisions. The Academies also encourage education and research, recognize outstanding contributions to knowledge, and increase public understanding in matters of science, engineering, and medicine.

Learn more about the National Academies of Sciences, Engineering, and Medicine at **www.national-academies.org.**

The **Transportation Research Board** is one of seven major programs of the National Academies of Sciences, Engineering, and Medicine. The mission of the Transportation Research Board is to increase the benefits that transportation contributes to society by providing leadership in transportation innovation and progress through research and information exchange, conducted within a setting that is objective, interdisciplinary, and multimodal. The Board's varied activities annually engage about 7,000 engineers, scientists, and other transportation researchers and practitioners from the public and private sectors and academia, all of whom contribute their expertise in the public interest. The program is supported by state transportation departments, federal agencies including the component administrations of the U.S. Department of Transportation, and other organizations and individuals interested in the development of transportation.

Learn more about the Transportation Research Board at **www.TRB.org**.

TRR
JOURNAL OF THE TRANSPORTATION RESEARCH BOARD

Article

Transportation Research Record
2018, Vol. 2672(28) 1–9

DOI: 10.1177/0361198118768528
journals.sagepub.com/home/trr
SAGE

Performance Assessment of Interlayers with Different Tack Coats by Considering Loading Types and Failure Modes

Soroosh Amelian[1] and Yong-Rak Kim[1]

Abstract

A monotonic direct shear test (DST) is the most commonly used method to evaluate bond strength of interlayers; however, it does not consider the actual repetitive traffic loading that pavement undergoes during its service life. This study aims to examine the link between various parameters obtained from a monotonic DST and the results of a cyclic loading test that more realistically assesses the interlayer performance with tack coats. Furthermore, to investigate the possible effect of the interface failure mode (shear vs. tensile) on the performance assessment of the tack coat interlayers, a simple indirect tensile test was conducted to determine the tensile strength of the interlayers. Field core samples extracted from a test section constructed in Nebraska were used for various laboratory tests. The field test section included 14 subsections containing various types of tack coat materials with different application methods. Permanent shear deformation and interface stiffness were selected as the failure criteria to characterize the interface resistance associated with cyclic load repetition. Shear strength, shear stiffness, and interlayer bond energy were chosen as the failure criteria for the monotonic DST. Results show that the interlayer shear strength from the monotonic DST had an acceptable correlation with the cyclic test results, and interlayer bond energy is a good predictor of the shear resistance of the tack coats with a high correlation with the cyclic test. The simple indirect tensile test was capable of differentiating interlayer performance with different types of tack coating.

Adequate bonding between lifts, and especially between the existing road surface and an overlay, is critical to ensure that the completed pavement structure behaves like a single unit with a sufficient level of strength while also preventing distress, such as delamination or slippage cracking. Therefore, reliable test methods are imperative for assessing tack coat quality. Bond failure at the interface is generally classified into three modes: shear failure, tensile failure, and mixed shear-tensile failure (*1,2*). In general, the test methods developed to evaluate the bond strength of an interface are categorized as monotonic loading tests and cyclic loading tests. The former method has been widely examined, whereas only a few studies have attempted to evaluate cyclic loading tests (*3*). Among the monotonic tests, the direct shear test (DST) appears to be the most widely used procedure to measure interface strength in asphalt concrete; thus, various DST devices have been developed (*4–7*). In relation to cycling loading tests, Romanoschi and Metcalf applied a testing configuration to determine the shear fatigue behavior of the interface, and they concluded that the cyclic test can be used to compare the durability of interfaces (*8*). Diakhaté et al. designed a fatigue test to propose an interface failure model. Based on their study, it was concluded that evaluating interface fatigue performance provides key information for assessing pavement durability (*9*). More recently, Safavizadeh and Kim developed a device called a double shear tester to study the cyclic performance of three-layered specimens. They found that the presence of tack coats improves the four-point bending beam fatigue life of specimens (*10*).

Because DST has a relatively simple configuration and is easy to perform by applying normal force, it has been approved and adopted as a national standard in many countries (*3*). In the United States, the Louisiana Interlayer Shear Strength Tester was developed during the National Cooperative Highway Research Program Project 9-40, and it has recently been approved by the American Association of State Highway and Transportation Officials (AASHTO) as a standard test to determine interlayer shear strength (ISS) as an indicator of asphalt pavement interlayer performance (*2,11*). Although many studies have assessed the capabilities and effectiveness of DST, there is still a significant lack of insight regarding the link between DST and a more realistic representation which considers the effects of repetitive traffic loads on interlayer damage. Recently, a state-of-the-art review conducted by Rahman et al. reported that there are no

[1]Department of Civil Engineering, University of Nebraska-Lincoln, Lincoln, NE

Corresponding Author:
Address correspondence to Yong-Rak Kim: yong-rak.kim@unl.edu

extensive studies in the literature investigating the link between monotonic and cyclic DST (*3*). Furthermore, although there could be a significant difference in the interface performance results depending on whether the test is performed in shear failure mode or tensile failure mode, the literature does not clearly state which type of failure is predominant in asphalt pavement interlayers.

Study Objectives and Scope

The primary purpose of this study is to examine the link between various parameters obtained from a monotonic DST and the results of a cyclic loading test, which more realistically assesses the interface performance. Furthermore, to investigate the possible effect of interface failure mode (shear vs. tensile) on the performance assessment of tack coat interlayers, a simple indirect tensile test (IDT) was conducted to determine the tensile behavior of the interlayers.

The scope of this research includes applying various types of tack coat materials with different application methods in a field test section. A total of 14 subsections in the field were considered in order to vary the type of tack coating used. The field cored samples were laboratory tested by considering different loading modes (monotonic vs. cyclic) and failure modes (shear vs. tensile). Test results were then used to examine the link between monotonic interlayer performance and cyclic performance and the similarities/differences in the interface resistance at different failure modes (shear vs. tensile).

Materials and Field Test Section

Because of seasonal variations in temperature and humidity levels in Nebraska, various types of tack coats were selected, including CSS-1h, CRS-2P, CFS-1, PG 64-22, and PG 64-22 plus a wax additive, namely a "trackless" tack coat. In comparison with CSS-1h (slow-setting type), CRS-2P, which is a rapid-setting emulsified asphalt, and CFS-1, which is a modified CSS-1 that provides a faster set, were selected to evaluate the difference in the breaking time in the field and to compare their bond behavior as tack coats. Throughout this research study, two main application rates were used in the emulsified tack coats: the typical application rate (0.08 gal/yd^2, 0.362 liter/m^2) used in most state projects and recommended by the specifications, and double the standard application rate (0.16 gal/yd^2, 0.724 liter/m^2) to identify the effect of using higher rates than those that are used in current typical tack coat practices. Whereas CRS-2P and CFS-1 were applied without dilution, two different dilution rates were applied for CSS-1h: the 50% dilution rate recommended by the State Department of Transportation specifications and a 30% dilution rate (i.e., 70% original emulsion plus 30% additional water). For asphalt binder tack coats, the standard application rate was achieved when visual observation confirmed full coverage of the pavement surface. The standard and double application rates were also used for the asphalt binder tack coats.

The construction of the pavement test section and the tack coating application process were conducted near Oconto, Nebraska (41°08 01.7 N, 99°45 29.7 W). The test section was constructed using two asphaltic layers. The first layer was placed and then compacted to the target thickness of 50 mm a few days before the tack coating stage. The dimensions of the test section were 7 m in width and approximately 28 m in length. The entire section was then subdivided using paint, marking it into 14 subsections that were 1.2 m long and 7 m wide. There was also a 60 cm gap between each subsection. For the emulsion tack coats, considering the dimension of subsections and specific gravity of emulsions, the weight of tack coat needed to obtain the target application rate was determined. The required amount of material was then acquired using a scale. The emulsion tack coats were then spread and distributed onto the subsections as shown in Figure 1. Even coverage of material was ensured by visual inspection during placement and distribution of emulsified asphalt. Trailer-mounted asphalt spray equipment was also used to spray the binder tack coats onto the relevant subsections. Thus, the first 8 of the 14 subsections were tack coated using emulsions; 4 subsections were tack coated with either PG 64-22 or trackless tack coat using spray equipment; 1 subsection (i.e., Subsection #14) was tack coated with trackless tack coat using the same method of application employed for the emulsions; and 1 subsection (i.e., Subsection #13) was left without any tack coat as a control. The type of tack coat used for each subsection along with final application rates applied are presented in Table 1. It is important to note that the main reason for not applying the same method of tack coat distribution used in actual pavement construction was to avoid any possible deviation from target application rates and/or non-uniform distribution of material, which could happen if the distribution truck was not perfectly calibrated/operated. Manually applying the tack coat materials to the subsections ensured that the exact amount of tack coat was applied and distributed uniformly. After all the tack coats were applied and the emulsion tack coats were broken, a 50 mm thick asphalt mixture was overlaid and compacted (Figure 1).

Sample Preparation

The coring process was conducted 1 day after the test section was overlaid. To avoid the possible effects of the haul truck's wheel on the performance of the tack coats, all the core samples were taken from a part of the section that was located away from the wheel path; then, core samples with a 150 mm diameter were extracted. A total of 258 cores, at least 18 per subsection, were obtained. All the cores were then collected, labeled with a paint pen, and transferred to the testing laboratory. Before testing each core, the core samples with a height above the target of 50 mm were cut using a wet masonry saw

Figure 1. Field implementation of tack coats, section paving, and samples coring.

Table 1. Application Rates Applied to the Test Section

Subsection number	Tack coat	Application rate diluted (gal/yd^2)**	Application rate undiluted (gal/yd^2)	Residual binder rate (gal/yd^2)
1	CSS-1h (50/50)	0.08	0.04	0.026
2	CSS-1h (50/50) × 2	0.16	0.08	0.052
3	CSS-1h (70/30)	0.08	0.056	0.036
4	CSS-1h (70/30) × 2	0.16	0.112	0.073
5	CRS-2P	N/A	0.1	0.065
6	CRS-2P × 2	N/A	0.16	0.104
7	CFS-1	N/A	0.08	0.052
8	CFS-1 × 2	N/A	0.16	0.104
9	PG 64-22	N/A	Standard Coverage	0.26
10	PG 64-22 × 2	N/A	Double	0.53
11	Trackless	N/A	Standard Coverage	0.26
12	Trackless × 2	N/A	Double	0.52
13	No Tack	N/A	0	0
14	Trackless × 2*	N/A	Double	0.69

Note: × 2*: Double application rate or coverage, 1 (gal/yd^2)** = 4.528 liter/m^2.

machine to a height of less than 50 mm. The interface in the cores was also marked to ensure the accuracy of the testing process.

Experimental Plan

Monotonic Direct Shear Test

A direct shear testing device manufactured by Pine Instrument Company (Grove City, PA) was used to assess the bond behavior of the interlayers. In the current study, a monotonic DST was performed by following the AASHTO TP 114 "Standard Method of Test for Determining the Interlayer Shear Strength (ISS) of Asphalt Pavement Layers" guidelines (*11*). A gap width of 1.3 cm between the moving and reaction frames was adopted. Based on a comprehensive literature review conducted by the authors, it was decided to perform the test without applying normal force to minimize the effect of aggregate interlocking at the interface, which could be a dominating source of interlayer strength, rather

Figure 2. (a) Cyclic loading test configuration and (b) IDT test sample preparation.

than the adhesion characteristics of the tack coats. Although removing normal force from the test does not simulate the actual pavement conditions under a vehicle's tire load, the AASHTO TP 114 guidelines do not require it; moreover, this approach has been reported to be an acceptable way to ensure quality control (*12*). The test was carried out at 25°C under displacement control mode at the rate of 2.54 mm/min. The ISS can then be computed as follows:

$$ISS = \frac{P_{\max}}{(\pi D^2 / 4)} \quad (1)$$

where P_{max} is the maximum load obtained from the test and D is the diameter of the specimen. The interlayer reaction tangential modulus k (N/m^3) can also be determined by diving the shear strength by the corresponding displacement.

Cyclic Direct Shear Test

To assess the interlayer behavior caused by repeated traffic loading, the cyclic DST was performed at 25°C using the same device that was used for the monotonic DST. Two linear variable differential transformers (LVDT) were mounted on the fixture to measure the average vertical displacement of the sample, as shown in Figure 2a. The test was conducted using servo-hydraulic UTM-25 kN equipment. Continuous sinusoidal loading at a frequency of 10 Hz in force-controlled mode was applied. A 50 N contact load was applied and maintained during the test, and 0.75 kN was selected as the peak force to target failure that occurred in approximately less than 150,000 cycles for all the samples. Both force and displacement were recorded during the test; a total of 25 data points were employed to plot each loading/unloading cycle. The average of the displacement measured by two LVDTs was then used to characterize the interlayer response caused by the cyclic loads.

Two main criteria were used to identify interlayer failure. The first criterion used permanent shear deformation (PSD), which was defined as the average accumulated shear deformation acquired from the two LVDTs. The PSD plot usually consists of three regions: primary, secondary, and tertiary. Initially, PSD grows fast; then, the rate decreases to a constant value in the secondary region. Finally, the tertiary region shows a dramatic increase in the rate of accumulated deformation. In the current study, the number of loading cycles at which the tertiary flow begins was selected as the failure point in the PSD graph. This point is typically referred to the flow number (F_N). Among the different relations proposed for identifying the F_N, the Francken relation is considered appropriate to determine the deformation rate and the corresponding F_N. The Francken relation is expressed as follows (*13*):

$$u_p = An^B + C\left(e^{Dn} - 1\right) \quad (2)$$

where u_p is the permanent shear deformation, n is the number of cycles, and A, B, C, and D are the fitting coefficients.

The F_N was then found by differentiating the permanent strain from the loading cycle curve and searching for the minimum value. Figure 3 depicts an example of the cyclic test results of PSD versus the loading cycles for subsection #11 and its corresponding curve fit based on the Francken equation. As seen in Figure 3, the equation precisely followed the experiment results.

The second criterion selected to characterize the interlayer failure resistance with cyclic loads was interlayer stiffness (IS), which is described as follows:

$$IS = \frac{\Delta F}{A \times \Delta u} \quad (3)$$

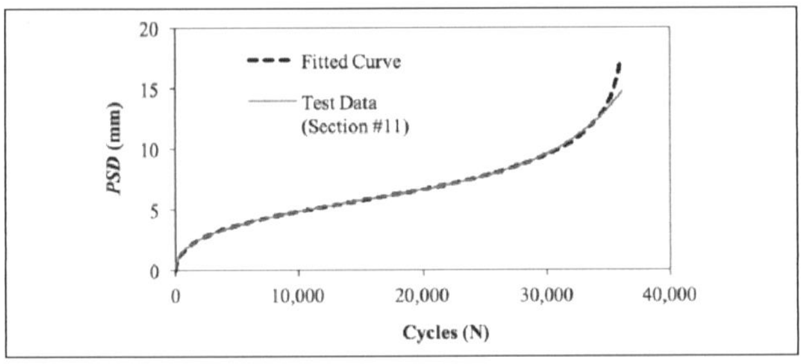

Figure 3. PSD Curve fitting based on Francken equation.

where ΔF is the amplitude of the applied shear force, Δu is the amplitude of the measured relative displacement, and A is the interface area. A 50% reduction in the initial IS was considered to be the second failure criterion.

Indirect Tensile Test

An IDT test was performed according to ASTM D6931 (*14*). Toward that end, 100 mm diameter cores were extracted from 150 mm field core samples so that the coring direction was perpendicular to the coring direction in the field (Figure 2b). The top and bottom of the cores were then cut so that the final height of specimens was 65 ± 5 mm (Figure 2*b*). The test was carried out at 25°C under a displacement rate of 50 mm/min.

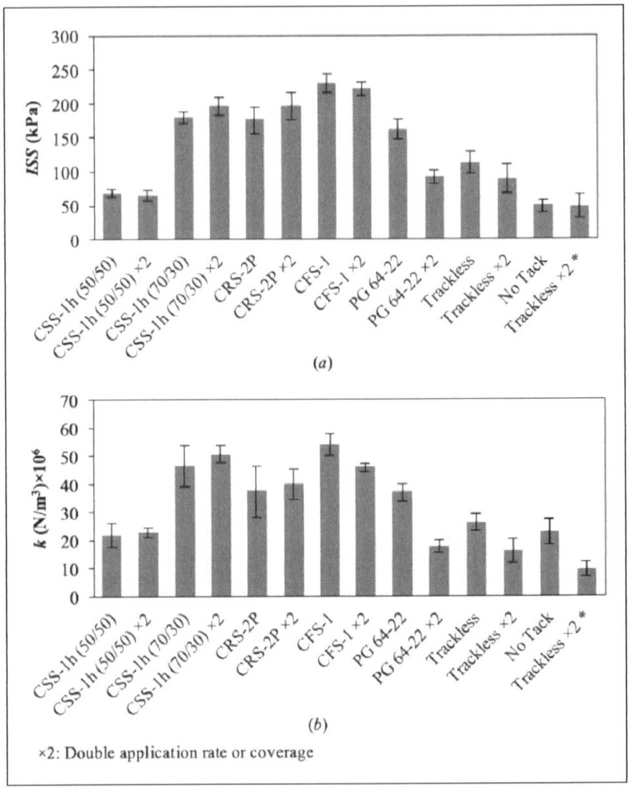

Figure 4. Results of monotonic DST: (a) shear strength and (b) tangential modulus.

Test Results and Analyses

Monotonic DST

Figure 4, *a* and *b* present the average ISS and interlayer reaction modulus (*k*-value), respectively. The average results of at least three replicates and the standard deviation, in the form of error bars, are depicted in Figure 4. Based on the results, CFS-1 showed the best performance in terms of shear strength. The highest ISS was found in the CFS-1 sample at the standard application rate, followed by CFS-1 with a double application rate. The lowest shear strength was also found in the subsection with no tack coat and subsection #14, shown as Trackless × 2^* (i.e., trackless tack coat with a double application rate applied without spray equipment). In addition, CSS-1h at 50% dilution (the most common practice in Nebraska) had a lower shear strength than the other tack coats. It should also be noted that, although similar application rates were used among cases, because different dilution rates were applied (ranging from 0% to 50%), the residual application rates among the cases are completely different (Table 1). Therefore, the relatively lower shear strength observed in the samples with CSS-1h at 50% dilution implies that both the residual binder application rate and the dilution rate could contribute to the performance of tack coats.

Whereas the increase in the application rate of the emulsion tack coats slightly increased the shear strength in the samples with CSS-1h at 30% dilution and CRS-2P, the shear strength decreased slightly from the standard to the double application rate in the samples with CFS-1; and it remained almost the same as the shear strength in the CSS-1h sample at 50% dilution. Thus, enhancing the application rate from a standard rate to a double application rate did not necessarily result in an increase in the shear strength at an intermediate temperature; this further implies that there might be an optimum application rate that yields the highest shear strength of emulsified asphalt interlayers, although evaluating the possible existence of an optimum application rate was not the scope of this study. For the asphalt binders, both the shear strength and interlayer reaction modulus decreased as a result of an increase in the application rate of the binder tack coats. This reduction is quite significant for the PG-64-22 sample, which suggests that using an excessive amount of binder as the tack coat could have adverse effects on the performance of the interlayer.

Cyclic DST

Figure 5a presents the average values of the test results in terms of failure Criterion I, which was defined as the number of cycles at which the lowest derivative of the PSD rate was obtained. The failure cycle is shown as F_N. The results show that the performance of the CFS-1 case at the standard

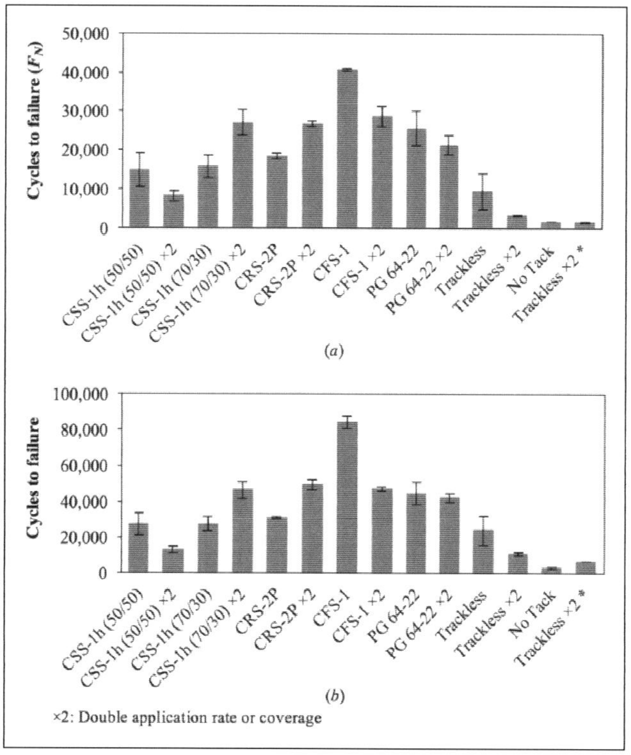

Figure 5. Results of cyclic loading test in terms of (a) Criterion I and (b) Criterion II.

application rate was noticeably better than the performance of the other cases, which is consistent with monotonic DST results. CSS-1h at 30% dilution, CRS-2P, CFS-1 all at the double application rates, and PG-64-22 with standard coverage exhibited a very similar performance; whereas the subsection without the tack coat and the subsection tack coated using a trackless tack coat with double coverage demonstrated the lowest resistance to cyclic loading. This is also in agreement with monotonic testing outcomes.

As mentioned earlier, the second failure criterion adopted to assess the interlayer performance of the subsections subjected to cyclic loading was 50% loss of initial IS. The cyclic test results revealed that, in many cases, particularly in the subsections with emulsion tack coats, the specimens reached 50% of the initial IS slightly ahead of complete failure, where complete debonding between two layers occurred. However, in other cases, particularly in the samples containing binder tack coats with higher application rates, neither 50% loss of the initial IS nor the complete debonding of layers occurred. This is mainly as a result of the relatively high viscosity of asphalt binder at 25°C, which reduces IS. This further induces a high amount of PSD at the interface without a significant loss of stiffness while still holding the two layers together. In this case, interlayer failure was determined to be the number of cycles at which PSD reached 20 mm. Although a PSD of 20 mm is an arbitrary value, it was considered to be an excessive deformation caused by repeated loading in the pavement.

Figure 5b depicts the average fatigue test results in terms of failure Criterion II, which was defined as the number of cycles needed to reach either 50% IS loss or a PSD of 20 mm, whichever came first. As shown in Figure 5b, CFS-1 at the standard application rate had the highest resistance. Similar to the results based on Criterion I, the samples without the tack coat and the samples with the trackless tack coat at the double application rate exhibited poor interlayer resistance. A comparison of the information presented in Figure 5, a and b show that both failure criteria were in generally good agreement for assessing interlayer performance under cyclic loading. Criterion I resulted in a lower number of cycles required to reach failure for all cases.

Link between the Monotonic DST and the Cyclic DST

Figure 6 presents a comparison of the ISS values obtained from the monotonic DST and the cyclic failure Criterion I, F_N, obtained from the cyclic DST. F_N was selected for this comparison. The results demonstrate that, although the performance ranking of the tack coats was not exactly the same for the monotonic ISS and the cyclic F_N, the agreement between the ranking performances of the tack coat layers from the two different loading modes was generally good. As seen in Figure 6, both testing approaches predicted that four subsections would have the best performance (i.e., Subsections #7, #8, #4, and #6). Similarly, both testing methods provided the same prediction for the cases with the lowest failure resistance (i.e., Subsections #13 and #14).

To quantitatively compare the monotonic DST and cyclic DST results, regression analysis was conducted. In addition to ISS, which is the most common parameter used in DST and the one that the AASHTO TP 114 guidelines recommend using to characterize the interlayer behavior of asphalt pavements, tangential modulus (k) from the monotonic DST was also correlated with the cyclic test results. Figure 7, a and b shows the linear regression results for the ISS values and the failure criteria from the cyclic tests. A higher correlation was found between the monotonic test results and Criterion I, with an R^2 value of 0.75 (Figure 7a), than with Criterion II, with an R^2 value of 0.67 (Figure 7b). Figure 7, c and d also depicts the correlation between monotonic tangential modulus and the cyclic failure criteria. Comparing Figure 7, a and b with Figure 7, c and d, the tangential modulus with R^2 values of 0.67 and 0.58 demonstrated a weaker correlation than the ISS. This implies that ISS could be considered to be a more desirable indicator than tangential modulus to characterize interlayer failure behavior because its results had a better correlation with the cyclic loading test that simulates repeated traffic loading in the pavement.

The initiation and evolution of interlayer damage caused by cyclic loading is strongly related to the amount of strain or deformation caused by loading on the pavement interlayer. Therefore, another parameter was further assessed: the

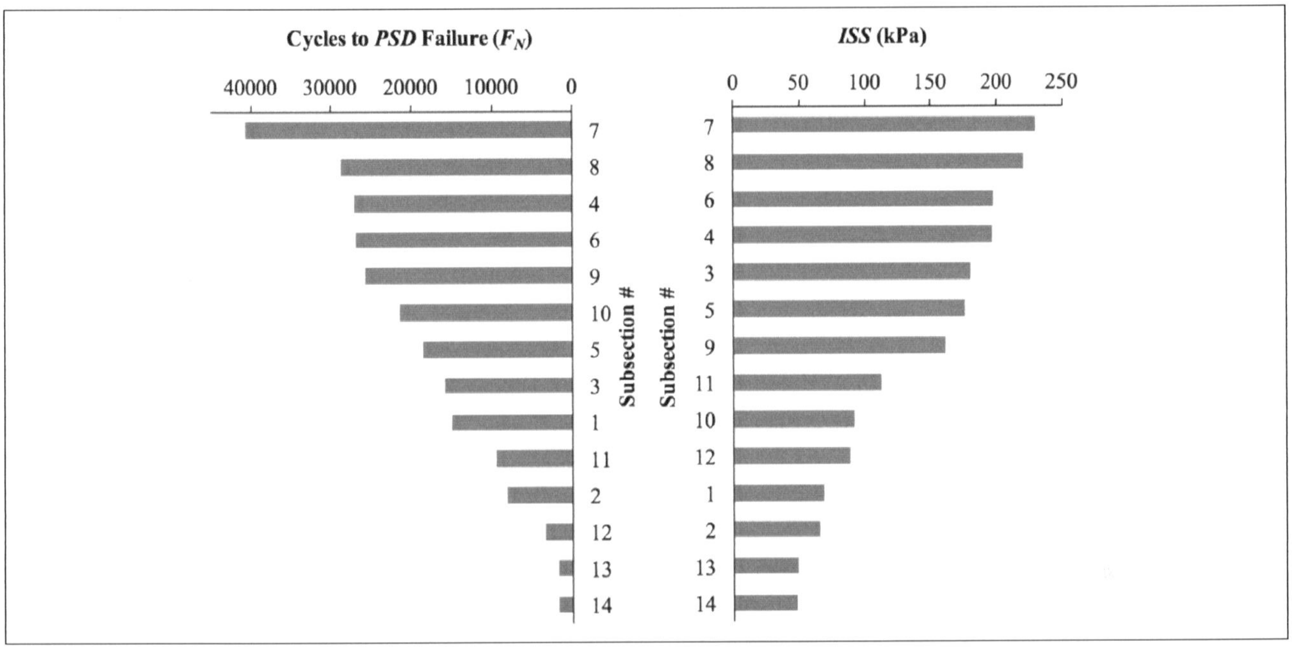

Figure 6. Comparison between monotonic and cyclic DST in performance ranking.

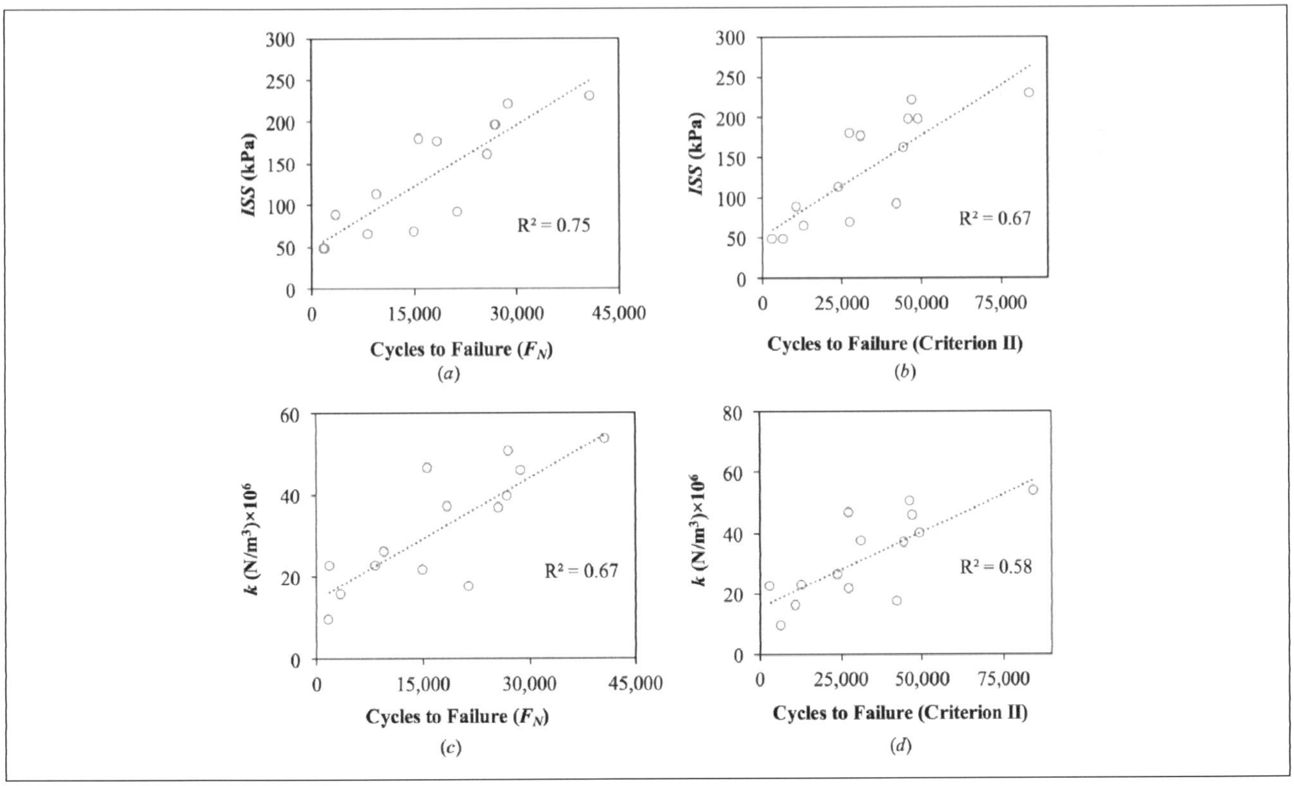

Figure 7. Correlation of monotonic ISS and *k*-value with cyclic loading test results: (a) ISS vs. Criterion I, (b) ISS vs. Criterion II, (c) k vs. Criterion I, (d) k vs. Criterion II.

area beneath the load-displacement curve until the peak from the monotonic DST. This parameter, which is denoted as the interlayer bond energy in this study, accounts for not only the maximum shear force in the interlayer but also the resulting associated displacement. It also represents the amount of energy dissipated during the stage in which the interlayer

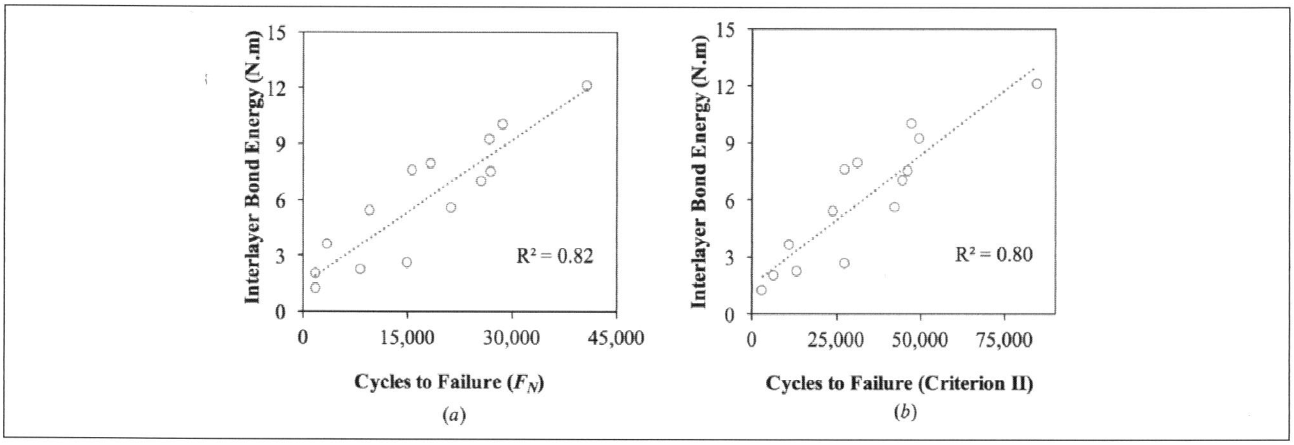

Figure 8. Correlation between interlayer bond energy and cyclic loading test results: (a) Criterion I and (b) Criterion II.

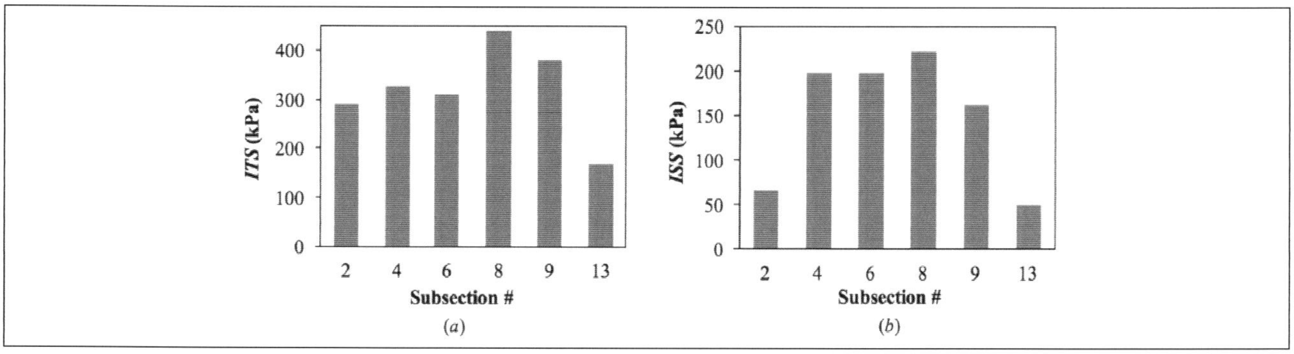

Figure 9. Comparison between (a) ITS from IDT test and (b) ISS from DST test.

resists the shearing. Thus, it is interesting to compare this parameter with the cyclic damage results. The correlation between the interlayer bond energy and the two different failure criteria from the cyclic testing is illustrated in Figure 8. As shown, with R^2 values of 0.82 and 0.80, for Criterion I and Criterion II, respectively, a stronger link was found with the bond energy than the ISS. This implies that the interlayer bond energy could be a good or better parameter to predict the resistance of asphalt interlayers subjected to repeated traffic loads.

Indirect Tensile Test

To examine the feasibility of a simple tensile test to capture the difference in the performance of the interlayers, an IDT test was performed on the specimens selected from six subsections. The selection of the subsections was based on the availability of a sufficient number of samples that could be tested. The following subsections were selected: CSS-1h (50% dilution), CSS-1h (30% dilution), CRS-2P and CFS-1 at the double application rate, PG 64-22 (standard coverage), and a subsection without any tack coat. Figure 9a shows the average indirect tensile strength (ITS) of the interlayers. The average ISS values for the same subsections are presented in Figure 9b for the sake of comparison. The IDT results show that the test was capable of differentiating the interlayer performance when different types of tack coating were used. Among the subsections tested, the highest tensile strength was attributed to the samples with CFS-1 at the double application rate (#8), whereas the samples without any tack coat (#13) exhibited the lowest tensile strength value. The aforementioned observation was consistent with the ISS results (Figure 9*b*), in which the samples from the same subsections showed the best and the poorest performance. However, the regression analysis between the IDT (tensile) and DST (shearing) outcomes resulted in an R^2 value of 0.65, which is considered a somewhat weak link between the two types of tests. Although a further investigation with more data is necessary to make any definite conclusion, the correlation between the two types of tests implies that interlayer testing of asphalt pavement on different modes of failure (i.e., shear vs. tensile) does not necessarily lead to the same performance prediction. In order to better understand which type of failure mode would be more appropriate for ranking the tack coat materials, more investigation is recommended by relating the laboratory experimental results to the field performance of asphalt pavement interlayers.

Summary and Conclusions

The study aimed to examine the relationship between a simple monotonic DST and a cyclic loading shear test for assessing the performance of asphalt pavement interlayers with different types of tack coat applications. The DST, which is one of the most widely used methods for assessing tack coat quality, was performed according to AASHTO TP 114. Several monotonic parameters: ISS, tangential modulus (k), and interlayer bond energy, were used in the DST, and two different indicators: PSD and IS, were adopted to assess failure in the cyclic loading mode. Both types of tests were then conducted on the field core samples acquired from a specially built test section in which 14 subsections with different types of tack coats and application rates were considered. Furthermore, this study investigated differences in the interlayer performance between shear and tensile failure modes. Thus, a simple IDT was adopted to capture the tensile resistance of the interlayers. From the results and analyses of data, the following conclusions can be made:

- The ISS values from the monotonic DST and the failure criteria from the cyclic loading test showed same performance prediction for the highest and the lowest shear resistance.
- Both of the failure criteria defined in the cyclic loading test demonstrated good agreement in predicting tack coat interlayer performance.
- An acceptable link was found between the ISS obtained from the monotonic DST and the PSD criterion (F_N) obtained from the cyclic loading test.
- Among the parameters measured using the monotonic DST method, interlayer bond energy (i.e., the area beneath the load-displacement curve until the peak), demonstrated the best correlation with the interlayer behavior under cyclic loading.
- The simple IDT was capable of differentiating tensile interlayer performance with different types of tack coating.

Acknowledgments

The authors gratefully acknowledge the financial support received from the Nebraska Department of Transportation (NDOT) for this study. In particular, the authors would like to thank Mr. Robert Rea and Mr. Matt Beran for their invaluable comments and discussions.

References

1. Raab, C., and M. Partl. Interlayer Shear Performance: Experience with Different Pavement Structures. *Proc., 3rd Eurasphalt and Eurobitumen Congress*, Vienna, Foundation Eurasphalt, Vol. 1, 2004, pp. 535–545.
2. Mohammad, L. N., M. A. Elseifi, A. Bae, N. Patel, J. Button, and J. A. Scherocman. *NCHRP Report 712: Optimization of Tack Coat for HMA Placement*. HRB, National Research Council, Washington, D.C., 2012.
3. Rahman, A., C. Ai, C. Xin, X. Gao, and Y. Lu. State-of-the-art Review of Interface Bond Testing Devices for Pavement Layers: Toward the Standardization Procedure. *Journal of Adhesion Science and Technology*, Vol. 31, 2017, pp. 109–126.
4. Leutner, R. Investigation of the Layered Composite with the Bituminous Superstructure. *Bitumen*, Vol. 3, 1979, pp. 84–91.
5. West, R. C., J. Zhang, and J. Moore. *Evaluation of Bond Strength between Pavement Layers*. NCAT Report No. 05–08. National Center for Asphalt Technology, Auburn, Ala., 2005.
6. Sholar, G. A., G. C. Page, J. A. Musselman, P. B. Upshaw, and H. L. Moseley. Preliminary Investigation of a Test Method to Evaluate Bond Strength of Bituminous Tack Coats. *Journal of the Association of Asphalt Paving Technologists*, Vol. 73, 2004, pp. 771–806.
7. Santagata, E., and F. Canestrari. Tensile and Shear Tests of Interfaces in Asphalt Mixes: A New Prospective on their Failure Criteria. *2nd International Symposium Highway Surfacing*, Ulster, 1994.
8. Romanoschi, S. A., and J. B. Metcalf. Characterization of Asphalt Concrete Layer Interfaces. *Transportation Research Record: Journal of the Transportation Research Board*, 2001. 1778: 132–139.
9. Diakhaté, M., A. Millien, C. Petit, A. Phelipot-Mardelé, and B. Pouteau. Experimental Investigation of Tack Coat Fatigue Performance: Towards an Improved Lifetime Assessment of Pavement Structure Interfaces. *Construction and Building Materials*, Vol. 25, 2011, pp. 1123–1133.
10. Safavizadeh, A., and Y. R. Kim. Mode II Fatigue and Reflective Cracking Performance of GlasGrid-Reinforced Asphalt Concrete under Repeated Loading. *Proc., 12th ISAP Conference on Asphalt Pavements*, Raleigh, North Carolina, Taylor and Francis, 2014, pp. 1893–1902.
11. AASHTO. *Standard Method of Test for Determining the Interlayer Shear Strength (ISS) of Asphalt Pavement Layers*. AASHTO TP 114, Washington, D.C., 2016.
12. Karshenas, A., S. H. Cho, A. Tayebali, M. Guddati, and Y. Kim. Importance of Normal Confinement to Shear Bond Failure of Interface in Multilayer Asphalt Pavements. *Transportation Research Record: Journal of the Transportation Research Board*, 2014. 2456: 170–177.
13. Francken, L. Pavement Deformation Law of Bituminous Road Mixes in Repeated Load Triaxial Compression. *Proc., 4th International Conference on the Structural Design of Asphalt Pavements*, Vol. 1, University of Michigan, Ann Arbor, 1977, pp. 483–496.
14. ASTM. *Standard Test Method for Indirect Tensile (IDT) Strength of Bituminous Mixtures*. ASTM D6931. West Conshohocken, Pa., 2012.

The Standing Committee on Surface Requirements of Asphalt Mixtures (AFK40) peer-reviewed this paper (18-01914).

Article

TRR
JOURNAL OF THE TRANSPORTATION RESEARCH BOARD

Transportation Research Record
2018, Vol. 2672(28) 10–21

DOI: 10.1177/0361198118756631
journals.sagepub.com/home/trr
SAGE

Conditioning and Testing Protocol Combinations to Detect Asphalt Mixture Damage

Rabeea W. Bazuhair[1], Carl V. Pittman[1], Isaac L. Howard[1], Walter S. Jordan III[2], James Michael Hemsley[2] Jr, and Gaylon L. Baumgardner[2]

Abstract

Asphalt mixes often have many ingredients that can interact with each other. When put into service, where there are multiple environmental effects, there are many interactions that need mixture testing. This paper's objective was to evaluate laboratory conditioning protocols coupled with subsequent property measurements for their ability to detect damage of asphalt mixtures in the southeastern U.S. climate (or similar climates). The investigation's focus is the property measurements themselves, and in particular how a given test can simultaneously assess multiple types of damage (i.e. oxidation, moisture damage, and freeze-thaw damage). While in service, mixtures can be damaged in multiple manners so laboratory conditioning protocols that expose specimens to multiple types of damage are needed as are test(s) that can detect these damages in a manner that can help assess performance during service. Four plant produced mixtures with all virgin ingredients were evaluated at intermediate temperatures with mixture and binder tests. The mixtures were well suited for such a comparison because they consisted of all virgin binder. Indirect tensile (IDT) strength did not relate to Cantabro Mass Loss (CML) or binder test results, which was concerning. Even more concerning was IDT's inability to respond to laboratory conditioning protocols that considered multiple environmental effects (i.e., oxidation, moisture, and freeze-thaw). CML results related to binder properties and were able to reasonably detect multiple types of environmental effects. As such, Cantabro testing is recommended over tensile strength for intermediate temperature mixture property assessments related to non-load associated environmental effects.

It has been understood for decades that environmental effects are a major contributor to asphalt pavement deterioration. Attempts to capture these effects are implemented in the American Association of State Highway and Transportation Officials' (AASHTO) practices in terms of material selection (e.g., R28), mix design (e.g., R30, T283, T324), and pavement design (*1,2*). In addition, there have been several studies attempting to capture environmental effects on mixture properties (e.g., *3–8*). There are many pavement distresses either caused or magnified by environmental effects, and different studies have tended toward solutions to a family of problems where oxidation, moisture damage (stripping or cohesive instability), and freeze-thaw (FT) are the primary environmental effects leading to the distresses. Fatigue cracking (bottom up or top down) and moisture damage due to cyclic pore pressures from traffic are usually the load-related distresses of primary concern since rutting resistance generally improves with environmental exposure absent moisture damage. Block cracking, stripping, weathering, raveling, and transverse cracking are usually the non-load related distresses of primary concern.

Modern asphalt mixes are often composed of many ingredients that can interact with each other, and when put into service, in which there are multiple environmental effects, there are many interactions that need mixture testing for a comprehensive assessment. This need is understood and several groups are pursuing conditioning and testing protocols to predict longer-term mixture performance. An abbreviated literature review is presented in the next section to provide an overview of some of the activities related to asphalt mixture conditioning and testing. Not considering the damage from multiple environmental effects simultaneously can lead to confounding results from binder and mixture properties. For example, oxidation of bitumen from aging leads to stiffening and brittleness, but not moisture damage, and conditioning of a mixture for a relatively brief period in warm to hot water

[1]Civil and Environmental Engineering, Mississippi State University, Mississippi State, MS
[2]Paragon Technical Services, Inc., Richland, MS

Corresponding Author:
Address correspondence to I. L. Howard: ilhoward@cee.msstate.edu

can affect binder adhesion to aggregates, but the binder may not be aged considerably. Both the aforementioned protocols could lead to, for example, the same tensile strength in mixture specimens. Binder aging characterization is more mature than mixture aging, and they overlap in some ways but diverge in other ways (*6, 9*).

Objectives and Scope

This paper's objective is to evaluate laboratory conditioning protocols coupled with subsequent property measurements for their ability to detect damage of asphalt mixtures in the southeastern U.S. climate (or similar climates). The investigation's focus is the property measurements themselves, and in particular how a given test can simultaneously assess multiple types of damage (i.e., oxidation, moisture damage, and free-thaw damage). While in service, mixtures can be damaged in multiple manners, so laboratory conditioning protocols that expose specimens to multiple types of damage are needed as are tests that can detect this damage in a manner that can help assess performance during service.

Durability and non-load associated cracking are two of the most pressing issues facing the paving industry in present day. Oxidation (aging), moisture damage (adhesive or cohesive), and FT all affect mixture durability, and they were evaluated individually and in combination in the conditioning protocols (CPs) described later in the paper. After conditioning, intermediate temperature non-load associated properties were assessed with binder, indirect tensile strength, and Cantabro testing. Companion data from the research team where field aging studies have been performed were used as a guide during analysis.

Abbreviated Literature Review

This section summarizes several studies that have made some contribution to mixture conditioning and mechanical testing or both for the purpose of detecting damage produced by environmental effects. These references do not include all investigations to date, and as such should not be considered a comprehensive and balanced assessment of the types of efforts occurring in the paving industry. The main purpose of this section is for general knowledge related to the type and extent of the work occurring in this area. The topics of mixture conditioning and subsequent mechanical testing are two of the most active current arenas in the paving industry.

One of the driving factors behind interest in mixture characterization is the ever-increasing design and production options. Table 1 summarizes some pertinent work (*4, 5, 7, 10–16*) related to mixture testing and subsequent mechanical testing. Examples of design and production options include, but are not limited to, warm mix additives (WMA) (e.g., *5, 8, 10, 13-15, 17, 18*), polymer modified binders (PMB) (*8, 10, 12, 18, 19*), crumb-rubber modified binders (CRMB) (e.g., *12, 17*), and high recycled material contents (e.g., *10*). Reasonably capturing the combined behavior of combinations of these options after environmental exposure, points to the need for mixture characterization. Oxidation by way of oven conditioning was commonly utilized in the efforts described in Table 1. On occasion, moisture CPs were observed in these studies (e.g., *4*), and none of these studies utilized FT. Table 1 should not be interpreted to mean that moisture conditioning is uncommon for asphalt (e.g., Doyle and Howard summarize several studies that condition specimens with moisture in various ways (*20*)), or that FT conditioning has not been investigated. FT cycles were considered in the development of AASHTO T283.

A test method capable of quantifying oxidation and moisture-related damages to asphalt mixtures is needed. As discussed in Isola et al. (*4*), and demonstrated in Smith (*6*) some test methods can reliably detect oxidation but not moisture-induced damage. Other studies have demonstrated that there is no relationship between tensile strength and moisture-related pavement distresses (*21, 22*). Cox et al. discuss several years' worth of Cantabro Mass Loss (CML) testing on dense graded asphalt to demonstrate the usefulness of mass loss (ML) values in detecting durability and brittleness (*23*). Therein, binder grade, binder type, binder content, aggregate type, recycled materials, air voids, dust-binder ratio, and aging were assessed in a favorable manner with CML data. One of the most useful attributes about CML testing is that distresses related to durability and brittleness all lead to increased ML values. To further this discussion, the remainder of this paper focuses on mechanical testing to characterize asphalt mixtures that have been exposed to distress mechanisms to simulate environmental effects.

Experimental Program

This paper is part of a multi-year effort to assess how environmental effects damage mixtures over time. Twenty mixes are considered in the overall effort, and to maintain consistency, the four mixtures evaluated herein are denoted M17 to M20. After presenting this paper's experimental program, a section is devoted to summarizing key findings from other parts of this work that are used to interpret the current findings. Unless stated otherwise, the field aging site and laboratory testing protocols were identical between companion efforts and this paper. Smith evaluated a full-scale test section over a 5-year period in which cores taken soon after construction and over time were used (in conjunction with laboratory CPs) to recommend laboratory protocols that could replicate time-dependent environmental effects (*6*). Hansen used some of the same binders as used herein within very simple mixtures produced with a single aggregate source and found that aggregate properties have meaningful

Table 1. Sampling of Recent Literature Related to Mixture Conditioning and Environmental Effects

Reference	Materials tested	Lab conditioning	Field aging	Testing	Major findings
Isola et al. (*4*)	12.5 mm NMAS with PG 67-22. Three mixes: granite, lime-treated granite, and limestone.	R30 STOA and LTOA with and without CPPC.	None.	SIDT- LMLC.	FE detected combined moisture and oxidation damage.
Yin et al. (*5*)	9.5 mm NMAS with PG 58-28: HMA, Evotherm, and Sasobit. 12.5mm NMAS with PG 70-22: HMA, Evotherm, and foam.	R30 and LTOA @ 60°C up to 16 weeks. STOA @ 135°C for HMA and 116°C for WMA.	Iowa 0–12 Mo. Texas 14 Mo.	M_r, IDT, and HLWT- PMFC and LMLC.	4–8 weeks LTOA @ 60°C simulated approx. one summer of service.
Howard and Doyle (7)	9.5 mm, 12.5 mm, and 19.0 mm NMAS with PG 67-22.	R30 and R30 STOA + 28 days @ 60°C.	Mississippi 12 Mo.	CML- PMLC.	R30 simulates under 1 year of field aging. 28 day oven at 60°C simulates approx. 1 year.
Yin et al. (*10*)	Various aggregate sources with binders ranging PG 76-28 to PG 64-22: HMA, foam, Sasobit, and Advera.	R30 and R30 STOA + 2 weeks @ 60°C or 3 days at 85°C.	Seven states in a range of climates.	M_r and HLWT- LMLC and PMFC.	M_r detected change in WMA type, aggregate absorption, and recycled materials, but not production temperature.
Elwardany et al. (*11*)	9.5 mm NMAS with PG 64-22: with and without SBS.	R30, LTOA @ 85°C of loose mix, PAV of loose mix and LMLC, and RTFO and PAV of binders.	None.	*E**, fatigue, CDT, FTIR, and DSR- LMLC, PMFC, conditioned and E/R binder.	LTOA of loose mix is recommended for new conditioning protocol.
Kumbargeni and Biligiri (*12*)	19mm NMAS with VG-10, VG-20, PMB-40, CRMB-60 binders.	R30, RTFO, and PAV.	None.	*E**, *Pen.*, softening point, η, and asphaltene extraction- conditioned and E/R binder.	Asphaltene content is an accurate indicator of binder oxidation.
Bankowski et al. (*13*)	SMA 8 and AC 16 P with 35/50 and 45/80-55: HMA and WMA.	STOA and LTOA @ 145°C for HMA and 125°C or 105°C for WMA.	None.	IDT, fatigue, low-temperature cracking, *G**, and SCB- LMLC and E/R binder.	ITSR changed less than 10% from unaged to STOA to LTOA mixes.
Safaei et al. (*14*)	9.5 mm NMAS with PG 70-22: Evotherm and foam.	R30 STOA for HMA, STOA @ 117°C for WMA, LTOA @ 85°C for 2 or 8 days.	None.	*E**, CDT, DSR, and fatigue- LMLC and E/R binder.	All test methods detected increase in oxidation damage.
Yin et al. (*15*)	9.5 mm NMAS with PG 58-28: HMA, Evotherm, and Sasobit. 12.5 mm NMAS with PG 70-22: HMA, Evotherm, and foam.	Different length of aging @ temperatures dependent on compaction temperature.	Iowa 0–6 Mo. Texas 0 Mo.	M_r and DSR- LMLC, PMLC, and PMFC.	WMA mixtures found to be more sensitive to conditioning temp. than conditioning time.
Harvey and Tsai (*16*)	19 mm NMAS granite with PG 58/64-10 and PG 58-28.	R30 STOA and LTOA @ 85°C for 3 or 6 days.	None	Fatigue- beams from LMLC slabs.	CSFB test detected aging, and changes in binder content and V_a.

Note: VG = Viscosity Grade; CPPC = Cyclic Pore Pressure Conditioning; SIDT = SuperPave Instrumented Indirect Tensile Test; *Mr* = Resilient Modulus; *HLWT* = Hamburg Loaded Wheel Tracking; *E** = Dynamic Modulus; CDT = Cyclic Direct Tension; E/R = Extracted & Recovered; η = Viscosity; *G** = Complex Modulus; SCB = Semi-Circular Bending; ITSR = Indirect Tensile Strength Ratio; FE = Fracture Energy; CSFB = Controlled Strain Flexural Beam; FTIR = Fourier Transform Infrared; LMLC = Lab Mixed-Lab Compacted; PMLC = Plant Mixed-Lab Compacted; PMFC = Plant Mixed-Field Compacted; IDT = Indirect Tensile; PAV = Pressure Aging Vessel; RTFO = Rolling Thin Film Oven; LTOA = Long-Term Oven Aging; STOA = Short-Term Oven Aging; PG = Performance Grade; NMAS = Nominal Maximum Aggregate Size; *Va* = Air Voids; DSR = Dynamic Shear Rheometer; DSR-LMLC = Dynamic Shear Rheometer- Lab Mixed Lab Compacted.

Table 2. Properties of Mixtures Tested

Mixture ID	M17	M18	M19	M20
NMAS (mm)	12.5	12.5	12.5	12.5
Binder Grade	PG 67-22	PG 67-22	PG 67-22	PG 67-22
WMT	None	Foam	0.5% Evotherm 3G™	1.5% Sasobit®
$T_{production}$ (°C)	143	132	132	132
P_b (%)	5.3	5.3	5.3	5.3
P_{be} (%)	4.5	4.5	4.5	4.5
VMA	14.3	14.3	14.3	14.3
Crushed Gravel (%)	25	25	25	25
Limestone (%)	60	60	60	60
Sand (%)	15	15	15	15
$P_{9.5mm}$ (%)	85	85	85	85
$P_{2.36mm}$ (%)	54	54	54	54
$P_{0.6mm}$ (%)	28	28	28	28
$P_{0.075mm}$ (%)	4.9	4.9	4.9	4.9

Note: NMAS = nominal maximum aggregate size; WMT = warm mix technology; $T_{production}$ = temperature at time of plant mixing; P_b = total binder content-mix mass basis; P_{be} = effective binder content-mix mass basis; VMA = voids in mineral aggregate; $P_{9.5mm}$ = percent passing a 9.5 mm sieve; $P_{2.36mm}$ = percent passing a 2.36 mm sieve; $P_{0.6mm}$ = percent passing a 0.6 mm sieve; $P_{0.075mm}$ = percent passing a 0.075 mm sieve.

effects on response to environmental effects (*9*). With these efforts as reference, specimens were laboratory compacted, subjected to a series of CPs, and thereafter mixture and binder property measurements were taken and used to assess how useful the measurements would be in terms of detecting environmental effects.

Materials and Specimens Tested

Properties of the four plant produced mixtures tested are provided in Table 2. They are well suited for this investigation because they do not contain any post-consumer products such as reclaimed asphalt pavement (RAP) or reclaimed asphalt shingles (RAS), and they do not contain hydrated lime. As such, relating binder properties to mixture properties is more straightforward and does not need to address binder to mixture offsets, as would be needed with typical mixtures. The Table 2 mixtures were plant produced for use by the Engineer Research and Development Center (ERDC) for accelerated airfield pavement testing (*24*), and during construction of their test sections a large sample was provided for this project (Figure 1*a*).

The plant produced material was stored in air-conditioned sealed metal buckets, and when ready for use, the buckets were heated to approximately 95°C so the material was just workable enough to split into pans. Once in pans, the mix was heated to between 116°C and 143°C depending on the mixture, but never more than the original mixing temperature, and compacted soon after achieving this temperature with a Superpave Gyratory Compactor (SGC). Specimens were compacted to AASHTO T331 based air void (V_a) levels of 8.0 ± 0.5%. Following compaction, specimens were either tested without conditioning or subjected to a conditioning protocol.

Laboratory Conditioning Protocols (CPs)

Eight laboratory CPs utilizing oven conditioning, hot water conditioning, FT cycles, or some combination, were utilized to simulate environmental effects (Figure 1, *b* to *d* and Table 3). Oven conditioning was performed in a forced draft oven with the specimen bottoms in direct contact with the oven shelves. Specimen sides and tops were not covered with enough space between specimens to allow air movement. To begin oven conditioning, room temperature specimens were transferred to a preheated oven. Following conditioning, specimens were cooled to room temperature in the oven with the oven turned off and the door open. Water conditioned specimens were first vacuum saturated to around 75% of their T166 measured V_a volume, then temporarily stored in room temperature water until transferred to a water bath preheated to 64°C. Transferring the room temperature specimens to the hot water bath caused the water to cool. Water temperature was returned to 64°C before conditioning time started. If specimens were not FT conditioned following water conditioning, then the water was drained away from the specimens at the end of the conditioning time, and the specimens were cooled to room temperature in air, then dried at room temperature for a minimum of 6 weeks prior to testing. For specimens that were subjected to FT cycles after water conditioning, the specimens were kept in the water bath while the water was cooled to room temperature overnight by using fans to circulate air over the water surface. Specimens were then transferred directly to a freezer

Figure 1. Photographs of experimental program.

Table 3. Laboratory Conditioning Protocols

Protocol ID	Oven (Figure 1*b*)	64°C Water (Figure 1*c*)	Freeze-Thaw (Figure 1*d*)
CP0	—	—	—
CP1	5 days @ 85°C	-	-
CP2	28 days @ 60°C	-	-
CP3	-	14 days	-
CP4	-	14 days	1 cycle
CP5	-	14 days	2 cycles
CP6	-	28 days	-
CP7	5 days @ 85°C	14 days	1 cycle
CP8	10 days @ 85°C	—	—

Note: Conditioning proceeded from oven to water to freeze-thaw.
CP0 = unconditioned control specimens; Dashes = a type of conditioning was not performed.

pre-chilled to –22°C. After 24 hours from the time the specimens were placed in the freezer, the freezer was turned off and specimens thawed with the freezer door closed to approximately 5°C, at which point the freezer door was opened. For specimens subjected to 2 FT cycles, the freezer door was closed after 30 minutes and the process was repeated. For single FT cycles, the specimens were returned to room temperature with the freezer door open before being removed from the freezer. FT specimens were also dried at room temperature for a minimum of 6 weeks prior to testing. Additional specimen conditioning details can be found in Smith (*6*).

Mixture Testing

CML testing was conducted on 15 cm × 11.5 cm specimens conditioned in air to 25°C and then subjected to 300 revolutions in an AASHTO T96 drum also conditioned to 25°C without steel spheres. Change in mass divided by initial specimen mass is referred to as ML; Figure 1, *e* and *f* are example photographs. Higher ML values indicate a mixture with greater susceptibility to durability and brittleness-related distresses. The CML test is described in detail in (*6, 23*).

Indirect tensile (IDT) testing was conducted on 10 cm × 6.3 cm specimens that had been conditioned in air to 25°C to measure tensile strength at failure (S_t). AASHTO T283 loading was used (Figure 1*g*). S_t may increase or decrease due to distresses related to durability or brittleness.

Binder Testing

Compacted and conditioned specimens had 1.3 cm slices taken off the top and bottom that were used to obtain binders for testing (Figure 1, *h* and *i*). Several identical specimens were sliced, and the collected slices had binder extracted and recovered as described in (*6*). Ethanol and toluene solvents were used alongside a centrifuge to separate aggregates. A second filterless centrifuge removed dust, and the binder was recovered with a Rotovapor® device.

Penetration (*Pen*) testing was performed according to ASTM D5 at 25°C (Figure 1*j*) in triplicate with the averaged values used herein. Dynamic Shear Rheometer (DSR) testing was performed at intermediate temperatures with an 8-mm plate and a 2-mm gap (referred to as DSR_8). The complex shear modulus (G^*) and phase angle (δ) were measured and used to calculate critical temperatures (T_c) as in (*6*). As per AASHTO M320, a maximum $G^*\sin\delta$ value of 5,000 kPa was taken as the failure criteria.

Key Findings from Companion Efforts

As mentioned in the Experimental Program section, data collected with the same laboratory CPs were available from companion efforts in which field data were also available. Howard and Doyle compared the damage due to environmental exposure in gyratory specimens that were field aged in Starkville, MS, for 1 year with exposure on all specimen sides to long-term R30 (i.e. CP1) and 28 days of conditioning in a 60°C oven (i.e. CP2) (*7*). CP1 and CP2 simulated less than 1 year and approximately 1year of field exposure based on CML testing, respectively. Smith conducted CP1 to CP7 on field cores collected soon after construction from a full-scale and un-trafficked test section in Columbus, MS (*6*). CML results of laboratory conditioned cores were compared to cores subjected to 0, 2, 3, 4, and 5 years of field aging prior to being taken. The field aging simulation times of (*6*) are provided in Figure 2*a*. Figure 2 is color coded for visual ease defining laboratory CPs by their most identifiable feature(s).

Gyratory compacted specimens were also field aged in Columbus, MS for 2 years within the same time frame as (*6*). These specimens were compacted from plant mixed material sampled during re-construction of the Columbus Air Force Base (CAFB) runway in July of 2013 (mixtures are referred to as $CAFB_{Surface}$ and $CAFB_{Base}$). Some of the compacted specimens were field aged with plastic sleeves exposing only their tops to simulate an in-service pavement, while others were exposed to several of the Table 3 laboratory CPs (Figure 2*b*). Figure 2*b* ranks results of CML testing on CAFB specimens from lowest to greatest ML. Linear ML regressions for the CAFB surface and base mixes were developed from the 0 to 2 years of environmental exposure and used to estimate the amount of field aging each CP simulated. This approach was assumed reasonable based on a statistically significant linear regression of ML data over the 5-year period that resulted in a 2.8% increase in ML per year, as documented by Smith (*6*). The CAFB slopes were lower than those detailed in Smith (*6*) at a 1.9% to 2.0% ML increase per year of field aging.

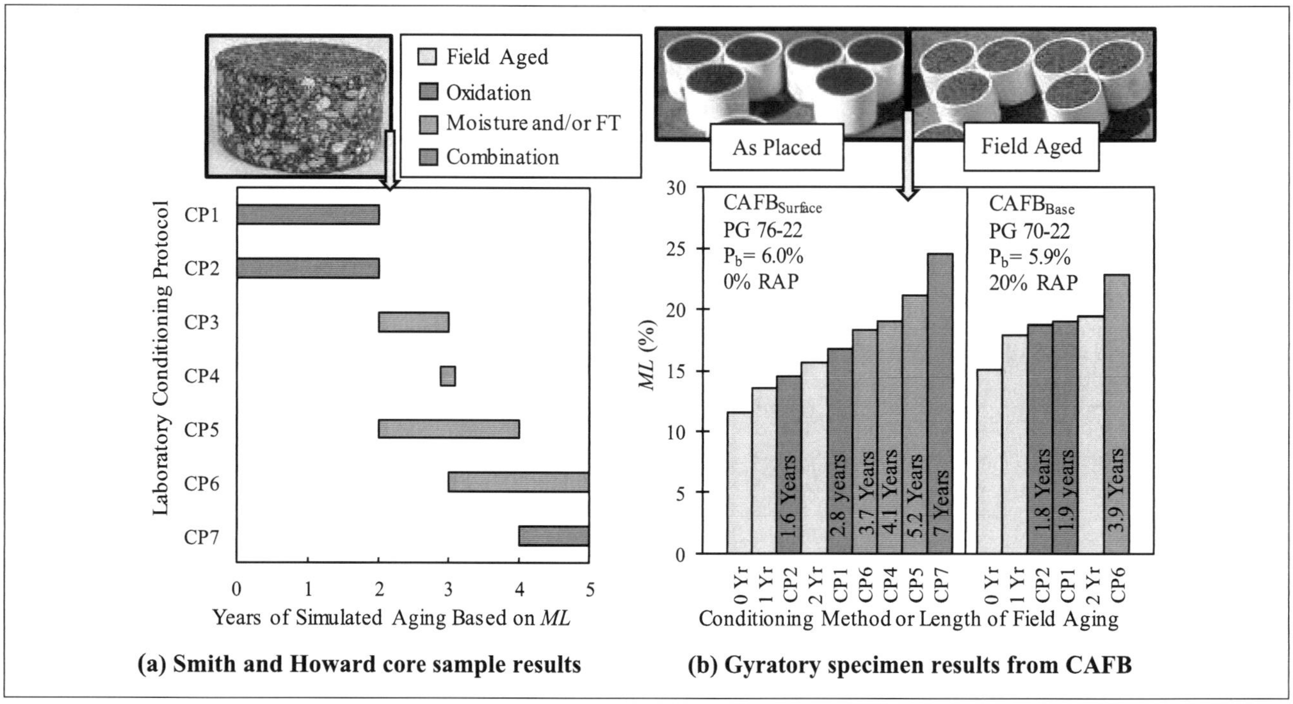

Figure 2. Companion laboratory conditioning and field aging ML results.

Results and Discussion

Test results are summarized in Table 4, and Figure 3 demonstrates the context within which these results are interpreted. In Figure 3, a mixture is designed and produced according to an acceptable method, and this process leads to a mixture with a CML and an S_t. This mixture can have properties anywhere in the spectrum of stiffness and durability. Once placed into a pavement, environmental effects begin to change the initial ML and S_t values. This paper's hypothesis is primarily shown in the Environmental Effects portion of Figure 3. If the mixture is placed into an oxidation-dominated environment (e.g., very hot and very dry), ML and S_t are prone to increase, but there is a limit to how much S_t can increase relative to ML (see Damage Envelope in Figure 3). At the other extreme, if the mixture is placed in an environment dominated by moisture and FT, ML is still prone to increase but S_t is prone to decrease. Combined environmental effects would be somewhere in between for S_t, but ML is still prone to increase. This hypothesis is based on observations from complimentary work (*6, 9*) and as supported by the data listed in Table 4, the case for Cantabro testing is solidified relative to tensile strength.

Figure 4 plots properties measured for each mix alongside linear regressions used to determine if there are significant relationships (i.e., greater than 95% confidence) between each property. For example, Figure 4*a* plots S_t from the fifth column of Table 4 on the vertical axis and ML from the sixth column of Table 4 on the horizontal axis. Based on Figure 4, *a* to *c*, *p*-values, ML did not correlate with S_t but did correlate with *Pen* and $T_c(DSR_8)$ with scatter (R^2 of 0.32 for *Pen* and 0.13 for $T_c(DSR_8)$). In addition, S_t did not correlate with *Pen* or $T_c(DSR_8)$ (Figure 4, *d* and *e*). The relationship between the two binder properties, *Pen* and $T_c(DSR_8)$, shown in Figure 4*f* was the strongest with a *p*-value of less than 0.01 and R^2 of 0.71. As such, further comparisons between mixture and binder properties rely only on *Pen*.

Figure 5 presents comparisons between the unconditioned control (CP0) and the eight CPs based on individual specimens (Table 4 mixture property values are averages of three specimens). For example, each point in Figure 5*a* represents two CML tested specimens in which one specimen was tested without conditioning (CP0) and one specimen was tested after CP1. For the equality plots shown, regressions (denoted with colored dashed lines) above the equality line indicate higher values measured after conditioning, values below the equality line indicate lower values measured after conditioning, and values on the equality line indicate no property change due to conditioning.

Table 5 presents statistical assessments to consider the effects of conditioning, represented by Δ(CP*i*–CP0) in which CP*i* represents any of the eight protocols, on ML and S_t. Arrows are used to respectively indicate increases or decreases in mixture properties. Comparison of means *t*-tests are provided in rows with no shading where Δ(CP*i*–CP0) was considered individually for each combination of mix, mixture property, and CP. For example, the value of 0.3 in column three of row one indicated a 0.3% ML increase in M17 specimens after CP1. Shaded rows in Table 5 provide results of multiple comparison *t*-groupings produced following an

Table 4. Test Results

CP	Mix ID	Pen (dmm)	T_c (DSR$_8$) (°C)	S_t (kPa)	ML (%)
CP0	M17	28	20.0	1,322	11.8
	M18	31	18.4	1,207	10.2
	M19	35	19.1	1,219	10.1
	M20	37	18.2	1,166	11.0
CP1	M17	28	18.5	1,375	12.1
	M18	33	18.4	1,385	12.8
	M19	34	18.9	1,478	12.2
	M20	27	19.7	1,390	14.5
CP2	M17	26	19.7	1,561	12.3
	M18	32	18.1	1,357	10.8
	M19	31	18.7	1,451	11.2
	M20	30	18.3	1,267	13.0
CP3	M17	25	18.7	1,391	15.6
	M18	31	18.0	1,069	13.5
	M19	37	17.7	1,350	14.5
	M20	28	19.1	1,199	16.5
CP4	M17	33	16.8	1,286	12.0
	M18	34	16.7	1,213	13.6
	M19	40	15.6	1,328	13.3
	M20	50[a]	11.9[a]	1,156	14.8
CP5	M17	30	18.3	1,387	14.0
	M18	38	16.1	1,180	12.8
	M19	30	19.2	1,413	14.7
	M20	29	18.2	1,295	14.6
CP6	M17	19	21.2	1,194	13.5
	M18	28	18.5	1,212	12.6
	M19	37	17.4	1,360	11.7
	M20	29	18.6	1,285	13.3
CP7	M17	18	21.8	1,388	16.2
	M18	22	19.4	1,285	15.1
	M19	24	20.9	1,426	16.8
	M20	22	19.7	1,238	17.9
CP8	M17	—	—	—	—
	M18	—	—	1,606	15.9
	M19	—	—	1,315	15.4
	M20	—	—	1,594	18.6

[a]Values believed to be outilers, and they were not included in analysis.

analysis of variance for both mixture properties based on the average Δ(CP*i*–CP0) for each CP for both mixture properties. For interpretation, all CPs assigned to a similar letter from *t*-grouping are statistically similar. For example, the average Δ(CP*i*–CP0) on ML for CP7 and CP8 were statistically comparable as both CPs were assigned to *t*-group "E."

Oxidation Effects

As represented in Figure 3, binder oxidation-dominated damage is expected to increase ML and S_t. Based on Figure 5 and Table 5, the oxidative CPs (i.e., CP1 and CP2) consistently increased ML and S_t at statistically equivalent magnitudes. The data presented here indicate that CP1 is slightly more damaging than CP2 based on coefficients from Figure 5 and average ML increases in Table 5. However, the effects are not statistically different based on Table 5 *t*-groups. This observation is like those found in Figure 2 in which there were little to no differences detected between ML for CP1 and CP2. Penetration results indicated a slight increase in oxidation due to conditioning; however, the values among CP0, CP1, and CP2 are within the multi-laboratory d2s limits of ASTM D5. Oxidation conditioning findings support the hypothesis in Figure 3 and do not differentiate ML from S_t (they were practically the same).

Moisture Effects

Figure 3 hypothesizes that moisture-induced damage would increase ML but decrease or have no effect on S_t. It is intuitive that an increased number of FT cycles would increase mixture damage, thus CP3, CP4, and CP5 would be expected to rank from least to most severe. in addition, CP6 would be expected to be more damaging than CP3 in all cases, as it is double the water exposure time. Figure 2 ranked the moisture conditioning protocols from least to greatest as CP3, CP4, CP5, CP6 (Figure 2*a*), and CP6, CP4, CP5 (Figure 2*b*). All moisture conditioning ranked more severe than oxidative-only conditioning for both Figure 2 data sets based on ML results. Data presented in Figure 5 and Table 5 indicate that CP6, CP4, CP5, and CP3 produce sequentially more damage based on ML results. S_t did not show the same trends in Figure 5 and Table 5 as they remained practically the same and were not statistically different than each other. Although the order of severity was counterintuitive, the chaining across Table 5 *t*-groups for ML results indicates that these protocols produced comparable damage in some senses. Penetration data ranked the moisture-dominated CPs from least to most severe as CP4, CP5, CP3, and CP6. Water conditioning findings support the hypothesis in Figure 3, and provide some evidence that ML has more potential than S_t.

Combined Effects of Oxidation and Moisture

Figure 3 hypothesizes that combined damage could be anywhere within the interior portion of the damage envelope, with a critical case being two or more damage mechanisms cancelling each other to produce the same S_t. Figure 3 also hypothesizes that multiple damage mechanisms would not cancel each other for Cantabro ML measurements. Logically, CP7 should be more severe than CP1, CP3, and CP4 since it is the combination of CP1 and CP4, which encompasses CP3. This trend is observed for ML in Figure 2 and supported by ML data in Table 5 and Figure 5. As hypothesized in Figure 3, S_t shows a practically negligible increase due to the combined conditioning mechanisms as seen in Table 5 and Figure 5. This observation is concerning regarding the

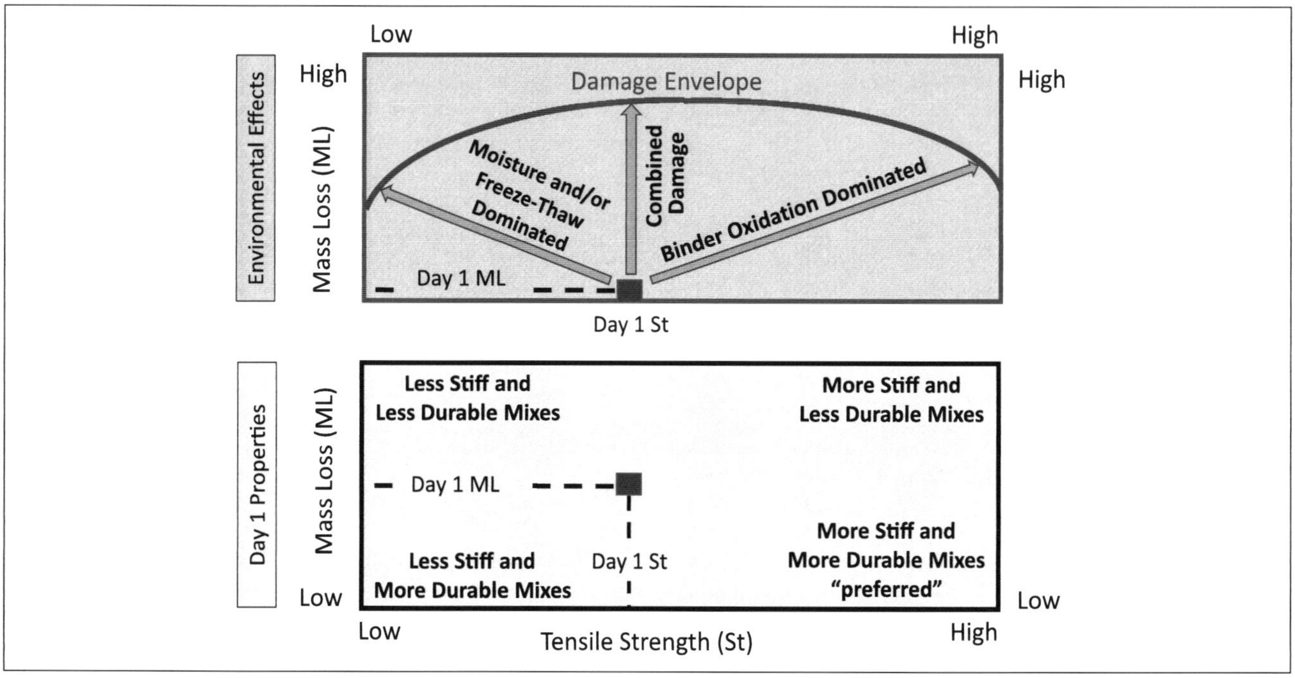

Figure 3. Mixture-Environmental effects space diagram.

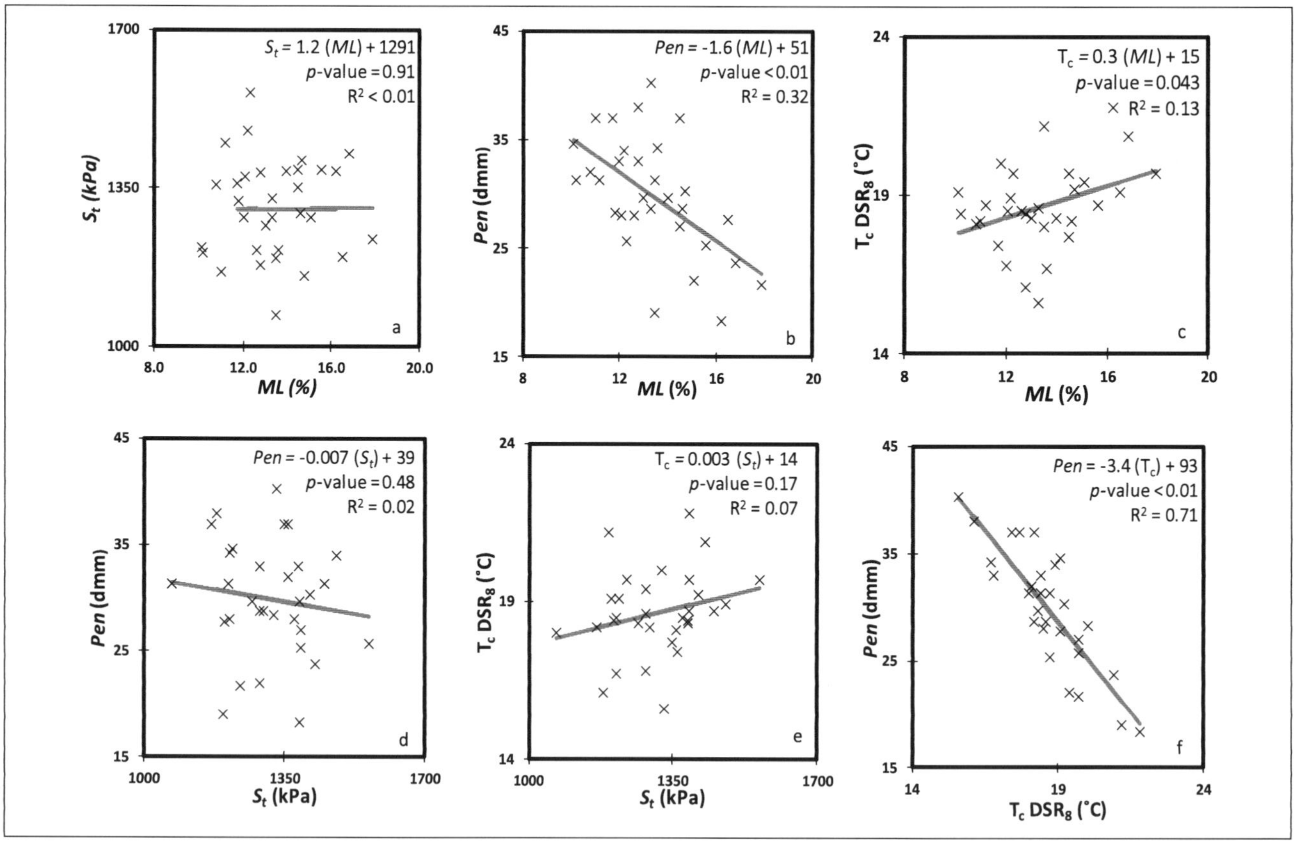

Figure 4. Relationship between test properties.

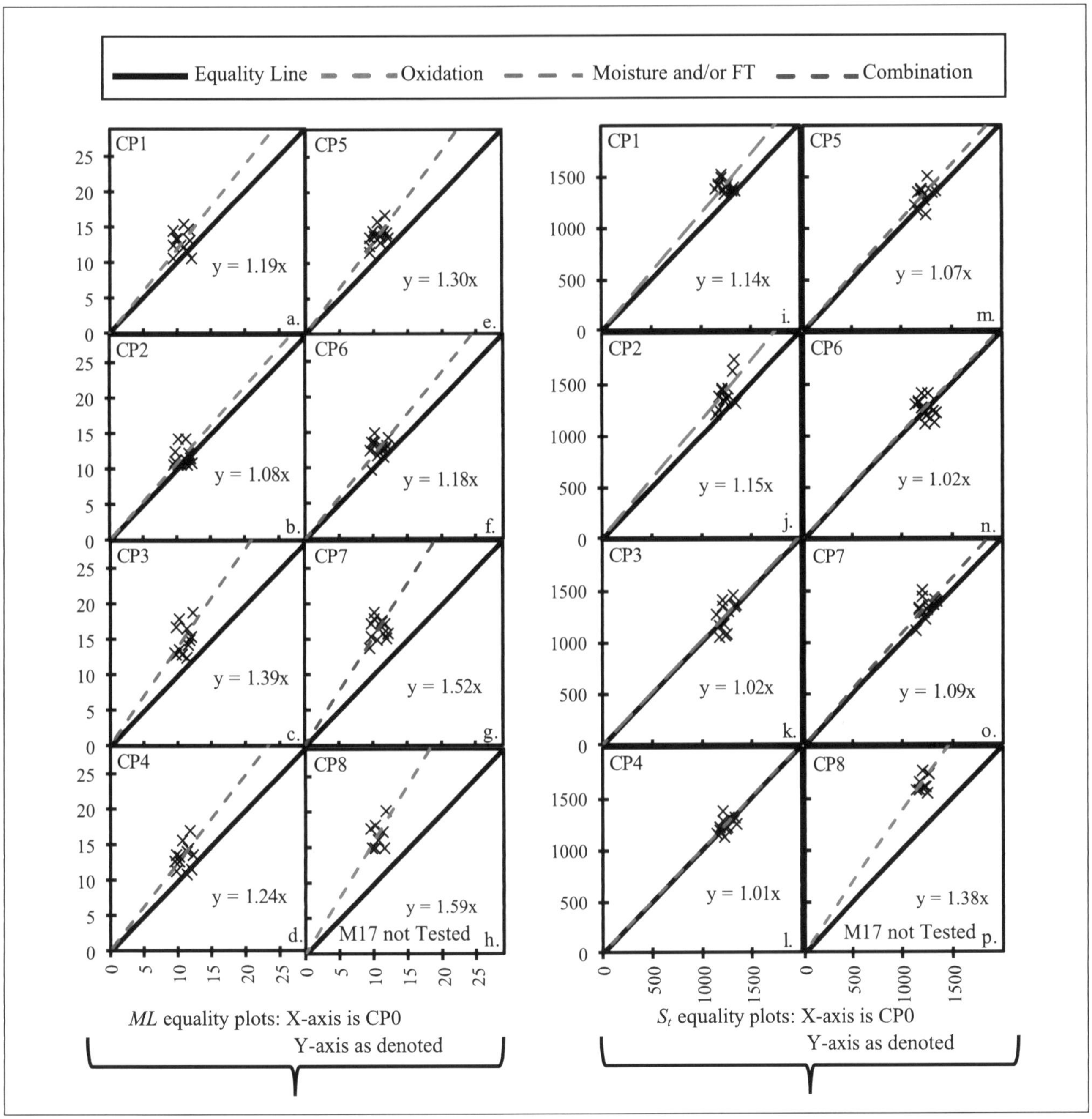

Figure 5. Equality plots for ML and S_t.

usefulness of tensile strength in detecting combined environmental effects.

To further evaluate the discrepancies between ML and S_t, CP8 was conducted with the expectation that it would produce increased damage relative to CP1 for both mixture properties. As per Table 5 and Figure 5, this expectation was validated. Comparing ML and S_t in the context of CPs 7 and 8 confirms that ML is a more suitable test than S_t for detecting combined environmental effects. ML observed high levels of damage for either heavily oxidized or combined oxidation, moisture, and FT conditioning, while S_t did not. Logically, CP7 does more damage than CP1, and S_t was unable to detect this behavior. Combined effects supported the Figure 3 hypothesis in a strong manner.

Conclusions and Recommendations

This paper's objective was to evaluate laboratory CPs coupled with subsequent property measurements for their ability to detect damage of asphalt mixtures in the southeastern US. To accomplish this objective, four plant produced mixtures with all virgin ingredients were evaluated at intermediate

Table 5. Statistical Assessments of Mixture Tests

CP	Mix ID	ML (%) Δ(CPi-CP0)	*p*-Value	Sig?	All Data *t*-Grouping	S_t (kPa) Δ(CPi-CP0)	*p*-Value	Sig?	All Data *t*-Grouping
CP1	M17	0.3↑	0.77	No	—	53↑	<0.01	Yes	—
	M18	2.6↑	0.07	No	—	178↑	<0.01	Yes	—
	M19	2.1↑	0.07	No	—	259↑	<0.01	Yes	—
	M20	3.5↑	0.01	Yes	—	224↑	<0.01	Yes	—
	all data	2.1↑	—	—	AB	179↑	—	—	A
CP2	M17	0.5↑	0.67	No	—	239↑	0.14	No	—
	M18	0.6↑	0.42	No	—	150↑	<0.01	Yes	—
	M19	1.1↑	0.16	No	—	232↑	<0.01	Yes	—
	M20	2.0↑	0.17	No	—	101↑	0.09	No	—
	all data	1.1↑	—	—	A	181↑	—	—	A
CP3	M17	3.8↑	0.12	No	—	69↑	0.13	No	—
	M18	3.3↑	0.01	Yes	—	-138↓	<0.01	Yes	—
	M19	4.4↑	0.02	Yes	—	131↑	0.02	Yes	—
	M20	5.5↑	<0.01	Yes	—	33↑	0.47	No	—
	all data	4.3↑	—	—	D	24↑	—	—	B
CP4	M17	0.2↑	0.81	No	—	-36↓	0.10	No	—
	M18	3.4↑	0.01	Yes	—	6↑	0.81	No	—
	M19	3.2↑	0.06	Yes	—	109↑	0.04	Yes	—
	M20	3.8↑	0.03	Yes	—	-10↓	0.75	No	—
	all data	2.7↑	—	—	BC	17↑	—	—	B
CP5	M17	2.2↑	<0.01	Yes	—	65↑	0.10	No	—
	M18	2.6↑	0.07	Yes	—	-27↓	0.63	No	—
	M19	4.6↑	<0.01	Yes	—	194↑	0.01	Yes	—
	M20	3.6↑	0.04	Yes	—	129↑	0.04	Yes	—
	all data	3.3↑	—	—	DC	90↑	—	—	AB
CP6	M17	1.7↑	0.03	Yes	—	-128↓	0.03	Yes	—
	M18	2.4↑	0.05	No	—	5↑	0.95	No	—
	M19	1.6↑	0.23	No	—	141↑	0.04	Yes	—
	M20	2.3↑	0.06	Yes	—	119↑	0.03	Yes	—
	all data	2.0↑	—	—	AB	34↑	—	—	B
CP7	M17	4.4↑	<0.01	Yes	—	66↑	0.01	Yes	—
	M18	4.9↑	<0.01	Yes	—	78↑	0.15	No	—
	M19	6.7↑	<0.01	Yes	—	207↑	0.01	Yes	—
	M20	6.9↑	<0.01	Yes	—	72↑	0.34	No	—
	all data	5.7↑	—	—	E	106↑	—	—	AB
CP8	M17	—	—	—	—	—	—	—	—
	M18	5.7↑	<0.01	Yes	—	399↑	<0.01	Yes	—
	M19	5.3↑	<0.01	Yes	—	96↑	<0.01	Yes	—
	M20	7.6↑	<0.01	Yes	—	428↑	<0.01	Yes	—
	all data[a]	6.2↑	—	—	E	308↑	—	—	A

[a]Δ(CP*i*–CP0) does not consider M17 from CP0.

temperatures with CML and IDT mixture tests that were supplemented by binder tests. Mixtures were exposed to a variety of laboratory CPs that included oxidation, moisture, FT, and combinations of all three. IDT strength did not relate to CML or binder test results, which was concerning. Even more concerning was IDT's inability to respond to laboratory CPs that considered multiple environmental effects. CML results related to binder properties and were able to reasonably detect multiple types of environmental effects. As such, Cantabro testing is recommended over tensile strength for intermediate temperature mixture property assessments related to non-load associated environmental effects.

Acknowledgments

The Mississippi Department of Transportation (MDOT) funded Project 106526, State Study 266, and State Study 270. The Engineer Research and Development Center (ERDC) of the U.S. Army Corps of Engineers supplied the mixtures tested. The Ergon Asphalt &

Emulsions Student Support Initiative in Construction Materials provided financial support for multiple students working on this effort. APAC Mississippi, Inc. (An Oldcastle® Company) produced the mixes. The staff at Paragon Technical Services, Inc. facilitated most binder testing and mentored students performing testing. Individuals deserving special thanks include Michael Bogue, Baxter Burns, Ben C. Cox, Jesse Doyle, Westin Graves, Brad Hansen, Amanda Hufft, Codrin Daranga, Alex Middleton, Drew Moore, Sonia Serna, Braden Smith, Scott Watson, and Ethan Whaley.

References

1. AASHTO Guide for Design of Pavement Structures. *American Association of State Highway and Transportation Officials*, Washington, D.C., 1993.
2. Mechanistic-Empirical Pavement Design Guide-A Manual of Practice-July 2008 Interim Edition. *American Association of State Highway and Transportation Officials*, Washington, D.C., 2008.
3. Copas, T. L., and H. A. Pennock. *NCHRP Synthesis 59: Relationship of Asphalt Cement Properties to Pavement Durability*. TRB, National Research Council, Washington, D.C., September 1979.
4. Isola, M., S. Chun, R. Roque, J. Zou, C. Kob, and G. Lopp. Development of and Evaluation of Laboratory Conditioning Procedures to Simulate Mixture Property Changes Effectively in the Field. *Transportation Research Record: Journal of the Transportation Research Board*, 2014. 2447: 74–82.
5. Yin, F., L. G. Caucalon, A. E. Martin, E. Arambula, and E. S. Park. Performance Evolution of Hot-Mix and Warm-Mix Asphalt with Field and Laboratory Aging. *Journal of Association of Asphalt Paving Technologists*, Vol. 83, 2014, pp. 109–114.
6. Smith, B. T. *Characterizing Short and Long Term Mixture Aging of a Full-Scale and Non-Trafficked Asphalt Test Section*. PhD dissertation. Mississippi State (Starkville), US, 2017.
7. Howard, I. L., and J. D. Doyle. Durability Indexes via Cantabro Testing for Unaged, Laboratory-Conditioned, and One-Year Outdoor-Aged Asphalt Concrete. Presented at 94th Annual Meeting of the Transportation Research Board, Washington, D.C., 2015.
8. Abbas, A. R., M. Nazzal, S. Kaya, S. Akinbowale, B. Subedi, M. S. Arefin, and L. A. Qtaish. Effect of Aging on Foamed Warm Mix Asphalt Produced by Water Injection. *Journal of Materials in Civil Engineering*, Vol 28, No.11, 2016. http://dx.doi.org/10.1061/(ASCE)MT.1943-5533.0001617.
9. Hansen, B. S. *Capturing Aggregate and Binder Interaction Effects on Aging via Mixture Testing for Single Aggregate Source Asphalt Mixtures*. MS thesis. Mississippi State (Starkville), US, 2017.
10. Yin, F., E. Arambula-Mercado, A. Epps-Martin, D. Newcomb, and N. Tran. Long Term Aging of Asphalt Mixtures. *Journal of Association of Asphalt Paving Technologists*, Vol. 85, 2016, pp. 1–31.
11. Elwardany, M. D., F. Y. Rad, C. Castorena, and Y. R. Kim. Evaluation of Asphalt Mixture Laboratory Long-Term Aging Methods for Performance Testing and Prediction. *Journal of Association of Asphalt Paving Technologists*, Vol. 85, 2016, pp. 35–75.
12. Kumbargeni, Y. S., and K. P. Biligiri. Rational Performance Indicators to Evaluate Asphalt Materials' Aging Characteristics. *Journal of Materials in Civil Engineering*, Vol. 28, No. 12, 2016. https://doi.org/10.1061/(ASCE)MT.1943-5533.0001681.
13. Bankowski, W., R. Horodecka, M. Gajewski, and K. Mirski. The Extended Assessment of Warm Mix Asphalts Durability. *Roads and Bridges*, Vol. 15, No. 2, 2016, pp. 157–173.
14. Safaei, F., J. Lee, L. Nascimento, C. Hintz, and Y. Kim. Implications of Warm-Mix Asphalt on Long Term Oxidative Aging and Fatigue Performance of Asphalt Binders and Mixtures. *Road Materials and Pavement Design*, Vol. 15, 2014,[1] pp. 45–61.
15. Yin, F., L. G. Cucalon, A. E. Martin, E. Arambula, A. Chowdhury, and E. S. Park. Laboratory Conditioning Protocols for Warm-Mix Asphalt. *Journal of Association of Asphalt Paving Technologists*, Vol. 82, 2013, pp. 177–212.
16. Harvey, J., and B. Tsai. Long-Term Oven-Aging Effects on Fatigue and Initial Stiffness of Asphalt Concrete. *Transportation Research Record 1590*, TRB, National Research Council, Washington, D.C., 1997, pp. 89–98. http://dx.doi.org/10.3141/1590-11.
17. Farshidi, F., D. Jones, and J. T. Harvey. *Warm-Mix Asphalt Study: Evaluation of Hot and Warm Mix Asphalt with Respect to Binder Aging*. Publication CA 13-2385A. California Department of Transportation, Sacramento, CA, 2013.
18. Behl, A., and S. Chandra. Aging Characteristics of Warm-Mix Asphalt Binders. *Journal of Materials in Civil Engineering*, Vol. 29, No. 10, 2017. http://dx.doi.org/10.1061/(ASCE)MT.1943-5533.0002013.
19. Anderson, D., and R. Bonaquist. *NCHRP Report 709: Investigation of Short-Term Laboratory Aging of Neat and Modified Asphalt Binders*. Transportation Research Board of the National Academies, Washington, D.C., 2012.
20. Doyle, J. D., and I. L. Howard. *Rutting and Moisture Damage Resistance of High Reclaimed Asphalt Pavement Warm Mixed Asphalt: Loaded Wheel Tracking vs. Conventional Methods. Road Materials and Pavement Design*, Special Issue from 88th Association of Asphalt Paving Technologists' Annual Meeting, Vol. 14, No. S2, 2013, pp. 148–172.
21. Zaniewski, J., and A. G. Viswanathan. *Investigation of Moisture Sensitivity of Hot Mix Asphalt Concrete*. Report Submitted to West Virginia Division of Highways, West Virginia University, Morgantown, WV, 2006.
22. Kanitong, K., and H. Bahia. Evaluation of HMA Moisture Damage in Wisconsin as it Relates to Pavement Performance. *International Journal of Pavement Engineering*, Vol. 9, No. 1, 2008, pp. 9–17.
23. Cox, B. C., B. T. Smith, I. L. Howard, and R. S. James. State of Knowledge for Cantabro Testing of Dense Graded Asphalt. *Journal of Materials in Civil Engineering*. Vol. 29, No. 10, 2017. http://dx.doi.org/10.1061/(ASCE)MT.1943-5533.0002020.
24. Rushing, J. F., M. Mejias-Santiago, and J. D. Doyle. *Comparing Production and Placement of Warm-Mix Asphalt to Traditional Hot-Mix Asphalt for Constructing Airfield Pavements*. Publication ERDC/GSL TR-13-35. Geotechnical and Structures Laboratory, Engineer Research and Development Center, Vicksburg, MS, 2013.

The Standing Committee on Surface Requirements of Asphalt Mixtures (AFK40) peer-reviewed this paper (18-06545).

Article

TRR
JOURNAL OF THE TRANSPORTATION RESEARCH BOARD

Transportation Research Record
2018, Vol. 2672(28) 22–30

DOI: 10.1177/0361198118759001
journals.sagepub.com/home/trr
SAGE

Development and Validation of a Model to Predict Interface Bonding Between Pavement Layers

Ramendra Das[1], Louay N. Mohammad[2], Mostafa Elseifi[1], Wei Cao[3], and Samuel B. Cooper, Jr.[3]

Abstract

During the planning of overlay construction activities, a contractor and/or state agency may be interested in selecting a tack coat material and application rate that would provide adequate interface bonding at the site. The objective of this study was to develop an interface shear strength (ISS) predictive model to estimate interface bonding between the hot-mix asphalt (HMA) overlay and the underlying pavement layer. To achieve this objective, 10 field rehabilitation projects were selected in six states that included 33 in-service test sections representing different traffic and climatic conditions. The overlay construction used different types of tack coat at various residual application rates. Field cores were obtained to measure ISS using a direct shear test device, referred to as the Louisiana Interlayer Shear Strength Tester (LISST). Influencing factors to be included in the model were pavement surface type and mean texture depth, residual application rate, and characteristics of tack coat residues. The experimental data were complemented with test results from a recently completed study for model development and validation. Two types of statistical regression models (linear and nonlinear) were evaluated. Results of the analysis showed that the predicted ISS obtained from the proposed models were in good agreement with the measurements. Further, the developed nonlinear model showed better prediction accuracy than the linear model. The nonlinear model can be used effectively for optimum selections of tack coats given the project conditions. The results presented herein were part of NCHRP Project 9-40A, *Field Implementation of the Louisiana Interface Shear Strength Test.*

Adequate interface bonding in a multilayered pavement system is a key component for satisfactory pavement performance. It has been widely acknowledged that if the interface cannot provide sufficient strength to withstand stresses from traffic and environmental loading, shear failure may occur at the interface, especially in areas of acceleration or deceleration. Lack of sufficient interface bonding between pavement layers may also cause delamination or de-bonding, which may accelerate the occurrence of other distresses, such as slippage cracking, fatigue cracking, shoving, rutting, and bleeding (*1–4*). Tack coat is a light application of asphalt, usually asphalt emulsion diluted with water, onto an existing, relatively nonabsorptive pavement surface (*5*). The purpose of using tack coat is to provide necessary adhesive bonding between adjacent pavement layers to ensure they behave as a monolithic system. The surface receiving tack coat can be hot-mix asphalt (HMA) or Portland cement concrete (PCC) pavement.

Selection of an optimum tack coat material and application rate is critical in the development of proper interface bond strength. However, the selection is currently made based on experience and/or engineering judgment. In addition, quality control and quality assurance testing of tack coat materials are rarely performed in the field, which may lead to premature pavement failure. The amount of tack coat material to be applied for residual asphalt content is a function of the existing pavement surface conditions (*4*). Excessive tack coat may weaken the interface by introducing a slipping plane and promote shear slippage as they act as a lubricant instead of a bonding agent. In contrast, insufficient tack coat may result in high tensile stresses at the bottom of the overlay owing to poor interface shear resistance (*6–8*).

[1]Department of Civil and Environmental Engineering, Louisiana State University, Baton Rouge, LA
[2]Department of Civil and Environmental Engineering, Louisiana State University, Louisiana Transportation Research Center, Baton Rouge, LA
[3]Louisiana Transportation Research Center, Louisiana State University, Baton Rouge, LA

Corresponding Author:
Address correspondence to Louay N. Mohammad: louaym@lsu.edu

A review of existing studies revealed that the interface bonding between pavement layers depends on several factors, such as properties of residual asphalt binder, pavement surface characteristics, application rates, mixture properties, temperature, and many other associated influencing factors (*1–4,9–16*). Many studies reported that polymer-modified emulsions have the potential to improve interface bond strength significantly, followed by unmodified emulsions and nontacked interfaces (*3,4,9–11*). In contrast, some studies showed that the improvement in interlayer adhesion property is not statistically significant when the performance of tacked interface is compared with the nontacked interface condition (*4,12*). Moreover, mechanical bonding between adjacent pavement layers is highly dependent on the characteristics of existing pavement surface before the construction of overlays. The surface micro- and macro-textures of the bottom layer play an important role in the development of proper interface bonding (*12,16*). In this regard, higher interlayer shear resistance can be related to higher surface roughness (*12*).

The construction of HMA overlays has been used widely as a cost-effective and swift pavement maintenance and rehabilitation (M&R) treatment. Therefore, proper design of asphalt overlays is critical for satisfactory serviceability and performance of the pavement system. With increasing need for reducing M&R costs and long-lasting pavement structures, the effectiveness of interface bonding between pavement layers should be determined by fundamental and/or simulative testing. However, determining interface bonding between pavement layers during the design phase is not feasible. This study presents the development of an interface shear strength (ISS) predictive model that can be used to estimate bonding between pavement layers. During the planning of overlay construction activities, a contractor and/or state agency may use the developed model in selecting a tack coat material and application rate that would provide adequate ISS given the project conditions.

Objective and Scope

The objective of this study was to develop an ISS predictive model to estimate interface bonding between HMA overlay and the underlying pavement layer. Two types of statistical models (linear and nonlinear) were evaluated using the most influential factors affecting interface bonding. To develop and validate the models, HMA field rehabilitation projects were identified across the United States to represent different traffic and climatic conditions. The HMA overlay construction used different types of tack coat materials on four types of pavement surfaces at various residual application rates. The experimental program in this study consisted of 10 field rehabilitation projects in six states that included 33 in-service test sections. For adequate model development and validation, data collected from these projects were complemented with the results from 17 comparable pavement sections from the recently completed NCHRP 9-40 study (*3*).

The statistical regression models were developed using the test results obtained from 50 test sections in total. The data were divided into 70% (35 sections) and 30% (15 sections) for model development and validation, respectively. The experimental data covered the following:

- four types of pavement surfaces (new HMA, existing HMA, milled HMA, and PCC);
- six types of emulsified tack coat materials (slow setting: SS-1, SS-1H, CSS-1H; rapid setting: CRS-1; and non-tracking: CBC-1H, NTSS-1HM);
- residual application rate ranging from 0.02 to 0.06 gal/yd^2; and
- mean pavement surface macro-texture depths between 0.63 mm and 2.14 mm.

Experimental Program

Research in this study was based on experiments conducted in a field environment followed by laboratory testing. Measurements of pavement surface macro-texture depths and tack coat application rates were performed before construction of overlays. Tack coat materials were sampled during construction to determine rheological properties of the residual asphalt binder. Subsequent to overlay construction, field cores were obtained and tested in the laboratory to determine ISS. The following presents a brief description of each experiment conducted in this study.

ISS Test

This test was conducted in accordance with AASHTO TP 114, *Standard Method of Test for Determining the Interlayer Shear Strength of Asphalt Pavement Layers*. A monotonic shear load was applied at a constant rate of 0.1 in./min (2.54 mm/min) until failure; see Figure 1*a*. A typical test result for the interface shear load versus displacement is shown in Figure 1*b*. Triplicate samples were employed at each condition to allow for statistical comparisons and the resulting coefficient of variation (COV) for ISS test results was less than 15% on average. The shear strength obtained on the interface was calculated as follows:

$$\mathrm{ISS} = \frac{P_{\mathrm{ult}}}{\frac{\pi D^2}{4}} \quad (1)$$

where

ISS = interface shear strength (psi);
P_{ult} = ultimate load applied to the specimen (lb.); and
D = diameter of the test specimen (in.).

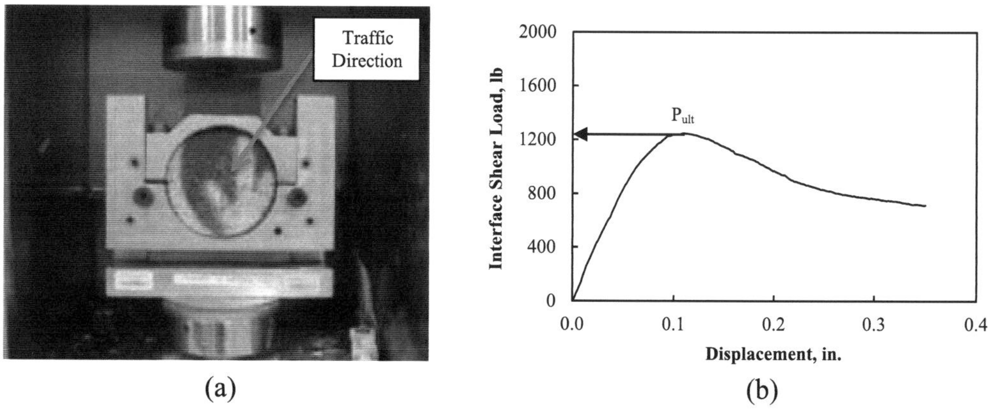

Figure 1. Interface shear strength test: (*a*) the LISST device; (*b*) typical testing curve of the interface shear load versus displacement.

Rheological Testing of Tack Coat Residues

Asphalt emulsion is composed of three basic ingredients: asphalt, water, and emulsifying agent. The residual asphalt content of the emulsified tack coats was obtained by performing the residue by evaporation test according to AASHTO T 59, *Standard Method of Test for Emulsified Asphalts*. With the obtained original residue, a suite of physical and mechanical tests were performed to determine the rheological properties. Penetration and softening point tests were performed according to AASHTO T 49, *Standard Method of Test for Penetration of Bituminous Materials*, and AASHTO T 53, *Standard Method of Test for Softening Point of Bitumen (Ring-and-Ball Apparatus)*, respectively. The apparent viscosity of the residual asphalt cement was measured at 135°C according to AASHTO T 316, *Viscosity Determination of Asphalt Binder Using Rotational Viscometer*. Performance grade (PG) of the residual asphalt binder was determined according to AASHTO M 320, *Standard Specification for Performance-Graded Asphalt Binder*. Three replicates were conducted in each test to allow for statistical comparisons.

Pavement Surface Texture Measurement

Pavement surface macro-texture depths as a measure of physical properties of the existing surfaces was determined before overlay construction in accordance with ASTM E 965, *Standard Test Method for Measuring Pavement Macrotexture Depth Using a Volumetric Technique*, a test commonly referred to as the sand patch test. Three tests were conducted within each test section. The mean texture depth (MTD) was calculated as follows:

$$\mathrm{MTD} = \frac{4V}{\pi D^2} \tag{2}$$

where

MTD = mean texture depth of pavement macro-texture (mm);

V = sample volume (mm^3); and

D = average diameter of the area covered by the material (mm).

Tack Coat Application Rate Measurement

Application of tack coat materials was performed immediately after calibration of the distributor truck and preparation of the existing surface. All emulsified tack materials were applied in the undiluted state. In some projects, two distributor trucks were utilized to avoid the use of different emulsion charges within the same distributor truck and to meet the construction schedule. During construction, the total application rate was measured in each project in accordance with ASTM D 2995 (Method A), *Standard Practice for Estimating Application Rate of Bituminous Distributors*, utilizing pre-weighed geotextile pads (12 ×12 in.) in both longitudinal and transverse directions. The residual application rate was calculated using the following equation:

$$\text{Residual application rate} = \text{percent residue} \times \text{total applicatiion rate} \tag{3}$$

Model Development Methodology

Table 1 presents the dependent and selected independent variables that were evaluated during the model development. The ISS obtained immediately after overlay construction was set as the dependent variable. The potential independent variables were selected as pavement type (i.e., HMA or PCC), pavement surface texture depth, residual application rate, and tack coat rheological properties (i.e., penetration, softening point, useful temperature interval, rotational viscosity, and high-temperature PG). It is noted that surface type and surface texture were identified as two distinct variables, because surface texture alone was not sufficient to

Table 1. Selected Variables in Characterizing Interface Bonding

Variable	Description
Interface shear strength (ISS), psi	Maximum interface shear stress
Pavement surface type (PT)	HMA and PCC
Residual application rate (RES), gal/yd^2	Measured in the field during construction
Mean texture depth (MTD), mm	
Useful temperature interval (UTI), °C	Rheological properties of tack coat materials
PG high temperature (HPG), °C	
Penetration (PEN) @ 25°C, 100g, 5 s, dmm	
Softening point (SFPT), °C	
Rotational viscosity (*RV*) @ 135°C, Pa.s	

Note: PT was selected as a discrete variable in the analysis, assuming 0 and 1 for Portland cement concrete and hot-mix asphalt surfaces, respectively. Differences in the hot-mix asphalt pavement surface type were addressed by MTD.

distinguish between HMA and PCC surfaces. The useful temperature interval was determined by taking the difference between the high-temperature grading and the low-temperature grading. The Statistical Analysis Software (SAS) was used to perform the statistical analyses (i.e., Pearson's correlation, coefficient of determination, and general linear model procedure). The stepwise variable selection method was used to identify the independent variables that were significant in predicting the dependent variable. Two statistical approaches were adopted for model development, namely multiple-linear regression analysis and nonlinear regression analysis. The former was performed using SAS while the latter was carried out using the Excel Solver tool. Subsequent to the model development, multicollinearity among independent variables was evaluated using collinearity diagnostics (i.e., variance inflation factor and tolerance value).

Pairwise Correlation Analysis

To avoid multicollinearity problems in the statistical predictive model, the test data for all independent variables were evaluated initially for potential correlation analysis. The Pearson's product-moment correlation coefficient or the linear correlation coefficient, r, is a convenient index of strength of the linear relationship between two variables (*17*). Its value varies from −1 to +1, with 0 indicating no relationship between two variables. The sign of the correlation coefficient indicates positive or negative relationship between two variables. In general, a large absolute value of the correlation coefficient (i.e., greater than 0.9) suggests potential multicollinearity. The Pearson's correlation coefficient, r, can be defined by the following equation:

$$r = \frac{n(\sum xy) - (\sum x)(\sum y)}{\sqrt{[n(\sum x^2) - (\sum x)^2][n(\sum y^2) - (\sum y)^2]}} \quad (4)$$

where

x = x variable;
y = y variable; and
n = number of variables.

Table 2 summarizes the results of the Pearson correlation analysis. As expected, the rheological properties (i.e., penetration, softening point, rotational viscosity, PG high temperature, and useful temperature interval) of tack coat materials were highly correlated with each other. Therefore, out of the five rheological variables, one or two that were most significantly related to the others would be sufficient in the development of the predictive model.

Identification of Significant Independent Variables

Analysis of Variance. An analysis of variance (ANOVA) between the dependent variable, ISS, and the independent variables was conducted using SAS. The ANOVA provided a p-value that helped to understand whether an independent variable was significant to the model or not. Independent variables that were not statistically significant in the model were removed. If, from the ANOVA analysis, a set of correlated variables are significant, then according to the pairwise correlation analysis, only one should be retained, and the selection can be made based on the p-value. Table 3 summarizes the results of the ANOVA analysis. It was observed that four independent variables (penetration, softening point, rotational viscosity, and useful temperature interval) were not statistically significant and were therefore excluded from further consideration in model development.

Correlations between ISS and Significant Independent Variables. Figure 2*a* illustrates the effect of pavement surface type on ISS obtained immediately after overlay construction. On average, the milled HMA surface provided the highest ISS followed by new HMA, existing HMA, and PCC surface types. However, use of the SS-1 tack coat material on milled HMA yielded an ISS value slightly below the minimum recommended ISS threshold value of 40 psi (*3*). Figures 2*b–d* present the correlations between the measured ISS and MTD, residual application rate (RES), and PG high temperature (HPG), respectively. As shown in these figures, in general an increasing trend of ISS can be noted for these three variables.

Multiple-Linear Regression Analysis

The objective of the multiple-linear regression (MLR) analysis was to model the linear relationship between the

Table 2. Summary of the Pearson Correlation Analysis

Variable	ISS	RES	MTD	PEN	SFPT	RV	HPG	UTI
ISS	1.00	0.76	0.66	−0.34	0.23	0.22	0.32	0.30
RES	0.76	1.00	0.34	−0.29	0.09	0.15	0.12	0.06
MTD	0.66	0.34	1.00	0.11	−0.13	−0.12	−0.08	0.05
PEN	−0.34	−0.29	−0.11	1.00	−0.87	−0.82	−0.88	−0.77
SFPT	0.23	0.09	−0.13	−0.87	1.00	0.93	0.95	0.77
RV	0.22	0.15	−0.12	−0.82	0.93	1.00	0.90	0.73
HPG	0.32	0.12	−0.08	−0.88	0.95	0.90	1.00	0.88
UTI	0.30	0.06	0.05	−0.77	0.77	0.73	0.88	1.00

Note: HPG = PG high temperature; ISS = interface shear strength; MTD = mean texture depth; PEN = penetration; RES = residual application rate; RV = rotational viscosity; SFPT = softening point; UTI = useful temperature interval.

Table 3. Summary of ANOVA Analysis

Variable	*F*-value	*p*-value
PT	62.24	<0.0001
RES	98.17	<0.0001
MTD	150.07	<0.0001
PEN	1.43	0.242
SFPT	1.31	0.263
RV	1.18	0.287
HPG	5.95	0.022
UTI	0.31	0.584

Note: HPG = PG high temperature; MTD = mean texture depth; PEN = penetration; PT = pavement surface type; RES = residual application rate; RV = rotational viscosity; SFPT = softening point; UTI = useful temperature interval.

dependent variable (Y) and the independent variables (Xs). In the development of the MLR model, identification of the appropriate independent variables significant in predicting the dependent variable is important to avoid overfitting and multicollinearity problems. This selection was made based on the result of stepwise regression and ANOVA analyses. The MLR model can be represented as follows:

$$Y = \beta_o + \beta_1 X_1 + \ldots\ldots + \beta_K X_K \quad (5)$$

where, β_0 is the regression coefficient for intercept, and β_i are coefficients for the independent variables X_1 through X_K.

Nonlinear Regression Analysis

Nonlinear regression (NLR) analysis is a statistical technique that is used to describe a nonlinear relationship between dependent and independent variables. NLR models are generally assumed to be parametric. The principle of this method is the same as linear regression analysis, which is to minimize the sum of squared error (SSE) between data and the fitted values (*18*). An iterative nonlinear least square fitting method was employed using the Excel Solver tool to develop the optimum NLR model. This method requires an initial estimate of the parameter values based on prior experience or reasonable guess of the function used to fit the data. The iteration is repeated by changing the parameter values and functions until the minimum possible value of SSE is achieved.

Multicollinearity Test

In regression analysis, it is possible that there is no pairwise correlation, but a combination of independent variables is correlated with some other combinations. The variance inflation factor (VIF) is the most common statistic used to identify this problem. VIF is a measure of inflation in the standard error associated with a particular weight caused by multicollinearity. A value of 5 or 10 has been typically recommended as the maximum level of VIF (*19*). Table 4 presents the multicollinearity test results. The VIF values for all independent variables were less than 5, which indicates no multicollinearity problem existed between or among the independent variables.

Result and Analysis

This section presents the model development using MLR and NLR analyses, followed by model validation.

Model Development

Two statistical regression models were developed to predict interface bonding between pavement layers based on the final list of significant independent variables. The MLR analysis was conducted using the PROC REG procedure in SAS, while the Excel Solver tool was used to develop the NLR model. The following two models are proposed to predict ISS based on the data considered in this study:

MLR model:

$$ISS = \beta_o + \beta_1 PT + \beta_2 \times RES + \beta_3 \times MTD + \beta_4 HPG \quad (6)$$

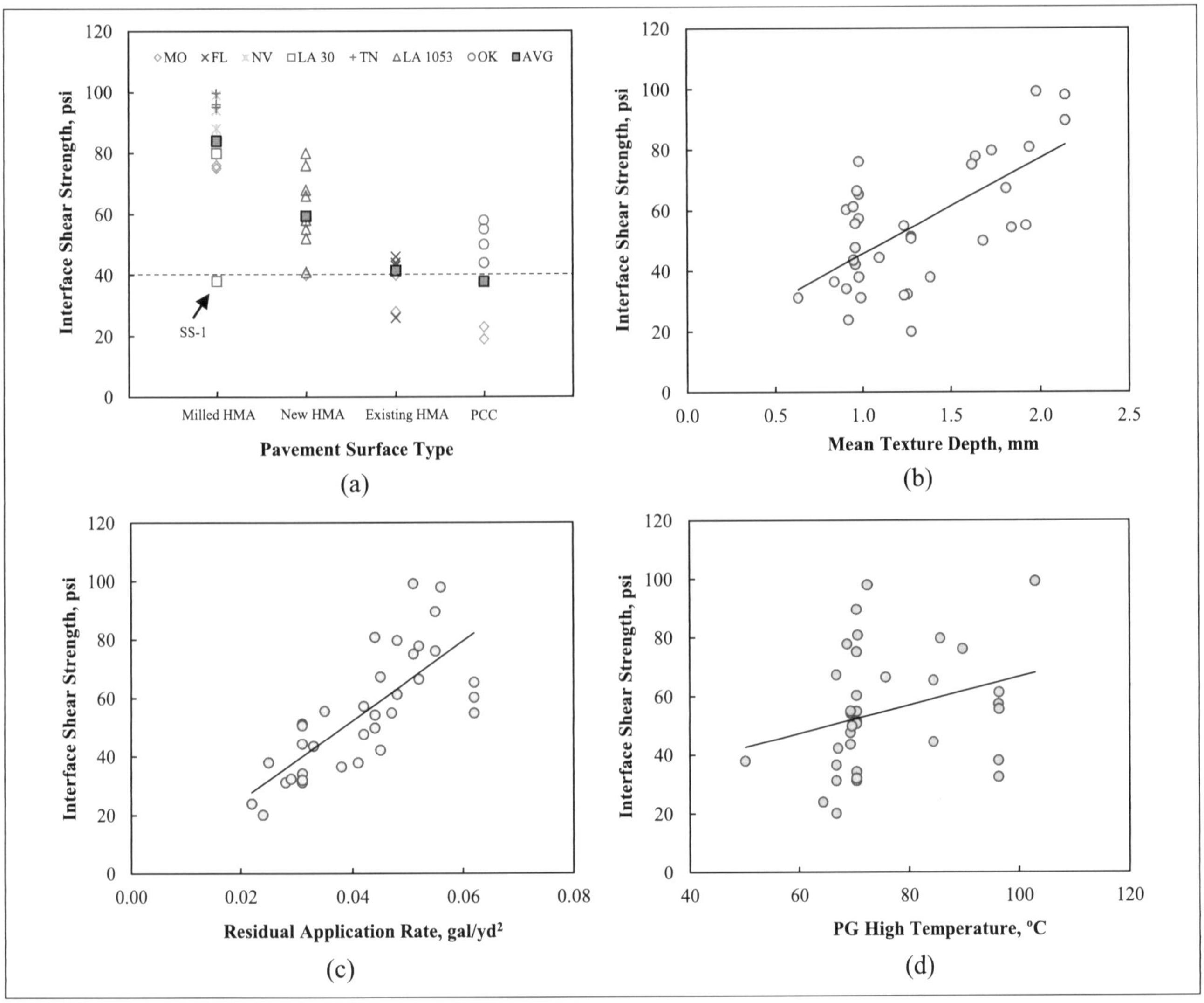

Figure 2. Correlations between ISS and (*a*) pavement surface type, (*b*) mean texture depth, (*c*) residual application rate, (*d*) high-temperature performance grade.

Nonlinear regression model:

$$ISS = \delta_1 \times \left[(PT + MTD)^3 + \delta_2 \times RES \times HPG \right] \quad (7)$$

where:

ISS = predicted ISS (psi);
PT = pavement surface type (i.e., 0 = PCC; 1 = HMA);
MTD = mean texture depth (mm);
RES = residual application rate (gal/yd^2);
HPG = PG high temperature (°C); and
β_{0-4}, and $\delta_{1,2}$ = model coefficients.

The ranges of values for parameters used for model development are as follows:

- RES is 0.02–0.06 gal/yd^2;
- MTD is 0.06–2.2 mm;
- HPG is 50–100°C.

Table 4. Multicollinearity Test Results

Variable	Tolerance	VIF
PT	0.94	1.06
RES	0.80	1.25
MTD	0.77	1.29
HPG	0.98	1.03

Note: HPG = PG high temperature; MTD = mean texture depth; PT = pavement surface type, RES = residual application rate; VIF = variance inflation factor.

Table 5 presents the estimated parameters and *p*-values for the MLR and NLR models. It can be seen that all model coefficients are significant at the 5% significance level. The root-mean-squared error (RMSE) was calculated as a comparative measure of goodness of fit. The RMSE value indicates the difference between the predictions and measurements, and was calculated via:

Table 5. Regression Model Parameters

Parameter	Multiple-linear regression					Nonlinear regression	
	β_0	β_1	β_2	β_3	β_4	δ_1	δ_2
Value	−69.5	17.9	897.4	27.7	0.49	1.55	8.17
Standard error	11.9	3.9	142.8	4.1	0.13	0.13	0.98
t-value	−5.9	4.6	6.3	6.7	3.8	12.22	8.34
p-value	<0.0001	<0.0001	<0.0001	<0.0001	0.0007	<0.0001	<0.0001

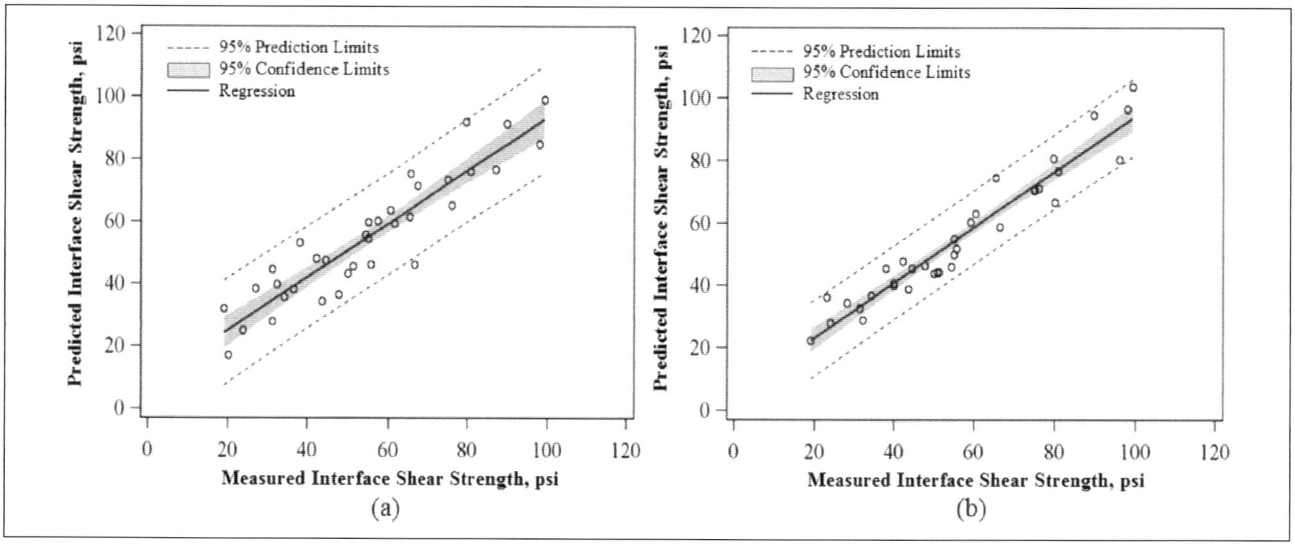

Figure 3. Development of the interface shear strength predictive model: (*a*) multiple-linear regression model; (*b*) nonlinear regression model.

$$\text{RMSE} = \sqrt{\frac{1}{n}\sum_{i=1}^{n}\left(y_i - \hat{y}_i\right)^2} \tag{8}$$

where y_i is measured value, $\hat{y}_i$ is predicted value, and n is number of observations.

Figure 3*a* illustrates the comparison between the measured and predicted ISS based on the developed MLR model. This figure shows a good correlation with an R^2 value of 0.85. The calculated RMSE was 8.9 psi. The range of measured ISS values used in model development was between 19 psi and 99 psi, which was a wide range encompassing both poorly and strongly bonded tack-coated interfaces. Based on the range of measured ISS and RMSE, the percentage error would range between 9% and 47%. The COV of the RMSE is 16.6% for the RMSE value of 8.9 psi, which is calculated by normalizing RMSE by the mean ISS (i.e., 53.6 psi) used in model development. This COV is deemed acceptable. However, if the minimum ISS value (i.e., 19 psi) is used to normalize RMSE, the obtained percentage error is 47%, which was considered high and may not be acceptable. Therefore, the NLR analysis was conducted for better prediction accuracy. Figure 3*b* illustrates the comparison between the predicted and measured ISS based on the developed NLR model. This model showed an improved R^2 value of 0.95 with a reduced RMSE of 4.8 psi.

Model Validation

As previously noted, 30% of the observations were used in the model validation phase as independent observations. Figure 4*a* presents the validation of the proposed MLR model by comparing the measured and predicted ISS values. There was an acceptable correlation with an R^2 value of 0.80 and an RMSE of 6.8 psi. The range of measured ISS values used for model validation was between 27 and 75 psi. Figure 4*b* presents validation of the proposed NLR model by comparing the measured and predicted ISS values. The NLR model showed an improved R^2 value of 0.95 with an RMSE value of 4.8 psi in the development phase and an R^2 of 0.84 with an RMSE of 6.2 psi in the validation phase. Based on the range of the measured ISS values and RMSE, the percentage error in validation ranged from 5% to 25%, which was a considerable reduction compared to the linear model and was deemed acceptable. Therefore, the NLR analysis provided a more suitable approach than linear regression, as better prediction accuracy was accomplished. The developed

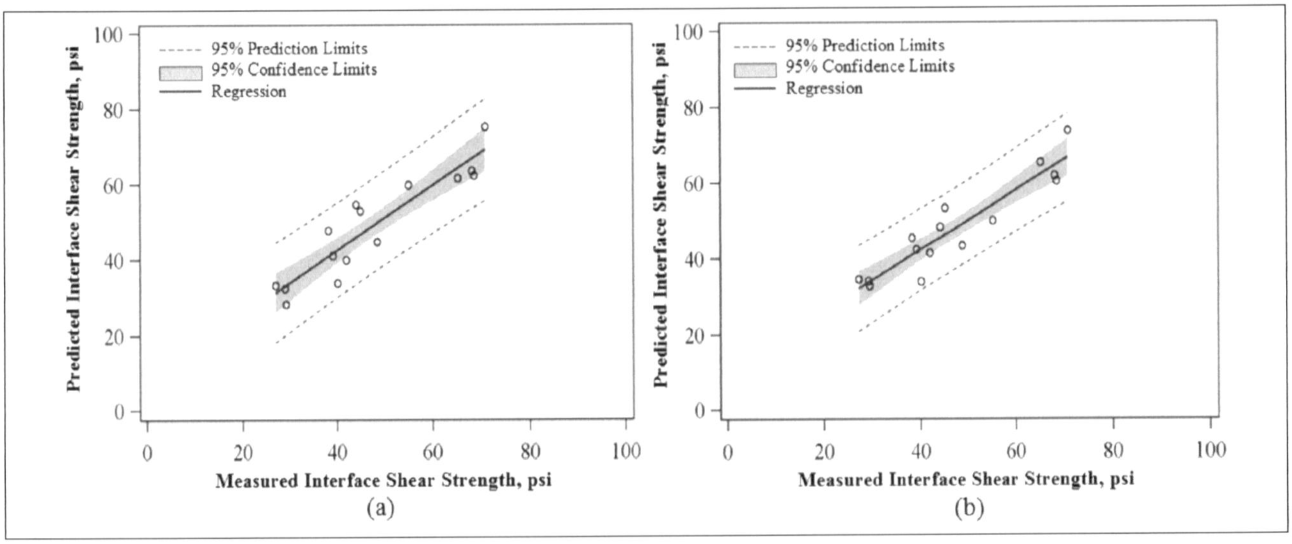

Figure 4. Validation of the interface shear strength predictive model: (*a*) multiple-linear regression model; (*b*) nonlinear regression model.

Table 6. Illustrative Applications of the Nonlinear ISS Predictive Model

Surface type	PT	MTD (mm)	Tack coat type	HPG (°C)	Predicted ISS
Milled HMA	1	1.77	NTSS-1HM	84	75
			SS-1H	72	69
			SS-1	50	58
New HMA	1	0.91	NTSS-1HM	84	53
			SS-1H	72	47
			SS-1	50	36
Existing HMA	1	0.97	NTSS-1HM	84	54
			SS-1H	72	48
			SS-1	50	37
PCC	0	1.49	NTSS-1HM	84	48
			SS-1H	72	42
			SS-1	50	30

HPG = PG high temperature; ISS = interface shear strength; MTD = mean texture depth; PT = pavement surface type; RES = residual application rate.

NLR model is recommended to be used for estimating the interface bonding during planning of overlay construction activities.

Illustrative Applications of the Nonlinear ISS Predictive Model

During the planning of overlay construction activities, a contractor and/or state agency may be interested in selecting a tack coat material and application rate that would provide adequate ISS at the site to ensure good pavement performance. Table 6 shows application of the developed nonlinear predictive model at 0.04 gal/yd^2 residual application rate to estimate ISS. The MTD was defaulted to the mean value for each pavement surface type measured in this study. Three types of tack coat materials were selected based on HPG: NTSS-1HM, SS-1H, and SS-1; the corresponding HPG values in Table 6 were also defaulted to the averaged measurements. As shown in the table, the NTSS-1HM tack coat with high HPG showed the highest interface bonding (i.e., 75 psi) for milled HMA surface type, while SS-1 with low HPG exhibited the minimum ISS (i.e., 30 psi) for PCC pavement surface.

Summary and Conclusions

This paper presents the development of two statistical models: multiple-linear regression model and nonlinear regression model to predict the interface bond strength between pavement layers. For this purpose, 10 field rehabilitation projects were selected in six states that included 33 in-service test sections. For adequate model development and validation, the experimental data were complemented with results from a recently completed study (NCHRP Project 9-40). The construction of HMA overlays involved use of different tack coat materials at various residual application rates. In the experimental program, pavement surface macrotexture depths and residual application rates were measured before overlay construction. Laboratory experiments on field-extracted cores and tack coat samples were performed to determine interface bonding and rheological properties of tack coat residues, respectively. All test results were carefully analyzed and critically evaluated for subsequent model development. The developed predictive model can be used by practitioners during the planning phase to select tack coat type and application rate, given the conditions in the field. Default values are provided to assist in the use of the predictive model in the field. In addition, caution should be exercised in using the ISS prediction model for values of parameters outside those considered in model development. The following conclusions can be drawn based on the findings of this study:

- ISS was largely dependent on the type of pavement surface receiving the tack coat (e.g., HMA versus PCC) and pavement surface texture (e.g., milled versus non-milled). In general, HMA surface showed better interface bonding than PCC surface.
- ISS was found to correlate well with pavement surface texture depth and tack coat residual application rate. In general, ISS was enhanced with increase in surface texture depth and residual application rate.
- The predicted ISSs obtained from the proposed models were in good agreement with the measurements.
- The developed nonlinear regression model showed better prediction accuracy than the linear model.

Acknowledgments

This paper is based on the experimental results of NCHRP Project 9-40A, *Field Implementation of the Louisiana Interface Shear Strength Test*. The authors acknowledge the assistance of the Technical Review Panel for NCHRP Project 9-40A and the Louisiana Department of Transportation and Development. The authors would like express their gratitude for the participation and cooperation of the state departments of transportation, industry associations, material suppliers, and contractors in providing the test sections, tack coat products, application equipment, and other associated supports throughout the course of the project.

References

1. Salinas, A., I. Al-Qadi, K. Hasiba, H. Ozer, Z. Leng, and D. Parish. Interface Layer Tack Coat Optimization. *Transportation Research Record: Journal of the Transportation Research Board*, 2013. 2372: 53–60.
2. Sholar, G. A., G. C. Page, J. A. Musselman, P. B. Upshaw, and H. L. Moseley. Preliminary Investigation of a Test Method to Evaluate Bond Strength of Bituminous Tack Coats. *Journal of Association of Asphalt Paving Technologists*, Vol. 73, 2004, pp. 771–806.
3. Mohammad, L. N., M. Elseifi, A. Bae, N. Patel, J. Button, and J. Scherocman. *NCHRP Report 712: Optimization of Tack Coat for HMA Placement*. Transportation Research Board of the National Academies, Washington, D.C., 2012.
4. Hachiya, Y, and K. Sato. Effect of Tack Coat on Binding Characteristics at Interface between Asphalt Concrete Layers. *Eighth International Conference on Asphalt Pavements*, Seattle, WA, Vol. I, 1997, pp. 349–362.
5. ASTM D8-02. *Standard Terminology Relating to Materials for Roads and Pavements, Annual Book of ASTM Standards*, Vol. 04.03, West Conshohocken, PA, 2004.
6. Asphalt Institute. *A Basic Asphalt Emulsion Manual*. 4th ed. Asphalt Institute, Lexington, KY, 2009.
7. Asphalt Institute. *The Asphalt Handbook*. 4th ed. Asphalt Institute, Lexington, KY, 1989.
8. Flexible Pavement of Ohio. *Proper Tack Coat Application*. Technical Bulletin, Flexible Pavement of Ohio, Dublin, OH, 2012.
9. Sangiorgi, C., A. C. Collop, and N. H. Thom. Laboratory Assessment of Bond Condition using the Leutner Shear Test. *Third International Conference on Bituminous Mixture & Pavements*, Thessaloniki, Greece, 2002, pp. 315–324.
10. Collop, A. C., M. H. Sutanto, G. D. Airey, and R. C. Elliott. Development of an Automatic Torque Test to Measure the Shear Bond Strength between Asphalt. *Construction and Building Materials*, Vol. 25, No. 2, 2011, pp. 623–629.
11. Johnson, E. N., K. C. Melissa, and J. Pantelis. *Tack Coat Testing: Measuring Field Bond Strength*. Report No. MN/RC 2015-25. Minnesota Department of Transportation, Research Services and Library, Saint Paul, MN, 2015.
12. Canestrari, F., G. Ferrotti, X. Lu, A. Millien, M. N. Partl, C. Petit, A. Phelipot-Mardelé, H. Piber, and C. Raab. Mechanical Testing of Interlayer Bonding in Asphalt Pavements. In *Advances in Interlaboratory Testing and Evaluation of Bituminous Materials*. Springer Dordrecht, 2013, pp. 303–360.
13. Uzan, J., M. Livneh, and Y. Eshed. Investigation of Adhesion Properties Between Asphaltic-Concrete Layers. *Proceeding of the Association of Asphalt Paving Technologists*, Vol. 47, 1978, 495–521.
14. Das, R., L. N. Mohammad, M. Elseifi, W. Cao, and S. B. Cooper Jr. Effects of Tack Coat Application on Interface Bond Strength and Short-Term Pavement Performance. *Transportation Research Record: Journal of the Transportation Research Board*, 2017. 2633: 1–8.
15. Mohammad, L. N., M. A. Raqib, and B. Huang. Influence of Asphalt Tack Coat Materials on Interface Shear Strength. *Transportation Research Record: Journal of the Transportation Research Board*, 2002. 1789: 56–65.
16. West, R. C., J. Zhang, and J. Moore. *Evaluation of Bond Strength Between Pavement Layers*. Report No. 05-08. National Center for Asphalt Technology, Auburn, AL, 2005.
17. Freund, R., W. Wilson, and P. Sa. *Regression Analysis*. Academic Press, New York, NY, 2006.
18. Brown, A. M. A step-by-step guide to non-linear regression analysis of experimental data using a Microsoft Excel spreadsheet. *Computer Methods and Programs in Biomedicine*, Vol. 65, No. 3, 2001, pp. 191–200.
19. Hair, J. F., W. C. Black, B. J. Babin, R. E. Anderson, and R. L. Tatham. *Multivariate Data Analysis*. Prentice Hall, Upper Saddle River, NJ, 1998.

The Standing Committee on Surface Requirements of Asphalt Mixtures (AFK40) peer-reviewed this paper (18-04194).

Article

TRR
JOURNAL OF THE TRANSPORTATION RESEARCH BOARD

Transportation Research Record
2018, Vol. 2672(28) 31–41

DOI: 10.1177/0361198118759064
journals.sagepub.com/home/trr
SAGE

Effects of Regular and Nano Sized Hydrated Lime Fillers on Fatigue and Bond Strength Behavior of Asphalt Mastic

Aditya Kumar Das[1] **and Dharamveer Singh**[1]

Abstract

The present study evaluates effects of regular sized hydrated lime (RHL) and nano sized hydrated lime (NHL) on fatigue and bond strength of asphalt mastic. The asphalt mastics were produced in the laboratory using AC-30 binder with different combinations of basalt–RHL, and basalt–NHL fillers. The dosages of RHL and NHL were selected as 0%, 5%, 10%, 15%, and 20% by weight of asphalt binder, and the percentage of basalt filler was adjusted accordingly. Filler to binder (F/B) ratio was selected as 0.8 (by mass ratio) for all mastic sample preparation. The fatigue damage behavior (number of cycles to fatigue damage/failure) of asphalt mastic was evaluated using a linear amplitude sweep (LAS) test. Further, the interfacial bond strengths of asphalt mastic and aggregate samples were evaluated using the bitumen bond strength (BBS) test. Overall test results indicate that mineralogy, surface area, and interaction properties of RHL and NHL fillers have a significant effect on fatigue, bond strength, and moisture damage performance of asphalt mastic. The results from the LAS test showed that NHL filler predominantly enhanced the fatigue life of asphalt mastic as compared with RHL filler. BBS test results imply that the contribution of NHL filler is significant over RHL filler in improving the bond strength and moisture damage resistance of asphalt mastic. Overall asphalt mastic with 20% NHL filler had better fatigue life, bond strength, and moisture damage performance over mastic with other percentages of RHL or NHL fillers.

Asphalt mix comprises three critical phases: asphalt, asphalt-mineral filler (asphalt mastic), and aggregates. Asphalt mastic contributes to in-service performance and durability of asphalt mix related to fatigue and bond strength at the aggregate surface interface (*1, 2*). Application of inert filler (originated from parent source of aggregate) and active filler (defined by reactive nature with binder such as hydrated lime (HL), cement, fly-ash, dolomite, etc.) plays an important role in enhancing the performance of asphalt mastic and mix (*3, 4*). Use of regular sized hydrated lime (RHL) as an active filler or modifier has a beneficial effect on the performance of the asphalt binder, mastic, and mixes due to its rugose surface with high specific surface area (SSA), antistripping, and antioxidant behavior (*5*). Many researchers have shown that addition of RHL improves the resistance of asphalt mastic and mixes to fracture growth, oxidative aging, rutting, fatigue, and moisture damage (*3, 6–8*). Some researchers have also reported that effectiveness of HL for asphalt mixes may be enhanced by reducing its size to nano level (*9, 10*). Nano size materials often possess better physicochemical properties that significantly differ from those of regular size bulk structured materials of the same elemental composition because of the high specific area and interaction potential (*11*). It has been reported that addition of nano sized hydrated lime (NHL) has positive effects on rutting and fatigue resistance performance of asphalt and mixes (*12, 13*). So far, limited studies are available on the potential application of NHL and its comparison with RHL. In addition, little attention is given on combined effects of either NHL or RHL with an inert filler, such as Basalt (B).

Therefore, this present study was motivated by the aim of addressing the effects of RHL and NHL, in combination with B filler, on fatigue and moisture damage performance of asphalt mastic. The asphalt mastics were produced in the laboratory using AC-30 binder with different combinations of basalt–RHL and basalt–NHL fillers. The dosages of RHL and NHL were selected as 0%, 5%, 10%, 15%, and 20% by weight of asphalt binder, and the percentage of basalt filler was adjusted accordingly. Filler to binder (F/B) ratio was selected as 0.8 (by mass ratio) for all mastic sample preparation. The fatigue damage behavior (number of cycles to fatigue damage/failure) of asphalt mastic was evaluated using the linear amplitude sweep (LAS) test, which is based

[1]Department of Civil Engineering, Indian Institute of Technology Bombay, Powai, Mumbai, India

Corresponding Author:
Address correspondence to Aditya Kumar Das: kumaradityanitr@gmail.com

Table 1. Conventional Properties of AC-30 Binder and Fillers

Binder	Penetration 0.1 mm at 25°C	Softening point (°C)	Ductility at 25°C, mm	Absolute viscosity at 60°C, poise
AC-30	46	48	>100	Min. 2400
Fillers	Particle diameter	SSA (m^2/g)	Specific gravity	Plasticity index
B	45 μm	9.2	2.78	Non-plastic
RHL	27 μm	2.15	2.19	<4
NHL	358 nm	43.17	2.15	<4

on the visco-elastic continuum damage (VECD) principle (*14*). The LAS test is an accelerated test method for analyzing fatigue behavior of asphalt mastic under a controlled strain condition. Further, the interfacial bond strengths of asphalt mastic and aggregate samples were evaluated using a pneumatic adhesion tensile testing instrument (PATTI) known as the bitumen bond strength (BBS) test (*2, 15*). This test is a reliable method for understanding the adhesive and cohesive behavior of asphalt mastic-aggregate systems (*15*). It is expected that the present study will help in developing a better understanding of the effects of RHL and NHL fillers on the performance of asphalt mastic.

Materials

Asphalt and Fillers

A control viscosity grade asphalt binder (AC-30) commonly used for the construction of flexible pavements in India was selected in this study. Three types of mineral fillers—Basalt filler (B), RHL and NHL—were used for the preparation of asphalt mastic samples. All the three fillers have particle-size distribution of 100% passing a 0.075 mm sieve. The NHL was produced from milling of RHL using a planetary ball mill by maintaining 250 rpm with a ball to powder ratio of 5:1 for a period of 10 h (*12*). The conventional test properties of AC-30 binder and all three fillers (B, RHL, and NHL) are shown in Table 1.

Characterization of Fillers

The average particle size was determined using light scattering (LS) graphs for B and RHL fillers, while dynamic light scattering (DLS) was used for NHL filler. LS and DLS methods are widely used to measure particle size of micron level and nano level materials, respectively. In addition, the SSA of fillers was measured using the Braunauer, Emmett, and Teller (1938) method. In addition, plasticity index and specific gravity (SG) tests on all three fillers were conducted along with scanning electron microscopy (SEM) imaging. Figure 1 shows the particle-size distribution and shape structure for the B, RHL, and NHL fillers. The average particle size of B, RHL, and NHL fillers was found to be 45 μm, 27 μm, and 358 nm, respectively. SEM images show that the surface texture of all the three fillers is flaky in appearance (Figure 1), which may contribute towards better bonding with asphalt. Furthermore, B and RHL fillers have lower SSAs of 9.2 m^2/g and 2.15 m^2/g, respectively (Table 1), but NHL has a high SSA of 43.17 m^2/g. Thus, NHL may have better interaction with asphalt particles and may possibly lead to improving the performance of asphalt mastic. The SG of B filler was high compared with RHL and NHL filler, indicating that B filler is denser than the others. All three fillers are approved for use as a typical mineral filler for asphalt mastic preparation (Ministry of Road Transport and Highways (MORTH) 5th revision, 2013, New Delhi, India).

Preparation of Asphalt Mastic Samples

In this study, filler to binder (F/B) ratio was selected as 0.8 (by mass ratio) for all mastic sample preparation, which is the average value of the F/B range of 0.6–1.2 (MORTH, 5th revision, 2013), and having a corresponding volume concentration of 22.5%. Asphalt mastics were produced in the laboratory using AC-30 binder with nine different combinations of basalt and regular sized hydrated lime fillers (B-RHL) and basalt and nano sized hydrated lime fillers (B-NHL). The dosages of RHL and NHL were selected as 0%, 5%, 10%, 15%, and 20% by weight of asphalt binder. The high dosages of RHL and NHL (20%) were chosen based on the equivalent percentage of HL used in the asphalt mix (1% to 2%). The nine different combinations of asphalt mastic are:

- Asphalt mastic with B filler: M80-0 (AC-30 + 80% B + 0% RHL);
- Asphalt mastic with RHL filler: M75-5RHL (AC-30 + 75% B + 5% RHL), M70-10RHL (AC-30 + 70% B + 10% RHL), M65-15RHL (AC-30 + 65% B + 15% RHL), M60-20RHL (AC-30 + 60% B + 20% RHL);
- Asphalt mastic with NHL filler: M75-5NHL (AC-30 + 75% B + 5% NHL), M70-10NHL (AC-30 + 70% B + 10% NHL), M65-15NHL (AC-30 + 65% B + 15% NHL), M60-20NHL (AC-30 + 60% B + 20% NHL).

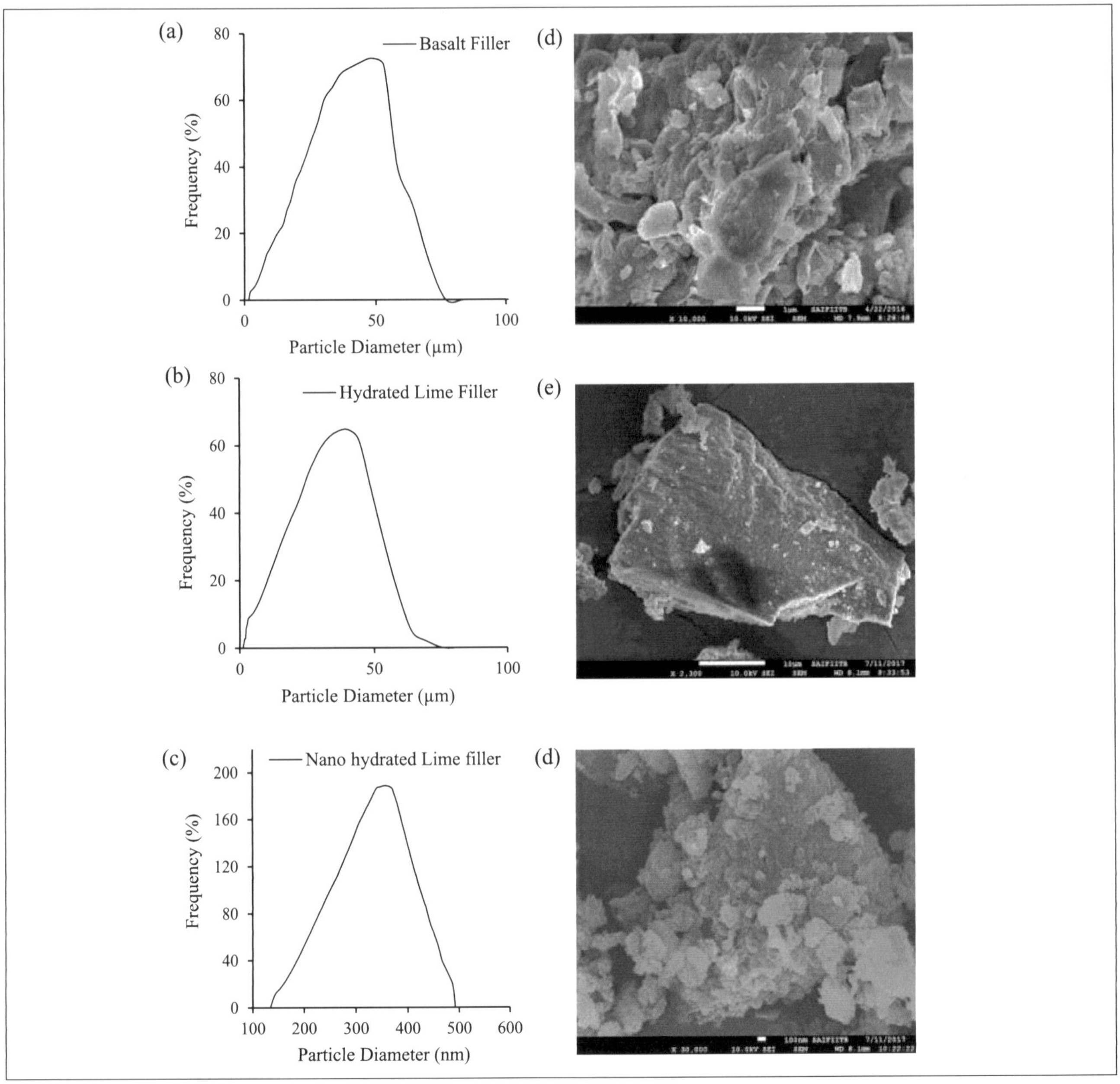

Figure 1. Particle-size distribution of fillers (*a*) B, (*b*) RHL, and (c) NHL, and SEM images of (*d*) B, (e) RHL, and (*f*) NHL.

MXX-YY represents the type of AC-30 mastic for different combinations of B-RHL and B and B-NHL fillers, where: XX is % of B (% by weight of binder) and YY is % of RHL or NHL (% by weight of binder). To first prepare the mastic samples, blends of RHL or NHL with B filler were made using the planetary ball mill for all the combinations. To achieve homogeneity in blending of fillers, the milling was executed by maintaining 250 rpm of speed with a ball to powder ratio of 1:1 for 10 min. Second, AC-30 binder was preheated in the oven for about 30 min at 150 ± 5°C to ensure its proper fluidity. Thereafter, different blends of B-RHL and B-NHL fillers were conditioned in an oven at 150°C to make sure that the fillers reached mixing temperature. Finally, AC-30 and B-RHL/B-NHL fillers were mixed using a high-shear mixer at a speed of 2000 rpm for 1 h at 150°C ± 5°C for the production of a homogenous mastic and to ensure uniform filler distribution in asphalt. Figure 2 shows the flow chart for the experimental program.

Experimental Investigation

Fatigue Performance using LAS Test

The LAS test method is intended to evaluate the resistance of asphaltic materials against cyclic load induced fatigue damage using VECD analysis. The VECD concept has been

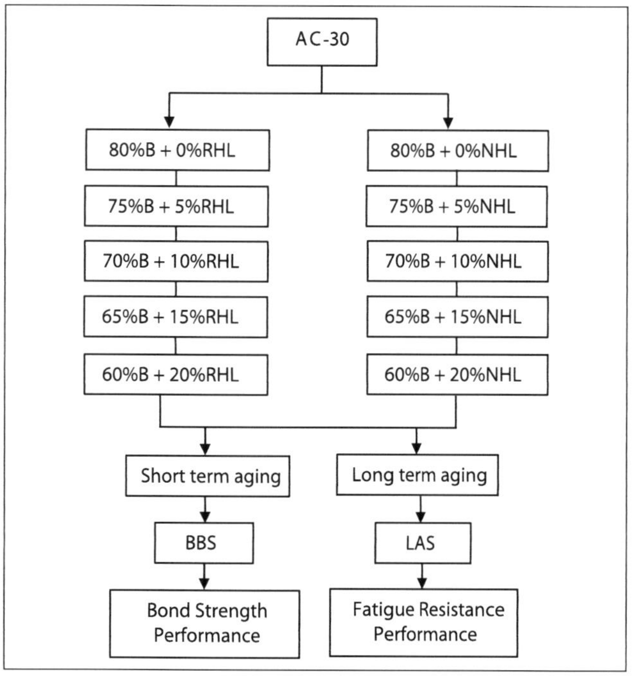

Figure 2. Flow chart for experimental program.

extensively used by various researchers in order to simulate the fatigue damage analysis of binders (*12, 16*). This accelerated test procedure systematically increases the load amplitude and has been reported to be a reliable test methodology for characterizing fatigue damage resistance (*14*). The LAS test is conducted using a dynamic shear rheometer (DSR) with the standard geometry of 8 mm parallel plates and 2 mm thickness gap as per AASHTO-TP-101. The damage accumulation in the sample is calculated using

$$D(t) \cong \sum_{i=1}^{n} \left[\pi \gamma_0^2 \left(C_{i-1} - C_i\right)\right]^{\alpha/1+\alpha} \left(t_i - t_{i-1}\right)^{1/1+\alpha} \quad (1)$$

where:

$C(t) = \frac{\left|G^*\right|(t)}{\left|G^*\right|_{initial}}$ = integrity parameter;

γ_0 = applied strain (%);

G^* = complex shear modulus in MPa;

$\alpha = 1/m$, where m is the slope of logarithmic plot between storage modulus and applied frequency;

t = testing time in seconds.

The relationship between $C(t)$ and $D(t)$ is developed by the curve fitting method using the power law expressed in

$$C(t) = C_0 - C_1 D^{C_2} \quad (2)$$

where C_1 and C_2 are coefficients of curve fitting equations. Further, damage value at failure point is being calculated. The value of $D(t)$ at failure, D_f, is defined as the $D(t)$ which corresponds to the reduction in initial $|G^*|$ at the peak shear stress, using

$$D_f = \left(\frac{C_0 - C\ at\ peak\ stress}{C_1}\right)^{1/C_2} \quad (3)$$

Finally, the number of fatigue cycles (N_f) is evaluated using

$$N_f = A\left(\gamma_{max}\right)^B \quad (4)$$

where A and B are regression parameters, and γ_{max} is the maximum expected binder strain for a given sample. A binder having a high value of N_f indicates superior resistance to fatigue cracking. First, the test procedure includes a strain amplitude of 0.1% for the frequency range from 0.1 to 30 Hz to calculate B, which indicates undamaged material properties. Second, parameter A is estimated based on the VECD concept and the test is conducted at a specified temperature using oscillatory shear loading in the strain-controlled mode of 10-Hz frequency. The samples are subjected to a loading pattern that involves a continuous oscillatory strain sweep. Loading is increased linearly from 1% to 30% over the sequence of 3,100 loading cycles. Peak shear stress, strain, δ (degree), and G^*(Pa) are measured at every 10 load cycles (1 s) (AASHTO-TP-101-14) (*16*). In this study, to access the fatigue damage behavior of asphalt mastics by simulating field conditions, asphalt mastic samples underwent a short-term aging condition using the thin film oven (TFO) method. The LAS test was conducted for short-term aged asphalt mastic containing RHL and NHL at 25°C.

BBS Test

The BBS test can be used for the evaluation of the bond strength of asphalt mastic or asphalt binder at the aggregate interface in dry and wet conditions (*1, 2, 15, 17*). The BBS device is composed of a portable pneumatic adhesion tester, pressure hose, piston, reaction plate and a metal pull-out stub (Figure 3). The evaluation of the BBS test is represented using three parameters as pull-off tensile strength (POTS), percentage loss in POTS (% loss), and mode of failure. The POTS can be calculated using

$$POTS = \frac{\left(BP \times A_g\right) - C}{A_{ps}} \quad (5)$$

where POTS is in kPa, BP is burst pressure in kPa, A_g is the contact area of gasket with reaction plate in mm^2

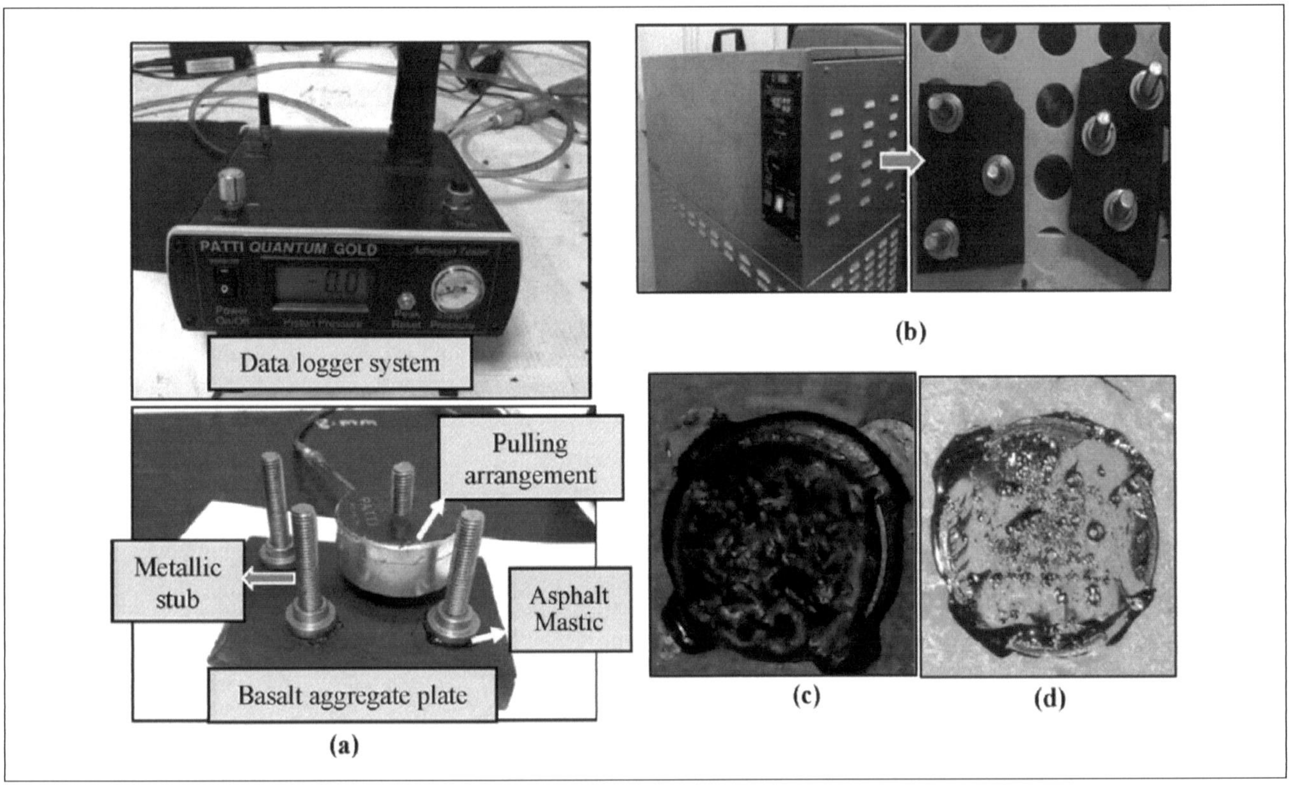

Figure 3. BBS test device: (*a*) BBS test set up with the typical sample arrangement, (*b*) water bath set up for moisture conditioning of sample, (*c*) typical cohesive failure surface, and (*d*) typical adhesive failure surface.

(2.009 mm^2), C is the piston constant (0.25), and A_{ps} is the area of the pull-off stub in square millimeters (0.4869 mm^2). To identify the potential effect of mastic on moisture damage resistance, % loss in bond strength is calculated using

$$\%\,Loss\ in\ bond\ strength = \frac{POTS_{dry} - POTS_{wet}}{POTS_{dry}} \quad (6)$$

where $POTS_{dry}$ is POTS measured in dry conditioning at ambient temperature and $POTS_{wet}$ is measured after conditioning in water for 24 h at 40°C. Finally, the mode of failure can be identified based on the visual inspection of the aggregate surface (Figure 3, *c* and *d*).

The BBS test was conducted on short term aged asphalt mastic specimens containing B-RHL and B-NHL for dry and wet conditions at ambient temperature (Figure 3, *a* and *b*). In this study, basalt aggregate was used for the preparation of aggregate samples, which underwent diamond saw cutting followed by polishing and then ultrasonic cleaning. A smooth surface was required so that friction would not make any contribution to the bond strength of asphalt. Five test replicates of mastic specimens were tested for both dry and wet conditions. The effect of moisture on bond strength of asphalt mastic was simulated by conditioning the whole specimen arrangements (Figure 3*b*) in a water bath at 40°C for 24 h.

Results and Discussion

Fatigue Damage using LAS Test

Influence of RHL and NHL on Fatigue Damage Parameters. Table 2 shows fatigue damage parameters for asphalt mastics containing RHL or NHL, obtained from the LAS test. The C_0 parameter for all asphalt mastics was found to be 1, indicating the highest integrity with no damage. Parameters α, A and τ_{max} were evaluated through damage analysis using the amplitude sweep test and are presented in Table 2. The fatigue damage resistance of asphalt is sensible to the parameters α and A. Parameter α indicates the rate of reduction in fatigue life with increase in strain amplitude (*18*). In addition, parameter A is directly proportional to N_f as per Equation 4. Thus, a lower value of α and a higher value of A are desirable for superior resistance to fatigue damage. A decrease in the α value with an increase in A can be seen for asphalt mastic with increase in RHL or NHL content. For example, asphalt mastic with addition of 5% RHL (i.e., M75-5RHL) and 5% NHL (i.e., M75-5NHL) decreased the α value by 5% and 6%, respectively, compared with base asphalt mastic (0% RHL) (i.e., M80-0). Similarly, the A value for M75-5RHL and M75-5NHL mastic was increased by 13% and 30%, respectively, relative to base asphalt mastic (0% RHL). Moreover, similar trends in α and A values were observed for asphalt mastics containing 10%, 15%, and 20% of RHL or NHL. Overall results imply that considering

Table 2. Fatigue Damage Parameters for Asphalt Mastics (LAS Results)

Mastic	C_0	C_1	C_2	α	A	B	τ_{max} (kPa)
M80-0RHL	1	0.448	0.17	2.538	135582	−4.553	375
M75-5RHL	1	0.364	0.209	2.407	153267	−4.813	481
M70-10RHL	1	0.309	0.257	2.378	188194	−4.755	523
M65-15RHL	1	0.3	0.261	2.344	249674	−5.077	533
M60-20RHL	1	0.37	0.205	2.276	313252	−4.688	551
M75-5NHL	1	0.274	0.264	2.381	177052	−4.356	561
M70-10NHL	1	0.212	0.348	2.256	212714	−4.203	574
M65-15NHL	1	0.292	0.25	2.178	361642	−4.553	593
M60-20NHL	1	0.307	0.257	2.101	567726	−4.88	732

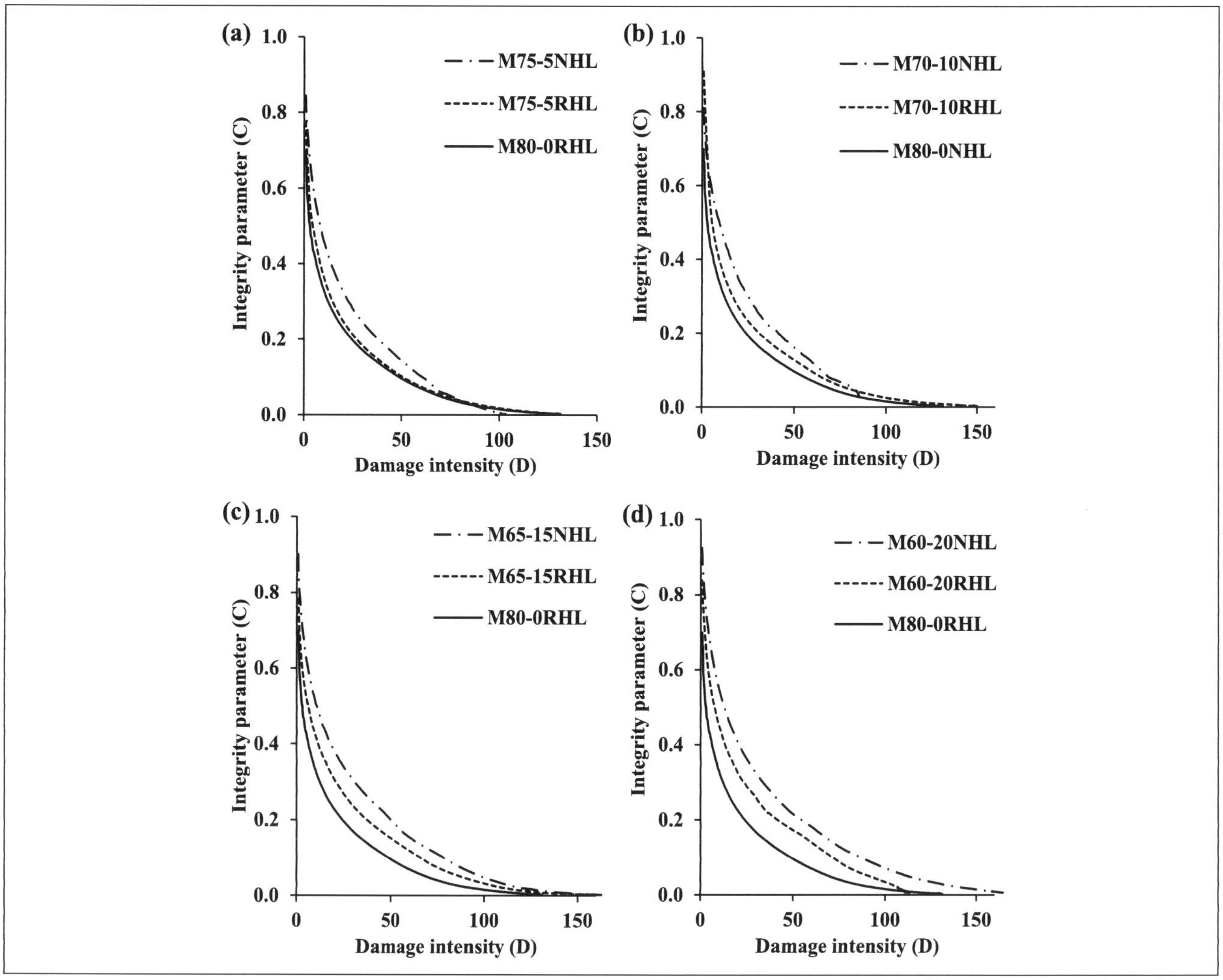

Figure 4. Integrity parameter versus damage intensity for asphalt mastics. Comparison of 0% RHL with (*a*) 5% of RHL and NHL, (*b*) 10% of RHL and NHL, (*c*) 15% of RHL and NHL, and (*d*) 20% of RHL and NHL.

similar amounts of dosage, NHL can contribute more effectively to improving the fatigue resistance of asphalt mastic as compared with RHL.

Likewise, the increase in the τ_{max} value for asphalt mastic was found to be from 481 kPa to 551 kPa with increase in RHL content from 5% to 20%, while increasing the NHL content from 5% to 20% showed an increase in τ_{max} values from 561 kPa to 732 kPa. Thus, the results indicate improved shear stress carrying capacity of asphalt mastic after inclusion of RHL or NHL. However, it can be observed that NHL can perform better to resist early fatigue as compared with RHL. Figure 4, *a–d* show that the rate of decrease of integrity level to asphalt mastic increases with increase in RHL or NHL content. It can also be observed that at any particular level of integrity, the damage to asphalt mastic was found to be highest in case of M60-20NHL (20% NHL) and lowest in case of base asphalt mastic. It has been reported that RHL particles can effectively remove the viscosity building components (resulting from oxidative aging) and thus increase the ability of the asphalt to dissipate stress build up in the asphalt matrix, which helps to increase the fatigue damage resistance (*19*). Hence, better performance in the fatigue behavior of asphalt mastic containing NHL may be thought to take place because of its high surface area and improved interaction with asphalt mastic.

Effect of RHL and NHL on Fatigue Life (N_f). The fatigue life N_f for asphalt mastics containing RHL and NHL was estimated at 2.5% of strain level using Equation 4. Figure 5*a* shows that asphalt mastic exhibited an increase in N_f with an increase in RHL or NHL dosages. It can be critically observed that with the addition of NHL, N_f of asphalt mastic increased by about or over 50% more than that for RHL. Such a finding signifies the potential role of NHL over RHL in improving the fatigue damage resistance of asphalt mastic, which was possibly because of an improvement in antioxidant nature. Moreover, the results indicate the establishment of a well-developed interaction between asphalt mastic and NHL, which possesses high SSA (43.17 m^2/g) and thus helps to reduce deleterious effects of oxidative aging of asphalt mastic (*6*). Figure 5, *b–e* show that, at any strain level, N_f of asphalt mastic increases with an increase in RHL or NHL content. More critically, it can be noticed that the difference in their fatigue life decreased with increases in strain level. A similar trend was observed for both cases of asphalt mastic containing RHL or NHL. Such a response in fatigue life describes the higher strain susceptibility performance of asphalt mastic containing NHL, especially for M60-20NHL (20% NHL) mastic.

BBS Test

Influence of RHL and NHL Fillers on POTS in Dry Condition. Figure 6*a* shows variations in pull-off tensile strength (POTS) for asphalt mastics containing RHL or NHL fillers in dry condition. It can be seen that the POTS value of asphalt mastic increased with an increased percentage of NHL or RHL. For example, the POTS value for asphalt mastic with 5% RHL (i.e., M75-5RHL) was found to be 2.58 MPa, which increased to 2.83 MPa for asphalt mastic with 20% RHL (i.e., M60-20RHL). Similarly, the POTS values for 5% NHL (i.e., M75-5NHL) and 20% NHL (i.e., M60-20NHL) were observed to be 2.64 MPa and 3.03 MPa, respectively. It can be seen that asphalt mastic with NHL showed higher POTS compared with RHL, indicating better bond strength performance of asphalt mastic with NHL, especially in the case of asphalt mastic containing 20% NHL. Furthermore, the rate of increase in POTS was higher for asphalt mastic containing NHL (slope: 3.08, Figure 6*a*) as compared with that of RHL (slope: 2.26, Figure 6*a*). Such an increasing response in POTS value implies that the contribution of NHL is predominant over RHL in enhancing the bond strength of asphalt mastic with the aggregate surface. It is also important to note that with the inclusion of NHL, the high SSA and potential interaction properties may have a key role in improving the bond strength of asphalt mastic to a superior level. Thus, the reduced size of HL can have a significant role in improving the performance of asphalt mastic.

Influence of RHL and NHL on POTS in Wet Condition. Figure 6*b* shows POTS of all combinations of asphalt mastic containing RHL or NHL, tested after 24 h of moisture conditioning (Figure 3*b*). The results showed that POTS values of asphalt mastic increase with an increase in RHL or NHL content in wet condition. The POTS value of asphalt mastic with 5% RHL (M75-5NHL) was found to be 2.21 MPa, which increased to 2.45 MPa after addition of 20% RHL. Similarly, with an increase in NHL content from 5% to 20%, the POTS values increase from 2.29 MPa to 2.55 MPa. It should be noted that, in wet condition, the POTS values of asphalt mastics containing RHL or NHL decreased as compared with those in a dry condition, which may be the result of moisture intrusion at the interface of mastic and aggregate surface. The rate of increase in POTS value for asphalt mastic containing NHL was higher (slope: 2.28, Figure 6*b*) than that for RHL (slope: 1.88, Figure 6*b*). Also, results signify that asphalt mastic with NHL had better bond strength and moisture damage resistance compared to mastic with RHL, especially in the case of 20% NHL. It is believed that RHL particles are involved in a cation exchange mechanism of agglomeration and pozzolanic reactions resulting in an alteration in properties of filler particles, which act as agglomerates and are held together by a pozzolanic cement. Thus, this makes the filler particles less susceptible to moisture by reducing hydrophilic nature (*6*). Hence, the superior performance of NHL in resisting moisture damage may be due to its high surface area and better interaction characteristics, which considerably helped to improve the hydrophobic behavior.

Influence of RHL and NHL on % Loss in Bond Strength. Figure 7 shows % loss in bond strength of asphalt mastics after wet conditioning, determined using Equation 6. The % loss in bond strength for asphalt mastic containing 0%, 5%, 10%, 15%, and 20% RHL was found to be 12.5%, 11%, 9.7%, 10.3%, and 9.6%, respectively. In addition, % loss in bond strength for asphalt mastic with 0%, 5%, 10%, 15%, and 20% NHL was 12.5%, 9.5%, 9.4%, 5.7%, and 4.5%, respectively. It can be seen that % loss in bond strength decreases with increase in RHL content from 0% to 10%; however, a

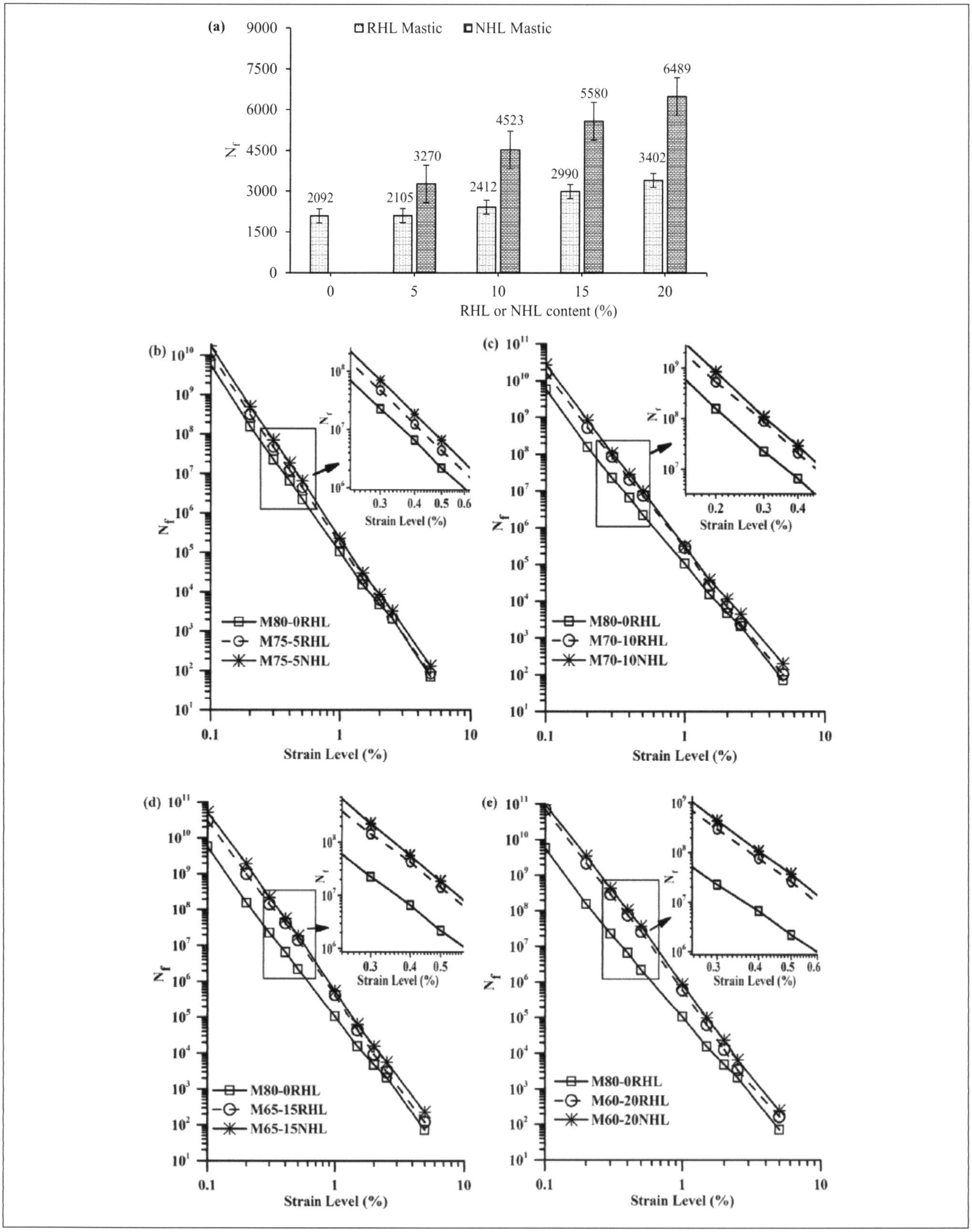

Figure 5. (*a*) Variations in N_f for asphalt mastics containing RHL or NHL. Plot of N_f versus strain level comparison of 0% RHL with (*b*) 5% of RHL and NHL, (*c*) 10% of RHL and NHL, (*d*) 15% of RHL and NHL, and (e) 20% of RHL and NHL.

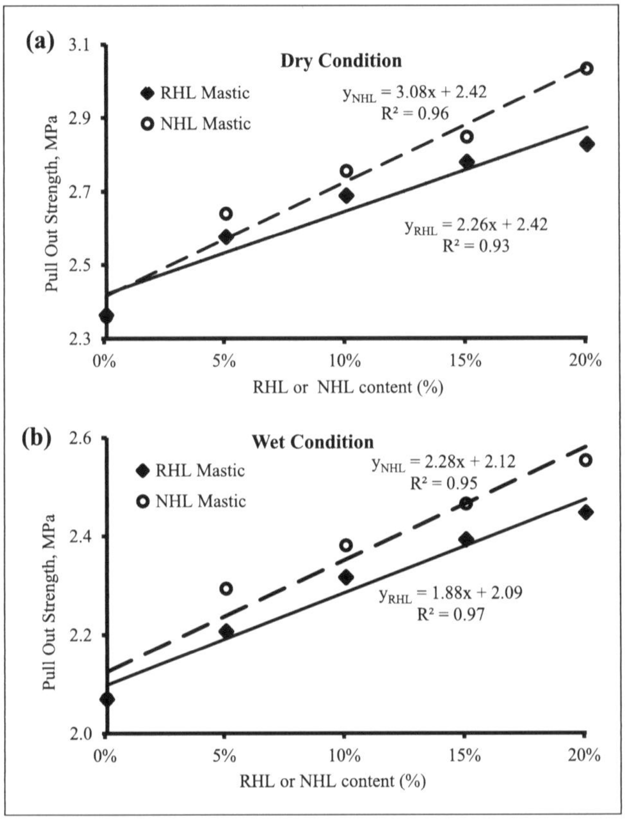

Figure 6. Effect of RHL and NHL on (*a*) POTS in dry condition and (*b*) POTS in wet condition.

further increase in RHL to 15% caused a marginal increase in % loss in bond strength. Such a resulting trend may be due to the difference in the decrease in POTS value of asphalt mastic (15% RHL) after 24 h of wet conditioning. Moreover, the addition of 20% RHL exhibited the lowest % loss in bond strength, which indicates superior resistance to moisture damage.

Likewise, with an increase in NHL content from 0% to 20%, the % loss in bond strength decreases from 12.5% to 4.5%. In this case, the % loss in bond strength for NHL was observed to be much lower than that for RHL (Figure 7*a*). These results showed that mineralogy, surface area, and interaction properties of NHL have a significant influence in improving bond strength and moisture damage resistance of asphalt mastic. However, asphalt mastic containing 20% NHL was observed to be superior in resisting bond failure by reducing the moisture intrusion at the interface of mastic and aggregate surface.

Effects of RHL and NHL on Mode of Failure. Mode of failure surface can only be detected by visual inspection of the aggregate surface (Figure 7*b*). The mode of failure can be characterized in two ways: cohesive and adhesive modes. Diffusion and migration of moisture in asphalt mastic at aggregate interface can cause of mode of failure to adhesion failure (*20*). Such failure can be defined as failure due to influence of moisture The BBS test showed that all asphalt mastics containing RHL and NHL had cohesive failure mode, which indicates that both RHL and NHL filler have the ability to resist moisture diffusion and migration at the interface of asphalt mastic and aggregate surface to some extent.

Conclusions

This laboratory study evaluates fatigue and moisture induced bond strength of AC-30 mastic prepared with various combinations of B and RHL or NHL fillers. The effects of RHL and NHL fillers on fatigue damage behavior of AC-30 mastic were evaluated using an LAS test. The bond strength and moisture damage performance of AC-30 mastic were evaluated using a BBS test. Based on the results, the following conclusions can be drawn.

- LAS test results showed that asphalt mastic containing NHL fillers resulted in better fatigue life as compared with RHL fillers. In contrast, M60-20NHL mastic (20% NHL) had the highest fatigue life, indicating less susceptibility to fatigue damage at intermediate temperature.
- The BBS test (dry condition) revealed that addition of NHL filler can be superior compared to RHL filler in improving bond strength performance of asphalt mastic. Among all combinations of B and RHL or NHL filler used in this study, it can be concluded that the 60% B + 20% NHL (M60-20NHL mastic) made the best contribution to enhancing the bond strength and moisture damage resistance of asphalt mastic.
- The mode of failure surface of all asphalt mastics containing RHL or NHL was cohesive in nature, indicating less susceptibility to moisture diffusion and migration at the interface of asphalt mastic and aggregate surface.
- The overall study implies that the combination of B and NHL filler performed significantly better than the combination of B and RHL filler in enhancing the fatigue life, bond strength, and moisture damage resistance of asphalt mastic.
- Moreover, the response of the asphalt mastics signifies that, both the RHL and NHL filler can improve fatigue, bond strength and moisture damage resistance of asphalt mastics. However, NHL was found to have better potential than RHL, which may be due to its mineralogy, surface area and improved interaction properties and is thought to take place as a result of reduction in particle size of RHL.

Understanding production time and cost of NHL is important for its use on a larger scale. The present study produced NHL in the laboratory, and it was observed that production

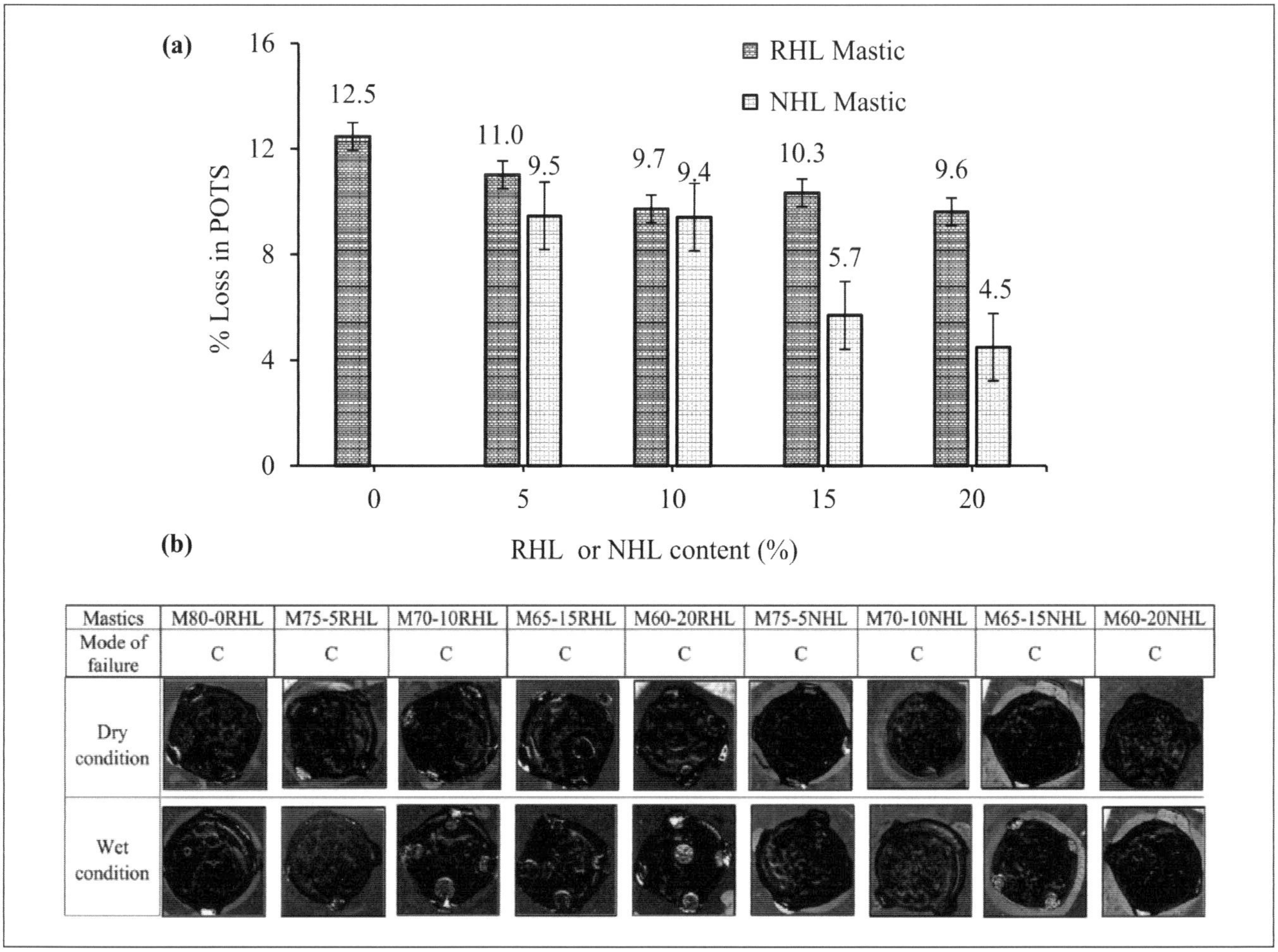

Figure 7. (*a*) Percent loss of bond strength of asphalt mastic after wet conditioning and (*b*) mode of failure surface after testing.

time is significant. However, it is expected that production time of NHL may be reduced once this technology is adopted by industry on a larger scale. Based on the findings of the present study, it was found that lower percentages of NHL filler can be more promising in improving fatigue, bond strength and moisture damage compared to RHL filler. Thus, considering performance benefits and future scenarios, it is expected that the cost of NHL may be either equal to or less than RHL. However, a future study and industry collaboration is required to have realistic data on production cost.

Acknowledgments

The authors acknowledge Prof. Prashant P. Date, Mr. Kanhu Charan Nayak from Department of Mechanical Engineering, IIT Bombay and Dr. Vishal R. Thombare, MCGM, Mumbai for their continuous support and help in collection and preparation of materials.

References

1. Apeagyei, A. K., J. R. A. Grenfell, and G. D. Airey. Moisture-Induced Strength Degradation of Aggregate–Asphalt Mastic Bonds. *Road Materials and Pavement Design*, Vol. 629, No. 15(sup1), 2014, pp. 239–262. https://doi.org/10.1080/14680629.2014.927951.
2. Chaturabong, P., and H. U. Bahia. Effect of Moisture on the Cohesion of Asphalt Mastics and Bonding with Surface of Aggregates. *Road Materials and Pavement Design*, Vol. 19, No. 3, 2016, pp. 741–753. https://doi.org/10.1080/14680629.2016.1267659.
3. Iwanski, M., and G. Mazurek. Hydrated Lime as the Anti-Aging Bitumen Agent. *Procedia Engineering*, Vol. 57, 2013, pp. 424–432. https://doi.org/10.1016/j.proeng.2013.04.055.
4. Cheng,Y.,J.Tao,Y.Jiao,Q.Guo,andC.Li.InfluenceofDiatomite and Mineral Powder on Thermal Oxidative Ageing Properties of Asphalt. *Advances in Materials Science and Engineering*, Vol. 2015, 2015, pp. 1–10. https://doi.org/10.1155/2015/947834.
5. Johansson, L. S. *Bitumen Ageing and Hydrated Lime.* Doctorial thesis. Institutionen för infrastruktur och samhallsplanering, 1998.
6. Little, D. N., J. A. Epps, and P. E. Sebaaly. *The Benefits of Hydrated Lime in Hot Mix Asphalt.* National Lime Association, 2006, pp. 1–79. https://doi.org/10.1.1.508.3233&rep=rep1&type=pdf.
7. Das, A. K., and D. Singh. Investigation of Rutting, Fracture and Thermal Cracking Behavior of Asphalt Mastic Containing

Basalt and Hydrated Lime Fillers. *Construction and Building Materials*, Vol. 141, 2017, pp. 442–452. https://doi.org/10.1016/j.conbuildmat.2017.03.032.
8. Verhasselt, A., and D. Puiatti. Effect of Hydrated Lime on Ageing Behaviour of Bituminous Mastics. *Proc., 3rd Eurasphalt and Eurobitume Congress, Foundation Eurasphalt*. Vol. 1, 2004, pp. 746–756.
9. Shen, J., Z. Xie, F. Xiao, and W. Fan. Evaluations of Nano-Sized Hydrated Lime on the Moisture Susceptibility of Hot Mix Asphalt Mixtures. *Applied Mechanics and Materials*, Vol. 174–177, 2012, pp. 82–90. https://doi.org/10.4028/www.scientific.net/AMM.174-177.82.
10. Diab, A., Z. You, and H. Wang. Rheological Evaluation of Foamed WMA Modified with Nano Hydrated Lime. *Procedia - Social and Behavioral Sciences*, Vol. 96, 2013, pp. 2858–2866. https://doi.org/10.1016/j.sbspro.2013.08.318.
11. Antonietti, M., and G. A. Ozin. Promises and Problems of Mesoscale Materials Chemistry or Why Meso? *Chemistry - A European Journal*, Vol. 10, No. 1, 2004, pp. 28–41. https://doi.org/10.1002/chem.200305009.
12. Kavussi, A., and P. Barghabany. Investigating Fatigue Behavior of Nanoclay and Nano Hydrated Lime Modified Bitumen Using LAS Test. *Journal of Materials in Civil Engineering*, Vol. 28, No. 3, 2016, pp. 1–7. https://doi.org/10.1061/(ASCE)MT.1943-5533.0001376.
13. Diab, A., and Z. You. Rheological Characteristics of Nano-Sized Hydrated Lime-Modified Foamed Warm Mix Asphalt. *Pavement Materials, Structures, and Performance*, 2014, pp. 79–89. https://doi.org/10.1061/9780784413418.009.
14. Martono, W., and H. Bahia. *Developing a Surrogate Test for Fatigue of Asphalt Binders*. Presented at 87th Annual Meeting of the Transportation Research Board, Washington, D.C., 2008.
15. Moraes, R., R. Velasquez, and H. U. Bahia. Measuring the Effect of Moisture on Asphalt-Aggregate Bond with the Bitumen Bond Strength Test. *Transportation Research Record: Journal of the Transportation Research Board*, 2011. 2209: 70–81..
16. Hintz, C., R. Velasquez, R. Associate, C. Johnson, S. Asphalt, and H. Bahia. Modificaton and Validation of the Linear Amplitude Sweep Test for Binder Fatigue Specification. *Transportation Research Record: Journal of the Transportation Research Board*, 2010. 2207: 99–106.
17. Zhang, J., G. D. Airey, and J. R. A. Grenfell. Experimental Evaluation of Cohesive and Adhesive Bond Strength and Fracture Energy of Bitumen-Aggregate Systems. *Materials and Structures*, Vol. 49, No. 7, 2016, pp. 2653–2667. https://doi.org/10.1617/s11527-015-0674-7.
18. Saboo, N., and P. Kumar. Performance Characterization of Polymer Modified Asphalt Binders and Mixes. *Advances in Civil Engineering*, Vol. 2016, 2016, pp. 1–12. https://doi.org/10.1155/2016/5938270.
19. Little, D. N., and J. C. Petersen. Unique Effects of Hydrated Lime Filler on the Performance-Related Properties of Asphalt Cements: Physical and Chemical Interactions Revisited. *Journal of Materials in Civil Engineering*, Vol. 17, No. 2, 2005, pp. 207–218. https://doi.org/10.1061/(ASCE)0899-1561(2005)17:2(207).
20. Cheng, D., D. Little, R. Lytton, and J. Holste. Surface Energy Measurement of Asphalt and Its Application to Predicting Fatigue and Healing in Asphalt Mixtures. *Transportation Research Record: Journal of the Transportation Research Board*, 2002. 1810: 44–53.

The Standing Committee on Surface Requirements of Asphalt Mixtures (AFK40) peer-reviewed this paper (18-03836).

Article

TRR
JOURNAL OF THE TRANSPORTATION RESEARCH BOARD

Transportation Research Record
2018, Vol. 2672(28) 42–51

DOI: 10.1177/0361198118787975
journals.sagepub.com/home/trr
SAGE

Effect of Synthetic Fiber State on Mechanical Performance of Fiber Reinforced Asphalt Concrete

Hossein Noorvand[1], Ramadan Salim[1], Jose Medina[1], Jeffrey Stempihar[1], and B. Shane Underwood[1]

Abstract

It has been recognized that there exists a potential benefit from using synthetic fibers to reinforce asphalt mixtures. In these mixtures, the state of the fibers may play an essential role in their reinforcement function. This study aims to quantify the state of synthetic fiber distribution for two different aramid fiber–based asphalt mixtures and then show the impacts of fiber dispersion on modulus, rutting, and fatigue performance of each asphalt mixture in comparison with one another and with respect to an equivalent non-reinforced asphalt mixture. Both a quantitative and qualitative assessment of aramid fibers distribution as well as state of fiber are investigated using a fiber extraction procedure and microscopy imaging, respectively. The results suggested that a higher level of micro-fibrillation as well as high distribution of aramid fibers improved the rutting resistance of asphalt mixtures, while the distribution level of aramid fibers and fibers state did not affect the modulus and fatigue. These results are specific to the mixture studied but provide the first objective and detailed study describing fiber state, fiber dispersion, and performance.

The use of reinforcing fibers in asphalt concrete (AC) has been documented in many studies (*1, 2, and the citations therein*). Some of these studies have shown that the addition of fiber in dense graded and open graded AC mixtures can change the dynamic modulus and phase angle of the mixture (*3*), relaxation ability and rutting resistance (*4–6*), and resistance to low-temperature cracking, fatigue cracking, and reflective cracking (*6–9*). More than half of state agencies in the United States use fibers in AC mixtures (*1*), but most use non-synthetic fibers to reduce drain down in open graded and stone mastic asphalt mixtures. Fibers commonly used for this purpose include cellulose, mineral fibers, and other plant-based materials. Synthetic polymeric fibers (i.e., polypropylene, polyester, aramid, and nylon), glass fibers, and recycled fibers are also used for the purposes of providing mechanical reinforcement and improving rutting and cracking in AC, referred to generically as mechanically fiber reinforced asphalt concrete (M-FRAC) (*1*).

Most studies in this area attempt to experimentally observe the effect of different fiber types on traditional mechanical measures. Previous studies have used indirect tensile tests, pull out tests, repeated load permanent deformation tests, triaxial shear strength tests, and uniaxial fatigue tests to quantify the benefits of synthetic fibers (*3, 6–8*). While published studies generally demonstrate positive benefits from using M-FRAC, it is not always the case, and when improvements are found they vary with respect to the types of fibers and the particular study (*10, 11*).

One possible reason for this inconsistency may be a lack of understanding in the basic, fundamental mechanisms that govern the mechanical response of these mixtures and the effect of fiber state with respect to these mechanisms. The essential role of synthetic fibers in M-FRAC is to act as reinforcement elements to enhance the tensile properties of the mixture, and the mechanisms that dictate their effectiveness are how easily the fibers pull out from the asphalt mastic and how easily the fibers break (*12*). Both mechanisms, and their ultimate effect on the composite performance are related to the specific state of fibers in the mixtures. The fiber state, inclusive of how the fibers are dispersed, where the fibers are located, and surface morphology directly affect these mechanisms. These effects have been recognized in previous studies, but little to no systematic study has been done to attempt to quantify and correlate the effect of fiber state to the performance improvements.

[1]Ira A. Fulton Schools of Engineering, School of Sustainable Engineering and the Built Environment, Arizona State University, Tempe, AZ

Corresponding Author:
Address correspondence to B. Shane Underwood:
shane.underwood@ncsu.edu

Objective

The specific objectives of this study are to measure the mechanical properties of two M-FRAC specimens with synthetic aramid fibers in different states, then to quantify the state of aramid fiber distribution in AC, and finally to examine the impacts of aramid fiber dispersion on laboratory modulus, rutting, and fatigue performance of asphalt mixtures.

Methodology

The experimental approach involved first conducting mechanical tests to evaluate the response and laboratory performance of M-FRAC mixtures and then evaluating the dispersion, distribution, and morphology of fibers.

Materials

A 12.5 mm Marshall mixture (PG 70-10, 5.2% binder) from Tucson, Arizona was used for this study. A total of three variations to this plant-produced mixture were used in this study: a non-reinforced mixture (control), a M-FRAC mixture with overall good dispersion of synthetic fiber (FA), and a M-FRAC mixture with overall poor dispersion of synthetic fiber (FB). Both M-FRAC mixtures contained synthetic aramid fibers at a dosage of 65.5 g per tonne of mix. The aramid fibers were 19 mm long and were introduced in bundles of approximately 3,000 individual fibers each. The fibers have a reported tensile strength between 2,700 and 3,000 MPa and are not affected by temperatures up to 426°C.

In the field, both fiber mixtures were added directly into the hot aggregates in the drum just before the application of hot binder at the dust port. Both manufacturers' products were added with the addition of forced air. Also, the fibers were mixed with hot aggregates for a few seconds before adding the hot binder. Differences in dispersion and distribution are achieved due to the use of different technologies in FA and FB. In FA, the aramid fibers are introduced along with polyolefin fibers while in FB the aramid fibers are coated in a thin membrane of wax, which melts as the fibers are mixed in. Theoretically, both FA and FB try to deliver fibers into the mixing process with high aspect ratios (ratio of diameter to length) so that early in the mixing process they distribute better and do not interfere with the process. Both technologies then rely on the fibers separating and dispersing throughout the mixture in a low aspect ratio state so that more volume of the mixture can be reinforced. In FA, the polyolefin fibers facilitate this transition by first carrying aramid bundles throughout the mixture. Then, as the polyolefin fibers melt they change shape to help pull aramids away from the bundles and also act to neutralize static charges that can cause aramids to agglomerate into cotton-ball-like structures. Aramids may also be dispersed by frictional forces between the aramid fibers and the aggregate. In FB, the wax keeps the aramids bundled until it melts, at which point the aramids are dispersed through frictional forces between the fibers and the aggregate. It is not known if the wax completely melts into the asphalt mixture or if as it melts it seeps deeper into the bundle so that dispersing the fibers requires continual agitation.

At the laboratory, the mixtures removed from the 19 L metal pail as received by heating them at 135°C for 90 minutes, and then samples were randomly taken from multiple pails. Also, representative samples from each mix were selected to perform extraction to check the equivalency of the gradation and binder content of the mixes so that any differences in the performance of the mixes could be attributed to differences in the fiber state. On average, the difference in percent passing on sieve sizes greater than 4.75 mm and less than 4.75 mm between the mixes was less than 2% and 0.5%, respectively. The binder content and the theoretical maximum specific gravities (Gmm) from the extraction were consistent across all three mixtures. Also, the Gmm of the mixtures was within the difference of two samples limit stated in AASHTO T209. Based on the results, the three mixtures appear to be consistent, with the exception that FA and FB mixtures contain fibers.

Mechanical Characterization

In total, three different mechanical experiments were conducted: dynamic modulus test, repeated load permanent deformation test, and uniaxial fatigue test. The target air void level for all tests was 5.5 ± 0.5%, which was the target air void content of the overlay surface course for the project. All specimens were compacted with the gyratory compactor, 150 mm in diameter and 170 mm in height. Specimens were then cut and cored to their final geometry.

Dynamic Modulus Test. Temperature and frequency sweep tests were performed per AASHTO T342 protocol except that the temperatures tested were 4.4, 21.1, and 37.8°C. Tests were performed on cylindrical specimens, 100 mm in diameter and 150 mm in height, and before testing they were instrumented with three linear variable displacement transducers (LVDTs) spaced at 120° intervals on the specimen surface. The readings from these LVDTs along with the applied forces were used to compute dynamic modulus, $|E^*|$, and phase angle, δ, at each test condition.

Repeated Load Permanent Deformation Test. The repeated load permanent deformation, also known as flow number test, was performed to evaluate the permanent deformation characteristics of the asphalt mixtures. The test procedure followed that specified in AASHTO TP79. The test was conducted under atmospheric conditions, a stress level of 160 kPa, and at 50°C. Three replicates were tested for each mixture. Tests were carried out on cylindrical specimens, 100 mm in diameter and 150 mm in height. The deformations were measured using LVDTs mounted to the surface of the sample.

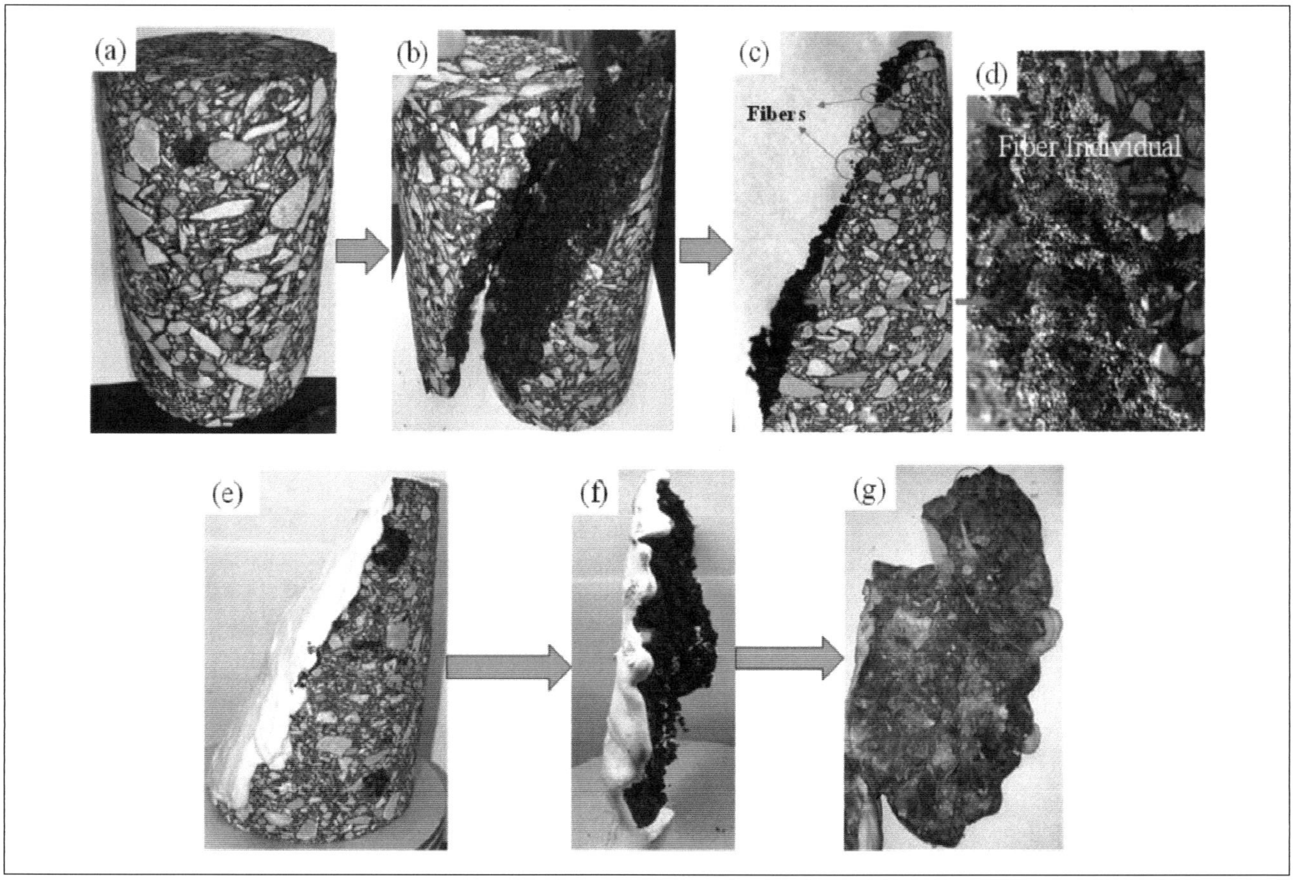

Figure 1. Summary of casting method to evaluate in-situ fiber state: (*a*) tested specimen, (*b*) splitting specimen along the diagonal, (*c*) half splitting specimen, (*d*) zoomed in photo of region identified in (*c*), (e) casting on split specimen, (*f*) separated asphalt mixtures from cast specimen, and (*g*) complete separation of asphalt mixture from cast specimen.

Uniaxial Fatigue Test. The uniaxial fatigue test was performed per AASHTO TP107 with the exception that the test specimens were 75 mm in diameter and 150 mm in height. A sinusoidal deformation along the long axis of a cylindrical test specimen is repeated until it fails. Tests were conducted in controlled actuator displacement mode at 10 Hz loading and 18°C. On-specimen measurements of displacement were recorded along with the applied load throughout the test. As per AASHTO TP107, the result of the test is a damage characteristic function used to predict the fatigue life of AC (*13*).

Fiber State

The state of the fibers is described based on three parameters: dispersion, distribution, and surface morphology. Dispersion refers to the process whereby the fiber bundles added to the mixture are dispersed in the mixture. Distribution represents how the dispersed fibers are spatially located through the mixture. Surface morphology characterizes the form and shape of fibers present in the mix.

Fiber Extraction. To evaluate dispersion, aramid fibers were extracted using a procedure developed at Arizona State University (ASU) (*14*), which is essentially the same as ASTM D2172 except that attention is given to observing, separating, and cleaning the fibers in the mixture after soaking the mixture and centrifuging off the solvent. The centrifugal extraction process has two limitations: (1) it can alter the state of the fibers, but studies at ASU (not shown here) reveal that the effect is generally negligible and does not alter the characteristics to a quantifiable amount and (2) separating and cleaning all individual fibers is tedious and time consuming, and in practice a grouping of 3–10 individual fibers (defined as a very small cluster) may be counted as individuals instead of clusters. The ASU test method describes the state of dispersed aramid fibers in AC as one of four conditions: bundles, agitated bundles, clusters, and individuals. A bundle is the condition of the fibers as supplied by the manufacturer (before introduction to AC). Their existence in the final mixture is undesirable as it indicates zero dispersion of the fibers. With only a little agitation, these bundles disperse some of the fibers, but the resulting agitated bundles may still be visibly similar to the original bundles. With sufficient agitation, the fibers will individually disperse throughout the mixture. These individual fibers are the most desirable condition since their existence suggests the formation of a

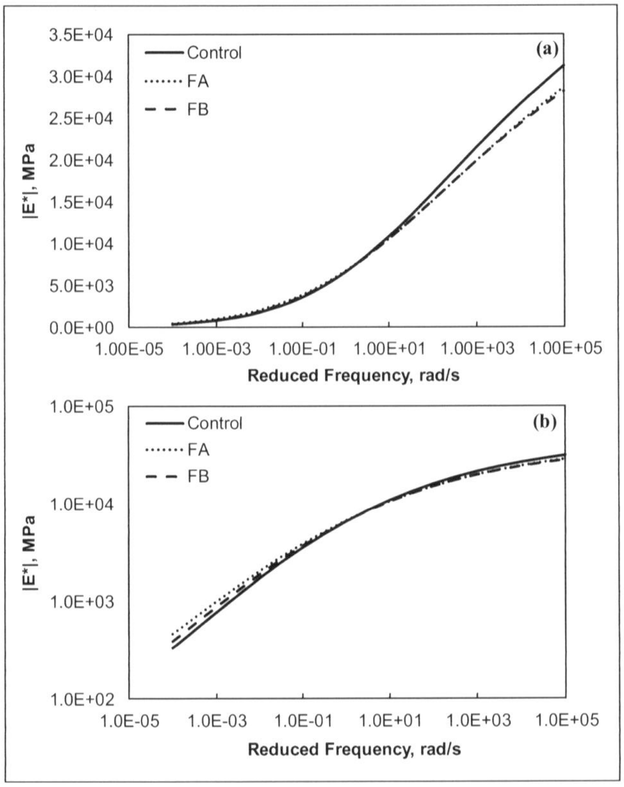

Figure 2. Comparison of $|E^*|$ master curves in: (*a*) semi-log scale and (*b*) log-log scale.

fibrous network throughout the mixture. As the condition transitions between agitated bundles and individuals, the fibers may form loosely connected clusters that provide a dense network in a relatively small volume of material.

After the asphalt binder is extracted, the fibers are separated into the aforementioned categories and the total aramid dosage rate is determined along with the relative proportions of fibers in each category. Of principal interest is the proportion of fibers in the individual state and Equation 1 defines the ratio that quantifies this proportion, known as aramid dispersion state ratio (*ADSR*). A larger *ADSR* signifies the existence of more fibers in the individual state. Supplemental calculations for fibers in the cluster and agitated bundle states are also performed. A test result is reported based on the average of two individual centrifugal extractions.

$$ADSR = F_I = \frac{M_I}{M_a} \times 100 \qquad (1)$$

where;
F_I = fibers in the individual state (%),
M_I = fibers in the individual state (g), and
M_a = total extracted mass before separation (g).

Failure Surface Casting. Castings of the failure surface of tested specimens were made to observe the distribution of the aramid fibers throughout the asphalt mixture. The process is performed on mechanically tested specimens shown in Figure 1*a*. First, mechanical test samples are heated to 60°C for 30 minutes and split along a diagonal line extending from the lower edge of the specimen to the upper opposite edge so that fiber orientations could be viewed with respect to both the horizontal and vertical directions, Figure 1*b*. Although the different dispersion states are evident on the surface of the split samples, Figure 1*c* and *d*, detailed comparisons are not possible because the fibers are coated with asphalt and hard to see. So, a thin layer of fluid plaster is poured on the fracture surface of samples and embed the surface fibers. Once the casting cures, Figure 1*e*, the samples are heated to 80°C to carefully separate as much mixture from the cast as possible, Figure 1*f*. Then the remaining samples are soaked in solvent to dissolve the asphalt and completely remove all remnants of the test specimen from the casting. Finally, photos are taken of the surface and evaluated, Figure 1*g*.

Microscopy Imaging. Original and extracted fibers were imaged using a Philips SEM-XL30 Environmental FEG scanning electron microscope (SEM) under the following conditions: environmental mode, secondary electron signal, and accelerating voltage of 15 kV. Fibers were also sputter coated with gold for 2 minutes to improve the electrical conductivity and resolution of the images.

Results

Mechanical Characterization

Dynamic Modulus Test. Data from the $|E^*|$ experiment is presented in the form of dynamic modulus master curves in Figure 2. The dynamic modulus data are shown in both log-log and semi-log scales so that any differences in the moduli at high and low temperatures, respectively, can be observed. It is seen that the moduli values for all three mixtures are very similar at the intermediate temperature. However, at lower temperatures (higher reduced frequencies) the mixes with aramid fibers yield lower moduli values and at higher temperature (lower reduced frequencies) they yield higher moduli. At these higher temperatures, the FA mix exhibits slightly higher average values than the FB mix. ANOVA and statistical *t*-test analyses were performed at a 95% confidence level. The ANOVA analysis was used to compare the difference between all means and evaluate the significance of the total differences and the *t*-test was performed to compare individual cases to the control mixture and individually between the fiber mixes. In all cases, the results indicated a lack of statistical significance in the moduli. Thus, while on-average differences are evident these effects are not significant.

Repeated Load Permanent Deformation Test. Figure 3 shows the accumulated permanent strain during the repeated load

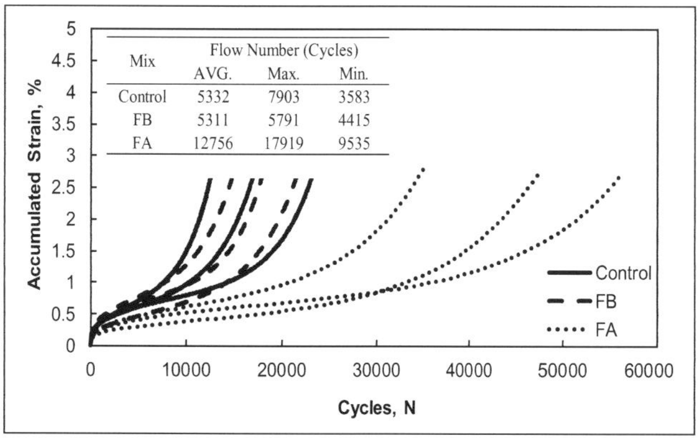

Figure 3. Accumulated strain during flow number test.

permanent deformation test for each of the mixtures. The resulting flow number values from this analysis are also shown as a tabular inset into this figure. Figure 3 shows that the FB and control mixture exhibit similar permanent strain response under the repeated loading, but that the FA mixture accumulates strain at a slower rate and undergoes more repetitions before reaching the onset of flow. In fact, the FA mixture exhibits an increase in average flow number of 139% compared with the control and FB aramid mixtures.

Axial Fatigue Test. The damage characteristic curve for the samples tested in this task is shown in Figure 4*a* for control, FA, and FB. These curves show that mixes with fibers have damage curves that are very similar to those of the control mixtures. These damage curves are fit to the analytical expression shown in Equation 2 and then used with Equation 3 to predict the fatigue life of the mixtures. More details on the analytical process to characterize the damage functions and to predict the number of cycles to failure can be found elsewhere (*13*). Figure 4*b* shows the failure envelopes for each mix type for constant strain–based analysis, which shows that the fiber reinforced samples have approximately twice the fatigue life at a fixed strain level with the FA mixture being slightly better than FB.

$$C = 1 - C_1 S^{C_2} \tag{2}$$

$$N_f = \frac{(f_{red})(2^{\alpha})\left(\frac{(1-C_f)}{C_1}\right)^{\frac{(\alpha-\alpha C_2+1)}{C_2}}}{(\alpha-\alpha C_2+1)(C_1C_2)^{\alpha}\left[(\varepsilon_{0,ppp})(|E^*|)\frac{\beta+1}{2}\right]^{2\alpha} K_1} \tag{3}$$

where;

C = pseudostiffness,
S = internal state variable to quantify damage,
C_1, C_2 = fitting coefficients,
N_f = predicted number of cycles to failure,
f_{red} = reduced frequency for the simulation condition, and
α = damage accumulation rate,
C_f = stiffness ratio at failure,
$\varepsilon_{0,pp}$ = peak-to-peak strain amplitude,
$|E^*|$ = dynamic modulus at the simulation condition,
β = wave shape factor (taken as 0 for this study), and
K_1 = loading shape factor (*13*).

Fiber State

Quantity and Condition of Aramid Fiber after Extraction 41. Table 1 summarizes the extraction results first in terms of the aramid dosage rate, which is simply the mass of the aramid fiber with respect to the total weight of the mixture. Both FA and FB contained similar dosage levels (within approximately 8% of one another) and were within 1% to 5% of the target rates (recall that the target fiber dosage rate was 65.5 g/tonne). The difference between replicates was higher in the FB mixture than the FA mixture. Also, the percentages of each of fiber states (*ADSR*, agitated bundles, and clusters)

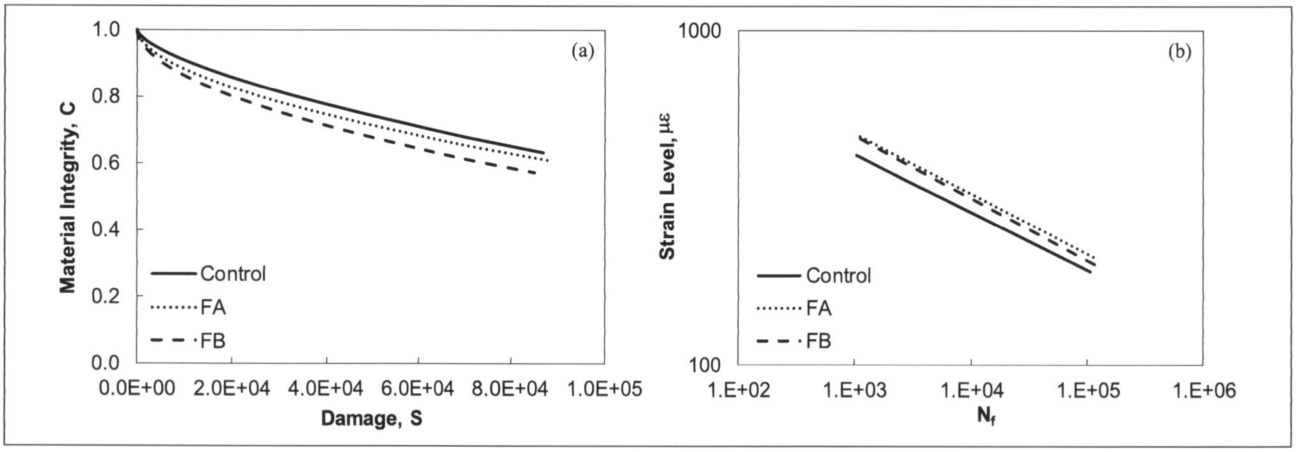

Figure 4. Results of fatigue evaluation: (*a*) damage characteristic curve and (*b*) predicted fatigue life relationships.

Table 1. Summary of Extracted Aramid Dosage and Fiber State

Specimen ID	Aramid dosage rate (g/tonne)/(oz/ton)	State of aramid (%)		
		Agitated bundle	Cluster	ADSR (individual)
FA-1	68.7/2.2	2.0	8.0	90.0
FA-2	65.6/2.1	0.0	13.0	87.0
Average FA	67.1/2.15	1.0	10.5	88.5
FB-1	50/1.6	69.0	12.0	19.0
FB-2	96.8/3.1	78.0	9.0	13.0
Average FA	71.8/2.3	73.5	10.5	16.0

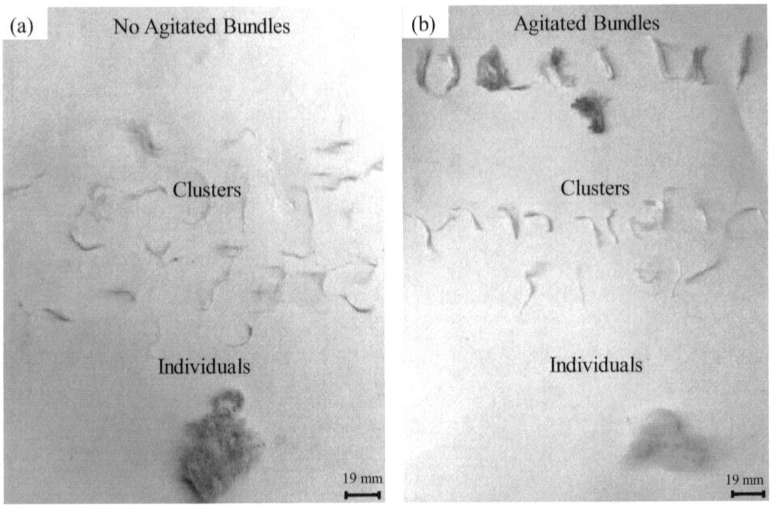

Figure 5. State of extracted aramid fibers from a single sample of (*a*) FA and (*b*) FB.

were determined. The FA mixtures had an average *ADSR* value of 88.5% while FB mixtures had an average *ADSR* of 16%. In the FB case, most of the fibers are dispersed in the agitated bundle condition (73.5% on average). These findings provide some support for the variation observed in the dosage rates of FB replicates because agitated bundles can lead to larger differences in sample-to-sample aramid mass since one or two additional or fewer bundles can alter the mass substantially.

Figure 5 also shows the clear difference between state of the fibers in the FA and FB mixtures after extraction. Figure 5*a* shows that only individuals and clusters exist in FA, and Figure 5*b* shows the presence of agitated bundles in FB and a much lower proportion of individuals than FA. It is important to mention that for the individual fibers the extraction and cleaning process result in the cotton-ball-like agglomeration as seen in these figures. This structure in no way represents how the individual fibers exist in the mixture.

Qualification of Distribution of Aramid Fibers. Images of the castings from the internal diagonal planes of both FA and FB mixtures are shown in Figures 6 and 7. A best effort was made to clearly capture fibers in each photo, and for further clarity of the fibers' state and location, the images were modified with stars, circles, and squares to show locations of fibers in the form of individuals, clusters, and bundles, respectively. A part of the photos in Figures 6*a* and 7*a* was enlarged to more clearly show the existence of individual fibers, as seen in Figures 6*b* and 7*b*. Overall, as the *ADSR* values suggest, a substantially greater number of individual fibers was observed in the casting of the FA mixture (Figure 6). Conversely, the casting of FB yielded a considerably greater number of agitated bundles (Figure 7). Additionally, it was recognized that individual fibers were evenly distributed throughout both FA and FB and they were oriented in all directions (horizontal, diagonal, and vertical). In this study, no attempt was made to quantify the relative proportion of individual fibers oriented in each direction, but it was observed that a larger proportion of the individual fibers were oriented with their long axis perpendicular to the compaction and testing direction.

With respect to distribution and orientation of bundles and clusters, it was observed that clusters and bundles were mostly distributed and located toward the edges of samples

Figure 6. Photos taken from the aramid fibers distributed on the surface fracture of each FA mix: (*a*) overall view of casting surface and (*b*) enlarged photo of central section.

Figure 7. Photos taken from the aramid fibers distributed on the surface fracture of each FB mix: (*a*) overall view of casting surface and (*b*) enlarged photo of central section.

and oriented either horizontally or vertically, unlike individual fibers that tend to be oriented mostly horizontally. It should be noted that the location of clusters and bundles toward the edges of the samples observed in this study might not necessarily be interpreted as the distribution behavior of clusters or bundles. Rather it could have been a random distribution of such fibers that happen to be located at the edges.

Fiber Surface Morphology. Representative results from the multiple images taken of aramid fibers before and after mixing are shown in Figures 8 and 9. Aramid are highly oriented aromatic polyamides, which presents challenges in creating fibers. The process of wet-spinning is typically adopted and yields a fiber with very high strength along the axis of the fiber, but made up of many layers bonded together with relatively low strength (*15*). As a result, during agitation the surface of the aramid fibers may fibrillate into strands that effectively increase the bonding area of the fiber resulting in increase of pull-out resistance.

Figures 8*c* and 9*b* show the surfaces of the clustered fibers and again they appear much the same with only slightly greater surface fibrillation. However, the images from individual condition showed a different morphology, Figures 8*d* and 9*c*. In these fibers, micro-fibrillation is apparent. Although it is not quantified, it also appears that the individual fibers from FA are more fibrillated than FB. This effect may occur because the fibers disperse more quickly in FA and are therefore agitated in the individual condition for a longer time compared with the wax coating on FB. The significance of this effect is that not all individual fibers have the same state even though they may be dispersed and distributed in the same way.

Figure 8*a* shows the fibers in their original condition and it is seen that each fiber is cylindrical and 10–14 μm long in diameter. Most of the fibers are smooth, with a few occasional micro-fibrillations (circled in red). These fibrillations are approximately 0.5–2 μm wide and 35–150 nm thick. In Figures 8*b*–*d* and 9*a*–*c*, small dots appear on the surface of the fibers, which are leftover dust from the extraction and cleaning process. Figures 8*b* and 9*a* show the surface of the agitated bundles, and it is seen that the fibers retain much of the same morphology as the unagitated bundles.

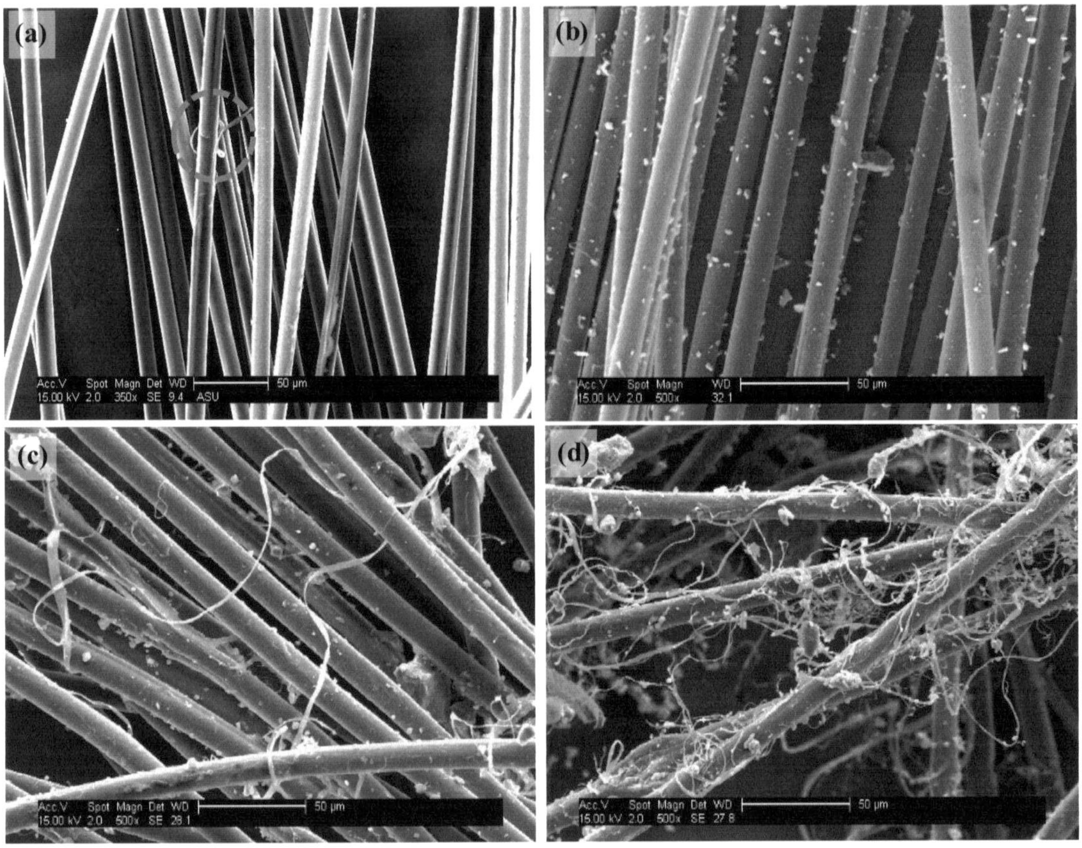

Figure 8. SEM images of aramid fibers of FA mix at different conditions: (*a*) original bundles (unagitated), (*b*) extracted agitated bundles, (*c*) extracted clusters, and (*d*) extracted individual fibers (all images at 500× magnification).

Discussion

In this study, the repeated load permanent deformation test showed improved performance in FA mixture over FB and control mixtures. Since both FA and FB mixtures contained aramid fibers at the same volumetric content but yielded different behaviors, it can be inferred that volumetric content alone is not a sufficiently descriptive metric to quantify fibers in M-FRAC. The fact that FA exhibits substantially less permanent strain accumulation and has a larger flow number, along with the fact that the fibers in FA were more dispersed, distributed, and fibrillated, may suggest that additional fiber metrics play a role in the fibers' effectiveness. From the improved permanent deformation test results and in light of the orientation findings, it is theorized that the fibers function, as they do in other composites, by bridging or reinforcing cracks. In this case horizontal cracks are developed during the flow number test. In this theory, it is postulated that a more dispersed and distributed fiber system acts to reinforce more flaws and thus more greatly improve the performance. It is also possible that the effectiveness of the fibers in bridging cracks is aided by micro-fibrillation of the fiber surfaces, although this is not verified in any way here. A higher level of this micro-fibrillation would mean a higher surface area of fibers in the matrix and a stronger overall bond.

Additional insights are gained by considering the fatigue tests. Here, both FA and FB showed improvement over the control, but there was no discernable difference between the two M-FRAC mixtures. The combined fatigue and permanent deformation behaviors are attributed to the same phenomenon, which is that the individual aramid fibers in these mixtures tended to orient perpendicular to compaction. In this configuration, the fibers are ideally oriented to reinforce cracks that form parallel to the loading direction (as in compression loading), but are not ideally oriented for reinforcing cracks developing perpendicular to the loading direction (as in tension loading). Since the agitated bundles are not well distributed, their effectiveness is reduced. While they may reinforce cracks, they are only located in a few places and so the overall role of this reinforcement on the continuum is apparently small. Stated differently, a well-distributed network of fiber reinforcement in the proper directions is necessary since failure can occur on any plane. A large volume of the material left unreinforced can dictate the overall performance of the material. This study does not identify a critical threshold of individual fibers needed to achieve an effective

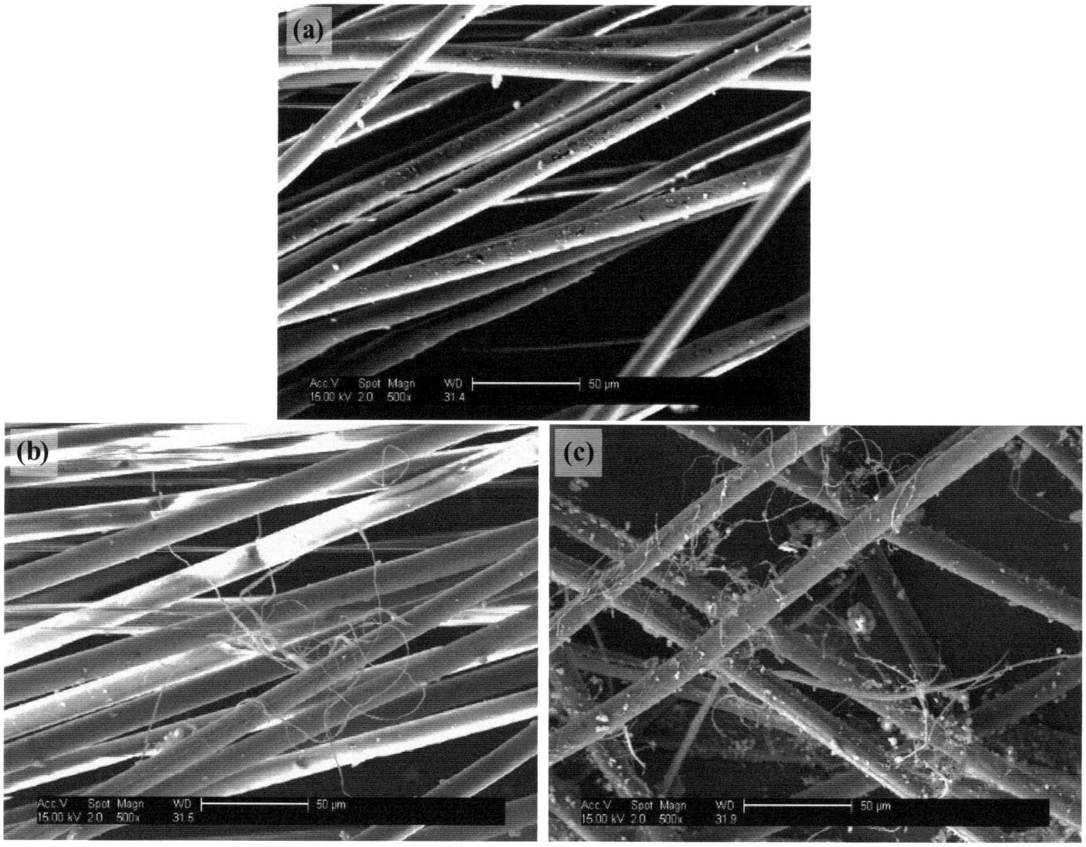

Figure 9. SEM images of extracted aramid fibers of FB mix at different conditions: (*a*) agitated bundles, (*b*) clusters, and (*c*) individual fibers (all images at 500× magnification).

network. It can only be stated that this threshold is somewhere between an *ADSR* of 16% (that from FB) and an *ADSR* of 88.5% (that from FA). Whether this network formation follows a simple linear, nonlinear, or percolation type relationship cannot be determined from the current data.

Finally, it should be stated that the orientations of the fibers with respect to the testing configuration in this study is a direct result of the laboratory fabrication procedure. The evaluation of field cores was beyond the scope of this study but would provide additional insights. Here the gyratory compactor is used to create specimens by compacting them along the same direction that the major principal stress is applied during tests. In the case of compression, these orientations generally align with pavement construction and vehicular loading. However, in tension these orientations differ since tension in a pavement is induced in a direction perpendicular to the compaction direction. Failure to consider these differences may lead to laboratory results that underestimate the true fatigue benefits from M-FRAC mixtures. Also, the study findings are isolated to the single mixture studied, and additional testing incorporating more careful documentation of fiber state is warranted in order to refine the theory further. What this study does demonstrate is that understanding M-FRAC benefits can be improved by considering the state of the fiber within the mixture.

Conclusion

This paper has investigated how the state of aramid fibers would influence the mechanical properties of asphalt mixtures. The following conclusions were reached in this study.

- Dynamic moduli values showed no statistical differences between the three mixtures.
- The FA mixture exhibited a 139% improvement in flow number compared with the control while the FB mixture had a similar performance to the control mixture.
- Fatigue simulations showed that both FA and FB mixtures had two times fatigue improvement compared with the control.
- Aramid fibers in FA mix were well dispersed and most of the fibers were in the individual state (*ADSR* of 88.5%) while FB fibers were poorly dispersed (ADSR of 16%).
- FA and FB specimens showed even distribution of the fibers in the individual fiber state but also a higher

number of individual fibers on the surface of FA specimens. Besides, it was observed that individual fibers tend to be oriented mostly horizontally.

- Aramid fibers in FA mix showed a higher degree of micro-fibrillation compared with aramid fibers in FB mix.
- The different behavior in fatigue and permanent deformation for each FA and FB was attributed to the orientations of the fibers with respect to the testing configuration. It was postulated that in the flow number test the fibers are ideally oriented to strengthen cracks that form parallel to the loading direction, whereas fibers are not preferably oriented for reinforcing cracks that develop perpendicular to the loading direction in the fatigue test.

Author Contributions

The authors confirm contributing to the paper as follows: study conception and design: Hossein Noorvand, B. Shane Underwood; data collection: Hossein Noorvand, Ramadan Salim, Jose Medina; analysis and interpretation of results: Hossein Noorvand, B. Shane Underwood; draft manuscript preparation: Hossein Noorvand, B. Shane Underwood, Jeffrey Stempihar. All authors reviewed the results and approved the final version of the manuscript.

References

1. Mcdaniel, R. *Fiber Additives in Asphalt Mixtures*. Transportation Research Board of the National Academies, Washington, D.C., 2015.
2. Abtahi, S. M., M. Sheikhzadeh, and S. M. Hejazi. Fiber-Reinforced Asphalt-Concrete – A Review. *Construction and Building Materials*, Vol. 24, No. 6, 2010, pp. 871–877.
3. Huang, H., and T. White. Dynamic Properties of Fiber-Modified Overlay Mixture. *Transportation Research Record: Journal of the Transportation Research Board*, 1996. 1545: 98–104.
4. Mcdaniel, R., and A. Shah. *Asphalt Additives to Control Rutting and Cracking*. Publication FHWA/IN/JTRP-2002/29. Joint Transportation Research Program, Indiana Department of Transportation and Purdue University, West Lafayette, 2003.
5. Behbahani, H., S. Nowbakht, H. Fazaeli, and J. Rahmani. Effects of Fiber Type and Content on the Rutting Performance of Stone Matrix Asphalt. *Journal of Applied Sciences*, Vol. 9, No. 10, 2009, pp. 1980–1985.
6. Kaloush, K. E., K. P. Biligiri, W. A. Zeiada, M. C. Rodezno, and J. X. Reed. Evaluation of Fiber-Reinforced Asphalt Mixtures Using Advanced Material Characterization Tests. *Journal of Testing and Evaluation*, Vol. 38, No. 4, 2010, pp. 400–411.
7. Tapkın, S. The Effect of Polypropylene Fibers on Asphalt Performance. *Building and Environment*, Vol. 43, No. 6, 2008, pp. 1065–1071.
8. Lee, S. J., P. J. Rust, H. Hamouda, Y. R. Kim, and R. H. Borden. Fatigue Cracking Resistance of Fiber-Reinforced Asphalt Concrete. *Textile Research*, Vol. 75, No. 2, 2005, pp. 123–128.
9. Maurer, D. A., and G. J. Malasheskie. Field Performance of Fabrics and Fibers to Retard Reflective Cracking. *Geotextiles and Geomembranes*, Vol. 8, No. 3, 1989, pp. 239–267.
10. Simpson, A. L, and M. C. Case Study of Modified Bituminous Mixtures: Somerset, Kentucky. *Proc., 3rd Materials Engineering Conference, San Diego, CA, ASCE*, New York, 1994, pp. 88–96.
11. Xu, Q., H. Chen, and J. A. Prozzi. Performance of Fiber Reinforced Asphalt Concrete Under Environmental Temperature and Water Effects. *Construction and Building Materials*, Vol. 24, No. 10, 2010, pp. 2003–2010.
12. Mobasher, B. *Mechanics of Fiber and Textile Reinforced Cement Composites*. CRC Press, Boca Raton, Fla., 2011.
13. Underwood, B., C. Baek, and Y. Kim. Simplified Viscoelastic Continuum Damage Model as Platform for Asphalt Concrete Fatigue Analysis. *Transportation Research Record: Journal of the Transportation Research Board*, 2012. 2296: 36–45.
14. Zeiada, W., J. Medina, S. Underwood, and J. Stempihar. *Fiber-Reinforced Asphalt Concrete Laboratory Procedure: Verification of Aramid Dispersion*. Forta Corporation, Grove City, PA 2017.
15. Hahn, C. Characteristics of P-aramid Fibers in Friction and Sealing Materials. *Journal of Industrial Textiles*, Vol. 30, No. 2, 2000, pp. 146–165.

The Standing Committee on Non-Binder Components of Asphalt Mixtures (AFK30) peer-reviewed this paper (18-05606).

Article

TRR
JOURNAL OF THE TRANSPORTATION RESEARCH BOARD

Transportation Research Record
2018, Vol. 2672(28) 52–58

DOI: 10.1177/0361198118790620
journals.sagepub.com/home/trr
SAGE

Utilizing Network and Project-Level Data to Evaluate Impact of Reclaimed Asphalt Pavement and Recycled Asphalt Shingles on Laboratory and Field Performance in Texas

Berenice Salaices[1], David Teutli[1], Imad Abdallah[1], Anjan Kumar Siddagangaiah[2], and Soheil Nazarian[1]

Abstract

The incorporation of reclaimed asphalt pavement (RAP) and recycled asphalt shingles (RAS) in asphalt mixes is encouraged as long as the modified mix exhibits equal or improved performance. Through the integration and data mining from several statewide databases, and extensive laboratory and field testing, this paper investigates the performance of asphalt mixes containing RAP, RAS, or both relative to similar mixes with virgin components. The integration of several databases permitted a network-level investigation of performance patterns. The reasonableness of the observed performance patterns was assessed using the results from the Hamburg wheel tracking tests (HWTT) and indirect tensile strength tests performed on specimens from several sites included in a network-level database. Furthermore, the retrieved network-level data were cross-referenced to the results from a historical performance database. At project level, HWTT and the overlay tester tests were performed on samples from several dozen sites to characterize laboratory performance. Field performance of these sites was also documented through periodic cracking and rutting surveys. From the network-level assessment, the impact of RAP, RAS, or both on the performance of the mixes was found to be mix-dependent. The combination of RAP and RAS seems to negatively affect cracking resistance. Further, dense-graded and Superpave mixes with RAP, RAS, or both exhibit higher cracking potential with minimal impact on rutting resistance. The project-level field performance evaluation shows that fatigue cracking propagates at a higher rate for sections with RAP and RAS than for sections with only RAP.

Recycled materials have been used in the construction of asphalt layers in the United States for a couple of decades. Reclaimed asphalt pavement (RAP) and recycled asphalt shingles (RAS) are two of the most common recycled materials used. The utilization of recycled materials in asphalt concrete (AC) mixes contributes to material conservation and cost savings. However, the benefits derived from the use of the recycled materials are controlled by the permitted uses and technological challenges such as production and construction. The degree of blending of the asphalt binder from the recycled materials with virgin asphalt is influenced by the mixing conditions such as time and temperature during plant production (*1*, *2*). To address such issues and to ensure sufficient blending, various additives such as rejuvenators and warm mix agents, are considered to improve or achieve the desired properties of the asphalt mixes with RAP and RAS (*3–5*).

Changes in the mechanical properties of the AC mixes due to the addition of RAP and RAS affect its resistance to rutting and fatigue cracking performance (*6*, *7*). As such, the use of RAP and RAS is promoted as long as the same or improved field performance is obtained. Over the years, RAP and RAS have been used with different types of mixes (e.g., dense-graded and gap-graded mixes). Different laboratory tests are currently being used to assess the asphalt mix performance. The Hamburg wheel tracking test (HWTT), indirect tensile

[1]Center for Transportation Infrastructure Systems, University of Texas at El Paso, El Paso, TX
[2]Department of Civil Engineering, Indian Institute of Technology Guwahati, Assam, India

Corresponding Author:
Address correspondence to Berenice Salaices:
bsalaicesgomez@miners.utep.edu

test (IDT) and overlay tester (OT), amongst others, are commonly used to delineate the asphalt mix resistance to rutting, moisture damage, and cracking. A substantial body of research has documented the resulting variations in the desired properties of the asphalt mixes with RAP and RAS (*1*, *6–8*). However, the information about the longevity of different mixes with and without RAP and RAS is limited, especially when structurally different mixes are involved. The methodology proposed by Rodriquez et al. to evaluate the longevity of asphalt mixes in terms of performance using in-service life served as a potential tool to address the above concern (*9*).

The objective of this study is to quantify the statewide impact of RAP and RAS on the performance of several asphalt mixes. The approach followed to meet this objective consisted of integrating three sources of data. First, at a network level, the available data from a statewide database were sorted to identify the different types of mixes with and without recycled materials and their corresponding laboratory results. Then, the performance lives of pavement sections with RAP and RAS in asphalt mixes from an integrated database were extracted. In the third stage, project-level data were used to evaluate the trends in laboratory and field performance. Finally, the observed patterns in the performance were collectively assessed to understand how the use of RAP and RAS affected the performance of asphalt mixes.

Methodology

The methodology adopted in this study is schematically represented in Figure 1. First, with significant effort at a network level, several sections stored in a quality control/quality assurance (QC/QA) database were merged, and integrated to link statewide data regarding the design characteristics of the asphalt mixes, the presence of recycled materials, and their respective HWTT and IDT results. The outcomes were then linked to an online platform called Performance Life of Mixes in Texas (PERMIT), developed by Rodriquez et al. (*9*). PERMIT estimates the longevity, serviceability state, performance information, and life predictions of thousands of road sections. The predicted performance lives, for virgin and recycled sections, were analyzed to assess the impact of RAP and RAS in asphalt mixes. In the next stage, a database developed by Walubita et al. that contained relevant information regarding the laboratory test results on mix properties, quality control data, and periodic field performance of about 100 test sections was used to verify the performance responses from the network level (*10*).

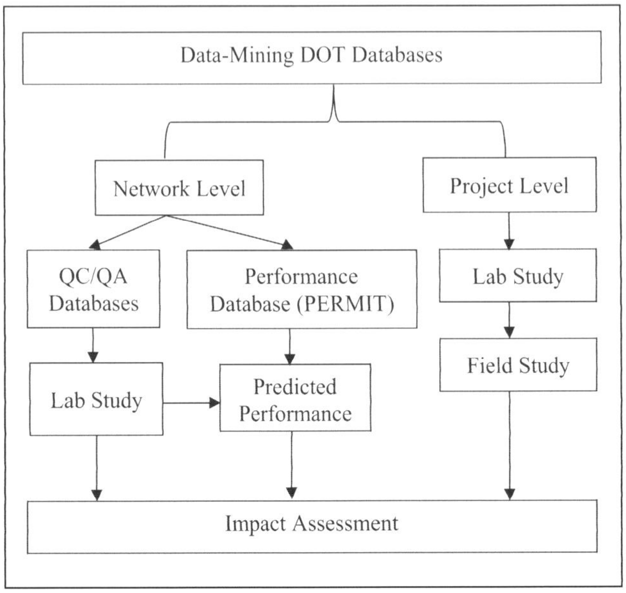

Figure 1. Schematic representation of methodology.

Data Mining

The statewide QC/QA database contains information regarding the design characteristics and laboratory results of several mixes designed in the laboratory. Each mix stored in the database can be identified by a unique sample ID, which is automatically generated by the corresponding form used. Therefore, multiple sample IDs can belong to one mixture. The forms used to extract the data were the hot mix asphalt (HMA) mix design form and the QC/QA design form (laboratory results). The database is structured in such a way that it is not possible to automatically extract information for one mixture from different forms. Each form saves its results in separate tabs of the database, and each field within the form is stored separately. Consequently, a linking process was established to integrate unique fields from each section of the database as well as descriptive fields shared to identify unique projects. By developing this linking process, mixes with virgin and recycled asphalt, and their corresponding laboratory results were cross-referenced. The linking process extracted results for all mixes designed between 2008 and 2015. After integration and data mining, 4,059 unique sample IDs were retrieved from the HMA mix design forms. These sample IDs contained information such as mix type, and the presence or absence of recycled materials in the asphalt mix. The outcomes were further referenced to the HWTT and IDT laboratory results. If an entry from the HMA mix design form did not meet the linking criteria or detected a blank field in the QC/QA design form, an entry could not be generated. With this linking criteria, the research team collected information for mixes that contained mix type, presence or absence of recycled materials, and laboratory results.

Table 1. Statewide Database Sample Size by Recycled Category

Category	Virgin mixes	Recycled asphalt		
		RAP	RAP with RAS	RAS
HWTT rut depth	830	541	704	43
IDT strength	952	588	751	44
Performance lives	349	340	318	20

These results were also referenced to the PERMIT database in order to assign a fatigue life prediction. This link of data was possible because the two databases contained similar unique keys such as mix type and control section job number. This undertaking was time consuming and challenging due to the structure of each database as well as issues like data gaps, formatting, and the extensive amount of information stored.

The sample sizes for the HWTT rut depth, IDT strength, and PERMIT results are presented in Table 1. The study focused on three prominent mixes, namely traditional dense-graded, stone matrix asphalt (SMA), and Superpave (SP) mixes. Different mixes within each mix type with different nominal maximum aggregate size (e.g., Type B, Type C, or Type D) were combined into their corresponding mix category to increase the sample size, and to consolidate the results. Predominantly, recycled mixes contained 15% or 20% RAP; whereas the combination of RAP and RAS ranged from 15% to 20% RAP with 5% RAS. Finally, 5% RAS was commonly used. Due to the low variation in the use of recycled materials, the mixes were divided into four different categories, virgin, RAP, RAP with RAS, and RAS in order to critically assess the impact of recycled materials.

Approximately 73% of the test sections from the project-level database contained data useful for the study. Dense-graded, SMA, and SP mixes were categorized as virgin, RAP, and RAP with RAS mixes. The RAS category was excluded from this portion of the study due to the low number of sections. The data retrieved from this database were the mix design characteristics, HWTT rut depth, OT parameters, and field performance results. Although, many design factors influence the impact of RAP and RAS in the mix performance, the research team decided to use the HWTT, IDT, and OT parameters to evaluate laboratory performance.

Results and Discussion

The integration of information from different databases was useful to evaluate the impact of RAP and RAS on the mix performance. The mix-specific network and project-level information, along with various performance measures from the laboratory and field were analyzed. First, the HWTT rut depth and IDT strength cross-plots were developed to assess the impact of RAP, RAS, or both using the network-level data for virgin and recycled mixes. Subsequently, the variations in the predicted lives of different asphalt mixes based on fatigue cracking deterioration were presented using the network-level data. Comparatively, the impact of RAP, RAP with RAS, and RAS on the fatigue cracking was analyzed using the OT cracking parameters contained in the project-level study. To demonstrate the impact of RAP and RAS the field performance of the mixes in terms of cracking was assessed using project-level data. Finally, the field and laboratory variations in the rutting resistance of the asphalt mixes with and without recycled materials were evaluated.

Network-Level Rutting and Tensile Strength

The rutting resistance and tensile strength of each asphalt mix from a large database were compiled with considerable effort. Highway agencies typically limit the RAP content from 15% to 20% and RAS content to 5%. An asphalt mix is considered failed if it does not comply with the HWTT number of passes to rut 0.5 in. (12.5 mm), whereas the recommended IDT strength should fall between 85 psi (586 kPa) to 200 psi (1,380 kPa).

The cross-plots of the median of the HWTT rut depths and the median of the IDT strengths compiled from the network-level database are shown in Figure 2 as a function of the mix type. In broad terms over the entire network, the medians of the three mix types are within the acceptable region. However, the error bars denote that some mix categories failed to pass the acceptance limits for HWTT and IDT. Based on the medians, the addition of RAP to dense-graded mixes marginally but negatively affects the rut resistance and has a minimal impact on IDT strength. On the other hand, the inclusion of the combination of RAP and RAS marginally improves IDT strength when compared with the corresponding virgin mixes, with virtually no impact on rutting performance.

The SMA virgin mixes exhibit the lowest IDT median strength when compared with the other two mixes evaluated. The addition of RAP to SMA mixes improves the rutting resistance, whereas it increases the tensile strength when compared with the virgin mixes. The presence of both RAP and RAS in SMA appears to profoundly influence the resistance to rutting, and IDT strength. The SP virgin mixes exhibit more potential for rutting when compared with the corresponding mixes with recycled materials. The addition of RAP to SP mixes improves rutting resistance with virtually no impact on IDT strength. The combination of RAP and RAS to SP mixes

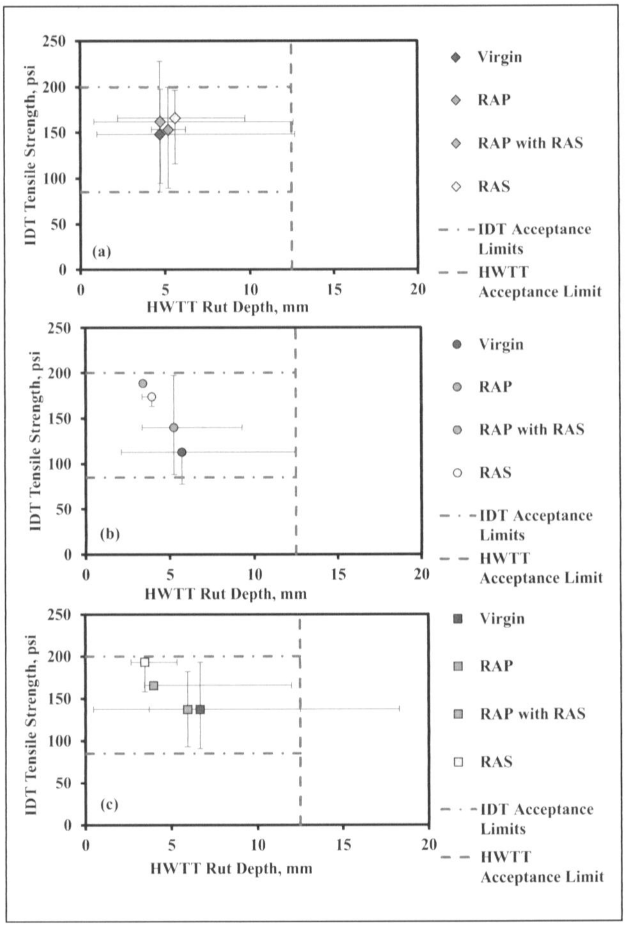

Figure 2. Median IDT tensile strength and HWTT rut depth interaction of mixes: (*a*) dense-graded; (*b*) SMA; (*c*) SP.

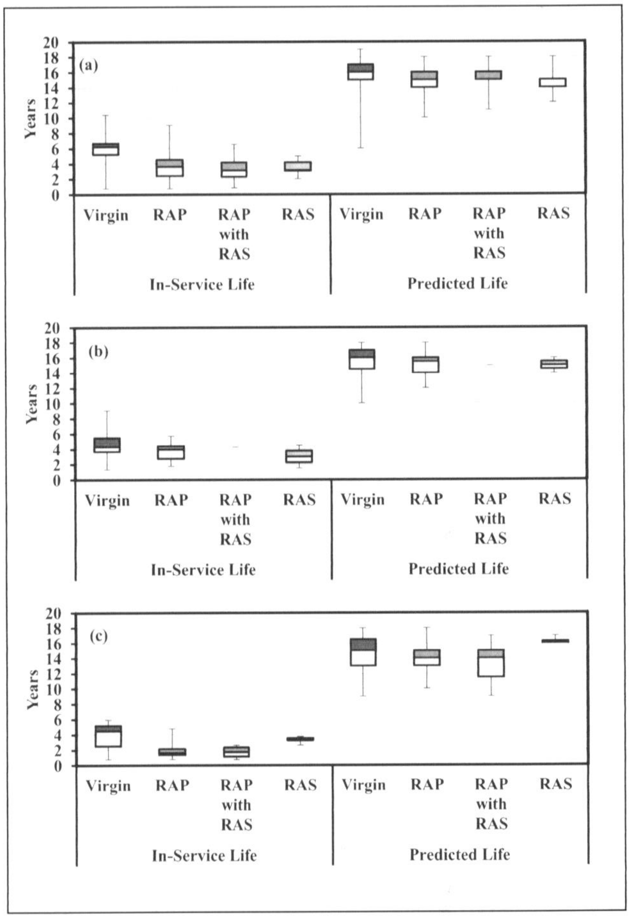

Figure 3. Fatigue cracking lives of mixes: (*a*) dense-graded; (*b*) SMA; (*c*) SP.

improves both rutting resistance and IDT strength. It is not possible to draw a specific conclusion on how the use of RAS affects rutting resistance and tensile strength for SMA (3 samples) and SP (8 samples) because of sample size limitations. In summary, the impact of the addition of RAP or RAS, or both RAP and RAS to mixes depends on the mix type. The addition of RAP, and RAP with RAS, usually positively affects rutting resistance and increases the tensile strength of the mixes.

Network-Level Fatigue Cracking Prediction

By linking the QC/QA database to PERMIT, the research team was able to predict the fatigue life of virgin and recycled mixes. PERMIT, which is available as an online tool, provides the in-service and predicted lives of the road sections. In-service lives refer to the age of the sections that have not experienced significant distress, maintenance, or rehabilitation during the study period. These sections are part of the existing network that had not failed at the time of the study. The predicted life refers to a fatigue cracking failure prediction calculated from the existing conditions (in-service lives). To calculate predicted life, PERMIT uses the progression of the annual fatigue cracking of the road section and generates a pavement performance curve for each section by fitting a Weibull survival function (*9*). PERMIT is capable of making a life prediction as long as the fatigue cracking is increasing over time. The time to reach a terminal value for fatigue cracking is calculated using the annual distress ratings, condition scores, and traffic data that are kept in another statewide performance database. Due to constraints in the data in the statewide database, only the fatigue cracking lives were estimated.

The results from PERMIT are summarized in Figure 3 as box plots showing the median, minimum, and maximum lives for the three mix types. Two sets of box plots are shown for each mix type, one depicting the in-service life and the other the predicted life. Each mix type was categorized as virgin, RAP, RAP with RAS, and RAS. All three mixes exhibit higher in-service and predicted fatigue lives for virgin mixes when compared with mixes with recycled materials. However, the predicted lives are less pronounced with a maximum difference of 1.5 years, when compared with the in-service lives. Again, mixes with RAS represent a very small sample size and cannot be assessed with enough confidence. Overall, results indicate a higher cracking resistance for virgin mixes when compared with mixes with content of recycled materials.

Laboratory Fatigue Cracking at a Project Level

Walubita et al. conducted laboratory performance tests (i.e., HWTT and OT tests) on loose mixes collected during paving, and cores obtained shortly after construction at more than 100 test sites (*10*). The OT test results were used to determine the susceptibility of asphalt mixes to fatigue cracking following the methodology proposed by Garcia, et al. (*11*). In that methodology, two parameters are used to evaluate the crack resistance of the mixes, namely the critical fracture energy as a surrogate for resistance to crack initiation, and crack progression rate for assessing the crack propagation properties. Figure 4 illustrates the interaction between these two parameters and their current acceptance limits for the three mix types. Each point represents the corresponding test results for one test site. No data is presented for the SMA and SP sections containing RAP with RAS because that combination was not used in any of the projects.

With crack progression rate (CPR) values that ranged from 0.3 to 1.5, most dense-graded mixes with RAP exhibited unacceptable performance for crack propagation. With CPR values in excess of 1.5, dense-graded mixes containing both RAP and RAS exhibited even higher potential for crack propagation as compared with the virgin and RAP mixes. All three SMA samples with RAP performed well in terms of cracking. Only two virgin samples were available for this type of mix: one passed and one failed to meet the acceptance limits. The cracking results for the virgin SP mixes for the most part satisfied the acceptance limits. In this case, the addition of RAP negatively affected the crack resistance as CPR increased. The dense-graded and SP virgin mixes appear to cluster closer to the OT acceptance limits, whereas a gradual movement toward higher CPR (higher potential for cracking) is observed for the asphalt mixes that contained RAP. The sections with dense-graded mix types containing a combination of RAP and RAS tend to be more crack prone as they exhibit higher CPR values. Based on these results, the addition of RAP to the SMA mixes appears to have a positive impact on cracking resistance.

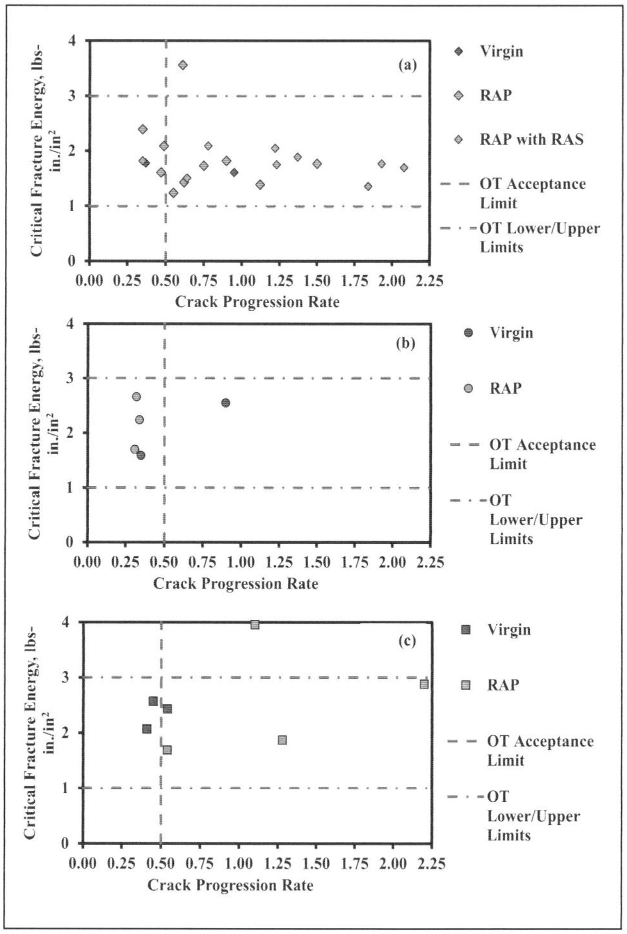

Figure 4. Evaluation of cracking characteristics of mixes: (*a*) dense-graded; (*b*) SMA; (*c*) SP.

Field Study at a Project Level

Walubita et al. also monitored the pavement performance of 500-ft test sections at nominally 6-month intervals for each of the mixes tested in the laboratory (*10*). These periodic visits involved visual crack survey and rut depth measurements among other functional and structural parameters. The test sections, which were 3- to 5-years old during the last visit, had not yet exhibited severe distresses. However, the available performance data were retrieved from the database to interpret the performance of the virgin and recycled mixes in terms of fatigue cracking and rutting.

The cross-plot of the field fatigue cracking along the test sections and OT CPR for different mix types is presented in Figure 5. The majority of the recycled asphalt sections are located outside of the OT acceptance limit, which indicates greater cracking potential. Only two sections with RAP within the OT acceptance limit have experienced fatigue cracking. On the other hand, seven sections with recycled materials that are outside the OT acceptance limit have experienced this type of distress.

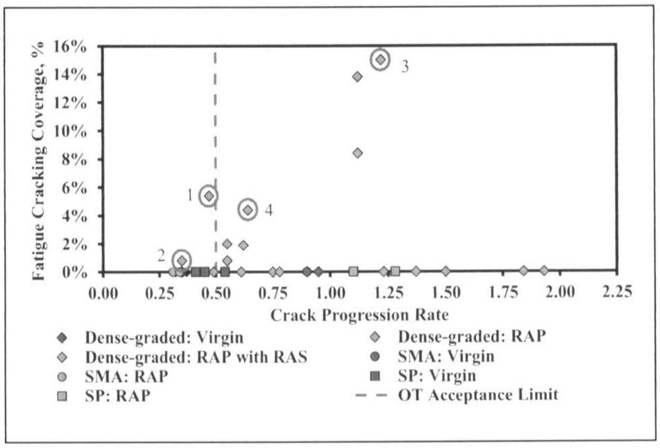

Figure 5. Relationship between field fatigue cracking and OT crack progression rate.

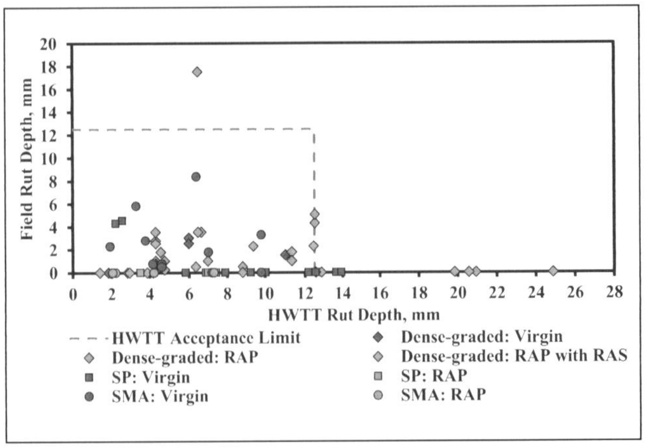

Figure 7. Relationship between field rutting and HWTT rut depth.

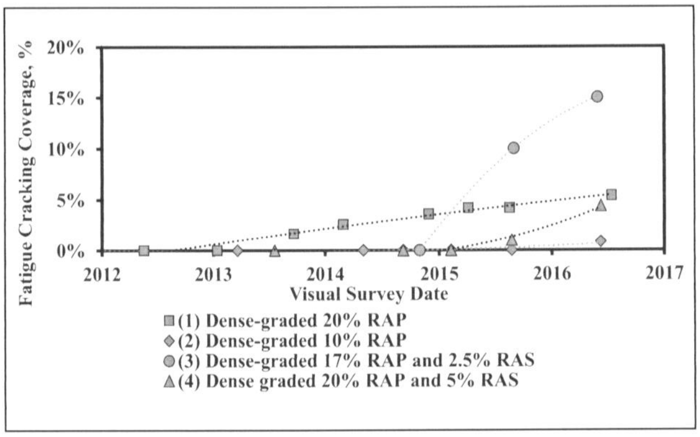

Figure 6. Progression of fatigue cracking for sections marked in Figure 5.

Four dense-graded sections, consisting of two sections with RAP and two sections with a combination of RAP and RAS that are marked in Figure 5 were selected to evaluate their yearly crack progression.

The deterioration of the pavements in terms of fatigue cracking over time for the selected mix types is presented in Figure 6. The dense-graded Mix 2 with 10% RAP exhibited less cracking as compared with Mix 1 that contained 20% RAP. Mix 3 and Mix 4 that contained both RAP and RAS show a higher rate of crack initiation as compared with the mixes with only RAP. The selected sections that exhibited fatigue cracking for asphalt mixes containing only RAP gradually increased over the years. On the other hand, sections with a combination of RAP and RAS have experienced considerable damage that propagated at a higher rate. The addition of RAS appears to directly impact the crack propagation susceptibility of asphalt mixes.

Figure 7 illustrates a cross-plot between the HWTT rut depths and field rut depths. The dense-graded mixes from the three categories (virgin, RAP, and RAP with RAS) exhibit similar field rutting. Only one dense-graded mix with RAP has experienced extensive field rutting. Mixes with RAP or RAP with RAS that exhibited higher HWTT rut depth in the laboratory have not yet exhibited field rutting. SMA and SP virgin mixes are perceived to have high rutting resistance. However, the results from the rut depth measurements indicate that the corresponding virgin mixtures rutted excessively in the field when compared with the corresponding sections with RAP. None of the SMA or SP sections with RAP has experienced rutting over this study period. These results indicate that the addition of RAP may improve the rutting resistance of these two mixes.

From the project-level investigation, the incorporation of RAP, RAS, or both to dense-graded and SP mixes negatively affected crack resistance. Therefore, the impact of recycled materials on crack resistance seems to be mix specific. The information from the project-level field studies demonstrate increased crack susceptibility for the dense-graded mixes designed with recycled materials. Comparative assessment of the field rutting and laboratory rutting data points toward improvement in the rutting resistance of the SMA and SP mixes with RAP, RAP with RAS, and RAS.

Observations, Summary and Conclusions

The utilization of recycled materials in AC mixes is controlled by various factors such as the mixture properties, expected performance, and lack of confidence in their long-term responses. This study presented a broad comparative analysis of the laboratory and field performance of virgin and recycled asphalt mixes. A systematic approach was adopted to integrate large databases that contained network-level and project-level information about the performance of mixes to assess the impact of

RAP and RAS use. Based on the available information, the performance of three different types of mixes (traditional dense-graded, SMA, and SP) with recycled asphalt were compared with conventional virgin mixes.

The addition of RAP and RAS to the asphalt mixes tends to improve rutting resistance. However, from a network-level point of view, the impact of RAP and RAS on rutting resistance and tensile strength for conventional dense-graded mixes was not substantial. On the other hand, the addition of RAP and RAS to the SMA and SP mixes improved their rutting potential and increased their tensile strengths. The network-level performance predictions from PERMIT indicate that the utilization of RAP and RAS was not substantially different when compared with virgin mixtures, presenting a maximum life difference of 1.5 years between the virgin and corresponding recycled mixes.

From the project-level investigations, the OT crack initiation and crack progression indices were used to delineate the mixture resistance to fatigue cracking. The dense-graded and SP mix types with RAP exhibited a greater potential for fatigue cracking, whereas the SMA mixes with RAP showed improved performance for fatigue cracking resistance. The results from the project-level field performance monitoring showed that fatigue cracking propagates at a higher rate for sections with RAP and RAS, than for the sections with only RAP. Finally, the project-level condition surveys indicated that the addition of RAP to the dense-graded mixes minimally affect their rutting resistance. Nonetheless, the addition of RAP to SMA and SP mixes improved field rutting.

Acknowledgments

The authors thank the Texas Department of Transportation for providing access to their state databases and for their support and guidance throughout this study.

Author Contributions

The authors confirm contribution to the paper as follows: study conception and design: BS, IA, SN; data collection: BS, DT, IA and AKS; analysis and interpretation of results: BS, IA, SN; draft manuscript preparation: BS, AKS, SN. All authors reviewed the results and approved the final version of the manuscript.

References

1. Bowers, B. F., J. Moore, B. Huang, and X. Shu. Blending Efficiency of Reclaimed Asphalt Pavement: An Approach Utilizing Rheological Properties and Molecular Weight Distributions. *Fuel*, Vol. 135, 2014, pp. 63–68.
2. Zhou, F., H. Li, R. Lee, T. Scullion, and G. Claros. Recycled Asphalt Shingle Binder Characterization and Blending with Virgin Binders. *Transportation Research Record: Journal of the Transportation Research Board*, 2013. 2370: 33–43.
3. Copeland, A., J. D'Angelo, R. Dongré, S. Belagutti, and G. Sholar. Field Evaluation of High Reclaimed Asphalt Pavement–Warm-Mix Asphalt Project in Florida Case Study. *Transportation Research Record: Journal of the Transportation Research Board*, 2010. 2179: 93–101.
4. Mallick, R. B., P. S. Kandhal, and R. L. Bradbury. Using Warm-Mix Asphalt Technology to Incorporate High Percentage of Reclaimed Asphalt Pavement Material in Asphalt Mixtures. *Transportation Research Record: Journal of the Transportation Research Board*, 2008. 2051: 71–79.
5. Yu, X., M. Zaumanis, S. Dos Santos, and L. D. Poulikakos. Rheological, Microscopic, and Chemical Characterization of the Rejuvenating Effect on Asphalt Binders. *Fuel*, Vol. 135, 2014, pp. 162–171.
6. Huang, S. C., A. T. Pauli, R. W. Grimes, and F. Turner. Ageing Characteristics of RAP Binder Blends - What Types of RAP Binders are Suitable for Multiple Recycling? *Road Materials and Pavement Design*, Vol. 15, 2014, pp. 113–145.
7. Li, X., M. O. Marasteanu, R. C. Williams, and T. R. Clyne. Effect of Reclaimed Asphalt Pavement (Proportion and Type) and Binder Grade on Asphalt Mixtures. *Transportation Research Record: Journal of the Transportation Research Board*, 2008. 2051: 90–97.
8. Zhao, S., S. N. Nahar, A. J. Schmets, B. Huang, X. Shu, and T. Scarpas. Investigation on the Microstructure of Recycled Asphalt Shingle Binder and its Blending with Virgin Bitumen. *Road Materials and Pavement Design*, Vol. 16, 2015, pp. 21–38.
9. Rodriguez, D., L. Torres, I. Abdallah, B. Haggerty, and S. Nazarian. Methodology for Determining Performance Life of Hot-Mix Asphalt Mixtures. *Transportation Research Record: Journal of the Transportation Research Board*, 2016. 2589: 110–118.
10. Walubita, L. F., T. Scullion, S. Lee, A.N. Faruk, S. Nazarian, and I. Abdallah. *Texas Flexible Pavements and Overlays, Year 5 Report. Complete Data Documentation.* 0-6658-3, Texas A&M Transportation Institute, Center for Transportation Infrastructure Systems, College Station, TX, 2017.
11. Garcia, V., A. Miramontes, J. Garibay, I. Abdallah, and S. Nazarian. *Improved Overlay Tester for Fatigue Cracking Resistance of Asphalt Mixtures.* TxDOT 0-6815-1, Center for Transportation Infrastructure Systems, The University of at Texas El Paso, El Paso, TX, 2016.

The Standing Committee on Non-Binder Components of Asphalt Mixtures (AFK30) peer-reviewed this paper (18-03729).

Article

TRR
JOURNAL OF THE TRANSPORTATION RESEARCH BOARD

Transportation Research Record
2018, Vol. 2672(28) 59–68

DOI: 10.1177/0361198118792998
journals.sagepub.com/home/trr
SAGE

Laboratory Evaluation of Mixture Type on Highly Modified Asphalt Mixtures in Virginia

Benjamin F. Bowers[1]

Abstract

The work presented attempts to address reflective cracking of asphalt-surfaced pavements through binder modification with a highly polymer (HP)-modified asphalt binder. Nine asphalt mixtures ranging from fine dense-graded mixtures to stone matrix asphalt (SMA) mixtures were investigated with conventional polymer modified binders and HP binder. The dynamic modulus test, overlay test (OT), and semi-circular bend (SCB) test were used to evaluate the mixtures. In the cracking tests, HP mixtures outperformed the conventionally modified control mixtures for the same mixture type. For HP mixtures, in general, SMA mixtures performed better in the cracking test than dense-graded mixtures. One of the dense-graded mixtures having larger nominal maximum aggregate size (NMAS) performed better than the mixture with a smaller NMAS, whereas the other having a larger NMAS was not significantly different in crack testing. Further, a discussion on the calculation of bulk specific gravity and percent air voids in a cut OT and SCB specimen using saturated surface dry or vacuum sealing methods is presented.

Whether covering an old, fatigued asphalt pavement or a jointed concrete pavement, reflective cracking in new asphalt overlays is a problem that plagues many departments of transportation (DOTs). Often DOTs do not have the budget for deep rehabilitation or total reconstruction, emphasizing the need to find a solution for reflective cracking that can be performed in the surface layers. Many different methods have been used to mitigate or prolong crack propagation, including the use of paving fabrics, stress-absorbing interlayers, fibers, and various binder additives. Binder additives are particularly intriguing from a constructability standpoint, especially if they do not significantly change asphalt mixture designs or plant operations from that of a conventional mixture. One natural way to achieve this is through polymer modification of the asphalt binder. There is extensive literature supporting the use of polymer modification for extending the service life of pavements (*1*). Increasing the polymer content of the binder may further extend the life of the asphalt pavement in harsh stress and strain environments by increasing the elasticity of the mixture. Highly modified binders, henceforth referred to as high polymer or "HP" binders, increase the styrene-butadiene-styrene (SBS) polymer content of the asphalt binder to roughly 7.5% from a conventional content of approximately 2.5% to 3.5%. There is much interest in whether or not such a highly modified binder may be able to improve the resistance to reflective cracking.

Research was performed at the National Center for Asphalt Technology that examined HP binders in the surface, intermediate, and base asphalt mixture layers. Comparisons were made with a conventional polymer modified binder with a 2.8% SBS content for the surface and intermediate layers. It was found that the HP binder reduced rut susceptibility, increased stiffness, and increased fatigue life versus the control (*2–4*). Mogawer et al. studied the impact of using HP binder in high-performance thin overlays, which is an overlay that is 1 in. or less in thickness (*5, 6*). The researchers' test matrix included zero and 40% Reclaimed Asphalt Pavement (RAP) mixtures, warm mix additives, and varied binder grades with and without HP additives. In most cases the authors observed increased stiffness, reduced rut susceptibility, a decrease in moisture susceptibility, and, interestingly, a reduction in crack performance in the overlay test (OT) (*5*). The second study by Mogawer et al. saw

[1]Virginia Transportation Research Council, Charlottesville, VA

Corresponding Author:
Address correspondence to Benjamin F. Bowers: bfbowers@auburn.edu

RAP contents between 0% and 25% as well as varying binder grades that included HP modification (*6*). The authors saw good crack performance through the OT and beam fatigue test and reported positive, though limited, field results at the time of publication.

In 2014 Virginia DOT (VDOT) began investigating HP mixtures as a potential solution to reflective crack issues in composite pavements. Bowers et al. documented the performance of a subdivision trial between an HP mixture and a control subdivision mixture and found equivalent or better performance in the HP mixture over that of the control mixture (*7*). This paper presents a natural progression of that work in which multiple mixture types that were placed over jointed concrete pavement throughout Virginia were investigated. Only the laboratory performance is presented here, with field performance to be presented following additional time in the field.

Objective

The objectives of this study are defined as follows: first, examine the difference in laboratory performance between mixtures having conventional polymer modified and HP asphalt binder. Second, determine if there are specific asphalt overlay mixture types (e.g., dense-graded surface mixture [SM] versus gap-graded stone matrix asphalt [SMA]) that perform superior to others when HP binder is used in the laboratory. The outcomes of this paper are intended to inform researchers and practitioners of the effects of HP binder on different asphalt mixture types in the laboratory and their expected impact on field performance. This information will lead to a better understanding of the appropriateness of using HP binder in asphalt overlay mixtures for crack mitigation. An additional investigation into the impact of bulk specific gravity (G_{mb}) and air void calculations for cut SMA specimens is also provided. This commentary is intended to encourage discussion around the best approach to measuring key properties of test specimens.

Methods

Sampling

All mixtures used in this study are plant produced. Asphalt mixture from each project was sampled and collected in boxes at the asphalt plant. The boxes were then returned to the Virginia Transportation Research Council (VTRC) laboratory for specimen preparation and testing. Boxes were heated until the mixture could be split into appropriate batches for making specimens. The batches were then covered with aluminum foil and heated until they reached mixing temperature, at which they were then compacted to the appropriately sized gyratory specimen.

Mixture Volumetrics, Gradation, and Binder Characterization

In order to determine the fundamental mix properties, volumetric analysis was performed on each mix sampled at the VTRC asphalt laboratory. Included in this analysis were the total asphalt content, gradation, bulk- and maximum theoretical (Rice) specific gravity (G_{mb} and G_{mm}), voids in the mineral aggregate (VMA), voids total in the mix or air voids (VTM), voids filled with asphalt (VFA), aggregate bulk and effective specific gravities (G_{sb} and G_{se}), dust to asphalt ratio, percent binder absorbed (P_{ba}), and effective binder content (P_{be}). Asphalt content was determined from AASHTO T308: *Determining the Asphalt Binder Content of Hot Mix Asphalt (HMA) by the Ignition Method*. The aggregate gradation was determined for the mixture after completion of the ignition furnace test. Binder grading was performed on tank material collected at the asphalt plant in accordance with AASHTO M 320: *Performance-Graded Asphalt Binder*.

Dynamic Modulus

The dynamic modulus tests were conducted on laboratory-produced specimens in accordance with AASHTO TP 79: *Standard Method of Test for Determining the Dynamic Modulus and Flow Number for Asphalt Mixtures Using the Asphalt Mixture Performance Tester*. Dynamic modulus specimens were fabricated in a gyratory compactor and were 150-mm tall and 100 mm in diameter. The specimen air void contents were 7.0 +/–0.5%. Three replicate dynamic modulus specimens were tested at 25, 10, 5, 1, 0.5, and 0.1 Hz at temperatures of 4.4, 21.1, 37.8, and 54.4°C. All mixture dynamic modulus master curves were generated from triplicate specimens. On occasion, one triplicate specimen deviated from the specification requirements. This was further investigated by examining the average master curve values with and without the specimen and, upon applying a test of reasonableness; it was decided to keep the specimen in the dataset.

Overlay Test

OT was generally conducted in accordance with the TX-248-F-09: *Test Procedure on a Universal Testing Machine* with a 100-kN load cell. A minimum of four specimens were tested for each mixture. Testing was performed at a temperature of 25 ± 0.5ºC. The failure point was defined as a 93% reduction from initial load, though the test was terminated at 1,200 cycles if a sample did not meet the

Table 1. Gradations of the Control and HP Mixtures after Ignition Furnace

		HP	Control		HP		Control		HP	
		SM-9.0	SM-12.5		SM-12.5		SMA-12.5	SMA-9.5	SMA-9.5	
Sieve No.	(mm)	A	B	C	D	E	F	G	H	I
2 in.	50.8	100.0	100.0	100.0	100.0	100.0	100.0	100.0	100.0	100.0
1 1/2 in.	37.5	100.0	100.0	100.0	100.0	100.0	100.0	100.0	100.0	100.0
1 in.	25.0	100.0	100.0	100.0	100.0	100.0	100.0	100.0	100.0	100.0
3/4 in.	19.0	100.0	100.0	100.0	100.0	100.0	100.0	100.0	100.0	100.0
1/2 in.	12.5	99.0	96.0	94.5	94.4	96.1	84.6	90.8	88.7	88.7
3/8 in.	9.5	93.0	84.9	84.3	84.0	86.9	62.3	72.4	70.7	69.6
No. 4	4.8	66.6	60.2	62.0	57.7	58.7	27.0	29.0	28.4	29.5
No. 8	2.4	46.7	42.7	43.7	38.1	35.8	18.2	20.3	19.2	20.7
No. 16	1.2	35.7	31.6	31.6	27.8	25.0	16.0	18.4	17.1	18.1
No. 30	0.6	26.3	21.8	21.8	19.6	18.3	14.5	16.9	15.5	16.0
No. 50	0.3	16.2	13.2	13.2	12.1	13.3	13.3	15.5	14.0	14.1
No. 100	0.2	8.4	7.2	7.5	7.1	9.2	12.0	13.9	12.4	12.5
No. 200	0.1	5.2	4.6	5.0	4.7	6.0	9.3	10.8	9.7	10.6

failure criteria. All tested specimens were within 7.0% ± 1.0% air voids as determined by AAHSTO T 269-14: *Percent Air Voids in Compacted Dense and Open Asphalt Mixtures* by way of AASHTO T166-16: *Bulk Specific Gravity (G_{mb}) of Compacted Hot Mix Asphalt (HMA) Using Saturated Surface-Dry Specimens*. A minimum of three replicates were considered for analysis after outliers were identified.

Semi-Circular Bend Test

The semi-circular bend (SCB) Illinois Flexibility Index Test (SCB I-FIT) was conducted in accordance with Illinois Test Procedure 405 (2016) using an instrumented load frame. Tests were conducted at ambient laboratory temperature (~21°C). All specimens, after notching, were within 7.0% ± 0.5% air voids as determined using AASHTO T 269-14 by way of the vacuum sealing method. One mixture, F, did not meet the air void requirements and had specimen air void contents between 8.7% and 9.3%. Analysis of results was conducted using the I-FIT (IL-SCB) analysis tool developed by the Illinois Center for Transportation and the University of Illinois Urbana-Champaign. A minimum of three replicates were considered for analysis after outliers were identified.

Results and Discussion

Mixture Volumetrics and Gradation

The gradation and volumetric properties of the mixtures examined in this study are presented in tables 1 and 2, respectively. It should be noted that Control SMA-F has a nominal maximum aggregate size (NMAS) of 12.5 mm, whereas all other SMAs have an NMAS of 9.5 mm. All control mixtures contained a polymer-modified PG76-22 binder, typical for Virginia interstate mixtures, whereas all HP mixtures were graded as a PG82-28. Each mixture contained 15% RAP with the exception of Mixture F, which contained 12% RAP. The conventional asphalt mixture-type naming mechanism within VDOT is referred to throughout this paper. The naming mechanism first provides the mixture-type acronym (SM or SMA) followed by the mixture's NMAS (9.0, 9.5, or 12.5 mm).

Dynamic Modulus

The master curves generated by the dynamic modulus test are presented in Figure 1, *a* and *b* for the surface and SMA mixtures respectively. The dynamic modulus master curves for the control SM-12.5 mixtures are nearly identical over all frequencies and temperatures. Overall, the HP mixtures are less stiff over all frequencies and temperatures than the control mixtures. Mixture A, an SM-9.0, is the least stiff at the lowest temperatures (on the right side of Figure 1) whereas nearly identical to Mixture E at the higher temperatures (on the left side of Figure 1). Mixture D, an SM-12.5, has higher stiffness values than all of the other HP mixtures but lower than that of the control. This may be caused by the lower P_{be} than all other HP mixtures at 4.82% (mixtures A [SM-9.0] and E [SM-12.5] are 5.73% and 5.45% respectively). The overall lower stiffness values may be attributed to the lower base binder grade as compared to the controls, which is reflected in the low temperature grade.

The master curves for the SMA specimens shown in Figure 1*b* exhibit similar results between the control and HP mixtures with mixtures F, G, and H all showing similar stiffness over the different temperatures and frequencies. Mixture I is less stiff overall, which may be attributed to the higher effective binder content than

Table 2. Volumetric Properties of all Mixtures

	HP	Control		HP		Control		HP	
	SM-9.0	SM-12.5		SM-12.5		SMA-12.5	SMA-9.5	SMA-12.5	
Property	A	B	C	D	E	F	G	H	I
Binder grade	82-28	76-22	76-22	82-28	82-28	76-22	76-22	82-28	82-28
Asphalt content, %	6.03	5.36	5.25	4.99	5.97	6.29	6.45	6.58	7.13
RAP content, %	15	15	15	15	15	12	15	15	15
Maximum theoretical specific gravity, G_{mm}	2.562	2.633	2.642	2.626	2.585	2.669	2.614	2.615	2.432
VTM, %	2.7	4.4	4.0	3.7	2.6	3.0	2.7	2.0	4.1
VMA, %	16.6	17.1	16.5	15.5	15.9	18.7	18.2	18.0	19.4
VFA, %	83.5	74.4	76.0	76.2	83.5	84.2	85.4	88.7	79.1
Fines to asphalt ratio	0.90	0.88	0.98	0.98	1.11	1.48	1.71	1.50	1.57
Mixture bulk specific gravity, G_{mb}	2.492	2.517	2.537	2.529	2.517	2.59	2.544	2.562	2.333
Aggregate bulk specific gravity, G_{sb}	2.808	2.874	2.879	2.844	2.816	2.987	2.91	2.919	2.688
Effective binder content, P_{be}, %	5.73	5.21	5.09	4.82	5.45	6.27	6.29	6.42	6.78

Note: VTM = voids in total mixture; VMA = voids in mineral aggregate; VFA = voids filled with asphalt.

the other mixtures at 6.78%; 0.36% higher than any other SMA. The observation that the lower base binder grade of the HP mixtures may be reflected in the lower stiffness values did not occur for the SMA mixtures, namely Mixture H. This may be due to the aggregate structure of the SMA, specifically in Mixture H, having an increased role than it does in a dense-graded mixture.

Overlay Test

OT was selected for crack evaluation due to the nature of the test as it relates to reflective cracking over jointed concrete, which is one of the primary uses of interest for the HP binder. OT data were processed for outliers by examining the interquartile range of the between-specimen cycles to failure for a given mixture. The first quartile (Q1) and third quartile (Q3) were identified, along with the interquartile range (IQR = Q3 – Q1). The upper limit is defined as Q3 + 1.5 IQR and the lower limit is Q1 – 1.5 IQR. Any specimens whose cycles to failure were outside the upper and lower limit were considered outliers and were removed from the analysis. Three specimens (one each from three unique mixtures) were removed as an identified outlier.

Figure 2 provides the results of the OT. The data were further processed to establish whether the mean cycles to failure between each mixture were significantly different. First, an *F*-test was employed to distinguish equal or unequal variance between each of the mixtures. A two-tailed two-sample *t*-test was then used with equal or unequal variance depending on the result of the *F*-test to determine whether the difference in means was or was not statistically significant. The results of the *t*-test ($\alpha = 0.05$) are presented in Table 3 and Table 4; the results of the *F*-test are omitted for brevity.

The statistical tests showed that for 83% of the mixture combinations the mean difference was statistically significant. In all mixture-type comparisons, the HP binder increased the OT-measured crack performance to a significant degree. Interestingly, the effect of using an SMA mixture versus a dense-graded surface mixture between control mixtures at conventional polymer loads was not significant with the exception of the B-G combination. For HP mixtures, however, the positive effect of using an SMA versus a surface mixture was statistically significant in all cases except for the E-H combination. The SM-9.0 HP mixture had a significantly better OT performance than all control mixtures, but lesser performance than all HP mixtures with the exception of Mixture D (SM-12.5 HP). Generally, with respect to HP mixtures, having a larger NMAS (9.0 to 12.5 mm) in dense-graded surface mixtures or using an SMA leads to better crack performance in the OT. One explanation for

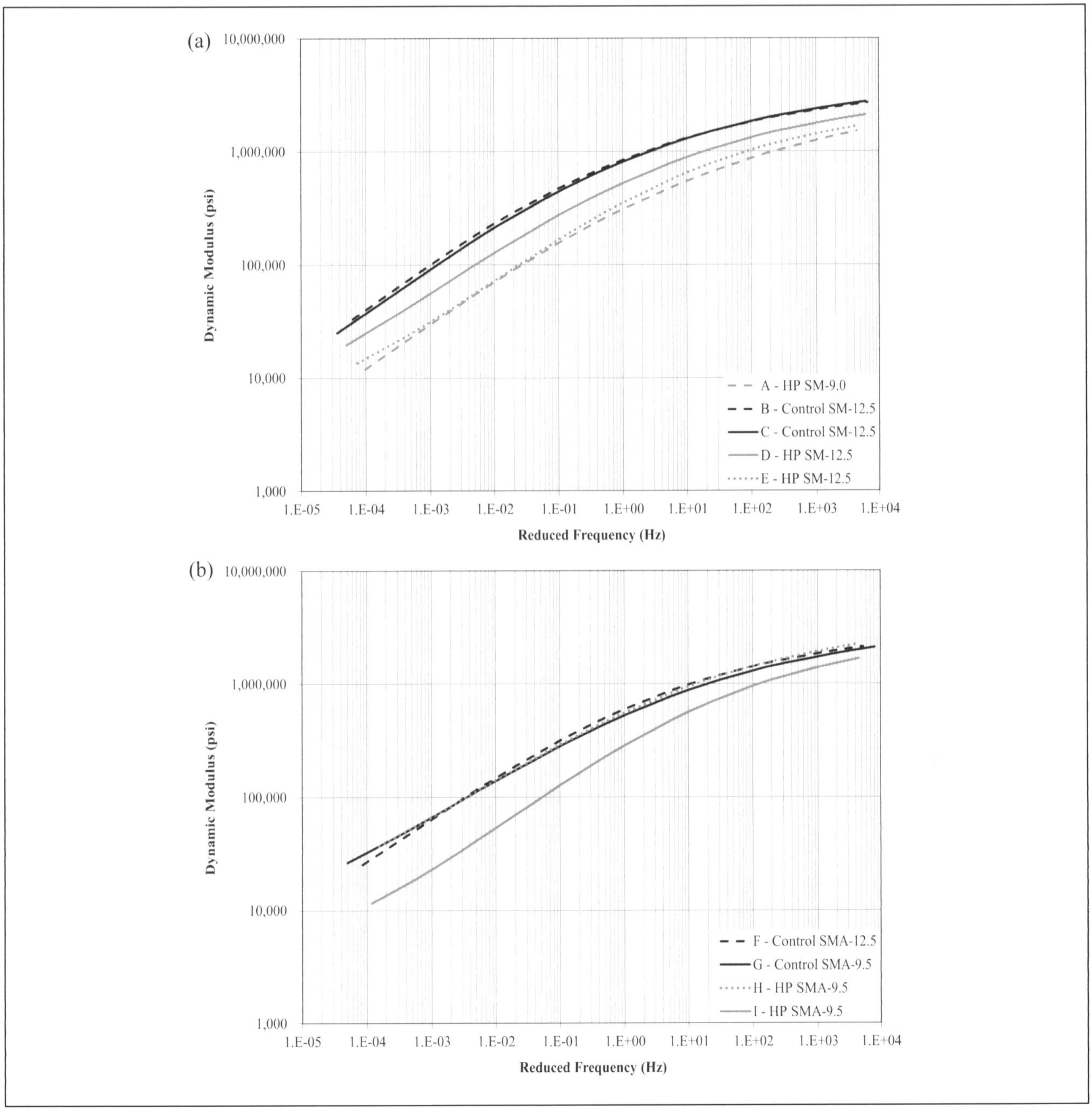

Figure 1. Dynamic modulus master curves for (*a*) the sampled SM, and (*b*) the sampled SMA.

the SMA outperforming other mixture is the higher effective binder content in conjunction with the higher polymer content. The HP SMAs had an average effective binder content of 6.6%, whereas the HP SM-12.5 and HP SM-9.0 mixtures had effective binder contents of 5.14% and 5.73% respectively. Although effective binder does not explain the difference in the HP SM-12.5 and HP SM-9.0 mixtures, it is possible that there is a smaller asphalt film thickness in the HP SM-9.0 mixture due to the finer gradation.

Semi-Circular Bend Test

The SCB I-FIT test was selected in this study as a fast, mid-range temperature cracking test, appropriate for Virginia's climate. The SCB I-FIT test results are shown in Figure 3. Due to the more limited number of samples, no statistical analysis was performed to remove outliers. Statistical testing was used in the form of an *F*-test and *t*-test ($\alpha = 0.05$), as previously described in the OT portion of this paper. All samples tested for Mixture B were identified by the I-FIT software as "brittle" and therefore

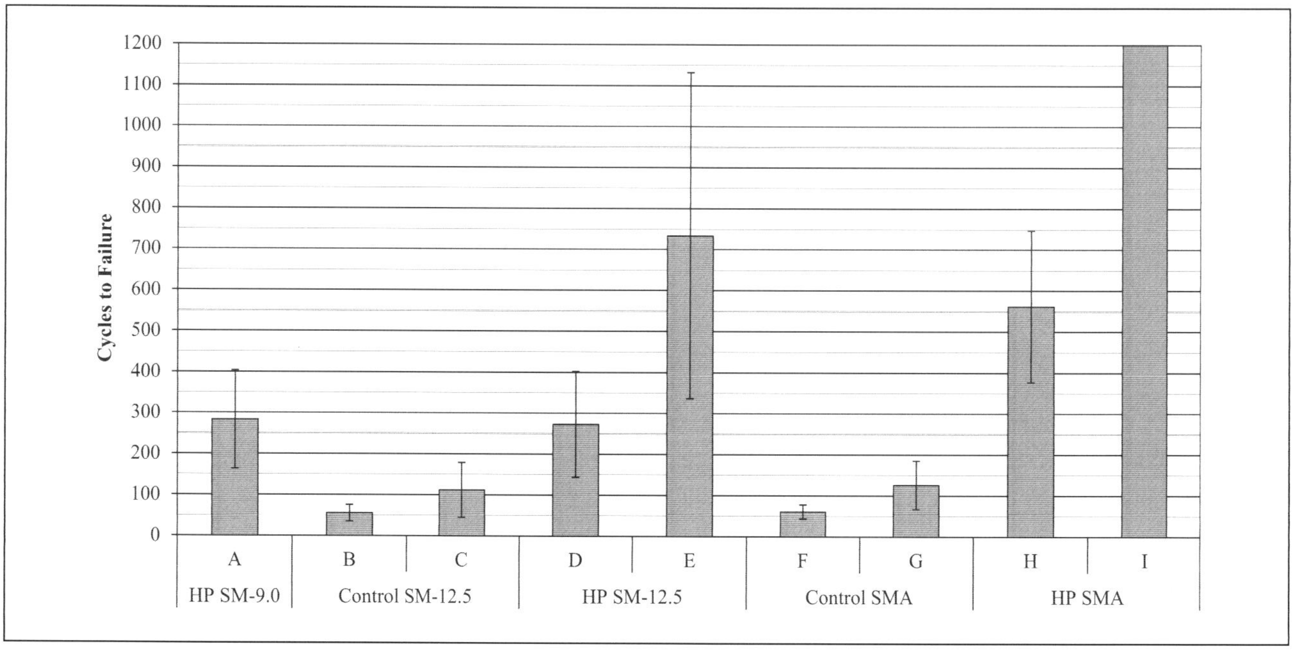

Figure 2. Overlay test results. Error bars represent + /– one standard deviation.

Table 3. T-test Results for the Overlay Test

	HP	Control		HP		Control		HP	
	SM-9.0	SM-12.5		SM-12.5		SMA-12.5	SMA-9.5	SMA-9.5	
Mixture	A	B	C	D	E	F	G	H	I
A	1								
B	*0.01*	1							
C	*0.02*	0.10	1						
D	0.89	*0.01*	*0.02*	1					
E	*0.04*	*0.01*	*0.01*	*0.04*	1				
F	*0.01*	0.72	0.25	*0.01*	*0.01*	1			
G	*0.03*	*0.04*	0.72	*0.05*	*0.01*	0.12	1		
H	*0.04*	*0.04*	*0.00*	*0.03*	0.51	*0.04*	*0.00*	1	
I	*0.00*	*0.00*	*0.00*	*0.00*	*0.03*	*0.00*	*0.00*	*0.03*	1

Note: Italics indicate the differences in means are statistically significant.

Table 4. *T*-test Results for the Semi-Circular Bend Test

	HP	Control		HP		Control		HP	
	SM-9.0	SM-12.5		SM-12.5		SMA-12.5	SMA-9.5	SMA-9.5	
Mixture	A	B	C	D	E	F	G	H	I
A	1								
B	*0.02*	1							
C	*0.01*	0.10	1						
D	0.56	*0.02*	*0.01*	1					
E	*0.01*	*0.00*	*0.00*	*0.02*	1				
F	0.22	*0.01*	*0.01*	0.36	0.17	1			
G	0.27	*0.00*	*0.01*	0.10	*0.00*	*0.04*	1		
H	*0.02*	*0.00*	*0.00*	*0.04*	0.80	0.22	*0.00*	1	
I	*0.01*	*0.02*	*0.00*	*0.01*	*0.01*	*0.01*	*0.03*	*0.00*	1

Note: Italics indicate the differences in means are statistically significant.

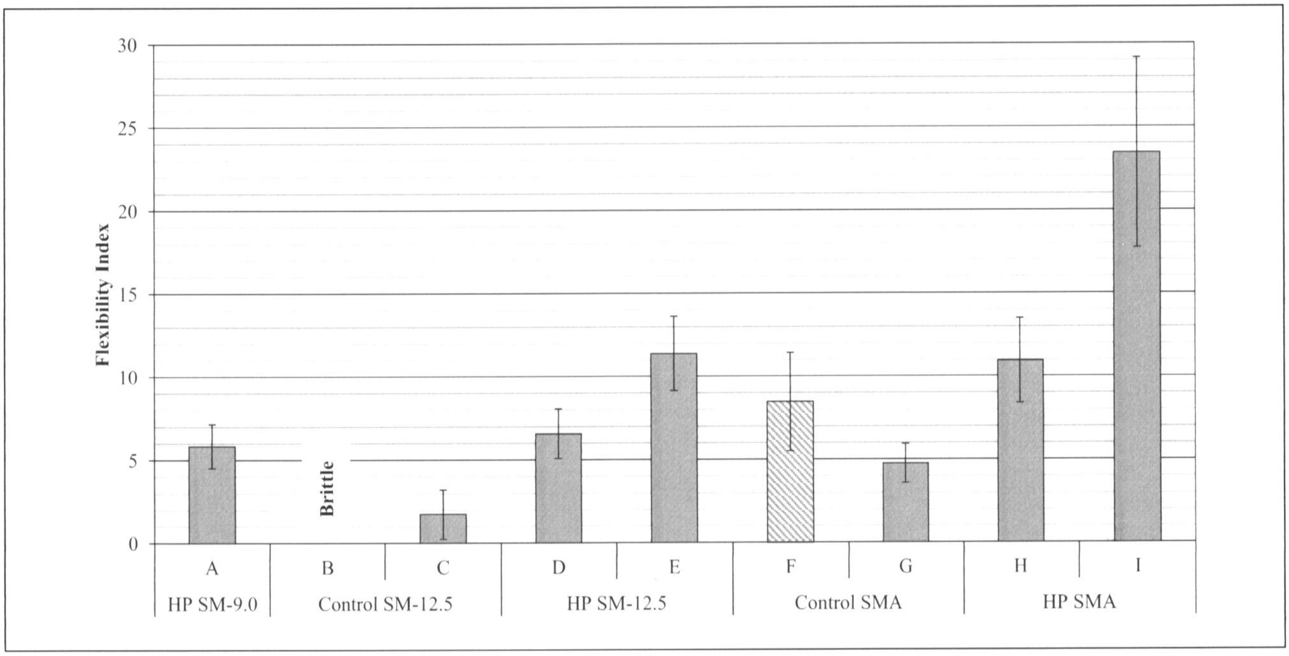

Figure 3. Semi-circular bend I-FIT test results. Error bars represent +/– one standard deviation. The hashed column, F, represents a specimen set with air voids outside of specification.

had a flexibility index of zero. Mixture C also had a single specimen identified as "brittle."

The statistical tests showed that for 75% of the mixture combinations the mean difference was statistically significant. In all mixture-type comparisons, with the exception of Control SMA-F and HP SMA-H, the HP binder increased the SCB-measured crack performance to a significant degree. For control mixtures, the effect of using an SMA mixture versus a dense-graded surface mixture was significant, with the SMA control having a higher flexibility index. Similar to the OT results, the positive effect of using an SMA versus a dense-graded surface mixture was significant for all HP mixtures except for the E-H combination. However, using HP binder in a dense-graded SM in one case outperformed a Control SMA (E-G). The SCB result from the SM-9.0 HP mixture was significantly higher than all control dense-graded mixtures, equivalent to control SMAs and HP SM-12.5-D, and lower than all remaining HP mixtures. Generally, with respect to HP mixtures, having a larger NMAS (9.0 to 12.5 mm) in dense-graded surface mixtures or using an SMA lead to better crack performance with respect to the flexibility index. The impact of the high effective binder and polymer contents in SMA may play a role for the flexibility index in the same way that it does for OT.

Crack Test Correlation Analysis

A correlation analysis was conducted between the mixture properties and the cracking test results, as provided in Table 5. The output of the correlation analysis is a correlation coefficient between −1.0 and 1.0. If there is a high positive correlation a value of 1.0 is the result. For an inverse correlation, −1.0 is the result. For no correlation a value of zero is the result. All values in between represent various degrees of positive or inverse correlation. For both crack tests there is an inverse relationship between the low temperature grade of the binder and the cycles to failure and flexibility index. There is also a correlation between the high binder grade and the laboratory cracking performance. In both cases, the OT sensitivity to binder grade change is stronger by 0.10. Interestingly, the OT result was more highly correlated to increasing binder content than the flexibility index. This may help explain why HP SMA Mixture I outperformed all mixtures in the OT test, as it had the highest total binder content of 7.13%. However, the finding that the correlation between the OT result and effective binder content is not stronger (0.48) was surprising. OT also had high inverse correlations between cycles to failure and Gmm, Gmb, and Gsb.

The flexibility index had a high inverse correlation to the design VTM whereas OT showed almost no correlation to design VTM. Although the analysis looks at the design VTM, it may be implied that specimen air voids have a more significant impact on laboratory performance in the I-FIT test than the OT, which is reflected in the allowable departure from target air voids for each specification (+/−1.0% for OT, +/−0.5% for SCB). The lack of correlation between design VTM and the

Table 5. Results of the Correlation Analysis between Mixture Volumetrics (Table 2) and Cracking Test Results (Figures 2 and 3)

Material/mixture property	Overlay test – cycles to failure	SCB test – flexibility index
Binder low grade	*–0.71*	*–0.61*
Binder high grade	*0.71*	*0.61*
Asphalt content, %	*0.62*	0.42
RAP content, %	0.31	–0.23
Maximum theoretical specific gravity, Gmm	*–0.87*	–0.08
Design VTM, %	–0.04	*–0.77*
VMA, %	0.30	0.03
VFA, %	0.15	*0.73*
Fines to asphalt ratio	0.32	0.36
Mixture bulk specific gravity, Gmb	*–0.78*	0.15
Aggregate bulk specific gravity, Gsb	*–0.77*	0.07
Effective binder content, Pbe, %	0.48	0.35

Note: VTM = voids in total mixture; VMA = voids in mineral aggregate; VFA = voids filled with asphalt. Correlation higher than +/–0.5 is italicized.

inverse correlation to Gmm, Gmb, and Gsb implies that less dense mixtures may perform better on the OT. However, it is well known in the asphalt paving community that increased density leads to better performance, thus this correlation is likely not causal. Finally, there was a moderately high correlation between VFA and the flexibility index. In general, it was surprising to find that there was a low correlation between laboratory cracking performance and the P_{be} and, in the case of the flexibility index, the total asphalt content. Future work will consider a multivariate analysis to determine whether volumetric properties can be used to predict the results of cracking tests.

Test Specimen Air Void Relationship

Due to the smooth faces of a cut specimen, especially with SMA specimens, it was not clear whether the saturated surface dry (SSD) method (AASHTO T166) or vacuum sealing method (AASHTO T331) should be used to calculate G_{mb}. G_{mb} is used to calculate percent air voids. An overlay specimen has an uncut face having an area of approximately 18.8 in.^{2} and an SCB specimen as an uncut area of approximately 37.7 in.^{2}. The influence of these uncut faces on determining G_{mb} using the SSD method is unknown, especially when considering a gap-graded mixture such as SMA. Thus, both SSD and vacuum sealing methods to calculate G_{mb} were evaluated on some of the SMA mixtures to investigate the relationship, as shown in Table 6 and Figure 4. The current method to calculate G_{mb} using the SSD procedure states that if the specimen absorption is greater than 2% the specimen should be tested by parafilm method or the vacuum sealing method. A recent study by Dukatz et al. recommended that this limit be changed to 1% (*8*).

Table 6 presents the air void results using the SSD and vacuum sealing methods for the SMA mixtures. The dataset includes data for specimens that were included in crack testing analysis and those that were not due to being outside of specification requirements. Only the SMA mixtures were evaluated because of the gap-graded nature of the mixture that can lead to higher water absorption levels and excess water loss in the SSD process. These changes can lead to a faulty G_{mb} measurement and thus an erroneous calculation of air voids in the specimen. For crack test analysis purposes the OT test relied on the SSD method, whereas the SCB test relied on the vacuum sealing method. This decision was made based on the SCB specimens having an uncut surface area approximately twice that of OT specimens. Mixture F for the OT test had the only specimen with an absorption level greater than 2.0%, meaning its G_{mb} should be measured using the vacuum sealing method. It should be noted that the SSD and vacuum methods were used on SCB specimens after the notch was cut into the specimen. In hindsight, this may have caused an artificial air void in the specimen worth between roughly 0.2% to 0.3%. Future researchers should consider performing the vacuum method prior to notching the specimen to collect more accurate information for the calculation of air voids.

One striking observation is the difference in air voids between specimens when measured using the vacuum method versus the SSD method. Of the 32 specimens measured with both methods, 20 of them had a difference in air voids between the vacuum method and SSD method of 1% or greater. Even if the difference caused by an artificial air void due to the notch in the SCB specimen in the vacuum method is accounted for, 16 of the specimens had a difference in air voids of greater than 1%. To further investigate these differences, the percent

Table 6. Air Voids Comparison for SSD and Corelok Method

	Overlay test				Semi-circular bend test			
Mixture #	SSD method	Absorption (%)	Vacuum method	Difference	SSD method	Absorption (%)	Vacuum method	Difference
F	7.6	*2.3*	*9.6*	2.0	7.2	2.0	*9.3*	2.1
	6.5	1.1	7.1	0.6	6.8	1.8	*8.7*	1.9
	6.8	1.7	*8.6*	1.7	7.4	1.6	*9.3*	2.0
	6.5	1.3	*7.7*	1.2	6.9	1.9	*8.9*	2.0
G	6.5	1.1	NT	NT	7.0	1.5	*8.3*	1.3
	7.3	2.0	NT	NT	*6.3*	1.0	7.3	1.0
	6.6	1.6	NT	NT	7.5	1.6	*8.8*	1.3
	5.8	1.0	NT	NT	*6.0*	0.9	6.9	0.9
	7.2	1.4	NT	NT	NT	NT	*7.8*	NT
	7.1	1.0	NT	NT	NT	NT	7.0	NT
	NT	NT	NT	NT	NT	NT	7.1	NT
	NT	NT	NT	NT	NT	NT	7.1	NT
H	6.4	0.9	6.9	0.5	*5.8*	0.6	6.6	0.8
	6.0	0.4	*5.9*	−0.1	*5.9*	0.7	6.8	0.9
	6.5	1.1	*8.3*	1.8	*6.2*	0.8	6.6	0.4
	5.9	0.7	6.8	0.9	*5.5*	1.1	6.8	1.3
	6.3	0.6	6.5	0.2	*5.9*	0.9	7.0	1.1
I	8.0	1.3	*8.6*	0.6	*5.9*	0.7	6.8	0.9
	7.8	1.8	*9.6*	1.7	6.5	0.9	7.5	1.0
	7.3	1.6	*8.5*	1.2	*5.3*	0.6	*6.4*	1.1
	7.4	1.2	7.9	0.4	*5.8*	0.5	7.3	1.5
	7.7	1.1	*8.7*	1.0	NT	NT	NT	NT
	7.3	1.1	*8.1*	0.8	NT	NT	NT	NT

Note: NT = not tested. Specimens in italics are outside of the specification limit.

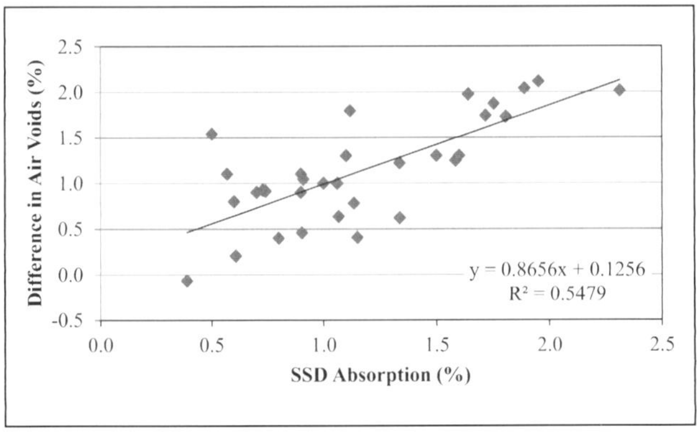

Figure 4. Relationship between SSD absorption and the difference in air voids (vacuum – SSD).

water absorption is plotted against the difference in air voids between methods in Figure 4. As expected, there is a linear trend between increasing water absorption and difference in air voids. At 1% absorption, the difference in air voids between methods is approximately 1%. This suggests that the limiting of SSD absorption to 1% as suggested by Dukatz et al. is warranted, and it could be argued that a more conservative value near 0.5% absorption is justified (*8*).

Findings and Conclusion

In an effort to better address reflective cracking, the laboratory cracking performance of multiple polymer modified mixtures with and without HP binder was investigated. The following findings and conclusions can be made on the basis of this research:

- The dense-graded surface mixtures having an HP binder are less stiff than the dense-graded control mixtures, as assessed by their dynamic modulus. One HP SMA mixture exhibited similar stiffness to the and control SMA mixtures, whereas one HP SMA mixture was less stiff overall.
- The OT and SCB tests yielded very similar results, with HP modified mixtures outperforming their controls for similar mixture types.
- Laboratory crack performance was either unchanged or improved as the NMAS increased in dense-graded mixtures, with the SMA providing the best performance overall.
- The OT correlated well with the binder low- and high-temperature grades, asphalt content, G_{mm}, G_{mb}, and G_{sb}. The SCB test correlated well with binder low- and high-temperature grades, design

VTM, and VFA. Interestingly, the correlation with P_{be} was only 0.48 for the OT and 0.35 for the SCB test.

- Consideration should be given to whether to measure cut SMA specimens using the SSD method or the vacuum sealing method. If using the vacuum sealing method for SCB specimens, it is recommended that the test is conducted prior to notching to avoid artificial air voids.
- The threshold percent water absorption at which the vacuum sealing method is required for G_{mb} and air void content calculations should be further investigated. It is likely that the 2.0% threshold (and even the 1.0% recommended by Dukatz et al. [*8*]) is too high.

Acknowledgments

The authors acknowledge the assistance of Bob Kluttz of Kraton Polymers; Cory Zimmerman of Superior Paving Corp.; Samuel Rosser and Tony Zhang of University of Virginia; Troy Deeds, Donnie Dodds, Benjamin Earl, Zach Robb, Jon Burch, Linda Evans, and Kevin McGhee of Virginia Transportation Research Council; Kim Snead, Ken Elliton, Sam DeLoche, and Guido Zaga, formerly of VTRC; Trenton Clark of Virginia Asphalt Association; personnel at Virginia Paving Corp, Lee Hy Paving, and Fort Myer Construction Company; and Rob Crandol, David Shiells, Tanveer Chowdhury, and Sungho Kim of VDOT.

Author Contributions

The author confirms sole contribution to the paper as follows: study conception and design; data collection; analysis and interpretation of results; and draft manuscript preparation. The author reviewed the results and approved the final version of the manuscript.

References

1. Von Quintus, H. L., J. Mallela, and M. Buncher. Quantification of Effect of Polymer-Modified Asphalt on Flexible Pavement Performance. *Transportation Research Record: Journal of the Transportation Research Board*, 2007. 2001: 141–154.
2. Timm, D. H., M. M. Robbins, J. R. Willis, N. Tran, and A. Taylor. *Field and Laboratory Study of High-Polymer Mixtures at the NCAT Test Track*. Interim Report 12-08. National Center for Asphalt Technology, Auburn, Ala., 2012.
3. Willis, J. R., A. Taylor, N. H. Tran, R. Kluttz, and D. H. Timm. Laboratory Evaluation of High Polymer Plant-Produced Mixtures. *Journal of the Association of Asphalt Paving Technologists*, Vol. 81, 2012, pp. 605–631.
4. Timm, D. H., M. M. Robbins, J. R. Willis, N. Tran, and A. Taylor. *Field and Laboratory Study of High-Polymer Mixtures at the NCAT Test Track*. Final Report. Report 13-03. National Center for Asphalt Technology, Auburn, Ala., 2013.
5. Mogawer, W. S., A. J. Austerman, R. Kluttz, and M. Roussel. High-Performance Thin-Life Overlays with High Reclaimed Asphalt Pavement Content and Warm-Mixture Asphalt Technology. *Transportation Research Record: Journal of the Transportation Research Board*, 2012. 2293: 18–28.
6. Mogawer, W. S., A. J. Austerman, R. Kluttz, and L. N. Mohammad. Development and Validation of Performance-Based Specifications for High-Performance Thin Overlay Mixture. In *Application of Asphalt Mixture Performance-Based Specifications*. Transportation Research Circular, No. E-C189. Transportation Research Board of the National Academies, Washington D.C., October 2014. pp. 52–68.
7. Bowers, B. F., S. D. Diefenderfer, and B. K. Diefenderfer. Laboratory Evaluation of a Plant-Produced High Polymer-Content Asphalt Mixture. *Transportation Research Record: Journal of the Transportation Research Board*, 2017. 2631: 144–152.
8. Dukatz, E., J. Haddock, K. Hall, J. Kliewer, C. Marek, J. Musselman, A. Regimand, R. West, G. Sholar, and N. Tran. *A Review of Aggregate and Asphalt Mixture Specific Gravity Measurements and Their Impacts on Asphalt Mix Design Properties and Mix Acceptance*. Report 12-06. National Center for Asphalt Technology, Auburn, Ala., 2009.

The Standing Committee on Non-Binder Components of Asphalt Mixtures (AFK30) peer-reviewed this paper (18-05313).

Article

TRR
JOURNAL OF THE TRANSPORTATION RESEARCH BOARD

Transportation Research Record
2018, Vol. 2672(28) 69–78

DOI: 10.1177/0361198118782023
journals.sagepub.com/home/trr
SAGE

Laboratory Observation and Evaluation of Asphalt Blends of Reclaimed Asphalt Pavement Binder with Virgin Binder using SEM/EDS

Yanxu Jiang[1], Xingyu Gu[1], Zhou Zhou[1], Fujian Ni[1], and Qiao Dong[1]

Abstract

In this paper, microscopic technique tests were carried out to observe and evaluate the degree of blending between reclaimed asphalt pavement (RAP) binder and virgin binder in hot mixed asphalt mixture. To this end, titanium dioxide (TiO_2) was selected as a tracer to tag virgin binder. Scanning electron microscope/energy dispersive spectrometer (SEM/EDS) experiments were conducted on compacted recycled asphalt mixtures and virgin asphalt mixtures. The element mass ratio of titanium over sulfur (Ti:S) was proposed as an quantitative indicator of blending ratio to accurately evaluate the degree of partial blending between RAP and virgin binders. The SEM/EDS images visually displayed the partial blending in high RAP mixtures. Different partial blending patterns were observed under different handling processes. The results of EDS tests indicated that with the increase of the RAP content, the blending degree of virgin and aged binder decreased rapidly, and the homogeneity of blended binder became weakened. In addition, aging process and recycling agent could improve the efficiency of RAP binder as it is blended with virgin one, and it should be noted that the inter-diffusion of old and new binders need enough time. This methodology provides a systemic approach to determine the degree of binder blending in RAP mixture.

Nowadays, reclaimed asphalt pavement (RAP) is commonly used in rehabilitation and maintenance of pavement due to its economic and environmental benefits. However, to assure the cracking properties and durability performance of recycled asphalt mixture, the proportion of RAP is strictly limited, usually below 30% (*1–3*). NCHRP Project 9-12, *Incorporation of Reclaimed Asphalt Pavement in the Superpave System*, has developed the recommendations for the use of RAP in asphalt mixtures. Those results were based on laboratory testing of asphalt mixtures performance and the fully blending of old–new binder was considered (*1–4*). However, with the high content of RAP in recycled mixes, partial blending between aged binder contained in RAP and virgin binder occurred and hence the aggregate in the final mixture was coated in inhomogeneous binder film (Figure 1) (*5–8*). As a result, the performance of final mixture cannot be ensured simply by changing the performance grade of virgin binder. Therefore, more recent research efforts focused on evaluating the blending ratio of RAP binder with virgin binder which will help in determining the virgin binder content to be added in the mixture.

To address this critical issue, several past researches have proposed a method based on performance tests to indirectly predict the blending ratio of virgin and RAP binder. Mogawer et al. obtained the dynamic modulus of recycled asphalt mixture through experiments and back-calculated the dynamic modulus based on the stiffness of recycled binder, and then evaluated the blending degree of the aged-virgin binder by comparing the differences of the dynamic modulus acquired through the above two methods (*9*). Bennert et al. have also adopted a similar approach by using the performance indicators of high temperature rutting factor $G^*/\sin\delta$ and fatigue factor $G^*\cdot\sin\delta$ (*10*). Due to the inaccuracy of back-calculation models such as the Hirch model, and the researches only being able to consider one performance index in most instances, this method cannot, after all, get the exact blending ratio of aged and virgin binder in recycled mixture.

Other researchers such as Huang et al. and Bowers et al. proposed an approach of staged extraction method (*11, 12*). Different layers of binder in RAP aggregate

[1]College of Transportation Engineering, Southeast University, Nanjing, China

Corresponding Author:
Address correspondence to Xingyu Gu: guxingyu1976@163.com

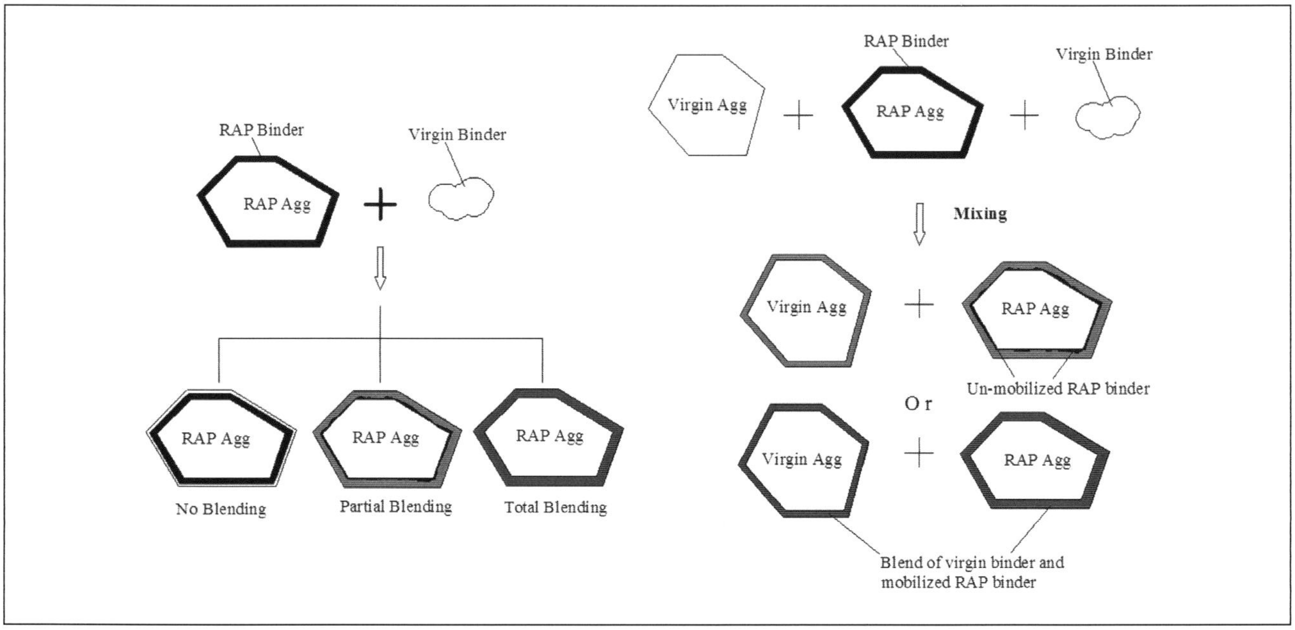

Figure 1. Different blending patterns in recycled asphalt mixtures.

surface was extracted in stages, and then the differences of performances between these binders was analyzed to evaluate the blending degree of old–new binder. Performance tests of binder vary, mainly including dynamic shear rheometer (DSR), Fourier Transform Infrared Spectroscopy (FTIR), and Gel Permeation Chromatography (GPC). However, staged extraction method relies on researchers to grasp the timing of stage and the amount of binder when extracting the binder, which is strongly subjective.

The two methods above could qualitative or semi-quantitative estimate the blending degree of aged and virgin binder. Recently more research has utilized microscopic testing technology to observe and analyze recycled asphalt mixtures. S.N. Nahar et al. firstly studied the microstructures of the 'blending zone' of RAP binder and virgin binder using atomic force microscopy (AFM) (*13*). Navaro et al. also adopted the microscopy to analyze this vital issue and furthermore, quantitative indicators were proposed to characterize the properties of the blend of binder and quantify the influence of production parameters, materials temperature and mixing time (*14*). A series work has done by Cavalli to investigate the 'black rock' effect and particle clustering phenomena in high RAP mixtures by using environmental scanning electron microscopy (ESEM) (*15*, *16*). Microscopic testing technology provides a new approach to investigate the binder blends in recycled asphalt mixtures. What's more, combined with macro performance testing, it would bring the results of microscopic testing from small length scales up to the engineering level (*17*).

Objective

- The objectives of this study include
- Developing better understanding of actual blending scenarios of RAP binder and virgin binder by utilizing scanning electron microscope (SEM)/energy dispersive spectrometer (EDS);
- Proposing a quantitative indicator of blending ratio to accurately evaluate the degree of partial blending between RAP and virgin binder and develop a model to quantify the amount of virgin binder added to recycled asphalt mixtures based on blending ratio; and
- Investigating the effects of recycling agent and aging process on the degree of blending of aged and virgin binder in recycled asphalt mixture.

Material and Sample Preparation

Tracer Binder

To trace virgin binder under SEM, titanium dioxide (TiO_2) powder was selected as a tracer to tag virgin binder and the mass content of the tracer was 20%. The preparation process of tracer binder was performed in two steps as follows: firstly, neat binder 70# was heated to 145°C and the TiO_2 power was added into the binder by a certain percentage. Next the mixture was sufficiently stirred with a modified drill for 20 minutes to obtain stable and homogeneous asphalt binder. Since the tracer binder was only used for microscopic observation, more attention was paid on its mixing liquidity. The results of

Table 1. Grading of RAP Aggregate, Virgin Mixture and Recycled Asphalt Mixtures

Mixture types	Passing %										Asphalt content %
	16	13.2	9.5	4.75	2.36	1.18	0.6	0.3	0.15	0.075	
RAP	100	99.0	93.0	70.1	51.6	35.8	22.9	12.6	8.1	4.8	4.67
Virgin mix	100	96.2	76.1	52.7	37.5	25.1	18.0	13.2	10.5	5.4	4.67
15% RAP	100	94.8	75.7	54.6	34.4	25.2	18.0	13.3	9.6	6.2	4.67
30% RAP	100	95.0	76.0	54.2	35.2	25.9	18.5	13.3	9.7	6.4	4.67
50% RAP	100	95.2	76.5	53.7	36.2	26.6	19.0	13.4	9.8	6.6	4.67
Specification	100	90~100	68~85	38~68	24~50	15~38	10~28	7~20	5~15	4~8	–

Table 2. Types of Engineered Mixtures under Different Producing Processes

Process condition	Engineered mixtures			
	Virgin mix	15% RAP	30% RAP	50% RAP
Normal condition	Mix 1	Mix 4	Mix 5	Mix 6
Short-term aging	Mix 2	Mix 7	Mix 8	Mix 9
Long-term aging	Mix 3	Mix 10	Mix 11	Mix 12
Recycling agent	—	—	Mix 13	Mix 14

135°C viscosity indicated that TiO_2 tracer binder had not shown the significant difference in 135°C viscosity compared with the original neat binder and the value of the viscosity was still in control.

Mixture Designs

The RAP was milled from the top layer of Huning Highway in Jiangsu Province of China and was prefractionated into two parts by using #4 (4.75 mm) sieve. Binder content and aggregate gradation were measured from laboratory tests. Test results are shown in Table 1. The mix design process of the tested mixtures was conducted according to the technical specifications for construction of highway asphalt pavements, JTG F40-2004 (*18*). The synthesis grading and specification requirements were presented in Table 1.

For microscopic observations, tracer binder was used as virgin binder. In order to distinguish RAP aggregate (dark basalt), light colored limestone was selected as virgin aggregate in recycled asphalt mixtures. It was supposed that aged binder was 100% recycled (aged and virgin was totally blended) in RAP mix, therefore, the optimal asphalt content was all 4.67%.

Samples for SEM Observations

In this study, 14 types of engineered mixtures (Mix 1–14) were produced (Table 2). Mix 1–3 were virgin asphalt mixtures. After mixing, the loose asphalt mixture was heated at the compaction temperature for 4 hours and 12 hours to simulate the short-term aging of the field mixing and transporting process and the long-term aging during pavement service, respectively (*19*, *20*). At the same time, use of the recycling agent was considered for high RAP mixture (Mix 13 and 14).

Flow chart presented in Figure 2 illustrates the mixing procedure. The virgin aggregate, virgin binder (tracer binder), and RAP were heated to the mixing temperature separately before mixing. For Mix 1–3, virgin aggregate and tracer binder were mixed at the mixing temperature in a mixing pot to produce virgin mixtures. For Mix 2–14, the heated RAP was first mixed with the virgin aggregate for 2 min. After the RAP binder was mobilized and transferred to the virgin aggregate, the tracer binder and recycling agent (if necessary) was added to mixing pot and mixed for 2 min. The mixing procedure was followed by the aging process. The normal group and the RA group were heated at compaction temperature for 45 min. The short-term aging and long-term aging groups were heated at compaction temperature for 4 h and 12 h, respectively.

After aging, 14 cylindrical specimens were compacted using the Superpave gyratory compactor at 145°C with the measurement of 100 mm (in diameter) by 63.5 mm (in height). The air voids of all the specimens were controlled to 4% ± 0.5%. Gyratory specimens were sliced into 1–1.5 mm slabs and cut into 25 mm^2 squares (Figure 3). The specimen surface was ground and polished, then sputter-coated with Au/Pd powder to make surface conductive for SEM observations (*21*).

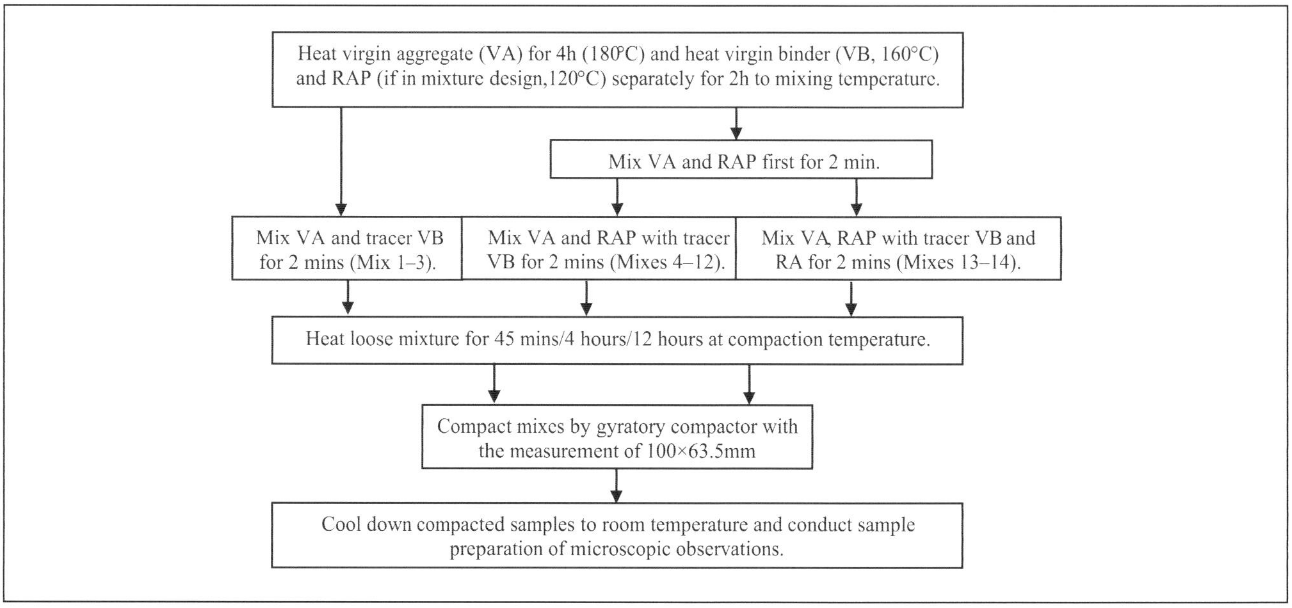

Figure 2. Flow chart of mixing procedure.

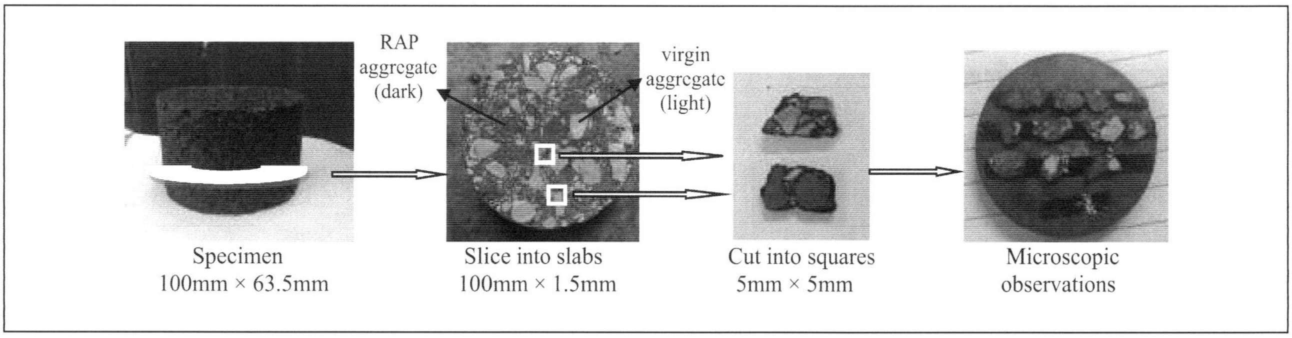

Figure 3. Sample preparation for microscopic observations.

Methodology

The SEM produces a virtual image of the sample from the signals generated by the high-energy electron beam scanning the sample. The electrons interact with the atoms and extra-nuclear electrons of samples and generate a variety of physical signals. By accepting the generated information, the image is enlarged and displayed to obtain the surface morphology of samples. This method can deliver valuable information for asphalt concrete as shown in earlier studies (*14–16*, *22*). The SEM used in this study was a FEI Inspect F50 (Figure 4). The main advantages of the SEM are the high stability and high brightness emission filament, which fulfill the requirements of high-resolution observation and analysis.

In this study, the SEM functions in a high vacuum environment capable of obtaining high resolution images and stable EDS and EBSD analysis. At the same time, the secondary electron detector was used, which detected extra-nuclear electrons which were bombarded out by incident beam electrons. It is extremely sensitive to the surface state of the samples and can therefore effectively show the surface morphology of the sample with higher resolution. Additionally, EDS measurements were performed. The EDS microanalysis gives specific information about the composition of the elements of the samples, in terms of quantity and distribution of the elements that allow their mapping (*22*).

Results and Discussion

Blending State of RAP Binder and Virgin Binder

In this study, EDS mapping experiment was carried out to distinguish the virgin asphalt by obtaining distribution of titanium element. Sulfur and titanium are the feature

Figure 4. FEI Inspect F50 scanning electron microscope (SEM).

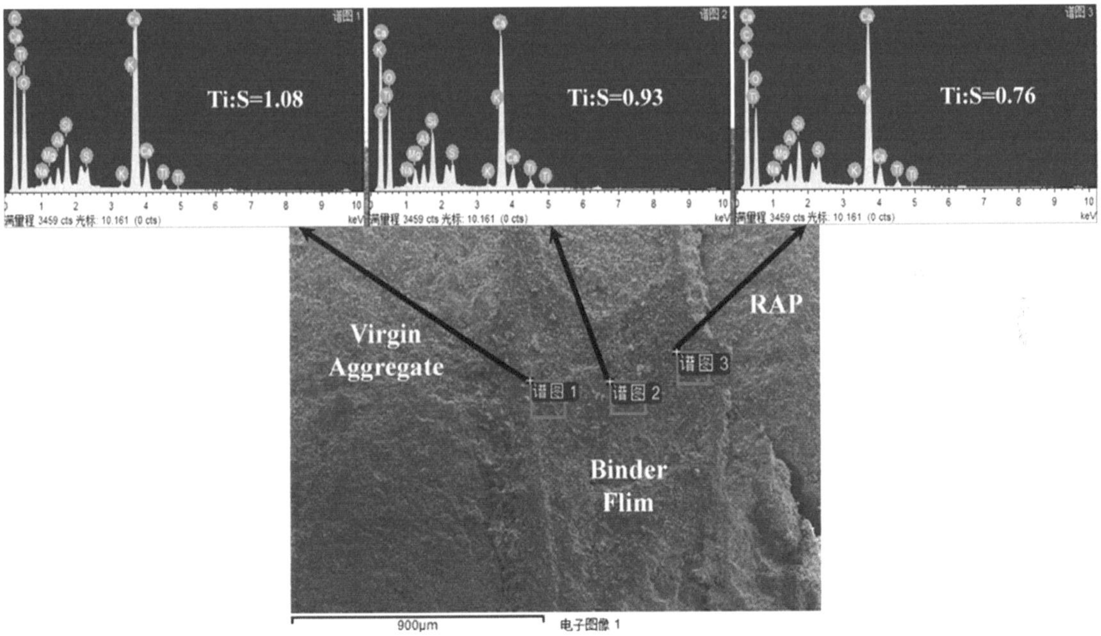

Figure 5. The EDS point scanning result of the binder film between virgin and RAP aggregates in Mix 5.

element of binder and tracer, respectively. The value of Ti:S (mass ratio) obtained from EDS tests can be a quantitative characterization of the ratio of tracer and binder. The EDS point scanning and line scanning of the binder film between virgin and RAP aggregates from Mix 5 and Mix 9, respectively, are presented in Figures 5 and 6. It can be seen in Figure 5, the more it closes to the RAP aggregates, the lower the mass ratio of Ti:S is, indicating that there are differences in the binder composition at these three locations. As shown in the line scanning element mapping (Figure 6), the peak content of titanium and sulfur is almost coincident in the area close to the virgin aggregate. However, on the surface of RAP aggregate, the content of titanium decreases significantly, while the content of sulfur still remains at peak level. It is speculated that the virgin binder (containing TiO_2) is not completely diffused to the surface of the RAP aggregate, and the partial blending of old–new binder does occur in these recycled mixes. What's more, it is likely that a certain amount of the aged binder coating on the RAP aggregate surface is simply not integrated with the virgin binder.

For the aim of understanding the actual blending state of virgin and RAP binder, the RAP aggregate (dark

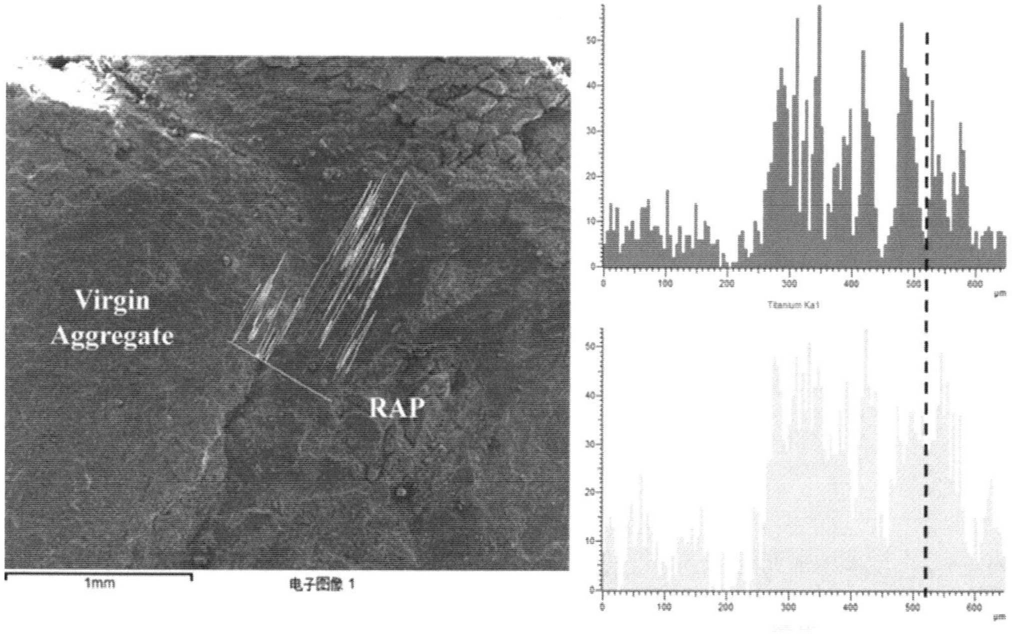

Figure 6. The EDS line scanning result of the binder film between virgin and RAP aggregates in Mix 9.

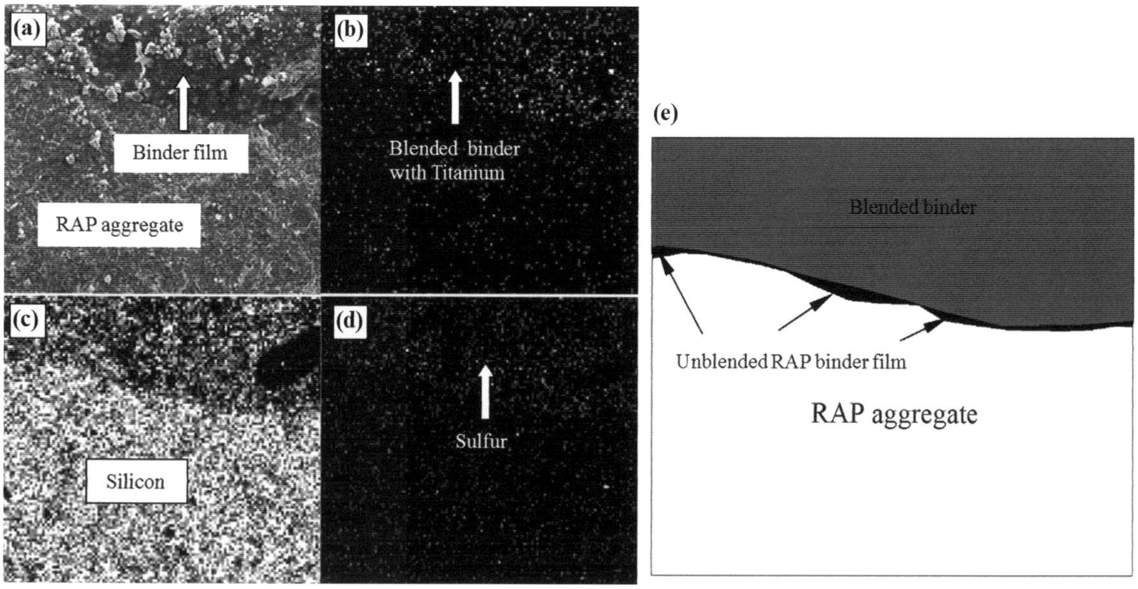

Figure 7. (*a*) SEM micrograph of a RAP aggregate from sample Mix 6 and EDS elements mapping of (*b*) titanium distribution, (*c*) silicon distribution, (*d*) sulfur distribution, and (e) schematic diagram of partial blending of virgin binder and RAP binder.

basalt) was select to perform SEM/EDS measurements. Figure 7 shows the SEM micrograph and relevant element mapping of a RAP aggregate sample obtained from Mix 6. Figure 7*b* presents the titanium distribution map which indicates the location of virgin binder. Figure 7*c* shows the silicon distribution map mainly consistent with the aggregates. Figure 7*d* shows the sulfur distribution map indicating location of all binder (including virgin and aged binder). Figure 7*e* illustrates the actual blend state of virgin and RAP binder. By subtracting the

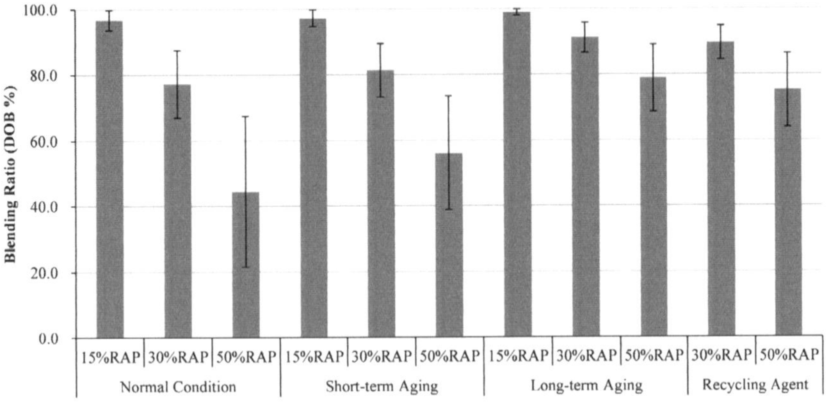

FIGURE 8

Figure 8. Blending ratio of virgin and RAP binder in all kinds of recycled mixtures.

silicon map from the titanium map one can obtain the RAP binder film which is not blended with virgin binder containing TiO_2.

It can be seen that in the high RAP mixes, some of the aged binder is coated on the RAP aggregate in its original form. It affects the homogeneity of the asphalt mastic seriously, which leads to the poor cracking performance of the recycled mixture. However, the EDS tests also found that the "sandwich" structure could hardly be observed in the recycled mixes under the condition of long-term aging process or adding recycling agent, indicating that there are no strict boundaries between the old and new binder. The gradual change of the blending state tends to be a common phenomenon.

Determination of the Degree of Partial Blending

The thickness of mastic (asphalt binder and fine aggregates) is approximately 100 μm, and in order to obtain accurate Ti:S (mass ratio) value in binder mastic coating in RAP and virgin aggregates, the EDS tests were performed in aggregates surface 30 μm range. Six replicates (three RAP aggregates and three virgin aggregates) were selected for each mixture (Mix 1–14). The initial Ti:S mass ratio ($i_{Ti:S}$) was obtained from EDS tests of virgin asphalt mixture (Mix 1–3). As the partial blending of virgin and aged binder occurs in recycled mixture, a portion of RAP binder is blended with virgin tracer binder, resulting in the change of Ti:S mass ratio of binder coating in virgin aggregates and RAP aggregates.

In this study, it is considered that the virgin aggregates are coated with homogeneous blended binder (Figure 1 Partial blending), so the final Ti:S mass ratio ($f_{Ti:S}$) of blended binder can be obtained by EDS test in the binder film of virgin aggregates. Furthermore, the blending ratio is calculated by the following formula to express the

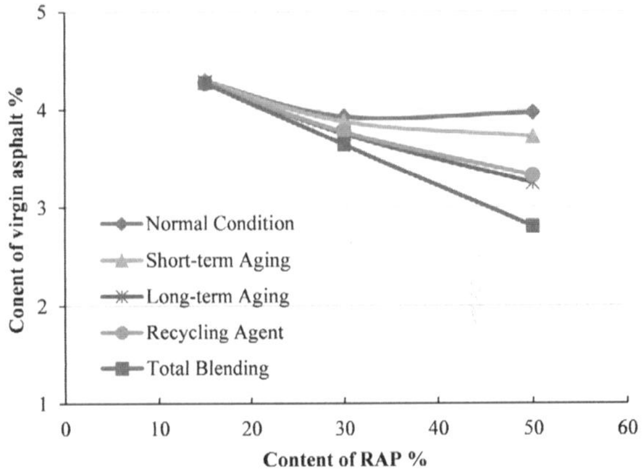

Figure 9. A model to quantify the amount of virgin binder added to recycled mixtures based on blending ratio.

degree of blending (DOB) of the virgin and aged binder. The blending ratio of tested recycled asphalt mixtures are shown in Figure 8. A model to quantify the amount of virgin binder added to recycled asphalt mixtures is developed based on blending ratio (see Figure 9).

$$\gamma = \text{blended aged asphalt}/\text{total aged asphalt}$$

$$= \frac{(\frac{i_{Ti:S}}{f_{Ti:S}} - 1)*(OAC - \alpha_1\beta_1 - \alpha_2\beta_2)*S_v/S_a}{\alpha_1\beta_1 + \alpha_2\beta_2} * 100\% \quad (1)$$

where

γ = blending ratio;

$i_{Ti:S}$ = initial value of Ti:S (mass ratio) in virgin binder;

$f_{Ti:S}$ = final value of Ti:S (mass ratio) in blended binder;

OAC = optimal asphalt content of asphalt mixture;

α_i = the proportion of RAP mass (two tranches: >4.75 mm and <4.75 mm);

β_i = the asphalt content of RAP (two tranches: >4.75 mm and <4.75 mm);

S_v = the content of sulfur in virgin binder;

S_a = the content of sulfur in aged binder;

From Figure 8, it can be seen that the aged binder almost all blended with the virgin one in recycled asphalt mixture containing 15% RAP, and was relatively stable, indicating the formation of homogeneous blended binder. With the increase of RAP content, the blending ratio decreases rapidly and one can see notable variance gradually comes into being especially in normal and short-term aging condition. It is demonstrated that the effective utilization rate of the aged binder becomes smaller with the increase of the RAP content, and the blended binder has greater variability in the different layers and different areas of the recycled asphalt mixture. This is due to the clump effect of RAP in recycled mixture, making the RAP unable to fully contact with virgin aggregates and binder to complete the transfer and diffusion of aged binder.

At the same time, short-term aging process seems to not make a qualitative improvement of the blending ratio. However, after long-term aging process, the blending ratio of virgin and aged binder increases significantly, and the variability correspondingly becomes smaller, especially in high RAP mixtures. It indicates that the blending process mainly occurs during the long-term use of the recycled mixture pavement. That is to say, the inter-diffusion of virgin and RAP binder needs enough time. It could be speculated that after the construction of the pavement, for a period of time the composition of the binder film between the virgin and RAP aggregates would gradually become consistent, and the gradual change of the blending state would also disappear. What's more, the use of recycling agent has also improved the efficiency of RAP binder as it is blended with virgin binder. Due to the improvement of the rheological properties and mobility of aged binder, the diffusion between the virgin and RAP binder has greatly accelerated.

Additionally, an important finding as noted in Figure 9 is that such great amount of virgin binder needs to be added in high RAP (50%) recycled mixture if the blending ratio is the only factor taken into consideration. In normal condition, it even exceeds in virgin binder to be added in the 30% RAP recycled asphalt mixture, which is obviously not economical. In the cases of long-term aging process and the addition of recycling agent, the amount of added virgin binder is close to the total blending circumstances. It could be appropriate to increase the amount of virgin binder to offset the partial blending of old–new asphalt in recycled mixture. It should be noted that there are apparent performance differences between the blended asphalt and virgin one. It is necessary to improve the properties of blended binder by other agents to ensure the performance of recycled asphalt mixture.

Conclusions

The presented study evaluated the blending efficiency of virgin and RAP binders in recycled asphalt mixture. Utilizing SEM/EDS microscopic technique and tracer method, the following conclusions can be drawn:

1. The partial blending of virgin and aged binder is commonly observed in recycled asphalt mixtures. In the high RAP mixes, a certain amount of the aged binder is coated on the RAP aggregate in its original form which affects the homogeneity of the asphalt mastic seriously, leading to the poor cracking performance of the recycled mixture.
2. In the cases of long-term aging process or adding recycling agent, there are no strict boundaries between the old and new binder. The gradual change of the blending state tends to be a common phenomenon in hot-mix asphalt including RAP.
3. With the increase of the RAP content in recycled mixture, the blending degree of virgin and aged binder decreases rapidly and the homogeneity of blended binder is weakened which may lead to the poor performance of the mixture.
4. The blending process of the virgin and RAP binder mainly occurs during the long-term use of the recycled mixture pavement. In other words, the inter-diffusion of RAP and virgin binder needs sufficient time.
5. Using recycling agent improved the efficiency RAP binder as it is blended with virgin binder. Because of the improvement of the rheological properties and mobility of aged binder, the diffusion between the virgin and RAP binder has greatly accelerated.
6. For high RAP content recycled asphalt mixture, it is ineffective and uneconomical to settle the influence of partial blending only by changing the addition of virgin binder. It is necessary to improve the utilization ratio of the aged binder and the properties of blended binder by recycling agents or other agents to ensure the performance of recycled asphalt mixture.

Acknowledgments

This work is sponsored by a project supported by the National Natural Science Foundation of China, Basic Research Plan, in

Jiangsu, China (No. BK20151413). The authors gratefully acknowledge the support.

Author Contributions

The authors confirm contribution to the paper as follows: study conception and design: Yanxu Jiang, Xingyu Gu, Zhou Zhou, Fujian Ni; data collection: Yanxu Jiang, Zhou Zhou; analysis and interpretation of results: Yanxu Jiang, Zhou Zhou; draft manuscript preparation: Yanxu Jiang, Qiao Dong. All authors reviewed the results and approved the final version of the manuscript.

References

1. McDaniel, R. S., H. Soleymani, R. M. Anderson, P. Turner, and R. Peterson. *NCHRP Report 9-12: Incorporation of Reclaimed Asphalt Pavement in the Superpave system*. Transportation Research Board of the National Academies, Washington, D.C., 1997.
2. West, R., J. R. Willis, and M. Marasteanu. *NCHRP Report 752: Improved Mix Design, Evaluation, and Materials Management Practices for Hot Mix Asphalt with High Reclaimed Asphalt Pavement Content*. Transportation Research Board of the National Academies, Washington, D.C., 2013.
3. Zhou, Z., X. Gu, Q. Li, F. Ni, and R. Yuan. Use of Rejuvenator, Styrene-Butadiene Rubber Latex, and Warm-Mix Asphalt Technology to Achieve Conventional Mixture Performance with 50% Reclaimed Asphalt Pavement. *Transportation Research Record: Journal of the Transportation Research Board*, 2016. 2575: 160–167.
4. Shirodkar, P., Y. Mehta, A. Nolan, E. Dubois, D. Reger, and L. Mccarthy. Development of Blending Chart for Different Degrees of Blending of RAP Binder and Virgin Binder. *Resources Conservation & Recycling*, Vol. 73, No. 4, 2013, pp. 156–161.
5. Chen, J. S., S. F. Chen, M. C. Liao, and S. W. Huang. Laboratory Evaluation of Asphalt Blends of Recycling Agents Mixed with Aged Binders. *Journal of Materials in Civil Engineering*, Vol. 27, No. 4, 2014, p. 04014143.
6. Shirodkar, P., Y. Mehta, A. Nolan, K. Sonpal, A. Norton, C. Tomlinson, E. Dubois, P. Sullivan, and R. Sauber. A Study to Determine the Degree of Partial Blending of Reclaimed Asphalt Pavement (RAP) Binder for High RAP Hot Mix Asphalt. *Construction & Building Materials*, Vol. 25, No. 1, 2011, pp. 150–155.
7. Colin, J. Experimental Investigation of the Homogeneity of the Blended Binder of a High Rate Recycled Asphalt. *Road Materials & Pavement Design*, Vol. 13, No. 3, 2012, pp. 566–575.
8. Zhao, S., B. Huang, X. Shu, and M. E. Woods. Quantitative Evaluation of Blending and Diffusion in High RAP and RAS Mixtures. *Materials & Design*, Vol. 89, 2016, pp. 1161–1170.
9. Mogawer, W., T. Bennert, J. S. Daniel, R. Bonaquist, A. Austerman, and A. Booshehrian. Performance Characteristics of Plant Produced High RAP Mixtures. *Road Materials & Pavement Design*, Vol. 13, No. supl, 2012, pp. 183–208.
10. Bennert, T., and R. Dongré. Backcalculation Method to Determine Effective Asphalt Binder Properties of Recycled Asphalt Pavement Mixtures. *Transportation Research Record: Journal of the Transportation Research Board*, 2010. 2179: 75–84.
11. Huang, B., G. Li, D. Vukosavljevic, X. Shu, and B. K. Egan. Laboratory Investigation of Mixing Hot-Mix Asphalt with Recycled Asphalt Pavement. *Transportation Research Record: Journal of the Transportation Research Board*, 2005. 1929: 37–45.
12. Bowers, B. F., B. Huang, X. Shu, and B. C. Miller. Investigation of Reclaimed Asphalt Pavement Blending Efficiency through GPC and FTIR. *Construction & Building Materials*, Vol. 50, No. 50, 2014, pp. 517–523.
13. Nahar, S. N., M. Mohajeri, A. J. Schmets, A. Scarpas, M. F. V. D. Ven, and G. Schitter. First Observation of Blending-Zone Morphology at Interface of Reclaimed Asphalt Binder and Virgin Bitumen. *Transportation Research Record: Journal of the Transportation Research Board*, 2013. 2370: 1–9.
14. Navaro, J., D. Bruneau, I. Drouadaine, J. Colin, A. Dony, and J. Cournet. Observation and Evaluation of the Degree of Blending of Reclaimed Asphalt Concretes using Microscopy Image Analysis. *Construction & Building Materials*, Vol. 37, 2012, pp. 135–143.
15. Cavalli, M. C., M. N. Partl, and L. D. Poulikakos. Measuring the Binder Film Residues on Black Rock in Mixtures with High Amounts of Reclaimed Asphalt. *Journal of Cleaner Production*, Vol. 149, 2017, pp. 665–672.
16. Bressi, S., M. C. Cavalli, M. N. Partl, G. Tebaldi, A. G. Dumont, and L. D. Poulikakos. Particle Clustering Phenomena in Hot Asphalt Mixtures with High Content of Reclaimed Asphalt Pavements. *Construction & Building Materials*, Vol. 100, 2015, pp. 207–217.
17. Rinaldini, E., P. Schuetz, M. N. Partl, G. Tebaldi, L. D. Poulikakos, E. Rinaldini, and G. Tebaldi. Investigating the Blending of Reclaimed Asphalt with Virgin Materials using Rheology, Electron Microscopy and Computer Tomography. *Composites Part B*, Vol. 67, 2014, pp. 579–587.
18. Research Institute of Highway Ministry of Transport. *Technical Specification for Construction of Highway Asphalt Pavements, JTG F40-2004*. China Communications Press, Beijing, 2011.
19. Wen, H., and K. Zhang. Investigation of Blending Mechanisms for Reclaimed Asphalt Pavement Binder and Virgin Binder in Laboratory-Produced Reclaimed Asphalt Pavement Mixtures. *Transportation Research Record: Journal of the Transportation Research Board*, 2016. 2575: 187–195.
20. Mandal, T., L. Cheng, P. Chaturabong, and H. Bahia (2017). Effects of Mixture Design Factors on Results of Semi Circular Bend (SCB-LSU) Test. Presented at 96th Annual Meeting of the Transportation Research Board, Washington, D.C., 2017.
21. Poulikakos, L. D., and M. N. Partl. Investigation of Porous Asphalt Microstructure using Optical and Electron

Microscopy. *Journal of Microscopy*, Vol. 240, No. 2, 2010, pp. 145–154.
22. Castorena, C., S. Pape, and C. Mooney. Blending Measurements in Mixtures with Reclaimed Asphalt: Use of Scanning Electron Microscopy with X-Ray Analysis. *Transportation Research Record: Journal of the Transportation Research Board*, 2016. 2574: 57–63.

The Standing Committee on Non-Binder Components of Asphalt Mixtures (AFK30) peer-reviewed this paper (18-01321).

Article

TRR
JOURNAL OF THE TRANSPORTATION RESEARCH BOARD

Transportation Research Record
2018, Vol. 2672(28) 79–87

DOI: 10.1177/0361198118780703
journals.sagepub.com/home/trr
SAGE

Effect of Partial Blending on High Content Reclaimed Asphalt Pavement (RAP) Mix Design and Mixture Properties

Shuai Yu[1], Shihui Shen[1,2], Xuli Zhou[3], and Xinghai Li[3]

Abstract

Reclaimed Asphalt Pavement (RAP) material, especially high RAP content mixture, is nowadays receiving more interest in asphalt pavement construction. When determining the optimum asphalt content (OAC) in RAP mixture design, RAP partial blending is sometimes neglected and total blending is assumed. Such an assumption influences the virgin binder content determination and can contribute to the poor fatigue and cracking performance of high RAP pavement in the field. This paper studied the impact of RAP partial blending on the OAC of mixtures with varied RAP content up to 40%. A Laboratory Hamburg Wheel Tracking Test, an IDT fracture test, and an SCB test were performed on mixtures designed by different methods to evaluate rutting and cracking performance. In particular, a partial blending design method was proposed to improve the mixture performance by adjusting the virgin binder content through the RAP blending ratio. It was found that the optimum virgin binder content decreased as the RAP content increased; the Superpave volumetric method required the lowest OAC while the partial blending method required the highest OAC. The rutting performance of the RAP mixtures considering partial blending was slightly lower than other mixtures. However, all RAP mixtures showed better rutting resistance than virgin mixtures. The cracking performance of the high RAP mixtures was significantly improved if the partial blending effect was considered in the design.

Usage of Reclaimed Asphalt Pavement (RAP) has become a continuous trend in pavement engineering with the rapid progress in efforts toward pavement rehabilitation and reconstruction. Because of the economic and environment benefits, high content RAP usage has been a focus of attention in recent years (*1*). Extensive research including, the NCHRP 09-46 project, has been carried out into the potential and practical approaches of applying high content RAP material in asphalt mixtures. Blending charts and other approaches were also developed to aid virgin binder grade selection to compensate the highly aged RAP binder. Relatively, the interaction between RAP binder and virgin binder was less understood, which could be one of the reasons for the poor field performance of some high RAP projects (*2*). Highly aged RAP binder, when blended with virgin asphalt binder at high temperatures, can result in number of performance issues including fatigue and thermal cracking, especially with more than 20% RAP content (*3–5*). The Federal Highway Administration showed that many construction projects still limited RAP content to around 20% for the sake of the pavement performance (*6*).

The design of asphalt mixture containing RAP, especially the determination of the optimum asphalt content (OAC), usually depended on the RAP content. The standards of the American Association of State Highway and Transportation Officials (AASHTO) suggested that for RAP usage up to 20%, no modification of mix design was needed (*7*, *8*). Practically, RAP binder was used to replace part of virgin binder to achieve material and cost reduction. It was believed that at low RAP content, the influence from RAP was minimal, and thus it was reasonable to use full RAP content reduction assuming that all RAP binder can be effective in coating aggregate. When the RAP content was more than 20%, in addition to implementing binder grade reduction (*7*, *8*), the OAC

[1]The Key Laboratory of Road and Traffic Engineering, Ministry of Education, Tongji University, Shanghai, China
[2]Department of Engineering, Penn State Altoona, Altoona, PA
[3]Beijing Key Laboratory of Highway Engineering Materials and Testing & Authentication Technology, Beijing Road Engineering Quality Supervision Station, Beijing, China

Corresponding Author:
Address correspondence to Shihui Shen: sshen2@tongji.edu.cn

must be re-designed according to the Superpave mixture design method.

When RAP was included in a hot mix asphalt (HMA) mixture, only part of the aged binders became soft enough in the mixing process to recoat the aggregates. Some hard RAP binders, as well as those absorbed RAP binders, might never become effective during the mixing and compaction stage. McDaniel (*9*) used RAP binder, RAP mixture, virgin binder and virgin aggregates to produce a series of mixtures to research their performance, and verified the existence of RAP partial blending in a high-temperature blending process. Huang (*10*) and Shirodkar (*11*) compared the differences between the binders around RAP aggregates and virgin aggregates after blending using gap gradation and stage extraction methods, and confirmed that RAP blending effectiveness changed with RAP content and virgin binder properties. A multistep extraction method was adopted by Liphardt (*12*) to assess the blending levels of binders in HMA containing RAP. According to the results from the Multiple Stress Creep Recovery (MSCR) test, there was no complete blending between RAP binder and virgin binder. A Gel Permeation Chromatograghy (GPC) technique was also introduced to study the RAP blending efficiency in HMA by Zhao et al. (*13*). A blending chart was built on RAP-virgin binder blends with GPC parameter Large Molecular Size Percentage (LMSP) and the relationship was found to be linear The relationship of %RAP and LMSP was then used to confirm the existence of RAP partial mobilization. Zhang (*14*) simulated the interaction of RAP material with virgin aggregate and asphalt binder during the plant mixing process using a discrete element method, and found that RAP binder transfer reached its maximum for 30% RAP, compared with the 10% and 50% RAP cases. Using a similar gap gradation approach, Yu and Shen (*15*) investigated blending effectiveness for various RAP content mixtures, and obtained the specific values of the RAP blending ratio based on different binder property parameters.

Neglecting partial blending effects might result in improper mix design and insufficient asphalt binder content in the RAP mixture, because less virgin binder will be required under the full blending assumption. Such insufficient asphalt binder content can be particularly problematic for high RAP mixtures, leading to higher cracking and raveling potential (*16–18*). Aiming to reduce the influence of the aged binder on the fatigue and cracking performance of the RAP mixtures, Boriack (*2*) investigated the effect of increasing virgin binder content in RAP mixtures on mix performance and compaction efforts. It was found that 0.5% more than optimum design binder significantly improves the RAP mix performance for both rutting resistance and fatigue cracking resistance, especially for low RAP content. Zhou (*19*) adopted the Superpave volumetric method to research the OAC differences of HMA containing varied RAP content. It was found that the optimum total binder (RAP binder and virgin binder) content increased with the increase of the RAP content. However, the cracking issues still occurred in these RAP mixtures. Balanced RAP mix design was then proposed to solve these issues. In this method, the mixture performance was evaluated at 98% theoretical maximum density and the final asphalt content was selected after optimizing the mix density, Hamburg wheel track test, and overlay test requirement.

In summary, although much research has been conducted into the complex RAP binder-virgin binder system, there is still merit in investigating how to reasonably determine the OAC of high RAP mixtures while, at the same time, ensuring acceptable field performance. This paper therefore applies the partial blending concept and the RAP blending ratio determined by Yu and Shen (*15*) to achieve the objectives of: (a) developing a method to determine the OAC of high RAP content mixtures by taking into account blending ratios; and (b) comparing three RAP mixture design methods for determining OAC and evaluating their effects on mixture properties.

High Rap Mixture Design

OAC Determination Based on Superpave Volumetric Method

The most conventional method for high RAP mixture OAC determination followed the Superpave Volumetric Mix Design method given by AASHTO M323. After selecting appropriate virgin binder grade and performing gradation design based on both virgin and RAP aggregates, the OAC of RAP mixture was determined to ensure mixtures designed to N_{design} gyration satisfying the Superpave specifications for V_a, VMA, VFA and other requirements (*20*).

OAC Determination Based on Total Blending Assumption

The main hypothesis of the total blending method is that all of the RAP binder is treated as active binder in the mixing process. Consequently, the OAC of RAP mixture can be the same as the OAC of the virgin HMA with no RAP, given that the two mixtures have the same gradation. Therefore, the amount of virgin binder needed for the RAP mixture, m_{vb}, can be calculated using Equation 1. This approach is very similar to the current practice of using RAP when the content of RAP is less than 20% of the total mixture.

$$m_{vb} = OAC \times m_{HMA} - Pb_{RAP} \times m_{RAP} \quad (1)$$

where, m_{vb} is the virgin binder content; m_{HMA} is the total mass of the RAP mixture; OAC is the optimum asphalt content of the virgin mixture that presents the same gradation as the RAP mixture; m_{RAP} is the mass of RAP materials; Pb_{RAP} is the asphalt content of the RAP materials.

OAC Determination Considering RAP Blending Ratio

The literature (*9–11, 21*) suggested partial blending always existed when RAP was mixed with virgin binder and aggregate. Yu and Shen (*15*) further used a parameter of a Blending Ratio (BR) and an approach of gap gradation to quantify the effectiveness of RAP binder at different RAP contents. In that approach, virgin coarse aggregates and fine RAP materials were used to produce the RAP mixture. The coarse aggregate mixture and fine aggregate mixture can thus be separated easily after the blending process, and the differences between the extracted CM (coarse mixture) binder and FM (fine mixture) binder can reflect the RAP blending effectiveness. Proportion binder, mixed RAP binder and virgin binder based on the proportion of RAP usage, was then produced to estimate the ideal full blending condition. Theoretically, the property of CM binder should always be between the property of virgin binder and the proportion binder; the closer the CM binder's position to the proportion binder's, the higher degree of blending it has. The specific value of BR can be determined by Equation 2,

$$BR = \frac{\ln(R'_c) - \ln(R_v)}{\ln(R_p) - \ln(R_v)} \times 100\% \quad (2)$$

where

R'_c = specific binder property parameters for the CM binder;

R_v = specific binder property parameters for the virgin binder;

R_p = specific binder property parameters for the proportion binder.

Although such values should vary depending on different RAP sources, binder grade and other factors, using those developed BR will provide an example to explain the influence of partial blending on OAC determination and RAP mixture performance, as compared with two other RAP design methods. Details of the BR determination according to various binder parameters and RAP content can be referred to reference (*15*).

The RAP blending ratio helps to distinguish how much aged binder in the RAP is involved in the blending process, and how much is absorbed or stuck on the RAP aggregate, behaving like a hard rock. Therefore, the optimum virgin binder content of the RAP mixture can be determined based on the OAC of virgin mixture and the RAP blending ratio using Equation 3. All parameters have the same meaning as Equation 1.

$$m_{vb} = OAC \times m_{HMA} - BR \times Pb_{RAP} \times m_{RAP} \quad (3)$$

Materials and Experiment

Focusing mainly on the methodology and general trends, a single source virgin limestone aggregate and PG 64-22 virgin binder were selected for the experiment. These materials were the same as the those used in Yu and Shen (*15*), so that the same BR can be used. For comparison purposes, no binder grade reduction was used including a 40% RAP mixture. Figure 1 presents the gradation of the RAP materials and the final RAP mixture,

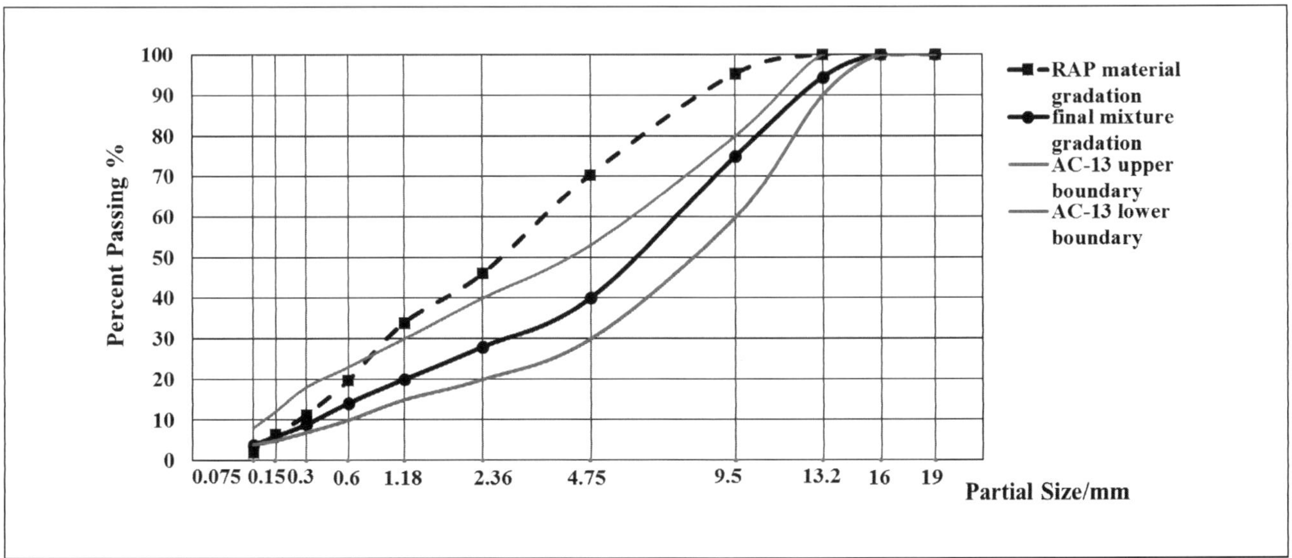

Figure 1. The gradation of RAP material and final mixtures.

Table 1. Summary of the Mixture Tests and Parameters

Experiment	HWTT	IDT	SCB
Specification	AASHTO T324-14	AASHTO T322-03	AASHTO TP124-16
Testing temperature	50°C water bath	25°C, 0°C	25°C
Testing condition	Cyclic wheel loading	50mm/min, 10mm/min	50mm/min
Specimen air void	7 ± 0.5%	4 ± 0.5%	7 ± 0.5%
Parameter	RRI	IDT strength Vertical failure deformation	Flexibility Index
Performance	Rutting	Cracking	Cracking

respectively, as well as the specification boundaries for the designed mixtures. The asphalt content of the RAP material was 4.8% following the binder extraction and recovery test method specified by AASHTO T164.

Before mixing, virgin aggregates were washed and dried in the oven overnight. The loss of fines during washing was accounted and added back for mix design. Virgin binder and virgin aggregates were heated to 175°C, and RAP material was heated to 110°C for less than 0.5 H to avoid further aging (*18*). To simulate the real field blending process, the following mixing approach was used: blend the pre-heated RAP materials and virgin aggregates first at 160°C temperature for 90 s, and then add appropriate amounts of virgin binder to the mixes for the another 90 s blending. Place the mixtures in the 135°C oven after blending for 2 H to simulate the short-term aging process. Compact the hot mixtures to specimens at different air voids for the mechanical testing.

Three RAP content mixtures, of 0%, 20%, and 40%, were prepared to study the influences of RAP content on the mixture performance. For the Superpave mix design approach, the control parameters according to local climate and traffic conditions include: $N_{initial}$ = 7, N_{design} = 75, N_{max} = 115, V_a @ N_{design} = 4%, VMA @ $N_{design} \geq$ 14%, VFA @ N_{design} = 65% - 80%, $\%G_{mm}$ @ $N_{initial}$ <89%, and $\%G_{mm}$ @ N_{max} <98% (*20*).

A Hamburg wheel track test (HWTT) at high temperature was performed to evaluate the mixture rutting performance, and an indirect tensile test (IDT) and a semi-circular bend test (SCB) at intermediate and low temperatures were performed to evaluate the mixture's cracking performance. These tests were selected because they were suggested by the literature as reasonable performance tests for asphalt mixtures (*16*, *21*, *22*). A summary of the mixture tests and parameters used in the evaluation is presented in Table 1. For each test, two replicate specimens were tested and their average values were used for data analysis. The Coefficient of Variance (COV) values of each test were checked to ensure the repeatability of the test results.

Table 2. Blending Ratio Base on Various Binder Parameters in Different RAP Content

RAP content	20%	40%	60%
$G^*/\sin\delta$	36.2%	83.0%	73.5%
$J_{nr0.1}$	37.4%	82.5%	78.0%
$J_{nr3.2}$	32.9%	84.0%	73.7%
$G^*\cdot\sin\delta$	21.2%	81.6%	63.3%
FE	20.9%	84.8%	71.4%
Average	30%	83%	72%

Results and Discussion

RAP Blending Ratio and OAC Determination

Table 2 presents the BR of RAP mixtures determined by Yu and Shen (*15*) for the same mixture type and RAP source used in this study. In particular, the blending ratios were determined from various binder properties, which were relatively consistent for either high or intermediate temperature parameters. Therefore, the average value of the blending ratio was used in the OAC determination for certain RAP content.

Using the three OAC determination methods discussed earlier, the OAC for the total binder content and the virgin binder content were obtained for each RAP content mixture, and the results are illustrated in Figures 2 and 3. As shown in Figure 2, the optimum total binder contents decreased when RAP was introduced. However, they stayed at a similar level for both 20% and 40% RAP content. The total blending method required the same amount of total OAC regardless of the RAP content, which was expected. The partial blending method suggested higher total binder content be used to compensate the amount of RAP binder that was ineffective during mixing. However, less virgin binder was actually needed due to the contribution from RAP binder. As shown in Figure 3, the least amount of virgin binder was needed if based on the Superpave design method, while the largest amount was needed if considering partial blending. In particular, significant virgin binder reduction was achieved no matter which design method was

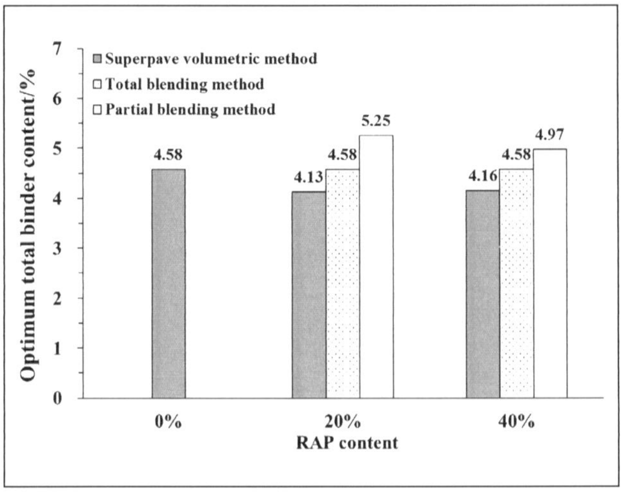

Figure 2. Optimum total binder content of different mixtures.

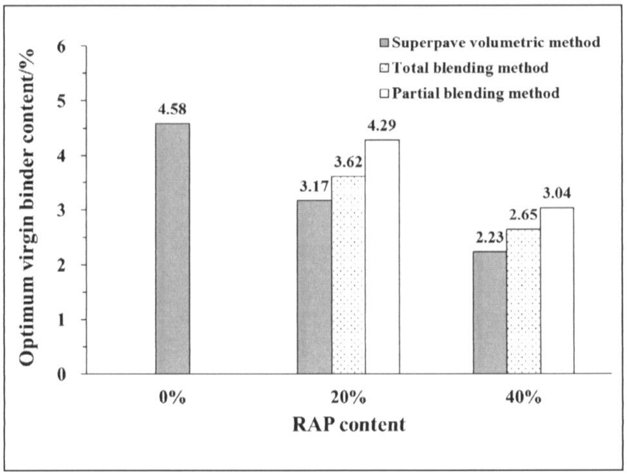

Figure 3. Optimum virgin binder content of different mixtures.

used; with the Superpave method giving the highest reduction and Partial blending method giving the lowest reduction. It was also worth noting that during mixing, the 40% RAP mixtures designed by the Superpave method appeared to be much drier than virgin mixture and other mixtures, indicating some concerns of low asphalt content.

Rutting Properties of RAP Mixture

An HWTT test in a 50°C water bath (AASHTO T324-14) was performed to compare the rutting properties of the RAP mixes. The specimen size for this test was $\phi 150 \times 62$ mm and its target air void content was 7 ± 0.5%. The test was set to stop at 20,000 wheel passes or 20 mm rut depth, whichever came first. The rutting resistance index (RRI), calculated by Equation 4, was used to

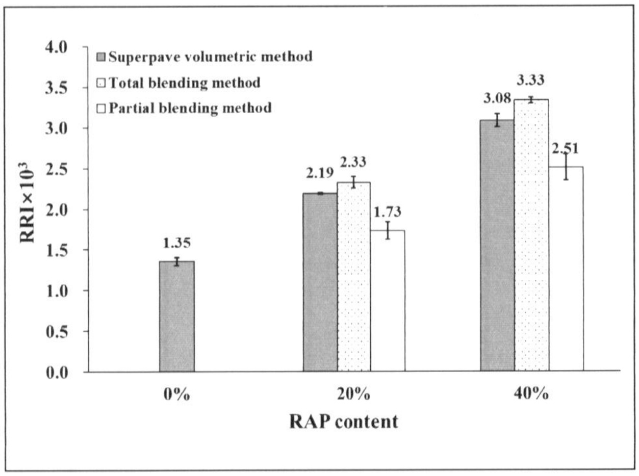

Figure 4. Rut resistance index of different mixtures.

compare the rutting potential of various mixtures. The RRI considered the number of passes and rut depth jointly, and was found to be a significant determinant of field rutting performance (*16*). A higher RRI value correlated with a lower rut depth value.

$$RRI = Number\ of\ passes \times \left(1 - \frac{rut\ depth}{25.4}\right) \tag{4}$$

Figure 4 summarizes the RRI values of different mixtures with error bars included (the error bar shows minimum and maximum values of the test results) and all of the data had COVs less than 10%. It was reasonable that better rutting properties occurred in the mixtures containing higher RAP content and lower asphalt content. The best rutting performance was achieved in the mixtures designed by the total blending method for both 20% and 40% RAP content. For 20% RAP content, the RRI value of mixtures designed by a partial blending method was much lower than the other two methods, indicating lower rutting resistance. However, adding 20% or 40% RAP always produced a more rut resistant mixture than 0% RAP mixture, regardless of the type of mix design method used for OAC determination. This finding was as expected because rutting generally was not a concern for high RAP mixtures.

Cracking Properties of RAP Mixture as Measured in IDT Test

To compare the cracking performance of the RAP mixtures designed by different approaches, IDT fracture tests at 25°C and 0°C were performed in accordance with AASHTO T322-03. The tests had specimen geometry of $\phi 150 \times 50$ mm, target air void content of 4 ± 0.5%, a line-loading rate of 50 mm/min for 25°C test, and 10 mm/min for 0°C test. The mixture's vertical failure

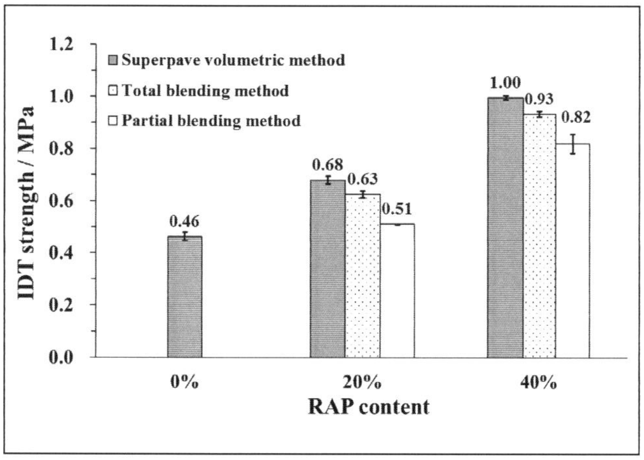

Figure 5. IDT strength at intermediate temperature.

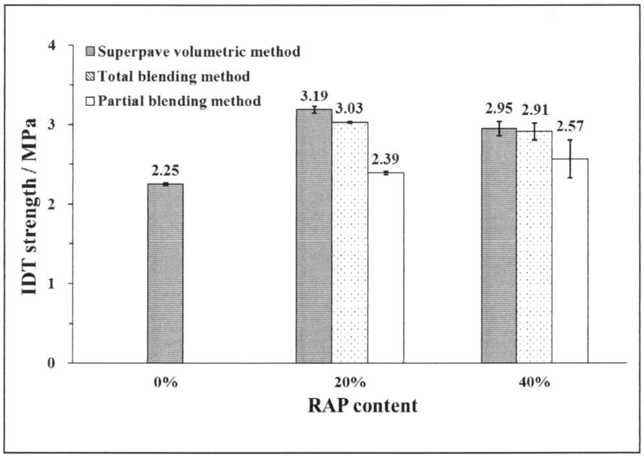

Figure 7. IDT strength at low temperature.

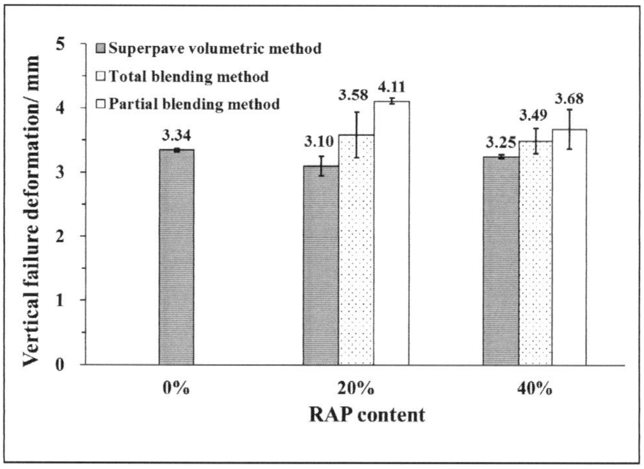

Figure 6. Vertical failure deformation at intermediate temperature.

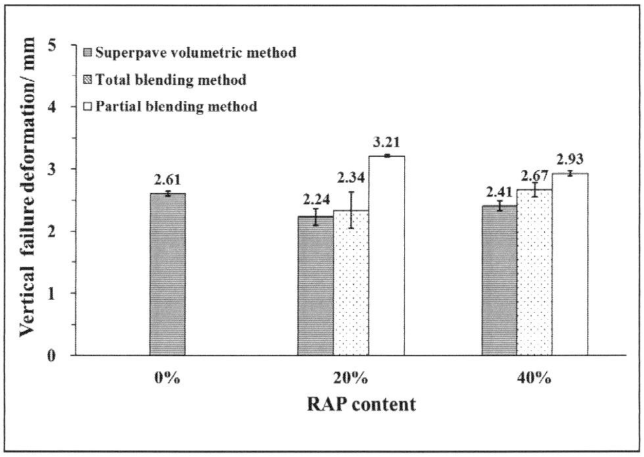

Figure 8. Vertical failure deformation at low temperature.

deformation and IDT strength value obtained from IDT was used to compare the mixture cracking performance. They were proposed by Shen et al. (*16*) as potentially significant determinants for wheel-path longitudinal cracking. A higher vertical failure deformation value and a lower IDT strength value correlated with less wheel-path longitudinal cracking in the field.

Figures 5–8 showed the IDT strength and vertical failure deformation values at intermediate temperature and low temperature with error bars indicated. All of the data had COV values less than 11%. It can be seen that the best cracking performance was achieved in partial blending mixtures for both 20% and 40% RAP content mixtures, as compared to two other design methods. The IDT strength and vertical failure deformation of 20% and 40% partial blending mixture were similar or better than the 0% RAP situation, showing clear cracking performance improvement due to having additional virgin binder added. It should be noted that here the 40% RAP mixture didn't use a lower PG grade binder and only focused on the effect of asphalt content on performance. In practice, reducing the PG grade of virgin binder should improve the overall cracking performance of high RAP mixtures, regardless of which design method is used.

Cracking Properties of RAP Mixture as Measured in SCB Test

SCB fracture tests at 25°C were performed in accordance with AASHTO TP124-16 to compare the cracking performance. The specimen size of the SCB test was ϕ150 × 55 mm semi-circular, the air void content was 7 ± 0.5%, and the line-loading rate was 50 mm/min. At the beginning of the experiment, a 0.1 ± 0.1 kN contact load with 0.05 kN/s loading rate was imposed first, and then the full load-displacement curve was recorded until the load dropped below 0.1 kN. The load-displacement

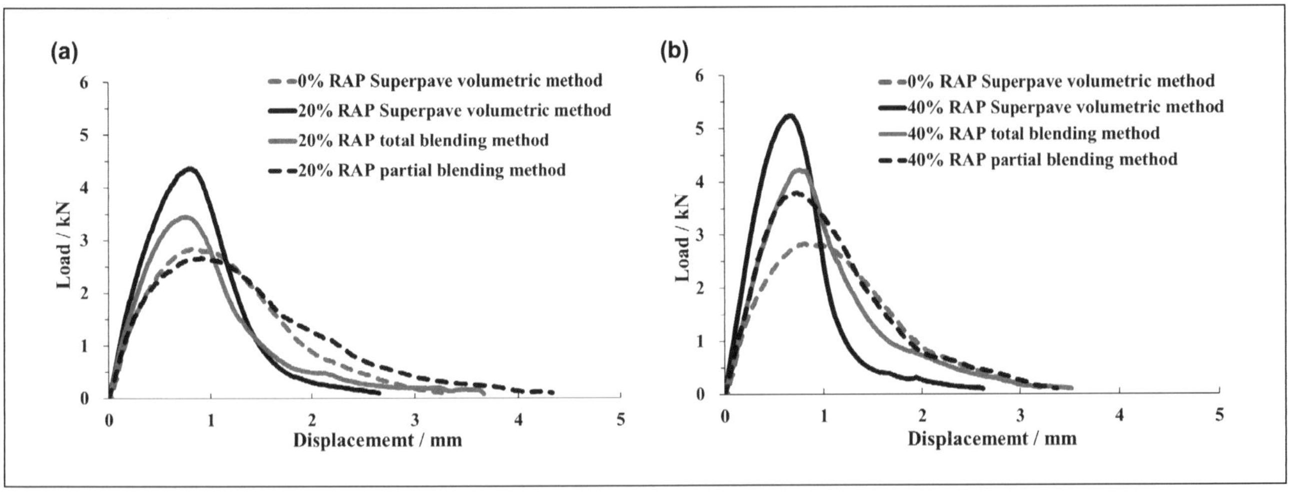

Figure 9. Load-displacement curves of SCB tests at intermediate temperature (*a*) for 20% RAP mixture and (*b*) for 40% RAP mixture.

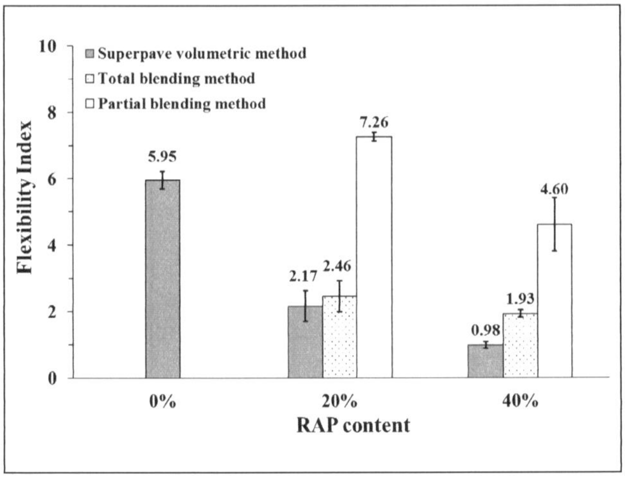

Figure 10. Flexibility index at intermediate temperature.

curve of varied mixes was presented in Figure 9. And the Flexibility Index (FI) was calculated using Equation 5.

$$FI = A \times \frac{W_f}{Area \times |m|} \times 10^6 \quad (5)$$

where, W_f is work of fracture, the area under the load-displacement curve; *Area* is the specimen ligament area; m is the inflection point determined on the load-displacement curve after the peak point. Factor A is used for unit conversion and scaling, A is equal to 0.01.

Figures 9 and 10 present the results of the load-displacement curve and flexibility index (FI) at intermediate temperature with error bars indicated. All data had COV values less than 20%, showing good repeatability. In particular, FI was derived from the fracture energy and slope of the load-displacement curve after a crack began to propagate, and mixtures with low FI values could have increased brittleness and reduced cracking resistance (*22*). It was clear that FI values decreased with the increase of RAP content. RAP mixtures designed by the partial blending method presented better cracking performance than the other two methods for both 20% and 40% RAP content. A 20% RAP mixture, considering the partial blending, even showed better cracking performance than the virgin mixture.

Conclusions and Recommendations

This paper has studied the effect of RAP partial blending on the OAC and the performance of RAP mixtures, especially for high RAP content mixtures. Considering the RAP blending ratio, partial blending design method was compared with two other RAP mix design methods including Superpave volumetric method and total blending method. The comparison was based on laboratory rutting and cracking performance tests for a virgin mixture, 20% RAP mixture, and 40% RAP mixture. Several conclusions can be drawn from this study:

- The optimum virgin binder content of RAP mixture designed by the Superpave volumetric method generally decreased with more RAP usage, but the total binder content including virgin binder and RAP binder remained at a similar level for different RAP content mixtures.
- The OAC value determined by the partial blending method was slightly higher than the other two methods because of RAP blending ratio considerations. The Superpave volumetric method determined the lowest OAC for the same gradation RAP mixture. Note no binder gradation reduction was considered here.

- The rutting performance of RAP mixture designed by the partial blending method was slightly worse than the RAP mixture designed by the conventional method. However, all RAP mixtures with 20% or 40% RAP content showed better rutting resistance than the virgin mixture.
- The cracking performance of the RAP mixture designed by the partial blending method was much better than the mixtures designed by the other two methods for both low and high RAP content. Low RAP content mixtures even presented better cracking performance than the virgin mixture.

Overall, the findings in this paper suggest that partial blending effects should be considered in the design of high RAP mixtures. Using the conventional Superpave mix design method to determine the OAC of high RAP mixtures might lead to insufficient virgin asphalt binder, as also noticed by other researchers when conducting balanced mix design (*19*). Low OAC might, at least partially, contribute to the current cracking performance issues of the high RAP mixtures. The laboratory mixing process also suggested that the mixture designed by the Superpave method appeared to be drier than 0% RAP virgin mixture as well as the mixture designed by partial blending method.

Acknowledgments

The authors would like to thank Jiaxing Jiahai Construction Co., Ltd. for supplying materials utilized in this study.

Author Contributions

The authors confirm contribution to the paper as follows: study conception and design: Shihui Shen; data collection: Shuai Yu and Xinghai Li; analysis and interpretation of the results: Shuai Yu, Shihui Shen, Xuli Zhou, and Xinghai Li; draft manuscript preparation: Shuai Yu and Shihui Shen. All authors reviewed the results and approved the final version of the manuscript.

References

1. Hansen K. R. and Copeland A., *2nd Annual Asphalt Pavement Industry Survey on Reclaimed Asphalt Pavement, Reclaimed Asphalt Shingles, and Warm-Mix Asphalt Usage: 2009–2011*. National Asphalt Pavement Association, Lanham, MD, 2013
2. Boriack, P., S. W. Katicha, and G. W. Fintsch. A Laboratory Study on the Effect of High RAP and High Asphalt Binder Content on the Stiffness, Fatigue Resistance and Rutting Resistance of Asphalt Concrete. Presented at 93rd Annual Meeting of the Transportation Research Board, Washington, D.C., 2014
3. Zhao, S., B. Huang, X. Shu, and M. E. Woods. Quantitative Evaluation of Blending and Diffusion in High RAP and RAS Mixtures. *Materials & Design*, Vol. 89, 2016, pp. 1161–1170.
4. Yu, B., X. Gu, M. Wu, and F. Ni. Application of a High Percentage of Reclaimed Asphalt Pavement in an Asphalt Mixture: Blending Process and Performance Investigation. *Road Materials & Pavement Design*, Vol. 18, No. 3, 2017, pp. 753–765.
5. Zhao, S., B. Huang, X. Shu, and M. Woods. Comparative Evaluation of Warm Mix Asphalt Containing High Percentages of Reclaimed Asphalt Pavement. *Construction & Building Materials*, Vol. 44, 2013, pp. 92–100.
6. Copeland, A. *Reclaimed Asphalt Pavement in Asphalt Mixtures: State of the Practice Publication*. FHWA-HRT-11-021. Turner-Fairbank Highway Research Center, Federal Highway Administration, 2011.
7. American Association of State Highway and Transportation Officials. *Mechanistic-empirical Pavement Design Guide: A Manual of Practice*, 2nd ed. American Association of State Highway and Transportation Officials, Washington, D.C., 2015.
8. Newcomb, B., E. Raybrown, and J. Epps. *Designing HMA Mixtures with High RAP Content: A Practical Guide*. Quality Improvement Series 124. National Asphalt Pavement Association, U.S. Department of Transportation, Federal Highway Administration, Washington, D.C., 2007, p. 36.
9. McDaniel, R. S., H. Soleymani, R. M. Anderson, P. Turner, and R. Peterson. *NCHRP Report 452: Recommended Use of Reclaimed Asphalt Pavement in the Superpave Mix Design Method*. Transportation Research Board of the National Academies, Washington, D.C., 2001.
10. Huang, B, G Li, D Vukosavljevic, X Shu, and K. Egan Brian. Laboratory Investigation of Mixing Hot-mix asphalt with Reclaimed Asphalt Pavement. *Transportation Research Record: Journal of the Transportation Research Board*, 2005. 1929: 37–45.
11. Shirodkar, P., Y. Mehta, A. Nolan, K. Sonpal, A. Norton, and C. Tomlinson. A Study to Determine the Degree of Partial Blending of Reclaimed Asphalt Pavement (RAP) Binder for High Rap Hot Mix Asphalt. *Construction & Building Materials*, Vol. 25, 2011, pp. 150–155.
12. Liphardt, A., P. Radziszewski, and J. Król. Binder Blending Estimation Method in Hot Mix Asphalt with Reclaimed Asphalt. *Procedia Engineering*, Vol. 111, 2015, pp. 502–509.
13. Zhao, S., B. Huang, X. Shu, and M. Woods. Quantitative Characterization of Binder Blending: How Much RAP-RAS Binder is Mobilized During Mixing? *Transportation Research Record Journal of the Transportation Research Board*, 2015. 2506: 72–80.
14. Zhang, K., H. Wen, and A. Hobbs. Laboratory Tests and Numerical Simulations of Mixing Superheated Virgin Aggregate with Reclaimed Asphalt Pavement Materials. *Transportation Research Record Journal of the Transportation Research Board*, 2015. 2506: 62–71.
15. Yu, S., S. Shen, C. Zhang, and X. Jia. Evaluation of the Blending Effectiveness of Reclaimed Asphalt Pavement Binder. *ASCE Journal of Materials in Civil Engineering*, Vol. 29, No. 12, 2017. DOI: 10.1061/(ASCE)MT.1943-5533.0002095.

16. Shen, S. H., S. Wu, W. Zhang, L. Mohammad, and B. Muhunthan. *NCHRP Report 843: Long-term Field Performance of Warm Mix Asphalt Technologies*. Transportation Research Board of the National Academies, Washington, D.C., 2017.
17. Randy, W., J. R. Wills, and M. Marasteanu. *NCHRP Report 752: Improved Mix Design, Evaluation, and Materials Management Practices for Hot Mix Asphalt with High Reclaimed Asphalt Pavement Content*. Transportation Research Board of the National Academies, Washington, D.C., 2013.
18. Zhao, S., B. Huang, and X. Shu. Investigation on Binder Homogeneity of RAP/RAS Mixtures Through Staged Extraction. *Construction & Building Materials*, Vol. 82, 2015, pp. 184–191.
19. Zhou, F., S. Hu, G. Das, R. Lee, T. Scullion, and G. Claros. Successful High Rap Mixes Designed with Balanced Rutting and Cracking Requirements. *Asphalt Paving Technology: Association of Asphalt Paving Technologists-Proceedings of the Technical Sessions*, Vol. 81, No. 1, 2012, pp. 477–505.
20. Brown, E. R., P. S. Kandhal, F. L. Roberts, Y. R. Kim, D. Y. Lee, and T. W. Kennedy. *Hot Mix Asphalt Materials, Mixture Design and Construction*. 3rd ed. National Asphalt Pavement Association Research and Education Foundation, Lanham, Maryland, 1996.
21. National Research Council. *NCHRP Report 673: A Manual for Design of Hot-Mix Asphalt with Commentary*. Transportation Research Board of the National Academies, Washington, D.C., 2011.
22. Ozer, H., I. L. Al-Qadi, J. Lambros, A. El-Khatib, P. Singhvi, and B. Doll. Development of the Fracture-based Flexibility Index for Asphalt Concrete Cracking Potential using Modified Semi-circle Bending Test Parameters. *Construction & Building Materials*, Vol. 115, 2016, pp. 390–401.

The Standing Committee on Non-Binder Components of Asphalt Mixtures (AFK30) peer-reviewed this paper (18-02942).

Article

TRR
JOURNAL OF THE TRANSPORTATION RESEARCH BOARD

Transportation Research Record
2018, Vol. 2672(28) 88–95

DOI: 10.1177/0361198118787936
journals.sagepub.com/home/trr
SAGE

Laboratory Performance of Asphalt Mixtures Containing Re-Refined Engine Oil Bottoms Modified Asphalt Binders

Taesun You[1], Yucheng Shi[2], Louay N. Mohammad[3], and Samuel B. Cooper, III[4]

Abstract

The increasing use of hardened oxidized asphalt binders in recycling materials such as recycled asphalt pavement (RAP) and recycled asphalt shingles (RAS) has increased the need to use re-refined engine oil bottoms (REOB) as a recycling agent. However, conflicting conclusions on the effect of REOB on mixture performance have been reported in the literature. The objective of this study was to characterize the laboratory performance of conventional asphalt mixtures and mixtures containing REOB modified asphalt binders. Four 12.5-mm asphalt mixtures, containing 0%, 5%, 10%, and 15% REOB modified asphalt binders, were designed to meet Superpave design criteria. A suite of laboratory tests was utilized to ascertain the mechanistic behavior of the mixtures considered. Mechanistic tests were performed to evaluate the high-, intermediate-, and low-temperature properties and moisture susceptibilities of laboratory produced mixtures using the Hamburg Loaded-Wheel Tester (LWT), the Semi-Circular Bending (SCB) test, and the Thermal Stress Restrained Specimen (TSRST), respectively. Results showed that the addition of REOB did not adversely affect the rutting and moisture susceptibility of the mixtures. Similar low-temperature mixture performance was observed with an REOB dosage up to 10%, while a mixture with 15% REOB showed significantly reduced low-temperature cracking performance. A decreasing trend in SCB J_c with an increasing REOB content was observed indicating a negative effect of REOB on intermediate-temperature cracking resistance.

Re-refined Engine Oil Bottoms (REOB), also known as Vacuum Tower Asphalt Extender (VTAE) is the non-distillable residuum from a vacuum tower during the re-refinery process of used engine oil (*1*). It has been implemented to reduce both the high and low temperatures in the Superpave Performance Grade (PG) system since the 1980s (*1*). The Federal Highway Administration (FHWA) reported that REOB was found in about 20% of over 1,000 asphalt binder samples collected from various parts of the U.S. (*2*). Recently, REOB has been used as a softening agent with the increasing use of recycled asphalt pavement (RAP) and recycled asphalt shingles (RAS). Hardened oxidized asphalt binder in RAP and RAS is required to be softened to increase cracking resistance.

A report from the Asphalt Institute (AI) entitled *The Use of REOB/VTAE in Asphalt* (*1*) provided a synthesis of the published literature on the performance of mixtures containing REOB/VTAE (*2–26*). Conflicting conclusions regarding the effect of REOB on both field and mixture performance were reported. Several studies showed an adverse effect of REOB (*13*, *14*, *20*), while others indicated a positive or negligible influence on field performance (*11*, *12*). Moreover, the effect of REOB on mixture performance was inconsistent (*6*, *23*, *26*).

Wielinski et al. (*27*) investigated the effect of REOB on the fatigue resistance of asphalt mixtures by conducting flexural beam fatigue tests. They reported that short-term and long-term aged asphalt mixtures containing 9% REOB showed equal or better fatigue performance as compared with mixtures without REOB.

Intermediate-temperature fatigue cracking of short-term and long-term aged asphalt mixtures was studied by Bennert et al. (*28*). Three fatigue tests, including overlay tester (OT), a semi-circular bend (SCB) test, and a flexural beam fatigue

[1]Louisiana Transportation Research Center, Louisiana State University, Baton Rouge, LA
[2]Department of Civil and Environmental Engineering, Louisiana State University, Baton Rouge, LA
[3]Department of Civil and Environmental Engineering, Engineering Materials Characterization Research Facility, Louisiana State University, Louisiana Transportation Research Center, Baton Rouge, LA
[4]Louisiana Transportation and Research Center, Baton Rouge, LA

Corresponding Author:
Address correspondence to Louay N. Mohammad: louaym@lsu.edu

test, were conducted with two types of asphalt binder (PG 64-22 and PG 58-28) and various REOB contents from two sources. It was observed that the REOB source marginally influenced the overall fatigue performance of asphalt mixtures. In addition, the fatigue resistance measured by OT and fatigue life obtained from the flexural beam fatigue test were inversely related. SCB test results did not show a direct trend on the fatigue cracking of asphalt mixtures containing REOB. Further, the research was extended to include low-temperature thermal cracking and high-temperature permanent deformation evaluation by Bennert et al. (*29*). It was concluded that there was no trend between the fracture temperature and REOB content (0%, 6%, 10%, and 20%) evaluated in this study. Furthermore, the addition of REOB did not affect the rutting potential of mixtures.

Ozer et al. (*30*) evaluated the effect of REOB on asphalt mixture performance including rutting susceptibility and low- and intermediate-temperature cracking resistance. Two REOB modifiers from different manufactures, varying REOB contents, and three mixture aging levels (i.e., short-term, long-term, and extended long-term) were considered during the Hamburg loaded-wheel track test (LWT) and SCB test. LWT results indicated no significant difference in rutting susceptibility between asphalt mixtures with and without REOB. A consistent reduction in fracture resistance was found with increasing dosage of both REOB products based on SCB test results.

A similar study was reported by Mogawer et al. (*31*). The effect of REOB on mixture performance after short- and long-term aging was investigated. The LWT, SCB, and Disc-shaped Compact Tension (DCT) tests were conducted to characterize moisture damage/rutting, overall cracking, and low-temperature cracking, respectively. The addition of REOB did not adversely affect the mixture performance (*31*).

Mensching et al. (*32*) evaluated various mixture aging methods, including short-term aging, long-term aging, and extended high-temperature oven aging, for REOB-modified mixtures, in terms of stiffness, damage, and thermal cracking. Among the three aging procedures, the extended high-temperature oven aging showed promise for characterizing the aging susceptibility of mixtures containing REOB.

The impact of REOB on dynamic modulus, moisture susceptibility, and fatigue and thermal cracking performance of asphalt mixtures containing various REOB contents was investigated by Li et al. (*33*). The results showed that mixtures with higher REOB content were more susceptible to moisture damage, while the cracking performance was inconsistent within the range of REOB tested (0%, 2.5%, 6%, and 15%). However, the test results indicated that high amounts of REOB (15% in this study) negatively affected mixture performance.

Cooper et al. (*34*, *35*) evaluated the laboratory performance of mixtures containing recycled asphalt shingles and REOB contents. The LWT, SCB test, and Thermal Stress Restrained Specimen (TSRST) were conducted to characterize the rutting susceptibility, intermediate-temperature cracking, and low-temperature performance of mixtures, respectively. The test results showed that REOB did not affect the rutting performance, but adversely influenced the intermediate-temperature cracking. In addition, a decreasing trend in low-temperature fracture performance was observed with an increasing REOB content.

Despite numerous efforts on performance of asphalt mixtures containing REOB, conclusions reported were inconsistent and sometimes conflicting. Some studies (e.g., *14, 16, 18*) indicate the adverse effect of the addition of REOB on mixture performance, while other research (e.g., *5–7, 31*) showed that mixtures containing REOB have equivalent or better performance as compared with those without REOB. With the increased potential of using REOB with higher recycled asphalt binder usage, there is a pressing need to investigate the effect of REOB and its dosage to various types of mixture performance across a wide range of temperatures.

Objectives and Scope

The objective of this study was to characterize the laboratory performance of conventional asphalt mixtures and mixtures containing REOB modified asphalt binders. Four 12.5-mm asphalt mixtures were designed to meet Superpave design criteria containing 0%, 5%, 10%, and 15% REOB modified asphalt binders. The Hamburg loaded-wheel test (LWT) was conducted to characterize the rutting and moisture susceptibility at high temperatures. Intermediate- and low-temperature cracking resistance was evaluated by a semi-circular bend (SCB) test and a thermal restrained specimen test (TSRST), respectively. Test results were statistically analyzed to group each mixture based on their mechanical performance. Furthermore, Superpave Performance Grading (PG) tests were performed for the asphalt binders considered. Triplicate specimens were used for each mixture test, except for the LWT test, in which two replicates were used. The air void content for each specimen was 7.0% ± 0.5%.

Experimental Design

Materials

Gravel and coarse natural sand aggregates commonly used in Louisiana were utilized in this study. Conventional asphalt mixture (0% REOB) was prepared using a styrene-butadiene-styrene (SBS) polymer-modified asphalt binder meeting Louisiana specifications for PG 70-22M (*36*). Three dosages of REOB (5%, 10%, and 15%) were used in the remaining three mixtures. The properties of REOB used in this study are:

- Cleveland Open Cup Flash by ASTM D 92 (*37*): 282°C (540 °F);
- API Gravity at 16°C (60 °F) by ASTM D 287 (*38*): 23 API;
- Pour Point by ASTM D 97 (*39*): −27°C (−17 °F);

Table 1. Job Mix Formula

Parameter	All mixtures
% G_{mm} at N_{ini}	88.8
% G_{mm} at N_{max}	97.0
Air Voids (%)	3.8
VMA (%)	13.8
VFA (%)	72.7
Total % AC	5.3
Metric (U.S.) sieve	Composite gradation blend
19.0 mm (3/4 in.)	100
12.5 mm (1/2 in.)	98
9.5 mm (3/8 in.)	85
4.75 mm (No. 4)	62
2.36 mm (No. 8)	43
1.18 mm (No. 16)	30
0.600 mm (No. 30)	23
0.300 mm (No. 50)	15
0.150 mm (No. 100)	9
0.075 mm (No. 200)	6.3

- Kinematic Viscosity at 100°C (212 °F) by ASTM D 445 (*40*): 80 cSt.

Mixture Design

Four Superpave asphalt mixtures utilizing a NMAS of 12.5 mm were designed and evaluated in this study. A Level 2 design ($N_{initial}$ = 8, N_{design} = 100, N_{final} = 160 gyrations) was performed in accordance with AASHTO R 35, "Standard Practice for Superpave Volumetric Design for Hot Mix Asphalt (HMA)," AASHTO M 323 (*41*), "Standard Specification for Superpave Volumetric Mix Design," and Section 502 of the 2006 Louisiana Standard Specifications for Roads and Bridges (*42*). Specifically, the optimum asphalt cement content was determined based on volumetric properties (VTM = 3.0%–5.0%, VMA ≥ 13%, VFA = 68%–78%) and densification requirements (%G_{mm} at $N_{initial}$ ≤ 89, %G_{mm} at N_{final} ≤ 98). The optimum asphalt binder content for all mixture is 5.3%. Table 1 presents the Job Mix Formula.

Laboratory Tests

Performance Grading was assessed for asphalt binders used according to AASHTO M 320 (*43*), "Standard Specification for Performance-Graded Asphalt Binder." Table 2 presents the mixture performance test factorial conducted for the asphalt mixtures evaluated in this study. Three tests, including LWT, SCB tests, and TSRST, were conducted to characterize the performance of asphalt mixtures. Cylindrical samples were fabricated using a Superpave gyratory compactor (SGC) with the exception of the TSRST. The laboratory prepared TSRST specimens were compacted into a rectangular slab, 260.8 mm wide by 320.3 mm long by 50 mm, using a linear kneading compactor. After compaction, the required beam specimens for the TSRST were obtained by sawing the rectangular slab to the correct dimensions as detailed in Table 2. The target air void for all specimens prepared in this study was 7.0% ± 0.5%.

Statistical Analysis

The mechanistic performance results of asphalt mixtures from the all tests were statistically analyzed using an analysis of variance (ANOVA) procedure provided in the Statistical Analysis System (SAS) 9.4 program (*47*). A multiple comparison procedure (Tukey test) with a confidence level of 0.05 was performed on the means. The groupings represent the mean for the test results reported by mixture type. The results of the statistical grouping are reported with letters: A, B, C, D, and so forth. Letter A was assigned to the highest mean followed by the other letters in appropriate order. A double (or more) letter designation, such as A/B (or A/B/C) indicates that the difference in the means is not clear-cut, and that the mean is close to either group in the analysis.

Results and Analysis

Performance Grade (PG) of Asphalt Binders

Figure 1 (*a*) presents the PG test results for the asphalt binders considered. It is noted that these binders did meet the Louisiana DOTD specifications for PG 70-22M (*36*). Further, the upper continuous PG temperature of the REOB modified binders (5%, 10%, and 15% modified REOB) was lower than that of the base binder (0% REOB). On the other hand, the addition of REOB improved the lower continuous PG temperature of the asphalt binders evaluated (i.e., a decreasing trend in the lower PG continuous temperature). Figure 1 (*b*) presents ΔT_c of all the binders. This parameter was used to

Table 2. Mixture Performance Tests

Tests	Protocols	Engineering properties	Specimen details	No. of specimens
SCB at 25°C	ASTM D 8044 (*44*)	Intermediate temperature: Fatigue cracking resistance	ϕ150 mm × 57 mm	4
TSRST	AASHTO TP 10 (*45*)	Low temperature: Thermal cracking resistance	50 ± 5 mm^2 × 250 ± 5 mm length	3
LWT at 50°C	AASHTO T 324 (*46*)	Rutting susceptibility and moisture resistance	ϕ150 mm × 60 mm	2

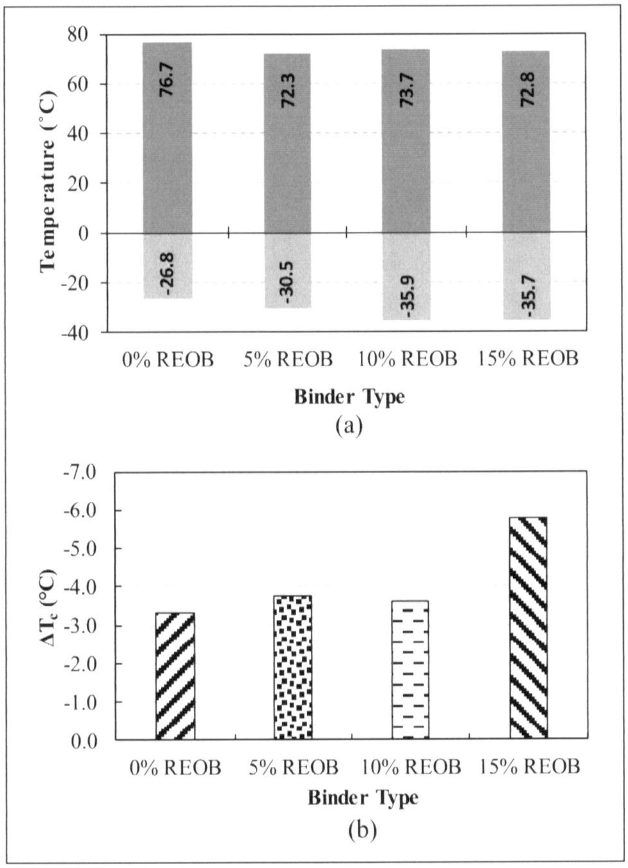

Figure 1. PG test results of asphalt binders: (*a*) continuous grading temperature, (*b*) ΔT_c of all binders.

ascertain binder cracking potential (e.g., *28, 29, 33*). The ΔT_c of asphalt binder containing 15% REOB had a negative effect on ΔT_c, while the other three binders (0%, 5%, and 15% REOB) had similar ΔT_c. This indicates that asphalt binder containing 15% REOB is more susceptible to low-temperature cracking than the other asphalt binders evaluated.

Rutting and Moisture Susceptibility

The rutting performance of the mixture was assessed using a Hamburg-type Loaded Wheel Tester (LWT). This test was conducted in accordance with AASHTO T 324, "Standard Method of Test for Hamburg Wheel-Track Testing of Compacted Hot Mix Asphalt (HMA)" (*46*). This test is considered a torture test that produces damage by rolling a 703 N (158 lb) steel wheel across the surface of a slab that is submerged in 50°C water for 20,000 passes at 56 passes per minute. The rut depth at 20,000 cycles was measured and used in the analysis. Louisiana specifies a maximum allowable rut depth of 6.0 mm at 20,000 passes at 50ºC (*36*). The average coefficient of variation was 14.0%.

Figure 2 presents the rut depth of the mixtures containing REOB modified asphalt binders. All mixtures met the Louisiana requirement and REOB did not adversely

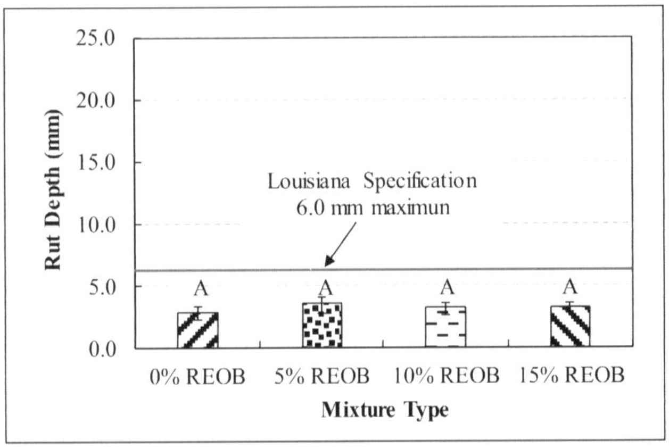

Figure 2. Hamburg loaded-wheel test results.

influence their rutting resistance. No stripping occurred in the all mixtures evaluated as measured by LWT. Moreover, there was no statistical difference among the mixtures tested in this study in terms of permanent deformation, Figure 2.

Intermediate-Temperature Cracking

The SCB test was conducted to characterize the intermediate-temperature cracking resistance of the mixtures evaluated with various REOB modified asphalt binders. This test was performed according to ASTM D 8044 "Standard Test Method for Evaluation of Asphalt Mixture Cracking Resistance using the Semi-Circular Bend Test (SCB) at Intermediate Temperatures" (*44*). Test specimens were long-term temperature-conditioned (5 days at 85°C) according to AASHTO R30 "Practice for Mixture Conditioning of Hot Mix Asphalt (HMA)" (*48*). This test characterizes the fracture resistance of HMA mixtures based on fracture mechanics principals, the critical strain energy release rate, also called the critical value of J-integral, or J_c. To determine the critical value of J-integral (J_c), semi-circular specimens with at least two different notch depths need to be tested for each mixture. In this study, three notch depths of 25.4 mm, 31.8 mm, and 38 mm were selected. Test temperature was selected to be 25°C. The semi-circular specimen is loaded monotonically until fracture failure under a constant cross-head deformation rate of 0.5 mm/min in a three-point bending load configuration. The load and deformation are continuously recorded and strain energy to failure (U) is the area under the loading portion of the load-deflection curves, up to the maximum load measured for each notch depth. Then, the critical value of J-integral (J_c) is determined by dividing a slope of a linear regression line in a strain energy-notch depth curve by a sample thickness as follows:

$$J_c = -\left(\frac{1}{b}\right)\frac{dU}{da} \quad (1)$$

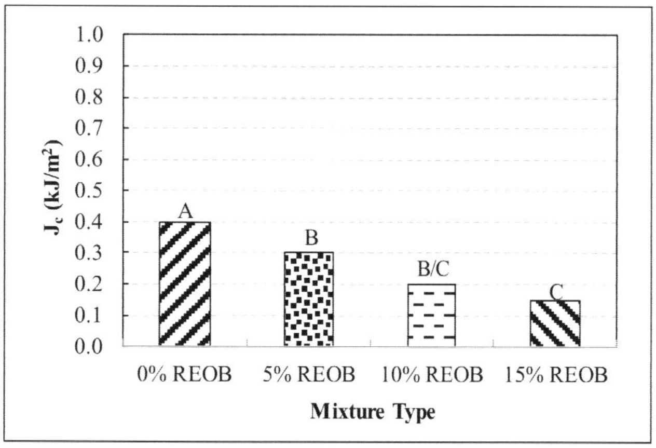

Figure 3. Semi-circular bend test results at 25°C.

where,

J_c = critical strain energy release rate (kJ/m^2),
b = sample thickness (m),
a = notch depth (m),
U = strain energy to failure (kJ), and
dU/da = change of strain energy with notch depth (kJ/m).

Figure 3 presents the critical strain energy (J_c) values for the mixtures containing various REOB contents. The average coefficient of variation was 8.9%. The control mixture exhibited significantly higher J_c value follow by mixtures containing 5%, 10%, and 15% REOB content, respectively. The decrease trend in J_c with an increasing REOB content is consistent with findings reported elsewhere (e.g., *30, 34, 35*). Thus, the use of REOB modified asphalt binders can adversely affect the intermediate-temperature cracking resistance measured by the SCB test as compared with the conventional mixture.

Low-Temperature Thermal Cracking

The low-temperature cracking performance was assessed using the Thermal Stress Restrained Specimen Tensile Strength (TSRST). TSRST is used to study changes in the asphalt binder's glass transition temperature and how it may affect the mixture performance in terms of low-temperature cracking. This test was conducted according to AASHTO TP 10, "Standard Test Method for Thermal Stress Restrained Specimen Tensile Strength (TSRST)" (*45*). Triplicate long-term aged specimens were 50 mm thick × 50 mm wide × 254 mm long. The specimen was glued at the ends to two aluminum platens. The test device then cooled the beam specimen at 10°C/hour from 5°C while restraining it from contracting. As the temperature drops, thermal stresses build up until the specimen fractures. Specimen temperature at failure, fracture strength, and the slope of the thermally induced stress curve were then determined. The average coefficient of variation was 7.1%.

Figure 4 (*a*) shows an improving TSRST low-temperature cracking resistance with an increase in REOB content up to 10%. This observation is consistent with the asphalt binder PG test results shown in Figure 1. This trend may be due to the softening of asphalt binders with higher REOB contents. However, it was observed that mixtures with 15% REOB were susceptible to low-temperature cracking, implying an adverse effect of high REOB dosage (≥15% in this study) to the TSRST low-temperature cracking resistance. The TSRST fracture strength of the mixtures with various REOB

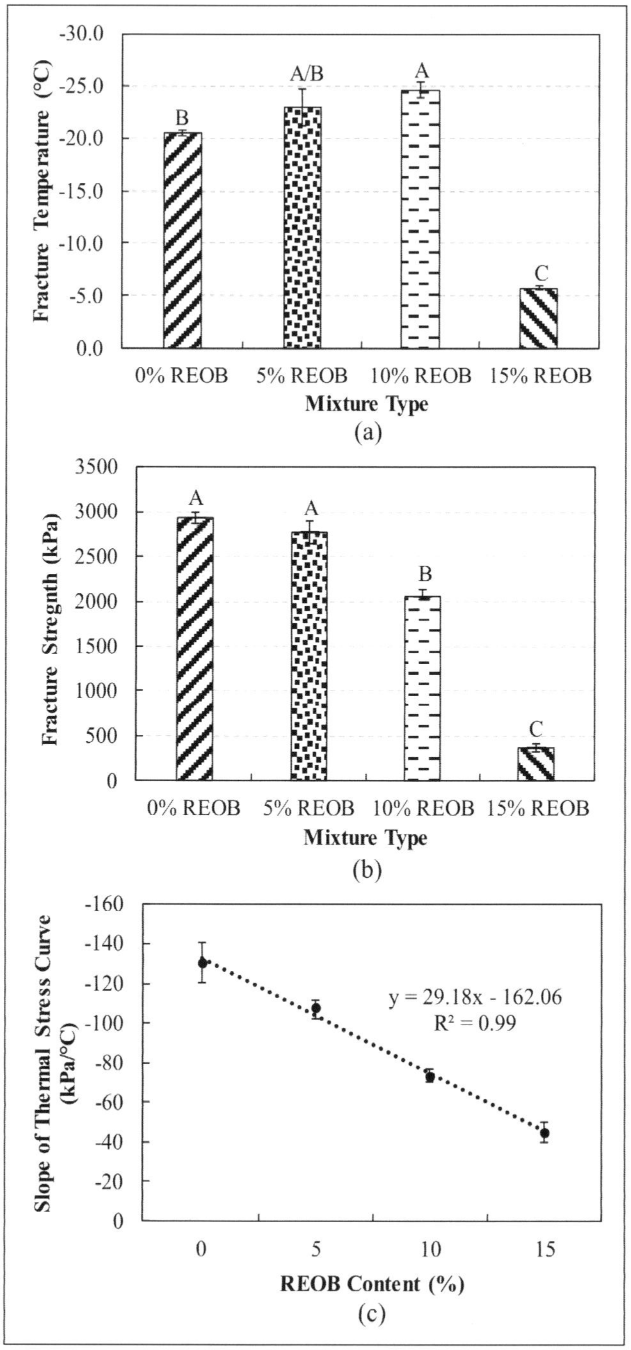

Figure 4. Thermal restrained specimen test results: (*a*) fracture temperature, (*b*) fracture strength, (*c*) slope of thermal stress curve.

Table 3. Summary of Test Results

	Rutting	Fracture, intermediate temp.	Fracture, low temp.
Mixture type	LWT	SCB	TSRST[a]
0% REOB	A[b]	A	B
5% REOB	A	B	A/B
10% REOB	A	B/C	A
15% REOB	A	C	C

[a]Fracture temperature was used.
[b]The letter A was assigned to the best performer followed by the other letters in appropriate order.

contents is shown in Figure 4 (*b*). The fracture strength decreased as the REOB content increased. This may be caused by the softer asphalt binder with the higher REOB content. It is noted that a sudden drop in fracture strength occurred for mixtures with 15% REOB. Figure 4 (*c*) presents the slope of thermal stress curve (i.e., ratio of a change in the thermal stress to a change in temperature) with an increase in REOB content. This slope decreased with an increase in REOB level due to the softening of the asphalt binders. This observation is consistent with the low temperature PG grade shown in Figure 1. In general, the low-temperature performance of mixtures containing REOB content up to 10% were similar, while mixture with 15% REOB dosage showed significantly reduced low-temperature mixture cracking resistance.

Summary of the Results

The tests evaluated in this study were selected to capture the laboratory performance of the asphalt mixtures prepared with various percentages of REOB additive. The laboratory test results were statistically analyzed and grouped using the analysis of variance (ANOVA) procedure provided in the Statistical Analysis System (SAS) program. A multiple comparison procedure with a significance level of 5% was performed for the means. The results of the statistical grouping were reported with the letters A, B, C, and so forth. The letter A was assigned to the best performer followed by the other letters in appropriate order. A double (or more) letter designation, such as A/B, indicates that the difference in the means is not clear-cut, and that the results could fall in either category.

Table 3 summarizes the statistical ranking of the laboratory test results for the four asphalt mixtures considered in this study. REOB did not adversely affect mixture rutting performance measured by LWT, while intermediate temperature cracking resistance of mixtures (i.e., SCB J_c) containing REOB decreased with an increase in REOB content. The TSRST low-temperature cracking performance of mixtures with REOB up to 10% (i.e., 0%, 5%, and 10% REOB) were statistically similar, but 15% REOB mixture showed significantly reduced low-temperature cracking resistance.

Summary and Conclusions

This study evaluated the laboratory performance of asphalt mixtures containing various REOB-modified asphalt binders. The Performance Grade (PG) asphalt binder tests and three laboratory tests, including the Hamburg loaded-wheel test (LWT), semi-circular bend (SCB) test, and thermal restrained specimen test (TSRST), were conducted to characterize rutting/moisture susceptibility, intermediate- and low-temperature cracking resistance, respectively. Four levels of REOB dosages (0%, 5%, 10%, and 15%) were considered and its effect on each mixture's peformance in a wide range of temperature was investigated. The test results were also statistically analyzed to ascertain the presence of significant differences among each mixture performance. Based on the results, the following conclusions can be drawn:

- Rutting performance of mixtures measured by LWT, containing various REOB contents (0%, 5%, 10%, and 15%) was comparable to that of the control mixture;
- Use of REOB did not adversely affect moisture susceptibility; no stripping was observed as measured from LWT test;
- A decreasing trend in SCB J_c with an increasing REOB content was observed indicating a negative effect of REOB on intermediate-temperature cracking resistance; and
- Similar TSRST low-temperature thermal cracking resistance was exhibited with an increase in REOB content up to 10%. However, mixtures containing 15% REOB showed a reduction of low-temperature cracking resistance as compared with other mixtures evaluated.

Acknowledgments

The authors would like to acknowledge the support of the Louisiana Transportation Research Center (LTRC). The contents of this paper do not necessarily reflect the official views or policies of the Louisiana Department of Transportation and Development or the Louisiana Transportation Research Center.

Author Contributions

The authors confirm contribution to the paper as follows: study conception and design: Mohammad, You; data collection: Shi, You;

analysis and interpretation of results: Mohammad, You, Cooper, draft manuscript preparation: Mohammad, You, Shi. All authors reviewed the results and approved the final version of the manuscript.

References

1. Asphalt Institute. *State-of-the-Knowledge: The Use of REOB/ VTAE in Asphalt*. Asphalt Institute, Lexington, KY, 2016.
2. Arnold, T. S., and A. Shastry. Analysis of Asphalt Binders for Recycled Engine Oil Bottoms by X-ray Fluorescence Spectroscopy. Presented at 94th Annual Meeting of the Transportation Research Board, Washington, D.C., 2015.
3. Zamhari, K. A., M. Hermadi, and W. Choy Fun. Characteristics of Rejuvenated Bitumen with Used Lubricating Oil as Rejuvenating Agent. *Proc., International Conference on Sustainable Infrastructure and Built Environment in Developing Countries*, Bandung, Indonesia, 2009.
4. Burke, K., and S. Hesp. Penetration Testing of Waste Engine Oil Residue Modified Asphalt Cements. *Proc., 1st Conference of Transportation Research Group of India (CTRG)*, Bangalore, India, 2011, pp. 7–10.
5. D'Angelo, J., K. Grzybowski, and S. Lewis. Asphalt Binder Modification with Re-refined Heavy Vacuum Distillation Oil. *Canadian Technical Asphalt Association*, Vancouver, Canada, No. 57, 2012, pp. 257–276.
6. D'Angelo, J. A., P. Eng, C. O. Officer, L. D'Angelo Consulting, V. Annandale, K. Grzybowski, S. Lewis, D. O. R. Sales, and R. Walker. Evaluation of the Performance Properties of Asphalt Mixes Produced with Re-refined Heavy Vacuum Distillate Bottoms. *Canadian Technical Asphalt Association*, St. John's, Newfoundland, Canada, No. 58, 2013. pp. 273–290.
7. DeDene, C. D., and Z.-P. You. The Performance of Aged Asphalt Materials Rejuvenated with Waste Engine Oil. *International Journal of Pavement Research and Technology*, Vol. 7, No. 2, 2014, pp. 145–152.
8. Golalipour, A. *Investigation of the Effect of Oil Modification on Critical Characteristics of Asphalt Binders*. PhD dissertation. The University of Wisconsin-Madison, 2013.
9. Golalipour, A., A. A. Engineer, and H. Bahia. Evaluation of Oil Modification Effect on Asphalt Binder Thermal Cracking and Aging Properties. *Canadian Technical Asphalt Association (CTAA)*, Winnipeg, Manitoba, Canada, 2014, pp. 345–374.
10. Herrington, P. Use of Rerefined Oil Distillation Bottoms as Extenders for Roading Bitumens. *Journal of Materials Science*, Vol. 27, No. 24, 1992, pp. 6615–6626.
11. Herrington, P., V. Dravitzki, C. Wood, and J. Patrick. Waste Oil Distillation Bottoms as Bitumen Extenders. *Road and Transport Research*, Vol. 2, No. 4, 1993, pp. 56–68.
12. Herrington, P., and P. Hamilton. *Recycling of Waste Oil Distillation Bottoms in Asphalt*. California Transit Association, Sacramento, CA, 1998.
13. Hesp, S., and H. Shurvell. Waste Engine Oil Residue in Asphalt Cement. *Proc., 7th International Conference on Maintenance and Rehabilitation of Pavements and Technological Control*, Auckland, New Zealand, 2012.
14. Hesp, S. A., and H. F. Shurvell. X-ray Fluorescence Detection of Waste Engine Oil Residue in Asphalt and Its Effect on Cracking in Service. *International Journal of Pavement Engineering*, Vol. 11, No. 6, 2010, pp. 541–553.
15. Jason, C., A. H. Wielinski, A. J. Kriech, and G. A. Hube. Laboratory Crack Testing of Hot Mix Asphalts Containing Vacuum Tower Asphalt Extender. *Proc., Sixtieth Annual Conference of the Canadian Technical Asphalt Association (CTAA)*, Winnipeg, Manitoba, Canada, 2015.
16. Johnson, K.-A., and S. Hesp. Effect of Waste Engine Oil Residue on Quality and Durability of SHRP Materials Reference Library Binders. *Transportation Research Record: Journal of the Transportation Research Board*, 2014. 2444: 102–109.
17. Oliveira, J. R., H. M. Silva, C. M. Jesus, L. P. Abreu, and S. R. Fernandes. Pushing the Asphalt Recycling Technology to the Limit. *International Journal of Pavement Research and Technology*, Vol. 6, No. 2, 2013, pp. 109–116.
18. Rubab, S., K. Burke, L. Wright, S. A. Hesp, P. Marks, and C. Raymond. Effects of Engine Oil Residues on Asphalt Cement Quality. *Canadian Technical Asphalt Association*, Québec City, Québec, Canada, No. 56, 2011, pp. 1–12.
19. Shurvell, H., S. Subramani, S. Hesp, S. Genin, and D. Scafe. Five-year performance review of a Northern Ontario pavement trial: validation of Ontario's double-edge-notched tension (DENT) and extended Bending Beam Rheometer (BRR) test methods. *Proc., Fifty-Fourth Annual Conference of the Canadian Technical Asphalt Association (CTAA)*, Moncton, NB, Canada, 2009.
20. Soleimani, A., S. Walsh, and S. Hesp. Asphalt Cement Loss Tangent as Surrogate Performance Indicator for Control of Thermal Cracking. *Transportation Research Record: Journal of the Transportation Research Board*, 2009. 2126: 39–46.
21. Uzarowski, L., G. MacDonald, J. Rizzo, G. Moore, and V. Henderson. Cold Winter and Early Asphalt Pavement Cracking Observed in Ontario. *Proc., Sixtieth Annual Conference of the Canadian Technical Asphalt Association (CTAA)*, Winnipeg, Manitoba, Canada, 2015.
22. Villanueva, A., S. Ho, and L. Zanzotto. Asphalt Modification with Used Lubricating Oil. *Canadian Journal of Civil Engineering*, Vol. 35, No. 2, 2008, pp. 148–157.
23. Wielinski, J., A. Kriech, G. Huber, A. Horton, and L. Osborn. Analysis of Vacuum Tower Asphalt Extender and Effect on Bitumen and Asphalt Properties. *Road Materials and Pavement Design*, Vol. 16, 2015, pp. 90–110.
24. Wielinski, J. C., A. J. Kriech, G. A. Huber, A. Horton, and L. V. Osborn. The Impact of Asphalt Blended with Re-refined Vacuum Tower Bottoms (RVTB) and Its Effect on HMA Mixture Performance. *Canadian Technical Asphalt Association*, Winnipeg, Manitoba, Canada, No. 59, 2014, pp. 402-426.
25. Wright, L., A. Kanabar, E. Moult, S. Rubab, and S. Hesp. Oxidative Aging of Asphalt Cements from an Ontario Pavement Trial. *International Journal of Pavement Research and Technology*, Vol. 4, No. 5, 2011, pp. 259–267.
26. Zaumanis, M., R. Mallick, and R. Frank. Evaluation of Rejuvenator's Effectiveness with Conventional Mix Testing for 100% Reclaimed Asphalt Pavement Mixtures. *Transportation Research Record: Journal of the Transportation Research Board*, 2013. 2370: 17–25.
27. Wielinski, J. C., A. J. Kriech, A. Horton, T. Reece, and G. A. Huber. Laboratory Evaluation of Vacuum Tower Asphalt Extenders Impact on HMA Fatigue Resistance and Asphalt Binder

Aging. Presented at 94th Annual Meeting of the Transportation Research Board, Washington, D.C., 2015.

28. Bennert, T., C. Ericson, D. Pezeshki, and R. Corun. Fatigue Performance of Re-Refined Engine Oil Bottom–Modified Asphalt: Laboratory Study. *Transportation Research Record: Journal of the Transportation Research Board*, 2016. 2574: 1–16.
29. Bennert, T., C. Ericson, D. Pezeshki, E. Haas, R. Shamborovskyy, and R. Corun. Laboratory Performance of Re-refined Engine Oil Bottoms (REOB) Modified Asphalt. *Association of Asphalt Paving Technologies Conference*, Indianapolis, IN, 2016.
30. Ozer, H., G. Renshaw, K. Hasiba, and I. L. Al-Qadi. *Evaluation of the Impacts of Re-Refined Engine Oil Bottoms (ReOB) on Performance Graded Asphalt Binders and Asphalt Mixtures*. Publication 0197-9191, Urbana, IL, 2016.
31. Mogawer, W. S., A. Austerman, I. L. Al-Qadi, W. Buttlar, H. Ozer, and B. Hill. Using Binder and Mixture Space Diagrams to Evaluate the Effect of Re-refined Engine Oil Bottoms on Binders and Mixtures After Ageing. *Road Materials and Pavement Design*, Vol. 18, 2017, pp. 154–182.
32. Mensching, D. J., A. Andriescu, C. DeCarlo, X. Li, and J. S. Youtcheff. Effect of Extended Aging on Asphalt Materials Containing Rerefined Engine Oil Bottoms. *Transportation Research Record: Journal of the Transportation Research Board*, 2017. 2632: 60–69.
33. Li, X., N. Gibson, A. Andriescu, and T. S. Arnold. Performance Evaluation of REOB-modified Asphalt Binders and Mixtures. *Road Materials and Pavement Design*, Vol. 18, 2017, pp. 128–153.
34. Cooper, S. B. Jr, L. N. Mohammad, and M. A. Elseifi. Laboratory Performance of Asphalt Mixtures Containing Recycled Asphalt Shingles and Re-Refined Engine Oil Bottoms. *Journal of Materials in Civil Engineering*, Vol. 29, No. 9, 2017, p. 04017106.
35. Cooper, S. B. Jr, I. Negulescu, S. S. Balamurugan, L. Mohammad, W. H. Daly, and G. L. Baumgardner. Asphalt Mixtures Containing RAS and/or RAP: Relationships Amongst Binder Composition Analysis and Mixture Intermediate Temperature Cracking Performance. *Road Materials and Pavement Design*, Vol. 18, 2017, pp. 209–234.
36. LA DOTD. *Louisiana Standard Specifications for Roads and Bridges*. State of Louisiana, Department of Transportation and Development, Baton Rouge, LA, 2016.
37. ASTM D 92. *Standard Test Method for Flash and Fire Points by Cleveland Open Cup Tester*. American Society for Testing and Materials, West Conshohocken, PA, 2016.
38. ASTM D 287. *Standard Test Method for API Gravity of Crude Petroleum and Petroleum Products (Hydrometer Method)*. American Society for Testing and Materials, West Conshohocken, PA, 2012.
39. ASTM D 97. *Standard Test Method for Pour Point of Petroleum Products*. American Society for Testing and Materials, West Conshohocken, PA, 2002.
40. ASTM D 445. *Standard Test Method for Kinematic Viscosity of Transparent and Opaque Liquids (and Calculation of Dynamic Viscosity)*. American Society for Testing and Materials, West Conshohocken, PA, 2006.
41. AASHTO T 323. *Standard Specification for Superpave Volumetric Mix Design. American Association of State Highway and Transportation Officials*, American Association of State Highway and Transportation Officials, Washington, D.C., 2013.
42. LA DOTD. *Louisiana Standard Specifications for Roads and Bridges*. State of Louisiana, Department of Transportation and Development, Baton Rouge, LA, 2006.
43. AASHTO M 320. *Standard Specification for Performance-Graded Asphalt Binder*. American Association of State Highway and Transportation Officials, American Association of State Highway and Transportation Officials, Washington, D.C., 2016.
44. ASTM D 8044. *Standard Test Method for Evaluation of Asphalt Mixture Cracking Resistance using the Semi-Circular Bend Test (SCB) at Intermediate Temperatures*. American Society for Testing and Materials, West Conshohocken, PA, 2016.
45. AASHTO TP 10. *Standard Test Method for Thermal Stress Restrained Specimen Tensile Strength*. American Association of State Highway and Transportation Officials, American Association of State Highway and Transportation Officials, Washington, D.C., 1993.
46. AASHTO T 324. *Standard Method of Test for Hamburg Wheel-Track Testing of Compacted Hot-Mix Asphalt (HMA)*. American Association of State Highway and Transportation Officials, American Association of State Highway and Transportation Officials, Washington, D.C., 2016.
47. SAS Institute. *SAS*. SAS Institute, Cary, NC.
48. AASHTO R 30. *Standard Practice for Mixture Conditioning of Hot-Mix Asphalt (HMA)*. American Association of State Highway and Transportation Officials, American Association of State Highway and Transportation Officials, Washington, DC, 2015.

The Standing Committee on Non-Binder Components of Asphalt Mixtures (AFK30) peer-reviewed this paper (18-04668).

Article

TRR
JOURNAL OF THE TRANSPORTATION RESEARCH BOARD

Transportation Research Record
2018, Vol. 2672(28) 96–106

DOI: 10.1177/0361198118773193
journals.sagepub.com/home/trr
SAGE

Wearing Course Mixtures Prepared with High Reclaimed Asphalt Pavement Content Modified by Rejuvenators

Stephan Büchler[1], Augusto Cannone Falchetto[1], Axel Walther[1], Chiara Riccardi[1], Di Wang[1], and Michael P. Wistuba[1]

Abstract

This paper presents an experimental investigation on the combined use of high content of reclaimed asphalt pavement (RAP) and rejuvenators, for producing asphalt mixtures for wearing courses. First, the new Binder-Fast-Characterization-Test (Bitumen Typisierungs Schnell Verfahren in German) BTSV method, recently proposed in Germany, is used to determine the amount of rejuvenator required to design recycled mixtures for a specific RAP source. Then, a set of seven asphalt mixtures for wearing course is prepared with different amounts of RAP and with three types of rejuvenators. The BTSV procedure is further applied to evaluate the rheological properties of the binder extracted from the mixtures. Resistance to permanent deformation, stiffness, fatigue, and low temperature behavior tests are then performed to determine the material response. It is observed that the use of recycled material and rejuvenators results in similar or better deformation resistance, higher stiffness and enhanced low temperature properties, with an improved fatigue behavior, also for high RAP content, when compared to mixtures prepared with virgin material. These findings suggest the possibility of using rejuvenators for field tests on wearing courses.

In the last decades, the economic and environmental needs for asphalt pavement with reduced construction costs and enhanced functional performance have promoted, within the concept of circular economy, an increasing use of reclaimed asphalt pavement (RAP). For this purpose, several studies have been carried out over the years to evaluate the influence of RAP content on the mechanical properties of asphalt mixtures. In some of these research efforts, an increase in complex modulus and stiffness was observed when RAP was included in the mix design (*1–5*). Moreover, such a stiffening phenomenon is associated with a decrease in the penetration grade (*6*) of the binder, for increasing amounts of RAP (*7*).

The additional embrittlement due to the aged binder restricts the amount of RAP in new asphalt mixtures. This may represent a serious constraint as in recent years, and most likely in the near future, increasing amounts of RAP will be included in pavement mixtures, ultimately reaching aging levels no longer tolerable for the design performance of the pavement. Rejuvenators can provide a valuable solution in the production of asphalt mixtures with high RAP content to modify and improve the characteristics of the aged and oxidized binder contained in the recycled material. For example, a decrease of the binder viscosity and resistance to oxidation can be promoted depending on the rejuvenator constituents (*8–11*). The conventional methods for rejuvenation of asphalt materials are based on reintegration of volatilized, oxidized missing components or similar constituents (replacement of dispersants or crude oil resins). This process can potentially lead to a recovery of the original ratio between maltenes and asphaltenes (*12*).

In this paper, the possibility of designing mixtures with high RAP content and rejuvenator is investigated. A novel test method recently introduced in Germany, known as Binder-Fast-Characterization-Test (Bitumen Typisierungs Schnell Verfahren in German) BTSV (*13, 14*), is introduced to detect the optimum amount of rejuvenator at the mix design level. Then, a set of wearing

[1]Department of Civil Engineering – ISBS, Technische Universität Braunschweig, Braunschweig, Germany

Corresponding Author:
Address correspondence to Augusto Cannone Falchetto: a.cannone-falchetto@tu-bs.de

course asphalt mixtures commonly used in Germany (*15*) were prepared with different amounts of RAP and different rejuvenators. Finally, the rheological response of the binder extracted from the mixtures as well as the mechanical performance of the mixtures were experimentally evaluated.

BTSV Method

In Germany, the design process to determine the allowed RAP content, which can be included in a recycled asphalt mixture without affecting the material performance, relies on the conventional softening point ring-and-ball (SPR&B) values of the extracted RAP and virgin binders (*16*). The SPR&B method is commonly used to obtain the properties of asphalt binder in the high temperature domain, and in the case of routine controls (*17*). In addition, the SPR&B is often used in Germany by road authorities and construction companies as a key indicator for low temperature behavior, as a high SPR&B value is linked with higher binder brittleness (*13*).

In recent years, with the increasing complexity of bituminous binders, numerous researchers have reported significant problems using the SPR&B method. The method does not work satisfyingly when polymer modification of the binder is involved, and when, in the frame of asphalt recycling, the binder properties have to be determined. Hence, there is a need for a new binder specification method which can be used for routine controls and that can easily replace the SPR&B procedure.

For this purpose, a simple method for asphalt binder characterization was recently proposed in Germany: the BTSV (*13, 14*). In this method, the principle of the SPR&B is transferred to the dynamic shear rheometer (DSR) (*18*) testing device, relying on the major findings of two different research efforts from Mattischek et al. (*19*) and Guericke (*20*). According to the latter, the SPR&B provides an estimation of the temperature when the binder condition changes from solid to liquid (*19*). On the other hand, binder deformation at constant load can be expressed in terms of viscosity from a rheological point of view. Within this framework, Mattischek et al. (*20*) found that SPR&B can be set equal to a zero-shear viscosity of $\eta_0 = 780$ Pa·s. Based on DSR oscillatory measurements, Guericke (*19*) equated SPR&B with a complex viscosity of $\eta^* = 1.3$ kPa·s at a frequency of $f = 0.0159$ Hz. While these findings are valid for plain binders, no consistent correlation of rheological parameters and SPR&B was found for polymer-modified material.

In the BTSV method, the DSR is used to soften the binder through a steady increase in temperature, by applying a temperature rate of $\Delta T = 1.2°C/min$, a constant shear stress of 500 Pa in oscillation mode, at a constant frequency $f = 10$ rad/s. In such a way, the binder is not experiencing any isothermal equilibrium, as set in conventional DSR measurements, but is subjected to a continuous increase in temperature as imposed during the SPR&B test. This shortens the overall test duration to approximately 60 minutes. Based on a large dataset on different binders, it was observed that the complex shear modulus $|G^*|$ at a temperature equal to SPR&B determines a mean value of $|G^*|$ of approximately 15 kPa (*13, 14*). Therefore, such a $|G^*|$ value is set as limiting threshold for the estimation of the material properties in the BTSV method. During the test, temperature, complex shear modulus $|G^*|$, and phase angle δ are recorded and reported against temperature (Figure 1).

As a result, two DSR output parameters are found (at $|G^*| = 15$ kPa), which represent the state of softening: the softening temperature (T_{BTSV}), and the corresponding phase angle (δ_{BTSV}). While the softening temperature T_{BTSV} directly replaces the SPR&B measurement, the phase angle δ_{BTSV} provides additional information on binder stiffness and possible degree of modification. Two replicates are commonly used to assess the BTSV parameters. The presented method has been successfully used for more than two years for differentiating the characteristics of asphalt binder in the high temperature range and it is currently part of the German technical standards (*14*). Figure 2 presents the BTSV plane (T_{BTSV}–δ_{BTSV}) with the data from approximately 80 tests on asphalt binders typically used in Germany.

Binder types are identified by individual boxes (domains). Plain binders are located in the top part of the plot, well differentiated by phase angle δ_{BTSV} remaining at a comparable level of about 80°. For polymer-modified binders, a lower phase angle is observed. A decrease in phase angle below 80° may, therefore, be

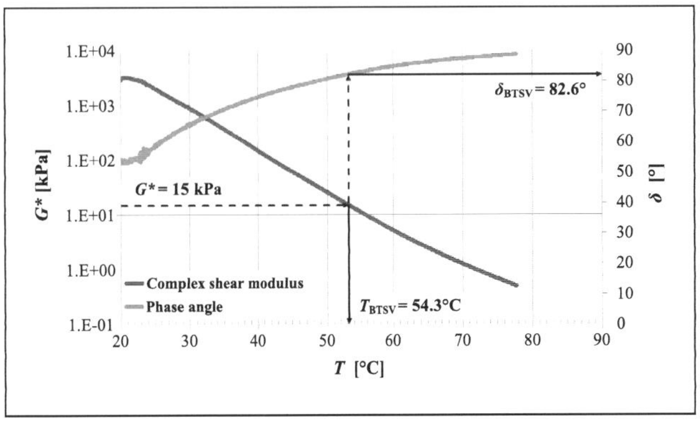

Figure 1. Complex shear modulus $|G^*|$ and phase angle δ as function of temperature: Example for determining softening temperature T_{BTSV} and corresponding phase angle δ_{BTSV}.

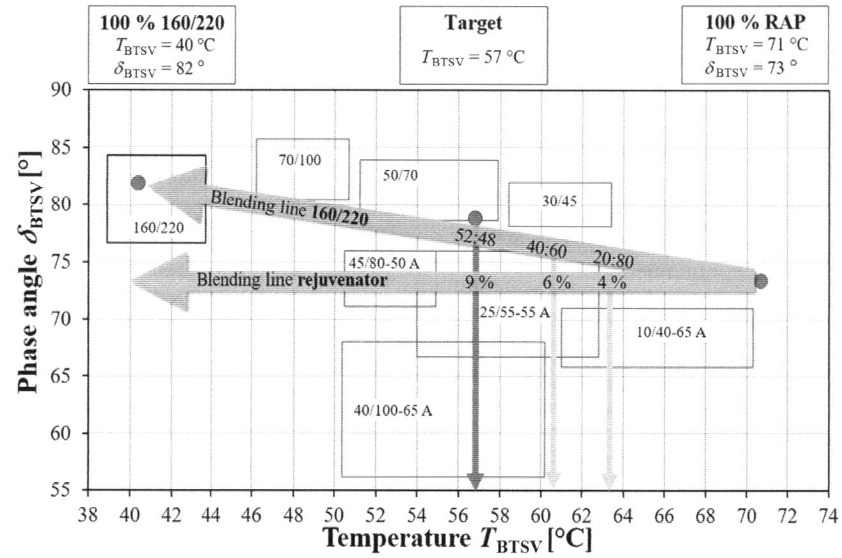

Figure 2. BTSV plane and example of binders blending in the context of recycling.

interpreted as an indicator for polymer modification. In the case of polymer-modified 25/55-55 A binder, overlapping with 10/40-65 A and 40/100-65 A binders occurs; this is most likely due to the inability of the conventional grading specification system, based on penetration and SPR&B values, to correctly discriminate different binders (*6, 16*). On the other hand, because of its simple grading system, the BTSV method can be easily and advantageously used to quickly check binder quality when delivered, since only a few grams of materials and a relatively short test of approximately 1 h are needed for characterization purposes.

The BTSV method and the BTSV plane can be further used for designing recycled asphalt mixtures. For example, Figure 2 shows how an aged polymer-modified binder extracted from RAP is blended with a virgin plain 160/220 binder. Depending on the proportions of aged and virgin binder the properties of the resulting blend will always lay along a "blending line" expressed by parameters T_{BTSV} and δ_{BTSV} and having a slope governed by the virgin binder. With increasing proportion of fresh binder, T_{BTSV} decreases, and δ_{BTSV} increases. From the proportions of RAP and virgin 160/220 binder, a 25/55-55 A binder can be targeted, while, for example, 45/80-50 A or 40/100-65 A do not belong to any possible blending. As shown in Figure 2, using a rejuvenator the softening temperature T_{BTSV} also decreases but the phase angle δ_{BTSV} will stay constant. Any target binder properties can be adjusted along an almost horizontal "blending line". This suggests a reduction in the susceptibility of the rejuvenated binder, while showing a considerable difference in the effect of oily rejuvenators with respect to soft binders.

Materials

In this research, a set of asphalt mixtures for wearing course was prepared according to a conventional mix design used in Germany (*15*) and identified as AC11 D S where AC 11 stands for asphalt concrete with a maximum aggregate size of 11 mm and D S stands for wearing course. The mixtures were prepared with different amounts of RAP (0%, 20% and 40% by mass) and using different rejuvenators. Then, the properties of the mixtures were compared to a reference mixture consisting of virgin materials (Gabbro aggregates and plain binder 50/70) without rejuvenators.

A total of 28 kg of each mixture was prepared with a target binder content of 6.1% by mass. RAP was heated at a temperature of 140°C for 4 h and then added as first component to the laboratory mixer. As a second component, the rejuvenator was poured into the mixer and it was mixed for 30 s to activate the aged RAP binder. Next, the virgin aggregate (including filler) was added to the mixture at a temperature of 175°C and mixed for 20 s, followed by the virgin binder for a mixing time of 180 s. Mixing was performed in a double-shaft paddle mixer, comparable to a mixer of a plant in a scale 1:8. The following (Table 1) rejuvenators were used in the mix design:

- Standard binder 160/220 (Total Bitumen Deutschland) that has a kinematic viscosity of 135 mm^2/s at 135°C,
- Rheofalt HP-EM (MT Merk Trading international GmbH) that has a viscosity of 2000 P/s at 22°C, and

Table 1. Composition and Laboratory Properties of the Asphalt Materials

Property		Unit	1	2	3	4	5	6	7
Sieve opening	0.063 mm	%	8.2	8.8	8.6	8.0	9.0	8.8	8.2
	0.125 mm	%	12.5	12.5	12.5	12.0	12.5	12.3	12.4
	0.25 mm	%	18.1	18.0	18.3	17.4	18.1	17.8	18.2
	1 mm	%	35.1	34.7	35.3	34.3	34.7	34.5	36.6
	2 mm	%	51.7	52.1	52.1	52.6	51.7	52.5	55.1
	5.6 mm	%	68.0	69.6	69.2	71.1	69.8	69.3	71.1
	8 mm	%	81.7	78.3	78.6	79.7	80.5	80.6	81.9
	11.2 mm	%	95.0	93.0	94.6	94.8	94.9	94.4	94.1
	16 mm	%	100	98.7	99.3	100	99.7	99.5	99.4
	22.4 mm	%		100	100		100	100	100
RAP		%	-	20	40	20	40	20	40
Rejuvenator		-	-	160/220	Rheofalt	Storflux			
		%	-	21.3	42.6	1.6	4.9	1.6	4.9
Fresh binder		%	100	59.0	16.4	78.7	54.1	78.7	54.1
Binder from RAP		%	-	19.7	41.0	19.7	41.0	19.7	41.0
Binder content		%	6.1	6.1	6.2	6.1	6.2	6.3	6.4
Softening Point		°C	53.8	57.0	56.6	55.2	56.0	56.2	55.8
T_{BTSV}		°C	58.3	57.8	57.0	59.1	58.6	58.3	58.9
δ_{BTSV}		°	77.7	76.4	76.4	77.0	76.8	76.0	75.0

- Storflux Nature (STORIMPEX AsphalTec GmbH) that has a kinematic viscosity of 38 mm^2/s at 40°C.

The last two rejuvenators have a natural origin. The Rheofalt HP-EM is derived from cashew nut shell oil, which contain mostly cardanol, a phenolic compound having a C_{15} unsaturated chain (see PCT International Publication nos. WO 2010/077141 and WO 2010/110651). The Storflux Nature is a vegetable oil that has a plant origin. It is a biogenic mixture of different fatty acid esters.

To determine the content of rejuvenator, to obtain a binder with the properties of a reference 50/70 penetration grade binder (*21*), a procedure based on the BTSV test was used. First, the T_{BTSV} of the extracted RAP binder was determined (100% RAP in Figure 2). Then, as shown in Figure 2, three different blends of fresh and RAP binders were prepared (20:80; 40:60; 54:46) and tested with the DSR to obtain the corresponding T_{BTSV}. The required content of virgin binder was determined in correspondence with the T_{BTSV} of the reference 50/70 binder (target point). When using a rejuvenator in place of a virgin binder (i.e., 160/220 binder), the dosage was increased in smaller dosage steps (4%, 6%, 9%) until the T_{BTSV} of the reference 50/70 binder was achieved. Using this method, the target content of fresh binder 160/220 was found to be 52% when 20% of RAP was used. That means 21.3 M.-% of fresh binder 160/220 was needed in the resulting mixture with 20 M.-% of RAP. A proportionally higher content of 42.6 M.-% was obtained for 40 M.-% of RAP.

When using Rheofalt and Storflux, the same dosage (8.1 M.-%) in relation to RAP binder content was reported. Therefore, mixtures with 20 M.-% of RAP needed 1.6 M.-% of rejuvenators in the resulting mixture. On the other hand, when 40 M.-% of RAP was included in the mix design, 4.9 M.-% for each rejuvenator was required. It must be noted that unlike in the case of the softer 160/220 binder used for mixture 2 and 3, the proportion of Rheofalt and Storflux are not in a ratio of 1:2 for 20% and 40% RAP binder. This may be due to the specific rejuvenator characteristics which do not follow a linear increment of the content for a proportional increase of RAP binder. In addition, it is interesting to observe that identical percentages of rejuvenator are required to achieve the same 50/70 T_{BTSV} target for both Rheofalt and Storflux when using 20% and 40% RAP binder. To verify that such a causality was not due to any experimental error, the binder was extracted from the mixtures prepared with the rejuvenators and the BTSV test performed once again. The T_{BTSV} obtained from the extracted binder contained in the rejuvenated mixtures confirmed the same percentages (1.6% and 4.9%) derived from the initial BTSV testing on the rejuvenated RAP binder both for Rheofalt and Storflux.

Table 1 shows the composition and properties of the seven AC 11 D S mixtures. The measured T_{BTSV} indicates that the binder properties obtained from extraction are comparable for the entire set of materials (1 to 7). The binders were extracted using the centrifuge method and were recovered using the rotatory evaporator, in accordance with EN12697-3 (*22*). With respect to T_{BTSV}, a reduction of 12°C is achieved for the extracted RAP

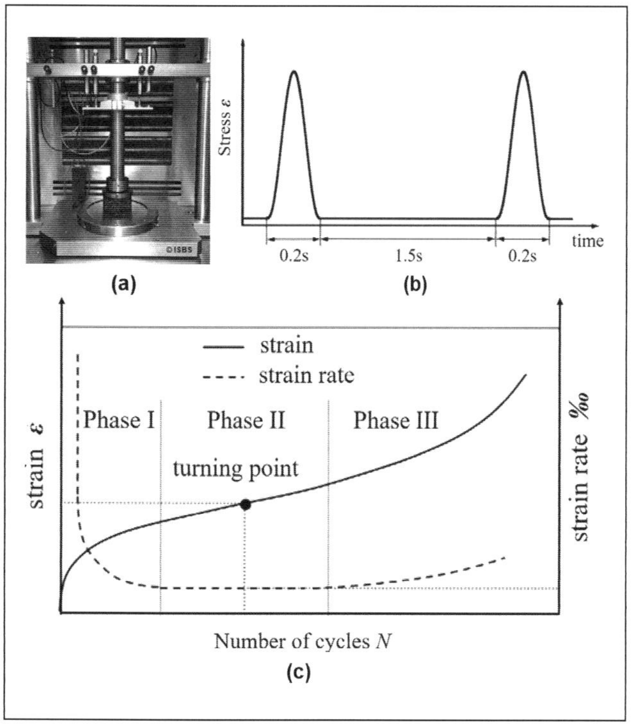

Figure 3. Uniaxial cyclic compression test: (*a*) Testing equipment, (*b*) load cycles, and (*c*) example of the results, creep curve with turning point and strain rate.

binder with all rejuvenators, starting from the initial 70°C down to approximately 58°C, which is typically observed for this plain 50/70 asphalt binder.

Testing

Asphalt Mixture Resistance to Permanent Deformation

Simple uniaxial cyclic compression tests were performed according to German specification (*23*) to determine the resistance of the recycled mixtures against permanent deformation. Cylindrical specimens with a diameter of 100 mm and height of 60 mm were cored from the asphalt mixture slabs and subjected to a haversine load of 0.35 MPa with a cycle duration of 0.2s for the loading phase and 1.2s for the rest period (see Figure 3).

At the end of each load cycle, the remaining strain [‰] is calculated and plotted as a creep curve in a diagram as function of the number of cycles (Figure 3). Results of the uniaxial creep test includes number of cycles at turning point, remaining strain and strain rate at the turning point. If no turning point is achieved, the values at the end of the test are used as results.

Stiffness Modulus

The stiffness modulus of the mixtures was determined according to the test method detailed in the European

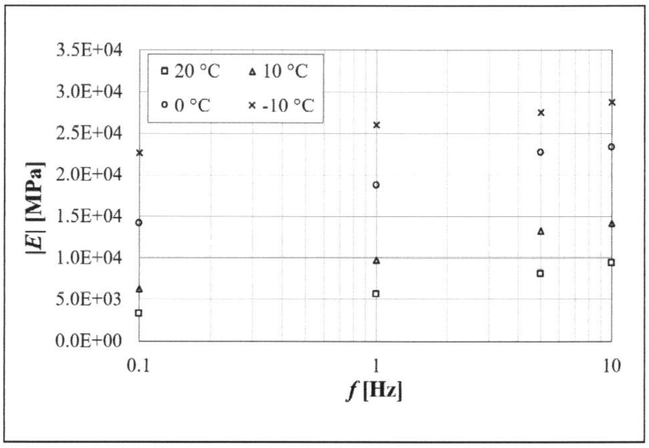

Figure 4. Stiffness modulus results ($T = -10$°C, 0°C, 10°C and 20°C) at different frequencies.

standard EN12697-26 (*24*), which relies on the cylindrical indirect tensile test mode (CIDT); at least four test temperatures (–10°C, 0°C, 10°C and 20°C) and three different loading frequencies in the range of 0.1 to 10 Hz, need to be used. Three replicates per testing conditions were used. Depending on test temperature and frequency, the stiffness modulus E was obtained as

$$E = [\Delta F \cdot (0.274 + \mu)]/[h \cdot \Delta u] \quad (1)$$

where
E = stiffness modulus,
ΔF = difference between the minimum and the maximum load,
μ = Posson's ratio,
h = specimen height, and
Δu = difference between the minimum and the maximum deformation.

Figure 4 shows an example of the measured stiffness moduli at each testing temperature T_{test} and frequency.

To generate the master curve of the stiffness modulus, the shift factor a_T can be derived from

$$\log \alpha_T = m \cdot (1/T - 1/T_0) \quad (2)$$

where the data obtained at frequency f can be expressed in terms of reduced frequency $f_R = a_T{*}f$.

Therefore, the master curve is approximated by a regression function such as the following sigmoid formula:

$$|E| = y_0 + \frac{w}{1 + e^{-\left(\frac{\log(f_R) - x_0}{z}\right)}} \quad (3)$$

where w is the absolute modulus and z is a material parameter.

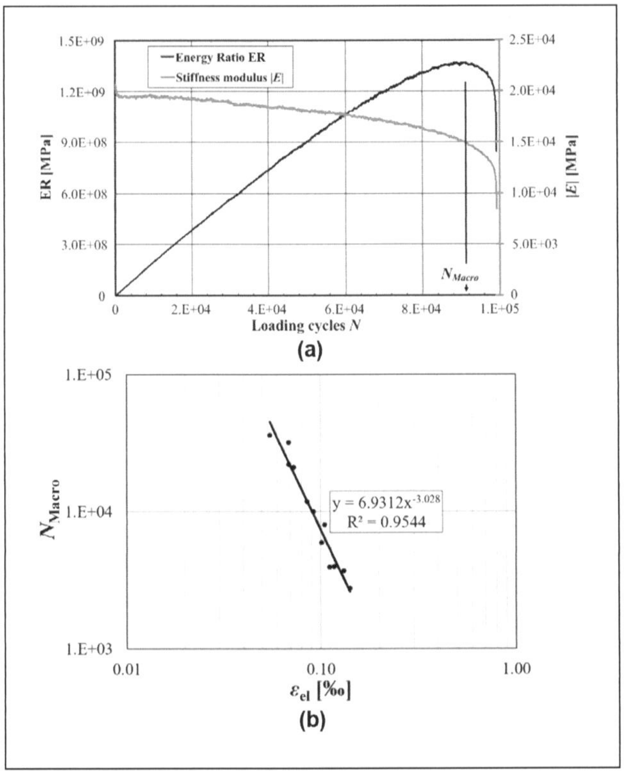

Figure 5. (*a*) Schematic to obtain the N_{macro} based on the maximum value of EN at f = 10 Hz, and (*b*) example of a fatigue function at T = 20°C and f = 10 Hz.

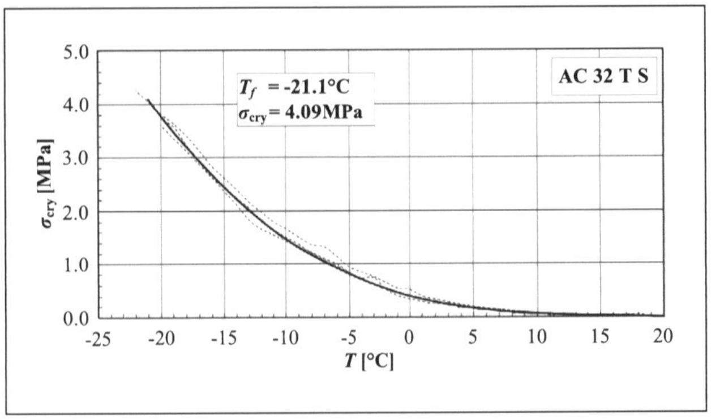

Figure 6. TSRST (*25*) example of cryogenic stress as function of temperature.

Fatigue Behavior

The fatigue behavior of asphalt mixture was analyzed by using cyclic laboratory tests on the same cylindrical geometry used for stiffness measurements (*25*) in the non-linear material domain. The force amplitude is imposed in such a way that with an increasing number of load cycles, the material stiffness modulus decreases quasi-linearly, indicating the occurrence of material damage. Figure 5 shows a typical fatigue response of the mixture in terms of decreasing stiffness modulus at $T = 20°C$ and $f = 10$ Hz (*26*). In addition, the figure provides the curve of the corresponding energy ratio ER obtained by multiplying the stiffness modulus $|E|_N$ by the corresponding number of load cycles N according to

$$ER(N) \sim N \cdot |E|_N \tag{4}$$

The peak of ER is used to determine failure due to fatigue, as this is associated with the beginning of the macro-crack, and identifies the allowable number of load cycles, N_{macro}, for the specimen (Figure 5*a*). At least nine single fatigue tests (three replicates per condition) must be performed at three different stress amplitudes to obtain the material's fatigue function. Based on the measured horizontal elastic initial strains ε_{el} and the corresponding N_{macro}, the fatigue function can be described by a power expression (Wöhler line) (*7*) as

$$N = C_1 \cdot \varepsilon^{C_2} \tag{5}$$

where C_1 and C_2 are fitting parameters. Figure 5*b* presents an example of fatigue curve.

Low Temperature Performance

Prismatic mixture specimens were subjected to thermal stress restrained specimen tests (TSRST) according to the EN12697-46 (*27*) standard for evaluating the low temperature performance behavior. During TSRST, the specimen is prevented from shrinking, while its temperature decreases starting from $T_{start} = 20°C$ with a constant cooling rate of $\Delta T = -10°C/h$ resulting in an increase of the tensile stress (known as cryogenic stresses) in the material. As in the previous tests, three replicates were used.

The test ends either at a minimum temperature of $T = -40°C$ or at failure when the cryogenic stress reaches the tensile strength of the asphalt mixture sample. The test result of TSRST is expressed by fracture temperature T_f [°C] and corresponding cryogenic stress σ_{cry} [MPa]. Figure 6 shows an example of the cryogenic stress evolution as function of temperature.

Results

Asphalt Mixture Resistance to Permanent Deformation

Figure 7 shows the results of the strain rate at the turning point or at the end of the uniaxial cyclic compression test (depending on the actual material response) for the entire set of mixtures.

A significant influence of the amount of RAP can be observed. It appears that the addition of RAP + rejuvenator positively influences the deformation resistance compared to the virgin mixture. An increase in RAP content

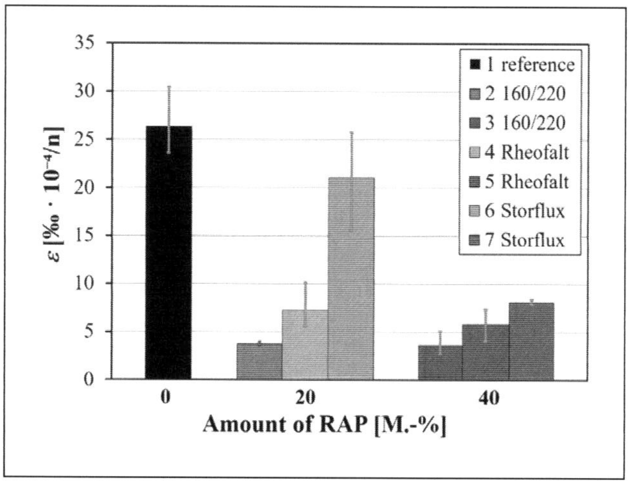

Figure 7. Strain rates at the turning point or at the end of the uniaxial cyclic compression test.

results in a similar or even decreasing strain rate, compared to the reference material, also when the rejuvenator is included in the mix design. With respect to the type of rejuvenator, the use of 160/220 results in the best deformation resistance followed by Rheofalt and Storflux.

Stiffness Modulus

Stiffness modulus was determined at different temperatures (20°C, 10°C, 0°C, and −10°C) and frequencies (0.1 Hz, 1.0 Hz, 5.0 Hz, and 10 Hz). Figure 8 shows the results at 10 Hz.

The value of the stiffness modulus of the recycled mixtures is comparable with that measured on the reference material (mixture 1). Nevertheless, it visually appears that when using Rheofalt and Storflux with a content of 40% of RAP (mixtures 5 and 7) lower stiffness values are observed. Figure 9 visualizes the resulting master curves at a frequency of 10 Hz.

Fatigue Behavior

In this section, the fatigue results are presented based on the analysis conducted on the fatigue measurements obtained using stress-controlled fatigue tests in the CIDT configuration. The number of loading cycles to fatigue failure N_{macro} (fatigue life) was plotted against the initial strain level ε_{el} at a loading frequency of 10Hz and at a test temperature of 20°C (Figure 10).

The coefficient of determination (R^2) is high for all mixtures and ranges from 90% to 98%. The fatigue functions of the wearing course mixtures exhibit an approximately parallel trend (Figure 10); this is confirmed by the comparable values of the slope parameter C_2. All mixtures prepared with rejuvenators show better fatigue properties than the control material, except mixture 6 (Storflux and 20% RAP) which presents fatigue behavior similar to mixture 1.

Low Temperature Performance

To evaluate the low temperature behavior of the set of asphalt mixtures investigated, TSRST (*25*) fracture temperature and corresponding cryogenic stress were selected for comparison purposes. Low fracture temperature and high cryogenic stress values indicate good material performance at low temperature. The experimental results are summarized in Figure 11.

With respect to fracture temperature, mixtures 1, 3, 6 and 7 show similar behavior. Using the Storflux rejuvenator (mixtures 6 and 7) fracture temperature comparable to the reference mixture was observed without any substantial impact of the amount of RAP. Also mixture

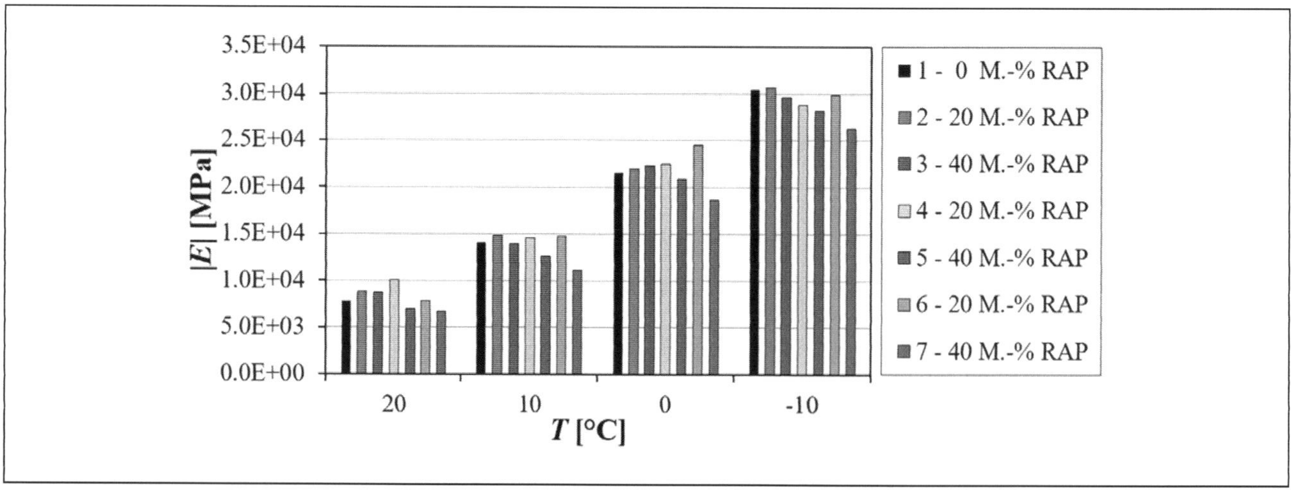

Figure 8. Stiffness modulus from CIDT test at T = 20°C, 10°C, 0°C, and −10°C, and f = 10 Hz.

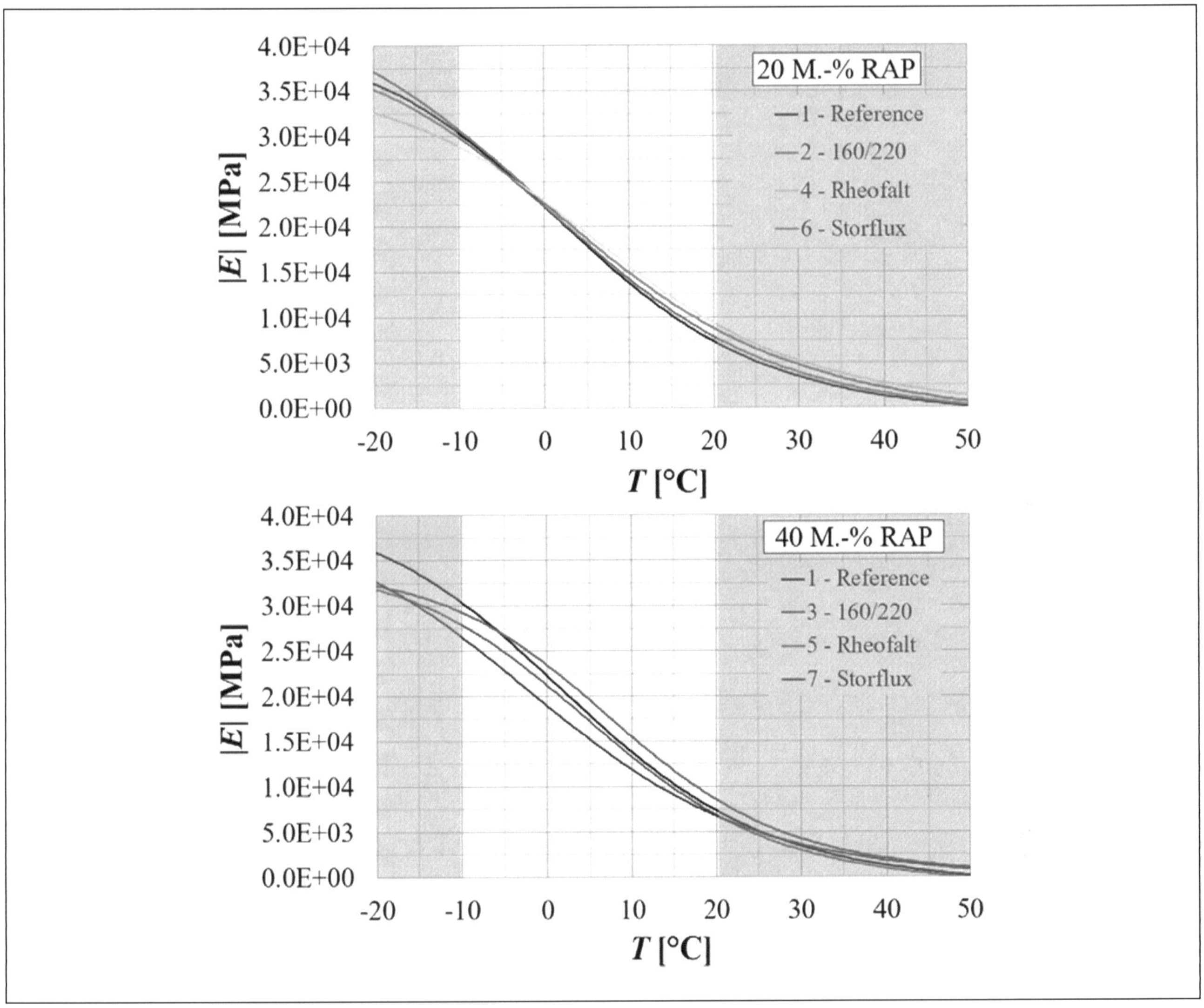

Figure 9. Master curves of stiffness moduli obtained from CIDT tests at f = 10 Hz.

3, prepared with a 160/220 binder as a rejuvenator and 40% of RAP, approaches the fracture temperature of the reference material. Using Rheofalt, higher fracture temperatures are observed for both RAP contents. Mixture 2, which includes a 160/220 binder as rejuvenator and only 20% of RAP, results in an unexpected high fracture temperature. The cryogenic stresses at fracture show comparable results for the entire set of mixtures. Mixture 4 presents the highest stress and fracture temperature; suggesting a relatively poor low temperature behavior. While for the specific result of mixture 4 it could be initially assumed that this is due to a low dosage of the rejuvenator (Rheofalt) associated with an erroneous material characterization, this hypothesis is not supported by the consistency of the other experimental results (permanent deformation, stiffness, and fatigue) and by the BTSV tests performed on the binder extracted from the rejuvenated mixtures (see materials section of this paper).

Summary and Conclusions

In this paper, the use of rejuvenators in combination with high RAP was investigated based on the analysis method BTSV, and then experimentally evaluated. A set of seven wearing course mixtures were produced using three amounts of RAP and three different types of rejuvenators. The properties of the extracted binder were checked with the BTSV method and similar BTSV temperature was observed. Resistance against deformation, mixture stiffness, fatigue behavior and resistance to low temperature cracking were next experimentally evaluated. It was found that rejuvenators and high amounts of RAP have a positive impact on deformation resistance, negligible impact on stiffness properties, improve the fatigue response, and result in comparable low temperature performance, except for a specific rejuvenator. Based on this case study the application of rejuvenators

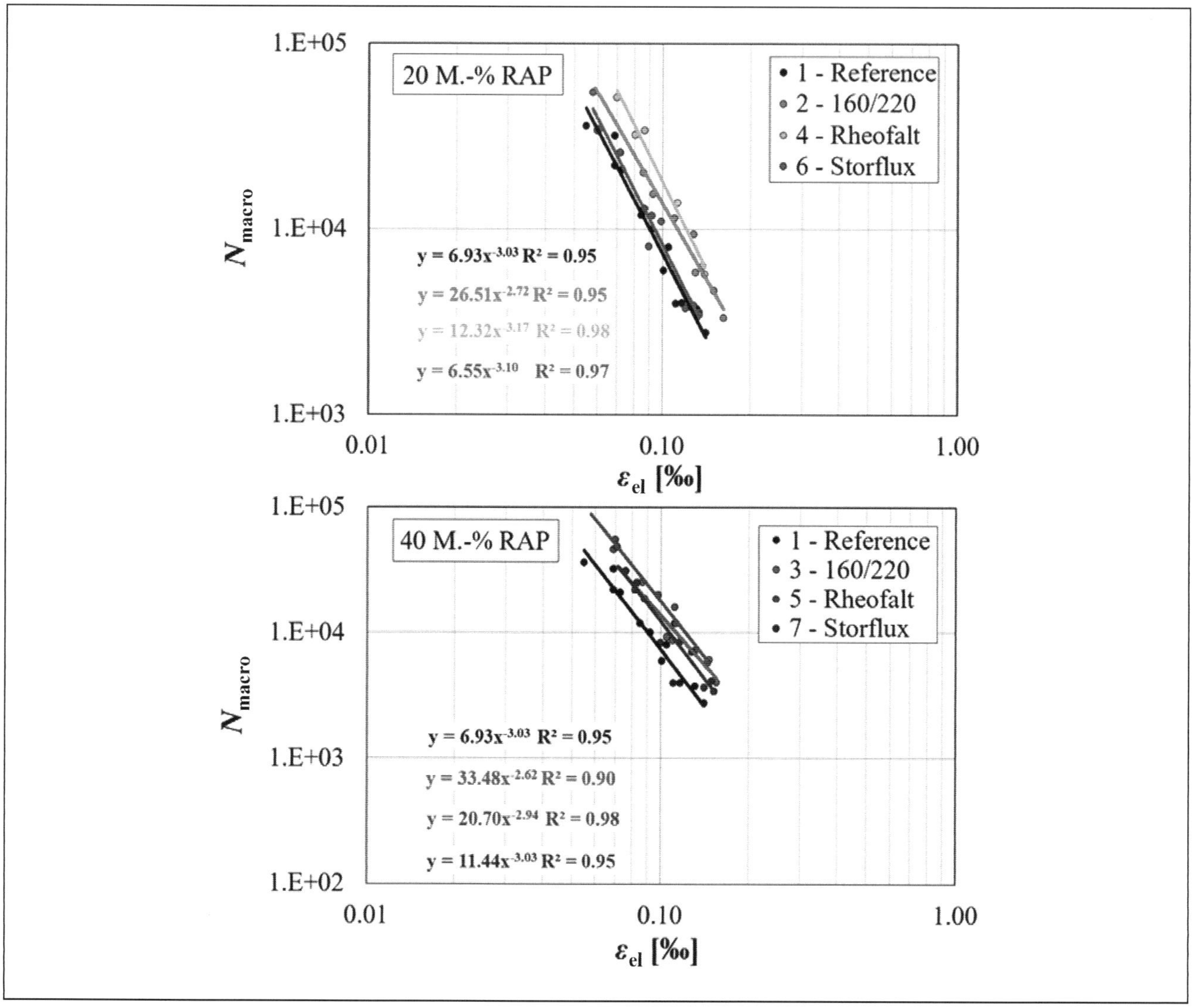

Figure 10. Loading cycles to fatigue failure N_{macro} vs initial strain level ε_{el} for three different loading amplitudes at f = 10 Hz and at a test temperature of 20°C.

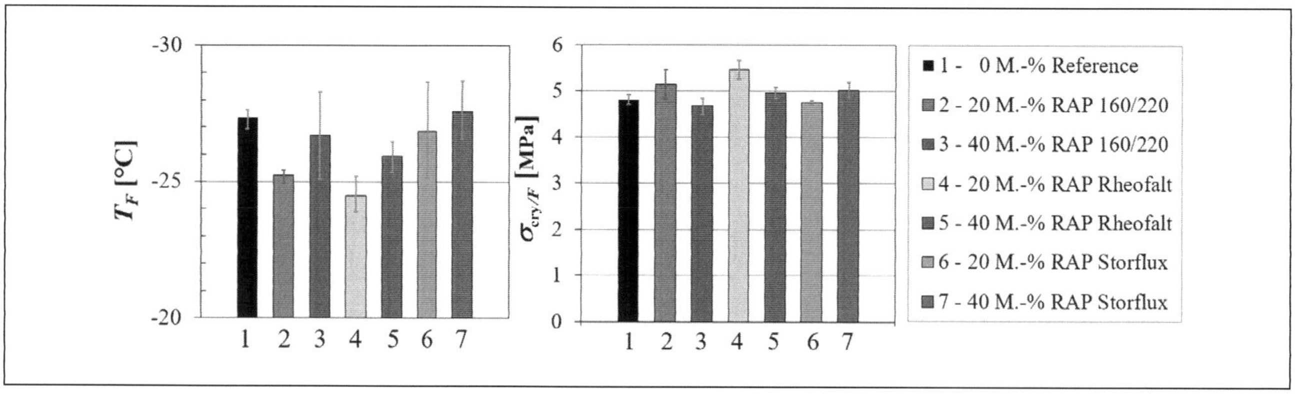

Figure 11. Fracture temperature and corresponding cryogenic stress from TSRST (25).

is recommended for field tests on wearing courses (7). Further research is needed to determine the response of the material in the case of long term aging when rejuvenators are used.

Author Contributions

Stephan Büchler conceived and designed the experiments; Augusto Cannone Falchetto wrote the paper; Axel Walther and Chiara Riccardi contributed to data analysis; Di Wang and Axel Walther provided additional recommendations; Michael P. Wistuba revised the paper.

References

1. Li, X., M. O. Marasteanu, R. C. Williams, and T. R. Clyne. Effect of RAP (Proportion and Type) and Binder Grade on the Properties of Asphalt Mixtures. *Transportation Research Record: Journal of the Transportation Research Board*, 2008. 2051: 92–97.
2. Cannone Falchetto, A., A. Montepara, G. Tebaldi, and M. O. Marasteanu. Microstructural and Rheological Investigation on the Use of Recycled Asphalt Material in Asphalt Mixtures at Low Temperature. *Construction and Building Materials*, Vol. 35, 2012, pp. 321–329.
3. Mangiafico, S., H. Di Benedetto, C. Sauzeat, F. Olard, S. Dupriet, L. Planque, and R. Van Rooijen. Effect of Reclaimed Asphalt Pavement on Complex Modulus and Fatigue Resistance of Bitumens and Asphalts. *Proc., 5th Eurasphalt & Eurobitume Congress*, Istanbul, Turkey. 2012.
4. Mangiafico, S., H. Di Benedetto, F. Sauzéat, S. Olard, F. Pouget, and L. Planque. Influence of Reclaimed Asphalt Pavement Content on Complex Modulus of Asphalt Binder Blends and Corresponding Mixes: Experimental Results and Modelling. *Road Materials and Pavement Design*, Vol. 14, 2013, pp. 132–148.
5. Moon, K. H., A. Cannone Falchetto, and J. W. Hu. Investigation of Asphalt Binder and Asphalt Mixture Low Temperature Creep Properties Using Semi Mechanical and Analogical Models. *Construction and Building Materials*, Vol. 53, 2014, pp. 568–583.
6. EN 12591. *Bitumen and Bituminous Binders - Specifications for Paving Grade Bitumens*. European Committee for Standardization, Brussels, Belgium, 2009.
7. Walther, A., and M. Wistuba. *Auswirkung der Wiederverwendung von Ausbauasphalt in Asphalttragschichten auf mechanische Eigenschaften*. Project Report, TU Braunschweig ISBS, 2014. (In German)
8. Guillaume, F. Die Wiederverwendung von bituminösem Mischgut in zentralen Mischanlagen. *La Technique Routière*, Vol. 1, 1987, pp. 56. (In German)
9. Holleran, G., T. Wieringa, and T. Tailby. *Rejuvenation Treatments for Aged Pavements. Proc., 8th Transit New Zealand and New Zealand Institute of Highway Technology (Nziht) Annual Conference*, Auckland, New Zealand, 2006.
10. Shen, J., S. Amirkhanian, and B. Tang. Effects of Rejuvenator on Performance-based Properties of Rejuvenated Asphalt Binder and Mixtures. *Construction and Building Materials*, Vol. 21, No. 5, 2007, pp. 958–964.
11. Scarpas, A., S. N. Nahar, A. J. M. Schmets, E. Schlangen, M. F. C. van de Ven, M. Shirazi, and G. Schitter. *Turning Back Time: Rheological and Microstructural Assessment of Rejuvenated Bitumen*. Transportation Research Board, Washington, D.C., 2014.
12. Sanders, P. *Review of Recycling and Rejuvenation Procedures*. SILVIA Project Report, SILVIA-TRL-016-02-WP4-240605. European Commission 2005.
13. Alisov, A., and M. Wistuba. Addressing Bitumen Softening Characteristics through Dynamic Shear Rheometer. *Proc., 7th International Conference of the European Asphalt Technology Association (EATA)*, Zürich, Switzerland, EATA publisher 2017.
14. AL BTSV (DSR). *Arbeitsanleitung zur Bestimmung des Verformungsverhaltens von Bitumen und bitumenhaltigen Bindemitteln im Dynamischen Scherrheometer (DSR) - Durchführung des BTSV (Bitumen-Typisierungs-Schnell-Verfahren)*. German Technical Specification (Draft), 2017. (In German)
15. TL Asphalt-StB. *Technische Lieferbedingung für Asphaltmischgut für den Bau von Verkehrsflächenbefestigung*. FGSV Verlag, 2013. (In German)
16. EN-1427. *Bitumen and Bituminous Binders - Determination of the Softening Point - Ring and Ball Method*. European Committee for Standardization, Brussels, Belgium, 2015.
17. Alisov, A., T. Hagner, and A. Walther. *Quantification of Polymer Content in Binder by modified MSCR-Test. Proc., 6th Eurasphalt & Eurobitume Congress*, Prag, 2016.
18. EN-14770. *Bitumen and Bituminous Binders - Determination of Complex Shear Modulus and Phase Angle - Dynamic Shear Rheometer (DSR)*. European Committee for Standardization, Brussels, Belgium, 2012.
19. Mattischek, H.-P., R. Sobczak, and J. Bleier. *Der Einsatz eines neuen Meßgerätes in der Bitumenprüfung - Korrelation der Nullviskosität mit dem Erweichungspunkt Ring und Kugel*. Bitumen 1/1996, S. 8-12. 1996. (In German)
20. Guericke, R. 100 Jahre Erweichungspunkt Ring und Kugel: Was Kommt Danach? [100 years Softening Point Ring and Ball-What Will Follow?] *Straße Und Autobahn*, Vol. 61, No. 7, 2010, 481–491. (In German)
21. EN 12591. *Bitumen and Bituminous Binders. Specification for Paving Grade Bitumens*. European Committee for Standardization, Brussels, Belgium, 2016.
22. EN 12697-3. *Bituminous Mixtures. Test Methods for Hot Mix Asphalt. Bitumen Recovery: Rotary Evaporator*. European Committee for Standardization, Brussels, Belgium, 2013.
23. EN 12697-25. *Bituminous Mixtures. Test Methods for Hot Mix Asphalt. Cyclic Compression Test*. European Committee for Standardization, Brussels, Belgium, 2016.
24. EN 12697-26. *Bituminous Mixtures. Test Methods for Hot Mix Asphalt. Stiffness*. European Committee for Standardization, Brussels, Belgium, 2012.
25. FGSV. *Arbeitsanleitung zur Bestimmung des Steifigkeits- und Ermüdungsverhaltens von Asphalten mit dem Spaltzug-Schwellversuch als Eingangsgröße in die Dimensionierung von Asphaltbefestigungen*. AL-SP – ASPHALT 09,

Forschungsgesellschaft für Straßen- und Verkehrswesen, Köln, 2009. (In German)
26. EN 12697-24. *Bituminous Mixtures. Test Methods for Hot Mix Asphalt. Resistance to Fatigue*. European Committee for Standardization, Brussels, Belgium, 2012.
27. EN 12697-46. *Bituminous Mixtures. Test Methods for Hot Mix Asphalt. Low Temperature Cracking and Properties by Uniaxial Tension Tests*. European Committee for Standardization, Brussels, Belgium, 2012.

The Standing Committee on Non-Binder Components of Asphalt Mixtures (AFK30) peer-reviewed this paper (18-05234).

Article

TRR
JOURNAL OF THE TRANSPORTATION RESEARCH BOARD

Transportation Research Record
2018, Vol. 2672(28) 107–118

DOI: 10.1177/0361198118787634
journals.sagepub.com/home/trr
SAGE

A Mechanical Approach to Quantify Blending of Aged Binder from Recycled Materials in New Hot Mix Asphalt Mixtures

Milad Zokaei Ashtiani[1], Walaa S. Mogawer[2], and Alexander J. Austerman[2]

Abstract

In the paving industry, there is increased interest in using recycled materials like recycled asphalt shingles (RAS) and reclaimed asphalt pavements (RAP) due to the valuable asphalt binder contained within them. The major concern with using these materials is that the binder they contain is highly aged, which could lead to reduced mixture durability. Therefore, a method is needed to quantify the extent to which the aged binders from these materials blend with virgin binder when producing mixtures in order to understand better their effects on mixture performance.

In this study, a new approach to quantify the amount of blending that occurs between aged RAS and RAP binders and a virgin binder was developed. Asphalt binders were extracted and recovered from RAS and RAP stockpiles and blended with a PG64-28 virgin binder in varying proportions. The master curves of these mixtures were constructed at 20°C. Asphalt mixtures containing different proportions of the same RAS and RAP stockpiles were then designed and the dynamic moduli of the mixtures were measured to construct mixture master curves at 20°C.

The binder master curves for each blending proportion were then substituted into a locally calibrated Hirsch model to predict the mixture master curves. Comparison of the measured and predicted mixture master curves suggested that the aged binder from RAS and RAP blends with the virgin binder less than 40% and 60%, respectively. Cracking tests were also conducted to validate the proposed degrees of blending from a mixture mechanical performance point of view.

Many state transportation agencies are interested in using recycled materials like recycled asphalt shingles (RAS) and reclaimed asphalt pavements (RAP) in new paving mixtures because they contain valuable asphalt binder. One of the main obstacles to utilizing RAS and/or higher amounts of RAP is that is not known how much of the aged asphalt binder contained with them will blend, or be activated, in a new mixture. Depending on this quantity of blending, the performance of the resulting mixture could be affected in terms of resistance to distresses. More blending of the aged material may result in a stiffer mixture which could be more prone to cracking but also more rut resistant. Since the asphalt binder used to produce RAS is a stiff air blown asphalt, it could have a more severe than the addition of RAP (*1*). Several studies have evaluated the blending in terms of whether it occurs or not. No studies have endeavored to quantify in a numerical sense how much blending is occurring. As an alternative, several studies have taken advantage of mechanical tests to evaluate the effect of using more recycled materials in mixtures with respect to distress (*2–5*).

Two terms have been used interchangeably with regard to blending. Binder contribution is a term used to describe the quantity of asphalt binder from RAS or RAP that participates as effective binder in a mixture design. Degree of blending, on the other hand, refers to the binder formed as a result of diffusion between two binders that have different rheological properties when combined (*6, 7*). Researchers have utilized different approaches to study the degree of blending, mainly focusing on whether blending occurs or not.

[1]Civil and Environmental Engineering Department, University of Washington, Seattle, WA
[2]Highway Sustainability Research Center (HSRC), Civil and Environmental Engineering Department, University of Massachusetts, Fall River, MA

Corresponding Author:
Address correspondence to Walaa S. Mogawer: wmogawer@umassd.edu

Table 1. Aggregate Gradation and Aggregate Specific Gravities

	Percent passing						
Sieve openings (mm)	Job mix formula (JMF)	12.5 mm coarse	9.5 mm coarse	Stone sand	RAP	MWSS	TOSS
19	100	100	100	100	100	100	100
12.5	95	87.1	99.9	100	99.4	100	100
9.5	80	27.8	91.2	100	93.3	100	100
4.75	49	5.4	19.7	97	63.1	99.8	99.8
2.36	32	2.1	3.3	71	45.4	98.6	99.3
1.18	22	1.8	2.6	46	34.7	80.7	81.7
0.6	15	1.7	2.4	30	27.1	55.7	67.7
0.3	11	1.6	2.2	17	20.2	46.1	52.0
0.15	7	1.5	2	8	13.6	36.0	42.5
0.075	4	1.0	1.5	4.4	8.1	26.2	30.3
G_{sb}	2.750	2.773	2.754	2.715	2.770	2.839	3.070
Binder content (%)	5.2	na	na	na	5.5	17.4	31.5

Note: G_{sb} = aggregate specific gravity; na = not applicable.

Previous research into blending has been conducted (*8–14*). An approach to studying the degree of blending, recommended by Bonaquist, was to utilize models which correlate the mixture dynamic modulus and binder shear modulus (*8*). As well as its simplicity, the Hirsch model was found to be more beneficial to link mechanical to rheological performance of asphalt mixtures when blending is concerned (*9, 10*). Extending the calculation from the Hirsch model to binder master curves, researchers were able to assess the degree of blending when either RAS or RAP was used (*5, 11*). However, none of the methods mentioned above aimed to quantify numerically the degree of blending between aged and virgin binder throughout the whole mixture. This study attempted to present a new approach to quantifying numerically the degree of blending.

Objectives

The objectives of this study were as follows: (a) propose a procedure which can be used to quantify the amount of binder from RAS or RAP that blends with a virgin binder, (b) validate the procedure using mixture tests that measure the cracking susceptibilities of mixtures, and (c) investigate the effects of RAS and RAP on cracking susceptibility.

Materials and Mixture Design

For this study, three aggregate stockpiles were used: 12.5 mm coarse aggregate, 9.5 mm coarse aggregate, and stone sand. RAP was also obtained from the same contractor as the aggregates. RAS was obtained from two sources: tear-off scrap shingles (TOSS) and manufacturer waste scrap shingles (MWSS).

A Superpave 12.5 mm mixture was designed and used in the study. Using the ignition oven test according to AASHTO T308, the binder content of TOSS, MWSS, and RAP was found to be 31.5%, 17.4%, and 5.5%, respectively. Table 1 shows the aggregate gradations and specific gravity of each stockpile. Optimum binder content for this mixture, using all virgin materials, was measured to be 5.2% by weight of the mixture. A performance grade PG64-28 binder was used as it is the standard specified binder grade in Massachusetts.

To achieve the objectives of this research, four different mixtures were developed: (a) mixture without recycled materials as control mix (control), (b) mixture containing 5% MWSS, (c) mixture containing 5% TOSS, and (d) mixture containing 15% RAP. Following recommendations from the literature, RAS was limited to 5% as that is the highest percentage that is allowed to be used in the production of hot mix asphalt without changing to a softer virgin binder grade (*12*).

Binder contributions were measured by first assuming 100% of the binder from RAS or RAP is effective. In cases in which the Superpave criteria were not met, changes to the binder contribution assumption were made so that the volumetric requirements for the 12.5-mm dense graded mixture were satisfied (AASHTO M323-13). All mixtures met Superpave criteria for 75-gyration mixtures at a traffic level of equivalent single axle load of 0.3 million to less than 3 million.

To replicate identical aggregate skeletons better, individual sieving was implemented to match the job mix formula (JMF) gradation for all four blends. The Superpave gyratory compactor (SGC) was used to fabricate all

Table 2. Mixture ID and Volumetric Properties

Mix ID	Volumetric properties				
	Maximum theoretical specific gravity (G_{mm})	Air voids (%)	Voids in mineral aggregates (VMA) (%)	Voids filled with asphalt (VFA) (%)	RAP/RAS binder contribution (%)
Control	2.576	3.7	14.2	74.1	na
5% MWSS	2.577	3.7	14.4	74.4	100
5% TOSS	2.555	3.7	15.5	76.0	70
15% RAP	2.554	3.6	14.3	76.6	100

Note: na = not applicable.

mixture samples unless otherwise specified. Volumetric properties for the mixtures are presented in Table 2. Following a procedure utilized in previous studies, dried RAS and RAP were added to heated aggregates 5 minutes and 2 hours prior to mixing, respectively, at 155°C and then conditioned at compaction temperature of 145°C. These mixing and compaction temperatures were specified by the supplier of the asphalt binder for the PG64-28 virgin binder used in this study.

Binder Testing

By means of a centrifuge and the rotary evaporator, asphalt binder from RAS and RAP were extracted and recovered using toluene as a solvent (ASTM D6847-02, ASTM D7906-14). Various proportions of virgin binder combined with recovered binder were considered that were equivalent to certain degrees of blending that we assumed might be present in the mixtures (see Table 3). Calculations of the equivalent degree of blending were completed by knowing the amount of virgin binder added to fabricate each mixture and the binder contents of RAS and RAP.

Multiplying the binder contents of RAS or RAP by the total weights of mixtures in which they were introduced results in the total quantity of binder attributed from recycled materials. Multiplying this total aged binder from RAS or RAP by the binder contributions, therefore, results in their effective binder contents. Subtraction of the optimum binder content of the control mix from the effective binder from RAS or RAP would result in the amount of virgin binder needed to add into mixtures. Assumptions on the degree of blending can then be made on how much the effective binder from RAS or RAP blends into the virgin binder. For example, 100% blending is the case in which all the binder from recycled materials is melted into the virgin binder, and 0% blending is equivalent to the virgin binder.

Recovered binders were heated to 170°C and then added to the virgin binder at 135°C, as suggested by Bonaquist (*12*). In order to ensure thorough blending, the blends of recovered and virgin binders were kept at 163°C for 90 minutes and stirred every 30 minutes. The blended binders were tested to measure their high, intermediate, and low continuous PG temperatures, and to construct master curves after aging in the pressure aging vessel. Table 3 also illustrates the continuous true PG of the blended binders. Because the RAP binder was not as stiff as the RAS binder, no blending proportion lower than 60% was assumed for 15% RAP mixture. Moreover, previous research studies indicated the degree of blending for RAP is believed to fall between 70% and 100% depending on the percentage of RAP used (*13*). Recovered RAS and RAP binders were also graded.

Methodology

To accomplish the first objective, extracted and recovered binders from each RAS and RAP were blended with a virgin binder in several assumed percentages which translate to different degrees of blending ranging from 0% to 100% for the mixtures. Each blended binder was tested in the dynamic shear rheometer to obtain a shear modulus (G*) data for subsequent construction of a binder master curve at a reference temperature of 20°C. This provided a set of G* master curves for each RAS and RAP blend showing the effects of assumed blending.

Next, for each mixture, a dynamic modulus (E*) master curve at a reference temperature of 20°C was obtained using the asphalt mixture performance tester (AMPT). The E* master curve of the control mixture was also measured and used to calibrate the Hirsch model. This calibrated Hirsch model was then used to estimate a set of E* master curves for each mixture from its corresponding set of (G*) master curves and various mixture properties. Each measured E* master curve was then compared with its corresponding set of estimated E* master curves to quantify degree of blending.

To validate the results of the degree of blending analysis and to assess the cracking susceptibilities of mixtures, the following tests were carried out: four-point flexural

Table 3. Performance Grading of Various Binder Contributions

Binder type	Binder blending proportions (%) Virgin	Recovered	Equivalent degree of blending in mixtures (%)	True high PG	True intermediate PG	True low PG	PG
Virgin	100	0	0	69.8	16.97	−29.5	64-28
MWSS	96.8	3.2	20	71.5	17.03	−28.4	70-28
	93.6	6.4	40	73.7	17.95	−28.2	70-28
	90.4	9.6	60	75.2	18.26	−27.0	70-22
	87.2	12.8	80	80.6	19.10	−25.7	76-22
	84	16	100	81.6	20.10	−25.3	76-22
	0	100	na	164.9[a]	40.8	NA	NA
TOSS	93.2	6.8	20	72.8	18.55	−27.8	70-22
	87.2	12.8	40	80.6	20.27	−26.4	76-22
	81.9	18.1	60	83.7	21.30	−24.3	82-22
	77.3	22.7	80	90.2	22.89	−23.4	88-22
	73.1	26.9	100	92.1	23.50	−18.5	88-16
	0	100	na	NA	48.8	NA	NA
RAP	90	10	60	72.0	18.32	−28.9	70-28
	85	15	100	72.6	18.67	−28.2	70-28
	0	100	na	92.3[a]	31.62	−13.9	88-10

Note: na = not applicable; NA = not available. 15% RAP Binder (100% Blending). 10% RAP Binder (60% Blending). 0% RAP Binder (0% Blending).
[a]High PG was determined assuming recovered binder was RTFO aged.

beam fatigue, semi-circular bending beam (SCB) and disc-shaped compact tension (DCT).

Asphalt Binder Master Curve (G)*

A master curve is a plot that summarizes the values of complex modulus (G*) at varying frequencies. Limited capabilities of experimental equipment to run cyclic loading tests at frequency ranges other than 0.01 to 100 rad/sec (or Hertz) have compelled researchers to make use of the time–temperature superposition principle. In that, if measurements of a material's response are made at varying temperatures, shifting the measured values to a reference temperature would cover a wider range of frequency. A functional form can then be fitted into the shifted data to form a single smooth plot of G* versus frequency (or equivalently temperature) known as a master curve.

For this study, frequency sweep tests from 0.01 to 100 rad/sec at temperatures of 10, 22, 34, 46, 58, and 70°C were used to construct master curves at reference temperature of 20°C. Arrhenius and modified Williams-Landel-Ferry by Kealble are two shift factor equations at colder and warmer temperatures, respectively, commonly used to construct master curves (Equations 1 and 2). Standard sigmoid function (Equation 3) with a higher asymptote has been shown to describe best the behavior of a binder master curve. A simultaneous solution of the shift factor equations and a functional form will result in the general shape of master curves (*14*). Figure 1 shows the master curves for all assumed blending proportions fabricated from mixing aged and virgin binder, as described earlier. From Figure 1, the effect of adding more recovered binder to virgin binder is evident as they have higher values of G* at the same frequencies.

$$Arrhenius : log a_T = a + b\left(\frac{1}{T} - \frac{1}{T_{ref}}\right) \quad (1)$$

$$Kealble\ Modification\ of\ WLF : log a_T = -\frac{C_1(T - T_{ref})}{C_2 + |T - T_{ref}|} \quad (2)$$

$$Standard\ Sigmoid : log G^* = \delta + \frac{\alpha}{[1 + \lambda e^{\beta + \gamma(log\omega)}]^{1/\lambda}} \quad (3)$$

Where
a_T = shift factor,
T = temperature (Kelvin),
T_{ref} = reference temperature (Kelvin),
ω = frequency (Hz), and
Other variables are constants.

Asphalt Mixture Master Curve (E)*

Specimens 150 mm in height and 100 mm in diameter were cored and cut from original samples fabricated in the SGC. Using AMPT and following AASHTO PP61-13, a haversine wave of loading is applied to measure the dynamic modulus which is the ratio of peak stress to peak strain. Each sample was tested at 4°C, 20°C, and 40°C at

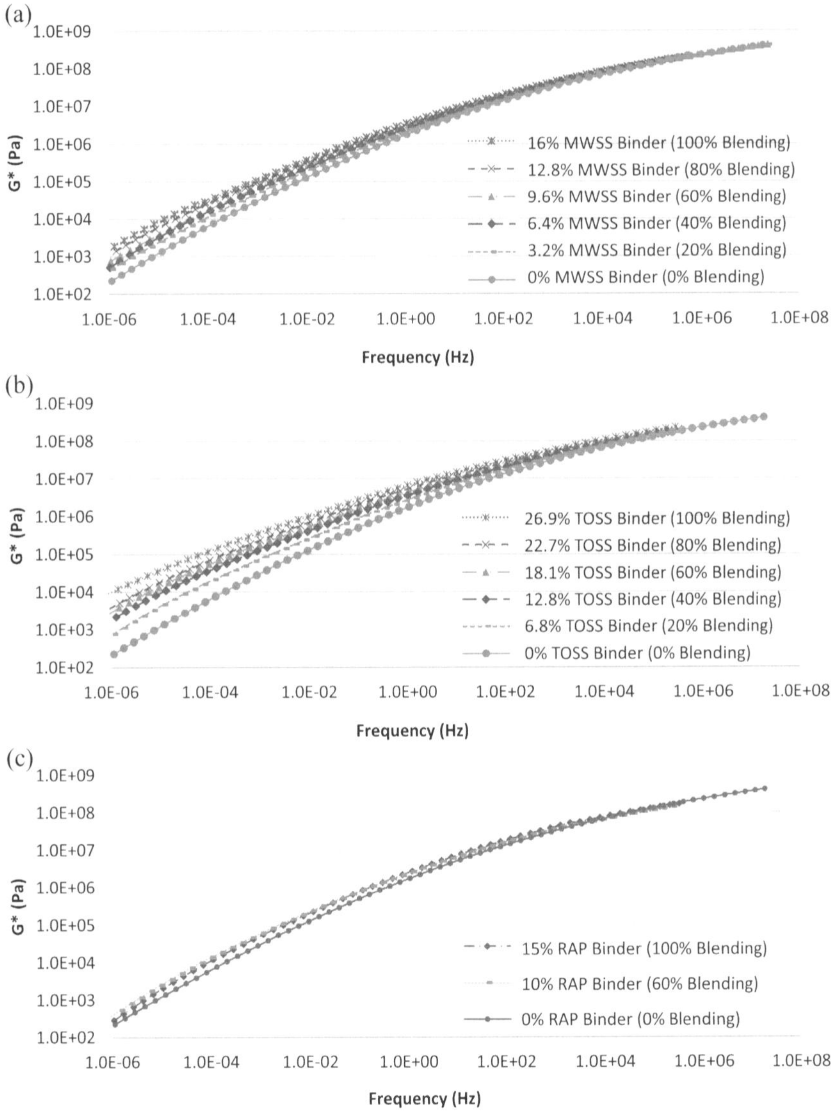

Figure 1. Asphalt binder master curves at 20°C for various degrees of blending of virgin binder and recovered binder from (*a*) MWSS, (*b*) TOSS, and (*c*) RAP.

frequencies of 10, 1, 0.1, and 0.01 Hz. Dynamic modulus data were then shifted at 20°C to create master curves for mixtures. The Arrhenius equation is thought to provide a good estimation of shift factors for asphalt mixtures (Equation 2). Equation 4 is also a general logistic sigmoid form to fit the shifted points into one smooth graph (Figure 2) (*14*). Variables in this equation are as described earlier. Figure 2 indicates that the stiffer binder from RAS caused the master curve of 5% RAS to be higher than 15% RAP, which itself is slightly higher than the control. This observation indicates that the effect of blending could be significant even for mixtures containing RAS.

$$logE^{*} = \delta + \frac{\alpha}{1 + e^{\beta + \gamma(log\omega)}} \quad (4)$$

Estimation of E* from Hirsch Model

Modulus is the link between load and deformation, on which all mechanical properties of a structure depend. For temperature-dependent viscoelastic materials like asphalt mixtures, the complex modulus is defined to account for the effect of loading time and temperature. Both aggregate skeleton and asphalt binder influence the modulus of mixtures. The Hirsch model represents a

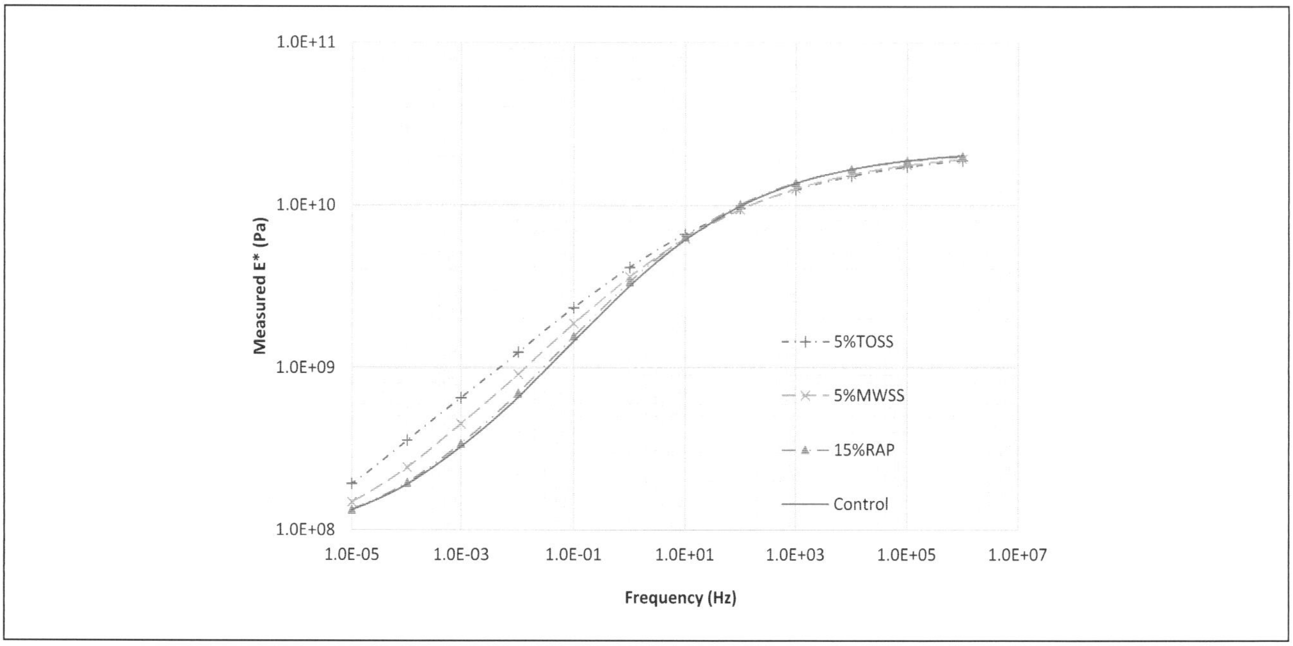

Figure 2. Asphalt mixture master curves at reference temperature of 20°C.

relationship between mixture dynamic modulus (|E*|) and asphalt binder shear modulus (|G*|) (Equations 5 and 6). This model is valid only if the same temperature and the same frequency are used for binder and mixture, and the model is most accurate at intermediate frequencies and temperatures (*9, 10*).

$$|E^*|_m = P_c\left[4,200,000\left(1-\frac{VMA}{100}\right)+3|G^*|_b\left(\frac{VMA\times VFA}{10,000}\right)\right] + \frac{(1-P_c)}{\frac{1-VMA/100}{4,200,000}+\frac{VMA}{3|G^*|_b(VFA)}} \quad (5)$$

$$P_c = \frac{\left(20+3|G^*|_b{}^{(VFA)}/_{(VMA)}\right)^{0.58}}{650+\left(3|G^*|_b{}^{(VFA)}/_{(VMA)}\right)^{0.58}} \quad (6)$$

Where

$|E^*|_m$ = asphalt mixture dynamic modulus, psi,
$|G^*|_b$ = asphalt binder complex shear modulus, psi,
VMA = voids in mineral aggregate,
VFA = voids filled with asphalt, and
P_c = aggregate contact volume.

Backcalculation of |G*| from the Hirsch equation is a method to determine whether blending takes place. Degree of blending has been qualitatively assessed for mixtures containing RAP in some studies. However, in most cases extraction and recovery has been made on mixtures that had undergone the dynamic modulus test (*5, 11, 15*). The main disadvantage of this approach is that the virgin and aged binder will dissolve together and form a homogenous binder that represents only a 100% blending scenario. Thus, attention has been drawn to making different ratios of recycled asphalt binder and virgin binder artificially to investigate the effect of several degrees of blending on the performance grade of asphalt binders (*12, 16*).

The original Hirsch model proposed by Christensen et al. made use of a database containing limited measured values of E* and G*. Moreover, no detailed information is provided by the developers of this model that can be used to evaluate its applicability to any specific mixture design. Level of binder aging, type of aggregates and gradation, mixing and conditioning temperatures of mixtures used to fabricate dynamic modulus samples, and volumetric properties of the mixtures are some of the factors that may affect the validity of the Hirsch model. In this sense, the original Hirsch model could be incapable of accurately estimating E* from G* for all mixtures. To overcome probable shortcomings of the original model, local calibration of the model was completed based on the control mixture and virgin binder used in this study. Equations 7 and 8 reformulate the Hirsch model to replace constant values with parametric variables.

$$|E^*|_m = P_c\left[a_1\left(1-\frac{VMA}{100}\right)+3|G^*|_b\left(\frac{VMA\times VFA}{10,000}\right)\right] + \frac{(1-P_c)}{\frac{1-VMA/100}{a_1}+\frac{VMA}{3|G^*|_b(VFA)}} \quad (7)$$

$$P_c = \frac{\left(a_2+3|G^*|_b{}^{(VFA)}/_{(VMA)}\right)^{a_4}}{a_3+\left(3|G^*|_b{}^{(VFA)}/_{(VMA)}\right)^{a_4}} \quad (8)$$

Since the dynamic modulus samples of the control mixture in this study were fabricated using only a virgin binder, substitution of binder master curve data into the Hirsch model should produce an accurate estimation of its associated mixture master curve. Calibration of the Hirsch model was completed based on the hypothesis that measured master curve for the control mixture must match the predictions from the model. Using the Solver tool in Microsoft Excel, the sum of squared errors between prediction and observation was minimized to optimize new coefficients of the Hirsch model (see Equations 9 and 10). This model was used for calculations of the E* values in the next section. Since a unique JMF was followed to design mixtures in this study, this local calibration is more accurate as it contains the properties of the aggregate structure (gradation and modulus) used to produce mixtures. Constants in the calibrated Hirsch model were not significantly different from the original model proposed by its developers which shows the strength of this formulation and its reliance on only one volumetric property of the mixture, VMA.

$$|E^*|_m = P_c\left[3,954,491\left(1-\frac{VMA}{100}\right)+3|G^*|_b\left(\frac{VMA\times VFA}{10,000}\right)\right] + \frac{(1-P_c)}{\frac{1-VMA/100}{3,954,491}+\frac{VMA}{3|G^*|_b(VFA)}}. \quad (9)$$

$$P_c = \frac{\left(307+3|G^*|_b{}^{(VFA)}/_{(VMA)}\right)^{0.834}}{10,159+\left(3|G^*|_b{}^{(VFA)}/_{(VMA)}\right)^{0.834}} \quad (10)$$

Degree of Blending Analysis and Results

Mixture dynamic modulus measurements along with the mixture volumetric properties satisfy the requirement for two of the Hirsch model inputs. To calculate the values of E* from the Hirsch model, for each mixture of 5% MWSS, 5% TOSS, and 15% RAP, all binder master curves were substituted into the model from Figure 1*a*, *b*, and *c*, respectively. As a result, mixture master curves were estimated based on different blending proportions. As explained previously, each blending proportion of virgin and aged binder was equivalent to an assumed degree of blending from a mixture perspective. Refer to Table 3 for details on equivalent degrees of blending for each blending proportion.

Figure 3 shows the results derived from substituting binder dynamic modulus (G*) into the Hirsch model to estimate E*. This figure also indicates the E* values measured from the AMPT device. This figure is limited to frequencies of 1 to 10 Hz to avoid using extrapolated values in the analysis. Mid-range frequency acts as an average value for the measurements carried out on the complex moduli. Therefore, to establish an average degree of blending through the mixture, only the frequencies that fall in a range in which all measurements are taken may result in the most precise region of a master curve. In addition, the Hirsch model seems to produce the most reliable results at the intermediate temperature regimes.

From Figure 3*a*, the experimentally measured master curve of the 5% MWSS mixture is plotted along with the predicted master curves for various degrees of blending. Figure 3*b* and *c* are also demonstrating the same concept but for 5% TOSS and 15% RAP mixtures, respectively. It was estimated that between 20% to 40% of binder from 5% MWSS was activated in the mixture, whereas less than 20% was activated for 5% TOSS. Activation of RAP binder was estimated at less than 60% for the 15% RAP mixture.

Several interpretations could be concluded from these results. First, the degree of blending is greater for RAP than for RAS, which was expected beforehand. Asphalt binder from RAP is less oxidized than RAS, and oxidation causes the stiffness of RAS binder to be much greater. Moreover, a higher binder content from RAS appears to result in a thicker binder film around its particles. Both of these may explain reasons for the higher degree of binder blending from RAP. Similarly, the degree of blending for TOSS is observed to be lower than for MWSS, which could be due to greater stiffness of the binder from TOSS compared with that from MWSS. The higher binder content of TOSS could also result in a thicker film of binder around stone particles which may contribute to its lower degree of blending.

Second, although having a lower degree of blending, RAS could be more detrimental than RAP in a mixture. Degree of blending by itself will not show the priority of one mixture over another, but the characteristics of the aged binder from RAS or RAP play a prominent role. Assume the degree of blending for 5% MWSS, 5% TOSS, and 15% RAP are rounded up to 40%, 20%, and 60%, respectively. In the same order, from

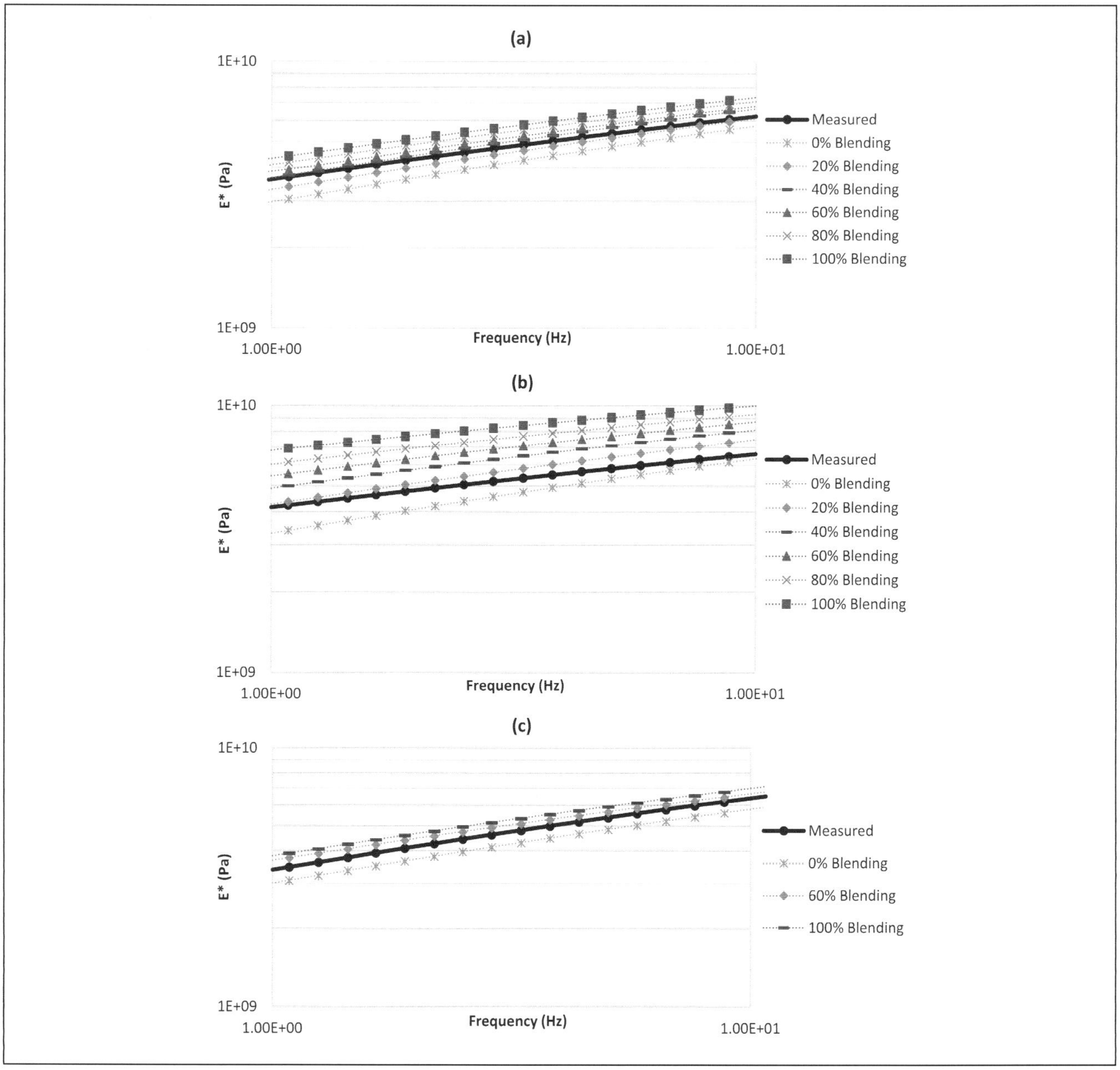

Figure 3. Degree of blending analysis for (*a*) 5% MWSS, (*b*) 5% TOSS, and (*c*) 15% RAP.

Table 3 and assuming a linear blending chart, their true performance grades would be PG73.7(17.95)-28.2, PG72.8(18.55)-27.8, and PG72(18.32)-28.9. Therefore, it appears that the effect of adding RAS to mixtures is more noticeable at low temperatures than at intermediate temperatures when compared with RAP. No significant difference is observable at intermediate temperatures; however, the effect of TOSS on fatigue properties of the mixture seems to be the most severe as the intermediate PG is higher. At high temperatures the addition of recycled materials improves the rutting resistance, as previously discussed by many researchers. Subsequent sections present mechanical cracking tests on the mixtures designed in this study as investigated to examine further the results of the methodology proposed in this paper.

Mixture Cracking Tests

Cracking of asphalt pavements due to cyclic traffic load and environmental circumstances has been under investigation by researchers in recent decades. With the use of additives and recycled materials, the mechanical behavior of pavement has become even more complicated. Cracks

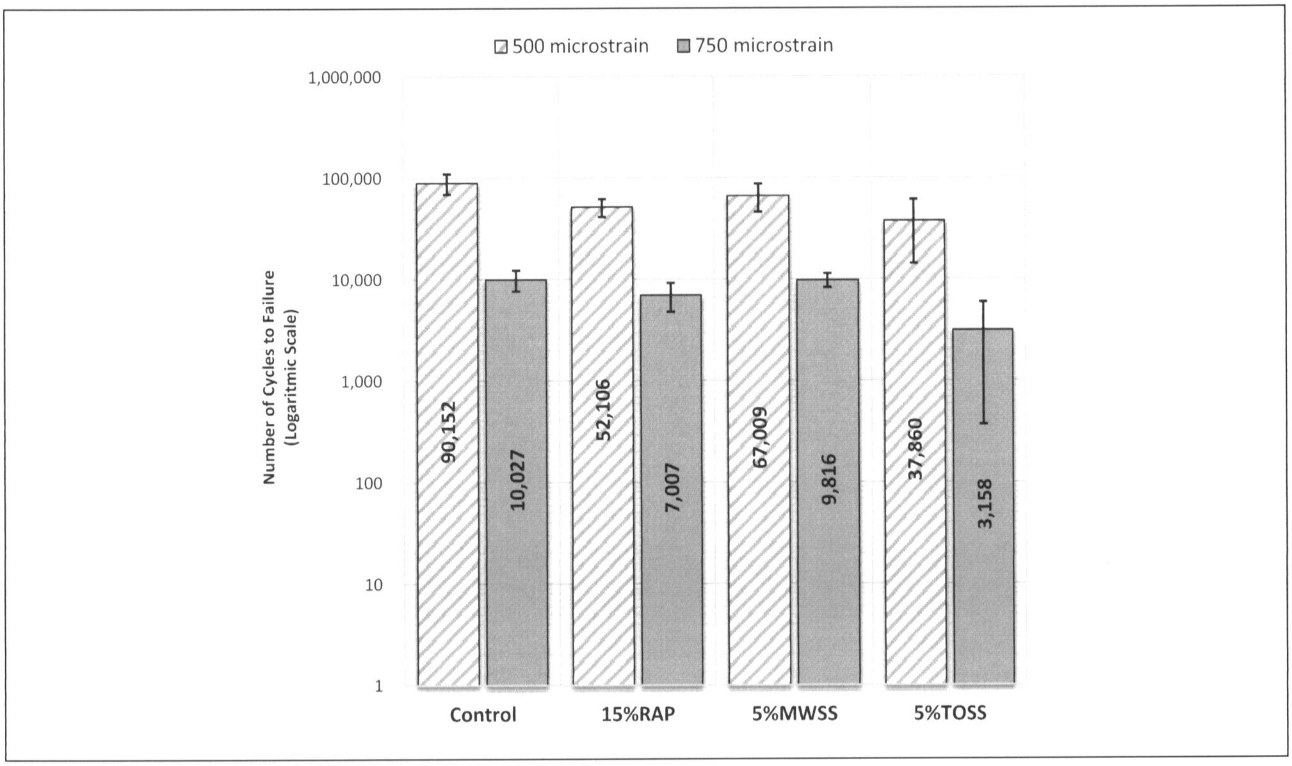

Figure 4. Flexural beam fatigue test results.

seem to initiate from binder bonds tearing apart at the weakest point in terms of tension, and then propagate mainly from the mastic phase of the mixture. Asphalt binder, in this sense, plays a major role in studying the cracking phenomenon. In addition, when including RAS or RAP into mixtures and due to partial blending, the PG of the resultant binder would differ from that of both the virgin binder and the aged binder. To shed more light on the effect of partial blending on the mechanical behavior of mixtures at intermediate and low temperatures, cracking performance tests were carried out in this study.

Crack Initiation: Four-Point Flexural Beam Fatigue Test

To investigate the fatigue cracking potential of the asphalt mixtures, a four-point beam fatigue test was carried out following ASTM D7460-10. The IPC Pressbox was used to compact slabs that were 450 mm × 150 mm × 150 mm. Testing beams of 380 mm × 63 mm × 50 mm were cut from the slabs. The target air voids of final beams were set to 7%. A strain-controlled mode was used to run the test at two strain levels of 500 and 750 micro strains and at temperature of 15°C. Flexural stiffness is the ratio of maximum tensile stress to maximum tensile strain and was computed for every 10 successive cycles.

The number of cycles to failure for the two strain levels is defined as the number in which flexural stiffness at the mid-span of the beam exceeds half of its initial value. Excessive brittleness due to aged binder from RAS or RAP influences the mixture's performance under fatigue loading. Figure 4 indicates a reduced number of cycles to failure for mixtures with RAS and RAP, which was as expected. The same trend was also observed for the proposed degrees of blending in the previous section. The intermediate true PG of the binders yielded the same conclusions as observed in the beam fatigue results in that the equivalent true PGs for the 5% MWSS, 5% TOSS, and 15% RAP mixtures were determined as 17.95°C, 18.55°C, and 18.32°C, respectively. Higher intermediate true PG thus implies that less resistance is expected due to cyclic fatigue loads. However, closeness of true intermediate PG for the proposed degrees of blending of RAS and RAP binder suggests that the behavior of their mixtures should not be significantly different. This in fact explains why the differences in number of cycles to failure for mixtures are insignificant.

Crack Propagation: Semi-Circular Bending Beam Test (SCB) and Disc-Shaped Compact Tension Test (DCT)

The capability of a mixture to withstand crack propagation is another parameter influenced mostly by binder

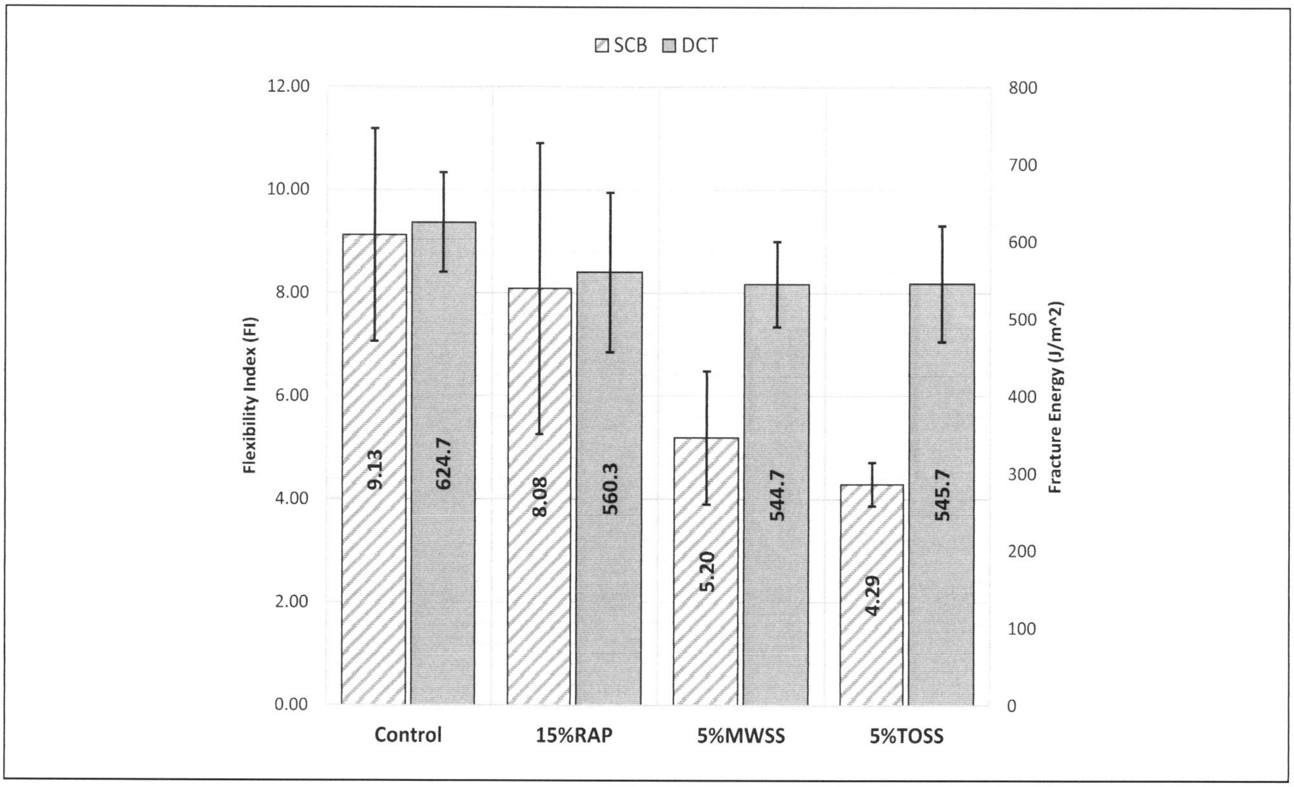

Figure 5. SCB and DCT test results.

properties. In the SCB test setup, a semi-circular sample with a thickness of 50 mm and target air void of 7% is notched at its center to artificially create a crack (AASHTO TP 105-13). Previous studies have found that an intermediate temperature of 25°C and a loading rate of 50 mm/min can depict differences with higher reliability (*3, 17*). The DCT test also evaluates the cracking potential of an asphalt mixture at low temperatures when a crack is already started. According to ASTM D7313-13, a disk-shaped specimen 50 mm in thickness and with 7% target air void is notched at one end and load is applied from two holes cored on each side of the notched crack to simulate a tension mode around the crack tip. A crack opening displacement rate of 0.017 mm/s at −18°C was used for this research.

Loading is then applied against the notch and causes the crack to propagate through the whole section of the specimen. The data collected are the load amplitude and resulting deformation. Fracture energy, which is the area under the load-displacement curve, has been measured as the only output of both the SCB and DCT tests. Nevertheless, researchers at Illinois Center for Transportation have recently developed a parameter called flexibility index (FI), believing that this index distinguishes better between mixtures. This index takes into account the post-peak behavior of mixtures by considering the slope of the load-deformation curve after its peak has reached.

Figure 5 illustrates the FI and fracture energy results for the SCB and DCT tests, respectively. A significant difference between the FI for control and 5% RAS mixtures may explain the adverse effects of adding RAS to mixtures. Despite a higher degree of blending between RAP and virgin binder, lower stiffness of RAP binder compared with RAS could have caused the RAP to be more resistant to crack propagation at intermediate temperatures. Lower fracture energies from DCT results for mixtures containing RAS or RAP further indicate a disadvantage of adding recycled materials at low temperatures. However, standard deviations of DCT test results show no statistically significant difference.

Nevertheless, one cannot neglect the fact that under-asphalting would also affect the results when RAS is added to a mixture. A combination of under-asphalting and partial blending can then justify the lower cracking resistance of mixtures produced with RAS and RAP. It is worthy of notice that as the binder PG system used in this study does not account for crack propagation, it is not yet understood how different degrees of blending of binders can affect this type of distress from the point of view of binder rheology. In other words, even knowing the exact degree of blending may not suffice to predict the behavior of mixtures in crack propagation.

Referring to the proposed degrees of blending and their true intermediate and low PG, similar conclusions could be drawn as for the beam fatigue test results. Running SCB at 25°C and DCT at −18°C should correlate with the true intermediate and low PG of the proposed degrees of blending. Continuous true PG of PG73.7(17.95)-28.2, PG72.8(18.55)-27.8, and PG72(18.32)-28.9 for 5% MWSS, 5% TOSS, and 15% RAP respectively can be compared with that for virgin binder which was determined as PG69.8(16.97)-29.5. Since no more than three measurements were made to calculate binder true PG, no statistical comparison of the temperatures is reasonable. However, negative effects coming from the TOSS binder are still evident and are more severe than the other two recycled materials. Crack propagation test results at both intermediate and low temperatures also support the hypothesis that the PG of the proposed degrees of blending could be used as an indication of the mixture's performance. Using higher percentages of recycled materials, on the other hand, may depict differences more clearly and is recommended for future research.

Summary and Conclusion

- In this study, a new method to quantify the amount of blending that occurs between aged RAS and RAP binders was developed. Local calibration of the Hirsch model was carried out by considering a control mixture containing only virgin materials. Asphalt binder master curves for assumed blending proportions were substituted into the calibrated model which results in an estimation of their corresponding asphalt mixture master curve.
- Using this method, it was estimated that around 20% to 40% of RAS binder from manufacturer's shingle waste was activated in the mixture, whereas less than 20% was activated for RAS binder from TOSS. Activation of RAP binder was estimated at 40% to 60%.
- Four-point beam fatigue, SCB, and DCT test configurations were implemented to investigate further the degree of blending results from a cracking perspective. In all cases a consistent trend was observed between the results of mixture cracking tests and the PG of the proposed degree of binder blending.

Author Contributions

The authors confirm contribution to the paper as follows: study conception and design: Ashtiani, Mogawer, and Austerman; data collection: Ashtiani, Mogawer, and Austerman; analysis and interpretation of results: Ashtiani, Mogawer, and Austerman; draft manuscript preparation: Ashtiani, Mogawer, and Austerman. All authors reviewed the results and approved the final version of the manuscript.

References

1. Alvergue, A., M. Elseifi, L. N. Mohammad, S. B. Cooper, and S. Cooper, III. Laboratory Evaluation of Asphalt Mixtures with Reclaimed Asphalt Shingle Prepared Using the Wet Process. *Journal of the Association of Asphalt Paving Technologists*, Vol. 83, 2014, pp. 197–218.
2. Ozer, H., I. L. Al-Qadi, A. I. Kanaan, and D. L. Lippert. Performance Characterization of Asphalt Mixtures at High Asphalt Binder Replacement with Recycled Asphalt Shingles. *Transportation Research Record: Journal of the Transportation Research Board*, 2013. 2371: 105–112.
3. Al-Qadi, I. L., H. Ozer, J. Lambros, A. E. Khatib, P. Singhvi, T. Khan, J. Rivera-Perez, and B. Doll. *Testing Protocols to Ensure Performance of High Asphalt Binder Replacement Mixes Using RAP and RAS*. Research Report No. FHWA-ICT-15-017. Illinois Center for Transportation, Urbana, IL, 2015.
4. Williams, R. C., A. Cascione, D. S. Haugen, W. G. Buttlar, R. A. Bentsen, and J. Behnke. *Characterization of Hot Mix Asphalt Containing Post-Consumer Recycled Asphalt Shingles and Fractionated Reclaimed Asphalt Pavement*. Iowa State University and Illinois State Toll Highway Authority, Ames, IA, 2011.
5. Foxlow, J. J., J. S. Daniel, and A. K. Swamy. RAP or RAS? The Differences in Performance of HMA Containing Reclaimed Asphalt Pavement and Reclaimed Asphalt Shingles. *Journal of the Association of Asphalt Paving Technologists*, Vol. 80, 2011, pp. 347–373.
6. Ozer, H., I. L. Al-Qadi, S. H. Carpenter, Q. Aurangzeb, G. L. Roberts, and J. Trepanier. Evaluation of RAP Impact on Hot-Mix Asphalt Design and Performance. *Journal of the Association of Asphalt Paving Technologists*, Vol. 78, 2009, pp. 317–351.
7. Shirodkar, P., Y. Mehta, A. Nolan, E. Dubois, D. Reger, and L. McCarthy. Development of Blending Chart for Different Degrees of Blending of RAP Binder and Virgin Binder. *Resources, Conservation and Recycling*, Vol. 73, 2013, pp. 156–161. https://doi.org/10.1016/j.resconrec.2013.01.018.
8. Konnur, S., S. V. Dinesh, and H. S. Lalithamba. An Experimental Investigation of Chemical and Rheological Properties of Virgin Bitumen Blended with Recovered RAP Binder. *Proceeding of International Conference on 11th Transportation Planning and Implementation Methodologies for Developing Countries (11th TPMDC)*, Transportation Systems Engineering, Mumbai, India, 2014.
9. Yousefi Rad, F., N. R. Sefidmazgi, and H. Bahia. Application of Diffusion Mechanism: Degree of Blending Between Fresh and Recycled Asphalt Pavement Binder in Dynamic Shear Rheometer. *Transportation Research Record: Journal of the Transportation Research Board*, 2014. 2444: 71–77.
10. Nahar, S. N., M. Mohajeri, A. J. M. Schmets, A. Scarpas, M. F. C. van de Ven, and G. Schitter. First Observation of Blending-Zone Morphology at Interface of Reclaimed Asphalt Binder and Virgin Binder. *Transportation Research*

Record: Journal of the Transportation Research Board, 2013. 2370: 1–9.
11. Zofka, A., M. O. Marasteanu, T. R. Clyne, X. Li, and O. Hoffmann. 2004. *Development of Simple Asphalt Test for Determination of RAP Blending Charts*. Report No. MN/RC - 2004-44, Minnesota Department of Transportation, St. Paul, MN, 2004.
12. Zhou, F., H. Li, R. Lee, T. Scullion, and G. Claros. Recycled Asphalt Shingle Binder Characterization and Blending with Virgin Binders. *Transportation Research Record: Journal of the Transportation Research Board*, 2013. 2370: 33–43.
13. McGraw, J., A. Zofka, D. Krivit, J. Schroer, R. Olson, and M. Marasteanu. Recycled Asphalt Shingles in Hot Mix Asphalt. *Journal of the Association of Asphalt Paving Technologists*, Vol. 76, 2007, pp. 235–274.
14. Al-Qadi, I. L., M. Elseifi, and S. H. Carpenter. *Reclaimed Asphalt Pavement - A Literature Review*. Research Report No. FHWA-ICT-07-001, Illinois Center for Transportation, Urbana, IL, 2007.
15. Bonaquist, R. New Approach for the Design of High RAP HMA. Presented at North East Asphalt User/Producer Group Meeting, Burlington, Vt., 2005. http://www.neaupg.uconn.edu/pdf/neaupg_oct2005_bonaquist.pdf.
16. Christensen, D. W., T. Pellinen, and R. F. Bonaquist. Hirsch Model for Estimating the Modulus of Asphalt Concrete. *Journal of the Association of Asphalt Paving Technologists*, Vol. 72, 2003, pp. 121–151.
17. Kim, R. Y., B. Underwood, M. Sakhaei Far, N. Jackson, and J. Puccinelli. *LTPP Computed Parameter: Dynamic Modulus*. Publication FHWA-HRT-10-035. Federal Highway Administration, Washington, D.C., 2011.
18. Booshehrian, A., W. S. Mogawer, and R. Bonaquist. How to Construct an Asphalt Binder Master Curve and Assess the Degree of Blending between RAP and Virgin Binders. *Journal of Materials in Civil Engineering*, Vol. 25, No. 12, 2013, pp. 1813–1821.
19. Bonaquist, R. *Effect of Recovered Binders from Recycled Shingles and Increased RAP Percentages Resultant Binder PG*. Report No. WHRP 11-13. Wisconsin Department of Transportation, 2011.
20. Shirodkar, P., Y. Mehta, A. Nolan, K. Sonpal, A. Norton, C. Tomlinson, E. Dubois, P. Sullivan, and R. Sauber. A Study to Determine the Degree of Partial Blending of Reclaimed Asphalt Pavement (RAP) Binder for High RAP Hot Mix Asphalt. *Construction and Building Materials*, Vol. 25, No. 1, 2011, pp. 150–155.
21. Rowe, G., G. Baumgardner, and M. Sharrock, Functional Forms for Master Curve Analysis of Bituminous Materials. In *Advanced Testing and Characterization of Bituminous Materials* (A. Loizos, M.N. Partl, T. Scarpas, I. L. Al-Qadi, ed.), CRC Press, London, 2009, Volume I, pp. 81–91.
22. Ozer, H., I. L. Al-Qadi, P. Singhvi, T. Khan, J. Rivera-Perez, and A. El-Khatib. Fracture Characterization of Asphalt Mixtures with RAP and RAS Using the Illinois Semi-Circular Bending Test Method and Flexibility Index. *Transportation Research Record: Journal of the Transportation Research Board*, 2016. 2575: 130–137.

The Standing Committee on Non-Binder Components of Asphalt Mixtures (AFK30) peer-reviewed this paper (18-02796).

Article

TRR
JOURNAL OF THE TRANSPORTATION RESEARCH BOARD

Transportation Research Record
2018, Vol. 2672(28) 119–129

DOI: 10.1177/0361198118786825
journals.sagepub.com/home/trr
SAGE

Effect of Binder Modification and Recycled Asphalt Pavement on the Performance of Permeable Friction Course

Mark J. Brum[1], Walaa S. Mogawer[2], Kevin D. Stuart[3], and Alexander J. Austerman[4]

Abstract

Permeable Friction Course (PFC) is a hot-mix asphalt that contains interconnecting voids that provide improved pavement surface drainage during rainfall. The objective of this study was to determine if PFC mixtures which incorporate reclaimed asphalt pavement (RAP) will provide performances that are similar to PFC mixtures which only use virgin materials and whether binder type will affect performance. Utilizing current design specifications, PFCs were designed with RAP contents of 0%, 15%, and 25% and four asphalt binders. These mixtures were subject to a barrage of tests which measured their ability to resist draindown, abrasion, fatigue cracking, rutting, and moisture damage. Other testing determined permeability, porosity and workability. Rutting testing employed the industry standard asphalt pavement analyzer. Fatigue cracking and moisture susceptibility testing utilized the semicircular bending test. It was determined that it is possible to design a PFC incorporating RAP which will have good performance. However, this will only be achieved when the proper materials are used. The RAP must be properly fractionated and a modified binder such as asphalt rubber or a highly modified asphalt binder (HiMA) must be used. The mixture tests combined with Life Cycle Cost Analyses demonstrated that a PFC with up to 15% RAP combined with an asphalt rubber or HiMA can provide good performance and be cost effective. Furthermore, the use of 25% RAP in combination with these binders was only limited by the asphalt rubber not meeting the specification for permeability.

A permeable friction course (PFC) is a hot-mix asphalt (HMA) which contains interconnecting voids that provide improved surface drainage during rainfall. These asphalt mixtures contain only a small portion of fine aggregate, creating a pavement with a relatively large percentage of air voids.

Asphalt binders that are modified with polymers are generally called polymer modified asphalts (PMA) by the HMA industry and are the most common type of modification, although asphalts have been modified in other ways. A common modifier is crumb rubber which is generally obtained from used tires (*1*). Although crumb rubber is a polymer, asphalts and mixtures incorporating crumb rubber are often given unique names such as "asphalt rubber" to distinguish them from other forms of PMA. Many agencies require the incorporation of a modified asphalt binder in their PFCs to increase performance.

Reclaimed asphalt pavement (RAP) has been widely used in the asphalt paving industry for decades. RAP is a readily available resource which has provided economic savings and environmental benefits. Recycling reduces the demand for non-renewable natural resources, both asphalt and aggregates, and thus, reduces the energy and emissions associated with the extraction and transportation of these resources (*2*).

Currently, PFC mixtures are designed with many different types of asphalt binders both modified and unmodified. RAP, on the other hand, has generally not been used in PFC, although in recent years, research has been performed regarding its use in PFCs (*3*). With the increasing amount of RAP that is becoming available as older pavements are removed, the use of RAP in PFCs should be encouraged. There is a strong possibility that the right combination of RAP and a modified binder may produce a sustainable PFC which will provide performances that are similar to what is presently being specified.

Objectives

The objectives of this study were to determine (1) if PFC mixtures incorporating RAP will provide performances similar to PFC mixtures fabricated using only virgin materials, and (2) whether the binder type used will affect PFC mixture performance.

[1]Research and Materials, Massachusetts Department of Transportation, Hopkinton, MA
[2]Highway Sustainability Research Center (HSRC), Civil and Environmental Engineering Department, University of Massachusetts, Fall River, MA
[3]Consultant, Phoenixville, PA
[4]Highway Sustainability Research Center (HSRC), University of Massachusetts, Fall River, MA

Corresponding Author:
Address correspondence to Walaa S. Mogawer: wmogawer@umassd.edu

Experimental Methodology

An experimental plan was first developed (Figure 1). The primary stages for the project were as follows:

1. Develop a control 12.5 mm PFC mixture in accordance with AASHTO Provisional Standard AASHTO PP 77 "Materials Selection and Mixture Design of Permeable Friction Courses."
2. Develop mixtures incorporating 0%, 15% and 25% RAP with four asphalt binders.
3. Evaluate mixture performance in terms of:
   - Draindown
   - Permeability and porosity
   - Durability
   - Workability
   - Rutting
   - Fatigue cracking
   - Moisture susceptibility
4. Determine the maximum amount of RAP that can be used in a PFC.
5. Develop cost analyses for the mixtures.

Material testing was completed following AASHTO, ASTM and other specifications as outlined in the subsequent sections. The aggregate testing and mixture designs were performed in accordance with AASHTO PP 77.

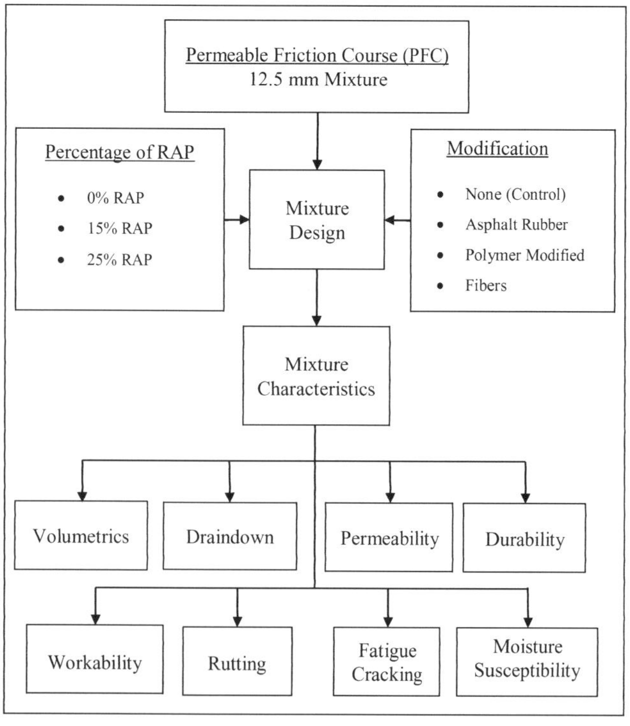

Figure 1. Experimental plan.

Materials

Aggregate

The aggregate was obtained from a Massachusetts quarry which supplies materials for HMA production. The aggregate properties were determined based on the requirements stipulated in AASHTO PP 77. Once aggregate testing was completed, the materials were selectively sieved into individual size fractions to allow for more precise and repeatable batching of mixture samples.

RAP

The RAP was obtained from a Rhode Island-based HMA producer. It was screened through a #4 sieve and the coarse fraction was retained. This allowed for the removal of approximately 7% dust passing the #200 sieve and provided the coarsest possible RAP that could be practically produced for plant production. This maximized the amount of RAP that could be used in the mixture.

Binder

Three asphalt binder options were chosen, and a fourth option did not change the PG compared to the control but was reported by the supplier to provide a more durable mixture.

Control. The control binder was a PG64-28 produced by a Massachusetts binder supplier. This is the standard PG binder grade used by the Massachusetts Department of Transportation (MassDOT) for conventional asphalt mixtures. Although this binder was not recommended by AASHTO PP 77, it has been used in the past on Massachusetts projects and would provide a baseline. The mixing temperature was 150°C and compaction temperature was 140°C.

Asphalt Rubber. The second binder used was an asphalt rubber which was also produced by a Massachusetts binder supplier. The base binder, which was a PG58-28, was blended with 15% vulcanized tire rubber. This met ASTM D6114 Standard Specification for Asphalt Rubber Binder (*4*). This is the modified binder that is currently required by MassDOT for PFCs. The mixing temperature was 177°C and compaction temperature was 157°C.

HiMA. The third binder was a highly modified asphalt (HiMA) produced by an asphalt polymer producer. HiMA is a special type of PMA where the polymer loading is more than twice that of ordinary PMA. The mixtures, however, are readily workable and easy to lay down in the field. Samples of PG58-28 from a local supplier were modified with 7% styrene-butadiene-styrene (SBS) polymer. The mixing temperature was 166°C and compaction temperature was 149°C.

Table 1. Trial Blend Gradations and Mixture Volumetric Properties

Sieve size	Blend 1 coarse	Blend 2 intermediate	Blend 3 fine	AASHTO PP 77 gradation specification
19.0 mm	100	100	100	100
12.5 mm	80	85	90	80–100
9.5 mm	35	42	48	35–60
4.75 mm	10	14	18	10–25
2.36 mm	5	7	8	5–10
1.18 mm	3	4	4	
0.600 mm	2	3	3	
0.300 mm	2	2	2	
0.150 mm	1	2	2	
0.075 mm	1.1	1.2	1.2	0–4
Binder	Control	Control	Control	n/a
Binder content (%)	6.2	6.2	6.2	n/a
G_{mm}	2.422	2.421	2.418	n/a
G_{mb}	1.953	1.992	2.007	n/a
Air voids (%)	19.4	17.7	17.0	n/a

Note: n/a = not applicable.

Fiber. The last binder option was not a reacted type of binder modification but consisted of utilizing the PG64-28 control binder with a proprietary synthetic high strength tensile aramid fiber blend formulated to reinforce asphalt mixtures. The fibers will not melt in an asphalt mixture and are designed to provide strength and durability at both high and low temperatures. The fibers were designed to a provide life cycle cost reductions by extending pavement life at conventional asphalt pavement thicknesses. The mixing and compaction temperatures were the same as for the PG64-28 control binder.

Stabilizing Fiber

Since PFC with relatively high binder contents are susceptible to draindown, a polyester stabilizing fiber called BoniFiber from Kapejo, Inc was also added when needed. These readily available fibers had been successfully used to stabilize a stone matrix asphalt in a previous project.

Mixture Designs

Trial blends were first developed. The mixture design process followed the requirements of Section 7 of AASHTO PP 77 which included:

1. A minimum asphalt binder content of 6.2% which is based on the combined aggregate bulk specific gravity;
2. Air void content of 18%–22%;
3. Adequate stone-on-stone contact with the voids in the mix (VCA_{mix}) less than the voids in the coarse-aggregate portion (VCA_{DRC});
4. Abrasion loss less than or equal to 15%;
5. Draindown at production temperature less than 0.3%.

Three trial blends meeting the PFC gradation specification band were selected (Table 1). One blend was coarse, the second intermediate, and the third fine. To be more conducive to the addition of RAP, each virgin blend utilized the addition of 1% dust. From experience this is the maximum amount that could be realistically utilized from an asphalt plant's baghouse during production. Designing a virgin mixture with extra dust would allow for the replacement of fines from the baghouse with the fines from the RAP.

The volumetric properties for each virgin control mix were determined (Table 1). Based on the results, the intermediate blend was chosen. Although the air voids were slightly below the 18% requirement, it would allow for more RAP to be added compared to the coarse blend and it was believed that the lack of voids would not detrimentally affect the permeability of the mixture.

The stone-on-stone contact for each mix was determined by comparing the air void content in the coarse aggregate (VCA), evaluated in both the dry rodded condition (VCA_{DRC}) and the compacted mixture (VCA_{mix}). AASHTO PP 77 requires that VCA_{mix} be less than the VCA_{DRC}. The ratio of VCA_{mix} to VCA_{DRC} was 0.9 for all the mixtures regardless of the binder used or the percentage of RAP which showed that adequate stone-on-stone contact was developed.

Mixture Performance Testing

Draindown

In accordance with AASHTO PP 77, draindown testing was performed following AASHTO T 305 (*5*). Stabilizing fibers were added to the mixtures in increasing amounts until draindown was below 0.3%. The asphalt rubber mixtures did not require any stabilizing fibers having 0.0% draindown for all three RAP contents, whereas, the control, HiMA, and fiber mixtures each required 0.6% stabilizing fiber. In conclusion, it is possible to produce a PFC with RAP which does not exhibit draindown although a stabilizing fiber might be needed.

Permeability and Porosity

To measure the ability to allow water to pass through it, each mixture was tested for permeability and porosity. No specific test method is provided in AASHTO PP 77, so permeability was tested in accordance with the Florida Department of Transportation's "Test for Measurement of Water Permeability of Compacted Asphalt Paving Mixtures" (*6*). A minimum permeability of 0.045 cm/s is required by MassDOT's OGFC-AR (*7*). All mixtures passed except for the two mixtures containing the control and asphalt rubber binders in combination with 25% RAP, having permeabilities of 0.040 and 0.042 cm/s, respectively. Perhaps, these deficiencies could be rectified with a slight change in gradation. Porosity was also measured on compacted gyratory specimens in accordance with ASTM D 7063 (*4*). There were no significant differences between the porosities.

In conclusion, it is possible to produce a PFC with RAP which has adequate permeability and porosity; however, the percentage of RAP might have to be limited to 15% using some asphalt binders.

Durability

AASHTO TP 108, also referred as the Cantabro test, was employed to assess the cohesion, bonding, and effects of traffic abrasion on the mixtures. An abrasion loss of 15% or less is usually required; however, AASHTO PP 77 does stipulate that some states have reported having successful PFC mixtures with up to 20% abrasion loss.

All mixtures passed the criteria of 15%. An analysis of variance (ANOVA) showed that both the binders and RAP contents provided significant differences with the mixtures without RAP performing better than those with RAP. Even so, since all mixtures passed, it is possible to produce a PFC with RAP which has adequate durability.

Workability

For mixtures with low workability, it can be more difficult to construct smooth pavements. If the compositional properties of a mixture, such as the aggregate physical properties and gradation are kept constant, then workability becomes a function of its binder properties (*8*). The workability of the PFCs was measured since the modified binders used in them tend to have higher viscosities, and thus, might decrease workability, although the mixing and compaction temperatures were adjusted to prevent this.

Compaction Energy Index. The compaction energy index (CEI) is defined as the area from the eighth gyration to 92% of maximum specific gravity (Gmm) under the gyratory compaction curve. It theoretically represents the energy that is applied by the roller to compact the mixture to the required density during construction (*9*). Mixtures with lower values of CEI would be preferred, however, too low a CEI might be an indication of a tender mix. It has been found that because PFCs have a higher air void content compared to a dense-graded mixture, a variation for the calculation of the CEI must be used (*10*). It was determined that the CEI should be represented by the area under the curve from the eighth gyration to the final gyration.

The CEI for each gyratory sample was determined and averages tabulated (Figure 2). All CEI were comparable to what Frigio et al. found and all were in the acceptable range of 50 to 1000 (*3*). An ANOVA showed that the binders provided significant differences but not RAP content. Even so, since all mixtures passed, it is possible to produce a PFC with RAP which has adequate workability.

Rutting

The asphalt pavement analyzer (APA) was used to evaluate rutting in accordance with AASHTO T 340 "Standard Method of Test for Determining Rutting Susceptibility of Hot-mix Asphalt (HMA) Using the Asphalt Pavement Analyzer" (*5*). The test temperature was 64°C (147°F) as specified by MassDOT. To be representative of what would be obtained in the field, the samples were compacted to the target air void content of 18 ± 0.5% instead of 7.0 ± 0.5%, as specified by AASHTO T 340. Based on research performed by NCAT, a rut depth of less than 5.5 mm was set as the acceptable threshold, which was determined from correlations between the APA and rutting provided by a full-scale test track (*11*).

The average rut depth for each mixture was determined (Figure 3). The use of 15% or 25% RAP had little effect on rut depth since all three RAP contents for each binder either passed or failed, except for the fiber mixture where 25% RAP provided a passing rut depth of 5.3 mm while the other two RAP contents failed. An ANOVA showed that there were significant differences between binder types. HiMA provided the lowest rut depths, followed by the asphalt rubber, and then the control and fiber mixtures whose rut depths were not significantly different and failed, except for the fiber mixture with 25% RAP. In conclusion, (a) the use of RAP only influenced rutting performance using the fiber mixture where 25% RAP increased rutting resistance, and (b) asphalt rubber or HiMA was generally needed to achieve adequate rutting resistance, which suggests that a RAP's ability to stiffen a mixture may not always be sufficient for achieving this.

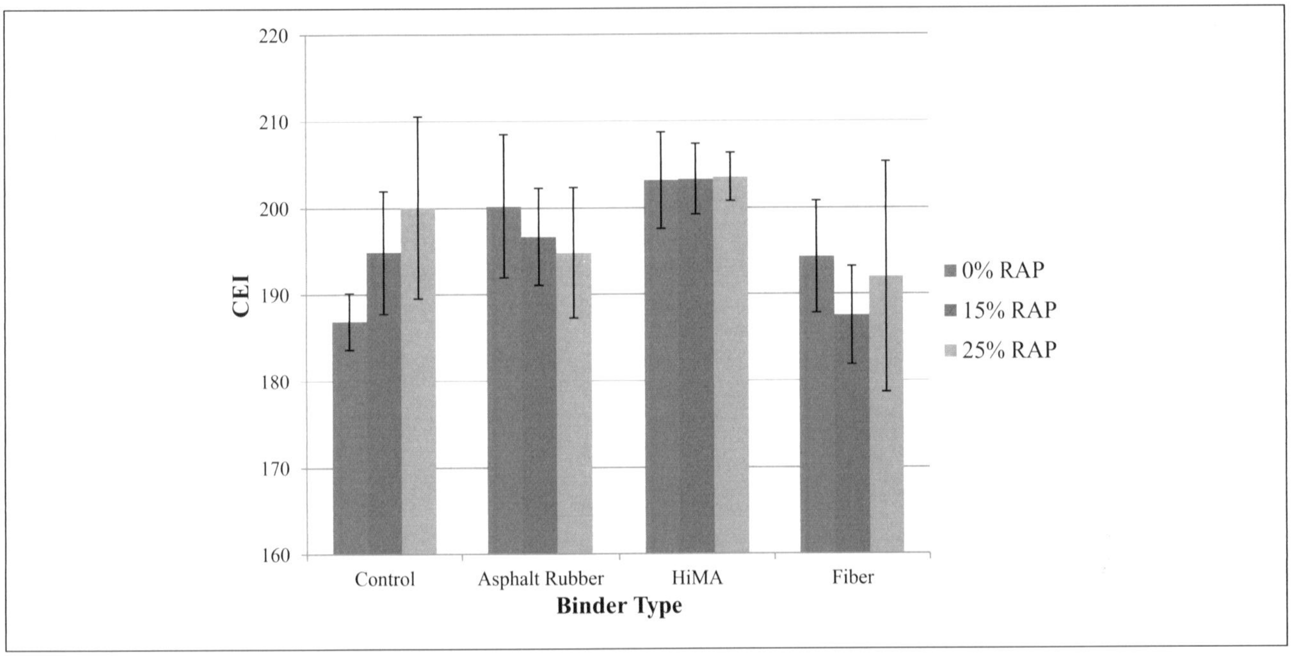

Figure 2. Compaction energy index (CEI).

Fatigue Cracking

The Flexibility Index Test (FIT) is a test method which utilizes the semicircular bend geometry (SCB) to determine the fracture cracking resistances of asphalt mixtures at intermediate temperatures. The provisional AASHTO test method TP 124 "Standard Method of Test for Determining the Fracture Potential of Asphalt Mixtures Using Semicircular Bend Geometry at Intermediate Temperature" was used in this study (*5*). To evaluate the effects of the binders and RAP on fatigue cracking using fracture energy (G_f) and flexibility index (FI), the FIT is performed at an intermediate temperature of 25°C and a loading rate of 50 mm/min.

The test was conducted using a custom-designed SCB fixture that was placed in a servo-hydraulic asphalt testing machine. A 97.8 kN load cell was used to measure the fracture load. The test was controlled by a linear variable differential transformer (LVDT) using a monotonic load applied along the vertical diameter of the specimen.

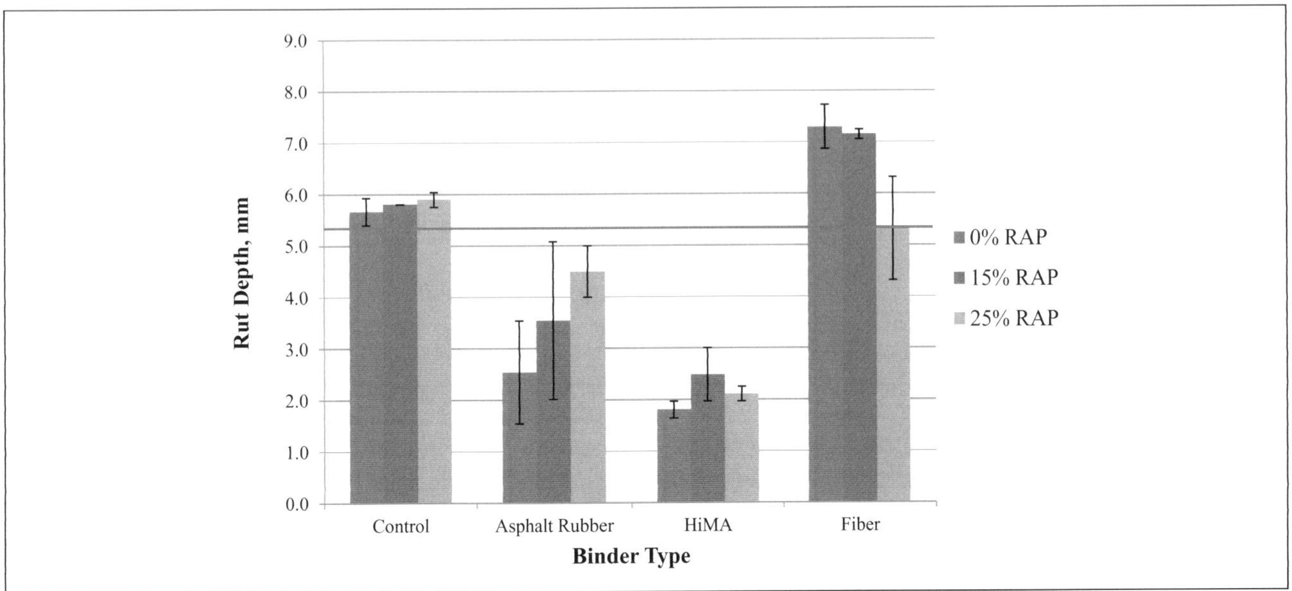

Figure 3. APA rut depths.

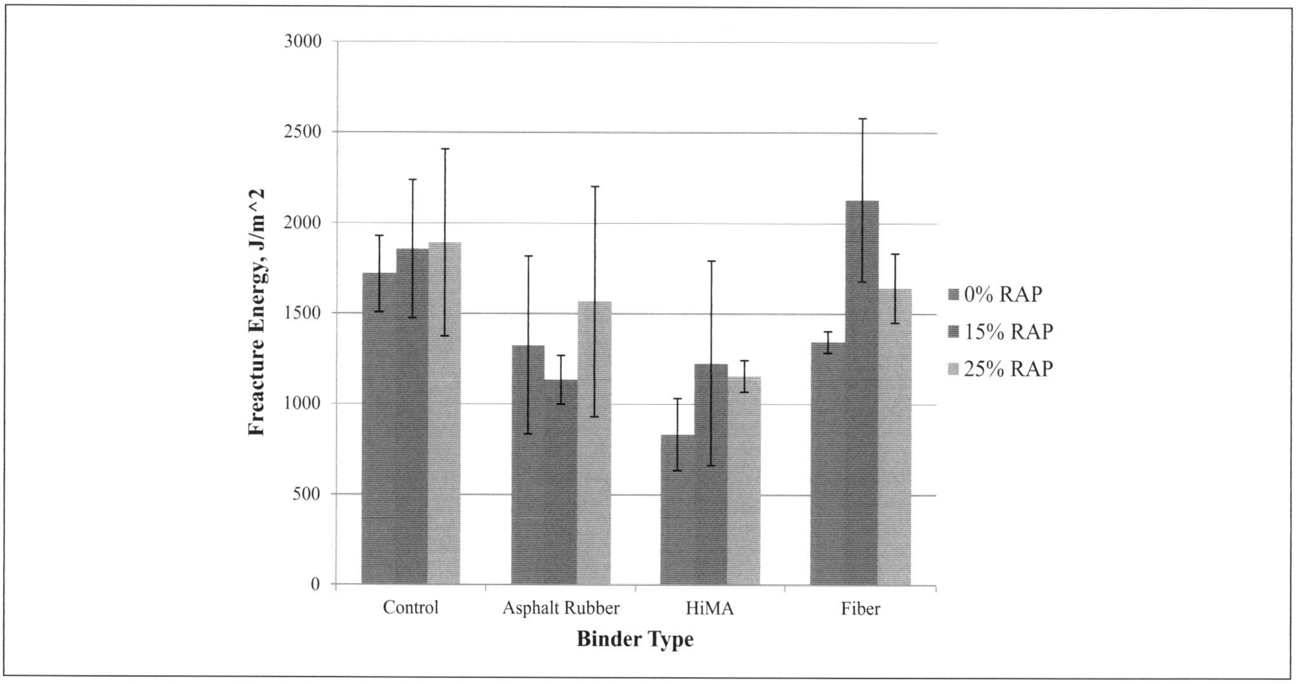

Figure 4. G_f of the mixtures.

Prior to specimen fabrication, each loose mixture was long-term oven-aged (LTOA) as outlined in a previous study by Braham et al. (*12*). The LTOA process proposed in this study was utilized to simulate actual field aging conditions. For this process, specimens of each mixture were batched, mixed, and aged at 135°C for 24 hours in a loose state. The specimens were then returned to compaction temperature and immediately compacted.

Five Superpave gyratory specimens were compacted to a height of 115 mm for each mixture. Then, one 50 mm slice was cut from the middle of each gyratory specimen. The slice was then cut in half across the diameter to yield two FIT specimens.

Fracture Energy (G_f). G_f is the energy required to create a unit surface area of a crack (*5*). It represents the overall energy that is required to fracture the specimen. Theoretically, an increase in G_f would indicate a better resistance to fatigue cracking.

The average G_f value for each mixture was measured (Figure 4). From ANOVA it was determined that there was a statistical difference between the RAP contents as well as for binder type. However, in terms of the individual binders in Figure 4, the use of 15% or 25% RAP either had no significant effect on G_f or increased it. The ANOVA showed that the control and fiber mixtures were not significantly different and generally had the higher average G_f while the asphalt rubber and HiMA mixtures were not significantly different and generally had the lower G_f.

In conclusion, the use of RAP either had no significant effect on G_f or it increased it, while the control and fiber mixtures performed better than the asphalt rubber and HiMA mixtures.

Flexibility Index (FI). FI is a primary parameter currently being used to characterize the fatigue cracking resistances of asphalt mixtures (*5*). The larger the FI, the more resistant the mixture is to fatigue cracking damage. A FI of less than 8.0 has been correlated with mixtures with poor field performance (*12*).

The average FI for each mix was determined (Figure 5). All mixtures had an FI of 8 or greater which met the threshold requirement. From ANOVA it was determined that the only significant difference was between HiMA and the other binders. HiMA provided higher FIs. However, if the FI threshold of 8 adequately predicts cracking in PFC mixtures, then all mixtures would not be susceptible to fatigue cracking. Based on this, RAP content and binder type had no significant effect on fatigue cracking performance as measured by FI.

Moisture Susceptibility

Based on the research performed by Frigio et al. (*3*), it has been hypothesized that moisture susceptibility can be determined using a ratio between dry versus wet conditioned SCB samples.

The G_f and FI ratios were determined using a combination of the FIT test, AASHTO T 283, and AASHTO PP 77. From the five gyratory specimens that were fabricated to produce the SCB samples, each half of the disc was separated into one of two subsets as required by T 283. One half was tested in the dry condition per the FIT test method and the other

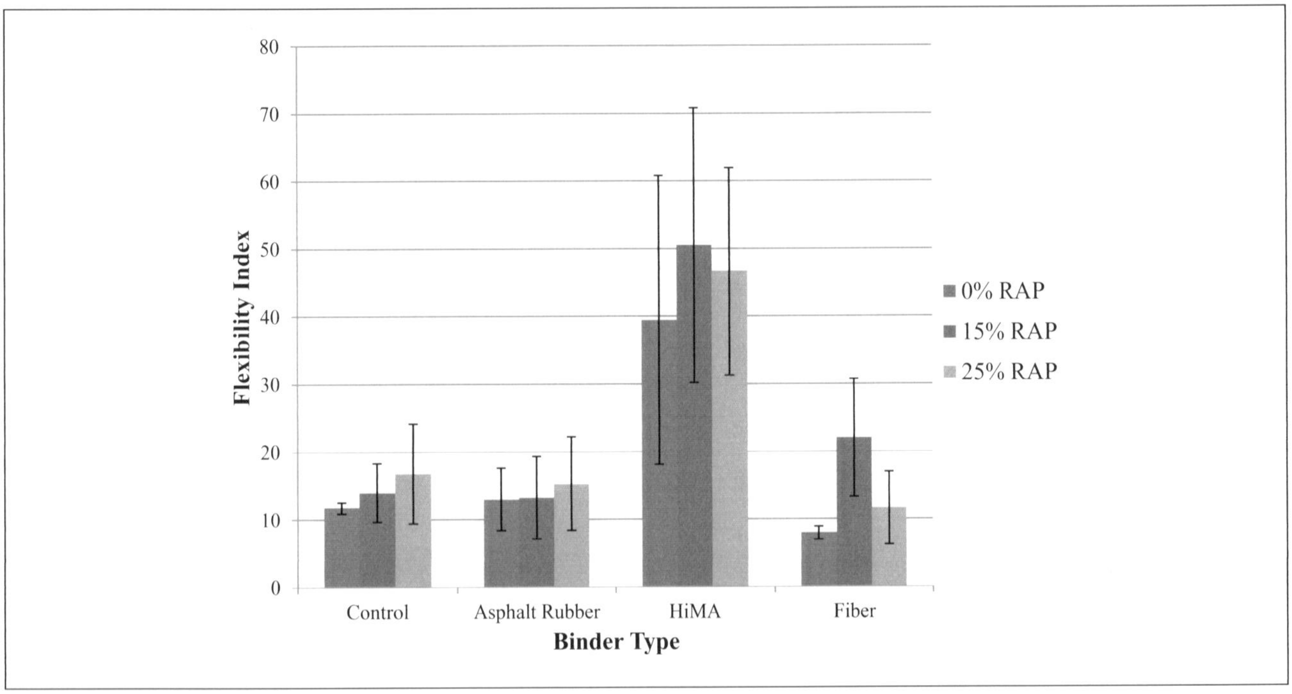

Figure 5. FI of the mixtures.

half were wet conditioned in accordance with AASHTO PP 77 Section 7.4.7.

The wet conditioned specimens were vacuum saturated at 87.8 kPa under water for 10 minutes. The vacuum saturation container was then submerged in a water bath and the saturated specimens were transferred underwater into a tub filled with water to ensure water did not escape from the specimen pores. The tub with the specimens was then transferred to a freezer at −18 ± 3°C for 24 hours. After the 24-hour freeze period, the samples were then transferred to a hot water bath at 60 ± 1°C for 24 hours. Lastly, the samples were brought back to 25 ± 0.5°C in an environmental chamber for 2 hours then tested in the SCB. After testing, each sample was fully separated along the crack and inspected for visual signs of moisture damage in the form of stripping.

The wet to dry ratios for G_f and FI were calculated using a modification to equation 6 from T 283.

$$\mathbf{G_f ratio} = \frac{\mathbf{G_f wet}}{\mathbf{G_f dry}} \tag{1}$$

$$\mathbf{FI\ ratio} = \frac{\mathbf{FIwet}}{\mathbf{FIdry}} \tag{2}$$

The average G_f and FI ratios for each mixture were calculated (Figure 6). The majority of mixtures had a G_f ratio and an FI ratio greater than one. This leads to the conclusion that most of the mixtures had no damage. For G_f, an ANOVA analysis did not show any statistically significant differences between the binders or RAP content. The same was found for the FI ratios. If the tensile strength ratio (TSR) threshold of 80% was to be applied to these G_f and FI ratios, all the mixtures would pass. The visual inspection of the samples also indicated little to no stripping in the aggregate. In conclusion, RAP content and binder type had no significant effect on moisture susceptibility because none of the mixtures exhibited significant moisture damage.

Performance Space Diagram

An interaction plot was constructed using the APA rut depth data at 8,000 passes and the FI values obtained from the FIT. Referred to as a performance space diagram (*13*), it can provide insight to make relative comparisons of mixtures. In Figure 7, the four quadrants define the overall resistance of each mixture to rutting and fatigue cracking.

- Stiff and Brittle: High rutting but low fatigue cracking resistance.
- Soft and Brittle: Low rutting and low fatigue cracking resistance.
- Soft and Flexible: Low rutting but high fatigue cracking resistance.
- Stiff and Flexible: High rutting and high fatigue cracking resistance.

The APA and dry conditioned FI values were plotted (Figure 7). The HiMA mixtures far outperformed the rest; lying deep in the "stiff and flexible" quadrant. The asphalt

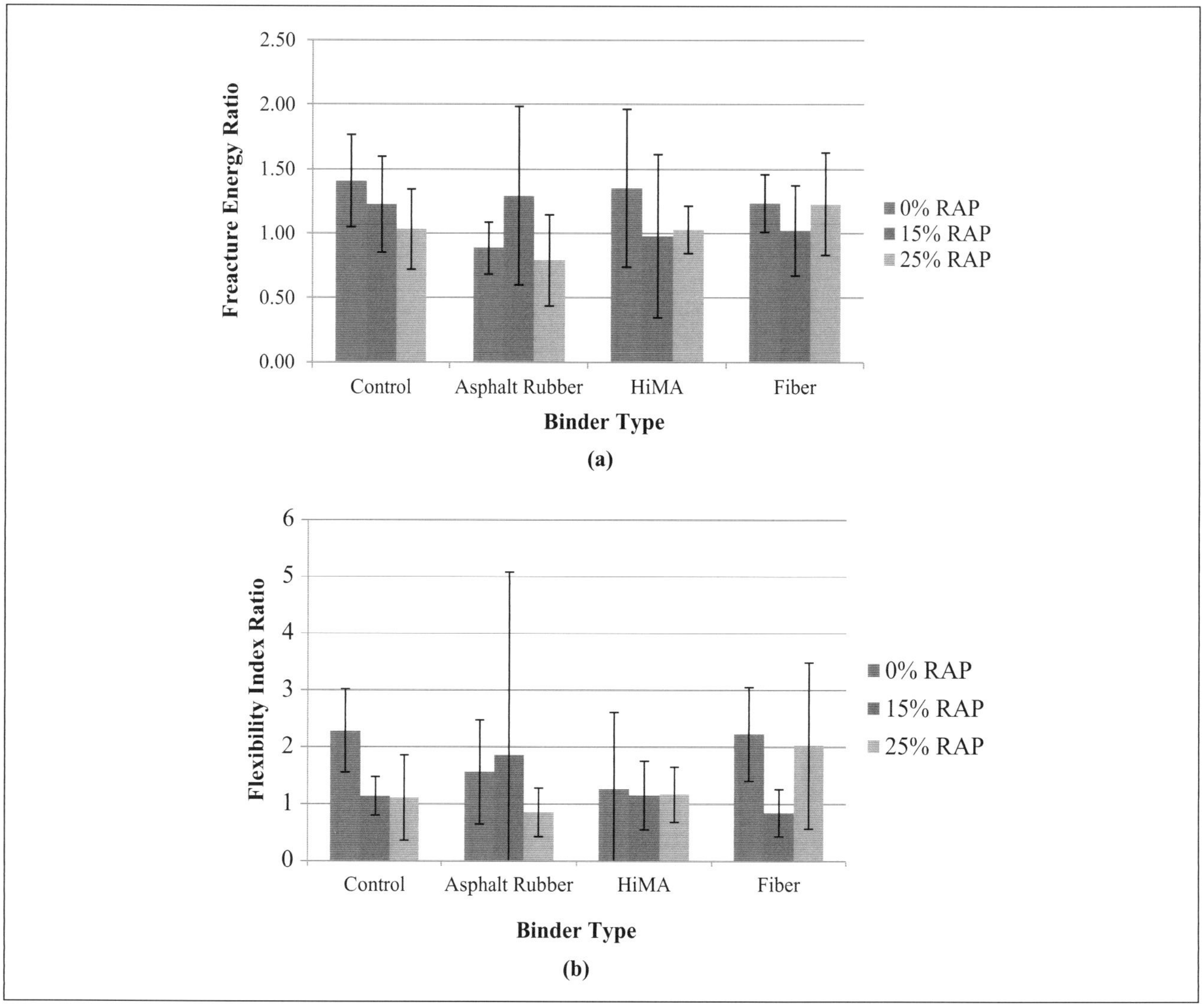

Figure 6. Moisture susceptibility: (*a*) G_f ratio and (*b*) FI ratio.

rubber mixtures and the fiber mixture with 25% RAP mixtures were the only other mixtures that would be considered acceptable. This indicates that to have a balanced PFC mixture, one may use RAP, but it will require the use of a modified binder such as asphalt rubber or HiMA to achieve the necessary balance between fatigue cracking and rutting resistance.

Cost Analysis

A two-part cost analysis was performed on the individual mixtures. The first step consisted of obtaining a price per ton for each mixture. The second step was to perform a life cycle cost analysis (LCCA) using each mixture as different alternatives.

Mixture Costs

To determine the difference in pricing between the individual mix designs, the cost per ton for each was obtained from various Massachusetts HMA producers familiar with the materials employed in the project. The producers were given each mix design. In return, the price per ton for each mix was provided. This price would be the overall cost to produce a mixture including the material and production costs. The costs were compiled and the average price for each mix was determined (Table 2).

The mixtures with the control binder were the least expensive as it utilized the least expensive binder. The HiMA was the most expensive due to the large percentage of polymer. The addition of RAP, even with its processing cost and limited binder content of 3.7%, did provide a cost saving between $2.00 and $6.30 per ton, depending on the binder type and percentage of RAP.

LCCA

An LCCA was performed for each mixture using the FHWA LCCA software Realcost 2.5 available on the FHWA website

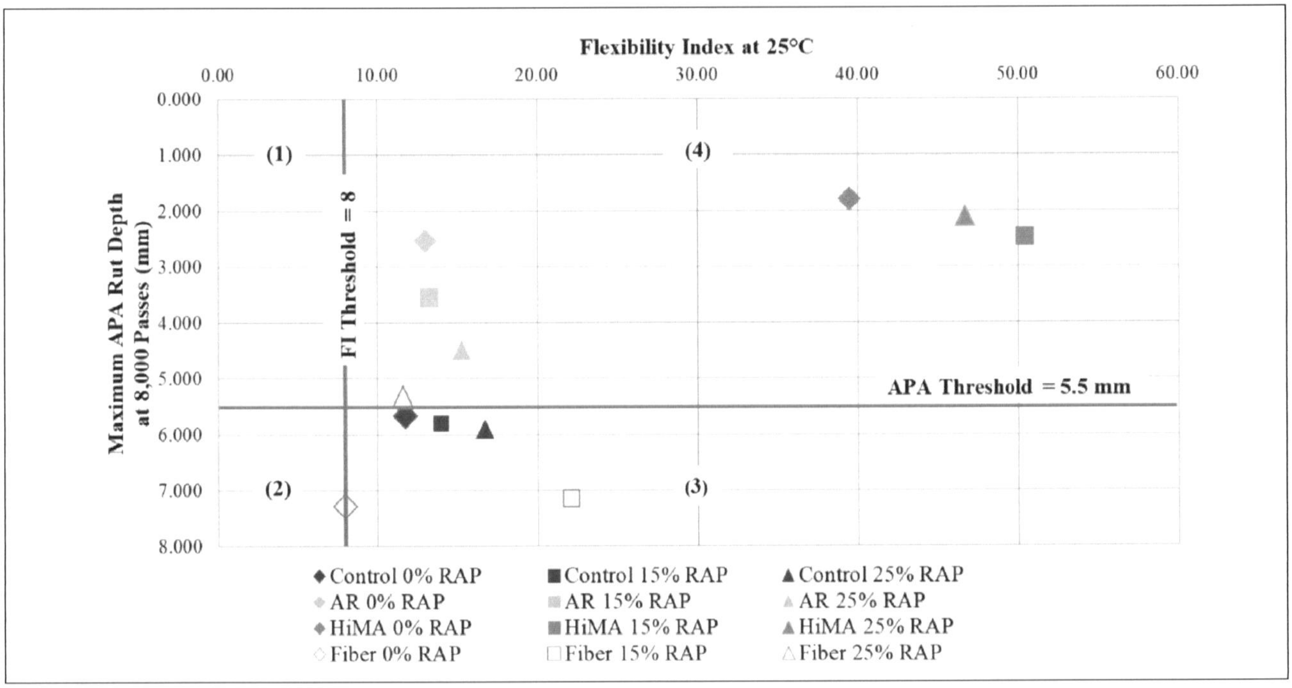

Figure 7. FI-APA performance space diagram.

(*14*). Using a MassDOT resurfacing contract for the project details portion of the software, a 20-year cost analysis was performed for each mixture. The contract referenced was a resurfacing contract for a portion of Interstate 91 in the towns of Deerfield, Hatfield, Northampton, and Whately. This project was chosen because of its relatively straightforward scope of work, which primarily focused on the resurfacing of the highway. The overall cost of the project was estimated to be $13,605,748 and required 25,960 tons of PFC. Based on the cost per ton for each mix, the overall project cost was recalculated (Table 3).

The equivalent uniform annual cost (EUAC) for each mixture was determined in Realcost 2.5 (Table 3). In simple terms, the EUAC is the "payment" required to fund the life cycle cost over the service life. The service life for each mixture was estimated based on the recommendations from MassDOT's Pavement Management Section. These recommendations were founded on the service lives of MassDOT's previously in-service PFC. The HiMA mixtures were expected to perform the longest, followed by the asphalt rubber mixtures, then the control and fiber mixtures which would be expected to perform the same. This was supported by the laboratory rutting tests performed in this study which was the controlling distress. In fact, the rutting tests suggest that the HIMA mixtures might last more than twice as long as the control and fiber mixtures. The performance space diagram in Figure 7 indicated that the mixtures with RAP would perform similar to the mixtures with no RAP. For that reason, the LCCA assumed that there would be no reduction in service life with the addition of RAP. The EUAC for each mixture and their overall ranking with regards to the overall project cost were then determined.

Based on the analyses in Table 3, if the RAP does not cause a reduction in service life, then the mixtures with the most RAP would be the most cost effective. The laboratory testing does suggest that this assumption is accurate. Also observed was that even though the HiMA has the higher upfront costs, over a 20-year analysis period, it is the most cost effective.

Table 2. Mixture Cost per Ton Comparison

Binder type	RAP (%)	Virgin binder (%)	RAP binder (%)	Cost per ton ($)
Control	0	6.2	0.0	76.90
	15	5.6	0.6	74.80
	25	5.3	0.9	73.50
Asphalt rubber	0	6.2	0.0	84.10
	15	5.6	0.6	81.20
	25	5.3	0.9	79.30
HiMA	0	6.2	0.0	96.30
	15	5.6	0.6	92.50
	25	5.3	0.9	90.00
Fiber	0	6.2	0.0	83.60
	15	5.6	0.6	81.60
	25	5.3	0.9	80.20

Conclusion

- Except for the permeabilities of the 25% RAP mixtures with the control and asphalt rubber binders, the addition of RAP was not a detriment to performance.

Table 3. Equivalent Service Life Comparison

Binder	RAP (%)	Construction cost (per $1000)	Service life (years)	EUAC ($)	% Difference from 0% RAP	Rank
Control	0	12746.47	10	1571.52	0	9
	15	12691.96	10	1564.80	−0.4	8
	25	12658.21	10	1560.64	−0.7	7
Asphalt rubber	0	12933.38	12	1401.29	−11.4	6
	15	12858.10	12	1393.13	−12.0	5
	25	12808.78	12	1387.79	−12.4	4
HiMA	0	13250.10	15	1219.69	−25.2	3
	15	13151.45	15	1210.61	−25.9	2
	25	13086.55	15	1204.63	−26.4	1
Fiber	0	12920.40	10	1592.97	+1.4	12
	15	12868.48	10	1586.57	+0.9	11
	25	12832.14	10	1582.09	+0.7	10

- The type of binder significantly affected performance. None of the three mixtures with the control PG64-28 binder met all performance criteria while all three mixtures with HiMA met all performance criteria. The asphalt rubber mixtures with 0% and 15% RAP and the fiber mixture with 25% RAP also met all criteria. The asphalt rubber mixture with 25% RAP failed the specification for permeability. The fiber mixtures with 0% and 15% RAP failed the rutting test and barely passed the rutting test using 25% RAP. Based on these results, it appears that a modified binder such as asphalt rubber or HiMA needs to be used in PFC containing RAP.
- Overall, the mixture tests demonstrated that a PFC with up to 15% RAP combined with asphalt rubber or HiMA can provide good performance. Furthermore, the use of 25% RAP was only limited by the asphalt rubber not meeting the specification for permeability.
- LCCA showed that if PFCs with RAP do have the same or better pavement service lives compared to virgin PFC, which was shown by the mixtures tests, then the RAP mixtures will be more cost effective.
- When the following steps are taken, it is possible to design a PFC with RAP which has good performance:
    - Using a modified binder such as asphalt rubber or HiMA.
    - Fractionating the RAP into coarse and fine stockpiles to accurately control the gradation of the aggregate blend.

Recommendation

It is recommended that pavement test sections using PFC with up to 15% RAP in combination with a modified binder such as asphalt rubber or HiMA be constructed and evaluated.

Author Contributions

The authors confirm contribution to the paper as follows: study conception and design: Brum, Mogawer; data collection: Brum; analysis and interpretation of results: Brum, Mogawer, Stuart, Austerman; draft manuscript preparation: Brum, Mogawer, Stuart, Austerman. All authors reviewed the results and approved the final version of the manuscript.

References

1. Walker, D. The Benefits of Modified Asphalts. *Asphalt: The Magazine of the Asphalt Institute*. Asphalt Institute, March 28, 2014. http://asphaltmagazine.com/the-benefits-of-modified-asphalts/. Accessed November 11, 2016.
2. West, R. *Best Practices for RAP and RAS Management*. Publication QIP 129. National Asphalt Pavement Association, Lanham, Md., 2015. https://www.asphaltpavement.org/PDFs/EngineeringPubs/QIP129_RAP_-_RAS_Best_Practices_lr.pdf. Accessed November 12, 2016.
3. Frigio, F., E. Pasquini, M. N. Partl, and F. Canestrari. Use of Reclaimed Asphalt in Porous Asphalt Mixtures: Laboratory and Field Evaluations. *Journal of Materials in Civil Engineering*. Vol. 27, No. 7, 2015. https://doi.org/10.1061/(ASCE)MT.1943-5533.0001182. Accessed September 22, 2014.
4. ASTM International. *Road and Paving Materials; Vehicle-Pavement Systems*. Vol. 04.03. ASTM International, West Conshohocken, Pa., 2014.
5. American Association of State Highway and Transportation Officials. *Standard Specifications for Transportation Materials and Methods of Sampling and Testing*. American Association of State Highway and Transportation Officials, Washington, D.C., 2016.
6. Florida Department of Transportation. *Measurement of Water Permeability of Compacted Asphalt Paving Mixtures*. FM 5-565. Florida Department of Transportation, Gainesville, 2015. http://www.fdot.gov/materials/administration/resources/library/publications/fstm/methods/fm5-565.pdf. Accessed May 20, 2015.
7. Massachusetts Department of Transportation. *Open Graded Friction Course Asphalt Rubber (OGFC-AR)*. Special Provision 450.2. Massachusetts Department of Transportation, Boston, 2015.
8. Gudimettla, J. Cooley, L. A., Jr., and E. R. Brown. *Workability of Hot Mix Asphalt*. NCAT Report 03-03. National Center for Asphalt Technology, Auburn, Ala., 2003.
9. Bahia, H. U., T. P. Friemel, P. A. Peterson, J. S. Russell, and B. Poehnelt. Optimization of Constructability and Resistance

to Traffic: A New Design Approach for HMA Using the Superpave Compactor. *Journal of the Association of Asphalt Paving Technologists*, 1998, Vol. 67, pp. 189–232.

10. Goh, S. W., and Z. You. Mechanical Properties of Porous Asphalt Pavement Materials with Warm Mix Asphalt and RAP. *Journal of Transportation Engineering*, Vol. 138, No. 1, 2012. https://doi.org/10.1061/(ASCE)TE.1943-5436.0000307. Accessed December 2, 2014.
11. West, R., D. Timm, R. Willis, B. Powell, N. Tran, D. Watson, M. Sakhaeifar, R. Brown, M. Robbins, A. V. Nordcbeck, F. L. Villacorta, X. Guo, and J. Nelson. *Phase IV NCAT Pavement Test Track Findings*. National Center for Asphalt Technology, Auburn, Ala., 2012.
12. Braham, A. F., W. G. Buttlar, T. R. Clyne, M. O. Marasteanu, and M. I. Turos. The Effect of Long-Term Laboratory Aging on Asphalt Concrete Fracture Energy. *Journal of the Association of Asphalt Paving Technologists*, Vol. 78, 2009, pp. 417–454.
13. Mogawer, W. S., and A. J. Austerman. *Evaluating the Performance of a Typical New England Asphalt Mixture Designed Using Asphalt Binders Modified with Re-refined Engine Oil Bottoms (REOB)*. University of Massachusetts Dartmouth – Highway Sustainability Research Center (HSRC), Fall River, Mass., 2016.
14. U.S. Department of Transportation Federal Highway Administration. *Life-Cycle Cost Analysis Software*. Federal Highway Administration, March 23, 2017. https://www.fhwa.dot.gov/infrastructure/asstmgmt/lccasoft.cfm. Accessed April 1, 2017.

The Standing Committee on Non-Binder Components of Asphalt Mixtures (AFK30) peer-reviewed this paper (18-01231).

Article

TRR
JOURNAL OF THE TRANSPORTATION RESEARCH BOARD

Transportation Research Record
2018, Vol. 2672(28) 130–142

DOI: 10.1177/0361198118786828
journals.sagepub.com/home/trr
SAGE

Investigating the Performances of Plant-Produced High-Reclaimed Asphalt Pavement Content Warm Mix Asphalts

Walaa S. Mogawer[1], Kevin Stuart[2], Alexander J. Austerman[1], and Ahmed A. Soliman[3]

Abstract

Laboratory-based studies have been conducted considering the incorporation of warm mix asphalt (WMA) technologies in paving mixtures both excluding and including reclaimed asphalt pavement (RAP). However, little research has been conducted to date that further advances the knowledge of WMA technologies in combination with high RAP contents beyond experiments using laboratory fabricated mixtures. The objective of this study was to expand on the existing laboratory research by investigating the moisture susceptibility, fatigue cracking and thermal cracking performances of plant-produced mixtures incorporating various WMA technologies and RAP contents. Nine 12.5-mm superpave mixtures were produced in two drum plants using three WMA technologies and three RAP contents. The three WMA technologies used were wax-based SonneWarmix™, chemical-based Evotherm®, and the Stansteel ACCU-SHEAR™ foaming process. Post production moisture in each mixture was negligible, which means that moisture had no effect on the laboratory mixture performance. The quality of blending analyses showed that the majority of the mixtures had poor blending, which would indicate that there should be performance issues with many of these mixtures. However, the quality of blending had no apparent effect on the laboratory mixture performance. It did not explain any poor performance or any difference in performance. Overall, some WMA with RAP contents up to approximately 50% RAP provided an acceptable laboratory performance. Fatigue cracking was the primary failure mode. Field trials or full-scale pavement accelerated tests with rigorous pavement monitoring are needed to confirm the findings of this study.

Many laboratory studies have been conducted concerning the use of warm mix asphalt (WMA) technologies in paving mixtures (*1–3*). These technologies allow for the production of paving mixtures at lower temperatures relative to hot mix asphalt (HMA). One concern with the lower production temperatures associated with WMA technologies, especially in mixtures containing high amounts of reclaimed asphalt pavement (RAP), is that they might not suitably dry all of the moisture out of the aggregates and the RAP which could lead to moisture susceptibility. In addition to this, some WMA technologies, such as those employing foaming, introduce water directly into the asphalt binder which could also lead to moisture susceptibility. Another concern which arises when using WMA with a high RAP content is whether the lower temperatures will provide sufficient blending between the virgin and aged asphalt binders. If blending is insufficient, the mixture may be under-asphalted and be susceptible to fatigue and thermal cracking. However, if blending is sufficient, the use of a high RAP content may increase mixture stiffness which may make the mixture susceptible to fatigue and thermal cracking unless it is compensated for by using a softer or modified asphalt binder. Counteracting these potential problems is the consideration that the use of lower production temperatures might decrease aging during production and improve the fatigue and thermal cracking performance.

Laboratory studies have researched the use of WMA with high amounts of RAP (*4*, *5*). However, little research has been conducted that advances the knowledge of WMA with high RAP contents beyond using laboratory prepared mixtures, which might not accurately represent plant-produced mixtures because of the unique production processes required when using a WMA technology in combination with a high RAP content.

[1]Highway Sustainability Research Center (HSRC), Civil and Environmental Engineering Department, University of Massachusetts Dartmouth, Fall River, MA
[2]Consultant, Phoenixville, PA
[3]Highway Sustainability Research Center (HSRC), University of Massachusetts Dartmouth, Fall River, MA

Corresponding Author:
Address correspondence to Walaa S. Mogawer: wmogawer@umassd.edu

Research Objective

The objective of this study was to expand on existing laboratory research on WMA containing RAP by investigating the moisture susceptibility, fatigue cracking and thermal cracking performances of plant-produced mixtures incorporating various WMA technologies and RAP contents.

Internet Based Survey

To assist in the development of the experimental plan for this study, two internet based surveys were distributed to members of state agencies and industries throughout the Northeast. One survey solicited feedback from state agency personnel while the other solicited feedback from industry personnel. The purpose of the state agency survey was to determine the specification limits for RAP usage, the WMA technologies approved for use, allowable RAP stockpile moisture contents, production and placement temperatures for both high RAP content and WMA mixtures, and approved moisture susceptibility tests for the mixtures. The purpose of the industry survey was to determine the typical RAP stockpile moisture contents, typical RAP properties, maximum permissible amounts of RAP used in production, the WMA technologies utilized, and production and placement temperatures for both high RAP content and WMA mixtures. Some highlights from the survey are as follows:

- Drum mix plants can produce mixtures with up to 50% RAP by weight of aggregate, whereas, batch plants are typically used to produce mixtures with only up to 20% RAP by weight of aggregate.
- The most commonly specified and used wax-based WMA technology is SonneWarmix™ while the most commonly specified and used chemical-based WMA technology is Evotherm®. Foaming/moisture-based processes are not common.
- RAP binder contents vary from 2% to 8% while RAP moisture contents vary from 0% to 7%. See Figure 1 for the distribution of responses.
- The survey indicated that HMA with high RAP contents had typical production temperatures of 143–168°C and compaction temperatures of 135–149°C. WMA mixtures had typical production temperatures of 135–168°C and compaction temperatures of 121–160°C.

Research Approach

To accomplish the objective, plant-produced mixtures incorporating three WMA technologies and three RAP contents were tested for laboratory performance in regards to the moisture susceptibility, fatigue cracking and thermal cracking. Two drum plants were used to produce the mixtures. The residual moisture content and the quality of blending between the virgin and aged RAP binders in each mixture were measured because they may be partially responsible for any differences in the laboratory mixture performance along with the WMA technology and RAP content. The experimental plan is shown in Figure 2.

WMA Technology Selection

An organic wax-based additive named SonneWarmix™, a chemical-based additive named Evotherm®, and a foaming process named Stansteel ACCU-SHEAR™ were selected from the Northeast Asphalt User Producer Group (NEAUPG) approved list of WMA technologies. SonneWarmix™ and Evotherm® were utilized at a dosage of 0.5% per weight of binder according to manufacturer's recommendations. The Stansteel ACCU-SHEAR™ foaming process utilized a water content of 3% by weight of binder.

Plant Produced Mixtures

Two Massachusetts approved 12.5-mm nominal maximum aggregate size (NMAS) superpave mixtures were selected. Each mixture design gradation was then modified to include the increasing amounts of RAP while keeping the job mix formula constant. Drum plants were selected to produce each mixture to accommodate RAP contents greater than 20%. As noted in the survey results, batch plants are typically limited to 20% RAP. The binder remained constant at each plant with no changes in the performance grade (PG) as the RAP content was increased. The binders used at both plants were a PG64-28 which have production temperatures around 160°C when used in HMA. No antistripping agents were used.

The first 12.5-mm mixture was a 100-gyration mixture produced using SonneWarmix™. Three mixtures were produced with RAP contents of 15% (low), 27.8% (medium) and 46.3% (high). The RAP content of 27.8% and 46.3% corresponded to 1.5% and 2.5% of the mixture virgin binder content being replaced, respectively. The binder content of the RAP was 5.4% and the moisture content of the RAP was 1.6%. The properties of these mixtures and production temperatures are shown in Table 1. The volumetric properties were measured after gyratory compaction. Based on the manufacturer recommendations, the production temperatures ranged from 133°C to 134°C.

The second 12.5-mm mixture was a 75-gyration mixture using Evotherm® and Stansteel ACCU-SHEAR™. Six mixtures, three for each WMA technology, were produced with RAP contents of 29% (medium), 39% (high) and 48% (high), which corresponded to 1.5%, 2.0% and 2.5% of the mixture virgin binder content being replaced, respectively. For the 29% and 39% RAP mixtures incorporating Evotherm®, the binder content of the RAP was 5.0% and the moisture content of the RAP was 2.9%. For the 48% RAP mixture incorporating Evotherm®, the RAP binder

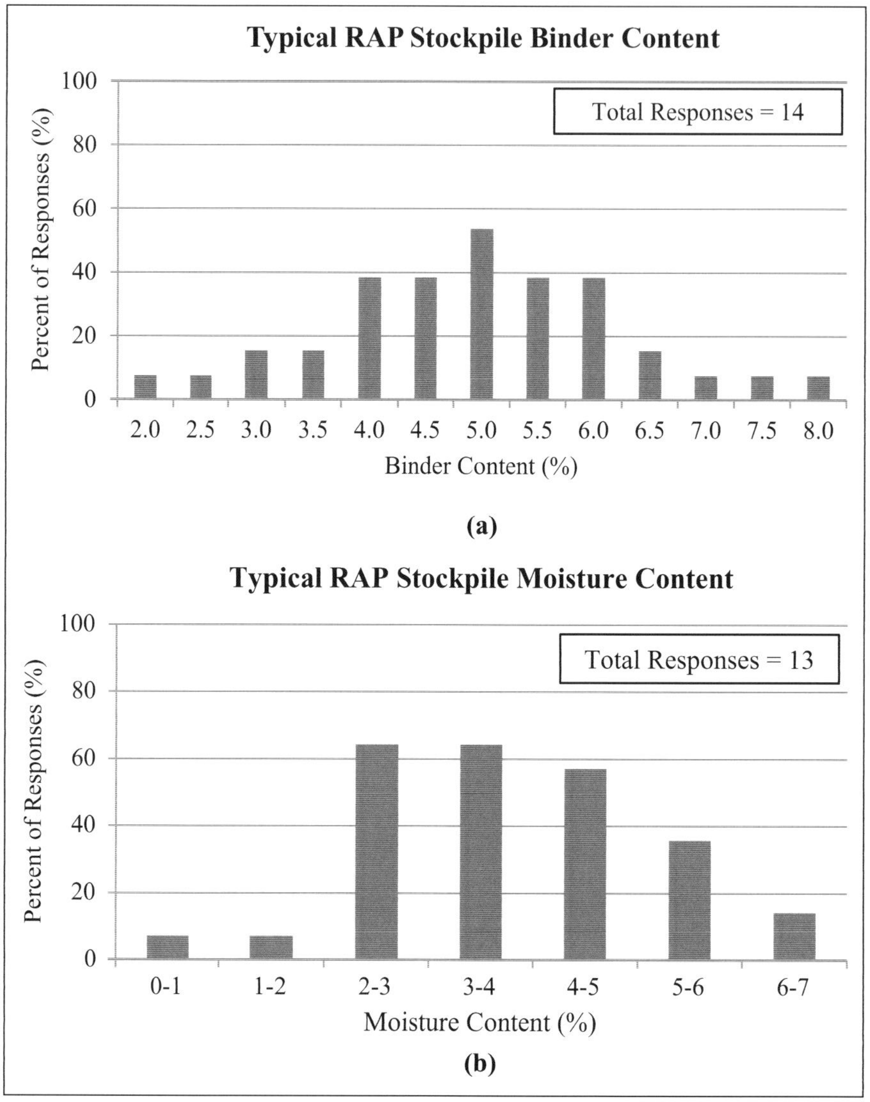

Figure 1. Industry survey results: (*a*) Typical RAP stockpile binder content and (*b*) Typical RAP stockpile moisture content.

content was 5.4% and the moisture content was 4.0%. For all three mixtures incorporating the Stansteel ACCU-SHEAR™ foaming process, the binder content of the RAP was 5.9% and the moisture content of the RAP was 4.1%. The properties of these mixtures and production temperatures are shown in Tables 2 and 3. The volumetric properties were measured after gyratory compaction. Based on the manufacturer recommendations, the production temperatures ranged from 135°C to 143 °C .

Finally, the MassDOT quality assurance specification only requires monitoring of air voids for acceptance testing during production. The mixture air voids must be within ±1.3% of the target 4%. The other superpave volumetric properties of the voids in mineral aggregate (VMA), voids filled with asphalt (VFA), and a dust-to-binder ratio are only required to be verified during the mixture design process and not during production. Thus, during production, only the air void percentage outside of the tolerance would trigger a corrective action by the contractor. Note that all the mixtures produced for this study met this air void requirements as shown in Tables 1–3.

Moisture Content

The moisture contents of all mixtures were measured immediately after production, with no silo storage, in accordance with the AASHTO T329 "Moisture Content of Hot Mix Asphalt (HMA) by Oven Method" (*6*). As shown in Tables 1–3, the moisture remaining after plant production in each mixture was negligible. This indicates that nearly all of the moisture in the RAP and aggregate was eliminated during plant production.

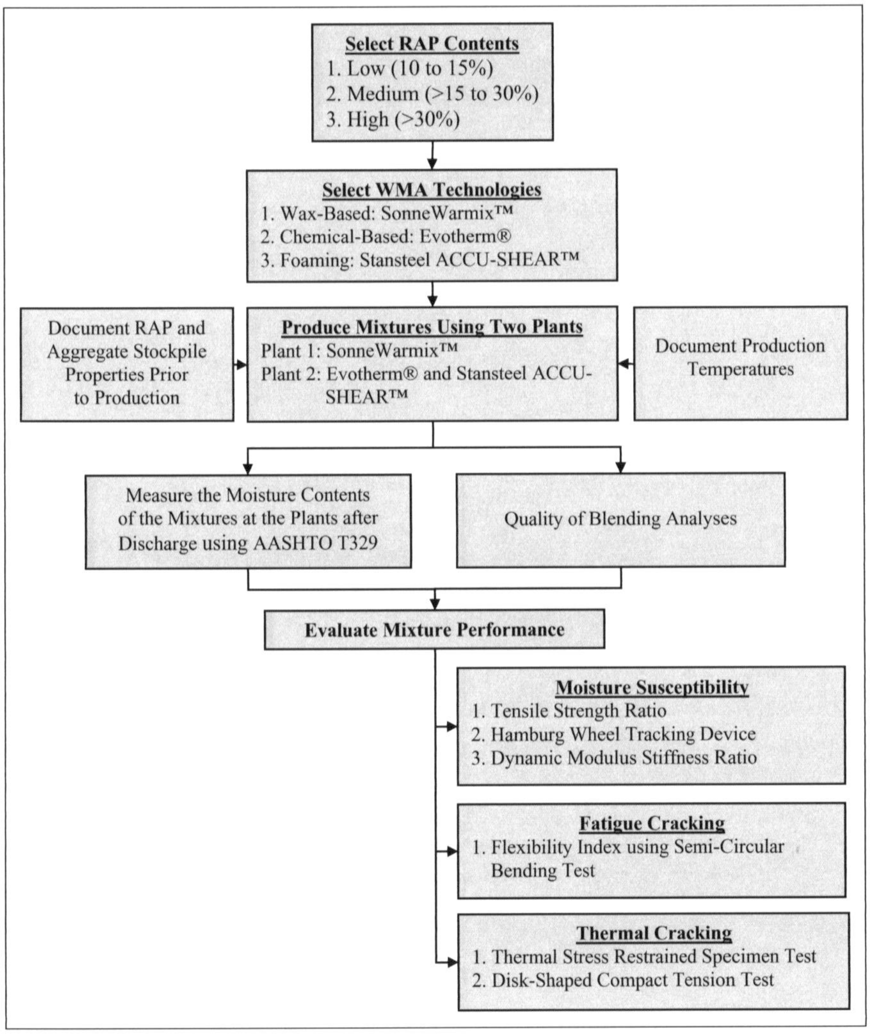

Figure 2. Experimental plan.

Laboratory Testing of Plant Produced Mixtures

Mixtures were collected during production and placed into buckets. These buckets were reheated in the laboratory to fabricate specimens for the performance evaluations. The same reheating procedure was used for each mixture and the amount of reheating was minimized. Each bucket was first heated at the post-production temperature for one hour with the lid loosely on it. After this, the lid was removed and the temperature was monitored until it reached 82ºC. The mixture was then removed from the oven and split and quartered manually to the appropriate test specimen weight in accordance with AASHTO T248 "Reducing Samples of Aggregate to Testing Size" (*6*). The spilt and quartered samples were placed back into the oven until the compaction temperature of 121°C was reached, which is the minimum compaction temperature noted for WMA mixtures from the survey responses. All specimens were compacted using a superpave gyratory compactor (SGC).

Quality of Blending Analysis

As lower WMA production temperatures might negatively impact the quality of blending between the virgin and aged RAP binders, and thus, possibly, might impact mixture performance, a blending analysis is critical. If no blending occurs, the resultant mixture will contain less than the targeted asphalt binder content which could lead to mixtures which are prone to cracking (*7–9*).

Factors that may affect the quality of blending are plant type, production temperature, mixing time, discharge temperature, RAP source and RAP properties. Several laboratory studies have researched the quality of blending. These studies illustrated that the quality of blending can be evaluated using the dynamic modulus (E*) of the mixtures and the complex shear modulus (G*) of the extracted and recovered binders (*7*, *8*).

To evaluate the quality of blending, the first step was to develop an E* master curve for each mixture in accordance with AASHTO PP-61 "Developing Dynamic Modulus

Table 1. Drum Plant #1: 12.5-mm Mixture Properties with 0.5% SonneWarmix™

Sieve size (mm)	% Passing by weight		
	15% RAP mixture	27.8% RAP mixture	46.3% RAP mixture
19.0	100	100	100
12.5	98.4	98.2	98.1
9.5	85.9	84.9	83.4
4.75	61.0	60.6	60.2
2.36	41.8	42.0	42.9
1.18	28.2	28.9	31.0
0.60	18.4	18.8	21.1
0.30	11.8	12.0	13.6
0.15	6.7	6.9	8.2
0.075	3.5	3.5	4.3
Total mixture binder content (%)	5.5	5.5	5.5
Mixture air voids (%)	4.1	3.4	2.8
Voids in mineral aggregate (VMA), %	14.0	14.2	13.7
Voids filled with asphalt (VFA), %	71.0	76.1	79.8
Production temperature (°C)	133°C	133°C	134°C
Mixture moisture content post production	0.16%	0.08%	0.10%

Note: RAP = reclaimed asphalt pavement.

Table 2. Drum Plant #2: 12.5-mm Mixture Properties with 0.5% Evotherm®

Sieve size (mm)	% Passing by weight		
	29% RAP mixture	39% RAP mixture	48% RAP mixture
19.0	100	100	100
12.5	97.4	98.5	99.0
9.5	84.1	85.0	88.2
4.75	59.7	59.7	63.5
2.36	47.1	46.7	50.5
1.18	39.9	39.1	38.2
0.60	32.9	32.1	34.0
0.30	21.4	20.9	22.2
0.15	9.8	9.8	11.3
0.075	4.1	4.2	4.8
Total mixture binder content (%)	5.0	5.0	5.0
Mixture air voids (%)	4.4	4.3	4.2
Voids in mineral aggregate (VMA), %	14.1	14.2	14.4
Voids filled with asphalt (VFA), %	69.0	69.6	71.1
Production temperature	138°C	135°C	143°C
Mixture moisture content post production	0.04%	0.03%	0.00%

Note: RAP = reclaimed asphalt pavement.

Master Curves for Asphalt Mixtures Using the Asphalt Mixture Performance Tester (AMPT)" (*6*). Three 170-mm tall specimens were compacted using the SGC. These specimens were then cored and cut to the final test specimen dimensions of 101 ± 0.5 mm in diameter and 150 ± 2 mm in height. The air voids were 7 ± 1%. Specimens were conditioned overnight at 4°C and then tested for E* at loading frequencies of 10, 1, and 0.1 Hz. The specimens were next conditioned at 20°C for a minimum of one hour and tested again at the same loading frequencies of 10, 1, and 0.1 Hz. Finally, the specimens were conditioned at 40°C between one and two hours and tested at loading frequencies of 10, 1, 0.1, and 0.01 Hz. For each mixture, the E* results were used to develop the master curve.

The second step was to extract and recover the binder from each mixture using toluene in accordance with ASTM

Table 3. Drum Plant #2: 12.5-mm Mixture Properties with ACCU-SHEAR™ Foaming

Sieve size (mm)	% Passing by weight		
	29% RAP mixture	39% RAP mixture	48% RAP mixture
19.0	100	100	100
12.5	97.4	98.5	99.0
9.5	84.1	85.0	88.2
4.75	59.7	59.7	63.5
2.36	47.1	46.7	50.5
1.18	39.9	39.1	38.2
0.60	32.9	32.1	34.0
0.30	21.4	20.9	22.2
0.15	9.8	9.8	11.3
0.075	4.1	4.2	4.8
Total mixture binder content (%)	5.0	5.0	5.0
Mixture air voids (%)	3.4	3.5	3.8
Voids in mineral aggregate (VMA), %	14.1	14.2	14.4
Voids filled with asphalt (VFA), %	76.2	75.6	73.5
Production temperature	138°C	135°C	143°C
Mixture moisture content post production	0.0%	0.0%	0.0%

Note: RAP = reclaimed asphalt pavement.

D2172/D2172M "Quantitative Extraction of Bitumen from Bituminous Paving Mixtures" (*10*) – Method A. The G* and phase angle (δ) of each binder was then measured using a dynamic shear rheometer in accordance with AASHTO T315 "Determining the Rheological Properties of Asphalt Binder Using Dynamic Shear Rheometer" (*6*) at 10, 22, and 34°C and strain levels of 0.1, 1, 1%, respectively, and at loading frequencies of 0.1, 1, 10, and 100 Hz.

For each mixture, the Hirsch model (*11*) was used to predict a mixture E* for a fully blended condition at each temperature and loading frequency from the binder G* using Equations 1 and 2:

$$E^* = P_c\left[4,200,000\left(1-\frac{VMA}{100}\right) + 3\times G^*\left(\frac{VFA\times VMA}{10,000}\right)\right] + \frac{1-P_c}{\left[\frac{\left(1-\frac{VMA}{100}\right)}{4,200,000}+\frac{VMA}{3\times VFA\times G^*}\right]} \quad (1)$$

$$P_c = \frac{\left(20+\frac{VFA\times 3G^*}{VMA}\right)^{0.58}}{650+\left(\frac{VFA\times 3G^*}{VMA}\right)^{0.58}} \quad (2)$$

In which:

E* = predicted asphalt mixture dynamic modulus, psi

G* = measured asphalt binder complex shear modulus, psi

P_c = aggregate contact volume

The measured E* master curve for each mixture was then utilized to determine E* at the temperatures of 10, 22, and 34°C and loading frequencies of 0.1, 1, 10, and 100 Hz used for the binder testing. For each temperature and loading frequency, the mean E* from the master curve and the predicted mean E* from the Hirsch model were statistically compared at a significance level of $\alpha = 0.05$ to determine if the blending was good or poor. If the means were significantly different, blending was poor.

The results are shown in Figure 3 with the shaded areas indicating poor blending. Note that this method does not directly measure the degree of blending as a percentage but is simply an indication of the quality of blending, either good or poor. Owing to this, whether there is good or poor blending for a particular mixture often varies with the temperature and loading frequency. When this is encountered, whether there is good or poor blending must be deduced from the overall results.

Figure 3 shows that blending was poor for the majority of mixtures. The 29% and 48% RAP mixtures with Evotherm® provided the best overall quality of blending, although why the quality of blending for them was better than for the 39% RAP content is not known. The reasons why the quality of blending for the 48% RAP mixture with ACCU-SHEAR™ foaming was so imprecise and why it was somewhat better than for the 29 and 39% RAP contents are also unknown.

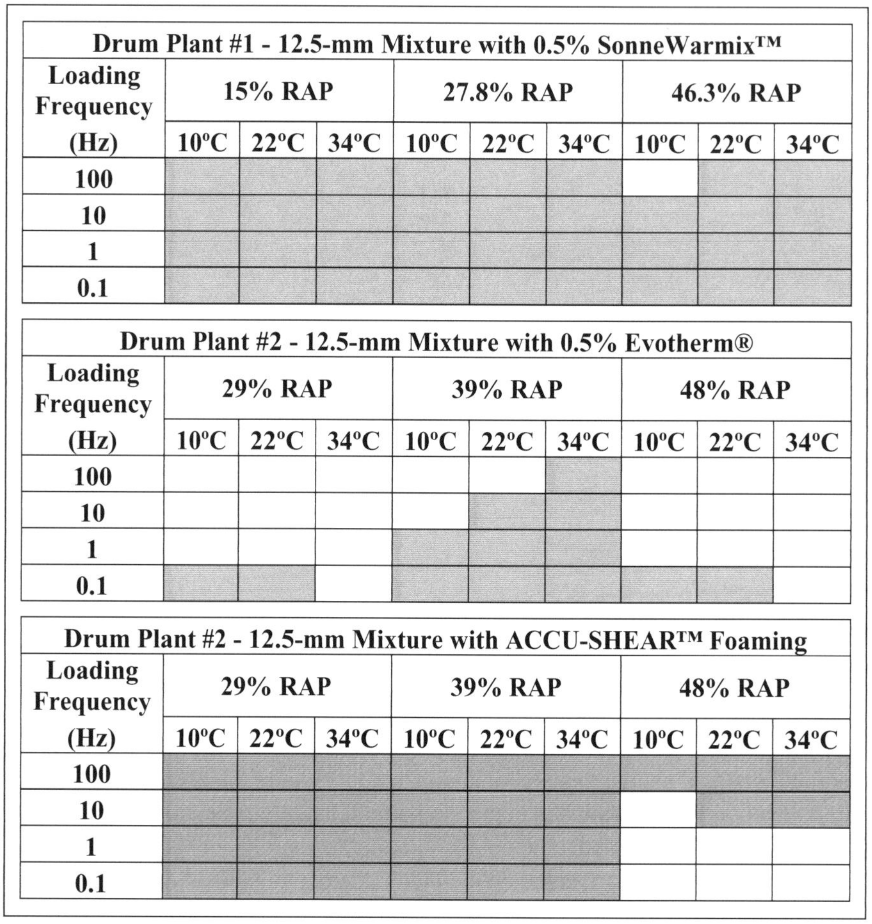

Figure 3. Quality of blending analysis.
Note: Shading indicates poor blending.

Mixture Performance: Moisture Susceptibility

The tensile strength ratio (TSR), Hamburg wheel tracking device (HWTD), and E* stiffness ratio (ESR) were used to evaluate the moisture susceptibilities of the mixtures. The TSR and HWTD were found to be the most common tests used to evaluate moisture susceptibility according to the internet survey.

Tensile Strength Ratio and Percent Area Stripped. A common method for evaluating the moisture susceptibilities of asphalt paving mixtures is the TSR performed in accordance with AASHTO T283 "Standard Method of Test for Resistance of Compacted Asphalt Mixtures to Moisture-Induced Damage" (*6*). Six replicate specimens per mixture were compacted using the SGC to a height of 95 mm and at an air void level of 7±1.0%. Three specimens were tested unconditioned while three were freeze thaw conditioned in accordance with AASHTO T283 - Section 10.3. The TSR is the ratio of the tensile strength of conditioned specimens to the tensile strength of the unconditioned specimens. The minimum passing TSR is 80%.

The TSR results are shown in Tables 4, 5 and 6. All mixtures passed except for the 39% RAP mixture with Evotherm® which had a borderline but failing TSR of 77.5%. However, this mixture easily passed the HWTD. None of the mixtures exhibited visual stripping that was more than 10% of the total area.

Hamburg Wheel Tracking Device Test. The HWTD test was conducted in accordance with AASHTO T324 "Hamburg Wheel-Track Testing of Compacted Hot Mix Asphalt (HMA)" (*6*). In this test, the rut depth versus the number of wheel passes is plotted to determine the stripping inflection point (SIP), which gives an indication of when the test specimen is beginning to exhibit moisture damage.

The SGC specimens were fabricated to an air void level of 7.0 ± 1.0% as required by AASHTO T324. The test temperature was 45°C which is required by the state agency where these mixtures were produced. Testing is terminated at 20,000 wheel passes or sooner if damage is high. Based on the internet survey, the passing criteria for the HWTD is a maximum rut depth of 12.5 mm at 20,000 passes combined with no SIP before 15,000 passes. These criteria are also

Table 4. Performance Test Results for Drum Plant #1: 12.5-mm Mixture with 0.5% SonneWarmix™

	15% RAP mixture	27.8% RAP mixture	46.3% RAP mixture
Moisture susceptibility tests			
AASHTO T283			
Tensile strength ratio (TSR)			
TSR (%)	92.0	88.7	93.9
Min. is 80%			
Area stripped (%)	<10%	<10%	<10%
AASHTO T324			
Hamburg wheel tracking device			
Stripping inflection point, (Wheel Passes)	11,766 **F**	None	None
Min. is 15,000			
Rut depth at 20,000 wheel passes (mm)	13.4 **F**	5.1	4.4
Max. is 12.5 mm			
Dynamic modulus stiffness ratio (ESR)			
Min. is 70 to 80%			
ESR, 10 Hz (%)	101.4	95.2	95.1
ESR, 1 Hz (%)	99.8	92.0	93.3
ESR, 0.1 Hz (%)	96.5	89.9	91.6
Thermal cracking tests			
AASHTO TP10			
Thermal stress restrained Specimen test (TSRST)			
Average failure temperature (°C)	−27.0	−27.3	−26.6
ASTM D7313			
Disk-shaped compact tension DC(T) test			
Average fracture energy (J/m^2)	499	518	474
Fatigue cracking test			
AASHTO TP124			
Semi-circular bending test (SCB)			
Flexibility index (FI)	10.6	16.2	15.8
Min is 8			

Note: RAP = reclaimed asphalt pavement; **F** = Failed the specification; Min. = Minimum; Max. = Maximum.

Table 5. Performance Test Results for Drum Plant #2: 12.5-mm Mixture with 0.5% Evotherm®

	29% RAP mixture	39% RAP mixture	48% RAP mixture
Moisture susceptibility tests			
AASHTO T283			
Tensile strength ratio (TSR)			
TSR (%)	82.2	77.5 **F**	85.9
Min. is 80%			
Area stripped (%)	<10%	<10%	<10%
AASHTO T324			
Hamburg wheel tracking device			
Stripping inflection point (wheel passes)	None	None	None
Min. is 15,000			
Rut depth at 20,000 wheel passes (mm)	2.8	2.1	2.3
Max. is 12.5 mm			
Dynamic modulus stiffness ratio (ESR)			
Min. is 70 to 80%			
ESR, 10 Hz (%)	90.4	77.8	92.3

(Continued)

Table 5. Continued

	29% RAP mixture	39% RAP mixture	48% RAP mixture
ESR, 1 Hz (%)	89.2	73.2	90.4
ESR, 0.1 Hz (%)	81.7	74.7	88.4
Thermal cracking tests			
AASHTO TP10			
Thermal stress restrained			
Specimen test (TSRST)			
Average failure temperature (°C)	−25.5	−26.2	−26.8
ASTM D7313			
Disk-shaped compact tension DC(T) test			
Average fracture energy (J/m^2)	504	562	562
Fatigue cracking test			
AASHTO TP124			
Semi-circular bending			
test (SCB)			
Flexibility index (FI)	6.1 **F**	5.4 **F**	3.7 **F**
Min. is 8			

Note: RAP = reclaimed asphalt pavement; **F** = Failed the specification; Min. = Minimum; Max. = Maximum.

Table 6. Performance Test Results for Drum Plant #2: 12.5-mm Mixture with ACCU-SHEAR™ Foaming

	29% RAP Mixture	39% RAP mixture	48% RAP mixture
Moisture susceptibility tests			
AASHTO T283			
Tensile strength ratio (TSR)			
TSR (%)	94.8	90.1	86.8
Min. is 80%			
Area stripped (%)	<10%	<10%	<10%
AASHTO T324			
Hamburg wheel tracking device			
Stripping inflection point (wheel passes)	None	None	None
Min. is 15,000			
Rut depth at 20,000 wheel passes (mm)	3.04	2.44	1.71
Max. is 12.5 mm			
Dynamic modulus			
stiffness ratio (ESR)			
Min. is 70 to 80%			
ESR, 10 Hz (%)	94.6	92.4	92.3
ESR, 1 Hz (%)	94.7	89.3	90.4
ESR, 0.1 Hz (%)	98.0	94.8	88.4
Thermal cracking tests			
AASHTO TP10			
Thermal stress restrained			
specimen test (TSRST)			
Average failure temperature (°C)	−26.9	−25.8	−23.0
ASTM D7313			
Disk-shaped compact tension DC(T) test			
Average fracture energy (J/m^2)	548	517	582
Fatigue cracking test			
AASHTO TP124			

(Continued)

Table 6. Continued

	29% RAP Mixture	39% RAP mixture	48% RAP mixture
Semi-circular bending test (SCB)			
Flexibility index (FI) Min. is 8	8.4	6.7 **F**	4.4 **F**

Note: RAP = reclaimed asphalt pavement; **F** = Failed the specification; Min. = Minimum; Max. = Maximum.

currently required by the state agency where these mixtures were produced.

The HWTD results are presented in Tables 4, 5, and 6. All mixtures passed except for the 15% RAP mixture with SonneWarmix™ which means an antistripping agent would be needed for this. The reason for its poor performance is unknown. It is hypothesized that the WMA technology might have softened the virgin asphalt binder and that the low RAP content might not have introduced enough stiff binder to counteract this softening effect. However, the mixture easily passed the other two tests for moisture susceptibility. All rut depths were low except for the mixture which failed from moisture damage.

Dynamic Modulus Stiffness Ratio. Previous studies showed that the ESR can also be used to evaluate moisture susceptibility (*12–14*). For each mixture, three E* specimens were compacted to a height of 170 mm. Test specimens were cored from them at a diameter of 101 ± 0.5 mm which were then cut to a height of 150 ± 2 mm. The air void level was 7 ± 1%. The specimens were conditioned at 20ºC for a minimum of two hours and then tested for E* at 20ºC and loading frequencies of 10, 1, and 0.1 Hz in accordance with AASHTO PP61 "Developing Dynamic Modulus Master Curves for Asphalt Mixtures Using the Asphalt Mixture Performance Tester (AMPT)"(*6*). The same specimens were then freeze thaw conditioned in accordance with AASHTO T283 - Section 10.3 with the exception that the vacuum was applied for 30 min regardless of the degree of saturation attained. The specimens were then tested in the AMPT to determine the E* once again at the same temperature of 20ºC and the same loading frequencies of 10, 1, and 0.1 Hz. The ESR is the ratio of the E* of the conditioned specimens to the E* of the unconditioned specimens which is computed at each frequency. A minimum ESR in the range of 70% to 80% is generally considered acceptable (*14*).

The ESR results are shown in Tables 4, 5 and 6. All ESR values were above 70%, suggesting good performance. Even so, the ESR may have further identified the 39% RAP mixture with Evotherm® to be slightly susceptible to moisture damage, having an ESR between 73 and 78%. The TSR for this mixture failed at 77.5%. The mixture easily passed the HWTD.

Mixture Performance: Fatigue Cracking

The resistances of the mixtures to fatigue cracking were evaluated using the Illinois flexibility index test (I-FIT) in accordance with AASHTO TP124 "Standard Method of Test for Determining the Fracture Potential of Asphalt Mixtures Using Semicircular Bend Geometry (SCB) at Intermediate Temperature." The SCB is controlled by a linear variable differential transformer (LVDT) using a monotonic load applied along the vertical diameter of the specimen at a displacement rate of 50 mm/min (*15*) and a test temperature of 25°C. SGC specimens were compacted to a height of 180 mm. Two 50-mm slices were cut from the middle of each specimen. Each slice was cut in half across the diameter to yield four specimens. The air void level was 7.0 ± 1.0%. Specimens were tested in the SCB and the fracture energy (G_f,) was calculated using the work of fracture method by finding the area under the load displacement curve and dividing it by the crack propagation area.

An important parameter obtained from SCB is the flexibility index (FI) which was developed to discriminate the fatigue cracking damage potential of asphalt mixtures. It accounts for both the G_f and the ductile nature of an asphalt mixture after the peak load through the slope M. The recommended minimum allowable FI is 8.0. The FI is calculated using Equation 3:

$$FI = A \times \frac{G_f}{|M|} \quad (3)$$

In which:

G_f = fracture energy, Joules/m^2

M = slope of the post-peak curve at the inflection point, kN/mm

A = unit conversion factor and scaling coefficient. The A in this study was 0.01.

The FI results are shown in Tables 4, 5 and 6. Four of the nine mixtures passed the test. An analysis of variance showed that the FI was a function of both the WMA technology and the RAP content with the Sonnewarmix™ mixtures having the highest FI. Although all three RAP contents using

Sonnewarmix™ had acceptable performances, the lowest RAP content provided the lowest FI, while for both Evotherm® and Stansteel ACCU-SHEAR™, the lowest RAP content provided the highest FI. For the latter two WMA technologies, the FI decreased with increasing RAP content which might be expected as the amount of aged asphalt increased with increasing RAP content. Neither the quality of blending in Figure 2 nor the differences in volumetrics in Tables 1, 2 and 3 explained this or any other difference in FI. Perhaps, it was because of the same mechanism which allowed the one WMA technology to provide better overall performance.

Mixture Performance: Thermal Cracking

Thermal Stress Restrained Specimen Test. Thermal cracking susceptibility was first measured using the thermal stress restrained specimen test (TSRST) in accordance with AASHTO TP10-93 (*16*) with the exception that SGC specimens were utilized.

In this test, an asphalt specimen is cooled at a constant rate of -10ºC/h while its length is held constant. As the specimen becomes colder, it tries to contract but the device restricts this contraction which results in the accumulation of thermal stresses. Eventually the thermal stresses exceed the tensile strength capacity of the specimen resulting in specimen fracture. The temperature at which this fracture occurs is the low cracking temperature of the mixture.

SGC specimens, which were 185 mm tall by 150 mm in diameter, were fabricated for each mixture. TSRST specimens were then cored and cut from them to a height of 160 ± 2 mm by 54 ± 0.5 mm in diameter. The air voids were 7 ± 1%.

The TSRST results are shown in Tables 4, 5 and 6. The TSRST cracking temperatures ranged from −23.0°C to −27.3°C. As a virgin PG64-28 binder was used in all mixtures, it might be expected that the mixtures would fail at a temperature equal to or colder than −28°C if there was no influence of the aged RAP binder. However, the TSRST often provides a temperature that is warmer than the low-temperature PG. For a low-temperature PG of −28°C, the TSRST usually provides a cracking temperature in the range of −26 to −28°C. Although the potential beneficial effects on thermal cracking performance because of the lower production temperatures associated with WMA were not realized in this study, it was concluded that the thermal cracking performances of the mixtures would be acceptable at temperatures down to −28°C except for one mixture. The 48% RAP mixture with foaming WMA potentially showed the effect of the aged RAP binder as it had the warmest cracking temperature of −23°C, whereas the rest of the mixtures were between −26 to −28°C.

Disk-Shaped Compact Tension. A second test utilized to evaluate thermal cracking performance was the disk-shaped compact tension (DC(T)) test. In this test, fracture energy is used to describe the thermal fracture resistance of a mixture. Specimens having a 150 mm height were fabricated, prepared, and tested in accordance with ASTM D7313 "Standard Test Method for Determining Fracture Energy of Asphalt-Aggregate Mixtures Using the Disk-Shaped Compact Tension Geometry" (*10*). As outlined in the method, a test temperature of −18ºC was utilized which is 10ºC warmer than the low-temperature PG of the asphalt binder used in the mixtures, which was −28°C. The crack mouth opening displacement rate was 0.017 mm/s.

Threshold values of DC(T) fracture energy have been developed based on field investigations of thermally-induced cracks under a FHWA pooled fund study (*17*). The thresholds for low, medium, and high traffic asphalt pavement mixtures were set at 400, 460, and 690 J/m^2, respectively. Mixtures from drum plant #1 were 100 gyration design mixtures which would be for high traffic, having a minimum criterion of 690 J/m^2, whereas, the mixtures from drum plant #2 were 75 gyration design mixtures which would be for medium traffic, having a minimum criterion of 460 J/m^2.

The DC(T) results are shown in Tables 4, 5 and 6. The mixtures from drum plant #1 did not meet the high traffic criteria but passed the medium traffic criteria. The mixtures from drum plant #2 passed the medium traffic criteria. Mixtures from drum plant #1 might exhibit thermal cracking at high traffic levels while all mixtures would be acceptable for medium traffic levels. A final comment is that the TSRST does not consider the traffic level. A mixture can be used at any traffic level as long as the pavement temperature will not go below the cracking temperature that the TSRST provides. Thus, the two tests might not always provide an identical conclusion.

Conclusions and Recommendations

- A major concern with using a WMA technology and RAP is the potential for residual moisture from the RAP to be present in the mixture after production because of the reduced production temperatures associated with WMA and also any residual moisture if a WMA foaming process is used. The residual moisture contents of the mixtures in this study after plant production were negligible, which would indicate that moisture had little to no effect on the laboratory mixture test results. The results indicate that moisture can be dried out of these types of mixtures by a drum plant. The moisture contents of the RAP stockpiles ranged from 1.6 to 4.1%.
- The quality of blending analyses showed that the majority of the mixtures had poor blending, which would indicate that there should be performance issues with many of these mixtures. However, the quality of blending had no apparent effect on the laboratory mixture performance. They did not explain any poor performance or any difference in laboratory

performance. Perhaps the method used to determine them is not accurate for these types of mixtures.

- The three moisture susceptibility tests generally indicated acceptable laboratory performance regardless of the WMA technology or RAP content. The only significant failure was provided by the low RAP content of 15% RAP in combination with one WMA technology when tested using the HWTD. However, the mixture easily passed the other two tests. The three tests for moisture susceptibility did not always agree with one another. All rut depths in the HWTD were low except for the mixture which failed from moisture damage.
- Four out of the nine mixtures had acceptable fatigue cracking laboratory performance based on FI with one WMA technology providing better overall performance. Although all three RAP contents using this WMA technology had acceptable performances, the lowest RAP content provided the lowest FI, while for other two WMA technologies, the lowest RAP content provides the highest FI. For the latter two WMA technologies, the FI decreased with increasing RAP content which might be expected as the amount of aged asphalt increased with increasing RAP content. Neither the quality of blending nor differences in volumetrics explained this or any other difference in FI. Perhaps, it was because of the same mechanism which allowed the one WMA technology to provide a better overall performance.
- The TSRST provided acceptable thermal cracking performances. The DC(T) showed that all mixtures would have acceptable thermal cracking performances for pavements with low or medium traffic levels.
- Overall, some WMA with RAP contents up to approximately 50% RAP provided acceptable laboratory performance. Field trials or full-scale pavement accelerated tests with rigorous pavement monitoring are needed to confirm the findings of this study. Studies are also needed to determine the applicability of the recommended minimum allowable FI of 8.0 to Northeast states.

Acknowledgments

The research data and results presented in this paper were part of a New England Transportation Consortium (NETC) project study entitled "Low Temperature and Moisture Susceptibility of RAP Mixtures with Warm Mix Technology." NETC is a funded research cooperative between the state DOTs of Connecticut, Maine, Massachusetts, New Hampshire, Rhode Island and Vermont.

Author Contributions

The authors confirm contribution to the paper as follows: study conception and design: Mogawer, Stuart, Austerman; data collection: Soliman; analysis and interpretation of results: Mogawer, Stuart, Austerman, Soliman; draft manuscript preparation: Mogawer, Stuart, Austerman, Soliman. All authors reviewed the results and approved the final version of the manuscript.

References

1. Bonaquist, R. *NCHRP Report 691: Mix Design Practices for Warm Mix Asphalt.* Transportation Research Board of the National Academies, Washington D.C., 2011.
2. Mogawer, W. S., A. J. Austerman, and H. U. Bahai. Evaluating Effect of Warm-Mix Asphalt Technologies on Moisture Characteristics of Asphalt Binders and Mixture. *Transportation Research Record: Journal of the Transportation Research Board*, 2011. 2209: 52–60.
3. Bennert, T., G. Reinke, W. S. Mogawer, and K. Mooney. Assessment of Workability and Compactability of Warm-Mix Asphalt. *Transportation Research Record: Journal of the Transportation Research Board*, 2010. 2180: 36–47.
4. Mogawer, W. S., A. J. Austerman, M. Louay, and M. Kutay. Evaluation of High RAP-WMA Asphalt Rubber Mixtures. *Road Materials and Pavement Design (RMPD)*, Vol. 14, Suppl. 2, 2013, pp. 129–147.
5. Mogawer, W. S., A. J. Austerman, R. Kluttz, and M. Roussel. High-Performance Thin-Lift Overlays with High Reclaimed Asphalt Pavement Content and Warm-Mix Asphalt Technology. *Transportation Research Record: Journal of the Transportation Research Board*, 2012. 2293: 18–28.
6. American Association of State Highway and Transportation Officials. *Standard Specifications for Transportation Materials and Methods of Sampling and Testing.* American Association of State Highway and Transportation Officials, Washington, D.C., 2015.
7. Bonaquist, R. *Laboratory Evaluation of Hot Mix Asphalt (HMA) Mixtures Containing Recycled or Waste Product Materials Using Performance Testing.* Publication FHWA-PA-2005-006+98-32(19). Pennsylvania Department of Transportation, Office of Planning and Research, 2005.
8. Shah, A., R. McDaniel, G. Huber, and V. Gallivan. Investigation of Properties of Plant-Produced RAP Mixtures. *Transportation Research Record: Journal of the Transportation Research Board*, 2007. 1998: 103–111.
9. Daniel, J., and J. Pochily. *Can We Make More Use of RAP? A Study of Extracted Binder Properties in NH, VT and Maine RAP Mixes.* http://www.neaupg.uconn.edu/pdf/09_Binder_Study_daniel-pochily.pdf. Accessed July 11, 2017.
10. American Society for Testing and Materials. *Annual Book of ASTM Standards.* Section 4, Volume 04.03, ASTM International, 2015.
11. Christensen, D., T. Pellinen, and R. Bonaquist. Hirsch Model for Estimating the Modulus of Asphalt Concrete. *Journal of the Association of Asphalt Paving Technologists*, Vol. 72, 2003, pp. 97–121.
12. Solaimanian, M., D. Fedor, R. Bonaquist, A. Soltani, and V. Tandon. Simple Performance Test for Mixture Damage Prediction in Asphalt Concrete. *Journal of the Association of Asphalt Paving Technologists*, Vol. 75, 2006, pp. 345–380.
13. Nadkarni, A., K. Kaloush, W. Zeiada, and K. Biligiri. Using Dynamic Modulus Test to Evaluate Moisture Susceptibility

of Asphalt Mixtures. *Transportation Research Record: Journal of the Transportation Research Board*, 2009. 2127: 29–35.

14. Mogawer, W. S., A. J. Austerman, E. Kassem, and E. Masad. Moisture Damage Characterization of Warm Mix Asphalt Mixtures. *Journal of the Association of Asphalt Paving Technologists*, Vol. 80, 2011, pp. 491–526.
15. Ozer, H., I. L. Al-Qadi, J. Lambros, A. Khatib, P. Singhvi, and T. Khan. Development of a Fracture-Based Flexibility Index for Asphalt Concrete Cracking Potential Using Modified Semi-Circle Bending Test Parameters. *Construction and Building Materials*, Vol. 115, 2015, pp. 390–401.
16. American Association of State Highway and Transportation Officials. *AASHTO Provisional Standards*. American Association of State Highway and Transportation Officials, Washington, D.C., 1993.
17. Marasteanu, M., K. H. Moon, E. Z. Teshale, A. C. Falchetto, M. Turos, W. Buttlar, E. Dave, G. Paulino, S. Ahmed, S. Leon, H. Bahia, A. Arshadi, H. Tabatabaee, J. Ojo, J. R. Velasquez, S. Mangiafico, C. Willaims, and A. Bus. *Investigation of Low Temperature Cracking in Asphalt Pavements National Pooled Fund Study -Phase II*. Final Report. Publication MN/RC 2012-23. Minnesota Department of Transportation, Saint Paul, Minn., 2012.

The Standing Committee on Non-Binder Components of Asphalt Mixtures (AFK30) peer-reviewed this paper (18-01235).

Article

TRR
JOURNAL OF THE TRANSPORTATION RESEARCH BOARD

Transportation Research Record
2018, Vol. 2672(28) 143–153

DOI: 10.1177/0361198118787940
journals.sagepub.com/home/trr
SAGE

Impact of Gyration Reduction and Design Specification Changes on Volumetric Properties of Virginia Dense-Graded Asphalt Mixtures

Stacey D. Diefenderfer[1], Benjamin F. Bowers[1], and Kevin K. McGhee[1]

Abstract

In 2015, the Virginia Department of Transportation (VDOT) proposed changes to their specification for asphalt mix design. The changes incorporated a reduction of design gyrations from 65 to 50; the addition of constraints on the No. 4 (4.75 mm) and No. 30 (600 μm) sieves; and adjustments to the production value for voids filled with asphalt and minimum voids in mineral aggregate. Before these modifications were fully adopted, a study was performed to assess the effect of the changes on mixture properties and laboratory performance. Eleven pairs of asphalt mixtures were evaluated; the pairs consisted of a typical VDOT 65-gyration mixture produced under the specification current at the time and a companion 50-gyration mixture designed and produced in accordance with the proposed specification. Mixtures were evaluated to determine the effect of the design parameters on volumetric properties, gradation, and permeability. The changes had very little effect on volumetric properties or gradation. Permeability effects were mixed, with 9.5 mm nominal maximum aggregate size mixtures requiring greater compaction to meet permeability requirements and 12.5 mm nominal maximum aggregate size mixtures showing improved permeability even at higher air-void contents. For the 50-gyration mixtures, core air voids were reduced, indicating the potential for increased durability attributable to improved densification in the field.

Periodically, the Virginia Department of Transportation (VDOT) updates their specifications for their asphalt mix design methodology to reflect advances in technology and address any current issues. VDOT adopted a modified version of the Superpave design methodology in 2000, in part to address the issue of rutting, which had been a widespread concern. This virtually eliminated rutting in Virginia, but concerns about the cracking durability of the Superpave mixtures have been widespread. As a result, a number of changes were made to the specification in an attempt to improve the durability of mixtures. When Superpave was adopted in 2000, VDOT decided to use compactive efforts of 65 and 75 gyrations in mix designs for different levels of traffic (*1*). In 2002, VDOT decided to require all mixtures to be designed with 65 gyrations as the compaction level (*2*). The number of gyrations was reduced to provide higher binder contents in an effort to increase durability. Stiffer binder types, instead of a higher compactive effort, were used to increase stability under higher traffic levels.

Despite these efforts, the perception of reduced durability for Superpave mixtures in Virginia remains. In 2003, Maupin found that the asphalt contents of a majority of nine evaluated mixtures could be increased in the laboratory up to 0.5% without compromising the integrity of the mixture (*3*). The increased asphalt content was used in plant-produced material designed to the optimal asphalt content in accordance with the specification in effect at the time. The laboratory evaluation indicated that, for most mixtures evaluated, the increased asphalt content would improve fatigue life. Laboratory rutting increased slightly, but rut depths remained well within the specification.

Maupin (*1*) undertook additional work in 2010 to address continued questions about mixture durability. Binder contents were evaluated to determine whether there were differences in quantities between VDOT's 9.5 mm Superpave surface mixture and two comparable Marshall mixtures. Results indicated that there was no significant difference in asphalt content. However, a laboratory evaluation determined that some mixtures may perform satisfactorily with increased asphalt content. Based on this finding, work

[1]Virginia Transportation Research Council, Charlottesville, VA

Corresponding Author:
Address correspondence to Stacey D. Diefenderfer:
stacey.diefenderfer@vdot.virginia.gov

Table 1. 9.5 mm NMAS Mix Designs

Property and gradation	Mix no. (binder type PG 64S-22)											
	C-65	C-50	D-65	D-50	F-65	F-50	I-65	I-50	J-65	J-50	K-65	K-50
Property												
Binder content, %	5.4	5.4	5.5	5.7	5.2	5.3	5.7	6.1	5.6	5.5	5.8	6.1
RAP content, %	30	30	26	26	25	30	30	30	30	30	25	25
Design VTM, %	4.0	4.0	3.7	3.7	3.6	4.0	4.0	3.9	3.6	4.0	4.1	3.8
Gradation												
1/2 in. (12.5 mm)	100	100	100	100	100	100	100	100	100	100	100	100
3/8 in. (9.5 mm)	93	93	94	94	94	94	98	99	91	90	97	97
No. 4 (4.75 mm)	61	60	63	63	62	61	64	85	60	60	69	69
No. 8 (2.36 mm)	39	41	44	44	40	40	47	46	38	38	47	47
No. 30 (600 μm)	NA	19	NA	NA[a]	NA	22	NA	21	NA	18	NA	22
No. 200 (75 μm)	5.4	5.4	5.6	5.6	5.4	5.3	5.2	5.2	6.3	6.9	5.6	5.6

Note: NMAS = nominal maximum aggregate size; RAP = reclaimed asphalt pavement; VTM = voids in total mix; NA = data not available.
[a]No data were provided, although the mix was designed to meet the specification.

continued to investigate means by which binder content could be increased for improved durability. In a laboratory study, Katicha and Flintsch investigated the rutting and cracking resistance of four plant mixtures modified to represent different gyration levels and binder contents (*4*). Rutting resistance was evaluated with the unconfined flow number test; cracking resistance was assessed with the indirect tensile strength test. Changes in binder content for the plant-produced mixtures resulted in mixed test responses, with some mixtures showing improved rutting resistance and some showing decreased resistance. For some mixtures it was possible to improve both rutting and cracking resistance by reducing the number of gyrations and thereby increasing binder content. Although results were encouraging, the study had one major limitation in that modifications were made to plant-produced material without redesigning the gradation; in practice, changes in design gyrations result in mixture gradation changes for mixture optimization rather than just the addition of binder. The current study was developed to assess proposed changes to the specification that resulted from the work of Katicha and Flintsch (*4*).

Purpose and Scope

This study investigated differences in volumetric properties and laboratory performance in mixtures to determine the effect of a proposed specification change. Eleven pairs of plant-produced mixtures were evaluated to determine whether the specification change requirements, which included the reduction from 65 to 50 design gyrations, for the percent passing the No. 4 (4.75 mm) and No. 30 (600 μm) sieves, and a change to the minimum voids in mineral aggregate (VMA), resulted in more durable mixtures. This paper reports the effect of the specification changes on mixture volumetric properties and compaction during construction.

Methods

Mixtures

The 11 pairs of plant-produced mixtures were designated with a letter (A to K) indicative of the mixture and a number (50 or 65) indicative of the design gyrations. Each pair was produced at the same plant. Mixtures were 9.5 mm or 12.5 mm nominal maximum aggregate size (NMAS) surface mixtures, and all contained reclaimed asphalt pavement (RAP) except Mixture B-65, which contained recycled asphalt shingles (RAS). Tables 1 and 2 summarize the mix designs. Mixtures were sampled from 10 asphalt plants throughout Virginia (mixtures E and K were sampled from the same plant). Figure 1 is a map of the paving locations.

Plant Sampling

Sampling of loose mixture was conducted at the plant during production. Because of the large amount of material sampled, approximately 0.5 tons per mixture, most samples were not collected from the truck bed. In most cases, to minimize delays in delivery of the mixture to the field, loose mixture was released from the silo into either a truck bed or a loader bucket and dumped on the ground for sampling to proceed in a safe manner. Care was taken during sampling to avoid any areas of apparent segregation or contamination resulting from the dumping process. Samples were stored in sample boxes in a climate controlled area until test specimens were manufactured.

Coring

Fifteen cores 6 in. (150 mm) in diameter were taken from each site for evaluation. Cores were taken from the center of the lane at approximately 100-ft intervals along the length of

Table 2. 12.5 mm NMAS Mix Designs

	Mix no. (binder type PG 64S-22 except as otherwise noted)									
Property and gradation	A-65	A-50	B-65	B-50	E-65	E-50	G-65[a]	G-50[a]	H-65	H-50
Property										
Binder content, %	5.3	5.5	6.0	6.1	5.2	5.6	5.9	6.1	5.2	5.4
RAP content, %	30	30	-	30	25	25	15	15	26	30
RAS content, %	-	-	2	-	-	-	-	-	-	-
Design VTM, %	4.0	3.9	4.0	4.1	3.2	4.2	4.0	3.5	4.0	4.0
Gradation										
3/4 in. (19.0 mm)	100	100	100	100	100	100	100	100	100	100
1/2 in. (12.5 mm)	96	95	97	96	98	98	96	96	97	98
3/8 in. (9.5 mm)	87	85	87	87	88	88	85	85	83	88
No. 4 (4.75 mm)	-	-	-	62	-	66	-	59	-	58
No. 8 (2.36 mm)	35	36	34	47	40	48	38	38	38	39
No. 30 (600 μm)	NA	NA[b]	NA	19	NA	23	NA	18	NA	21
No. 200 (75 μm)	6.5	6.7	6.5	6.8	5	5	5.6	5.6	5.6	5.4

Note: NMAS = nominal maximum aggregate size; RAP = reclaimed asphalt pavement; RAS = recycled asphalt shingles; VTM = voids in total mix; NA = no data provided.
[a]Binder type was PG 64E-22.
[b]No data were provided, although the mix was designed to meet the specification.

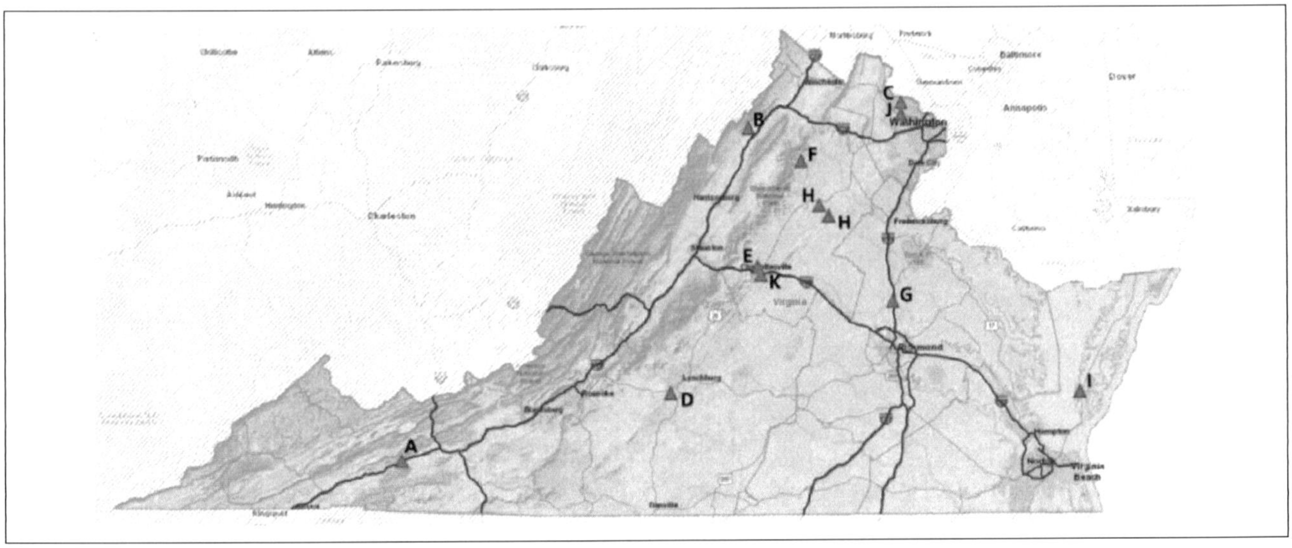

Figure 1. Geographic distribution of paving locations.

the pavement site to minimize the long-term effects of coring. This length included tonnage adjacent to the plant-sampled loose mixtures. With the exception of Site G, cores were collected at the time of construction; Site G cores were collected within approximately 2 weeks of construction.

Volumetric Analysis

Volumetric analyses were performed to determine fundamental mixture properties.

Properties included maximum theoretical specific gravity (G_{mm}), bulk specific gravity (G_{sb}), voids in total mix (VTM), voids filled with asphalt (VFA), and VMA.

Asphalt Content and Gradation

Asphalt content was determined using the ignition furnace in accordance with AASHTO T 308, Determining the Asphalt Binder Content of Hot-Mix Asphalt by the Ignition Method. Sieve analysis was performed on the aggregate in accordance with AASHTO T 30, Mechanical Analysis of Extracted Aggregate.

Core Air Voids

Air-void contents were determined in accordance with AASHTO T 269, Percent Air Voids in Compacted Dense and Open Asphalt Mixtures.

Table 3. Volumetric Properties for 9.5 mm NMAS Mixtures

	Mix no.											
Volumetric property	C-65	C-50	D-65	D-50	F-65	F-50	I-65	I-50	J-65	J-50	K-65	K-50
Asphalt content, %	5.38	5.44	5.81	5.50	5.13	5.30	5.37	5.94	5.57	5.74	5.71	6.13
Rice specific gravity, G_{mm}	2.724	2.695	2.533	2.527	2.697	2.663	2.514	2.513	2.723	2.702	2.565	2.545
VTM, %	4.8	3.7	4.3	4.3	5.2	4.7	4.7	3.5	4.0	3.8	3.4	2.5
VMA, %	17.6	16.9	17.4	16.7	17.2	17.4	16.8	17.0	18.1	18.3	15.6	15.8
VFA, %	73.0	78.0	75.3	73.9	69.6	72.8	72.1	79.4	78.0	79.0	78.5	84.1
FA ratio	1.16	1.20	1.07	1.21	1.10	1.08	1.12	0.85	1.19	1.13	1.09	1.00
Mixture bulk specific gravity, G_{mb}	2.594	2.595	2.424	2.417	2.556	2.537	2.396	2.425	2.614	2.598	2.479	2.481
Aggregate effective specific gravity, G_{se}	3.004	2.971	2.783	2.760	2.955	2.922	2.738	2.764	3.015	2.998	2.820	2.815
Aggregate bulk specific gravity, G_{sb}	2.980	2.954	2.764	2.741	2.928	2.908	2.725	2.748	3.015	2.998	2.771	2.766
Absorbed binder content, P_{ba}, %	0.28	0.20	0.25	0.26	0.32	0.17	0.18	0.22	0.00	0.00	0.65	0.65
Effective binder content, P_{be}, %	5.11	5.25	5.57	5.26	4.83	5.14	5.20	5.74	5.57	5.74	5.10	5.52
Effective film thickness, F_T, microns	9.0	8.8	9.8	8.8	8.7	9.0	9.3	11.6	9.7	10.1	8.9	9.7

Note: VTM = voids in total mix; VMA = voids in mineral aggregate; VFA = voids filled with asphalt; FA ratio = fines/asphalt ratio.

Table 4. Volumetric Properties for 12.5 mm NMAS Mixtures

	Mix no.									
Volumetric property	A-65	A-50	B-65	B-50	E-65	E-50	G-65	G-50	H-65	H-50
Asphalt content, %	5.55	5.49	5.56	5.64	5.99	6.87	6.20	6.58	5.27	5.52
Rice specific gravity, G_{mm}	2.501	2.513	2.626	2.627	2.569	2.556	2.484	2.486	2.749	2.749
VTM, %	3.6	6.6	3.0	6.1	1.9	1.6	5.4	4.9	2.7	4.5
VMA, %	16.2	18.7	15.1	18.6	15.6	17.4	18.5	19.3	16.1	16.9
VFA, %	77.6	64.7	80.1	67.3	87.6	90.6	72.7	74.5	83.2	73.5
FA ratio	1.49	1.26	1.22	1.59	0.97	0.93	0.93	1.28	1.10	1.04
Mixture bulk specific gravity, G_{mb}	2.411	2.347	2.547	2.467	2.520	2.514	2.358	2.364	2.674	2.626
Aggregate effective specific gravity, G_{se}	2.731	2.743	2.890	2.895	2.840	2.869	2.740	2.761	3.030	3.046
Aggregate bulk specific gravity, G_{sb}	2.718	2.729	2.835	2.860	2.806	2.835	2.719	2.737	3.020	2.986
Absorbed binder content, P_{ba}, %	0.18	0.19	0.69	0.44	0.44	0.43	0.29	0.33	0.11	0.68
Effective binder content, P_{be}, %	5.38	5.31	4.91	5.22	5.57	6.47	5.92	6.27	5.16	4.88
Effective film thickness, F_T, microns	8.8	10.0	8.8	7.3	11.2	12.3	11.6	9.8	9.0	9.2

Note: VTM = voids in total mix; VMA = voids in mineral aggregate; VFA = voids filled with asphalt; FA ratio = fines/asphalt ratio.

Permeability

Permeability testing was performed on cores in accordance with Virginia Test Method 120, Method of Test for Measurement of Permeability of Bituminous Paving Mixtures Using a Flexible Wall Permeameter (*5*).

Results and Discussion

Volumetric and gradation analyses were performed for the plant mixture sampled during production. Results are shown in tables 3 and 4 for the 9.5 mm NMAS and 12.5 mm NMAS mixtures, respectively. Each 65-gyration mixture was a typical mixture produced by the contractor under VDOT's existing specification. The companion 50-gyration mixture was redesigned to meet the proposed specification. Tables 5 and 6 summarize the design requirements for each gyration level.

Figure 2 shows the 50-gyration results plotted versus the 65-gyration results for asphalt content, effective asphalt content, film thickness, VMA, and VFA for each mixture. Asphalt content, effective asphalt content, and VMA were slightly higher for the majority of the 50-gyration mixtures. The results for film thickness and VFA were more mixed, with four mixtures showing reductions in film thickness and VFA.

Paired two-sample Student's *t*-test analyses were performed to assess the significance of the differences, as summarized in Table 7. With the exception of asphalt content, the differences in properties were not statistically significant. However, the statistical evaluation does not consider the

Table 5. Design Volumetric Requirements for 65- and 50-Gyration Mixtures

No. of gyrations (ndesign)	Mix type	VTM (%) production	VFA (%) design	VFA (%) production[a]	Minimum VMA (%)	Fines/asphalt ratio[b]
65	SM-9.5A	2.0–5.0[c]	73–79	68–84	15	0.6–1.2
65	SM-9.5D	2.0–5.0[c]	73–79	68–84	15	0.6–1.2
65	SM-9.5E	2.0–5.0[c]	73–79	68–84	15	0.6–1.2
65	SM-12.5A	2.0–5.0[c]	70–78	65–83	14	0.6–1.2
65	SM-12.5D	2.0–5.0[c]	70–78	65–83	14	0.6–1.2
65	SM-12.5E	2.0–5.0[c]	70–78	65–83	14	0.6–1.2
50	SM-9.5A	3.0-6.0[d]	na	65–80	16	0.7–1.3
50	SM-9.5D	3.0–6.0[d]	na	65–80	16	0.7–1.3
50	SM-9.5E	2.0–5.0[d]	na	70–85	16	0.7–1.3
50	SM-12.5A	3.0–6.0[d]	na	65–80	16	0.7–1.3
50	SM-12.5D	3.0–6.0[d]	na	65–80	16	0.7–1.3
50	SM-12.5E	2.0–5.0[d]	na	70–85	16	0.7–1.3

Note: VTM = voids in total mix; VFA = voids filled with asphalt; VMA = voids in mineral aggregate; na = no requirement.
[a]During production of an approved job mix, the VFA must be controlled within these limits.
[b]Fines-asphalt ratio is based on effective asphalt content.
[c]Asphalt content should be selected at 4.0 % air voids.
[d]Asphalt content should be selected at 4.0% air voids for A and D mixes and at 3.5% for E mixes.

Table 6. Design Gradation Requirements for 65- and 50-Gyration Mixtures (% Passing Sieve)

Design gyrations	Mix type	Sieve size						
		3/4 in.	1/2 in.	3/8 in.	No. 4	No. 8	No. 30	No. 50
65	SM-9.5	na	100	90–100	na	38–67	na	na
65	SM-12.5	100	95–100	90 max.	na	34–50	na	na
50	SM-9.5	na	100	90–100	58–80	38–67	23 max.	na
50	SM-12.5	100	95–100	90 max.	58–80	34–50	23 max.	na

practical aspects of differences in properties (such as acceptable variability), which may be assessed using the acceptance tolerance as described in the VDOT *Road and Bridge Specifications* (*2*). In this case, only the asphalt content is measured for acceptance; the process tolerance is defined as ±0.60% from the design binder content for a single test. Practically speaking, this means that a single test may vary from the design content by up to 0.60% without being considered unacceptably "different" and subject to penalty. Investigation of the volumetric properties presented in tables 3 and 4 shows that only one mixture pair (Mixture E) had a difference in asphalt content between the 65-gyration and 50-gyration mixtures that exceeded 0.60%; thus, the remaining pairs of mixtures have asphalt contents that fall within the tolerance to be considered the same asphalt content.

Gradation Analysis

Figure 3 presents the gradations of the 9.5 mm NMAS mixtures. Figure 3 shows that most of the 9.5 mm NMAS mixtures had generally similar gradations. Mixtures K-65 and K-50 deviated from the other gradations at the 4.75 mm and 2.36 mm sieves and were finer at those sieves. Mixtures J-65 and J-50 were coarser at the 9.5 mm sieve and followed that trend until converging with the other gradations at the 0.30 mm sieve. The other differences in the gradations generally occurred between the 2.36 mm and 0.60 mm sieves.

Figure 4 shows the gradations of the 12.5 mm NMAS mixtures. These gradations were similar except for Mixture B-50, which was finer in the range of the 4.75 mm through 0.30 mm sieves.

Tables 8 and 9 present the detailed gradation results. Although some of the test values did not meet the design requirements, all values met the tolerance for acceptance. Overall, it is clear that the gradations did not change to a great degree between the 65- and 50-gyration mixtures. This is further illustrated in figures 4 and 5, which show the difference in gradation at each sieve between the 50- and 65-gyration 9.5 mm NMAS and 12.5 mm NMAS mixtures, respectively.

Figures 5 and 6 show the difference in percent of aggregate passing each sieve between the 65- and 50-gyration mixtures. For each sieve, positive values indicate that there is more material passing for the 65-gyration mixture than for the 50-gyration mixture, with negative values signifying the opposite. Figure 5 shows that the maximum change in

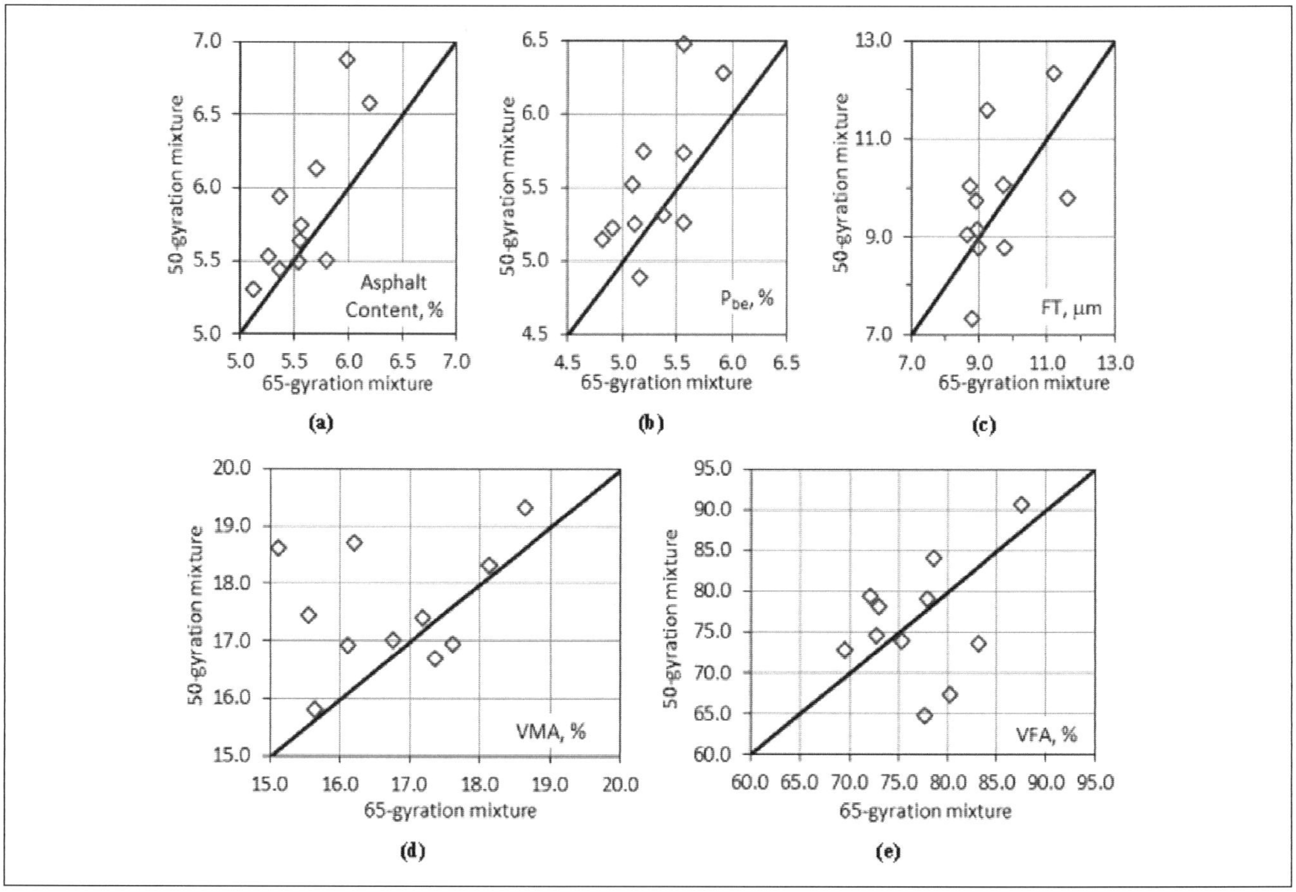

Figure 2. Differences in (*a*) asphalt content, (*b*) effective binder content, (*c*) film thickness, (*d*) VMA, and (*e*) VFA between 50-gyration and 65-gyration mixture pairs.

Table 7. Paired Two-Sample Student's *t*-test Analyses of Properties (α = 0.05)

Property	*P*(T <= *t*) two-tail	Outcome
Asphalt content	0.034	Significant difference
Effective asphalt content	0.063	No significant difference
Film thickness	0.684	No significant difference
VMA	0.075	No significant difference
VFA	0.699	No significant difference

percent passing any sieve for the 9.5 mm NMAS mixtures was approximately 4%. Mixture I showed a positive difference in percent passing for all sieves smaller than the 9.5 mm sieve, indicating that the mixture coarsened. Changes in mixtures C and D were mixed, although both followed a similar trend. Mixtures F-50 and J-50 became finer at most sieves. Mixture K showed very little change, and it should be noted that mixtures K-65 and K-50 were produced with the same mix design, so the changes seen are indicative of production variability.

Figure 6 shows that changes in gradation were greater for the 12.5 mm NMAS mixtures than for the 9.5 mm NMAS mixtures, with a maximum change in percent passing of approximately 8%. Mixture B-50 had increases in the percent passing for the 50-gyration mixture (a negative chage on the graph), indicating that the mixture became finer. Mixtures A-50 and G-50 became slightly coarser. Mixture H showed minimal changes in gradation, as did Mixture E. Again, as with Mixture K, mixtures E-65 and E-50 were produced with the same mix design, so changes are indicative of production variability.

Field Air Voids

Air voids were measured for each field core collected. It was noted in the field that the same compaction effort was applied to each pair of mixtures. A histogram was constructed by

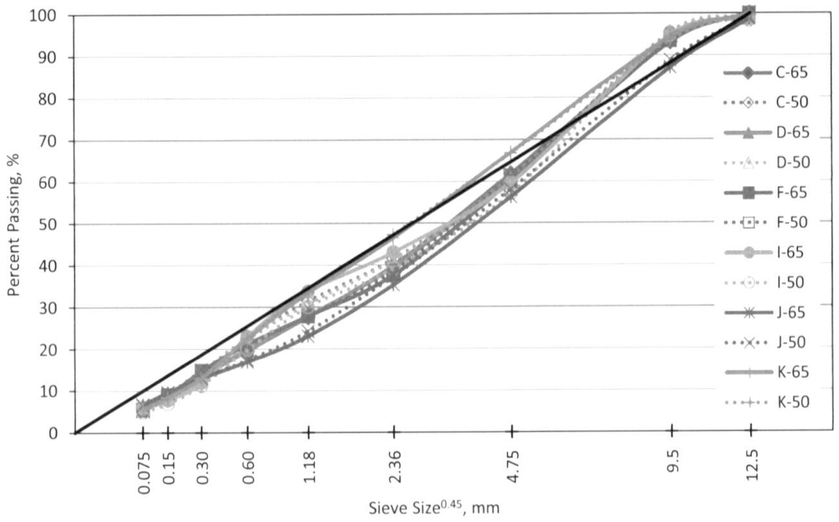

Figure 3. Gradations for 9.5 mm NMAS mixtures.

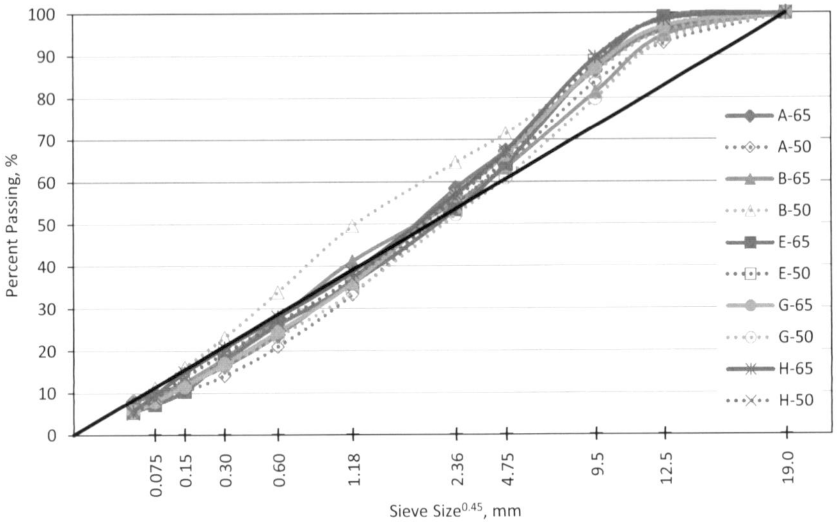

Figure 4. Gradations for 12.5 mm NMAS mixtures.

Table 8. Gradations for 9.5 mm NMAS Mixture (% Passing Sieve)

	Mix no.												
Sieve size	C-65	C-50	D-65	D-50	F-65	F-50	I-65	I-50	J-65	J-50	K-65	K-50	Design requirement
1/2 in. (12.5mm)	100.0	98.8	100.0	99.5	100.0	99.9	98.7	98.3	98.7	99.0	99.8	99.5	100
3/8 in. (9.5mm)	93.2	95.6	93.5	95.5	93.5	94.1	94.8	95.4	87.2	88.9	94.5	93.8	90–100
No. 4 (4.75mm)	61.8	60.0	60.9	60.0	59.9	61.2	59.6	57.1	56.3	58.1	66.9	66.7	58–80[a]
No. 8 (2.36mm)	39.9	41.5	39.5	41.1	37.6	38.6	43.0	39.2	35.3	37.3	46.5	47.2	38–67
No. 16 (1.18mm)	27.7	31.5	27.9	30.8	27.9	28.0	33.8	30.1	23.1	24.3	33.4	33.7	
No. 30 (600 μm)	19.6	22.7	19.9	22.3	21.1	21.1	23.0	19.9	16.9	17.4	22.4	22.5	23 max.[a]
No. 50 (300 μm)	13.4	14.6	13.7	14.6	14.7	14.8	11.8	11.1	13.0	13.1	13.0	12.9	
No. 100 (150 μm)	8.9	8.7	8.9	9.0	8.7	9.0	7.7	6.9	9.6	9.5	8.3	8.3	
No. 200 (75 μm)	5.9	6.3	6.0	6.3	5.3	5.5	5.8	4.9	6.6	6.5	5.6	5.5	

[a]Requirements for No. 4 and No. 30 sieves apply only to 50-gyration mixtures.

Table 9. Gradations for 12.5 mm NMAS Mixture (% Passing Sieve)

	Mix no.										
Sieve size	A-65	A-50	B-65	B-50	E-65	E-50	G-65	G-50	H-65	H-50	Design requirement
3/4 in. (19.0 mm)	100.0	100.0	100.0	100.0	100.0	100.0	100.0	100.0	100.0	100.0	100
1/2 in. (12.5 mm)	96.0	93.0	94.8	95.7	99.0	98.6	96.9	93.7	98.4	98.6	95–100
3/8 in. (9.5 mm)	86.5	83.8	81.3	86.1	87.6	88.6	87.2	79.7	89.6	89.7	90 max.
No. 4 (4.75 mm)	58.5	54.7	55.1	64.6	53.3	54.7	56.8	52.2	57.1	56.1	58–80[a]
No. 8 (2.36 mm)	36.8	33.3	41.2	49.6	35.9	36.3	36.0	33.7	38.6	37.2	34–50
No. 16 (1.18 mm)	23.8	21.0	27.6	33.8	26.1	26.3	24.5	23.3	28.0	26.7	
No. 30 (600 μm)	16.3	14.1	18.2	23.1	17.7	18.1	16.9	16.5	20.7	19.4	23 max.[a]
No. 50 (300 μm)	12.2	10.4	12.4	16.1	10.4	11.0	11.6	11.5	14.8	13.7	
No. 100 (150 μm)	9.7	8.2	8.6	11.2	7.2	7.9	7.9	8.0	9.5	8.5	
No. 200 (75 μm)	8.0	6.7	6.0	8.3	5.4	6.0	5.5	5.7	5.7	5.1	

[a]Requirements for No. 4 and No. 30 sieves apply only to 50-gyration mixtures.

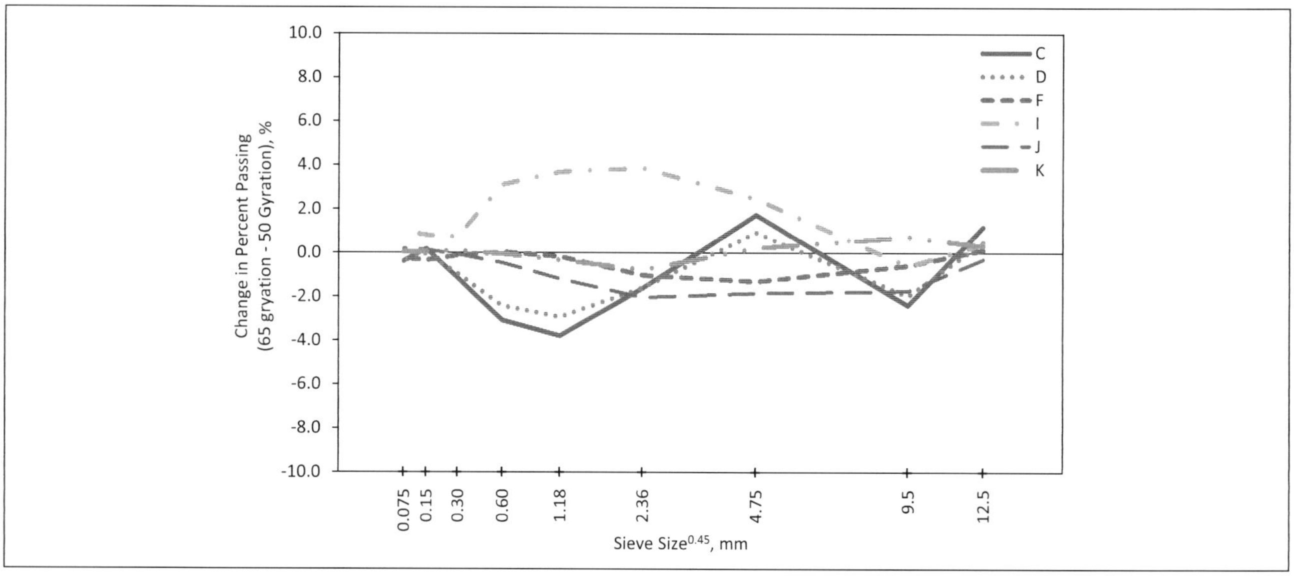

Figure 5. Change in gradation by sieve for 9.5 mm NMAS mixtures.

rounding each measurement to the nearest percent and is shown in Figure 7. From the histogram, the 50-gyration mixtures showed an improvement in compactability, as the peak of the curve is shifted toward the lower air-void contents. This is a positive outcome, as it is well documented that increased in-place density has a direct effect on the extension of pavement service life (*6*). Correlation analyses were conducted to assess potential relationships among mixture gradation, asphalt content, and average void content; no significant relationships were found.

Permeability

The results from laboratory permeability testing of field cores are shown in Figure 8. The 165 65-gyration field cores and the 166 50-gyration field cores were tested. The specimens with zero permeability are included in the figure. Sixteen percent of the 65-gyration cores and 37% of the 50-gyration cores had zero permeability. Correlation analyses were conducted to assess potential relationships among mixture gradation, asphalt content, and average permeability; no significant relationships were found.

In addition, trend lines were determined for the permeability results for each mixture. The data used to determine the trend lines did not include the zero permeability data points, as the exponential trend most appropriate for permeability data cannot incorporate those points. These expressions were used to predict the air voids at 150 x 10^5 cm/s permeability, which is the limit used for VDOT's permeability specification (*4*). Table 10 presents the results. The air voids at failing permeability were reduced for the 50-gyration mixtures in comparison to the 65-gyration mixtures for four of the five

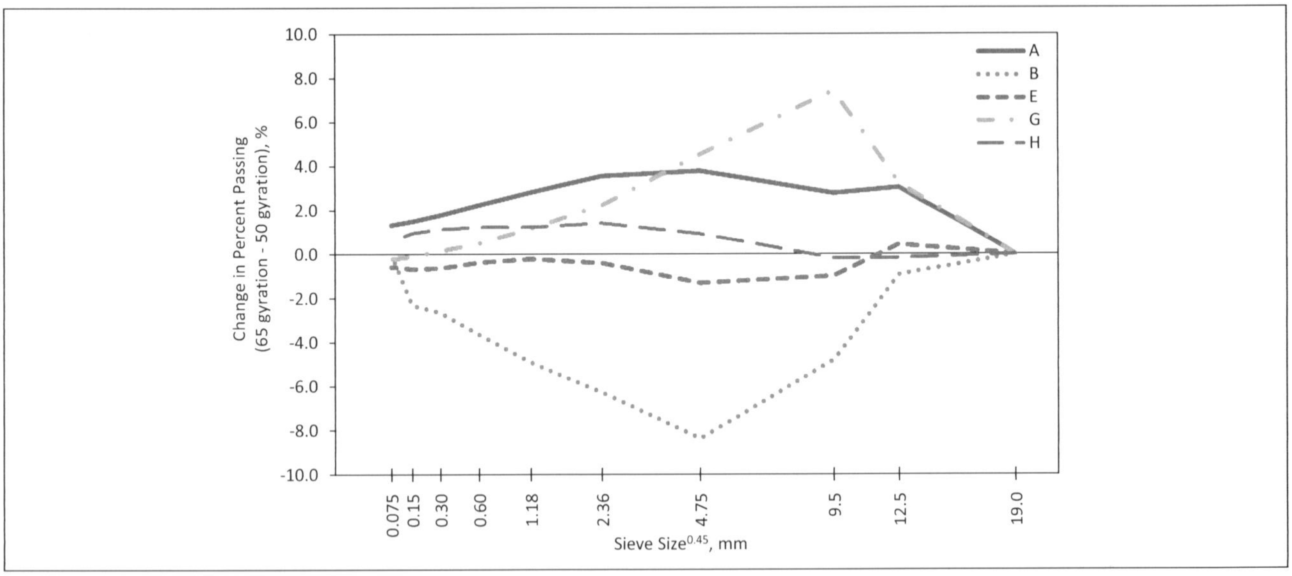

Figure 6. Change in gradation by sieve for 12.5 mm NMAS mixtures.

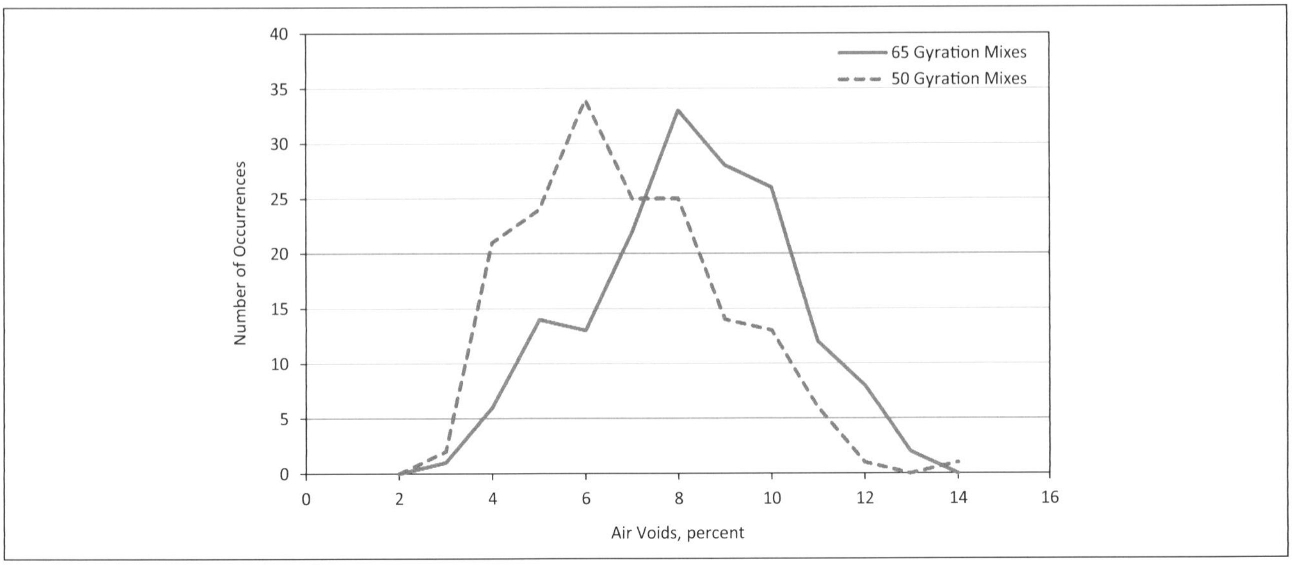

Figure 7. Air-void contents of field cores.

9.5 mm NMAS mixtures evaluated. This indicates that the 9.5 mm NMAS 50-gyration mixtures must be densified further than the 65-gyration mixtures to meet the specification. The opposite trend was seen for the 12.5 mm mixtures. In this case, the 50-gyration mixtures met the permeability requirement at higher void contents than the 65-gyration mixtures, indicating that these mixtures show improved permeability at higher void contents.

Findings and Conclusions

Eleven pairs of dense-graded asphalt surface mixtures were assessed to evaluate the effect of changes to VDOT's specification, including the requirements for VFA production, minimum VMA, and fines/asphalt ratio; the reduction in design gyrations from 65 to 50; and the addition of requirements for percent passing the No. 4 and No. 30 sieves. The findings were as follows:

- Changes in the VDOT specification resulted in slight changes to mixture volumetric properties. Asphalt contents changed significantly as determined by the Student *t*-test; however, the increase or decrease in asphalt content was mixture dependent and was not practically significant, as determined by the process tolerance for mixture acceptance. No statistically

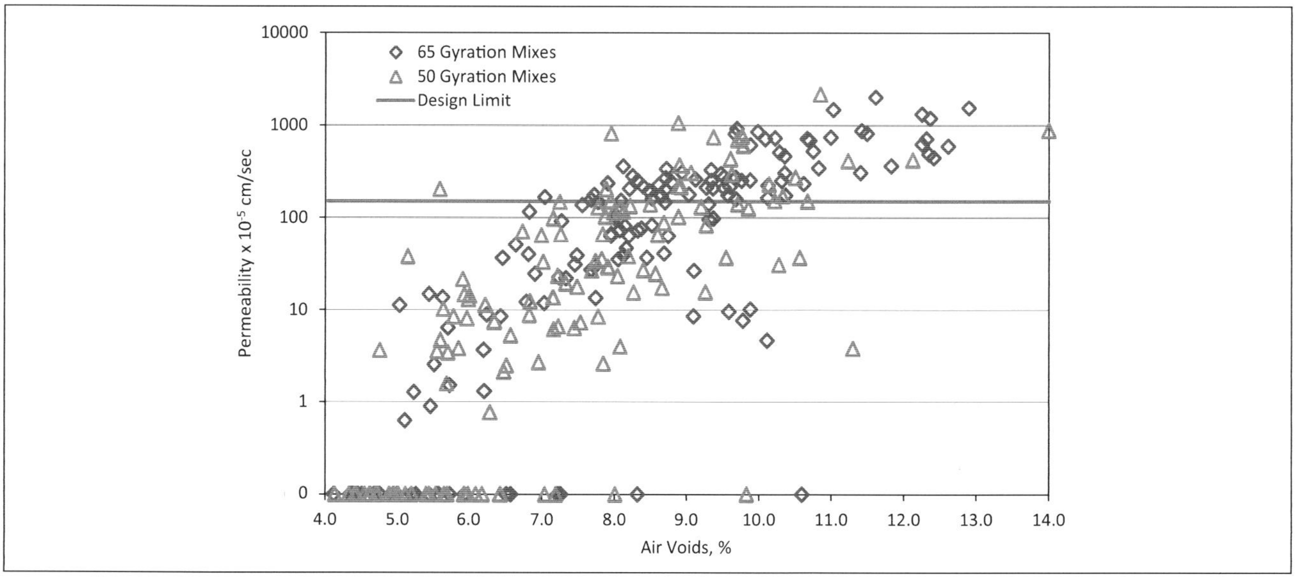

Figure 8. Permeability results for cores.

Table 10. Predictions of Void Contents at Failing Permeability

NMAS	Mixture	Number of cores	Cores having 0 permeability	Permeability trend line	R^2	Predicted air voids at 150 x 10^5 cm/s permeability[a]
9.5 mm	C-65	15	14	na[b]	na	na
	C-50	16	10	na[b]	na	na
	D-65	15	1	$y = 0.5935e^{0.48x}$	0.1904	11.5
	D-50	15	7	$y = 0.002e^{1.2911x}$	0.7425	8.7
	F-65	15	0	$y = 0.0266e^{0.9218x}$	0.7756	9.4
	F-50	15	0	$y = 0.5048e^{0.7124x}$	0.5885	8.0
	I-65	15	0	$y = 0.0694e^{0.6452x}$	0.5192	11.9
	I-50	15	9	$y = 0.0175e^{0.8637x}$	0.7861	10.5
	J-65	15	2	$y = 0.0668e^{0.806x}$	0.4198	9.6
	J-50	15	5	$y = 0.0036e^{1.2749x}$	0.8971	8.3
	K-65	15	0	$y = 0.4383e^{0.5724x}$	0.8846	10.2
	K-50	15	0	$y = 0.154e^{0.6668x}$	0.8729	10.3
12.5 mm	A-65	15	1	$y = 0.0332e^{0.923x}$	0.6018	9.1
	A-50	15	4	$y = 0.0318e^{0.7776x}$	0.2398	10.9
	B-65	15	9	$y = 0.0029e^{1.2804x}$	0.6481	8.5
	B-50	15	11	$y = 4.1924e^{0.3448x}$	0.1699	10.4
	E-65	15	0	$y = 0.0808e^{0.8658x}$	0.7816	8.7
	E-50	15	3	$y = 0.1392e^{0.7987x}$	0.6929	8.7
	G-65	15	0	$y = 0.3294e^{0.7287x}$	0.8427	8.4
	G-50	15	3	$y = 0.1082e^{0.5683x}$	0.7602	12.7
	H-65	15	0	$y = 1.7805e^{0.564x}$	0.8068	7.9
	H-50	15	10	na[b]	na	na

Note: na = not available.
[a]VDOT requires that mixtures meet a maximum predicted permeability of 150 x 10^{-5} cm/s at 7.5% air voids (*4*).
[b]Trend lines and predictions could not be determined because of the lack of data points.

significant differences were found for effective asphalt content, binder film thickness, VMA, or VFA.

- Mixture gradations changed only slightly with the additional requirements for percent passing the No. 4 and No. 30 sieves. With the exception of one mixture, the 9.5 mm NMAS mixtures were slightly finer, and the effect on the 12.5 mm NMAS mixtures was mixed.

- The changes resulted in an improved ability of the 50-gyration mixtures to compact in the field. Although the same compaction effort was used for each pair of mixtures, the field cores for the 50-gyration mixture had lower air-void contents than those for the 65-gyration mixtures.
- Permeability generally improved for the 50-gyration mixtures, as a higher percentage of field cores had zero permeability. The 9.5 mm NMAS 50-gyration mixtures required greater field densification to meet permeability requirements than did the 65-gyration mixtures. The 50-gyration 12.5 mm NMAS mixtures met permeability requirements at higher field air-void contents than the 65-gyration mixtures.

In conclusion, although the specification changes did not result in practically significant increases in asphalt content, the results indicated that the changes to the specification had positive outcomes, particularly as related to field density. This is expected to improve the durability of Virginia's dense-graded surface mixtures.

In order to improve mixture durability by increasing asphalt content, more substantial changes to specifications are necessary. Addressing mixture volumetric properties is likely to be the least efficient manner, as the low-bid system of procurement does not provide an incentive for increased asphalt contents since the binder is the most costly mixture component. The use of performance-based specifications requiring mixtures to meet performance criteria may lead to improved durability if the performance tests and criteria are appropriately chosen.

Acknowledgments

The authors thank the Virginia Transportation Research Council (VTRC) and the FHWA for their support of this research. The authors thank John Burch, Troy Deeds, Donnie Dodds, Ben Earl, Zack Robb, and Jennifer Samuels of VTRC and Ken Elliton, Mark Potter, and Guido Zaga, formerly of VTRC, for their efforts. The assistance of personnel from VDOT's Bristol, Culpeper, Hampton Roads, Lynchburg, Northern Virginia, Richmond, and Staunton districts was greatly appreciated as were the efforts of Trenton Clark, Virginia Asphalt Association, and all participating contractors. Appreciation is also extended to Linda Evans and André Surles of VTRC for editorial and graphics assistance, respectively.

Author Contributions

The authors confirm contribution to the paper as follows: study conception and design: SD, KM; data collection: SD, KM, BB; analysis and interpretation of results: SD, BB, KM; draft manuscript preparation: SD, BB. All authors reviewed the results and approved the final version of the manuscript.

References

1. Maupin, G. W., Jr. *Investigation of Optimized Mixture Design for Superpave Surface Mixtures*. Publication VTRC 11-R5. Virginia Transportation Research Council, Charlottesville, 2010.
2. Virginia Department of Transportation. *Road and Bridge Specifications*. Richmond, 2002.
3. Maupin, G. W., Jr. *Additional Asphalt to Increase the Durability of Virginia's Superpave Surface Mixtures*. Publication VTRC 03-R15. Virginia Transportation Research Council, Charlottesville, 2003.
4. Katicha, S. W., and G. W. Flintch. *Improving Mixture Durability through Design Gyrations, Air Voids, and Binder Content*. Publication VTRC 16-R17. Virginia Transportation Research Council, Charlottesville, 2016.
5. Virginia Department of Transportation, Materials Division. *Virginia Test Methods*. Richmond, 2013.
6. Tran, N., P. Turner, and J. Shambley. *Enhanced Compaction to Improve Durability and Extend Pavement Service Life: A Literature Review*. Publication 16-02R. National Center for Asphalt Technology, Auburn, Ala., 2016.

The Standing Committee on Surface Requirements of Asphalt Mixtures (AFK40) peer-reviewed this paper (18-04756).

Article

TRR
JOURNAL OF THE TRANSPORTATION RESEARCH BOARD

Transportation Research Record
2018, Vol. 2672(28) 154–166
© National Academy of Sciences: Transportation Research Board 2018
Article reuse guidelines:
sagepub.com/journals-permissions
DOI: 10.1177/0361198118784377
journals.sagepub.com/home/trr
SAGE

Influence of Recycled Asphalt Pavement on Interfacial Energy and Bond Strength of Asphalt Binder for Different Types of Aggregates

Ayyanna Habal[1] and Dharamveer Singh[1]

Abstract

The bond between aggregate and asphalt binder depends on many factors, such as chemistry of materials, nature of aggregates, and quality of asphalt binder. The present study evaluates the effects of recycled asphalt pavement (RAP) on the bonding behavior of asphalt binder for acidic (granite and quartzite) and basic (basalt and limestone) aggregates. AC30 binder was blended with different proportions of RAP binder (15%, 25%, and 40%). The interfacial energy and bond strength of 16 combinations (four asphalt binders × four aggregates) of aggregate-asphalt binder system were evaluated based on a surface free energy (SFE) approach and binder bond strength (BBS) test, respectively. One of the strong motivations of the present study was to evaluate how interfacial energy parameters calculated based on the SFE of an aggregate-asphalt binder system may correlate with the outcome obtained from a BBS test. The results showed that AC30 binder modified with RAP binder had a good performance for basic aggregates compared with acidic aggregates. Basic aggregates (limestone and basalt) showed cohesive failure before and after conditioning, indicating minimal effects of water on interfacial bond strength. However, acidic aggregates (granite and quartzite) showed cohesive failure in dry conditions and adhesive failure after wet conditioning, implying detrimental effects of water on interfacial bond strength for these aggregates. The failure pattern of aggregates in cohesion was supported by a reduction in cohesion energy of asphalt binder after the addition of RAP binder. Correlation between interfacial energy parameters (obtained from SFE) and BBS strength depends on aggregate types and failure patterns.

It is essential to ensure a good bond between aggregate and asphalt binder to have a moisture resistant mix. Usually, indirect test methods such as indirect tensile strength ratio, Hamburg wheel tracking, and static and boiling are used to evaluate bonding or moisture resistance potential of an aggregate-asphalt binder system (*1*, *2*). However, none of the test methods address expected failure patterns (cohesive or adhesive) or the role of chemistry of materials. The bond between aggregate and asphalt binder depends on many factors, such as the chemistry of materials, the nature of aggregates, and the quality of the asphalt binder. For example, a selected asphalt binder may act differently with granite, limestone, basalt and quartzite, and sandstone aggregates. Further, the response of asphalt binder toward an aggregate may alter after the addition of a modifier or additives (e.g., anti-stripping, anti-aging, recycled asphalt binder, polymer, etc.). Thus, the role of surface chemistry and type of aggregate and asphalt binders cannot be denied to ensure a good bond. Recently two new test protocols have been developed to address the bonding behavior of asphalt binder and aggregates: (i) the implementation of a surface free energy (SFE) concept to quantify cohesion and adhesion bond energy of different aggregate-asphalt binder systems (*1*–*6*), (ii) the use of a binder bond strength (BBS) test to measure bond strength between asphalt binder and aggregates (*7*). The BBS test can help to identify types of failure (cohesion and adhesion) of an asphalt-aggregate system before and after water conditioning. The BBS test is repeatable and reproducible and effectively measures the effect of moisture conditioning (*7*).

Theoretically SFE seems to be a promising approach to evaluate the compatibility of the aggregate-asphalt binder; however, it was emphasized this approach should be validated based on the mechanical performance of

[1]Department of Civil Engineering, Indian Institute of Technology Bombay, Powai, Mumbai, India

Corresponding Author:
Address correspondence to Ayyanna Habal: ayyanna.habal@gmail.com

compacted asphalt mixes. Hamedi and Nejad (*1*) correlated four different energy parameters derived based on the SFE of aggregate-asphalt binder with tensile strength ratio (TSR) of compacted mix. They reported that energy parameters had a good correlation with TSR results. Similarly, many studies by Arabani et al. (*3*), Bhasin et al. (*8*), and Howson et al. (*9*) showed that the energy parameters obtained using the SFE concept can have a good correlation with laboratory performance parameters of asphalt mixes, namely, dynamic modulus, resilient modulus ratio, fatigue ratio, and work of fracture. In another study, Moraes et al. (*10*, *11*) showed that a good relationship exists between BBS and SFE test methods (*11*). However, as per the authors' knowledge, limited studies have been conducted correlating SFE parameters of an aggregate-asphalt binder system with strength parameters obtained from a BBS test, which provides a direct measurement of bond strength and type of failures. Therefore, one of the strong motivations of the present study was to evaluate how interfacial energy parameters calculated based on the SFE of an aggregate-asphalt binder system may correlate with an outcome obtained from a BBS test. Further, the present study addresses the effects of recycled asphalt pavement (RAP) binder on the bonding behavior of asphalt binder for four different types of aggregates (granite, limestone, basalt, and quartzite). These four aggregates were selected considering their acidic and basic nature. Obviously, RAP binder being aged and stiff may alter the chemical composition and properties of virgin asphalt binder, which can influence bonding behavior. The authors could not find a study that brings both SFE and BBS tests together for a binder containing RAP.

In the present study, AC30 binder was blended with different proportions of RAP binder (0, 15, 25, and 40%). The first phase of the work was to estimate the SFE of aggregate and asphalt binders selected in the study. The SFE components of asphalt binders were measured using the Wilhelmy plate (WP) method. Likewise, the SFE components of basalt, limestone, granite, and quartzite aggregates were measured using a dynamic vapor sorption (DVS) device. A total of 16 combinations of aggregate-asphalt binder systems were selected (four asphalt binder × four aggregates). The interfacial energy of all 16 combinations of aggregate-asphalt binder system was calculated based on SFE components. The second phase of the present work was to measure the bond strength of all 16 combinations of aggregate-asphalt binder in dry and wet conditions using a BBS test. Bond strength ratio (BSR) was calculated for each selected combination based on wet and dry bond strength measured through a BBS test. Also, this paper presents a comparison between bond strength obtained through BBS- and SFE-based interfacial energy parameters. It is expected that the outcome of this study will help with understanding the effects of RAP binder on bond strength, interfacial energy, and failure pattern with different types of aggregates.

Objectives

The objectives of the present study were to:

1. Evaluate the effects of RAP binder and aggregate type on the interfacial energy parameters of aggregate-asphalt binder systems using the SFE approach
2. Evaluate pull-off tensile strength (POTS) and failure pattern of basalt, limestone, granite, and quartzite aggregates for asphalt binder containing RAP binder in dry and wet conditions using a BBS test
3. Compare POTS of different aggregate-asphalt binder combinations obtained through a BBS test with interfacial energy parameters estimated using the SFE approach

Materials

Virgin asphalt binder (AC30), RAP binder, and four different types of aggregates (basalt, limestone, granite, and quartzite) were considered in the study. It was found that calcareous aggregates such as basalt and limestone are basic in nature with predominant calcite and least silica content. Likewise, siliceous aggregates, namely, granite and quartzite, are acidic in nature with high silica content (*12*). The binder was extracted from RAP material using centrifuge extraction and rotary evaporator methods. AC30 binder was blended with different proportions of RAP binder (0, 15, 25, and 40%) at 165 ± 5°C, and 1,000 rpm for 1 hour. Further, the chemical changes of these asphalt binders were studied by Fourier transform infrared (FTIR) spectroscopy method. A total of 16 combinations of aggregate-asphalt binder systems were prepared in this study.

Methodology and Experimental Plan

First, the SFE components of asphalt binders with and without RAP binder, and different types of aggregates were measured. Later, the bond strength of all 16 asphalt binder-aggregate combinations in dry and wet conditions was measured using a BBS test. The detailed description of the test methods and an experimental plan is discussed below.

Surface Free Energy

The SFE of a solid is defined as the magnitude of work required to create a unit area of new surface in vacuum (*13*). The total SFE of any material is divided into three components, namely, Lifshitz-van der Waals component (γ^{LW}), Lewis acid component (γ^{+}), and Lewis base component (γ^{-}). The total SFE (γ) can be obtained using Equation 1. The work of cohesion (W_{BB}) for asphalt binder and work of adhesion (W_{AB}) between asphalt binder and aggregates can be estimated using an SFE parameter by Equations 2 and 3, respectively. Similarly, work of adhesion in the presence of water, also known as work of debonding (W^{wet}_{ABW}), can be estimated using Equation 4. A higher work of adhesion may imply a stronger bond between asphalt binder and aggregate and vice versa (*2*).

$$\gamma = \gamma^{LW} + \gamma^{+-} = \gamma^{LW} + 2\sqrt{\gamma^{+}\gamma^{-}}. \quad (1)$$

$$W_{BB} = 2\gamma_{B}. \quad (2)$$

$$W_{AB} = \gamma_{li}(1 + cos\theta_{i}) = 2\sqrt{\gamma_{A}^{LW}\gamma_{B}^{LW}} + 2\sqrt{\gamma_{A}^{+}\gamma_{B}^{-}} + 2\sqrt{\gamma_{A}^{-}\gamma_{B}^{+}}. \quad (3)$$

$$W^{wet}_{ABW} = \gamma_{AW} + \gamma_{BW} - \gamma_{AB} \quad (4)$$

Where, subscripts A, B, and W indicate aggregate, binder, and water, respectively.

γ_{A}^{LW} *and* γ_{B}^{LW} = Lifshitz-van der Waals component of aggregate and binder, respectively. γ_{A}^{+} *and* γ_{B}^{+} Lewis acid component of aggregate and binder, respectively, and, γ_{A}^{-} *and* γ_{B}^{-} Lewis base component of aggregate and binder, respectively.

γ_{AW}, γ_{BW}, and γ_{AB} are interfacial energy between aggregate-water, asphalt binder-water, and aggregate-asphalt binder, respectively.

Interfacial energy between two materials, i and j (γ_{ij}.) can be determined using Equation 5.

$$\gamma_{ij} = \gamma_{i} + \gamma_{j} - 2\sqrt{\gamma_{i}^{LW}\gamma_{j}^{LW}} - 2\sqrt{\gamma_{i}^{+}\gamma_{j}^{-}} - 2\sqrt{\gamma_{i}^{-}\gamma_{j}^{+}} \quad (5)$$

Energy Parameters

In the present study, energy parameters, EP_1 (Equation 6) and EP_2 (Equation 7) suggested by Hamedi and Ned (*1*) were considered for quantifying the bonding behavior of all 16 combinations of asphalt binder-aggregate system. The surface energy of water is higher than that of bitumen, and therefore it tries to replace bitumen from aggregate. Therefore, the interfacial energy of adhesion between aggregate-binder (γ_{AB}) and water-aggregate (γ_{WA}) can be used for judging the stripping potential. The difference between these two interfacial energies (γ_{AB} - γ_{WA}) indicates the stripping potential. As this difference is greater, the aggregate is more likely to strip in the presence of water. EP_1 is the ratio of (γ_{AB}-γ_{WA}) to γ_{AB} (Equation 6). A lower value of EP_1 implies better compatibility of an asphalt-aggregate system and vice versa (*1*).

$$EP_{1} = \frac{\gamma_{AB} - \gamma_{WA}}{\gamma_{AB}} \quad (6)$$

Similarly, another energy parameter, EP_2 (Equation 7), can also be used as a tool to detect the compatibility of aggregate-asphalt binder systems. EP_2 is the ratio of interfacial energy of adhesion between aggregate-binder (in dry conditions) γ_{AB}^{dry} to the difference between interfacial energy of adhesion in dry and wet conditions (γ_{AB}^{dry}- γ_{AB}^{wet}). A higher value of EP_2 implies better compatibility of the aggregate-asphalt binder system and vice versa (*1*).

$$EP_{2} = \frac{\gamma_{AB}^{Dry}}{\gamma_{AB}^{Dry} - \gamma_{AB}^{Wet}} \quad (7)$$

Measurement of SFE of Asphalt Binders

The SFE components of AC30 control binder with different RAP binder content (0%, 15%, 25%, and 40%) were determined using a contact-angle based approach. First, the contact angle of selected asphalt binders was measured with three probe liquids (distilled water, glycerol, and formamide) using the WP method. Thereafter, the SFE components were calculated using acid-base theory (Equation 3) (*13*). Five replicate samples of each asphalt binder were tested for contact angle. More details on sample preparation, selection of probe liquid, and test parameters used to find the SFE of asphalt binders can be found in other studies (*1, 6, 14*). Table 1 summarizes the SFE of all asphalt binder samples.

Measurement of SFE of Aggregates

In the present study, the SFE of aggregates was measured using the DVS method. Three probe vapors (n-Octane, ethyl acetate, and chloroform) were passed into a sealed chamber containing a dry aggregate sample. At different partial pressures, probe vapor is allowed to be absorbed by the dry aggregate sample at 25ºC. At each partial pressure, the amount of probe vapor absorbed was measured by a high accuracy microbalance connected to the chamber. From this data, adsorption isotherm can be plotted for all three probe vapors. Through adsorption isotherm, the SFE components of aggregate can be calculated. A detailed discussion on measurement of aggregate SFE by DVS method can be found in the literature (*2*). Table 1 summarizes the SFE of all the aggregates considered in this study and a detailed discussion on results is provided later in this paper.

Table 1. SFE Components of Asphalt Binders and Aggregates

Material	SFE components, mJ/m^2					Cohesion energy (mJ/m^2)
	γ^{LW} (Non-Polar)	γ^{+} (Acid)	γ^{-} (Base)	γ^{AB} (Polar)	γ^{Total}	
Asphalt binders						
AC30 (control)	13.99	0.00	11.21	0.00	13.99	27.98
AC30 + 15% RAP	4.50	1.52	7.79	6.88	11.38	22.75
AC30 + 25% RAP	3.88	3.15	1.22	3.91	7.79	15.58
AC30 + 40% RAP	1.33	4.05	9.55	12.44	13.77	27.54
Aggregates						
Basalt	89.5	232.9	134.4	353.7	443.3	–
Limestone	84.0	51.4	122.7	158.8	242.8	–
Granite	159.6	22.1	53.3	68.7	228.2	–
Quartzite	142.1	10.6	4.20	13.3	155.4	–

Binder Bond Strength Test

The test was conducted as per AASHTO T 361-16 (*15*) to measure POTS between asphalt binder and aggregate. POTS can be calculated using Equation 8. The test can also help to identify the type of failure (cohesion and adhesion) in asphalt-aggregate systems before and after water conditioning (Figure 1*e*). Researchers have suggested POTS can be an indicator of moisture damage in an asphalt-aggregate system (*7*, *16*). Therefore, POTS was measured on conditioned and unconditioned samples to check the influence of moisture. Conditioning was done at 40° for 24 hours. Further, BSR was calculated by considering the ratio between POTS of conditioned and unconditioned samples (Equation 9).

$$POTS = \frac{(BP^{*}A_g) - C}{A_{ps}} \tag{8}$$

$$BSR = \frac{POTS_{wet}}{POTS_{dry}} \tag{9}$$

Where, A_g = contact area of gasket with reaction plate (mm^2); BP = burst pressure (kPa); A_{ps}= area of pull stub (mm^2); C = piston constant, POTS for conditioned samples referred as $POTS_{wet}$ and for unconditioned samples $POTS_{dry}$.

Aggregate Sample Preparation. Figure 1*a* depicts the aggregate sample preparation procedure for a BBS test. For this method, first, aggregate samples were cut in sizes of approximately 100 mm × 100 mm to accommodate at least five pull-off stubs. After cutting, surface roughness was controlled through lapping at different sized silicon carbide grits. Polished substrates were cleaned in an ultrasonic bath at 60ºC for 60 minutes. Then the samples were dried in an oven at 110ºC for 24 hours.

Asphalt Sample and Stub Preparation. First, pull-off stubs and aggregate substrates were heated at 150ºC for 30 minutes to remove the moisture. Thereafter, aggregate substrates were kept at 60ºC for preparing BBS test samples. Molten asphalt binder was poured into silicon molds (fabricated to accommodate the required volume) (Figure 1*b*). The aggregate substrate and stubs were removed from the oven and the asphalt binder sample was placed immediately on the surface of the stub for 10 s. Then, the stub with the asphalt binder sample was pressed firmly on the aggregate surface until it reached the surface (Figure 1*c*). For testing unconditioned samples, dry samples were left at room temperature for 24 hours. For wet conditioning, samples were first kept at room temperature for 1 hour then submerged in a water bath at 40ºC for 24 hours. Testing was performed using Pneumatic Adhesion Tensile Testing Instrument (PATTI) and accessories at a calibrated loading rate of 100 psi/sec (Figure 1*d*).

FTIR Spectroscopy

The FTIR spectroscopy technique was used to evaluate changes in the chemical composition of the control binder after the addition of RAP binder. For this test, asphalt and tetrachloroethylene solvent solution was prepared (0.75 gm:10 ml ratio). IR spectrum was passed through prepared samples to get FTIR spectra. For organic materials like asphalt binder, the FTIR spectra were obtained for a wavelength range of 4,000 cm^{-1} to 400 cm^{-1}. Aging in asphalt binder can be captured by observing the peaks at 1,700 cm^{-1} (C = O), which shows the presence of ketones and carboxylic acids. Likewise, a peak at 1,030 cm^{-1} indicates aging due to sulfoxide (S = O) (*17*). For the purpose of quantitative analysis, aging and stiffness indicator indices ($I_{C=O}$ and $I_{S=O}$) suggested by Liu et al. (*17*) were calculated using Equations 10 and 11, respectively.

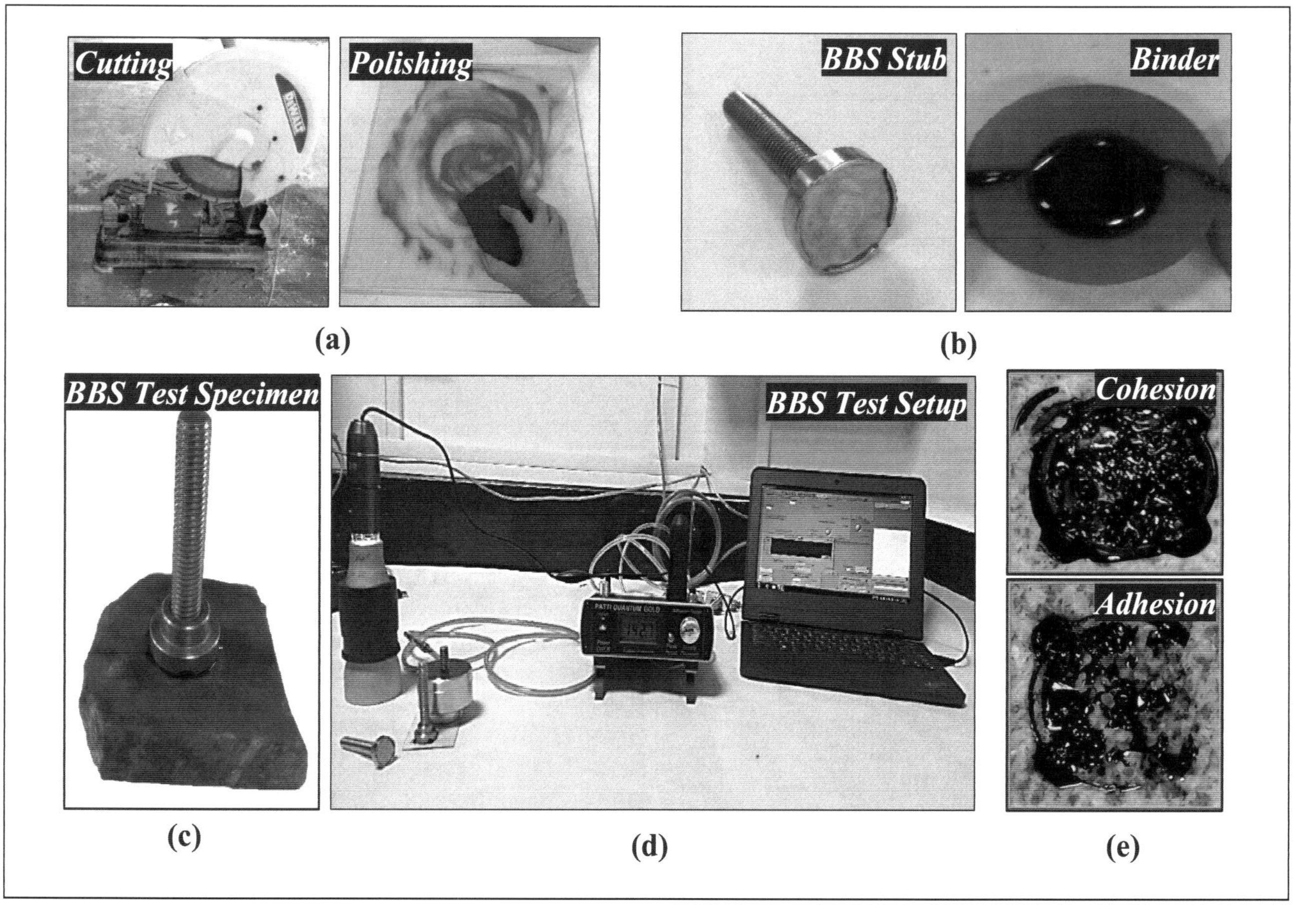

Figure 1. (*a*) Aggregate sample preparation, (*b*) stub and bitumen sample, (*c*) samples for BBS test, (*d*) BBS test set up, and (*e*) Failure pattern.

$$I_{C=O}\frac{\text{Area around 1700 cm}^{-1}}{\text{Area around 1460 cm}^{-1}\text{and Area around 1375 cm}^{-1}}. \quad (10)$$

$$I_{S=O}\frac{\text{Area around 1030 cm}^{-1}}{\text{Area around 1460 cm}^{-1}\text{and Area around 1375 cm}^{-1}} \quad (11)$$

Results and Discussions

Chemical Composition of Asphalt Binders

Figure 2, *a* and *b*, presents the FTIR spectra and indices, respectively, for AC30 asphalt binder after the addition of 15%, 25%, and 40% RAP binder. It can be seen that RAP binder (100% RAP) shows highest peaks at 1,700 cm^{-1} and 1,030 cm^{-1} wavenumbers (indicating highly aged), followed by AC30 control binder with 40%, 25%, 15%, and 0% RAP binder (Figure 2*a*). The arrangement of peaks shows the difference in aging between RAP binder and AC30 control binder. Interestingly, relatively small peaks observed at 1,700 cm^{-1} and 1,030 cm^{-1} for AC30 control binder may be attributed to the sample preparation process and short-term aging. Figure 2*b* shows that both the indices ($I_{C=O}$ and $I_{S=O}$) increased with the addition of RAP binder. For example, $I_{C=O}$ for AC30 binder increased from 0.06 to 0.10, 0.13, and 0.12 after the addition of 15%, 25%, and 40% RAP binder, respectively. This validates the visual observations (arrangement of peaks for different RAP content) made in the FTIR spectral trend (Figure 2*a*).

SFE of Asphalt Binders

The total SFE, LW, and base SFE components of AC30 control binder decreased with the addition of RAP binder (Table 1). For example, the LW component of AC30 control binder decreased from 13.99 to 4.50, 3.88, and 1.33 mJ/m^2 after the addition of 15%, 25%, and 40% RAP binder, respectively. However, the polar and acid SFE component of AC30 binder increased with the addition of RAP binder. This may be because RAP binder is basically aged binder, which comprised a higher amount of polar carboxylic acids compared with the control binder (*18*). An increment in carbonyl functional group after

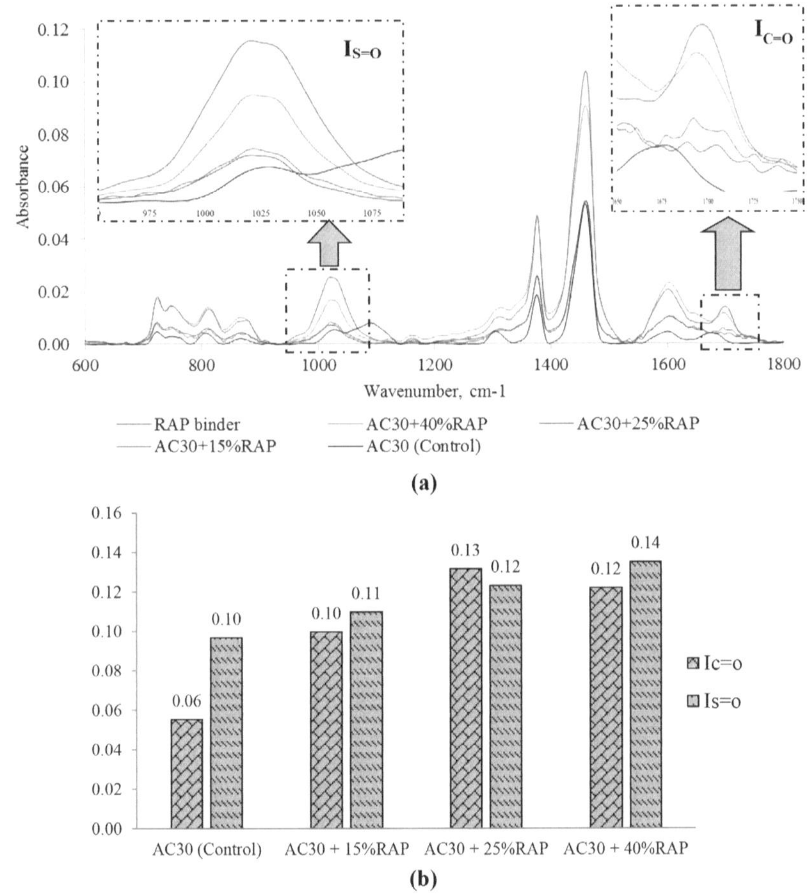

Figure 2. (*a*) FTIR spectra and (*b*) FTIR indices for AC30 binder with different RAP binder content.

the addition of RAP binder was evident from the observation made in the FTIR test conducted in this study.

The total SFE is a function of acid, base, and non-polar components (Equation 1). It was observed that the acid component of the control binder increased with the addition of RAP. However, the basic component of the control binder decreased with the addition of RAP up to 25%, and then it increased after the inclusion of 40% RAP. Thus, after addition of 40% RAP, basic and acidic components showed an opposite trend, which influenced the total SFE (Equation 1) of the control binder. Therefore, total SFE did not yield a specific trend with an increase in RAP content. It seems control binders behave differently after the addition of 40% RAP (especially the basic components of the binder), and this may be attributed to the interaction level of aged RAP binder with the control binder. Similar trends in SFE components were also observed in the literature for control binders with the addition of RAP binder (*4*). Overall, it can be noted that the chemistry of the control binder is altered due the addition of RAP binder, which is reflected in terms of changes in the SFE components of the control binder. However, a detailed chemical investigation needs to be conducted to gain more insight into the SFE of RAP blended binders.

SFE of Aggregates

Table 1 summarizes the SFE components of different types of aggregates considered in this study. Basic aggregates (basalt and limestone) had high polar and high basic SFE components compared with acid aggregates (granite and quartzite). The total SFE of basalt aggregate was higher (443.3 mJ/m^2) followed by limestone (242.8 mJ/m^2), granite (228.2 mJ/m^2), and quartzite (155.4 mJ/m^2). Basalt aggregate was found to be highly polar with a polar SFE component of 353.7 mJ/m^2 and quartzite was found to be least polar with a polar SFE component of 13.3 mJ/m^2. The non-polar component of acid aggregates (i.e., granite and quartzite) was higher than that of basic aggregates (i.e., basalt and limestone). A broad range of SFE component values indicates a

Table 2. Interfacial Energy and Energy Parameters for Different Asphalt-Aggregate Combinations

		Interfacial energy, mJ/m^2			Energy parameters	
		γ^{AS}	γ^{SW}	γ^{ASW}		
Asphalt binder	Aggregate	Asphalt-stone	Stone-water	Asphalt-stone-water	EP_1	EP_2
AC30 (Control)	Basalt	284.3	156.5	−109.7	0.45	0.72
AC30 + 15% RAP		300.8	156.5	−120.5	0.48	0.71
AC30 + 25% RAP		339.0	156.5	−149.3	0.54	0.69
AC30 + 40% RAP		294.2	156.5	−113.4	0.47	0.72
AC30 (Control)	Limestone	140.2	45.7	–76.4	0.67	0.65
AC30 + 15% RAP		148.0	45.7	–78.5	0.69	0.65
AC30 + 25% RAP		159.3	45.7	–80.5	0.71	0.66
AC30 + 40% RAP		146.5	45.7	–76.5	0.69	0.66
AC30 (Control)	Granite	116.3	61.8	–36.4	0.47	0.76
AC30 + 15% RAP		141.8	61.8	–56.2	0.56	0.72
AC30 + 25% RAP		150.0	61.8	–55.0	0.59	0.73
AC30 + 40% RAP		154.4	61.8	–68.3	0.60	0.69
AC30 (Control)	Quartzite	58.4	63.3	23.0	–0.08	1.65
AC30 + 15% RAP		93.0	63.3	–5.9	0.32	0.94
AC30 + 25% RAP		101.8	63.3	–5.3	0.38	0.95
AC30 + 40% RAP		113.3	63.3	–25.7	0.44	0.82

significant difference in surface chemistry and mineralogy of selected aggregates.

Energy Parameters Based on SFE of Asphalt Binder and Aggregate

The SFE components of asphalt binders and aggregates reported in Table 1 were used to calculate the interfacial energies and energy parameters using Equations 6 and 7. Table 2 enumerates interfacial energies and energy parameters for all combinations of asphalt binder-aggregates used in the present study.

Interfacial Energy between Asphalt Binder and Aggregate in Dry Conditions (γ^{AS}). In general, γ^{AS} increased with the addition of RAP binder for all types of aggregates. For example, the γ^{AS} value for AC30-quartzite aggregate was found to be 58.4 mJ/m^2, which increased to 93.0, 101.8, and 113.3 mJ/m^2 with the addition of 15%, 25%, and 40% RAP binder, respectively. Nevertheless, for basic aggregates (basalt and limestone), γ^{AS} increased up to the addition of 25% RAP binder, and then decreased with higher RAP binder content (40% RAP). On the other hand, for acid type aggregates (granite and quartzite), γ^{AS} value continued to increase with the addition of RAP binder. The basalt aggregate had the highest γ^{AS} for AC30 with and without RAP binder, followed by limestone, granite, and quartzite aggregates. This may because surface chemistry of Lewis acids and bases in the case of an acidic asphalt and basic aggregate favor adhesion, and a good bond between basic aggregate and an acidic asphalt can be achieved (*19*). It would be interesting to check if γ^{AS} can have a correlation with a strength parameter obtained from a BBS test.

Interfacial Energy between Asphalt Binder and Aggregate in Wet Conditions (γ^{ASW}). Interfacial energy in the presence of water (γ^{ASW}) was calculated for all 16 combinations of aggregate-asphalt binder system. A higher γ^{ASW} value indicates a greater chance of the asphalt binder being stripped from the aggregate. In general, γ^{ASW} increased with the addition of RAP binder for all types of aggregates, indicating degradation of the bond between binder-aggregate systems in the presence of water. For example, the magnitude of γ^{ASW} for AC30-basalt combination was found to be 109.7 mJ/m^2, which increased to 120.5, 149.3, and 113.4 mJ/m^2 with the addition of 15%, 25%, and 40% RAP binder, respectively. Nevertheless, for basic aggregates (basalt and limestone), γ^{ASW} increased up to the addition of 25% RAP binder, and then decreased with higher RAP binder content (40% RAP). On other hand, for acid type aggregates (granite and quartzite), the γ^{ASW} value showed a mixed trend with the addition of RAP binder content. In general, quartzite aggregate had the least γ^{ASW} for AC30 with and without RAP binder, followed by granite, limestone, and basalt aggregates.

Energy Parameters (EP_1 and EP_2). Two energy parameters (EP_1 and EP_2) were calculated for the 16 asphalt binder-aggregate combinations using Equations 6 and 7, respectively. As mentioned earlier, a lower value of EP_1 and a higher value of EP_2 indicates better bonding potential of asphalt binder-aggregates considering both dry and wet conditions. In general, EP_1 increased, while EP_2 decreased

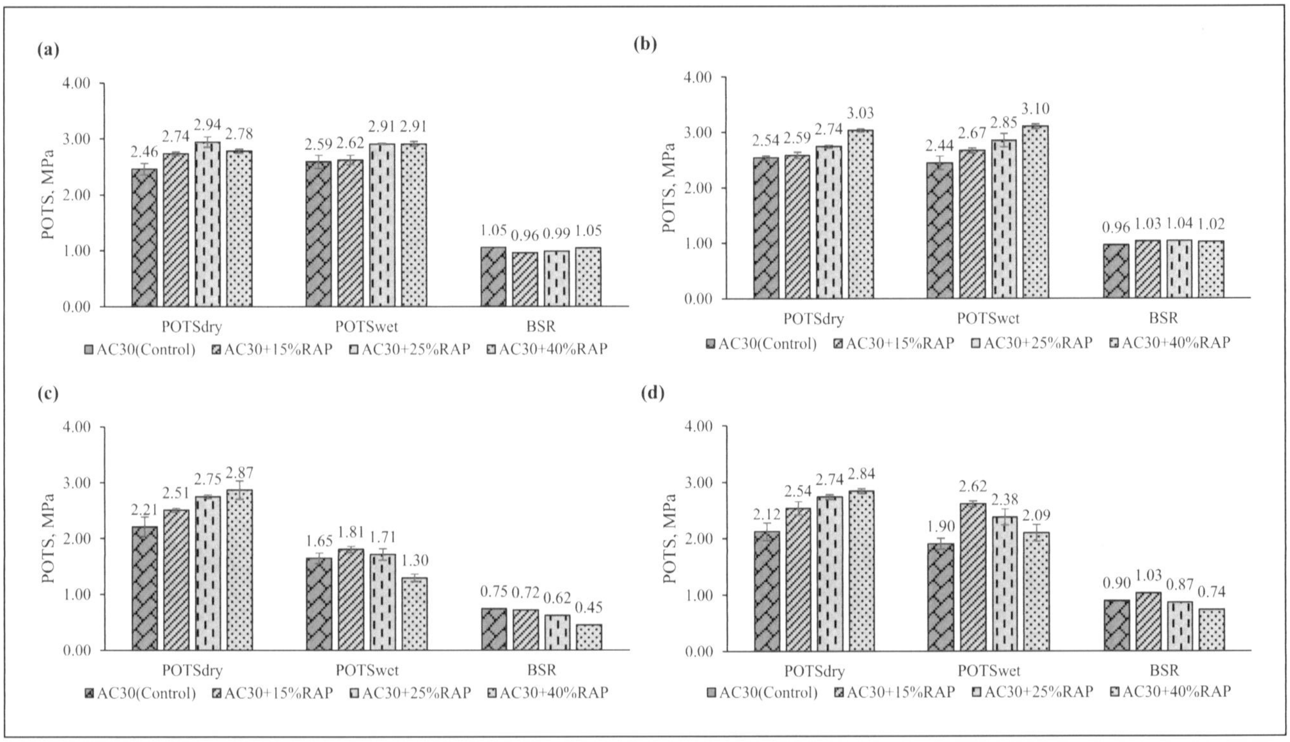

Figure 3. BBS of AC30 with and without RAP binder for (*a*) basalt, (*b*) limestone, (*c*) granite, and (*d*) quartzite.

with the addition of RAP binder to control binder for all types of aggregates. For example, the EP_1 value for AC30-basalt combination increased from 0.45 to 0.48, 0.54, and 0.47 after the addition of 15%, 25%, and 40% RAP binder, respectively. This indicates that the compatibility of AC30 control binder decreased with the addition of RAP binder. A similar trend was observed for other aggregates as well. Nevertheless, for basic aggregates (basalt and limestone), EP_2 did not differ much with the addition of RAP binder to AC30 control binder. This implies that the addition of RAP binder may not affect the performance of AC30 binder with basic aggregates. On other hand, for acid type aggregates (granite and quartzite), EP_2 value continued to decrease with the addition of RAP binder, indicating a negative effect of RAP binder on acidic aggregates. This may be because carboxylic acids in bitumen readily bond with siliceous aggregate surfaces but these bonds are easily displaced by water (*12*). An increase in carboxylic acids with the addition of RAP binder was evident from the FTIR test conducted in the study. It is to be noted that the EP_1 for AC30-quartzite combination was found to be an erroneous value of -0.08 (EP cannot be negative) and was not considered for comparing with the BBS results.

Asphalt Binder Bond Strength

The POTS value measured in dry and wet conditions for four different aggregates (basalt, limestone, granite, and quartzite) and AC30 containing different percentages of RAP binder 0%, 15%, 25%, and 40%) is presented in Figure 3. The error bars shown in Figure 3 represent the standard deviation (variability) of five replicate samples tested in the present study.

$POTS_{dry}$. $POTS_{dry}$ value increased with the addition of RAP binder irrespective of the nature of the aggregates. For example, $POTS_{dry}$ for control AC30-quartzite aggregate was measured to be 2.12 MPa, which increased to 2.54, 2.74, and 2.84 MPa with the addition of 15%, 25%, and 40% RAP binder, respectively (Figure 3*d*). A similar trend can also be seen for limestone, granite, and basalt aggregates (Figure 3). An increase in bond strength may be attributed to the stiffer or viscous nature of AC30 binder after the addition of RAP binder (*20*). This can also be supported by indices, $I_{C=O}$ and $I_{S=O}$ for AC30 binder, increased after the addition of RAP binder, confirming the increment of binder stiffness and aging. Therefore, Bahia et al. (*20*) suggested performing the BBS test at iso-stiffness temperature. Interestingly, the magnitude of increment in bond strength varied from aggregate to aggregate, indicating that the type of aggregate plays an important role. For control AC30 binder, basic aggregates (basalt and limestone) performed better in dry conditions compared with acidic aggregates (granite and quartzite). This trend changes after the addition of RAP binders, and acidic aggregates showed a

relatively similar performance to that of basic aggregates, indicating that RAP addition is favorable for acidic aggregates in dry conditions. The observations made in this study are in line with the literature findings for aged binder (*16*, *20*).

$POTS_{wet}$. For basic aggregates, the increment was consistent. For example, $POTS_{wet}$ for AC30-limestone was increased from 2.44 to 2.67, 2.85, and 3.10 MPa after the addition of 15%, 25%, and 40% RAP binder, respectively. Whereas, for acidic aggregates (granite and quartzite), $POTS_{wet}$ value increased for the addition of 15% RAP binder, and then after a higher percentage of RAP (i.e., 25% and 40%) decreased $POTS_{wet}$ value. Even though $POTS_{wet}$ value decreased, it was still higher than the control binder for quartzite aggregate (2.09MPa > 1.90MPa). However, granite aggregate showed a $POTS_{wet}$ value lower than the control binder for 40% RAP binder (1.30 < 1.65MPa), indicating that very high RAP binder content leads to degradation of bond strength in acidic aggregate and is relatively more severe for granite aggregate. Also, acidic aggregates (granite and quartzite) showed the influence of moisture conditioning by showing reduced POTS value on conditioning. This may be because carboxylic acids in bitumen readily bond with siliceous aggregate surfaces but these bonds are easily displaced by water (*12*).

BSR. $POTS_{dry}$ and $POTS_{wet}$ values indicate the strength of an asphalt-aggregate system in dry and wet conditions, respectively. However, in reality, both dry and wet conditions are prevalent simultaneously. Therefore, in order to consider the combined effect of wet and dry conditions, BSR was calculated (Equation 9). For AC30 control binder, BSR for basalt aggregate was the highest, followed by limestone, quartzite, and granite. However, after the addition of RAP binder, the order of BSR was found to be limestone, basalt, quartzite, and granite aggregates. Limestone had the highest BSR with the addition of RAP binder, and granite had the least. For basic aggregates, BSR value did not vary much with the addition of RAP binder, indicating that addition of RAP binder had a minimal effect on the performance of basic aggregates. However, for acidic aggregates, BSR value decreased with an increase in RAP binder content (except for quartzite-AC30 + 15% RAP combination). Overall the results showed that the addition of RAP binder increased moisture-induced sensitivity in acidic aggregate combinations. This may be because the acidic aggregates (granite and quartzite) considered in this study are siliceous in nature. Carboxylic acids in asphalt binder readily bond with siliceous aggregate surfaces and are easily displaced by water (*12*). It can be concluded that caution should be taken while using RAP for acidic aggregates.

Failure Mechanism. After BBS tests, the failure pattern (cohesive or adhesive) was captured for all 16 asphalt-aggregate systems. The failure was classified as cohesive failure (C) if more than 50% area asphalt binder retained to an aggregate substrate. Likewise, it was classified as an adhesive failure (A), when failure had occurred at an interface, exposing the aggregate surface more than 50% of the area. In some cases, failure may occur partly in cohesion and partly in adhesion and was thus classified as a C/A (*7*). Table 3 summarizes the failure type, and photos represent all the replicate samples for different combinations of asphalt-aggregates in dry and wet conditions.

The results showed that cohesive failure (C) was dominant in dry conditions for all 16 combinations of asphalt binder-aggregates. This signifies that adhesive strength between binder-aggregate was stronger than cohesive strength of asphalt binder. The cohesive energy of control AC30 binder was 27.98 mJ/m^2 (Table 1), which decreased with the addition of RAP binder. This implies that if combination of control AC30-aggregate fails in cohesive, then obviously other asphalt-aggregate combinations with RAP would also fail in cohesion. And this hypothesis is validated through the nature of failure observed in the present study. On other hand, in case of wet conditioned samples, the failure pattern changed from C to A failure for acidic aggregates (i.e., granite and quartzite). This indicates the significant influence of moisture on asphalt binder bonding at interface for acidic aggregates. This trend can also be supported by diminishing POTS value in wet conditions for acidic aggregates (Figure 3). The trends observed in this study are in line with reported literature (*7*, *20*). Surprisingly, the failure pattern remains C for basic aggregates (i.e., basalt and limestone) even after conditioning. The fact can also be validated through minimal change in POTS value after conditioning for basic aggregates (Figure 3). The failure pattern for AC30-quartzite and AC30 + 15% RAP-limestone was found to be C/A (Table 3). Therefore, these combinations were not considered for correlation.

Relationship between Interfacial Energy and Asphalt Binder Bond Strength

Energy parameters (EP_1 and EP_2) obtained from an SFE approach to strength parameters (i.e., $POTS_{dry}$, $POTS_{wet}$, and BSR) measured from BBS tests were compared for all 16 combinations of aggregate-asphalt binder. First, the correlation between $POTS_{dry}$ and EP_1; $POTS_{dry}$ and EP_2; BSR and EP_1; BSR and EP_2 were tried for all 16 combinations. However, a weak correlation (R^2 <0.25) was observed between these selected pairs of parameters. The possible reason for not showing any correlation may be because the failure encountered

Table 3. Failure Types for Different Asphalt-Aggregate Combinations from BBS Test

Aggregate	Conditioning	AC30 (Control)		AC30+15% RAP		AC30+25% RAP		AC30+40% RAP	
		FP	Photo	FP	Photo	FP	Photo	FP	Photo
Basalt	Dry	C	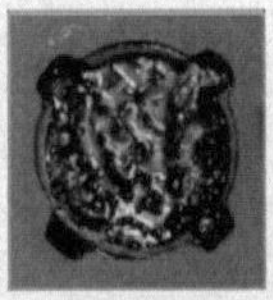	C	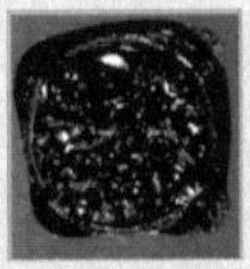	C		C	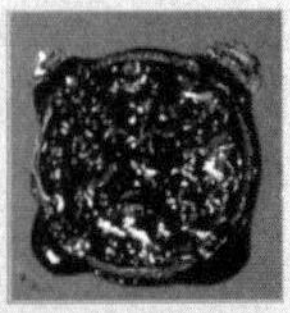
	Wet	C	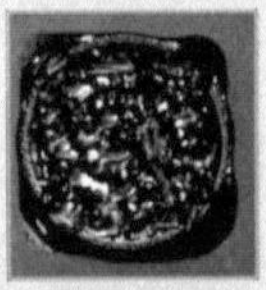	C	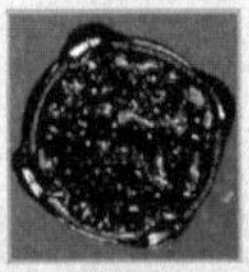	C	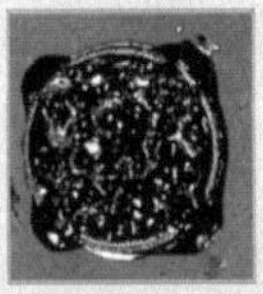	C	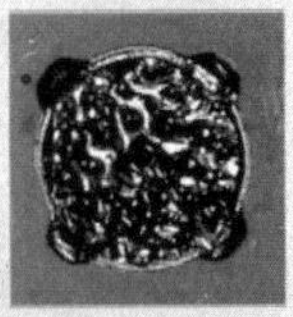
Limestone	Dry	C		C	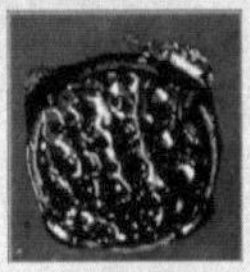	C	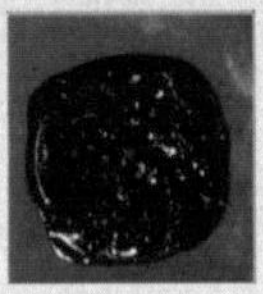	C	
	Wet	C	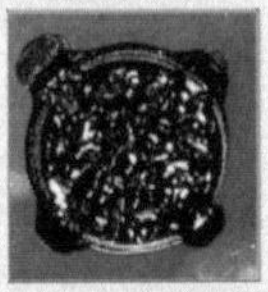	C/A	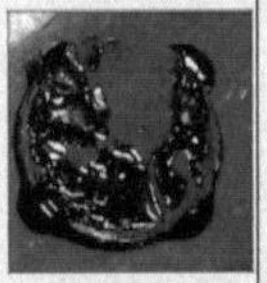	C	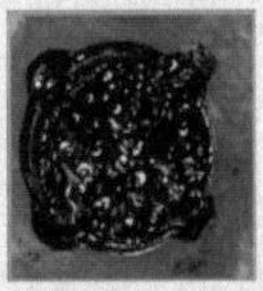	C	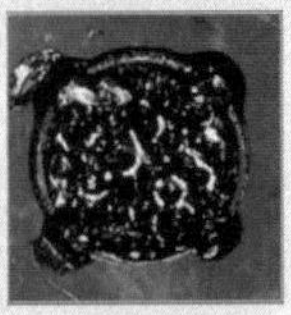
Granite	Dry	C	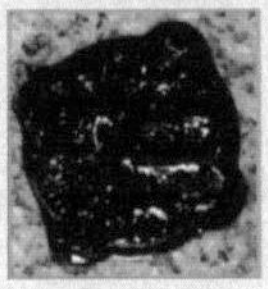	C	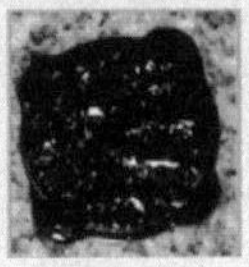	C	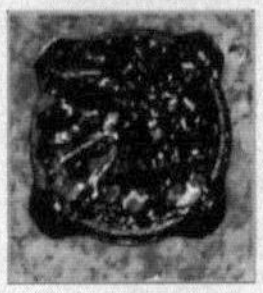	C	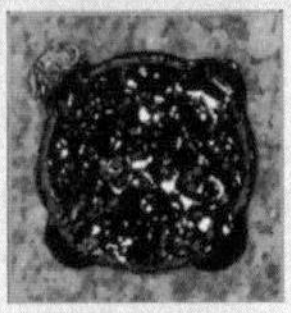
	Wet	A	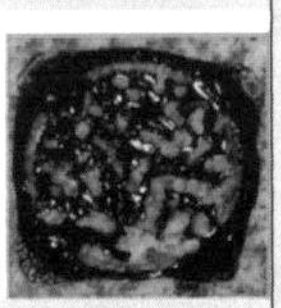	A	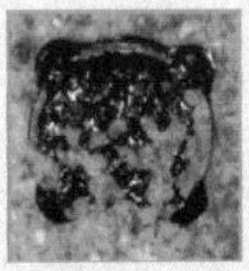	A	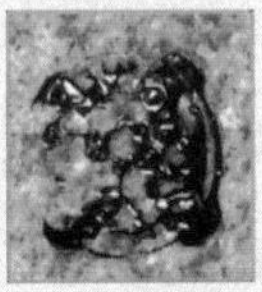	A	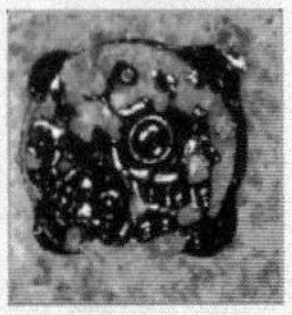
Quartzite	Dry	C	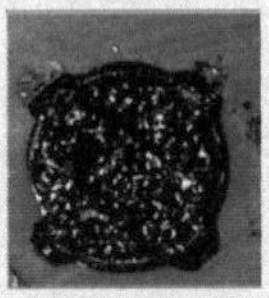	C		C	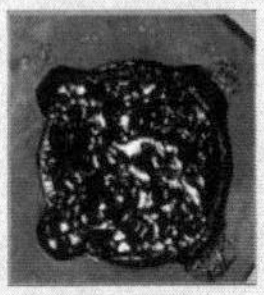	C	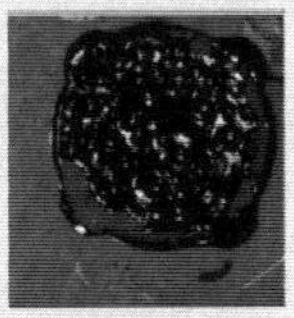
	Wet	C/A	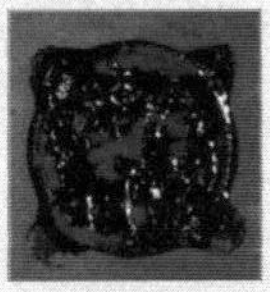	A	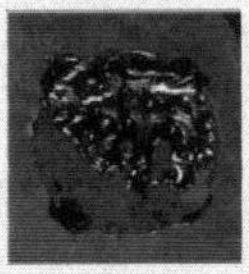	A	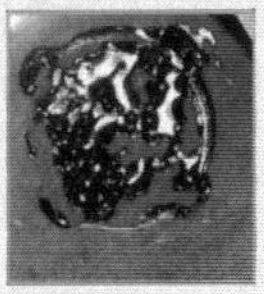	A	

Note: FP = failure pattern; C = cohesive failure; A = adhesive failure; and C/A = cohesion-adhesion.

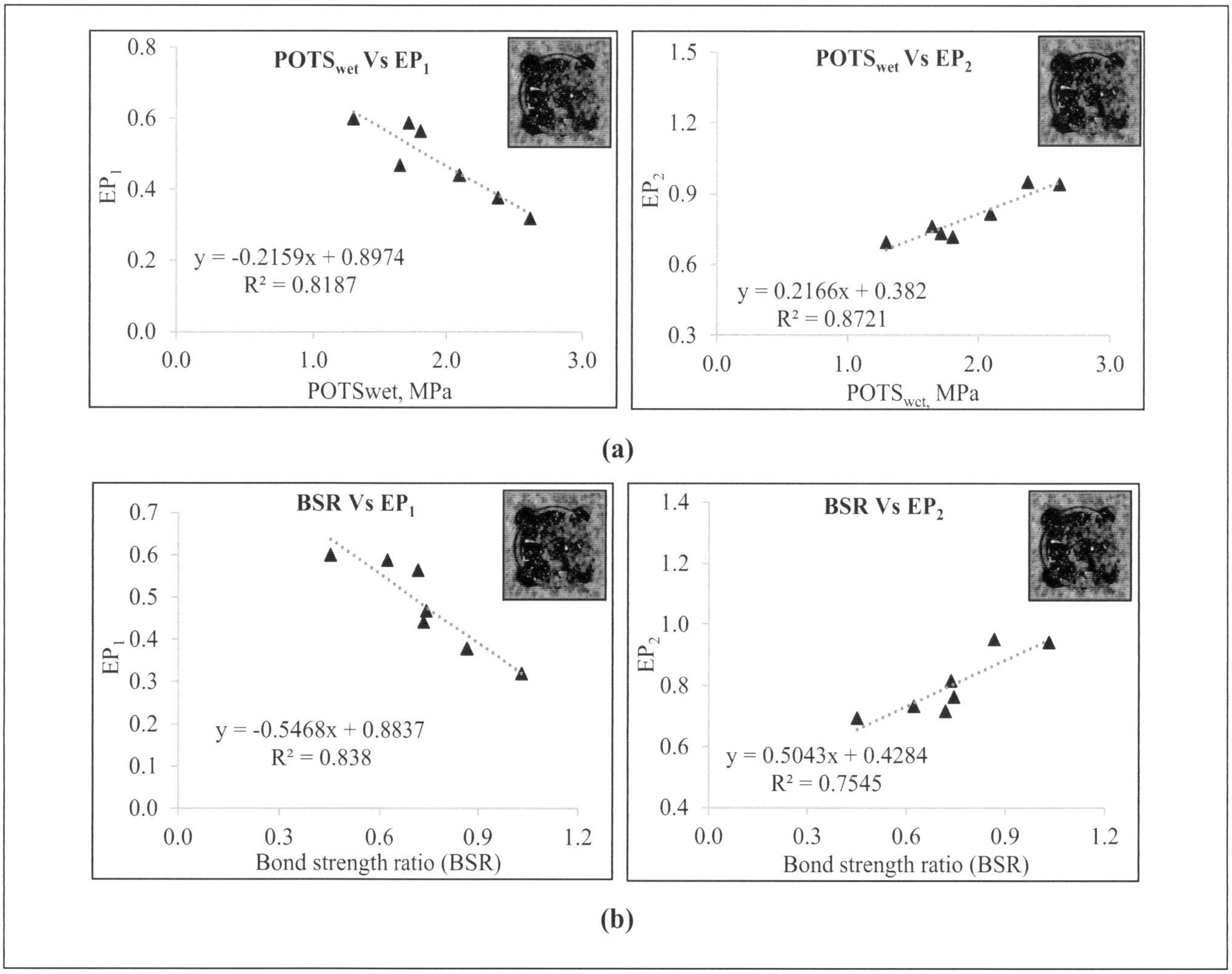

Figure 4. Correlation for acidic aggregates: (*a*) $POTS_{wet}$ and energy parameters (EP_1 and EP_2) and (*b*) BSR and energy parameters (EP_1 and EP_2).

was within asphalt phase, that is, in cohesion, and estimation of EP_1 and EP_2 (Equations 6 and 7) parameters takes into account interfacial energy between both asphalt binder and aggregates (Equation 5). Also, EP_1 and EP_2 consider the presence of moisture, which is not considered for $POTS_{dry}$.

Results from BBS tests showed (Table 3) that, for dry conditions, all asphalt-aggregates (both acidic and basic) samples failed in cohesion, indicating a weak cohesive energy of asphalt binder. Thus, checking the correlation between $POTS_{dry}$ and EP may not be logical in case of cohesive failure. Similarly, for conditioned samples, BBS tests showed that failure was adhesive in nature for acidic aggregates, while it was cohesive for basic aggregates. In this case, similar correlations were tried for both acidic and basic aggregates separately. However, no correlation was observed for basic aggregates, because all samples failed in cohesion.

Interestingly, a good correlation (R^2 = 0.8187 and 0.8721) was observed between $POTS_{wet}$ and energy parameters (EP_1 and EP_2) for acidic aggregates (Figure 4*a*). It can be noted that all the samples of these combinations exhibited adhesive failure in wet conditions. These correlations seem to be more logical as failure was encountered at the interface (involving aggregate and binder simultaneously). Energy parameters derived from an SFE approach also considered interfacial adhesion energies between asphalt binder and aggregate in wet and dry conditions.

Similar to $POTS_{wet}$, BSR also showed good correlation (R^2 = 0.838 and 0.7545) with both the energy parameters for acidic aggregates that portrayed adhesive failure (Figure 4*b*). However, as expected, basic aggregates did not show any correlation with BSR. This may be because all these samples exhibited cohesive failure.

Conclusion and Recommendations

The effects of RAP binder on bond strength behavior of AC30 control binder was evaluated for four different types of aggregates using interfacial energy and bond strength parameters. The critical conclusions drawn based on the selected combinations of aggregate-asphalt binder are listed below:

- In general, energy parameters EP_1 increased and EP_2 decreased after the addition of RAP binder to control binder for all types of aggregates, showing a reduction in compatibility of control binder with the addition of RAP binder. Results showed that the bonding potential of basic aggregates was less affected compared with acidic aggregates.
- $POTS_{dry}$ value of the control binder increased after addition of RAP binder for all types of aggregates, which may be attributed to an increase in stiffness of the control binder after the addition of RAP binder.
- BSR and $POTS_{wet}$ values showed that the addition of RAP binder increases the performance of control AC30 binder for basic aggregates. Therefore, usage of RAP can be suggested for basic aggregates (basalt and limestone). However, addition of higher RAP binder content degrades the bond strength of acidic aggregates (granite and quartzite) in the presence of water. Also, BSR demonstrated the effect of moisture conditioning on acidic aggregate. Based on the selected combinations of aggregate-asphalt binder used in the present study, it is recommended that caution should be exercised while using RAP modified binder with an acidic type of aggregates.
- For control AC30 binder, BSR was highest for basalt aggregate, followed by limestone, quartzite, and granite. However, after the addition of RAP binder, the order of BSR was found to be limestone, basalt, quartzite, and granite aggregates. For basic aggregates, BSR value did not vary much with the addition of RAP binder, indicating that the addition of RAP had a minimal effect on the performance of basic aggregates. However, for acidic aggregates, BSR value decreased with an increase in RAP binder content.
- Basic aggregates (limestone and basalt) selected in the present study showed cohesive failure before and after conditioning, indicating a minimal effect of water on these aggregates. However, acidic aggregates (granite and quartzite) showed cohesive failure in dry conditions, and adhesive failure in wet conditions, implying detrimental effects of the presence of water for these aggregates. The failure pattern in cohesion was supported by reduction in cohesion energy of asphalt binder after addition of RAP binder.
- The analyses showed that correlation between POTS and EP was not found for cohesive types of failure, particularly for basic aggregate samples. Remarkably, a good correlation was observed between energy parameters (EP_1 and EP_2) and bond strength parameters ($POTS_{wet}$ and BSR) for samples that failed in adhesion, particularly acidic aggregates.

It is recommended to conduct detailed chemical investigation and comprehensive lab and field performance tests to gain more insight into changes in SFE of asphalt binder and validate the applicability of SFE and BBS test methods presented in this paper.

Author Contributions

The authors confirm contributing to the paper as follows: study conception and design: Ayyanna Habal; data collection: Ayyanna Habal; analysis and interpretation of results: Ayyanna Habal, Dharamveer Singh; draft manuscript preparation: Ayyanna Habal, Dharamveer Singh. All authors reviewed the results and approved the final version of the manuscript.

References

1. Hamedi, G. H., and F. Moghadas Nejad. Using Energy Parameters Based on the Surface Free Energy Concept to Evaluate the Moisture Susceptibility of Hot Mix Asphaltc *Materials and Pavement Design*, Vol. 16, No. 2, 2015, pp. 239–255.
2. Little, D. N., and A. Bhasin. *Using Surface Energy Measurements to Select Materials for HMA Pavements.* NCHRP Project 9-37. Transportation Research Board, Washington D.C., 2006.
3. Arabani, M., H. Roshani, and G. H. Hamedi. Estimating Moisture Sensitivity of Warm Mix Asphalt Modified with Zycosoil as an Antistrip Agent Using Surface Free Energy Method. *Journal of Materials in Civil Engineering*, Vol. 24, No. 7, 2012, pp. 889–897.
4. Ghabchi, R., D. Singh, and M. Zaman. Evaluation of Moisture Susceptibility of Asphalt Mixes Containing RAP and Different Types of Aggregates and Asphalt Binders Using the Surface Free Energy Method. *Construction and Building Materials*, Vol. 73, 2014, pp. 479–489.
5. Hossain, Z., B. Bairgi, and M. Belshe. Investigation of Moisture Damage Resistance of GTR-Modified Asphalt Binder by Static Contact Angle Measurements. *Construction and Building Materials*, Vol. 95, 2015, pp. 45–53.
6. Habal, A., and D. Singh. Moisture Damage Resistance of GTR-Modified Asphalt Binders Containing WMA Additives Using the Surface Free Energy Approach. *Journal of Performance of Constructed Facilities*, Vol. 31, No. 3, 2017, p. 4017006.
7. Moraes, R., R. Velasquez, and H. Bahia. Measuring the Effect of Moisture on Asphalt-Aggregate Bond with the

Bitumen Bond Strength Test. *Transportation Research Record: Journal of the Transportation Research Board*, 2011. 2209: 70–81.
8. Bhasin, A., D. N. Little, K. L. Vasconcelos, and E. Masad. Surface Free Energy to Identify Moisture Sensitivity of Materials for Asphalt Mixes. *Transportation Research Record: Journal of the Transportation Research Board*, 2007. 2001: 37–45.
9. Howson, J., E. Masad, D. Little, and E. Kassem. Relationship between Bond Energy and Total Work of Fracture for Asphalt Binder-Aggregate Systems. *Road Materials and Pavement Design*, 2012.
10. Moraes, R. *Validation of the Bitumen Bond Strength Test Using Surface Energy Measurements*. MSc thesis. University of Wisconsin-Madison, 2011.
11. Moraes, R., R. Velasquez, and H. Bahia. Using Bond Strength and Surface Energy to Estimate Moisture Resistance of Asphalt-Aggregate Systems. *Construction and Building Materials*, Vol. 130, 2017, pp. 156–170.
12. Hefer, A., and D. Little. *Bitumen-Aggregate Systems and Quantification of the Effects of Water on the Adhesive Bond*. Report No. ICAR/505-1, Texas Transportation Institute, the Texas A&M University, 2005.
13. Van Oss, C. J., M. K. Chaudhury, and R. J. Good. Interfacial Lifshitz-van Der Waals and Polar Interactions in Macroscopic Systems. *Chemical Reviews*, Vol. 88, No. 6, 1988, pp. 927–941.
14. Habal, A., and D. Singh. Comparison of Wilhelmy Plate and Sessile Drop Methods to Rank Moisture Damage Susceptibility of Asphalt - Aggregates Combinations. *Construction and Building Materials*, Vol. 113, 2016, pp. 351–358.
15. AASHTO T 361-16. *Standard Method of Test for Determining Asphalt Binder Bond Strength by Means of the Asphalt Bond Strength (ABS) Test*. AASHTO, Washington D.C., 2016.
16. Aguiar-Moya, J. P., J. Salazar-Delgado, A. Baldi-Sevilla, F. Leiva-Villacorta, and L. Loria-Salazar. Effect of Aging on Adhesion Properties of Asphalt Mixtures with the Use of Bitumen Bond Strength and Surface Energy Measurement Tests. *Transportation Research Record: Journal of the Transportation Research Board*, 2015. 2505: 57–65.
17. Liu, G., E. Nielsen, J. Komacka, G. Leegwater, and M. Van De Ven. Influence of Soft Bitumens on the Chemical and Rheological Properties of Reclaimed Polymer-Modified Binders from The "old" surface-Layer Asphalt. *Construction and Building Materials*, Vol. 79, 2015, pp. 129–135.
18. Siddiqui, M. N., and M. F. Ali. Studies on the Aging Behavior of the Arabian Asphalts. *Fuel*, Vol. 78, 1999, pp. 1005–1015.
19. Robertson, R. E. Chemical Properties of Asphalt S and Their Relationship to Pavement Performance. *Strategic Highway Research Program - National Research Council*, 1991, p. 41.
20. Bahia, H., R. Moraes, and R. Velasquez. The Effect of Bitumen Stiffness on the Adhesive Strength Measured by the Bitumen Bond Strength Test. *Proc., 5th Eurasphalt & Eurobitume Congress*, 13–15 June 2012, Istanbul, Turkey.

The Standing Committee on Surface Requirements of Asphalt Mixtures (AFK40) peer-reviewed this paper (18-06664).

Article

TRR
JOURNAL OF THE TRANSPORTATION RESEARCH BOARD

Transportation Research Record
2018, Vol. 2672(28) 167–177

DOI: 10.1177/0361198118796967
journals.sagepub.com/home/trr
SAGE

Effect of Mixing Time and Temperature on the Homogeneity of Asphalt Mixtures Containing Reclaimed Asphalt Pavement Material

Jiantao Wu[1], Quan Liu[1,2], Yu Wang[1], Jun Chen[1], Dawei Wang[2], Linlin Xie[1], and Cadnel Ago[1]

Abstract

Laboratory-produced reclaimed asphalt pavement (RAP) material, new aggregate, and tinted new binder were mixed (RAP proportion: 40%) with 12 different combinations of mixing temperatures and times, based on which circular specimens of various thicknesses were fabricated and then subjected to multi-direction indirect tensile stiffness modulus (ITSM) testing and color image analysis. Statistical calculations including average value, coefficient of variation, standard deviation, and range were carried out to investigate the effects of different mixing conditions on the homogeneity of asphalt mixtures containing RAP. The results show that the deterioration of homogeneity of asphalt mixtures containing RAP was mainly caused by the asynchronous breaking of clusters. The mixing temperature was decisive in determining the resistance of clusters to breaking, and for mixing conditions chosen in this study, the resistance increases with the increase of mixing temperature. Short mixing time might lead to a phenomenon of "momentary homogeneity," in which clusters were not broken and the blending degree between aged and new binder was low.

In the past two decades, reclaimed asphalt pavement (RAP) has been increasingly used in new paving asphalt mixes because of the significant cost, resource, and environmental benefits this brings. As the raw materials (i.e., virgin mineral aggregate and asphalt binder) for asphalt mixtures are non-renewable resources, their availability declines gradually and their prices increase dramatically. For instance, it was reported that the 2006 price ($400 per ton) of asphalt binders in the United States increased by almost 50% over the 2005 price ($250 per ton) (*1*). Therefore, the use of RAP not only reduces the consumption of invaluable natural resources, but also effectively cuts the cost of purchasing and transporting raw materials for asphalt mixtures. In addition, as the used asphalt was historically disposed of in landfills, the current high re-use rate of old asphalt (*2*, *3*) fundamentally reduces the area of landfill disposal and its related environmental impacts.

Because of the advantages listed above, paving asphalt has long been the most recycled material in many countries (*4*). However, it should be pointed out that the average RAP content in new asphalt mixtures is only 10–20%, even though specifications allow higher percentages (*5–8*). One of the main concerns about utilizing a high percentage of RAP in asphalt mixtures is that the degree of blending and diffusion between the RAP and new binder is not well understood, and may significantly affect the mix design and result in unexpected field performance. Therefore, many studies have been performed to quantify the degree of blending and/or diffusion between the RAP and neat binders and the effects on pavement performance (*9–13*). In addition, extensive studies have recently been carried out by several researchers (*14–16*) to explore the phenomenon of RAP clusters. Bressi et al. defined the clustering phenomenon as the agglomeration of RAP particles formed either before or during the new mix phase (*15*). It was stated by Bressi et al. that the formation of RAP clusters could prevent the RAP and new binders from contacting and increase the heterogeneity of asphalt mixtures, which would in turn decrease the degree of blending between RAP and new binders (*17*).

[1]College of Civil and Transportation Engineering, Hohai University, Nanjing, China
[2]Institute of Highway Engineering Aachen, RWTH Aachen, Germany

Corresponding Author:
Address correspondence to Jiantao Wu: Jiantao.wu@hhu.edu.cn

In the authors' opinion, the homogeneity of asphalt mixtures containing RAP is one of the decisive factors that affects the degree of blending and/or diffusion between RAP and new binders, and the RAP cluster—which is mainly caused by the agglomeration of small-size RAP particles (*15*)—is just one of the reasons reducing the homogeneity of asphalt mixtures containing RAP. In fact, all RAP particles, including those that do not form clusters, can significantly increase the heterogeneity of asphalt mixtures containing RAP, as a result of the sticky RAP binders coating them and the insufficient mixing times (mostly shorter than 1 min). Therefore, investigating the homogeneity of asphalt mixtures containing RAP is as important as the blending studies.

The objective of this study is to investigate the effects of mixing time and temperature on the homogeneity of asphalt mixtures containing a relatively high percentage RAP components. Asphalt mixtures containing 40% RAP were mixed at different mixing condition combinations of three temperatures and four durations. Multi-direction indirect tensile stiffness modulus (ITSM) tests and color image analyses were applied to cylindrical and disc specimens, respectively, based on which statistical analyses were carried out to characterize the homogeneities of different mixtures.

Materials and Experimental Program

Materials

Bitumen. In this study, a PG70-22 bitumen was used in both the RAP production and the mixing of asphalt mixtures containing RAP. The only difference was that the binder used in the new mixing was tinted with 30% (by mass of bitumen) iron oxide powder. The empirical parameters of the virgin PG70-22 bitumen and the tinted binder are shown in Table 1.

Table 1. Empirical Parameters of PG70-22 Bitumen and the Tinted Binder

Parameters	Binders	
	PG70-22 bitumen	Tinted binder
Penetration @25°C (dmm)	68	52
R&B softening point (°C)	49.2	49.7
Ductility @15°C (cm)	158.4	26.4

Mineral Aggregate and Filler. Basalt aggregates and limestone fillers were collected from the local market. Their combined objective gradation curve, which satisfies the gradation for AC-13 from specification, is shown in Figure 1.

RAP Preparation. To prepare enough RAP materials for the following studies, AC-13 asphalt mixtures with the objective gradation of aggregate (as shown in Figure 1) and 4% binder content were first mixed and short-term aged (at 135°C for 4 h) in accordance with the Strategic Highway Research Program (SHRP). After that, the loose mixtures were compacted into cylindrical specimens (150 mm diameter, 115 ± 5 mm height and 4.5% target air void) using a gyratory compactor. The compacted specimens were then SHRP long-term aged (at

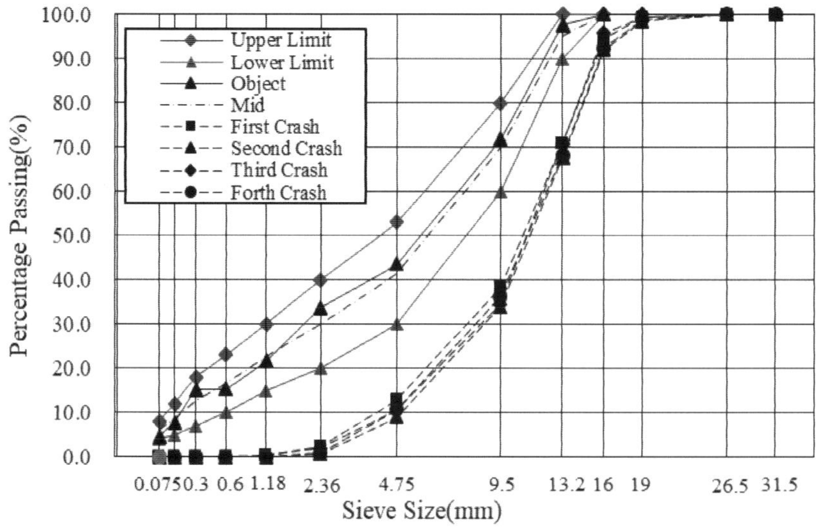

Figure 1. Grading curves for aggregates and crushed RAP.

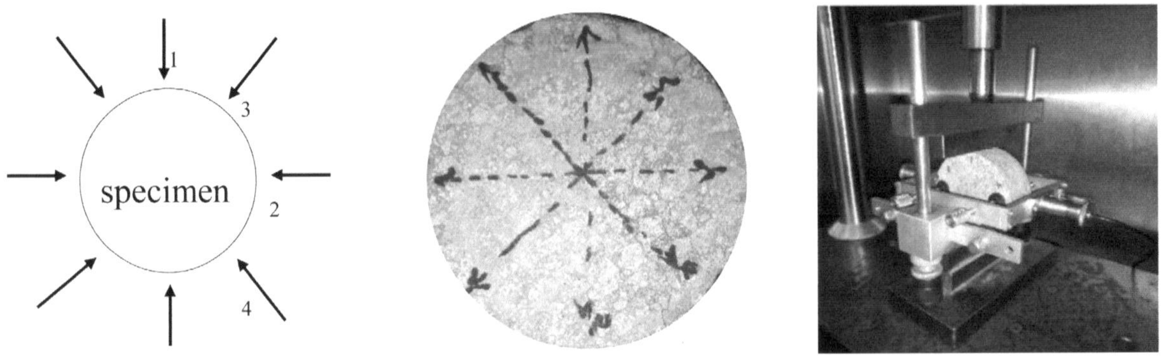

Figure 2. Multi-direction ITSM test.

85°C for 5 days), followed by at least 1 day conditioning at room temperature. Finally, the specimens were heated at 135°C for 4 h and manually crushed. In total, 55 specimens were aged, crushed, and mixed together to produce sufficient RAP materials. Four samples were randomly taken from the crushed RAP and subjected to gradation tests, with the results also shown in Figure 1. It can be seen that the grading results are very close to each other.

Experimental Program

Fabrication of Specimens Containing RAP. In this study, crushed RAP, new aggregate, and virgin binder (tinted) were mixed at 90, 110, and 130°C using a mechanical mixer. Before mixing, the crushed RAP was preheated at the mixing temperature for 2 h, the new aggregate was preheated at 150°C for 8 h, and the virgin binder was preheated at 135°C for 2 h. After preheating, the mixing was done by blending crushed RAP (40%), virgin aggregate, and binder in the mechanical mixer which was maintained at the planned mixing temperature (90, 110, or 130°C) for 30, 60, 120, or 240 s. Therefore, there are 12 different combinations of mixing temperature and time adopted in this study.

For each mixing condition combination, four cylindrical specimens (with 150 mm diameter and 115 ± 5 mm height) were compacted using a gyratory compactor. The specimens were conditioned at room temperature for 1 day, and then cored into cylindrical specimens with diameter of 100 mm. After that, two specimens from each group were cut and trimmed at both ends to fabricate two cylindrical specimens with height of 40 mm, and another two specimens were sliced into 10 disc specimens with thickness of 20 mm.

Multi-Direction ITSM Tests. Instead of the standard two-direction ITSM test, the cylindrical specimens were subjected to ITSM tests in four directions at 20°C using UTM-25, as shown in Figure 2. This is so that eight ITSM values for asphalt mixtures mixed under one combination of temperature and time can be used to mechanically demonstrate the material homogeneity.

Color Image Analysis. High-resolution digital pictures were taken for both sides of each disc specimen using a digital camera with the lighting and exposure conditions carefully controlled. It was found that the red-tinted binder was easier to identify when the specimen was wetted, and therefore water was sprayed before pictures were taken. In total, 240 pictures of 120 disc specimens were taken, and Figure 3 shows the disc specimens and an example picture.

In order to identify the distribution of the red-tinted new binder, image binarization based on image gray level was first processed. However, it was found that it was extremely difficult to distinguish the red-tinted new binder from the black RAP binder, as the gray levels of these two colors are very close. Therefore, a RGB color image analysis technique was employed and further developed to quantify the distribution of new binder.

It is well known that the RGB color model is an additive color model in which all colors can be reproduced by adding red (R), green (G), and blue (B) light together at various component rates. Based on this principle, the color of every point in an image can be quantitatively defined by the three components and represented as a point in a vector space. In this study, a computer program was written with MATLAB software so that the image processing which includes image reading, smoothing, segmentation, and binarization can be done more efficiently. According to this program, a representative red color was first selected as a reference, and its position in the RGB vector space was quantified by values of the three components. As the color of every point in the

Figure 3. Picture of disc specimens and example image.

specimen image can also be expressed as a point in the RGB vector space, the Euclidean distance between the points of the target color and reference color can be calculated by Equation 1. A threshold value of $D(z, m)$ was then set to identify the distribution of red color (new binder) within end surfaces of disc specimens, based on which more detailed analyses were carried out:

$$D(z, m) = z - m = \sqrt{(z_R - m_R)^2 + (z_G - m_G)^2 + (z_B - m_B)^2} \quad (1)$$

where $D(z, m)$ = Distance between the point of target color (z) and reference color (m) in RGB vector space; z_R, z_G, z_B = values of red, green, and blue components of target color (z) in the RGB color model; m_R, m_G, m_B = values of red, green, and blue components of the reference color (m) in the RGB color model.

Experimental Plan. In order to address the objective of this study more clearly, the experimental plan was developed as shown in Figure 4.

Results and Discussion

Homogeneity Analysis Based on ITSMs

Coupling Effect of Mixing Time and Temperature on Stiffness Homogeneity. Table 2 presents the results of ITSM tests in four different directions for each specimen. As stated above, for each combination of mixing temperature and time, eight ITSM values from eight different directions of two specimens were obtained. The mean value, standard deviation (*SD*) and coefficient of variation (*CoV*) of the ITSMs were calculated for each combination, with the calculation results also included in Table 2.

Figure 5*a* shows the average ITSM values of different groups. It can be seen that, in general, the stiffness of asphalt mixture containing RAP increases gradually with the increase of mixing temperature and mixing time. This is a reasonable result as it is well known that either a higher mixing temperature or a longer mixing duration will result in more significant age hardening to both the RAP and new binders. In addition, the increased mixing temperature and time can also lead to stronger blending between RAP and new binders, which means more RAP binder is contributing to the viscoelastic properties of asphalt mixtures. The result in Figure 5*a* also indicates that a homogeneity study based on *SD* values may be misleading to a certain extent, as the average ITSMs of different groups differ from each other. Therefore, the *CoV*, which is defined as the ratio of *SD* to the mean value, is being mainly used to show the effects of mixing temperature and time on the stiffness homogeneity of asphalt mixtures containing RAP material.

Figure 5*b* demonstrates the influences of different combinations of mixing temperature and time on the *CoVs* calculated from the eight ITSMs of each group. It can be clearly observed that the *CoV* varies significantly as the mixing condition changes, which indicates that the mixing temperature and time can considerably affect the mechanical homogeneity of asphalt mixtures containing RAP materials. It has to be noted that, different from the average stiffness results shown in Figure 5*a*, the relationship between mixing conditions and *CoV* results is not monotonic. It can clearly be seen that the relatively larger *CoVs* (representing poor homogeneity) generally occur at the intermediate levels of mixing temperatures and/or times in this study, revealing a complicated mechanism of mixing conditions affecting the stiffness homogeneity of asphalt mixtures containing RAP.

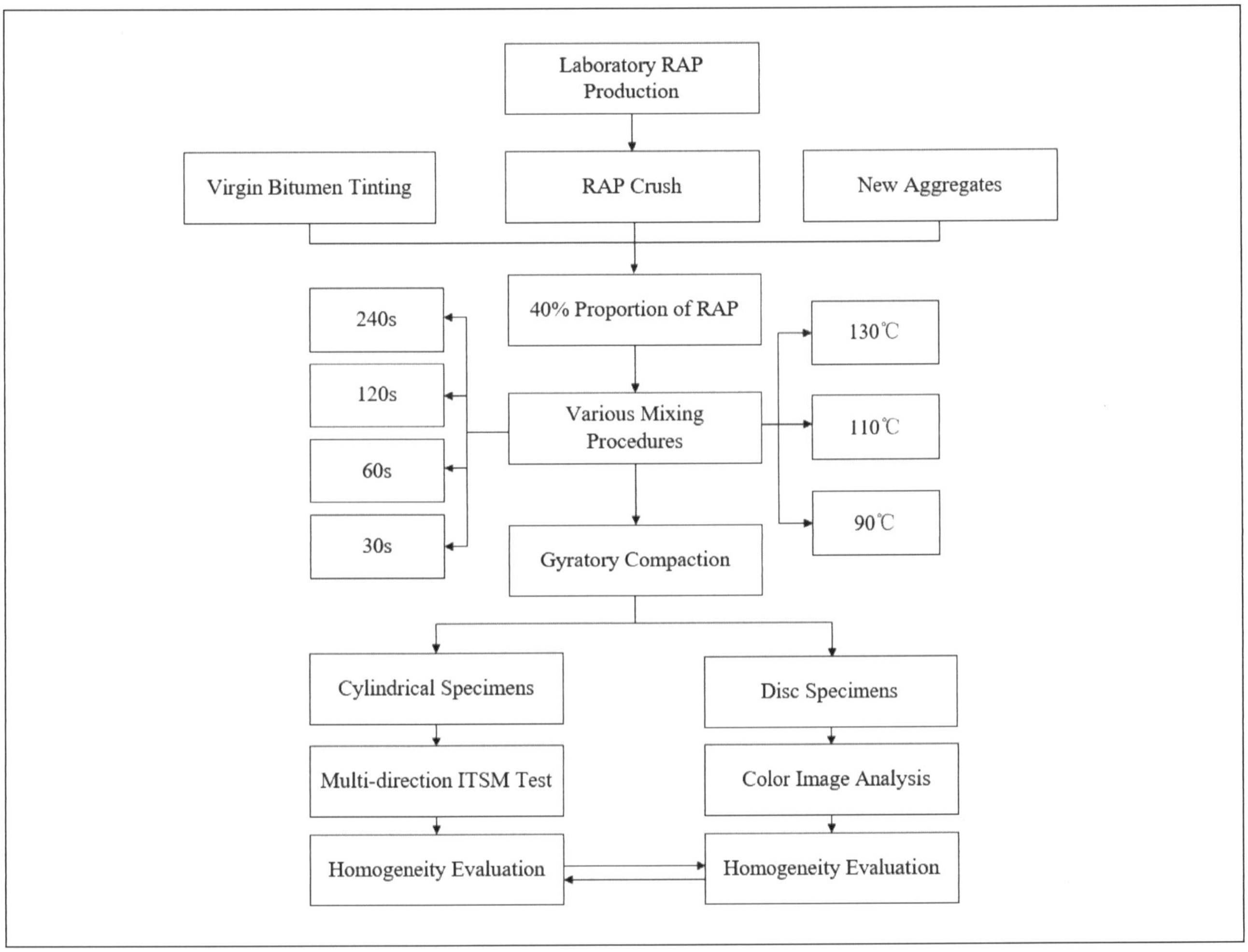

Figure 4. Experimental plan.

Effect of Mixing Time on Stiffness Homogeneity. To provide a better understanding of how mixing conditions affect the stiffness homogeneity of asphalt mixtures containing RAP, it is essential to quantitatively identify the effect of a single factor (either the mixing time or the mixing temperature) by fixing the other factor. Figure 6, *a–c*, presents the average (with *SD*), maximum, and minimum stiffness results of specimens fabricated after different mixing times at 90, 110, and 130°C. It can be seen that, at all the temperatures studied in this research, the results of *SD* and range (difference between maximum and minimum stiffness at certain mixing condition) first increase then decrease with the increase of mixing time, which indicates that at the three mixing temperatures selected in this study, the stiffness homogeneity of the specimens containing RAP first deteriorated and then improved as the mixing time increased.

The formation of clusters during RAP preheating and new mixture mixing and their asynchronous breaking during mixing are considered as the main causes of the above phenomenon. Clustering was clearly observed after RAP preheating and therefore, in this study, the lumps were briefly crushed by hand after preheating. However, it was found that clusters formed by the agglomeration of small-size RAP particles were difficult to break. In addition, these small RAP particles were observed to cluster again very easily after manual breaking. Therefore, it is believed that these clusters acted as aggregate particles in new mixtures before they were crushed by the mechanical mixer. Based on the results shown in Figure 6, it is highly likely that, at the mixing temperatures selected in this study, 30 s of mechanical mixing was not long enough to break the clusters, resulting in a momentary stiffness homogeneity of the mixtures containing RAP. After that, the clusters at different positions and with different resistance to the mechanical breaking work were crushed asynchronously, which led to a deterioration of stiffness homogeneity of the mixtures containing RAP. As the mixing time extended, most clusters were crushed and a stable stiffness homogeneity achieved.

Figure 7 presents the plot of the *CoVs* calculated from ITSMs versus the mixing times at different mixing

Table 2. Results of ITSM Tests (@20°C)

Temperature (°C)	Time (s)	Group code	Direction	ITSM (MPa)			SD (MPa)	CoV (%)
				Specimen 1	Specimen 2	Mean		
90	30	90c–30s	1	5806	5582	5206	452	9
			2	4835	5149			
			3	4253	5416			
			4	5203	5404			
	60	90c–60s	1	3743	5418	5730	1379	26
			2	6664	6174			
			3	6452	8157			
			4	5385	3843			
	120	90c–120s	1	7751	6149	5961	1007	18
			2	5505	4035			
			3	6791	5997			
			4	5446	6010			
	240	90c–240s	1	8652	8185	7794	411	6
			2	7494	7435			
			3	7549	7477			
			4	7579	7982			
110	30	110c–30s	1	5353	4553	4871	484	11
			2	4955	4769			
			3	4499	5670			
			4	5123	4043			
	60	110c–60s	1	7007	5107	5611	1365	26
			2	6311	4563			
			3	7849	3610			
			4	6167	4272			
	120	110c–120s	1	6440	4775	6051	1532	27
			2	9149	7145			
			3	4202	5398			
			4	6658	4639			
	240	110c–240s	1	6721	8856	7492	779	11
			2	7686	8404			
			3	6300	7434			
			4	7313	7220			
130	30	130c–30s	1	5918	6081	5304	551	11
			2	5690	5373			
			3	4456	4984			
			4	4618	5314			
	60	130c–60s	1	5126	6934	6115	625	11
			2	5382	6292			
			3	6179	6839			
			4	5651	6518			
	120	130c–120s	1	5133	6084	6534	1390	23
			2	5239	9839			
			3	7123	6352			
			4	6534	5970			
	240	130c–240s	1	7782	7698	7074	584	9
			2	6079	7608			
			3	7332	6493			
			4	6864	6732			

temperatures. It can be seen that for all the mixing temperatures (90, 110, and 130°C) the *CoV* varied considerably with mixing time, illustrating that the mixing time can significantly affect the stiffness homogeneity of asphalt mixtures containing RAP. In addition, the *CoVs* corresponding to mixing times of 30 s and 240 s are much smaller than those with mixing times of 60 s and 120 s, which further demonstrates the above-mentioned phenomena of "momentary homogeneity" and "stable homogeneity." Special attention should be paid to the change in *CoVs* from 30 s to 60 s for different mixing temperatures. It is shown that for the mixing temperature of 130°C, the *CoV* with mixing time of 60 s remained at the same level as the *CoV* with mixing

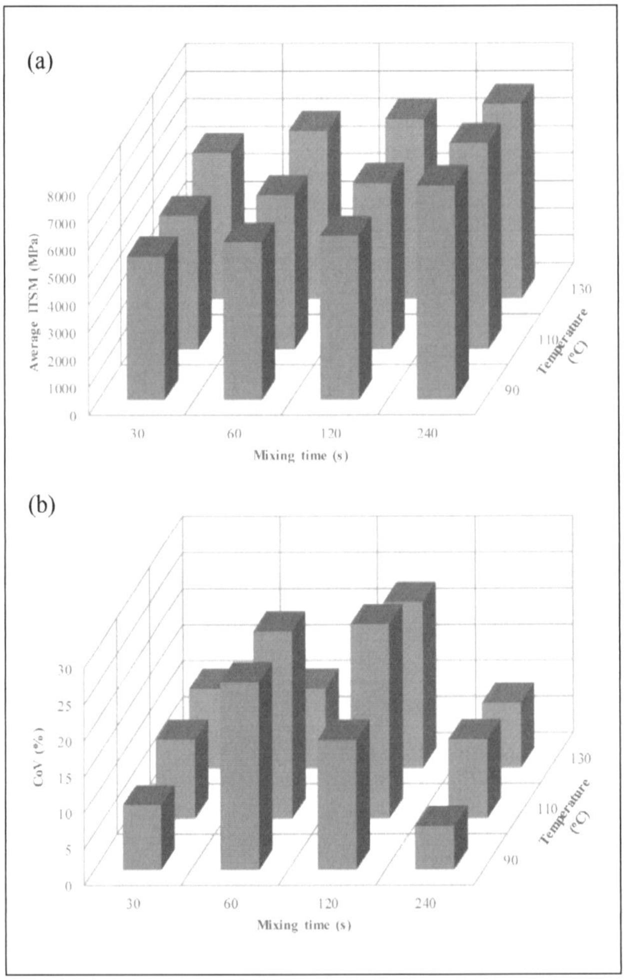

Figure 5. The average ITSM and CoV versus mixing temperature and time: (*a*) ITSM results; (*b*) CoVs.

time of 30 s. As stated above, the deterioration of the stiffness homogeneity of the mixtures containing RAP is believed to mainly result from the asynchronous crushing of RAP clusters during the mechanical mixing. Therefore, this implies that it took longer than 60 s for the mechanical mixer to break the clusters at the mixing temperature of 130°C. In contrast, for the mixing temperatures of 90 and 110°C, obvious increase of *CoVs* can be observed when mixing time extends from 30 s to 60 s, revealing that at these two mixing temperatures clusters started to be broken between mixing time of 30 s and 60 s. Therefore, the mixing temperature is considered to be decisive in determining the resistance of clusters to breaking, and within the mixing conditions of this study, the resistance increases with the increase of mixing temperature.

To further evaluate the effects of mixing time on mixture homogeneity, third-degree polynomial trendlines were added for the data of different temperatures (also shown in Figure 7). Although the trendlines are not able to represent the real relationship between *CoV* and mixing time, informative observation can still be made with respect to the predicted mixing time where the peak value of *CoV* appears. It is shown in Figure 7 that the predicted mixing time corresponding to the maximum value of *CoV* increases with the increase of mixing temperatures, which further supports that the resistance of the clusters to the breaking work from mechanical mixer increased with the increase of mixing temperature.

An interesting observation can also be made by combining the results shown in Figures 6 and 7. From Figure 6, it can be seen that for mixing temperatures of 90 and 110°C, the increase rate of average ITSMs between mixing times of 120 s and 240 s is higher than that between mixing times of 60 s and 120 s. In Figure 7, horizontal secant lines going through the points of *CoVs* with 240 s mixing time were added to the trendlines of mixing temperatures 90 and 110°C. This is so that the approximate time point for the mixtures to achieve "stable homogeneity" can be displayed. Based on this, the concept of a cluster-breaking period is defined as the period from the time point when clusters began to break to the time point when the mixtures initially achieved "stable homogeneity." During the cluster-breaking period, increasing amount of aged binder from inside clusters would be available for blending with new binder and the total gradation would be improved, both of which can accelerate the increase of ITSMs of mixtures containing RAP. After the cluster-breaking period, the blending between the aged and new binder continued until complete blending was accomplished. In terms of a mixing temperature of 130°C, as its cluster-breaking period started later than 90 and 110°C and was roughly on the point of achieving "stable homogeneity" at mixing time of 240 s, the acceleration of its ITSM increase as a result of binder blending and gradation improvement did not appear.

Effect of Mixing Temperature on Stiffness Homogeneity. Figure 8 shows the plot of average, maximum, and minimum ITSMs and *CoVs* versus mixing temperatures with different mixing times. From this figure, it can clearly be seen that, with a certain mixing time, the mixing temperature is shown to have no significant effect on the average ITSM values. With regards to the *CoV* results (homogeneity), the influence of mixing temperature varies depending on the mixing temperature. For mixing times of 30 s and 240 s (as shown in Figure 8, *a* and *d*), the change of mixing temperature does not result in dramatic differences in *CoV* results. In contrast, considerable difference can be observed among the *CoVs* at different mixing temperatures for the 60 s and 120 s mixing times. In light of the discussion in the above subsection, the main reason for this is that mixing times of 60 s and 120 s were still

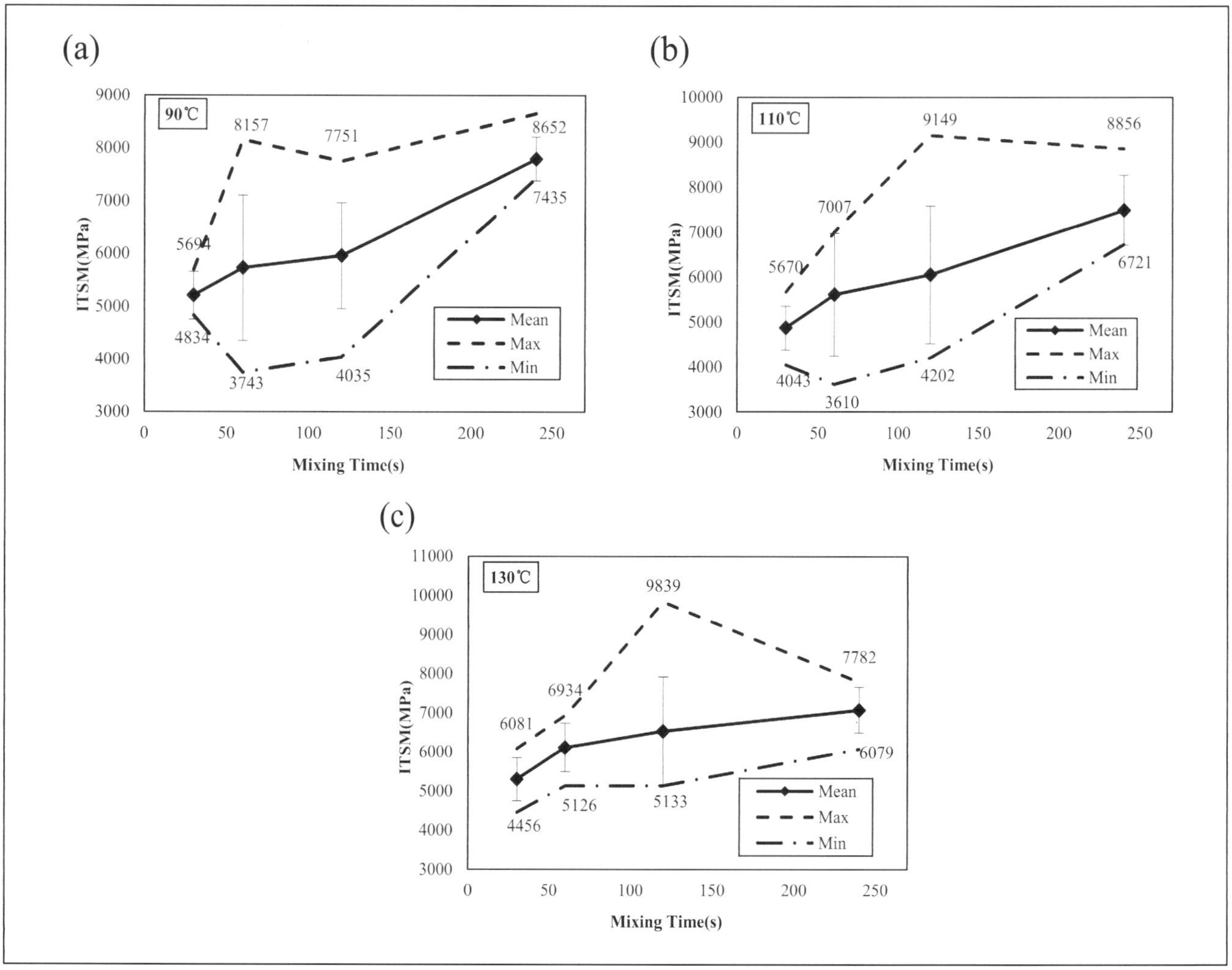

Figure 6. The average, maximum and minimum stiffness values versus mixing time at different mixing temperatures: (*a*) ITSM results at 90°C; (*b*) ITSM results at 110°C; (*c*) ITSM results at 130°C.

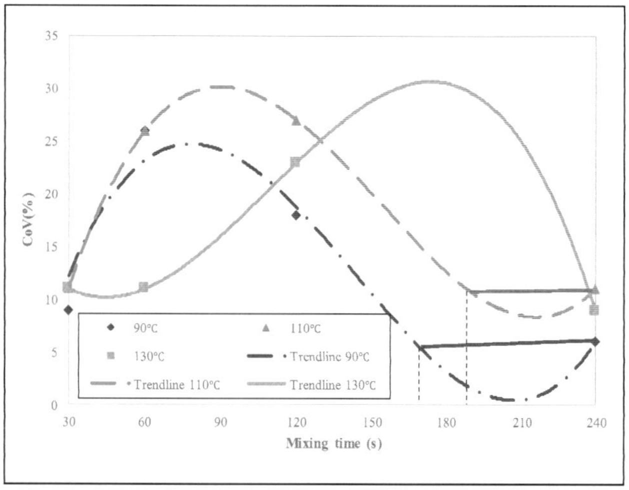

Figure 7. The plot of CoV value versus mixing time at different mixing temperatures.

within the cluster-breaking period. During this period the mixing temperature was decisive in determining the resistance of clusters to the breaking work of mechanical mixer, which in turn affected the proportion of crushed clusters and therefore the homogeneity of the asphalt mixtures containing RAP.

Homogeneity Analysis Based on Color Images

Although the MATLAB program developed in this study improved the efficiency of the image processing (i.e., image reading, smoothing, segmentation, and binarization), the program configuration was still extremely time consuming. In addition, the homogeneity analysis based on stiffness above has shown that, within the mixing conditions used in this study, the effects of mixing time on homogeneity of asphalt mixtures containing RAP was

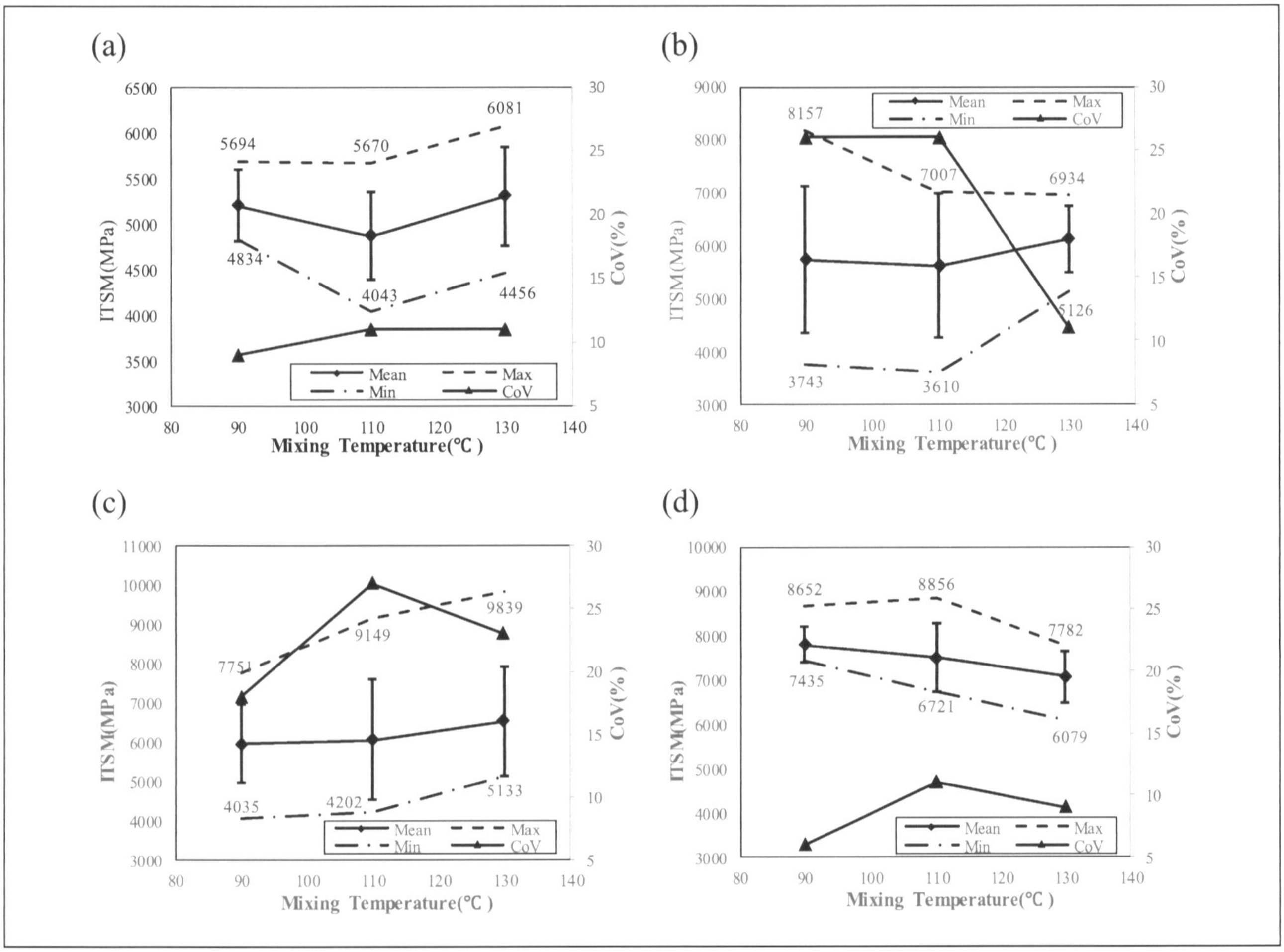

Figure 8. ITSM results and CoVs versus mixing temperatures with different mixing times: (*a*) 30 s; (*b*) 60 s; (*c*) 120 s; (*d*) 240 s.

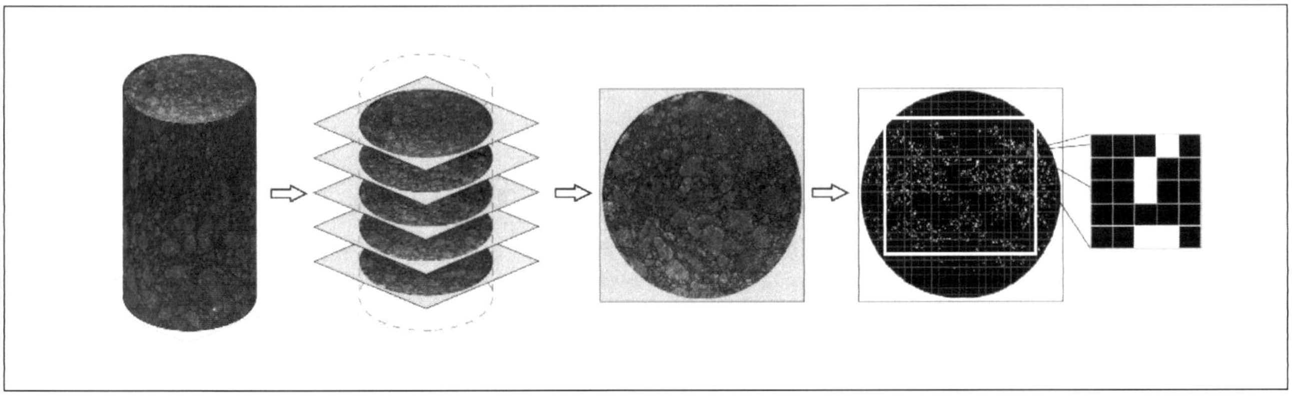

Figure 9. Image analysis based on pixel counting.

more dominant than the mixing temperature. Therefore, in this paper, image analysis was carried out only for specimens fabricated via mixing conditions at 110°C for 30, 60, 120, and 240 s.

Forty binary images of 20 specimens were obtained using the specially written MATLAB program. After this, in order to make use of these images to analyze the homogeneity of the asphalt mixtures, another program written with Python was developed. In this program, a 40 × 40 pixel mesh filter was first built and then used to progressively scan a square area of every binary image, based on which the number of white pixels (representing

Table 3. Pixel Number SDs for Different Surfaces of Different Mixing Times (@ 110°C)

Surface no.	30 s SD	30 s Average	60 s SD	60 s Average	120 s SD	120 s Average	240 s SD	240 s Average
1	12.87	16.62	19.22	19.66	18.61	17.37	19.27	17.17
2	16.28		17.93		21.30		14.01	
3	15.89		22.74		18.19		16.04	
4	14.34		15.29		20.26		16.80	
5	16.47		16.01		13.68		16.31	
6	16.69		24.35		18.96		14.49	
7	22.93		20.56		18.83		17.91	
8	18.57		18.69		16.32		19.40	
9	16.00		18.85		13.48		19.47	
10	16.16		22.92		14.05		17.95	

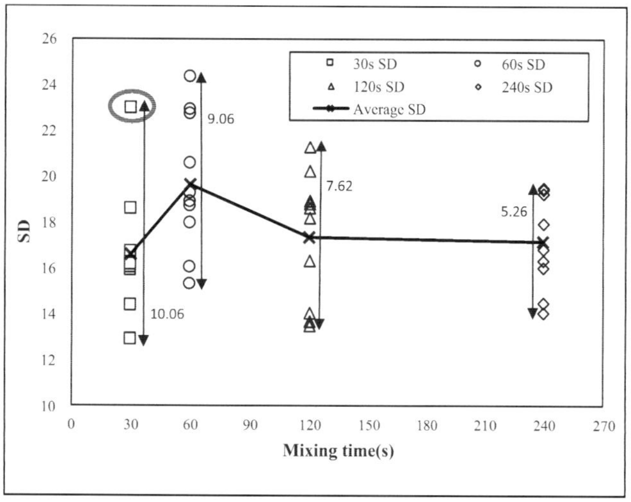

Figure 10. Effects of mixing time on pixel number SDs for different surfaces (@ 110°C).

the new red binder) within each of the 40 × 40 area was counted and the standard deviation (*SD*) was calculated with the scanning results. The scanning program is graphically shown in Figure 9.

The calculated *SD* results are presented in Table 3 and Figure 10. The mixing time of 60 s is shown to have the highest value of average *SD*, and the average *SDs* of other three mixing times are at a similar level. This indicates that, based on the average *SDs* of different mixing times, the homogeneity of asphalt mixtures containing RAP firstly deteriorated and then improved as the mixing time increased, which correlates very well with the observations made from the multi-direction stiffness analyses (as shown in Figures 6 and 7).

The *SD* range (difference between maximum and minimum *SDs*) of each mixing time was also calculated and listed in Figure 10. Different from the average *SD* results, the *SD* range is observed to decrease with the increase of mixing time, which indicates that the homogeneity of mixture continuously improved with the increase of mixing time. However, it should be noted that the large *SD* range for mixing time of 30 s is mainly because of an irregular maximum value (as shown in Figure 10), and therefore observations made with these data included are considered to be misleading.

Conclusions

Based on the results and discussion of this study, the following conclusions can be drawn:

1. Both the multi-direction ITSM test and the color image analysis technique were proved to be capable of effectively characterizing the homogeneity property of asphalt mixtures containing RAP. However, for the color image analysis technique, it is suggested that a more advanced tinting technique with more distinguishing color and fewer negative effects on mechanical/rheological properties of binders be used in future study.
2. The deterioration of homogeneity of asphalt mixtures containing RAP was mainly caused by the asynchronous breaking of clusters.
3. For the mixing conditions adopted in this study, it was shown that the effects of mixing time on the homogeneity of asphalt mixtures containing RAP was more significant than mixing temperature. However, the mixing temperature was decisive in shaping the influencing law of mixing time on the homogeneity.
4. The mixing temperature was decisive in determining the resistance of clusters to breaking work. For the mixing conditions chosen in this study, the resistance increases with the increase of mixing temperature, and therefore it took a longer mixing time to start and complete the break of clusters at higher mixer temperature.

5. Special attention should be paid to the phenomenon of "momentary homogeneity," which was detected and defined in this study. The "momentary homogeneity" was a result of short mixing time and most of the clusters were not broken, resulting in poor blending between aged and new binders.

Acknowledgments

The authors would like to acknowledge the financial support provided by The National Natural Science Foundation of China (No.51108157), and The Fundamental Research Funds for the Central Universities (Project No. 2015B17214 and 2015B17014).

Author Contributions

The authors confirm contribution to the paper as follows: study conception and design: JW, QL, JC; data collection: QL, YW, LX; analysis and interpretation of results: DW, CA; draft manuscript preparation: JW. All authors reviewed the results and approved the final version of the manuscript.

References

1. Apeagyei, A. K., T. M. Clark, and T. M. Rorrer. Stiffness of High-RAP Asphalt Mixtures: Virginia's Experience. *Journal of Materials in Civil Engineering*, Vol. 25, No. 6, 2013, pp. 747–754.
2. Rinaldini, E., P. Schuetz, M. N. Partl, G. Tebaldi, and L. D. Poulikakos. Investigating the Blending of Reclaimed Asphalt with Virgin Materials using Rheology, *Electron Microscopy and Computer Tomography. Composites: Part B*, Vol. 67, 2014, pp. 579–587.
3. European Asphalt Pavement Association. *Asphalt in Figures [online]*, 2012. http://eapa.org/usr_img/Asphalt%20in%20figures%20Version%2022-12-2011.pdf.
4. Federal Highway Administration and the U.S. Environmental Protection Agency. *A Study of the Use of Recycled Paving Material*. FHWA-RD-93 and EPA/600/R-93/095, Washington, D.C., 1993.
5. Copeland, A. *Reclaimed Asphalt Pavement in Asphalt Mixtures: State of the Practice*. Publication FHWA-HRT-11-021. Turner-Fairbank Highway Research Center, Federal Highway Administration, 2011.
6. Swiertz, D., E. Mahmoud, and H. Bahia. Estimating the Effect of Recycled Asphalt Pavements and Asphalt Shingles on Fresh Binder, Low-Temperature Properties without Extraction and Recovery. *Transportation Research Record: Journal of the Transportation Research Board*, 2011. 2208: 48–55.
7. Huang, S., A. T. Pauli, and R. W. Crimes. Aging Characteristics of RAP Binder Blends: What Types of RAP Blenders Are Suitable for Multiple Recycling? *Journal of the Association of Asphalt Paving Technologist*, Vol. 83, 2014, pp. 297–343.
8. Zaumanis, M., and R. B. Mallick. Review of Very High-Content Reclaimed Asphalt Use in Plantproduced Pavements: State of the Art. *International Journal of Pavement Engineering*, Vol. 16, No. 1, 2015, pp. 39–55.
9. Carpenter, S., and J. Wolosick. Modifier Influence in the Characterization of Hot-Mix Recycled Material. *Transportation Research Record: Journal of the Transportation Research Board*, 1980. 777: 15–22.
10. Potter, J., and J. Mercer. Full-Scale Performance Trials and Accelerated Testing of Hot-mix Recycling in the UK. *Proc., 8th International Conference on Asphalt Pavements*, International Society for Asphalt Pavements, Seattle, WA, 1997, pp. 593–607.
11. Huang, B., G. Li, D. Vukosavljevic, X. Shu, and B. K. Egan. Laboratory Investigation of Mixing Hot-Mix Asphalt with Reclaimed Asphalt Pavement. *Transportation Research Record: Journal of the Transportation Research Board*, 2005. 1929: 37–45.
12. Al-Qadi, I. L., M. A. Elseifi, and S. H. Carpenter. *Reclaimed Asphalt Pavement: A Literature Review*. Report Number FHWA-ICT-07-001. Illinois Center for Transportation, Springfield, IL, 2007.
13. Shirodkar, P., Y. Mehta, A. Nolan, K. Sonpal, A. Norton, C. Tomlinson, E. Dubois, P. Sullivan, and R. Sauber. A Study to Determine the Degree of Partial Blending of Reclaimed Asphalt Pavement (RAP) Binder for High RAP Hot Mix Asphalt. *Construction and Building Materials*, Vol. 25, No. 1, 2011, pp. 150–155.
14. Navaro, J., D. Bruneau, I. Drouadaine, J. Colin, A. Dony, and J. Cournet. Observation and Evaluation of the Degree of Blending of Reclaimed Asphalt Concretes using Microscopy Image Analysis. *Construction and Building Materials*, Vol. 37, 2012, pp. 135–143.
15. Bressi, S., A. G. Dumont, and M. Pittet. Cluster Phenomenon and Partial Differential Aging in RAP Mixtures. *Construction and Building Materials*, Vol. 99, 2015, pp. 288–297.
16. Bressi, S., M. C. Cavalli, M. N. Partl, G. Tebaldi, A. G. Dumont, and L. D. Poulikakos. Particle Clustering Phenomena in Hot Asphalt Mixtures with High Content of Reclaimed Asphalt Pavements. *Construction and Building Materials*, Vol. 100, 2015, pp. 207–217.
17. Bressi, S., M. Pittet, A. G. Mumont, and M. N. Partl. A Framework for Characterizing RAP Clustering in Asphalt Concrete Mixtures. *Construction and Building Materials*, Vol. 106, 2016, pp. 564–574.

The Standing Committee on Surface Requirements of Asphalt Mixtures (AFK40) peer-reviewed this paper (18-04587).

Article

TRR
JOURNAL OF THE TRANSPORTATION RESEARCH BOARD

Transportation Research Record
2018, Vol. 2672(28) 178–189

DOI: 10.1177/0361198118787960
journals.sagepub.com/home/trr
SAGE

Performance of Virginia's Early Foamed Warm Mix Asphalt Mixtures

Stacey D. Diefenderfer[1]

Abstract

The Virginia Department of Transportation began allowing the use of warm mix asphalt (WMA) in 2008. Although several WMA technologies were investigated prior to implementation, foamed WMA was not. This study evaluated the properties and performance of foamed WMA placed during the initial implementation of the technology to determine whether the technology had performed as expected. Six mixtures produced using plant foaming technologies and placed between 2008 and 2010 were identified and subjected to field coring and laboratory testing. Coring was performed in 2014, resulting in pavement ages from 4 to 6 years. Three comparable hot mix asphalt (HMA) mixtures were cored at 5 years for comparison. Cores were evaluated for air-void contents and permeability and were subjected to dynamic modulus, repeated load permanent deformation, and overlay testing. In addition, binder was extracted and recovered for performance grading. Similar properties were found for the WMA and HMA mixtures. One WMA mixture had high dynamic modulus and binder stiffness, but overlay testing did not indicate any tendency for premature cracking. All binders had aged between two and three performance grades above that specified at construction. WMA binders and one HMA binder aged two grades higher, and the remaining two HMA binders aged three grades higher, indicating a likely influence on aging of the reduced temperatures at which the early foamed mixtures were typically produced. Overall results indicated that foamed WMA and HMA mixtures should be expected to perform similarly.

Warm mix asphalt (WMA) is produced by incorporating additives into asphalt mixtures or by foaming asphalt binder to allow production and placement of the mixture at temperatures below the production temperatures of conventional hot mix asphalt (HMA). Benefits such as reduced plant emissions, improved compaction in the field, extension of the paving season into colder weather, longer haul distances, reduced opening time to traffic, and reduced energy consumption at the plant may be realized with different applications.

Research on the uses and benefits of WMA has been promising, and many states, including Virginia, now allow several WMA technologies. However, knowledge with regard to the long-term effect of these technologies on mixture performance is limited. In NCHRP Project 09-49, Martin et al. found no evidence from an examination of in-service pavements that WMA is more susceptible to moisture damage than HMA, although laboratory results indicated potential issues (*1*). In NCHRP Project 09-49A, the long-term performance of pavements constructed using various WMA technologies was comparable with that of the companion HMA pavement sharing similar pavement structure, climate, and traffic conditions (*2*). In Virginia, the use of some form of WMA has become nearly ubiquitous, comprising more than 75% of asphalt mixture production reported in 2014 (*3*). The Virginia Department of Transportation (VDOT) is thus interested in evaluating the lifetime performance of in-state WMA installations to determine their performance as compared to that of conventional HMA. In the past several years, more research has been initiated to investigate the performance of WMA. Although the majority of these studies were laboratory studies, the purpose of a select few was to evaluate the performance of WMA in the field. Most of the studies were limited to initial performance, as WMA has not been in use long enough for its lifetime performance to be evaluated.

VDOT's initial WMA trials were constructed in 2006 and documented in several studies (*4–6*). These trials assessed the Sasobit additive and Evotherm DAT technology and determined based on initial construction and early performance data that WMA should perform similarly to HMA. After the trials, VDOT developed a special provision for the use of WMA and in 2009 incorporated the use of approved WMA technologies in their specifications. At the time the trials were constructed, plant foamed WMA had not been introduced; however, by 2010, plant foaming was increasingly used to produce WMA because of its cost-competitiveness with HMA.

[1]Virginia Transportation Research Council, Charlottesville, VA

Corresponding Author:
Address correspondence to Stacey D. Diefenderfer:
stacey.diefenderfer@vdot.virginia.gov

Table 1. Mixture Sites

Mixture	Year	VDOT district	County	Route	Mileposts	Length	Mixture type
A	2008	Lynchburg	Pittsylvania	US 29 SBL	0.97-4.45	3.48 mi	SM-9.5D
B	2009	Staunton	Frederick	US 522 NBL	0.0-1.52	1.52 mi	SM-12.5D
C	2010	Staunton	Frederick	SR 7 WBL	0.0-3.64	3.64 mi	SM-12.5D
D	2010	Staunton	Rockingham	US 33 WBL	30.01-37.71	7.70 mi	SM-12.5A
E	2010	Lynchburg	Campbell	SR 128	0.16-0.69	0.53 mi	SM-9.5D
F	2010	Lynchburg	Pittsylvania	US 29 NBL	4.42-7.72	3.30 mi	SM-9.5D
HMA 1	2006	Culpeper	Rappahannock	SR 211 EBL	5.22-5.72	0.50 mi	SM-9.5A
HMA 2	2006	Staunton	Highland	US 220 NBL	2.41-2.91	0.50 mi	SM-12.5A
HMA 3	2006	Hampton Roads	York	SR 143 NBL	5.87-6.17	0.30 mi	SM-9.5D

Note: VDOT = Virginia Department of Transportation; SBL = southbound lane; NBL = northbound lane; WBL = westbound lane; HMA = hot mix asphalt; EBL = eastbound lane.

Purpose and Scope

The purpose of this study was to document the material properties of early foamed WMA sections constructed in Virginia and investigate the long-term performance of the sections to validate the expectation of equivalent performance. Six mixtures produced using plant foaming technologies and placed between 2008 and 2010 were identified and subjected to field coring and laboratory testing. Coring was performed in 2014 when pavement ages ranged from 4 to 6 years.

Information collected included the documentation of sections constructed with plant foaming technologies. Sites were visited and the general condition of the sections was observed to assess performance. Cores were collected to verify mixture properties such as in-place density and binder grade and content. The cores were also tested to determine modulus values and evaluate resistance to rutting.

No "control" HMA pavements were constructed for direct comparison with the foamed WMA pavements. For the purpose of comparison, three HMA mixtures with known properties and performance were cored at 5 years of age to provide average age reference points for material properties.

Methods

Evaluation of Sections

Sites. Site information and the designation for each mixture are provided in Table 1. Mix design information is provided in Table 2 for all mixtures. Figure 1 shows the design gradation of each mixture. Production information was not available.

Coring. Ten cores 6 in. (150 mm) in diameter were taken from each foamed WMA pavement section for evaluation. Six cores 6 in. (150 mm) in diameter were taken from each HMA section for evaluation. Core sites were randomized along the length and width of the section.

VDOT Pavement Management System (PMS) Data. Distress data for the foamed WMA sites were extracted from VDOT's PMS. VDOT's maintenance division acquires and maintains the results of an annual condition survey of all interstates, all primaries, and approximately 20% of secondary pavements. The survey collects and summarizes detailed distress data for each 0.1 mi of right-lane or principal direction pavement surface. Condition is reported on a scale from 0 to 100, completely failed to new or like new, respectively. The overall section rating, the composite condition index (CCI), is the lower of two ratings that summarize the load-related and non-load-related distresses for a pavement.

Laboratory Evaluation

Core Air Voids. Air-void contents were determined in accordance with AASHTO T 269.

Permeability. Permeability testing was performed on cores in accordance with Virginia Test Method 120, Method of Test for Measurement of Permeability of Bituminous Paving Mixtures Using a Flexible Wall Permeameter (*7*).

Dynamic Modulus Test. Dynamic modulus tests were performed using an asphalt mixture performance tester (AMPT) with a 25 to 100 kN loading capacity in accordance with AASHTO T 342. Tests were performed on 38-mm-diameter × 110-mm-tall specimens cored horizontally from field cores (*8*, *9*). Three testing temperatures (4.4°C, 21.1°C, and 37.8°C) and six testing frequencies (ranging from 0.1 Hz to 25 Hz) were used. Tests were conducted from the coldest temperatures to the warmest temperatures. At each temperature, tests were performed from the highest to the lowest frequency. Load levels were selected in such a way that at each temperature–frequency combination, the applied strain was in the range of 75 to 125 microstrains. All tests were conducted in the uniaxial mode without confinement. Stress versus strain values were captured continuously and used to calculate dynamic modulus. Dynamic modulus was computed automatically using IPC $|E^*|$ software. The results for each temperature–frequency combination for each mixture type are reported for three replicate specimens.

Table 2. Mix Design Information for Plant Foamed WMA and HMA Mixtures

	Mixture								
Mix design	A	B	C	D	E	F	HMA 1	HMA 2	HMA 3
Sieve size, % Passing									
19 mm	100	100	100	100	100	100	100	100	100
12.5 mm	100	96	96	96	100	100	100	96	100
9.5 mm	94	84	84	86	96	93.5	92	86	93
4.75 mm	64	NA	NA	NA	64	58	60	NA	61
2.36 mm	46	40	38	35	NA	43	43	34	44
0.075 mm	5.8	6.5	6.5	5.6	5.8	5.2	5.7	6	5
Volumetric properties									
VTM	3.5	4.0	NA	3.2	4.0	3.8	NA	NA	NA
VFA	78	NA	NA	NA	NA	NA	NA	NA	NA
Asphalt content	5.8	5.5	5.5	5.7	5.8	5.6	5.5	5.2	5.7
VMA	15.9	NA	NA	NA	NA	NA	NA	NA	NA
FA ratio	1.0	NA	NA	NA	NA	NA	NA	NA	NA
G_{mm}	2.436	2.613	NA	2.423	2.420	2.529	NA	NA	NA
Other properties									
RAP content	25%	25%	25%	15%	26%	26%	12%	10%	20%
Binder grade	PG 64-22	PG 64-22	PG 64-22	PG 64-22	PG 64-22	PG 64-22	PG 64-22	PG 64-22	PG 70-22

Note: WMA = warm mix asphalt; HMA hot mix asphalt; NA = data not available; VTM = voids in total mix; VFA = voids filled with asphalt; VMA = voids in mineral aggregate; FA ratio = fines/aggregate ratio; RAP = recycled asphalt pavement.

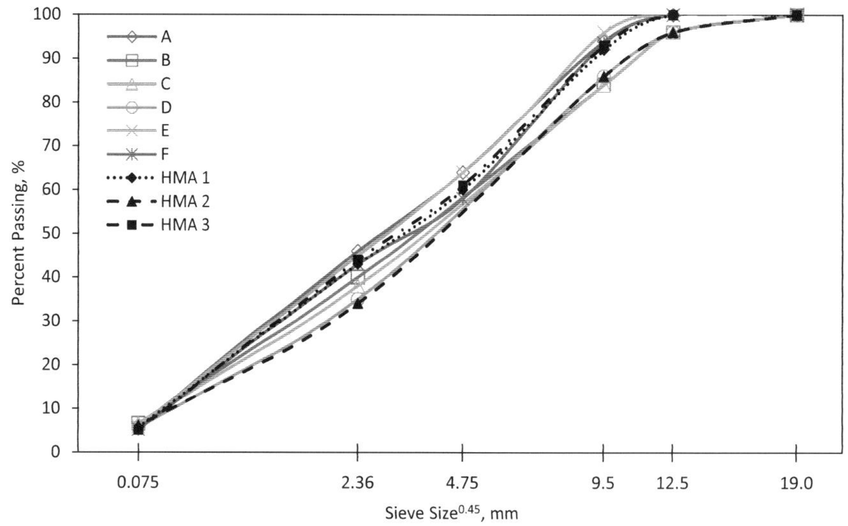

Figure 1. Design mixture gradations.
Note: HMA = hot mix asphalt.

Repeated Load Permanent Deformation (RLPD) Test. The repeated load permanent deformation (RLPD) test was used to evaluate rutting resistance. An AMPT with a 25 to 100 kN loading capacity was used. Testing was performed on 38-mm-diameter × 110-mm-tall specimens cored horizontally from field cores. Testing was conducted on specimens previously tested for dynamic modulus. Tests were conducted at 54°C based on LTPPBind software that represents the 50% reliability maximum high pavement temperature at sites in central Virginia. A repeated haversine axial compressive load pulse of 0.1 s every 1.0 s was applied to the specimens. The tests were performed in the confined mode using a confining stress of 10 psi (68.9 kPa) and a deviator stress of 70 psi (483 kPa). The tests were continued for 10,000 cycles or a permanent strain of 10%, whichever came first. During the test, permanent strain (εp) versus the number of loading cycles was recorded automatically, and the results were used to estimate the flow number, which was determined numerically as the cycle number at which the strain rate is at a minimum, based on the Francken model.

The slope and intercepts of the secondary portion of the cycle versus permanent strain curve were estimated in log-log space by applying a power-law function:

$$\varepsilon_p = aN^b \quad (1)$$

where
ε_p = permanent strain,
a = intercept in log-log space,
N = cycle number, and
b = slope in log-log space.

The slope of this line is indicative of the plastic strain rate of the mixture. A higher slope relative to other mixtures indicates a mixture that is likely to be more susceptible to rutting. The intercept of the line is indicative of the plastic strain at cycle $N = 1$ (*10*, *11*).

No preconditioning load was used during testing; however, Gibson and Li reported that the initial RLPD testing performed in the laboratory testing for NCHRP Project 09-30A used a preconditioning stage consisting of the application of 100 cycles of 10% of the deviator stress (*12*). Gibson and Li conducted an analysis to determine a method wherein virtual preconditioning could be applied to test results obtained without preconditioning to approximate the change in response predicated by the actual preconditioning cycles. This virtual preconditioning consisted of the removal of the first 51 cycles of data and the subsequent re-zeroing of the 52nd cycle to approximate the response curve after preconditioning cycles have been performed. Except for the determination of flow number, all analyses in the current study were conducted after virtual preconditioning was applied.

Overlay Test. The Texas overlay test was performed to assess the susceptibility of each mixture to cracking (*13*). Testing was performed using a universal testing machine with a loading capacity of 25 to 100 kN on field cores having a 150-mm diameter and varying thickness. Testing was performed at 25°C ± 0.5°C. Loading was applied for 1,200 cycles or until a reduction of 93% or more of the maximum load was reached.

Binder Extraction and Recovery. Extraction of binder from cores was performed in accordance with AASHTO T 164, Method A, using n-propyl bromide as the solvent. Binder was recovered from the solvent using the Rotavap recovery procedure specified in AASHTO T 319.

Binder Testing. Binder grading was performed in accordance with AASHTO M 320.

Results and Discussion

Visual Assessment

Visual assessment of the foamed WMA mixtures indicated no issues with performance. Mixture B had been overlaid at an age of 6 years during the summer prior to coring and thus the pre-overlay condition could not be seen. HMA mixtures were generally performing as expected.

PMS Data

Distress data extracted from VDOT's PMS for sections containing the foamed WMA mixtures and HMA mixtures are summarized in tables 3 and 4, respectively.

Section B had a CCI of 59 prior to the overlay placed in 2015. This appeared to be due primarily to rutting; a warehouse distribution center and shopping center located at the north end of this section were likely contributors. Sections A, F, and HMA 1 had the next lowest CCIs: 69, 72, and 72 at ages 8, 6, and 5 years, respectively. Section A had considerable cracking. All other sections had CCIs of 80 or greater. Ride quality for all sections was acceptable, with six of the nine sections having an International Roughness Index (IRI) below 95 and the remaining three sections having an IRI below 115. Section HMA 1 had the lowest IRI at 60; this was due to pavement structural issues and was not a function of the surface course performance.

Core Air Voids

Air voids were measured for all cores from all foamed WMA and HMA mixtures. Table 5 details the results. Cores from mixtures B and D had very low average void contents, at 2.97% and 1.94%, respectively, and the remaining mixtures had average void contents from 5% to 9%. The average air-void content and standard deviation for all WMA specimens were 5.9% and 2.8%, respectively; those for HMA specimens were 6.3% and 1.5%, respectively. A paired Student's *t*-test assuming unequal variance indicated that the air-void contents of the WMA and HMA specimen sets were not significantly different.

Permeability results for all tested cores along with the predicted permeability at 7.5% air voids are shown in Table 6. Cores from mixtures HMA 1 and HMA 2 were not tested. A plot of the permeability results (Figure 2) indicated that all results with the exception of those for Mixture D were below the specification requirement of 150×10^{-5} cm/s at predicted permeability at 7.5% air voids.

Dynamic Modulus Test

Dynamic modulus testing was performed on small-scale specimens cored horizontally from the road cores. Table 7 provides the results. Ideally, specimen sets should be ±0.5% of the target air-void content, which was the average air-void

Table 3. PMS Distress Data for Foamed WMA Sections

Data	Section					
	A	B[a]	C	D	E[b]	F
Year paved	2008	2009	2010	2010	2010	2010
Year evaluated	2016	2014	2016	2016	2015	2016
Segment length, mi	2.99	1.56	1.81	7.65	0.28	3.03
Average IRI, left wheel path	77	112	78	78	112	80
Average IRI, right wheel path	91	111	81	85	116	74
Average IRI	*84*	*112*	*80*	*82*	*114*	*77*
Transverse cracking severity 1, avg. ft/mi	5176.2	0	16.6	442.0	146.4	871.3
Transverse cracking severity 2, avg. ft/mi	114.7	0	0	4.3	0	0
Longitudinal cracking severity 1, avg. ft/mi	254.4	0	0	83.7	3.6	11.9
Longitudinal cracking severity 2, avg. ft/mi	10.0	0	0	2.6	0	0
Longitudinal joint cracking severity 1, avg. ft/mi	292.5	0	286.9	1637.4	0	271.2
Longitudinal joint cracking severity 2, avg. ft/mi	0	0	0	59.6	0	0
Alligator cracking severity 1, avg. ft^2/mi	4516.5	37.3	59.1	1124.0	42.9	1070.3
Alligator cracking severity 2, avg. ft^2/mi	1326.0	33.4	1.7	189.4	0	14.2
Alligator cracking severity 3, avg. ft^2/mi	0	0	0	0.3	0	0
Patching area, wheel path, avg. ft^2/mi	25.4	0	3.32	0	0	0
Patching area, non-wheel-path, avg. ft^2/mi	0.7	0	0	0	0	0
Delamination area, avg. ft^2/mi	7.4	0	0	0	0	0
Bleeding severity 1, avg. ft^2/mi	0	30.8	0	0	0	0
Bleeding severity 2, avg. ft^2/mi	0	62.9	0	0	0	0
Average rut depth, in.	0.10	0.29	0.16	0.12	0.17	0.08
Total average rutting per mile, in./mi	1.03	2.97	1.71	1.22	1.79	0.87
Average load distress rating	69	59	94	87	94	75
Average non-load-distress rating	79	100	98	85	98	73
Average composite condition index	69	59	93	82	94	72

Note: PMS = pavement management system; WMA = warm mix asphalt; IRI = International Roughness Index.
[a]Section B was overlaid in 2015 at an age of 6 years. PMS data were collected prior to overlaying.
[b]Section E condition data were available only from 2015. These data were collected in the westbound travel lane; coring of Section E was performed in the eastbound passing lane.

content of the cores in this study. However, because of the inherent variability of cored field specimens, meeting the ideal was not always possible. An effort was made to choose field cores with air-void contents as similar as possible for use in preparing the dynamic modulus specimens; despite this, variability in air voids was greater than desired within some specimen sets. Table 7 also summarizes pertinent mixture properties that influenced the dynamic modulus.

Dynamic modulus mastercurves, created using a reference temperature of 70°F (21°C), are shown in Figure 3 for all mixtures.

Figure 3*a* shows results on a log-log scale, which emphasizes the differences in modulus for mixtures below approximately 0.1 Hz, corresponding to slow loading frequencies, which are analogous to higher temperatures. Mixture A is the stiffest mixture at frequencies below 0.1 Hz, and Mixture C is the least stiff at frequencies below 0.1 Hz. Figure 3*b* shows results on a log-linear scale to emphasize differences in modulus at higher frequencies that are masked by a log-log scale. This indicates that Mixtures HMA 1 and HMA 2 are stiffer at frequencies above 0.1 Hz and that mixtures C and E are least stiff. Of interest, Mixture A, although being stiffer at lower frequencies, had a considerably lower slope than the other mixtures, and so its stiffness was less relative to that of the other mixtures with increasing frequency of load application. Practically speaking, this indicates a stiffer (more rut-resistant) mixture at higher temperatures that is also less stiff (and likely to be crack-resistant) at low temperatures. However, this mixture showed considerable cracking in the field, according to PMS data. All other mixtures maintained similar rankings in terms of stiffness throughout the range of test frequencies.

RLPD Test

Figure 4 shows the averaged RLPD data for each mixture after virtual preconditioning was applied, as described previously. All further analysis was conducted after virtual preconditioning was applied. Figure 4 clearly shows grouping among the average response of the mixtures. Mixture A accumulated the lowest strain. Mixtures B, F, and HMA 3 were grouped, as were mixtures C, D, E, HMA 1, and HMA

Table 4. PMS Distress Data for HMA Sections

Data	Section		
	HMA 1	HMA 2	HMA 3
Year paved	2006	2006	2006
Year evaluated	2011	2011	2011
Segment length, mi	0.5	0.5	0.3
Average IRI, left wheel path	53	92	96
Average IRI, right wheel path	67	101	97
Average IRI	60	96	97
Transverse cracking severity 1, avg. ft/mi	24.8	10.8	850
Transverse cracking severity 2, avg. ft/mi	0.8	0	53.3
Longitudinal cracking severity 1, avg. ft/mi	13	3.2	1.1
Longitudinal cracking severity 2, avg. ft/mi	5.8	0.4	3.3
Longitudinal joint cracking severity 1, avg. ft/mi	0	0	26.7
Longitudinal joint cracking severity 2, avg. ft/mi	0	0	0
Alligator cracking severity 1, avg. ft^2/mi	447.4	91.2	234.4
Alligator cracking severity 2, avg. ft^2/mi	94.2	20.4	7.8
Alligator cracking severity 3, avg. ft^2/mi	0	0	0
Patching area, wheel path, avg. ft^2/mi	0.2	392.8	0
Patching area, non-wheel-path, avg. ft^2/mi	0.2	340.4	0
Delamination area, avg. ft^2/mi	0	0	0
Bleeding severity 1, avg. ft^2/mi	0	0	0
Bleeding severity 2, avg. ft^2/mi	0	0	0
Average rut depth, in.	0.09	0.11	0.08
Total average rutting per mile, in./mi	0.86	1.14	0.77
Average load distress rating	72	87.4	96
Average non-load-distress rating	94	92.4	80
Average composite condition index	72	87	80

Note: PMS = pavement management system; HMA = hot mix asphalt; IRI = International Roughness Index.

Table 5. Road Core Air Voids, %

Specimen	Mixture								
	A	B	C	D	E	F	HMA 1	HMA 2	HMA 3
1	8.4	5.5	6.7	3.1	8.3	10.3	5.6	7.1	6.4
2	8.2	3.7	6.5	1.7	7.6	9.9	5.9	6.9	4.4
3	7.9	4.8	6.7	2.1	8.6	11.3	3.3	6.1	6.1
4	8.1	1.8	6.6	1.2	8.5	10.2	6.2	9.4	8.2
5	7.9	2.2	5.1	1.5	7.0	9.6	4.2	6.8	7.2
6	6.6	1.5	6.5	2.2	7.0	9.1	7.0	4.8	8.0
7	7.9	1.7	6.4	2.0	6.0	8.3			
8	8.2	2.9	6.8	1.1	6.3	6.4			
9	5.8	2.3	6.4	3.1	7.6	7.1			
10	9.2	3.4	6.3	1.5	8.0	7.6			
Average	7.83	2.97	6.41	1.94	7.49	8.97	5.36	6.85	6.72
SD	0.96	1.37	0.49	0.73	0.88	1.57	1.36	1.50	1.41
Average			5.9					6.3	
SD			2.8					1.5	

Note: HMA = hot mix asphalt; SD = standard deviation.

2. Groupings were not influenced by volumetric properties or by nominal maximum aggregate size. Rutting was found to be an issue only with Mixture B, which was overlaid after only 6 years of service; the rutting was likely to have been

Table 6. Permeability of Road Cores (Permeability × 10^{-5} cm/s)

	Mixture								
Specimen	A	B	C	D	E	F	HMA 1	HMA 2	HMA 3
1	17.34	1.10	6.82	8.68	1.30	3.80	NA	NA	0.00
2	12.21	0.00	8.28	27.90	3.03	6.24			0.00
3	4.29	1.39	2.44	6.53	1.49	5.83			0.00
4	3.03	0.00	4.23	1.62	1.52	1.57			6.30
5	14.30	0.00	5.68	1.47	2.80	3.93			0.00
6	1.21	0.00	4.31	7.31	2.00	1.60			0.89
7	8.90	0.00	4.33	4.24	0.57	0.51			
8	4.27	0.00	2.06	0.00	0.83	0.00			
9	0.43	0.00	6.46	19.13	0.80	1.64			
10	25.36	0.00	6.36	2.63	2.01	0.00			
Predicted permeability at 7.5% air voids	3.37	na[a]	na[b]	843.25	1.44	1.05	NA	NA	na[a]

Note: HMA = hot mix asphalt; NA = data not available; na = not applicable.
[a]Insufficient data points with permeability ≠ 0 to fit trendline.
[b]Air voids do not span sufficient range to fit trendline appropriately.

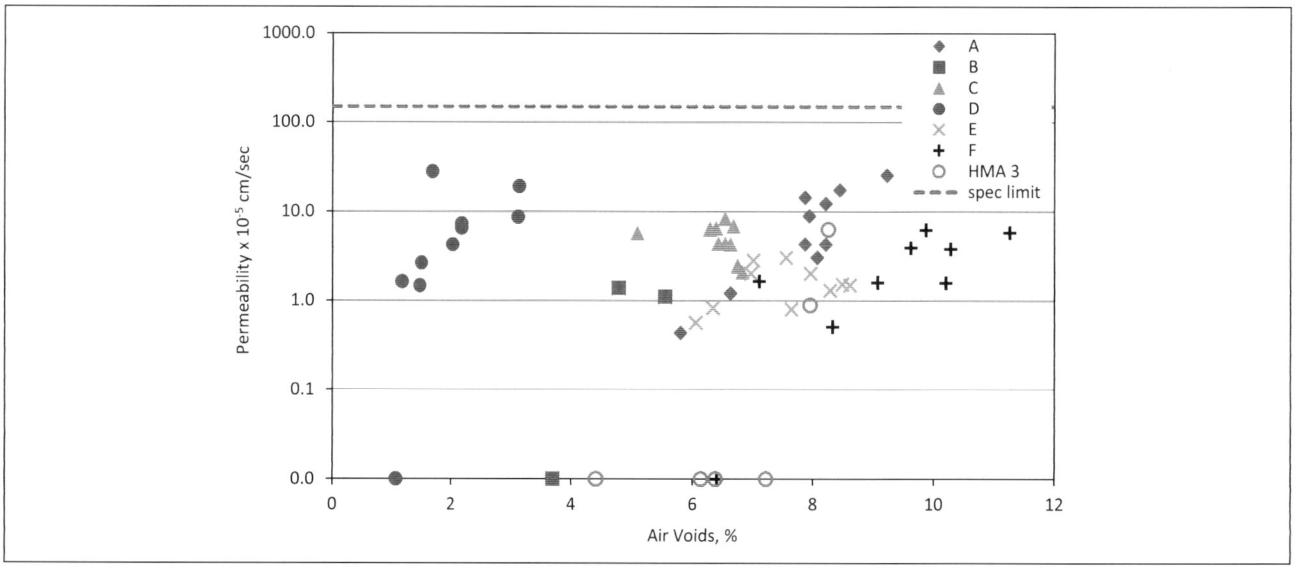

Figure 2. Road core permeability results. Specification requires a maximum of 150 × 10^{-5} cm/s permeability at 7.5% air voids.
Note: HMA = hot mix asphalt.

accelerated by an increase in truck traffic as a result of a warehouse distribution center built mid-service.

Overlay Test

Overlay test results were not available for the HMA mixtures as the test was not available when those mixtures were sampled. Replicate tests were performed on each WMA mixture, with five replicate cores tested for each mixture. Interquartile outlier analysis was applied such that any result more than 1.5 times the interquartile region was discarded. Only one test result was discarded; one core from Section A was an outlier. Figure 5 summarizes the results. The results indicated that all foamed WMA mixtures should perform acceptably, as all mixtures reached 1,200 load applications with the exception of Mixture C, which exceeded 800 cycles. The difference in response between Mixture C and the other mixtures was not seen in the PMS data, where Mixture C, along with Mixture E, had the highest load distress rating.

Binder Testing

Binders were extracted and recovered from cores collected for each mixture and were tested after recovery and after

Table 7. Dynamic Modulus and RLPD Specimen Void Contents and Mixture Properties

	Mixture								
Specimen	A	B	C	D	E	F	HMA 1	HMA 2	HMA 3
1	7.9	5.5	6.4	1.7	7.6	8.3	5.6	6.9	6.4
2	7.9	3.7	6.4	2.2	7.0	7.1	5.9	6.1	6.1
3	6.6	4.8	6.3	2.2	7.6	7.6	4.2	6.8	7.2
Average	7.45	4.67	6.36	2.01	7.40	7.66	5.22	6.60	6.57
Standard deviation	0.72	0.93	0.07	0.29	0.35	0.62	0.91	0.43	0.57
All cores average	7.83	2.97	6.41	1.94	7.49	8.97	5.36	6.85	6.72
NMAS, mm	9.5	12.5	12.5	12.5	9.5	9.5	9.5	12.5	9.5
Binder type					PG 64-22				
Asphalt content, %	5.8	5.5	5.5	5.7	5.8	5.6	5.5	5.2	5.7
RAP content, %	25	25	25	15	26	26	12	10	20

Note: RLPD = repeated load permanent deformation; HMA = hot mix asphalt; NMAS = nominal maximum aggregate size; RAP = reclaimed asphalt pavement.

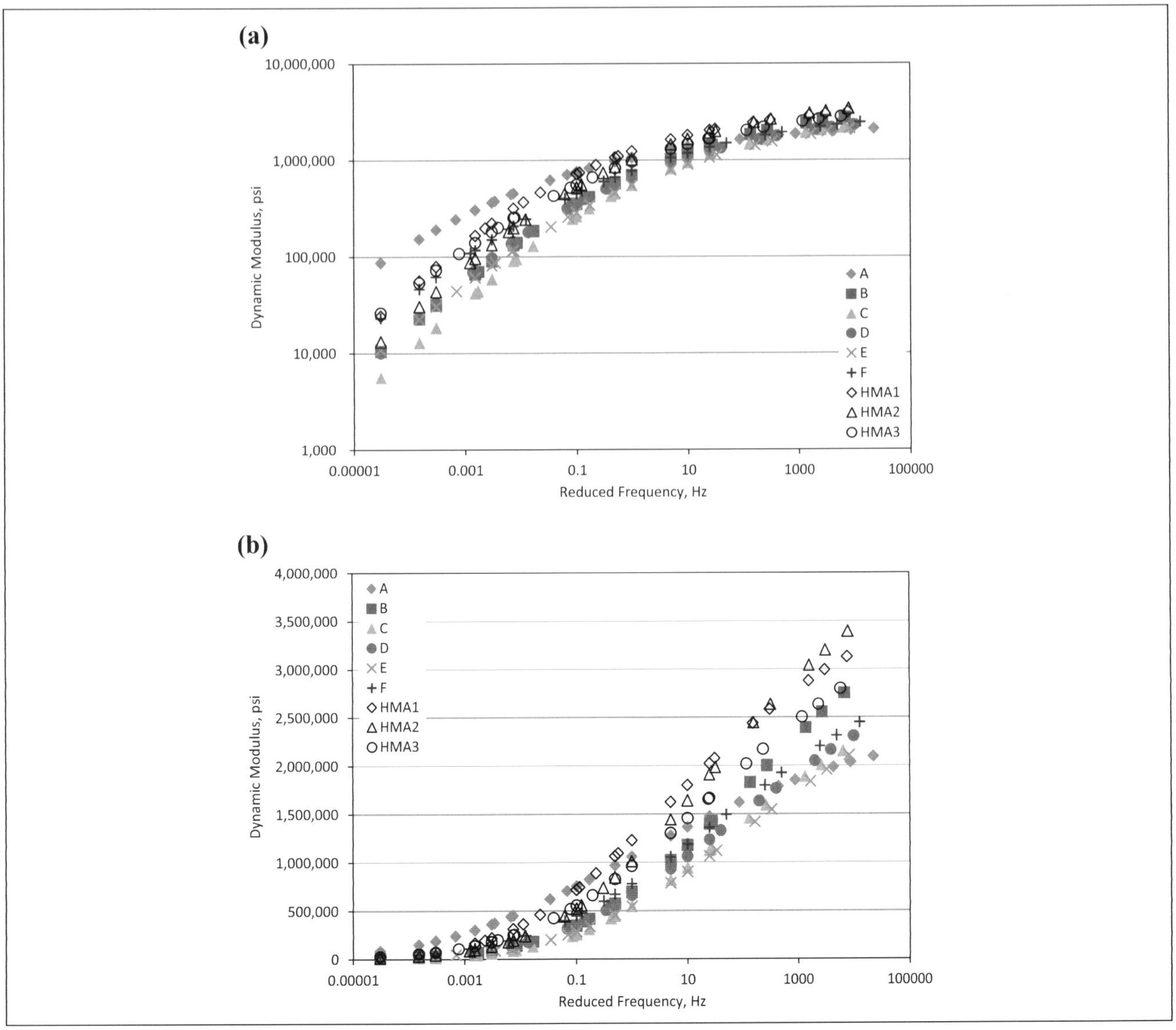

Figure 3. Comparison of dynamic modulus mastercurves for road cores using (*a*) log and (*b*) linear scales.
Note: HMA = hot mix asphalt.

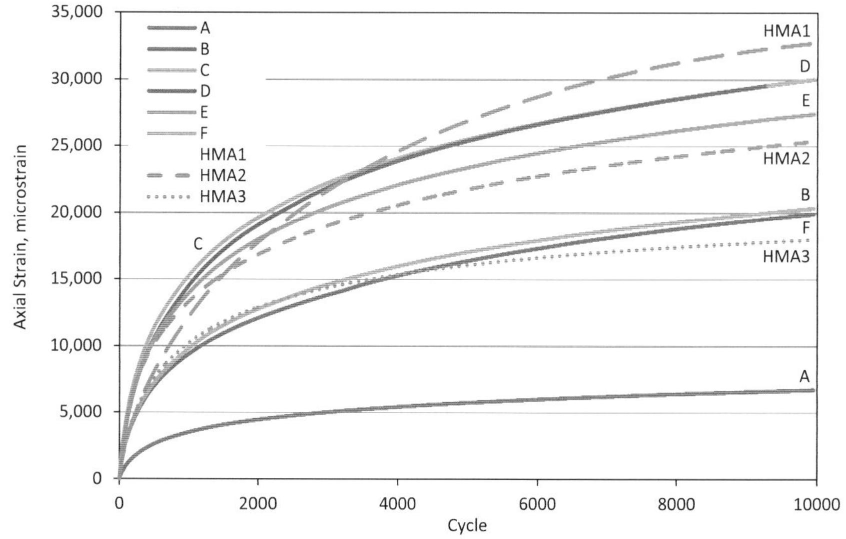

Figure 4. RLPD mixture averages after virtual preconditioning.
Note: HMA = hot mix asphalt.

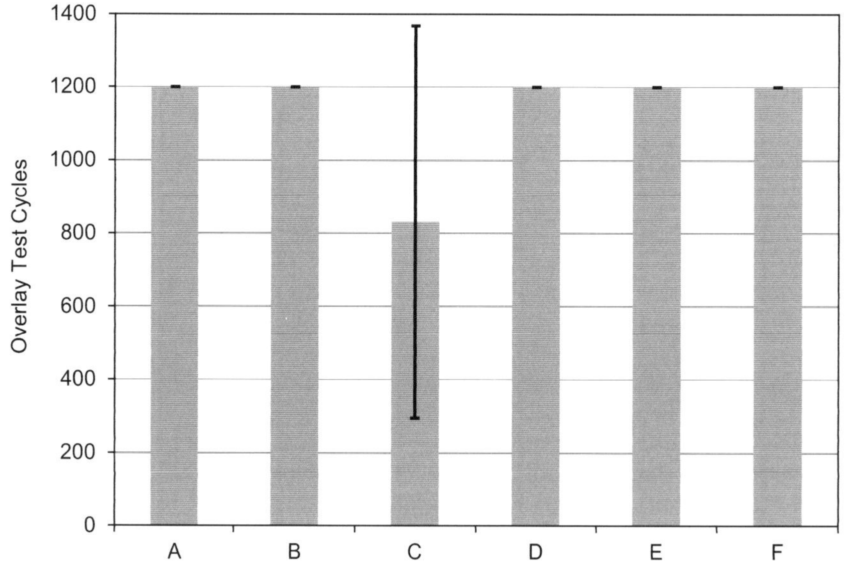

Figure 5. Overlay test results for WMA cores. I-bars indicate one standard deviation.
Note: WMA = warm mix asphalt.

aging in a pressure aging vessel. Table 8 summarizes the performance grading parameters for each binder from the foamed WMA sites. Data were not available for Mixture F. Table 9 summarizes the performance grading parameters for each HMA binder.

With the exception of Binder A, foamed binders graded to a PG 76-16 binder. It is clear that in-service aging is not well simulated by laboratory aging, as all of the foamed WMA mixtures were produced with virgin PG 64-22 binders and varying RAP contents from 15% to 26%. Even Binder D, which was produced with PG 64-22 binder and 15% RAP, aged to a PG 76-16 binder after 5 years in service. Binder A was exceptionally stiff with a high failure temperature of 94.1°C and a low failure temperature of −3.0°C; the failure temperatures fell outside those established in AASHTO M 320, so this binder did not meet any performance grade. It is unclear why Binder A was so stiff; RAP content may have been an influence but this is unlikely as four additional foamed

Table 8. Performance Graded Binder Properties of Foamed WMA Sites[a]

	Site					
Property	A	B	C	D	E	F
Dynamic shear, 10 rad/s, specification: G*/sin delta > 2.20 kPa						
Recovery G*/sin delta, 76°C		2.222	2.358	2.753	4.294	NA
Recovery G*/sin delta, 82°C	9.27	1.098	1.144	1.321	2.037	
Recovery G*/sin delta, 88°C	4.464					
Recovery G*/sin delta, 94°C	2.213					
Recovery G*/sin delta, 100°C	1.112					
Recovery G*, 76°C		2.207	2.344	2.738	4.252	
Recovery G*, 82°C	9.132	1.095	1.14	1.318	6.026	
Recovery G*, 88°C	4.426					
Recovery G*, 94°C	2.203					
Recovery G*, 100°C	1.109					
Recovery phase angle, 76°C		83.39	83.83	84.12	81.95	
Recovery phase angle, 82°C	80.09	85.30	85.67	85.92	84.16	
Recovery phase angle, 88°C	82.45					
Recovery phase angle, 94°C	84.43					
Recovery phase angle, 100°C	85.93					
Recovery failure temperature, °C	94.12	76.08	76.57	77.83	81.38	
Dynamic shear, 10 rad/s, specification: G* sin delta < 5000 kPa						
PAV G* sin delta, 25.0°C		6066	6381			NA
PAV G* sin delta, 28.0°C		4312	4495	5593	6589	
PAV G* sin delta, 31.0°C			3112	4146	4712	
PAV G* sin delta, 40.0°C	7294					
PAV G* sin delta, 45.0°C	4105					
PAV G*, 25.0°C		9.09E + 06	9.44E + 06			
PAV G*, 28.0°C		6.14E + 06	6.30E + 06	8.28E + 06	9.98E + 06	
PAV G*, 31.0°C			4.16E + 06	5.88E + 06	6.75E + 06	
PAV G*, 40.0°C	1.01E + 07					
PAV G*, 45.0°C	5.33E + 07					
PAV phase angle, 25.0°C		41.84	42.54			
PAV phase angle, 28.0°C		44.60	45.49	42.51	41.30	
PAV phase angle, 31.0°C			48.40	44.81	44.26	
PAV phase angle, 40.0°C	46.26					
PAV phase angle, 45.0°C	50.37					
PAV failure temperature, °C	43.29	26.07	27.06	29.12	30.47	
Creep stiffness, 60 s, specification: stiffness < 300 MPa and m-value > 0.300						
Stiffness, 0°C	262					NA
M-value, 0°C	0.250					
Stiffness, −6°C		117	126	140	168	
M-value, −6°C		0.338	0.318	0.313	0.304	
Stiffness, −12°C	900	251	262	288	313	
M-value, −12°C	0.164	0.294	0.287	0.276	0.263	
ΔT_c, °C	−8.3	−2.2	−3.6	−4.2	−5.0	
Stiffness failure temperature, °C	−11.3	−23.4	−23.1	−22.3	−21.6	
M-value failure temperature, °C	−3.0	−21.2	−19.5	−18.1	−16.6	
Performance grade	–	76-16	76-16	76-16	76-16	NA

Note: WMA = warm mix asphalt; PAV = pressure aging vessel; NA = data not available; – = grade does not conform to grades specified in AASHTO M 320.

WMA mixtures were produced using PG 64-22 binder and RAP contents of 25% to 26% and only Mixture A had the drastic increase in stiffness. The behavior seen with recovered Binder A was also supported by the mixture testing, as Mixture A showed generally higher dynamic modulus across most of the range of testing and resistance to deformation in the RLPD analysis. Of interest, Mixture A did not show any increased tendency toward cracking in the overlay test that would be

Table 9. Performance Graded Binder Properties of HMA Sites

	Site		
Property	HMA 1	HMA 2	HMA 3
Dynamic shear, 10 rad/s, specification: G*/sin delta > 2.20 kPa			
Recovery G*/sin delta, 76°C	6.408	4.410	5.832
Recovery G*/sin delta, 82°C	3.056	2.061	2.809
Recovery G*/sin delta, 88°C	1.510		1.396
Recovery G*, 76°C	6.308	4.371	5.737
Recovery G*, 82°C	3.029	2.051	2.780
Recovery G*, 88°C	1.503		1.389
Recovery phase angle, 76°C	79.90	82.37	79.59
Recovery phase angle, 82°C	82.37	84.49	82.03
Recovery phase angle, 88°C	84.43		84.11
Recovery failure temperature, °C	84.83	81.48	84.14
Dynamic shear, 10 rad/s, specification: G* sin delta < 5000 kPa			
PAV G* sin delta, 28.0°C		6454	5197
PAV G* sin delta, 31.0°C	5418	4590	3776
PAV G* sin delta, 34.0°C	3868	3203	
PAV G*, 28.0°C		9.61E + 06	7.53E + 06
PAV G*, 31.0°C	7.90E + 06	6.47E + 06	5.26E + 06
PAV G*, 34.0°C	2.38E + 06	4.30E + 06	
PAV phase angle, 28.0°C		42.18	43.65
PAV phase angle, 31.0°C	43.33	45.18	45.89
PAV phase angle, 34.0°C	45.99	48.15	
PAV failure temperature, °C	31.72	30.21	28.36
Creep stiffness, 60 s, specification: stiffness < 300 MPa and m-value > 0.300			
Stiffness, –6°C	179	177	148
M-value, –6°C	0.313	0.308	0.327
Stiffness, −12°C	362	349	294
M-value, −12°C	0.264	0.255	0.278
ΔT_c, °C	−2.8	−3.8	−2.9
Stiffness failure temperature, °C	−20.4	−20.7	−22.2
M-value failure temperature, °C	−17.6	−16.9	−19.3
Performance grade	82-16	76-16	82-16

Note: HMA = hot mix asphalt; PAV = pressure aging vessel.

expected from a binder this stiff. This is particularly surprising given the ΔT_c value for Binder A in Table 8 and the quantity of cracking seen in the PMS data. ΔT_c is calculated as the difference between the critical temperature for stiffness and m-value for a binder and is an indicator of non-load-related cracking susceptibility (*14*).

Binders for mixtures HMA 1 and HMA 3 graded to a PG 82-16 binder; that for Mixture HMA 2 graded to a PG 76-16 binder but was nearly a PG 82-16 binder. In general, the decreased aging of the foamed binders may be due to slightly reduced production temperatures.

Conclusion

- Results from this study verified that foamed WMA mixtures should be expected to perform similarly to HMA mixtures. Although air-void contents and permeability were different among individual mixtures, the results were within expected ranges for most mixtures. The dynamic modulus, RLPD, overlay test results, and PMS performance data all indicated trends of similar performance, with the exception of Mixture A, which had a different behavioral trend than that of the other mixtures.
- Binder grading indicated that foamed WMA mixtures may age slightly less than HMA mixtures; this is likely to be because of the slightly lowered production temperatures. However, all mixtures, regardless of virgin binder grade and RAP content, aged to at least a PG 76-16 binder, one to two grades higher than the binder grade specified for design.

Acknowledgments

The authors thank the Virginia Transportation Research Council (VTRC) and the FHWA for their support of this research. The authors also thank Troy Deeds, Donnie Dodds, and Ben Earl, of

VTRC, and Sam Deloach, Ken Elliton, Shane Moomaw, and Kim Snead, formerly of VTRC, for their outstanding efforts in sample collection and testing. Kevin Wright of VTRC is also acknowledged for his assistance in implementation efforts. Appreciation is also extended to Linda Evans of VTRC for her editorial assistance. The authors also appreciate the project review panel for their expertise and guidance: Bill Bailey, Sungho Kim, Haroon Shami, Chaz Weaver, and Chung Wu of VDOT; Ben Bowers and Brian Diefenderfer of VTRC; and Trenton Clark of the Virginia Asphalt Association.

Author Contributions

The author confirms sole contribution to the paper as follows: study conception and design; data collection; analysis and interpretation of results; and draft manuscript preparation. The author reviewed the results and approved the final version of the manuscript.

References

1. Martin, A. E., E. Arambula, F. Yin, L. Garcia, A. Cucalon, R. Chowdhury, J. Lytton, C. Epps, C. Estakhri, and E. S. Park. *NCHRP Report 763: Evaluation of the Moisture Susceptibility of WMA Technologies*. Transportation Research Board of the National Academies, Washington, D.C., 2014.
2. Washington State University, Pennsylvania State University–Altoona, and Louisiana Transportation Research Center. *NCHRP Report 843: Long-Term Field Performance of Warm Mix Asphalt Technologies*. Transportation Research Board of the National Academies, Washington, D.C., 2017.
3. Hansen, K. R., and A. Copeland. *Asphalt Pavement Industry Survey on Recycled Materials and Warm-Mix Asphalt Usage: 2014*. Information Series 138, 5th ed. National Asphalt Pavement Association, Lanham, Md., 2015.
4. Diefenderfer, S. D., and A. J. Hearon. *Laboratory Evaluation of a Warm Asphalt Technology for Use in Virginia*. Publication VTRC 09-R11. Virginia Transportation Research Council, Charlottesville, 2008.
5. Diefenderfer, S. D., and A. J. Hearon. *Performance of Virginia's Warm Asphalt Trial Sections*. Publication VTRC 10-R17. Virginia Transportation Research Council, Charlottesville, 2010.
6. Diefenderfer, S. D., K. K. McGhee, and B. M. Donaldson. *Installation of Warm Mix Asphalt Projects in Virginia*. Publication VTRC 07-R25. Virginia Transportation Research Council, Charlottesville, 2007.
7. Virginia Department of Transportation Materials Division. *Virginia Test Methods*. Richmond, 2013. http://www.virginiadot.org/business/resources/materials/bu-mat-vtms.pdf. Accessed April 20, 2017.
8. Bowers, B. F., B. K. Diefenderfer, and S. D. Diefenderfer. Evaluation of Dynamic Modulus in Asphalt Paving Mixtures Utilizing Small-Scale Specimen Geometries. *Journal of the Association of Asphalt Paving Technologists*, Vol. 84, 2015, pp. 497–526.
9. Diefenderfer, B. K., B. F. Bowers, and S. D. Diefenderfer. *Asphalt Mixture Performance Characterization Using Small-Scale Cylindrical Specimens*. Publication VTRC 15-R26. Virginia Center for Transportation Innovation and Research, Charlottesville, 2015.
10. Khosravifar, S., I. Haider, Z. Afsharikia, and C. W. Schwartz. Application of Time-Temperature Superposition to Develop Master Curves of Cumulative Plastic Strain in Repeated Load Permanent Deformation Tests. *International Journal of Pavement Engineering*, Vol. 16, No. 3, 2015, pp. 214–223.
11. Von Quintus, H. L., J. Mallela, R. Bonaquist, C. W. Schwartz, and R. L. Carvalho. *NCHRP Report 719: Calibration of Rutting Models for Structural and Mix Design*. Transportation Research Board of the National Academies, Washington, D.C., 2012.
12. Gibson, N., and S. Li. *Performance Characterization of Oklahoma Asphalt Mixtures*. Turner-Fairbank Highway Research Center, FHWA, McLean, Va., 2013.
13. Texas Department of Transportation. *Test Procedure for Overlay Test Tex-248-F. Effective Dates: January 2009–October 2013*. http://ftp.dot.state.tx.us/pub/txdot-info/cst/TMS/200-F_series/archives/248-0109.pdf. Accessed June 30, 2018.
14. Anderson, R. M., G. N. King, D. I. Hanson, and P. B. Blankenship. Evaluation of the Relationship Between Asphalt Binder Properties and Non-Load Related Cracking. *Journal of the Association of Asphalt Paving Technologists*, Vol. 80, 2011, pp. 615–664.

The Standing Committee on Critical Issues and Emerging Technologies in Asphalt (AFK10) peer-reviewed this paper (18-04779).

Article

TRR
JOURNAL OF THE TRANSPORTATION RESEARCH BOARD

Transportation Research Record
2018, Vol. 2672(28) 190–199

DOI: 10.1177/0361198118781377
journals.sagepub.com/home/trr
SAGE

Chemo-Rheological Study of Hardening of Epoxy Modified Bituminous Binders with the Finite Element Method

P. Apostolidis[1], X. Liu[1], C. Kasbergen[1], M.F.C. van de Ven[1], G. Pipintakos[1], and A. Scarpas[1,2]

Abstract

The chemical irreversible hardening of epoxy modified bitumen is affected by various physical factors and the successful application of this technology is directly linked with full understanding of chemo-rheological material characteristics. This study proposes a model to describe the material viscosity evolution during hardening of epoxy modified bitumen. The findings from numerical analyses performed to assess the mechanical response of epoxy modified bituminous binders are presented. Information of the chemical interaction of epoxy within a bituminous matrix was collected and all the influential factors have been determined. The proposed chemo-rheological model accounting for the polymerization of the epoxy in the bitumen was formulated and the sensitivity of material parameters, such as activation energy, reaction order and extent of hardening reaction until the gel point of epoxy modified binders, was demonstrated. Results of the analyses suggest that lower levels of activation energy increase the degree of hardening and the rate of viscosity development. By decreasing the hardening reaction until the gel point the achieved viscosity of epoxy modified bitumen was increased showing the importance of gel reaction extent on material viscosity evolution. The numerical studies have shown also that the polymerization rate in the epoxy modified bitumen is highly dependent on the temperature under various (non-) isothermal conditions. Also, the polymerization rate should be considered through all the material curing processes to avoid unwanted variations in the mechanical properties.

Due to the challenge to minimize the public expenditures and the road administration costs for repairing and re-paving the transport infrastructure networks, epoxy modified bituminous binders have captured the interest of road authorities as an important alternative to conventional binder technologies. Nowadays, bituminous materials modified with epoxy-type resins are used mainly for pavements in tunnels, steel deck bridges and at intersections of heavy duty roads to provide enhanced performance without major repairs for more than 30–40 years (*1–5*).

Recent studies suggest a successful industrial transfer of this technology to pavements but contractors still face serious issues during mixing, transport and pavement construction phase. Two main issues are for example the accelerated and the slow chemical hardening of material before compaction and during the early traffic period, respectively (*6*). Moreover, production plants need to be modified to allow proportional in-line feeding of epoxy modifiers into the pugmill (i.e., through a static mixer before adding them to the drum). The chemical reaction rate of epoxy with the bituminous phase is influenced by several factors such as the added hardening agents, the working temperature, mixing time in-plant, transport and compaction of the mix on the road. Higher mixing temperatures accelerate the hardening reactions of epoxies in bituminous mixes and lower the time window for material transport and compaction. As a consequence of the above, it becomes obvious that it is crucial to have full control over the whole production chain of epoxy modified pavement structures.

In comparison with bituminous mixes, the processing of epoxies and epoxy modified bituminous composites is more complicated and more difficult to control because of the chemical reactions that take place. The chemical hardening of epoxy modified bitumen due to crosslinking of linear pre-polymers and the polymeric network formation from monomers is a complex process that involves the continuous chemical alteration and interaction of epoxy resin with the

[1]Section of Pavement Engineering, Faculty of Civil Engineering and Geosciences, Delft University of Technology, Delft, the Netherlands
[2]Department of Civil Infrastructure and Environmental Engineering, Khalifa University of Science and Technology, Abu Dhabi, United Arab Emirates

Corresponding Author:
Address correspondence to P. Apostolidis: p.apostolidis@tudelft.nl

bituminous binder. Within this framework, the development of rheological properties plays an important role to characterize the mix workability and performance before and after construction. To optimize the processing and the properties of the epoxy-bituminous materials it is necessary to understand the chemo-rheological relationship which varies at different temperatures and time as polymerization reactions proceed.

This study proposes a chemo-rheological numerical model able to describe the mechanical response of epoxy modified bituminous binder during material chemical hardening. The proposed chemo-rheological modeling approach accounting for the epoxy polymerization within the bituminous matrix and the sensitivity of the physical model parameters on the mechanical response of the matrix were investigated by performing numerical analyses.

Epoxy Modified Bituminous Binders

Bituminous binders are characterized by their high chemical complexity containing various molecular types. Bitumen is mostly classified based on differences in their solubility and polarity into two major groups; asphaltenes and maltenes. The fraction of asphaltenes consists of highly condensed planar and heteroatom polar groups, polar aromatic rings and large amounts of heteroatom polar functional groups. The maltenes fraction is sub-divided in saturates, aromatics and resins groups (*7*). On the other hand, the epoxy-type systems are copolymers which consist typically of two liquid components; the monomers or short chain prepolymers with epoxide groups at their ends named epoxy resins, and the co-monomers which react with the epoxides named hardening agents. These two components are combined and crosslinking chains of molecules are formed through the polymerization process. When these components are mixed with bituminous binders they produce paving materials with enhanced durability and increased longevity.

Incorporation of Epoxy Resins into Bituminous Binders

Initially, epoxy resins were added in bituminous binders to minimize the thermal susceptibility and permanent deformation of asphalt concrete mixes since the thermoset nature of epoxies results in excellent rut resistant binders. Later on, it was realized that the polymeric structure of epoxy within the bituminous matrix enhanced the fatigue, moisture and oxidative degradation resistance of bituminous mixes (*1–6*). These materials cannot be re-melted and their application in pavements has been predominantly for high-performance solutions, which does not need any special construction equipment.

However, epoxy modified bituminous binders are thermodynamically immiscible blends appearing with phase separation issues, mainly at elevated temperatures. Experimental observations showed that, in the epoxy modified bitumen, part of the binder is substituted with epoxy and an archipelago of polymeric chains is dispersed in the continuous bituminous-rich phase as polymerization proceeds. The structure of these chain networks dramatically influences the hardening crosslinking density, the characteristics of crosslinking chains (i.e., the size) and, consequently, the final product properties. For the material preparation, rapid epoxy-bitumen blending, and addition of extra compatibility agent can assist in producing homogeneous thermoset dispersion in toughening bituminous blends. The most preferred is formed by epichlorhydrin and bisphenol A-based epoxy resin (*2*, *4*, *6*) which needs a mixing temperature of 110–120°C and a time available of 57–94 minutes transported from the production plant to field (*6*).

Incorporation of Hardening Agents into Bituminous Binders

Typically, the hardening agents react with the epoxide groups of resins to produce covalent bonds. Based on the type of hardening agent used, the epoxies are classified as amine- and acid-type resins. The application of epoxy resins is a well-established field in several industries and the performance of epoxy-hardening agents is extensively studied. For pure epoxies, acid anhydrides are preferred above the organic acids, which are less soluble in resins and sometimes release water during the polymerization process. Also, acid anhydrides are more often applied in epoxies than the traditional amine-based agents since they provide higher compatibility and better thermal stability with epoxies, leading to a rapid viscosity reduction when mixed (*8*, *9*).

In the paving industry, acids and anhydride acids are commonly used with a typical completed reaction after 4h at 120°C. It is reported in the literature that methyl tetrahydropthalic of cyclic molecular structure, methyl hexahydropthalic and tung oil anhydrides improve the compatibility of bitumen and epoxy. They help to produce materials with enhanced strength and heat resistance (*10–12*). However, the high volume of rigid benzene groups in the epoxy-methyl tetrahydropthalic system could limit its applicability. Polymerized fatty acid agents have been synthesized successfully and added to the system together with tung oil maleic tribasic acid agent to increase the stability between the components (*13*). Moreover, stable modified binders of short polymeric chains are produced when epoxy modified bitumen is hardened with maleated bituminous agents (*14*).

Also, amine-type hardeners have been used for epoxy-bitumen blends. For example, octadecylamine, an aliphatic amine, has shown a very quick hardening effect (1h at 150°C or 3d at 60°C) but resulted in an unstable epoxy-bitumen with restrictions for the amount of bituminous binder to be used (*15*). This had direct effect on the total cost since the limited relative volume ratio of the bitumen causes not only a more expensive but also a less viscoelastic material. On the other hand, a polyetheramine was used with longer chains and polar ether groups to increase the stability between epoxy and bitumen without phase separation and to provide an inexpensive hardening option with sufficient hardening rates (1h at 160°C or 3d at 60°C) for paving applications (*16*). It is believed that the performance of this hardener is related to the light networks crosslinked within the bituminous matrix.

Motivation and Objectives

In contrast to the wax fraction and the crystallization-related physical thermo-reversible hardening in pure bituminous binders (*17*, *18*), the chemical irreversible hardening of epoxy-bituminous materials is influenced by different factors. Apart from the influence of isothermal conditioning on material stiffening for both physical and chemical hardening, the evaluation of reaction order, frequency and activation energy of hardening is of great importance to predict the hardening phenomena and to reach optimal cure. Therefore, an appropriate chemo-rheological model which can describe the reaction processes to avoid unwanted variations in the mechanical properties (i.e., complex modulus, phase angle, viscosity) of epoxy modified bituminous binders is required. The model should be capable of simulating not only the temperature–time process of various operations but also the influence of different resins, agents and other additives on properties to design a better material.

The objective of this study is to introduce a procedure to predict the hardening of epoxy modified binders by taking into account their time–temperature dependency. Also, the proposed procedure can be used as a tool to understand the chemical hardening of blends and to calculate the mechanical properties of epoxy modified bitumen throughout the polymerization reactions phase. Since it is crucial to identify the influential parameters of the processes, the model has been deployed to simulate the reaction temperatures and the chemical conversion of these materials.

Modeling of Epoxy Modified Bitumen Hardening

Extensive researches (*19–22*) have been conducted to develop models for describing the chemo-rheological behavior of epoxies with the phenomenological model rather than a mechanistic one when the hardening process is very complex. Among others, the autocatalytic models have been successfully applied to various thermosets. Also, models developed based on *n*-th order kinetics can describe sufficiently the chemical hardening of modified thermoplastic blends of relatively low amount of thermosets (*23*, *24*). Since previous investigations have shown no significant effect on the extent of the epoxy hardening after blending with bitumen under the same conditions (*16*), the exothermic reaction is considered as a one-step *n*-th order reaction with negligible secondary reactions in this study.

The exothermic polymerization reaction rate of an epoxy-bitumen blend is a temperature dependent reaction and the conversion is initiated via heating. Hence, the ability of the epoxy modified bitumen to harden at a sufficient time depends on the temperature of initial blending in-plant production, material transport, laying and compaction. To simulate the reaction, the heat transfer and the viscosity evolution mechanisms of the complex chemical hardening processes of epoxy modified bituminous binders, the governing equations are given in the following sub-sections.

Governing Hardening Kinetics Equation

In polymerization process of epoxy modified bitumen, the kinetic model allows to predict the chemical conversion or hardening degree α and the reaction rate as a function of time and temperature as:

$$\frac{\partial \alpha}{\partial t} = K(T) \cdot f(\alpha) \tag{1}$$

where $K(T)$ is a temperature-dependent parameter, which is described by an Arrhenius equation, and $f(\alpha)$ is a function of conversion. Also, the shape of the $f(\alpha)$ shows the reaction performance of the system and is determined by the *n*-th order kinetics.

Therefore, the rate of change of hardening degree is described as:

$$\frac{\partial \alpha}{\partial t} = k_0 \exp\left(-\frac{E_\alpha}{RT}\right) \cdot (1-\alpha)^n \tag{2}$$

where α is the hardening degree, k_0 is the pre-exponential kinetic factor, E_α is the activation energy or the energy barrier to be overtaken to begin the reaction, R is the universal gas constant and n is the reaction order upon the hardening mechanism.

During chemical hardening of pure epoxies under isothermal conditions, the total cure of the resin is not always achieved, due to the fact that the glass transition temperature of epoxy increases progressively during hardening. The crosslinked molecules do not flow easily when the conditioning temperature of epoxy becomes the full polymerization temperature and, thus, their mobility lowers. Therefore, the Equation 2 is not always valid for modeling the epoxy resins. However, in the case of epoxy modified bituminous mixes, it is assumed to be a sufficient function for describing the chemical reactions of these materials. An illustration of crosslinked molecules of epoxy within the bituminous matrix is given in Figure 1*a*.

Governing Heat Transfer Equations

The governing equation of the transient heat conduction within the epoxy modified bitumen is described by:

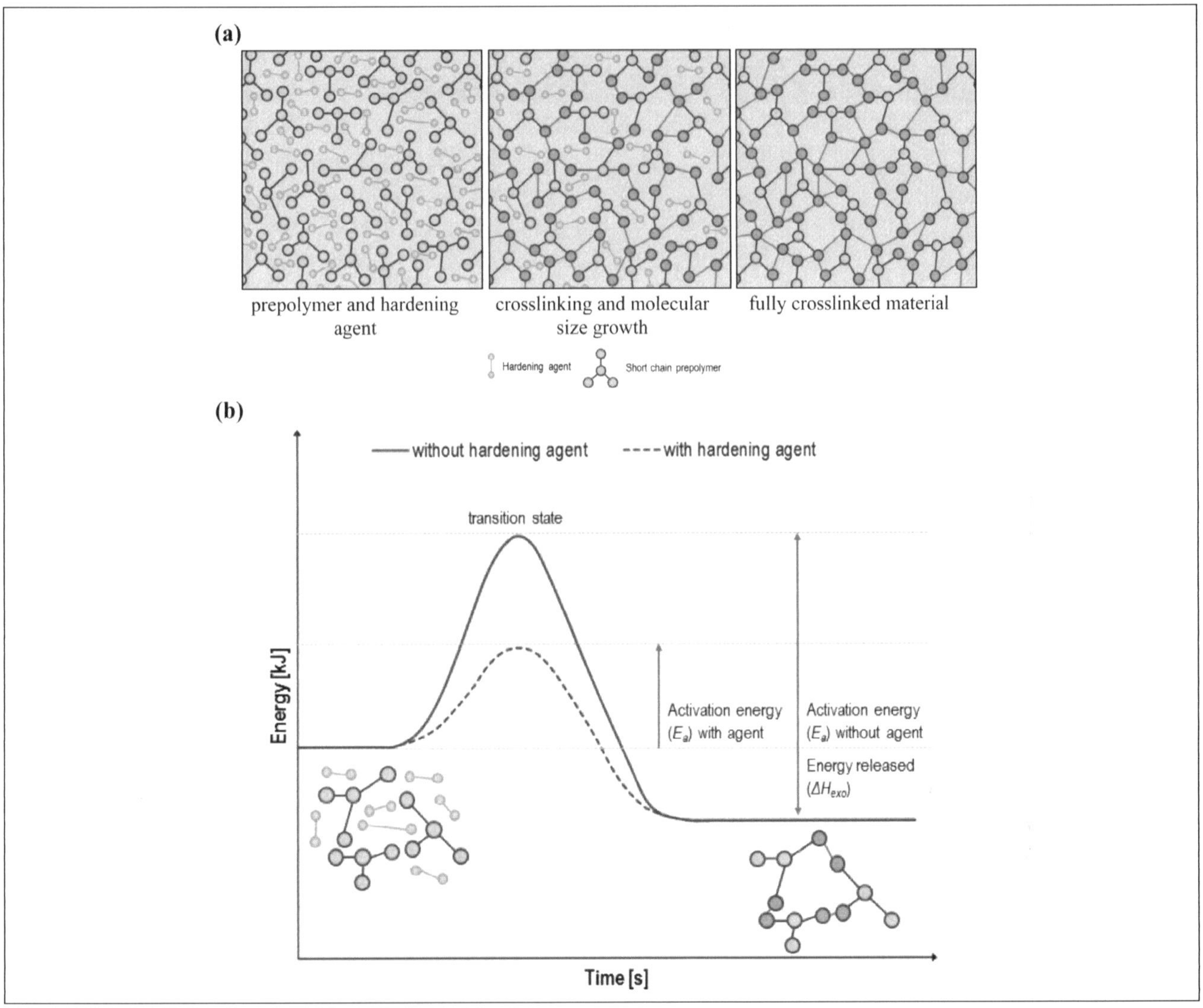

Figure 1. Crosslinking of epoxy modified bitumen: (*a*) stepwise polymerization of epoxy in the bituminous matrix; (*b*) exothermic chemical reaction diagram.

$$\rho c_p \nabla T - \nabla \cdot (k \nabla T) = Q \tag{3}$$

where ρ is the mass density of epoxy modified asphalt mastic, k denotes the thermal conductivity, c_p is the heat capacity, T is the temperature, Q represents the exothermic heat source (kW/m^3). It is assumed that the convection and radiation heat do not have important impact on the energy balance of the system.

As the resin turns into a networked microstructure and the epoxy modified bitumen hardens, heat is released. The heat releasing rate is proportional to the consumption rate of reactive elements in the epoxy binder. Assuming no heat flow, the volumetric heat source Q accounts the exothermic hardening effect and is described by:

$$Q = \rho \Delta H_{exo} \frac{\partial \alpha}{\partial t} \tag{4}$$

where ΔH_{exo} is the exothermic reaction heat. The reaction rate gradients generated by the temperature gradients affect the temperature profile in the hardening material. The exothermic reaction diagram is shown in Figure 1*b*.

Chemo-Rheological Equation

The chemical hardening prediction of the epoxy modified bitumen was the ultimate scope of this study. On molecular level, the viscosity of the epoxy-bitumen is affected by (i) the increasing temperature, which increases molecular mobility, and (ii) the molecular size growth and the subsequent crosslinking of molecules. When the average molecular weight rises towards infinity the viscosity of epoxy-bitumen becomes infinite as well and the gel point is reached (*25*). In both network-forming polymerization and the crosslinking of polymeric chains, the gel point is expressed as an extent of

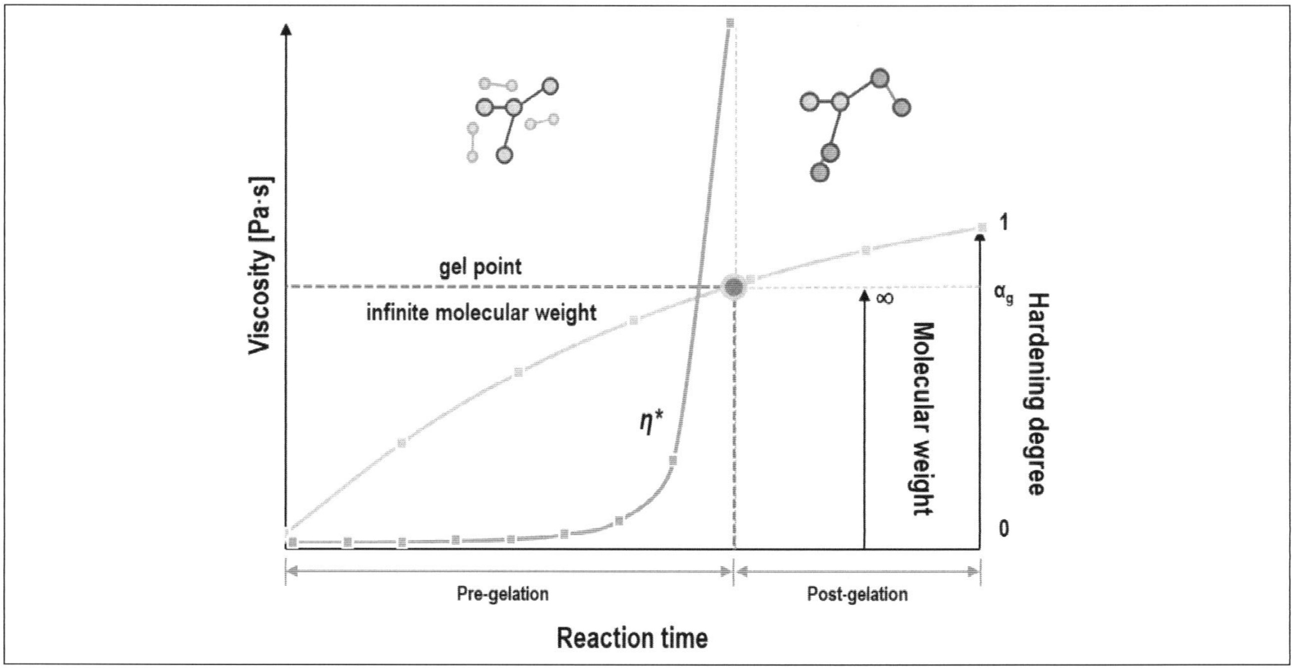

Figure 2. Schematic of variation of epoxy modified bitumen properties during isothermal chemical hardening.

chemical reaction. At the gel point a solid (network) material spanning the entire system is formed. Despite the solidification of epoxy which occurs at any stage of reaction, gelation of epoxy is a critical event throughout the crosslinking process that determines the maximum available time before the completion of pavement construction. Epoxy gelation happens at a distinct extent of hardening reactions, as shown in Figure 2.

To predict the chemo-rheological performance of epoxy modified bitumen as function of time and temperature, the model considers the viscosity changes due to epoxy irreversible hardening as:

$$\eta^*(\mathrm{T},\alpha)=\eta^*_g\cdot\exp\left[-\frac{C_1\left(T-T_g(\alpha)\right)}{C_2+T-T_g(\alpha)}\right]\cdot\left(\frac{\alpha_g}{\alpha_g-\alpha}\right)^{n_r} \quad (5)$$

where α_g is the extent of reaction at the gel point, $T_g(\alpha)$ is the glass transition temperature of the uncured material which is a function of the hardening degree, η^*_g is the complex viscosity at the glass transition temperature, C_1 and C_2 are material-dependent and temperature-independent constants obtained from the rearranged William-Landel-Ferry (WLF) equation and n_r is a material-dependent constant. Equation 5, which is a combination of the WLF equation (*26*) and a conversion term, was obtained by others (*27*) and describes the chemo-rheology of an epoxy-bituminous blend as function of time and temperature.

For the determination of the chemo-rheological model, several criteria can be used to characterize the glass transition temperature in epoxy resins and epoxy modified bitumen in general. The crossover point between the storage and loss modulus curves, the point where the loss tangent becomes frequency independent, and the inflection point of the modulus curve are some of the methods used to determine the glass transition temperature. However, unlike pure bitumen, epoxy modified binders undergo reactions during hardening leading to polymerization of the insoluble resinous part of modified bitumen and the slight reduction in the glass transition temperature. The decrease of glass transition temperature may be caused by the limited polar interaction of epoxy resin with asphaltene groups of bitumen. Typically, the glass transition temperature of many epoxy systems is defined using empirical DiBenedetto equation (*28*). It should be mentioned that hardening agents are capable of increasing both the glass transition temperatures of bituminous and resinous phase in the epoxy modified bitumen.

According to previous study (*15*), the activation energy of epoxy-bitumen was marginally lower (46–49 kJ/mol) and subsequently released less heat than the pure epoxy (50 kJ/mol) showing negligible impact of bitumen on chemical hardening reactions, when part of epoxy was substituted by bitumen. However, in cases where part of the bitumen was substituted with epoxy, the difference of activation energies of pure epoxy and epoxy modified binders will be higher; 78 kJ/mol and 65 kJ/mol for epoxy and epoxy modified binders, respectively (*13*). The contact opportunities between epoxy and hardening agent are reduced by bitumen and the chemical reactions are inhibited leading to higher energy needs.

Numerical Simulation of Hardening

For the qualitative analyses, one three-dimensional cubic finite-element mesh of E10-3 m geometry and of 16,200

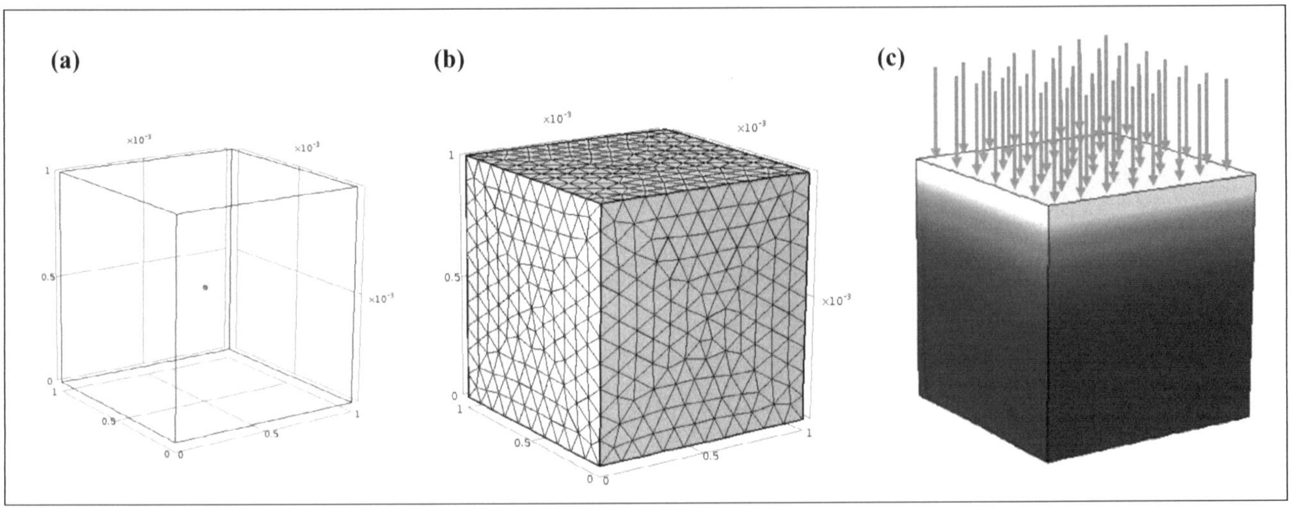

Figure 3. Finite element cubic geometry: (*a*) location of output point in the center of cube; (*b*) mesh; (*c*) direction of applied thermal field at the top boundary.

elements was created. The proposed chemo-rheological model of epoxy modified bituminous blends suitable for paving applications was implemented in COMSOL Multiphysics to simulate the viscosity evolution during chemical hardening as a measure of reaction progress. By varying the different parameters insight is gained into the effect of each of them on the overall hardening process of epoxy modified binders. For these analyses, the density (1600 kg/m3), heat capacity (920 J/kg·K) and thermal conductivity (0.45 W/m·K) were assumed to be constant (*29*). The exothermic reaction heat (ΔH_{exo}) was considered constant (100 kJ/kg) as well. Additionally, it is assumed that the glass transition temperature and the gelation of the unreacted epoxy-bituminous binder occurs at a fixed hardening degree (HD) and, since through processing of epoxy modified blend before compaction the material is not fully cured. The viscosity at the glass transition temperature was assumed to be 1 Pa·s. The imposed thermal field at the top boundary of the finite element cube is shown in Figure 3 with the rest boundaries to be thermally insulated. The model predictions are demonstrated at a point located in the cubic center.

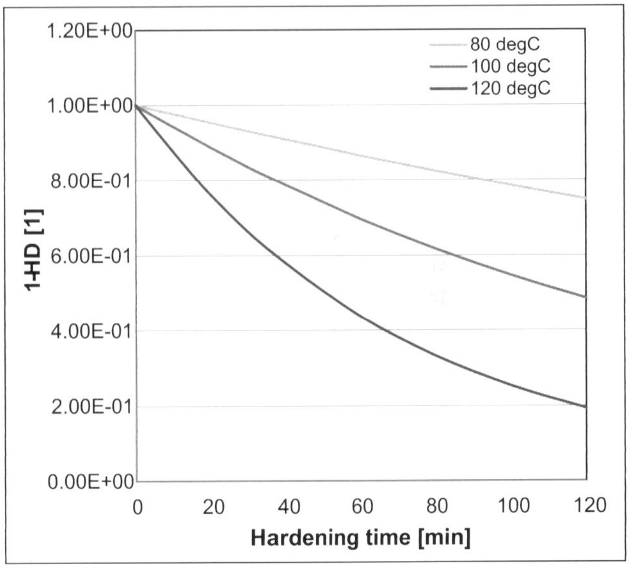

Figure 4. Typical isothermal hardening of epoxy modified bitumen for different heating temperatures (E_a = 50 kJ/mol, n = 1, k_0 = 100 1/s).

Results and Discussion

To apply epoxy modified binders for pavement construction successfully, it is necessary to understand the reaction kinetics of epoxy modified bituminous mixes for various processing scenarios from the plant production to transport, laying, and compaction of these materials. Continuous monitoring of epoxy reactions in bitumen is highly important to obtain the desired end product performance. The hardening of epoxy-bituminous blends involves heat transfer, cure kinetics phenomena, and mechanical changes. This section presents the numerical predictions of HD and viscosity, which is very sensitive to molecular changes at the early material hardening stages of epoxy-bituminous binder, blends with various characteristics and under different operational conditions. The findings of the numerical analyses are discussed in the following subsections.

Influence of Activation Energy and Reaction Parameters on Hardening Degree

To assess the influence of reaction kinetics on the hardening caused by epoxy polymerization in bituminous binder, the activation energy, the hardening reaction order and the kinetic factor were varied. For the current analyses, values of the kinetic parameters as in previous researches were utilized. From Figure 4, the typical HD evolution at different temperatures is shown.

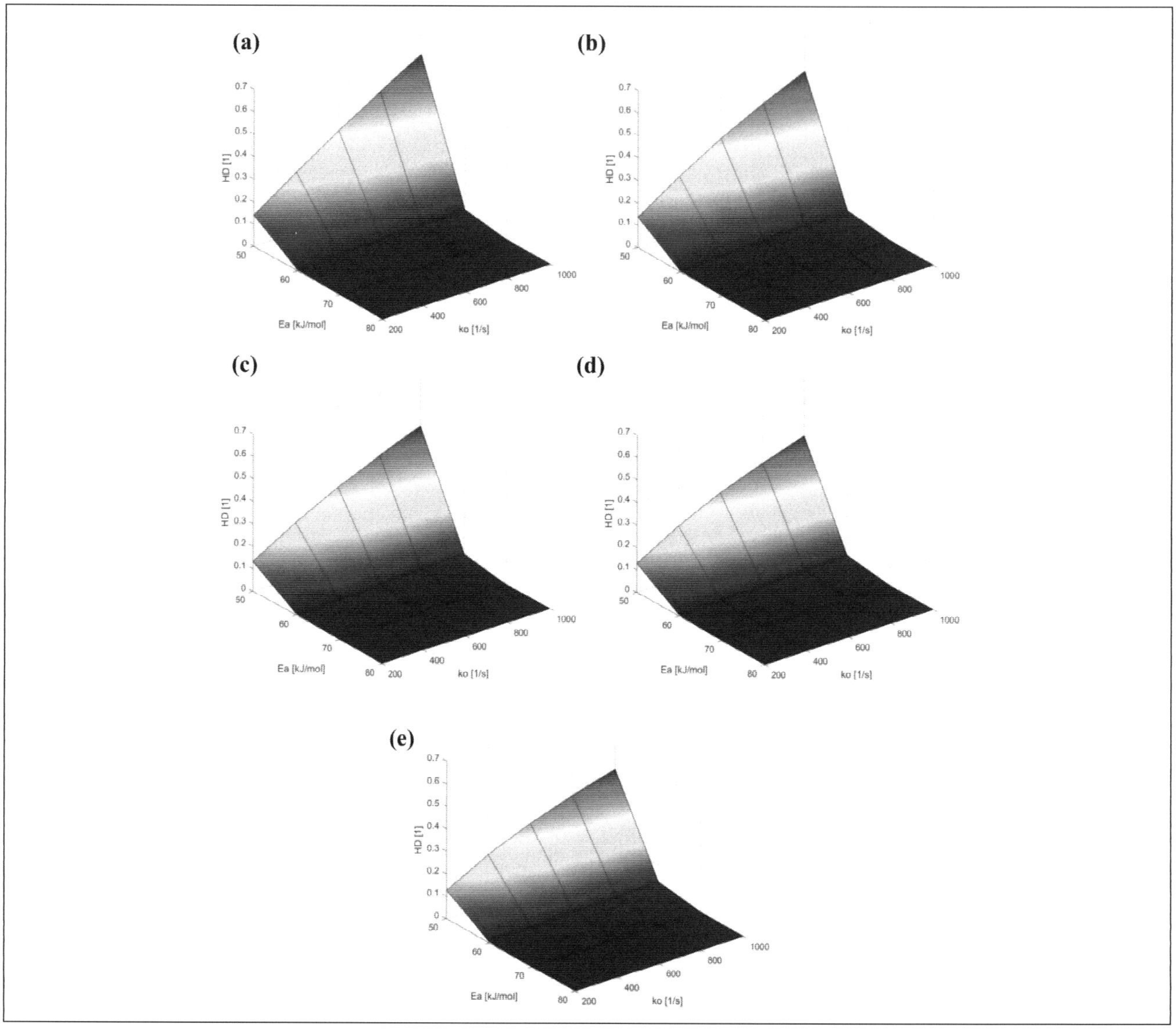

Figure 5. Variation of hardening degree for different activation energies and kinetic factors with reaction orders of (*a*) 0.4, (*b*) 0.8, (*c*) 1.2, (*d*) 1.6, and (e) 2.0 (T = 100°C and 120 min).

The numerical results after decreasing the activation energy E_α from 80 kJ/mol to 50 kJ/mol, subjected to isothermal conditions (100°C) and with varying the hardening reaction order *n* (0.2 to 2, step of 0.4) and the kinetic factor k_0 (200 to 1000 1/s, step of 200 1/s) are presented in Figure 5.

It is observed that activation energy has a strong impact on HD compared to the other parameters (Figure 5), which can be explained by the influence on the polymerization occurrence and crosslinking of epoxy in the modified binder. As the activation energy decreases, HD increased remarkably. For example, the increase of HD was observed to become 3.94E-01 from 1.72E-02 when the activation energy changes from 60 kJ/mol to 50 kJ/mol, for *n* 0.4 and k_0 600, respectively. This performance is also apparent for higher levels of activation energy and it is related with the type of epoxy components (i.e., hardening agent and epoxy resin) and the chemical interaction between them. A more reactive agent can be attributed to lower the energy demands to trigger the epoxy polymerization. Since bitumen is substituted by epoxy through the modification, increasing the amount of epoxy within the bituminous matrix can increase the contact opportunities of the thermoset phase and results a subsequent increase of crosslinking density. Based on these results, this phenomenon leads to higher epoxy percentages but lower activation energy levels.

Influence of Temperature and Reaction Extent on Viscosity Development under Isothermal Conditions

The duration at high temperatures of the blend from plant to field will affect the achieved viscosity. To investigate this

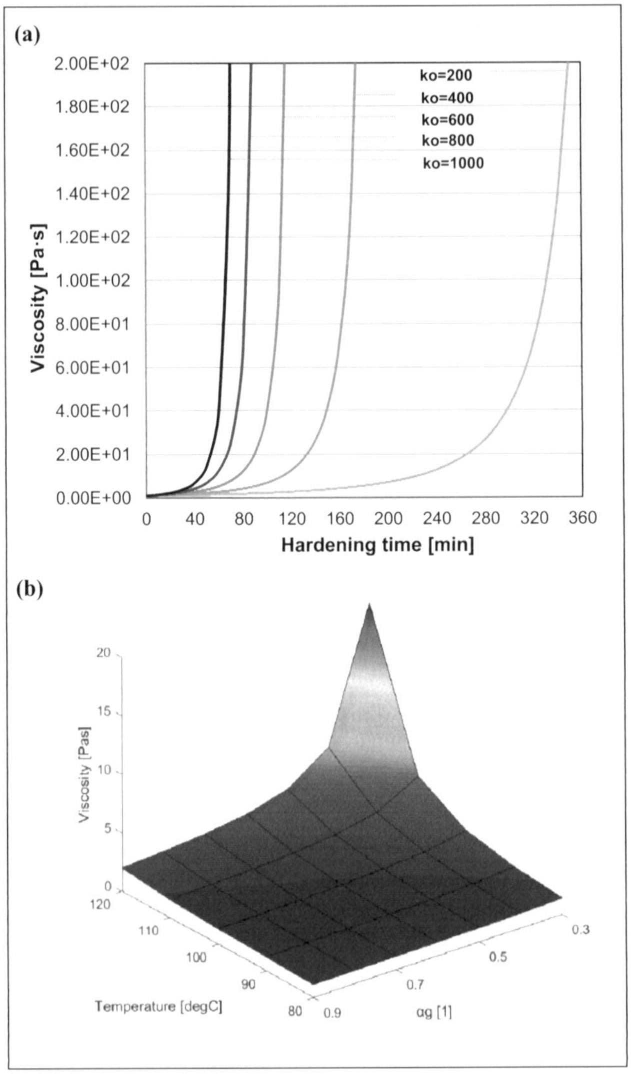

Figure 6. Predicted viscosity development of epoxy modified bitumen for (*a*) different k_0 and (*b*) different α_g at various temperatures (Ea = 50 kJ/mol, n = 0.2, k_0 = 100 l/s).

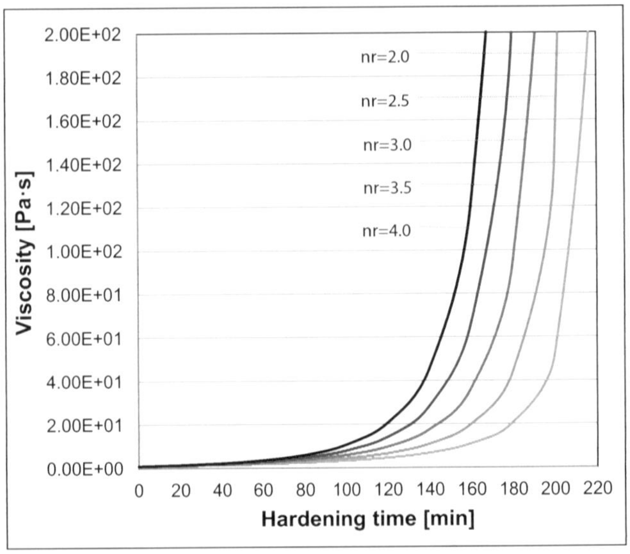

Figure 7. Predicted viscosity development of epoxy modified bitumen under isothermal heating conditions for different n_r at T = 100°C (E_a = 50 kJ/mol, n = 0.2, k_0 = 100 l/s, α_g = 0.6).

different isothermal analyses were considered, because of the importance for simulating real practical scenarios. The parameter of chemical reaction extent at the gel point (α_g) was considered to be constant and the numerical analyses were performed with varying this parameter from 0.3 to 0.9 under defining a heat flux related to warm and half-warm mixed production conditions (e.g., 80–120°C with step of 10°C) at the boundary. Also, it should be pointed out that normally the reaction kinetics do not change during the occurrence of gelation of epoxy-type blends.

The impact of kinetic factor on viscosity development appears in the isothermal hardening curves of Figure 6*a* showing that the highly reactive blends harden faster. Figure 6*b* demonstrates the influence of temperature on viscosity achieved after 120 min of hardening. Increase in the predicted viscosity after 120 min hardening was achieved with increasing temperature. The computed results illustrate the role of applied temperatures on material polymerization and on viscosity at isothermal conditions since the material reached a higher HD when the conditioning temperature was higher.

The trend above was also observed in previous practical studies (*5*). The epoxy modified bituminous mixes were produced and compacted at a temperature range substantially lower than for conventional hot-mix asphalt mixes resulting in a decrease of the material production energy and reduction of carbon footprint of the whole pavement manufacturing chain. Therefore, apart from the influence of physical reaction parameters, the rate of polymeric network formation is also a function of temperature which assists in the speed of hardening activation of epoxy in the bituminous matrix. Moreover, a similar trend was observed for α_g, as a material parameter linked with the gel point of epoxy, on building the viscosity of epoxy modified binders (Figure 6*b*). For example, for all the predefined applied temperatures, the achieved viscosity was higher when α_g decreases from 0.9 to 0.3. As a result, if the intermolecular interactions between the epoxy-bitumen components produce a composite with lower value of hardening extent until the material gelation, then the initiation of hardening can be caused with minimum energy demands and the material can be fully cured at a higher rate.

In Figure 7, the influence of the constant n_r is depicted showing the importance of this parameter on providing realistic numerical predictions. The rate of viscosity development was found to increase with increasing value of n_r from 2 to 4 with step of 0.5 for the initially un-reacted epoxy modified bitumen having as consequence a faster hardening. Thus, for the ranking of the individual epoxy components and the blend of epoxy with bitumen, it is important to identify experimentally these factors since these contribute vitally on material structural development and on the shape

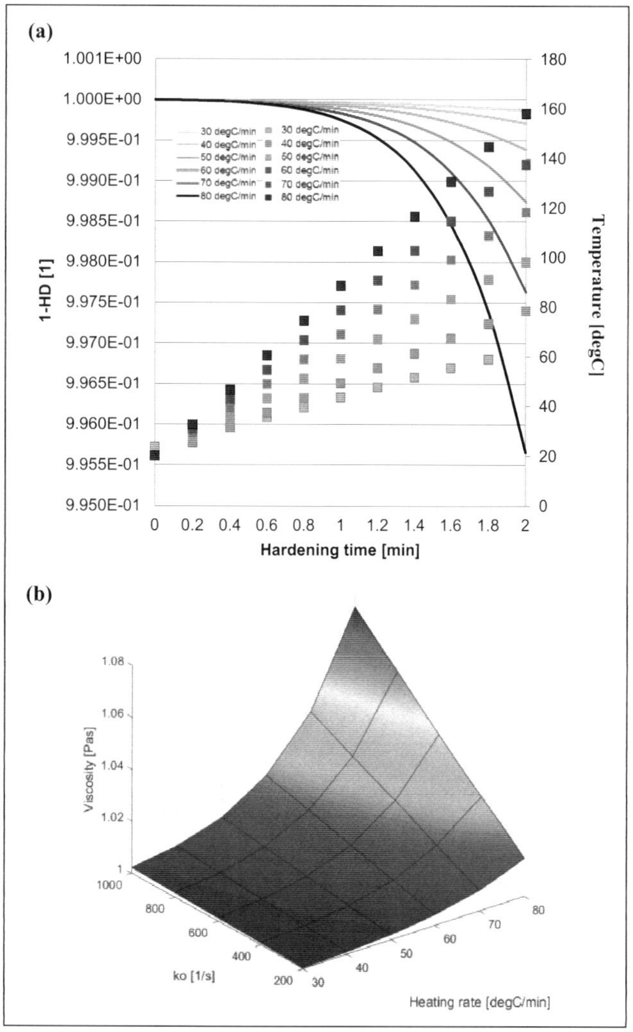

Figure 8. Non-isothermal chemical hardening of epoxy modified bitumen: (*a*) predicted hardening degree development of epoxy modified bitumen (E_a = 50 kJ/mol, n = 0.2, k_0 = 200 1/s, α_g = 0.6); (*b*) predicted viscosity with various heating rates and kinetic factor k_0 after 2 min hardening.

of chemo-rheological plots. The observed trends of predicted viscosity coincide with the similar trends of others (*5*).

Influence of Heating Rate on Viscosity Development under Non-Isothermal Conditions

Next, due to the fact that the current model was developed to characterize the performance of material under the usual operating conditions in a hardening cycle (mixing plant to construction field), it is also important to study the evolution of the viscosity at non-isothermal heating mode. Epoxy modified bitumen was analyzed from 30 to 80°C/min with 10°C/min step and the evolution of hardening degree under these applied heating rates is presented in Figure 8*a*. The rising temperature activates reaction between epoxy and hardening agent forming polymeric chains quicker. However, it should be noted that high heating rates can cause undesired temperature differences and irregularities in epoxy chemical hardening. In Figure 8*b*, the predicted viscosity at the first 120 sec of hardening under different heating rates and kinetic factors k_0, as a very influential reaction parameter regarding previous analyses, is demonstrated. Higher heating rates and k_0 increase the material achieved viscosity. These predicted results can be attributed to the trend of consuming more energy to result in quicker material hardening, something which can increase the risk of unwanted issues within the blend as well, such as variation in porosity or epoxy distribution in bituminous phase. Although the material was studied at different heating rates, the viscosity development was non-linear and related with the applied thermal field, which is a common trend identified in all studied cases of these analyses.

Conclusions and Future Work

This paper presents the hardening process and the subsequent predictions of viscosity evolution of epoxy modified bitumen via performing numerical analyses. These analyses showed the impact of all the model parameters on chemical hardening of epoxy blends. Particularly, results of the analyses suggest that lower levels of activation energy increase the degree of hardening and the rate of viscosity development. In addition, the performed numerical sensitivity analyses have shown that with the increase of temperature the material viscosity increased, whereas the polymerization rate was highly dependent on the temperature. By decreasing the hardening reaction until the gel point the achieved viscosity of epoxy modified bitumen, after a certain isothermal hardening period, was increased which shows the importance of gel reaction extent on material viscosity evolution.

In the near future, the proposed model will be verified through differential scanning calorimetry (DSC) tests at (non-) isothermal conditions. All the reactions should be evaluated to quantify the temperature dependence of kinetic parameters under different heating modes. The glass transition of epoxy modified bituminous binders and their behavior need to be measured to plan sufficiently the field operations mainly because the manufacturing and the delivering conditions of the mixes have direct impact on long-term pavement performance. Once the DSC studies are developed, they can be used to optimize performance-related details of epoxy modified bituminous pavements design.

Author Contributions

The authors confirm contribution to the paper as follows: study conception and design: P. Apostolidis, A. Scarpas; data collection: P. Apostolidis; analysis and interpretation of results: P. Apostolidis, X. Liu, C. Kasbergen, M.F.C. van de Ven, G. Pipintakos, A. Scarpas; draft manuscript preparation: P. Apostolidis. All authors reviewed the results and approved the final version of the manuscript.

References

1. Widyatmoko, I., B. Zhao, R. C. Elliott, and W. G. Lloyd. Curing Characteristics and the Performance of Epoxy Asphalts. *Proc., 10th International Conference on Asphalt Pavements*, Quebec City, Canada, August 2006, pp. 496–505.
2. Herrington, P., and D. Alabaster. Epoxy Modified Open-graded Porous Asphalt. *Road Materials and Pavement Design* Vol. 9, No. 3, 2008, pp. 481–498.
3. Xiao, Y., M. F. C. van de Ven, A. A. A. Molenaar, Z. Su, and F. Zandvoort. Characteristics of Two-component Epoxy Modified Bitumen. *Materials and Structures* Vol. 44, No. 3, 2010, pp. 611–622.
4. Wei, J., and Y. Zhang. Study on the Curing Process of Epoxy Asphalt. *Journal of Testing and Evaluation*, Vol. 40, No. 7, 2012, pp. 1–8.
5. Yin, C., H. Zhang, and Y. Pan. Cracking Mechanism and Repair Techniques of Epoxy Asphalt on Steel Bridge Deck Pavement. *Transportation Research Record: Journal of the Transportation Research Board*, 2016. 2550: 123–130.
6. International Transport Forum. *Long-life Surfacings for Roads: Field Test Results*. ITF Research Reports, OECD, Paris, France, 2017.
7. Redelius, P. G. Solubility Parameters and Bitumen. *Fuel*, Vol. 79, 2000, pp. 27–35.
8. Harper, C. A. *Handbook of Plastics, Elastomers and Composites*. McGraw-Hill, New York, 1996.
9. Odian, G. *Principles of Polymerization*. John Wiley &Sons, New York, 1981.
10. Yu, J. Y., P. L. Cong, and S. P. Wu. Laboratory Investigation of the Properties of Asphalt Modified with Epoxy Resin. *Journal of Applied Polymer Science*, Vol. 113, 2009, pp. 3557–3563.
11. Luo, S., Q. Lu, and Z. Qian. Performance Evaluation of Epoxy Modified Open-graded Porous Asphalt Concrete. *Construction and Building Materials*, Vol. 76, 2015, pp. 97–102.
12. Zhou, W., H. Zhao, J. Wen, F. C. Cai, N. Ma, and T. Jiang. Study on Effects of Flexible Curing Agent on the Structure and Performance of Epoxy-asphalt. *Journal of Wuhan University of Technology*, Vol. 33, No. 7, 2011, pp. 28–31.
13. Li, S., K. Huang, X. Yang, M. Li, and J. Xia. Design, Preparation and Characterization of Novel Toughened Epoxy Asphalt Based on a Vegetable Oil Derivative for Bridge Deck Paving. *Royal Society of Chemistry*, Vol. 4, 2014, pp. 44741–44749.
14. Kang, Y., F. Wang, and Z. M. Chen. Reaction of Asphalt and Maleic Anhydride: Kinetics and Mechanism. *Chemical Engineering Journal*, Vol. 164, 2010, pp. 230–237.
15. Yin, H., H. Jin, C. Wang, Y. Sun, Z. Yuan, H. Xie, Z. Wang, and R. Cheng. Thermal, Damping, and Mechanical Properties of Thermosetting Epoxy-Modified Asphalts. *Journal of Thermal Analysis and Calorimetry*, Vol. 115, 2014, pp. 1073–1080.
16. Kang, Y., Q. Wu, R. Jin, P. Yu, and J. Cheng. Rubber-Like Quasi-Thermosetting Polyetheramine-Cured Epoxy Asphalt Composites Capable of Being Opened to Traffic Immediately. *Scientific Report*, Vol. 6, 2016, p. 18882.
17. Anderson, D. A., D. W. Christensen, H. U. Bahia, R. Dongre, M. G. Sharma, C. E. Antle, and J. Button. *SHRP-A-369: Binder Characterization and Evaluation, Vol. 3: Physical Characterization*. TRB, National Research Council, Washington, D.C., 1994.
18. Anderson, D. A., and M. O. Marasteanu. Physical Hardening of Asphalt Binders Relative to Their Glass Transition Temperatures. *Transportation Research Record: Journal of the Transportation Research Board*, 1999. 1661: 27–34.
19Kamal, M. R., and S. Sourour. Kinetics and Thermal Characterization of Thermoset Cure. *Polymer Engineering and Science*, Vol. 13, No. 1, 1973, pp. 59–64.
20. Ryan, M. E., and A. Dutta. Kinetics of Epoxy Cure: A Rapid Technique for Kinetic Parameter Estimation. *Polymer*, Vol. 20, 1978, pp. 203–206.
21Halley, P. J., and M. E. Mackay. Chemorheology of Thermosets-An Overview. *Polymer Engineering and Science*, Vol. 36, No. 5, 1996, pp. 593–609.
22. Yousefi, A., and P. G. Lafleur. Kinetic Studies of Thermoset Cure Reactions: A Review. *Polymer Composites*, Vol. 18, No. 2, 1997, pp. 157–168.
23. Yang, Y., H. Fujiwara, T. Chiba, and T. Inoue. Morphology Development in a Thermoset/Thermoplastic Blend: DAP/PPE System via Apparent Two-step Spinodal Decomposition. *Polymer*, Vol. 39, No. 13, 1998, pp. 2745–2750.
24.Blanco, I., G. Cicala, O. Motta, and A. Recca. Influence of a Selected Hardener on the Phase Separation in Epoxy/Thermoplastic Polymer Blends. *Journal of Applied Polymer Science*, Vol. 94, 2004, pp. 361–371.
25. Flory, P. J. Molecular Size Distribution in Three Dimensional Polymers. II. Trifunctional Branching Units. *Journal of the American Chemical Society*, Vol. 63, No. 11, 1941, pp. 3091–3096.
26. Williams, M. L., R. F. Landel, and J. D. Ferry. The Temperature Dependence of Relaxation Mechanisms in Amorphous Polymers and Other Glass-forming Liquids. *Journal of the American Chemical Society*, Vol. 77, No. 14, 1955, pp. 3701–3707.
27. Castro, J. M., and C. W. Macosko. Kinetics and Rheology of Typical Polyurethane Reaction Injection Molding Systems. *Society of Polymer Engineers Technical Paper*, Vol. 26, pp. 434–438.
28DiBenedetto, A. T. Prediction of the Glass Transition Temperature of Polymers: A Model Based on the Principle of Corresponding States. *Journal of Polymer Science, Part B: Polymer Physics*, Vol. 25, No. 9, 1987, pp. 1949–1969.
29.Apostolidis, P., X. Liu, A. Scarpas, C. Kasbergen, and M. F. C. van de Ven. Advanced Evaluation of Asphalt Mortar for Induction Healing Purposes. *Construction and Building Materials*, Vol. 126, 2016, pp. 9–25.

The Standing Committee on Critical Issues and Emerging Technologies in Asphalt (AFK10) peer-reviewed this paper (18-02446).

Article

TRR
JOURNAL OF THE TRANSPORTATION RESEARCH BOARD

Transportation Research Record
2018, Vol. 2672(28) 200–210

DOI: 10.1177/0361198118780702
journals.sagepub.com/home/trr
SAGE

Comparison of the Relative Long-Term Field Performance among Various Warm Mix Asphalt (WMA) Pavements

Weiguang Zhang[1], **Shihui Shen**[2,3], **and Shenghua Wu**[4]

Abstract

The use of warm mix asphalt (WMA) has increased greatly over the past decade because it provides potential to lower energy demands, reduce emissions, and increase haul distances. However, questions remain about the long-term performance of WMA pavements, including whether different WMA technologies can provide the same field performance. In this paper, field rutting and cracking (transverse and longitudinal) performance data for chemical, organic, and foaming WMA technologies are compared pairwise. Thirty-four WMA-WMA pairs from 13 field test roads are included, which cover different pavement ages, traffic volumes, pavement structures, WMA technologies, and four climate zones across the United States. For the WMA-WMA pairs that exhibited statistically different performance, their material properties were investigated further to identify potential factors that could be used to characterize such field differences. The results indicate that, generally, no statistical performance differences are evident between the WMA-WMA pairs. However, it is worth noting that over the longer term, there are individual projects whose organic pavement sections showed statistically higher amounts of cracking than the chemical and foaming sections. Based on the laboratory testing results on field cores, it is suggested that the greater amounts of transverse cracking exhibited by organic pavements may be caused by organic's lower mixture fracture work density and lower binder failure strain values compared with those of the chemical and foaming sections. The greater amounts of longitudinal cracking exhibited by the organic pavements may be caused by organic's higher mixture indirect tensile strength, lower mixture fracture work density, and lower binder fracture energy values compared with those of the chemical and foaming sections.

Warm mix asphalt (WMA) has been used widely as a sustainable pavement technology to lower greenhouse gas emissions and energy consumption. Compared with traditional hot mix asphalt (HMA), WMA technologies can reduce the mixing and compaction temperatures by approximately 17°C to 28°C, depending on the selected WMA type (*1*). Most previous WMA studies have focused on comparing HMA and WMA with respect to material properties and field performance, whereas a direct comparison between the WMA technologies should also be of interest to agencies when selecting the most appropriate WMA technology.

Several studies have compared laboratory test results among different WMA technologies with regard to resistance to permanent deformation, fatigue cracking, low-temperature cracking, and moisture damage. Based on the use of plant/laboratory-mixed laboratory-compacted specimens, Hamburg wheel tracking test results indicate that Sasobit has better rutting resistance than Evotherm and foaming WMA (*2*, *3*), whereas the rutting resistance of Evotherm and foaming are comparable (*4*).

Based on the elastic component of the complex shear modulus, referred to as $G^* sin\delta$, and the low-temperature performance grade (PG), researchers (*5*, *6*) have reported that foaming WMA and Sasobit show similar fatigue resistance, but the addition of Sasobit lowers resistance to thermal cracking. Based on plant/laboratory-mixed laboratory-compacted specimens, Evotherm 3G and foamed asphalt have been found to be comparable in terms of fracture resistance (*4*). In contrast, Sasobit mixtures show either similar or lower fracture resistance than Evotherm and foamed asphalt (*2–4*, *7*).

Researchers cannot find agreement with regard to moisture resistance among the various WMA technologies. Utilizing the tensile strength ratio as an evaluation parameter, Sargand et al. (*8*) noted that Sasobit and Evotherm field

[1]School of Transportation Engineering, Southeast University, Nanjing, China
[2]Rail Transportation Engineering, Pennsylvania State University, Altoona, PA
[3]Tongji University, Shanghai, China
[4]Department of Civil, Coastal, and Environmental Engineering, University of South Alabama, Mobile, AL

Corresponding Author:
Address correspondence to Shihui Shen: szs20@psu.edu

cores have better moisture resistance than cores taken from foamed sections. However, in another field study that also was based on tensile strength ratio tests, Sasobit showed higher moisture susceptibility than Evotherm 3G and foamed mixtures (*9*). In the National Cooperative Highway Research Program (NCHRP) Reports 763 (*10*) and 779 (*11*), the authors indicate that they found no evidence of moisture damage in field pavements, and thus, WMA technologies should perform equally well in resisting moisture-induced damage.

The field performance comparisons among WMA pavements are based primarily on falling weight deflectometer tests, accelerated pavement tests (APTs), and short-term field performance. The falling weight deflectometer test results indicate that Sasobit pavement sections have either identical or significantly lower stiffness values than Evotherm 3G and foamed asphalt (*8*, *12*). APT results indicate that an Evotherm 3G test section typically experienced greater rut depths than Sasobit and foamed sections (*12*). A comparison of six field projects in Florida found no practical differences in cracking, ride, and rut depth between foamed and Evotherm sections (*13*). A paved road in Connecticut showed that one year after construction, the Sasobit section had significantly deeper ruts and more transverse cracks than the foaming section (*7*). A short-term (less than 3-year) field performance study conducted in Texas found no evidence of rutting or cracking for all the tested sections that consisted of Sasobit, Evotherm, or foaming asphalt, and thus, the WMA technologies were considered to perform equally well in the short-term.

In summary, the literature review indicates that limited comparisons among WMA technologies are based primarily on laboratory tests and short-term field pavement performance evaluations. In some cases, WMA field performance was compared at distant pavement sites, and thus, the results could have been confounded by other factors, such as variations in the pavement structure, subgrade conditions, and pre-overlay conditions of the existing pavements. Therefore, a thorough understanding of WMA technologies calls for comprehensive studies of the long-term performance of WMA pavements based on a relatively large sample population of field experiments so that a statistically sound comparison of WMA field performance can be conducted and conclusive findings of specific performance characteristics (transverse cracking, wheel-path longitudinal cracking, and rutting) can be proposed. In this study, the term "long-term" refers to pavement service years up to 100 months (over 8 years).

Methodology and Scope

The WMA performance comparisons presented in this study are based on field distress data collected during the NCHRP 9-49A study, as detailed in NCHRP Report 843 (*14*). Twenty-eight projects were selected for this study, including five newly constructed projects and 23 in-service projects that reflect different pavement ages, traffic volumes, pavement structures, WMA technologies, and four climate zones across the United States.

Two rounds of distress surveys were conducted in 2012/2013 and 2014/2015 for all the in-service projects that were considered in this study. For the newly constructed pavements, only the second-round distress survey was being undertaken; the first survey was not conducted because they were not opened to traffic at that point. Within each WMA project, three 200-ft test sections were selected to measure the distresses. The same 200-ft test sections were used for both the first- and second-round distress surveys. The distresses measured include rutting and cracking (transverse and longitudinal) in accordance with the LTPP Distress Survey Manual (*15*).

This study's field performance comparisons among the WMA technologies include transverse crack length, wheel-path longitudinal crack length, and rut depth. The crack lengths of the WMA pavements were compared based on statistical *t*-test results at a significance level of 0.05. The null hypothesis was that there was no significant difference in crack length among the tested WMA technologies. The rut depths were compared based on a practical criterion, a 1.6 mm threshold value, because the manual measurement of a rut depth has a precision of 1.6 mm. For each WMA-WMA pair, three comparison results are possible: WMA > WMA, WMA = WMA, or WMA < WMA.

Within each project, the different WMA pavements were constructed physically near to each other at the same time, and thus, they share similar traffic, climate, and pavement structures. Therefore, the authors assumed that any field performance differences among the WMA technologies were caused by differences in their material properties. Any such differences also were assessed in this study to determine the laboratory asphalt and mixture properties that could distinguish between any differences in distress type.

Data Collection

The collected data include project information and field pavement performance information about 34 WMA-WMA pairs from 13 projects. Although HMA sections were also available for the research project, they are not included because the focus of this study is on comparisons between any WMA-WMA pair. Field performance comparisons between HMA and WMA can be found elsewhere (*16–18*). Table 1 provides a summary of the project information that includes the WMA type, climate, pavement age, pavement structure, and annual average daily truck traffic (AADTT). The WMA technologies are divided into three categories: chemical (including Evotherm and LEA), organic (including Sasobit), and foaming (including foaming, Double Barrel Green [DBG], Aquablack [AQB], Advera, Gencor, and Aspha-min). Two types of foaming technology, water-based

Table 1. Project Information

Project	Climate zone	WMA type	Age by 1st round	Age by 2nd round	Pavement structure, mm	AADTT
CO I-70	DF	chemical, foaming, organic	5	7	55+275+0	3,000
MT I-15	DF	chemical, foaming, organic	0	2	51+178+411[a]	833
NE US 14	DF	chemical, foaming	4	6	38+178+102	270
CA 3a	DNF	chemical (2 types), foaming	NA[b]	NA[b]	61+61+399	NA
CA 3b	DNF	chemical, foaming (2 types), organic	NA[b]	NA[b]	61+61+399	NA
IA US 34	WF	chemical, organic	0	3	38+305+203	703
MO Hall	WF	chemical, foaming, organic	5	8	51+298+38	1,470
OH SR 541	WF	chemical, foaming, organic	6	8	32+173+229	10
PA SR 2006	WF	foaming, organic	3	5	36+129+102	100
LA US 61	WNF	chemical, organic	0	3	38+406+152	4,779
LA 116	WNF	chemical, organic	3	5	38+51+216	578
TN SR 46	WNF	chemical, foaming (2 types), organic	5	7	25+97+152	133
TX FM 324	WNF	chemical (2 types), foaming, organic	4	7	31+152+254	209

Note: DF =dry freeze; DNF = dry non-freeze; WF = wet freeze; WNF = wet non-freeze.
[a]Corresponds to overlay + existing pavement + unbound layer.
[b]CA 3a and CA 3b were tested using APTs for a short-term duration (less than 1 year).

and water-containing were both included. The water-based foaming technology refers to those that used 100% water such as Astec DBG and AquaBlack technologies. The water-containing foaming technology refers to the zeolite or particle that contains 20% water such as Asphamin and Advera technologies. More project information, such as mix designs, can be found elsewhere (*14*).

Field Distress

Transverse crack lengths were measured manually within the three 200-ft test sections of the WMA pavements in both the inside and outside wheel paths. Figure 1*a* shows an example of field transverse cracking. In most cases, transverse cracks in the existing asphalt layer overlapped the transverse cracks in the asphalt overlay and initiated from the top of the pavement, as shown in Figure 1*b*. For such transverse cracking type, the existing crack did not reflect to the pavement surface. Instead, a new crack initiated and propagated downward to match the existing crack. Detailed explanation of such cracks can be found elsewhere (*19*). In the case of no existing cracks in the existing asphalt layer, the transverse crack also initiated from the top, as shown in Figure 1*c*. Both thermal cracking and reflective cracking can occur in the form of transverse cracking in an asphalt overlay. The fact that transverse cracks in the WMA layers often overlapped with transverse cracks in the existing pavement indicates that the transverse cracks could be a combination of thermal cracking and reflective cracking.

With regard to fatigue cracking, this study focused on wheel-path longitudinal cracking, which is in general fatigue-related. Similar to the measurements for transverse cracking, crack lengths were measured within the three 200-ft test sections of the WMA pavements inside the wheel paths. Figure 1*d* shows an example of a longitudinal crack, and Figure 1*e* shows that a crack initiated from the top and stopped within the asphalt layer.

Rut depths were measured within the three 200-ft test sections of the WMA pavements every 50 feet for both the inside and outside wheel paths. Figure 1*f* illustrates the rut depth measurement process performed in the field.

Summary of Field Performance

Field Transverse Cracking Identification

Figure 2 presents a summary of the field transverse cracking measurements based on climatic zone. Note that 10 out of 13 projects are listed in the figure because the other three projects did not show transverse cracking. As shown, by the time the first-round survey was conducted, transverse cracks were observed in the MO Hall St. and OH SR 541 projects. Minor cracks were observed in the CO I-70, NE US 14, and TN SR 46 pavement sections. The field cores indicate that transverse cracks in the MO Hall St. and OH SR 541 projects initiated from the pavement surface and propagated downward. Thus, these cracks can be categorized into thermal cracks or surface-initiated reflective cracks (*19*). Additionally, the MO Hall St. and OH SR 541 projects were the oldest (8 years old) of all the projects. As specified by Shen et al. (*14*), a higher pavement age (i.e., an older pavement) corresponds to a greater amount of field transverse cracking for both traditional HMA and WMA pavements. By the time the second-round survey was implemented (2 to 3 years later), more pavement sections in the MT I-15, IA US 34, and LA US 61 projects were showing new transverse cracks. Cracks that were observed in the first round had grown with various degrees of severity.

Figure 1. Field distress and field cores taken at the tip of cracks: (*a*) transverse cracking; (*b*) overlay cracking with transverse cracking in existing asphalt layer; (c) overlay cracking only without transverse cracking in existing asphalt overlay; (*d*) wheel-path longitudinal cracking; (e) field core taken from tip of a wheel-path longitudinal cracking; (*f*) rut depth measurements.

Field Wheel-Path Longitudinal Cracking Identification

Figure 3 provides a summary of the field measurements for projects that showed longitudinal cracks. Similar to the transverse cracking results, the MO Hall St. and OH SR 541 projects showed the most cracking at the time the first-round survey was conducted. By the time the second-round survey was conducted, very few new cracks had developed except for the chemical WMA pavements in CO I-70, organic WMA in PA SR 2006, and foaming WMA in TN SR 46. Figure 3 also shows that the cracks observed from the first round had grown with various degrees of severity and that the pavement sections in the TN SR 46 project showed the most significant increase in the amount of cracking. One possible reason for this outcome is the very thin overlay that was

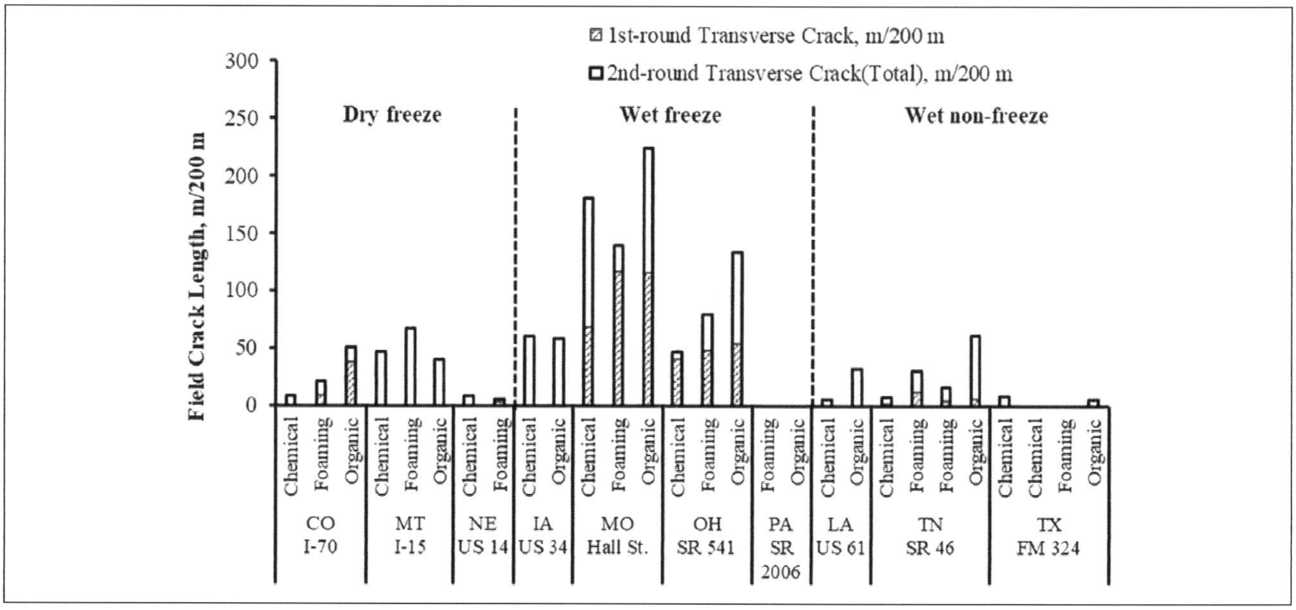

Figure 2. Summary of field transverse cracking.

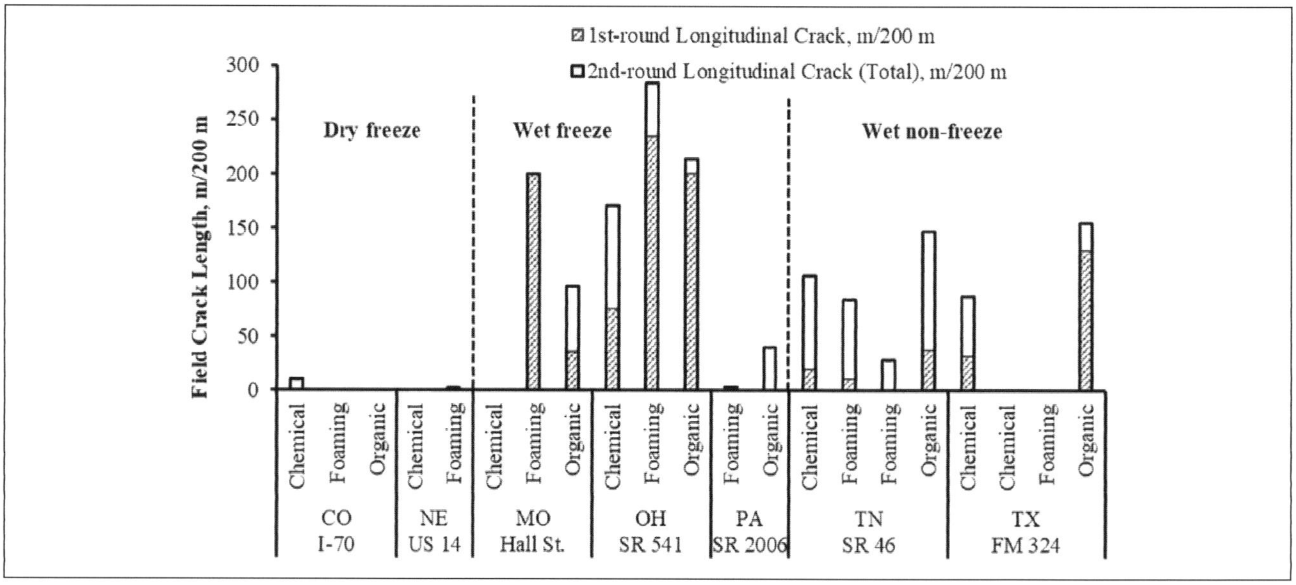

Figure 3. Summary of field longitudinal cracking.

applied in the TN SR 46 project (19 mm to 31 mm). It was found that thinner pavement structures could potentially correspond to worse longitudinal cracking resistance than thicker pavements (*17*).

Field Rut Depth Identification

Figure 4 presents a summary of the field rut depths for projects that showed measurable rut depth during the second-round distress survey. The CO I-70 project shows the greatest rut depths, which may have been caused by the soft binder used (PG 58-28), the high truck traffic level (3,000 AADTT) experienced, and the studded tires. The CA projects were tested using APTs that were terminated when a threshold value of 12.5 mm was reached. However, this termination of test did not affect the comparisons among the WMA technologies.

Comparison among WMA Technologies

Based on the field performance data, statistical *t*-tests with a significance level of 0.05 were conducted to check for any

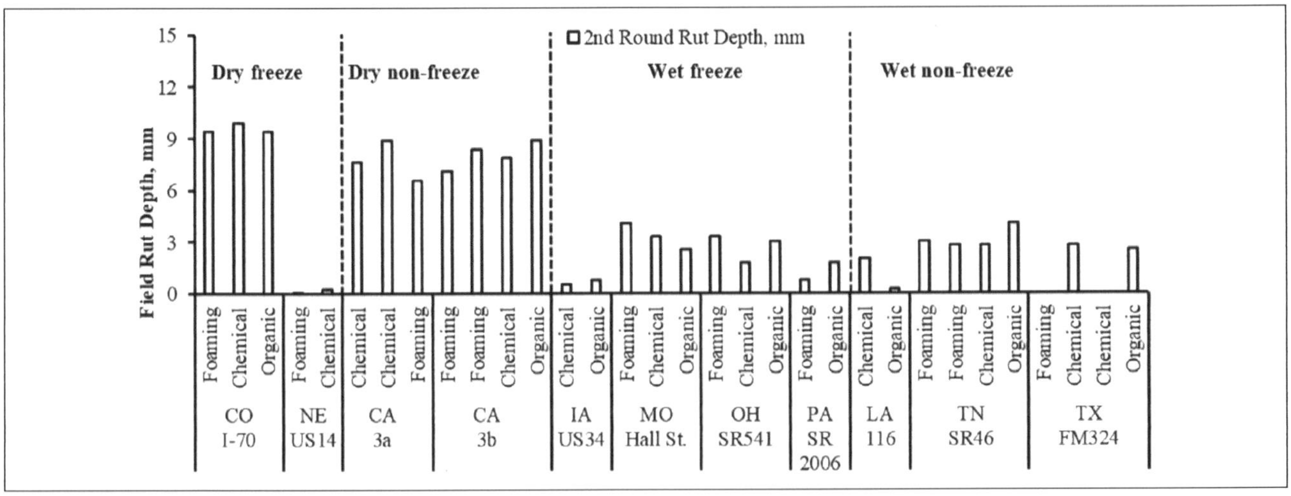

Figure 4. Summary of field rut depths.

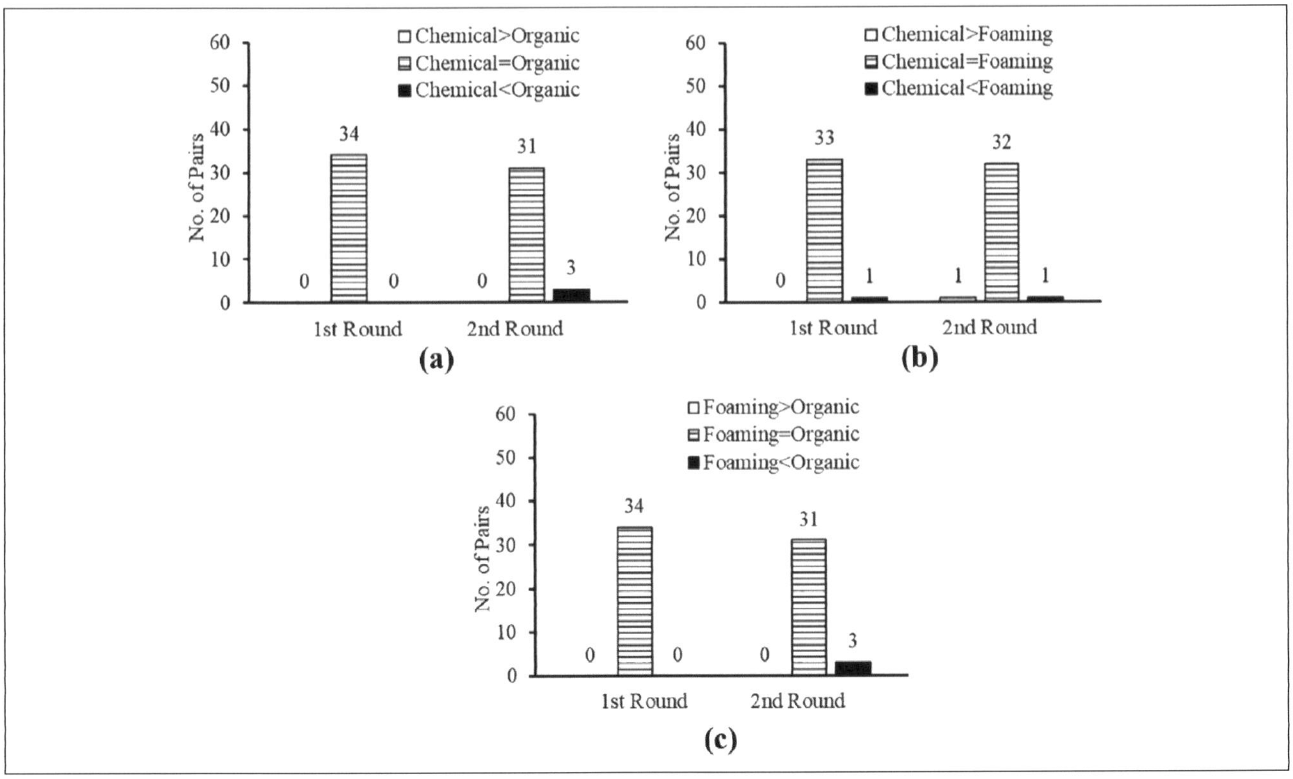

Figure 5. Transverse cracking comparisons among WMA pavements in terms of crack length based on distress survey results: (*a*) comparison between chemical and organic WMA technologies; (*b*) comparison between chemical and foaming WMA technologies; (*c*) comparison between foaming and organic WMA technologies.

statistical differences between each WMA-WMA pair with regard to cracking performance. The criterion that was used to differentiate rutting performance is 1.6 mm.

Field Transverse Cracking Comparison

Figure 5 presents a summary of the statistical comparisons between the WMA-WMA pairs. Note that "chemical > organic" indicates the crack length of chemical WMA pavements is greater than organic pavements; "chemical = organic" indicates the crack length of chemical WMA pavements is not statistically different from organic pavements; "chemical < organic" indicates the crack length of chemical WMA pavements is smaller than organic pavements. The other legends have similar meanings. Based on the results from the first-round distress survey, no performance

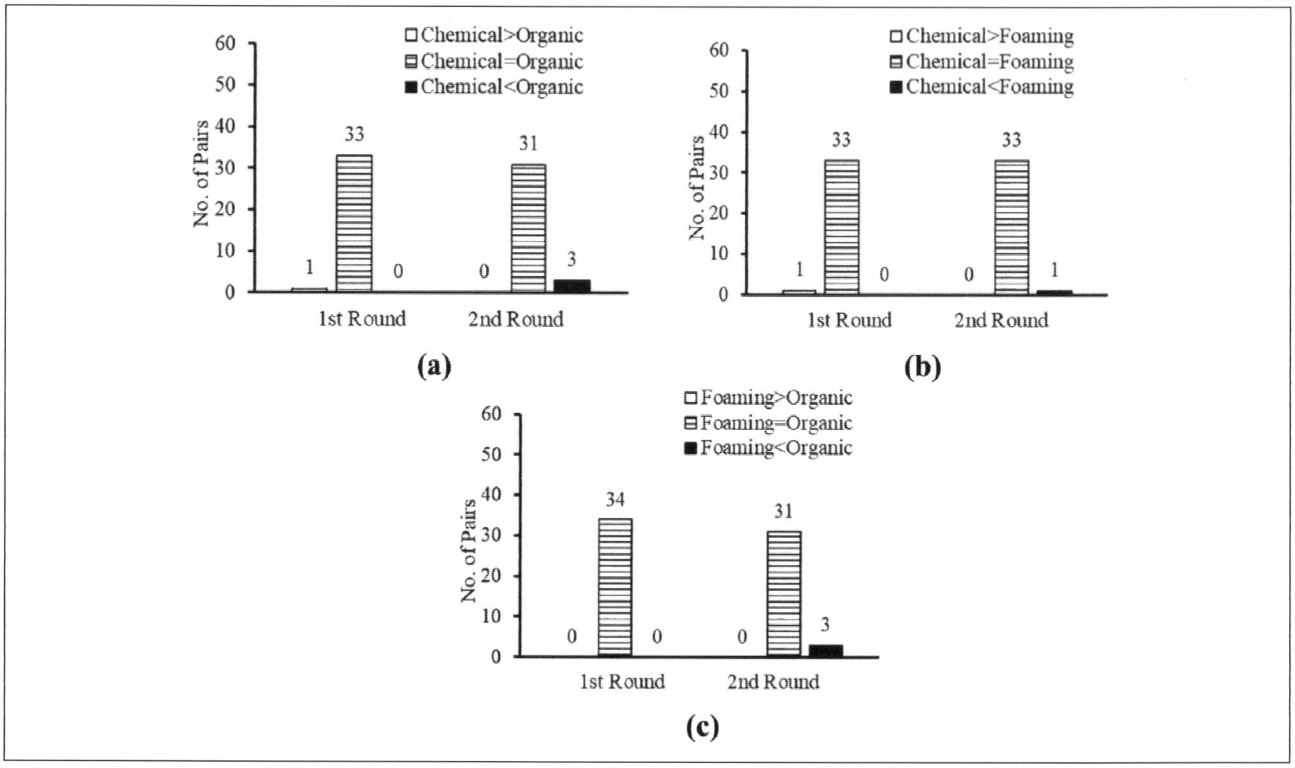

Figure 6. Wheel-path longitudinal cracking comparisons among WMA pavements in terms of crack length based on distress survey results: (*a*) comparison between chemical and organic WMA technologies; (*b*) comparison between chemical and foaming WMA technologies; (*c*) comparison between foaming and organic WMA technologies.

difference between any WMA-WMA pair was observed. The pavements that contained the three respective WMA types (chemical, foaming, and organic) exhibited similar field performance in terms of transverse cracking. It is noted that all the projects were no more than 6 years old by the time the first distress survey was conducted.

Further comparisons among the WMA technologies were made based on the second-round transverse cracking survey (2 to 3 years after the first-round survey). As Figure 5 shows, in general, the three WMA technologies are comparable. However, Figure 5, *a* and *c* indicate that the three organic modified WMA pavements from the CO I-70, OH SR 541, and TN SR 46 projects have significantly longer transverse crack lengths than their companion WMA pavements (chemical and foaming), whereas Figure 5*b* shows that the chemical and foaming WMA pavements experienced the same resistance to transverse cracking. These findings are different from the first-round analysis results. Given that the three pavements had aged for 7 years or longer by the time the second-round distress survey was conducted, it is possible that aging might be more critical to the organic modified WMA pavement than to the other two WMA technologies. However, further study is needed to evaluate such subtle differences between the WMA technologies.

Field Longitudinal Cracking Comparison

Figure 6 presents results from the comparisons of wheel-path longitudinal cracking for each WMA-WMA pair. Based on the first-round distress survey results, the three WMA technologies show similar field performance in terms of wheel-path longitudinal cracking. It is noted that all the project pavements were no more than 6 years old at the time the first distress survey was conducted.

Further comparisons were made based on the second-round distress survey results. Similar to the transverse cracking results, three WMA-WMA pairs showed that the organic modified WMA pavements (OH SR 541, TN SR 46, and PA SR 2006) had significantly longer longitudinal cracks than their companion WMA pavements (chemical and foaming), as shown in Figure 6, *a* and *c*, whereas Figure 6*b* shows that the chemical and foaming WMA pavements have similar resistance to longitudinal cracking. The relatively high pavement age (7 years or older) of all three pavements plus the ultra-thin overlay thickness in the TN SR 46 project (which made it easier to age) suggest that the effect of aging could be more prominent in organic WMA pavements than in chemical and foaming WMA pavements. Again, further study is needed to evaluate such subtle differences between the WMA technologies.

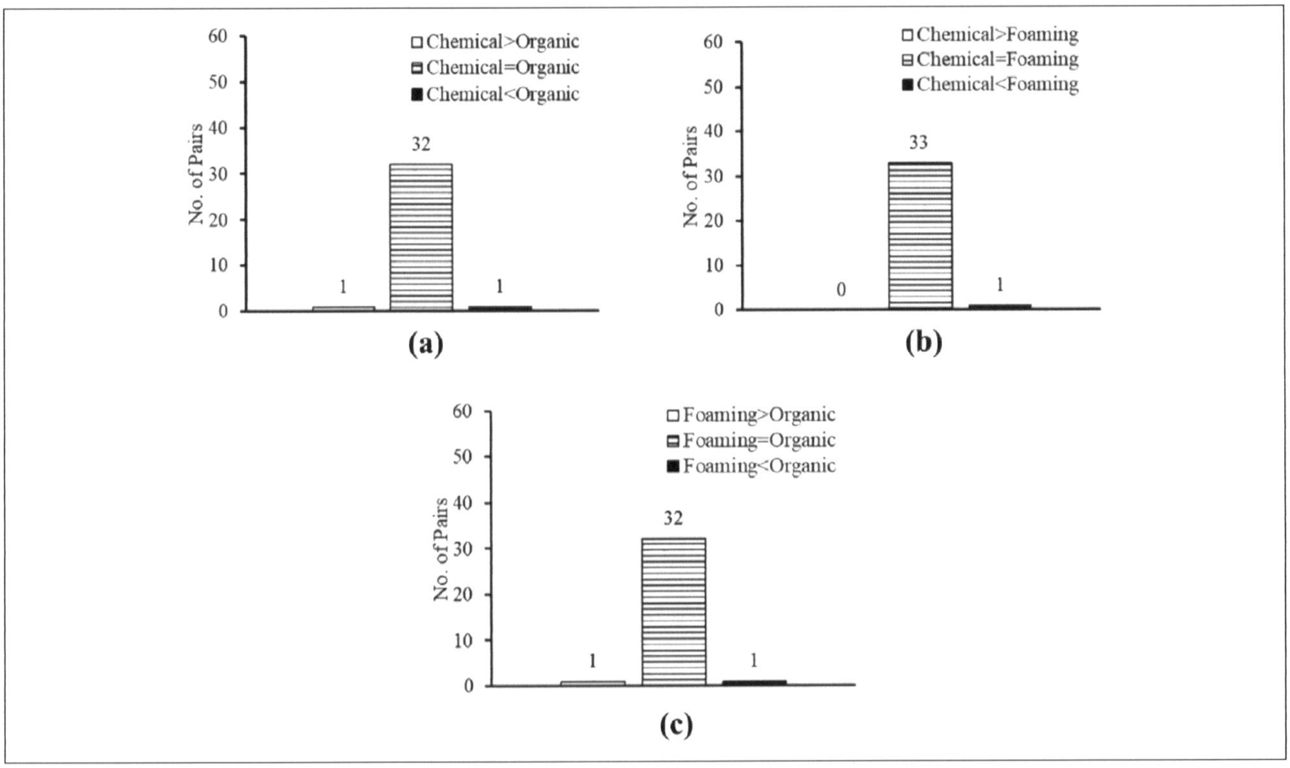

Figure 7. Rut depth comparisons among WMA pavements based on second-round distress survey results: (*a*) comparison between chemical and organic WMA technologies; (*b*) comparison between chemical and foaming WMA technologies; (*c*) comparison between foaming and organic WMA technologies.

Field Rut Depth Comparison

Figure 7 presents comparisons of the rutting performance of the WMA pavements based on the second-round survey results. The pavements with respectively different WMA technologies generally show comparable rutting performance.

Material Properties Comparisons

The analysis results indicate that most of the WMA-WMA pairs are not statistically different from each other in terms of field cracking and rutting performance. However, the organic WMA sections from the CO I-70, OH SR 541, and TN SR 46 projects showed worse transverse cracking resistance than the chemical and foaming pavements. Also, the organic WMA sections from the OH SR 541, TN SR 46, and PA SR 2006 projects showed inferior longitudinal cracking resistance. Such common behavior is worth further exploration and, therefore, the above projects were evaluated to see what material properties are significant for identifying the performance difference.

As aforementioned, each WMA-WMA pair was compared within the same project and it was assumed that any field performance differences were caused by differences in the material properties. Because numerous material properties (more than 50) were tested in the NCHRP 9-49A project, this study considered only the variables that correlated with field cracking performance to some extent, as identified elsewhere (*16*, *17*, *20*, *21*).

Table 2 lists each variable for both asphalt binder and asphalt mixture properties that was evaluated using the statistical *t*-test. The table lists the test results only from the CO I-70, OH SR 541, PA SR 2006, and TN SR 46 projects. Note that the variables tested at 20°C were used to characterize longitudinal cracking, and the parameters tested at lower temperatures (5°C and lower) were used to characterize transverse cracking. All the mixture properties were tested using field cores drilled during the first-round distress survey. The asphalt binders were obtained by extracting and recovering the field cores. The extraction process followed Method A in AASHTO T 164, and the recovery procedure was carried out in accordance with AASHTO R 59.

Results indicated that the material properties of fracture work density (tested at −10°C) and binder failure strain (tested at 5°C) were the two most critical factors that correlated with field performance. Mixture fracture work density is the area of the load-vertical deformation curve divided by the volume of the specimen, as obtained from indirect tensile (IDT) testing (*22*). Binder failure strain is the shear strain that corresponds to the peak shear stress, as obtained from DSR testing. Higher mixture fracture work density and binder failure strain values are potentially correlated with

Table 2. Material Properties from Projects in Which Organic Pavements Show More Field Cracking Than the Chemical and Foaming Sections

	CO I-70			OH SR 541			PA SR 2006		TN SR 46			
Variable	Chemical	Organic	Foam	Chemical	Organic	Foam	Organic	Foam	Chemical	Organic	Foam1[a]	Foam2
In-place air voids, %	3.6	4.1	3.6	5.9	6.3	6.5	6.0	8.2	7.1	8.5	6.1	6.7
Percentage passing No. 200 sieve, %	8.0	7.4	7.3	5.6	4.9	0.6	6.1	4.7	5.8	5.1	6.1	6.0
Mix fracture energy, MPa, 20°C	14.3	14.1	17.2	7.0	6.6	4.8	6.1	8.4	1.8	3.5	3.6	6.7
Mix tensile strength, MPa, 20°C	1.4	1.3	1.2	1.4	2.1	1.9	2.6	2.7	1.7	2.0	1.9	2.0
Mix VFD, mm, 20°C	2.3	2.1	2.5	1.5	1.2	1.1	2.0	1.6	1.5	0.9	1.0	1.5
Mix HFS, mm, 20°C	0.013	0.013	0.018	0.006	0.004	0.003	0.003	0.004	0.001	0.002	0.002	0.004
Mix FWD, kPa, 20°C	113.6	111.6	121.0	89.0	91.4	79.3	144.3	149.5	81.4	52.6	55.0	91.8
Binder maximum stress, kPa, 20°C	238.3	346.7	264.3	2,159.3	2,137.9	2,992.8	1,320.0	895.0	4,998.2	3,446.5	4,086.0	4,119.3
Binder fracture energy, kPa, 20°C	335.0	529.5	519.5	15,241.6	14,300.2	15,492.4	1,652.2	1,364.9	3,975.7	2,528.5	9,036.0	5,291.0
Mix fracture energy, MPa, −10°C	31.7	21.8	35.4	5.6	5.9	8.7	2.1	1.3	0.7	1.6	2.0	1.8
Mix tensile strength, MPa, −10°C	4.4	3.7	4.5	2.3	1.9	2.6	3.6	2.9	2.5	2.4	2.6	3.1
Mix FWD, kPa, −10°C	0.1	0.1	0.1	0.2	0.1	0.5	0.04	0.03	0.02	0.02	0.03	0.02
Mix creep compliance, 1/GPa, 0°C, 100 s	0.5398	0.5982	0.7132	0.1475	0.1603	0.1740	0.1707	0.1402	0.1518	0.1181	0.1442	0.2333
Mix dynamic modulus, MPa, −10°C, 10 Hz	17,692	15,052	18,915	15,450	14,086	14,421	18,628	19,559	17,729	16,817	18,461	17,561
Binder low PG, °C	−27.6	−27.0	−28.4	−16.2	−7.4	−5.8	−19.6	−17.5	−6.4	−12.5	−13.0	−14.4
Binder BBR m-value, −6°C	0.463	0.432	0.483	0.302	0.228	0.274	0.313	0.350	0.231	0.274	0.290	0.307
Binder fracture energy, kPa, 5°C	2,308.4	1,507.9	2,631.7	8,739.2	8,286.6	3,742.0	4,019.2	2,965.7	278.7	860.2	3,202.2	3,739.0
Binder failure strain, 5°C	1.6	1.7	1.2	1.3	0.8	1.6	1.4	1.5	0.1	0.1	0.4	0.5

Note: VFD = vertical failure deformation; HFS = horizontal failure strain; FWD = fracture work density; BBR = bending beam rheometer.
[a]TN SR46 project used two types of foaming.

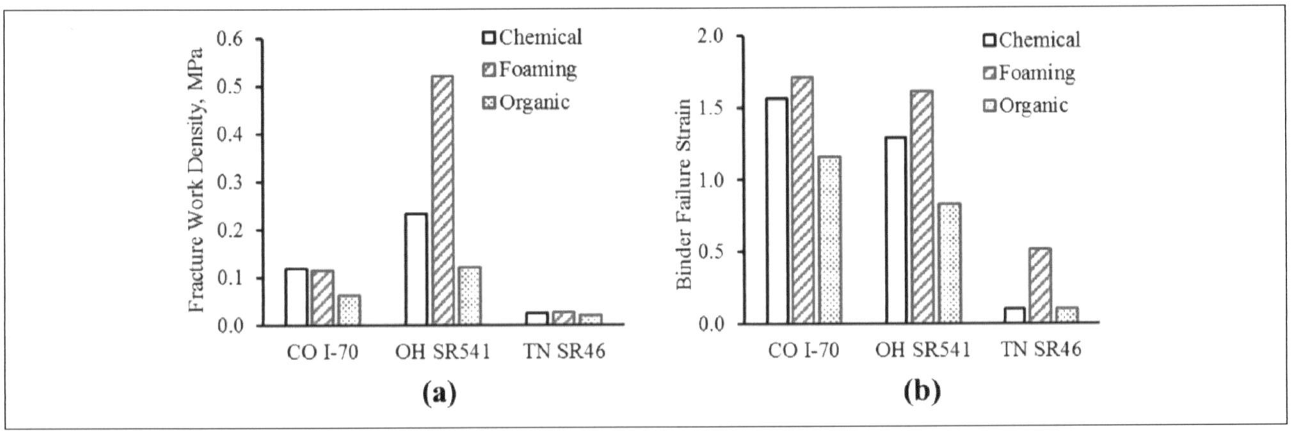

Figure 8. Material properties comparison among WMA technologies: (*a*) mixture fracture work density and (*b*) binder failure strain.

better cracking resistance in the field pavements, as specified elsewhere (*16*, *20*). As shown in Figure 8, as well as the statistical analysis results, the fracture work density and the binder failure strain properties of the organic pavement section are not statistically higher than the chemical and the foaming pavement sections, which corresponds to higher cracking in the field. The rest parameters that are potentially correlated with field transverse or longitudinal crack were also checked, such as low-temperature PG, whereas no obvious trend was observed.

The material properties obtained from the OH SR 541, TN SR 46, and PA SR 2006 projects were tested for longitudinal cracking. Compared with the chemical and foaming sections, the organic pavements showed comparable or higher mixture tensile strength values, comparable or lower mixture fracture work density values, and comparable or lower binder fracture energy values, indicating a greater potential for large amounts of cracking, as specified in the literature (*17*, *20*). Here, mixture tensile strength indicates the strength of a specimen under tension, as obtained from IDT tests. Binder fracture energy is defined as the area underneath the shear stress versus shear strain curve up to the peak shear stress. Regarding the trend between mixture tensile strength and longitudinal cracking, Shen et al. (*17*) indicate that, for pavements with high pavement ages (i.e., 7 years or older), overlay mixtures with high IDT strength could be related to high stiffness and brittleness and therefore would be prone to longitudinal top-down cracking.

Summary of Findings and Conclusions

This study compared WMA technologies in terms of field performance, including transverse cracking, top-down longitudinal cracking, and rut depth. The comparisons were based on two field surveys; the first survey targeted pavement ages of 0 to 6 years, and the second survey targeted pavement ages of 2 to 8 years. The study considered chemical, organic, and foaming WMA technologies; 34 WMA-WMA pairs from 13 field projects were included for analysis. The performance differences between the chemical and organic, chemical and foaming, and organic and foaming WMA technologies were analyzed. The material properties that could cause these field performance differences among these WMAs also were investigated. Based on the analysis results, the following conclusions can be drawn:

- Generally, no statistical differences were found among the WMA-WMA pairs. The pavements that contained the three WMA technologies (chemical, foaming, and organic) showed similar field performance in terms of transverse cracking and longitudinal top-down cracking.
- During the second-round survey, a few (three) organic modified WMA pavements showed higher amounts of transverse and longitudinal cracking than their companion WMA pavements (chemical and foaming). Given that the three organic WMA pavements with higher amounts of cracking had aged for at least 7 years by the time the second-round distress survey was conducted, it is possible that aging is a critical factor for organic WMA pavement crack resistance.
- Comparisons of the rut depths between the WMA-WMA pairs in the second-round survey indicate that the pavements with different WMA technologies generally show comparable rutting performance.
- The three WMA-WMA comparisons in which organic showed more transverse cracking were possibly because of its lower mixture fracture work density and lower binder failure strain values compared with the chemical and foaming pavements.
- The three WMA-WMA comparisons in which organic showed more longitudinal cracking were possibly because of its higher mixture tensile strength, lower mixture fracture work density, and lower binder fracture energy values compared with the other WMA technologies.

Acknowledgments

This study was sponsored by the National Cooperative Highway Research Program project 09-49A. The authors would like to acknowledge and thank Dr. Ed Harrigan of the NCHRP staff and panel members for their assistance. Thanks also go to the entire NCHRP 09-49A research project team members and other partner universities and highway agencies for data collection and generous help.

Author Contributions

The authors confirm contributing to the paper as follows: study conception and design: Shihui Shen, Weiguang Zhang, and Shenghua Wu; data collection: Weiguang Zhang and Shenghua Wu; analysis and interpretation of the results: Weiguang Zhang, Shenghua Wu, and Shihui Shen; draft manuscript preparation: Weiguang Zhang, Shihui Shen, and Shenghua Wu. All authors reviewed the results and approved the final version of the manuscript.

References

1. Bower, N., H. Wen, S. Wu, K. Willoughby, J. Weston, and J. DeVol. Evaluation of the Performance of Warm Mix Asphalt in Washington State. *International Journal of Pavement Engineering*, Vol. 17, 2016, pp. 423–434.
2. Estakhri, C., J. Button, and A. Alvarez. *Field and Laboratory Investigation of Warm Mix Asphalt in Texas*. FHWA/TX-10/0-5597-2. Texas Department of Transportation, Texas, 2010.
3. Bennert, T., A. Maher, and R. Sauber. Influence of Production Temperature and Aggregate Moisture Content on the Initial Performance of Warm-Mix Asphalt. *Transportation Research Board: Journal of the Transportation Research Board*, 2011. 2208: 97–107.
4. Leng, Z., A. Gamez, and I. Al-Qadi. Mechanical Property Characterization of Warm-Mix Asphalt Prepared with Chemical Additives. *Journal of Materials in Civil Engineering*, Vol. 26, No. 2, 2014, pp. 304–311.
5. Lee, S., S. Amirkhanian, N. Park, and K. Kim. Characterization of Warm Mix Asphalt Binders Containing Artificially Long-term Aged Binders. *Construction and Building Materials*, Vol. 23, 2009, pp. 2371–2379.
6. Akisetty, C., S. Lee, and S. Amirkhanian. Laboratory Investigation of the Influence of Warm Asphalt Additives on Long-term Performance Properties of CRM Binders. *International Journal of Pavement Engineering*, Vol. 11, No. 2, 2010, pp. 153–160.
7. Bernier, A., A. Zofka, R. Josen, and J. Mahoney. Warm-Mix Asphalt Pilot Project in Connecticut. *Transportation Research Board: Journal of the Transportation Research Board*, 2012. 2294: 106–116.
8. Sargand, S., M. D. Nazzal, A. Al-Rawashdeh, and D. Powers. Field Evaluation of Warm-Mix Asphalt Technologies. *Journal of Materials in Civil Engineering*, Vol. 24, No. 11, 2012, pp. 1343–1349.
9. Porras, J., E. Hajj, P. Sebaaly, S. Kass, and T. Liske. Performance Evaluation of Field-Produced Warm-Mix Asphalt Mixtures in Manitoba, Canada. *Transportation Research Board: Journal of the Transportation Research Board*, 2012. 2294: 64–73.
10. Martin, A. E., E. Arambula, F. Yin, L. G. Cucalon, A. Chowdhury, R. Lytton, J. Epps, C. Estakhri, and E. S. Park. *Evaluation of the Moisture Susceptibility of WMA Technologies*. NCHRP Report 763. Transportation Research Board, Washington, D.C., 2014.
11. West, R., C. Rodezno, G. Julian, B. Prowell, B. Frank, L. Obsborn, and T. Kriech. *Field Performance of Warm Mix Asphalt Technologies*. NCHRP Report 779. Transportation Research Board, Washington, D.C., 2014.
12. Mejias, M., J. Doyle, and J. F. Rushing. Accelerated Pavement Testing of Warm-Mix Asphalt for Heavy-Traffic Airfields. *Transportation Research Board: Journal of the Transportation Research Board*, 2014. 2456: 11–20.
13. Choubane, B., S. Chun, H. Lee, P. Upshaw, J. Greene, and A. Nazef. *Evaluation of Warm-mix Asphalt (WMA) Performance in Florida*. FL/DOT/SMO/13-558. Florida Department of Transportation, State Materials Office: Gainesville, FL, 2013.
14. Shen, S., S. Wu, W. Zhang, L. Mohammad, and B. Muhunthan. *Long-Term Field Performance of Warm Mix Asphalt Technologies*. NCHRP Report 843. Transportation Research Board, Washington, D.C., 2017.
15. Miller, J., and W. Bellinger. *Distress Identification Manual for the Long-Term Pavement Performance Program*. FHWA-HRT-13-092. Federal Highway Administration, McLean, VA, 2014.
16. Zhang, W., S. Shen, P. Basak, H. Wen, S. Wu, A. Faheem, and L. N. Mohammad. Development of Predictive Models for Initiation and Propagation of Field Transverse Cracking. *Transportation Research Record: Journal of the Transportation Research Board*, 2015. 2524: 92–99.
17. Shen, S., W. Zhang, L. Shen, and H. Huang. A Statistical Based Framework for Predicting Field Cracking Performance of Asphalt Pavements: Application to Top-Down Cracking Prediction. *Construction and Building Materials*, Vol. 116, 2016, pp. 226–234.
18. Wen, H., S. Wu, L. N. Mohammad, W. Zhang, S. Shen, and A. Faheem. Long-Term Field Rutting and Moisture Susceptibility Performance of Warm-Mix Asphalt Pavement. *Transportation Research Record: Journal of the Transportation Research Board*, 2016. 2575: 103–112.
19. Shen, S., W. Zhang, H. Wang, and H. Huang. Numerical Evaluation of Surface-Initiated Cracking in Flexible Pavement Overlays with Field Observations. *Road Materials and Pavement Design*, Vol. 18, 2017, pp. 221–234.
20. Wu, S., H. Wen, W. Zhang, S. Shen, A. Faheem, and L. N. Mohammad. Long-term Field Transverse Cracking Performance of Warm-Mix Asphalt Pavement and Its Significant Material Property. *Transportation Research Record: Journal of the Transportation Research Board*, 2016. 2576: 109–120.
21. Zhang, W., S. Shen, S. Wu, and L. N. Mohammad. Long-term Field Aging of Warm Mix and Hot Mix Asphalt Binders. *Transportation Research Record: Journal of the Transportation Research Board*, 2017. 2632: 140–149.
22. Wen, H. Use of Fracture Work Density Obtained from Indirect Tensile Testing for the Mix Design and Development of a Fatigue Model. *International Journal of Pavement Engineering*, Vol. 14, 2013, pp. 561–568.

The Standing Committee on Critical Issues and Emerging Technologies in Asphalt (AFK10) peer-reviewed this paper (18-02967).

Article

TRR
JOURNAL OF THE TRANSPORTATION RESEARCH BOARD

Transportation Research Record
2018, Vol. 2672(28) 211–222

DOI: 10.1177/0361198118781389
journals.sagepub.com/home/trr
SAGE

Evaluation of Marshall Stability Design Principles: Applied to Cold In-Place Recycling

Ben C. Cox[1] and Isaac L. Howard[2]

Abstract

The objective of this paper is to document limitations and robustness issues with Marshall stability (MS) testing for cold in-place recycling (CIR) mix designs. This paper focuses efforts on MS because many departments of transportation are using this test for CIR design, and the data presented in this paper supports abandonment for any type of CIR mix design. Unconfined compressive strength (UCS) and indirect tensile strength data is used throughout the paper for benchmarking, as both tests are used in various capacities for in-place recycling designs. Portland cement and asphalt emulsion binders were assessed in this paper. In comparison with UCS design curves with cement, MS results for emulsion were not as rational and often did not respond to emulsion content changes in a logical manner. MS repeatability was poor for the materials evaluated in this paper. Replicate design curves ranged from all points below a design threshold (i.e., all emulsion dosage levels failed to meet the design criteria) to all points above a design threshold (i.e., all emulsion dosage levels successfully met the design criteria). The typical minimum MS criteria (5.56 kN) was exceeded without any emulsion for specimens oven conditioned for 2 h at 40°C. A notable limitation of the data presented in this paper was that retained MS (i.e., assessment of moisture conditioning effects) was not performed comprehensively throughout the work.

In-place recycling materials, namely cold in-place recycling (CIR) and full-depth reclamation (FDR), can be stabilized with asphalt emulsion, portland cement, or many other additives. The state-of-practice often utilizes different mix design methods depending on the stabilization agent. For example, projects intending to use portland cement usually rely on unconfined compressive strength (UCS) as their primary design criteria, whereas projects intending to use asphalt emulsion often rely on Marshall stability (MS). Indirect tensile (IDT) testing is also utilized to varying degrees for CIR design.

The ideal scenario for CIR going forward would be to perform designs in a consistent manner that is independent of the stabilization material being utilized (e.g., *1*). To do so, mechanical property test methods need to be robust enough to provide useful information about multiple types of stabilization agents. However, before a mechanical property test can be robust enough to accommodate multiple stabilization materials, it needs to be able to properly assess at least one stabilization material. To this end, the objective of this paper is to document limitations and robustness issues with MS testing for CIR mix designs. This paper selected MS because many departments of transportation (DOTs) are using this test for CIR design, and the data presented supports abandonment of MS for any type of CIR design. UCS and IDT data are used to benchmark MS throughout this presentation.

This paper first reviews pertinent information related to MS, alongside existing design practices for emulsion-stabilized and cement-stabilized CIR. That review clearly shows nearly all DOT methods using MS (alongside several complimentary tests) to select emulsion dosages, whereas cement dosages are determined largely via UCS. CIR's utilization as a rehabilitation technique began during a period in which Marshall principles governed most DOT asphalt paving, and as such Marshall principles found their way into CIR. Data presented in this paper, however, shows that MS testing is not optimal for CIR moving forward.

Literature Review

Marshall Stability

The Marshall method was first introduced for asphalt around 1940 by Bruce Marshall of the Mississippi Highway Department and was furthered by the U.S. Army Corps of Engineers. Over the next 50 years, the Marshall method became the most widely used design method in existence. Collectively, Waterways Experiment Station (*2*, *3*), Marshall (*4*), Corps of Engineers (*5*), White (*6*), Kandhal and Koehler (*7*), Roberts et al. (*8*) and Leahy and McGennis (*9*)

[1]U.S. Army Engineer Research and Development Center, Vicksburg, MS
[2]Civil and Environmental Engineering, Mississippi State University, Mississippi State, MS

Corresponding Author:
Address correspondence to Ben C. Cox: benjamin.c.cox@usace.army.mil

summarize its development and use through the years. MS and flow tests were key aspects of the method.

Design asphalt content was generally selected based on bulk density, air voids (V_a), voids in mineral aggregate, stability, and flow (*3*, *8*). Method specifics varied, but the two most common were, first, the National Asphalt Pavement Association method in which design asphalt content was selected to yield 4% V_a alongside values within specification limits for the other properties (*8*) and, second, the Asphalt Institute method in which design asphalt content was taken as the average of the asphalt contents at 4% V_a, maximum density, and maximum stability so long as all other properties were within specification limits at this average asphalt content (*7–8*). Specification criteria, particularly for stability but also for flow, varied significantly as each state adapted the method for their local materials and needs. In 1985, potentially at the height of the Marshall method when 75% of state DOTs were using it, minimum 60°C MS requirements ranged from 2.24 to 8.90 kN (500 to 2,000 lbs) (*7*).

Throughout the 1980s, the belief that the Marshall method had outlived its usefulness grew (*7–9*). More fundamental approaches were desired in place of Marshall's relatively empirical philosophies such as impact compaction, which was not believed to fully represent field compaction, and the stability test, which is a surrogate for field stability (*7–9*).

Ironically, CIR using modern milling machines did not become customary until around 1980 (*1*). Perhaps given the Marshall method's simplicity and familiarity alongside its continued use through nearly the end of the 20th century, Marshall design principles, namely the use of Marshall compaction (now mostly replaced by gyratory compaction) and MS, were carried over to CIR mix design as shown in the subsequent section. Marshall principles became further integrated into CIR design practices following the *AASHTO-AGC-ARTBA Task Force 38* report published in 1998, which recommended either Marshall or Hveem procedures, of which Marshall was the predominant method at the time (*10*).

State of CIR Mix Design Practice

Table 1 provides a summary of existing DOT emulsion CIR mix design practices. Table 1 covers several of the major components of each mix design such as moisture content and compaction requirements; however, the design binder content selection tests are of most importance to this paper.

At large, Table 1 practices are generally based on specification recommendations of Thomas and Kadrmas (*20*) who selected the emulsion content at 5.56 kN MS as long as it satisfied 70% or greater retained MS (RMS) after moisture conditioning, 2% or less mass loss in the ASTM D7196 raveling test (RT), and an AASHTO T322 critical cracking temperature (T_{crit}) lower than the LTPPBind 98% reliability low performance grade temperature at the top of the CIR layer. Although some states' requirements vary in different aspects, all require MS in some form. Most require a minimum AASHTO T245 MS of 5.56 kN (1,250 lbs) at 40°C and use this criteria as the primary means to select a design emulsion content so long as other criteria are also met at that emulsion content.

For comparison, Cox and Howard (*1*) summarize existing portland cement mix design practices for five states for CIR and FDR (most cement-stabilized in-place recycling is FDR however, at least with respect to binder content selection, FDR and CIR are similar). The states summarized were Alabama, California, Mississippi, Pennsylvania, and New York (*21–25*). States with cement specifications all utilize Proctor compaction (standard and modified) and one of two curing protocols (moist cured in a sealed bag at 21°C for 7 days or sealed in a bag at 40°C for 7 days). Similar to Table 1 generally using MS as the primary mix design test, states with cement specifications all use UCS as the sole means to select cement contents. While UCS testing of cement-stabilized CIR is not the focus of this paper, comparing emulsion mix design methods to cement better illustrates concerns with MS.

CIR Marshall Stability Reference Values

Table 2 provides a collection of MS and RMS values from literature largely covering the time period since CIR literature first became available around 1980. Not every reference including Marshall testing is provided in Table 2. For example, testing details (e.g., temperature) were not provided in some references.

Throughout the years, several test temperatures have been utilized with significant MS effects. Though this is a fairly crude assessment, MS in Table 2, on average and for a distribution of emulsion contents (sometimes including secondary binders such as cement or hydrated lime), is 12.1, 9.1, and 4.4 kN at 25, 40, and 60°C, respectively. At 40°C (most common Table 1 test temperature), all Table 2 MS values meet or exceed the typical MS criteria of 5.56 kN. It is important to note that, in some cases, RMS was less than 70% for lower emulsion contents, meaning it would govern the design emulsion content selection.

It is interesting that, during this work, references published in 2000 and more recently were reviewed to determine a distribution of the most common mix tests used. In all, 78 references were identified that included mix tests. Of those, only 11 presented MS results, and all but four of those were published by authors outside the United States (e.g., China, Iran, Poland, Switzerland) where the Marshall method is more prevalent at present. Three of the four references within the United States were by the same research group that, interestingly, was presenting a new mix design methodology for foamed asphalt CIR that departed from MS testing.

In contrast, IDT strength (S_t) results were presented in 49 of those 78 references whereas resilient modulus (M_r) and elastic modulus (E^*) were included in 21 and 10, respectively. This simple assessment suggests a trend that for nearly the last two decades, research has shifted away from MS while DOTs, on

Table 1. Existing Emulsion Mix Design Methods

State	Designation	Mixing and compaction MC	Compaction	Curing	Density	Design binder content selection tests [test criteria in brackets]
California (*11*)	LP-8	Expected *MC* during milling, typ. 1.5% to 2.5%	75-blow Marshall or 30-gyration SGC	60°C to constant mass (in 16 to 48 h)	T209[a] G_{mm}, T166[a] G_{mb}, report V_a	1. T245 MS [5.56 kN min at 40°C] 2. T245 RMS [70% min at 40°C after VS and 24-h soak] 3. D7196[a] RT [2% max, 20-gyr, cured at 21°C for 4 h]
Illinois (*12*)	S.P. LR 400-5	*MC* needed for emulsion dispersion	30-gyration SGC	NS	T209[a] G_{mm}, T331[a] or T166[a] G_{mb}, report V_a	1. T245[a] MS [5.56 kN min, NS] 2. T245[a] RMS [70% min, NS] 3. RT [2% max, at 10°C]
Iowa (*13*)	I.M. 504 App. B	1.5% *MC*	30-gyration SGC	60°C for 48 h	T166[a] G_{mb}	1. T245[a] MS [4.45 kN min at 40°C] 2. T245[a] RMS [70% min at 40°C after VS and 24-h soak] 3. D7196[a] RT [2% max, 20-gyr, cured at 10°C for 4 h] 4. T322 T_{crit} [−20°C max]
Kansas (*14*)	C.M. Part V-5.3.4	Expected *MC* during milling, typ. 1.5% to 2.5%	30-gyration SGC	60°C to constant mass (in 16 to 48 h)	T209[a] G_{mm}, T166[a] G_{mb}, report V_a	1. T245[a] MS [4.45 kN min at 40°C] 2. T245[a] RMS [70% min at 40°C after VS and 24-h soak] 3. D7196[a] RT [2% max, 20-gyr, cured at 21°C for 4 h] 4. T322[a] T_{crit} [less than LTPPBind 98% reliability low temp at top of CIR layer]
Mississippi (*15*)	S.P. 907-425-1	T180 Proctor-determined *OMC*	30-gyration SGC	NS	NS	1. MT-59 boil test [60% min] 2. MT-63 S_t and TSR [310 kPa and 55%] 3. Marshall quotient [61.3 kN/cm]
Montana (*16*)	S.P. 405-2	Expected *MC* during milling	30-gyration SGC	60°C to constant mass (in 16 to 48 h)	T209[a] G_{mm}, T166[a] G_{mb}, report V_a	1. T245 MS [5.56 kN min at 40°C] 2. T245 RMS [70% min at 40°C after VS and 24-h soak] 3. D7196 RT [5% max, 20-gyr, cured at 21°C or 10°C for 4 h] 4. T322 T_{crit} [−31°C max]
New York (*17*)	Proposed Spec. as per (18)	Expected *MC* during milling, typ. 1.5% to 4.5%	75-blow Marshall or 30-gyration SGC	60°C to constant mass (in 16 to 48 h)	T209 G_{mm}, T166 G_{mb}, report V_a	1. T245 MS [5.56 kN min at 40°C] 2. T245 RMS [70% min at 40°C after VS and 24-h soak] 3. D7196 RT [5% max, 20-gyr, cured at 10°C for 4 h at 50% RH] 4. T322 T_{crit} [−20°C max]
Texas (*18*)	S.S. 3254	Expected *MC* during milling, typ. 1.5% to 2.5%	35-gyration SGC	60°C to constant mass (in 16 to 48 h)	T209[a] G_{mm}, T166[a] G_{mb}, report V_a	1. T245[a] MS [11.12 kN min at 40°C] 2. T245[a] RMS [70% min at 40°C after VS and 24-h soak] 3. D7196 RT [2% max, 20-gyr, cured at 10°C for 4 h at 50% RH] 4. T322 T_{crit} [report only] 5. Tex-226-F S_t [276 kPa min] 6. Tex-242-F HLWT [5,000 < $P_{12.5}$ < 15,000]
Virginia (*19*)	S.P. S315X01-1214	T180 Proctor-determined *OMC*	75-blow Marshall or 30-gyration SGC	60°C to constant mass	NS	**1.** T245 *MS* [5.56 kN min at 40°C] **2.** T245 RMS [70% min at 40°C after VS and 24-h soak] **3.** D7196 RT [2% max, 20-gyr, cured at 10°C for 4 h at 50% RH] **4.** T322 T_{crit} [less than LTPPBind 98% reliability low temp at top of CIR layer]

Note: G_{mb} = bulk specific gravity; G_{mm} = maximum specific gravity; HLWT = Hamburg loaded wheel tracker; MC = moisture content; MS = Marshall Stability; NS = not specified; OMC = optimum MC; RH = relative humidity; RMS = retained moisture content; RT = raveling test; SGC = Superpave gyratory compactor; S_t = indirect tensile strength; T_{crit} = critical cracking temperature; TSR = tensile strength ratio; V_a = air voids; VS = vacuum saturated.

[a]Equivalent AASHTO, ASTM, or state test method used; the most common standard method was always reported in Table 1 for ease of comprehension.

Table 2. Cited Emulsion-Stabilized CIR Marshall Stability Values

Reference	Year	Test temp (°C)	Field/Lab	Binders studied	MS (kN)	RMS (%)
Mamlouk and Ayoub (*26*)	1983	25	Lab	2.5e, 3.7e	11.1, 7.6	NA
Dudley et al. (*27*)	1987	25	Lab	5.2e, 6.8e, 8.3e	15.7, 21.5, 13.4	NA
			Lab	5.8e, 6.3e, 7.8e	9.0, 8.9, 9.9	NA
Forsberg et al. (*28*)	2002	40	Lab	1.3e, 2e	9.3 to 9.4	42 to 50
			Lab	2e, 2.7e, 3.4e	7.2 to 9.7	66 to 92
Yan et al. (*29*)	2009	40	Lab	3e2c	6.6 to 13.4	89 to 114
			Lab	3.5e2c	6.7 to 11.8	86 to 89
			Lab	4e2c	5.8 to 12.9	77 to 99
			Lab	4.5e2c	10.7	97
			Lab	5e2c	5.6	115
Dudley et al. (*27*)	1987	60	Lab	5.2e, 6.8e, 8.3e	2.4, 1.9, 2.0	NA
			Lab	5.8e, 6.3e, 7.8e	0.2, 0.5, 0.5	NA
Steward (*30*)	1987	60	Lab	1.5e	1.3	NA
Cohen (*31*)	1989	60	Lab	5e	2.1	NA
			Field	5e	1.1 to 2.6 (7 days to 12 months)	NA
Scholz et al. (*32*)	1991	60	Lab	1e	2.7 to 5.2	NA
			Field	1e	3.1 to 8.1 (3 to 36 months)	NA
			Lab	1.9e	3.8 to 5.8	NA
			Field	1.9e	2.3 to 7.1 (3 to 36 months)	NA
Niazi and Jalili (*33*)	2009	60	Lab	3.5e	8.3	57
			Lab	3.5e2c	12.7	83
			Lab	3.5e2HL	11.8	77
			Lab	3.5e2HLS	11.6	69

Note: Binder blends identified by number (dosage percentage) and letter (binder type) designations for emulsion (e), cement (c), hydrated lime (HL), and hydrated lime slurry (HLS). For example, 3.5e2c implies 3.5% emulsion with 2% cement. NA = not available.

the other hand, still predominantly utilize Marshall stability as their primary strength/stability indicator.

In a similar manner, several agencies other than state DOTs allow mix designs to be based on either MS or another test (typically S_t), perhaps suggesting the phasing out of MS among those agencies. For example, AASHTO recently released provisional standards MP31 and PP86 for CIR design that allow either an IDT (minimum 310 kPa S_t and 70% TSR) or MS (minimum 5.56 kN MS and 70% RMS) protocol. Either protocol requires a 7% maximum RT mass loss. The Asphalt Recycling and Reclamation Association also allows both MS and S_t protocols (*34*). Others such as the Federal Lands Highways and Wirtgen procedures use S_t protocols (*35*, *36*). The second edition of the Basic Asphalt Recycling Manual states there is not a nationally accepted method from which to perform cold recycling designs (*37*). MS (T245) and S_t (T283) are listed as the primary methods, with mention of Hveem stability (T246), rutting (T324 or T340), and dynamic modulus – flow number (TP79) as either rarely used or potential methods for future designs.

Experimental Program

Materials Tested

Three reclaimed asphalt pavement (RAP) materials (Table 3) were tested, which were sampled from locations up to around 150 mi apart. R1 and R2 were obtained during Mississippi DOT (MDOT) CIR projects on U.S. Highways 49 (US-49) and 45Alt (US-45Alt), respectively, and were tested at their as-received (A/R) gradations. The US-49 and US-45Alt projects, being about 150 mi apart, were deemed sufficient to represent two distinct materials. R3 was obtained from a Mississippi asphalt producer's RAP stockpile, meaning it was reclaimed in more traditional manners (e.g., milling). R3 was tested at two fabricated gradations. First, R3 was tested at R1's A/R gradation, which, based on a review of 28 CIR gradations (*38*), is relatively fine for CIR. Second, R3 was tested at a coarse gradation (GC) that approximated the coarser-end CIR gradations in Cox and Howard (*38*). All RAP materials were dried and sieved into multiple sizes for controlled batching.

One engineered emulsion (CIR-EE) was used throughout the US-49 CIR construction project (US-49 design was 4%

Table 3. RAP Properties

Category		CIR-milled		Traditionally milled	
RAP ID		R1	R2	R3	
Source		US-49	US-45Alt	Lowndes Co. RAP stockpile	
Gradation ID		A/R	A/R	A/R[a]	GC
Percent passing (Bulk RAP)	25 mm	100	100	100	100
	19 mm	100	99	100	93
	12.5 mm	94	89	94	78
	9.5 mm	85	73	85	65
	4.75 mm	55	41	55	34
	2.36 mm	38	25	38	21
	0.075 mm	1.5	4.9	1.5	0.8
Percent passing (Extracted Agg.)	25 mm	99	100	100	100
	19 mm	98	100	99	100
	12.5 mm	96	97	99	98
	9.5 mm	89	87	95	90
	4.75 mm	63	58	71	61
	2.36 mm	48	42	52	42
	0.075 mm	9.9	11.3	10.1	8.8
$P_{AC\text{-}T164}$[b] (%)		4.8	5.1	5.6	5.6
RAP G_{mm}[c]		2.447	2.374	2.373	2.373
Aggregate NMAS		12.5	12.5	9.5	9.5
Aggregate FAA (%)		41	—	43	43

Note: Washed gradations for bulk RAP and extracted aggregates determind by ASTM C117 and C136. Fine aggregate angularity (FAA) determined by ASTM C1252 Method C. NMAS = nominal maximum aggregate size.
[a]R3 as-received (A/R) gradation is that which matches the R1 gradation at controlled sieve sizes.
[b]Asphalt content by extraction (AASHTO T164). A blend of 85% toluene and 15% ethanol was used.
[c]Maximum theoretical specific gravity determined via Corelok® vacuum sealing. See Cox and Howard (*39*) for details.

emulsion with 1% hydrated lime) and was also used throughout the work presented in this paper. CIR-EE was laboratory-produced in small batches by Paragon Technical Services, Inc. (PTSi) and classified as a CSS-1h with a T59 residue of 63.3% and a T200 pH of 2.38. All emulsion CIR specimens included 1% hydrated lime as an anti-stripping agent. Cement-stabilized CIR specimens were stabilized with a Type I portland cement.

Test Methods

Emulsion CIR specimens were mixed in a bucket mixer for two minutes following the predominant Table 1 mixing guidelines. Water was added at the beginning of mixing (6% based on Cox et al. [*40*]), emulsion (heated to 60°C) was added at 60 s, and hydrated lime was added at 90 s. Specimens (100 mm diameter by 63.5 mm height) were immediately compacted to 30 gyrations, extruded from molds, and 60°C oven cured to constant mass (most common Table 1 compaction and curing requirements). Specimens were weighed every 2 h until percent mass loss (P_{ML}) in a 2-h period was less than 0.05% provided the cure time was 16 to 48 h. Specimens were then cooled for 12 to 24 h following typical Table 1 specifications.

After cooling, specimen G_{mb} was measured via AASHTO T331 (CoreLok® method) then T166 (saturated surface dry method). G_{mm} for each blend was determined according to Cox and Howard (*39*).

MS specimens were oven conditioned to 40°C for 2 h prior to testing per Table 1 specifications. MS testing was conducted largely in accordance with AASHTO T245 and ASTM D6927 (replaced D1559 – now withdrawn but commonly referenced in DOT specifications). Specimens were loaded on a Marshall load frame at 50 mm/min within 30 s of removing from the oven, and D6927 thickness corrections were applied where applicable.

RMS specimens, in place of 2-h oven conditioning, were vacuum saturated following T283 protocols to 55% to 75% saturation and soaked in a 25°C water bath for 23 h then in a 40°C water bath for 1 h to bring specimens to test temperature (total of 24 h soaking). Specimens were tested for stability immediately when removed from the final water bath, and RMS was reported as a percentage of MS. A small set of specimens were tested, which were placed in the 40°C water bath for only 30 min to investigate any differences between the two conditioning methods since some Table 1 specifications require 30 min rather than an hour. No meaningful differences were observed, and 1 h was selected for testing here.

IDT specimens were tested similarly to MS specimens except they were tested at 25°C (i.e., not oven conditioned) and were tested using an IDT fixture in the same load frame.

Though a smaller component of this paper, cement CIR specimens were bucket mixed; however, UCS specimens (101.6 mm diameter by 116.4 mm height) were either standard Proctor compacted (3 layers, 25 blows/layer) or

Table 4. Test Plan

	Emulsion-stabilized CIR							Cement-stabilized CIR				
								UCS[h]				IDT S_t[i]
								Std. Proctor		Mod. Proctor		30-gyr SGC
Material	MS[a]	RMS[b]	MS rep.[c]	MS var.[d]	MS curing effects[e]	IDT S_t[f]	IDT S_t var.[g]	25°C	40°C	25°C	40°C	25°C
R1(A/R)	✓[j]	✓	✓	✓	✓	✓	✓	✓	✓	✓	✓	✓
R3(A/R)	✓	✓	NA	NA	NA	✓	NA	✓	NA	NA	✓	✓
R3(GC)	✓	✓	NA	NA	NA	✓	NA	✓	NA	NA	✓	✓

Note: Binders dosed as a percentage of dry RAP mass. Water dosed as a percentage of solids (i.e., RAP, emulsion residue, cement, and/or hydrated lime) and was total water (i.e., batched water plus emulsion water).
[a]MS: conventional MS vs. emulsion investigation at 0%, 1.5%, 2%, 2.5%, 3%, 3.5%, 4%, 5%, and 6% emulsion.
[b]RMS: retained MS checks at 4% emulsion (chosen because 4% emulsion was the US-49 design binder content).
[c]MS rep. (repeatability): three MS vs. emulsion (1%, 2%, 3%, 4%, 5%, and 6%) curves over a 7-month period with testing occurring in Dec 2014, Feb 2015, and Jul 2015 with different emulsion samples.
[d]MS var. (variability): 15-replicate MS variability investigation at 2.5% and 4% emulsion contents.
[e]MS curing effects: MS vs. cure time investigation from 16 to 48 h at 2.5% emulsion.
[f]Emulsion IDT S_t: conventional S_t vs. emulsion investigation at 0%, 1.5%, 2%, 2.5%, 3%, 3.5%, 4%, 5%, and 6% emulsion.
[g]S_t var. (variability): 15-replicate S_t variability investigation at 2.5% and 4% emulsion contents.
[h]UCS: conventional UCS vs. cement investigation at 3%, 4%, 5%, and 6% cement for a given combination of compaction method (Std. or Mod. Proctor) and curing method (25°C curing room or 40°C sealed bag).
[i]Cement IDT S_t: conventional S_t vs. cement investigation at 3%, 4%, 5%, and 6% cement when SGC compacted to 30 gyrations and cured in the 25°C curing room.
[j]For R1(A/R) MS testing at 0% emulsion, five replicates were tested.

modified Proctor compacted (5 layers, 25 blows/layer). IDT specimens were compacted like emulsion specimens (30 gyrations, 100 mm diameter, 63.5 mm height). Cement specimens were cured in either a moist curing room (~23°C, 100% RH) or in a sealed plastic bag at 40°C. G_{mb} was measured via T331, and G_{mm} was determined following (*39*). UCS specimens were tested in general accordance with ASTM D1633 at 1.27 mm/min and 25°C. IDT specimens were tested identically to emulsion-stabilized specimens.

Test Plan

Table 4 summarizes all testing performed. For more conventional testing (e.g., standard MS versus emulsion content design curves), all RAP materials and gradations were tested except for R2(A/R), which was used for several verification points. For supplemental testing (testing not typical of DOT mix designs, such as variability and repeatability testing), only R1(A/R) was tested as it was RAP from an actual CIR project. In most cases, three replicates were tested per unique material and blend; 15 replicates were tested during variability investigations. In all, Table 4 corresponds to 171 MS tests, 105 emulsion S_t tests, 96 UCS tests, and 36 cement S_t tests.

Results

Conventional Design Practices (MS and UCS)

Figure 1 presents design curves for emulsion and cement CIR materials including the most prevalent Table 1 criteria that were 5.56 kN (1,250 lb) for MS and 2,068 to 3,447 kPa (300 to 500 psi) for UCS. Both curves for cement-stabilized CIR (Standard Proctor, 25°C and Modified Proctor, 40°C) are shown in Figure 1.

For emulsion CIR, emulsion contents of 1.5% to 4% were initially evaluated; all yielded MS values considerably greater than 5.56 kN with the exception of R1(A/R) at 4% emulsion. There were two issues with this result. First, even the lowest emulsion content tested (1.5%) exceeded the minimum threshold considerably; second, true optimum emulsion contents, or peaks, were not clear in most cases.

From an economics standpoint, the optimum emulsion content would be that which just meets the minimum MS criteria, yet the 1.5% emulsion content exceeded the minimum threshold by a considerable margin, suggesting a more economical design is possible (absent the other testing requirements in Table 1). Also, from a material mechanics standpoint, properties should increase with emulsion content to a point as cohesion increases and then decrease once the emulsion phase increases such that its behavior dominates. Other than perhaps R1(A/R), this trend was not observed from 1.5 to 4% emulsion in Figure 1.

As a result, Figure 1 was broadened by testing additional emulsion contents of 0% (i.e., unstabilized, without emulsion or hydrated lime) and 5% and 6%, extreme values that would not likely be considered in practice but bounded the problem. At the higher emulsion contents, MS trended downward though it remained above 5.56 kN for R3(A/R) and R3(GC). Perhaps of greater interest is at the 0% emulsion content where MS remained well above 5.56 kN for all but

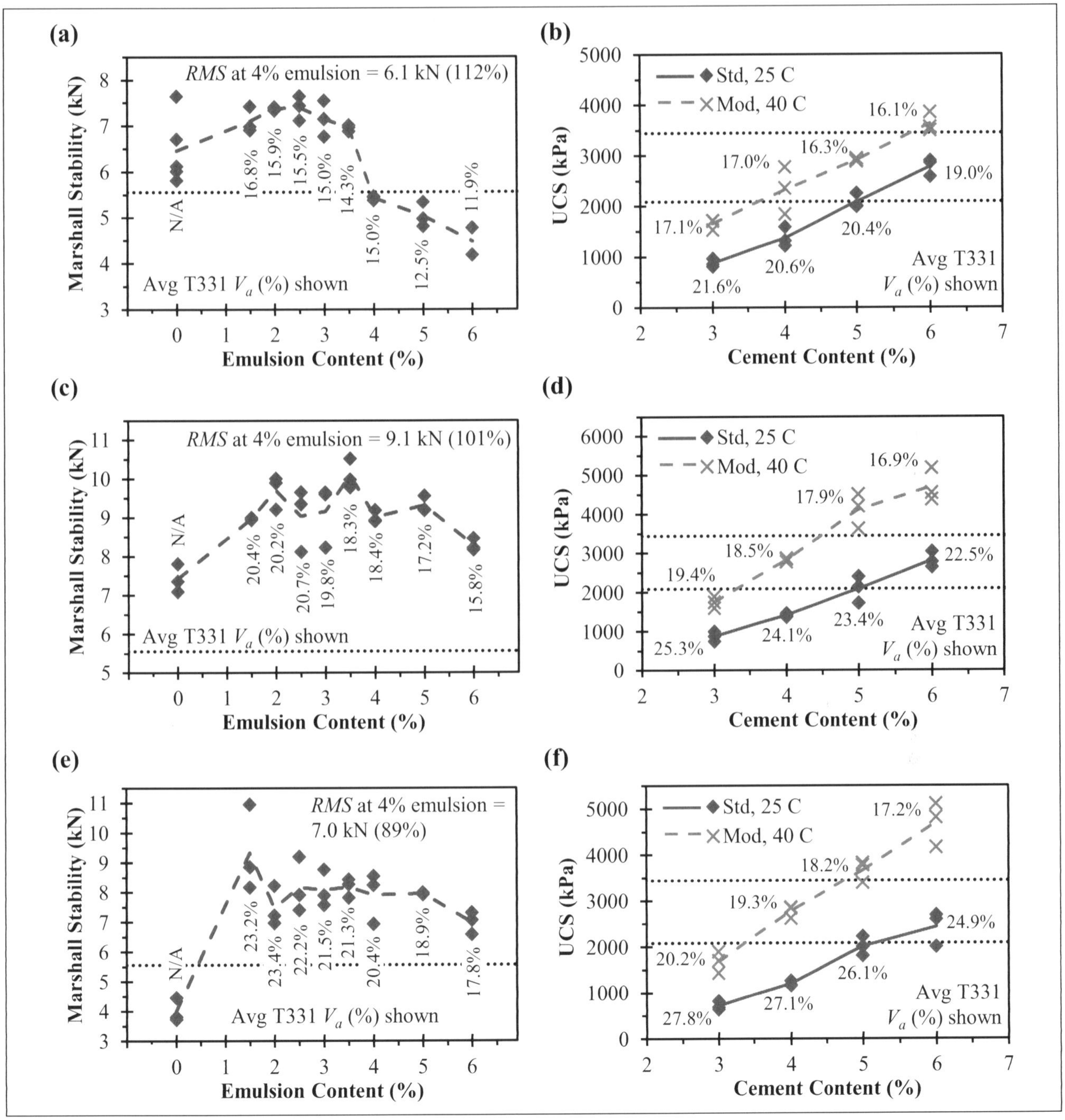

Figure 1. Conventional CIR mix design results: (*a*) R1(A/R) emulsion-stabilized MS; (*b*) R1(A/R) cement-stabilized UCS; (*c*) R3(A/R) emulsion-stabilized MS; (*d*) R3(A/R) cement-stabilized UCS; (*e*) R3(GC) emulsion-stabilized MS; (*f*) R3(GC) cement-stabilized UCS.

R3(GC). It is not greatly surprising that R3(GC) stands alone in this regard given its much coarser gradation. A limited amount of R2(A/R) was available, so 0% emulsion testing was performed with it as well to incorporate another RAP material obtained directly from a CIR project. Its average *MS* was 6.9 kN. This finding that the MS threshold was generally met without any stabilization additive is exceptionally concerning, and its implications are examined further in the discussion section.

Compared with the emulsion CIR mixtures in which most reasonable emulsion contents were not clearly distinguished from each other using MS, the cement CIR results in Figure 1 present a stark contrast in terms of variability, trends, and resolution between cement contents. This is true for either of the two compaction and curing protocols. As stated, UCS of cement CIR is not the focus of this paper; however, it is useful to present the type of clean, logical, and discernable results that are desired from a mix design test. It is also useful in showing that the RAP materials themselves are not responsible for the poor MS results since they produced, albeit with cement instead of emulsion, reasonable UCS results.

Supplemental Testing with Conventional Design Tests (MS and UCS)

Based on results in the previous section, further testing considered variability, a range of cure times in the allowable 16- to 48-h period, and repeatability. A minor amount of cement CIR testing was also conducted to further support UCS's rationality.

MS variability testing at 2.5% and 4% emulsion yielded average MS values of 6.8 kN and 5.5 kN and coefficients of variation (COVs) of 9.5% and 8.4%, respectively. In comparison, four airfield hot mix asphalt materials (mix details provided in [1]) compacted to 4% and 7% V_a and unaged or field-aged 2 years (16 total combinations) were tested at either 3 or 9 replicates with COVs ranging from 2.2% to 13.1% with an average COV of 7.2%. Thus, for an in-place recycled material, the degree of variability obtained does not seem unreasonable.

A small experiment was conducted to determine MS sensitivity within the typical 16- to 48-h curing window as some states do not specify the cure-to-constant-mass requirement. MS increased fairly linearly ($R^2 = 0.56$) from 4.7 to 5.7 kN from 16 to 48 h, respectively (figure not shown for brevity). Increasing at a rate of approximately 0.25 kN per 8 h, this finding, though less related to the MS test itself than curing protocols, is notable.

Figure 2*a* shows R1(A/R) MS repeatability results with unpredictable behaviors from one curve to the next even though specimen preparation and testing were carefully controlled. The only variation between each trial was that a new batch of emulsion was used; however, those batches, being laboratory-produced, exhibited insignificant property ranges. There is no obvious pattern to Figure 2*a* results, and a wide range of MS values were observed from below the threshold (not a single Trial 2 specimen exceeded 5.56 kN) to well above. Air voids (Figure 2*b*) were considered to investigate any influence on Figure 2*a* results; V_a's seemed reasonable except for Trial 3 at 1% and 2% emulsion. A simple V_a adjustment factor (observed V_a divided by average V_a) at each emulsion content was used to normalize MS results (Figure 2*c*). While approximate, it illustrates that V_a differences likely only explain a small part of the Figure 2*a* repeatability issue. Overall, MS sensitivity in Figure 2*a* presents considerable cause for concern.

As in the previous section, Figure 2*d* serves to contrast UCS against MS. In Figure 2*d*, the compaction and curing combinations were interchanged, and the results were predictable, clean, logical, and more desirable. Essentially, for a given compaction method, 40°C curing increased UCS slightly compared with 25°C curing, and for a given curing method, modified Proctor compaction increased UCS relative to standard Proctor compaction. Compaction had more pronounced effects on UCS than curing method.

Alternative Design Test (IDT)

Figure 3 presents IDT results for emulsion-stabilized and cement-stabilized CIR showing a 310 kPa S_t threshold (more conservative than Mississippi and Texas specifications in Table 1). In comparison to Figure 1, cement CIR S_t trends are similar visually (lower magnitude) to UCS trends.

Emulsion CIR S_t trends, though they do not necessarily resemble perfect curves, are much more reasonable than Figure 1 MS results for several reasons. First, S_t values both above and below the threshold were obtained, enabling selection of reasonable design emulsion contents (between 1.5% and 3% for the 310 kPa threshold). Second, curves were less erratic with greater resolution between emulsion contents. In general, S_t increased linearly around 90 to 120 kPa per 1% emulsion ($R^2 = 0.87$ to 0.94) in the range of more reasonable emulsion contents tested (i.e., 1.5% to 4%).

Regarding S_t variability, COV for emulsion CIR S_t at 2.5% and 4% emulsion was 9.7% and 6.0%, respectively, which was not meaningfully lower on average than for MS. Comparisons with the four airfield asphalt mixtures (same testing layout) were again conducted with COVs ranging from 0.8% to 8.3% and averaging 3.7%. The asphalt S_t COVs were noticeably lower than for asphalt MS or CIR S_t (which were similar). Even so, COVs in the 5% to 10% range for in-place recycled materials is reasonable. S_t repeatability testing was not explicitly conducted; however, other S_t testing on 150 mm diameter R1(A/R) specimens conducted later for a different purpose yielded average S_t values of 283, 365, and 400 kPa for 2%, 3%, and 4% emulsion, respectively, compared with 287, 381, and 455 kPa (Figure 3*a*).

Discussion of Results

As discussed in the Introduction, a mechanical property test used for mix design must be robust enough to adequately characterize at least one stabilization material. Data presented here suggests MS is not sufficiently robust for CIR design with emulsion, especially when compared with UCS for cement CIR. Being able to meet the minimum MS criteria without any stabilization additive (likely just with relivening of the RAP binder during curing) challenges the intent of following a mix design. The premise behind a mix design is that the mixture will require some percentage of binder to improve its properties. Though 0% emulsion likely would not pass other requirements (such as 70% RMS) (note that insufficient RMS testing is a limitation of this work), the possibility that the first and primary test in a design method may indicate no binder is needed is not rational. MS repeatability was also concerning.

IDT S_t produced reasonable results for both cement and emulsion CIR. The use of such a test presents a more ideal scenario for CIR design in which a single, consistent design protocol is utilized regardless of the stabilization material used. Under MDOT State Study 250 (SS250), this type of universal CIR mix design framework suitable for

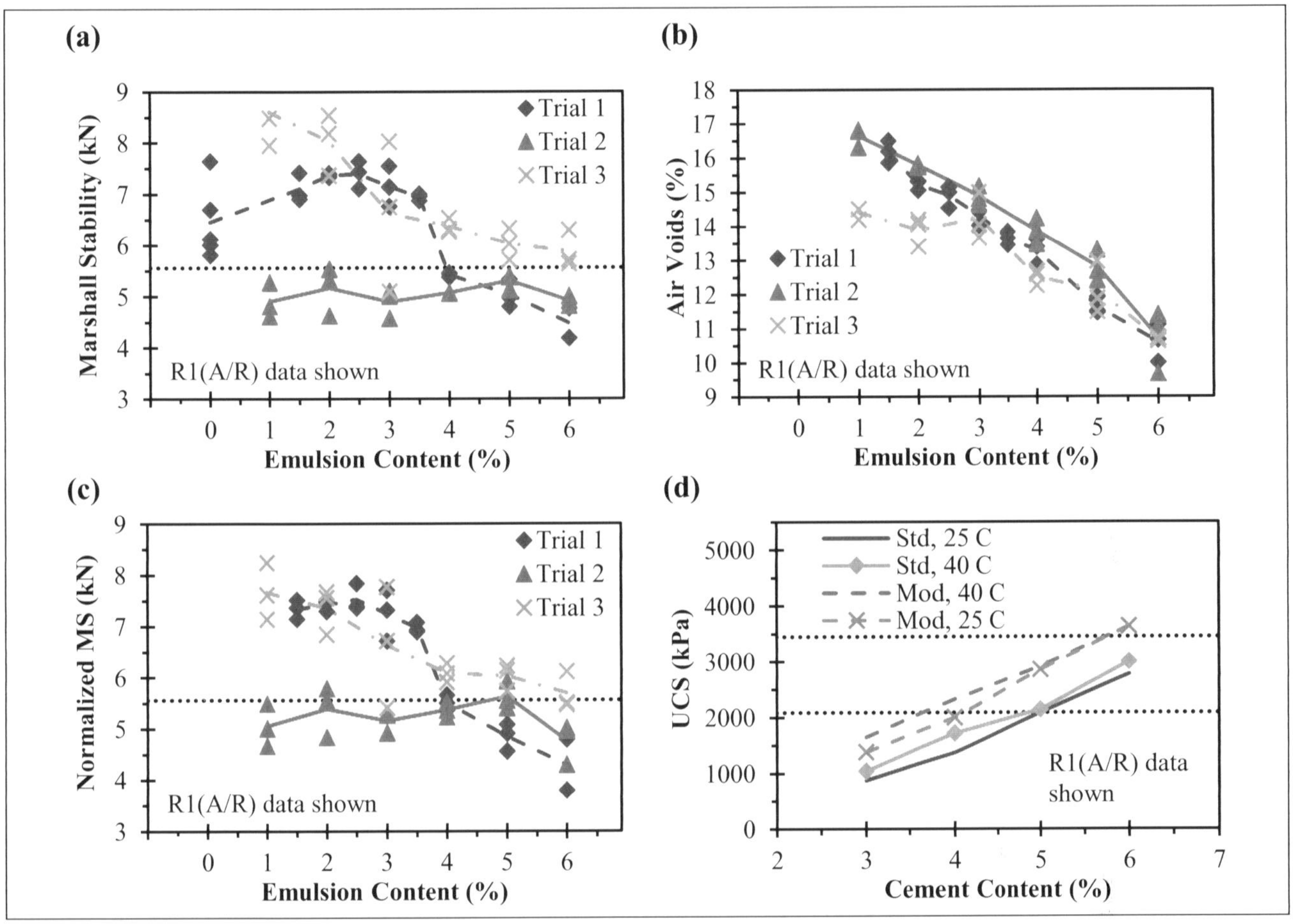

Figure 2. Supplemental CIR results: (*a*) MS repeatability; (*b*) air voids for MS repeatability curves; (*c*) MS repeatability normalized by V_a; (d) compaction and curing effects on UCS.

bituminous, cementitious, or combined stabilization blends was investigated (*1*). Within the framework developed, all specimens, irrespective of stabilization additives, are identically mixed, compacted, cured, and tested.

While S_t results presented in this paper provide a simple illustration of a test that is robust enough to provide useful information about different stabilization materials, relying solely on intermediate temperature S_t for design can lead to problems such as brittle mixtures. For this reason, the SS250 mix design approach incorporates a balanced testing regime consisting of a high-temperature rutting test and a low-temperature instrumented IDT fracture test. This universal mix design approach can facilitate various stabilization materials (e.g., cement, emulsion, or a combination) and would enable an agency to allow competition between different stabilization approaches within a consistent framework where a balance between rutting and cracking can be maintained.

Table 5 presents additional R1(A/R) data from other SS250 tests for 2%, 3%, and 4% emulsion. The purpose of Table 5 is two-fold. First, it further emphasizes the need to abandon MS. It shows that the issues observed with MS were not a result of the materials (RAP or emulsion) but the test itself as other tests did not experience the same issues. Unlike MS, there are many other tests from the high-temperature to the low-temperature spectrum that respond logically to emulsion content, similarly to the S_t results presented here. This presents a compelling argument for adopting a new testing regime in place of MS. Second, it provides, with more tests than would be utilized in an actual mix design, an example of the design approach recommended from SS250, which includes high-temperature rutting, intermediate-temperature strength, and low-temperature fracture resistance tests to produce mix designs with balanced performance.

Summary and Conclusions

The objective of this paper was to present data demonstrating that MS is not a suitable emulsion CIR mix design test. Key conclusions and observations supporting this objective are:

- In comparison to UCS design curves (cement), MS results (emulsion) were not as rational and often did not respond to emulsion content changes in a logical manner.
- The typical minimum MS criteria of 5.56 kN (1,250 lb) was exceeded without stabilization additives

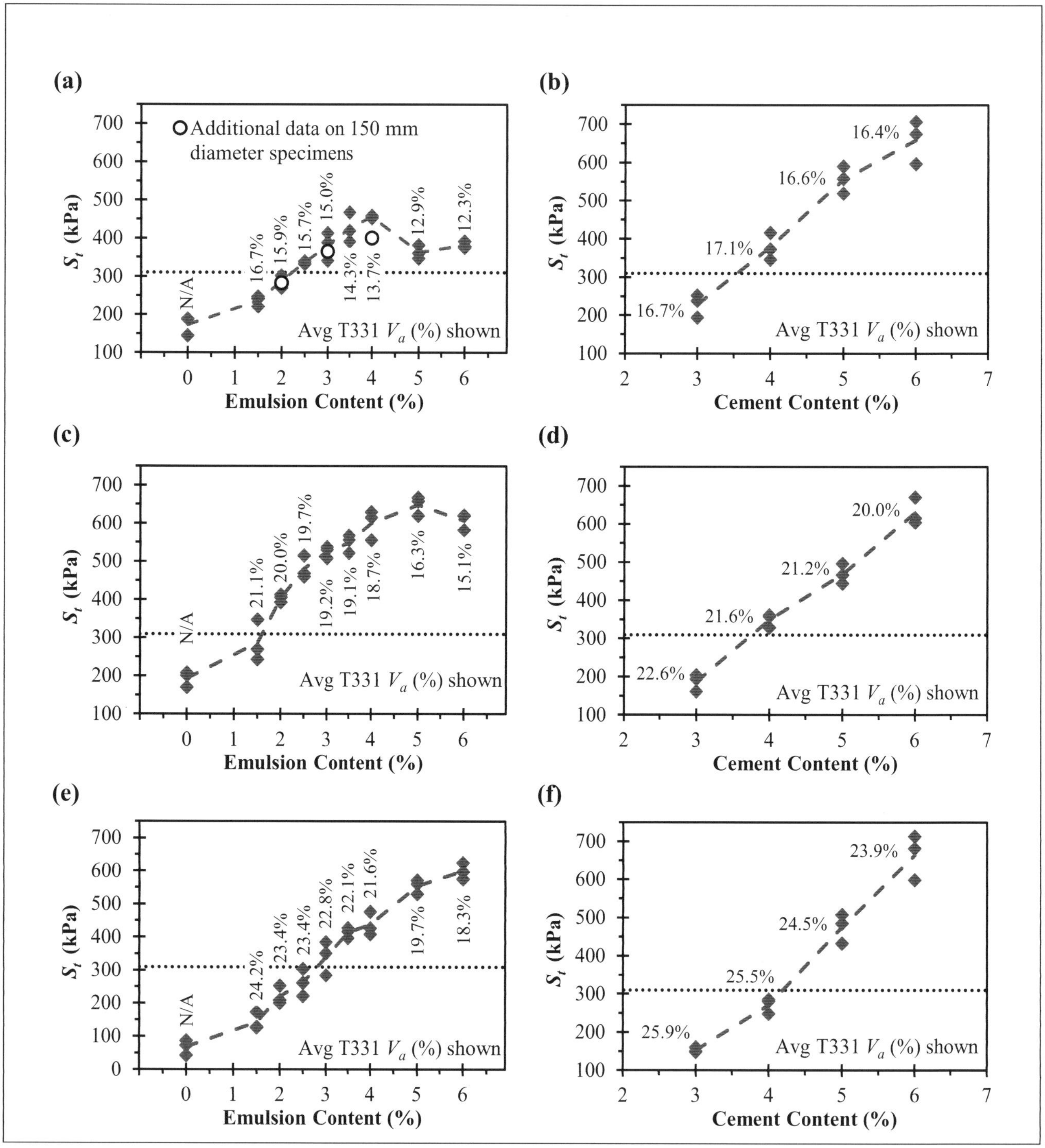

Figure 3. IDT results: (*a*) R1(A/R) emulsion-stabilized S_t; (*b*) R1(A/R) cement-stabilized S_t; (*c*) R3(A/R) emulsion-stabilized S_t; (*d*) R3(A/R) cement-stabilized S_t; (e) R3(GC) emulsion-stabilized S_t; (*f*) R3(GC) cement-stabilized S_t.

for specimens oven conditioned for 2 h at 40°C, but not soaked in water. Retained MS was not consistently performed, and might have provided more insight.

- MS repeatability was poor for the materials tested here. Three replicate design curves ranged from all points below the 5.56 kN threshold (all emulsion contents failed) to all points above the threshold (all passed).
- For agencies that continue conventional CIR design (single binder type), it is recommended that MS be abandoned for emulsion CIR design.

Table 5. Additional Test Results for Balanced Mix Design Framework

Emulsion content (%)	APA rut depth (mm)	PURWheel no. passes to 12.5 mm rut depth	S_t (kPa)	FE (kJ/m^3)	FE (kJ/m^3)	T_{crit} (°C)
	64°C dry	64°C wet	20°C	20°C	−10°C	0 to −20°C
2	4.9	3,250	283	0.29	0.13	−17.8
3	6.7	1,200	365	0.59	0.23	−18.7
4	8.4	270	400	0.87	0.39	−21.6

Note: APA = asphalt pavement analyzer; PURWheel = submerged rubber-tire loaded wheel test; FE = IDT fracture energy; T_{crit} = AASHTO T322 critical cracking temperature.

Acknowledgments

MDOT-funded State Study 250, and Ergon Asphalt & Emulsions and Paragon Technical Services, Inc. also provided financial or other support. Permission to publish was granted by the Director of the Geotechnical and Structures Laboratory, U.S. Army Engineer Research and Development Center.

Author Contributions

The authors confirm contribution to the paper as follows: study conception and design, data collection, analysis and interpretation of results, and draft manuscript preparation: B.C.C. and I.L.H. All authors reviewed the results and approved the final version of the manuscript.

References

1. Cox, B. C., and I. L. Howard. *Cold In-Place Recycling Characterization Framework and Design Guidance for Single or Multiple Component Binder Systems*. Final Report No. FHWA/MS-DOT-RD-15-250-Volume 2, Mississippi Department of Transportation, Jackson, MS, 2015.
2. Waterways Experiment Station. *Investigation of Stability of Asphalt Paving Mixtures*. Interim Report No. 1, War Department, Corps of Engineers, U.S. Army, Waterways Experiment Station, Vicksburg, MS, 1944.
3. Waterways Experiment Station. *Investigation of the Design and Control of Asphalt Paving Mixtures*. Technical Memorandum No. 3-254, Department of the Army, Corps of Engineers, Waterways Experiment Station, Vicksburg, MS, 1948.
4. Marshall, B. G. *The Marshall Method for the Design & Control of Bituminous Paving Mixtures*. 3rd Revision, Marshall Consulting and Testing Laboratory, 1949.
5. Corps of Engineers. *Symposium on Asphalt Paving Mixtures*. Research Report No. 7-B, Highway Research Board, National Research Council, Washington, D.C., 1949.
6. White, T. D. Marshall Procedures for Design and Quality Control of Asphalt Mixtures. *Journal of the Association of Asphalt Paving Technologists*, Vol. 54, 1985, pp. 265–284.
7. Kandhal, P. S., and W. S. Koehler. Marshall Mix Design Method: Current Practices. *Journal of the Association of Asphalt Paving Technologists*, Vol. 54, 1985, pp. 284–303.
8. Roberts, F. L., P. S. Kandhal, E. R. Brown, D. Y. Lee, and T. W. Kennedy. *Hot Mix Asphalt Materials, Mixture Design, and Construction*, 2nd ed. National Asphalt Pavement Association Research and Education Foundation, Lanham, MD, 1996.
9. Leahy, R. B., and R. B. McGennis. Asphalt Mixes: Materials, Design, and Characterization. *Journal of the Association of Asphalt Paving Technologists*, Vol. 68A, 1999, pp. 70–127.
10. AASHTO-AGC-ARTBA. *Report on Cold Recycling of Asphalt Pavements*. Task Force No. 38 Report. AASHTO-AGC-ARTBA Joint Committee, Washington, D.C., 1998.
11. California Department of Transportation. *Method of Test for Determining the Percent of Emulsified Recycling Agent to Use for Cold Recycling of Asphalt Concrete*. Lab Procedure No. 8. California Department of Transportation, Sacramento, CA, 2005.
12. Illinois Department of Transportation. *Special Provision for Cold In-Place Recycling (CIR) with Emulsified Asphalt*. Special Provision LR 400-5. Illinois Department of Transportation, Springfield, IL, 2012.
13. Iowa Department of Transportation. *Mix Design Method for CIR with Engineered Emulsion*. Matls IM 504 Appendix B. Iowa Department of Transportation, Ames, IA, 2008.
14. Kansas Department of Transportation. *Mix Design Procedures for CIR (Cold in Place Recycling) Material*. Kansas Department of Transportation Construction Manual, Part V, Section 5.3.4. Kansas Department of Transportation, Topeka, KS, 2014.
15. Mississippi Department of Transportation. *Roadbed Reclamation with Emulsified Asphalt*. Special Provision 907-425-1. Mississippi Department of Transportation, Jackson, MS, 2010a.
16. Montana Department of Transportation. *Cold In-Place Recycling (Partial Depth) Mixture Design*. Special Provision 405-2. Montana Department of Transportation, Helena, MT, 2015.
17. Cross, S. A., E. R. Kearney, H. G. Justus, and W. H. Chesner. *Cold-In-Place Recycling in New York State*. Report C-06-21. New York State Department of Transportation, Albany, NY, 2010.
18. Texas Department of Transportation. *Cold In-Place Recycling of Asphalt Concrete Pavement*. Special Specification 3254. Texas Department of Transportation, Austin, TX, 2004.
19. Virginia Department of Transportation. *Special Provision for Cold In-Place Recycling (CIR)*. Special Provision S315X01-1214. Virginia Department of Transportation, Richmond, VA, 2014.

20. Thomas, T., and A. Kadrmas. Performance-Related Tests and Specifications for Cold In-Place Recycling: Lab and Field Experience. Presented at 82nd Annual Meeting of the Transportation Research Board, Washington, D.C., 2003.
21. Alabama Department of Transportation. *Laboratory Design of Soil-Cement and Full-Depth Reclamation Mixes*. ALDOT Procedure ALDOT-416. Alabama Department of Transportation, Montgomery, AL, 2009.
22. California Department of Transportation. *Full Depth Reclamation Using Cement*. Caltrans Division of Maintenance, California Department of Transportation, Sacramento, CA, 2013.
23. Mississippi Department of Transportation. *Roadbed Reclamation with Portland Cement*. Special Provision 907-499-1. Mississippi Department of Transportation, Jackson, MS, 2010b.
24. Morian, D. A., M. Solaimanian, B. Scheetz, and S. Jahangirnejad. *Developing Standards and Specifications for Full-Depth Pavement Reclamation*. Report FHWA-PA-2012-004-090107. Pennsylvania Department of Transportation, Harrisburg, PA, 2012.
25. New York State Department of Transportation. *Geotechnical Engineering Manual: Design and Construction Guidelines for Full Depth Reclamation of Asphalt Pavement*. GEM-27. New York State Department of Transportation, Albany, NY, 2014.
26. Mamlouk, M. S., and N. F. Ayoub. Evaluation of Long-Term Behavior of Cold Recycled Asphalt Mixture (Abridgment). *Transportation Research Record: Journal of the Transportation Research Board*, 1983. 911: 64–66.
27. Dudley, S. W., K. Majidzadeh, and K. Kaloush. The Recycling of Cold-Mix, In-Place Asphalt for Low-Volume Roads in Ohio. *Transportation Research Record: Journal of the Transportation Research Board*, 1987. 1106: 163–172.
28. Forsberg, A., E. Lukanen, and T. Thomas. Engineered Cold In-Place Recycling Project: Blue Earth County State Aid Highway 20, Minnesota. *Transportation Research Record: Journal of the Transportation Research Board*, 2002. 1813: 111–123.
29. Yan, J., F. Ni, Z. Tao, and J. Jia. *Development of Asphalt Emulsion Cold In-Place Recycling Specifications*. Geotechnical Special Publication No. 190, American Society of Civil Engineers, Reston, VA, 2009, pp. 49–55.
30. Steward, D. A. Cold, In-Place Asphalt Pavement Recycling. *Transportation Research Record: Journal of the Transportation Research Board*, 1987. 1106: 172–178.
31. Cohen, E., A. Sidess, and G. Zoltan. Performance of a Full-Scale Pavement Using Cold Recycled Asphalt Mixture. *Transportation Research Record: Journal of the Transportation Research Board*, 1989. 1228: 88–93.
32. Scholz, T., D. F. Rogge, R. G. Hicks, and D. Allen. Evaluation of Mix Properties of Cold In-Place Recycled Mixes. *Transportation Research Record: Journal of the Transportation Research Board*, 1991. 1317: 77–89.
33. Niazi, Y., and M. Jalili. Effect of Portland Cement and Lime Additives on Properties of Cold In-Place Recycled Mixtures with Emulsion. *Construction and Building Materials*, Vol. 23, 2009, pp. 1338–1343.
34. Asphalt Recycling and Reclaiming Association. *Basic Asphalt Recycling Manual*, 1st ed. Asphalt Recycling and Reclaiming Association, Glen Ellyn, IL, 2001.
35. Federal Lands Highway. *Standard Specifications for Construction of Roads and Bridges on Federal Highway Projects*. FP-14. US Department of Transportation, Federal Highway Administration, Federal Lands Highway, Washington, D.C., 2014.
36. Wirtgen. *Cold Recycling: Wirtgen Cold Recycling Technology*, 1st ed. Wirtgen GmbH, Windhagen, Germany, 2012.
37. Asphalt Recycling and Reclaiming Association. *Basic Asphalt Recycling Manual*, 2nd ed. Report FHWA-HIF-14-001. Asphalt Recycling and Reclaiming Association and Federal Highway Administration, 2015.
38. Cox, B. C., and I. L. Howard. *Cold In-Place Recycling and Full-Depth Reclamation Literature Review*. White Paper Number CMRC WP-13-1. Mississippi State University, Construction Materials Research Center, 2013. http://www.cee.msstate.edu/downloads/(2013)CoxandHoward-CMRCWP13-1-LitReviewofCIRandFDR.pdf
39. Cox, B. C., and I. L. Howard. Vacuum Sealing-Based Volumetric Density Measurement Approach for Cold In-Place Recycling. *Transportation Research Record: Journal of the Transportation Research Board*, 2014. 2444: 11–19.
40. Cox, B. C., I. L. Howard, and R. Battey. In-Place Recycling Moisture-Density Relationships for High-Traffic Applications. *Proc., IFCEE 2015 (GSP 256)*, San Antonio, TX, 2015, pp. 349–358.

The Standing Committee on Critical Issues and Emerging Technologies in Asphalt (AFK10) peer-reviewed this paper (18-03446).

Article

TRR
JOURNAL OF THE TRANSPORTATION RESEARCH BOARD

Transportation Research Record
2018, Vol. 2672(28) 223–234

DOI: 10.1177/0361198118790836
journals.sagepub.com/home/trr
SAGE

Comparing Pressure Aging Vessel Time to Field Aging of Binder as a Function of Pavement Depth and Time

Braden T. Smith[1], Isaac L. Howard[2], Walter S. Jordan III[3], Codrin Daranga[3], and Gaylon L. Baumgardner[3]

Abstract

The AASHTO R28 pressure aging vessel (PAV) binder conditioning protocol has been heavily used for over two decades as part of purchasing specifications. Many paving industry changes have occurred since PAV was implemented in the early 1990s. This paper provides discussion with respect to a controlled test section in which the exact binders from construction were subjected to multiple PAV times and compared with properties observed after 2 or 4 years of field aging. The data suggests that 2 and 4 years of field aging were best simulated by 21 h and 45 h of PAV time for binders extracted from the top 1.3 cm of the pavement surface. For binders extracted from 5.0 to 6.3 cm below the pavement surface (i.e., at depth), 2 and 4 years of field aging were best simulated by 3 and 14 h of PAV conditioning, respectively. These results indicate that a PAV rate of 11 h per year of simulation seemed adequate for binders aged on the pavement surface, and a PAV rate of 3 h per year of simulation seemed most reasonable for binders aged at depth. Considering these rates, the current 20-h protocol in R28 seems to simulate roughly 2 years of aging at the pavement surface and about 6.5 years at depth. The amount of field aging simulation at depth is within the range of 5 to 10 years of aging suggested by R28.

The asphalt industry actively pursues methods to simulate environmental factors to characterize durability over time. This pursuit has led to the increased use of mixture tests to characterize paving materials in many cases, but mixture tests are not ideal for use in specifying asphalt binder products. Ideally, binder specifications would provide: 1) a competitive, unbiased, and non-subjective means for classifying asphalt binder products, and 2) a direct method for considering changes binders experience in production and service after leaving a producer. The asphalt industry currently utilizes the rolling thin film oven (RTFO) protocol, AASHTO T240, to simulate short-term binder property changes that occur during plant mixing and construction. The pressure aging vessel (PAV) method, AASHTO R28, uses 20 h of conditioning to consider longer term binder property changes. These protocols have been used together for the performance grading (PG) system since the early 1990s with few changes.

With increased emphasis on pavements after environmental exposure, there have been many recent efforts to improve binder conditioning methods used to consider the effects of aging. This pursuit is evidenced in current activities of the Federal Highway Administration (FHWA) asphalt binder expert task group (ETG), an ongoing National Cooperative Highway Research Program (NCHRP) study, and efforts of multiple state departments of transportation. These efforts have collectively considered binder conditioning protocol changes to better characterize changes that occur in asphalt binders as there have been vast industry transitions in the past quarter century (and in particular in the past decade).

A field aged binder dataset in which construction occurred in a controlled setting and the exact binders were sampled during paving and evaluated after laboratory conditioning is rare. This paper presents findings from an investigation in which a full-scale test section was constructed in November of 2011 and evaluated for binder property changes through November 2015. The exact binders used in the plant-mixed asphalt were sampled at the mix plant the day of construction, and cores were collected periodically to characterize changes

[1]Hunt Refining Company, Tuscaloosa, AL
[2]Civil and Environmental Engineering, Mississippi State University, Mississippi State, MS
[3]Paragon Technical Services, Inc., Richland, MS

Corresponding Author:
Address correspondence to Isaac L. Howard: ilhoward@cee.msstate.edu

to binder properties with consideration given to account for the effect of pavement depth on binder aging.

Size and Scale Factors

When changes to a test protocol as predominant as R28 are considered, there are multiple size and scale factors that should be considered. The next three subsections discuss size and scale factors for the R28 protocol from the standpoints of: binder conditioning complexities, PAV method investigations, and effects that changing R28 might have on the asphalt industry at large.

Binder Conditioning Complexities

Although binder conditioning protocols are necessary to provide a competitive, unbiased, and non-subjective means for classifying asphalt binder products, like R28 does, multiple complexities limit the ability of conditioning protocols to directly consider all changes that binders experience as they leave a refinery, are handled by an asphalt mix plant, and are ultimately used in a pavement.

Many asphalt mixture properties have been linked to immediate effects and longer term effects on asphalt binder aging (*1–4*), which increases stiffness and decreases resistance to cracking. Adsorption of binder portions onto aggregate surfaces has been linked to binder property changes (*5–7*). Incorporation of reclaimed asphalt pavement (RAP) has been linked to changes in composite binder properties (*8–9*). Hydrated lime typically increases binder stiffness initially with increased resistance to binder aging (*10–14*). Construction effects, such as plant mixing temperature, have also been linked to short-term binder property changes (*15*).

Once placed, exposure to environmental effects varies in two primary ways: 1) climate to climate variation, and 2) depth of placement below a pavement surface. AASHTO M320 considers varying climates by specifying different R28 temperatures for varying binder grades. However, climate variations have been documented within distances that a single refinery can service (*2*). Referring to the effects of pavement depth, most have demonstrated that binder hardening is more severe at the pavement surface with severity decreasing non-linearly with pavement depth (*16–19*). Occasionally, studies have suggested uniform aging with depth (*20*). Exposure caused by pavement depth makes selecting binder conditioning protocols more complex as identical binders produced on the same day could be placed in different environments. For example, one shipment of PG 67-22 could be placed in a 2.5-cm-thick overlay on a rural low volume road whereas the next shipment could be placed in a lower layer of new pavement construction, such as the Interstate 269 project near Memphis, TN. The Interstate 269 pavement design is 36-cm thick (with multiple mixtures placed over a 19 mm nominal maximum aggregate size (NMAS) mix with PG 67-22). Environmental exposure for these PG 67-22 binders is substantially different, and the illustration made in this paragraph represents the current paving market in north Mississippi.

PAV Studies

The R28 protocol was developed as part of Project A-002A of the Strategic Highway Research Program (SHRP). The original protocol was recommended in Anderson et al. (*21*) and remains, for the most part, unchanged in R28-12. The early 1990s was a time in which asphalt mixes were typically placed in thicker layers, used more asphalt binder, and used less recycled material (e.g., RAP). As such, the magnitude of change that a binder property is expected to undergo after leaving the possession of an asphalt supplier has increased. The previous example of north Mississippi with Interstate 269 and thin overlays is a realistic example of very different aging potentials, but it is not the current norm nationally. Most markets have more surface or near-surface work than new construction. As such, binders are in a higher aging-prone zone of pavements.

SHRP Project A-002A, claimed that "The test [PAV] accounts for temperature effects but is not intended to account for mixture variables such as air voids, aggregate type, and aggregate adsorption" (*21*). It should not be forgotten that any binder-only conditioning protocol will have these same concessions. Further, a review of original PAV publications indicated where the field validation considered the effects of temperature and core density, but the documents do not mention thickness or depth of binder samples (*3*, *21*, *22*). As previously mentioned, industry transitions in pavement layer thickness have most likely increased the amount of environmental exposure that asphalt binders experience during service.

Over the past two decades, multiple groups have considered changes to longer term binder conditioning methods. The FHWA asphalt binder ETG assembles industry experts to provide timely responses to pressing industry concerns. Binder aging is one such concern addressed by the binder ETG. Long-term PAV aging and related methods have been recurring topics of concern to the binder ETG since inception of the PG grading specification. Most recently, the binder ETG has appointed task groups to evaluate extended PAV aging for prediction of non-load related pavement cracking and effects of recycled materials on long-term binder performance. Based on a review of the Transportation Research Board database of research in progress, ongoing investigations relative to longer term aged binder characterization are being conducted by NCHRP (Project 09-61), Texas

DOT (Contract 0-6925), and New Hampshire DOT (Contract 26962O). Farshidi et al. demonstrated approximately equal rheological properties of four PG 64-16 asphalts (one control blend, and three warm mix technologies) that were PAV conditioned or field aged in 19-mm NMAS mixes after 4 years of aging in California (*23*). Two examples for which increased PAV times have been recommended or implemented are the Asphalt Institute (*24*) and Australasian Road Transport and Traffic Agencies AGPT/T194, which was recently implemented as a 72-h PAV specification.

Effects of Changing PAV Protocols

Multiple effects could occur to the industry at large if the R28 protocol were changed or replaced. Logistically, asphalt refineries would assume higher risks if conditioning methods for acceptance were increased from 20 h. Changes to conditioning protocols to simulate later age properties would also force changes in binder formulations. All other factors being equal, harsher laboratory conditioning environments would most likely lead to softer base asphalt cements with increased modification rates to simultaneously meet early age and later age properties.

Although changes to the R28 protocol could produce effects within the industry at large, there are some factors that a single conditioning protocol cannot address. A single conditioning protocol cannot account for variations in environmental exposure due to climate or pavement depth, as mentioned previously. Further, binder protocols cannot account for the effects of ingredient interaction after binders leave a refinery (e.g., RAP, REOB, reclaimed asphalt shingles [RAS], GTR, hydrated lime, aggregates, etc.). With increased potential for ingredient interaction effects, there is a need to consider the effects of environmental exposure within mixtures during design, as demonstrated in companion research (*25*). However, binder conditioning protocols like R28 will most likely remain invaluable parts of specifications for purchasing asphalt binder products.

Experimental Program

The Figure 1 test section was constructed by Howard et al. (*25*) and evaluated for field aging in Smith (*26*). The test section used three asphalt mixes (M14, M15, and M16) that were produced from a single 160°C target mixing temperature design. Four strips of each mix (twelve strips in total) were placed in a single lift that was thicker than normal (7 cm on average). A pavement using a single lift of this thickness was ideal for considering effects of pavement depth on binder aging. All mixes used a PG 67-22 base asphalt produced in Vicksburg, MS. M14 used the PG 67-22 without modification. M15 foamed the same PG 67-22 as M14 with 2% moisture by binder mass prior to mixing. M16 used the same PG 67-22 binder, but 0.5% M1 Evotherm 3G^{TM} was added. This paper refers to the binder used in M14 and M15 as B1 (PG 67-22 without additive) and to the M16 binder as B2 (PG 67-22 with additive). After construction, the test section was closed to traffic and left in place to consider environmental effects on M14 to M16. Smith et al. ultimately monitored the section over 5 years, and this paper used 48 binder samples recovered from 144 cores (*26*). Table 1 provides properties of the cores selected from the six test strips considered for binder aging (strips 1, 3, 5, 7, 9, and 10).

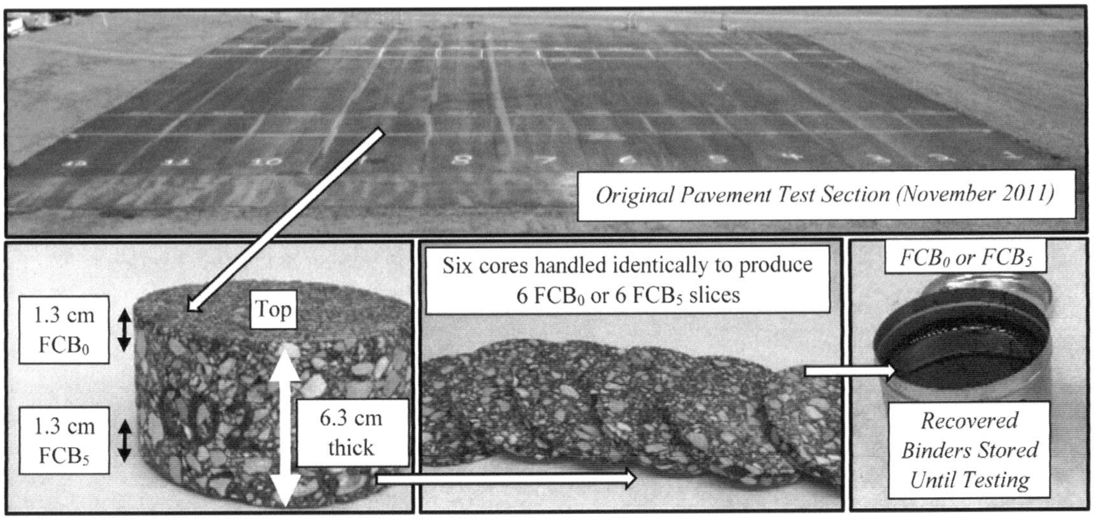

Figure 1. Experimental program photographs.

Table 1. Field Core Properties

Binder	B1 (PG 67-22 without additive)				B2 (PG 67-22 with additive)	
Mix (modification)	M14 (none)		M15 (foamed 2%)		M16 (0.5% M1 Evotherm 3G™)	
Mix temperature at plant	164°C		153°C		148°C	
Strip (See Howard et al. [24] for explanation)	1	3	5	7	9	10
Haul time (h)	1.0	5.8	1.1	5.6	1.0	5.7
0-year V_a average (%)	9.2	9.1	8.0	9.5	8.7	11.2
2-year V_a average (%)	9.2	9.3	7.9	9.0	8.5	11.1
4-year V_a average (%)	8.9	9.3	7.6	9.0	8.6	11.0
Acceptable V_a range (%)	8.5–10.0	8.5–10.0	7.5–9.0	8.3–9.8	8.0–9.5	10.3–11.8

Note: V_a based on AASHTO T331 bulk mixture specific gravity (G_{mb}).

The mix design had a total binder content (P_b) of 5.4%, absorbed content based on mix mass ($P_{ba(mix)}$) of 0.8%, and voids in mineral aggregate of 14.1%. Aggregate contents consisted of 39% crushed gravel, 35% limestone, 15% RAP, 10% sand, and 1% hydrated lime. The nominal maximum aggregate size (NMAS) was 12.5 mm, and the percentages of aggregates finer than 2.36 mm (no. 8), 0.60 mm (no. 30), and 75 μm (no. 200) were 36%, 19%, and 6%, respectively.

Binder Sampling Approaches

Binder samples were collected using two approaches: 1) as-received binder (ARB) samples were collected from binder supply tankers the morning of mixing, and 2) field core binder (FCB) samples were extracted from cores.

As-Received Binder (ARB) Samples. ARB samples were tested in eight conditions. The original binder (OB) was tested as received without binder conditioning (i.e., no RTFO, no PAV) for a conversational reference, and the remaining seven conditions evaluated the ARB binders after RTFO conditioning per AASHTO T240-13 and PAV conditioning for 0, 10, 20, 30, 40, 60, and 80 h. All PAV conditioned binders were RTFO conditioned as per AASHTO T240 prior to PAV conditioning, which was performed at 100°C and standard pressure conditions as per AASHTO R28. These values are denoted PAV_0 to PAV_{80}. This paper relies on the assumption that T240 is reasonable for simulation of construction effects and focuses on longer term property change.

Field Core Binder (FCB) Samples. FCB samples were extracted from slices removed from the pavement surface (i.e., FCB_0) or from 5 cm below the pavement surface (i.e., FCB_5). The process of collecting FCB samples from cores is demonstrated in Figure 1 in which a core was collected from the test section, initially sliced from the bottom to 6.3 cm to determine V_a using AASHTO T331. Then, 1.3-cm slices were removed from the tops (FCB_0) and bottoms (FCB_5) of six comparable cores (based on V_a, test strip, and age or laboratory condition). Central portions of cores were used for other purposes (*26*). Binder was collectively extracted from these groups of six slices. Field-aged binders were evaluated after 0, 2, and 4 years of field aging, and one mixture conditioning protocol from Smith (*26*) was conducted on cores collected within a few days of construction. The mixture conditioning protocol (referred to as CP4) was conducted on 6.3-cm tall cores by submerging cores in 64°C water for 14 days followed by one freeze-thaw cycle. The same process documented in Figure 1 was applied to CP4 cores. The FCB extraction process is described in Smith (*26*) and is summarized as follows. Extractions relied primarily on toluene for an initial 45-min soak, a variable number of 250-ml toluene washes, and a minimum of three 250-ml washes using an 85% toluene and 15% ethanol blend by volume. All extractions were completed in 6 to 8 washes after initial soaking, and recoveries used a Büchi Rotavapor R-114.

Binder Test Methods

Multiple binder test methods were selected to consider property changes over field-aging time and PAV time. These methods were chosen to consider: 1) penetration at 25°C as a property understood in most markets, 2) chemistry changes in Fourier transform infrared spectroscopy (FTIR), and 3) the methods in AASHTO M320. In essence, the intention was to measure high, low, and intermediate temperature rheological properties alongside chemistry changes.

Penetration (Pen). *Pen* testing was conducted at 25°C in 3-oz binder tins per ASTM D5. Tests were conducted in triplicate while submerged in water. All *Pen* values are reported as averages.

Fourier Transform Infrared Spectroscopy (FTIR). FTIR testing was performed using a Nicolet 380 FTIR analyzer. The carbonyl index (CI) and sulfoxide index (SI) were calculated by dividing absorbance peak heights at wave counts of 1,700 cm^{-1} and 1,031 cm^{-1} respectively, by the absorbance peak height at 1,375 cm^{-1}. For interpretation, the CI and SI were added together for aging considerations, and samples with higher CI + SI combinations indicate more aging had occurred.

Dynamic Shear Rheometer (DSR). Dynamic shear rheometer (DRS) testing was conducted with an Anton Paar SmartPave Plus 301, as per AASHTO T315, to determine critical temperatures (T_c) for 25-mm (DSR_{25}) and for 8-mm (DSR_8) plates based on complex shear modulus (G^*) and phase angle (δ) requirements. DSR_{25} testing was conducted with a 1.0-mm gap to determine $T_c(DSR_{25})$ in which $G^*/\sin\delta$ was 2.2 kPa, and DSR_8 was tested with a 2.0-mm gap to determine $T_c(DSR_8)$ in which $G^*\sin\delta$ was 5.0 MPa. All DSR T_c values were determined using Equation 1.

Bending Beam Rheometer (BBR). Bending beam rheometer (BBR) testing determined T_c for each binder sample based on stiffness (S) of 300 MPa or *m*-value of 0.300 using a Cannon TE-BBR Thermoelectric BBR. These critical behaviors are described by BBR_S and BBR_m. $T_c(BBR_S)$ was determined using Equation 1 whereas $T_c(BBR_m)$ used Equation 2.

$$T_c(BBR_S\ DSR_8,\ \text{or}\ DSR_{25}) = T_1 + (T_1 - T_2)\left(\frac{\log_{10}(p_c) - \log_{10}(p_1)}{\log_{10}(p_1) - \log_{10}(p_2)}\right) \quad (1)$$

$$T_c(BBR_m) = T_1 + (T_1 - T_2)\left(\frac{p_c - p_1}{p_1 - p_2}\right) \quad (2)$$

where

T_c = critical temperature,
T_1 = test temperature of passing test,
T_2 = test temperature of failure test,
p_c = critical property (2.2 kPa, 5.0 MPa, 300 MPa, or 0.300)
p_1 = measured property of passing test,
p_2 = measured property of failure test.

Limitations on the difference between $T_c(BBR_m)$ and $T_c(BBR_S)$, as defined between Equation 3, have been suggested by some as a durability parameter in virgin binders not blended with RAP or RAS binders. Minimum ΔT_c values of $-2.5°C$ and $-5°C$ have been suggested by Anderson et al. and Rowe (*27*, *28*), respectively, as indicators to consider preventative maintenance or take immediate action.

$$\Delta T_c = T_c(BBR_S) - T_c(BBR_m) \quad (3)$$

Results

Binder test results are provided in tables 2 to 5. Individual measurements to reach final results are provided in Smith (*26*). Figure 2 plots relationships between laboratory-conditioned ARB data and PAV time (PAVi) where "i" denotes the hours of PAV time.

Table 2. Penetration (*Pen*) Results

		Field core samples				As-received samples							
Sample description	Short term[a]	0 yr	2 yr	4 yr	CP4	OB	PAV_0	PAV_{10}	PAV_{20}	PAV_{30}	PAV_{40}	PAV_{60}	PAV_{80}
B1													
ARB	na	—	—	—	—	63	37	25	21	18	16	11	10
B2													
ARB	na	—	—	—	—	57	35	24	20	17	15	12	11
M14													
FCB_0	1 h (S1)	23	15	10	23	—	—	—	—	—	—	—	—
FCB_0	6 h (S3)	27	16	10	22	—	—	—	—	—	—	—	—
FCB_5	1 h (S1)	29	41	17	27	—	—	—	—	—	—	—	—
FCB_5	6 h (S3)	25	21	19	21	—	—	—	—	—	—	—	—
M15													
FCB_0	1 h (S5)	28	25	11	24	—	—	—	—	—	—	—	—
FCB_0	6 h (S7)	49	17	11	22	—	—	—	—	—	—	—	—
FCB_5	1 h (S5)	50	45	18	21	—	—	—	—	—	—	—	—
FCB_5	6 h (S7)	29	30	16	21	—	—	—	—	—	—	—	—
M16													
FCB_0	1 h (S9)	48	16	10	20	—	—	—	—	—	—	—	—
FCB_0	6 h (S10)	26	15	10	18	—	—	—	—	—	—	—	—
FCB_5	1 h (S9)	36	25	22	23	—	—	—	—	—	—	—	—
FCB_5	6 h (S10)	28	22	20	20	—	—	—	—	—	—	—	—

Note: — = data not measured.
[a]Short-term aging (aging times rounded to the nearest hour with test strip [S] indicated); not applicable (na) for B1 and B2 since short-term conditioning included within each property.

Table 3. FTIR Indices (CI + SI) Results

Sample description	Short term[a]	Field core samples				As-received samples							
		0 yr	2 yr	4 yr	CP4	OB	PAV_0	PAV_{10}	PAV_{20}	PAV_{30}	PAV_{40}	PAV_{60}	PAV_{80}
B1													
ARB	na	—	—	—	—	0.16	0.21	0.32	0.36	0.42	0.46	0.54	0.59
B2													
ARB	na	—	—	—	—	0.54[b]	0.18	0.28	0.35	0.37	0.46	0.53	0.63
M14													
FCB_0	1 h (S1)	0.57	0.49	0.80	0.60	—	—	—	—	—	—	—	—
FCB_0	6 h (S3)	0.47	0.69	0.76	0.62	—	—	—	—	—	—	—	—
FCB_5	1 h (S1)	0.56	0.56	0.69	0.60	—	—	—	—	—	—	—	—
FCB_5	6 h (S3)	0.57	0.80[b]	0.61	0.56	—	—	—	—	—	—	—	—
M15													
FCB_0	1 h (S5)	0.56	0.68	0.88	0.64	—	—	—	—	—	—	—	—
FCB_0	6 h (S7)	0.56	0.77	1.03	0.69	—	—	—	—	—	—	—	—
FCB_5	1 h (S5)	0.61	0.53	0.56	0.64	—	—	—	—	—	—	—	—
FCB_5	6 h (S7)	0.60	0.60	0.75	0.65	—	—	—	—	—	—	—	—
M16													
FCB_0	1 h (S9)	0.57	0.67	0.70	0.61	—	—	—	—	—	—	—	—
FCB_0	6 h (S10)	0.56	0.69	0.84	0.62	—	—	—	—	—	—	—	—
FCB_5	1 h (S9)	0.60	0.55	0.59	0.64	—	—	—	—	—	—	—	—
FCB_5	6 h (S10)	0.58	0.57	0.60	0.57	—	—	—	—	—	—	—	—

Note: — = data not measured.
[a]Short-term aging (aging times rounded to the nearest hour with test strip (S) indicated); not applicable (na) for B1 and B2 as short-term conditioning included within each property.
[b]Value measured believed to be erroneous.

Analysis of Results

The analysis conducted in this paper began with some fundamental understandings. First, the investigation focuses on one climate region. As such, the conclusions drawn for this test section are not necessarily applicable to all climate regions, but it would be well suited to couple with data from other climate regions. Second, many factors relative to the construction approach for this test section were considered in Smith and Howard (*29*) in which the same test section was considered from a mixture perspective. Findings in Smith and Howard (*29*) suggested that haul time effects between the strips considered in this paper were secondary effects. Thus, construction effects were considered on a mix basis and haul time effects were not considered directly.

PAV Trends

Figure 2 shows predictable trends between PAVi and binder properties ($R^2 > 0.90$ in all cases), and most relationships between PAVi and binder properties were linear. The only exception to linear behavior was *Pen* (Figure 2*a*). The observation of linear behaviors in Figure 2, *b–f* is not unusual as behaviors of binder properties vs PAV time in Glover et al. and Anderson et al. (*20, 21*) indicated that some binders changed properties in a linear fashion whereas others did not. Ongoing discussions within the FHWA binder ETG also suggest that linear forecasting of binder properties with respect to PAV time could be appropriate in some cases. The best fit line for Figure 2*a* was exponential when considering PAV times of 0 to 80 h, but a linear fit of all Figure 2*a* data between 10 and 80 h resulted in an R^2 of 0.91 and a linear decrease in *Pen* of 0.2 dmm/h. Based on trendline slopes in Figure 2, the effects of PAVi were more pronounced in B1 than in B2 for $T_c(DSR_{25})$, $T_c(BBR_m)$, and ΔT_c. In other words, the additive decreased the rate of aging.

Binder Conditioning Complexity Considerations

Column 4 of Table 6 determined mixture offsets to collectively estimate the effects of utilizing B1 or B2 binders in mixtures M14 to M16. Mixture offsets were defined by subtracting properties of ARB samples after RTFO conditioning from average FCB sample properties with no longer term aging. For example, the M14 *Pen* offset of −11 dmm was determined by averaging FCB values without longer term aging from M14 (23, 27, 29, and 25 dmm) and subtracting the RTFO conditioned ARB *Pen* for B1 (37 dmm). Had the Figure 1 test section been constructed to evaluate PAV effects instead of demonstrating emergency paving capabilities of asphalt, large quantities of all raw materials would have been obtained and a systematic and comprehensive offset between virgin binder and mixture would

Table 4. Dynamic Shear Rheometer Critical Temperature: $T_c(DSR_{25})[T_c(DSR_8)]$

Sample description	Short-term[a]	Field core samples: 0 yr	2 yr	4 yr	CP4	As-received samples: OB	PAV_0	PAV_{10}	PAV_{20}	PAV_{30}	PAV_{40}	PAV_{60}	PAV_{80}
B1													
ARB	na	—	—	—	—	61.1[13.8]	70.4[17.6]	77.6[20.6]	83.3[22.8]	88.5[24.6]	93.2[26.2]	101.2[28.5]	112.0[32.1]
B2													
ARB	na	—	—	—	—	62.6[14.7]	70.7[17.7]	77.7[21.1]	82.6[23.4]	87.4[24.9]	90.9[26.1]	98.0[28.6]	105.8[31.5]
M14													
FCB_0	1 h (S1)	79.3[20.8]	87.5[25.4]	94.1[28.0]	79.6[19.8]	—	—	—	—	—	—	—	—
FCB_0	6 h (S3)	80.1[19.5]	90.0[25.4]	96.4[28.3]	82.0[19.8]	—	—	—	—	—	—	—	—
FCB_5	1 h (S1)	75.7[17.8]	74.7[15.9]	84.7[23.3]	77.0[17.9]	—	—	—	—	—	—	—	—
FCB_5	6 h (S3)	79.6[19.5]	83.2[23.1]	83.8[21.6]	82.5[21.0]	—	—	—	—	—	—	—	—
M15													
FCB_0	1 h (S5)	73.3[16.0]	81.6[20.1]	92.6[28.4]	78.9[19.7]	—	—	—	—	—	—	—	—
FCB_0	6 h (S7)	71.9[14.5]	87.1[23.9]	95.0[29.6]	80.6[20.5]	—	—	—	—	—	—	—	—
FCB_5	1 h (S5)	70.8[16.7]	72.2[16.1]	82.6[23.2]	78.0[19.8]	—	—	—	—	—	—	—	—
FCB_5	6 h (S7)	77.3[18.9]	76.2[17.9]	85.6[23.8]	80.0[21.2]	—	—	—	—	—	—	—	—
M16													
FCB_0	1 h (S9)	70.2[12.5]	85.8[23.8]	93.7[29.6]	79.3[20.9]	—	—	—	—	—	—	—	—
FCB_0	6 h (S10)	76.5[18.9]	88.9[26.1]	96.0[30.8]	81.4[21.9]	—	—	—	—	—	—	—	—
FCB_5	1 h (S9)	74.4[17.4]	75.7[17.7]	78.8[20.3]	77.6[20.1]	—	—	—	—	—	—	—	—
FCB_5	6 h (S10)	77.2[18.9]	81.4[23.8]	81.0[21.8]	79.5[20.5]	—	—	—	—	—	—	—	—

Note: — = data not measured.

[a]Short-term aging (aging times rounded to the nearest hour with test strip [S] indicated); not applicable (na) for B1 and B2 as short-term conditioning included within each property; each data column represents Tc(DSR25)[Tc(DSR8)] in °C.

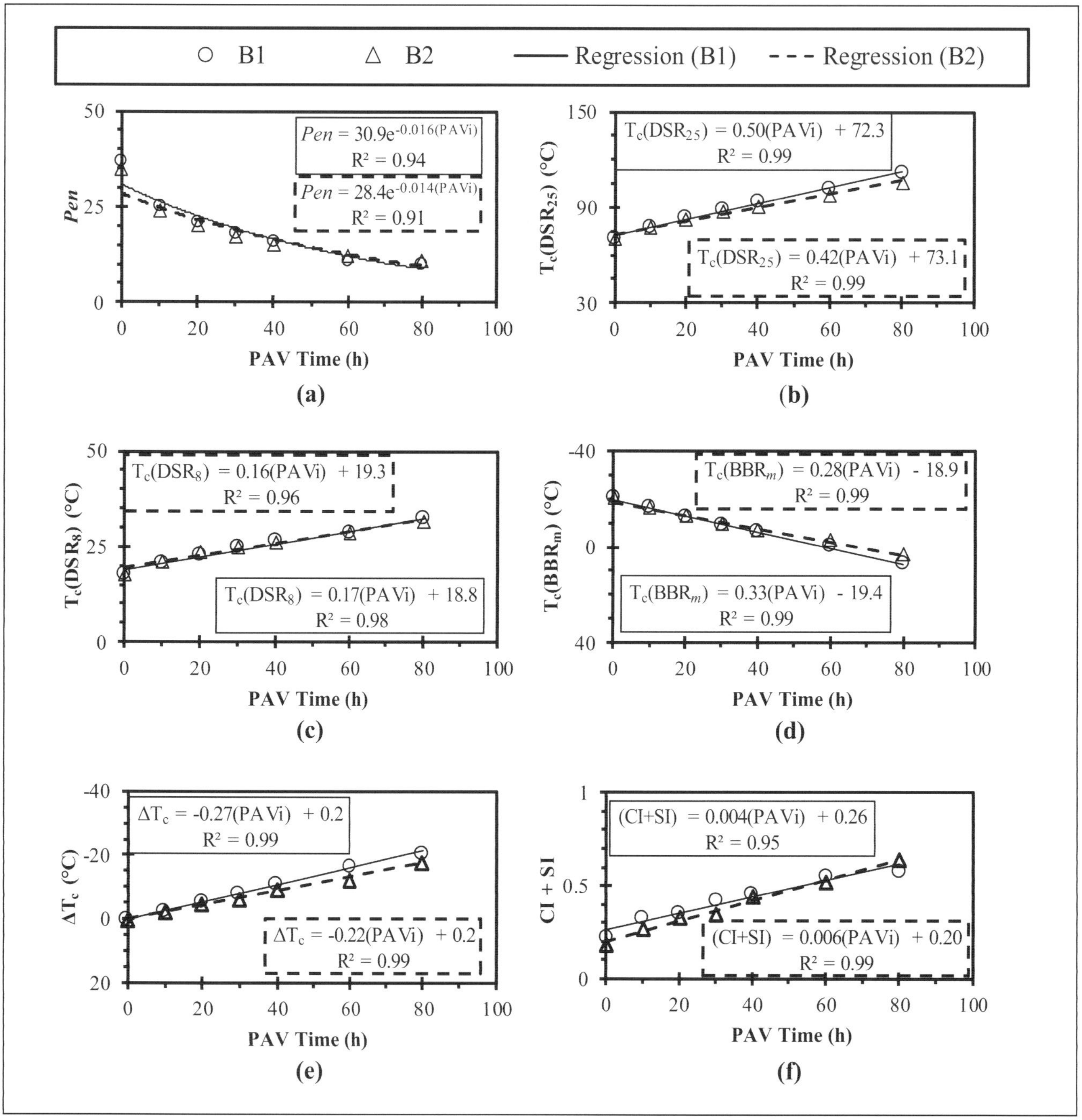

Figure 2. Pressure aging vessel time behaviors for ARB: (*a*) Pen vs. PAVi, (*b*) $T_c(DSR_{25})$ vs PAVi, (*c*) $T_c(DSR_8)$ vs PAVi, (*d*) T_c(BBRm) vs PAVi, (e) ΔT_c vs PAVi, (*f*) CI + SI vs PAVi.

have been measured. As this was not possible years later, the approximate (but believed reasonable by the authors) offset was performed between RTFO aged ARB and FCB prior to field aging. The primary value for using this test section to assess PAV conditioning remains that the test section was constructed some time ago, virgin binder was sampled from tankers a few minutes before mix production, and the test section was carefully monitored for years.

Comparing PAVi to Field Aging

The mixture offsets determined in the previous section were used to uniformly shift Figure 2 equations and produce the PAVi equations in the fifth column of Table 6. For example, the −11 dmm offset for M14 *Pen* was added as a constant with *Pen* = $30.9e^{-0.016(PAVi)}$ to produce *Pen* = $30.9e^{-0.016(PAVi)}$ – 11 before solving for PAVi as shown in the first row of column five in Table 6. These equations

Table 6. PAV Times (PAVi) in Hours to Simulate Binder Changes in Mixtures

Binder property	Mix	Figure 2 regression	Mixture offset	Regression plus offset solved for PAVi (modified Figure 2 regression)	Test Strip	2-year PAVi FCB_0	2-year PAVi FCB_5	4-year PAVi FCB_0	4-year PAVi FCB_5	CP4 PAVi FCB_0	CP4 PAVi FCB_5
Pen	M14	$Pen = 30.9e^{-0.016(PAVi)}$	−11	PAVi = ln[(*Pen* + 11)/	1	11	0	24	6	0	0
(dmm)				30.9](−1/0.016)	3	8	0	24	2	0	0
	M15	$Pen = 30.9e^{-0.016(PAVi)}$	+ 2	PAVi = ln[(*Pen*-2)/	5	18	0	77	41	21	30
				30.9](−1/0.016)	7	45	6	77	49	27	30
	M16	$Pen = 28.4e^{-0.014(PAVi)}$	−1	PAVi = ln[(*Pen* + 1)/	9	37	6	68	15	22	12
				28.4](−1/0.014)	10	41	15	68	22	29	22
$T_c(DSR_{25})$	M14	$T_c(DSR_{25})$ = 0.50(PAVi) + 72.3	+ 8.3	PAVi = ($T_c(DSR_{25})$–80.6)	1	14	0	27	8	0	0
(°C)				(1/0.50)	3	19	5	32	6	3	4
	M15	$T_c(DSR_{25})$ = 0.50(PAVi) + 72.3	+ 2.9	PAVi = ($T_c(DSR_{25})$–75.2)	5	13	0	35	15	7	6
				(1/0.50)	7	24	2	40	21	11	10
	M16	$T_c(DSR_{25})$ = 0.42(PAVi) + 73.1	+ 3.9	PAVi = ($T_c(DSR_{25})$–77.0)	9	21	0	40	4	5	1
				(1/0.42)	10	28	10	45	10	10	6
$T_c(DSR_8)$	M14	$T_c(DSR_8)$ = 0.17(PAVi) + 18.8	+ 1.8	PAVi = ($T_c(DSR_8)$–20.6)	1	28	0	44	16	0	0
(°C)				(1/0.17)	3	28	15	45	6	0	2
	M15	$T_c(DSR_8)$ = 0.17(PAVi) + 18.8	−1.1	PAVi = ($T_c(DSR_8)$–17.7)	5	14	0	63	32	12	12
				(1/0.17)	7	36	1	70	36	16	21
	M16	$T_c(DSR_8)$ = 0.16(PAVi) + 19.3	−0.8	PAVi = ($T_c(DSR_8)$–18.5)	9	33	0	69	11	15	10
				(1/0.16)	10	48	33	77	21	21	13
$T_c(BBR_m)$	M14	$T_c(BBR_m)$ = 0.33(PAVi)–19.4	+ 0.2	PAVi = ($T_c(BBR_m)$ + 19.2)	1	15	0	29	14	2	0
(°C)				(1/0.33)	3	14	6	34	5	2	4
	M15	$T_c(BBR_m)$ = 0.33(PAVi)–19.4	−4.0	PAVi = ($T_c(BBR_m)$ + 23.4)	5	15	0	42	23	13	14
				(1/0.33)	7	27	5	48	30	16	17
	M16	$T_c(BBR_m)$ = 0.28(PAVi)–18.9	−1.8	PAVi = ($T_c(BBR_m)$ + 20.7)	9	21	0	46	5	11	6
				(1/0.28)	10	28	15	52	9	13	8
ΔT_c	M14	$\Delta T_c = -0.27(PAVi) + 0.2$	−2.9	PAVi = (ΔT_c + 2.7)	1	10	0	17	8	3	0
(°C)				(−1/0.27)	3	9	3	18	3	3	1
	M15	$\Delta T_c = -0.27(PAVi) + 0.2$	−2.2	PAVi = (ΔT_c + 2.0)	5	11	0	19	3	4	3
				(−1/0.27)	7	12	0	24	15	6	3
	M16	$\Delta T_c = -0.22(PAVi) + 0.2$	−2.7	PAVi = (ΔT_c + 2.5)	9	12	0	23	0	5	0
				(−1/0.22)	10	15	2	25	0	2	2
CI + SI	M14	(CI + SI) = 0.004(PAVi) + 0.26	+ 0.33	PAVi = ((CI + SI)–0.59)	1	0	0	52	25	2	2
				(1/0.004)	3	25	—	42	5	7	0
	M15	(CI + SI) = 0.004(PAVi) + 0.26	+ 0.37	PAVi = ((CI + SI)–0.63)	5	12	0	62	0	2	2
				(1/0.004)	7	35	0	100	30	15	5
	M16	(CI + SI) = 0.006(PAVi) + 0.20	+ 0.40	PAVi = ((CI + SI)–0.61)	9	10	0	15	0	0	5
				(1/0.006)	10	14	0	39	0	2	0

were used to determine the amount of PAVi necessary to produce the binder properties experienced after field aging (2 years or 4 years) and CP4 as reported in tables 2 to 5. The methods conducted in Table 6 resulted in negative PAVi values in some cases. When negative PAVi values were calculated, PAVi was reported as zero considering the understanding that binder properties do not improve after field aging or soaking in 64°C water.

Discussion

Table 7 summarizes the Table 6 analysis using both material properties and test strips. The final row of Table 7 summarizes the Table 6 analysis considering an average of all data. The third through tenth columns present ranges of PAVi values determined for age and depth combinations. Columns 11 to 14 present linear regressions for which average PAVi values at the surface (i.e., columns five and nine) or for non-surface material (i.e., columns seven and eleven) were forced through the origin. Based on slopes provided for surface and non-surface materials, 11.1 h of PAVi and 3.1 h of PAVi produced property changes comparable to that of 1 year for surface and non-surface binders, respectively.

When considering PAV rates determined for the test section as divided by binder property, there was considerable variation between rates determined for each property. For comparison, intermediate temperature properties [*Pen* and $T_c(DSR_8)$] required more PAVi than average at both surface and non-surface materials. Low [$T_c(BBR_m)$ and ΔT_c] and high ($T_c(DSR_{25})$ temperature properties required less than average PAVi, and chemical

Table 7. Summary of PAV Times (PAVi) in Hours to Simulate Field Aging

Category	Data included	2-year PAVi				4-year PAVi				Linear regression through origin (RTO)			
		FCB_0 (surface)		FCB_5 (non-surface)		FCB_0 (surface)		FCB_5 (non-surface)		FCB_0 (surface)		FCB_5 (non-surface)	
		Range	Avg.[1]	Range	Avg.[1]	Range	Avg.[1]	Range	Avg.[1]	Slope[2]	R^2	Slope[2]	R^2
Data sorted by binder properties	$T_c(DSR_{25})$	13–28	20	0–10	3	27–45	37	4–21	11	9.4	0.99	2.5	0.84
	$T_c(DSR_8)$	14–48	31	0–33	8	44–77	61	6–36	20	15.3	0.99	4.8	0.96
	Pen	8–45	27	0–15	5	24–77	56	2–49	23	13.9	0.99	5.1	0.79
	$T_c(BBR_m)$	14–28	20	0–15	4	29–52	42	5–30	14	10.4	0.99	3.2	0.86
	ΔT_c	9–15	12	0–3	1	17–25	21	0–15	5	5.4	0.96	1.1	0.78
	CI + SI	0–35	16	0–0	0	15–100	52	0–30	10	12.0	0.88	2.5	0.60
Data sorted by Table 1 test strip	Strip 1	0–28	13	0–0	0	17–52	32	6–25	13	7.7	0.96	2.6	0.60
	Strip 3	6–28	17	0–15	6	18–45	33	2–6	5	8.3	0.99	1.6	0.00
	Strip 5	11–18	14	0–0	0	19–77	50	0–41	19	11.4	0.85	3.8	0.60
	Strip 7	12–45	30	0–6	2	24–100	60	15–49	30	15.0	0.99	6.2	0.66
	Strip 9	10–37	22	0–6	1	15–69	44	0–15	6	11.0	0.99	1.3	0.74
	Strip 10	14–48	29	0–33	13	25–77	51	0–22	10	13.1	0.96	3.3	0.00
All data	All data	0–48	21	0–33	3	15–100	45	0–49	14	11.1	0.99	3.1	0.88

Note: Avg. = average. Slope is PAVi (hours) divided by field age, or t_F (years), for example, 9.4 means 1 year of field aging was replicated by 9.4 h of PAV conditioning.

properties (CI + SI) required about average PAVi times for surface and non-surface materials. Though the PAV rates to simulate binder property change varies between properties, considering a single PAV condition for all properties is more practical for purchasing specifications.

PAV rates also varied when comparing rates from test strip to test strip. However, there were multiple variables that could have affected the rates between test strips (e.g., air voids, mixing temperature, and haul time) (*29*). Regression analysis between Table 7 PAV rates and average air void levels from Table 1 indicated that there was no significant relationship between air voids and the amount of PAVi necessary to simulate field aging for the test section considered. Surface materials had an R^2 of 0.06 and *p*-value of 0.63, and non-surface materials had an R^2 of 0.00 with a *p*-value of 0.92. Although air void effects were likely for this dataset, interaction of the factors described in Smith and Howard (*29*) prevented the authors from considering air voids directly.

Referring to average values from Table 7, the PAVi to produce average property changes as seen after 2 or 4 years of field aging was 21 h and 45 h, respectively, for binder collected from the pavement surface. For non-surface materials, the same field ages were best simulated by 3 h and 14 h of PAVi. Considering these values and the average PAV rates from Table 7, the 20-h PAV time specified in AASHTO R28 would be likely to simulate approximately 2 years of binder property change at the pavement surface and 6.5 years of property change for non-surface materials.

Conclusion

Based on the analysis conducted in this paper, the PAV times that most appropriately simulated 2 years and 4 years of field aging were 21 h and 45 h for binders collected from the top 1.3 cm of the pavement, whereas 3 h and 14 h best simulated 2 years and 4 years of field aging for binders collected from 5 cm to 6.3 cm below the pavement surface. To simulate 1 year of field aging, 11 h of PAV time was required for the pavement surface, and 3 h was required for materials at depth. These statements correspond with the 20-h protocol in R28 simulating approximately 2 years at the pavement surface and 6.5 years at depth for the test section considered. The authors recommend these values be considered alongside investigations from other climatic regions and current marketplace factors (e.g., more severe overall aging environments in present day than when the PAV was developed) as the industry moves forward with aging discussions that affect standards and specifications.

Acknowledgments

The Mississippi Department of Transportation (MDOT) funded Project 106526, State Study 266, and State Study 270, which, among other items helped support the lead author while a doctoral student at Mississippi State University. Test section construction was performed by APAC Mississippi supported by Ergon Asphalt and Emulsions and funded as part of Project 70015 by the Southeast Region Research Initiative through Oak Ridge National Laboratory (ORNL). Individuals deserving special thanks are Alex Middleton (MDOT), Griffin Sullivan (MDOT), James Williams (MDOT), Benjamin Thomas (ORNL), Mike Bogue (APAC), and Dwayne Boyd (APAC). Special thanks to the staff of Paragon Technical Services, Inc. (specifically Scott Watson, Mike Hemsley, Sonia Serna, and Dwayne Johnson) for hosting binder testing. Multiple Mississippi State University students assisted this work in some way: Amanda Hufft, Rabeea Bazuhair, Carl Pittman, Ethan Whaley, Robert Moore, Brad Hansen, and Westin Graves. The Ergon Asphalt and Emulsions Student Support Initiative in Construction Materials funded some of these students while working.

Author Contributions

The authors confirm contribution to the paper as follows: study conception and design: all authors; data collection: all authors; analysis and interpretation of results: all authors; draft manuscript preparation: BTS, ILH, GLM. All authors reviewed the results and approved the final version of the manuscript.

References

1. Krchma, L. C. Symposium on Hardening of Asphalt Pavements. *Journal of the Association of Asphalt Paving Technologists*, Vol. 27, 1958, pp. 171–231.
2. Kemp, G., and G. Sherman. Significant Studies on Asphalt Durability: California Experience. *Transportation Research Record: Journal of the Transportation Research Board*, 1984. 999: 36–40.
3. Petersen, J. C., R. E. Robertson, J. F. Branthaver, P. M. Harnsberger, J. J. Duvall, S. S. Kim, D. A. Anderson, D. W. Christiansen, and U. H. Bahia. *Binder Characterization and Evaluation Volume 1*. SHRP-A-367. Strategic Highway Research Program, National Research Council, Washington, D.C., 1994.
4. Won, J., A. Chowdhury, and C. Glover. Field Aging of Unmodified Asphalt Binder in Three Texas Long-Term Performance Pavements. *Transportation Research Record: Journal of the Transportation Research Board*, 2008. 2051: 15–22.
5. Bell, C. A., and D. Sosnovske. *Aging: Binder Validation*. Publication SHRP-A-384. Strategic Highway Research Program, National Academy of Sciences, Washington, D.C., 1994.
6. Baek, C., B. Underwood, and Y. R. Kim. Effects of Oxidative Aging on Asphalt Mixture Properties. *Transportation Research Record: Journal of the Transportation Research Board*, 2012. 2296: 77–85.
7. Copas, T. L., and H. A. Pennock. *NCHRP Synthesis 59: Relationship of Asphalt Cement Properties to Pavement Durability*. Transportation Research Board of the National Academies, Washington, D.C., 1979.
8. Huang, S., and T. F. Turner. Aging Characteristics of RAP Blend Binders: Rheological Properties. *Journal of Materials in Civil Engineering*, Vol. 26, No. 5, 2014, pp. 966–973.
9. Jacques, C., J. S. Daniel, T. Bennert, G. Reinke, A. Norouzi, C. Ericson, W. Mogawer, and Y. R. Kim. Effect of Silo Storage Time on the Characteristics of Virgin and Reclaimed Asphalt Pavement Mixtures. *Transportation Research Record: Journal of the Transportation Research Board*, 2016. 2573: 76–85.
10. Plancher, H., E. Green, and J. Petersen. Reduction of Oxidative Hardening of Asphalts by Treatment with Hydrated Lime – A Mechanistic Study. *Journal of the Association of Asphalt Paving Technologists*, Vol. 45, 1976, pp. 1–24.
11. Edler, A. C., M. M. Hattingh, V. P. Servas, and C. P. Marais. Use of Aging Tests to Determine the Efficacy of Hydrated Lime Additions to Asphalt in Retarding its Oxidative Hardening. *Journal of the Association of Asphalt Paving Technologists*, Vol. 54, 1985, pp. 118–139.
12. Lesueur, D., and D. Little. Effect of Hydrated Lime on Rheology, Fracture, and Aging of Bitumen. *Transportation Research Record: Journal of the Transportation Research Board*, 1999. 1661: 93–105.
13. Zou, J., R. Roque, G. Lopp, M. Isola, and M. Bekoe. Impact of Hydrated Lime on Cracking Performance of Asphalt Mixtures with Oxidation and Cyclic Pore Pressure. *Transportation Research Record: Journal of the Transportation Research Board*, 2016. 2576: 51–59.
14. Huang, S., J. C. Petersen, R. E. Robertson, and J. F. Branthaver. Effect of Hydrated Lime on Long-Term Oxidative Aging Characteristics of Asphalt. *Transportation Research Record: Journal of the Transportation Research Board*, 2002. 1810: 17–24.
15. Lund, J. W., and J. E. Wilson. Evaluation of Asphalt Aging in Hot Mix Plants. *Journal of the Association of Asphalt Paving Technologists*, Vol. 53, 1984, pp. 1–18.
16. Wright, P. H., and R. J. Paquette. Hardening of Hot Bituminous Mixes During the Hauling Process. *Highway Research Record*, 1966. 132: 11–28.
17. Coons, R. F., and P. H. Wright. An Investigation of the Hardening of Asphalt Recovered from Pavements of Various Ages. *Journal of the Association of Asphalt Paving Technologists*, Vol. 37, 1967, pp. 510–528.
18. Mirza, M. W., and M. W. Witczak. Development of a Global Aging System for Short and Long Term Analysis of Asphalt Cements. *Journal of the Association of Asphalt Paving Technologists*, Vol. 64, 1995, pp. 393–430.
19. Luo, X., F. Gu, and R. L. Lytton. Prediction of Field Aging Gradient in Asphalt Pavements. *Transportation Research Record: Journal of the Transportation Research Board*, 2015. 2507: 19–28.
20. Glover, C. J., R. R. Davison, C. H. Domke, Y. Ruan, P. Juristyarini, D. B. Knorr, and S. H. Jung. *Development of a New Method for Assessing Asphalt Binder Durability with*

Field Validation. Publication FHWA/TX-05/1872-2, Texas Department of Transportation, 2005.
21. Anderson, D., D. W. Christensen, H. U. Bahia, R. Dongre, M. G. Sharma, C. E. Antle, and J. Button. *Binder Characterization and Evaluation. Volume 3: Physical Characterization*. Report SHRP-A-369. Strategic Highway Research Program, Washington, D.C., 1994.
22. Bahia, H. U., and D. A. Anderson. *The Pressure Aging Vessel (PAV): A Test to Simulate Rheological Changes Due to Field Aging*. ASTM STP 1241. American Society for Testing and Materials, West Conshohocken, Pa., 1995, pp. 67–88.
23. Farshidi, F., D. Jones, and J. T. Harvey. *Warm-Mix Asphalt Study: Evaluation of Hot and Warm Mix Asphalt with Respect to Binder Aging*. Report UCPRC-RR-2013-02. University of California Pavement Research Center, 2013.
24. *State-of-the-Knowledge: The Use of REOB/VTAE in Asphalt*. Information Series No. 235. Asphalt Institute, Lexington, Ky., 2016.
25. Howard, I. L., B. A. Payne, M. Bouge, S. Glusenkamp, G. L. Baumgardner, and J. M. Hemsley. *Full Scale Testing of Hot-Mixed Warm-Compacted Asphalt for Emergency Paving*. Publication SERRI 70015-011. U.S. Department of Homeland Security, Washington, D.C., 2012.
26. Smith, B. T. *Characterizing Short and Long Term Mixture Aging of a Full-Scale and Non-Trafficked Asphalt Test Section*. Dissertation. Mississippi State University, 2017.
27. Anderson, M., G. King, D. Hanson, and P. Blankenship. Evaluation of the Relationship Between Asphalt Binder Properties and Non-Load Related Cracking. *Journal of the Association of Asphalt Paving Technologists*, Vol. 80, 2011, pp. 615–661.
28. Rowe, G. M. Prepared Discussion of Evaluation of the Relationship between Asphalt Binder Properties and Non-load Related Cracking. *Journal of the Association of Asphalt Paving Technologists*, Vol. 80, 2011, pp. 649–662.
29. Smith, B. T., and I. L. Howard. Short and Longer Term Effects of Short Term Aging on Asphalt Mixture Properties. *Journal of Materials in Civil Engineering*, Vol. 30. No. 3, 2018, p. 04018005.

The Standing Committee on Critical Issues and Emerging Technologies in Asphalt (AFK10) peer-reviewed this paper (18-04401).

Article

TRR
JOURNAL OF THE TRANSPORTATION RESEARCH BOARD

Transportation Research Record
2018, Vol. 2672(28) 235–243

DOI: 10.1177/0361198118793316
journals.sagepub.com/home/trr
SAGE

Modeling Shear Viscosity of Asphalt through Nonequilibrium Molecular Dynamics Simulation

Yongjie Ding[1,2], Baoshan Huang[3], and Xiang Shu[2]

Abstract

This paper investigates a quantifiable relationship between the chemical composition and the shear viscosity of the asphalt binder through the nonequilibrium molecular dynamics (NEMD) simulation. The fix Deform and fix NVT/SLLOD ensembles were used to perform a shear effect on the model. The relationship between the shear viscosity and the microstructure characteristics of the asphalt model was studied. The results indicated that the asphalt model exhibited shear thinning as the strain rate increases. The effects of energy and chemical structure (molecular weight, aromatic carbon and heteroatom percentage) on shear viscosity were studied from the perspective of microstructure. The results showed that the larger fused aromatic ring core caused a nonlinear relationship between viscosity and strain rate. The viscosity of the asphalt model increased with the molecular weight, the percentage of aromatic carbon and heteroatoms. The molecular weight had a greater effect on viscosity than aromatic carbon and heteroatom percentage. Considering energy variations with strain rate and chemical structure, the NEMD method can be used to predict the behavior of the asphalt molecules under shear strain at a certain level.

As a widely used viscoelastic material for building pavements and roofs, asphalt is generally graded based on physical properties. However, in some cases, due to differences in chemical composition, the same grade of asphalt behaves differently (*1*). Until recently, there was still limited knowledge of the chemical factors that affect physical properties. A better understanding of the relationship between chemical composition and the physical properties of asphalt (such as rheological) can provide useful insights into the design and modification of asphalt for long-lasting pavements.

Viscosity is a critical property that relates to the performance of asphalt on pavement, which depends on the rate of loading and temperature. Since Van der Poel (*2*) introduced the concept of asphalt stiffness as a function of temperature and time of loading frequency, several models have been proposed to characterize asphalt's viscoelasticity. One popular form of presentation for viscoelasticity is the plot of master curves using the shifting procedure developed by Williams et al. (*3*).

The fundamental factors that govern the viscoelasticity of asphalt relate to the microstructure parameters down to the molecular level, so a better understanding of the intrinsic relationship between the viscosity and molecular structural behavior is critical for designing better material and for advancing the state of knowledge concerning the fundamental characteristics of asphalt molecules (*1*, *4*, *5*).

Combining the knowledge of statistical mechanics and thermodynamics, molecular dynamics (MD) simulation can solve the full statistical mechanics of realistic systems numerically. MD is based on the analysis of atom dynamics and interactions. It can predict the structural and thermodynamic properties of relatively complex materials (*6*).

Although multiple works (*7–10*) have been conducted to investigate the structure and properties of asphalt binder using molecular simulation, limited research has been found on the rheology property simulation (*11*). MD can be used to predict viscosity using either equilibrium or nonequilibrium methods. Typical equilibrium MD (EMD) methods use either a Green-Kubo integral or

[1]College of Chemical Engineering, State Key Laboratory of Heavy Oil Processing, China University of Petroleum, Qingdao, China
[2]Department of Civil and Environmental Engineering, The University of Tennessee, Knoxville, TN
[3]School of Transportation Engineering, Tongji University, Shanghai, China. Department of Civil and Environmental Engineering, The University of Tennessee, Knoxville, TN

Corresponding Author:
Address correspondence to Baoshan Huang: bhuang@utk.edu

Einstein relation to relate fluctuations of off-diagonal elements of the pressure tensor to viscosity (6). Zhang and Greenfield (8) used EMD simulation to estimate the temperature-dependent change in zero shear asphalt viscosity. The results show that the viscosity can only be calculated at a high temperature (443 K) because the EMD simulation of viscosity relies on the prediction of molecular rotation radius, which is difficult to achieve in the time scale of molecular simulation (10^{-12} s) because asphalt molecules cannot rotate significantly (7). In addition, since shear stress is common in pavements, the shear viscosity seems to be more critical to the performance of the asphalt.

The development of shear flow nonequilibrium molecular dynamics (NEMD) avoids the high computational costs associated with EMD simulations (12–15), which makes it possible to measure the viscosity accurately because the shear strain applied to the model can force the molecular rotation to accelerate. Molecular modeling also allows the study of microstructure characteristics of asphalt under conditions that are difficult to perform in real experiments. Therefore, the NEMD simulation of asphalt may help to understand the basic properties of asphalt in the shear flow by observing its flow dynamics at atomic time and length scales.

Objective and Scope

The objective of the present study was to investigate a quantifiable relationship between the chemical composition and the shear viscosity of the asphalt binder by NEMD simulation. The shear behavior of asphalt molecule was examined from an energy point of view, such as how the energy performed and correlated with the viscosity quantities in vastly different flow fields. To correlate chemical character to viscosity, a series of asphaltic models were constructed based on the variation in molecular weight and aromatic carbon percentage as well as heteroatom percentage. Qualitative discovery is expected for the understanding of chemical effects on the shear viscosity of asphalt molecules. Understanding the contribution of chemical properties to shear viscosity will contribute to the modification and evaluation of asphalt.

Simulation Details

Basic Asphaltic Model

Based on solubility and chromatography, asphalt components can be divided into three groups: asphaltene, resin, and oil. In this study, two asphalt models proposed by Zhang and Greenfield (16) were used as the basic model. The model consists of three components, including asphaltenes, resins, and oil. The weight ratio of asphaltene, resin, and oil is selected to be about 20-20-60 (16). The concentration of components was selected based on the alkane: aromatic carbon ration reported by Storm (17). Resin and oil were represented with n-C_{22} and 1,7-dimethylnaphthalene, respectively. The research of Kowalewski indicates that n-C_{22} is the highest concentration alkane in the asphalt sample they studied (18). The number of aromatic rings and side chains makes 1,7-dimethylnaphthalene intermediate between oil and asphaltenes.

The asphaltene structure is from studies by Artok (19) and Groenzin (20). The asphaltene 1 molecule consists of many fused aromatic rings with short aliphatic branches, while asphaltenes 2 have less fused aromatic rings and are surrounded by longer aliphatic chains. They represent two typical asphaltenes, and are able to compare the effects of short vs long aliphatic chains and small vs big fused aromatic rings (16).

Modified Asphaltic Model

The two basic asphalt models used above are a far from accurate quantification of the chemical structure of the asphalt because asphalt is a mixture of millions of organic ingredients. According to the literature (21), the molecular weight (M), and the aromatic carbon percentage (A), as well as the heteroatom percentage (H) are three major factors that may affect the mechanical properties of asphalt. In this study, the basic model used above was modified to generate a series of models with different molecular weights, aromatic carbon, and heteroatoms percentages. The shear viscosity of the modified model was calculated and the influence of the chemical composition was compared.

Table 1 lists the molecular distributions of the modified model. The different asphaltene/resin ratios were used to represent models with different molecular weights. The sulfur atom of the asphaltene molecule was deleted to eliminate the effect of the heteroatom. A new resin molecule with the same carbon atom and oil (shown in Table 2) was established for the aromatic carbon percentage. Substituting a percentage of the resin instead of the oil can produce a different aromatic carbon content model. To compare the effects of heteroatoms, four novel resin molecules (shown in Table 2) with similar structure but different heteroatom content were constructed. The asphalt model was then produced by using different resin molecules. Table 3 shows the average molecular weight, aromatic carbon and heteroatom percentages of models.

NEMD Method

The steady state NEMD method was used to calculate the viscosity of the asphalt model using an open source

Table 1. Distribution of Molecules in Modified Asphaltic Model

System	Molecules	Number of molecules					
		Model No. 1	Model No. 2	Model No. 3	Model No. 4	Model No. 5	Model No. 6
M	Asphaltene	0	1	2	3	4	5
	Resin	27	22	17	12	7	2
	Oil	41	41	41	41	41	41
A	Asphaltene	1	1	1	1	1	1
	Resin	100	80	60	40	20	0
	Oil	0	10	20	30	40	50
H	Asphaltene	1	1	1	1	1	1
	Resin-S	50	40	30	20	10	0
	Resin-N	50	40	30	20	10	0
	Resin-1	0	10	20	30	40	50
	Resin-2	0	10	20	30	40	50
	Oil	10	10	10	10	10	10

Note: M = molecular weight; A = aromatic carbon percentage; H = heteroatom percentage.

Table 2. Structure of Molecules Used in Modified Asphaltic Model

Molecular weight	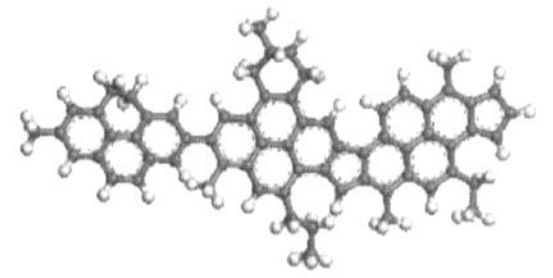 Asphaltene		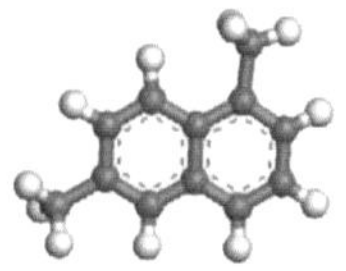 Resin	
Aromatic carbon percentage	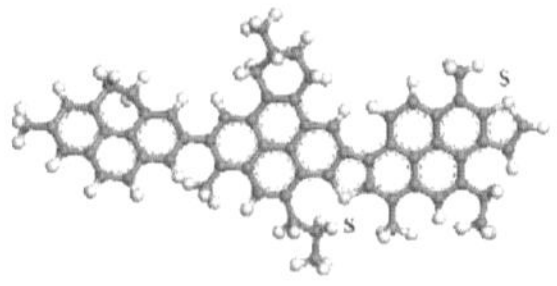 Asphaltene		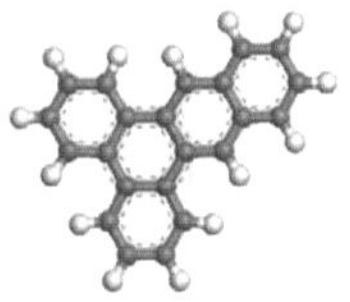 Resin	
Heteroatom percentage	 Resin-N	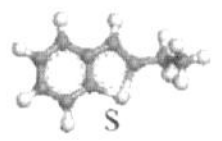 Resin-S	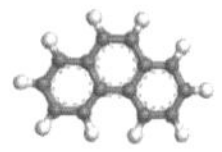 Resin-1	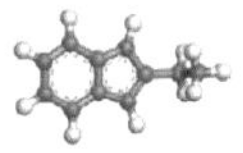 Resin-2

code of Large-scale Atomic/Molecular Massively Parallel Simulator (LAMMPS) (*13*, *22*). The fix Deform and fix NVT/SLLOD ensembles were used to perform a shear effect on the model. The SLLOD method and sliding-brick boundary conditions were applied in the simulation. The SLLOD method specifies the shear field by manipulating the simulated boundary and measures the viscosity by the appropriate nondiagonal component of the pressure or stress tensor (*23*) as shown in (*24*)

$$\dot{\boldsymbol{q}}_{ia} = \frac{\boldsymbol{p}_{ia}}{m_{ia}} + \boldsymbol{q}_{ia} \cdot \nabla \mathbf{u} \quad (1)$$

$$\dot{\boldsymbol{p}}_{ia} = \boldsymbol{F}_{ia} - \boldsymbol{p}_{ia} \cdot \nabla \mathbf{u} - m_{ia} q_{ia} \cdot \nabla \mathbf{u} \cdot \nabla \mathbf{u} - \zeta \boldsymbol{p}_{ia} \quad (2)$$

A thermostat is required to remove the energy injected by viscous heating (*25*). The Nose method was used to maintain a constant temperature (*26*).

In Equations 1 and 2, m_{ia}, $\boldsymbol{p}_{ia}$, $\boldsymbol{q}_{ia}$, and $\boldsymbol{F}_{ia}$ are the mass, momentum, position, and force vector of atom *a* of molecule *i*, respectively, and $\nabla \mathbf{u}$ is the velocity gradient tensor. The term $m_{ia}\boldsymbol{q}_{ia} . \nabla \mathbf{u} . \nabla \mathbf{u}$ vanishes since (*24*)

Table 3. Chemical Structure Parameters of Modified Asphaltic Model

	Molecular weight	Aromatic carbon percentage	Heteroatom percentage
Model No. 1	249	0.05	0.00
Model No. 2	266	0.24	0.02
Model No. 3	285	0.42	0.05
Model No. 4	306	0.61	0.07
Model No. 5	331	0.80	0.09
Model No. 6	360	0.99	0.11

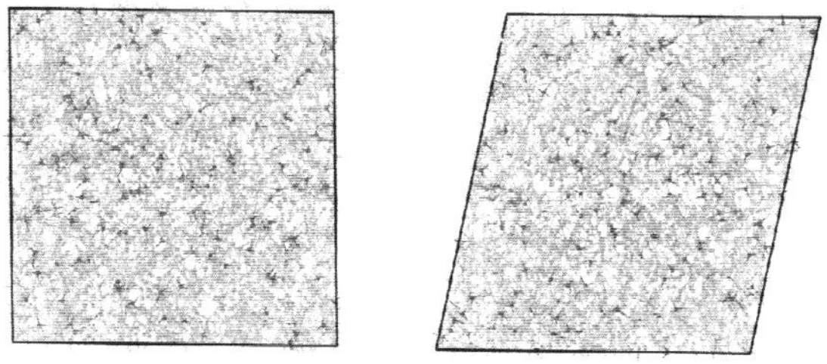

Figure 1. Asphalt model under shear force.

$$\nabla \mathbf{u} = \begin{pmatrix} 0 & 0 & 0 \\ \dot{\gamma} & 0 & 0 \\ 0 & 0 & 0 \end{pmatrix} \quad (3)$$

where $\dot{\gamma}$ is the applied shear rate. The other terms appearing in Equations 1 and 2 are defined as (*24*)

$$\dot{\zeta} = \frac{p_\zeta}{Q} \quad (4)$$

$$p_\zeta = \sum_i \sum_a \frac{\boldsymbol{p}_{ia}^2}{m_{ia}} - DNk_BT \quad (5)$$

$$Q = DNk_BT\tau^2 \quad (6)$$

The N, V, and T are total number of atoms, volume of the system, and absolute temperature, respectively. The $k_\mathbf{B}$ and D are the Boltzmann constant and the dimensionality of the system. The ζ, p_ζ and Q are the coordinate- and momentum-like variables and inertia mass of the Nose thermostat (*24*).

In this study, molecules were first placed on a lattice. Next, molecular dynamics in the canonical ensemble (NVT) was applied for 20,000 fs for the equilibration of the model. Then, NEMD shear and equilibrate were turned on by introducing a deformation with a certain rate (shown in Figure 1) for 50,000 fs. The simulation continued for another 500,000 fs for data gathering. The viscosity was calculated using

$$Viscosity = \frac{-dM}{Vel\,gradient} \quad (7)$$

where dM = momentum flux in the y-direction; and $Vel\,gradient = \frac{dVelX}{dY}$, which means the change in x-velocity over a distance dY in the y-direction.

An explicit atom representation of each molecule was used in the simulations. Parameters were taken from the cvff force field (*27*). Each aromatic ring was maintained close to planar using improper torsion angles. The periodic boundary condition was used in all simulations. The time step was 1 fs. The cut-off distance for interaction was always 1 nm. The simulation temperature was 298 K. The molecular model was assembled in a cube box of 4–5 nm in length.

Newton High Performance Computer Cluster (Newton HPC) located at the University of Tennessee in Knoxville was employed to perform the computation. The Newton HPC has a peak performance of 30 Tflop/s and storage resources of 150 TBytes of high-performance Lustre storage for use by computational jobs.

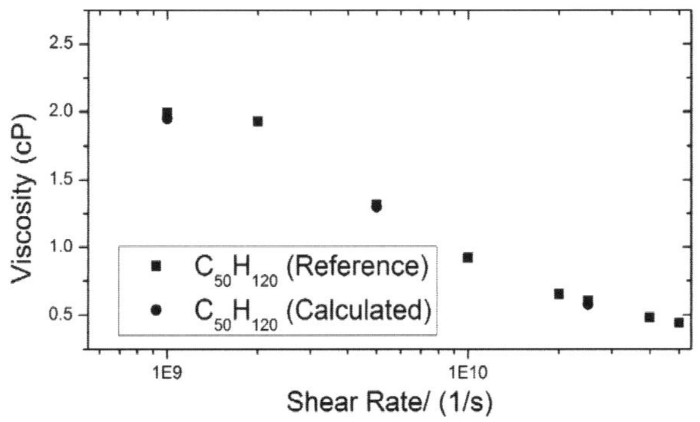

Figure 2. Comparison of calculated and reference (*24*) viscosities of $C_{50}H_{120}$ (real units rather than reduced units are used for the horizontal axis).

Results and Discussion

Shear Viscosity Estimation for Ethylbenzene and n-Decane

The simulation method was first verified by known shear viscosity data from literature (*15*, *24*), which selected $C_{50}H_{120}$ as a representative molecule. Figure 2 compares the calculated and reference viscosity of $C_{50}H_{120}$ under different shear rates. It can be seen that the calculated viscosities were quite close to that of the literate (*24*), indicating that the NEMD method could accurately calculate the viscosity of materials containing carbon hydrogen structure.

Shear Viscosity Estimation for Basic Asphaltic Model and Components

Figure 3 shows the calculated viscosities of the basic asphaltic model and its components at different strain rates. Viscosity decreased with an increasing strain rate, indicating shear thinning behavior for all three components and the entire asphalt model. Shear thinning is a term used for rheology to describe non-Newtonian fluids with reduced viscosity when subjected to shear strain. Furthermore, at a lower shear rate, the curves of resin

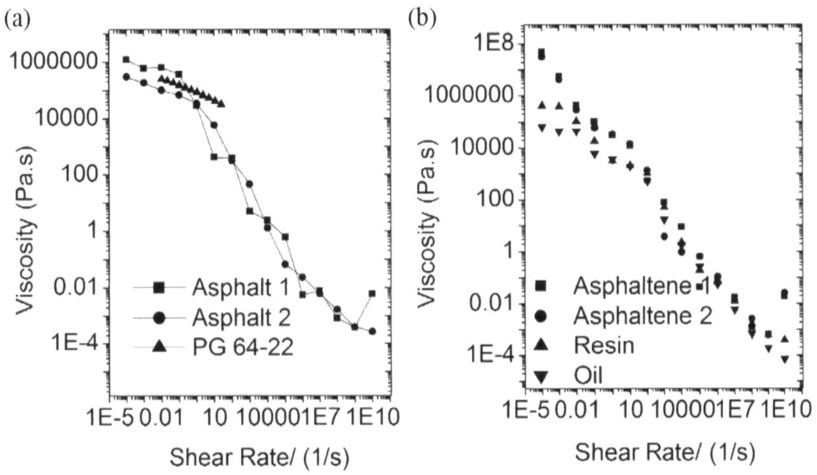

Figure 3. Viscosities of basic models' components at different shear rates: ((*a*) basic asphaltic model and (*b*) components).

and oil showed a plateau, indicating that there was a nonshear viscosity. However, the plateau did not develop on the curves of asphaltene under a low shear rate. The trend of viscosity with the strain rate changed from decreasing to increasing at high strain rates for asphalt 1 and asphaltenes 1 and 2, which was contradictory to the reality. This shows that the NEMD method could not predict the viscosity of these three models at high strain rates.

The results from a laboratory Dynamic Shear Rheometer (DSR) test gave a complex viscosity range from 10^4 Pa·s to 10^5 Pa·s at the strain rates ranging from $0.01\,s^{-1}$ to $25\,s^{-1}$ at 298 K for a typical asphalt used in Tennessee, USA. As shown in Figure 3, the viscosity of the PG 64-22 asphalt was between the calculated viscosities of the two basic models, indicating that the model constructed in this study was similar to the actual asphalt. However, as the strain rate increased, the calculated viscosity decreased faster than the experimental results. This may be because the loading pattern of the DSR was not exactly the same as the loading pattern in the NEMD simulation, resulting in a difference between the calculated viscosity and the experimental viscosity. DSR apply a cyclic torque to cause an angular rotational strain on asphalt, and the simulated strain in this study was parallel to the x-axis. It should also be noted that the molecular structure and composition percentage of the asphalt model and the PG 64-22 adhesive used in this study did not match exactly, so the calculated viscosity was different from the experimental results.

The curves of asphaltenes 1 and 2 exhibited a nonlinear trend around the strain rates from $1{,}000\,s^{-1}$ to $100{,}000\,s^{-1}$ and even increased at a higher strain rate rather than decreased, while asphalt 1 also appeared in this trend. The common feature of asphalt 1 and asphaltenes 1 and 2 was their high consistency aromatic ring content, which indicates that fused aromatic rings had a significant effect on the asphalt shear thinning behavior.

Mean Pressure on xy-Plane for Basic Asphaltic Models

The pressure on the xy-plane was used to calculate the viscosity because the deformation was applied on the xy-plane of the model. Figure 4 shows the mean pressure on the xy-plane at different strain rates. In general, the mean pressure was very similar for asphalt 1 and asphaltenes 1 and 2. After a plateau at low and intermediate strain rates, the pressure then increased sharply at high strain rates. The sharp increase may explain the reason for the failure of the NEMD method to calculate the viscosities of asphalt 1 and asphaltenes 1 and 2 at high strain rates. For asphalt 2, resin and oil, at low strain rates, there was a plateau in the mean pressure profile. Then the pressures began to increase slowly at intermediate strain rates.

Energy Distribution of Basic Asphaltic Models

The stress tensor can be decomposed into two parts, dynamics and potential parts from a microscopic perspective. The potential energy is defined by the sum of all interactions between all the particles in the system, which plays a leading role in determining the average pressure (*28*). Kinetic energy is determined by the velocity of all particles and can be used to determine whether the model is in a steady state since it is forced by the system temperature. In this study, the characteristics of the potential energy relative to the strain rate were studied and compared with the kinetic energy distribution.

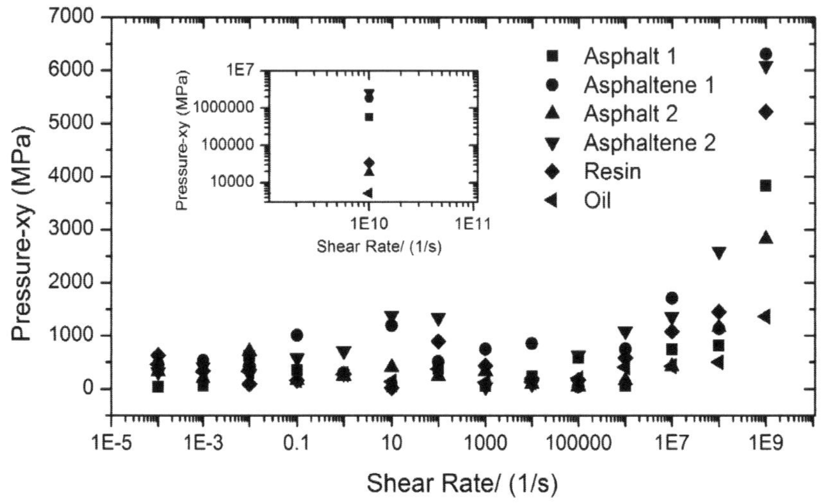

Figure 4. Comparison of mean pressure on xy-plane.

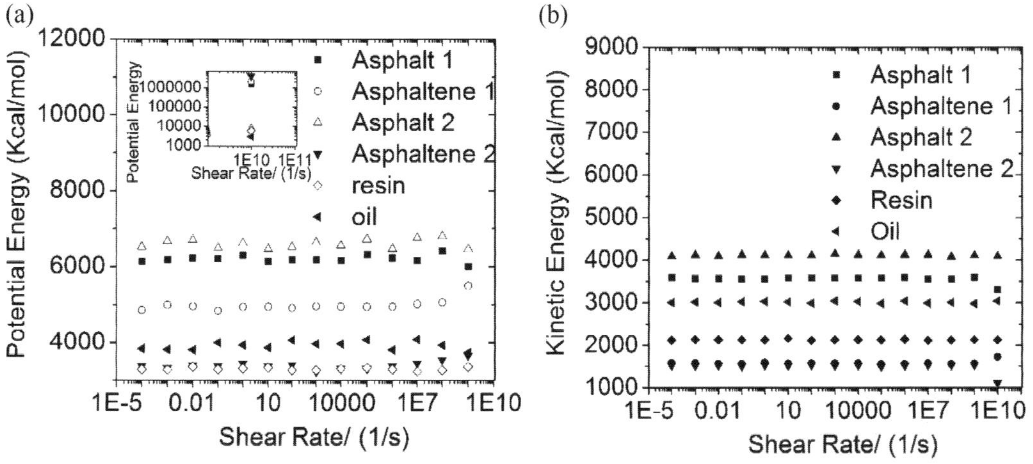

Figure 5. (*a*) Potential energy and (*b*) Kinetic energy vs. strain rate.

Figure 5*a* shows the potential energy at different strain rates. Except for oil, the overall trend of these energy values was very similar and showed a sharp increase at high strain rates, whereas the energy value of oil exhibited a sharp decline at very high strain rates. Compared with other models, the oil did not contain an aromatic ring. Thus, these phenomena can be explained by considering two different effects of the chemical structure of the aliphatic chain and aromatic rings. For the aliphatic chain, the torsion angle maintained more trans-conformation as the strain rate increased. This increased the area of interaction between adjacent molecules and reduced the potential energy. For aromatic rings, their twist angles were less than the twist angle of the aliphatic chain and did not have a trans-conformation at low strain rates. With the increase of strain rate, molecular collision (atomic interaction, promote molecular separation) played an increasingly important role, and ultimately dominated the expansion effect. Therefore, at a high strain rate, the potential energy increased dramatically.

Figure 5*b* shows the kinetic energy at different strain rates. The overall trend of these energy values remained flat as the strain rate increased, indicating that the model is in a steady state. However, the values of asphalt 1 and asphaltenes 1 and 2 showed an increase or decrease at very high strain rates. These phenomena can be explained by considering the energy variation factor at high strain rates (see Table 4). It can be seen that the kinetic energy

Table 4. Coefficient of Variation of Energy at High Strain Rates

Coefficient of variation	Asphalt I	Asphaltene I	Asphalt 2	Asphaltene 2	Resin	Oil
Potential energy (kcal/mole)	0.018	0.060	0.023	0.031	0.064	0.043
Kinetic energy (kcal/mole)	0.103	0.496	0.022	0.243	0.012	0.016

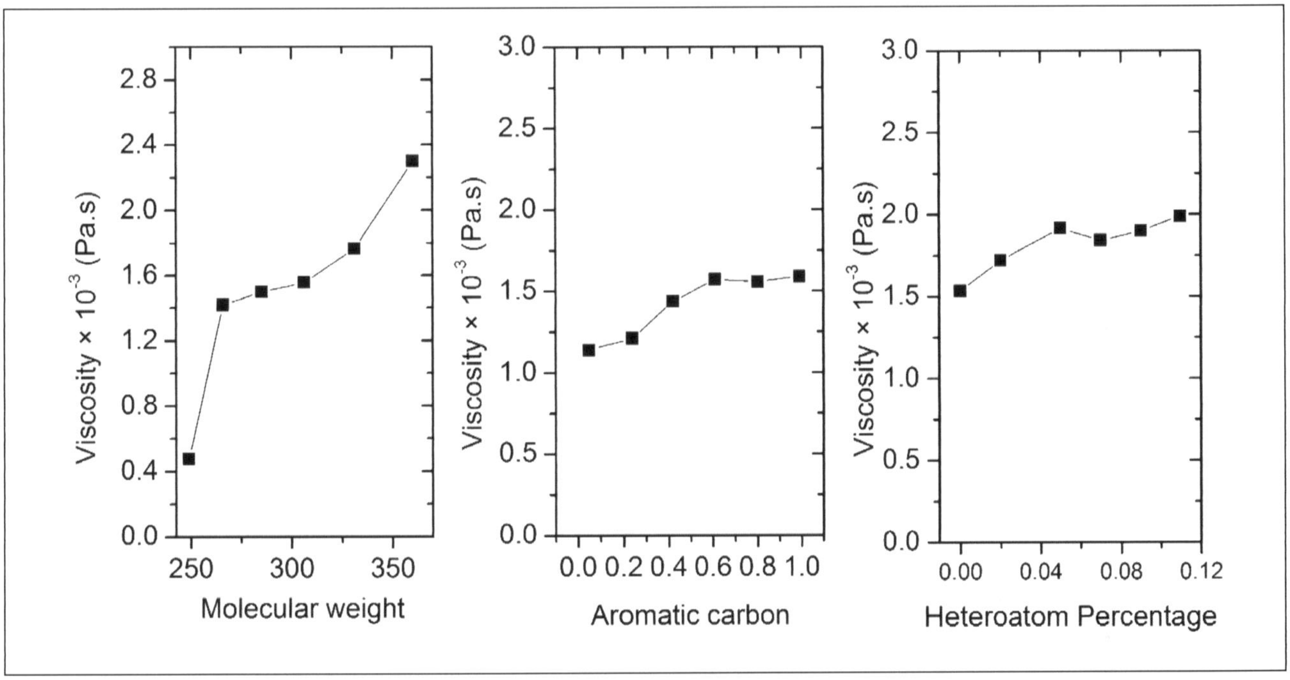

Figure 6. Viscosity of modified asphaltic models.

variation of these three models was much higher than that of other models, indicating that these three models were not as stable as other models at high strain rates. This may explain why the the NEMD method could not accurately predict the viscosity of these three models at high pollution rates.

Shear Viscosity Estimation for Modified Asphaltic Model

Figure 6 shows the calculated viscosities and mean pressure at the strain rate of 10^8 s^{-1} for modified asphaltic models. The viscosities increased with molecular weight, aromatic carbon percentage, and heteroatom percentage. The molecular weight appeared to have a stronger effect on viscosity than aromatic carbon percentage and heteroatom percentage. As shown in Figure 6, a nonlinear relationship was observed between the calculated viscosity and molecular weight, aromatic carbon percentage and heteroatom percentage, which can be attributed to the nonlinear relationship between viscosity and chemical structure. The shear viscosity of the modified asphalt model showed significant dependence on the chemical structure parameters, which indicates that the predicted viscosity was sensitive to model structures.

Conclusions

In this study, NEMD simulations were applied to two basic asphalt models and 18 modified models to investigate the effect of chemical structure on the viscosity of asphalt. Moreover, the energy variations with strain rate were analyzed to determine the behaviors of asphaltic molecules under shearing effect. Based on the results, the following conclusion can be drawn:

1) The viscosity of the asphalt model calculated by the NEMD method was close to the experimental results. For the asphalt model and its composition, the shear thinning behavior with increased strain rate was observed. In addition, the large fused aromatic ring core caused the nonlinear relationship between viscosity and strain rate.

2) The results of pressure and energy showed that the aromatic ring content affected the behavior of molecules in shear flow, which leads to the behavior of the asphalt 1 model being more similar to the asphaltene model.
3) The energy changed sharply at the high strain rate, indicating that the NEMD method had a strain rate limit. The kinetic energy of models with high aromatic carbon content showed higher variability than other models, indicating that these models were not as stable as other models at high strain rates.
4) The results of the modified model show that their viscosity increased with increasing molecular weight, aromatic carbon percentage and heteroatom percentage. Moreover, the molecular weight had a greater effect on viscosity than aromatic carbon percentage and heteroatom percentage.

Although the method proposed in this study was suitable for the research of base asphalt, the promotion to modified asphalt needs to be further investigated. First, the modifier such as SBS can significantly affect the intermolecular structure of asphalt (8), so the asphalt model should be adjusted based on the intermolecular interaction between SBS and asphalt molecules. Second, because the molecular weight of SBS (>100,000) is beyond the scale range of the conventional molecular dynamic method, other approaches such as a course-grained model can be introduced into the model development of modified asphalt.

Acknowledgments

This study was financially supported by the National Natural Science Foundation of China (No. 51578415).

Author Contributions

The authors confirm contribution to the paper as follows: study conception and design: Baoshan Huang; data collection: Yongjie Ding; analysis and interpretation of results: Yongjie Ding; draft manuscript preparation: Yongjie Ding, Xiang Shu. All authors reviewed the results and approved the final version of the manuscript.

References

1. Altgelt, K., and O. Harle. The Effect of Asphaltenes on Asphalt Viscosity. *Industrial & Engineering Chemistry Product Research and Development*, Vol. 14, No. 4, 1975, pp. 240–246.
2. Poel, C. D. V. A General System Describing the Visco-Elastic Properties of Bitumens and Its Relation to Routine Test Data. *Journal of Chemical Technology and Biotechnology*, Vol. 4, No. 5, 1954, pp. 221–236.
3. Williams, M. L., R. F. Landel, and J. D. Ferry. The Temperature Dependence of Relaxation Mechanisms in Amorphous Polymers and Other Glass-Forming Liquids. *Journal of the American Chemical Society*, Vol. 77, No. 14, 1955, pp. 3701–3707.
4. Lin, M.-S., J. M. Chaffin, M. Liu, C. Glover, R. Davison, and J. Bullin. The Effect of Asphalt Composition on the Formation of Asphaltenes and their Contribution to Asphalt Viscosity. *Fuel Science and Technology International*, Vol. 14, No. 1–2, 1996, pp. 139–162.
5. Pauli, A. T., and J. F. Branthaver. Relationships between Asphaltenes, Heithaus Compatibility Parameters, and Asphalt Viscosity. *Petroleum Science and Technology*, Vol. 16, No. 9–10, 1998, pp. 1125–1147.
6. Allen, M. P., and D. J. Tildesley. *Computer Simulation of Liquids*. Oxford University Press, 1989.
7. Sun, D., G. Sun, X. Zhu, F. Ye, and J. Xu. Intrinsic Temperature Sensitive Self-Healing Character of Asphalt Binders Based on Molecular Dynamics Simulations. *Fuel*, Vol. 211, 2018, pp. 609–620.
8. Zhang, L., and M. L. Greenfield. Relaxation Time, Diffusion, and Viscosity Analysis of Model Asphalt Systems Using Molecular Simulation. *The Journal of Chemical Physics*, Vol. 127, No. 19, 2007, p. 194502.
9. Ding, Y., B. Huang, X. Shu, Y. Zhang, and M. E. Woods. Use of Molecular Dynamics to Investigate Diffusion between Virgin and Aged Asphalt Binders. *Fuel*, Vol. 174, 2016, pp. 267–273.
10. Ding, Y., B. Tang, Y. Zhang, J. Wei, and X. Cao. Molecular Dynamics Simulation to Investigate the Influence of SBS on Molecular Agglomeration Behavior of Asphalt. *Journal of Materials in Civil Engineering*, Vol. 27, No. 8, 2013, p. C4014004.
11. Pauli, A. T. *Chemomechanics of Damage Accumulation and Damage-Recovery Healing in Bituminous Asphalt Binders*. Delft University of Technology, 2014.
12. Hess, S. Rheological Properties Via Nonequilibrium Molecular Dynamics: From Simple Towards Polymeric Liquids. *Journal of Non-Newtonian Fluid Mechanics*, Vol. 23, 1987, pp. 305–319.
13. Morriss, G. P., and D. J. Evans. *Statistical Mechanics of Nonequilbrium Liquids*. ANU Press, Canberra, 2007.
14. Kröger, M. *Models for Polymeric and Anisotropic Liquids*. Springer Science & Business Media, Berlin, 2005.
15. Cui, S., S. Gupta, P. Cummings, and H. Cochran. Molecular Dynamics Simulations of the Rheology of Normal Decane, Hexadecane, and Tetracosane. *The Journal of Chemical Physics*, Vol. 105, No. 3, 1996, pp. 1214–1220.
16. Zhang, L., and M. L. Greenfield. Analyzing Properties of Model Asphalts Using Molecular Simulation. *Energy & Fuels*, Vol. 21, No. 3, 2007, pp. 1712–1716.
17. Storm, D., J. Edwards, S. DeCanio, and E. Sheu. Molecular Representations of Ratawi and Alaska North Slope Asphaltenes Based on Liquid-and Solid-State NMR. *Energy & Fuels*, Vol. 8, No. 3, 1994, pp. 561–566.
18. Kowalewski, I., M. Vandenbroucke, A. Huc, M. Taylor, and J. Faulon. Preliminary Results on Molecular Modeling of Asphaltenes Using Structure Elucidation Programs in

Conjunction with Molecular Simulation Programs. *Energy & Fuels*, Vol. 10, No. 1, 1996, pp. 97–107.
19. Artok, L., Y. Su, Y. Hirose, M. Hosokawa, S. Murata, and M. Nomura. Structure and Reactivity of Petroleum-Derived Asphaltene. *Energy & Fuels*, Vol. 13, No. 2, 1999, pp. 287–296.
20. Groenzin, H., and O. C. Mullins. Molecular Size and Structure of Asphaltenes from Various Sources. *Energy & Fuels*, Vol. 14, No. 3, 2000, pp. 677–684.
21. Kennedy, T. W., R. J. Cominsky, E. T. Harrigan, and R. B. Leahy. *Hypotheses and Models Employed in the SHRP Asphalt Research Program.* Report SHRP-A/WP-90-008. Strategic Highway Research Program, National Research Council, Washington, D.C., 1990.
22. Plimpton, S., P. Crozier, and A. Thompson. LAMMPS-Large-Scale Atomic/Molecular Massively Parallel Simulator. *Sandia National Laboratories*, Vol. 18, 2007, p. 43.
23. Evans, D. J., and G. Morriss. Nonlinear-Response Theory for Steady Planar Couette Flow. *Physical Review A*, Vol. 30, No. 3, 1984, p. 1528.
24. Kim, J., D. J. Keffer, M. Kröger, and B. Edwards. Rheological and Entanglement Characteristics of Linear-Chain Polyethylene Liquids in Planar Couette and Planar Elongational Flows. *Journal of Non-Newtonian Fluid Mechanics*, Vol. 152, No. 1, 2008, pp. 168–183.
25. Tenney, C. M., and E. J. Maginn. Limitations and Recommendations for the Calculation of Shear Viscosity Using Reverse Nonequilibrium Molecular Dynamics. *The Journal of Chemical Physics*, Vol. 132, No. 1, 2010, p. 014103.
26. Evans, D. J., and B. L. Holian. The Nose–Hoover Thermostat. *The Journal of Chemical Physics*, Vol. 83, No. 8, 1985, pp. 4069–4074.
27. Lifson, S., A. Hagler, and P. Dauber. Consistent Force Field Studies of Intermolecular Forces in Hydrogen-Bonded Crystals. 1. Carboxylic Acids, Amides, and the C: O. cntdot.. cntdot.. cntdot. H-Hydrogen Bonds. *Journal of the American Chemical Society*, Vol. 101, No. 18, 1979, pp. 5111–5121.
28. Kröger, M., W. Loose, and S. Hess. Rheology and Structural Changes of Polymer Melts Via Nonequilibrium Molecular Dynamics. *Journal of Rheology (1978-present)*, Vol. 37, No. 6, 1993, pp. 1057–1079.

The Standing Committee on Asphalt Binders (AFK20) peer-reviewed this paper (18-06028).

Article

TRR
JOURNAL OF THE TRANSPORTATION RESEARCH BOARD

Transportation Research Record
2018, Vol. 2672(28) 244–255

DOI: 10.1177/0361198118777630
journals.sagepub.com/home/trr
SAGE

Analysis of Asphalt Oxidation by Means of Accelerated Testing and Environmental Conditions

Rafael E. Villegas-Villegas[1], Alejandra Baldi-Sevilla[1], José Pablo Aguiar-Moya[1], and Luis Loria-Salazar[1]

Abstract

The chemical changes that occur to asphalt with oxidation include an increase in oxygen content, as well as the unsaturation of the molecules. This change raises the polarity and stiffness of the material. As a consequence, the elastic response increases, altering its rheology. These transformations are essential to determine pavement performance during its service life. However, the characterization of these chemical and rheological processes cannot be completed because of the inefficiency of current procedures that simulate asphalt oxidation. For this reason, the objective of this study is to characterize asphalt oxidation fundamentally and to relate the observed changes in the materials' mechanical response. To achieve this goal, a representative group of asphalt samples has been exposed to environmental oxidation, and alternatively to thermal and ultraviolet aging in the laboratory. The samples were characterized chemically and rheologically before the start of the experiment. It was possible to find a correlation between the content of specific chemical species in the material and their mechanical behavior at low and intermediate temperatures. In addition, the present study helps to understand the oxidation phenomena, and helps verify the ineffectiveness of traditional aging techniques, so that they can be modified to simulate the environmental process better.

Asphalt oxidation occurs in two stages: (1) during the extraction, production of the asphalt mixture, and placement of the upper layer of the pavement, and (2) during the service life of the pavement (*1*). The reactions that occur during these stages are dependent on the chemical composition of the asphalt (*2*, *3*). However, increases in the amount of oxygenated functional groups, double bonds, and aromaticity are expected (*4–7*).

Specifically, oxidation promotes the formation of aromatic structures and unsaturations (carbon-carbon double bonds), which increase the rigidity of the asphalt. Such chemical changes can be monitored by Fourier transform infrared spectroscopy (FTIR) (*8–10*). Also, after oxidation, there is an increment in polarity of molecules because of the presence of oxygen and double bonds in the molecules of the material, which in turn increase the number and magnitude of intermolecular interactions. As a consequence, the elastic component of the asphalt is expected to become dominant as the oxidation in the binder advances: this is usually accompanied by a reduction in the viscous component of the material. Therefore, the correct interpretation of the contribution of the elastic and viscous components to the asphalt rheology is useful in the characterization and prediction of its performance. For example, a high contribution of the elastic component could represent an increase in the susceptibility of the binder to the fracture, which in turn is related to the energy dissipation capacity of the material: a rigid (oxidized) asphalt is expected to be less efficient in energy dissipation by flow, since its molecules will have reduced mobility, compared with a less rigid (non-oxidized) asphalt. In other words, an oxidized asphalt dissipates energy through the fracture.

Asphalt oxidation is characterized by the oxygenation of molecules, as well as by the increase in the number of double bonds and aromaticity (*11*). These transformations alter the rheological and mechanical responses of the material by increasing its fragility and susceptibility to fracture (*12*). In the laboratory, asphalt oxidation is

[1]National Laboratory of Materials and Structural Models (LanammeUCR), University of Costa Rica, San José, Costa Rica

Corresponding Author:
Address correspondence to Alejandra Baldi-Sevilla: alejandra.baldi@ucr.ac.cr

simulated with thermal treatments performed in specially designed ovens, such as the rolling thin-film oven (RTFO) and the pressure aging vessel (PAV). In these procedures, asphalt is subjected to high temperatures, continuous airflow, and pressure (*13*). For example, after being exposed to the PAV, the asphalt sample is expected to oxidize to the equivalent of 7–10 years of in-field service (*14*). However, several studies suggest that thermal treatments do not achieve complete oxidation of asphalt as they ignore the contribution of ultraviolet (UV) radiation (photo-oxidation) (*15*). Consequently, photo-oxidation has been proposed to complement and improve this type of simulation. An example of such a procedure consists in subjecting the asphalt that has been previously oxidized by the conventional thermal method, to UV radiation. Although this additional treatment has proved to be a more accurate approximation of oxidation, it is still an incomplete simulation because several environmental factors that could contribute to oxidation are not considered, such as water, impurities in the air, and radiation of the entire electromagnetic spectrum. Consequently, the present study intends to analyze the chemical changes that occur to asphalt under environmental conditions and to relate them to the mechanical consequences of oxidation in the material. In addition, based on the results, the efficiency of the accelerated aging treatments (RTFO, PAV) commonly applied in the laboratory will be analyzed (*16*).

Objective

The objective of this study is to analyze the chemical and rheological transformations that occur in the asphalt binder after environmental oxidation. In addition, and based on the previous, the efficiency of accelerated aging procedures commonly applied in the laboratory (such as PAV) in simulating oxidation of asphalt binder were analyzed.

Methodology

Materials

The materials analyzed in this study are neat binders denoted as Costa Rica 1 (PG 64-22), Costa Rica 2 (PG 70-25), Guatemala 1 (PG 64-22), Guatemala 2 (PG 64-22), Nicaragua (PG 64-25), El Salvador (PG 64-22), and Bolivia (PG 64-22), according to their origin country.

Oxidation of Samples

The asphalts were subjected to thermal aging in RTFO and subsequently oxidized in the PAV, based on Superpave specifications. The Costa Rica 2 asphalt was also subjected to three and five consecutive PAV cycles, to induce a more severe oxidation in the material.

In the case of environmental oxidation, a fixed amount of Costa Rica 2 asphalt was poured into non-stick molds to produce a 1 mm-thick film, which has been chosen to simulate the film thickness of binder that coats the aggregate in an asphalt mixture. The films were exposed to the weather for two years in order to subject the binder to environmental oxidation (Figure 1). The samples were kept to 1200 MASL (meters above the sea level), and the temperature ranged between 20 and 27°C during such period.

Samples were removed from these conditions and analyzed monthly. An asphalt sample was maintained as a control. It is necessary to clarify that RTFO-treated asphalt was used in the preparation of the films, to simulate the short-term aging that occurs before the binder is placed on the pavement layer.

Asphalt Characterization

Chemical changes induced by oxidation were determined by quantifying the particular signal area of the infrared spectrum of the samples. In the case of the environmental-oxidized films, infrared spectra were performed on the exposed and non-exposed surfaces of each film. The quantified signals were the following: oxygenated compounds (3500 cm^{-1}, hydroxyl, 1700 cm^{-1}, carbonyl and 1030 cm^{-1}, sulfoxide), aromatic and unsaturated compounds (containing C=C) (3050 cm^{-1} and 1600 cm^{-1} respectively), and saturated compounds (containing C–H) (2800, 2900 cm^{-1}).

The rheological analysis consisted of two stages. First was a temperature sweep (5–64°C) at controlled strain (1%) and a frequency of 10 rad/s, which allows finding the region where the asphalt becomes susceptible to fatigue. For the second stage, after finding the fatigue-susceptible region, the behavior of the samples was evaluated under controlled stress (1 MPa) at a frequency of 10 rad/s and 10°C, to ensure that the asphalt will show fatigue susceptibility. In both tests, the 8 mm geometry was used for all samples.

Results and Discussion

Chemical Transformations in Asphalt during Oxidation

The signal area of a chemical species in an infrared spectrum is directly related to the concentration of such species (*17*). Therefore, this technique was used to quantify the extent of oxidation in the proposed asphalt samples. The results are listed in Table 1.

Environmental Oxidation. Based on Table 1, the exposed surface of the asphalt films was found to be significantly more oxidized than the RTFO sample (corresponding to the initial condition of the films). The data show an

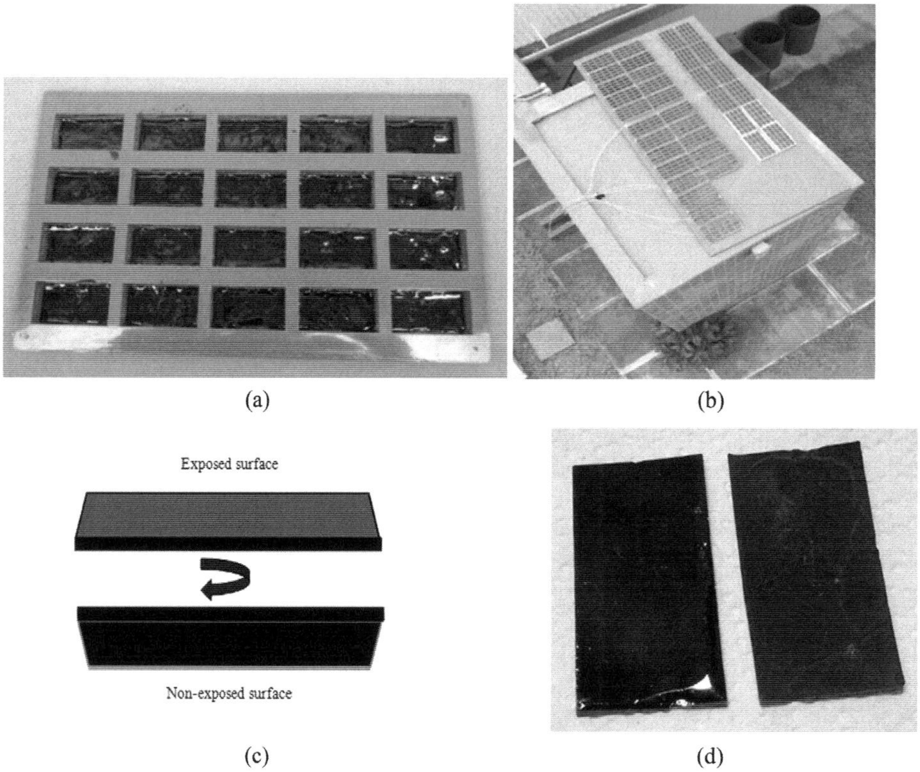

Figure 1. *(a)* Asphalt samples placed in non-stick molds, *(b)* location of samples for environmental oxidation, *(c)* diagram of exposed/non-exposed surface of asphalt films, and *(d)* asphalt film after six months of environmental oxidation (non-exposed [left] and exposed [right] surfaces).

increase in oxygenated functional groups, such as hydroxyl, carbonyl, and sulfoxide. In addition, an increase in carbon-carbon double bonds and aromatization was observed, which in turn comes with the consequent decrease of aliphatic carbon-hydrogen bonds. These changes may increase the rigidity of the material, as double bonds and aromatic molecules are structures with less flexibility than those composed of carbon-carbon single bonds. On the other hand, the non-exposed surface shows an increase in oxygenated compounds mainly, which points out that oxygen penetrates the film by reacting with the asphalt along the film and producing such compounds, perceptible even on the unexposed surface. In contrast, UV radiation, responsible for the formation of unsaturations in the material, does not appear to penetrate the film sufficiently as no significant increase in the signal of the C–C double bond or the signal corresponding to aromatic compounds is observed.

Comparison of Thermal and Environmental Oxidative Aging. Comparing the oxidation of samples exposed to the weather and those treated in PAV, a significant difference was observed in the hydroxyl, carbonyl, sulfoxide, aromaticity, and carbon-carbon double bonds. The previous is accompanied by a corresponding reduction in carbon-hydrogen bonds, being the environmental-oxidized samples the most affected. This finding suggests that the surface of asphalt exposed to environmental conditions during a short period undergoes a more severe oxidation than the asphalt subjected to thermal + pressure treatment, which intends to simulate up to 10 years of environmental conditions. On the other hand, the unexposed surface of the asphalt films experienced a slight increase in aromaticity and unsaturations. However, no significant increase in the number of oxygen compounds was observed. The data from Table 1 indicate that the thermal treatment (PAV) is not able to completely simulate asphalt oxidation: it is observed that PAV promotes the formation of oxygenated compounds but is insufficient in simulating the production of unsaturations and aromatic compounds in the material. The fact that the PAV does not simulate the environmental oxidation of binders has been highlighted in prior studies (*18*). To justify such observation, it is necessary to quantify the signal areas of the asphalt in original condition (non-oxidized) and compare them with the signal areas

Table 1. Area of the Infrared Signals Related to Asphalt Oxidation

Asphalt Costa Rica 2	Wavelength (cm^{-1}) 3500[a]	3050	2900	2800	1700	1600	1030
Original	0.19 ± 0.07[b]	0.029 ± 0.002	8.37 ± 0.71	2.76 ± 0.19	0.010 ± 0.006	0.76 ± 0.035	0.107 ± 0.038
RTFO	0.16 ± 0.04	0.028 ± 0.003	8.61 ± 0.15	2.83 ± 0.04	0.019 ± 0.005	0.75 ± 0.01	0.194 ± 0.004
PAV (1 cycle)	0.38 ± 0.09	0.028 ± 0.002	8.36 ± 0.21	2.73 ± 0.07	0.129 ± 0.005	0.66 ± 0.07	0.672 ± 0.040
PAV (3 cycles)	0.54 ± 0.15	0.027 ± 0.003	8.07 ± 0.43	2.64 ± 0.14	0.323 ± 0.060	0.66 ± 0.03	1.025 ± 0.065
PAV (5 cycles)	0.39 ± 0.09	0.025 ± 0.004	7.10 ± 1.00	2.33 ± 0.34	0.399 ± 0.054	0.64 ± 0.08	0.920 ± 0.140
Outdoors 6 months (exposed surface)	5.25 ± 0.83	0.043 ± 0.008	4.10 ± 1.10	1.31 ± 0.37	2.960 ± 0.450	1.52 ± 0.29	1.388 ± 0.204
Outdoors 6 months (non-exposed surface)	0.33 ± 0.04	0.033 ± 0.002	8.60 ± 0.20	2.83 ± 0.05	0.073 ± 0.011	0.80 ± 0.02	0.543 ± 0.022
Outdoors 12 months (exposed surface)	4.89 ± 1.16	0.027 ± 0.004	3.89 ± 0.74	1.23 ± 0.25	1.006 ± 0.289	2.10 ± 0.42	1.446 ± 0.217
Outdoors 12 months (non-exposed surface)	0.68 ± 0.21	0.033 ± 0.005	8.46 ± 0.15	2.79 ± 0.06	0.123 ± 0.022	0.97 ± 0.05	0.879 ± 0.057
Outdoors 24 months (exposed surface)	2.52 ± 0.24	0.017 ± 0.002	3.42 ± 0.85	1.12 ± 0.28	0.286 ± 0.067	1.36 ± 0.10	1.910 ± 0.245
Outdoors 24 months (non-exposed surface)	0.70 ± 0.22	0.038 ± 0.004	8.10 ± 0.30	2.66 ± 0.10	0.148 ± 0.047	1.02 ± 0.08	1.198 ± 0.245

[a]Average of seven replicas.
[b]Confidence interval 95%.

of the thermally treated asphalt. The results show that the only significant difference is the increase in oxygen compounds after PAV treatment. The differences can be attributed to the fact that the severe conditions of air pressure and high temperature to which the asphalt is subjected to during the PAV force the oxygenation reactions. Furthermore, it should be emphasized that the oxidation procedure in PAV does not consider the photo-oxidation of the asphalt, which promotes the formation of double bonds and aromatic compounds.

In addition, the asphalt was subjected to three and five consecutive aging cycles in the PAV. The areas of the signals of the spectra corresponding to these samples were quantified, and the results were quantitatively compared with those of the environmental-oxidized samples. It was found that three and five consecutive PAV cycles are not sufficient to produce the amount of carbonyls and sulfoxides formed after exposure of the asphalt to environmental conditions. This suggests that not even five consecutive PAV cycles correctly simulate the oxidation experienced by asphalt film outdoors. These observations reinforce the idea that heat treatments oxidize the binder only through the incorporation of oxygen into its molecules. In this way, subjecting the material to this type of treatment before mechanical evaluation could compromise the correct characterization of the asphalt. As a result, information regarding the performance of the binder throughout its service life might be lost. The following sections will address this issue.

Rheological Consequences of Oxidative Aging

The viscoelastic behavior of polymeric materials is governed by the variation of the magnitude of viscous and elastic components with temperature. Figure 2*a* illustrates this behavior, in which four regions are distinguished: glassy region, transition region, rubbery region, and terminal region. The movement and final arrangement of the molecules define the mechanical characteristics of the polymer in the different regions of the viscoelastic behavior. In the glassy region, molecular motion is reduced: there are only vibrational or short translational movements. In this region, the deformations are associated with elongation and shortening of the bonding distance and bonding angles. The transition region is, as the name implies, a transition from the vitreous region of restricted movement to the rubbery region, where more complex configurations occur. In this region, there are translational and configurational movements. The increase in temperature promotes the mobility and flexibility of the molecules, and therefore, they begin to behave as a fluid. In the rubbery region, the molecular chains begin to move in a more rapid and coordinated way. This rapid mobility causes deformation

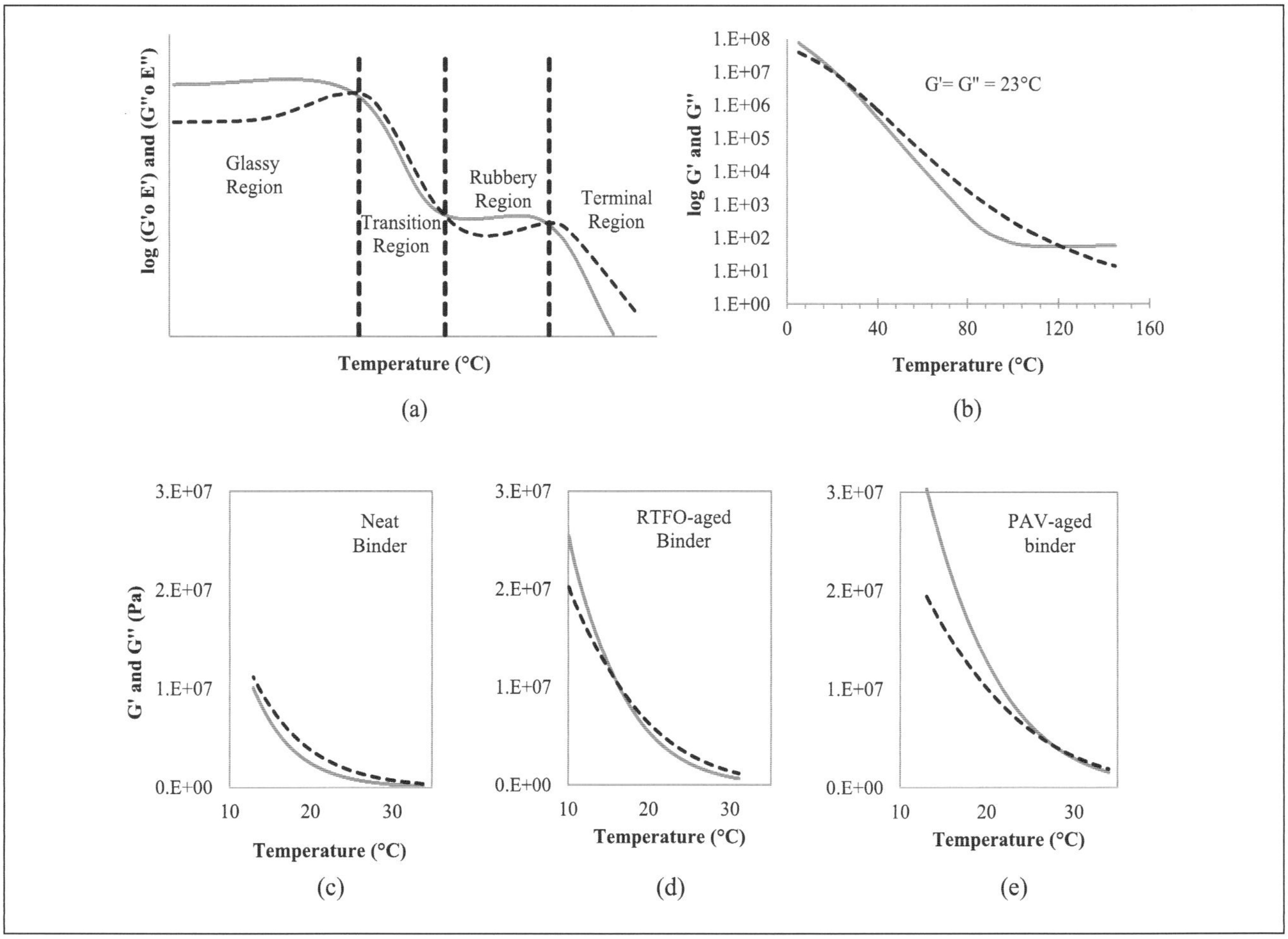

Figure 2. Variation of elastic component, G′ (continuous line) and viscous component G″ (dotted line) with temperature in (*a*) polymeric materials, (*b*) environmental-oxidized asphalt, (*c*) neat binder, (*d*) RTFO-aged binder, and (*e*) PAV-oxidized binder.

and relaxation to occur almost instantaneously. In this region, the material modulus varies little with temperature. Finally, in the terminal region, the molecules have a greater freedom of movement and the secondary bonds break and re-establish. At this point, unrecoverable deformations occur as a result of long-range configurational changes (*19–21*).

To assess the mechanical behavior of asphalt binder, a sweep of the elastic and viscous components was performed at different temperatures (5–145°C) for the Costa Rica 2 asphalt exposed to environmental aging for six months. The sweep was performed at a frequency of 10 rad/s and a strain of 1%. The results are shown in Figure 2*b*.

According to Figure 2*b*, it is possible to observe in the asphalt a behavior similar to a polymeric material. However, it is important to note that because of the heterogeneous composition of the asphalt, the regions shown in Figure 2*a* could overlap or elongate in the case of the binder. Therefore, in asphalt, several phase transitions can be found, such as melting and glass transitions, covering a wide temperature range. In other words, these transitions do not occur at specific points in the asphalt, as is the case in homogeneous materials.

The importance of this analysis is that the location of the intersection points between G′ and G″ (23°C in Figure 2*b*) determines the behavior of the asphalt at a given temperature. For example, the first point where G′ = G″ in Figure 2*b* indicates the boundary between the glassy and transition regions on the asphalt. Consequently, it is expected that below this temperature, the material behaves rigidly: the material becomes more elastic and increases the probability of fatigue cracking. In this sense, and considering the polymeric behavior of asphalt (*16*), it is expected that the temperature at which G′ = G″ can be an important indicator of oxidation in the asphalt: an oxidized binder exhibits a G′ = G″ point at a higher temperature, as compared with a non-oxidized one.

Table 2. Crossover Temperature of Asphalt Samples at Different Oxidation Levels

	T (°C) (G″ > G′)			
Asphalts	Original	RTFO	PAV	Outdoors
Guatemala 1	<10°C	12.1	22.5	–
Guatemala 2	<10°C	13.1	25.1	–
Nicaragua	<10°C	13.2	22.5	–
Bolivia	<10°C	17.2	27.9	–
Costa Rica 1	<10°C	12.5	25.5	–
Costa Rica 2	<10°C	16.4	27.7	–
Costa Rica 2 (3 PAV cycles)	–	–	45.1	–
Costa Rica 2 (5 PAV cycles)	–	–	63.4	–
Costa Rica 2 (6 months outdoors)	–	–	–	23.4
Costa Rica 2 (1 year outdoors)	–	–	–	30.3
Costa Rica 2 (2 years outdoors)	–	–	–	33.8

The previous observation is illustrated after monitoring the elastic and viscous components for the Costa Rica 2 asphalt samples at different levels of oxidation. In the analysis, a temperature sweep from 5 to 34°C was used. In the case of non-oxidized asphalt (Figure 2*c*), only the transition region was identified, indicating that the rheology of such material is governed by the viscous component (G″) at these temperature ranges. However, as aging progresses, the elastic component (G′) becomes more important: it was found that the crossover temperature in the RTFO-aged asphalt is significantly lower than for the PAV-treated asphalt (Figure 2, *d* and *e*, respectively). Therefore, for the laboratory-treated samples the elastic component governs the mechanical response of the asphalt material aged in RTFO below 16.4°C, and for the asphalt aged in PAV, below 27.5°C. This implies that the binder is more susceptible to fracture at temperatures at which the elastic component dominates. In these regions the molecular motion is reduced, a condition that becomes more important in oxidized samples.

Asphalt samples from different countries in Latin America were used to extend the analysis of the Costa Rica 2 asphalt and corroborate the findings from Figure 2, *c* through *d*. The analyzed samples consisted of neat, RTFO, and PAV-aged binders, in order to relate the location of the crossover point to the oxidation level in the samples. Furthermore, by comparison, the Costa Rica 2 sample was analyzed after being exposed to 6, 12 and 24 months of environmental oxidation. The results are shown in Table 2.

Based on Table 2, the crossover temperature of the viscous and elastic components is below 10°C for the non-aged asphalt samples. On the other hand, RTFO-aged samples show crossover temperatures higher than 10°C. It is worth noting that at this level of thermal aging an increase in the amount of sulfoxides in the material and the formation of the carbonyl in a smaller amount is expected. This trend can be observed in Table 3, which illustrates the content of these chemical species for the asphalts under different aging and oxidation conditions.

Table 3. Carbonyl (1700 cm^{-1}) and Sulfoxides (1030 cm^{-1}) Signal Areas for Selected Asphalts

	Wave length (cm^{-1})	
Asphalt	1700	1030
Costa Rica 2 original	0.010	0.107
Costa Rica 2 RTFO	0.019	0.194
Costa Rica 2 PAV	0.129	0.672
Nicaragua original	0.020	0.173
Nicaragua RTFO	0.048	0.201
Nicaragua PAV	0.133	0.760
Bolivia original	0.001	0.161
Bolivia RTFO	0.018	0.192
Bolivia PAV	0.153	0.515

When the samples are subjected to thermal oxidation in the PAV, the crossover temperatures exceed 19°C, which is accompanied by a significant increase in the number of carbonyls and sulfoxides, as expected for this type of oxidized sample. It must also be noted that for highly oxidized binders (five consecutive PAV cycles) the crossover point occurs at a temperature at which deformation is expected. This is consistent with the data in Table 3.

It is important to note that the crossover temperature of asphalt placed outdoors for one year period exceeds the value corresponding to PAV-oxidized asphalts, which is intended to simulate the oxidation of asphalt after ten years of service. The previous can be explained by the chemical analysis summarized in Table 1, where it was observed that the oxidation of the asphalt in the PAV is incomplete compared with the environmental oxidation. Photo-oxidation of asphalt produces rigid structures with greater polarity, which could generate an increase in rigidity of the material as the level of environmental

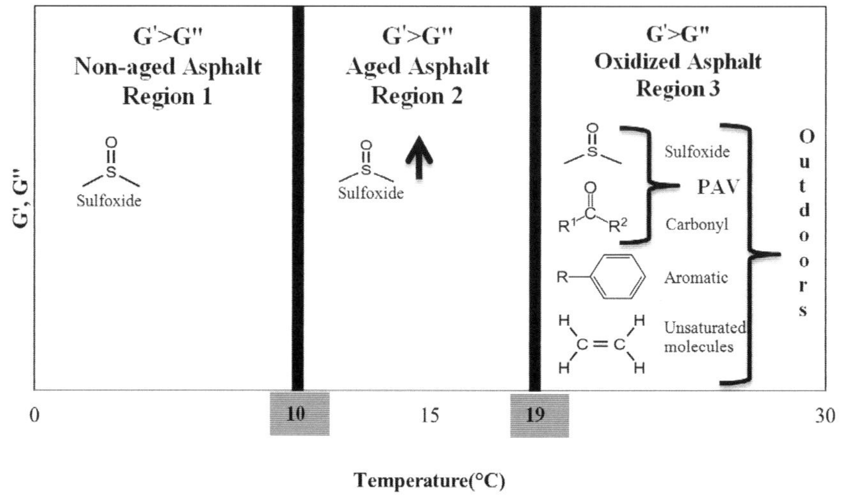

Figure 3. Rheological evaluation of the level of aging and oxidation of asphalt and its relationship with the chemical composition. Here "aged" refers to short aged binder (RTFO) and "oxidized" refers to a more severe oxidized binder (i.e., PAV).

oxidation advances. This would explain the significant increase in crossover temperature of environmental-oxidized films compared with thermal aging. The former suggests that aging in PAV is a less severe method in simulating the oxidative conditioning of the asphalts, as photo-oxidative reactions are not accounted for: valuable information is lost regarding the chemical composition of the binder and its final mechanical response.

Relationship between FTIR and Dynamic Shear Rheometer (DSR) in Oxidized and Aged Asphalts

Figure 3 summarizes the findings from Tables 2 and 3 and shows the relationship between the rheology and the chemical composition of asphalts.

Based on Figure 3 and data in Table 2, asphalts in non-aged condition were found to exhibit a crossover temperature ($G' = G''$) below 10°C (Region 1). At temperatures below 10°C, the elastic component (G') dominates over the viscous component (G''). A material with lower crossover temperature is expected to have a longer service life at intermediate temperatures. On the other hand, asphalts having a crossover temperature between 10 and 19°C generally have a higher sulfoxide content than a non-aged asphalt (Tables 2, 3). Therefore, it is expected that the asphalts with a crossover temperature within this range can be short-term aged. Consequently, materials within this range of crossover temperatures can oxidize at a higher rate, mainly in asphalt samples with crossover temperatures near the top of the range. Finally, if the crossover temperature is above 19°C, there is a high probability that the asphalt has been oxidized. This is because samples that exhibit this behavior show the presence of chemical species such as carbonyls and an important content in sulfoxides, as is the case of PAV-aged samples. For samples exposed to environmental conditions, an increase in aromatic and unsaturated molecules, accompanied by the increase in carbonyls and sulfoxides is observed.

The most important contribution of Figure 3 is that it relates the chemical composition of the asphalt binder to its mechanical response. It should be noted that rheological tests do not replace chemical analyses. However, in the specific case of the findings of Figure 3, a simple rheological analysis can facilitate the material selection.

It should be noted that, in addition to the previously analyzed asphalts, LanammeUCR has a database of almost 20 years of asphalts from different countries and, in general, the consistency of Figure 3 has been observed.

Stress-Controlled Fatigue Test

Historically, the most common method to evaluate fatigue of asphalt binders in the DSR is to apply a 10% strain at a 10 rad/s frequency, using an 8 mm geometry with a sample thickness of 2 mm. The temperature to be used is at the discretion of the user (ASTM D 2872). The most commonly used failure criterion is to consider that a 50% reduction of the initial modulus is indicative of fatigue damage. Figure 4 compares the results of the previous fatigue test under strain-controlled conditions (Figure 4, *a, b*) and under stress-controlled conditions (Figure 4, *c, d*) for Costa Rica 2 PAV-oxidized sample.

Based on Figure 4*a* (strain-controlled), it is emphasized that the behavior of the material be log-linear as

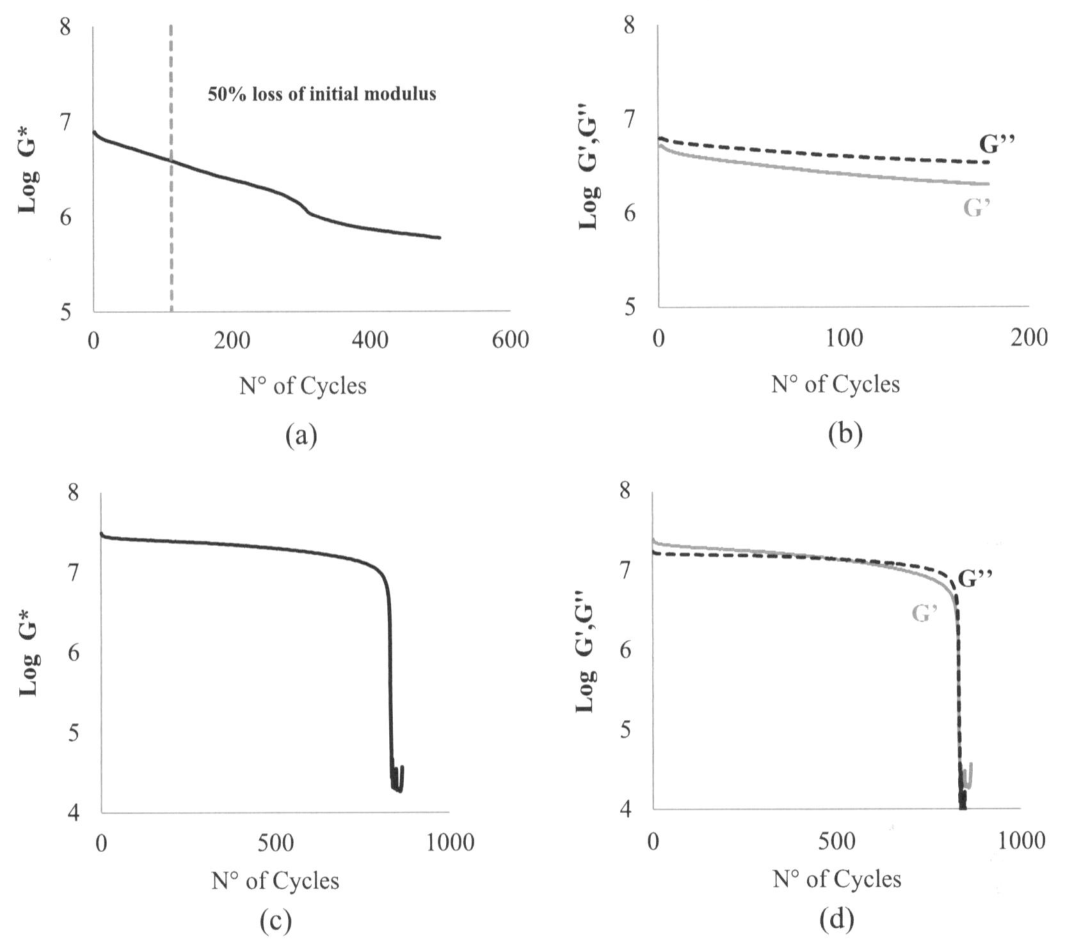

Figure 4. (*a*) Strain-controlled fatigue test (dotted line denotes a 50% loss in initial modulus), (*b*) behavior of elastic (gray) and viscous (black) components, (*c*) stress-controlled fatigue test, and (*d*) behavior of its elastic (gray) and viscous (black) components for PAV-oxidized Costa Rica 2 in DSR.

evidenced by its correlation coefficient. In the region where the material loses 50% of its modulus, there is still a high degree of linearity in its response: this shows that with each load application the modulus decreases linearly, suggesting that after each deformation cycle the material relaxes and enters flow. On the other hand, as shown in Figure 4*c* (stress-controlled), the material exhibits a small decrease in its modulus as a result of a deformation of its structure, but after approximately 800 load cycles a significant decrease in the modulus is observed, evidencing fatigue in the asphalt binder. Therefore, according to the results obtained, the fatigue phenomenon at intermediate temperatures is a process in which the material deforms slightly in the initial cycles of the test, and the presence of fatigue can be associated with an abrupt drop in the modulus after which it does not recover again.

As observed in the strain-controlled test (Figure 4*b*), the viscous component (G″) dominates over the elastic component (G′) from the start of the test, and this behavior is accentuated as the number of cycles increases. This phenomenon shows that the asphalt binder tends to be susceptible to deformation throughout the test. On the other hand, under stress-controlled conditions (Figure 4*c*), the response of the material is entirely different: the modulus drops abruptly, and the elastic and viscous components compete to dominate the behavior. Figure 4*d* shows that during the first load cycles the elastic component dominates the behavior of the asphalt. However, as the load cycles increase, there is an increasing influence of the viscous component and, therefore, the drop in modulus observed might be occurring as a result of damage in the sample, until it breaks.

The behavior of the phase angle was also monitored through the stress-controlled test (Figure 5*a*), and it was observed that at the beginning of the test the phase angle is below 45°, indicating that the elastic component (G′)

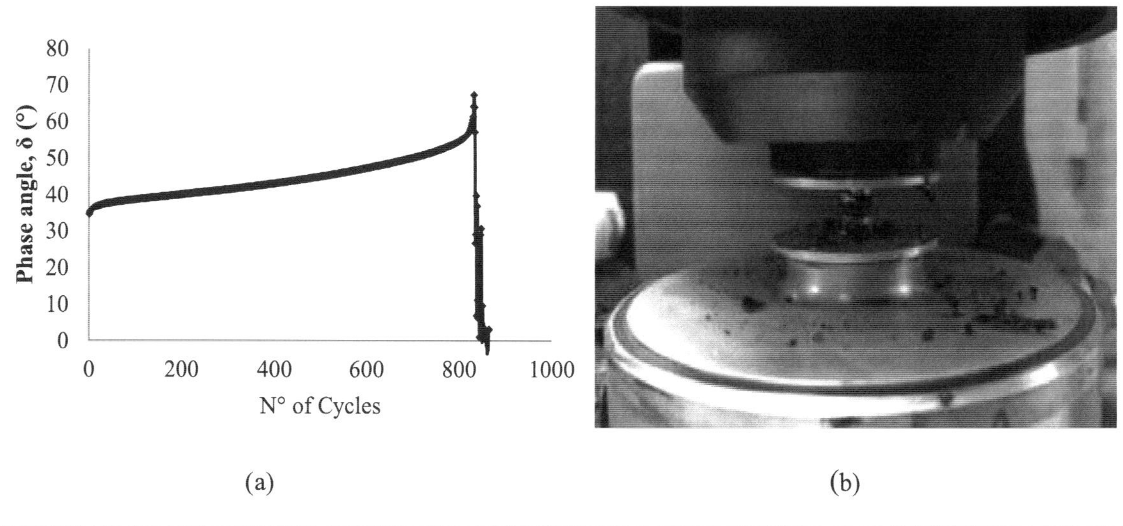

Figure 5. (*a*) Phase angle response during the stress-controlled fatigue test for Costa Rica 2 asphalt and (*b*) observed failure condition.

dominates over the viscous response (G″). As the number of load cycles increases the asphalt undergoes deformation, which is associated with the increase in the phase angle before failure occurs. The observed deformation is associated with the free volume of the binder molecules which, at such temperature, have enough space to move and tend to deform before the structure collapses. From the 800th cycle, this parameter drops abruptly, which confirms that there is a fatigue failure in the material. As shown in Figure 5*b*, the samples tested under stress-controlled conditions experienced an evident fracture, and their structure is completely broken upon failure.

Comparison between Environmental Conditions and Multiple PAV Cycles on Asphalt Oxidation

Fatigue on samples of asphalt Costa Rica 2 exposed to multiple PAV cycles and environmental oxidation were tested according to the previously described methodology. The obtained results are shown in Figure 6.

At this point, it is important to recall that according to the chemical analysis, the oxidation process caused by the PAV is incomplete compared with environmental oxidation. As shown in Figure 6, the oxidation achieved with the PAV is less drastic than 6 months of environmental oxidation. Consequently, this incomplete PAV oxidation means that the binder is more fatigue resistant than that exposed to environmental conditions. On the other hand, it was observed that the sample exposed for 24 months undergoes an increase in its modulus, but its resistance to fatigue cycles is reduced by an increase in its fragility.

A feasible solution to obtain a more drastic and realistic oxidation of asphalt could be to increase the number PAV cycles to which the material is subjected. Figure 6 shows the effect of increased PAV cycles on fatigue life for Costa Rica asphalt 2, evaluated under stress-controlled conditions. As shown in Figure 6, as the asphalt is subjected to continuous PAV cycles, its modulus increases. In the case of the samples subjected to one, two, and three consecutive PAV cycles, the asphalt becomes increasingly resistant to fatigue, which is evidenced by the increasing resistance to load cycles. On the other hand, even if the modulus of the samples oxidized after four and five PAV cycles increases, the material becomes fragile and can withstand a lower number of load cycles compared with the less oxidized material. Therefore, trying to simulate the environmental oxidation of binders by increasing the number of PAV cycles should be done with caution, given that the increase in one, two, or three PAV cycles could favor the fatigue resistance of the asphalt. The same analysis was performed using several binder sources to validate the findings for the sample Costa Rica 2. The results are shown in Table 4.

Conclusion

The thermal oxidation simulated in the PAV is achieved by increasing the content of oxygenated molecules in the material. However, it was found that environmental oxidation is a more severe process. Therefore, it is concluded that laboratory PAV aging is deficient in simulating the chemical oxidation of asphalt because it does not consider the effect of UV radiation, which causes unsaturation and aromaticity in the material. UV radiation might be more severe than RTFO and PAV treatments. This

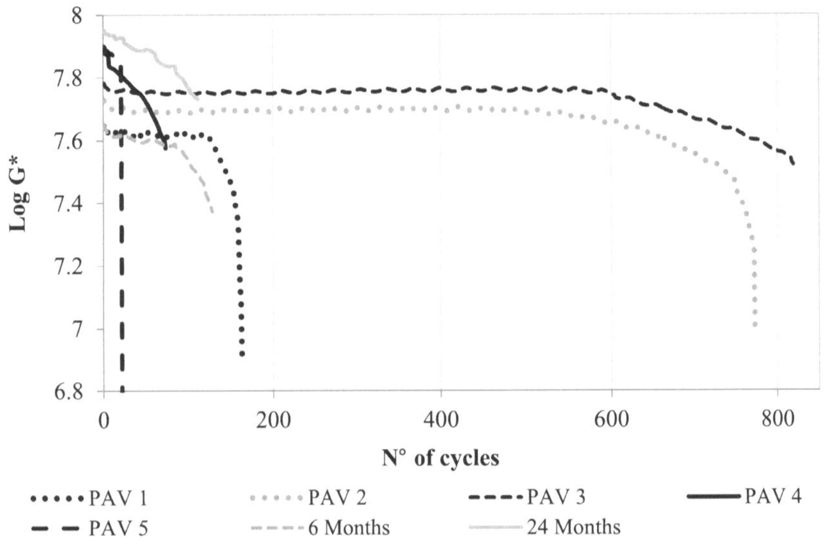

Figure 6. Comparison of stress-controlled fatigue test results for Costa Rica 2 samples oxidized under environmental conditions during 6 and 24 months (the curve of one cycle PAV-oxidized sample is shown for a simpler comparison), up to five PAV cycles.

Table 4. Stress-Controlled Fatigue Analysis Results

Asphalt	Condition	Initial modulus (MPa)	Failure modulus (MPa)	Number of cycles
Costa Rica 2	PAV 1	44.3	22.4	158
	PAV 2	53.4	22.8	762
	PAV 3	60.6	33.2	820
	PAV 4	76.6	37.7	73
	PAV 5	79.5	70.1	21
	6 months outdoors	44.9	23.6	128
	2 years outdoors	89.4	54.0	111
Bolivia	PAV 1	22.5	2.04	239
	PAV 2	26.3	2.62	381
	PAV 3	29.4	2.99	607
	PAV 4	31.7	7.53	36
	PAV 5	34.3	10.7	22
El Salvador	PAV 1	32.0	11.4	283
	PAV 2	47.9	20.4	893
	PAV 3	55.3	31.3	178
	PAV 4	57.7	42.0	15
	PAV 5	68.5	58.3	7
Nicaragua	PAV 1	33.3	12.4	539
	PAV 2	39.1	13.3	780
	PAV 3	42.2	28.3	1060
	PAV 4	44.5	33.7	153
	PAV 5	69.8	34.2	109

demonstrates the ineffectiveness of thermal treatments in simulating the oxidative aging that occurs to asphalt binders under environmental conditions. However, it must be recognized that oxidizing the material under environmental conditions for a complete characterization is a time-consuming process. Therefore, the results obtained serve as starting point for the calibration of an aging method that more realistically simulates the oxidation of the binder. In this regard, it is necessary to note that relying on multiple PAV cycles must be carefully done since it was found that two and three consecutive PAV cycles increment the fatigue resistance of the material and still do not emulate the photo-oxidative reactions that occur in the field. In summary, it is proposed to incorporate UV radiation, as well as humidity and temperature, in such an accelerated laboratory method, so that they can be adjusted to the climatic conditions of each region. The authors are currently working on an improved methodology that can simulate more efficiently the complete oxidation of the asphalt binder.

On the other hand, the fatigue behavior of binders is governed not only by the modulus but also by fragility. The previous was observed with fatigue testing: the fatigue test performed under stress-controlled conditions is a better approximation to the actual fatigue process and is more reliable in representing fatigue in the material, compared with the test performed at strain-controlled conditions. This happens because with the stress-controlled condition, a rapid modulus decrease can be observed and the sample fails when tested in the DSR (Figure 5*b*).

Finally, the data also showed a good relationship between the crossover temperature and the oxidation level of the asphalt. This results in an important tool for the evaluation of asphalt binders, especially when the equipment required for chemical analysis is not readily available.

Author Contributions

The authors confirm equal contribution to the paper. All authors reviewed the results and approved the final version of the manuscript.

References

1. Huang, S., and W. Grimes. Influence of Aging Temperature on Rheological and Chemical Properties of Asphalt Binders. *Transportation Research Record: Journal of the Transportation Research Board*, 2010. 2179: 39–48.
2. Lu, X., and U. Isacsson. Effect of Aging on Bitumen Chemistry and Rheology. *Construction and Building Materials*, Vol. 16, No. 1, 2002, pp. 15–22.
3. Branthaver, J. F., J. C. Petersen, R. E. Robertson, J. J. Duvall, S. S. Kim, P. M. Harnsberger, T. Mill, E. K. Ensley, F. A. Barbour, and J. F. Schabron. *Binder Characterization and Evaluation, SHRP-A-368: Chemistry*. Transportation Research Board, National Research Council, Washington, D.C., 1993.
4. Lamontagne, J., P. Dumas, V. Mouillet, and J. Kister. Comparison by Fourier Transform Infrared (FTIR) Spectroscopy of Different Aging Techniques: Application to Road Bitumens. *Fuel*, Vol. 80, No. 4, 2001, pp. 483–488.
5. Mouillet, V., J. Lamontagne, F. Durrieu, J. P. Planche, and L. Lapalu. Infrared Microscopy Investigation of Oxidation and Phase Evolution in Bitumen Modified with Polymers. *Fuel*, Vol. 87, No. 7, 2008, pp. 1270–1280.
6. Yao, H., Z. You, L. Li, S. W. Goh, C. H. Lee, and Y. K. Yap. Rheological Properties and Chemical Analysis of Nanoclay and Carbon Microfiber. *Construction and Building Materials*, Vol. 38, 2013, pp. 327–337.
7. Yang, X., Z. You, and J. Mills-Beale. Asphalt Binders Blended with a High Percentage of Biobinders: Aging Mechanism Using FTIR and Rheology. *Journal of Materials in Civil Engineering*, Vol. 27, No. 4, 2015, pp. 04014157.
8. Loría-Salazar, L. G. Evaluación de asfaltos modificados en laboratorio con distintos polímeros. *Infraestructura Vial*, Vol. 17, 2007, pp. 46–53.
9. Liu, M., K. M. Lunsford, R. R. Davidson, C. J. Glover, and J. A. Bullin. The Kinetics of Carbonyl Formation in Asphalt. *American Institute of Chemical Engineers Journal*, Vol. 42, No. 4, 1996, pp. 1069–1076.
10. Lui, M., M. A. Ferry, R. R. Davidson, C. J. Glover, and J. A. Bullin. Oxygen Uptake as Correlated to Carbonyl Growth in Aged Asphalts and Asphalt Corbett Fractions. *Industrial & Engineering Chemistry Research*, Vol. 37, No. 12, 1998, pp. 4669–4674.
11. Yang, X., Z. You, and J. Mills-Beale. Asphalt Binders Blended with a High Percentage of Biobinders: Aging Mechanism Using FTIR and Rheology. *Journal of Materials in Civil Engineering*, Vol. 27, No. 4, 2015, pp. 1–11.
12. *Transportation Research Circular E-C140: A Review of the Fundamentals of Asphalt Oxidation*. Transportation Research Board of the National Academies, Washington, D.C., 2009. http://onlinepubs.trb.org/Onlinepubs/circulars/ec140.pdf.
13. Bahia, H., and D. Anderson. The Pressure Aging Vessel (PAV): A Test to Simulate Rheological Changes Due to Field Aging. In *Physical Properties of Asphalt Cement Binders, STP18189S* (J. Hardin, ed.), ASTM International, West Conshohocken, PA, 1995, pp. 67–88.
14. Wu, S., L. Panga, L. Mo, J. Qiua, G. Zhu, and Y. Xiao. UV and Thermal Aging of Pure Bitumen-comparison Between Laboratory Simulation and Natural Exposure Aging. *Road Materials and Pavement Design*, Vol. 9, 2008, pp. 103–113.
15. Durrieu, F., F. Farcas, and V. Mouillet. The Influence of UV Aging of Styrene/Butadiene/Styrene Modified Bitumen: Comparison Between Laboratory and on Site Aging. *Fuel*, Vol. 86, No. 10–11, 2007, pp. 1446–1451.
16. Polacco, G., J. Stastna, D. Biondi, and L. Zanzotto. Relation Between Polymer Architecture and Nonlinear Viscoelastic Behavior of Modified Asphalts. *Current Opinion in Colloid & Interface Science*, Vol. 11, No. 4, 2006, pp. 230–245.
17. Walas, S. M. *Reaction Kinetics for Chemical Engineers: Butterworths Series in Chemical Engineering*. Butterworth-Heinemann, Newton, 2013.
18. Qin, Q., J. F. Schabron, R. B. Boysen, and M. J. Farrar. Field Aging Effect on Chemistry and Rheology of Asphalt Binders and Rheological Predictions for Field Aging. *Fuel*, Vol. 121, 2014, pp. 86–94.

19. Brinson, H. F., and L. C. Brinson. *Polymer Engineering and Viscoelasticity: An Introduction*, 2nd ed. Springer US, New York, 2015.
20. Rosen, S. *Fundamental Principles of Polymeric Materials*, 2nd ed. John Wiley & Sons, Inc., New York, 1993.
21. Padmarekha, A., and J. Murali Krishnan. Viscoelastic Transition of Unaged and Aged Asphalt. *Journal of Materials in Civil Engineering*, Vol. 25, No. 12, 2013, pp. 1852–1863.

The Standing Committee on Asphalt Binders (AFK20) peer-reviewed this paper (18-04684).

Article

TRR
JOURNAL OF THE TRANSPORTATION RESEARCH BOARD

Transportation Research Record
2018, Vol. 2672(28) 256–265

DOI: 10.1177/0361198118791913
journals.sagepub.com/home/trr
SAGE

Low-Temperature Vacuum Drying Procedure for Rapid Asphalt Emulsion Residue Recovery

Haritha Malladi[1], Meron Asnake[1], Andrew LaCroix[2], and Cassie Castorena[1]

Abstract

Asphalt emulsions are used extensively in tack coats and preservation surface treatments. The current specifications for asphalt emulsion residue recovery in AASHTO PP 72 are based on low-temperature evaporative drying. The shortest residue recovery procedure included in AASHTO PP 72 requires 6 h of conditioning in an oven to dry an emulsion. The ability to recover emulsion residues efficiently is of great interest given the ongoing efforts to develop emulsion performance-graded specifications based on residual binder properties. In this study, a rapid, vacuum drying technology was evaluated for asphalt emulsion residue recovery. The procedure enables the recovery of sufficient residual binder for dynamic shear rheometer (DSR) testing within 20–40 min. Five emulsions of different classifications were evaluated using both the vacuum drying procedure and the AASHTO PP 72 procedures. The vacuum drying procedure leads to similar water loss to the AASHTO PP 72 procedures. Based on the temperature-frequency sweep and multiple stress creep and recovery (MSCR) test results, the vacuum-dried residues are softer and more viscous than residues recovered using the AASHTO PP 72 procedures. Fourier transform infrared spectroscopy (FTIR) was performed to identify if oxidation levels could explain the observed in rheological trends. However, the results suggest no clear trend in the oxidation levels of residues recovered using AASHTO PP 72 compared with the rapid vacuum procedure. Future work is necessary to infer which method best reflects residual binders placed in the field.

Asphalt emulsions are used extensively in preservation surface treatments. In recent years, significant effort has been placed on establishing a performance-graded (PG) specification for asphalt emulsions used in preservation surface treatments (*1–3*). Because the water in the emulsion will leave the system through evaporation and absorption upon placement (*4*), the majority of performance parameters under consideration for specification are based on residual binder properties. Consequently, emulsion residue recovery is a critical component of future emulsion PG specifications.

Several methods have been employed for emulsion residue recovery, including (1) distillation, (2) low-temperature evaporative, and (3) vacuum techniques. Distillation methods (e.g., ASTM D6997 [*5*]) require high temperatures (163–260°C) that greatly exceed the temperatures that the emulsions will be exposed to in the field. In addition, the high temperatures used in distillation residue recovery can degrade polymer and latex-modified residues (*6*).

Consequently, the majority of efforts in recent years have focused on evaporative methods that utilize much lower temperatures. AASHTO PP 72 (*7*) includes provisions for two evaporative residue recovery procedures. Procedure A involves the placement of a roughly 2 mm thick film of emulsion on a silicone sheet that is conditioned at 25°C for 24 h followed by an additional 24 h of conditioning at 60°C. The relatively large sample thickness allows a substantial amount of residue to be recovered for testing. However, the procedure has been deemed impractical because of the time requirements. AASHTO PP 72 Procedure B (*7*) was developed to improve the efficiency of emulsion residue recovery at low temperature. In AASHTO PP 72 Procedure B (*7*), a wet film applicator is used to apply a 381 μm thick layer of emulsion to a silicone sheet that is conditioned in an oven at 60°C for 6 h. AASHTO PP 72 Procedure B (*7*) is

[1]Department of Civil, Construction, and Environmental Engineering, North Carolina State University, Raleigh, NC
[2]InstroTek, Inc., Raleigh, NC

Corresponding Author:
Address correspondence to Cassie Castorena: cahintz@ncsu.edu

the method currently recommended by the Federal Highway Administration Emulsion Task Force for the recovery of emulsion residues (*2*).

ASTM D7403 specifies the recovery of emulsion residues using a vacuum oven (*8*). The procedure is conducted at 135°C, which, like the distillation method, greatly exceeds temperatures that the emulsions will be exposed to in the field. However, several recent studies have proposed low-temperature vacuum methods for emulsion residue recovery. Reinke et al. proposed that the application of a vacuum of 12–15 kPa for 3 h at 60°C to 315 μm thick films of emulsion can be used to recover residual binders (*9*). Islam et al. reduced pressures to 0.5–6 kPa in a vacuum oven at 60°C and successfully recovered residues from emulsion samples prepared in accordance with AASHTO PP 72 Procedure A (*7*) (i.e., 2 mm thick films) in 3 h (*10*). Reinke et al. proposed the use of the CoreDry®, which is a rapid asphalt drying system, as an alternative means to low-temperature vacuum residue recovery (*11*). A maximum of 1 h was required for the recovery of residual binders. Reinke et al. (*11*) used 380 μm thick films of emulsion that were subjected to 35–40 cycles of vacuum–release. Vacuum levels ranged from 1.3 to 26.7 kPa with a maximum temperature of 70°C. Reinke et al. concluded that the CoreDry® method of residue recovery is effective, but stated that additional research is needed to confirm the findings and optimize the procedure (*11*). In addition, they noted that the quantity of residue recovery was limited to 2–3 g and suggested that future research was necessary to increase the residue yielded by the procedure if both 25 and 8 mm dynamic shear rheometer (DSR) tests are to be conducted for PG determination.

Objectives

The objectives of this study are to:

(1) Refine the low-temperature vacuum method for emulsion residue recovery to improve efficiency in terms of time requirements and residue yield.
(2) Compare the properties of residues recovered using the low-temperature vacuum process with those recovered following AASHTO PP 72 (*7*).

Low-Temperature Vacuum Rapid Emulsion Residue Recovery Procedure

The vacuum drying apparatus used in this study to recover emulsion residues is described in ASTM D7227 (*12*). This device has also been used extensively to remove moisture from compacted asphalt mixture specimens. The vacuum drying process was modified for emulsion residue recovery by the manufacturer.

The application of a vacuum reduces the boiling temperature of water. Consequently, moisture is removed efficiently by using a vacuum pump and a cold trap system within the drying system. Within the vacuum chamber used in this study, a hot plate at the bottom of the chamber is set to a temperature of 60°C. A vacuum is applied and the absolute pressure is monitored throughout the drying process and used to detect when the sample is dry. Experimentally, it has been found that the asphalt emulsion is dry when the absolute pressure is maintained below 1.33 kPa (10 mm Hg). The drying procedure consists of subjecting the emulsion to several vacuum–release cycles. The system automatically monitors the pressure, and when the pressure is maintained below 1.33 kPa it detects that the emulsion is dry and stops the process.

Emulsion samples are applied on circular silicone mats with 150 mm diameter using a Bird-type wet film applicator to produce emulsions with a 0.381 mm thickness. Five mats are prepared for each emulsion sample and stacked vertically on a pan holder adapted from the one used in the pressure aging vessel. The mats, sample pan holder, and film applicator are shown in Figure 1. The pan holder with the mats is placed inside the chamber of the vacuum drying unit. Based on the five emulsions evaluated here, drying only requires several cycles of vacuum–release and is completed within 20–40 minutes. By using five mats, approximately 10 g of residue can be recovered, which is sufficient for performing both 25 and 8 mm DSR tests.

Experimental Plan

Materials

Five plant-produced emulsions were evaluated in this study to encompass a broad range of emulsion types: CRS-2, CRS-2L, CRS-1H, CQS-1H, and SS-1.

Figure 1. Vacuum apparatus emulsion film applicator and sample pan holder for rapid residue recovery.

Residue Recovery Procedures

Three residue recovery methods were evaluated: the rapid low-temperature vacuum drying procedure described above and AASHTO PP 72 Procedures A and B (*7*). While following AASHTO PP 72 Procedure A (*7*), 18–20 g of asphalt emulsion was applied evenly across a 100 × 100 mm (0.01 m^2) square-shaped area on a silicone mat; this resulted in a wet film thickness of approximately 2 mm. Emulsion samples prepared following AASHTO PP 72 Procedure B (*7*) were spread on a silicone mat using a Bird-type wet film applicator to achieve a wet emulsion film thickness of 0.381 mm.

For ease of reporting, AASHTO PP 72 Procedures A and B (*7*) are referred to as "48 h oven dry" and "6 h oven dry", respectively, in the results section. The rapid low-temperature vacuum drying process is referred to as "vacuum dry."

Emulsion Residue Evaluation

To compare the relative effectiveness of the different residue recovery methods at removing water, the percent water loss was determined by comparing the mass of samples before and after residue recovery. Weight measurements were conducted directly after residue recovery. All other laboratory tests were performed within 24 h of recovering the residue. Caution was taken to avoid heating the recovered residues before testing to prevent any further water loss and/or aging.

To quantify if the residues obtained from the three residue recovery methods differed, linear viscoelastic properties were characterized using temperature-frequency sweeps using the 8-mm parallel plate geometry in an Anton Paar MCR 302 DSR. Temperature-frequency sweep testing was conducted over the temperature range of 5–50°C and frequency range of 0.1–30 Hz. Temperature-frequency sweep results were used to generate dynamic shear modulus and phase angle master curves.

Multiple stress creep recovery (MSCR) tests were conducted in an Anton Paar MCR 302 DSR on all recovered residues as per the standard procedure described in AASHTO T 350 (*13*). MSCR tests were performed at 67°C to reflect the proposed high-temperature emulsion PG temperature in Raleigh, NC (*3*). The results were used to determine the non-recoverable creep compliance at a stress level of 3.2 kPa. A minimum of two replicates were performed for the temperature-frequency sweep and MSCR tests.

Attenuated total reflectance (ATR) Fourier transform infrared (FTIR) spectroscopy was performed on all emulsion residues using a Burker Optics ALPHA FTIR spectrometer to evaluate the oxidation levels of recovered residues. FTIR-ATR simultaneously collects absorbance data with in a wide spectral range (400–4000 cm^{-1}). In this study, the ATR spectra were collected using 64 scans at a resolution of 4 cm^{-1} with a minimum of three replicates. A baseline correction was applied to all spectral data acquired. For each residual binder, all replicates under different conditions were normalized to the same absorbance value of 2 at wave number 1455 cm^{-1}. This wave number was selected for normalization because absorbance at this wave number is not affected by the level of oxidation or presence of water. Changes in the carbonyl and sulfoxide absorbance peaks were tracked at wave numbers 1702 and 1032 cm^{-1}, respectively. The sum of absorbance values at these two wavenumbers were used to assess oxidation levels.

Results

Water Loss

The weights of the original emulsion samples and the recovered residues were recorded and used to calculate the water loss as a percentage of the original emulsion weight. The water loss results are shown in Figure 2. Because the vacuum dry process involves five mats, the weights were separately recorded for each mat, resulting in a range of values. The average water loss across all five mats was calculated and reported. Average water loss from a minimum of two replicates was also reported for both 48 and 6 h oven dry procedures. The error bars shown in Figure 2 represent the range of observed water loss values. No trend of water loss was observed based on mat location within the sample pan holder. Generally, it can be seen that the percentage water loss obtained from the three residue recovery procedures are similar for all emulsion types. The error bars indicate relatively

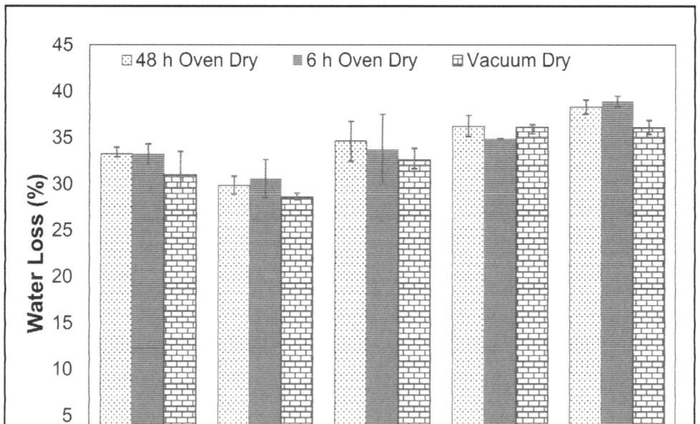

Figure 2. Percent water loss by weight of emulsion using the three residue recovery procedures.
Note: Error bars indicate range of values observed.

Table 1. Connecting Letters Report for Tukey HSD Test on % Water Loss Values

Level	Connecting letters	Mean
48 h oven dry	A	33.62
6 h oven dry	A	33.01
Vacuum dry	A	32.86

low variability in the water loss from each mat in the vacuum dry procedure.

It is noted that the average water loss values for the vacuum dry procedure are slightly lower than the values for the AASHTO PP 72 (*7*) procedures for four out of the five emulsions investigated. Therefore, the Tukey honest significant difference (HSD) test (α level 0.05) was used to detect statistically significant differences in mean water loss results obtained from the three procedures (*14*). In the Tukey HSD test, pairwise comparisons of the mean percentage water loss from each recovery procedure were performed to detect statistically significant differences. With three recovery procedures, the total number of pairwise comparisons is three (48 h oven dry–6 h oven dry, 48 h oven dry–vacuum dry, 6 h oven dry–vacuum dry).

The Tukey HSD test generates a connecting letters report, which is summarized in Table 1. Mean percent water loss values are listed in descending order. A letter is shown next to each mean value. Means that do not share a letter are significantly different. All three residue recovery procedures have the same connecting letter listed next to them, showing that the average % water loss values are not significantly different from each other.

The results of Tukey HSD are shown as a visual comparison of mean water loss percentage in Figure 3. Scatterplots of percent water loss are shown for each recovery procedure on the left. Circles drawn on the right represent confidence intervals of mean percent water loss for each residue recovery procedure. The degree of intersection of these circles represents how different these means are. As can be seen, the confidence interval circles are nested, indicating no statistically significant difference in percent water loss between the three residue recovery procedures. Thus, the results suggest that the rapid low-temperature vacuum drying procedure is as successful as the AASHTO PP 72 (*7*) oven-drying procedures in removing water from asphalt emulsions.

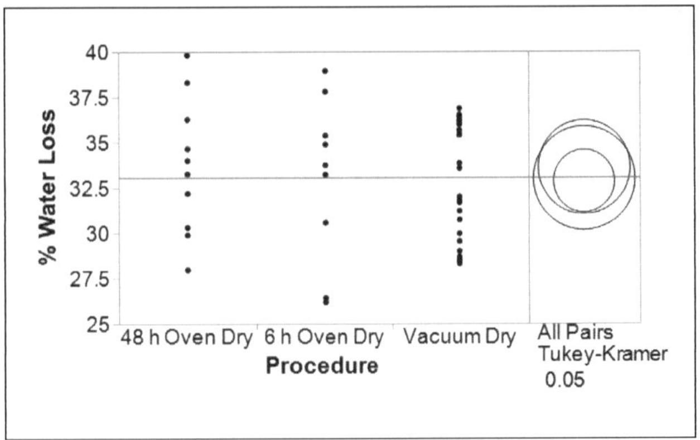

Figure 3. Tukey HSD comparison circles plot for shear modulus from different drying procedures with emulsion type as blocking.

Rheology

Linear Viscoelastic Mastercurves. The dynamic shear modulus and phase angle mastercurves corresponding to residues recovered using the three methods evaluated are shown in Figures 4 and 5, respectively. In general, it can be visually observed that the vacuum dry residual binder samples have slightly lower dynamic shear moduli and higher phase angle values than the residual binders recovered using the AASHTO PP 72 (*7*) procedures.

A statistical analysis was conducted to determine if the dynamic shear moduli of the residues recovered using the three procedures differ significantly. Dynamic shear modulus at 10 rad/s for all test temperatures (50, 35, 20, and 5°C) were considered as the response variables. The analysis was performed separately at each test temperature. Recovery procedure was considered a factor with three levels (48 h oven dry, 6 h oven dry, and vacuum dry). Emulsion type was considered as a blocking factor to control variability caused by difference in emulsion properties.

Interaction plots of the dynamic shear modulus results to show the relative effects of the drying procedure and emulsion type are shown in Figure 6. The interaction plots were drawn separately for each test temperature. In these plots, the dynamic shear modulus values are shown along the y-axis with three residue recovery procedures represented on the x-axis. Variation of mean shear modulus is shown for each emulsion type. From Figure 6, it can be noted that in general, the 48 h oven dry samples have the highest modulus values and the vacuum dry samples have the lowest modulus values at all test temperatures.

The Tukey HSD test was used to detect statistically significant differences in mean dynamic shear modulus results obtained from the three procedures (*14*). The Tukey HSD connecting letters report for mean dynamic shear modulus is summarized in Table 2. At each test temperature, mean dynamic shear modulus values are listed in descending order along with the connecting letters. As noted in the previous section, means that do not share a letter are significantly different. At 50, 35, and 20°C, mean dynamic shear modulus from all pairs of recovery procedures significantly differ from each other.

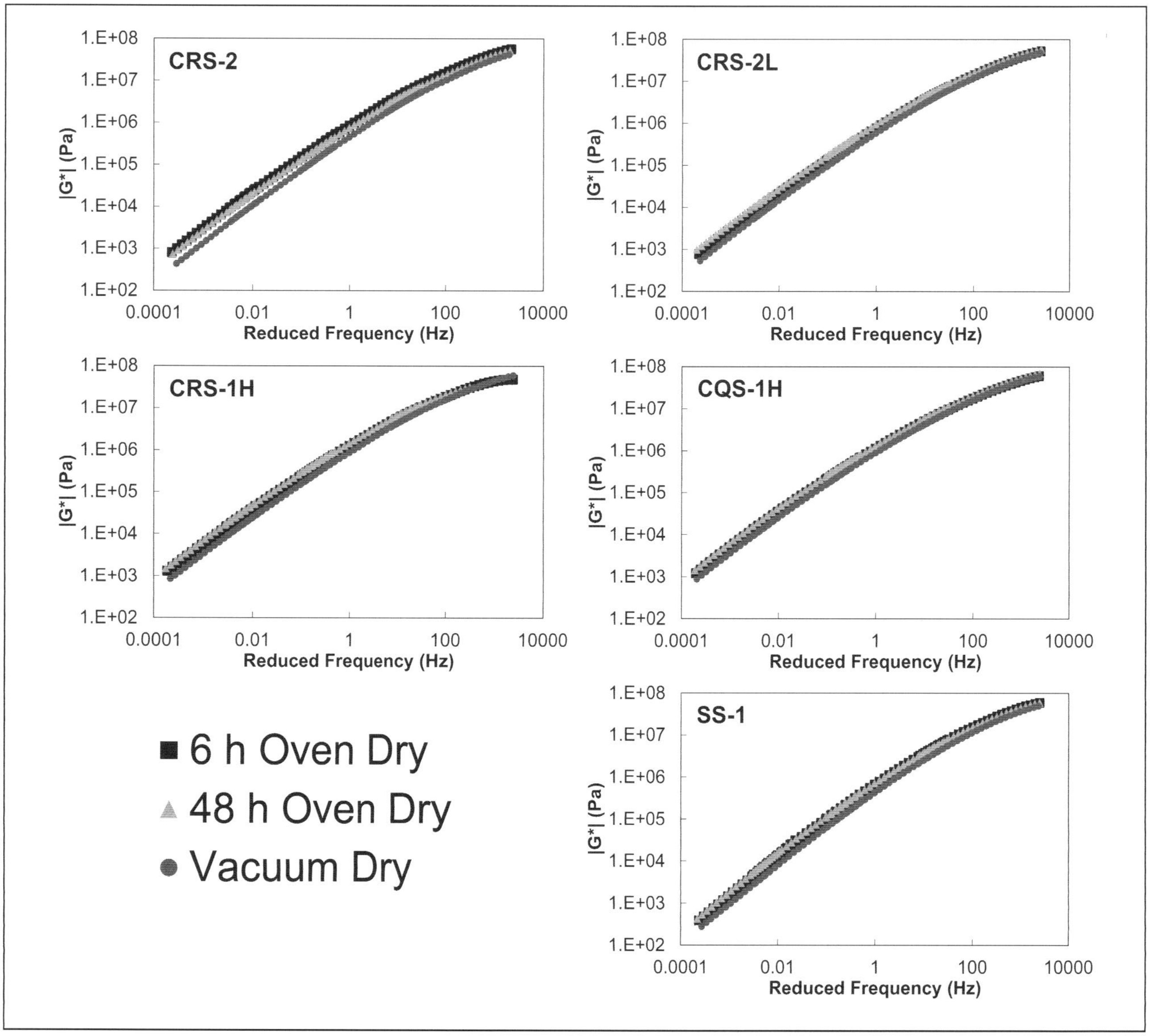

Figure 4. Dynamic shear modulus master curves for each emulsion type for all three residue recovery procedures.

At 5°C, the two AASHTO PP 72 (*7*) residues are not significantly different from each other, whereas the vacuum dry residue is significantly different from both of them.

Scatterplots and comparison circles for mean dynamic shear modulus are shown in Figure 7. At each test temperature, scatterplots of dynamic shear modulus values are shown for each recovery procedure on the left. Circles drawn on the right represent confidence intervals of mean dynamic shear modulus for each residue recovery procedure after controlling for block-to-block variability (based on emulsion type). As noted before, the degree of intersection of these circles represents how different these means are. The confidence interval for vacuum dry residues does not intersect with the ones for oven-dried residues at any temperature. Based on the Tukey HSD results, it can be concluded that the dynamic shear modulus of vacuum dry residues are significantly lower than both oven-dried residues.

MSCR Results. The non-recoverable creep compliance (J_{nr}) values corresponding to the 3.2 kPa stress level of the MSCR test are summarized in Figure 8. The error bars shown in Figure 8 represent the standard error of the test results. For all emulsion types, the J_{nr} values of the vacuum dry residues are significantly higher than those for residues obtained from the AASHTO PP 72 (*7*) procedures, indicating a softer response which is consistent with the findings of temperature-frequency sweep testing findings.

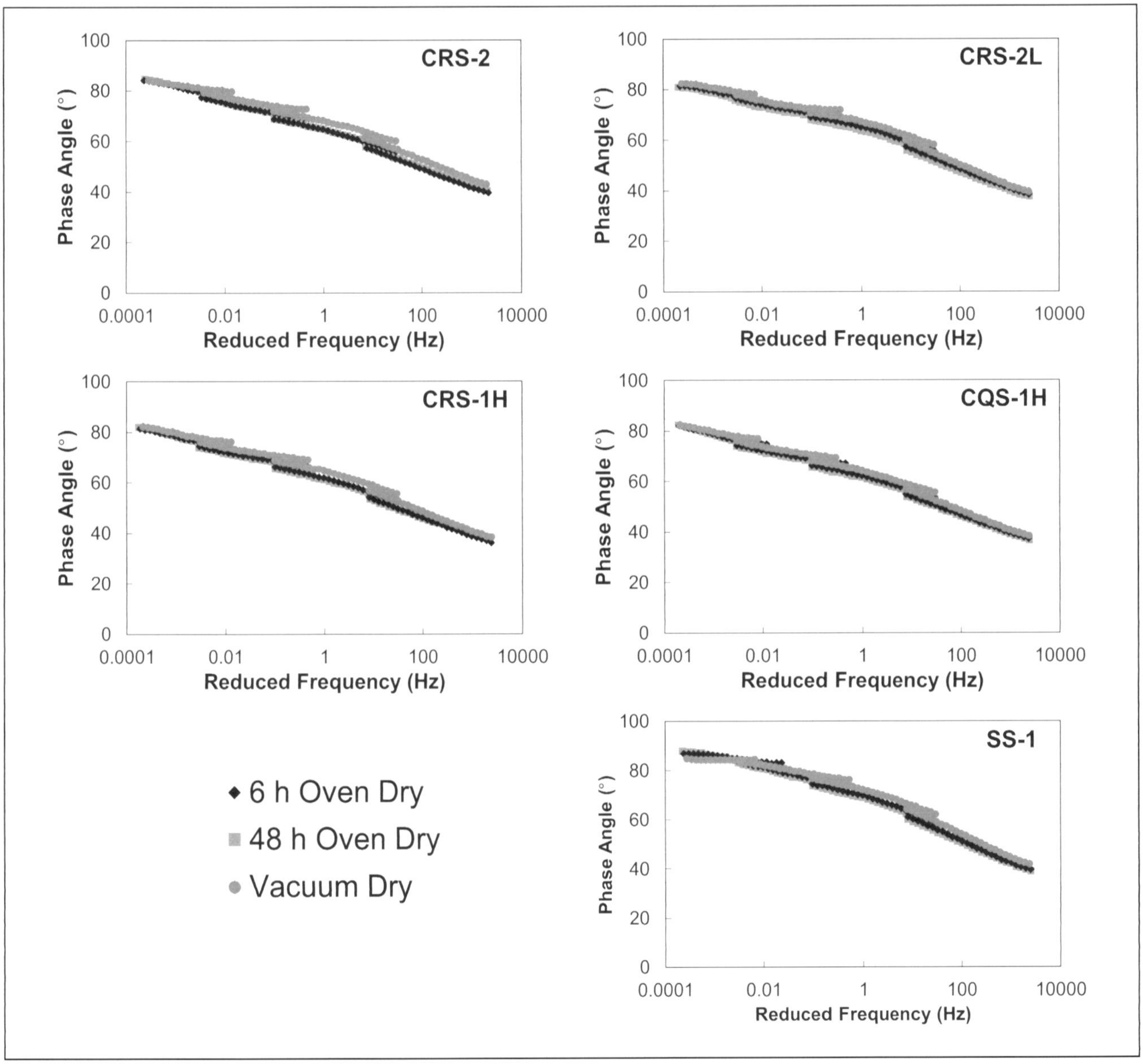

Figure 5. Phase angle master curves for each emulsion type for all three residue recovery procedures.

Discussion of Rheological Results. The rheological property results demonstrate that the residues recovered from the vacuum dry procedure are softer and more viscous than those recovered using the AASHTO PP 72 (*7*) procedures. There are several factors that may lead to differences in the results of the vacuum dry and AASHTO PP 72 (*7*) procedures. The presence of residual water could lead to a softer and more viscous response. However, the results presented in Figure 2 suggest generally comparable water loss for all residue recovery procedures. Furthermore, whereas the CQS-1H emulsion exhibited an average percentage water loss from the vacuum dry procedure that exceeded the water loss from the AASHTO PP 72 procedures, the vacuum dry residual binder demonstrated a softer and more viscous response than the residual binders recovered using the AASHTO PP 72 (*7*) procedures. In addition, Islam et al. demonstrated that rheological results of emulsion residue are not sensitive to small amounts of residual water (*10*). Thus, the presence of residual water does not explain the observed trends.

It is possible that the application of a vacuum could remove light oils from the binder. However, the removal of oils would result in stiffening, which does not match the observed trends. In addition, there was no visual evidence that the vacuum drying process removes anything other than moisture from the emulsion samples. Therefore, it is concluded that the removal of oils does not occur during the vacuum dry procedure.

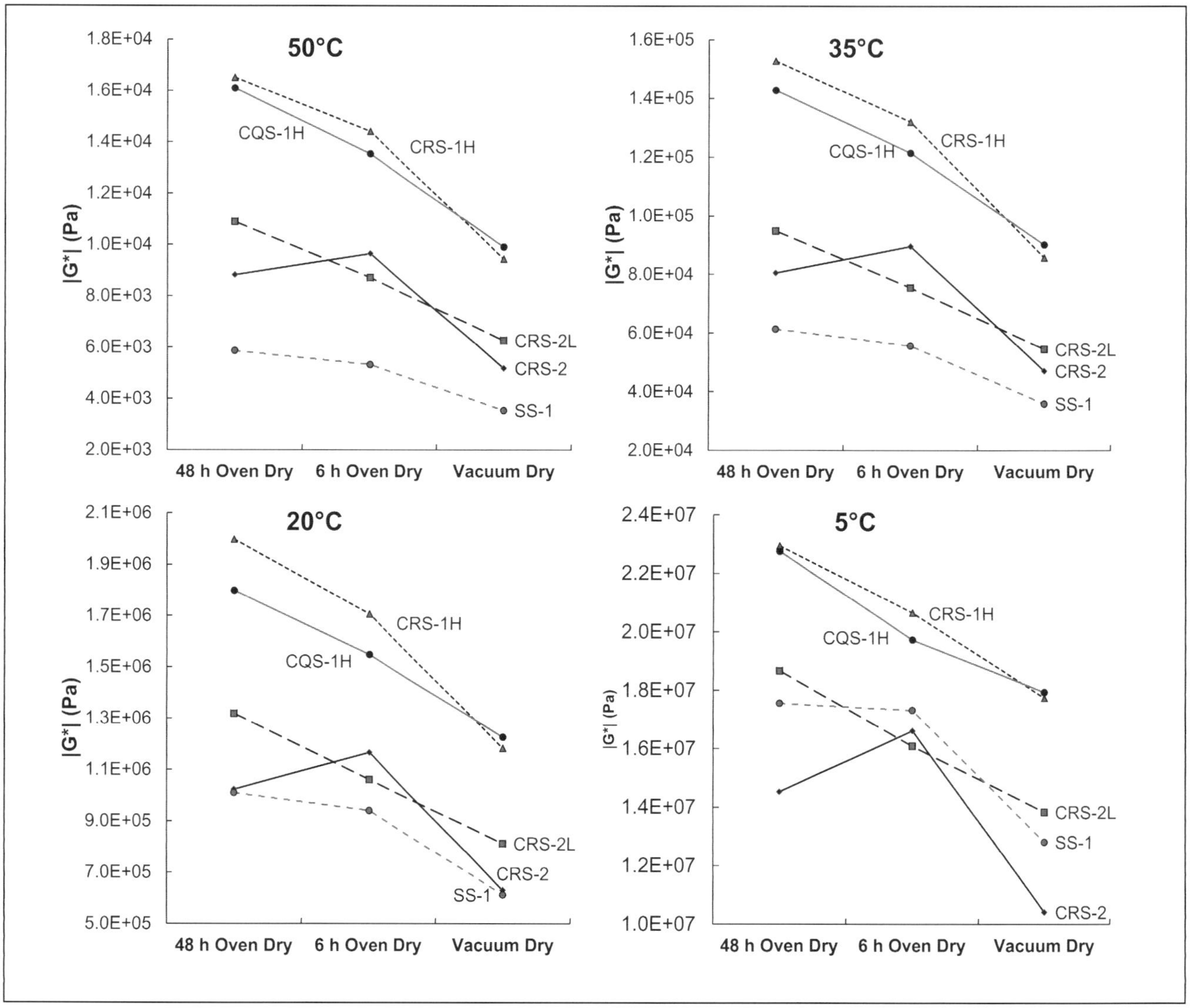

Figure 6. Interaction plots of shear modulus between residue recovery procedures and emulsion types at different test temperatures and 10 rad/s.

Table 2. Connecting Letters Report for Tukey HSD test on Dynamic Shear Modulus Values

Temperature = 50°C				
Level	Connecting letters			Mean
48 h oven dry	A			1.16E + 04
6 h oven dry		B		1.03E + 04
Vacuum dry			C	6.86E + 03

Temperature = 35°C				
Level	Connecting letters			Mean
48 h oven dry	A			1.06E + 05
6 h oven dry		B		9.48E + 04
Vacuum dry			C	6.26E + 04

Temperature = 20°C				
Level	Connecting letters			Mean
48 h oven dry	A			1.43E + 06
6 h oven dry		B		1.28E + 06
Vacuum dry			C	8.92E + 05

Temperature = 5°C				
Level	Connecting letters			Mean
48 h oven dry	A			1.90E + 07
6 h oven dry	A			1.80E + 07
Vacuum dry		B		1.50E + 07

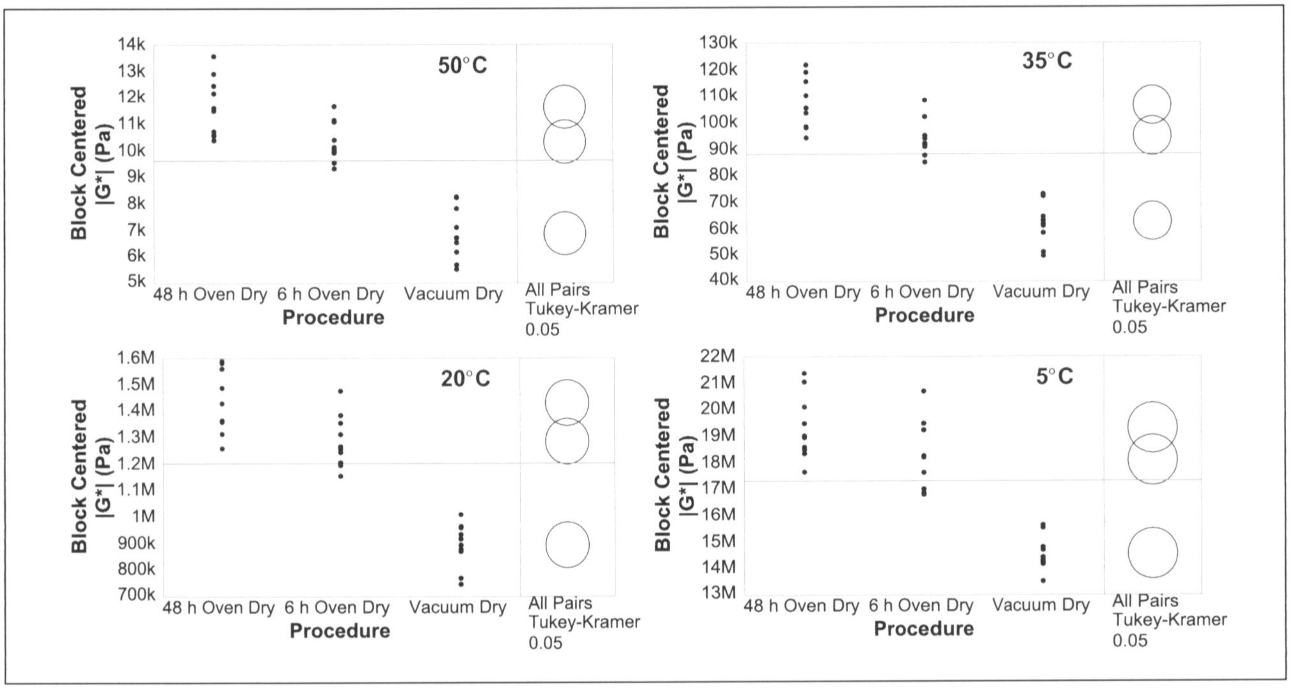

Figure 7. Tukey HSD comparison circles plot for shear modulus from different drying procedures with emulsion type as blocking factor.

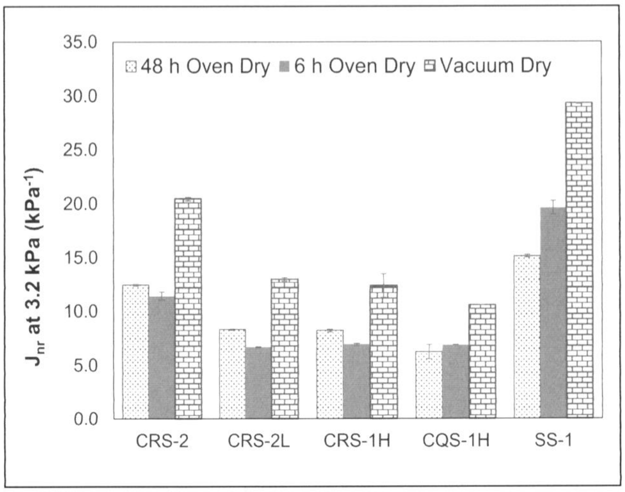

Figure 8. J_{nr} values at 3.2 kPa stress obtained from MSCR tests at 67°C.

Oxidation of the binder during the residue recovery procedure could lead to a stiffer and more elastic response. The vacuum drying process takes significantly less time to complete compared with the AASHTO PP 72 (*7*) procedures. Thus, the conditioning of the emulsion at 60°C for extended time in the AASHTO PP 72 (*7*) procedures could lead to oxidation of the binder that does not occur in the vacuum dry procedure, which could explain the observed trends. Reinke et al. (*11*) reported that residues recovered from AASHTO PP 72 (*7*) procedure B have higher stiffness than the base binder used for emulsion production. Hanz et al. also reported that significant stiffening occurs during evaporative residue recovery (*15*). Islam et al. demonstrated that residues recovered using a 48 h vacuum procedure were softer than residues recovered using AASHTO PP 72, which was attributed to differences in oxidation levels (*10*). However, findings in these past studies were not confirmed by an evaluation of binder chemistry. To evaluate the relative oxidation levels of the recovered residues here, FTIR-ATR testing was conducted.

Chemistry

Figure 9 shows the sum of the carbonyl and sulfoxide absorbance peaks for all residues. Carbonyls and sulfoxides constitute the chemical species that are produced upon the oxidation of asphalt binders. The error bars shown in Figure 9 represent the standard error of the test results. Based on the FTIR data, there is no clear trend in the relative oxidation level of emulsion residues recovered using the three residue recovery methods. Given the overlap in the error bar ranges for the results of the different residue recovery methods for a given emulsion type, the oxidation imposed by the three residue recovery methods is similar for most the emulsions evaluated. Significant differences observed follow no clear trend among different emulsion types. Consequently, the FTIR results do not explain the differences between the rheological results of the residues recovered using the different procedures.

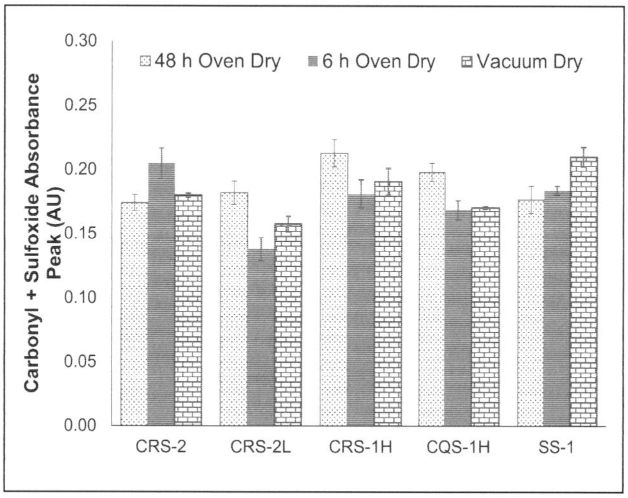

Figure 9. Sum of carbonyl and sulfoxide absorbance peaks for all recovered emulsion residues.

Given that the literature suggests that the AASHTO PP 72 (*7*) procedures lead to stiffening of the binder (*11*, *15*), it is suggested that future research be conducted to further investigate the mechanism of stiffening imposed in evaporative residue recovery relative to the vacuum recovery procedure. In addition, it is recommended that an investigation of the chemical and rheological properties of residual binders recovered from the field be evaluated to identify the residue recovery procedure best reflects the residual binders placed in the field.

Conclusions and Recommendations

The following conclusions and recommendations are drawn from this study:

1. The low-temperature vacuum drying technology is a rapid method of recovering residues from asphalt emulsions that requires less than 1 h to complete including sample preparation time. This is significantly faster than the two current evaporative oven-drying procedures recommended by AASHTO PP 72 (*7*), which involve a drying time of 6 or 48 h.
2. The percentage water loss measured in the vacuum drying process and the two AASHTO PP 72 (*7*) procedures are statistically equivalent.
3. Rheological tests show that vacuum drying residues are consistently softer and more viscous than the oven-dried residues.
4. FTIR-ATR results did not reveal any clear trend in the relative oxidation level of residual binders recovered using the low-temperature vacuum procedure and the AASHTO PP 72 (*7*) procedures.
5. It is recommended that future work be conducted to explain the differences in the rheological results of residues recovered using the vacuum method to those recovered using AASHTO PP 72 (*7*) to infer which best reflects residual binders placed in the field.

Acknowledgments

The authors would like to acknowledge and thank Brian Cancio and Shawn Byram of Hammaker East Emulsions in Greensboro, NC, and the staff at Asphalt Emulsion Inc. in Dunn, NC for providing the emulsions evaluated in this study. The authors also acknowledge and thank Michael Elwardany for his help with FTIR testing and analysis.

Author Contributions

The authors confirm contribution to the paper as follows: study conception and design: Cassie Castorena, Andrew LaCroix; data collection: Haritha Malladi, Meron Asnake; analysis and interpretation of results: Haritha Malladi, Meron Asnake, Cassie Castorena; draft manuscript preparation: Haritha Malladi, Cassie Castorena. All authors reviewed the results and approved the final version of the manuscript.

References

1. King, G., A. Kadrma, D. Salomon, A. J. Hanz, P. Johannes, P. Chaturabong, H. U. Bahia, Y. R. Kim, J. Adams, J.-H. Im, J. Lee, M. J. Farrar, Q. Qin, R. W. Grimes, J.-P. Planche, R. Boysen, J. Loveridge, S. Salmans, A. O. Cookman, G. Reinke, M. Ryan, S. Engber, and D. Herlitzka. *Transportation Research Board E-Circular 182: Progress Toward Performance-Graded Emulsified Asphalt Specifications*. Transportation Research Board of the National Academies, Washington, D.C., 2013.
2. Vijaykumar, A., A. Epps-Martin, and E. Arambula. Revision and Further Validation of Surface Performance-Graded (SPG) Specification for Chip Seal Binders. *Transportation Research Record: Journal of the Transportation Research Board*, 2013. 2370: 44–52.
3. Kim, Y. R., J. Adams, C. Castorena, M. Illias, J. Im, H. Bahia, and P. Johannes. *NHCRP Report 837: Performance-Related Specifications for Asphalt Binders Used in Preservation Surface Treatments*. Transportation Research Board of the National Academies, Washington, D.C., 2017.
4. Salomen, D. R. *Transportation Research Board E-Circular 102: Asphalt Emulsion Technology*. Transportation Research Board of the National Academies, Washington, D.C., 2006.
5. American Society for Testing and Materials. *Standard Test Method for Distillation of Emulsified Asphalt*. ASTM D 6997, West Conshohocken, PA, 2012.
6. Kadrmas, A. *Transportation Research Board E-Circular 182: Asphalt Emulsion Residue Recovery Update*. Transportation Research Board of the National Academies, Washington, D.C., 2013.

7. American Association of State Highway and Transportation Officials. *Standard Practice for Recovering Residue from Emulsified Asphalt Using Low-Temperature Evaporative Techniques*. AASHTO PP 72, Washington, D.C., 2013.
8. American Society for Testing and Materials. *Standard Test Method for Determination of Residue of Emulsified Asphalt by Low-temperature Vacuum Distillation*. ASTM D 7403, Washington, D.C., 2015.
9. Reinke, G., M. Ryan, S. Engberm, and D. Herlitzka. *Transportation Research Board E-Circular 182: Techniques for Accelerating Recovery of Asphalt Emulsion Residues at 60°C by Using Thin Film Procedures and Test Methods Suited to the Limited Amount of Residue Obtained to Characterize the Recovered Binder*. Transportation Research Board of the National Academies, Washington, D.C., 2013.
10. Islam, M., S. Ashani, N. Wasiuddin, and W. King. Effects of Curing Time, Temperature, and Vacuum Pressure on Asphalt Emulsion Residue Recovered by Vacuum Drying Method. *Journal of Testing and Evaluation*, Vol. 43, No. 5, 2015, pp. 1134–1145.
11. Reinke, G., S. Engber, D. Herlitzka, S. Veglahn, and K. Quillin. Evolution of Techniques for Obtaining Emulsion Residues at 60°C Using Thin Film Procedures and Test Methods Suited to the Limited Amount of Residue Obtained to Characterize the Recovered Binder. Presented at the ISAET Symposium, Washington, D.C., 2012.
12. American Society for Testing and Materials. *Standard Practice for Rapid Drying of Compacted Asphalt Mixture Specimens Using Vacuum Drying Apparatus*. ASTM D 7227-7227M, West Conshohocken, PA, 2017.
13. American Association of State Highway and Transportation Officials. *Standard Method of Test for Multiple Stress Creep and Recovery (MSCR) Test of Asphalt Binder Using a Dynamic Shear Rheometer (DSR)*. AASHTO T 350, American Association of State Highway and Transportation Officials, Washington, D.C., 2016.
14. Tukey, J. Comparing Individual Means in the Analysis of Variance. *Biometrics*, Vol. 5, No. 2, 1949, pp. 99–114.
15. Hanz, A., Z. Arega, and H. U. Bahia. Rheological Behavior of Emulsion Residues Produced by Evaporative Recovery Method. *Transportation Research Record: Journal of the Transportation Research Board*, 2010. 2197: 102–108.

The Standing Committee on Asphalt Binders (AFK20) peer-reviewed this paper (18-02848).

Article

TRR
JOURNAL OF THE TRANSPORTATION RESEARCH BOARD

Transportation Research Record
2018, Vol. 2672(28) 266–276

DOI: 10.1177/0361198118790133
journals.sagepub.com/home/trr
SAGE

Low-Temperature Emulsion Performance-Graded Specification for Chip Seals

Cassie Castorena[1], Mohammad Ilias[2], Javon Adams[3], and Y. Richard Kim[1]

Abstract

Chip seals are applied to existing roadways to slow deterioration and improve pavement surface conditions without increasing the pavement's structural capacity. Chip seals typically are comprised of asphalt emulsion and aggregate. Although it is well known that the properties of asphalt emulsion are critical to chip seal performance, the current specifications for these binders are empirical in nature and are not based on the properties that relate directly to chip seal performance. This study proposes emulsion performance-graded specifications that are designed to mitigate aggregate loss at low temperatures. To develop the proposed specifications, the research team executed an experimental plan to identify the residual binder properties that relate to chip seal aggregate loss at low temperatures. This research considered both fracture mechanics and rheology-based binder properties. Dynamic shear moduli, evaluated at critical phase angle values, demonstrated the strongest correlation to chip seal aggregate loss at low temperatures. Critical phase angle values were varied as a function of the low-temperature climatic emulsion performance grade of interest. This study also established preliminary specification limits for the dynamic shear modulus values based on critical chip seal performance thresholds.

Chip seals are applied to existing roadways to slow deterioration and improve the pavement's surface condition without increasing its structural capacity. It is well known that the properties of asphalt emulsion are critical to chip seal performance. Therefore, emulsion performance-graded (EPG) specifications that specify the asphalt emulsion properties that relate directly to performance will improve the emulsion selection process for a given chip seal to prevent premature failure. The urgent need for EPG specifications is clear in the responses to surveys conducted by Johnston and King (*1*) in which pavement practitioners in the United States were asked about the need for such specifications.

EPG specifications for chip seals should address aggregate loss and bleeding, which are the two critical distresses found in chip seals. Bleeding occurs primarily at high temperatures whereas aggregate loss occurs at intermediate and low temperatures. At intermediate temperatures, aggregate loss resistance is driven by the compatibility between the aggregate and emulsion (*2*). However, the compatibility between the aggregate and emulsion cannot be addressed in an emulsion purchase specification alone. Based on laboratory observations, aggregate loss is due primarily to the cohesive fracture within the residual binder at low temperatures (*3*).

This study seeks to develop recommendations in future EPG specifications that address low-temperature aggregate loss. The study investigated both fracture mechanics parameters and rheological parameters that have been proposed previously as indicators of cracking potential. The strength of the relationship between low-temperature chip seal aggregate loss and the binder properties, equipment requirements, time requirements, and repeatability were used to select a low-temperature parameter for future EPG specifications. Also, this study established preliminary specification limits for the selected binder parameter based on the values of the binder properties that correspond to critical chip seal aggregate loss performance thresholds.

[1]Department of Civil, Construction, and Environmental Engineering, North Carolina State University, Raleigh, NC
[2]Fugro Roadware, Inc., Austin, TX
[3]College of Engineering, Office of Academic Affairs, North Carolina State University, Raleigh, NC

Corresponding Author:
Address correspondence to Cassie Castorena: cahintz@ncsu.edu

Objectives

The objectives of this study are to:

1. Identify the residual binder properties that relate to the low-temperature aggregate loss performance of chip seals.
2. Develop low-temperature specifications for emulsion residues used in chip seals to address aggregate loss.

Experimental Plan

The North Carolina State University research team executed a laboratory experimental plan to develop the proposed low-temperature EPG specifications. Although the original intent was for the findings to relate to field performance, relating the material properties directly to field performance was deemed impractical due to the potentially confounding effects of various material application rates and exposure conditions.

Materials

The asphalt emulsions evaluated in this study are commercial-grade emulsions that have been used in chip seals throughout the United States. These emulsions are expected to resist significant performance problems under the climatic and traffic conditions for which they were designed. The inclusion of such emulsions ensures that the proposed specifications are not so restrictive that they preclude the use of asphalt emulsion formulations that have performed well in existing projects. This study evaluated two unmodified emulsions and four modified emulsions from four producers, as detailed below:

- CRS-2-NC
- CRS-2-F
- CRS-2L-C
- CRS-2P-A
- CRS-2L-NC
- CRS-2L-F

The letters after the emulsion class correspond to the producer of the emulsion and are coded to preserve anonymity.

Residual binders were recovered using AASHTO PP 72 Method B (*4*) before testing. The residues were tested without long-term aging because most low-temperature aggregate loss is expected to take place during the first winter following construction, before significant oxidation is expected.

The chip seal mixture specimens were fabricated using a performance-based mix design and a sample fabrication procedure that closely simulates the field construction process (*5*). Each chip seal specimen was designed such that the residual asphalt content was 67% after curing to achieve 50% aggregate embedment. Granite 78M aggregate was used for all the fabricated specimens with an aggregate application rate of 15.5 lb/yd^2, which was found to yield a single stone coverage.

Test Methods and Parameters

Chip Seal Mixture Vialit Test. Chip seal low-temperature aggregate loss was measured in this study using a modified version of the standard Vialit test (*6*) conducted at −16°C, −22°C, and −28°C. The Vialit test uses both gravity and impact to measure raveling resistance by subjecting an inverted chip seal sample affixed to a steel plate to the strike of a 500 g steel ball that is dropped from a fixed height. The weight of the specimen before and after testing is used to calculate the percentage of the aggregate loss. The Vialit test was developed to quantify the aggregate retention at intermediate temperatures. It was found in this study that the chip seal materials debonded from the Vialit plate when the standard test was conducted at a low temperature. To improve bonding, the Vialit plate surface was roughened and the plate thickness was increased to 6.35 mm based on the recommendations of Jordan and Howard (*7*). The Vialit drop height was lowered from the standard 50 cm to 12.5 cm to further mitigate debonding (*8*).

Chip seal samples were prepared in the laboratory using the ChipSS aggregate spreader (*5*), which contains a spreader box that allows aggregate to be applied at a controlled rate in a manner that mimics field construction. The samples were fabricated in a temperature-controlled greenhouse, with the emulsion preconditioned to 60°C. The Vialit test specimens were compacted and cured following ASTM D7000 (*9*). After curing, the specimens were inverted and a brush was used to remove loose aggregate. Subsequently, the samples were weighed and conditioned to the proper test temperature for four hours based on the recommendations of Jordan and Howard (*7*). The specimens were removed from the temperature chamber and tested within 30 seconds to avoid any significant increase in the temperature. Five replicate tests were conducted for each emulsion at each test temperature.

It should be noted that the Vialit test was chosen as the aggregate loss test for this evaluation of binders over an accelerated wheel loading test, that is, the one-third scale model mobile load simulator (MMLS3) test, due to the Vialit test's superior ability to differentiate between binder types (*10*).

Residual Binder Single-Edge Notched Beam Test. Residual binder fracture mechanics parameters were obtained from single-edge notched beam (SENB) tests conducted at −16°C, −22°C, and −28°C using a minimum of three replicates. The SENB test subjects a beam (made of binder) with a notch at its midpoint to constant displacement rate loading. For this study, the SENB tests were conducted using standard bending beam rheometer (BBR) beams (6.25 mm × 12.5 mm × 102 mm) with a 3-mm notch at the specimen center (*11*). These beams were loaded at a rate of 0.1 mm/s until failure in an electro-magnetic test system using air as the cooling medium to mimic field conditions. In a standard BBR test, ethanol is used as the cooling medium. However, Marasteanu et al. (*12*) demonstrated that the fracture resistance of asphalt is greatly diminished in ethanol compared with air.

Fracture energy and fracture toughness were both evaluated for the proposed low-temperature specifications. These parameters provide an indication of the ability of a cracked material to resist fracture. Fracture energy and fracture toughness were calculated using the force and displacement results obtained from the SENB tests. Fracture energy is calculated as the area under the force versus displacement curve up to the point of maximum force (*P*). Fracture toughness (K_{IC}) is calculated using Equations 1 and 2 (*11*):

$$K_{IC} = \frac{PL}{BW^{3/2}} f\left(\frac{a}{W}\right) \quad (1)$$

where P = maximum force (N), L = beam length (102 mm), a = notch depth (0.3 mm), W = beam width (12.5 mm), B = beam depth (6.25 mm), and:

$$f\left(\frac{a}{W}\right) = \frac{3\left(\frac{a}{W}\right)^{1/2}\left[1.99 - \frac{a}{W}\left(1 - \frac{a}{W}\right)\left(2.15 - 3.93\left(\frac{a}{W}\right) + 2.7\left(\frac{a}{W}\right)^2\right)\right]}{2\left(1 + 2\frac{a}{W}\right)\left(1 - \frac{a}{W}\right)^{3/2}} \quad (2)$$

Residual Binder Dynamic Shear Rheometer Test. Dynamic shear rheometer (DSR) temperature-frequency sweep tests were conducted using the 8-mm parallel plate geometry to obtain the residual binder dynamic rheological properties. These tests were conducted at temperatures ranging from 5°C to 15°C and frequencies ranging from 0.1 to 30 Hz using a minimum of two replicates. The test results were used to construct dynamic shear modulus ($|G^*|$) and phase angle (δ) mastercurves.

The dynamic rheological properties evaluated include: the dynamic shear modulus ($|G^*|$), phase angle (δ), $|G^*|\cdot sin\delta$, the Glover-Rowe parameter (*G-R*), and the crossover modulus ($|G^*|_c$). The parameter $|G^*|\cdot sin\delta$ was evaluated because it is used in the current hot binder performance-graded specifications to evaluate performance at an intermediate temperature. These specifications were developed under the assumption that pavement fatigue cracking is strain-controlled and that minimizing energy dissipation would improve pavement cracking resistance (*13*). The parameters $|G^*|$, δ, and $|G^*|\cdot sin\delta$ were predicted at the corresponding chip seal mixture test temperatures of −16°C, −22°C, and −28°C at a frequency of 12 Hz. The loading frequency of 12 Hz was selected based on the anticipated load pulse time at the pavement surface when subjected to 30 mph vehicular traffic (*14*). Snow plows induce the most severe aggregate loss at low temperatures, and 30 mph represents a typical snow plow speed.

Rowe proposed the G-R parameter (i.e., $|G^*|\cdot cos^2\delta/sin\delta$) as an indicator of binder ductility (*15*). Anderson et al. (*16*) demonstrated that the G-R parameter correlates with pavement durability (block cracking) when evaluated at 15°C and 0.0008 Hz (0.005 rad/s). Here, the research team evaluated the G-R parameter under these conditions recommended by Anderson et al. (*16*) and at the mixture performance test temperatures (−16°C, −22°C, and −28°C) and 12 Hz frequency that correspond to the chip seal mixture conditions.

The research team also evaluated the $|G^*|_c$ as a potential low-temperature specification parameter. The $|G^*|_c$ value is the $|G^*|$ value that corresponds to a δ of 45°. The $|G^*|_c$ is closely related to a mastercurve shape parameter, the *R*-value. The *R*-value is equal to the logarithmic difference between the glassy $|G^*|$ value and crossover modulus value and is proportional to the width of the relaxation spectrum of an asphalt binder (*17*). For most binders, the glassy modulus value is approximately 10^9 Pa (*16*). Thus, the *R*-value is inversely related to the asphalt binder $|G^*|_c$ value. Rowe et al. demonstrated that the binder *R*-value is highly correlated with strain-controlled asphalt mixture beam fatigue test results, with higher *R*-values (i.e., lower $|G^*|_c$ values) corresponding to better fatigue resistance (*18*). However, Rowe (*19*) proposed that a low binder *R*-value (and thus, higher $|G^*|_c$ value), is desirable for resisting age-induced durability (block cracking) and raveling in asphalt pavements. These $|G^*|_c$ values were always within the measured range of data acquired from frequency sweep tests at temperatures ranging from 5°C to 15°C. To evaluate a similar property under the low-temperature conditions at which the mixture tests were conducted, the $|G^*|$ values that correspond to a δ value of 5° were predicted and analyzed.

Residual Binder BBR Test. BBR stiffness (*S*[60]) and the m-value (*m*[60]) at 60 seconds of creep loading were measured in this study in accordance with AASHTO T 313 (*20*) using a minimum of two replicates. The current PG specifications assume that a lower stiffness value will result in less thermal stress accumulation on cooling in pavements and that a higher m-value will lead to a greater ability to relax such accumulated thermal stress, thereby reducing the thermal cracking potential. In addition, the research team evaluated creep stiffness at 8 seconds of loading (*S*[8]) as recommended for low-temperature chip seal binder specification by Vijaykumar et al. (*21*). Lastly, the team investigated the *delta T critical* (ΔT_c), which is equal to the difference between the low-temperature critical PG specification temperatures for *S*(60) and *m*(60), based on the results of Anderson et al. (*16*) who demonstrated a strong correlation between this parameter and pavement durability (block cracking).

Results

Relationships between Chip Seal Mixture Aggregate Loss and Residual Binder Properties

Residual Binder Fracture Mechanics Properties. Figure 1 presents relationships between chip seal aggregate loss and residual binder (*a*) fracture energy and (*b*) fracture toughness. Each data point in Figure 1 represents the average for multiple specimens at a given temperature. The figure demonstrates that the Vialit mixture tests were able to capture the aggregate loss difference between the modified and unmodified emulsions, as the modified binders show a clear reduction in aggregate loss. However, the binder fracture energy results shown in Figure 1*a* for the modified and unmodified binders overlap significantly and show poor correlation with the aggregate loss of the chip seal mixture.

The results presented in Figure 1*b* demonstrate that binder fracture toughness and chip seal aggregate loss are related. However, although a relationship is evident, the data show a significant amount of spread with some unexpected trends. The expectation was that an increase in binder fracture toughness with a change in temperature would correspond to a decrease in chip seal aggregate loss for a given binder. However, three of the six residual binders evaluated did not follow this expected trend, thereby prohibiting the reliable use of fracture toughness as a parameter in specifications.

Residual Binder Rheology.

Temperature-Dependent Creep Properties. Figure 2 presents relationships between chip seal aggregate loss and the BBR parameters evaluated: (a) *S*(60), (b) *m*(60), and (c) *S*(8). Each data point in Figure 2 represents the average for multiple specimens at a given temperature. The BBR parameters do not reflect the performance of modified versus unmodified binders and, thus, do not correlate with chip seal mixture aggregate loss, thereby precluding the use of BBR parameters in specifications.

Temperature-Dependent Dynamic Rheological Properties Figure 3 presents relationships between chip seal aggregate loss and (a) $|G^*|$, (b) δ, (c) $|G^*|\cdot sin\delta$, and (d) the G-R parameter, all evaluated at 12 Hz frequency. Each data point in the figure represents the average for multiple specimens at a given test temperature. The results demonstrate that the dynamic rheological properties presented in Figure 3 are comparable for the modified and unmodified binders. Therefore, the rheological properties shown

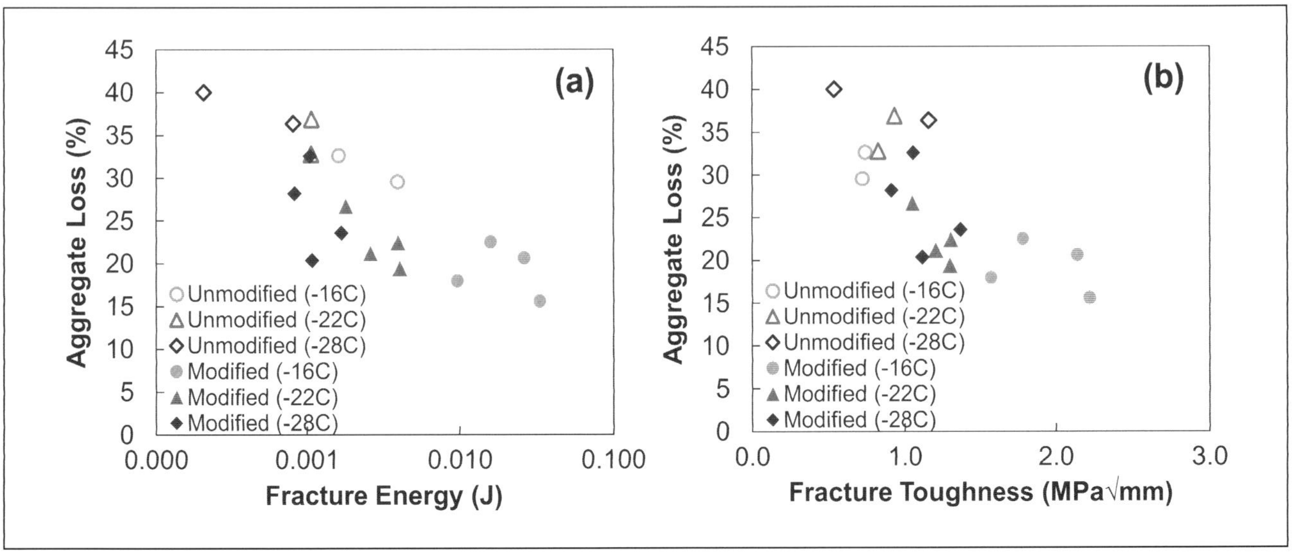

Figure 1. Relationships between chip seal mixture aggregate loss and residual binder (*a*) fracture energy and (*b*) fracture toughness.

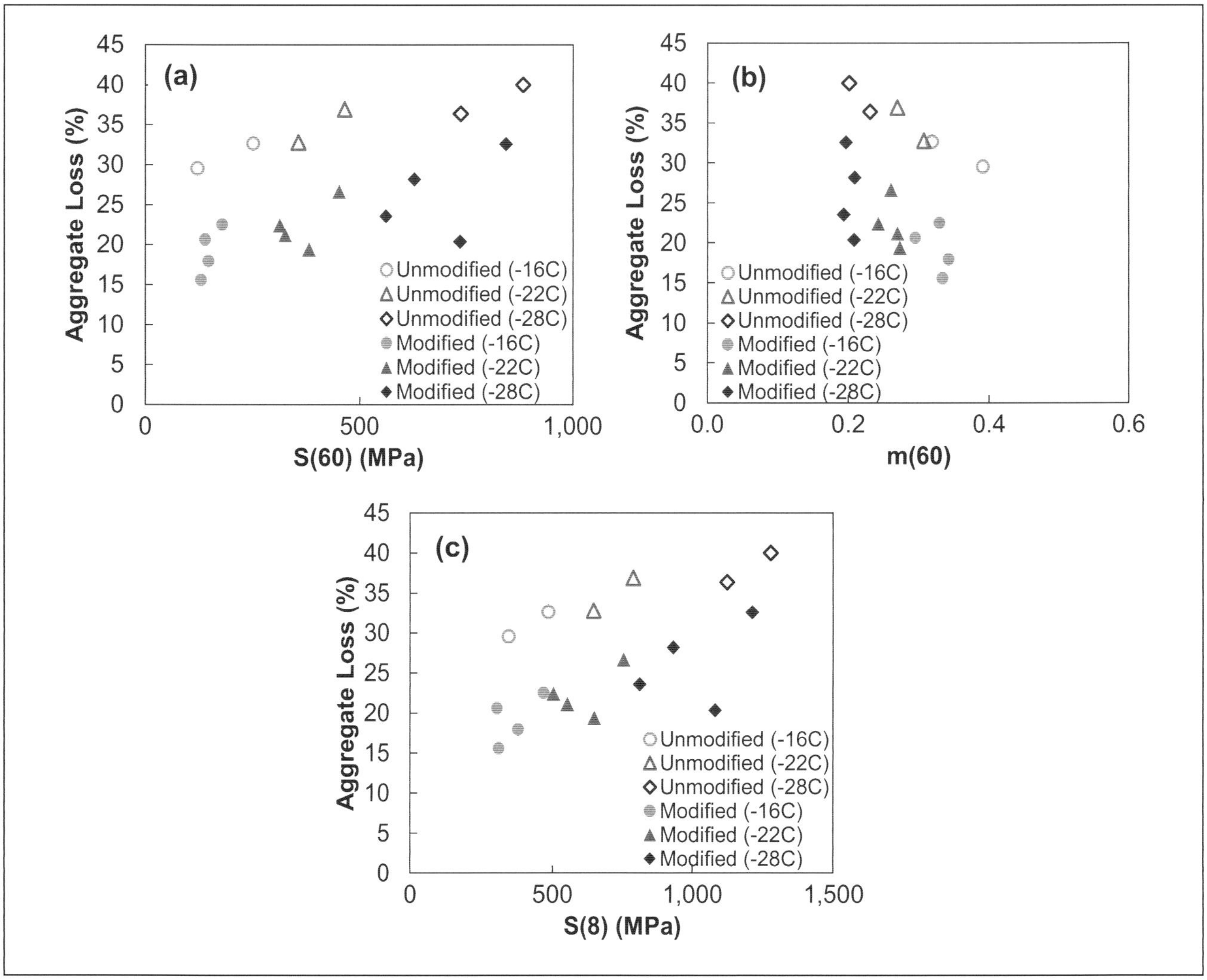

Figure 2. Relationships between chip seal mixture aggregate loss and residual binder: (*a*) *S*(60); (*b*) *m*(60); (*c*) *S*(8).

in Figure 3 are not deemed appropriate for specification purposes.

Temperature-Independent Rheological Parameters Related to Brittleness and Durability. Figure 4 presents the relationships between the chip seal mixture aggregate loss and (a) ΔT_c, (b) the G-R parameter at 15°C and 0.0008 Hz, (c) $|G^*|_c$, and (d) $|G^*|$ at δ of 5°. Note that, although mixture aggregate loss varies as a function of temperature, the binder parameters presented in Figure 4 do not; consequently, a separate data series is included for each mixture test temperature. Each data series includes the same residual binder properties but different aggregate loss values, corresponding to the different chip seal test temperatures. Excluding the results for the CRS-2L-F emulsion, the results presented in Figure 4 show that ΔT_c, $|G^*|_c$, and $|G^*|$ at δ of 5° are all related to chip seal aggregate loss at each temperature evaluated. Note that the aggregate loss of chip seal mixtures that contain CRS-2L-F emulsion does not vary significantly with temperature, unlike the other materials, which may explain the outlier behavior. The results presented in Figure 4*b* demonstrate that the G-R parameter at 15°C and 0.0008 Hz is not related to chip seal aggregate loss.

Figure 4*a* indicates that a lower ΔT_c value corresponds to better aggregate loss resistance. These results suggest that if a binder's stiffness value is lower for a given m-value, then better aggregate loss resistance is expected. Figure 4*c* shows that a lower residual binder $|G^*|_c$ value leads to less aggregate loss. This finding suggests that a higher *R*-value will improve performance, which matches the trends reported by Rowe et al. (*18*) for strain-controlled fatigue cracking in asphalt mixtures, thus also suggesting that aggregate loss in chip seals may be a strain-controlled phenomenon. However, the finding that a lower $|G^*|_c$ value is advantageous contradicts Rowe's findings regarding age-induced durability surface cracking in pavements, as Rowe reported that a lower

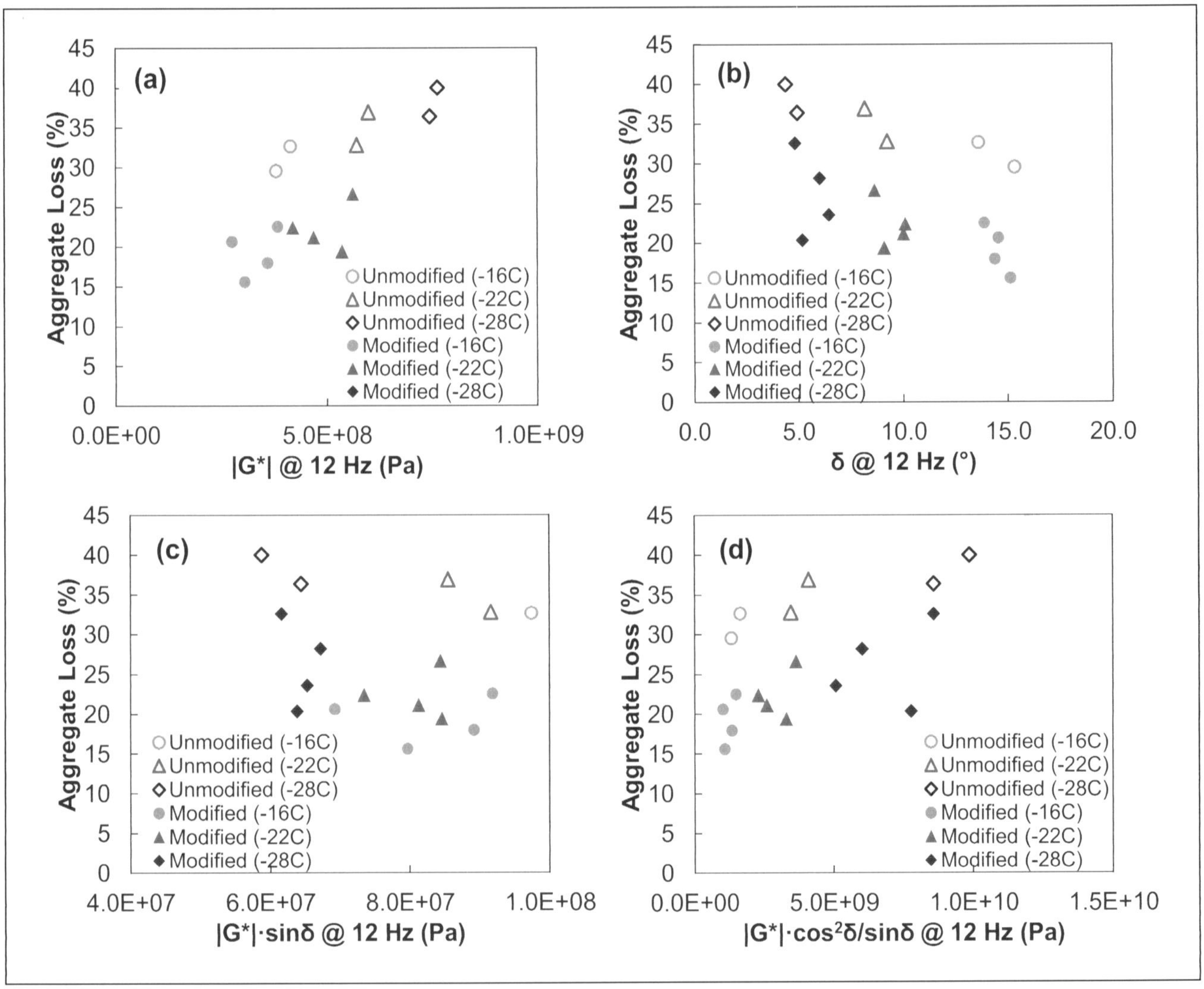

Figure 3. Relationships between chip seal mixture aggregate loss and residual binder: (*a*) $|G^*|$; (*b*) δ; (*c*) $|G^*|\cdot sin\delta$; (*d*) Glover-Rowe parameter.

R-value is advantageous (*19*). The $|G^*|_c$ value of a binder decreases with aging, and thus, findings here suggest that binders become less prone to aggregate loss as they age; this should be confirmed by future research.

The relationship between chip seal aggregate loss and $|G^*|$ at δ of 5°, shown in Figure 4*d*, closely resembles the relationship between chip seal aggregate loss and $|G^*|_c$, shown in Figure 4*c*. Thus, the results indicate that there is no advantage in relating binder and mixture results by using a critical phase angle condition that corresponds to low temperatures. For all of the binders evaluated, the crossover point occurred within the range of the measured data acquired from temperature-frequency sweep tests at temperatures ranging from 5°C to 15°C. Thus, the $|G^*|_c$ values were determined via interpolation of the data, whereas significant extrapolation was required to predict the $|G^*|$ value for a δ of 5°. DSR testing is not possible below 0°C for many rheometers, and thus, the ability to use an intermediate temperature measurement to control low-temperature performance is advantageous.

The m-value represents the slope of the logarithmic stiffness versus log time plot. The slope of the $|G^*|$ mastercurve is approximately equal to $\delta/90$ (*22*), and thus, the m-value and δ are related. The ΔT_c value is analogous to evaluating the *S*(60) at a given *m*(60) with a lower ΔT_c value, implying a lower *S*(60) for a given *m*(60). Thus, given the relationship between the m-value and δ, the finding that a lower ΔT_c value is better for aggregate loss resistance is consistent with the finding that a lower $|G^*|$ value for a given δ value is better for aggregate loss resistance. The DSR test is advantageous over the BBR test because the DSR requires smaller quantities of residue than a BBR. Therefore, the $|G^*|_c$

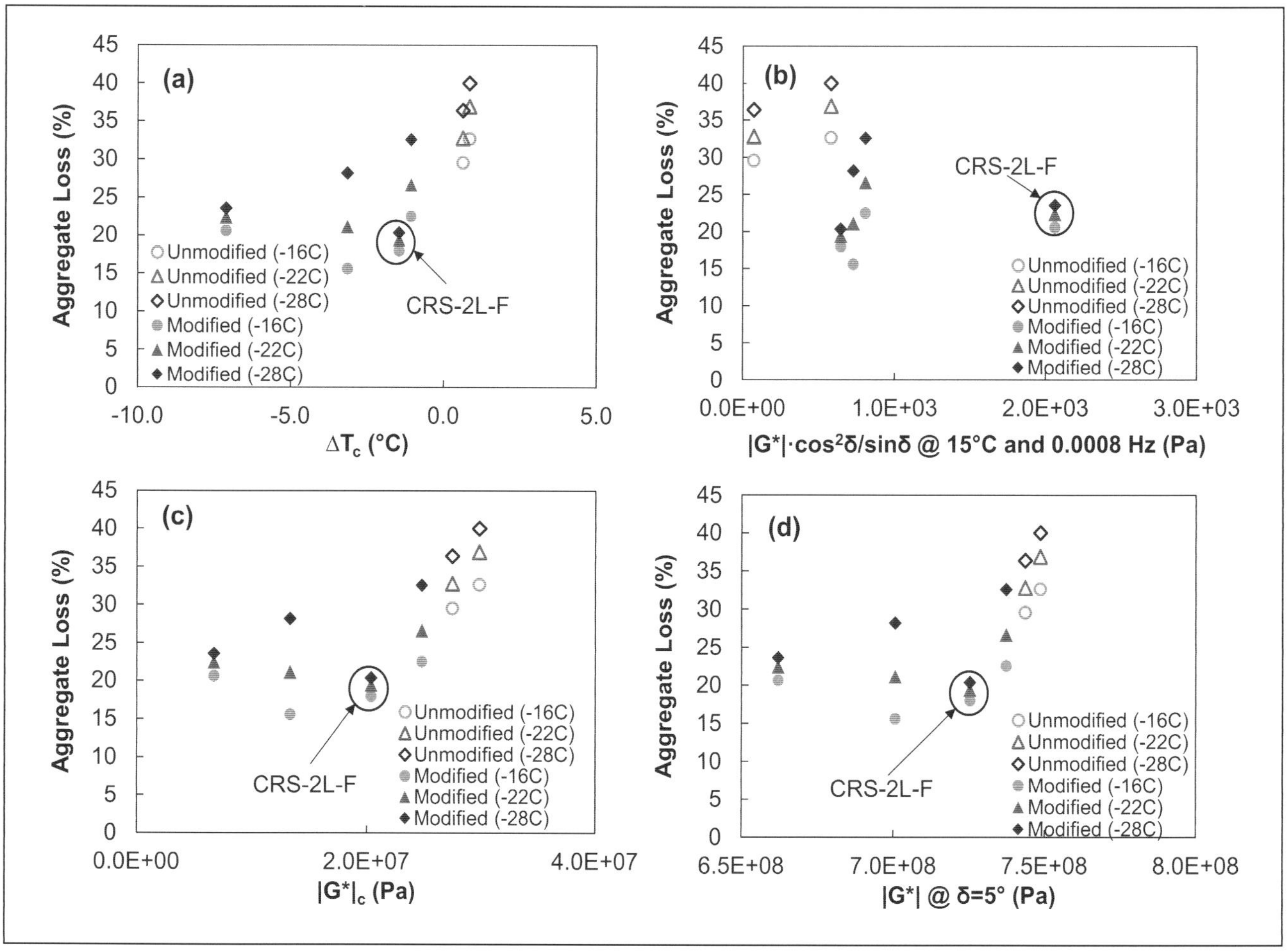

Figure 4. Relationships between chip seal mixture aggregate loss and residual binder: (*a*) ΔT_c; (*b*) G-R parameter at 15°C and 0.0008 Hz; (*c*) $|G^*|_c$; (*d*) $|G^*|$ at δ of 5°.

was selected as the parameter to use in developing the proposed low-temperature EPG specifications.

Proposed Specification Development

A preliminary EPG low-temperature specification was established based on the selection of the $|G^*|_c$ as the specification parameter. Specification development involved establishment of low-temperature climatic grades, establishment of traffic classifications, extension of the $|G^*|_c$ parameter for use in EPG determination, and the establishment of preliminary specification limits.

Low-Temperature Climatic Grades. Chip seals are placed on the surface of pavements. Therefore, the emulsion climatic PG for a given location should be determined based on the pavement surface temperature. High-temperature climatic PG requirements for a given location are based on a depth of 20 mm (*23*). Based on pavement-depth relationships, Kim et al. proposed that the high-temperature climatic EPG thresholds should be shifted 3°C higher than the PG thresholds (*3*).

The low-temperature grade in the current PG specifications is defined as the annual one-day minimum pavement surface temperature. Therefore, the depth correction that is required for a high temperature does not apply to the low-temperature grade determination. However, it would be simplest to shift both the high- and low-temperature grades from the PG specifications to those of the EPG specifications for simplicity (e.g., PG 64-22 versus EPG 67-19). To this end, in this study, the research team analyzed temperature data from the Long-Term Pavement Performance program database to determine the grade requirements for a given location if the grade thresholds were increased by 3°C. When determining the grade requirement for a given location, the continuous grade was rounded (conservatively) to the nearest available grade. For example, if the low-temperature continuous grade was determined to be −18°C, the PG low-temperature grade requirement

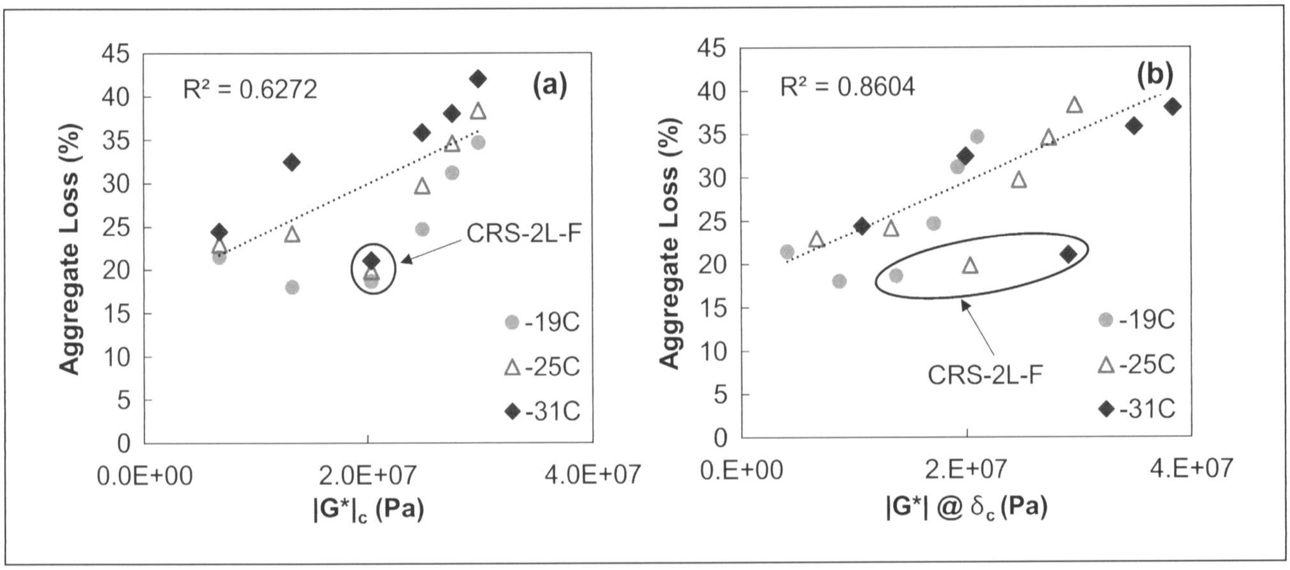

Figure 5. Chip seal mixture aggregate loss versus residual binder $|G^*|$ at δ_c corresponding to (*a*) $\delta = 45°$ and (*b*) δ_c as a function of temperature.

would be −22°C. The corresponding EPG low-temperature grade requirement would be −19°C if the available low-temperature PGs were increased by 3°C. As a result of this rounding, most locations would have an EPG low-temperature grade that is 3°C higher than its corresponding PG low-temperature grade. Furthermore, chip seals are designed for limited life, and thus, the chance of the temperature falling below a low-temperature grade selected based on the 98% reliability one-day minimum pavement temperature is quite low.

Table 1. Critical Phase Angle Values for Chip Seal Residual Binders

Low-temperature emulsion PG (°C)	δ_c (°)
−7	54
−13	51
−19	48
−25	45
−31	42
−37	39

Traffic Classifications. Chip seal aggregate loss at low temperatures is dependent on traffic level. Therefore, in this study, specification limits are defined similar to the performance grading in AASHTO M 332 (*24*), including high, medium, and low traffic levels. A survey of emulsion professionals, a literature review (*10*, *25*), and the authors' experience have led to the following recommendations for the following traffic classifications, categorized according to average annual daily traffic (AADT) volume:

- Low traffic: 0−500 AADT
- Medium traffic: 501−2,500 AADT
- High traffic: 2,500−20,000 AADT

Proposed Specification Development. The underlying concept used in performance-graded specifications is the existence of a temperature-independent relationship between the binder properties and mixture performance. This relationship allows the use of a single specification limit for various test temperatures as a function of climatic conditions, which is the basis for the current PG specifications. The crossover modulus is a unique property of a given binder. To adapt the $|G^*|_c$ concept for incorporation into the EPG specification framework, the research team determined critical phase angle (δ_c) values as a function of chip seal test temperature to produce a temperature-independent relationship between the mixture low-temperature performance and corresponding $|G^*|$ values. Critical phase angle values were adjusted to yield the highest correlation between aggregate loss and $|G^*|$ values. For the analysis, aggregate loss was estimated at the three most common EPG low temperatures by fitting a functional relationship to the measured aggregate loss versus temperature data using least squares regression.

Figure 5*a* shows the relationship between aggregate loss and the $|G^*|_c$. Figure 5*b* shows the relationship between aggregate loss and the $|G^*|$ when the δ_c values are adjusted as a function of the chip seal aggregate loss temperature. Table 1 presents a summary of the δ_c values

Table 2. Proposed Low-Temperature EPG Specification Limits

EPG specification test	Performance parameter	Traffic level	Temperature	EPG specification limit
DSR frequency sweep	Maximum $\|G^*\|$ @ δ_c	Low	5°C and 15°C	<30 MPa
		Medium		<20 MPa
		High		<12 MPa

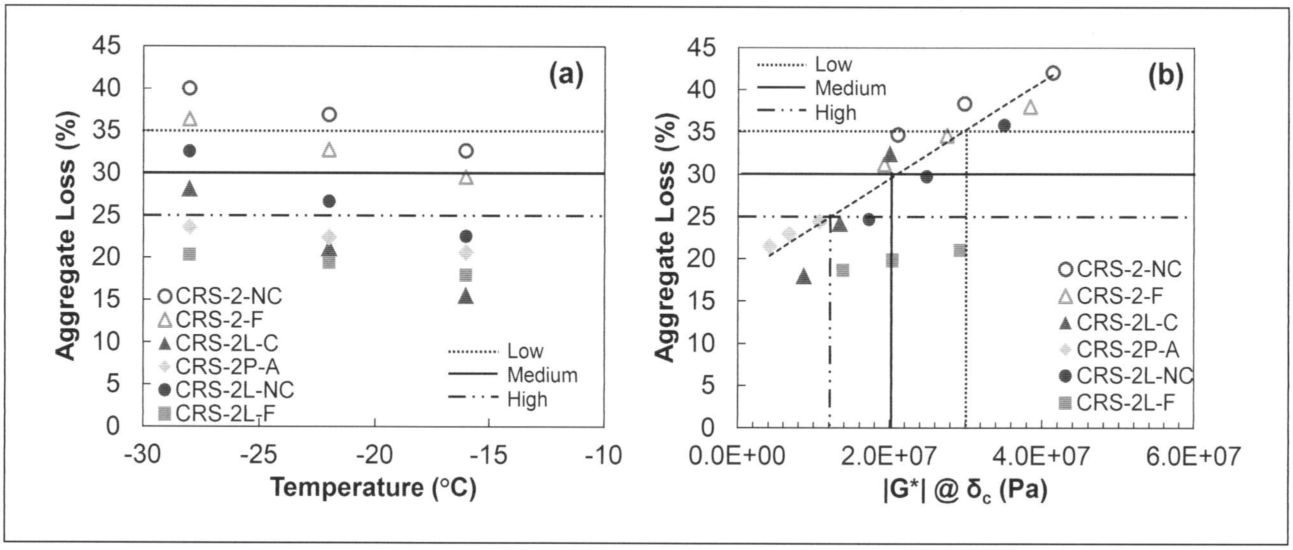

Figure 6. Low-temperature aggregate loss for chip seal mixtures (a) as a function of temperature and (b) versus residual binder $|G^*|$ at δ_c values.

for the expected range of EPG low temperatures. The δ_c value that corresponds to –25°C remains 45°. However, the δ_c values at other temperatures differ. These results demonstrate good agreement ($R^2 = 0.86$) between the aggregate loss and $|G^*|$ values irrespective of the mixture test temperature, with the exception of CRS-2L-F.

The research team established preliminary binder specification limits that are based on the values of the $|G^*|$ at δ_c values that correspond to the critical chip seal aggregate loss performance thresholds. Trends in the test results combined with *a priori* knowledge of the emulsions' performance were used to establish critical performance thresholds for a chip seal mixture's Vialit aggregate loss at low temperatures. Figure 6*a* presents the resultant relationship between chip seal mixture aggregate loss and temperature. The aggregate loss threshold for low traffic was set to 35% based on the known success of CRS-2-NC emulsion on low traffic roadways in a –19°C region. To determine the high traffic threshold, the team employed trends in aggregate loss with temperature. As shown in Figure 6*a*, the CRS-2L-NC emulsion experiences a significant increase in aggregate loss when the temperature is decreased from −19°C to −25°C. Similarly, the CRS-2L-C emulsion experiences an abrupt increase in the sensitivity of aggregate loss to temperature when the temperature drops below −25°C. In both instances, the aggregate loss before the observed abrupt loss of raveling resistance with the decrease in temperature is 25%. In addition, CRS-2P-A emulsion has been used successfully in −25°C regions on medium traffic roadways and passes this criterion, further validating its reasonableness. The research team selected the medium traffic aggregate loss threshold simply as the average of the high and low traffic thresholds. Then, the research team defined the residual specification limits based on the relationships between aggregate loss and the $|G^*|$ values that correspond to the δ_c values and the established allowable aggregate loss thresholds for low, medium, and high traffic volumes, as shown in Figure 6*b*. The maximum $|G^*|$ thresholds for low, medium, and high traffic volumes are 30, 20, and 12 MPa, respectively, and are summarized in Table 2.

Table 3 provides a summary of the corresponding PGs of the emulsions evaluated. The CRS-2-NC emulsion can be used only for low traffic levels in −19°C and −25°C regions. The results match intuition because this binder is currently used successfully on low traffic roadways in a −19°C region. With the exception of CRS-2-F at −19°C,

Table 3. Summary of Low-Temperature EPGs of Chip Seal Emulsions Evaluated

	Low-temperature emulsion PG		
Emulsion	−19°C	−25°C	−31°C
CRS-2-NC	Low	Low	FAIL
CRS-2-F	Medium	Low	FAIL
CRS-2L-C	High	Medium	Medium
CRS-2P-A	High	High	High
CRS-2L-NC	Medium	Low	FAIL
CRS-2L-F	Medium	Low	Low

medium and high traffic volumes require the use of a modified emulsion. Based on the known benefits of modification, these results are reasonable. Note that only one of the emulsions evaluated (CRS-2P-A) passes the high traffic criterion at both −25°C and −31°C. CRS-2P-A is the only emulsion evaluated that is currently used in a −25°C or colder region, and therefore, the performance grading results appear reasonable.

Conclusions and Recommendations

The following conclusions and recommendations are drawn from this study:

1. Based on analysis of both rheological and fracture properties combined with practical considerations, the most promising residual binder property evaluated for use in future EPG specifications to address low-temperature aggregate loss is the crossover modulus. To adapt the crossover modulus for use in EPG specifications, the critical phase angle values at which the dynamic shear modulus is evaluated should be varied as a function of the EPG low temperature. The dynamic shear modulus values that correspond to the critical phase angle values can be determined using DSR frequency sweep tests at 5°C and 15°C.
2. A maximum limit should be placed on dynamic shear modulus values at critical phase angle values to preclude premature failure due to low-temperature aggregate loss.
3. The effects of long-term oxidative aging on asphalt emulsion aggregate loss performance at low temperature should be investigated in future work. If aging is found to be significant, the specifications should be refined to include the evaluation of residue performance for long-term aged conditions.
4. The proposed specification limits should be refined by field validation in future work.

Acknowledgments

The authors would like to acknowledge the financial support from the National Cooperative Highway Research Program (NCHRP) under project NCHRP 9-50.

Author Contributions

The authors confirm contributing to the paper as follows: study conception and design: Cassie Castorena, Javon Adams, Richard Kim; data collection: Mohammad Ilias; analysis and interpretation of results: Cassie Castorena, Mohammad Ilias, Javon Adams, Richard Kim; draft manuscript preparation: Cassie Castorena, Richard Kim. All authors reviewed the results and approved the final version of the manuscript.

References

1. Johnston, J. B. and G. King. *Using Polymer Modified Emulsions in Surface Treatments.* Federal Lands Highway Interim Report, 2008.
2. Adams, J., C. Castorena, J. H. Im, M. Ilias, and Y. R. Kim. Addressing Raveling Resistance in Chip Seal Specifications. *Transportation Research Record, Journal of the Transportation Research Board*, 2017. 2612: 39–46.
3. Kim, Y. R., J. Adams, C. Castorena, M. Illias, J. Im, H. Bahia, and P. Johannes. *NHCRP Report 837: Performance-Related Specifications for Asphalt Binders Used in Preservation Surface Treatments.* Transportation Research Board of the National Academies, Washington, D.C., 2017.
4. American Association of State Highway and Transportation Officials. *Standard Practice for Recovering Residue from Emulsified Asphalt Using Low-Temperature Evaporative Techniques.* AASHTO PP 72. American Association of State Highway and Transportation Officials, 2013.
5. Kim, Y. R. and J. Adams. *Development of a New Chip Seal Mix Design Method.* Publication HWY-2008-04. N.C. Department of Transportation, 2011.
6. BS EN. *Surface Dressing - Test Method - Part 3: Determination of Binder Aggregate Adhesivity by the Vialit Plate Shock Test Method.* BS EN 12272-3. European Standards, 2003.
7. Jordan, W. S., and I. L. Howard. Applicability of Modified Vialit Adhesion Test for Seal Treatment Specifications. *Journal of Civil Engineering and Architecture*, Vol. 5, No. 3, 2011, pp. 215–223.
8. Ilias, M. *Development of Performance -Related Specification for Fresh Emulsions Used for Surface Treatments and Performance Study of Chip Seals and Microsurfacing.* PhD dissertation. North Carolina State University, Raleigh, NC, 2015.
9. American Society for Testing and Materials. *Standard Test Method for Sweep Test of Bituminous Emulsion Surface Treatment Samples.* ASTM D 7000. American Society for Testing and Materials, 2011.
10. Im, J. K. *Performance Evaluation of Chip Seals for High Volume Roads Using Polymer-Modified Emulsions and Optimized Construction Procedures.* PhD dissertation. North Carolina State University, 2013.

11. Velasquez, R. A., H. A. Tabatabaee, and H. U. Bahia. Low-temperature Cracking Characterization of Asphalt Binder by Means of the Single Edge Notch Bending (SENB) Test. *Journal of the Association of Asphalt Paving Technologists*, Vol. 80, 2011, pp. 583–614.
12. Marasteanu, M. O., A. C. Falchetto, M. Turos, and J. Le. *Development of a Simple Test to Determine the Low-temperature Strength of Asphalt Mixtures and Binders*. Highway IDEA Project 151 Final Report. National Research Council, Washington, D.C., 2012.
13. Hicks, R. G., F. N. Finn, C. L. Monishmith, and R. B. Leahy. Validation of the SHRP Binder Specification Through Mix Testing. *Journal of the Association of Asphalt Paving Technologists*, Vol. 62, 1993, pp. 565–614.
14. Brown, S. F. Determination of Young's Modulus for Bituminous Materials in Pavement Design. *Highway Research Record*, Vol. 431, 1973, pp. 38–49.
15. Rowe, G. Evaluation of the Relationship between Asphalt Binder Properties and Non-load Related Cracking. *Journal of the Association of Asphalt Paving Technologists, Prepared Discussion*, Vol. 80, 2011, pp. 649–663.
16. Anderson, R. M., G. N. King, D. I. Hanson, and P. B. Blankenship. Evaluation of the Relationship between Asphalt Binder Properties and Non-load Related Cracking. *Journal of the Association of Asphalt Paving Technologists*, Vol. 80, 2011, pp. 615–663.
17. Anderson, D. A., D. W. Christensen, H. U. Bahia, R. Dongre, M. G. Sharma, C. E. Angle, and J. Button. *SHRP A-369: Binder Characterization and Evaluation Volume 3: Physical Characterization*. Transportation Research Board of the National Academies, Washington, D.C., 1994.
18. Rowe, G. M., M. J. Sharrock, and S. Raposo. Cracking and Linear Viscoelastic Binder Properties. *Proc., 8th RILEM International Conference on Mechanisms of Cracking and Debonding in Pavements*, Vol. 13, 2016, pp. 3–8.
19. Rowe, G. M. Linear Visco-Elastic Binder Properties and Asphalt Pavement Cracking. *Proc., 11th Conference on Asphalt Pavements for Southern Africa (CAPSA15)*, Sun City, South Africa, 2015.
20. American Association of State Highway and Transportation Officials. *Standard Method of Test for Determining the Flexural Creep Stiffness of Asphalt Binder Using the Bending Beam Rheometer (BBR)*. AASHTO T 313. American Association of State Highway and Transportation Officials, 2012.
21. Vijaykumar, A., A. Epps-Martin, and E. Arambula. Revision and Further Validation of Surface Performance-Graded (SPG) Specification for Chip Seal Binders. *Transportation Research Record, Journal of the Transportation Research Board*, 2013. 2370: 44–52.
22. Anderson, D. A., and M. O. Marasteanu. Continuous Models for Characterizing Linear Viscoelastic Behavior of Asphalt Binders. In *International Society for Asphalt Pavements Workshop on Asphalt Binders and Mastics*, Madison, WI, 2011.
23. Mohseni, A., S. H. Carpenter, and J. D'Angelo. Development of Superpave High-Temperature Performance Grade (PG) Based on Rutting Damage (With Discussion and Closure). *Journal of the Association of Asphalt Paving Technologists*, Vol. 74, No. 1, 2005, pp. 197–254.
24. American Association of State Highway and Transportation Officials. *Standard Specification for Performance-Graded Asphalt Binder Using Multiple Stress Creep and Recovery Test*. AASHTO M 332. American Association of State Highway and Transportation Officials, 2016.
25. Gransberg, D. D. and D. M. B. James. *NCHRP Synthesis Report 342: Chip Seal Best Practices*. Transportation Research Board of the National Academies, Washington, D.C., 2005.

The Standing Committee on Asphalt Binders (AFK20) peer-reviewed this paper (18-01198).

Article

TRR
JOURNAL OF THE TRANSPORTATION RESEARCH BOARD

Transportation Research Record
2018, Vol. 2672(28) 277–289

DOI: 10.1177/0361198118797205
journals.sagepub.com/home/trr
SAGE

Impact of Recycled Materials and Recycling Agents on Asphalt Binder Oxidative Aging Predictions

Sara Pournoman[1], Elie Y. Hajj[2], Nathan Morian[3], and Amy Epps Martin[4]

Abstract

The overall objective of this study was to evaluate the influence of selected recycling agents (RAs) and recycled materials on the development of cracking potential with respect to oxidative aging. Given the complex nature of varying base asphalt binders, recycled materials, whether recycled asphalt pavement (RAP), reclaimed asphalt shingles (RAS), or both, and the complexity of their combined interactions with recycling agents, standard evaluation protocols for binder grading and evaluation may be insufficient. The binder blend aging predictions or oxidation modeling evaluation was introduced as a means to evaluate the combined influence of both binder oxidation kinetics and resulting rheological changes on the measured cracking potential of the various binder blends—that is, Glover–Rowe (G-R) parameter—driven by temperature estimation modeling over simulated in-service durations at example geographic locations. This evaluation has demonstrated the importance of adequate characterization of the specific materials being used in conjunction with selection of the correct dose of the appropriate recycling agent to ensure sufficient resistance to cracking and embrittlement of proposed material combinations. The combined influence of all the interested components did not always add up to the sum of the individual parts, nor are the measured interactions consistent with increased levels of oxidation. Therefore, the prevailing conclusion of the study as a whole indicated that material-specific evaluations are needed to identify the complex interactions taking place within the material combinations of interest, but also multiple levels of aging at appropriate intervals may be necessary for comprehensive characterization.

Among many types of recycled materials that are routinely used in asphalt mixtures in the United States, recycled asphalt pavement (RAP) materials are the most common in the asphalt industry. The use of reclaimed asphalt shingles (RAS) in asphalt mixtures has recently received more attention since this type of material contains increased quantities of binder, usually 20–30% binder by total weight (*1*). Several studies have shown that mixtures containing 10–30% RAP, commonly used in the US asphalt industry, are performing as well as virgin asphalt mixtures (i.e., 0% RAP). However, numerous state highway departments of transportation and contractors are continually looking to incorporate larger amounts of recycled materials into asphalt mixtures without sacrificing the fatigue and other cracking resistance characteristics of the mixture (*2–10*).

In this regard, the addition of recycling agents (RAs) to the recycled blends as a restoration tool has been recently explored by several researchers. Although true restoration and softening mechanisms are not fully understood, several investigations have been conducted that indicate the effectiveness of RA depends on the uniform dispersion of the RA within the mixture, as well as diffusion of the RA into the aged binder (*11*). In addition, not all types of RAs have a similar influence on the chemical or physical properties of an aged binder (*12–15*). In selecting an RA for a specific material combination, it is crucial to determine the best type and dose of the RA that provides the highest restoration level. However, it is almost impossible to state which type of RA is best for a particular binder without first testing

[1]Dynatest North America Inc., Ventura, CA
[2]Department of Civil and Environmental Engineering, University of Nevada, Reno, NV
[3]Nevada Department of Transportation, Carson City, NV
[4]Department of Civil Engineering, Texas A&M University, College Station, TX

Corresponding Author:
Address correspondence to Nathan Morian: nmorian@dot.nv.gov

the impact of the RA on the properties of the specific aged recycled binder, including compatibility and temperature susceptibility (*16–19*). While low dose are not always able to restore the recycled binder blend to that of the virgin binder, the use of higher dosage of RAs have been found to soften the aged binders, which is beneficial for cracking resistance. However, there is a potential for these high dose to negatively impact the mixture resistance to permanent deformation. Therefore, the dose should be carefully optimized in this range (*20*).

Attempts to simplify the dose selection or to follow traditional blending and binder grading protocols have identified the need for more comprehensive aging evaluations owing to oxidation level-specific variations in the effectiveness of certain blended materials. Thus, robust evaluations necessarily must include varied aging levels or explore rates of oxidation to provide comprehensive findings. To better understand the aging behavior of the recycled material, many efforts have been conducted in the asphalt industry using several oxidation models, including the Global Aging System (GAS), the Western Research Institute (WRI) model, and the Texas A&M methodology, which are all summarized elsewhere (*21*). The latter two are much more complicated when compared to the GAS model, which consists of regression relationships based on viscosity. The WRI procedure is a dual-mechanism kinetics model (*22*, *23*) that primarily characterizes the oxidation products through the development of ketone and sulfoxide functional groups. To incorporate the respective pavement temperature and associated oxidation models, this study was conducted following protocols more closely aligned with the Texas A&M methodology. This methodology builds upon previous work with this particular model system, including aging protocols, testing procedures, and analysis tools (*21*, *24*, *25*).

The experimental matrix used in this study produced parameters and measures specifically to utilize the Texas A&M aging prediction methodology, determine the aging characteristics of the binder blends evaluated, and finally predicted the aging of the binder blends in a simulated pavement. These aging predictions readily permit the observations of the material-specific physical and rheological behavior of the complex interactions taking place within the binder blends, leading to practical correlations with mixture characteristics and field performance measures.

Research Objective

The overall objective of this study was to evaluate the influence of RAs and recycled materials on the development of cracking potential with respect to oxidation of various binders at different geographic locations. The binder blend aging predictions or oxidation modeling evaluation was introduced as a means to evaluate the combined influence of both binder oxidation kinetics and resulting rheological changes on the measured cracking potential of the various binder blends—that is, Glover–Rowe (G-R) parameter—driven by temperature estimation modeling within a given pavement section at selected geographic locations.

This effort partially stemmed from the highly material-specific influence of multiple components observed through various modes of laboratory testing to evaluate the level of aging and rheological characterization of multiple binder blends and components. The study included neat and modified asphalt binders, RAP and RAS, as well as two types of RAs to evaluate the complex interactions of these varied components as a function of simulated in-service aging conditions.

Prediction Methodology

The oxidative aging of the respective materials was modeled as the evolution of carbonyl functional groups, represented as carbonyl area (CA), with time for the asphalt binder at a specific location. These predictions were accomplished using diffusion-based oxidative aging models developed at Texas A&M University (*26*). The inputs for the aging model are summarized as follows:

- predicted hourly pavement temperatures at selected depths within the asphalt concrete (AC) layer;
- asphalt binder aging kinetics, activation energy (E_a), and hardening susceptibility (HS);
- asphalt binder initial carbonyl at the beginning of constant-rate aging; and
- average representative air void radius and effective aging distance (e.g., asphalt binder film).

In summary, the CA values at various sublayers within the asphalt binder film were predicted by determining the oxygen partial pressure at each sublayer and the oxidative aging kinetics of the asphalt binder. The oxygen partial pressures were estimated from a numerical solution of a partial differential equation with age-dependent diffusivity. The age-dependent diffusivity was defined by relating the asphalt binder low shear viscosity (LSV) to the hardening susceptibility properties (HS and m) of the binder. Details of the aging model can be found in the literature (*24*, *26*). The aging model was simplified by modeling specific locations within the pavement structure, such as oxidation levels at the pore or air void interface rather than all points throughout the binder film. In this case, the model required pavement temperature profile, asphalt binder kinetics parameters (E_a and A), and

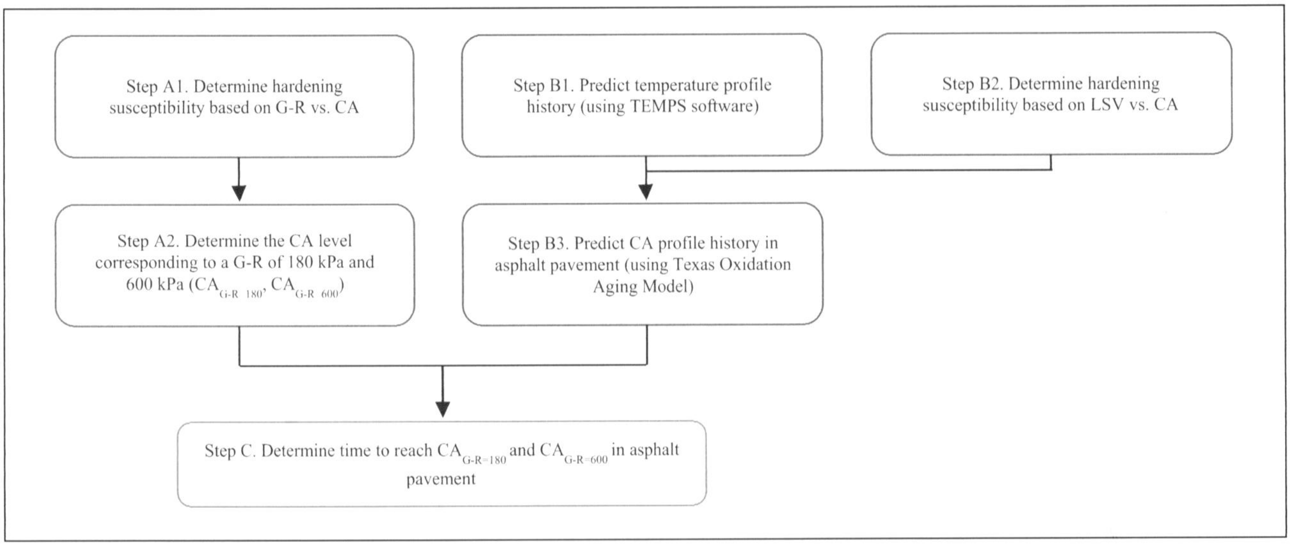

Figure 1. Asphalt pavement life prediction outline.

CA_{Tank}. Numerical computation of the aging model was performed using the finite control volume method (FCVM) in implicit time scheme. Oxidative aging kinetics and hardening parameters of the model were determined through measurements of CA and LSV of the asphalt binder after being aged at various combinations of time and temperature. The CA of the asphalt binder was measured by Fourier transform infrared (FT-IR) spectroscopy, and the LSV was determined from the complex modulus master curve, which was measured using a dynamic shear rheometer (DSR) (*21*). A general outline of the overall process is provided in Figure 1.

The use of this procedure permits the relative comparison of binder blends evaluated for an estimated time to reach the two predefined G-R thresholds (i.e., 180 kPa and 600 kPa) for the onset and significant cracking levels, respectively, as a function of in-service time at a given environment or physical location. While these are hypothetical aging simulations utilizing measured chemical and rheological binder properties in a particular environmental condition, they are valuable comparisons to make between respective binder blends. The detailed steps of the methodology are provided as follows.

Temperature Profile Prediction

Accurate pavement temperature data at various depths of the AC layer are required to estimate the oxidation level of the binder blend over time. A model has been employed to predict pavement temperatures at various pavement and sublayer depths (up to 3 m below the pavement surface) over a predefined analysis period. The development of the proposed model was primarily derived from efforts conducted at Texas A&M University (*27*) and Arizona State University (*28*). The model utilizes hourly air temperature, solar radiation, wind speed, and thermal diffusivity properties of the pavement layer materials (i.e., conductivity, density, and heat capacity), as well as monthly variable pavement surface radiation properties (i.e., albedo, emissivity, and absorption coefficients). In this particular study, the hourly air temperature and wind speed data were obtained from the Enhanced Integrated Climatic Model database, with synchronized solar radiation data gathered from the National Solar Radiation Database. Additional advancements of the prediction methodology resulted in numerical calculations of the proposed heat transfer model using FCVM in a fully implicit scheme (*24*). The improved time-dependent surface boundary condition accounts for a variation in the surface radiation properties, allowing for a better simulation of heat transfer in the pavement and thus more realistic predictions of pavement temperatures. Details of the model components, calculation approach, and validation results are available in the literature (*29*). This model has been optimized and resulted in a stand-alone calculation tool called the Temperature Estimate Model for Pavement Structure (TEMPS) (*30*). Figure 2 presents an example of predicted hourly pavement temperatures for one year of analysis (June 2014 to May 2015) at respective depths within the AC layer for Matterhorn Blvd. in Reno, Nevada.

Once the environmental inputs to the oxidative aging prediction model were determined, they were combined with material-specific aging properties, such as oxidation kinetics and HS, discussed in the following sections, to model the overall extent of aging within the pavement structure throughout the in-service lifespan.

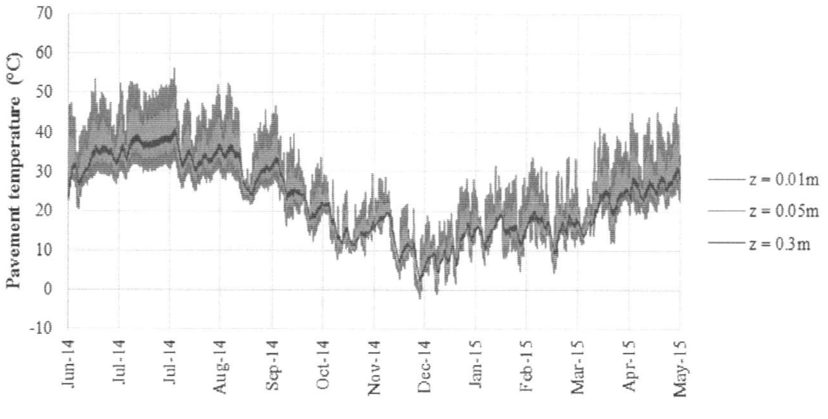

Figure 2. Predicted pavement temperature at various depths of asphalt mixture surface layer in Reno, NV environment.

Oxidation Kinetics

The primary method for evaluating the oxidation rate is through the measurements of carbonyl functional groups relative to each oxidation temperature at specific durations. The CA was calculated as the area, in arbitrary units, between the attenuated total reflectance (ATR) infrared absorption spectrum and the magnitude of the absorption at 1,524 and 1,820 cm^{-1}, used as a baseline. This area was integrated from 1650–1820 cm^{-1} wavenumbers. The CA values were normalized to conduct the analysis based on carbonyl growth (i.e., CA_g), or the oxygen uptake from the initial starting point of the evaluation. This was done in part to account for the respective levels of carbonyl functional groups present in the blended materials, which could be drastically different between the respective base binders, additives, and RAs. It also permits the analyses to initiate from a common point of reference, such as CA_{Tank}. The normalized CA values were calculated by subtracting the CA at the unaged level (i.e., CA_{Tank}) from the measured CA at each of the respective aging levels. An example of typical oxidation kinetics relationships is presented in Figure 3*a*, fit to the overall model function presented by Equation 1.

$$CA_g = M^*\left(1 - e^{-k_f t}\right) + k_c t \quad (1)$$

where CA_g is the carbonyl area growth ($CA - CA_0$); CA is the carbonyl area; CA_0 is the original or tank CA measurement; M is the initial jump, magnitude of fast rate reaction in terms of CA_g; k_f is the fast rate of CA growth; k_c is the slow or constant rate of CA growth; and t is time in days.

Figure 3*a* presents an example of a PG 64-28 base binder with the CA_g measurements and predicted curves resulting from the oxidative aging prediction model indicated by single symbols and dashed lines, respectively. The CA_g was highly dependent on the aging temperature, with separate fast and constant oxidation rates observed within each temperature path.

The duration of the fast rate was much shorter than the constant rate as the temperature increased; this phenomenon was consistent with the chemistry principle described by the Arrhenius reaction function presented in Equation 2. The oxidation kinetics parameters are defined by the fast and constant oxidation rates, k_f and k_c, respectively, as presented in Equation 1.

$$r_{CA} = AP^{\alpha}e^{-E_a/RT} \quad (2)$$

where r_{CA} is the rate of carbonyl area growth (CA_g), either k_f or k_c; A is the pre-exponential factor; P is the absolute oxygen pressure during oxidation (atm); α is the reaction order with respect to oxidation pressure; E_a is activation energy (kJ/mol); R is the ideal gas constant (8.3144621 L/mol·°K); and T is temperature (°K).

Comparisons in the current study were based on the constant oxidation rates (k_c) as a function of the inverse of the aging temperature multiplied by the gas constant, R, as represented in Figure 3*b*. This parameter (i.e., k_c) can be considered as a direct indication of the temperature dependency of the long-term oxidation rate (e.g., units of CA_g/day), understood to provide the most significant contribution to long-term aging of asphalt materials.

Hardening Susceptibility

One of the most significant parameters in characterizing the binder oxidation properties is HS, which is defined as the relationship between the binder rheology and aging. Historically, the LSV and CA from the FT-IR measurements were used to characterize the binder stiffness and aging, respectively.

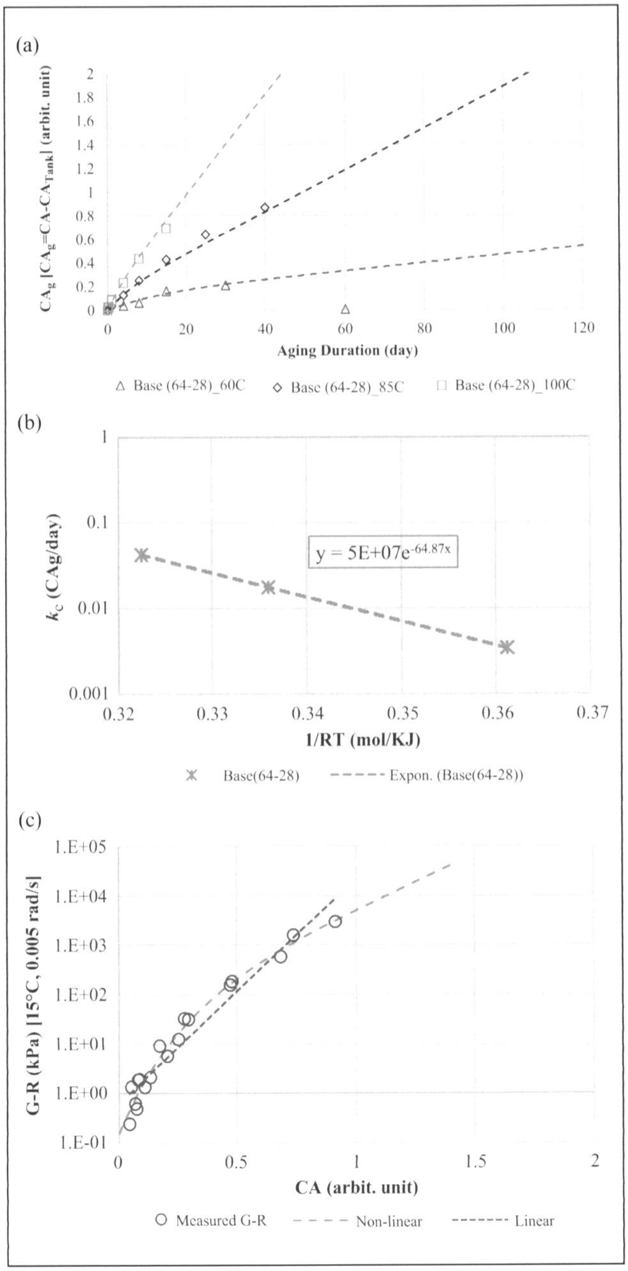

Figure 3. (*a*) Oven aging; (*b*) oxidation kinetics; and (*c*) hardening susceptibility of the PG 64-28 base binder for the G-R parameter.

Low Shear Viscosity

LSV has been identified as a rheological indicator of asphalt binder to represent the capability of asphalt mix to resist shear deformation at high temperatures. To determine LSV, the complex viscosity (G*) is plotted as a function of testing frequency. This plot creates a clear plateau in complex viscosity with lower frequencies. It should be mentioned that the LSVs in the current study were measured at a temperature of 60°C and a reduced frequency of 0.001 rad/s (*21*).

The corresponding HS is a linear relationship between the LSV and CA, mathematically defined by Equation 3.

$$\ln LSV = HS*CA + m \tag{3}$$

where: LSV is the low shear viscosity of the asphalt binder; HS is the hardening susceptibility, slope of relationship; CA is the carbonyl area, arbitrary units – unitless; and m is the intercept of the $\log\eta_0^*$ and CA relationship.

Glover–Rowe Parameter

Another rheological parameter becoming more commonplace is known as the Glover–Rowe (G-R) parameter, which was originally defined by Glover et al. (*31*) as the DSR function $(G'/(\eta'/G'))$ and reformulated by Rowe (*32*) in a discussion of Anderson et al. as the G-R parameter:

$$G\text{-}R = G'/(\eta'/G')/\delta = G^*(\cos\delta)^2/\sin\delta \tag{4}$$

where, G^* is the complex dynamic shear modulus; G' is the storage or elastic shear modulus; η' is the storage dynamic viscosity; $\eta' = G''/\omega$; and δ is the phase angle in degrees.

Each of the G-R components were determined at 0.005 rad/s and 15°C, largely based on the observed correlations with ductility and thus cracking resistance, as well as binder oxidation levels (*33*). Kandhal (*34*) proposed a damage zone between cracking onset and significant cracking thresholds, where cracking likely begins owing to brittle rheological behavior defined by low ductility values of 5 cm to 3 cm, respectively. These ductility measures have later been correlated to the G-R thresholds of 180 and 600 kPa, respectively (*31*). The G-R parameter captures both rheological parameters needed to characterize binder viscoelastic behavior: stiffness (G^* at high and intermediate temperatures) and phase angle (δ at high and intermediate temperatures).

In this study, the HS relationships were also determined with respect to the G-R parameter as a binder stiffness parameter. However, in this case not all the HS relationships were verified to have a linear relationship (especially in the case of polymer-modified binders). Similar to the two-phase kinetic relationships described previously, separate fast and constant HS rates were also observed in the G-R hardening susceptibility plots; therefore, a nonlinear, two-phase equation was developed to mathematically characterize the relationship between the G-R parameter and CA_g. Equation 5 represents the most recent update of the equation. Also, Figure 3*c* indicates the binder HS measures and predictions for PG 64-28 with both the linear and nonlinear equations, in which the latter equation seems to represent a more robust fit.

Table 1. Recycling Agents Properties

Category	Description	ID
Aromatic extracts	Refined crude oil products with polar aromatic oil components	A1
Tall oil	Plant extracts including long-chain organic acids, resin acids, and fatty acids	T1

Table 2. Recycled Materials Identification

Category		Description	Source
RAP		Recycled asphalt pavement	TX
RAS	MWAS	Manufactural waste asphalt shingles	TX
	TOAS	Tear-off asphalt shingles	TX

$$\ln(\text{G-R}) = M\left(1 - e^{-k_f' CA_g}\right) + k_c' CA_g + \ln(\text{G-R})_0 \quad (5)$$

where: G-R is the Glover–Rowe parameter (kPa) at 15°C and 0.005 rad/s; G-R_0 is the initial Glover–Rowe parameter (kPa) at 15°C and 0.005 rad/s; CA_g is the carbonyl area growth; k_f' is the fast rate of G-R growth; and k_c' is the constant rate of G-R growth.

Experimental Plan

The asphalt binder blends in this study include two unmodified binder sources, PG 64-22 (TX) and PG 64-28 (NH), and one SBS modified (NV), which originated as a subset of the oxidation studies conducted as part of the NCHRP 9-58 project. As a consequence of that effort, the binders evaluated as part of this effort also include binders that have been treated with a variety of RAs and recycled materials including RAP and RAS. Tables 1 and 2 provide a summary of the evaluated RAs and recycled materials, respectively.

Detailed information on the nomenclature of the binder blends is presented in Table 3. To clarify, the RAs in Table 3 are categorized as either rPGL or FLD; rPGL is an abbreviation for restored low PG grade, denoting the dose of the RA that restored the binder low PG grade; the FLD designation indicates the dose used in the field project if it were not the determined rPGL value. The calculated RA dose are represented as binder replacement in the evaluated binder blends.

Forced-Draft Oven Aging

The binder blends were aged to determine the oxidation kinetics as a function of time and temperature when exposed to free atmospheric air. Accordingly, 1 mm thick asphalt binder films covering 140 mm diameter pressure aging vessel pans were placed in forced-draft ovens under free atmospheric air pressure at the specific temperatures and durations shown in Table 3.

At their respective aging conditions, replicate binder samples were evaluated by DSR testing at multiple temperatures to determine the rheological parameters utilized in this study. The rheological measures were determined by conducting isothermal frequency sweeps at multiple temperatures ranging from 4°C to 100°C, depending on the particular binder. The isotherms were then shifted into master curves of dynamic shear modulus ($|G^*|$) and phase angle (δ), utilizing the Rhea software package (*21*, *35*).

Oxidative Aging Prediction Model Inputs

To determine the inputs for the oxidative aging prediction model, the hourly pavement temperature profile was determined at the pre-selected depth of 0.01 m, utilizing the TEMPS software. Then, the oxidative aging kinetics including the activation energy (E_a) and pre-exponential factor (AP^α) were determined for all binder blends at each respective aging temperature through the correlation between the CA values and the aging durations. The LSV-based HS plots are employed in the TX oxidative aging prediction model for the three base binders and their respective blends.

The third input for the oxidative aging prediction model is the initial CA of the asphalt binder at the beginning of model prediction. In this study, it was considered appropriate to conduct the CA prediction from the approximate aging level of the plant-produced material, that is, after short-term aging. Therefore, the CA predictions were conducted with the RTFO-aged binder as the CA prediction starting point (CA_{RTFO}).

Lastly, the average representative air void radius and effective aging distance (e.g., asphalt binder film) was assumed to be 0.65 and 1 mm, respectively, based on research by Han (*27*) using X-ray computed tomography

Table 3. Binder Blend Identification and Aging Experimental Plan

Binder ID	Base binder type	RAP (RAP binder ratio)	RAS (RAS binder ratio)	RA	Dose by replacement (%)	Forced-draft oven aging: Temperature (°C) 60	85	100
Base (64-22)	TX PG64-22	–	–	–	–	1, 4, 8, 15, 30, 60, 100 days	1, 4, 8, 15, 25, 40 days	2 hr, 6 hr, 1, 4, 8, 15 days
Base w/T1 @ TX FLD[1] (64-22/2.65%)		–	–	$T1^{FLD}$	2.65			
Recycled w/T1 @ TX FLD (64-22/2.65%/0.25 TX RAP, 0.25 TX MWAS)		TX (0.10)	TX MWAS (0.20)	$T1^{FLD}$	2.65			
Base (64-28)	NH PG64-28	–	–	–	–			
Base w/T1 @ TX FLD (64-28/2.65%)		–	–	$T1^{FLD}$	2.65			
Base w/A1 (64-28/6%)		–	–	A1	6			
Recycled w/T1 @ rPGL[2] (64-28/12.5%/0.25 TX RAP, 0.25 TX TOAS)		TX (0.25)	TX TOAS (0.25)	$T1^{rPGL}$	12.5			
Recycled w/A1 @ rPGL (64-28/6%/0.40 TX RAP)		TX (0.40)	–	$A1^{rPGL}$	6			
Base (64-28P)	NV PG64-28P	–	–	–	–			
Base w/T1 @ TX FLD (64-28P/2.65%)		–	–	$T1^{FLD}$	2.65			
Recycled w/T1 @ rPGL (64-28P/11%/0.25 TX RAP, 0.25 TX TOAS)		TX (0.25)	TX TOAS (0.25)	$T1^{rPGL}$	11			

Note: [1] = Field dosage; and [2] = Restored low PG grade dosage.

techniques and image processing. The values used in this study were within the range of observed values for a typical asphalt mixture.

A summary of the input details for the utilized binders from the experimental plan are shown in Table 4, noting the temperature-dependent values of k_c are not individually input to the oxidative aging prediction but are combined into the oxidation kinetics (e.g., E_a and pre-exponential factor, as noted in Equation 2). Additional details and a more robust description of the modeling and inputs may be found in the literature (*25*).

Results and Discussion

Modeling of Binder Oxidation

Using the oxidative aging prediction model (*24*, *36*) with the corresponding material and environmental inputs discussed in the previous section, CA values were predicted for all the experimental binder blends. A summary of the CA_g predictions for the blended binders representing 0.01 m below the pavement surface in Tyler, TX with the PG 64-22, Durham, NH for the PG 64-28, and Reno, NV for the PG 64-28P binder blends are provided in Figure 4. As an interim summary of Figure 4, it can readily be observed that the different base binders in combination with their respective geographic locations resulted in different levels of oxidation, shown as CA_g. In addition, it can further be observed that within each base binder group, the differing components added to the respective base binders did not always have a consistent influence. For instance, the addition of the T1 RA at the TX FLD dose had a fairly minor influence on the PG 64-22 and PG 64-28 binders, but there was a substantial decrease in the CA_g rate with the polymer-modified PG 64-28P. By similar comparison, the addition of the T1 to the PG 64-28 had practically no influence on the oxidation rate, while the A1 RA did have a measurable influence, noting the higher rPGL dose rate as well as the differing chemical makeup of the respective RAs.

Additional observations noted the reduction in CA_g with both of the blends containing RAP and tear-off asphalt shingles (TOAS), but an increase in the oxidation with the PG 64-22 that included RAP and manufactural waste asphalt shingles (MWAS). Perhaps of greater influence is the difference in the dose rates of the respective RAs, with the PG 64-28 and PG 64-28P receiving rPGL dose while the PG 64-22 blend was at a lower FLD dose.

While these findings may be beneficial for observing the chemical interactions of the respective materials and

Table 4. Oxidative Aging Model Parameters for the Evaluated Asphalt Binders

Binder ID		E_a (kJ/(mol·°K))	Pre-exponential Factor, AP^{α}	HS (1/CA)	m (ln(poise))	CA_{RTFO} (arbit. unit)	k_c Temperature (°C) 60	85	100
TX	Base (64-22)	75	1.129E + 09	11.31	46.91	0.323	0.002	0.013	0.035
	Base w/T1 @ TX FLD (64-22/2.65%)	76	1.357E + 09	13.28	0.12	0.637	0.002	0.012	0.034
	Recycled w/T1 @ TX FLD (64-22 / 2.65%/0.25 TX RAP, 0.25 TX MWAS)	64	4.049E + 07	11.61	0.29	0.857	0.003	0.016	0.038
NH	Base (64-28)	65	5.189E + 07	9.46	201.22	0.131	0.003	0.018	0.042
	Base w/T1 @ TX FLD (64-28/2.65%)	69	2.354E + 08	8.92	6.55	0.420	0.004	0.022	0.056
	Recycled w/T1 @ rPGL (64-28/12.5%/0.25 TX RAP, 0.25 TX TOAS)	70	1.400E + 08	8.43	1.00e-05	2.446	0.001	0.009	0.022
	Base w/A1 (64-28/6%)	70	3.700E + 08	8.16	123.22	0.107	0.004	0.021	0.053
	Recycled w/A1 @ rPGL (64-28/6%/0.40 TX RAP)	73	8.101E + 08	5.93	75.43	0.679	0.003	0.020	0.052
NV	Base (64-28P)	73	5.931E + 08	6.53	997.61	0.067	0.002	0.015	0.041
	Base w/T1 @ TX FLD (64-28P/2.65%)	79	4.709E + 09	8.90	22.91	0.383	0.002	0.013	0.038
	Recycled w/T1 @ rPGL (64-28P/11%/0.25 TX RAP, 0.25 TX TOAS)	86	5.045E + 10	6.12	0.01	2.112	0.002	0.013	0.04

Note: E_a = activation energy; AP^{α} = pre-exponential factor; HS = hardening susceptibility; m = hardening function constant; CA_{RTFO} = the CA after rolling thin-film oven (RTFO) aging.

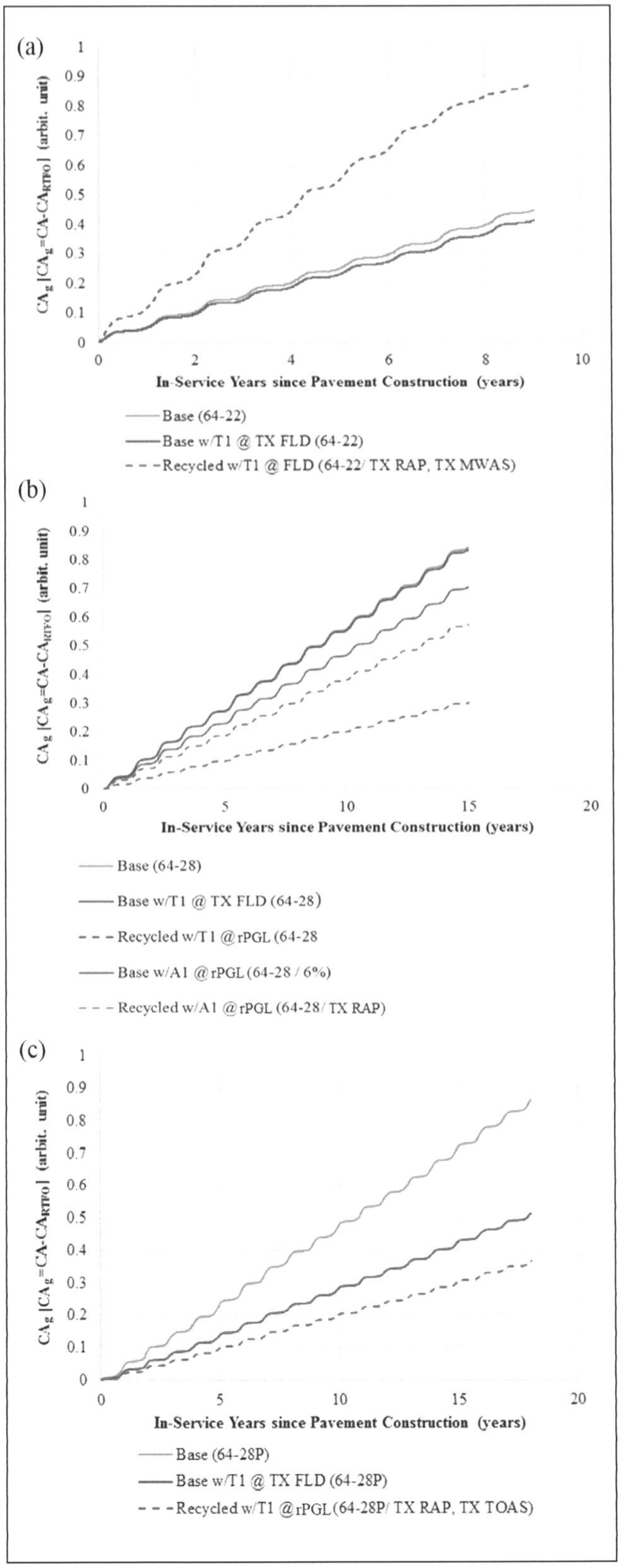

Figure 4. Carbonyl area prediction in asphalt mixture surface layer for binder blends in respective environments.

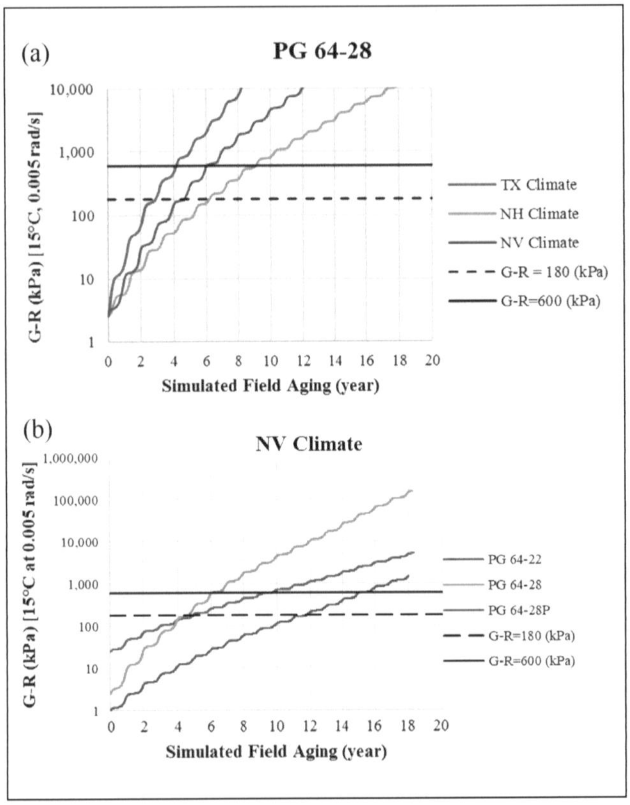

Figure 5. G-R prediction for *a*) PG 64-28 base binder in TX, NH, and NV climates and *b*) base binders in NV climate.

support the selection of the appropriate dosage for a given blend of materials, this analysis will focus on observations of the combined influence of these oxidation predictions with the physical characterization of the G-R parameter described by Equations 3 and 5.

Oxidation Modeling into Glover–Rowe Analysis

Figure 5 presents example plots of the progression of G-R along the modeled in-service duration for the example base binders. Figure 5*a* depicts the increase in the modeled G-R of the PG 64-28 base binder in the three simulated environments (Texas, New Hampshire, and Nevada). Figure 5*b* presents the G-R development of each of the three base binders, but all subjected to the NV climate. These example depictions provide a summary of the methodology used to establish the field age to reach G-R limits presented in Figure 6.

Implementing the step-by-step procedure summarized in Figure 1 demonstrates the simulated in-service years starting at construction to reach the onset and significant cracking limits of G-R = 180 kPa and 600 kPa, respectively. Logically, all the binders took more simulated in-

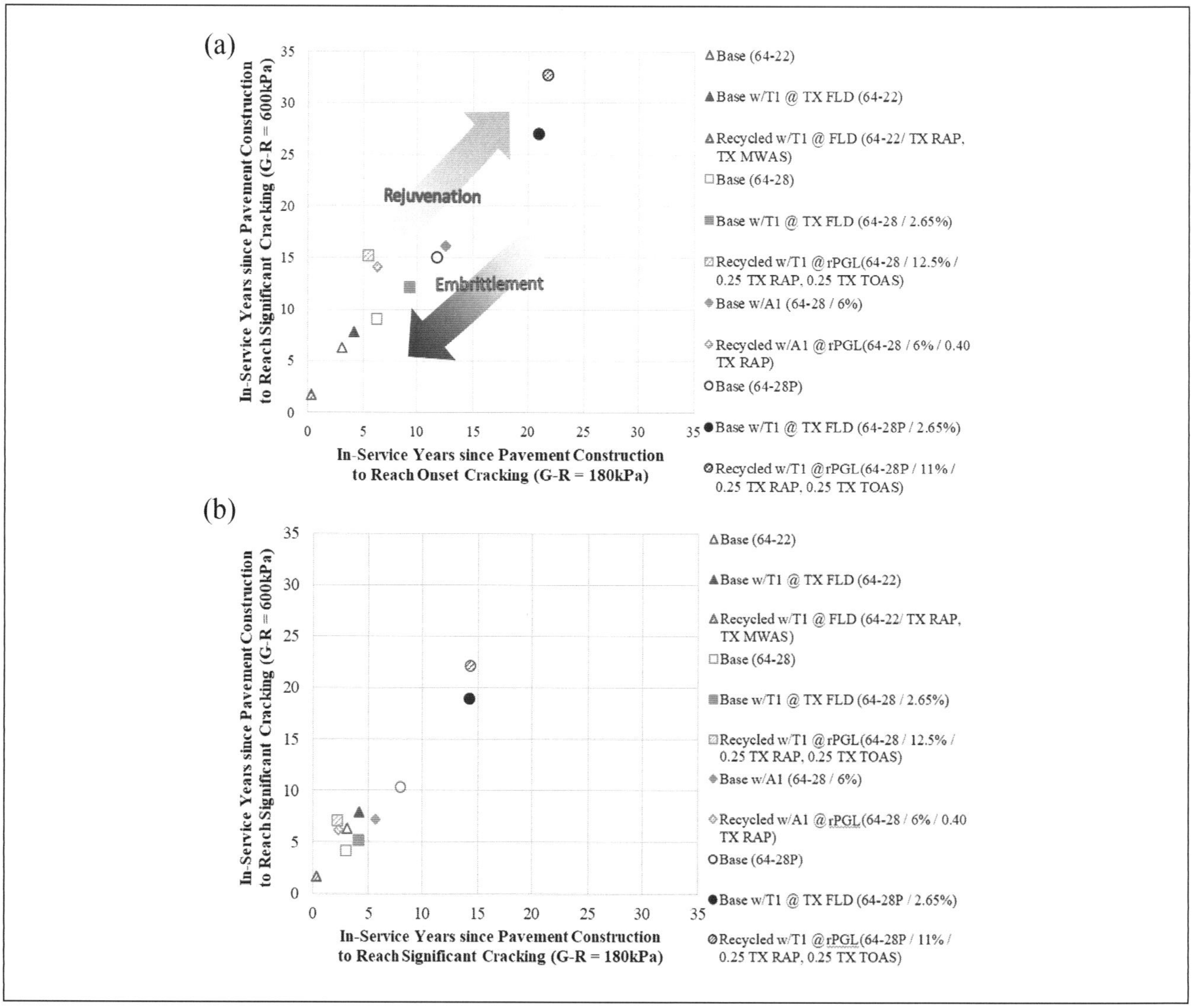

Figure 6. Overall comparison of the in-service years to reach the G-R threshold, including all binder blends at (*a*) respective locations and (*b*) Tyler, TX environment.

service years to reach the significant cracking level compared to the onset level. In addition, the effect of the RAs and recycled materials can readily be observed from Figure 6*a*, noting that the base binders are signified with hollow markers as the point of reference within each binder type. The binder blends with RAs and without recycled material are shown in solid, while those with both modifications are represented by hatched data points. The blends with RAs only (solid points) showed a clear increase in the predicted life, or delay in the onset of cracking, compared to the base binder (hollow points). Thus, both RAs, T1 and A1, had a beneficial effect on the base binder lifetime, with the larger doses resulting in more substantial benefits of the RA, noting the PG 64-28 blend with A1 at 6% dose rate. However, this consistent trend did not continue with the inclusion of the recycled material, which tended to counteract the RA, as expected.

The simulated oxidative aging predictions generally provided the following conclusions for the unmodified PG 64-22 blends. The addition of the recycled material at the field dose of 2.65% to the PG 64-22 base binder significantly reduced the predicted in-service duration. In other words, 2.65% was not a large enough dose to permit full restoration of the binder properties including 0.30 RBR (reclaimed asphalt binder ratio) (0.20 RBR of TX RAP and 0.10 RBR of TX MWAS).

The recycled blend with the PG 64-28 base binder at the rPGL dose of 12.5% and 0.50 RBR (0.25 TX RAP and 0.25 TX TOAS) nearly presented a restoration to the base binder by the onset cracking and an improvement in significant cracking limits. Similarly, the recycled blend with A1 at the rPGL dose of 6% and 0.40 RAPBR (RAP binder ratio) represented a slightly better restoration as compared to the base binder. These two observations imply that using the RAs at the rPGL dose can potentially have a positive influence on restoring the binder properties.

The 64-28P recycled blend contained 0.50 RBR, including 0.25 of RAP and 0.25 of TOAS and 11% (rPGL) of T1. The rPGL dose of the T1 along with the high RBR ratio of 0.5 not only restored the binder blend life prediction to the base PG 64-28P, but extended it by almost 6 years.

For the sake of comparison, the in-service duration to reach the G-R thresholds for PG 64-22 and PG 64-28 binder blends were both predicted in the Tyler, TX environment. The overall direction of the results in Figure 6*b* was observed to be similar to those in which the binder-specific locations were implemented for the binder blends. It is noteworthy to emphasize all the G-R parameters were calculated at a constant temperature of 15°C and a constant frequency of 0.005 rad/s.

Acknowledging the differences in crude source and realistic discrepancies in the geographic location of usage with each binder, side-by-side comparisons are potentially limited while they still add to the comprehension of the overall evaluation technique. However, a substantial reduction in the pavement life was noted in the harsh Nevada climate, implying better performance of the binder blends in the binder-specific locations, as may be expected. The resulting time simulations of the base binders culminate in the influence of the determined k_c rates combined with the relative differences in the HS relationships. Specifically, it should be noted that there was a minor reduction in the k_c terms of the PG 64-28P binder relative to the other two base binders (particularly at the lower temperatures, i.e., higher 1/RT values). Similar k_c terms for the PG 64-28 and PG 64-22 base binders yielded modest G-R duration simulations when compounded by the minor discrepancies noted in the HS between the two. These conditions were further exemplified by the lower HS for the PG 64-28P base binder, resulting in the longest simulated G-R limit durations of the three base binders.

Summary and Conclusions

In summary, the binder blend oxidative aging predictions were helpful in identifying the influence of RAs at the respective dose and recycled materials on the cracking potential of the various binders evaluated at specific geographic locations. The results provide a robust evaluation of the overall oxidation progression of the blended materials, albeit fully blended in the laboratory. By evaluating the combined influence of both the chemical changes (kinetics) and the progression of physical characteristics (rheology), the combined influence of such complicated and often unpredictable outcomes of such compositions may be evaluated.

The results indicate that using the recycled materials along with the RAs at the rPGL dose restored the binder blend properties to those of the base binder, though not always at the lower field dose. However, discrepancies were noted between initial levels of brittle behavior (G-R of 180 kPa) and the onset of severe cracking potential indicated by G-R levels over 600 kPa. This finding confirms the need for a robust RA dose selection methodology, being proposed under the NCHRP 9-58 (*37*, *38*) project, which in general showed the need for a significantly higher rPGL RA dose to restore the PG of the base binder when compared to the low dose used in the field for the example field blends.

This evaluation has demonstrated the importance of adequate characterization of the specific materials being used in conjunction with selection of the correct dose of the appropriate RA to ensure sufficient resistance to cracking and limit the embrittlement of proposed combinations. The combined influence of all components do not always add up to the sum of the individual parts, nor are the measured interactions consistent with increased levels of aging. The complex interactions and unpredictable influences make more robust evaluations and modeling of this type valuable tools in a system of evaluation techniques focused on recycled and rejuvenating materials in asphalt pavements. This finding is particularly true when considering that certain mixtures show beneficial usage of RAs initially, but lesser influence after varying degrees of aging. Other blended materials show quite the opposite trend, with similar results initially and discernable influences noted after longer aging durations.

Therefore, the prevailing conclusion of the study as a whole indicates that material-specific evaluations are needed to identify the complex interactions taking place within the material combinations of interest, but also multiple levels of aging at appropriate intervals may be necessary for comprehensive characterization.

Acknowledgments

The authors would like to thank Dr. Charles Glover and Dr. Geoff Rowe for their advice and contribution to asphalt binder modeling and characterization.

Author Contributions

The authors confirm contribution to the paper as follows: study conception and design: AEM, EH, NM; data collection: SP;

analysis and interpretation of results: AEM, SP, EH, NM; draft manuscript preparation: SP and NM. All authors reviewed the results and approved the final version of the manuscript.

References

1. Zhou, F., J. W. Button, and J. A. Epps. *Best Practice for Using RAS in HMA*. Report No. FHWA/TX-12/0-6614-1. Texas Transportation Institute, College Station, TX, 2012.
2. Shah, A., R. S. McDaniel, G. A. Huber, and V. L. Gallivan. Investigation of Properties of Plant-Produced Reclaimed Asphalt Pavement Mixtures. *Transportation Research Record: Journal of the Transportation Research Board*, 2007. 1998: 103–111.
3. Li, X., M. O. Marasteanu, R. C. Williams, and T. R. Clyne. Effect of RAP (Proportion and Type) and Binder Grade on the Properties of Asphalt Mixtures. *Transportation Research Record: Journal of the Transportation Research Board*, 2008. 2051: 90–97.
4. Hajj, E. Y., P. E. Sebaaly, and R. Shresta. Laboratory Evaluation of Mixes Containing Recycled Asphalt Pavement (RAP). *International Journal of Road Materials and Pavements Design*, Vol. 10, No. 3, 2009, pp. 495–517.
5. West, R., A. Kvasnak, N. Tran, B. Powell, and P. Turner. Testing of Moderate and High RAP Content Mixes: Laboratory and Accelerated Field Performance at the National Center for Asphalt Technology Test Track. *Transportation Research Record: Journal of the Transportation Research Board*, 2009. 2126: 100–108.
6. Hussain, A., and Q. Yanjun. Evaluation of Asphalt Mixes Containing Reclaimed Asphalt Pavement for Wearing Courses. *International Conference on Traffic and Transportation Engineering (ICTTE 2012)*, Hong Kong, China, 2012, pp. 43–48.
7. Valdés, G., F. Pérez-Jiménez, R. Miró, A. Martínez, R. Lee, and R. Botella. Experimental Study of Recycled Asphalt Mixtures with High Percentages of Reclaimed Asphalt Pavement (RAP). *Construction and Building Materials*, Vol. 25, 2011, pp. 1289–1297.
8. Mogawer, W., T. Bennert, J. Daniel, R. Bonaquist, A. J. Austerman, and A. Booshehrian. Performance Characteristics of Plant Produced High RAP Mixtures. *Journal of the Road Materials and Pavement Design*, Vol. 13, 2012, pp. 183–208.
9. Mogawer, W., A. Austerman, L. Mohammad, and M. E. Kutay. Evaluation of High RAPWMA Asphalt Rubber Mixtures. *Road Materials and Pavement Design*, Vol. 14, 2013, pp. 129–147.
10. West, R., J. R. Willis, and M. Marasteanu. *NCHRP Report 752: Improved Mix Design, Evaluation, and Materials Management Practices for Hot Mix Asphalt with High Reclaimed Asphalt Pavement Content*. Transportation Research Board of the National Academies, Washington, D.C., 2013.
11. Tran, N. H., A. Taylor, and R. Willis. *Effect of Rejuvenator on Performance Properties of HMA Mixtures with High RAP and RAS Contents*. NCAT Report 12-05. National Center for Asphalt Technology, Auburn, AL, 2012.
12. Little, D. N., R. J. Holmgreen, and J. A. Epps. Effect of RAs on the Structural Performance of Recycled Asphalt Concrete Materials. *Journal of the Association of Asphalt Paving Technologists*, Vol. 50, 1981, pp. 32–63.
13. Lin, P., T. L. Wu, C. Chang, and B. Chou. Effects of RAs on Aged Asphalt Binders and Reclaimed Asphalt Concrete. *Materials and Structures*, Vol. 44, No. 5, 2011, pp. 911–921.
14. Mogawer, W. S., A. Booshehrian, S. Vahidi, and A. J. Austerman. Evaluating the Effect of Rejuvenators on the Degree of Blending and Performance of High RAP, RAS, and RAP/ RAS Mixtures. *Road Materials and Pavement Design*, Vol. 14, 2013, pp. 193–213.
15. Zaumanis, M., R. Mallick, and R. Frank. Evaluation of Different RAs for Restoring Aged Asphalt Binder and Performance of 100% Recycled Asphalt. *Materials and Structures*, Vol. 48, 2014, 2475–2488.
16. Zaumanis, M., R. Mallick, and R. Frank. Determining Optimum Rejuvenator Dose for Asphalt Recycling Based on Superpave Performance Grade Specifications. *Construction and Building Materials*, Vol. 69, 2014, pp. 159–166.
17. Epps, J. A., D. N. Little, R. J. Holmgreen, and R. L. Terrel. *NCHRP Report 224: Guidelines for Recycling Pavement Materials*. Transportation Research Board of the National Academies, Washington, D.C., 1980.
18. Holmgreen, R. J., J. A. Epps, D. N. Little, and J. A. Button. *RAs for Recycled Bituminous Binders: Executive Summary*. Report No. FHWA-RD-82-122. Texas Transportation Institute, Texas A&M University System, Arlington, TX, 1982.
19. O'Sullivan, K. A. *Rejuvenation of Reclaimed Asphalt Pavement (RAP) in Hot Mix Asphalt Recycling with High RAP Content*. MSc thesis. Worcester Polytechnic Institute, University of Worcester, Worcester, England, United Kingdom, 2011.
20. Arámbula-Mercado, E., F. Kaseer, A. Epps Martin, F. Yin, and L. Garcia Cucalon. Evaluation of Recycling Agent Dosage Selection Methods for Asphalt Mixtures with High RAP and RAS Contents. *Construction and Building Materials*, Vol. 158, 2018, pp. 432–442.
21. Morian, N. E. *Influence of Mixture Characteristics on the Oxidative Aging of Asphalt Binders*. PhD dissertation. University of Nevada, Reno, NV, 2014.
22. Petersen, J. C., and P. M. Harnsberger. Asphalt Aging, Dual Oxidation Mechanism and Its Interrelationships with Asphalt Composition and Oxidative Age Hardening. *Transportation Research Record: Journal of the Transportation Research Board*, 1998. 1638: 47–55.
23. Petersen, J. C., and R. Glaser. Asphalt Oxidation Mechanisms and the Role of Oxidation Products on Age Hardening Revisited. *Road Materials and Pavement Design*, Vol. 12, No. 4, 2011, pp. 795–819.
24. Alavi, M. Z. *Comprehensive Methodologies for Analysis of Thermal Cracking in Asphalt Concrete Pavements*. PhD dissertation. University of Nevada, Reno, NV, 2014.
25. Pournoman, S. *Oxidative Aging of Binders with High Recycled Asphalt Materials*. MSc thesis. University of Nevada, Reno, NV, 2017.
26. Han, R., X. Jin, and C. J. Glover. Oxygen Diffusivity in Asphalts and Mastics. *Petroleum Science and Technology*, Vol. 31, No. 15, 2013, pp. 1563–1573.

27. Han, R. *Improvement to a Transport Model of Asphalt Binder Oxidation in Pavements: Pavement Temperature Modeling, Oxidation Diffusivity in Asphalt Binders and Mastics, and Pavement Air Void Characterization.* PhD dissertation, Texas A&M University, Arlington, TX, 2011.
28. Gui, F. Q., J. Wu, F. Chen, X. Liao, X. Hu, Z. Zhang, and Z. Wang. Inactivation of Polyphenol Oxidases in Cloudy Apple Juice Exposed to Supercritical Carbon Dioxide. *Food Chemistry*, Vol. 100, 2007, pp. 1678−1685.
29. Alavi, M. Z., and E. Y. Hajj. Effect of Cooling Rate on the Thermo-Volumetric, Thermo-Viscoelastic, and Fracture Properties of Asphalt Mixtures. *Asphalt Pavements*, Vol. 1, 2014, pp. 405–416.
30. Asphalt Research Consortium. *TEMPS Software (Version 0.0.1)*, 2014. http://www.arc.unr.edu/Software.html#TEMPS.
31. Glover, J. C., R. R. Davison, C. H. Domke, Y. Ruan, P. Juristyarini, D. B. Knorr, and S. H. Jung. *Development of a New Method for Assessing Asphalt Binder Durability with Field Evaluation.* Report No. FHWA/TX/05-1872-2. Texas Transportation Institute, College Station, TX, 2005.
32. Rowe, G. M. Prepared Discussion for the AAPT Paper by Anderson et al.: Evaluation of the Relationship between Asphalt Binder Properties and Non-Load Related Cracking. *Journal of the Association of Asphalt Paving Technologists*, Vol. 80, 2011, pp. 649–662.
33. Ruan, Y., R. R. Davison, and C. J. Glover. An Investigation of Asphalt Durability: Relationships between Ductility and Rheological Properties for Unmodified Asphalts. *Petroleum Science and Technology*, Vol. 21, 2003, pp. 231–254.
34. Kandhal, P. S. Low-Temperature Ductility in Relation to Pavement Performance. In *ASTM STP 628: Low-Temperature Properties of Bituminous Materials and Compacted Bituminous Paving Mixtures* (Marek, C. R., ed.), American Society for Testing and Materials, Philadelphia, PA, 1977.
35. RHEA. *Rhea, Rheology Analysis Software, version 1.2.1*, Abatech, Inc., Blooming Glen, PA, 2011.
36. Han, R., X. Jin, and C. J. Glover. Oxygen Diffusivity in Asphalts and Mastics. *Petroleum Science and Technology*, Vol. 31, 2013, pp. 1563–1573.
37. Epps Martin, M., F. Zhou, E. Arambula, E. S. Park, A. Chowdhury, F. Kaseer, J. C. Munoz, E. Hajj, J. Daniel, and C. Glover. *The Effect of Recycling Agents on Asphalt Mixtures with High RAS and RAP Binder Ratios, Phase I Interim Report.* Transportation Research Board, Washington, D.C., 2015, http://onlinepubs.trb.org/onlinepubs/nchrp/docs/NCHRP09-58_PhI_InterimReport.pdf.
38. Epps Martin, M., F. Zhou, E. Arambula, E. S. Park, A. Chowdhury, F. Kaseer, F. Yin, J. C. Munoz, A. Rose, E. Hajj, J. Daniel, and C. Glover. *NCHRP 09-58: The Effect of Recycling Agents on Asphalt Mixtures with High RAS and RAP Binder Ratios, Phase II Interim Report.* Transportation Research Board of the National Academies, Washington, D.C., 2016.

The Standing Committee on Asphalt Binders (AFK20) peer-reviewed this paper (18-04761).

The contents of this manuscript reflect the views and opinions of the authors and do not necessarily reflect those of AASHTO, National Cooperative Highway Research Program (NCHRP), the project panel, or the Transportation Research Board, the National Academies, or the program sponsors.

Article

TRR
JOURNAL OF THE TRANSPORTATION RESEARCH BOARD

Transportation Research Record
2018, Vol. 2672(28) 290–300

DOI: 10.1177/0361198118781378
journals.sagepub.com/home/trr
SAGE

Study of Asphalt Binder Fatigue with a New Dynamic Shear Rheometer Geometry

Panos Apostolidis[1], Cor Kasbergen[1], Amit Bhasin[2], Athanassios Scarpas[1,3], and Sandra Erkens[1]

Abstract

With the effort to precisely predict the lifetime of asphalt binders and subsequently optimize their utilization in a more economical way, the objective of this study was to introduce a new methodology to improve the fatigue characterization of asphalt binders through a new dynamic shear rheometer (DSR) sample testing geometry. Initially, numerical analyses were performed to study the geometry-related issues of a standard DSR sample on time sweep tests, and to assist in the effort to increase understanding of the DSR damage phenomena of asphalt samples. On the basis of these numerical analyses, a new testing geometry, the parallel hollow plate, was developed and its test results compared with the standard sample testing geometry. A single type of asphalt binder was assessed using amplitude sweep tests. The obtained results demonstrated a significant difference between the fatigue of the two sets of DSR sample geometries. On the basis of these, time sweep tests were conducted for the same sample geometries and the results demonstrated that the new testing geometry yields material response consistency under different loading conditions. The lifetime prediction of the standard parallel plates showed a significant difference with the newly developed DSR sample testing geometry by overestimating the total number of cycles until asphalt binder failure. The new testing geometry allowed the isolation of the damaged area of asphalt binder by localizing the shear stresses in the samples' periphery.

Owing to the extremely complex nature of asphalt binders, it is difficult for infrastructure designers to accurately predict the lifetime of the pavement structures. Taking into account the higher and heavier traffic on the highways in the last decades, implementation of new asphalt binders for the highway network has increased remarkably resulting in higher initial costs for pavement construction. Furthermore, difficulties have been identified in estimating serviceability and planning maintenance operations during the pavements' service life. Various parameters affect the performance prediction of asphalt binders and instead of making progress in the testing techniques, the challenge of precisely characterizing binder fatigue life still needs to be addressed.

Fatigue damage as one of the main asphalt binders' distress modes can be described as the material degradation process because of repeated loading by which the micro-cracks grow and the coalescence to macro-cracks. Typically, fatigue in asphalt mixes is studied by subjecting the test material to some form of cyclic stresses at a lower level than the ultimate strength and then determining the relative change in their mechanical properties, such as stiffness and strength. Therefore, having a test method that can predict the mechanical degradation of material, will allow comprehension of the exact damage mechanisms in detail and subsequently to optimize the utilization of asphalt binders in a more feasible way.

However, asphalt binder fatigue characterization is not uncharted territory for the paving industry. Several laboratory studies have been conducted to provide understanding of the degradation mechanism due to repeated stresses, and ranking of the binders' susceptibility to resist these stresses. Unfortunately, the results from the tests do not precisely predict the field performance of asphalt mixes. As a result, the need to improve testing

[1]Section of Pavement Engineering, Faculty of Civil Engineering and Geosciences, Delft University of Technology, Delft, the Netherlands
[2]Department of Civil, Architectural and Environmental Engineering, University of Texas at Austin, Austin, TX
[3]Department of Civil Infrastructure and Environmental Engineering, Khalifa University of Science and Technology, Abu Dhabi, United Arab Emirates

Corresponding Author:
Address correspondence to P. Apostolidis: p.apostolidis@tudelft.nl

methods for the quality control of asphalt binders to ensure accurate lifetime estimations has been increasing. The dynamic shear rheometer (DSR) has been introduced for use in fatigue characterization of binders (*1–3*). However, a satisfactory link between the measured binder fatigue response using DSR and the potential field material performance over a range of various operational conditions is still under investigation.

Within this framework, a study has been initiated to evaluate a potentially more appropriate DSR fatigue-testing method. The new parallel hollow plate (PHP) system was designed and developed with an outer diameter of 25 mm, as with the standard geometry of DSR parallel plates (PP), but with a concentric hollow area of 19 mm diameter and 0.1 mm depth. After filling the inner hollow area with a silicon paper, the new testing system was used to explore the impact of the mechanical performance of asphalt binders. For the selected new geometry system, after carrying out assessment of the repeatability of the test results, different dynamic shear measurements were conducted to evaluate the material's response. The experimental results demonstrated the significant variations on binder performance at low and high cyclic torque level tests between the new and the standard DSR apparatus. This comparison underlines the significance of geometry for DSR plates for a more accurate material characterization, and the upcoming need to minimize the geometry-related issues by localizing the shear damage in the tested material.

Fatigue in Asphalt Binder

Fatigue damage in asphalt is the degradation of the material due to repeated loading by which the cracks grow and the material loses its capability to resist more loads. Significant effort has been spent on evaluating asphaltic materials' fatigue life and thus several methods have been developed through this process. These methods differ mainly in terms of the fatigue damage approaches and testing configurations, such as the sample's geometry, loading conditions, and so forth. Here, emphasis is given to assessing the fatigue performance of asphalt binders and, for this reason, state-of-the-art DSR utilization as a fatigue characterization tool is discussed.

Fatigue Damage Approaches

The fatigue life of asphalt binders has been thoroughly examined, and several approaches such as energy-related, mechanistic, and phenomenological approaches have been utilized to evaluate the material's response under cyclic load repetitions and to determine the remaining life of the material.

Among energy-related approaches, the energy ratio as a function of the number of cycles and the complex shear modulus for the different controlled modes has been applied as a fatigue life criterion (*4*). In the stress-controlled mode particularly, the fatigue life of the material is defined as the point when the energy ratio reaches its peak in the relationship of energy ratio versus the number of cycles. On the other hand, in the strain-controlled mode, the fatigue life is defined as the number of load cycles at which the slope of the energy ratio deviates from a straight line.

Another energy approach is the dissipated energy ratio, which is defined as the ratio of the difference between the dissipated energy for the successive load cycles to the dissipated energy of the previous cycles (*5*, *6*). The dissipated energy ratio is the area inside the hysteric loop (*7*, *8*), and the fatigue life of the material is considered as the transition point where the dissipated energy ratio starts to increase rapidly from an approximately constant value (*6*). Similarly, the dissipated strain energy approach has been used by converting the actual strain to an equivalent pseudo-strain in order to remove the viscoelastic contribution (*2*, *9*), and to quantify the damage manifestation using mechanistic approaches such as continuum damage and fracture mechanics (*10–12*).

Finally, phenomenological approaches are the most frequently used to define fatigue life. One example of such an approach is the determination of fatigue as the number of cycles when the complex modulus decreases to 10% and 50% of the initial complex modulus for stress- and strain-controlled testing modes, respectively (*13*, *14*). However, the failure criterion of 50% complex modulus reduction is irrelevant to the damage accumulation as this value is arbitrary and varies at different loading modes. Others consider the fatigue life of asphalt as the point at which the stress level changes rapidly (*15*), however this approach is sensitive to the test loading conditions. In this study, the total number of fatigue cycles until complete failure of the sample or the end of test is used as the fatigue life criterion (*16*).

DSR Fatigue Damage Characterization

DSR is commonly used as standard performance testing equipment to characterize the viscoelastic properties of asphalt binders (*17–19*). Additionally, to evaluate the fatigue damage mechanism and to predict the fatigue life in asphalt binders, DSR has been utilized using the oscillatory time sweep (TS) test (*1–3*). Previous researchers have demonstrated that damage initiates at the outer periphery of the material and propagates through the sample with increasing numbers of loading cycles. Thus, damage results in a reduction of the radius of the test sample. Specialized imaging techniques have been used to demonstrate fatigue damage during DSR testing and the obtained images clearly demonstrate non-uniform damage with fractures at the outer edge of the testing plates and an intact center (*2*, *3*, *20*).

Others who studied the phenomena of fatigue with DSR have shown damage propagation as hairline cracks propagating toward the center accompanied by modulus decrease (*21*). The fatigue damage mechanism does not include internal damage because the edge fracture is dominant, especially in oscillatory TS tests (*5*). However, these are not the only issues that are encountered with the standard DSR test methods using a PP: the accuracy of the complex modulus is limited since the generated radial stress field is non-linear. Many aspects of DSR fatigue characterization are elaborated with approximations and extrapolations analogous to how Ptolemy used epicycles to explain the planets' movements around the earth. The urgency to improve fatigue-testing methods and the asphalt binders' quality lies in the need to resolve the inaccurate use and interpretation of DSR and to link the DSR-measured response of binders with field pavement performance. In the following section, numerical analyses are presented that study the geometry-related effects of DSR sample testing on fatigue damage. Furthermore, the numerical simulations of fatigue damage will assist in the effort to increase understanding of the damage phenomena of asphalt samples during DSR TS tests and to further optimize the testing configurations for obtaining more realistic material properties.

Numerical Simulation of DSR Fatigue Damage

Model Parameters Determination

A damage model was developed to illustrate the damage distribution of asphalt binder during a DSR TS test. The material parameters that were required as an input were modeled based on a linear viscoelastic response. The complex modulus values of asphalt binder were determined from frequency sweep tests in the standard PP DSR system. These tests were carried out over a temperature and frequency range from –10°C to 60°C and from 0.1 to 100 Hz, respectively. Instrument compliance was measured and accounted for in these measurements. The asphalt binder used was a commonly applied binder for porous asphalt mixes in Dutch roads, the penetration grade 70/100 unmodified bitumen. By employing the frequency-temperature superposition principle, the master curve in the frequency domain was defined (reference temperature of 20°C).

Continuum Damage Model

After determining the material parameters, with the Prony series coefficients (G_∞, G_i and ρ_i) obtained by fitting the experimental data with the storage modulus, the relaxation modulus could be expressed in the time domain as follows:

$$G(t) = G_\infty + \sum_{i=1}^{n} G_i e^{-t/\rho_i} \quad (1)$$

where

$G(t)$ is the shear relaxation modulus in time domain,
t is the loading time,
G_∞ is the long-time equilibrium modulus,
G_i are the spring constants in the generalized Maxwell model,
ρ_i are the relaxation times, and
n is the number of Maxwell components in the generalized model.

If it is assumed that the Poisson's ratio of binder is time independent and that the material is isotropic, the following expression that relates the $G(t)$ to $E(t)$ can be written as

$$E(t) = 2 \cdot G(t) \cdot (1 + \nu) \quad (2)$$

where $E(t)$ is the relaxation modulus and ν is the Poisson's ratio.

In continuum mechanics, the damage is defined as a function of any micro-mechanical change that develops in a homogeneous continuum media. To include damage in the above-described material model, the following damage evolution equation was proposed based on total dissipated energy as

$$\xi(t) = 1 - exp(-k \cdot W(t)^r) \quad (3)$$

where

ξ is damage degradation of asphalt binder,
t is time,
W is the total dissipated energy, and
k and r are damage rate parameters.

In incremental form Equation 4 can be written as

$$\xi(t + \Delta t) = 1 - (1 - \xi(t)) \cdot exp(-k \cdot (W(t + \Delta t)^r - W(t)^r)) \quad (4)$$

where Δt is the time increment. If the value of ξ is zero it indicates no damage and if the value of ξ is one it resembles full damage.

The total energy dissipation W can be computed in incremental form as

$$W(t+\Delta t) = W(t) + \sum_{i=1}^{n} \int_{t}^{t+\Delta t} S_i^{eff}(\tau) : \dot{E}(\tau) d\tau \quad (5)$$

$$S_i^{eff}(\tau) = (1 - \xi(\tau)) \cdot S_i(\tau) \quad (6)$$

where

i is the index of the Maxwell component,

n is the number of Maxwell components,

τ is the time integration variable,

S_i is the second Piola-Kirchhoff stress in the i-th Maxwell component,

S_i^{eff} is the effective or remaining second Piola-Kirchhoff stress in the i-th Maxwell component after damage has been taken into account, and

$\dot{E}$ is the total Lagrange-Green strain rate.

Using the midpoint integration rule Equation 5 can be simplified to

$$W(t+\Delta t) = W(t) + \sum_{i=1}^{m} \left(\frac{S_i^{eff}(t+\Delta t) + S_i^{eff}(t)}{2} \right) (E(t+\Delta t) - E(t)) \quad (7)$$

Numerical Implementation

The CAPA 3D system was utilized. Three user-defined 3D finite-element (FE) meshes were created to study the damage distribution and the localization of asphalt sample deterioration in a sinusoidal (oscillating) loading mode during a TS DSR test (Figure 1). The first FE mesh representing the standard DSR geometry of 2,400 cubic finite elements was developed. This DSR geometry comprises the two PPs in which the asphalt binder is located in-between, with the top plate being subjected to torsion and the bottom plate being fixed. Similarly, the second FE mesh of DSR geometry with a ring as the top plate and an inner and outer diameter of 19 mm and 25 mm, respectively, was created of 2,200 finite elements. This configuration was called a one ring-type testing system. A third mesh, a two rings-type testing geometry, comprising two rings instead of solid plates of 2,000 finite elements was generated.

To assess the fatigue damage behavior under the same applied torque (0.245 Nm), the load level was converted to shear stress (τ) based on the testing geometries. According to the elastic torsional theory, the shear stress (τ) calculations for plate-type and ring-type testing geometries (Figure 2) are given in the following equations,

$$\tau_{plate-type} = \frac{2T}{\pi R_0^3} \quad (8)$$

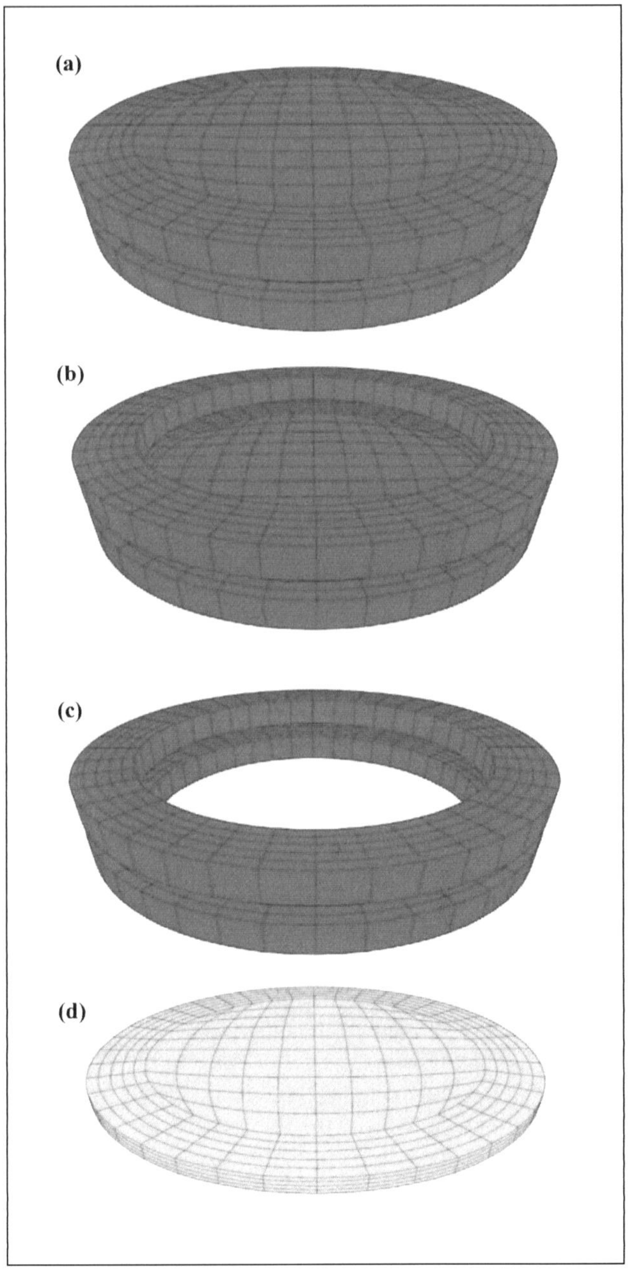

Figure 1. Three-dimensional meshes of (*a*) standard plate-type, (*b*) one ring-type, (*c*) two rings-type DSR sample testing geometries, and (*d*) the testing sample.

$$\tau_{ring-type} = \frac{TR_0}{\left(\frac{\pi (R_0^4 - R_i^4)}{2} \right)} \quad (9)$$

where

T is torque,

R_0 is the outer radius, and

Ri is the inner radius of the plate.

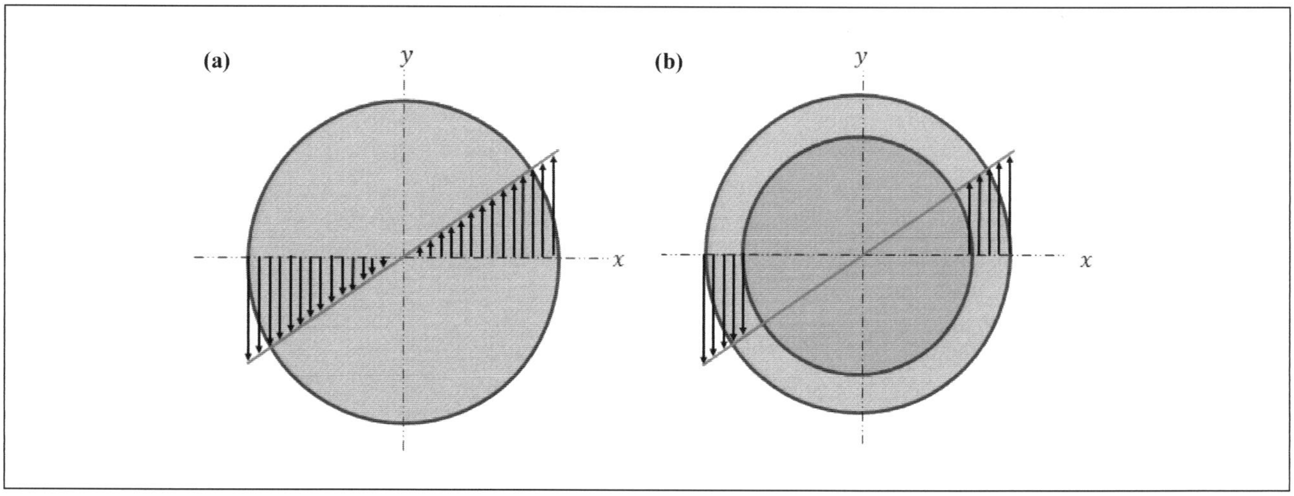

Figure 2. Shear stress distribution on (*a*) standard plate-type, and (*b*) one ring-type sample geometry.

Numerical Predictions. In Figure 3, the damage distribution within the specimen was obtained after subjecting the standard plate-type model to a torque of 0.245 Nm at 10 Hz frequency. The results from this analysis demonstrate that the material degradation during a PP DSR TS test differs across the sample's radius. Specifically, the top part of Figure 3 visualizes the damage progress over time for the first six TS cycles. With increased loading, it is apparent that the damage, as reflected by the different colors in the figure, is concentrated in the outer periphery. Plotting the damage values versus time gives the bottom graph in Figure 3, where the damage increases more rapidly in the points closer to the sample's periphery. As can be observed, the damage of the inner area of binder is not the same as the edge or at locations close to the edge. The damage rate shows the inner part of the testing binder is not affected significantly by the torsional-induced damage of the plates. Therefore, these results corroborate the previously mentioned mechanism of damage initiation at the outer periphery of the sample and the almost intact center during a DSR fatigue test (*5*).

Figure 4 compares the performance of the standard DSR geometric configuration and of the two ring-type geometries. The new geometries show a higher magnitude of damage localized on the ring area than the plate-type sample geometry for a given number of loading cycles (bottom of Figure 4). This difference is explained by the way the area that resists the applied torque is limited in the ring-type geometry compared with the standard system. Additionally, the impact of the top rotating part on the shear stress field and the subsequent damage propagation generated by the applied torque across the sample's radius is shown in Figure 5. For the ring-type geometries, the stress flow field appeared only on the outer periphery of the sample with very limited inward stress propagation for the one ring-type and two rings-type sample geometries, respectively. The edge damage phenomenon to the ring-type geometry occurs earlier than the plate-type sample geometry on account of the higher stress magnitude.

Additionally, the stress and damage difference across the sample's thickness at three different points at a certain time period is demonstrated in Figure 6. It is obvious that the standard geometry shows significant variation in damage across the sample's thickness at all these points. The one ring-type geometry has a bit less damage at the same location than the two rings-type testing configuration. All these predicted results reinforce recent studies on the lack of accuracy of the standard DSR sample testing geometry and the limitations of this system on providing true material properties (*22–25*).

Improving DSR Fatigue Damage Characterization

On the basis of the evidence from past research and the predicted results from implementing the previously described continuum damage model, the main objective of this part of the study was to introduce a new methodology to accurately characterize the fatigue performance of asphalt binders through a new DSR testing system. Different dynamic shear measurements were performed to assess the material response by using the standard PP and the newly developed PHP configuration. The ability of the new geometry to characterize asphalt binder fatigue was also evaluated.

Test Methods

The standard DSR sample geometry is the PP with smooth polished surfaces and a typical diameter of 25

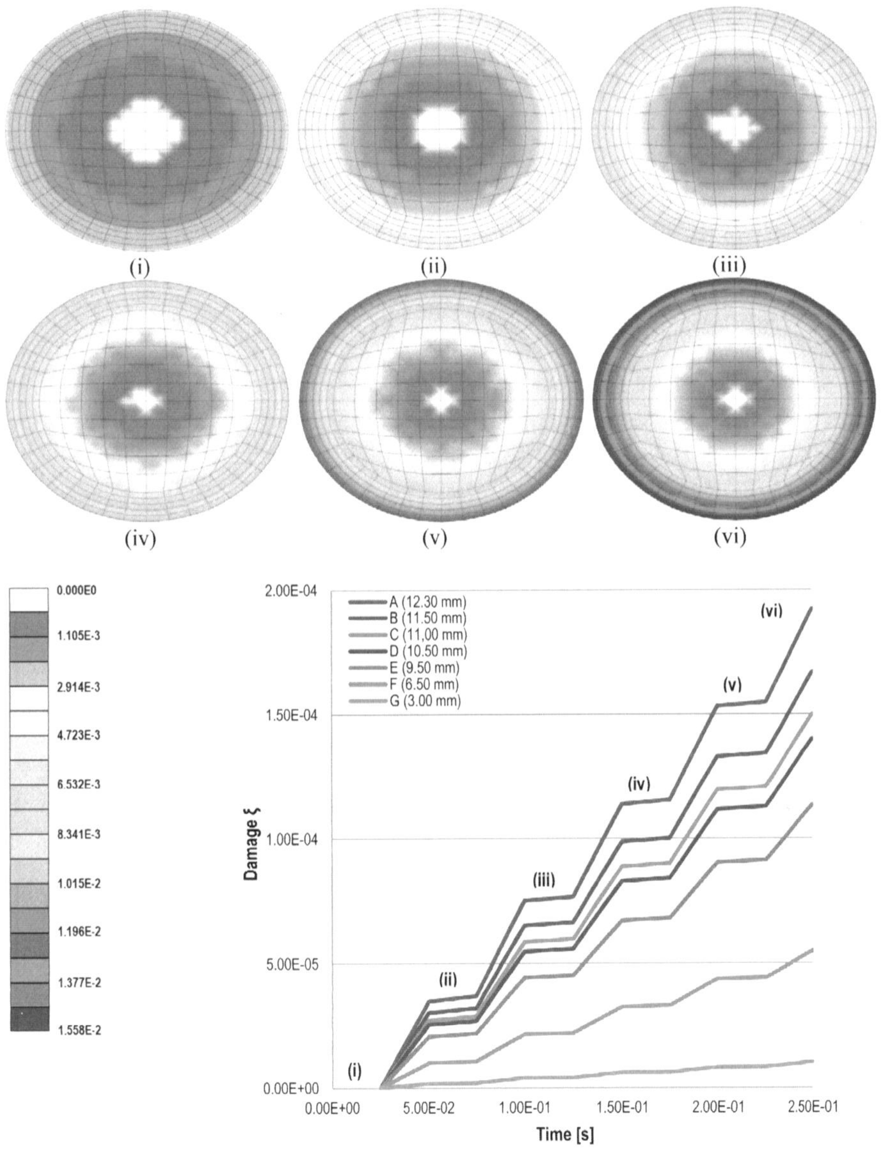

Figure 3. Predicted development of damage along the radius of the standard DSR sample testing geometry.

mm. A new sample testing geometry was designed and manufactured on the basis of the previous numerical analyses. Similar to the one ring-type geometry, the new PHP sample geometry has an outer diameter of 25 mm with a concentric hollow space of 19 mm diameter and 0.1 mm depth. The testing procedure is shown in Figure 7. The DSR setup was utilized for testing with the conventional PP and the new PHP, both with a 1 mm gap in accordance with the Superpave specifications, and obtaining the material response. After filling the inner hollow space of the PHP with a silicon paper, the new testing system was used to explore the impact of the mechanical performance of asphalt binder. A zero gap between the upper and lower plates was established and after reaching it, a 1 mm gap was set by moving the plates apart.

Amplitude Sweep Measurements. To obtain a dynamic material response for a very short loading time, a varying torque signal is applied with a fixed sinusoidal oscillatory frequency. In this study, a cyclic strain-controlled torque was applied throughout the test causing a constant rotational strain. These DSR experiments resulted in

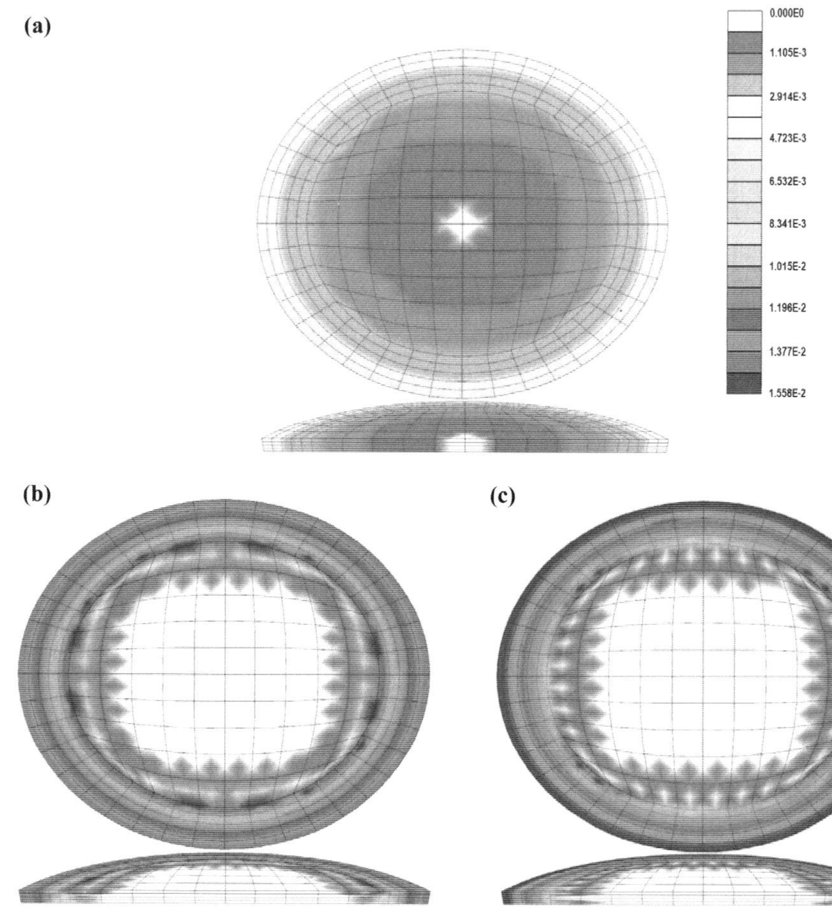

Figure 4. Simulation of damage distribution of: (*a*) standard plate-type, (*b*) one ring-type and (*c*) two rings-type DSR sample testing geometries at the end of the analyses.

amplitude sweep results for the two different sample geometries at 35°C for further comparison. Furthermore, these results were used to determine the linear viscoelastic range and a level of applied torque of 10 Hz frequency for conducting the TS studies in the latter step.

Time Sweep Measurements. The material damage manifests as a decrease in the complex modulus and an increase in the phase angle in the asphalt binder. In this study, the damage was quantified as the reduction in the complex modulus measured during the cyclic loading test with DSR. The TS torque-controlled loading mode was used to evaluate the binder fatigue life and the performance difference between the two sample testing geometries. During these tests, the samples were subjected to a sinusoidal loading mode with a fixed frequency of 10 Hz at 35°C.

Test Results

Amplitude Sweep Results. For the selected geometries, after assessing the repeatability of the test results, different dynamic shear measurements were conducted to evaluate the material response using an amplitude sweep test. Figure 8 depicts the variation in viscoelastic properties versus applied torque at 10 Hz frequency and 35°C. The effect of the new testing geometry is also demonstrated. The torque amplitude was increased in small amounts instead of large steps in each cycle. From the data, it can be observed that the complex modulus dropped and the phase angle increased first when the material was tested using the PHP configuration. The limited area in the outer periphery of the PHP caused quicker degradation than the PP system when the torque was increased. Thus, it is obvious that the material degradation rate is a func-

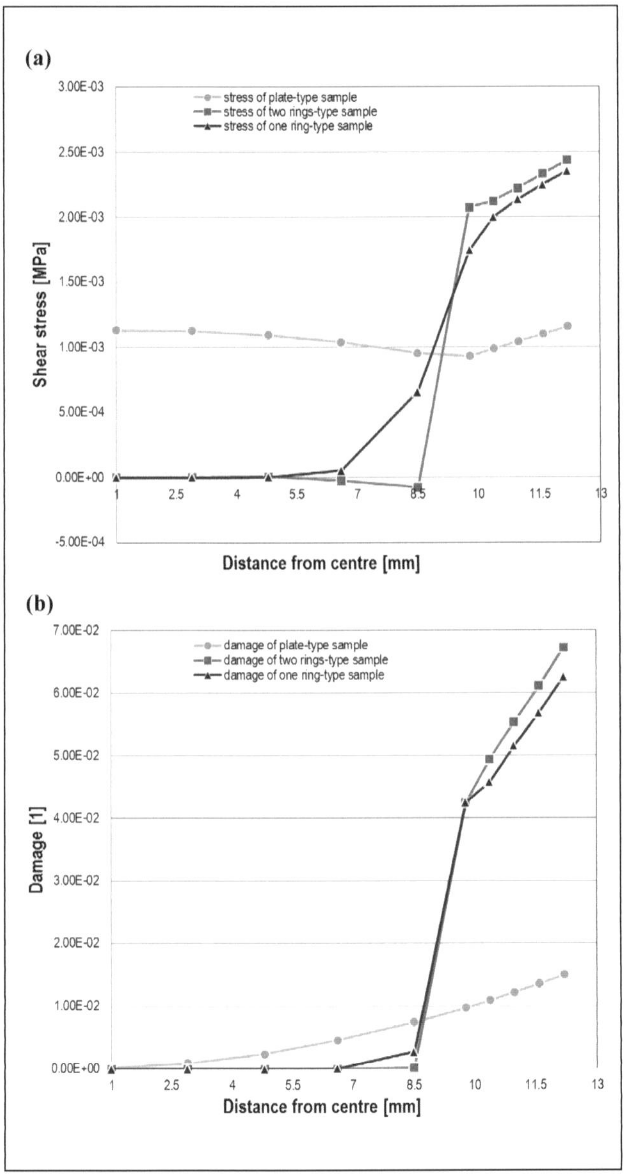

Figure 5. Predicted (*a*) stress and (*b*) damage distribution over the sample's radius of different DSR sample testing geometries at the end of the analyses.

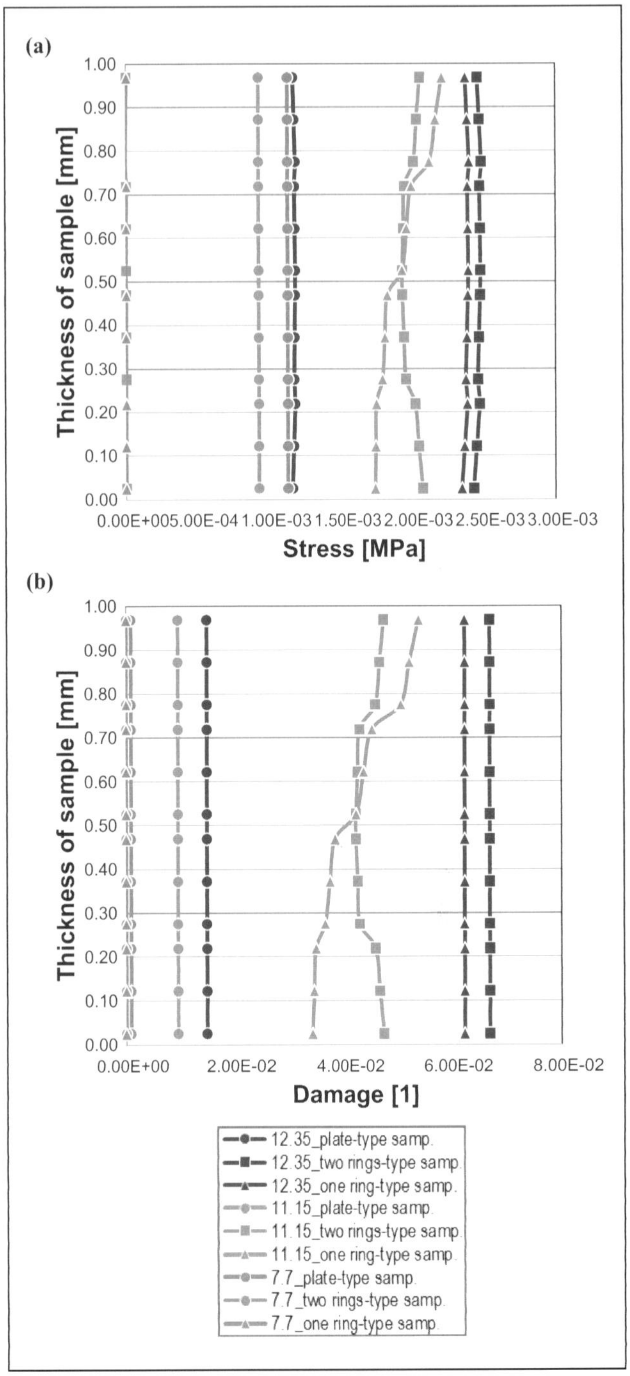

Figure 6. Predicted stress and damage distribution over the sample's height at different points over the sample's radius of the DSR sample testing geometries at the end of the analyses.

tion of the damaged area for an amplitude sweep test and subsequently of the testing geometry.

Time Sweep Results. The fatigue life of asphalt binder is influenced by various factors such as temperature, loading level, and frequency. In this study, the testing was done at 35°C in which the initial complex shear modulus was 0.5 MPa. Very different fatigue performances were observed between PP and PHP geometries. Figure 9 demonstrates the complex shear modulus versus the number of cycles from applying a torque level of 50 mNm. As expected, with increasing numbers of fatigue cycles, the complex modulus of the PHP dropped first as the tested area was limited, indicating the faster occurrence of damage. In addition, the shear modulus reported using the PP geometry is actually an average of the damaged periphery and the intact core.

Figure 10 shows the fatigue life curves for PP and PHP DSR geometries. Here, the most commonly applied

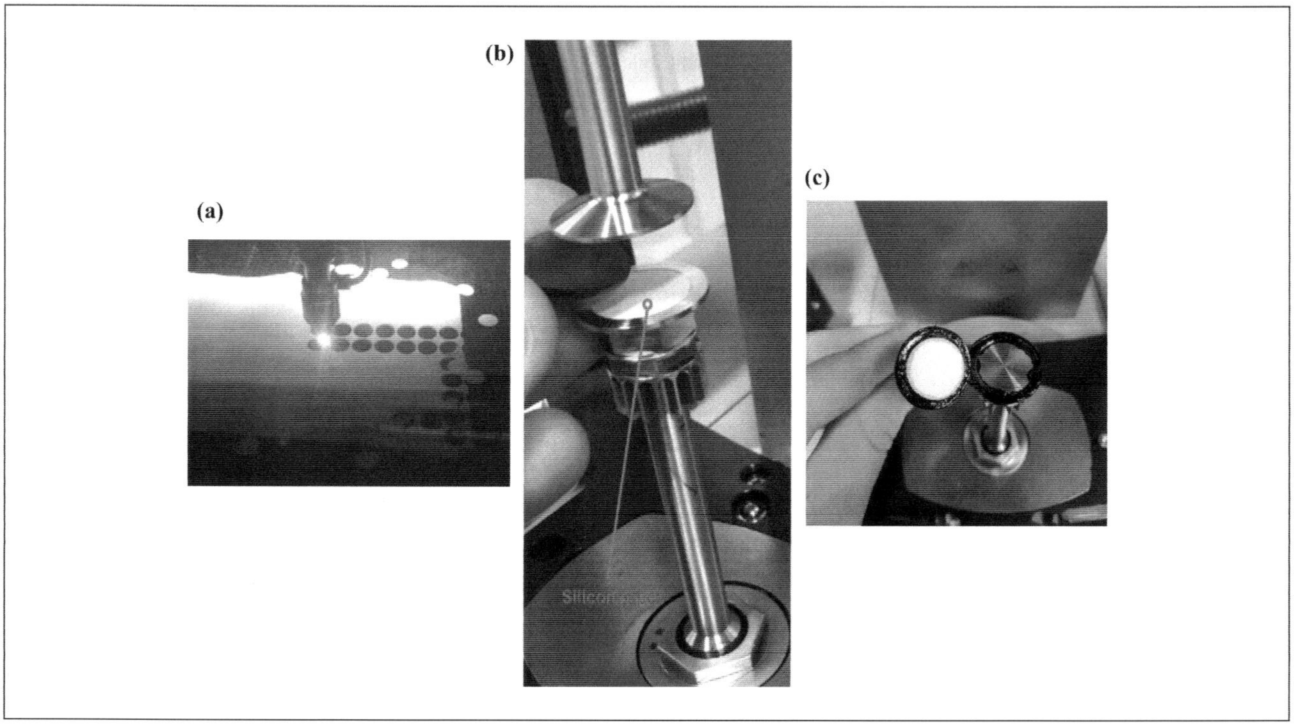

Figure 7. PHP DSR sample testing system: (*a*) laser cutting of silicon paper, (*b*) sample placed on the PHP DSR plates, and (*c*) view of top plate after test completion.

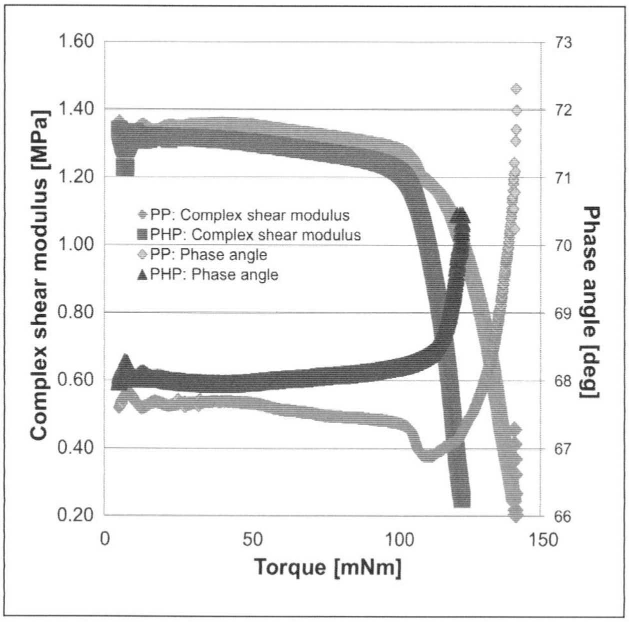

Figure 8. Amplitude sweep results. Rheological properties versus torque for the different sample testing geometries.

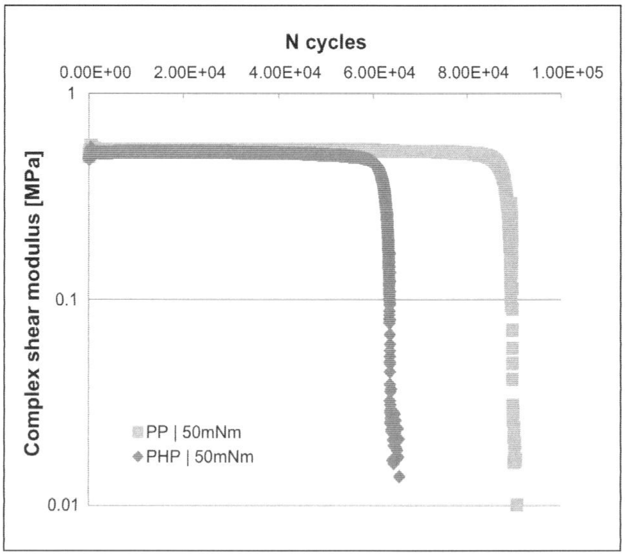

Figure 9. Complex shear modulus versus number of cycles of different sample testing geometries.

fatigue life criterion is considered to be the number of loading cycles at which the complex shear modulus reaches its lowest value. As failure happened only at the sample's periphery in the PHP system, PHP appeared to result in a shorter binder fatigue life for different applied torque levels. The propagation of the micro-cracks from the edges to the internal area of the sample using the standard geometry produced more cycles in the TS tests. The TS results of the newly developed sample testing geometry indicate the importance of characterizing

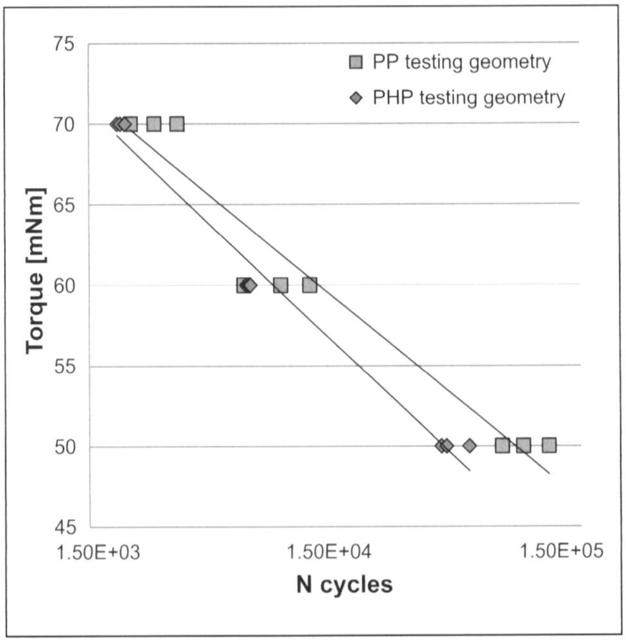

Figure 10. Fatigue life curves of different sample testing geometries.

fatigue performance accurately. According to these results, the fatigue resistance offered by the PP in a TS test was influenced as an artifact of the geometry. However, in addition to the various models that are utilized to successfully predict the fatigue life of material, precise testing to obtain accurate material properties should be a priority.

Summary of Findings and Future Work

From the perspective of pavement design, it is important to be able to predict the fatigue life of an asphalt binder as a result of cyclic loading over time. This study proposed a new testing geometry to more accurately predict the binder's fatigue life. On the basis of analyses and test results collected in this study, it can be stated that a less geometry-dependent measurement of fatigue damage was achieved using the newly developed DSR configuration, showing the importance of using precise testing systems for accurate material performance predictions.

The damage continuum model that was developed to demonstrate the non-uniform damage distribution of asphalt binder subjected to sinusoidal loads with the standard sample geometry showed that the damage was localized in the sample's periphery, leaving the center intact. The visualization of the concentration of damage during the fatigue testing with DSR was used as evidence to manufacture a new testing configuration with an inner hollow space in the center of the bottom plate. After conducting TS experiments using PP and PHP configurations, the fatigue life predictions of the two geometries showed a significant difference, with the edge damage phenomenon happening earlier for the PHP than the PP. The very different observed fatigue performances were derived from the way the new sample testing geometry allowed isolation of the material damage by localizing the shear stresses in the sample's periphery.

Further study is needed to maximize the damage by increasing the diameter of the inner hollow space, and the test loading and environmental conditions should be expanded to provide more realistic fatigue predictions. Moreover, extensive experimental programs are required to be performed to develop transferring functions to convert the results of the new geometry to the results derived from the standard DSR geometry for modified and unmodified binders.

Author Contributions

The authors confirm contribution to the paper as follows: study conception and design: P. Apostolidis, A. Scarpas, A. Bhasin; data collection: P. Apostolidis, C. Kasbergen; analysis and interpretation of results: P. Apostolidis, A. Bhasin, A. Scarpas S. Erkens; draft manuscript preparation: P. Apostolidis. All authors reviewed the results and approved the final version of the manuscript.

References

1. Kim, Y. R., D. N. Little, and R. L. Lytton. Use of Dynamic Mechanical Analysis to Evaluate the Fatigue and Healing Potential of Asphalt Binders in Sand Asphalt Mixtures. *Journal of the Association of Asphalt Paving Technologists*, Vol. 71, 2002, pp. 176–205.
2. Masad, E., V. T. F. Castelo Branco, D. N. Little, and R. Lytton. A Unified Method for the Analysis of Controlled-Strain and Controlled-Stress Fatigue Testing. *International Journal of Pavement Engineering*, Vol. 9, No. 4, 2008, pp. 233–246.
3. Hintz, C., and H. Bahia. Simplification of Linear Amplitude Sweep Test and Specification Parameter. *Transportation Research Record: Journal of the Transportation Research Board*, 2013. 2370: 10–16.
4. Van, D., and W. Visser. Energy Approach to Fatigue for Pavement Design. *Journal of the Association of Asphalt Paving Technologists*, Vol. 46, 1977, pp. 1–40.
5. Keentok, M., and S. Xue. Edge Fracture in Cone-Plate and Parallel Plate Flows. *Rheological Acta*, Vol. 38, 1999, 321–348.
6. Ghuzlan, K. A., and S. H. Carpenter. Energy-Derived, Damage-Based Failure Criterion for Fatigue Testing. *Transportation Research Record: Journal of the Transportation Research Board*, 2000. 1723: 141–149.
7. Ghuzlan, K. A., and S. H. Carpenter. Fatigue Damage Analysis in Asphalt Concrete Mixtures using the Dissipated Energy Approach. *Canadian Journal of Civil Engineering*, Vol. 33, No. 7, 2006, pp. 890–901.

8. Daniel, J. S., W. Bisirri, and Y. R. Kim. Fatigue Evaluation of Asphalt Mixtures using Dissipated Energy and Viscoelastic Continuum Damage Approaches. *Journal of the Association of Asphalt Paving Technologists*, Vol. 73, 2004, pp. 557–583.
9. Bhasin, A., V. T. Castelo Branco, E. Masad, and D. N. Little. Quantitative Comparison of Energy Methods to Characterize Fatigue in Asphalt Materials. *Journal of Materials in Civil Engineering*, Vol. 21, No. 2, 2008, pp. 83–92.
10. Lee, H. J., J. S. Daniel, and Y. R. Kim. Continuum Damage Mechanics Based Fatigue Model of Asphalt Concrete. *Journal of Materials in Civil Engineering*, Vol. 12, 2000, pp. 105–112.
11. Park, S. W., Y. R. Kim, and R. A. Schapery. A Viscoelastic Continuum Damage Model and its Application to Uniaxial Behavior of Asphalt Concrete. *Mechanics of Materials*, Vol. 24, 1996, pp. 241–255.
12. Si, Z., D. Little, and R. Lytton. Characterization of Microdamage and Healing of Asphalt Concrete Mixtures. *Journal of Materials in Civil Engineering*, Vol. 14, No. 6, 2002, pp. 461–470.
13. The American Association of State Highway and Transportation Officials. *Method for Determining the Fatigue Life of Compacted Hot Mix Asphalt (HMA) Subjected to Repeated Flexural Bending*. TP8-94. AASHTO Provisional Standards, Washington, D.C., 2002.
14. Kim, Y. R., H.-J. Lee, and D. N. Little. Fatigue Characterization of Asphalt Concrete Using Viscoelasticity and Continuum Damage Theory. *Journal of the Association of Asphalt Paving Technologists*, Vol. 66, 1997, pp. 521–569.
15. Himeno, K., T. Watanabe, and T. Maruyama. Estimation of Fatigue Life of Asphalt Pavement. *Proc., Sixth International Conference on the Structural Design of Asphalt Pavements*, University of Michigan, 1987, pp. 272–289.
16. Tayebali, A. A., G. M. Rowe, and J. B. Sousa. Fatigue Response of Asphalt-Aggregate Mixtures. *Journal of the Association of Asphalt Paving Technologists*, Vol. 61, 1992, pp. 333–360.
17. Bahia, H. U., D. I. Hanson, M. Zeng, H. Zhai, M. A. Khatri, and R. M. Anderson. *NCHRP Report 459: Characterization of Modified Asphalt Binder in Superpave Mix Design*. Transportation Research Board of the National Academies, Washington, D.C., 2001.
18. *Performance Graded Asphalt Binder Specification and Testing*. Superpave Series No. 1 (SP-1). Asphalt Institute, Lexington, Ky., 1996.
19. Bahia, H. U., D. Perdomo, and P. Turner. Applicability of Superpave Binder Testing Protocols to Modified Binders. *Transportation Research Record: Journal of the Transportation Research Board*, 1997. 1586: 16–23.
20. Anderson, D. A., Y. M. Le Hir, M. O. Marasteanu, J. P. Planche, D. Martin, and G. Gauthier. Evaluation of Fatigue Criteria for Asphalt Binders. *Transportation Research Record: Journal of the Transportation Research Board*, 2001. 1766: 48–56.
21. Soenen, H., and B. Eckmann. Fatigue Testing of Bituminous Binders with a Dynamic Shear Rheometer. *Proc., 2nd Eurasphalt and Eurobitumen Congress*, Barcelona, Spain, 2000.
22. Carreau, P. J., D. C. R. De Kee, and R. P. Chhabra. *Rheology of Polymeric Systems-Principles and Applications*. Hanser/Gardner Publications, Cincinnati, 1997.
23. Motamed, A., and H. U. Bahia. Influence of Test Geometry, Temperature, Stress Level, and Loading Duration on Binder Properties Measured using DSR. *Journal of Materials in Civil Engineering*, Vol. 23, No. 10, 2011, pp. 1422–1432.
24. Shan, L., S. Tian, H. He, and N. Ren. Internal Crack Growth of Asphalt Binders during Shear Fatigue Process. *Fuel*, Vol. 189, 2017, pp. 293–300.
25. Underwood, B. S. A Continuum Damage Model for Asphalt Cement and Asphalt Mastic Fatigue. *International Journal of Fatigue*, 82, 2016, pp. 387–401.

The Standing Committee on Asphalt Binders (AFK20) peer-reviewed this paper (18-02473).

Article

TRR
JOURNAL OF THE TRANSPORTATION RESEARCH BOARD

Transportation Research Record
2018, Vol. 2672(28) 301–310

DOI: 10.1177/0361198118772726
journals.sagepub.com/home/trr
SAGE

Effects of Light-Activated Self-Healing Polymers on the Rheological Behaviors of Asphalt Binder Containing Recycled Asphalt Shingles

Sharareh Shirzad[1], Marwa M. Hassan[2], Max A. Aguirre[1], Samuel Cooper, Jr.[3], and Ioan I. Negulescu[4]

Abstract

A new generation of ultraviolet (UV) light-activated, self-healing polymers was evaluated with the aim to enhance the elastic recovery of the binder and to increase its self-healing abilities. This study had three main objectives: (a) to develop an optimized synthesis procedure for producing light-activated self-healing polymers, (b) to examine the thermal stability of the prepared self-healing polymers, and (c) to evaluate the effect of self-healing polymers on the rheological properties of asphalt binder containing binder extracted from recycled asphalt shingles (RAS). Fourier transform infrared (FT-IR) spectroscopy analysis confirmed the successful synthesis of UV-activated polymers in the laboratory. In addition, thermogravimetric analysis showed that the materials produced achieved the required thermal stability at high temperature. Measuring the viscosity of different binder blends with and without RAS and with and without self-healing polymers revealed that the additive decreased the viscosity of the binder blends containing RAS, thereby providing blends with a better workability. Furthermore, rheological results showed that the rutting resistance of the binder blends containing RAS was improved by increasing the percentage of self-healing polymer. Results also showed improved rheological behaviors at low service temperature with 5% self-healing polymer and with exposure to UV light.

Pavement performance is highly dependent on the rheological properties of asphalt binder, to the extent that, with time, the degradation of asphalt rheological properties can lead to the failure of the pavement (*1*). During its service life, asphalt pavement is subjected to oxidation due to a reaction with atmospheric oxygen. The age hardening of asphalt pavement, caused by asphalt oxidation, leads to a brittle asphalt binder that not only starts to initiate micro-cracks but also results eventually in the deterioration and failure of the pavement (*2*). On the other hand, rehabilitation of damaged asphalt pavement produces large amounts of aged asphalt waste. The use of recycled asphalt waste, such as recycled asphalt shingles (RAS) and/or reclaimed asphalt pavement (RAP) as a partial replacement for virgin materials may significantly reduce consumption of virgin materials, negative economic impacts, and negative environmental effects of asphalt pavement reconstruction. However, there exists a serious challenge to this approach due to the fact that the asphalt binder extracted from recycled materials has been subjected to severe oxidation; as a result it becomes a hardened and brittle binder. Use of this material in new asphalt pavement construction may affect the performance and service life of asphalt pavement by increasing its susceptibility to cracking.

Polymers, which may be defined as large chains of repetitive small molecules, can be added to asphalt mixtures to improve the performance of pavement and thus prolong its service life. Polymer-modified binders show an increased resistance to rutting, thermal cracking, and fatigue, as well as a decrease in stripping and temperature susceptibility. Use of polymers can also lead to a greater elastic recovery, a higher softening point, an

[1]Department of Construction Management, Louisiana State University, Baton Rouge, LA
[2]Department of Construction Management and Industrial Engineering, Louisiana State University, Baton Rouge, LA
[3]Louisiana Transportation Research Center, Baton Rouge, LA
[4]LSU AgCenter and Department of Chemistry, Louisiana State University, Baton Rouge, LA

Corresponding Author:
Address correspondence to Marwa M. Hassan: marwa@lsu.edu

increased cohesive strength, and substantial ductility (*3*). Furthermore, asphalt binder possesses self-healing properties, which may be defined as the ability to recover its original properties. The self-healing concept can be introduced as a solution to the cracking of asphalt pavement, especially those pavements containing recycled materials. Self-healing of asphalt may be compared to the self-healing of injured skin. Injured skin self-heals due to nutrient supplies existing in the body that repair the damaged parts. In asphalt pavement, self-healing properties may be used to repair the damaged area by closing the cracks, stopping crack propagation, and eventually enhancing the performance of the pavement.

In the present study, a new generation of ultraviolet (UV) light-activated, self-healing polymers is evaluated to enhance elastic recovery of the binder and to increase its self-healing abilities. The propagation of micro-cracks, due to aging and excessive loading, causes the chemical breakage of polymer bonds, which in turn produces free radicals. The free radicals produced subsequently recombine through exposure to UV light and thus close the micro-cracks (*4*).

Objectives and Scope

This study had three main objectives: (a) to develop an optimized synthesis procedure for producing light-activated self-healing polymers, (b) to examine the thermal stability of the prepared self-healing polymers, and (c) to evaluate the effects of light-activated self-healing polymers on the rheological properties of asphalt binder with and without RAS (in the following, RAS refers to binder extracted from recycled asphalt shingles).

The synthetic degree of produced self-healing polymers was examined using Fourier transform infrared (FT-IR) spectroscopy, while the thermal stability was analyzed using thermogravimetric analysis (TGA). Furthermore, the effect of light-activated self-healing polymers on the rheological properties of asphalt binder was quantified by blending virgin binder with RAS, with and without the synthesized self-healing polymer.

Background

Recycled Asphalt Shingles

Application of recycled materials as a partial replacement for virgin binder and aggregate may lead to a significant reduction in the required amount of virgin materials, thereby decreasing the overall cost and environmental impacts of asphalt mixture production. One of the sources of recycled asphalt material is recycled asphalt shingles, which is a waste product from the roofing industry. These shingles consist of approximately 28% binder and 58% mineral aggregate, together with mineral filler and glass fibers. Previous studies show that each year, 11 million tons of shingles are generated in the USA, with only 20% of that amount reused (*5*). The main challenge for the asphalt industry in using recycled binder is related to the oxidized nature of the binder. Due to production characteristics and the aging process, the asphaltene-to-maltene ratio of the binder tends to increase (*6*). Aging of the binder invariably leads to a stiff and brittle binder, which may exhibit low deformation characteristics, thereby causing various distresses such as fatigue and thermal cracking (*7*). There are also concerns with the level of blending between virgin and recycled binders, with regard to the contribution of recycled binder to the total asphalt content of the mixture. These challenges have rendered the use of recycled materials, especially binder extracted from shingles, to be limited to a low content. For instance, the use of such recycled binder is not allowed in all states of the USA; in most of the states where it may be used, the material is limited to 5% by weight of the mixture (*8*).

Polymer Applications in Asphalt Pavement

Addition of polymers, natural or synthetic, was initiated in 1843, and has continued to increase during the last several decades (*9*). Polymers are added to the asphalt binder in order to improve the performance of asphalt pavement and to increase its service life. A survey of U.S. state agencies showed that 47 states use modified binders at present; 35 of these states would accept an increase in the amount of modifier in their binders (*10*). Previous studies (*11–14*) show that styrene–butadiene–styrene (SBS), styrene–butadiene rubber (SBR), Elvaloy, ethylene vinyl acetate (EVA), rubber, and polyethylene are the most commonly used asphalt modifiers. Furthermore, studies show that the addition of polymers is useful in increasing asphalt pavement's resistance to rutting, thermal cracking, and fatigue cracking, as well as a reduction of stripping and temperature susceptibility. The application of polymers can also result in a greater elastic recovery, a higher softening point, an increased cohesive strength, and substantial ductility (*3*).

Self-Healing of Asphalt Binder

Since the early 2000s, the self-healing property of asphalt pavement has received considerable attention as a means of addressing the challenges related to the application of recycled binder in asphalt pavement construction. The self-healing property of asphalt may be defined as the recovery of the original properties of the asphalt which have been damaged. However, the self-healing property of asphalt binder is highly dependent on crack width, rest period, and temperature (*15*). Self-healing of the asphalt binder can be slow at ambient temperatures and under continuous traffic loads. Different mechanisms may be used to increase the self-healing capabilities of asphalt binder. Self-healing mechanisms may be divided into two groups, extrinsic and intrinsic. Examples of extrinsic self-healing mechanisms are microcapsule and vascular network self-healing mechanisms, whereby the damage is repaired by releasing the preserved healing agent into different containers once the crack has appeared (*16–19*). The second group of self-healing materials presents an intrinsic self-healing material, which holds the ability to heal repetitively after the appearance of damages/cracks, with no need for any other healing agent or catalyst. Photochemical self-healing polymers are an example of intrinsic self-healing materials, and these are used as the self-healing mechanism in this research (*16*).

Light-Activated Self-Healing Polymer

UV light-activated, self-healing polyurethane, used in this study, was introduced by Ghosh et al. (*4*). The self-healing material is developed by combining polyurethane and oxetane-substituted chitosan (OXE-CHI) into a cross-linked polymer of oxetane-substituted chitosan-polyurethane (OXE-CHI-PUR). This new polymer has the ability to self-repair upon exposure to UV light by means of remodeling the damaged network. During this self-healing process, dormant oxonium ions are activated to react with accessible macromolecular ends, which result from the damage or cracks.

The main components of light-activated self-healing polymer are chitosan (CHI), oxetane (OXE) and polyurethane (PUR). These components were selected based on their functionality. Chitosan was selected to provide UV light sensitivity, oxetane was chosen to provide a constraint in the four-member ring, and PUR was selected to provide mechanical integrity as well as network heterogeneity, and to facilitate the cleavage of the oxetane ring. During the appearance of micro-cracks in asphalt pavement containing light-activated self-healing polymer, different chemical bonds in OXE-CHI-PUR will be broken. Consequently, a breakage of the bonds generates unstable, free radicals. These radicals will recombine through UV light exposure; during the recombination, the radicals will repair damaged areas of the asphalt binder (*4*).

Experimental Program

Test Materials

The experimental program encompassed a straight binder (PG 67-22) and post-consumer waste roofing shingles with a 20% binder content, which were provided by a local contractor. Asphalt binder was extracted from the shingles based on AASHTO T 164 and using trichloroethylene as a solvent. Afterwards, the solvent (trichloroethylene) was removed based on the procedure described in AASHTO R 59. The binder extracted from the shingles (RAS) was then added to selected asphalt binder blends by 5% weight of the binder.

Procedure for Preparation of Light-Activated Self-Healing Polymer

The chemicals required for production of OXE-CHI are chitosan (CHI), sodium hydroxide (NaOH) beads, isopropanol alcohol, and oxetane (OXE). In addition, dimethyl sulfoxide (DMSO), dibutyltin dilaurate (DBTDL), polyethylene glycol (PEG), and hexamethylene diisocyanate (HDI) were used for production of OXE-CHI-PUR.

Chitosan was added into 150 g of sodium hydroxide solution and stirred for 48 hours at 0°C; the solution was then refrigerated at 0°C for another 48 hours. In the following step, 50 ml of pre-cooled isopropyl alcohol was added to the thawed solution and stirred for one hour. After adding pre-cooled 3-chloro-3-methyl oxetane (OXE) into the mixture, the temperature was raised to 80°C and stirred for 12 hours. The prepared solution was then filtered and washed with methanol to provide a neutral pH. Next, the solution was dried at 60°C for 12 hours to remove the excess methanol. The oxetane-substituted chitosan macromonomer (OXE-CHI) was polymerized by dispersing OXE-CHI in dimethyl sulfoxide (DMSO) with a pH = 6.8 and an exposure to UV radiation for 20 minutes. The product was then washed with methanol and dried. A non-acidic route was selected for the OXE-CHI macromonomer production in order to preserve the functionality of the amino group, as well as to maintain the un-opened ring structure of the OXE (*4*).

Polymerized OXE-CHI was then dispersed in DMSO and sonicated at 25°C for 12 hours, followed by stirring at 80°C for 48 hours. The final, self-healing polymer was generated by reacting HDI with dispersed OXE-CHI and PEG under a nitrogen atmosphere at 25°C for 10 minutes (*4*).

Characterization of Light-Activated Self-Healing Polymer

Fourier Transform Infrared (FT-IR) Spectroscopy. During FT-IR spectroscopy, an infrared radiation was passed through the self-healing polymer samples; the infrared spectrum was achieved by determining the fraction of the incident radiation absorbed at a specific energy. The produced OXE-CHI and OXE-CHI-PUR cross-linked networks were characterized by comparing obtained FT-IR spectra of CHI with OXE-CHI, as well as HDI with OXE-CHI-PUR.

Table 1. Composition of Binder Blends

Binder blend	Binder type	RAS (by weight of binder)	Self-healing polymer (by weight of binder)	UV exposure
67CO	67–22	–	–	–
67PG5P	67–22	5%	–	–
67PG5P1SHP	67–22	5%	1%	1 h, 24 h, 48 h
67PG5P3SHP	67–22	5%	3%	1 h, 24 h, 48 h
67PG5P5SHP	67–22	5%	5%	1 h, 24 h, 48 h

Thermal Stability. The procedure for preparation of an asphalt mixture may include heating at up to 163°C for two hours. Thus, the produced self-healing polymer should have the thermal strength needed to resist this temperature. The thermal stability of light-activated self-healing polymer was examined by thermogravimetric analysis (TGA) using a Universal V4.3A TA Instruments thermobalance. Results obtained from the TGA analysis demonstrated changes in the properties of the self-healing polymer, regarded as a function of increasing temperature with a constant heating rate (20°C/min).

Effect of Light-Activated Self-Healing Polymer on Rheological Properties of Asphalt Binder

To assess the effects of light-activated self-healing polymer on the rheological properties, asphalt binder blends consisting of a binder mixed with or without RAS and self-healing polymer were prepared in the laboratory. The prepared binder extracted from RAS was added as 5% of the weight of the binder, and self-healing polymer was added to the binder at three different dosage rates by weight of the binder; the results were compared with virgin asphalt binder (Table 1). In order to achieve a uniform distribution of self-healing polymer in the binder blends, a mechanical stirrer with a high-shear rate of 3,600 rpm for 30 minutes was used. Prepared blends were characterized by laboratory rheological tests (using rotational viscometer, dynamic shear rheometer, and bending beam rheometer) and by comparing the Superpave performance grade (PG) of the modified binder blends with the unmodified binder. Rolling thin film oven (RTFO) and pressure aging vessel (PAV) were used respectively, for short term and long term aging of prepared binder blends. Furthermore, prepared samples (i.e., the dynamic shear rheometer and bending beam rheometer test samples) were exposed to three different durations of UV light (1 h, 24 h, and 48 h) to examine the effect of various UV exposures on the performance of the binder blends containing self-healing polymer.

The viscosity of the binder blends, with or without RAS, and with or without self-healing polymer, were measured using a rotational viscometer (RV), based on AASHTO T 316 (*20*). During the test, the torque required for a constant rotational speed of a spindle submerged in the binder was measured and then converted to represent the viscosity of the binder. The RV test was conducted at high temperature in order to simulate the manufacturing and construction temperature; the results relate to the workability required for pumping and mixing of the binder.

A dynamic shear rheometer (DSR) test was performed based on AASHTO T 315, in order to characterize the viscous and elastic behaviors of asphalt binders at intermediate to high temperatures (*21*). DSR was conducted on both unaged and RTFO-aged samples to measure the complex shear modulus (G*) and phase angle (δ) of each sample. The complex shear modulus is an indication of the binder's resistance to deformation, while the phase angle represents the lag between applied stress and resulted strain. $G^*/\sin\delta$, which is the elastic portion of complex modulus, should be large where rutting is of the greatest concern. The higher the G*, the more resistant the binder is to permanent deformation; with a lower δ, the binder shows improved elastic properties. The bending beam rheometer (BBR) test was performed based on AASHTO T 313-06 (*22*) to evaluate the performance of prepared binder blends at low service temperature. Parameters measured using BBR are creep stiffness and creep slope (m-value). Moreover, the useful temperature interval (UTI) of the binder blends was calculated as the range between the minimum and maximum temperature where the binder blends is expected to have adequate performance.

A multiple stress creep recovery (MSCR) test of asphalt binder was performed, based on AASHTO TP 70, in order to examine the properties at high service temperature. and accordingly to assess the rutting potential of the prepared asphalt binder blends, with or without self-healing polymer. The MSCR test measures the percent recovery, which indicates the elastic response in an asphalt binder; the non-recoverable creep compliance, in turn, is an indicator of the asphalt binder's resistance to permanent deformation under a repeated load. MSCR was performed at two stress levels, 0.1 kPa and 3.2 kPa, with 10 creep and recovery cycles tested at each stress level (*23*).

Results and Analysis

Light-Activated Self-Healing Polymer Characterization

Fourier Transform Infrared (FT-IR) Spectroscopy. The first step of preparing the self-healing polymer reacted

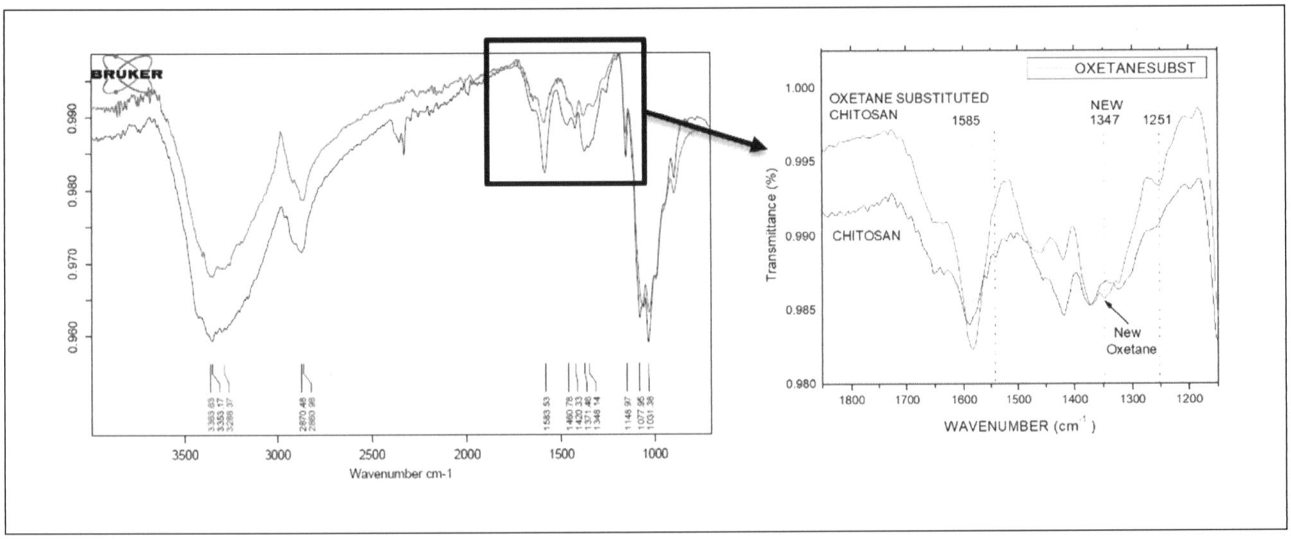

Figure 1. FT-IR spectra of CHI and OXE-CHI.

chloromethyl oxetane (OXE-cl) with chitosan in order to attach the OXE ring to the -OH group of CHI at C6 position of the CHI. In this study, OXE-CHI macromonomers were produced with a molar ratio of (OXE-CHI = 1:1). The FT-IR spectra of CHI and OXE-CHI, presented in Figure 1, confirmed the mentioned reaction.

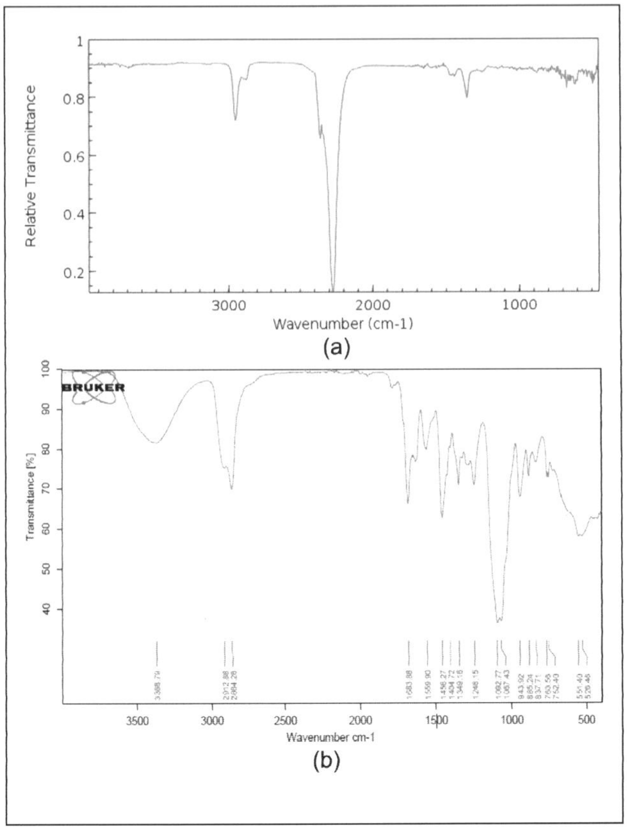

Figure 2. FT-IR spectra of (*a*) HDI and (*b*) OXE-CHI-PU.

Characteristic bands of CHI, observed on both the FT-IR spectra of CHI and OXE-CHI, may be listed as: 1030 cm^{-1} and 1070 cm^{-1} corresponding to the C-O bond; 1148 cm^{-1} corresponds to the ether bond of CHI; 1371 cm^{-1} and 1420 cm^{-1} correspond to C-C and CH bonds; while 1583 cm^{-1} corresponds to Amide I. However, a new peak was observed at 1348 cm^{-1}, but only on the FT-IR spectra of OXE-CHI (Figure 1). This new peak corresponds to C-CH3 of OXE, thereby confirming the successful reaction of OXE and CHI.

In the next step, macromonomers of OXE-CHI were used to produce cross-linked networks of OXE-CHI-PUR. PUR networks were produced based on an adjusted stoichiometric of reactive groups of HDI, PEG, and OXE-CHI; respectively, NCO, OH and NH2 (*4*). During this reaction, amidogens of CHI react with an isocyanate group of HDI, thus producing a carbamide of self-healing polymer. Due to this reaction, the band at 2260 cm^{-1}, which corresponds to the isocyanate group of HDI (Figure 2*a*), was removed in the FT-IR spectra of OXE-CHI-PUR; the band at 1616 cm^{-1} was added, showing the presence of carbamide.

Thermal Stability. Figure 3 presents the TGA of the OXE-CHI-PUR cross-linked networks produced. As shown in the plot, the weight loss of sample was less than 10% in the first 200°C. Weight loss was observed between 181°C and 476°C, which is higher than the preparation temperature of asphalt mixture.

Effect of Light-Activated Self-Healing Polymer on Rheological Properties of Asphalt Binder

Five binder blends, with or without RAS, and with or without self-healing polymer, were prepared and tested

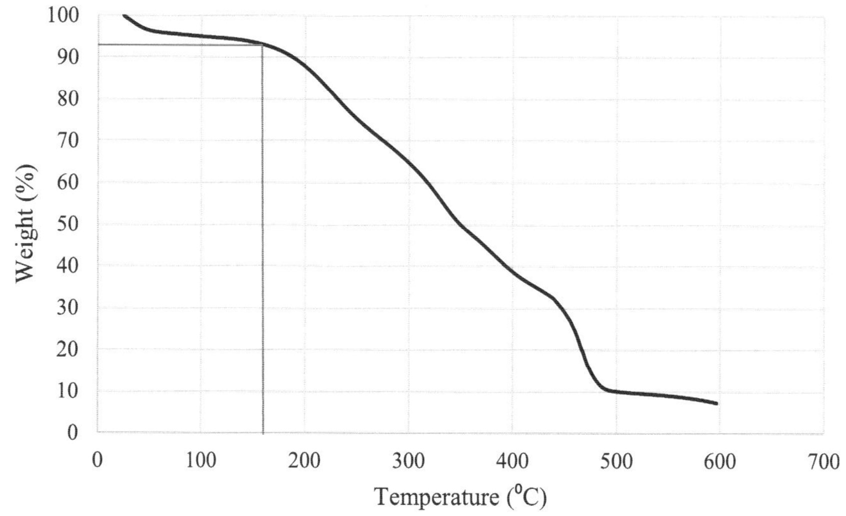

Figure 3. TGA result of OXE-CHI-PUR.

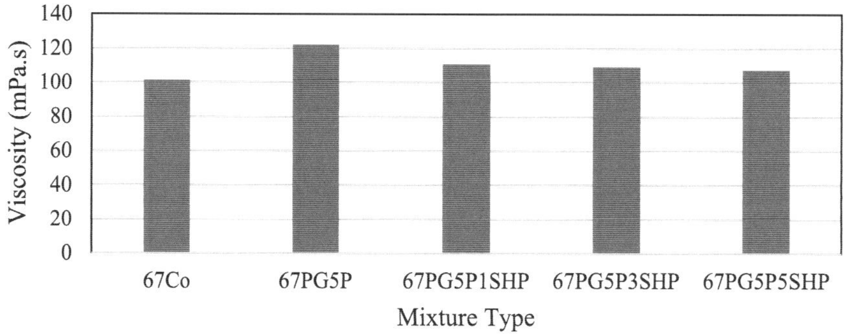

Figure 4. Measured viscosity of binder blends.

in this study in order to examine the effects of OXE-CHI-PUR, a cross-linked network, on the rheological properties of asphalt binder compared with an unmodified binder.

Rotational Viscometer (RV). The viscosity of the five prepared binder blends was measured at 135°C using an RV. The results are presented in Figure 4. The viscosity of the unmodified binder (i.e., 67CO) was 101 mPa.s, compared with a greater viscosity of 122 mPa.s for the binder blend containing 5% RAS (i.e., 67PG5P). However, a decrease was observed in the viscosity of the binder blends containing the self-healing polymer. The viscosity of the binder blends containing 1%, 3%, and 5% self-healing polymers by weight of binder (i.e., 67PG5P1SHP, 67PG5P3SHP, and 67PG5P5SHP) was 110 mPa.s, 109 mPa.s and 107 mPa.s, respectively. The measurements showed that the addition of self-healing polymer reduced the viscosity of the binder blends containing RAS. However, all exceeded the viscosity of the unmodified binder.

Dynamic Shear Rheometer (DSR). The effects of self-healing polymer on the rheological properties of the five prepared samples were tested at intermediate and high temperature using DSR and at low temperature using BBR. The results, presented in Figure 5, show that the addition of RAS to the unmodified binder caused an increase in the complex shear modulus (G* value) of the unaged samples from 1.92 for an unmodified binder (67CO) to 2.72 for a binder blend containing 5% RAS (i.e., 67PG5P). Furthermore, the G* value of the binder blend

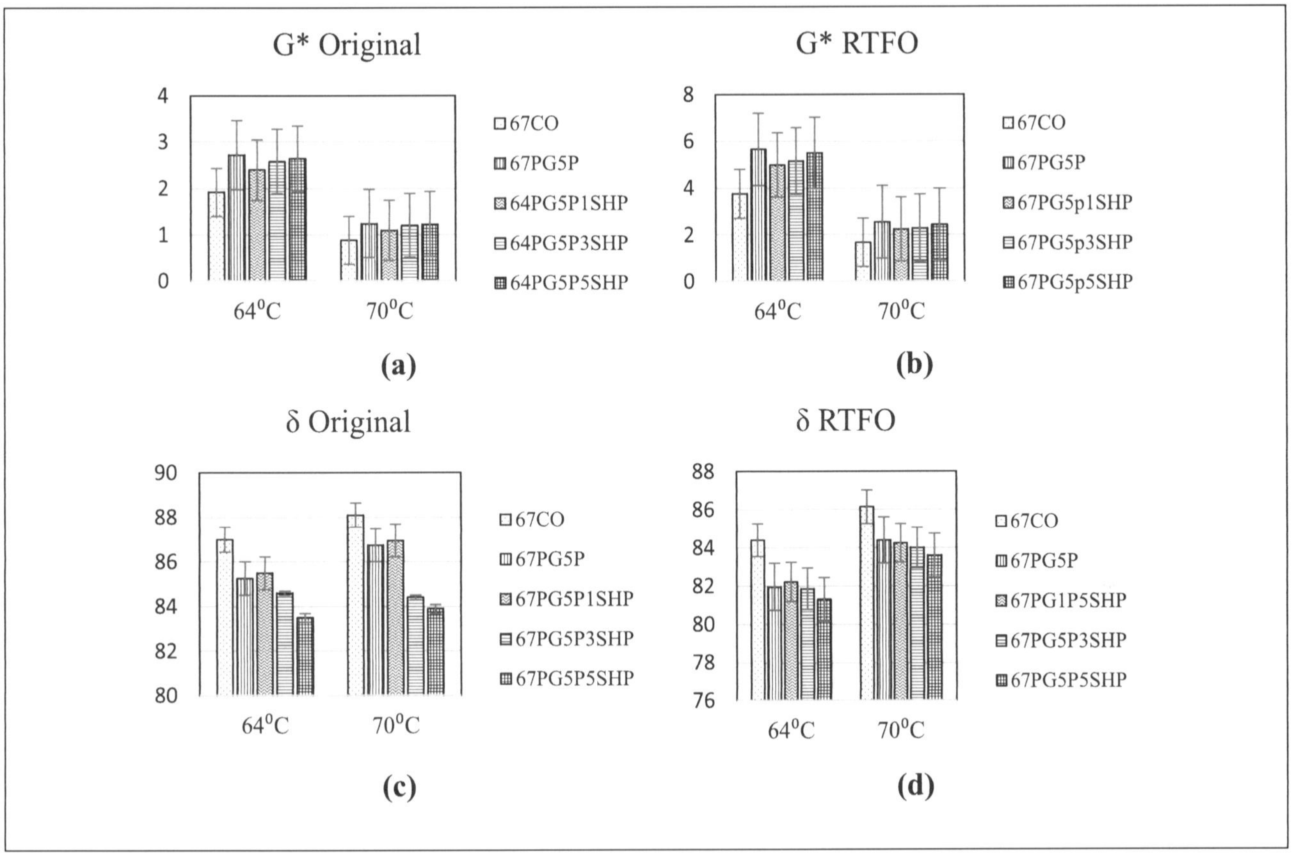

Figure 5. Measured complex modulus (G*) and phase angle (δ) of binder blends: (*a*) complex modulus of original blends, (*b*) complex modulus of RTFO blends, (*c*) phase angle of original blends, (*d*) phase angle of RTFO blends.

containing 5% RAS (i.e. 67PG5P) was decreased to 2.40, 2.58, and 2.64 for binder blends containing 1%, 3%, and 5% self-healing polymer (i.e., 67PG5P1SHP, 67PG5P3SHP, and 67PG5P5SHP), respectively. The phase angle (δ) of the unmodified binder (i.e., 67CO) for an unaged sample was measured at 87º at 67°C. This value decreased to 85.3º when 5% RAS was added to the unmodified binder blend (i.e., 67PG5P). The phase angle also decreased to 85.5º, 84.6º, and 83.5º due to the addition of 1%, 3%, and 5% self-healing polymer to the binder blends (i.e., 67PG5P1SHP, 67PG5P3SHP, and 67PG5P5SHP). The lower phase angle values from the addition of the self-healing polymer may be an indication of improvement in the elastic behavior of the binder blends containing RAS and self-healing polymer. The same behavior was also observed for RTFO samples and at different temperatures (i.e., 67ºC, 70ºC, and 76ºC).

G*/sinδ of the binder blends (Table 2) was measured as an indicator of the susceptibility to rutting of different asphalt binder blends. The G*/sinδ value of the unmodified binder (i.e., 67CO) increased from 1.92 to 2.73 due to the addition of RAS. When 1% self-healing polymer was added to the blend containing an unmodified binder and RAS (i.e., 67PG5P1SHP), the G*/sinδ value decreased to 2.39. The blends containing 3% and 5% self-healing polymer (i.e., 67PG5P3SHP and 67PG5P5SHP) exhibited a higher G*/sinδ value of 2.60 and 2.66, compared with the blend with 1% self-healing polymer.

Bending Beam Rheometer (BBR). Low temperature properties of the prepared binder blends were examined using BBR. The results presented in Table 2 show that the addition of RAS caused an increase in the stiffness and a decrease in the m-value of the binder blends. Furthermore, an increase in the stiffness and a decrease in the m-value of the binder blends were observed as the result of self-healing polymer modification.

Performance Grade (PG). The final performance grade is presented in Table 2. In summary, the addition of RAS caused an increase in the high temperature grade of the binder blend (from 64 to 70), which is an indication of a stiffer binder. G*/sinδ values were reduced by incorporation of self-healing polymers (from 71.2 to 70.2, 70.3 and

Table 2. Rheological Test Results of Binder Blends

Test	Specification	Temperature	67CO	67PG5P	67PG5P1SHP	67PG5P3SHP	67PG5P5SHP
			Test on Original Binder				
DSR ($G^*/\sin\delta$)	>1.0 kPa	64°C	1.92 kPa	2.73 kPa	2.39 kPa	2.60 kPa	2.66 kPa
DSR ($G^*/\sin\delta$)	>1.0 kPa	70°C	0.88 kPa	1.24 kPa	1.10 kPa	1.21 kPa	1.24 kPa
DSR ($G^*/\sin\delta$)	>1.0 kPa	76°C	–	0.61 kPa	0.53 kPa	0.61 kPa	0.63 kPa
			Test on RTFO				
DSR ($G^*/\sin\delta$)	>2.2 kPa	64°C	3.75 kPa	–	5.02 kPa	5.15 kPa	5.49 kPa
DSR ($G^*/\sin\delta$)	>2.2 kPa	70°C	1.68 kPa	2.55 kPa	2.26 kPa	2.30 kPa	2.45 kPa
DSR ($G^*/\sin\delta$)	>2.2 kPa	76°C	–	1.21 kPa	1.08 kPa	1.09 kPa	1.15 kPa
			Test on RTFO + PAV				
DSR ($G^*.\sin\delta$)	<5000 kPa	25°C	3920 kPa	4885 kPa	4150 kPa	4110 kPa	4170 kPa
DSR ($G^*.\sin\delta$)	<5000 kPa	22°C	5795 kPa	6925 kPa	5855 kPa	5850 kPa	5950 kPa
BBR-S	<300 MPa	−6	92 Mpa	98 Mpa	110 Mpa	110 Mpa	115 Mpa
BBR-S	<300 MPa	−12	197 Mpa	190 Mpa	228 Mpa	218 Mpa	228 Mpa
BBR-S	<300 MPa	−18	351 Mpa	360 Mpa	401 Mpa	397 Mpa	401 Mpa
BBR m-value	>0.3	−6	0.374	0.357	0.363	0.370	0.365
BBR m-value	>0.3	−12	0.312	0.302	0.306	0.312	0.309
BBR m-value	>0.3	−18	0.268	0.264	0.265	0.265	0.259
PG	—	—	**64–22**	**70–22**	**70–22**	**70–22**	**70–22**

Table 3. Rheological Test Results of Binder Blends Exposed to UV Light

		Binder blend	PG	Continuous grading	UTI	ΔTc
	-	67Co	64–22	67.9-23.6	91.5	−2.8
		67PG5P	70–22	71.2–22.3	93.5	−4
		67PG5P1SHP	70–22	70.2–24.9	95.1	−2
		67PG5P3SHP	70–22	70.3–25.2	95.5	−1.7
		67PG5P5SHP	70–22	70.8–24.9	95.7	−1.8
UV Light Exposure	**1 h**	67PG5P1SHP	70–22	70.9–25.7	96.6	−1.3
		67PG5P3SHP	70–22	70.9–25.8	96.7	−0.6
		67PG5P5SHP	70–22	72–25.7	97.7	−0.2
	24 h	67PG5P1SHP	70–22	71.5–24.8	96.3	−0.6
		67PG5P3SHP	70–22	71.6–25.6	97.2	−0.6
		67PG5P5SHP	70–22	71.8–25.7	97.5	−0.5
	48 h	67PG5P1SHP	70–22	72–25.3	97.3	−1.6
		67PG5P3SHP	70–22	71.5–25.4	96.9	−0.8
		67PG5P5SHP	70–22	72–25.4	97.4	−0.4

70.8 respectively by application of 1%, 3%, and 5% of self-healing polymer), but the decrease was insufficient to regain the original high temperature grade of the unmodified binder. Furthermore, changes in low temperature grade caused by addition of RAS and self-healing polymer were not significant enough to alter the low temperature grade of the binder blends.

Results from tests performed on binder blends exposed to UV light are presented in Table 3. An increase in the continuous high temperature grade of the blends was observed with the increase in the duration of UV exposure. Furthermore, delta Tc was calculated for all binder blends as the difference between the critical stiffness temperature and the m-value critical temperature. As a result of the addition of RAS, delta Tc was increased from −2.8 (67CO) to −4.0 (67PG5P). However, the addition of 1%, 3%, and 5% of self-healing polymer decreased delta Tc from −4.0 to −2.0, −1.7 and −1.8, respectively. These values were further reduced with UV light exposure.

Multiple Stress Creep Recovery (MSCR). The percent recovery, the non-recoverable creep compliance (J_{nr}), and the percent difference between J_{nr} were measured using MSCR, in order to evaluate the rutting susceptibility of the prepared binder blends at high temperature. The test was conducted at 67°C and the results are presented in Table 4. The percent recovery of the samples increased by increasing the percentage of self-healing polymers. A

Table 4. MSCR Test Results

Blends type	% Recovery		J_{nr} (1/kPa)		J_{nr} Difference
	0.1 kPa	3.2 kPa	0.1 kPa	3.2 kPa	
67CO	1.62%	−0.50%	3.47	3.77	8.65%
67PG5P	4.86%	0.58%	2.32	2.60	12.31%
67PG5P1SHP	4.33%	0.27%	2.70	3.00	12.85%
67PG5P3SHP	5.17%	0.56%	2.46	2.81	14.09%
67PG5P5SHP	6.69%	0.81%	2.25	2.64	17.12%
67PG5P1SHP-1h UV exposure	5.76%	0.71%	2.22	2.53	13.92%
67PG5P3SHP-1h UV exposure	5.40%	0.69%	2.31	2.61	13.52%
67PG5P5SHP-1h UV exposure	6.25%	1.50%	1.80	2.11	16.88%
67PG5P1SHP-24h UV exposure	6.38%	0.91%	2.12	2.42	14.95%
67PG5P3SHP-24h UV exposure	6.18%	0.93%	2.13	2.45	14.60%
67PG5P5SHP-24h UV exposure	6.12%	0.79%	2.23	2.59	15.73%
67PG5P1SHP-48h UV exposure	6.33%	1.14%	1.96	2.24	13.87%
67PG5P3SHP-48h UV exposure	6.80%	1.01%	2.07	2.41	15.99%
67PG5P5SHP-48h UV exposure	6.52%	0.90%	2.24	2.61	16.66%

recovery of 4.9% was observed for a binder blend containing 5% RAS (i.e., 67PG5P). The addition of 1% self-healing polymer (i.e., 67PG5P1SHP) decreased the recovery value to 4.3%. However, this value was increased to 6.7% for a binder blend with 5% self-healing polymer (i.e., 67PG5P5SHP). Additionally, a decrease was observed in the non-recoverable creep compliance (J_{nr}) of the samples. The measured J_{nr} of the binder blends with 1%, 3%, and 5% self-healing polymer was 3.00, 2.81, and 2.64, respectively. The results obtained from the MSCR test indicate an improvement in the rutting resistance of the binder blends with the use of 5% self-healing polymer.

Summary and Conclusions

An innovative, light-activated, self-healing polymer was synthesized in the laboratory by means of a photocatalytic-based chemical method. The synthetic degree of produced self-healing polymers was examined using FT-IR spectroscopy, while the thermal stability was analyzed by means of TGA. In addition, the effect of light-activated self-healing polymer on the rheological properties of asphalt binder was evaluated by blending virgin binder with RAS with and without self-healing polymers. Based on the results of this study, the following conclusions may be drawn:

- The FT-IR analysis confirmed the successful synthesis of cross-linked networks of OXE-CHI-PUR polymer in the laboratory. In addition, TGA results showed that the produced polymers achieved the required thermal stability for resisting the high temperature during asphalt mixture's production processes.
- The results of the rotational viscosity test demonstrated that the addition of RAS to an unmodified binder could lead to an increase in the viscosity of the binder blend. However, a reduction in the measured viscosity of binder blends containing RAS was observed using self-healing polymers. This decrease, caused by the incorporation of self-healing polymer, may improve the workability of binder blends containing stiff, recycled materials.
- DSR test results showed that G* and G*/sinδ values of binder blends containing only RAS decreased with the addition of 1% self-healing polymer. These values increased by increasing the percentages of self-healing polymer to 3% and 5%. An increase in G* and G*/sinδ and a decrease in δ of binder blends was observed with self-healing polymer when compared with the binder blends containing only RAS.
- Performance grading results showed an increase in high temperature grade of the binder blends containing RAS and binder blends containing both RAS and self-healing polymer. However, the low temperature grade was the same for all tested binder blends.
- The difference between the critical stiffness temperature and the m-value critical temperature (Delta Tc) showed an improvement in performance at low service temperature for samples exposed to UV light, with binder blend containing 5% self-healing polymer showing the best results.
- Based on the results from the MSCR test, a binder blend containing 5% RAS and 5% self-healing polymer showed the highest percent recovery among the tested binder blends, indicating that this binder blend showed the most desirable characteristics against rutting.

Further research is recommended to measure the mechanistic properties of asphalt mixtures containing RAS/RAP and self-healing UV-activated polymer in the laboratory.

Acknowledgments

This study is part of project 17BSLU02 funded by the Transportation Consortium of South-Central States (Tran-SET), UTC and the Louisiana Transportation Research Center (LTRC). The authors also acknowledge the laboratory support provided by the LTRC and the TGA work in the frame of the USDA S-1041 program.

References

1. Petersen, J. C., P. M. Harnsberger, and R. E. Robertson. Factors Affecting the Kinetics and Mechanisms of Asphalt Oxidation and the Relative Effects of Oxidation Products on Age Hardening. *American Chemical Society Division of Fuel Chemistry, Preprints*, Vol. 41, No. 4, 1996, pp. 1232–1244.
2. Petersen, J. C., and P. M. Harnsberger. Asphalt Aging: Dual Oxidation Mechanism and Its Interrelationships with Asphalt Composition and Oxidative Age Hardening. *Transportation Research Record: Journal of the Transportation Research Board*, 1998. 1638: 47–55.
3. Yildirim, Y. Polymer Modified Asphalt Binders. *Construction and Building Materials*, Vol. 21, No. 1, 2007, pp. 66–72.
4. Ghosh, B., V. Chellappan, and W. M. Urban. Self-Healing Inside a Scratch of Oxetane-Substituted Chitosan-Polyurethane (OXE-CHI-PUR) Networks. *Journal of Materials Chemistry*, Vol. 21, No. 38, 2011, pp. 14473–14486.
5. Oldham, D., E. Fini, and F. Chailleux. Application of a Bio-Binder as a Rejuvenator for Wet Processed Asphalt Shingles in Pavement Construction. *Construction and Building Materials*, Vol. 86, 2015, pp. 75–84.
6. Ongel, A., and M. Hugener. Impact of Rejuvenators on Aging Properties of Bitumen. *Construction and Building Materials*, Vol. 94, 2015, pp. 467–474.
7. Yaghoubi, E., M. R. Ahadi, M. Alijanpour Sheshpoli, and H. Jahanian Pahlevanloo. Evaluating the Performance of Hot Mix Asphalt with Reclaimed Asphalt Pavement and Heavy Vacuum Slops as Rejuvenator. *International Journal of Transportation Engineering*, Vol. 1, No. 2, 2013, pp. 115–124.
8. Mogawer, W. S., A. Booshehrian, S. Vahidi, and A. J. Austerman. Evaluating the Effect of Rejuvenators on the Degree of Blending and Performance of High RAP, RAS, and RAP/RAS Mixtures. *Road Materials and Pavement Design*, Vol. 14, No. S2, 2013, pp. 193–213.
9. Thompson, D. C., and A. J. Hoiberg (ed.). *Bituminous Materials: Asphalt Tars and Pitches*. Robert Krieger Publishing Co, New York, 1979.
10. Bahia, H. U., D. Perdomo, and P. Turner. Applicability of Superpave Binder Testing Protocols to Modified Binders. *Transportation Research Record: Journal of the Transportation Research Board*, 1997. 1586: 16–23.
11. Sargand, S. M., and S. S. Kim. Performance Evaluation of Polymer Modified and Unmodified Superpave Mixes. *Proc., 2nd International Symposium on Maintenance and Rehabilitation of Pavements and Technological Control*, Auburn, AL, 2001.
12. Sebaaly, P. E., G. M. Bazi, and Y. Vivekanathan. *Evaluation of New Pavement Technologies in Nevada*. Report No: RDT-03-034, Interim Report, Report No. 13AX-1, University of Nevada, 2003.
13. Newcomb, D. Limit the Strain at the Bottom of an Asphalt Pavement, and What Do You Get?: A Perpetual Pavement. *Hot Mix Asphalt Technology*, Vol. 8, No. 6, 2003, pp. 30–32.
14. Partl, M. N., and J. K. Newman. Flexural Beam Fatigue Properties of Airfield Asphalt Mixtures Containing Styrene–Butadiene Based Polymer Modifiers. *Proc., 6th International Rilem Symposium*, Zurich, Switzerland, 2003, pp. 357–363.
15. Qiu, J., V. Ven, S. Wu, and J. Yu. Investigation of Self Healing Behaviour of Asphalt Mixes Using Beam on Elastic Foundation Setup. *Materials and Structures*, Vol. 45, No. 5, 2012, pp. 777–791.
16. Hia, L., V. Vahedi, and P. Pasbakhsh. Self-Healing Polymer Composites: Prospects, Challenges, and Applications. *Polymer Reviews*, Vol. 56, No. 2, 2016, pp. 225–261.
17. Aguirre, M. A., M. Hassan, S. Shirzad, W. H. Daly, and L. N. Mohammad. Micro-Encapsulation of Asphalt Rejuvenators Using Melamine-Formaldehyde. *Construction and Building Materials*, Vol. 114, 2016, pp. 29–39.
18. Shirzad, S., M. Hassan, M. A. Aguirre, S. Cooper, I. Negulescu, and L. N. Mohammad. Microencapsulated Sunflower Oil for Rejuvenation and Healing of Asphalt Mixtures. *Journal of Materials in Civil Engineering*, Vol. 29, No. 9, 2017, 04017147.
19. Shirzad, S., M. Hassan, M. A. Aguirre, S. Cooper, W. H. Daly, and L. N. Mohammad. Evaluation of Sunflower Oil as a Rejuvenator and Its Microencapsulation as a Healing Agent. *Journal of Materials in Civil Engineering*, Vol. 28, No. 11, 2016, 04016116.
20. AASHTO. *Standard Method of Test for Viscosity Determination of Asphalt Binder Using Rotational Viscometer*. AASHTO T 316-13, Washington, D.C., 2013.
21. AASHTO. *Determining the Rheological Properties of Asphalt Binder Using a Dynamic Shear Rheometer (DSR)*. AASHTO T 315-12, Washington, D.C., 2016.
22. AASHTO. *Standard Method of Test for Determining the Flexural Creep Stiffness of Asphalt Binder Using the Bending Beam Rheometer (BBR)*. AASHTO T 313-06, Washington, D.C., 2016.
23. AASHTO. *Standard Method of Test for Multiple Stress Creep Recovery (MSCR) Test of Asphalt Binder Using a Dynamic Shear Rheometer (DSR)*. AASHTO TP 70, Washington, D.C, 2013.

The Standing Committee on Asphalt Binders (AFK20) peer-reviewed this paper (18-02638).

TRR
JOURNAL OF THE TRANSPORTATION RESEARCH BOARD

Article

Transportation Research Record
2018, Vol. 2672(28) 311–323

DOI: 10.1177/0361198118792999
journals.sagepub.com/home/trr
SAGE

Developing Simple Binder Indices for Cracking Resistance of Asphalt Binders at Intermediate and Low Temperatures

Raquel Moraes[1] and Hussain Bahia[1]

Abstract
Oxidative aging causes hardening of asphalt binders and, consequently, contributes to deterioration of asphalt pavements. Non-load related cracking of asphalt pavements (i.e., transverse and block cracks) is related to original properties and hardening of the asphalt binder. In recent years, researchers have proposed new indices derived from Superpave Performance Graded (PG) testing to identify changes in asphalt cracking susceptibility with aging. These indices include the parameter $G'/(\eta'/G')$ and the difference between continuous low temperature binder grade measured via Superpave creep stiffness and m-value (ΔT_c). This study aims to develop interrelationships between tests allowing choice selection for the determination of simpler parameters that could be used for specification tests of asphalt binders. Two simpler asphalt binder indices for cracking resistance are investigated by means of Dynamic Shear Rheometer (DSR), Bending Beam Rheometer (BBR) and Single Edge Notched Bending (SENB) tests results. At intermediate temperatures, the slope of the $|G^*|$-frequency curve obtained from a simple frequency sweep test is proposed as an alternative approach to directly calculate the durability parameter $G'/(\eta'/G')$. At low temperatures, results indicated a direct correlation between failure energy at fracture obtained from SENB and ΔT_c. To confim the validity of these indices and the changes at molecular level, Gel Permeation Chromatograph (GPC) results are presented to indicate that asphalts with higher content of large molecular size (LMS) molecules are likely to crack. Limits for specifcations for the slope of $|G^*|$-frequency curve and ΔT_c can be derived based on testing a wide range of binders and field experience.

In 1987, the Strategic Highway Research Program (SHRP) was established to improve the performance and durability of roads in the United States, and the Superpave (Superior Performing Asphalt Pavements) asphalt binder grading system was a product of this research effort. The Superpave Performance Graded (PG) asphalt binder specification was developed to relate binder properties to field performance, and allow the selection of an appropriate binder for a specific climatic condition.

The long-term pavement performance is a function of traffic load and volume, material properties, construction practices, and environmental factors. Pavement deterioration manifests itself in several common types of distress including rutting, fatigue cracking, low temperature cracking, reflective cracking, aging, raveling, and stripping. The conventional materials used in the asphalt mixture may perform satisfactorily relative to one distress type, but fail prematurely relative to the others. Asphalt production, however, is not one of the main profit-generating processes in the refining industry, since most refineries in the United States deal with asphalt as a by-product of crude fractionation (*1*). Therefore, the production of better-performing asphalts is not considered one of the common strategies in planning refining practices.

As the asphalt binder is responsible for the visco-elastic behavior characteristic of asphalt pavements, it plays a large part in determining many aspects of road performance, particularly resistance to permanent deformation and cracking (*2*). In general, the proportion of any induced strain in asphalt that is attributable to viscous flow, i.e., non-recoverable, increases with both loading time and temperature. Non-load related cracking of asphalt pavements (i.e., transverse and block cracks) is related to oxidation and hardening of the asphalt binder. Transverse cracks are usually formed because of thermal movement. It may occur because of shrinkage of the binder surface due to low temperatures or asphalt binder

[1]Department of Civil and Environmental Engineering, Modified Asphalt Research Center, University of Wisconsin-Madison, Madison, WI

Corresponding Author:
Address correspondence to Raquel Moraes: moraes@uwalumni.com

hardening. Block craking is typically caused by an inability of the asphalt binder to expand and contract with temperature cycles because of aging, as the binder starts to lose its durability.

Researchers have made significant efforts to classify the cracking resistance of asphalt binders using an index parameter. Kandhal et al. in 1977 (*3*) noted that the decrease in low-temperature ductility is an important factor as asphalt binder ages. Following this work and by using a mechanical-empirical relationship with observed cracking, Glover et al. (*4*) suggested a parameter relating storage modulus (G′) and dynamic viscosity (η′) to ductility at 15°C and 0.005 rad/s.

Anderson et al. in 2011 (*5*) investigated the relation between ductility and binder properties to non-load associated cracking for airport pavements. The findings of the study identified the Glover parameter (G′/[η′/G′]), the fatigue parameter *B*, and the difference between the continuous low-temperature binder grade measured via the Superpave creep stiffness and m-value (ΔT_c) as parameters to identify changes in cracking susceptibility with aging.

Rowe in 2011 (*6*) proposed modifications of the Glover parameter by introducing the Glover-Rowe (G-R) parameter ($G^*[\cos\delta]^2/\sin\delta$), which focuses attention on complex modulus ($|G^*|$) and phase angle (δ) of asphalt binders at 15°C and a frequency of 0.005 rad/s. On his proposal, Rowe ignored the frequency term expressing the G-R parameter purely in terms of $|G^*|$ and δ, allowing users to plot the ductility-based failure planes in Black Space diagrams.

In addition to these efforts, the American Association of State Highway and Transportation Officials (AASHTO) TP 101 "Estimating Damage Tolerance of Asphalt Binders Using the Linear Amplitude Sweep" was proposed in 2012 (*7, 8*) and used in various studies to estimate fatigue cracking resistance.

In the present study, a simpler asphalt binder index for cracking resistance at intermediate temperature is investigated. At low temperatures, failure energy at fracture obtained from the Single Edge Notched Bending (SENB) test, relaxation properties obtained from ΔT_c, and changes at the molecular level by using Gel Permeation Chromatograph (GPC) are investigated. The study aims to either confirm the validity of the cracking indices proposed by previous researchers, or suggest a replacement of these values with simpler parameters that could be used for specification tests of asphalt binders.

Background

Asphalt Chemistry

The chemical composition of asphalts cannot be defined exactly, because it is extremely complex and variable, depending on the source of the crude oil from which the asphalt originates and on modification induced by treatments in the refinery or during the "in service" life. Based on differences in solubility and polarity, the asphalt can be separated into four main chemical classes, designated as saturates (S), aromatics (A), resins (R), and asphaltenes (A).

Asphalt is often described as a colloidal dispersion of asphaltenes into an oily matrix constituted by saturates, aromatics, and resins. It has been established that asphaltenes are stabilized in crude oils by natural resins which are surfactant-like agents. The action of resins on the asphaltenes aggregation and precipitation processes is thought to be important. Due to their molecular constitution, asphaltenes and resins have a mutual intrinsic effect on the stability of molecular self-assembling, either in the form of asphaltene-resin association (which promotes re-dispersion) or asphaltene-asphaltene association (which promotes precipitation) (*9*). It is important to mention that both asphaltenes and resins are responsible for the visco-elastic properties of the asphalt at ambient temperatures (*2*).

Aging Process in Asphalts. The aging process in asphalt pavements may be detrimental when excessive hardening or stiffening is observed, subjecting the pavement to cracking, either under traffic load or thermal stress. There are two types of aging in asphalt: (i) physical aging caused by temperature, and (ii) oxidative aging caused by air or oxygen. Physical aging is a reversible phenomenon that takes place as a consequence of cooling or quenching of amorphous materials from melt temperatures to below the glass-transition temperature. Oxidative aging is irreversible hardening of asphalts. During the oxidative aging process of asphalt binders, the concentration of polar functional groups increases, resulting in an immobilization of molecules through intermolecular association (*10*). Therefore, the molecules or molecular agglomerates lose sufficient mobility to flow past one another under thermal or mechanical stress. This results in embrittlement of the asphalt, making it more susceptible to fracturing or cracking and resistant to healing (*11*).

According to the literature (*12, 13*), thermo-oxidative aging increases the content of large asphalt molecules and decreases the content of small molecules, leading to an increase in the molecular weight of the asphalt binders. Because the various asphalt fractions have different reactivities toward oxidation, during oxidative aging a reduction of aromatics content is observed along with an increase in the content of resins and asphaltenes (*13*). Oxidation causes the oils (i.e., maltene fraction) to convert to resins, and the resins to convert to asphaltenes (*14*). Changes in the content of saturates are almost negligible due to their inert nature. Because of this change in the asphaltenes-maltene ratio that occurs during oxidation, the stiffness property of the asphalt is affected.

Asphalt Modification to Improve Its Performance

For several decades, blending different types of asphalt was the applicable solution to obtain improvements in the asphalt binder properties. In recent years, the modification of the asphalt binder to enhance its performance at both high and low temperatures and under traffic loading has served as one of the cost-effective engineering solutions to allow the construction of superior pavements (*15–17*).

Polymer Modified Asphalts. Over the years, polymer modification of asphalt binders for pavements has found increasing usage, since demands made upon roads increase year by year. This type of modification provides increased viscosity at high pavement temperatures, resulting in a reduced tendency for permanent deformation (i.e., rutting). At the same time, polymers can provide increased ductility at low pavement temperatures and thus a reduced tendency for thermal cracking (*18*). Other advantages of polymer modification are improved resistance to abrasion as well as to oxidation and aging, and the possibility of applying a thinner pavement in comparison to the one where a conventional binder is used.

Blends of asphalt with polymers form multiphase systems. Such systems contain a phase rich in polymer, a phase rich in asphaltenes not adsorbed by the polymer, and a phase formed by maltenes (*19*). As a result, the properties of the modified asphalt binders depend on the concentration and the type of polymer used for modification. The main point in polymer modification is, therefore, the formation of a thermodynamically unstable but kinetically stable mixture, in which the polymer is partially swollen by the lighter asphalt components (*20*).

Although great advances have been achieved in the field of asphalt polymer modification, there are still various drawbacks which are limiting its future developments, such as higher costs, polymer low aging resistance and poor storage stability.

Rejuvenators. With increased demand and limited aggregate and binder supply, hot mix asphalt (HMA) producers have begun using reclaimed asphalt pavement (RAP) as a valuable component in HMA. Recycled roofing shingles (RAS) are also identified as a viable replacement or supplement to RAP due to their relatively high asphalt binder content and the presence of hard mineral aggregates and fibrous filler materials (*18*).

RAP and RAS are useful alternatives to virgin materials because they reduce the use of virgin aggregate and the amount of virgin asphalt binder required in the production of HMA. However, the hard, oxidized nature of reclaimed asphalt binder is a major concern when incorporating recycled materials into hot mix asphalt mixtures, since the stiff binder that is present in RAP/RAS would cause premature fatigue and low temperature cracking failures (*20, 21*).

In general, there are two types of chemical additives that can be added to pavements containing RAP/RAS: rejuvenating agents and softening agents. The primary difference between a rejuvenating agent and a softening agent is that a rejuvenating agent will restore the chemical structure of aged asphalt via the restoration of the original binder asphaltenes to maltenes ratio (*22*), while a softening agent will reduce the overall viscosity of the binder.

For optimal rejuvenation of the aged binder, consideration should be given not only to the viscosity reducing capacity of the rejuvenator, but also to its chemical composition. Furthermore, the degree of diffusion of the rejuvenator into the aged binder is of the utmost importance since it will allow chemical changes to take place, affecting the properties of the recycled asphalt mixture.

Researchers have used a wide range of oils to enhance binder performance, specifically at low service temperature (*23, 24*). Concerns exist, however, with respect to oil modification as it relates to high temperature performance and aging susceptibility. Softening effects of oil modification can detrimentally affect high temperature performance and increase the rate of oxidative aging.

Hybrid Modification. Hybrid modification is a term that is applied when two different types of modifiers are combined to produce an improved modification. The main interest in hybrid modification of binders is to capture the properties of the different modifiers.

Materials and Testing Procedure

Materials

Four unmodified asphalt binders and six modified asphalt binders were selected for investigation in this study. The modification type, content and the final performance grades for each selected binder are shown in Table 1. The target factor for the selection of both oil and polymer content was the required climatic needs found in the location of the project. It should be noted that oil type A is a bio-oil while the type B is a refined waste oil.

To validate lab findings, heavily aged binders from the Pooled Fund TPF-5 (302) project (*25*) were also included in the experimental matrix of this study. Two asphalt binders commonly used in the Mid-West region of the United States were selected to produce 'artificial'

Table 1. Performance Grade of the Selected Unmodified and Modified Binders

	Modifier	High temp. (Unaged & Rolling Thin-Film Oven (RTFO) aged)	Intermediate temp. (Pressure Aging Vessel (PAV) aged)	Low temp. true grade (PAV aged), (°C)		PG grade
Asphalt binder	Type and content (%)	True grade (°C)	True grade (°C)	S	m-value	High-low temp.
C (Russian crude source)	N/A	61.8	18.4	−25.8	−26.3	58-22
D (Russian crude source)	N/A	62.5	15.9	−28.9	−28.7	58-28
E (Russian crude source)	N/A	69.9	12.6	−34.2	−33.1	70-28
G (Russian crude source)	N/A	52.3	14.5	−28.4	−30.0	52-28
M-A (Venezuelan crude source)	5% Oil-A	58.3	10.8	−29.9	−32.2	58-28
M-B (Russian crude source)	10% Oil-B + 2% Plastomer	58.5	6.7	−38.1	−38.5	58-34
M-C (Russian crude source)	8% Oil-B + 3% Elastomer	59.7	7.9	−36.6	−37.6	58-34
M-D (Russian crude source)	8% Oil-B + 2% Plastomer	62.5	6.6	−38.4	−38.1	58-34
M-F (Venezuelan crude source)	8% Oil-B + 4% Plastomer	59.7	7.9	−36.8	−39.8	58-34
M-G (Russian crude source)	8% Oil-B + 5% Elastomer	59.2	9.1	−38.7	−39.9	58-34
A-RAP$_1$	N/A	90.1	N/A	−15.4	−12.2	88-12
A-RAP$_2$	N/A	105.3	N/A	−21.4	−14.6	100-12

reclaimed asphalt pavement (RAP) materials: an unmodified (neat) PG 64-22 binder (A-RAP$_1$), and a polymer modified asphalt (PMA) PG 70-28 binder (A-RAP$_2$) collected from the member state Wisconsin. The artificial RAP materials were produced by exposing the virgin binders to extended cycles of aging in the Pressure Aging Vessel (PAV) (40 and 60 hours for A-RAP$_1$ & A-RAP$_2$, respectively). The A-RAP materials (A-RAP$_1$ & A-RAP$_2$) were blended with polymer modified asphalts sampled from partner states. The two PMAs are WI 70-28 and KS 64-34, which were nominated PMA$_1$ and PMA$_2$, respectively. The combinations of PMA/A-RAP prepared for this study are: 100%/0%, 80%/20%, 60%/40%, and 0%/100%. The combinations of PMA/A-RAP were prepared using a low shear mixer for one 1 h and 30 min at a temperature of 150°C.

The asphalt binder modified with 5% of oil-A (sample M-A) was prepared using a high shear mixer for 30 min at temperature of 150°C. The asphalt binders modified with oil and elastomer (samples M-C and M-G) were prepared using a high shear mixer for 60 min at temperature of 180°C. After 1 hr of mixing, the cross-linker (Sulfur) was added, and the shear mixing was kept for another 3 h. The samples were then cured overnight at 145–150°C prior further testing. The asphalt binders modified with oil and plastomer (samples M-B, M-D, and M-F) were prepared using a low shear mixer for 60 min at temperature of 150°C.

Methods

A summary of the test methods selected for testing materials in this study is provided in Table 2.

Linear Amplitude Sweep (LAS) Test. The Linear Amplitude Sweep test is a method proposed to evaluate asphalt binder fatigue at intermediate temperature using the Dynamic Shear Rheometer (DSR) (*8*). The LAS test procedure was performed using 8 mm parallel plate geometry. Before testing, all samples were aged in RTFO followed by PAV aging for 20 h at 100°C. The test consists of two steps: a frequency sweep and an amplitude sweep. In the first part of the LAS procedure, an initial 100 cycles is applied at small strain (0.1%) to determine undamaged linear viscoelastic properties. The second part of the procedure consists of ramping strain amplitude, beginning at 0.1% and ending at 30% applied strain, over 3,100 cycles of loading at 10 Hz. Once the strain sweep is applied to the sample, damage accumulation can be then determined through Viscoelastic

Table 2. Selected Testing for Materials Evaluation

Property	Test type	Standard	Testing conditions
PG Grading	Dynamic shear rheometer (DSR)	AASHTO M 320 (*26*) and associated laboratory testing procedures	@ High PG and Intermediate PG temperature
Aging	Rolling thin-film oven (RTFO)	AASHTO T 240 (*27*)	@ 325°F (163°C) for 85 minutes
	Pressure aging vessel (PAV)	AASHTO R 28 (*28*)	@ 2.10 MPa and 100°C for 20 hours
Intermediate temperature cracking resistance	Linear amplitude sweep (LAS)	Proposed revised version of the AASHTO TP 101-12 (*7*)	Frequency sweep Amplitude sweep Tested @ Intermediate PG temperature
Low temperature cracking resistance	Bending beam rheometer (BBR)	AASHTO T 313 (*29*)	@ Low PG temperature
	Single edge notched beam (SENB)	AASHTO T XXX-11 (*30*)	@ Low PG temperature
Molecular size distribution (MSD)	Gel permeation chromatography (GPC)	—	Flow rate of 1 mL/min @ 40°C

Continuum Damage (VECD) analysis, resulting in the fatigue power law damage model (Equation 1) and the corresponding coefficients, *A* and *B*.

$$N_f = A(\gamma_{max})^{-B} \quad (1)$$

N_f is the traffic volume failure criteria and defines the number of cycles to fatigue failure at a user-defined damage level. γ_{max} is the maximum tensile strain expected in the binder phase under traffic loading, which will be a function of pavement structure. *A* is the LAS power-law parameter representing the intercept at 1% strain. *B* is the LAS power-law parameter representing the slope of the N_f-strain curve. The logarithmic slope of the storage modulus (G′[ω]) as a function of angular frequency is used to calculate the damage accumulation and the parameter *B*.

Single Edge Notched Bending (SENB) Test. The SENB system was developed for measuring asphalt binder, mastic, and mortar cracking resistance at low service temperatures (*31, 32*). The SENB test uses a modified Bending Beam Rheometer (BBR) with the addition of a loading motor that controls the displacement rate during testing, a load cell with a higher capacity than the regular BBR, and a modified beam placement fixture. In the SENB analysis, the failure energy (G_f) is defined as the total area under the entire load-deflection (*P-u*) curve, divided by the area of the ligament, as shown in Equation 2. The displacement at failure, u_f, is also reported from this test procedure as a characterization parameter.

$$G_f = \frac{W_f}{A_{lig}} = \frac{\int P du}{A_{lig}} \quad (2)$$

where W_f is work of failure, G_f is failure energy, *P* and *u* are the load and displacement measured by the SENB, and A_{lig} is the area of the ligament. In this study, SENB tests were performed at a constant displacement rate of 0.01 mm/s, at a low temperature PG grade of each asphalt binders. Before testing, all samples were aged in RTFO followed by PAV aging for 20h at 100°C.

Gel Permeation Chromatograph (GPC). In this study, the Gel Permeation Chromatograph was used to measure the molecular size distribution of unmodified and modified asphalt binders. In this method, the asphalt binder is dissolved in a solvent and is injected into the GPC. The injected sample travels through a series of columns which separates the sample based on molecular size. The larger molecular size particles exit the columns first and are detected by the GPC system. The smaller molecular size particles travel into the pores of the columns and, therefore, have longer retention times. A molecular size distribution (analogous to a sieve analysis on a smaller scale) is then obtained. For the GPC analysis of the asphalt binders selected in the study, the detector used was a differential refractometer, which is a concentration sensitive detector that measures the difference in refractive index (dRI) between the eluent in the reference side, and the sample + eluent in the sample side. The GPC columns used were PolyPore 5 μm mixed, 300 mm × 7.5 mm. The asphalt samples were prepared in concentrations of 5 mg/2 mL. The flow rate was 1 mL/min and the testing temperature was 40°C. The tetrahydrofuran (THF) solvent was selected for the GPC analysis since previous researchers had successfully selected it for GPC applications in asphalt (*13*). To better perform the analysis of the asphalt binders, each GPC chromatogram was divided into 13 equal elution time areas, and the molecules eluted during the first third of the elution period were classified as large molecular size (LMS), those eluted during the second third as medium molecular size (MMS), and those eluted in the last third as having small molecular size (SMS) (*13*).

Determination of Limits at Intermediate and Low Temperature to Cover Durability of Asphalts

Correlation between $G'/(\eta'/G')$ and Fatigue Parameter B

Research conducted by Glover et al. (*4*) indicated that asphalt ductility results of unmodified binders obtained at 15°C and 1 cm/min correlated well with $G'/(\eta'/G')$ for different binders aged at different conditions. On further investigation, the same authors concluded that if ductility at 15°C and 1 cm/min can be used as an indication of durability then so can $G'/(\eta'/G')$ at 15°C, 0.005 rad/s. Considering that this low frequency is not accessible to most instruments except through data taken at several temperatures, and over a range of several orders of magnitude in frequency, the same researchers followed the time-temperature superposition principle and shifted this correlation to measurements at 44.7°C and 10 rad/s.

In this present study, an alternative approach is considered to directly calculate the durability parameter $G'/(\eta'/G')$ developed by Glover et al. (*4*). DSR test results collected from the LAS frequency sweep procedure were used to obtain the storage modulus (G') and loss modulus (G'') of each asphalt binder at fixed frequency of 10 rad/s. To calculate $G'/(\eta'/G')$ at this frequency value, an interpolation of the testing results for frequency *versus* G' and G'' was perfomed for each binder. From the determined values of storage modulus (G') and loss modulus (G''), the durability parameter was then calculated for each binder at its intermediate performance grade temperature and 10 rad/s.

Since,

$$\eta' = G''/\omega \quad (4)$$

Therefore,

$$G'/(\eta'/G') = (G')^2/(G'').\omega \quad (5)$$

The proposed approach does not require manipulations using time-temperature superposition. Therefore, measured values can be plotted among each other without the use of shift factor and sigmoidal fit functions.

Figure 1*a* shows the absolute value of the LAS *B* parameter for the unmodified and modified binders selected in this study. Figure 1*b* shows the absolute value of the LAS *B* parameter for the artificially recycled binder, prepared in a laboratory under heavy aging in PAV (40 and 60 h for A-RAP_1 & A-RAP_2, respectively). Please note that the fatigue parameter *B* (slope parameter) is derived from the frequency sweep procedure, that is conducted before the strain sweep procedure in the LAS test.

Anderson et al. (*5*) showed a correlation between the absolute value of the fatigue parameter *B* and the durability parameter $G'/(\eta'/G')$, calculated at 15°C and 0.005 rad/s from a temperature-frenquency sweep produced mastercurve. Figure 1*c* shows the correlation between the absolute value of the fatigue parameter *B* and the durability parameter $G'/(\eta'/G')$ calculated at 10 rad/s and at the intermediate performance grade temperature of each binder. Because the fatigue parameter *B* simply represents the slope of the $|G^*|$-frequency curve at a given temperature, it is expected that it will relate to $G'/(\eta'/G')$, indicating a materials response to oxidative aging.

As can be seen in Figure 1*c*, given the fairly good correlation indicated by $R^2 = 0.76$, it does appear that the absolute value of the LAS *B* parameter (measured at the binders' intermediate PG temperature) is related to $G'/(\eta'/G')$ parameter (calculated from LAS frequency sweep results collected at intermediate PG temperature and 10 rad/s). Therefore, the frequency sweep step of LAS test could be used as a substitute for the "temperature-frequency sweep test with mastercurve determination" for calculation of the durability parameter $G'/(\eta'/G')$. From the results it can be see that this new approach works for unmodified, modified, and artificial recovered asphalt binders, which is very positive since the correlation found by Glover et al. (*4*) only included unmodified binders.

According to Glover et al. (*4*), a value of $G'/(\eta'/G')$ larger than 0.003 MPa/s at 15°C, 0.005 rad/s (i.e., ductility of about 3 cm at 15°C, 1 cm/min) should be taken as a critical value for pavement binder failure, while a value of 0.0009 MPa/s (ductility of about 5 cm) should be taken as an indication that the binder is approaching a value that will lead to pavement cracking. Please note that the authors did not consider the climate influence while setting these limit values. Anderson et al. (*5*) showed with actual field data that the threshold values identified by Glover et al. for the $G'/(\eta'/G')$ parameter at 15°C and 0.005 rad/s require more validation work. Therefore, setting limits for LAS fatigue parameter *B* based on the 0.0009 MPa/s at 15°C and 0.005 rad/s was not performed, and instead a correlation with the 10 rad/s frequency used in current specifications was investigated.

To propose a new limit for the $G'/(\eta'/G')$ parameter calculated with the alternative approach suggested in this paper (i.e., from LAS frequency sweep results collected at intermediate PG temperature and 10 rad/s) a value for the LAS *B* parameter was selected from a field study performed in Wisconsin (*33*). In the mentioned study, the

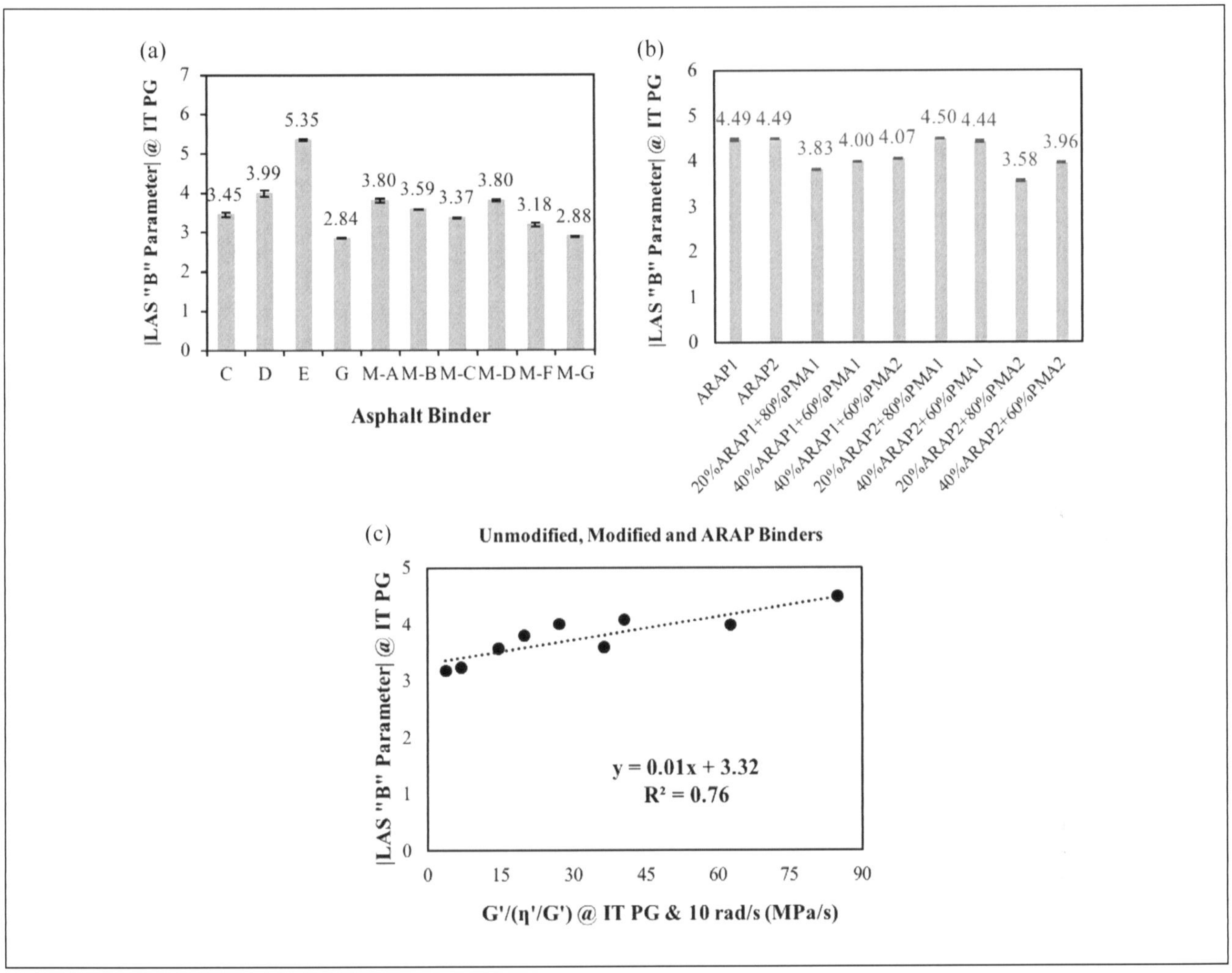

Figure 1. (*a*) Absolute value of the fatigue parameter *B*, for unmodified and modified binders. (*b*) Absolute value of the fatigue parameter *B*, for recycled binders (unmodified and modified with polymers). (*c*) Comparison of absolute value of LAS parameter *B* to G′/(η′/G′) calculated from LAS frequency sweep results collected at intermediate PG temperature and 10 rad/s.

researchers found that the LAS test related very well to field performance when performed at the required Superpave intermediate temperature grade of the project location, and the fatigue life parameter (N_f) was determined at the peak stress. To a specific project section, it was also found that a value of the LAS *B* parameter approximately = −3.43 corresponds to 29%–50% of cracked segments (Table 3). Considering this range of cracked segments as a maximum limit of allowed cracking, the curve in Figure 1*c* can be used to calculate the limit value for the G′/(η′/G′) parameter, at intermediate temperature and 10 rad/s. Thus, the G′/(η′/G′) parameter value at cracking limit is 11.0 MPa/s, with this limit being suggested for binders with up to 50% of RAP. Please note that Figure 1*c* shows the binders that presented LAS *B* parameter between −3.17 and −4.5, to mimic the values obtained in the field and to include the heavily aged binders from recycled pavements.

Table 3. LAS *B* Parameter and Field Performance Information (*33*)

Project ID	Cracked segments	LAS B parameter @ IT PG
1190-14-70	67%	−3.39
9049-09-70	14%	−3.61
1130-12-71	100%	−3.06
9140-07-70	0%	−3.17
1177-10-70	29%	−3.41
6590-00-70	50%	−3.44
31-20-06-70	100%	−3.34

Correlation between ΔT_c and Fatigue Parameter B

Anderson et al. (*5*) introduced the ΔT_c parameter as a means of indexing the non-load associated cracking potential of asphalt binders:

$$\Delta T_c = T_{c,S} - T_{c,m} \quad (6)$$

where,

ΔT_c = Difference in critical low temperature PG grade,

$T_{c,S}$ = Critical low temperature grade predicted using the BBR Stiffness (S), and

$T_{c,m}$ = Critical low temperature grade predicted using the BBR m-slope (m-value).

In this study, Equation 6 was followed for the ΔT_c calculation of the selected binders. Please note that, for all binders, the critical low temperature grade predicted using the BBR is listed in Table 1, for stifiness and m-value. ΔT_c results were correlated with the $G'/(\eta'/G')$ parameter calculated with the alternative approach suggested in this paper and described above (i.e., from LAS frequency sweep results collected at intermediate PG temperature and 10 rad/s).

As can be seen in Figure 2*a*, a strong correlation indicated by $R^2 = 0.87$ is seen for the relationship between ΔT_c and the $G'/(\eta'/G')$ parameter (calculated from LAS frequency sweep results collected at intermediate PG temperature and 10 rad/s). As ΔT_c decreases and becomes more negative, indicating that the binder has lost some of its relaxation properties and is not able to recover to its initial performing stage, higher values of the durability parameter $G'/(\eta'/G')$ are obtained. Therefore, both parameters appear to have potential to relate to binder durability, being used as a surrogate quantifying the loss of relaxation properties as an asphalt binder ages.

Figure 2*b* shows the relationship between the LAS *B* parameter (measured at the binders' intermediate PG temperature) and ΔT_c. As can be seen, a very good correlation is indicated by $R^2 = 0.72$. Please note that the binders included in Figure 2*b* presented LAS *B* parameter between −3.17 and −4.5, to mimic the values obtained in the field and to include the heavily aged binders. Binders with ΔT_c out of the allowable range (i.e., −5 to 2.5°C) were removed before the comparison.

A limit to reduce the risk of crack initiation was set at $\Delta T_c = -2.5$°C by Anderson et al. (*5*), at which point a preventive maintenance was suggested to avoid the pavement reaching a critical stage. The same authors also suggested a limit of ΔT_c minimum of −5.0°C to be used as an indication of the need for immediate pavement rehabilitation. The AASHTO PP78 "Design Considerations When Using Reclaimed Asphalt Shingles (RAS)" in Asphalt Mixtures includes a criteria for ΔT_c minimum of −5.0°C, meaning that a recovered and aged binder with a ΔT_c of −6.0°C is not acceptable. Based on the minimum limit of $\Delta T_c = -5.0$°C, the curve in Figure 2*b* can be used to calculate the limit value for the LAS *B* parameter. Thus, the calculated LAS *B* parameter value at cracking limit is −4.69. Please note that this value of LAS *B* parameter = −4.69 was not found in the field study (*33*), even when considering project sections with 100% of cracked segments (Table 3). However, two considerations are important to explain this value: (1) the field data did not include sections containing RAP; and (2) the graph in Figure 2*b* includes 100% RAP binder, which in reality does not happen in the field.

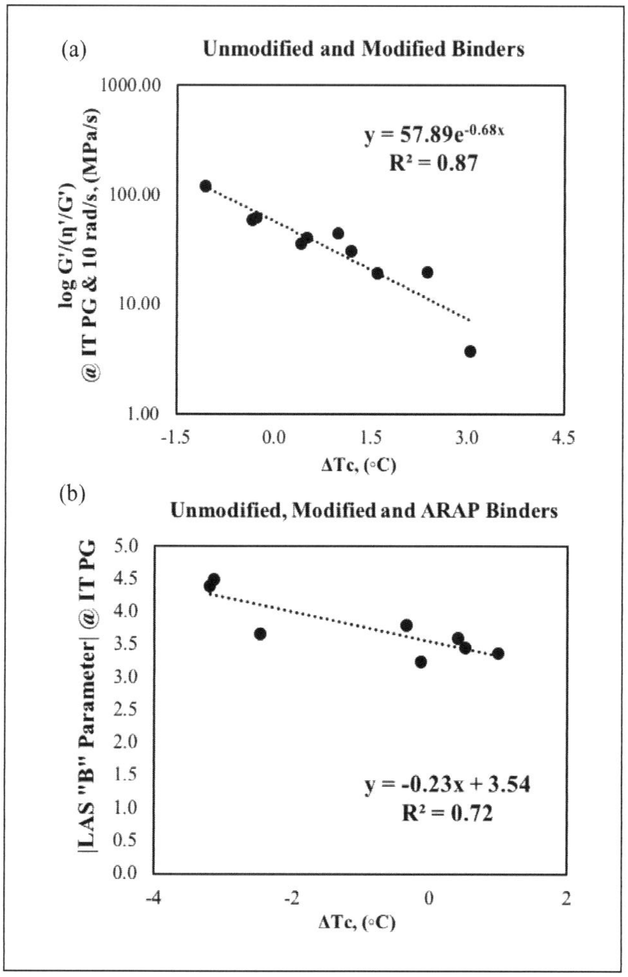

Figure 2. (*a*) Log $G'/(\eta'/G')$ *versus* ΔT_c, for unmodified and modified binders. (*b*) Absolute value of LAS parameter *B versus* ΔT_c, for unmodified, modified and artificially recycled binder.

Determination of Limits at Low Temperature to Cover Fracture of Asphalts

Correlation between $G'/(\eta'/G')$ and Fracture Energy

Research studies performed by Marasteanu et al. (*34*) showed that the SENB test results correlate well with the observed field thermal cracking pavement condition index (PCI), indicating the high promise of using this binder fracture test as an unmodified and modified binder low temperature characterization test to complement the current BBR test. According to the authors, the SENB test is capable of clearly discriminating between modified

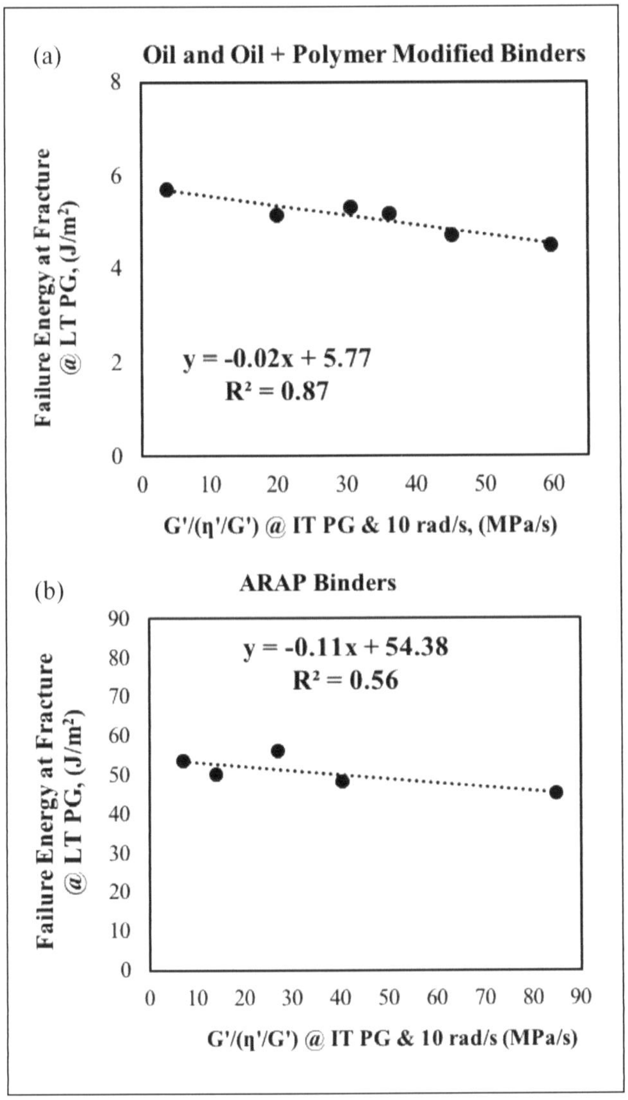

Figure 3. (*a*) Correlation between failure energy at fracture and $G'/(\eta'/G')$ for modified binders. (*b*) Correlation between failure energy at fracture and $G'/(\eta'/G')$ for artificially aged binder.

and unmodified binders in a repeatable fashion. Because unmodified asphalt binders generally have very similar failure stresses, the SENB fracture energy parameter (G_f) of the modified binder will be the focus of this section.

Figure 3*a* shows a correlation between failure energy at fracture (calculated from SENB at the low temperature performance grade) and $G'/(\eta'/G')$ parameter (calculated from LAS frequency sweep results measured at intermediate PG temperature and 10 rad/s). Given the very good $R^2 = 0.87$, it clearly appears that both parameters are directly correlated. Figure 3*b* shows that the correlation still remains even when considering severely aged materials (i.e., RAP), with these materials showing much higher values of failure energy at fracture.

The results in Figure 3 clearly indicate that the SENB fracture energy parameter (G_f) is strongly and inversely correlated with the durability parameter $G'/(\eta'/G')$. This observation seems logical since the durability parameter increases with aging due to increased $|G^*|$ while the fracture energy decreases due to brittleness.

Therefore, higher values of fracture energy which have been correlated to a lower propensity for cracking can be considered as an indication of higher duralibity of the binders, or vice-versa. This good correlation between rheological measurements and fracture properties can lead to a complete characterization of the effect of aging on binder properties and cracking.

Correlation between ΔT_c and Fracture Energy

Figure 4*b* shows the correlation between failure energy at fracture (calculated from SENB at the low temperature performance grade) and the ΔT_c parameter. The fairly good correlation indicated by $R^2 = 0.62$ shows that both parameters appear to quantify the loss of relaxation properties as an asphalt binder ages. Higher values of fracture energy (i.e., lower propensity for cracking) are seen at higher positive values of ΔT_c. To confirm that this correlation is in fact related to the molecular differences between binders, Gel Permeation Chromatography (GPC) analyses were performed and are going to be discussed next.

Correlation between ΔT_c and Molecular Size Distribution

In a previous study, Moraes et al. (*13*) showed that changes in the colloidal structure of asphalt binders due to changes in the degree of association of the different asphalt fractions after aging are reflected in the molecular size distribution as obtained by Gel Permeation Chromatography (GPC). The authors divided the GPC chromatogram into 13 equal elution time areas, and the molecules eluted during the first third of the elution period were classified as large molecular size (LMS), those eluted during the second third as medium molecular size (MMS), and those eluted in the last third as having small molecular size (SMS) (Figure 4*a*).

By applying the same approach, the present study divided into three molecular size ranges the GPC chromatogram of a select group of binders. To simulate the same aging conditions that the binders faced before the low temperature testing performed in this study, all samples were aged in RTFO followed by PAV aging for 20 h at 100°C before being analyzed through GPC.

Figure 4*c* shows the relationship between the large molecular size (LMS) molecules and ΔT_c. As can be seen, a good explanation for the ΔT_c can be obtained

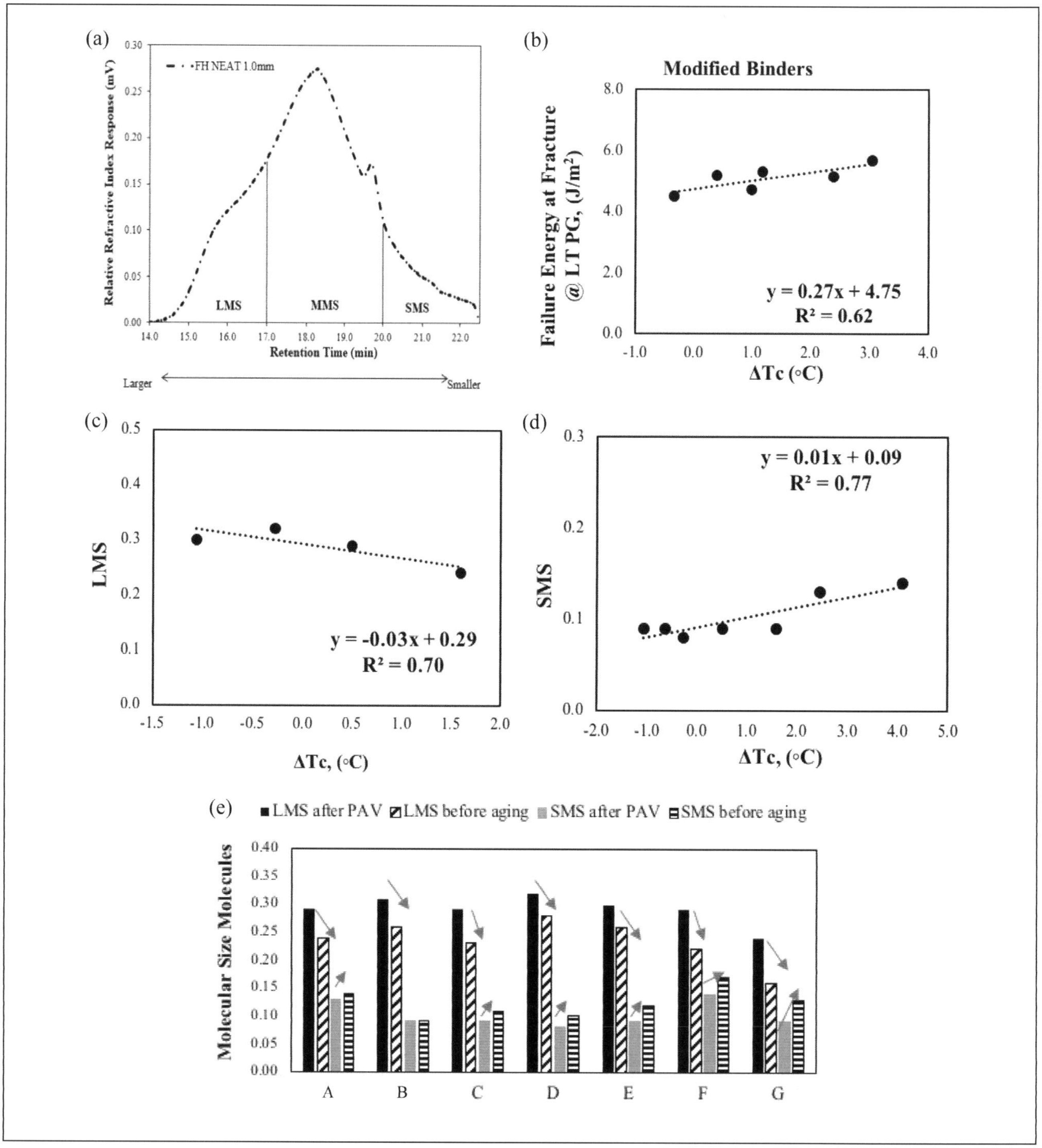

Figure 4. (*a*) Typical GPC chromatogram of an asphalt binder showing three elution time fractions (*13*). (*b*) Correlation between failure energy at fracture (calculated from SENB at the low temperature performance grade) and ΔT_c. (*c*) Correlation between LMS and ΔT_c. (*d*) Correlation between SMS and ΔT_c. (e) Molecular size distribution of asphalt binders before and after oxidative aging.

when looking to the molecular distribution of the binders, since aging results in a shift toward the higher molecular weight molecules. Therefore, asphalts with higher content of LMS are likely to crack. Figure 4*d* validates this finding, since it shows that asphalt binders with higher content of small molecular size (SMS) molecules are less prone to crack, as showed by higher positive values of ΔT_c.

To summarize, Figure 4*e* confirms the above findings showing that an increase in large molecular sized molecules at the expense of the small molecular sizes happens as the asphalt ages.

The changes in the colloidal structure of the asphalt binders due to changes in the asphalt fractions (i.e., asphaltenes and maltenes) upon aging are clearly reflected in the molecular weight distribution as obtained by GPC. The LMS is the percentage of high molecular weight molecules in a binder. In asphalt binders, since oxidative aging causes formation of more polar molecules at the expense of the lower weight molecules, aging results in a shift toward the higher molecular weight molecules. Therefore, the transformation of the molecular size distribution seems to be the reason behind the changes in the performance properties of asphalts.

Summary and Conclusions

This study was conducted to evaluate various durability parameters and the effects of oxidation on the properties of asphalt binders, measured at intermediate and low pavement temperatures. Based on the results collected using modified testing and analysis procedures to cover durability and cracking of binders, it is found that several useful interrelationships exist between new parameters, allowing a choice when selecting simpler measures that could be used for specification tests. The following points summarize the main findings of this study.

- The current DSR procedure for determination of the durability parameter $G'/(\eta'/G')$ is time consuming since it requires a temperature-frequency sweep test with mastercurve generation. Also, the suggested testing temperature of 15°C is not applicable for specific binder's grades selected for diverse climates.
- An alternative approach is proposed to directly calculate the durability parameter $G'/(\eta'/G')$ developed by Glover et al. (*4*). In the proposed approach, $G'/(\eta'/G')$ is calculated from DSR test results using the frequency sweep step of the LAS test procedure, performed at intermediate performance grade temperature and 10 rad/s.
    - The suggested approach allows a quicker (30-minute test) determination of the durability parameter, and utilizes a small amount of material (8 mm DSR sample). In addition, the test is performed at the actual intermediate performance grade temperature of the binder.
    - With this approach, the $G'/(\eta'/G')$ paramenter can be calculated without mastercurve data, and disregarding complicated shifts.
- By applying the LAS *B* parameter = −3.43, collected from a field sections showing between 29%–50% of cracked segments, a limit value for the $G'/(\eta'/G')$ at intermediate temperature and 10 rad/s was determined. The calculated $G'/(\eta'/G')$ parameter value at cracking limit is 11.0 MPa/s, with this limit being suggested for binders with up to 50% of RAP.
- Failure energy at fracture (calculated from SENB at the low temperature performance grade) and $G'/(\eta'/G')$ parameter (calculated from LAS frequency sweep results collected at intermediate PG temperature) were found to inversely correlate very well. Higher values of fracture energy, which have been correlated to a lower propensity for cracking, can be an indication of higher durability of the binders, or vice-versa. This good correlation between rheological measurements and fracture properties can lead to a complete characterization of the effect of aging on binder properties and cracking.
- Failure energy at fracture also showed a correlation with the ΔT_c parameter. Thus, both parameters appear to quantify the loss of relaxation properties as an asphalt binder ages. Higher values of fracture energy (i.e., a lower propensity for cracking) are seen at higher positive values of ΔT_c.
- From the GPC results it was found through a correlation between the large molecular size (LMS) molecules and ΔT_c that asphalts with higher content of large molecular size (LMS) molecules are likely to crack. It was also found that asphalt binders with higher content of small molecular size (SMS) molecules are less prone to crack, as shown by higher positive values of ΔT_c. Therefore, the transformation of the molecular size distribution seems to be the reason behind the changes in the performance properties of asphalts.

It is recommended that the durability parameter $G'/(\eta'/G')$, calculated from a simple frequency sweep at intermediate temperature performance grade, and the ΔT_c be used to control the cracking potential at intermediate and low temperatures, respectively. The limits should be based on testing binders with and without recycled materials, and binder additives (e.g., oils and polymers).

Author Contributions

The authors confirm contribution to the paper as follows: study conception and design: Raquel Moraes, Hussain Bahia; data collection: Raquel Moraes; analysis and interpretation of results: Raquel Moraes, Hussain Bahia; draft manuscript preparation: Raquel Moraes, Hussain Bahia. All authors reviewed the results and approved the final version of the manuscript.

References

1. Bahia, H. Critical Evaluation of Asphalt Modification Using Strategic Highway Research Program Concepts. *Transportation Research Record: Journal of the Transportation Research Board*, 1995. 1488: 82–88.

2. Read, J., and D. Whiteoak. *The Shell Bitumen Handbook*. 5th ed. Thomas Telford Publishing, 2003.
3. Kandhal, P. S. *Low-Temperature Ductility in Relation to Pavement Performance. Low-Temperature Properties of Bituminous Materials and Compacted Bituminous Paving Mixtures, ASTM STP 628* (Marek, C. R., ed.), American Society for Testing and Materials, 1977, pp. 95–106.
4. Glover, C., R. R. Davison, C. H. Domke, Y. Ruan, P. Juristyarini, D. B. Knorr, and S. H. Jung. *Development of a New Method for Assessing Asphalt Binder Durability with Field Validation*. Publication FHWA/TX-05/1872-2, Report 0-1872-2, U.S. Department of Transportation, 2005.
5. Anderson, M., G. N. King, D. I. Hanson, and P. B. Blankenship. Evaluation of the Relationship between Asphalt Binder Properties and Non-Load Related Cracking. *Journal of the Association of Asphalt Paving Technologists*, Vol. 80, 2011, pp. 615–664.
6. Rowe, G. M. Evaluation of the Relationship between Asphalt Binder Properties and Non-Load Related Cracking (Prepared Discussion). *Journal of the Association of Asphalt Paving Technologists*, Vol. 80, 2011, pp. 615–664.
7. AASHTO TP 101. *Estimating Damage Tolerance of Asphalt Binders Using the Linear Amplitude Sweep*. American Association of State Highway and Transportation Officials, 2014.
8. Hintz, C., R. Velasquez, Z. Li, and H. Bahia. Effect of Oxidative Aging on Binder Fatigue Performance. *Journal of Association of Asphalt Paving Technologists*, Vol. 80, 2011, pp. 527–548.
9. Ortega-Rodrıguez, A., S. A. Cruz, A. Gil-Villegas, F. Guevara-Rodrıguez, and C. Lira-Galeana. Molecular View of the Asphaltene Aggregation Behavior in Asphaltene-Resin Mixtures. *Energy & Fuels*, Vol. 17, 2003, pp. 1100–1108.
10. Ruan, Y., R. R. Davison, and C. J. Glover. Oxidation and Viscosity Hardening of Polymer-Modified Asphalts. *Energy & Fuels*, Vol. 17, 2003, pp. 991–998.
11. Petersen, J. C. *Transportation Research Circular E-C140: A Review of the Fundamentals of Asphalt Oxidation - Chemical, Physicochemical, Physical Property, and Durability Relationships*. Transportation Research Board of the National Academies, Washington, D.C., 2009.
12. Tarefder, R. A., and I. Arisa. Molecular Dynamic Simulations for Determining Change in Thermodynamic Properties of Asphaltene and Resin Because of Aging. *Energy & Fuels*, Vol. 25, 2011, pp. 2211–2222.
13. Moraes, R., and H. Bahia. Effect of Mineral Filler on Changes in Molecular Size Distribution of Asphalts during Oxidative Ageing. *Road Materials and Pavement Design*, Vol. 16, 2015, pp. 55–72.
14. Dehouche, N., M. Kaci, and K. A. Mokhtar. Influence of Thermo-Oxidative Aging on Chemical Composition and Physical Properties of Polymer Modified Bitumens. *Construction and Building Materials*, Vol. 26, 2012, pp. 350–356.
15. Khattak, M. J., and G. Y. Baladi. Engineering Properties of Polymer-Modified Asphalt Mixtures. *Transportation Research Record: Journal of the Transportation Research Board*, 2014. 1638: 12–22.
16. Bahia, H., D. I. Hanson, M. Zeng, H. Zhai, and M. A. Khatri. *NCHRP Report 459: Characterization of Modified Asphalt Binders in Superpave Mix Design*. Transportation Research Board of the National Academies, Washington, D.C., 2001.
17. Collins, J. H., M. G. Bouldin, R. Gelles, and R. A. Berker. Improved Performance of Paving Asphalt by Polymer Modification. *Journal of Association of Asphalt Paving Technologists*, Vol. 60, 1991, pp. 43–79.
18. Rostler, F. S., and R. M. White. Composition and Changes in Composition of Highway Asphalts, 85-100 Penetration Grade. *Journal of Association of Asphalt Paving Technologists*, Vol. 31, 1962, pp. 35–89.
19. Stastna, J., L. Zanzotto, and O. J. Vacin. Viscosity Function in Polymer-Modified Asphalts. *Journal of Colloid and Interface Science*, Vol. 259, 2003, pp. 200–207.
20. Polacco, G., A. Muscente, D. Biondi, and S. Santini. Effect of composition on the properties of SEBS modified asphalts. *European Polymer Journal*, Vol. 42, No. 5, 2006, pp. 1113–1121.
21. McDaniel, R. S., and R. M. Anderson. *NCHRP Report 452: Recommended Use of Reclaimed Asphalt Pavement in the Superpave Mix Design Method: Technician's Manual*. Transportation Research Board of the National Academies, Washington, D.C., 2001.
22. Shen, J., S. Amirkhanian, and J. A. Miller. Effects of Rejuvenating Agents on Superpave Mixtures Containing Reclaimed Asphalt Pavement. *Journal of Materials in Civil Engineering*, Vol. 19, No. 5, 2007, pp. 376–384.
23. D'Angelo, J. A., K. Grzybowski, S. Lewis, and R. Walker. *Evaluation of the Performance Properties of Asphalt Mixes Produced with Re-refined Heavy Vacuum Distillate Bottoms*. Canadian Technical Asphalt Association, November, 2013.
24. Golalipor, A. *Investigation of the Effect of Oil Modification on Critical Characteristics of Asphalt Binders*. PhD thesis, University of Wisconsin-Madison, Madison, Wisconsin, 2013.
25. Bahia, H, E. Lyngdal, R. Varma, D. Sweirtz, and P. Teymourpour. *Modified Binder (PG +) Specifications and Quality Control Criteria*. Wisconsin Highway Research Program (WHRP) 0092-14-22, Transportation Pooled Fund Program (TPF-5/302), 2016.
26. AASHTO M 320. *Standard Specifications for Transportation Materials and Methods of Sampling and Testing, Part 2B*. American Association of State Highway and Transportation Officials, 2016.
27. AASHTO T 240. *Effect of Heat and Air on a Moving Film of Asphalt (Rolling Thin-Film Oven Test)*. American Association of State Highway and Transportation Officials, 2013.
28. AASHTO R 28. *Accelerated Aging of Asphalt Binder Using a Pressurized Aging Vessel (PAV)*. American Association of State Highway and Transportation Officials, 2009.
29. AASHTO T 313. *Standard Method of Test for Determining the Flexural Creep Stiffness of Asphalt Binder Using the Bending Beam Rheometer (BBR)*. American Association of State Highway and Transportation Officials, 2005.
30. AASHTO T XXX. *Determining the Fracture Properties of Asphalt Binders Using the Single Edge Notched Bending Test*. American Association of State Highway and Transportation Officials, 2011.

31. Bahia, H., R. Velasquez, H. Tabatabaee, and S. Puchalski. *The Role of Asphalt Binder Fracture Properties in Thermal Cracking Performance of Mixtures and Pavements*. Canadian Technical Asphalt Association, 2012.
32. Velasquez, R., H. Tabatabaee, and H. Bahia. Low Temperature Cracking Characterization of Asphalt Binders by Means of the Single-Edge Notch Bending (SENB) Test. *Journal of the Association of Asphalt Pavement Technologists*, Vol. 80, 2011, pp. 583–614.
33. Bahia, H., H. Tabatabaee, and T. Mandal. *Field Validation of Wisconsin Modified Binder Selection Guidelines – Phase II*. Wisconsin Highway Research Program Project: 0092-13-02 Final Report. December 2013.
34. Marasteanu, M., W. Buttlar, H. Bahia, and C. Williams. *Investigation of Low Temperature Cracking in Asphalt Pavements*. National Pooled Fund Study - Phase II, Minnesota Department of Transportation Research Services, 2012.

The Standing Committee on Asphalt Binders (AFK20) peer-reviewed this paper (18-05364).

Article

TRR
JOURNAL OF THE TRANSPORTATION RESEARCH BOARD

Transportation Research Record
2018, Vol. 2672(28) 324–336

DOI: 10.1177/0361198118782029
journals.sagepub.com/home/trr
SAGE

Nonrecoverable Behavior of Polymer Modified and Reclaimed Asphalt Pavement Modified Binder under Different Multiple Stress Creep Recovery Tests

Zhou Zhou[1], Xingyu Gu[1], Jiwang Jiang[1], Fujian Ni[1], and Yanxu Jiang[1]

Abstract

Chinese transportation agencies usually use the styrene–butadiene–styrene (SBS) modified binder and rejuvenators when applying higher reclaimed asphalt pavement (RAP) in surface courses. However, the rutting performance of RAP binder containing SBS binder and rejuvenators remains a problem. In order to better evaluate the rutting performance of asphalt binder, recently, a multiple stress creep recovery (MSCR) test (AASHTO T 350) was proposed and approved. This paper critically reviewed the AASHTO T 350 test method and AASHTO M 332 specification by carrying out a series of MSCR tests. A total of 18 types of SBS modified and modified RAP binders were tested. The effect of a rejuvenator on the rutting performance was analyzed. Three different MSCR tests were conducted considering different stress levels and creep recovery cycles, including test conditions of AASHTO T 350. The results indicated that AASHTO T 350 can grade the binders based on the nonrecoverable creep compliance and stress sensitivity parameters. Since SBS modified binders always have lower J_{nr} values at lower stress levels, higher stress levels and more creep recovery cycles are recommended to better evaluate the rutting resistance of binder. The addition of RAP decreased the J_{nr} values while the addition of rejuvenator had the opposite effect. The added SBS modified binder and rejuvenator have a negligible effect on the percentage recovery. It is suggested that a suitable rejuvenator dosage and virgin binder type should be chosen carefully to guarantee the rutting performance of RAP binder.

Rutting is a major distress in asphalt pavement. Poor quality asphalt binder could be one of the main causes of rutting failure of asphalt pavement. The rutting performance of asphalt binders is characterized in the laboratory using Superpave rutting factor ($G^*/\sin\delta$) (*1*). However, the rutting factor has been proved to be inadequate for characterizing the high-temperature behavior of asphalt binders, especially for polymer modified binders (*2*). Recently, a specification based on the multiple stress creep recovery (MSCR) test (AASHTO T 350) was proposed to address the shortcomings of the Superpave high-temperature binder specifications (*3, 4*). The MSCR test method is applied on the rolling thin film oven (RTFO)-aged asphalt binder in a typical dynamic shear rheometer (DSR) machine. The test conducts 20 creep and recovery cycles at a 0.1 kPa stress level, immediately followed by another 10 cycles of creep and recovery at a stress level of 3.2 kPa. The MSCR test determines the nonrecoverable creep compliance (J_{nr}) and the percentage of recovery (%R) of the asphalt binder, which have been proved to have good correlations with field rutting resistance (*5–8*). While the J_{nr} parameter at 3.2 kPa and comparison of J_{nr} at 0.1 kPa and 3.2 kPa have been suggested to have better correlations with the rutting performance of asphalt mixes in the field and laboratory, several studies reported that MSCR results at higher stress levels could better correlate to mixture rutting performance (*9, 10*). A 3.2 kPa stress level considered in the MSCR test protocol could be less than stress experienced in the field (*11*).

There is a trend toward utilizing more reclaimed asphalt pavement (RAP) in asphalt mixtures because of economic and environmental benefits in recent years. However, the use of RAP in surface courses is relatively low in China. Chinese transportation agencies usually use styrene–butadiene–styrene (SBS) modified binder in the construction of surface courses of highways and its performance grade is often PG70-XX or

[1]School of Transportation, Southeast University, Nanjing, P.R. China

Corresponding Author:
Address correspondence to Xingyu Gu: guxingyu1976@163.com

Table 1. Superpave Test Results of Blended Binders with Different Rejuvenator Dosages

	High temp.			Low temp. PAV (20 h)		
Tested binder	Origin	RTFOT	Inter. temp.	S/ MPa	m-slope	PG grade
RAP	84.1	–	22.1	−16.7	−15.1	84–15
RAP+2%RA	77.0	76.6	22.6	−20.1	−19.2	76–19
RAP+4%RA	73.9	73.0	20.4	−23.2	−22.6	73–23
RAP+6%RA	69.8	69.1	14.2	−26.1	−25.8	69–25

Note: – = data not applicable.

PG76-XX (*12–15*). And when using high RAP (usually higher than 25%) in the surface course, the rejuvenator is often used to soften the RAP binder (*16*). However, transportation agencies are reluctant to use rejuvenator due to potential rutting damage causes by the rejuvenator, which may excessively soften the aged binder (*17–19*). So far, limited studies research the effect of rejuvenator on the rutting performance of aged binder containing virgin SBS modified asphalt binder. Therefore, this paper evaluates the rutting performance of reclaimed asphalt binder and SBS modified binder containing a rejuvenator.

Objectives

The main objectives of this study are as follows:

(1) Evaluate the effect of rejuvenator on the rutting performance of RAP binder containing virgin SBS modified binder;
(2) Evaluate the effects of high stress levels and more creep recovery cycles on the nonrecoverable response of SBS modified and RAP binder;
(3) Compare the nonrecoverable behavior of SBS modified and RAP binder under different MSCR tests.

Materials and Experimental Plan

Materials

SBS modified binder. In this study, two types of PG70-22 SBS modified binder and four types of PG76-22 SBS modified binder were collected from three sources, and PG 64-XX neat binder was collected from one source.

Reclaimed asphalt pavement binder. Reclaimed asphalt pavement (RAP) was collected from the top layer of the Huning Highway in Jiangsu Province of China. Binders were extracted and recovered from the RAP mixes in accordance with AASHTO T 164 and ASTM D 5404, respectively. A rejuvenator, which is commonly used in China, was then blended with the recovered binder to soften and recover the performance of the aged binder (*20*). The rejuvenator dosage was determined based on the performance grading test according to AASHTO T 315. The dosage was selected as 2%, 4%, and 6% by the weight of the RAP binder.

The performance results of blended binders with different rejuvenator dosages are shown in Table 1.

From Table 1, it can be seen that a dosage rate of 2% of the rejuvenator by the weight of RAP binders could be used to restore the continuous grade of the RAP binder from 84.1–15.1 to 76.6–19.2. When the dosage increases to 4%, the continuous grade changes to 73.0–22.6. Based on the results, this paper produced three groups of blended rejuvenated binder:

(1) Group I: The recovered aged binder was mixed with PG70-22 SBS modified binder (named SBS I in this paper) to produce blended binders containing 15%, 30%, 40%, and 50% of RAP binder by the weight of the virgin binder. These are named RAP15-70, RAP30-70, RAP40-70, and RAP50-70.
(2) Group II: The recovered aged binder was first blended with 4% dosage by the weight of the RAP binder and then mixed with PG70-22 SBS modified binder (SBS I) to produce blended binders containing 15%, 30%, 40%, and 50% of rejuvenated binder by the weight of the virgin binder. These are named RAP15-4%RA-70, RAP30-4%RA-70, RAP40-4%RA-70, and RAP50-4%RA-70.
(3) Group III: The recovered aged binder was first blended with 2% dosage by the weight of the RAP binder and then mixed with PG76-22 SBS modified binder (named SBS V in this paper) to produce blended binders containing 15%, 30%, 40%, and 50% of rejuvenated binder by the weight of the virgin binder. These are named RAP15-2%RA-76, RAP30-2%RA-76, RAP40-2%RA-76, and RAP50-2%RA-76.

Experimental Plan

Superpave Testing. DSR tests were conducted in accordance with AASHTO T 315. The DSR test method was conducted to measure rheological properties of unaged and short-term aged asphalt binder using parameters of the dynamic shear modulus (G*) and phase angle (δ).

MSCR Tests. The MSCR tests were conducted using a DSR device: the MCR 102 model from Anton Paar. The samples were measured in a geometry of 25 mm in diameter and 1 mm in thickness. The binder samples were tested after being RTFO aged.

In this study, the MSCR tests were conducted using three different methods:

- **Method A**: this method was in accordance with AASHTO T 350, which consists of 20 cycles of "1s creep period" and "9s recovery period" at 0.1 kPa stress level, immediately followed by another 10 cycles of creep and recovery at 3.2 kPa.
- **Method B:** this method was first conducted at 0.025 kPa with 20 cycles of creep and recovery period and then conducted at 0.05kPa, 0.1kPa, 0.2kPa, 0.4kPa, 0.8kPa, 1.6kPa, 3.2kPa, 6.4kPa, 12.8kPa, and 25.6kPa with 10 cycles of creep and recovery period.
- **Method C:** this method was conducted for 10 cycles of creep and recovery cycles under three stress levels of 0.1, 3.2, and 12.8 kPa, and 30 cycles in total.

In Jiangsu Province of China, the maximum temperature of pavement surface varies from 60°C~70°C with a 98% reliability (*21*). Taking account of the temperatures, the MSCR testing temperatures for PG64-22 binder were 64°C; those for PG 70-22 binders were 64°C and 70°C, and those for PG76-22 binders were 64°C, 70°C, and 76°C. The average results of three replicates are reported in this paper.

Results and Discussion

Superpave Test Results

The DSR test results of the SBS modified and RAP binders at various testing temperatures are shown in Table 2. It can be found that: for Group I binders, the high-temperature grade increases with the increasing RAP content, indicating that the addition of RAP is expected to improve the rutting performance of the asphalt binder. For Group II and Group III binders, they have a similar high-temperature performance grade, including seven types of PG 70-22 binders and 11 types of PG 76-22 binders.

MSCR Test Results

Classification of Asphalt Binders by AASHTO M 332 Standard. AASHTO M 332 specifications provided $J_{nr3.2}$ value instead of $G^*/\sin\delta$ on the basis of traffic condition. There are four different equivalent single axle load (ESAL) based traffic levels included in the specification: standard (S) (less than 10 million ESALs), heavy (H) (between 10 and 30 million ESALs), very heavy (V) (greater than 30 million ESALs), and extreme (E) (greater than 30 million ESALs plus standing traffic). Upper limits on the $J_{nr3.2}$ value decrease with each successive traffic level: 4.5 kPa^{-1} for the S level, 2 kPa^{-1} for the H level, 1 kPa^{-1} for the V level, and 0.5 kPa^{-1} for the E level. Stress sensitivity is another important MSCR parameter for prediction of the performance of polymer modified binders at a high stress level. The condition of stress sensitivity is that the difference between J_{nr} at 3.2 kPa and 0.1 kPa stress level at high target testing temperature must be no more than 75% of J_{nr} at 0.1 kPa stress level (*22*).

Table 3 shows a summary of the test results obtained from all binders. Based on the data, the following observations could be found:

(1) $J_{nr3.2}$: Figure 1 shows the $J_{nr,3.2}$ value of all tested binders. Because of the addition of RAP, the $J_{nr3.2}$ values of Group I are less than those of PG 70-22 SBS modified binders (SBS I and II), which means RAP can improve the rutting performance of asphalt binder. And as the percentage of RAP increases, the $J_{nr3.2}$ value decreases. For Group II, the addition of rejuvenator has negative effects on the RAP binder compared with Group I. For PG 70-22 binders, at 64°C, the $J_{nr3.2}$ value of four binders is less than 1 kPa^{-1} and three binders is less than 2 kPa^{-1}. When the testing temperature is 70°C, four binders is less than 2 kPa^{-1} and three binders is less than 4.5 kPa^{-1}. For PG 76-22 binders, at 64°C, the $J_{nr3.2}$ value of 10 binders is less than 0.5 kPa^{-1} and one binder is less than 1 kPa^{-1}. At 70°C, nine binders is less than 1 kPa^{-1} and two binders is less than 2 kPa^{-1}. While at 76°C, two binders is less than 2 kPa^{-1} and nine binders is less than 4.5 kPa^{-1}. For rejuvenated RAP binders with similar high-temperature grade (PG 70-22 or PG 76-22), they have similar $J_{nr3.2}$ values at 64, 70, and 76°C. It should be noted that binders with a higher-temperature grade did not have lower $J_{nr3.2}$ values. It seems that a high-temperature grade is not always consistent with $J_{nr3.2}$ values.

(2) $J_{nr,diff}$: A very interesting phenomenon is found when comparing the $J_{nr,diff}$ values of these binders. For all 7 PG 70-22 binders, at 64°C, three binders could be graded as PG 64H-XX, three binders could be graded as PG 64V-XX, and one binder could be graded as PG 64E-XX. For the same binders, at 70°C, two virgin SBS modified binders cannot pass the limits of $J_{nr,diff,}$ so these binders cannot be graded. Besides, three binders can be graded as PG 70S-XX and the rest can be graded as PG 70H-XX. For 11 PG 76-22 binders, at 64°C, the $J_{nr,diff}$ value of one binder cannot meet the requirement, and one binder could be graded as PG 64V-XX. The remaining binders can be graded as PG 64E-XX. For the same tested binders, at 70°C, three binders cannot be

Table 2. DSR Test Results for SBS Modified Binders and RAP Binders

Binder type	Sample condition	$G^*/\sin\delta$ (kPa) 64°C	70°C	76°C	82°C	Minimum temperature corresponding to 1.0 or 2.2kPa(°C)	True PG grade
Neat binder	Unaged	1.450	0.685	–	–	67.0	PG 65-xx
	RTFO	2.814	1.278	–	–	65.9	
SBS I	Unaged	–	1.540	0.734	–	73.5	PG 73-xx
	RTFO	–	2.973	1.692	–	73.2	
SBS II	Unaged	–	2.052	0.937	–	75.5	PG 74-xx
	RTFO	–	3.395	1.926	–	74.6	
SBS III	Unaged	–	–	1.031	0.588	76.3	PG 76-xx
	RTFO	–	–	2.202	1.303	76.0	
SBS IV	Unaged	–	–	1.134	0.651	77.4	PG 76-xx
	RTFO	–	–	2.284	1.370	76.4	
SBS V	Unaged	–	–	1.141	0.654	77.4	PG 76-xx
	RTFO	–	–	2.376	1.412	76.9	
SBS VI	Unaged	–	–	1.622	0.897	80.9	PG 78-xx
	RTFO	–	–	2.666	1.662	78.5	
RAP15-70	Unaged	–	1.810	1.002	–	76.0	PG 75-xx
	RTFO	–	3.692	2.054	–	75.3	
RAP30-70	Unaged	–	–	1.209	0.712	78.2	PG 76-xx
	RTFO	–	–	2.352	1.290	76.7	
RAP40-70	Unaged	–	–	1.326	0.810	79.5	PG 77-xx
	RTFO	–	–	2.586	1.412	77.6	
RAP50-70	Unaged	–	–	1.458	0.902	80.7	PG 79-xx
	RTFO	–	–	3.275	1.762	79.9	
RAP15-2%RA-76	Unaged	–	–	1.400	0.832	79.8	PG 77-xx
	RTFO	–	–	2.493	1.460	77.4	
RAP30-2%RA-76	Unaged	–	–	1.390	0.832	79.9	PG 77-xx
	RTFO	–	–	2.560	1.469	77.6	
RAP40-2%RA-76	Unaged	–	–	1.422	0.778	79.5	PG 78-xx
	RTFO	–	–	2.785	1.560	78.4	
RAP50-2%RA-76	Unaged	–	–	1.426	0.789	79.6	PG 78-xx
	RTFO	–	–	2.852	1.569	78.6	
RAP15-4%RA-70	Unaged	–	1.571	0.708	–	73.4	PG 72-xx
	RTFO	–	2.881	1.631	–	72.9	
RAP30-4%RA-70	Unaged	–	1.345	0.696	–	72.7	PG 72-xx
	RTFO	–	2.781	1.543	–	72.4	
RAP40-4%RA-70	Unaged	–	1.348	0.711	–	72.8	PG 71-xx
	RTFO	–	2.554	1.389	–	71.5	
RAP50-4%RA-70	Unaged	–	1.352	0.724	–	72.9	PG 71-xx
	RTFO	–	2.501	1.385	–	71.3	

graded, two binders could be graded as PG 70H-XX, and the remaining binders could be graded as PG 70V-XX. At 76°C, only five binders can be graded as PG 76S-XX and the remaining binders cannot be graded. It can be seen from Table 3 that most of the virgin SBS modified binders cannot meet the $J_{nr,diff}$ requirement in AASHTO M 332. For example, SBS VI binder has the minimum $J_{nr3.2}$ values at 64, 70, and 76°C. However, $J_{nr,diff}$ values of this binder at 64, 70, and 76°C are too large to be graded at these three temperatures. For RAP binders, only two rejuvenated RAP binders cannot pass the $J_{nr,diff}$ requirement at 76°C, which indicates that the addition of rejuvenators may be harmful to the rutting performance of asphalt binder.

(3) %R: AASHTO M 332 proposed a method to detect the polymer in the asphalt binders (*4*). The polymer modification curve adopted in AASHTO M 332 was developed based on the testing data. Binders with %R above the polymer modification curve indicated a good elastomeric behavior. Figure 2 shows that all virgin SBS modified binders were in the passing

Table 3. AASHTO T 350 test results for all tested binders

Binder type		Testing temp. (°C)	%Recovery (0.1 kPa)	%Recovery (3.2 kPa)	$R_{diff(0.1-3.2)}$,%	$J_{nr(0.1kPa)}$, 1/kPa	$J_{nr(3.2kPa)}$, 1/kPa	$J_{nr,diff(3.2-0.1)}$,%	Stress sensitivity (AASHTO M332)	MSCR grade
Neat binder		64	2.29	–0.87	137.82	3.249	3.668	12.89	YES	PG 64S-xx
SBS I		64	63.30	46.84	26.00	0.415	0.641	54.55	YES	PG 64V-xx
		70	57.63	32.03	44.43	0.863	1.531	77.45	NO	–
SBS II		64	67.60	53.71	20.55	0.313	0.488	55.95	YES	PG 64E-xx
		70	69.72	45.82	34.28	0.566	1.095	93.63	NO	–
SBS III		64	74.58	63.37	15.04	0.223	0.337	51.16	YES	PG 64E-xx
		70	69.92	49.14	29.72	0.480	0.871	81.75	NO	–
		76	58.80	28.22	52.00	1.149	2.288	99.05	NO	–
SBS IV		64	82.73	72.13	12.81	0.161	0.272	68.67	YES	PG 64E-xx
		70	81.37	61.46	24.47	0.313	0.678	116.65	NO	–
		76	74.05	36.75	50.37	0.756	2.075	174.52	NO	–
SBS V		64	79.07	71.00	10.21	0.179	0.258	44.32	YES	PG 64E-xx
		70	74.79	59.35	20.64	0.383	0.653	70.62	YES	PG 70V-xx
		76	67.13	38.39	42.80	0.803	1.696	111.22	NO	–
SBS VI		64	93.66	88.00	6.04	0.055	0.110	101.91	NO	–
		70	92.17	80.76	12.38	0.109	0.285	161.52	NO	–
		76	85.38	45.65	46.53	0.339	1.578	365.46	NO	–
Group I	RAP15-70	64	55.76	38.66	30.68	0.428	0.632	47.75	YES	PG 64V-xx
		70	46.71	24.53	47.47	1.050	1.657	57.86	YES	PG 70H-xx
	RAP30-70	64	50.58	34.49	31.82	0.396	0.558	40.79	YES	PG 64V-xx
		70	41.16	20.14	51.07	0.937	1.421	51.60	YES	PG 70H-xx
		76	31.03	8.57	72.38	2.236	3.490	56.11	YES	PG 76S-xx
	RAP40-70	64	49.04	34.14	30.38	0.347	0.472	35.83	YES	PG 64E-xx
		70	38.97	19.21	50.70	0.911	1.334	46.44	YES	PG 70H-xx
		76	29.10	8.40	71.12	1.839	2.778	51.06	YES	PG 76S-xx
	RAP50-70	64	46.86	34.26	26.89	0.266	0.341	28.12	YES	PG 64E-xx
		70	36.77	18.82	48.81	0.693	0.965	39.34	YES	PG 70V-xx
		76	26.94	7.55	71.99	1.576	2.323	47.41	YES	PG 76S-xx
Group II	RAP15-4%	64	60.05	42.72	28.86	0.508	0.781	53.68	YES	PG 64V-xx
	RA-70	70	51.73	26.51	48.74	1.144	1.940	69.66	YES	PG 70H-xx
	RAP30-4%	64	50.47	31.78	37.02	0.699	1.056	51.24	YES	PG 64H-xx
	RA-70	70	45.08	20.73	54.00	1.392	2.260	62.33	YES	PG 70S-xx
	RAP40-4%	64	42.38	24.27	42.74	0.838	1.216	44.99	YES	PG 64H-xx
	RA-70	70	34.77	12.96	62.73	1.868	2.864	53.31	YES	PG 70S-xx
	RAP50-4%	64	40.06	21.18	47.13	0.899	1.305	45.13	YES	PG 64H-xx
	RA-70	70	32.82	10.70	67.40	2.017	3.127	55.06	YES	PG 70S-xx
Group III	RAP15-2%	64	70.07	58.87	15.97	0.234	0.338	44.57	YES	PG 64E-xx
	RA-76	70	63.78	45.00	29.45	0.524	0.861	64.23	YES	PG 70V-xx
		76	55.90	26.38	52.81	1.203	2.323	93.03	NO	–
	RAP30-2%	64	63.71	53.12	16.62	0.256	0.345	34.98	YES	PG 64E-xx
	RA-76	70	58.01	40.94	29.43	0.567	0.860	51.71	YES	PG 70V-xx
		76	52.84	26.07	50.66	1.188	2.146	80.62	NO	–
	RAP40-2%	64	57.64	44.13	23.45	0.305	0.425	39.28	YES	PG 64E-xx
	RA-76	70	53.48	33.83	36.74	0.639	0.998	56.36	YES	PG 70V-xx
		76	45.07	19.06	57.72	1.384	2.365	70.94	YES	PG 76S-xx
	RAP50-2%	64	55.67	42.00	24.57	0.296	0.408	37.72	YES	PG 64E-xx
	RA-76	70	52.26	32.62	37.58	0.617	0.955	54.73	YES	PG 70V-xx
		76	37.99	13.15	65.39	1.574	2.649	68.36	YES	PG 76S-xx

Note: temp. = temperature.

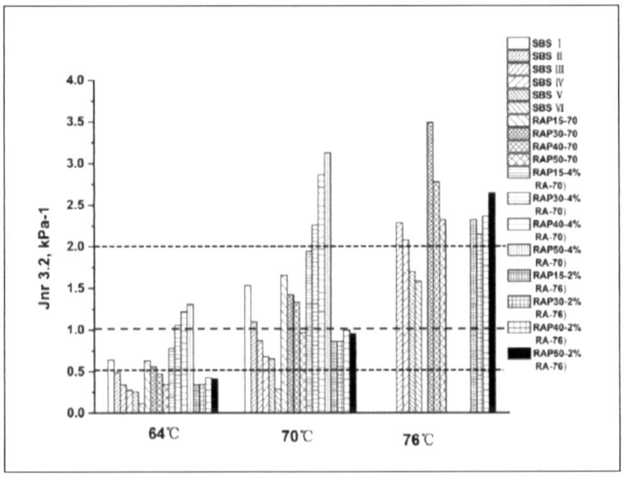

Figure 1. $J_{nr,3.2}$ for all tested binders.

zone. RAP binders plotted nearly at or below the polymer curve while just a few types of binders were plotted above the polymer modification curve. The addition of virgin SBS modified binder does not significantly affect the %R results. It is interesting that the rejuvenating RAP binder with 2% rejuvenator and PG 76-22 SBS modified binder is in the passing zone whereas the 4% content of rejuvenator and PG 70-22 SBS modified binder improves the %R results of RAP binder slightly.

Nonrecoverable Behavior under Different Stress Levels. The nonrecoverable creep compliance values of all binders under multiple stress levels from 0.025 to 25.6 kPa at different temperatures in Method B were calculated as shown in Table 4. A plot of the nonrecoverable compliance J_{nr} against creep stress levels at high-temperature grade of each binder is shown in Figure 3. Based on Table 4 and Figure 3, J_{nr} is weakly dependent on stress up to the level of 3,200 Pa at high-temperature grade of the binders and when the stress level is more than 3,200 Pa, an increasing dependence is observed with increasing stresses. As the stress level increases, nonlinear behavior of J_{nr} is observed. The J_{nr} parameter could indicate the boundary of the linear viscoelastic behavior in the tested binders. It is interesting that the J_{nr} values under different stress levels of the binder present different trends. At 64°C, the addition of RAP binder reduces J_{nr} while the addition of rejuvenator has the opposite effect. At 70°C, for PG 70-22 binders, when adding 15% RAP binder, the J_{nr} values were higher than the virgin binder under stress levels of 25 Pa to 3,200 Pa. The J_{nr} value of 15% RAP binder was less than the virgin SBS modified binder at stress

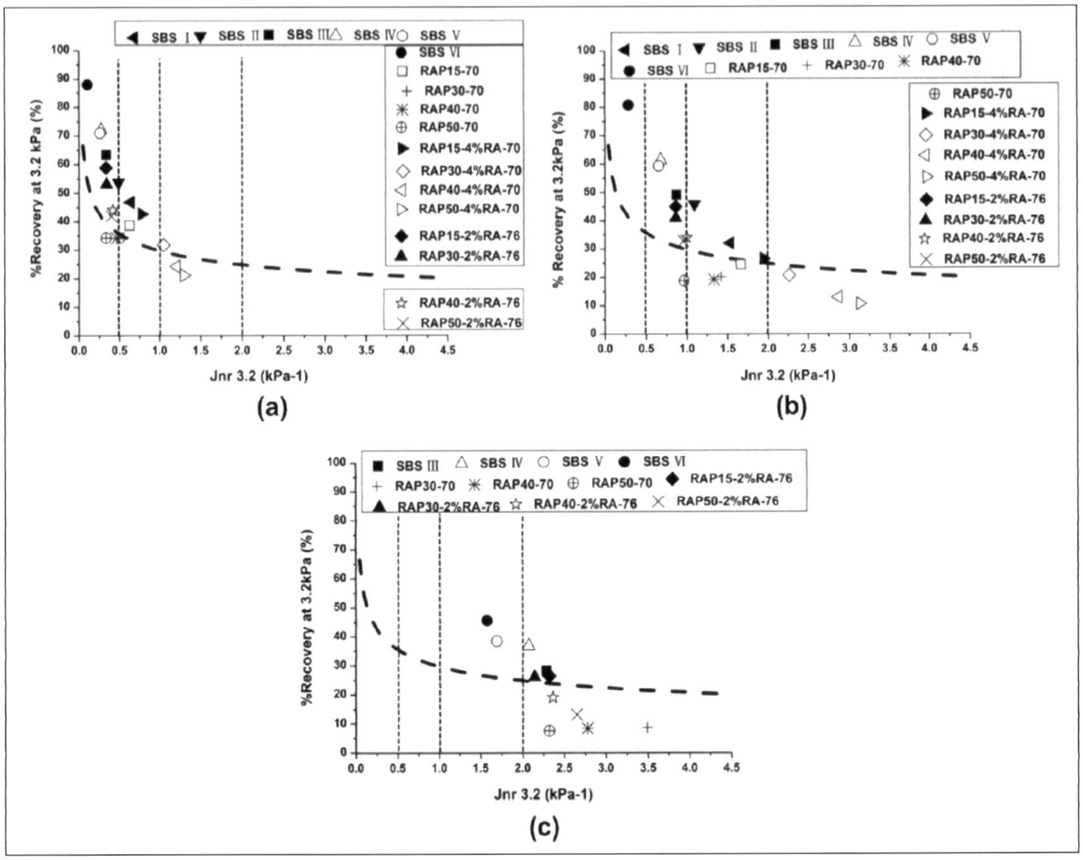

Figure 2. MSCR %R vs J_{nr} at 3.2 kPa for all tested binders: (*a*) 64°C, (*b*) 70°C, and (*c*) 76°C.

Table 4. J_{nr} Values at Each Stress Level for All Tested Binders

Binder type		Testing temp. (°C)	J_{nr} at each stress level (Pa), 1/kPa										
			25	50	100	200	400	800	1,600	3,200	6,400	12,800	25,600
Neat binder		64	3.377	3.320	3.312	3.667	3.406	3.473	3.617	3.813	4.158	7.460	69.700
SBS I		64	0.455	0.483	0.482	0.521	0.481	0.517	0.594	0.719	0.878	1.130	1.651
		70	0.785	0.835	0.847	0.947	0.939	1.117	1.354	1.625	2.054	2.827	4.206
SBS II		64	0.322	0.341	0.339	0.365	0.331	0.336	0.372	0.469	0.627	0.837	1.316
		70	0.660	0.702	0.692	0.740	0.699	0.794	0.999	1.298	1.722	2.540	3.972
SBS III		64	0.206	0.217	0.217	0.221	0.222	0.232	0.256	0.314	0.435	0.600	0.947
		70	0.485	0.521	0.524	0.574	0.547	0.615	0.745	0.933	1.220	1.791	2.916
		76	1.091	1.135	1.133	1.301	1.344	1.622	1.978	2.477	3.448	5.016	8.462
SBS IV		64	0.177	0.200	0.200	0.215	0.194	0.191	0.207	0.266	0.383	0.523	0.830
		70	0.426	0.463	0.457	0.490	0.456	0.512	0.655	0.863	1.138	1.695	2.974
		76	0.873	0.920	0.909	1.024	1.082	1.391	1.767	2.241	3.195	4.933	8.866
SBS V		64	0.177	0.201	0.202	0.219	0.197	0.194	0.205	0.237	0.342	0.517	0.824
		70	0.375	0.414	0.413	0.443	0.408	0.449	0.537	0.702	0.992	1.533	2.563
		76	0.829	0.882	0.879	0.992	1.021	1.241	1.515	1.933	2.790	4.229	7.182
SBS VI		64	0.043	0.069	0.071	0.075	0.069	0.065	0.064	0.072	0.137	0.357	0.986
		70	0.109	0.138	0.139	0.147	0.134	0.141	0.183	0.335	0.701	1.543	4.165
		76	0.306	0.333	0.323	0.363	0.407	0.677	1.176	1.677	2.751	5.646	11.760
Group I	RAP15-70	64	0.409	0.426	0.430	0.508	0.460	0.494	0.560	0.644	0.727	0.872	1.157
		70	0.997	1.022	1.033	1.180	1.164	1.302	1.472	1.655	1.939	2.454	3.346
	RAP30-70	64	0.401	0.411	0.413	0.455	0.426	0.458	0.508	0.579	0.657	0.784	0.999
		70	0.818	0.838	0.846	0.952	0.914	0.995	1.114	1.261	1.456	1.776	2.259
		76	2.058	2.085	2.108	2.434	2.344	2.608	2.941	3.362	3.950	4.825	6.935
	RAP40-70	64	0.339	0.344	0.342	0.376	0.350	0.369	0.404	0.456	0.525	0.617	0.790
		70	0.744	0.752	0.753	0.839	0.789	0.846	0.940	1.067	1.232	1.523	1.928
		76	1.852	1.871	1.887	2.152	2.063	2.265	2.548	2.927	3.394	4.127	5.666
	RAP50-70	64	0.317	0.325	0.326	0.360	0.332	0.346	0.373	0.421	0.489	0.594	0.795
		70	0.734	0.738	0.737	0.817	0.763	0.809	0.893	1.019	1.186	1.463	1.923
		76	1.637	1.657	1.673	1.883	1.784	1.930	2.162	2.467	2.875	3.574	5.024
Group II	RAP15-4% RA-70	64	0.465	0.496	0.502	0.556	0.530	0.597	0.693	0.812	0.955	1.216	1.714
		70	1.055	1.117	1.122	1.287	1.311	1.512	1.743	2.007	2.464	3.247	4.645
	RAP30-4% RA-70	64	0.693	0.725	0.729	0.813	0.783	0.864	0.977	1.105	1.263	1.552	2.039
		70	1.464	1.514	1.541	1.796	1.782	1.999	2.254	2.551	3.042	3.826	5.327
	RAP40-4% RA-70	64	0.822	0.838	0.837	0.935	0.906	0.992	1.115	1.247	1.428	1.733	2.263
		70	1.709	1.789	1.826	2.116	2.104	2.373	2.685	3.044	3.612	4.516	6.545
	RAP50-4% RA-70	64	0.903	0.929	0.940	1.061	1.018	1.110	1.239	1.397	1.585	1.911	2.511
		70	1.847	1.895	1.938	2.230	2.204	2.444	2.720	3.050	3.564	4.424	6.424
Group III	RAP15-2% RA-76	64	0.242	0.259	0.258	0.281	0.258	0.266	0.295	0.348	0.443	0.581	0.839
		70	0.476	0.510	0.510	0.559	0.531	0.595	0.703	0.843	1.065	1.488	2.263
		76	1.162	1.182	1.182	1.359	1.393	1.628	1.909	2.314	3.065	4.234	6.614
	RAP30-2% RA-76	64	0.258	0.277	0.279	0.304	0.280	0.290	0.318	0.372	0.465	0.595	0.835
		70	0.549	0.573	0.570	0.625	0.589	0.639	0.737	0.880	1.097	1.471	2.107
		76	1.251	1.279	1.283	1.466	1.460	1.663	1.926	2.280	2.905	3.901	5.925
	RAP40-2% RA-76	64	0.254	0.271	0.274	0.303	0.280	0.293	0.324	0.375	0.449	0.555	0.743
		70	0.616	0.639	0.641	0.712	0.680	0.749	0.856	0.988	1.172	1.520	2.117
		76	1.332	1.388	1.405	1.615	1.598	1.798	2.048	2.372	2.906	3.750	5.445
	RAP50-2% RA-76	64	0.301	0.307	0.305	0.334	0.308	0.321	0.354	0.407	0.475	0.573	0.753
		70	0.704	0.723	0.724	0.806	0.766	0.837	0.942	1.077	1.256	1.581	2.093
		76	1.420	1.462	1.480	1.695	1.663	1.856	2.119	2.459	3.012	3.854	5.553

Note: temp. = temperature.

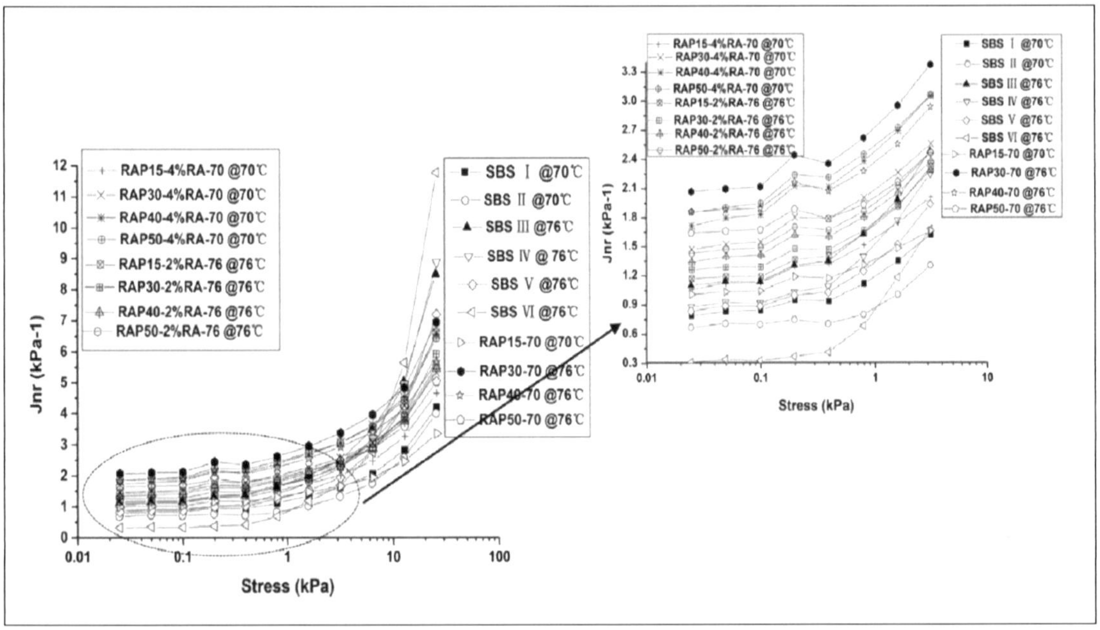

Figure 3. Plot of the J_{nr} values of the tested binders at high-temperature grade over a wide range of stress values.

levels greater than 3,200 Pa. When adding more RAP binder, the J_{nr} value reduces gradually. It seems that higher stress levels could better evaluate rutting performance of asphalt binder. The addition of rejuvenator increases the J_{nr} value of binders. At 76°C, for RAP binders with and without rejuvenator, the J_{nr} values were greater than those of virgin SBS modified binders when the stress level was less than 3,200 Pa, whereas the contrast trend occurred when the stress level exceeded 3,200 Pa. Generally, the SBS modified binders always have lower J_{nr} values in lower stress levels but higher J_{nr} in higher stress levels, which also can be seen based on $J_{nr,diff}$ at different stress levels.

Figure 4 shows $J_{nr,diff}$ at 3.2 kPa and 0.1 kPa ($J_{nr,diff(3.2-0.1)}$), $J_{nr,diff}$ at 6.4 kPa and 0.1 kPa ($J_{nr,diff(6.4-0.1)}$), $J_{nr,diff}$ at 12.8 kPa and 0.1 kPa ($J_{nr,diff(12.8-0.1)}$) and $J_{nr,diff}$ at 25.6 kPa and 0.1 kPa ($J_{nr,diff(25.6-0.1)}$) of tested binders. It is observed that the $J_{nr,diff}$ of RAP binders are less than those of SBS modified binders. For example, SBS VI binder has the minimum J_{nr} when the stress levels are in the range of 25 Pa to 400 Pa. After that, its J_{nr} increased dramatically causing maximum $J_{nr,diff}$ in the tested binders (419.20%, 751.70%, 1,647.99%, and 3,540.87% for $J_{nr,diff(3.2-0.1)}$, ($J_{nr,diff(6.4-0.1)}$, $J_{nr,diff(12.8-0.1)}$, and $J_{nr,diff(25.6-0.1)}$). The maximum $J_{nr,diff}$ leads to the maximum J_{nr} in 25,600 Pa, which suggests that SBS VI binder may have poor rutting performance at higher stress. Consequently, $J_{nr,diff}$ is also an important parameter for predicting the rutting performance of asphalt binder even though the binder has lower J_{nr} at lower stress levels.

Test results of Method B indicate that J_{nr} values at higher stress levels may be more appropriate to rank the rutting performance of binders. In this paper, J_{nr} value at 25.6 kPa is recommended to predict the high-temperature performance of asphalt binder.

Effects of Increased Stress Levels and More Cycles. Table 5 shows the result for all tested binders in Method C. Figure 6 illustrates J_{nr} values at three stress levels for all tested binders at three temperatures. As can be seen in Table 4 and Figure 6, binders show higher J_{nr} values compared with Method A (AASHTO T 350), which means the resistance to deformation of the binder decreases as more creep and recovery cycles are added. As with Method A and Method B, the addition of RAP binder only reduces the J_{nr} values. At 64°C and 70°C, the rejuvenated RAP binder with an addition of PG 70-22 SBS binder and 4% rejuvenator has larger J_{nr} values compared with other binders, indicating that these binders may have poor rutting performance. Virgin PG 70-22 binders

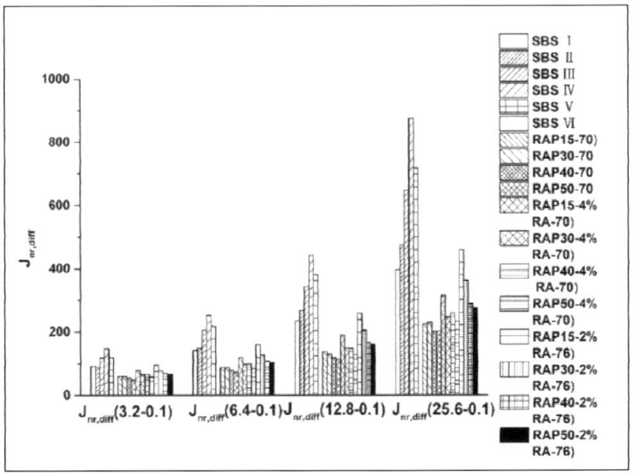

Figure 4. Percentage difference in nonrecoverable compliance of tested binders depending on stress level (SBS VI binder was not included).

Table 5. Method C Test Results for All Tested Binders

Binder type		Testing temp. (°C)	$J_{nr(0.1kPa)}$, 1/kPa	$J_{nr(3.2kPa)}$, 1/kPa	$J_{nr(12.8kPa)}$, 1/kPa	$J_{nr,diff(3.2–0.1)}$,%	$J_{nr,diff(12.8–0.1)}$,%	%R (0.1 kPa)	%R (3.2 kPa)	%R (12.8 kPa)
Neat Binder		64	3.396	3.721	18.117	9.57	433.48	2.36	–0.79	–17.11
SBS I		64	0.971	1.049	1.046	8.03	7.72	61.53	46.92	17.71
		70	1.892	2.216	2.503	17.12	32.29	56.40	31.00	–4.60
SBS II		64	0.906	0.961	0.984	6.07	8.61	69.01	59.38	24.30
		70	1.744	1.955	2.239	12.10	28.38	63.78	41.81	–2.32
SBS III		64	0.685	0.723	0.719	5.55	4.96	73.14	64.40	33.64
		70	1.385	1.523	1.655	9.96	19.49	66.02	47.21	5.31
		76	2.451	2.902	3.694	18.40	50.71	57.77	28.10	–13.15
SBS IV		64	0.758	0.778	0.770	2.64	1.58	79.42	76.51	38.01
		70	1.454	1.554	1.772	6.88	21.87	76.22	59.99	6.78
		76	2.391	2.800	3.849	17.11	60.98	69.46	37.13	–13.70
SBS V		64	0.684	0.707	0.687	3.36	0.44	77.98	74.37	40.24
		70	1.307	1.443	1.565	10.41	19.74	72.45	57.17	8.91
		76	2.101	2.401	3.092	14.28	47.17	67.52	39.95	–10.62
SBS VI		64	0.639	0.685	0.679	7.20	6.26	90.96	89.50	57.94
		70	1.120	1.260	1.641	12.50	46.52	86.38	74.69	14.99
		76	1.813	2.422	4.407	33.59	143.08	81.53	46.24	–14.22
Group I	RAP15-70	64	0.925	1.006	1.045	8.76	12.97	53.43	37.18	13.93
		70	1.838	2.124	2.421	15.56	31.72	46.67	23.42	–3.58
	RAP30-70	64	0.705	0.758	0.832	7.52	18.01	47.83	35.53	13.46
		70	1.436	1.633	1.917	13.72	33.50	38.70	18.72	–1.89
		76	2.597	3.228	4.112	24.30	58.34	30.60	8.04	–7.33
	RAP40-70	64	0.563	0.596	0.681	5.86	20.96	45.81	32.29	14.33
		70	1.170	1.319	1.672	12.74	42.91	36.55	17.99	3.92
		76	2.235	2.713	3.393	21.39	51.81	28.31	7.42	–8.47
	RAP50-70	64	0.453	0.473	0.576	4.42	27.15	44.95	32.83	14.18
		70	0.982	1.090	1.490	11.00	51.73	34.11	17.16	3.42
		76	1.912	2.293	3.250	19.93	69.98	25.60	6.64	–0.06
Group II	RAP15-4% RA-70	64	1.266	1.390	1.453	9.79	14.77	55.84	37.78	7.83
		70	2.261	2.616	3.112	15.70	37.64	48.93	23.67	–6.48
	RAP30-4% RA-70	64	1.066	1.186	1.303	11.26	22.23	46.78	26.33	8.16
		70	2.079	2.440	2.856	17.36	37.37	37.81	15.94	–5.40
	RAP40-4% RA-70	64	1.355	1.521	1.709	12.25	26.13	45.30	25.77	3.80
		70	2.679	3.212	4.023	19.90	50.17	36.09	12.31	–7.47
	RAP50-4% RA-70	64	1.494	1.692	1.933	13.25	29.38	40.37	21.01	0.49
		70	2.814	3.445	4.437	22.42	57.68	30.63	9.31	–4.85
Group III	RAP15-2% RA-76	64	0.639	0.671	0.680	5.01	6.42	68.97	60.23	33.51
		70	1.220	1.345	1.491	10.25	22.21	62.91	44.72	7.07
		76	2.198	2.614	3.338	18.93	51.87	55.65	26.93	–10.54
	RAP30-2% RA-76	64	0.654	0.690	0.720	5.50	10.09	62.63	52.38	26.30
		70	1.215	1.374	1.606	13.09	32.18	53.37	33.67	3.82
		76	2.283	2.783	3.758	21.90	64.61	44.11	17.77	–5.99
	RAP40-2% RA-76	64	0.598	0.637	0.688	6.52	15.05	56.65	43.97	21.73
		70	1.185	1.343	1.581	13.33	33.42	47.16	27.20	1.92
		76	2.317	2.829	3.788	22.10	63.49	39.07	14.19	–5.13
	RAP50-2% RA-76	64	0.564	0.601	0.692	6.56	22.70	50.09	37.25	16.71
		70	1.148	1.300	1.597	13.24	39.11	44.03	24.13	1.75
		76	2.337	2.847	3.809	21.82	62.99	34.12	11.14	–3.98

Note: temp. = temperature.

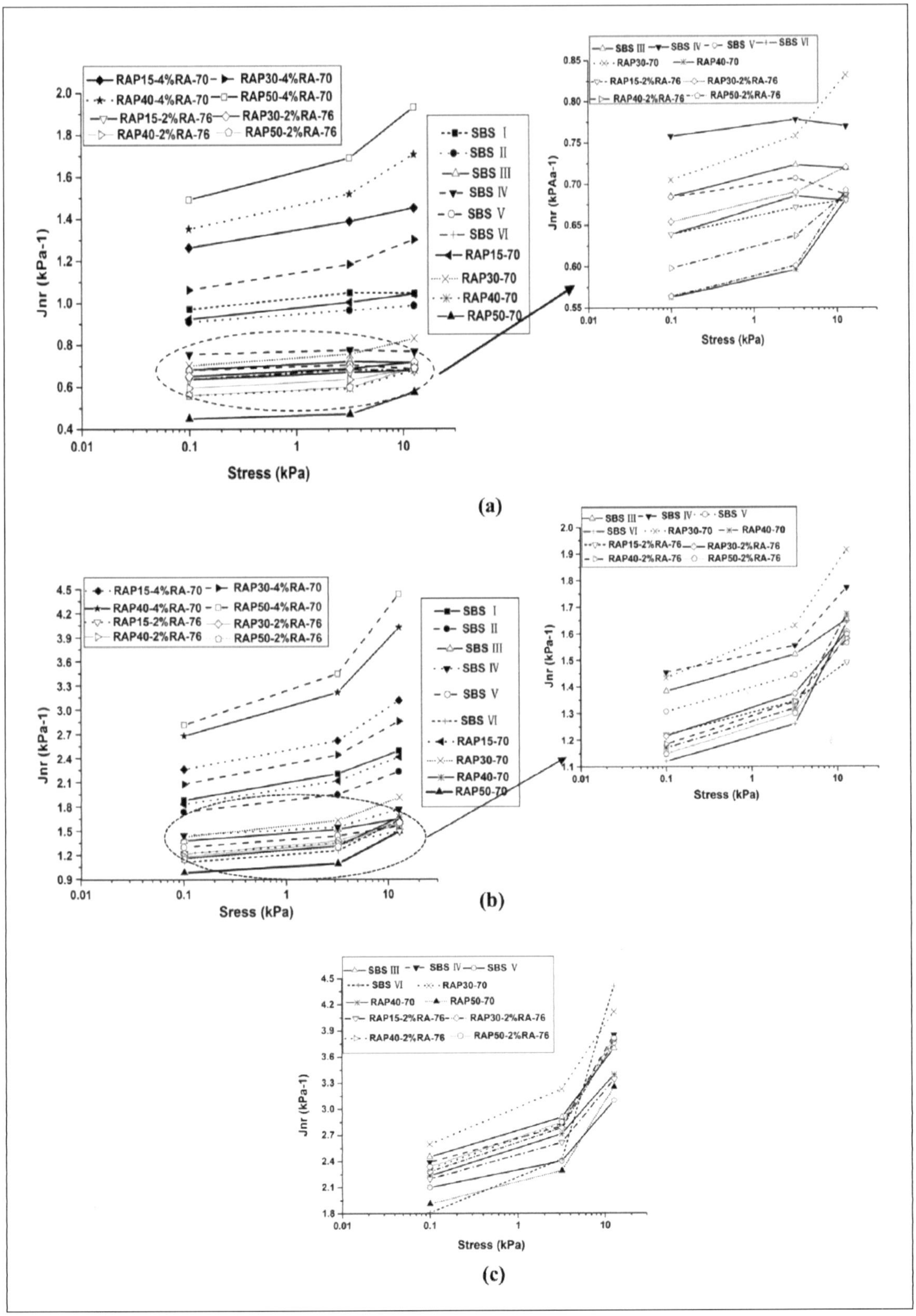

Figure 5. J_{nr} values at three stress levels for tested binders: (*a*) 64°C, (*b*) 70°C, and (*c*) 76°C.

and RAP 15-70 binder have similar J_{nr} values at 64°C and 70°C, as those for PG 76-22 binders. For PG 76-22 binders, at 76°C, J_{nr} values sharply increase at 12.8 kPa, which suggests that the rutting resistance of these binders decreases rapidly. If just accounting for J_{nr} values at 0.1 and 3.2 kPa, SBS VI binder may have the best rutting performance. However, it has the largest J_{nr} value at 12.8kPa in the PG 76-22 binders, which means the rutting performance of this binder may be doubtful at higher stress levels. To address this issue, $J_{nr,diff}$ is an important indicator. Figure 6 shows the $J_{nr,diff}$ between stress levels of 0.1 and 3.2 kPa and between stress levels of 0.1 and 12.8 kPa. From Figure 6, the $J_{nr,diff}$ values are less than those in Method A. Higher stress levels and high-temperature grades achieve higher $J_{nr,diff}$, which can make it easier to discriminate rutting performance of asphalt binder compared with J_{nr} values and $J_{nr,diff}$ values at lower stress levels. Generally, PG 70-22 SBS binder and 4% rejuvenator has bigger $J_{nr,diff}$ values while two types virgin PG 70-22 SBS binder have similar $J_{nr,diff}$ values. For PG 76-22 binders, except for SBS VI binder, other binders have similar $J_{nr,diff}$ values. SBS VI binder has the biggest $J_{nr,diff}$ values among all tested binders. This suggests that the $J_{nr,diff}$ parameter should be considered separately from the J_{nr} parameter.

In terms of %R results, %R values decrease with testing temperatures. All binders show negative %R values at high-temperature grade at 12.8 kPa, meaning that binder has been damaged using Method C. In addition, RAP binders have lower %R results compared with virgin SBS binder, which suggests that SBS binder and rejuvenator have a poor effect on the elastic recovery performance.

In Method C, it can be found that RAP binders containing more than 30% RAP without rejuvenator or RAP binders rejuvenated with 2% rejuvenator and PG 76-22 binder have lower J_{nr} values at 12.8 kPa compared with virgin SBS binders, indicating that adding a higher stress level (12.8 kPa) and increasing the number of cycles would be helpful to rank the binder rutting performance if only considering the J_{nr} parameter.

Comparison of three methods. By comparing the results of these three test methods, the following observations could be found:

(1) The addition of RAP binder decreased the J_{nr} value whereas rejuvenator has the opposite effect.
(2) The addition of SBS modified binder and rejuvenator have a negligible effect on the %R results based on the AASHTO T 350 and Method C results.
(3) The rejuvenating RAP binders with PG 70-22 binder and 4% rejuvenator had larger J_{nr} values at 64°C and 70°C in all three methods compared with other binders.
(4) PG 76-22 SBS binders showed different nonrecoverable behavior in all three methods. In AASHTO T 350 and Method B tests, these binders have lower J_{nr} values compared with rejuvenated RAP binders at lower stress levels. When a higher stress level was applied, the J_{nr} values also increased, especially for SBS VI binder. To address this concern, the $J_{nr,diff}$ parameter is a useful index for rutting performance predicting other than J_{nr}.
(5) Unlike J_{nr} results in Method A and Method B, RAP binders had lower J_{nr} values and $J_{nr,diff}$ values than virgin SBS modified binders in Method C.
(6) For RAP binders, rutting performance is good when more than 30% RAP was added without rejuvenator could make good. When adding rejuvenator, 2% dosage of rejuvenator and PG 76-22 binder may be suitable to guarantee the binder high-temperature deformation resistance of RAP binder.
(7) The rankings by different methods were different. In AASHTO T 350, the $J_{nr,diff}$ parameter must be considered separately from the J_{nr} indicator. Many SBS modified binders have lower J_{nr} values when applying a lower stress level. However, the J_{nr} of these binders may sharply increase at a higher stress level. This phenomenon was confirmed in Method B. The

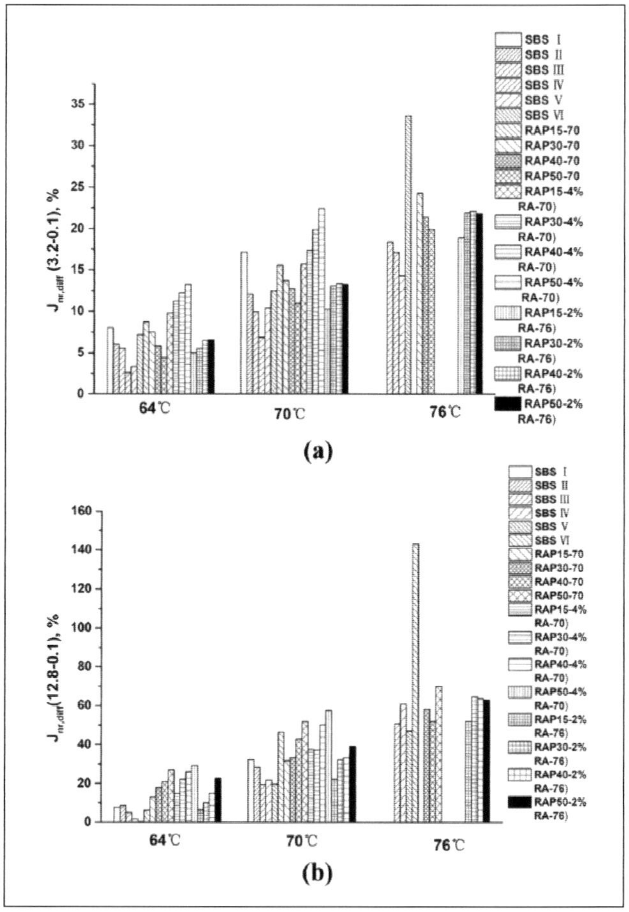

Figure 6. $J_{nr,diff}$ values for all tested binders: (*a*) $J_{nr,diff\ (3.2–0.1)}$ and (*b*) $J_{nr,diff\ (12.8–0.1)}$.

J_{nr} value at 25.6 kPa may be a suitable parameter to predict the rutting performance of asphalt binder. In Method C, adding more creep recovery cycles and a higher stress level (12.8 kPa) help to make a fair ranking if just accounting for J_{nr} value at 12.8 kPa.

Summary and Conclusion

Based on the work presented in this paper, the following conclusions and recommendations could be obtained:

(1) This paper critically reviewed the AASHTO T 350 test method and the AASHTO M 332 specification for prediction of rutting performance of asphalt binder. It was found that most SBS modified binders and two rejuvenated RAP binders in this study cannot meet the $J_{nr,diff}$ limits, although their J_{nr} values can pass the specification in AASHTO M 332. SBS binders distribute above the polymer curve while most RAP binders are nearly at or below the polymer curve.

(2) By using the 11 stress levels MSCR test, higher stress levels and higher temperatures are helpful for predicting the rutting performance of binder since the J_{nr} values of some SBS binders increase with increasing stress levels.

(3) Increasing the creep recovery numbers and adding a higher stress level (12.8kPa) help to rank SBS modified and RAP binders.

(4) The addition of RAP binders decreased J_{nr} values while the addition of rejuvenator increased the J_{nr} value. The addition of SBS modified binder and rejuvenator have a negligible effect on the %R results.

Acknowledgments

This work is sponsored by a project supported by Basic Research Plan (Natural Science Foundation) in Jiangsu Province of China (Grant No. BK20151413). The authors gratefully acknowledge their support.

Author Contributions

The authors confirm contributing to the paper as follows: study conception and design: Zhou Zhou, Xingyu Gu, Fujian Ni; data collection: Zhou Zhou, Yanxu Jiang; analysis and interpretation of results: Zhou Zhou; draft manuscript preparation: Zhou Zhou, Jiwang Jiang. All authors reviewed the results and approved the final version of the manuscript.

References

1. Golalipour, A. *Modification of Multiple Stress Creep and Recovery Test Procedure and Usage in Specification*. University of Wisconsin-Madison, 2011, pp. 64–66.
2. D'Angelo J., and R. Dongré. Superpave binder specifications and their performance relationship to modified binders[C]// *Proceedings of Canadian Technical Asphalt Association*. Vol. 47, 2002, pp. 91–103.
3. Behnood, A., M. Modiri Gharehveran, and A. M. Roshandeh. High-temperature Properties of Asphalt Binders: Comparison of Multiple Stress Creep Recovery and Performance Grading Systems. *Transportation Research Record: Journal of the Transportation Research Board*, 2016. 2574: 131–143.
4. Rahaman, M. Z., Zahid, H., and Musharraf Z. *Nonrecoverable Compliance and Recovery Behavior of Polymer-Modified and Reclaimed Asphalt Pavement-Modified Binders in Arkansas*. No. 17-01422. 2017.
5. D'Angelo, J. A. The Relationship of the MSCR Test to Rutting. *Road Materials and Pavement Design*, Vol. 10, 2009, pp. 61–80.
6. Ling, C., A. Arshadi, and H. Bahia. Importance of Binder Modification Type and Aggregate Structure on Rutting Resistance of Asphalt Mixtures Using Image-based Multi-scale Modelling. *Road Materials and Pavement Design*, Vol. 18, No. 4, 2017, pp. 785–799.
7. D'Angelo, J. Current Status of Superpave Binder Specification. *Road Materials and Pavement Design*, Vol. 10, 2009, pp. 13–24.
8. Wasage, T. L. J., J. Stastna, and L. Zanzotto. Rheological Analysis of Multi-stress Creep Recovery (MSCR) Test. *International Journal of Pavement Engineering*, Vol. 12, No. 6, 2011, pp. 561–568.
9. Jafari, M., A. Babazadeh, and S. Aflaki. Effects of Stress Levels on Creep and Recovery Behavior of Modified Asphalt Binders with the Same Continuous Performance Grades. *Transportation Research Record: Journal of the Transportation Research Board*, 2015. 2505: 15–23.
10Kataware, A. V., and D. Singh. A Study on Rutting Susceptibility of Asphalt Binders at High Stresses Using MSCR Test. *Innovative Infrastructure Solutions*, Vol. 1, No. 2, 2017, pp. 1–9.
11.Golalipour, A., H. U. Bahia, and H. A. Tabatabaee. Critical Considerations Toward Better Implementation of the Multiple Stress Creep and Recovery Test. *Journal of Materials in Civil Engineering*, Vol. 29, No. 5, 2016: 04016295.
12.Xu, J., S. Huang, and Y. Qin. Asphalt Pavement Recycling in Mainland China. *Application of Reclaimed Asphalt Pavement and Recycled Asphalt Shingles in Hot-Mix Asphalt 51*, 2014.
13.Yu, B., X. Gu, M. Wu, and F. Ni. Application of a High Percentage of Reclaimed Asphalt Pavement in an Asphalt Mixture: Blending Process and Performance Investigation. *Road Materials and Pavement Design*, 2016, pp. 1–13.
14. Copeland, A. *Reclaimed Asphalt Pavement in Asphalt Mixtures: State of the Practice*. No. FHWA-HRT-11-021. 2011.
15.Li, Q., X. Chen, G. Li, and S. Zhang. Fatigue Resistance Investigation of Warm-mix Recycled Asphalt Binder, Mastic, and Fine Aggregate Matrix. *Fatigue & Fracture of Engineering Materials & Structures*, 2017.
16.Guo, N., Z. You, Y. Zhao, Y. Tan, and A. Diab. Laboratory Performance of Warm Mix Asphalt Containing Recycled Asphalt Mixtures. *Construction and Building Materials*, Vol. 64, 2014, pp. 141–149.
17. Zaumanis, M., and R. B. Mallick. Review of Very High-content Reclaimed Asphalt Use in Plant-produced Pavements: State of

the Art. *International Journal of Pavement Engineering*, Vol. 16, No. 1, 2015, pp. 39–55.

18. Zhou, Z., X. Gu, Q. Li, F. Ni, and R. Yuan. Use of Rejuvenator, Styrene-Butadiene Rubber Latex, and Warm-Mix Asphalt Technology to Achieve Conventional Mixture Performance with 50% Reclaimed Asphalt Pavement. *Transportation Research Record: Journal of the Transportation Research Board*, 2016. 2575: 160–167.
19. Mogawer, W. S., A. J. Austerman, R. Kluttz, and S. Puchalski. Using Polymer Modification and Rejuvenators to Improve the Performance of High Reclaimed Asphalt Pavement Mixtures. *Transportation Research Record: Journal of the Transportation Research Board*, 2016. 2575: 10–18.
20. Li, J., F. Ni, J. Jin, and Z. Zhou. A Comparison of Rejuvenator and Sryrene–butadiene Rubber Latex Used in Hot In-place recycling. *Road Materials and Pavement Design*, 2016, pp. 1–15.
21. Li, Q., H. Yang, F. Ni, X. Ma, and L. Luo. Cause Analysis on Permanent Deformation for Asphalt Pavements Using Field Cores. *Construction and Building Materials*, Vol. 100, 2015, pp. 40–51.
22. Stevens, R., J. Stempihar, B. S. Underwood, and P. Pal. Evaluation of Multiple Stress Creep and Recovery (MSCR) Data for Arizona. *International Journal of Pavement Research and Technology*, Vol. 8, No. 5, 2015, pp. 337–345.

The Standing Committee on Asphalt Binders (AFK20) peer-reviewed this paper (18-02313).

Article

TRR
JOURNAL OF THE TRANSPORTATION RESEARCH BOARD

Transportation Research Record
2018, Vol. 2672(28) 337–348

DOI: 10.1177/0361198118781376
journals.sagepub.com/home/trr
SAGE

Rheological Behavior and Its Chemical Interpretation of Crumb Rubber Modified Asphalt Containing Warm-Mix Additives

Haopeng Wang[1], Xueyan Liu[1], Panos Apostolidis[1], and Tom Scarpas[1,2]

Abstract

The microstructure and chemical composition of asphalt binders have a significant effect on their rheological properties and, therefore, their performance as road paving binders. This study aims to investigate the effects of warm-mix asphalt (WMA) additives, organic type and chemical type, on the rheological properties and chemical internal structure of base asphalt and crumb rubber modified asphalt (CRMA). A set of dynamic shear rheometer (DSR) tests was conducted to obtain the rheological parameters (e.g., complex viscosity, complex modulus, phase angle) of asphalt binders. The flow activation energy was calculated from Arrhenius equation based on viscosity data to rank the thermal susceptibility. Black diagrams and master curves of complex modulus and phase angle were utilized to analyze the rheological properties. The molecular weight distributions of asphalt binders were inverted from the phase angle master curve to evaluate the molecular weight characteristics. It was found that the the addition of crumb rubber into base asphalt improves the rheological properties of enhanced modulus and elasticity. Organic and chemical types of WMA additives have different chemo-physical effects on both base asphalt and CRMA. Phase angle inversion method provides a powerful tool to monitor the molecular structure change and, therefore, the chemo-physical interactions of asphalt binders induced by modifications. Finally, there is a good correlation between flow activation energy and molecular weight.

Recycling crumb rubber from end-of-life tires into asphalt paving has been applied for several decades for its tremendous economic and environmental benefits. It is reported that the incorporation of crumb rubber modifier (CRM) into asphalt binders can improve the overall performance of asphalt pavements (*1*, *2*), such as improved aging and oxidation resistance, greater resistance to fatigue/thermal cracking and rutting, lower noise generation, higher skid resistance, and so forth. The above-improved performance of rubberized asphalt pavement relies on the interaction of CRM with asphalt. Depending on different interaction parameters (temperature, time and mixing rate, etc.), rubber-asphalt interaction is generally related to two mechanisms (*3*, *4*): (*a*) swelling of CRM particles in asphalt matrix through absorbing the light aromatic oils of asphalt, and (*b*) degradation (devulcanization and depolymerization) of CRM through the release of its components into the liquid phase of asphalt.

The improvement of the rheological properties of the rubberized asphalt binders has been observed after the interaction of CRM with asphalt (*5*). However, due to the high viscosity of rubberized binders, high mixing and compaction temperatures are required to achieve desirable workability and density of asphalt mixtures. In addition, the consequent high emissions and compromised work conditions during the construction process of rubberized asphalt pavement have been criticized (*6*). Warm-mix asphalt (WMA) technologies are developed to substantially decrease the production and placement temperatures of hot-mix asphalt (HMA) through mechanisms of viscosity reduction and lubricity enhancement (*7*–*9*). The coupling of WMA technology with rubberized asphalt is supposed to be a sustainable paving technology with the merits of energy conservation,

[1]Section of Pavement Engineering, Faculty of Civil Engineering and Geosciences, Delft University of Technology, Delft, The Netherlands
[2]Department of Civil Infrastructure and Environmental Engineering, Khalifa University of Science and Technology, Abu Dhabi, the United Arab Emirates

Corresponding Author:
Address correspondence to Haopeng Wang: haopeng.wang@tudelft.nl

environmental protection, performance optimization and durability extension (*10*). WMA products can be categorized into three main types (*11*): foaming processed, organic (wax-based) additives, and chemical additives. With the addition of new chemicals, the microstructure and molecular characteristics of original asphalt binders can be alternated. The different characteristics of WMA additives determine the different effects on the rheological properties of asphalt binder. It is vital to investigate the influence of WMA additives on the rheological behaviors and chemical structures of rubberized binders before applying them in the mixture level.

Chemical Interpretation from Rheology

Asphalt is commonly accepted as a multi-disperse colloidal system, where high molecular weight (MW) asphaltene micelles are peptized by resins in low molecular weight maltenes, with rheological behavior resembling that of a low-molecular-weight polymer (*12*). The chemical composition and microstructure of asphalt binder has a significant effect on the material rheological properties and, therefore, on its performance as a road paving binder.

A large amount of research works has been done to link the chemical nature of asphalt with its physical behavior, and thus to explain the mechanical performance. Typically, the microstructure and composition of asphalt can be determined through classical chemical characterization techniques, such as atomic force microscopy (AFM) to obtain surface topography, gel permeation chromatography (GPC) to obtain molecular weight distribution, thermogravimetric analysis (TGA) for thermal analysis, Fourier transform infrared (FTIR) spectroscopy to identify functional groups, and so forth (*13*). However, these chemical characterization methods need exclusive instruments, special training, and complex sample preparation. More importantly, due to the chemo-physical complexity of certain asphalt products (modified or emulsified), it is not suitable to use all the above techniques in every case (*14*, *15*). Recently, a few attempts were made to obtain the molecular weight distribution (MWD) and activation energy (E_a) for the flow of asphalt through rheological tests.

Relationship between Rheology and Flow Activation Energy

Viscosity is a measure of material's resistance to flow and deformation. The concept of the flow activation energy of asphalt describes the minimum energy barrier which asphalt molecules must overcome for flow to occur (*16*). It can be used to differentiate asphalt binders and to characterize the thermal susceptibility of the asphalt binders (*17*). The temperature dependence of the viscosity of asphalt can be described by an Arrhenius equation (Equation 1).

$$\eta^* = Ae^{E_a/RT} \tag{1}$$

where A is a pre-exponential parameter, R is the universal gas constant (8.314 J/[mol·K]), E_a is the activation energy (kJ/mol), T is temperature in degrees K, and η^* is the complex viscosity at zero or low shear rate (Pa·s). The activation energy can be calculated based on known paired data of temperature and zero-shear or low-shear viscosity (*18*).

Since zero shear viscosity (ZSV) is a theoretical concept and it is impossible to measure the absolute value directly, various models were developed to extrapolate the ZSV value (*19*). However, it was found that highly modified asphalt binders exhibit extreme high viscosity gradients at low frequencies, resulting in unrealistically high ZSV values using model predictions (*20*). It was also found in this study that CRMA binders and wax-based additive modified binders yielded unpractical ZSV values. Therefore, the low shear viscosity at 0.01 rad/s was used as a replacement to calculate the activation energy. In addition, the E_a values can also be obtained through fitting shift factors using the Arrhenius equation during the process of constructing master curves (*21*). Others reported similar E_a values using a shift factor-based Arrhenius equation as the viscosity-based Arrhenius equation did (*21*, *22*). This is mainly because different rheological parameters are interrelated.

Relationship between Rheology and Molecular Weight Distribution

It is well known that asphalt is composed of low (aromatic oils), medium (resins), and high (asphaltenes) molecular weight constituents and the incorporation of new additives (e.g., polymers and chemical surfactants) into base asphalt will inevitably change the proportions of these components. Consequently, the MWD alters due to the various interactions. Since MWD is associated with the internal structure as well as the structure-sensitive properties (e.g., rheological properties) of asphalt binders, comparing the MWDs of base and modified binders is a useful method to evaluate the chemo-physical interactions between base asphalts and additives. It is known that the linear viscoelastic parameters of polymer solutions and melts, as well as asphalt, are strong functions of molecular weight. Zanzotto et al. (*23*) inverted the fractional model of complex modulus to generate the MWD of regular and modified asphalt and compared with the MWDs obtained by GPC test. It was found that the MWDs calculated from rheological data are in good agreement with the corresponding GPC data. In subsequent research, Zanzotto et al. (*14*) found the phase angle to be more sensitive to molecular weight changes than the complex modulus. In addition, they noted that GPC is more sensitive to low molecular weights while the rheological inversion yields a higher resolution in a high molecular weight band. It is noticeable that solvents used to dilute asphalt in GPC test potentially change the microstructure of asphalt, resulting in an erroneous representation of the MWDs of the undiluted asphalt. On the other hand, in case of materials containing insoluble components (e.g., rubber) in common solvents, the use of the chromatographic method is not feasible. Considering the above two limitations of GPC tests for asphaltic materials, in terms of AR binders containing WMA chemicals, deducing MWDs from linear viscoelastic parameters seems a more promising alternative.

Before deriving the MWDs from the linear viscoelastic properties, two important definitions should be known, namely, differential molecular weight distribution (DMWD) and cumulative molecular weight distribution (CMWD). DMWD is a molecular weight distribution function (or a probability density function) of molecular weight. CMWD is an integral of DMWD from zero to a specified molecular weight. The relationship between DMWD and CMWD can be described using molecular weight as follows:

$$w(M) = \frac{d f_c(M)}{d \log M} \quad (2)$$

or

$$f_c(M) = \int_{-\infty}^{\log M'} w(M) d \log M \quad (3)$$

where M = molecular weight; $w(M)$ = molecular weight distribution function or probability density function, it represents the relative amount of different molecular weights; $f_c(M)$ = the cumulative weight fractions of molecules from zero up to a specified molecular weight (M'). Based on the double reptation mixing rule (*24*), the phase angle can be related to the MWD through the following relationship:

$$\delta(\omega) = \int_0^{\infty} w(M')c(M', \omega)dM' \quad (4)$$

where $\delta(\omega)$ = phase angle as a function of angular frequency; $w(M')$ = weight distribution function; $c(M', \omega)$ = monodisperse phase angle of a relaxation unit. With the assumption that the monodisperse phase angle is proportional to the Heaviside step function H:

$$c(M', \omega) \sim 1 - H(M' - M) \quad (5)$$

Equation 5 explains that, at a given frequency, if the specified molecular weight (M') is smaller than the corresponding molecular weight (M), it will relax and make no contribution to the viscoelastic phase angle (*25*). A simple power law was used to relate the molecular weight domain to the frequency domain (ω = crossover frequency):

$$M = \kappa \omega^{-\alpha} \quad (6)$$

The calibrated values for constants in Equation 6 based on test data were $\log \kappa = 2.544$, $\alpha = 0.06768$ at reference temperature of 0°C (*14*). Substituting Equations 5 and 6 into Equation 4, the following relationship can be obtained for normalized molecular weight:

$$\delta(x) \sim \int_0^{M = \kappa 10^{-\alpha x}} w(M')dM' \quad (7)$$

where $x = \log \omega$. Comparing Equations 3 and 7, it can be found that both cumulative weight fractions and phase angle are functions of molecular weight and there exists a link between cumulative weight fractions and phase angle in a logarithmic scale. Since phase angle is also a function of frequency, if the power law relationship between molecular weight and frequency is known, then the phase angle can be plotted as a function of molecular weight. Considering the constraint and boundary conditions of the cumulative weight fractions:

when $M \to 0$, $\delta(M) = 0, f_c(M) = 0$;
when $M \to \infty$, $\delta(M) = 90^{\circ}, f_c(M) = 1$.

The relationship between cumulative weight fractions and phase angle can be written as:

$$f_c(M) = \frac{1}{90}\delta(M) \quad (8)$$

Therefore, with the phase angle master curve and the calibration of Equation 6, the molecular weight distribution can be obtained through Equations 2 and 8.

Based on above derivations, one should note that rheological properties actually depend on the apparent MWDs that include the associations of various molecular weight constituents rather than on the MWDs of an individual molecule. Therefore, the MW obtained in this study means the apparent MWs of different molecular associations. Figure 1 summarizes the flow diagram of the steps and representative graphs involved in the conversion of linear viscoelastic data into an MWD, which is called the phase angle inversion method in this study.

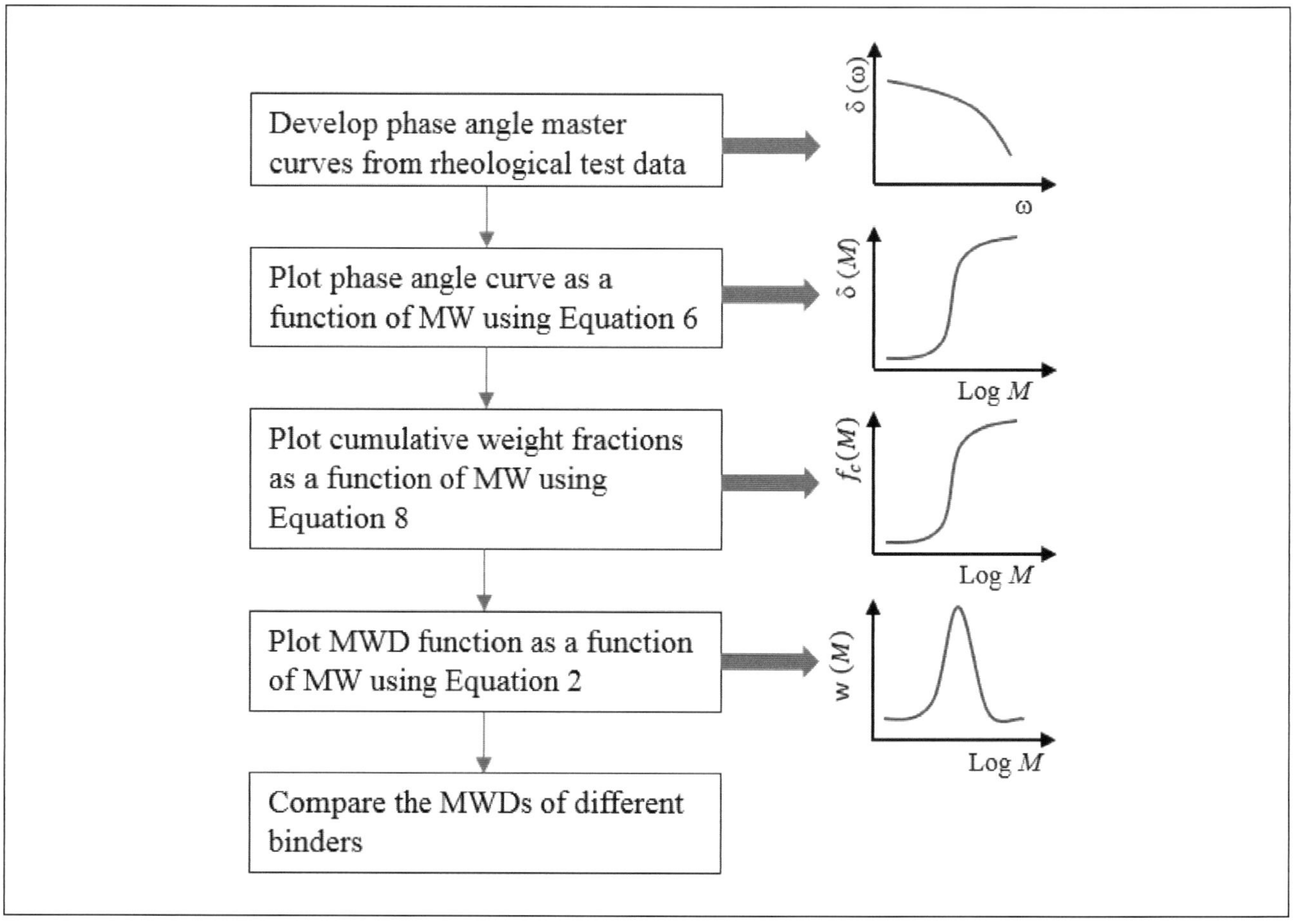

Figure 1. Flow chart of converting phase angle data to MWD data.

Objective

Based on above considerations, the main objectives of this study are:

- to investigate the effects of WMA additives on the rheological properties of base asphalt and CRMA;
- to obtain flow activation energies and molecular weight distributions (MWD) of asphalt binders based on rheological properties; and
- to build up the potential relationship between flow activation energy and molecular weight.

Materials and Methods

Raw Materials

Base asphalt of 70/100 penetration grade commonly used in the Netherlands was provided by NYNAS. The SARA fractions of the base asphalt are 7% for saturates, 51% for aromatics, 22% for resins, and 20% for asphaltenes, respectively. The CRMs, supplied by RUMAL, were produced by ambient grinding of scrap truck tires. The CRMs have particle sizes ranging from 0 to 0.5 mm and consist of about 55% rubber, 30% carbon black, and 15% other impurities. Two types of typical non-foaming WMA additives, wax-based product **W** and chemical-based product **C**, were utilized in this study. Additive W is a synthetic microcrystalline wax that is free from sulfur and other impurities and Additive C is a liquid chemical package of products, such as surfactants, polymers, additives, anti-stripping agents, and so forth.

Binder Preparation

The CRMA was produced in the laboratory by blending 18% CRMs by the weight of base asphalt based on the previous studies (*1*) and trial mixing. Manual stirring for 5 min was applied to pre-distribute CRM in the base asphalt, then the blend was mixed using a Silverson high shear mixer at 180°C and 6000 rpm for 30 min. During the laboratory mixing process, the mixing head was immersed into the hot bitumen to avoid a vortex which may involve the potential oxygen. Each WMA additive was incorporated into both base asphalt and CRMA with the same dosage. The percentages of WMA additives W and C were 2.0% and 0.6%, which were determined based on manufacturers' recommended dosage and preliminary tests. Therefore, a total of six types asphalt binders were prepared in this study, namely 70/100, 70/100-W, 70/100-C, CRMA, CRMA-W, and CRMA-C. To solely investigate the effect of WMA additives on the rheological properties of binders, all binder samples were tested in fresh states without artificial ageing.

Rheological Measurements

Dynamic shear rheometer (Anton Paar) was utilized to obtain the rheological parameters (complex shear modulus, phase angle and complex viscosity) of different asphalt binders. Frequency sweep tests were carried out from 0.01 to 100 rad/s over a temperature range of 10°C ~ 50°C with an increment of 10°C. Before the frequency sweep tests, strain amplitude sweep tests were conducted to identify the linear viscoelastic (LVE) range of different binders and thus to guarantee that the frequency sweep tests were undertaken within the binder's LVE region of response.

Results and Discussion

Complex Viscosity

Figure 2 plots the complex viscosities of different asphalt binders through dynamic shear rheometer (DSR) measurements. Viscosities at various combinations of temperature and frequency showed similar trends and are, therefore, omitted here. It can be clearly seen that the viscosity is dependent on temperature and shear rate (frequency) which decreases as the temperature and/or frequency increases. Figure 2*a* shows that at 50°C, the base asphalt 70/100 with and without additive C both behave like Newtonian fluids, whose viscosities are almost independent of shear rates. In contrast, 70/100-W and all CRMA based binders exhibit non-Newtonian behaviors as the viscosities increase significantly as the shear rate decreases. It was also found that CRMA presents much higher viscosity than base asphalt.

The addition of chemical WMA additive has an insignificant effect on the viscosity of both base and CRMA binders at a certain shear rate. However, organic additive increases the viscosities of both base and CRMA binders significantly, which seems contradictory to the viscosity-reduction effect of wax-based additive W at high construction temperatures (*26*). It should be noted that additive W is a synthetic wax with high molecular hydrocarbon chains, and its melting point is around 90°C. Therefore, when tested at 50°C, additive W in asphalt binders will crystallize to form a lattice structure and, therefore, stiffen the binders. This phase transition characteristic of additive W also explains the improved deformation resistance of binders containing W at service temperatures. In addition, through comparing the slopes of complex viscosity versus temperature curves in Figure 2*b*, CRMA based binders and 70/100-W are less thermal susceptible than base binder 70/100.

From Rheology to Flow Activation Energy

Figure 3 presents the plot of logarithmic viscosity as the ordinate and reciprocal temperature as the abscissa. The E_a values can be calculated based on the slope of $ln\ (\eta^*)$-1/T curves with known R value according to Equation 1. As can be seen from Figure 3, the E_a value obtained for base asphalt is much higher than CRMA. Chemical-based additive C had insignificant effects on the activation energies of both base asphalt and CRMA, while wax-based additive W significantly decreased the activation energies. Activation energy can be used as an indicator for characterizing the thermal susceptibility of asphalt binders. Binders with lower activation energy were found to be less susceptible to the temperature changes (*16*, *17*, *22*). It should be emphasized that asphalt with higher E_a values do not necessarily mean it has higher viscosities. The calculated activation energy is highly dependent on the testing temperatures at which the material can be in different physical states. Jamshidi et al. (*27*) found E_a values of asphalt blends at different phases (liquid, semiliquid, or solid) have opposite variation laws. Therefore, base asphalt, which has lower viscosity and higher activation energy, is more vulnerable to temperature change in the testing temperature range. From this point of view, the incorporation of CRM or additive W into base binders can improve the thermal susceptibility and deformation resistance within service temperature range, which coincides with the results of this study and previous research findings (*26*, *28*).

Complex Modulus and Phase Angle

Black Diagram. A black diagram is a graph plotting complex shear modulus (G^*) versus phase angle (δ) obtained from frequency sweep tests. This type of representation of test data eliminates the frequency and temperature and allows us to compare the viscoelastic response of bituminous materials without manipulating the raw data through time-temperature superposition principle. Generally, a black diagram is a useful tool in identifying possible discrepancies in test data, in verifying time-temperature equivalence and the thermo-rheological simplicity (*29*) of test samples, and in identifying different types of asphalt binders.

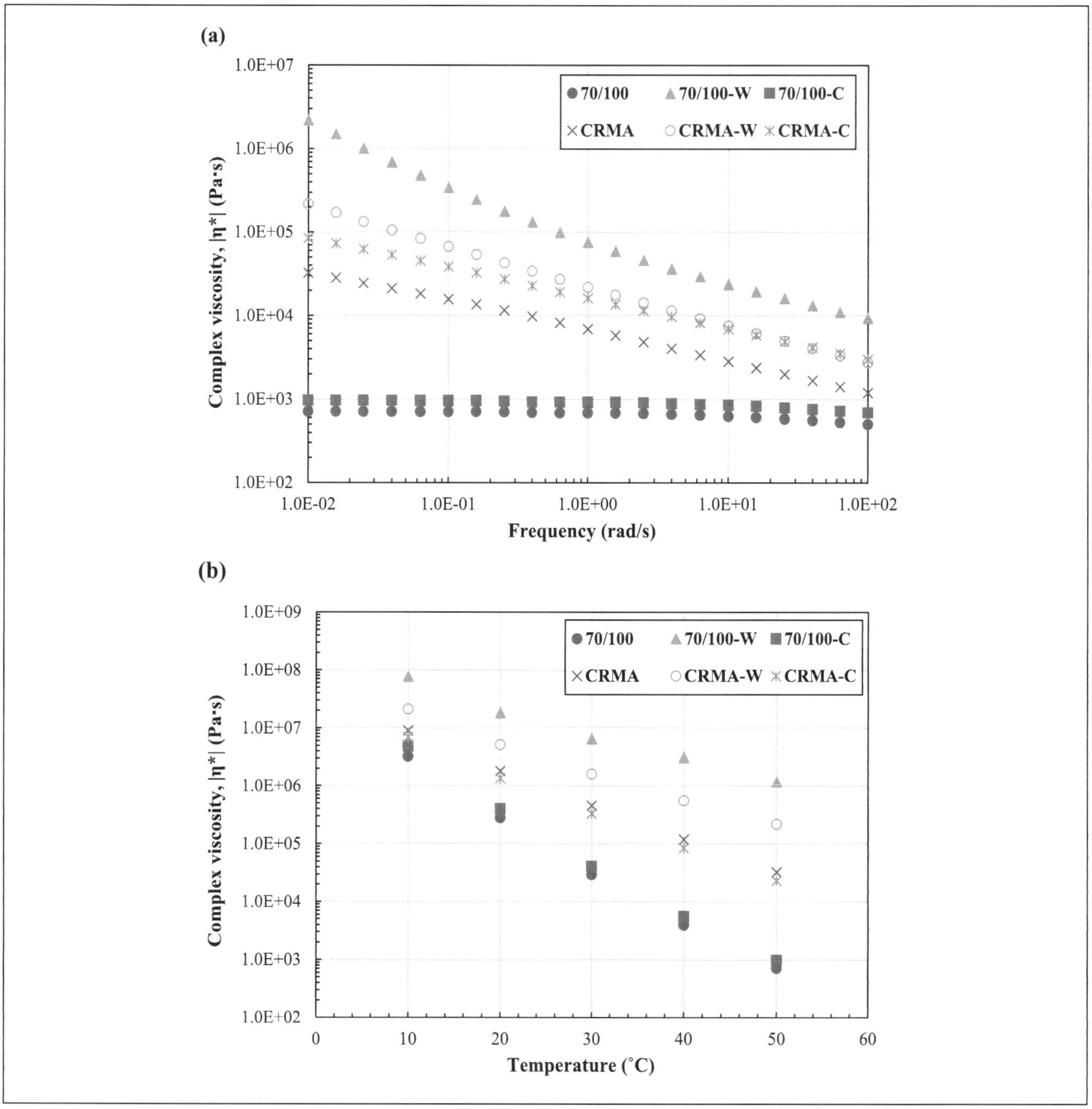

Figure 2. Complex viscosity of different binders versus (*a*) frequency at 50°C; (*b*) temperature at 0.01 rad/s.

Figure 4 shows the representation of dynamic data in black space. It can be clearly seen that the circled data of CRMA binder is out of line compared with the rest of the data. This unconformity indicates instrument compliance errors at the particular temperatures and frequencies. After checking the broad raw data (e.g., shear strain, shear stress, torque), it was found that the torques corresponding to the circled data exceeded the maximum torque value (230 mN·m) of the DSR instrument. Therefore, these erroneous data were omitted before generating an accurate master curve. Except for the limited erroneous data, all the dynamic data produced smooth curves with slight scattering, indicating that the asphalt binders can be considered as thermo-rheologically simple materials. Compared with the base binder and 70/100-C, the addition of CRM and additive W to the base binders results in a shifting of the rheological data toward a lower phase angle (left), which means more elastic behaviors. This could be due to the polymer network of CRM and the crystalline structure of W

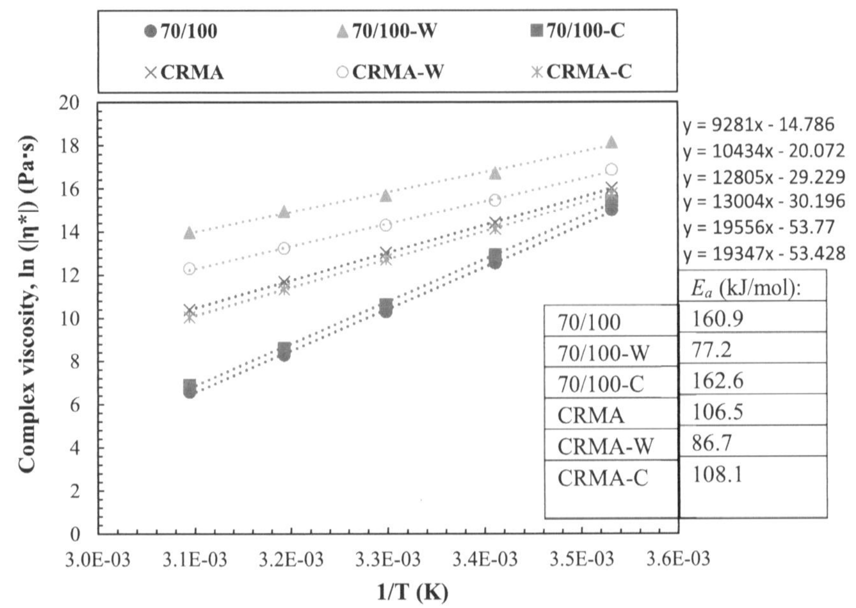

Figure 3. Temperature dependence of low-shear viscosity and calculation of activation energy.

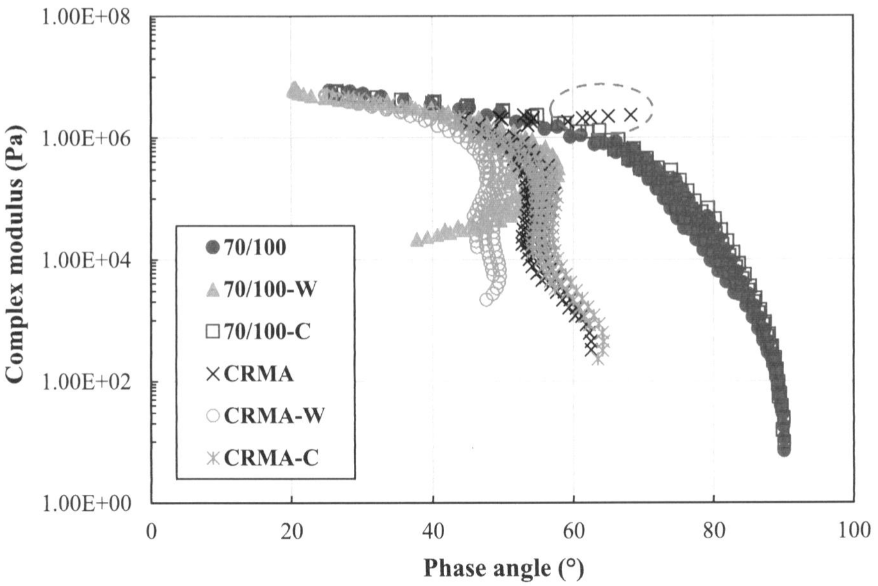

Figure 4. Black diagram of different asphalt binders.

formed within the asphalt matrix. Since different asphalt binders show different curve patterns in black space, the black diagram can also be used to differentiate materials. Unlike the base binder and 70/100-C, the black diagrams of all CRMA based binders show an inverse "S" pattern which is typically found in rubberized asphalt binders (*30*). The single modification by W changes the curve in black space into an inverse "C" pattern which is similar to that of SBS modified binder (*29*). At the phase angle of around 57°, there is evidence of phase transition of wax existing in the 70/100-W binder.

Master Curves of Complex Modulus and Phase Angle. Master curves are developed from the measured linear

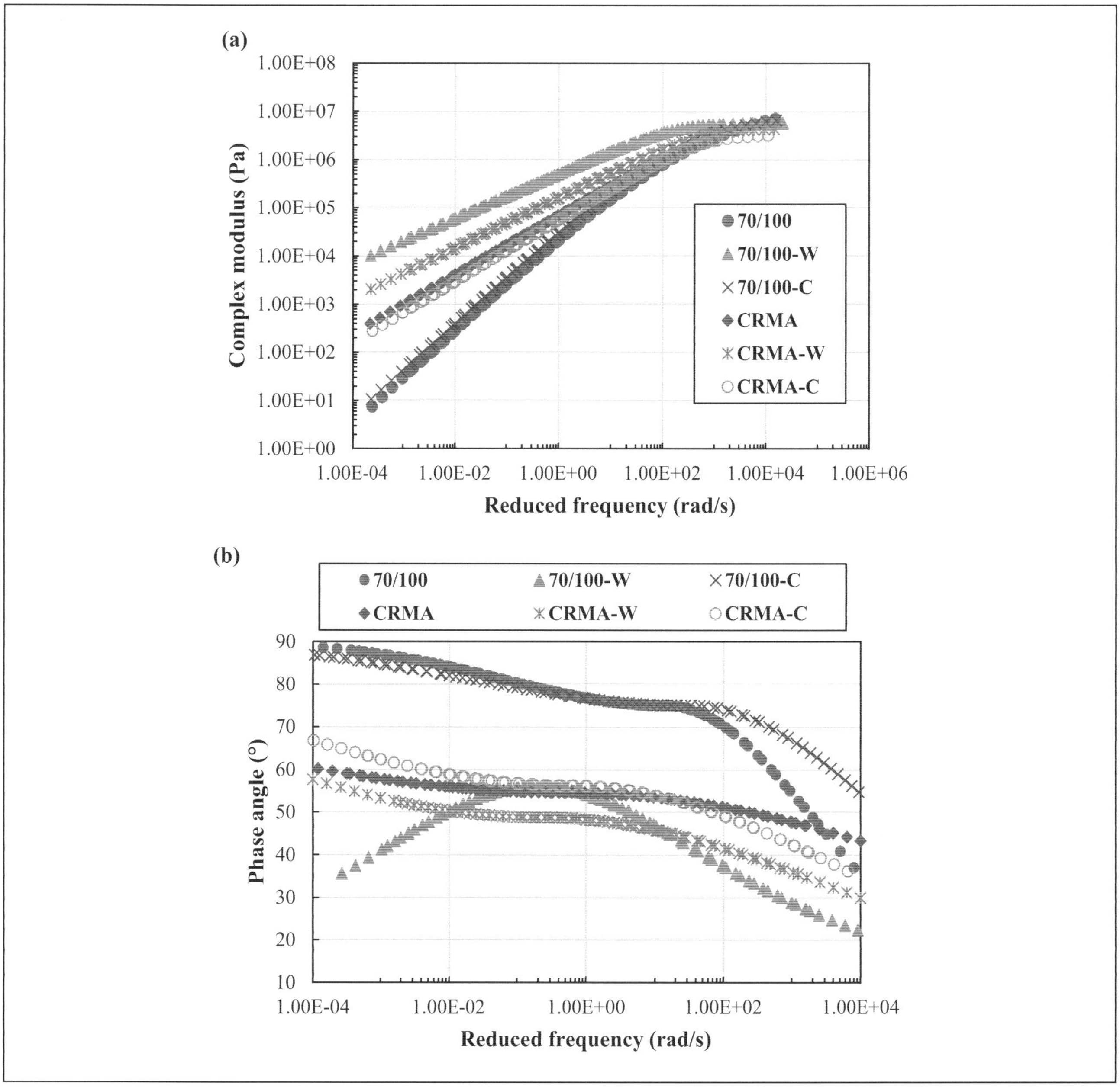

Figure 5. Master curve of (*a*) complex modulus and (*b*) phase angle at 30°C.

viscoelastic data using the time-temperature superposition principle at a reference temperature. In the present study, a modified Christensen-Anderson-Marasteanu (CAM) model (*31*) and Williams-Landel-Ferry (WLF) equation for shift factors fitting were used to develop complex modulus and phase angle master curves.

The master curves of complex modulus and phase angle at a reference temperature of 30°C are shown in Figure 5. Similar effects of CRM and WMA additives on the rheological properties were observed as previous analysis. From Figure 5*a*, all the modification of base asphalt increased the complex modulus at relatively low frequencies (elevated temperatures). However, the complex modulus of all binders merged together at high frequencies (low temperatures), where the base asphalt is relatively stiff and the rheological properties of modified binders are dominated by the characteristics of base asphalt. From Figure 5*b*, the modifications of base asphalt by CRM and additive W showed marked reductions in phase angle. The addition of additive C had insignificant effects on phase angle at relatively low frequencies but increased the phase angle at high frequencies due to the softening effect of the unique chemicals, which may be beneficial to the low-temperature performance. The presence of phase angle plateaus at intermediate frequencies for CRMA based binders indicates the three-dimensional rubber molecular networks or entanglements in the modified binders. The presence of a phase angle peak of 70/100-W again verifies the phase transition behavior of wax in the binder at intermediate frequencies (temperatures). To sum up, the CRM and additive W enhanced the stiffness and elasticity of base asphalt, which is derived from the physical cross-linking of molecules into the three-dimensional network and the unique crystalline lattice structure at the service temperature range, respectively.

Furthermore, the phase angle master curves are less uniform than that of complex modulus. Measurements of phase angle are more sensitive to the chemical structure change and thus the modification of asphalt than the complex modulus. This finding has driven the work of linking phase angle to the chemical characteristics of asphaltic materials.

From Rheology to Molecular Weight Distribution

After obtaining the CMWD curve ($f_c(M)$ versus $\log M$), the DMWD curve ($w(M)$versus $\log M$) is determined by numerical differentiation according to Equation 2 with the help of mathematical tools to guarantee convergence. The DMWD curves of different asphalt binders are presented in Figure 6. In general, Figure 6 clearly shows the internal structural changes of modified asphalt binders. From Figure 6*a*, the phase angle inversion method gives similar number-average molecular weight of around 800 Da for base asphalt as reported in the literature using GPC test (*14*, *15*, *32*). This comparable result verifies the robustness of the inversion method.

In terms of the effect of WMA additives on the MWDs of asphalt binders, it seems the MWD of base asphalt was not affected significantly by the additive C. The addition of W into base asphalt shifted the molecular weight to the heavier direction. It is known that additive W is a kind of microcrystalline wax with high molecular weight hydrocarbons in the range C_{40}~C_{120}. Besides the molecular population of average MW ($\approx$ 1000 Da), a new molecular population of average MW ($\approx$ 3000 Da) came up due to the modification of W. Unlike the effects of WMA additives on base asphalt in Figure 6*a*, neither of the WMA additives changed the bimodal nature of MWDs of CRMA binders in Figure 6*b*. However, they both increased the relative amounts of molecular weights corresponding to the peak distribution. Comparing the MWDs of base asphalt and CRMA, it can be found that the unimodal MWD of base asphalt changed to bimodal MWD due to the interactions of CRM. This MWD change provides evidence that the internal structure of base asphalt was altered due to the addition of CRM.

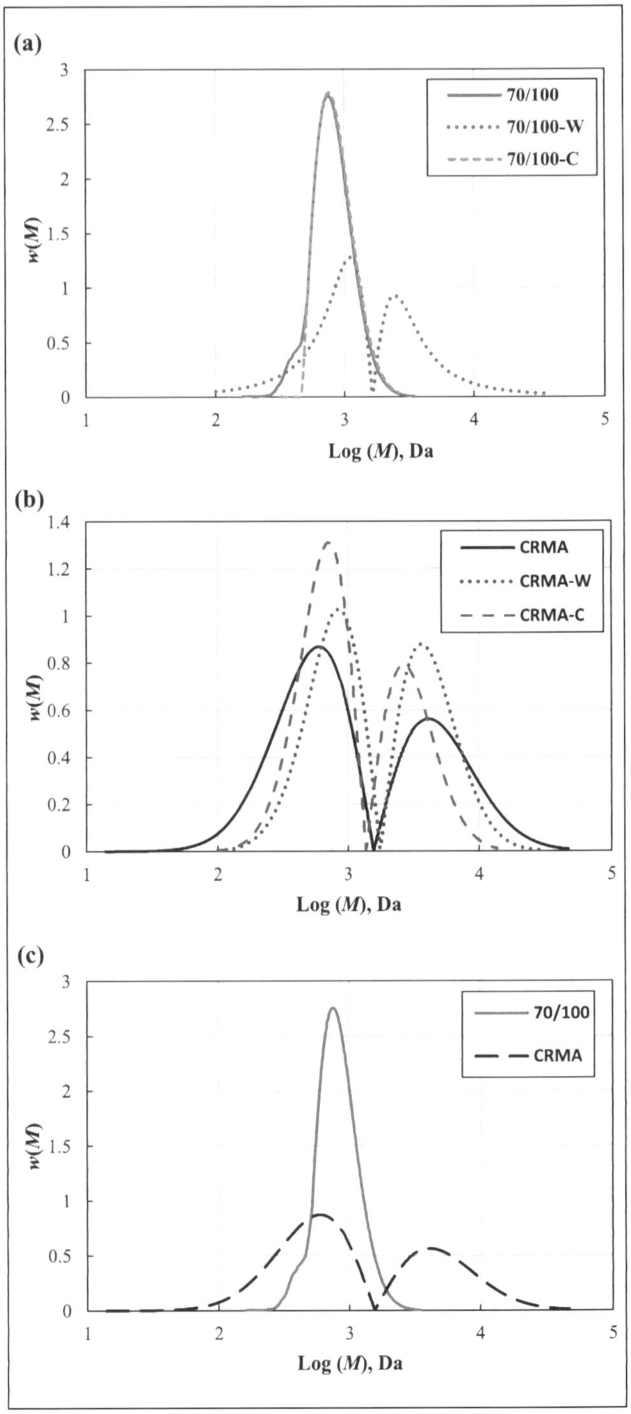

Figure 6. Molecular weight distribution comparisons between (*a*) base asphalt and warm-mix base binders, (*b*) CRMA and warm-mix CRMA binders, and (*c*) base asphalt and CRMA.

The large cross-linking molecules of rubber and the potential released polymer components from rubber make contributions to the increase in the average molecular weight of base asphalt.

In principle, the valley between two peaks of the MWDs should be smooth for real cases. The appearance

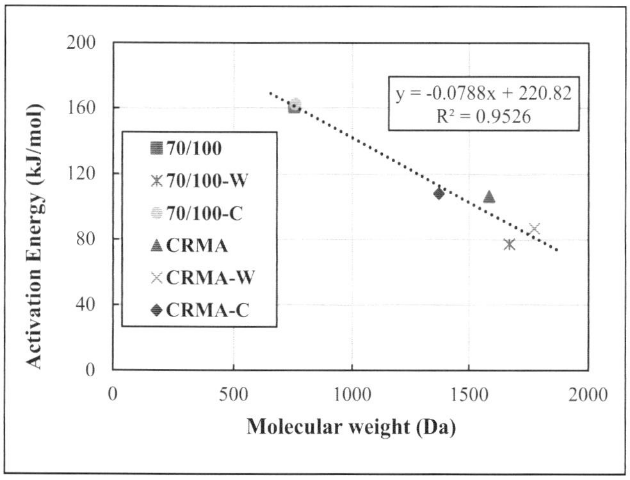

Figure 7. Relationship between molecular weight and flow activation energy.

of the sharpness in this case is mainly due to the inherent limitations of the phase angle master curve model and numerical differentiation error. Based on the above analysis, it can be concluded that the phase-angle inversion method has sufficient sensitivity to distinguish the structural changes of different binders and give comparable molecular weight distributions. However, the accuracy of this inverse method relies on the calibration of Equation 6 and phase angle master curve. More works need to be done on these two aspects to interpret the real molecular structure of asphaltic materials.

Relationship between Flow Activation Energy and Molecular Weight

The number-average molecular weight for each type of asphalt calculated from the MWD in Figure 6, as well as the flow activation energy, is summarized in Figure 7. It shows a strong linear relationship between molecular weight and flow activation energy. The flow activation energy decreases as the molecular weight increases at a constant shear rate in the tested temperature region. Contradictory results were also presented in the literature. García-Morales et al. (*18*) found that polymer addition leads to a reduction in E_a, while Ait-Kadi et al. (*22*) found that the presence of polymer in asphalt results in an E_a increase. Also, Collins and Metzger (*33*) found that the flow activation energy based on viscosities at constant shear rate decreases as the molecular weight increases. They also found that the influence of molecular weight change on the activation energies obtained in the high temperature region is different from that in the low temperature range. Fundamentally, activation energy obtained from rheological tests relies on testing temperature, loading mode (constant shear strain rate or constant shear stress) and the material flow mechanism. The comparison of E_a values tested at different conditions may result in discrepant conclusions. The introductions of CRM and microcrystalline wax into base asphalt increase the molecular weight, and simultaneously provide cross linking and micro crystallization to the asphalt structure, resulting in a harden microstructure. This modification changes the rubbery plateau zone and the corresponding transition temperature between the terminal zone and the rubbery zone of master curves, and finally changes the flow mechanism in different temperature ranges. With limited experimental data, it is speculated that an increase in molecular weight has similar effects on flow activation energy as a decrease in transition temperature, which means decreasing activation energy (*33*). However, more fundamental work needs to be done to verify this phenomenon.

Conclusions

Warm-mix asphalt technology has been used in the rubberized asphalt paving industry to alleviate the issues of high construction temperature and high emissions stemming from the high viscosity of CRMA binders. Based on the rheological tests and their chemical interpretation, the following conclusions can be drawn:

- The incorporation of CRM into base asphalt dramatically increases the complex viscosity. It also improves the rheological properties of base asphalt with enhanced stiffness and elasticity provided by the cross-linking polymer network. Wax-based WMA additive increases the complex viscosity and complex modulus, and decreases the phase angle of asphalt due to the stiffening effect of the unique microcrystalline lattice structure at service temperatures. The chemical type of the additive has an insignificant effect on the rheological properties at the in-service temperatures.
- Asphalt binders modified by either CRM or wax-based additive have decreased flow activation energies, indicating lower thermal susceptibility.
- The phase-angle inversion method provides adequate and comparable molecular weight distributions, and this simply implemented method offers new perspectives in the interpretation of rheological data to monitor the change in the molecular structure of asphalt. Both CRM and the wax-based additive increased the average molecular weight of asphalt. The chemical-based additive had an insignificant effect on the molecular weight distribution of asphalt.
- There is a strong linear relationship between flow activation energy and molecular weight. The

activation energy decreases as the molecular weight increases at a constant shear rate in the tested temperature region.

For future research, more fundamental work needs to be done to further explain the mechanism of flow activation energy. GPC tests for various types of asphalt binders at both unaged and aged states should be conducted to produce a more accurate calibration equation for the phase-angle inversion method.

Acknowledgments

The authors acknowledge RUMAL for providing testing materials. The corresponding author would like to thank the financial support from China Scholarship Council.

Author Contributions

The authors confirm contribution to the paper as follows: study conception and design: Xueyan Liu, Tom Scarpas; data collection: Haopeng Wang; analysis and interpretation of results: Haopeng Wang, Panos Apostolidis; draft manuscript preparation: Haopeng Wang. All authors reviewed the results and approved the final version of the manuscript.

References

1. Lo Presti, D. Recycled Tyre Rubber Modified Bitumens for Road Asphalt Mixtures: A Literature Review. *Construction and Building Materials*, Vol. 49, 2013, pp. 863–881.
2. Wang, T., F. Xiao, S. Amirkhanian, W. Huang, and M. Zheng. A Review on Low Temperature Performances of Rubberized Asphalt Materials. *Construction and Building Materials*, Vol. 145, 2017, pp. 483–505.
3. Abdelrahman, M. A., and S. H. Carpenter. Mechanism of the interaction of asphalt cement with crumb rubber modifier. *Transportation Research Record: Journal of the Transportation Research Board*, 1999. 1661: 106–113.
4. Ghavibazoo, A., M. Abdelrahman, and M. Ragab. Mechanism of Crumb Rubber Modifier Dissolution into Asphalt Matrix and Its Effect on Final Physical Properties of Crumb Rubber-Modified Binder. *Transportation Research Record: Journal of the Transportation Research Board*, 2013. 2370: 92–101.
5. Ragab, M., M. Abdelrahman, and A. Ghavibazoo. Performance Enhancement of Crumb Rubber-Modified Asphalts Through Control of the Developed Internal Network Structure. *Transportation Research Record: Journal of the Transportation Research Board*, 2013. 2371: 96–104.
6. Farshidi, F., D. Jones, and J. T. Harvey. *Warm-Mix Asphalt Study: Evaluation of Rubberized Hot- and Warm-Mix Asphalt with Respect to Emissions.* Research Report: UCPRC-RR-2013-03. University of California Pavement Research Center, Davis, Calif., 2013.
7. Baumgardner, G. L., and G. R. Reinke. Binder Additives for Warm Mix Asphalt Technology. *Journal of the Association of Asphalt Paving Technologists*, No. 82, 2013, pp. 685–709.
8. Bennert, T., G. Reinke, W. Mogawer, and K. Mooney. Assessment of Workability and Compactability of Warm-Mix Asphalt. *Transportation Research Record: Journal of the Transportation Research Board*, 2010. 2180: 36–47.
9. Hanz, A., A. Faheem, E. Mahmoud, and H. Bahia. Measuring Effects of Warm-Mix Additives. *Transportation Research Record: Journal of the Transportation Research Board*, 2010. 2180: 85–92.
10. Wang, H., X. Liu, P. Apostolidis, and T. Scarpas. Review of Warm Mix Rubberized Asphalt Concrete: Towards a Sustainable Paving Technology. *Journal of Cleaner Production*, Vol. 177, 2018, pp. 302–314.
11. Rubio, M. C., G. Martínez, L. Baena, and F. Moreno. Warm Mix Asphalt: An Overview. *Journal of Cleaner Production*, Vol. 24, 2012, pp. 76–84.
12. Lesueur, D. The Colloidal Structure of Bitumen: Consequences on the Rheology and on the Mechanisms of Bitumen Modification. *Advances in Colloid and Interface Science*, Vol. 145, No. 1–2, 2009, pp. 42–82.
13. Zofka, A., M. Chrysochoou, I. Yut, C. Johnston, M. Shaw, S.-P. Sun, J. Mahoney, S. Farquharson, and M. Donahue. *SHRP2 Report S2-R06B-RR-1: Evaluating Applications of Field Spectroscopy Devices to Fingerprint Commonly Used Construction Materials.* Transportation Research Board of the National Academies, Washington, D.C., 2013.
14. Zanzotto, L., J. Stastna, and S. Ho. Molecular Weight Distribution of Regular Asphalts From Dynamic Material Functions. *Materials and Structures*, Vol. 32, No. 3, 1999, pp. 224–229.
15. Themeli, A., E. Chailleux, F. Farcas, C. Chazallon, and B. Migault. Molecular Weight Distribution of Asphaltic Paving Binders From Phase-Angle Measurements. *Road Materials and Pavement Design*, Vol. 16, No. sup1, 2015, pp. 228–244.
16. Salomon, D., and H. Zhai. Ranking Asphalt Binders by Activation Energy For Flow. *Journal of Applied Asphalt Binder Technology*, Vol. 2, No. 2, 2002, pp. 52–60.
17. Haider, S. W., M. W. Mirza, A. K. Thottempudi, J. Bari, and G. Y. Baladi. Characterizing Temperature Susceptibility of Asphalt Binders Using Activation Energy for Flow. In *Transportation and Development Institute Congress 2011: Integrated Transportation and Development for a Better Tomorrow*, 2011, pp. 493–503.
18. García-Morales, M., P. Partal, F. J. Navarro, F. Martínez-Boza, C. Gallegos, N. González, O. González, and M. E. Muñoz. Viscous Properties and Microstructure of Recycled Eva Modified Bitumen. *Fuel*, Vol. 83, No. 1, 2004, pp. 31–38.
19. Biro, S., T. Gandhi, and S. Amirkhanian. Determination of Zero Shear Viscosity of Warm Asphalt Binders. *Construction and Building Materials*, Vol. 23, No. 5, 2009, pp. 2080–2086.
20. De Visscher, J., and A. Vanelstraete. Practical Test Methods for Measuring the Zero Shear Viscosity of Bituminous Binders. *Materials and Structures*, Vol. 37, No. 5, 2004, pp. 360–364.

21. Partal, P., F. Martınez-Boza, B. Conde, and C. Gallegos. Rheological Characterisation of Synthetic Binders and Unmodified Bitumens. *Fuel*, Vol. 78, No. 1, 1999, pp. 1–10.
22. Ait-Kadi, A., B. Brahimi, and M. Bousmina. Polymer Blends for Enhanced Asphalt Binders. *Polymer Engineering & Science*, Vol. 36, No. 12, 1996, pp. 1724–1733.
23. Zanzotto, L., J. Stastna, and K. Ho. Characterization of Regular and Modified Bitumens Via Their Complex Modulus. *Journal of Applied Polymer Science*, Vol. 59, No. 12, 1996, pp. 1897–1905.
24. Tsenoglou, C. Molecular Weight Polydispersity Effects on the Viscoelasticity of Entangled Linear Polymers. *Macromolecules*, Vol. 24, No. 8, 1991, pp. 1762–1767.
25. Tuminello, W. H. Molecular Weight and Molecular Weight Distribution From Dynamic Measurements of Polymer Melts. *Polymer Engineering & Science*, Vol. 26, No. 19, 1986, pp. 1339–1347.
26. Jamshidi, A., M. O. Hamzah, and Z. You. Performance of Warm Mix Asphalt containing Sasobit®: State-of-the-Art. *Construction and Building Materials*, Vol. 38, 2013, pp. 530–553.
27. Jamshidi, A., M. O. Hamzah, Z. Shahadan, and A. S. Yahaya. Evaluation of the Rheological Properties and Activation Energy of Virgin and Recovered Asphalt Binder Blends. *Journal of Materials in Civil Engineering*, Vol. 27, No. 3, 2015.
28. Attia, M., and M. Abdelrahman. Enhancing the Performance of Crumb Rubber-Modified Binders Through Varying the Interaction Conditions. *International Journal of Pavement Engineering*, Vol. 10, No. 6, 2009, pp. 423–434.
29. Airey, G. D. Use of Black Diagrams to Identify Inconsistencies in Rheological Data. *Road Materials and Pavement Design*, Vol. 3, No. 4, 2011, pp. 403–424.
30. Celauro, B., C. Celauro, D. Lo Presti, and A. Bevilacqua. Definition of a Laboratory Optimization Protocol for Road Bitumen Improved with Recycled Tire Rubber. *Construction and Building Materials*, Vol. 37, 2012, pp. 562–572.
31. Bahia, H. U., D. I. Hanson, M. Zeng, H. Zhai, M. A. Khatri, and R. M. Anderson. *NCHRP Report 459: Characterization of Modified Asphalt Binders in Superpave Mix Design*. Transportation Research Board of the National Academies, Washington D.C., 2001.
32. Domin, M., A. Herod, R. Kandiyoti, J. W. Larsen, M. Lazaro, S. Li, and P. Rahimi. A Comparative Study of Bitumen Molecular-Weight Distributions. *Energy & Fuels*, Vol. 13, No. 3, 1999, pp. 552–557.
33. Collins, E., and A. Metzger. Polyvinylchloride Melt Rheology II—The Influence of Molecular Weight on Flow Activation Energy. *Polymer Engineering & Science*, Vol. 10, No. 2, 1970, pp. 57–65.

The Standing Committee on Asphalt Binders (AFK20) peer-reviewed this paper (18-02433).

Article

TRR
JOURNAL OF THE TRANSPORTATION RESEARCH BOARD

Transportation Research Record
2018, Vol. 2672(28) 349–358

DOI: 10.1177/0361198118784138
journals.sagepub.com/home/trr
SAGE

Evaluation of Chemical and Rheological Aging Indices to Track Oxidative Aging of Asphalt Mixtures

Farhad Yousefi Rad[1], Michael D. Elwardany[2], Cassie Castorena[3], and Y. Richard Kim[3]

Abstract

Oxidative age hardening in asphalt binder leads to embrittlement. Embrittled asphalt is prone to fatigue and thermal cracking. Therefore, the ability to predict asphalt binder oxidative age hardening within a pavement throughout its service life could inform improved pavement material selection, design, and maintenance practices. Studying the evolution of oxidative aging requires the use of key properties to track oxidation levels, termed aging index properties (AIPs) here. The objective of this study is to identify suitable rheological and chemical AIPs to track oxidation levels in asphalt materials. A wide range of laboratory and field aged materials were evaluated in this study. A range of chemical AIPs determined by Fourier transform infrared spectroscopy (FTIR) absorbance peaks and areas were evaluated based on their correlation with laboratory aging duration. Rheological AIPs were evaluated based on the strength of their relationship to the chemical changes induced by oxidation. The rheological AIPs evaluated included the dynamic shear modulus, zero shear viscosity, Glover-Rowe parameter, and crossover modulus. The chemical AIP evaluation that most strongly correlated with laboratory aging duration is the carbonyl plus the sulfoxide absorbance peaks. The results indicate that both the dynamic shear modulus and Glover-Rowe parameter constitute rheological AIPs that relate directly to the chemical changes induced by oxidation.

Asphalt binder is an organic end product that reacts with oxygen in the environment. The oxidation of asphalt binder is a physicochemical reaction that causes permanent hardening and reduces the flexibility of the asphalt binder. Consequently, oxidation increases a pavement's susceptibility to cracking under traffic loading and thermal stress. Thus, given the importance of asphalt binder oxidation to pavement performance, it is imperative to define practical aging index properties (AIPs) that can be used effectively to determine oxidation levels. The AIPs then can be used to calibrate kinetics diffusion models and, by extension, be used within pavement performance prediction models. For an AIP to be effective, it must be easy to quantify and must relate directly to oxidative aging in an asphalt binder. AIPs generally are divided into two main categories: chemical functional groups and rheological parameters. Although numerous AIPs have been proposed in the literature, a comprehensive paper that examines the chemistry-based and rheology-based AIPs that best correlate to binder oxidation levels is currently missing.

The chemistry-based AIPs provide the most direct indicators of oxygen uptake and can be measured using Fourier transform infrared spectroscopy (FTIR) techniques. Carbonyls and sulfoxides are the two primary chemical functional groups that are produced on oxidation of asphalt binder. The carbonyl chemical functional group has long been used to indicate the level of oxidation within asphalt binder (*1*). The ketone functional group is the major component of the carbonyl infrared absorption region. Ketone formation increases the polarity of the associated aromatic ring components within an asphalt binder and consequently increases its viscosity (*1*, *2*). The production of sulfoxides by the oxidation of asphalt has received less attention in the past. However, Petersen and Glaser conducted a thorough study to understand the effect of sulfoxide formation on oxidative age hardening in asphalt binders (*2*). Their work demonstrated that the sulfoxides produced during oxidation have a significant impact on viscosity, especially for high sulfur content asphalts. Chemical AIPs have been represented by both absorbance peaks (e.g., *1*, *2*) and areas (e.g., *3–6*).

Different rheology-based AIPs also have been introduced and include, for example, the crossover modulus, low shear viscosity, and so on (*1*, *7*, *8*). Rheological AIPs are advantageous over chemical AIPs because they do not require

[1]Anton Paar, Ashland, VA
[2]Western Research Institute, Laramie, WY
[3]Department of Civil, Construction, and Environmental Engineering, North Carolina State University, Raleigh, NC

Corresponding Author:
Address correspondence to Y. Richard Kim: kim@ncsu.edu

Table 1. Summary of Different Aging Trials

Material source	Binder	Aging temperature (°C)	Aging duration
NC	PG 64-22	70	10, 12, 35 days
		85	8, 10, 15 days
		95	10, 17, 26 days
FHWA ALF control	PG 70-22	70	10, 35 days
		95	12, 14, 18 days
FHWA ALF-SBS-LG	PG 70-28 (SBS-modified)	70	10, 35 days
		85	8, 10, 20 days
		95	8, 10, 18, 22 days
SHRP AAD-1	PG 58-28	70	10, 18, 21 days
		95	10, 15, 27 days
SHRP AAG-1	PG 58-10	70	10, 18 days
		95	10, 15, 22, 27 days
WesTrack	PG 64-22	70	12, 18 days
		95	10, 15, 20 days

infrared spectroscopy devices and can be correlated more directly to the mechanical properties of an asphalt mixture than chemical AIPs (*9*). However, a comprehensive study to determine the rheological properties that relate most directly to the chemical changes caused by oxidation is currently missing in the literature.

Objectives

The objectives of this study are to:

- Evaluate different chemical AIPs based on their sensitivity to oxidation.
- Evaluate the relationship between rheological AIPs and the chemical changes induced by oxidation.
- Identify suitable chemical and rheological AIPs for tracking the oxidation of asphalt.

Test Method and Analysis Procedure

Table 1 presents details about the six asphalt mixtures evaluated and the aging trials conducted in this study. The materials were selected to encompass a wide range of binder types and different aggregates across the United States. All of the mixtures were subjected to laboratory loose mixture aging and were fabricated using original, component materials.

To evaluate the changes in the chemical and rheological AIPs in terms of oxidation, the loose mixtures were subjected to laboratory aging at multiple temperatures and for various durations in an oven under ambient pressure. To prepare the loose mixtures for long-term aging, the component materials were combined according to the original mix design and subjected to short-term aging at 135°C for 4 hours in accordance with AASHTO R 30 (*10*). A mix design was not available for the SHRP AAD-1 and SHRP AAG-1 binders. Therefore, these SHRP binders were combined with the FHWA ALF (Federal Highway Administration Accelerated Load Facility) aggregate according to the FHWA ALF mix design.

The short-term aged loose mixture was spread in a single layer in metal trays such that the thickness of the mix was equal to or less than the largest aggregate size in the mix. The thin layer of loose mixture was used to ensure homogeneous exposure to the heat and oxygen during the laboratory aging process. Moreover, it was assumed that using a thin layer of asphalt loose mix would expedite oxidative aging in the binder film, as most of the mixture was in direct contact with circulating hot air. Three aging temperatures, 70°C, 85°C, and 95°C, were used for long-term aging loose mixtures. The 95°C is the maximum temperature without inducing changes in chemical and rheological AIPs' correlations based on previous studies (*1*, *11*). The 85°C is the temperature used in AASHTO R30, and 70°C is roughly the maximum temperature that pavements reach in the U.S. Samples were collected at different time intervals for binder extraction and recovery and subsequent AIP testing to track the oxidation levels. The micro-extraction and recovery method developed by the Western Research Institute (WRI) (*12*) was used to obtain the binder samples from the aged loose mixtures.

Micro-Extraction and Recovery

Qualitative micro-extraction and recovery of the asphalt binder from the asphalt mixtures was accomplished following a proposed AASHTO specification prepared by the WRI (*12*). This procedure uses a solvent mixture of toluene and ethanol (85:15) for extraction and recovery. The mixture sample size is limited to 200 g to produce approximately 10 g of asphalt binder per extraction, which is adequate for both the FTIR with attenuated total reflectance (FTIR-ATR) sampling technique and dynamic shear rheometer (DSR) testing. To prevent further aging of the binder sample during the extraction and recovery procedure, a rotary evaporator

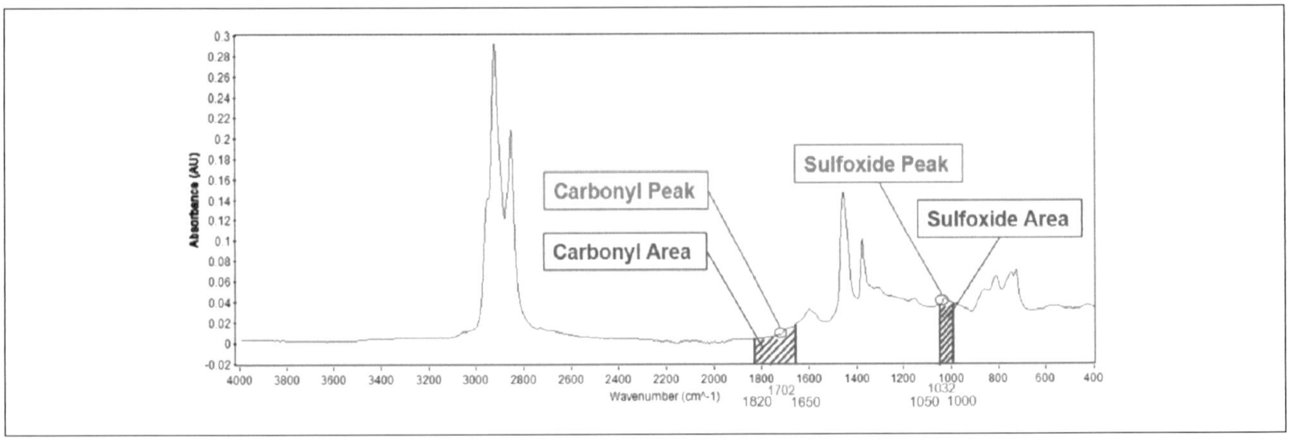

Figure 1. Example of IR spectrum data for asphalt binder and chemical AIPs calculation.

balloon was subjected to vacuum pressure of 80.0 ± 0.7 kPa (600 ± 5 mm of mercury) under nitrogen gas. The extraction and recovery process consists of the application of temperatures ranging from 70°C to 170°C at different stages. FTIR-ATR testing was conducted following extraction and recovery to ensure that no residual solvent was evident. The presence of toluene was tracked by the absorbance peaks at wavenumbers 730 and 694 cm^{-1}.

Measurement of Aging Index Properties (AIPs)

The chemical and rheological AIPs were measured using FTIR-ATR spectrometry and DSR tests, respectively. The micro-extraction and recovery process was conducted to collect asphalt binder from loose mixtures for the AIP measurements. Summaries of the FTIR and DSR procedures follow.

FTIR-ATR Spectroscopy. FTIR-ATR spectroscopy allows absorbance data to be collected within a wide spectral range (400–4000 cm^{-1}). Peaks in the spectra can be used to detect the presence of chemical functional groups. In this study, the ATR spectra were collected using 64 scans at a resolution of 4 cm^{-1} with a minimum of two replicates. For each binder, all replicates under different conditions were normalized to the same absorbance value at wave number 1375 cm^{-1}. This wave number was selected for normalization because absorbance is not affected by the level of oxidation. The chemical AIPs evaluated in this study are the carbonyl area, carbonyl plus sulfoxide area (C+S area), and carbonyl plus sulfoxide peaks (C+S peaks). Changes in the carbonyl and sulfoxides peaks were tracked at wave numbers 1702 and 1032 cm^{-1}, respectively. The carbonyl areas were determined according to the area under the FTIR absorbance curves from wave numbers 1650 to 1820 cm^{-1}. The sulfoxide areas were determined according to the area under the FTIR absorbance curve from wave numbers 1000 to 1050 cm^{-1}. The trapezoidal rule was used to determine numerically the area under the band between the specified wave number ranges. It should be noted that the absorbance peaks are direct measurements taken from the FTIR spectrum band, whereas the area-based measurements require numerical integration under the absorbance spectrum. The absorbance peak-based AIP is the easiest AIP to determine and involves less error or bias with regard to the method of calculation compared with the area-based AIPs. Figure 1 represents an example of an asphalt binder's infrared (IR) data and the wavelengths at which chemical AIPs are measured.

Dynamic Shear Rheometer Procedure. Frequency sweep testing was conducted at frequencies ranging from 0.1 to 30 Hz at multiple temperatures (5°C, 20°C, 35°C, 50°C, and 64°C) in an ARG-2 DSR from TA Instruments. The tests were conducted using an 8-mm parallel plate spindle with a strain amplitude of 1%. The four rheological AIPs that were calculated using the DSR are described in the following.

Dynamic Shear Modulus (G).* The dynamic shear modulus (G^*) is a property that can be measured directly and used to track the rheological changes caused by oxidation. The use of the G^* as an AIP is advantageous because it is a direct output of the DSR and is the most commonly used parameter that relates the physical properties of asphalt binder to asphalt pavement performance. This study found that oxidative age hardening affects the G^* most significantly at high temperatures or low frequencies, which is demonstrated by the asphalt binder mastercurves presented in Figure 2. The results shown in Figure 2 correspond to a single-source asphalt binder (FHWA ALF-SBS) that had been conditioned to produce various age levels. The figure shows that at lower reduced frequencies (and, correspondingly, higher temperatures), the differences among the G^* values are most significant. Therefore, the G^* value at 64°C and 10 rad/s was selected as one of the rheological AIPs to evaluate here, which corresponds to a typical measure determined in Superpave Performance Grading.

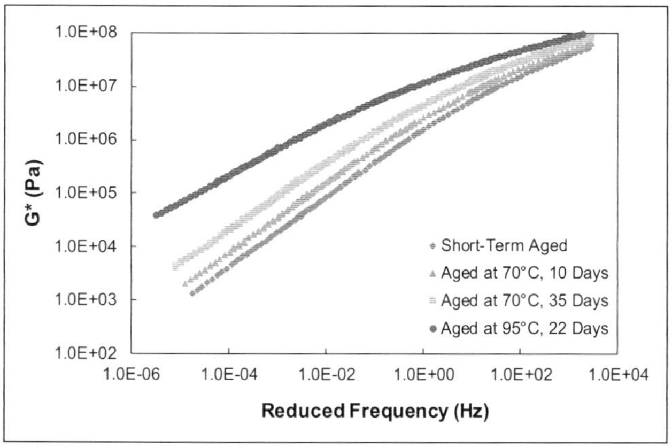

Figure 2. Comparison between dynamic shear modulus (G^*) mastercurves for asphalt binders extracted and recovered from FHWA ALF-SBS asphalt mixture aged at different levels.

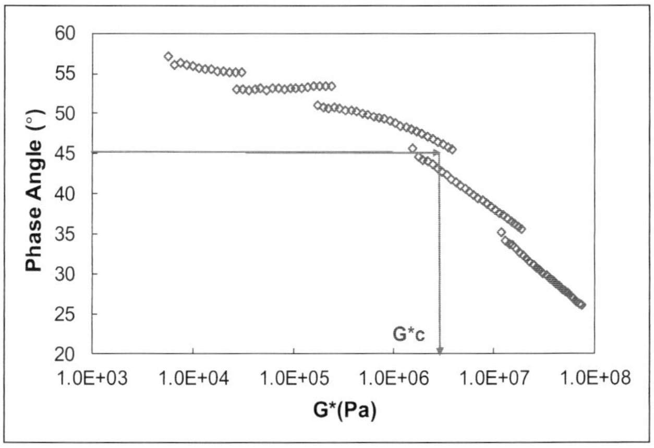

Figure 3. Crossover modulus (G^*c) values for binders extracted and recovered from FHWA ALF-SBS loose mix after 20 days of aging at 85°C.

Crossover Modulus (G^*c). The crossover modulus (G^*c) is defined as the G^* value that corresponds to the reduced frequency at which the storage modulus (G') and loss modulus (G'') mastercurves cross (i.e., where the phase angle equals 45°). The crossover modulus is a parameter that is found in the widely-applied Christensen-Anderson (CA) model (*13*) and has been proposed as an AIP to study the evolution of oxidative aging in asphalt binders (*8*). Several different methods have been proposed for calculating the crossover modulus (e.g., data smoothing, interpolation, and curve fitting) (*8*). In this study, the method developed by Farrar et al. (*8*) was applied. This method determines the crossover modulus as follows:

1. Examine the data in black space and remove any spurious points. Such points are particularly troublesome at the highest test frequencies.
2. Use the Christensen-Anderson-Marasteanu (CAM) model to pre-smoothen the G^* data and obtain mastercurves for both the G^* and phase angle.
3. Sort the measured data points according to reduced frequency in ascending order.
4. Apply a polynomial fit of the black space plot (phase angle versus logarithm of G^*) to the as-measured data in the neighborhood of 45° (40°–50°).
5. Interpolate the polynomial fit for the crossover modulus by finding the value of G^* when the phase angle is exactly 45°.

Figure 3 presents the crossover modulus values estimated from a black space diagram for the binder extracted and recovered from the FHWA ALF-SBS loose mixture after 20 days of conditioning at 85°C. As shown in Figure 3, the binder is not perfectly thermorheologically simple, thus leading to differences in the data obtained from different isotherms. Therefore, the estimation of the crossover modulus value requires interpolation between the data that correspond to different isotherms within the black space diagram. Consequently, the determination of the crossover modulus may be subject to more variability than is the case for parameters that are direct measurements (e.g., G^*).

Oxidative aging leads to a decrease in the crossover modulus (G^*c) value. To achieve a positive relationship with the aging duration, the inverse of the log of the crossover modulus ($1/\log G^*c$) is used as a rheological AIP (*8*). Farrar et al. reported that a linear relationship is found between the oxygen uptake and $1/\log G^*c$ for a given binder (*8*).

Zero Shear Viscosity (ZSV). Zero shear viscosity (ZSV) also was considered as a rheological AIP in this study. ZSV is defined as the viscosity when the shear rate reaches zero (*14*, *15*). ZSV can be determined from creep or frequency sweep testing. In this study, frequency sweep test results were used to estimate ZSV values using the method proposed by Marasteanu and Anderson (*16*). To apply their method, the CAM model, (i.e., Equation 1), was fitted to the frequency sweep test results. If the parameter m in the CAM model is larger than 1, the CAM model is shifted to a reference temperature of 60°C by dividing the crossover modulus value by the time–temperature shift factor that corresponds to 60°C.

$$G^* = \frac{G_g^*}{[1+(f/f_c)^k]^{m/k}} \quad (1)$$

where

G^*_g = glassy modulus (Pa),

f_c = location parameter with dimensions of frequency (rad/s),

f = reduced frequency (rad/s),

m = CAM model shift factor, and

k = CAM model parameter.

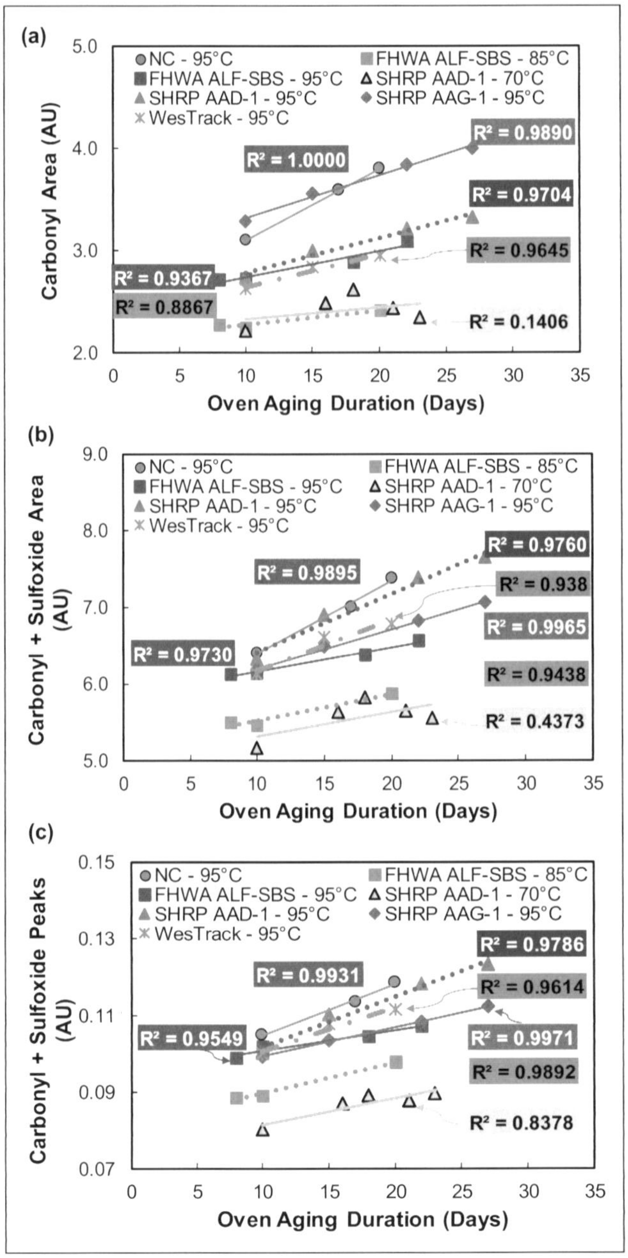

Figure 4. Sensitivity of different chemical AIPs to aging duration: (*a*) carbonyl area; (*b*) C+S area; (*c*) C+S peaks.

Then, the Cox-Merz rule, shown in Equation 2, is applied to convert the G^* value to viscosity. The ω is the frequency.

$$G^*/\omega = \text{Viscosity} \tag{2}$$

Lastly, the limit of the function, as the frequency goes to zero, is used to obtain the ZSV, as shown in Equation 3.

$$\text{ZSV} = \frac{G_g^*}{\omega_{c,\text{refT}=60°\text{C}}}\left(\frac{(m-1)^{m-1}}{m^m}\right)^{1/k} \tag{3}$$

where $\omega_{c,\,\text{refT}=60°\text{C}}$ is the crossover frequency at the reference temperature (rad/s).

If $m < 1$, then the CAM model is refitted to the data with $m = 1$, and the same algorithm is applied.

Glover-Rowe (G-R) Parameter. The Glover-Rowe (G-R) parameter was also evaluated in this study as a rheological AIP. The G-R parameter was proposed by Rowe based on his analysis of Glover et al.'s (*17*) work on low temperature ductility (*18*). Glover et al. (*17*) used the Maxwell mechanical analog model to derive a rheological parameter that is related to ductility, that is, $G'/(\eta'/G')$, termed the "Glover parameter." This parameter has been proposed by Glover et al. (*17*) as an indicator of pavement cracking potential. Rowe et al. (*19*) simplified Glover's parameter to arrive at

$$\frac{G^*(\cos\delta)^2}{\sin\delta} \tag{4}$$

where

G^* = dynamic shear modulus measured at 15°C and 0.005 rad/s (Pa) and

δ = phase angle measured at 15°C and 0.005 rad/s (°).

Testing at 0.005 rad/s is not possible. Therefore, the G^* and phase angle values at a reduced frequency that corresponds to 15°C and 0.005 rad/s were determined from the G^* and phase angle mastercurves constructed from temperature–frequency sweep DSR testing and frequency–temperature shift factors.

Results and Discussion

Evaluation of Chemical AIPs

The chemical AIPs were evaluated based on their correlation to the aging duration. Asphalt materials exhibit relatively similar kinetics, with an initial fast reaction period (also known as the spurt), followed by a slower reaction period that has an approximately constant rate (*2, 4, 9, 20–22*). The focus here is long-term aging. Therefore, the chemical AIPs were evaluated within the constant rate period because this period corresponds to higher age levels.

Figure 4 presents comparisons of the three chemical AIPs evaluated (i.e., carbonyl area, C+S area, and C+S peaks) in arbitrary units (AU) with respect to aging duration (note that S stands for sulfoxide). The rate of oxidation is temperature-dependent. Therefore, the chemical AIPs measured from long-term aging at different temperatures for a given mixture are plotted separately. All of the data included in the analysis of the chemical AIPs correspond to aging trials in which extracted and recovered binders were analyzed for at least three aging durations to compare the correlations. Conditions with fewer than three aging conditions were excluded from the evaluation. As shown in Figure 4, generally all three chemical AIPs are strongly correlated with the aging duration. However, it

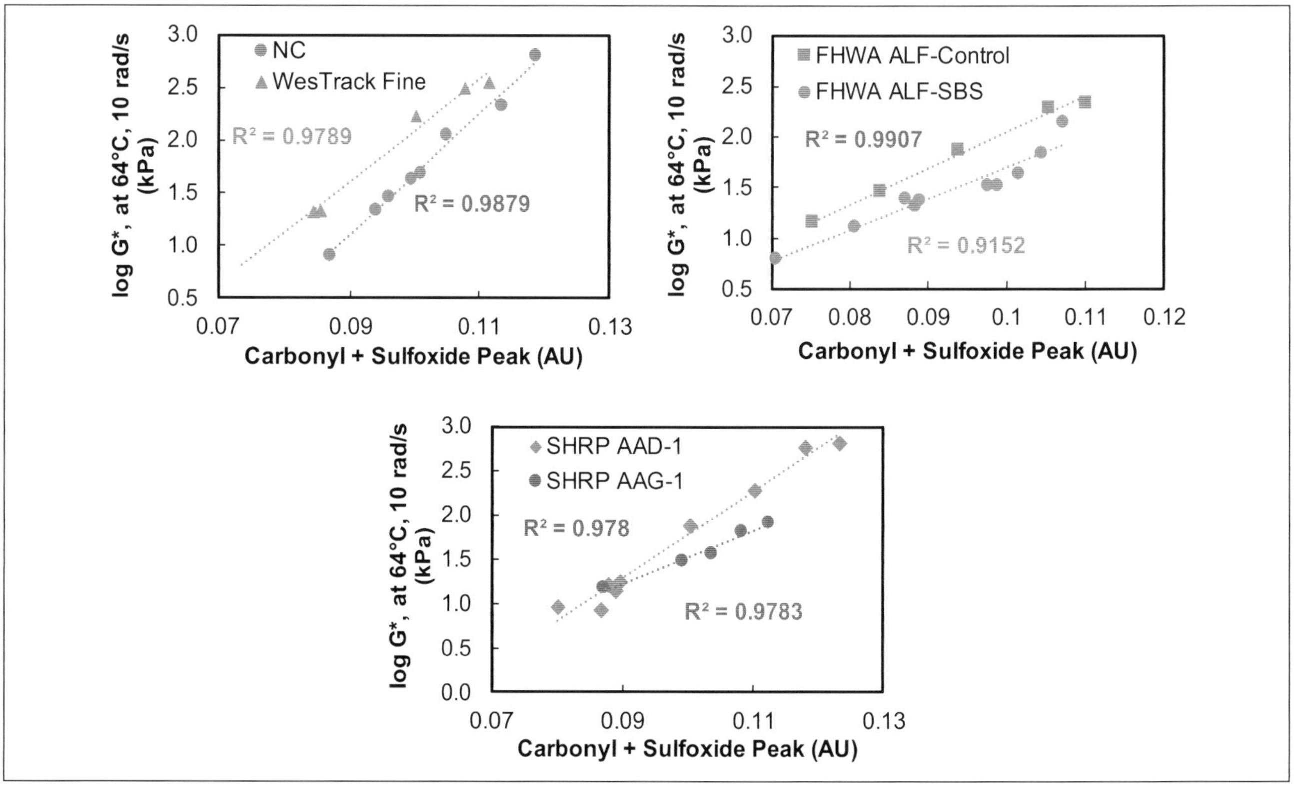

Figure 5. Correlations between log G^* at 64°C, 10 rad/s and C+S peaks for six mixtures.

can be noted that the effect of the sulfoxide functional group on the chemical AIP-based aging rates is of great importance. Although the C+S area and C+S peaks show very similar aging rates and rankings, the carbonyl area versus aging duration shows a different ranking of the materials. The literature indicates that sulfoxides have a significant effect on rheology, and thus this observation suggests that it is important to consider the sulfoxide functional group when tracking oxidative aging. The C+S peaks versus aging duration graph exhibits the overall highest R^2 values. Furthermore, the C+S peaks can be calculated using the direct output of the ATR-FTIR data, whereas the calculation of the C+S area requires numerical integration under the infrared spectrum. Therefore, C+S peaks is an easy parameter to calculate and is not sensitive to the method chosen for the calculation. Thus, in this study C+S peaks was selected as the most promising chemical AIP identified and was used in the subsequent analysis of the rheological AIPs.

Evaluation of Rheological AIPs

The correlation between the rheological AIPs and C+S peaks for each mixture was used to evaluate the rheological AIPs. Data that correspond to different aging temperatures ranging from 70°C to 95°C were included in the evaluation. Past studies have demonstrated that the relationship between chemistry and rheology is not affected by aging temperature if the temperature is below 100°C (*1, 11*). Moreover, the linear relationships between chemical parameters and aging duration presented in Figure 4 suggests that the aging durations are past the spurt region of aging and can, therefore, be studied in a linear regime. Therefore, the data for multiple aging temperatures are included for each mixture evaluated.

Figure 5 presents the correlations between log G^* at 64°C and 10 rad/s and C+S peaks for all of the mixtures evaluated. Each data point corresponds to a different aging duration or aging temperature. A log scale of the rheological parameters was used to produce a linear relationship to C+S peaks. The results demonstrate that G^* at 64°C and 10 rad/s is highly correlated with the chemical changes induced by oxidation for all of the mixtures evaluated.

Figure 6 presents the correlations between 1/log G^*c and C+S peaks. The results demonstrate that the crossover modulus is highly correlated with the chemical changes caused by oxidation. However, for certain highly aged mixtures, including the NC mix aged at 95°C for 20 days, FHWA ALF-SBS aged at 95°C for 28 days, and SHRP AAD-1 aged at 95°C for 22 and 27 days, the G^*c could not be estimated because of calculation difficulties and high variability. These difficulties and variability are caused by differences between the results from different isotherms in the black space, which led to inconsistent crossover modulus values from two test replicates. In addition, Figure 6 shows that the slopes of the 1/log G^*c versus C+S peaks lines are very low for many of the mixtures evaluated. Thus, although a good correlation exists,

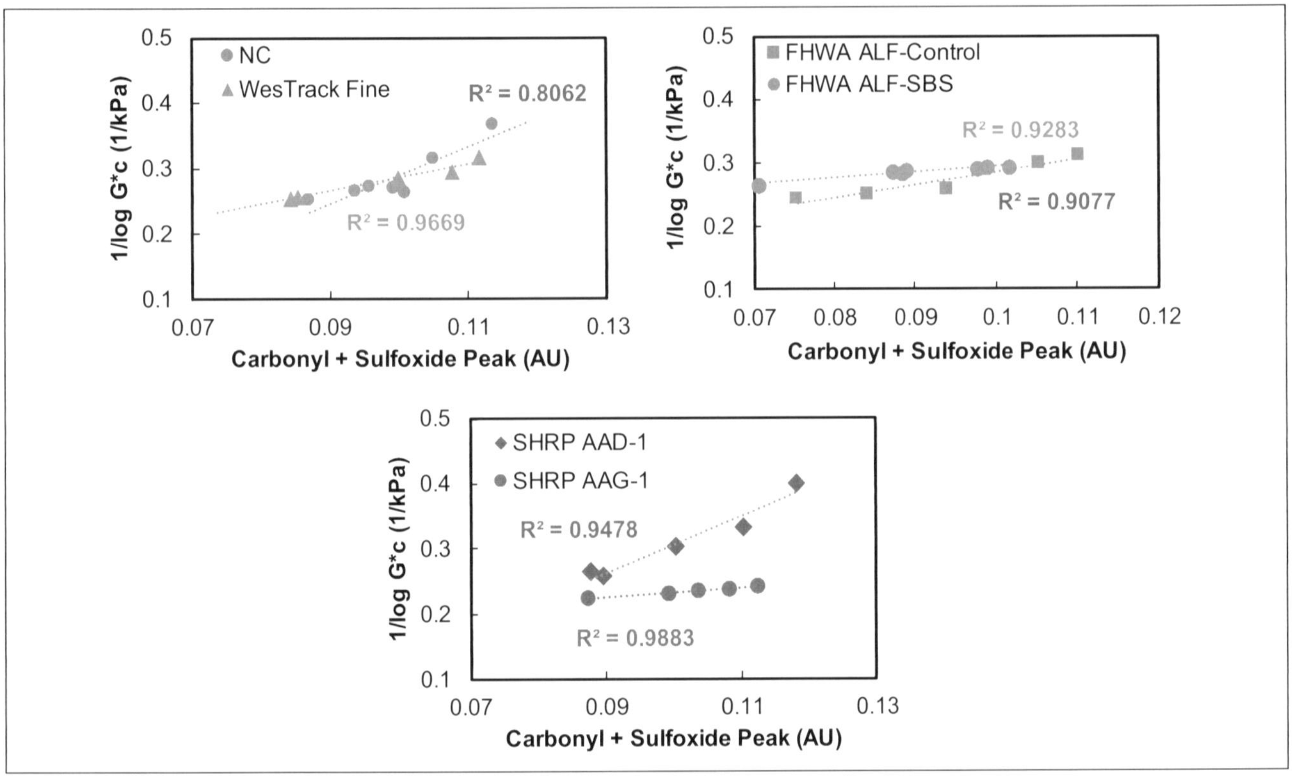

Figure 6. Correlations between log 1/G^*c and C+S peaks for six mixtures.

the crossover modulus changes relatively little as the C+S peaks values change, thereby indicating that the crossover modulus is not very sensitive to oxidation.

Figure 7 presents the correlation between the ZSV at 60°C and C+S peaks for each mixture evaluated. It is noted that relatively significant differences among the ZSV values for different test replicates were observed for some asphalt binder samples, which was attributed to the variability involved in the extrapolation process to estimate the ZSV from G^* mastercurves. The same data that caused the variability in the ZSV values were used to calculate other rheological AIPs and resulted in considerably less variability, indicating the weakness of the ZSV parameter as a reliable AIP. These problematic data were not included in Figure 7. The results in Figure 7 demonstrate that ZSV is highly correlated with C+S peaks for all of the mixtures evaluated with the exception of FHWA ALF-SBS. Note that the FHWA ALF-SBS mixture is the only mixture in this study that contains polymer-modified binder.

Figure 8 presents the correlation between the G-R parameter and C+S peaks for each of the mixtures evaluated. The results demonstrate that the G-R parameter is highly correlated with the chemical changes that result from oxidation for all of the mixtures evaluated.

Based on the results presented, the most promising rheological AIPs for tracking oxidation levels are G^* at 64°C and 10 rad/s and the G-R parameter. The ZSV and crossover modulus values proved problematic in some instances. The problems associated with ZSV and the crossover modulus are related to the interpolation and extrapolation of the data. In addition, the crossover modulus is less sensitive to the chemical changes caused by oxidation compared with the other rheological parameters evaluated. With the exception of the crossover modulus, all of the rheological AIPs evaluated correspond to low reduced frequencies. Low reduced frequencies are achieved by evaluating the properties either at a high temperature or at slow rates (i.e., low frequencies). Although the G-R parameter was evaluated at 15°C, it was evaluated at a very low frequency (0.005 rad/s) and, therefore, constitutes a relatively low reduced frequency within the mastercurve. For example, the condition of 15°C and 0.005 rad/s is equivalent to approximately 45°C and 10 rad/s based on typical binder time–temperature shift factors. Therefore, the G-R parameter also represents high-temperature characteristics of the asphalt binder.

Figure 9 presents the correlations between log G^* 64°C and 10 rad/s and log G-R. Figure 9 shows that G^* at 64°C and 10 rad/s and the G-R parameter are highly correlated, indicating that either can be used effectively as an AIP. The G-R parameter requires frequency sweep testing at multiple temperatures and the fitting of both a mastercurve and time–temperature shift factor function to the rheological data because testing at 0.005 rad/s is not possible. In contrast, G^* at 64°C and 10 rad/s is a single-point measurement that can be acquired directly

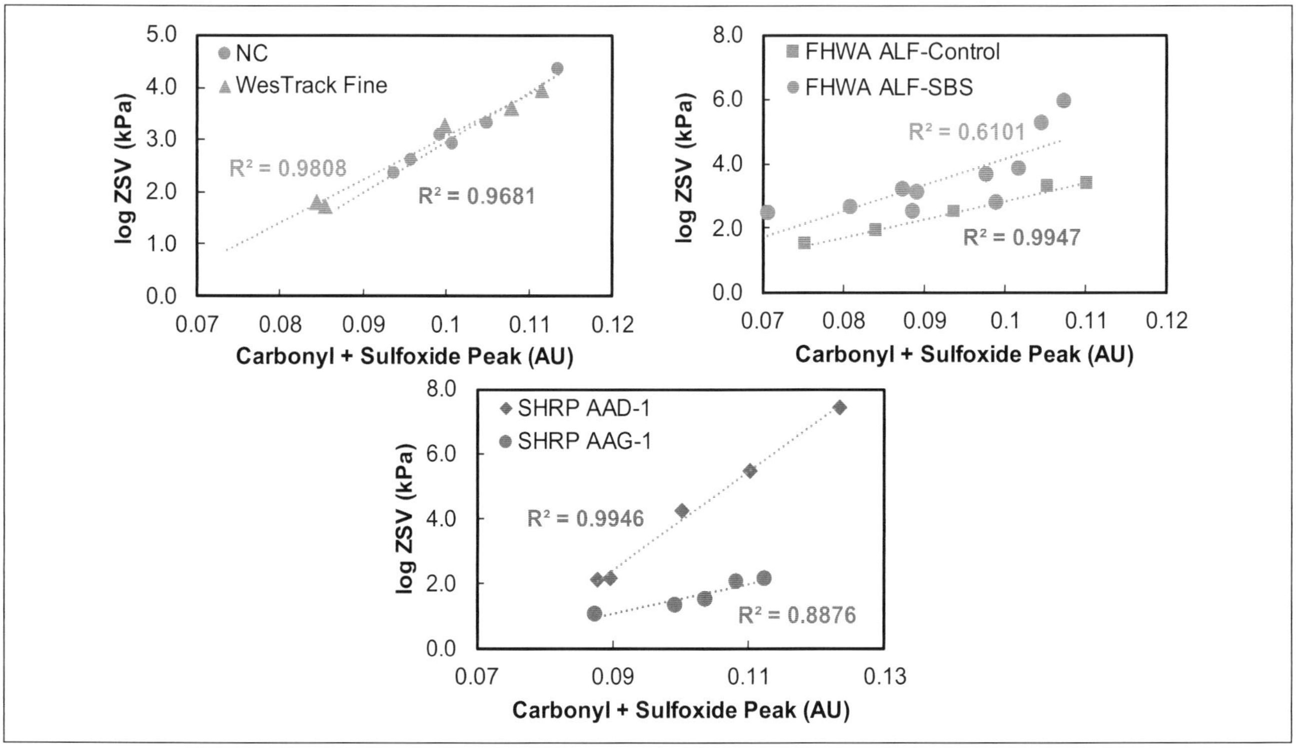

Figure 7. Correlations between log ZSV at 60°C and C+S peaks for six mixtures.

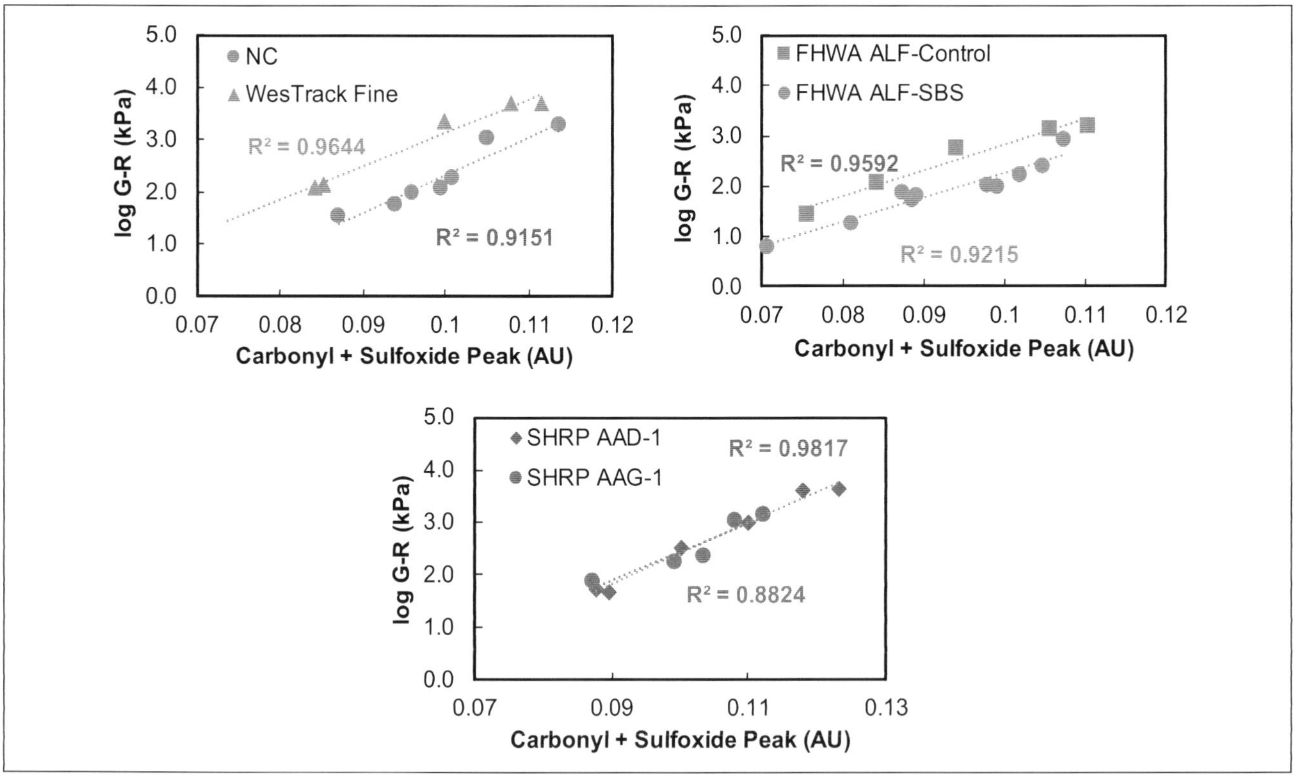

Figure 8. Correlations between log G-R and C+S peaks for six mixtures.

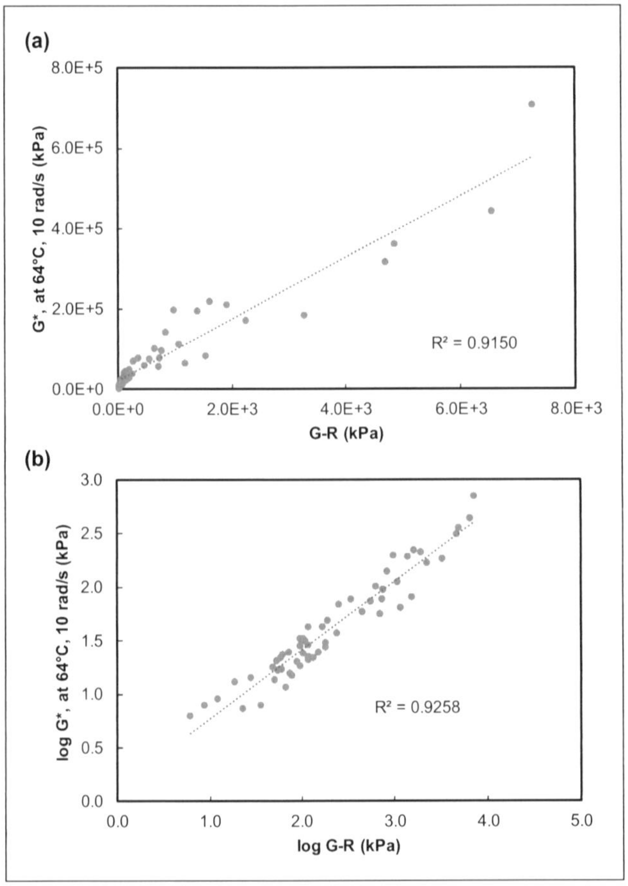

Figure 9. Correlation between G^* at 64°C and 10 rad/s and *G-R* parameter in (*a*) arithmetic scale and (*b*) log-log scale.

from testing at a single temperature and frequency. Therefore, the G^* at 64°C and 10 rad/s constitutes the most efficient and effective rheological AIP of those evaluated.

Conclusions and Recommendations

The main conclusions and recommendations drawn from this study are:

- Comparing the oxidation rates of different materials based on the combined effects of carbonyl and sulfoxide can lead to significantly different conclusions than when considering carbonyl alone.
- C+S peaks is an effective chemical AIP for tracking oxidation and can be obtained directly from FTIR-ATR spectrometry data.
- The two most promising rheological AIPs for tracking oxidation levels are G^* at 64°C and 10 rad/s and the G-R parameter.
- The ZSV and crossover modulus AIPs are problematic in some instances because of the required interpolation or extrapolation (or both) of the data used to determine the AIP values.
- The G^* at 64°C and 10 rad/s and the G-R parameter are highly correlated, indicating that either can be used effectively as an AIP. However, the G^* at 64°C and 10 rad/s is a more direct output of DSR.

Author Contributions

The authors confirm contribution to the paper as follows: study conception and design: Y. Richard Kim, Cassie Castorena, Farhad Yousefi Rad, Michael Elwardany; data collection: Farhad Yousefi Rad, Michael Elwardany; analysis and interpretation of results: Farhad Yousefi Rad, Michael Elwardany, Cassie Castorena, Y. Richard Kim; draft manuscript preparation: Farhad Yousefi Rad, Michael Elwardany, Cassie Castorena, Y. Richard Kim. All authors reviewed the results and approved the final version of the manuscript.

References

1. Petersen, J. C. *Transportation Research Circular E-C140: A Review of the Fundamentals of Asphalt Oxidation: Chemical, Physicochemical, Physical Property, and Durability Relationships*. Transportation Research Board of the National Academies, Washington, D.C., 2009.
2. Petersen, J. C., and R. Glaser. Asphalt Oxidation Mechanisms and the Role of Oxidation Products on Age Hardening Revisited. *Road Materials and Pavement Design*, Vol. 12, No. 4, 2011, pp. 795–819.
3. Han, R. *Improvements to a Transport Model of Asphalt Binder Oxidation in Pavements: Pavement Temperature Modeling, Oxygen Diffusivity in Asphalt Binders and Mastics, and Pavement Air Void Characterization*. Doctoral dissertation. Texas A&M University, College Station, TX, 2011.
4. Jin, X., R. Han, Y. Cui, and C. J. Glover. Fast-Rate–Constant-Rate Oxidation Kinetics Model for Asphalt Binders. *Industrial & Engineering Chemistry Research*, Vol. 50, No. 23, 2011, pp. 13373–13379.
5. Lamontagne, J., P. Dumas, V. Mouillet, and J. Kister. Comparison by Fourier Transform Infrared (FTIR) Spectroscopy of Different Ageing Techniques: Application to Road Bitumens. *Fuel*, Vol. 80, No. 4, 2001, pp. 483–488.
6. Zhang, F., J. Yu, and J. Han. Effects of Thermal Oxidative Ageing on Dynamic Viscosity, TG/DTG, DTA and FTIR of SBS- and SBS/Sulfur-Modified Asphalts. *Construction and Building Materials*, Vol. 25, No. 1, 2011, pp. 129–137.
7. Underwood, B. S., M. S. Sakhaei Far, and Y. R. Kim. Using Limited Purchase Specification Tests to Perform Full Viscoelastic Characterization of Asphalt Binder. *ASTM Journal of Testing and Evaluation*, Vol. 38, No. 5, 2010, pp. 558–566.
8. Farrar, M. J., T. F. Turner, J. P. Planche, J. F. Schabron, and P. M. Harnsberger. *Evolution of the Crossover Modulus with Oxidative Aging*. Transportation Research Board, Washington, D.C., 2013.
9. Glaser, R., J. Schabron, T. Turner, J. P. Planche, S. Salmans, and J. Loveridge. Low-Temperature Oxidation Kinetics of Asphalt Binders. *Transportation Research Record: Journal of the Transportation Research Board*, 2013. 2370: 63–68.
10. American Association of State Highway and Transportation Officials (AASHTO). *Standard Practice for Mixture*

Conditioning of Hot Mix Asphalt. AASHTO R 30, Washington, D.C., 2010.

11. Yousefi Rad, F., M. D. Elwardany, C. Castorena, and Y. R. Kim. Investigation of Proper Long-Term Laboratory Aging Temperature for Performance Testing of Asphalt Concrete. *Construction and Building Materials*, Vol. 147, 2017, pp. 616–629.
12. Farrar, M. J., R. W. Grimes, S. Wiseman, and J. P. Planche. *Asphalt Pavement: Micro-Sampling and Micro-Extraction Methods, Fundamental Properties of Asphalts and Modified Asphalts III*. Quarterly Technical Report, Federal Highway Administration (FHWA), Contract No. DTFH61-07-D-00005, 2015.
13. Christensen, D. W., and D. A. Anderson. Interpretation of Dynamic Mechanical Test Data for Paving Grade Asphalt Cements (with Discussion). *Journal of the Association of Asphalt Paving Technologists*, Vol. 61, 1992, pp. 67–116.
14. Binard, C., D. Anderson, L. Lapalu, and J. P. Planche. Zero Shear Viscosity of Modified and Unmodified Binders. *Proc., 3rd Eurasphalt & Eurobitume Congress*, Vienna, Vol. 2, 2004.
15. Biro, S., T. Gandhi, and S. Amirkhanian. Determination of Zero Shear Viscosity of Warm Asphalt Binders. *Construction and Building Materials*, Vol. 23, No. 5, 2009, pp. 2080–2086.
16. Marasteanu, M. O., and D. A. Anderson. Determination of Asphalt Binders Viscosity from Other Rheological Parameters. *Proc., Ninth International Conference on Asphalt Pavements*, Copenhagen, Denmark, 2002.
17. Glover, C. J., R. R. Davison, C. H. Domke, Y. Ruan, P. Juristyarini, D. B. Knorr, and S. H. Jung. *Development of a New Method for Assessing Asphalt Binder Durability with Field Validation*. Federal Highway Administration (FHWA) Report, No. FHWA/TX-05/1872-2, 2005.
18. King, G., M. Anderson, D. Hanson, and P. Blankenship. Using Black Space Diagrams to Predict Age-Induced Cracking. *Proc., 7th RILEM International Conference on Cracking in Pavements*, Springer, the Netherlands, 2012, pp. 453–463.
19. Rowe, G. M., G. King, and M. Anderson. The Influence of Binder Rheology on the Cracking of Asphalt Mixes in Airport and Highway Projects. *Journal of Testing and Evaluation*, Vol. 42, No. 5, 2014, pp. 1–10.
20. Petersen, J. C. A Dual, Sequential Mechanism for the Oxidation of Petroleum Asphalts. *Petroleum Science and Technology*, Vol. 16, No. 9–10, 1998, pp. 1023–1059.
21. Petersen, J. C., and P. M. Harnsberger. Asphalt Aging: Dual Oxidation Mechanism and Its Interrelationships with Asphalt Composition and Oxidative Age Hardening. *Transportation Research Record: Journal of the Transportation Research Board*, 1998. 1638: 47–55.
22. Petersen, J. C., P. M. Harnsberger, and R. E. Robertson. Factors Affecting the Kinetics and Mechanisms of Asphalt Oxidation and the Relative Effects of Oxidation Products on Age Hardening. Preprints of Papers, American Chemical Society, Division of Fuel Chemistry, Vol. 41, No. CONF-960807, 1996.

The Standing Committee on Asphalt Binders (AFK20) peer-reviewed this paper (18-03234).

Article

TRR
JOURNAL OF THE TRANSPORTATION RESEARCH BOARD

Transportation Research Record
2018, Vol. 2672(28) 359–371

DOI: 10.1177/0361198118790639
journals.sagepub.com/home/trr
SAGE

Selecting a Laboratory Loose Mix Aging Protocol for the NCAT Top-Down Cracking Experiment

Chen Chen[1], Fan Yin[1], Pamela Turner[1], Randy C. West[1], and Nam Tran[1]

Abstract

The objective of this study was to select a laboratory loose mix aging protocol for the NCAT top-down cracking (TDC) experiment. Field aging of asphalt mixtures was characterized using cumulative degree-days (CDD), which was defined as the accumulation of daily high temperature throughout mixtures' service life. Performance data from over 80 existing pavements showed that TDC typically initiated after a critical field aging condition of approximately 70,000 CDD. A laboratory experiment was conducted to select an aging protocol that was representative of this critical CDD. Materials used in the experiment were from five projects in Michigan, Washington, and Alabama. Four loose mix aging protocols were evaluated in terms of their effects on the rheological and oxidation properties of asphalt binders. Results from the dynamic shear rheometer (DSR), bending beam rheometer (BBR), and Fourier transform infrared spectroscopy (FT-IR) tests showed that the 24-h, 135°C protocol yielded the most significant level of asphalt aging, followed by the 12-h, 135°C protocol, 5-day, 95°C protocol, and 6-h, 135°C protocol, respectively. No significant difference in the oxidation–hardening relationship of asphalt binders was observed for mixes aged at 95°C versus 135°C. Among the four aging protocols, the 5-day, 95°C protocol was most representative of 70,000 CDD of field aging. Additionally, DSR and FT-IR results indicated that loose mix aging of 8-h at 135°C and 5 days at 95°C were likely to achieve an equivalent aging level; thus, the 8-h, 135°C protocol was recommended as an alternative protocol to simulate 70,000 CDD of field aging.

Top-down cracking (TDC) has been widely reported as a common type of distress in asphalt pavements. Different from traditional bottom-up cracking, TDC initiates at the surface of the asphalt pavement and progresses downward to the bottom of the asphalt layers. TDC is typically longitudinal, with surface crack widths around 3–4 mm and decreasing with pavement depth (*1*, *2*). Over the past few decades, several factors have been identified that contribute to the development of TDC. These factors include high bending-induced surface tension and shear-induced near-surface tension caused by tire–pavement interactions, age hardening of asphalt binder, and lower stiffness of upper hot-mix asphalt (HMA) layers (*3*). Despite several cracking tests being currently available for evaluating the cracking potential of asphalt mixtures, there is not a consensus within the asphalt pavement industry on the most appropriate method to address TDC.

In 2015, the National Center for Asphalt Technology (NCAT) partnered with the Minnesota Department of Transportation's Road Research facility (MnROAD) for a national-level pavement-cracking study. The overall objective of the study is to validate laboratory cracking tests by establishing correlations between test results and measured cracking in real pavements using real loading conditions. Test sections were built on the NCAT Test Track and MnROAD to evaluate TDC and thermal cracking tests, respectively. Seven mixtures with a wide range of cracking potentials were included in the experiment that focused on TDC evaluation (hereinafter referred to as the NCAT TDC experiment). Each mixture was constructed as the 1.5-inch surface layer on the NCAT Test Track on top of highly polymer modified base and binder layers. The base and binder layers were intentionally designed to be highly elastic to mitigate the

[1]National Center for Asphalt Technology, Auburn, AL

Corresponding Author:
Address correspondence to Fan Yin: f-yin@auburn.edu

occurrence of bottom-up cracking (*4*, *5*). The laboratory experiment consisted of five cracking tests: energy ratio, Texas overlay, NCAT-modified overlay, semi-circular bend (SCB), and Illinois flexibility index (I-FIT) tests. Considering the importance of age hardening of asphalt mixtures in the development of TDC, both laboratory-mixed laboratory-compacted (LMLC) and plant-mixed laboratory-compacted (PMLC) specimens were short-term conditioned and long-term aged before testing.

The standard practice for laboratory long-term oven aging (LTOA) per AASHTO R 30 is to condition a compacted specimen for 5 days at 85°C. This protocol, which was initially developed in a pre-SHRP study by Bell et al. (*6*), is anticipated to simulate field aging of asphalt pavements over 7–10 years of service. However, the National Cooperative Highway Research Program (NCHRP) project 09-52 reported that this protocol was only representative of approximately one or two years of field aging in warmer and colder climates, respectively (*7*). The conclusion was obtained based on the resilient modulus (M_R) and Hamburg wheel tracking test (HWTT) results of LMLC specimens and field cores for over 40 different asphalt mixtures. Similar findings were also reported by Islam et al. (*8*) and Howard and Doyle (*9*) in bending beam fatigue (BBF) and Cantabro tests, respectively.

In addition to conditioning compacted specimens, LTOA can be conducted by aging asphalt loose mix before compaction. The latter protocol typically yields a more severe level of aging following increased exposure of asphalt mix to oxygen and elevated temperature. The loose mix aging process can be further accelerated using a higher temperature without concerns of specimen distortion (i.e., changes in specimen air voids and geometry) (*10*). Over the past few years, several loose mix aging protocols have been used by different researchers. Braham et al. (*11*) conditioned the loose mix at 135°C for 24 h and evaluated its effect on mixture fracture energy from the disk-shaped compact tension (DCT) test. In another study conducted by Reinke et al. (*12*), loose mixes from three MnROAD test sections were conditioned at 135°C for 12 and 24 h before being compacted for mixture testing. NCHRP project 09-54 evaluated loose mix aging at multiple temperatures, ranging from 70°C to 135°C (*13*). It was reported that, for certain asphalt binders, there was a significant change in the relationship between binder rheology and chemistry when the aging temperature increased from 95°C to 135°C. A reduction in mixture fatigue resistance was also observed for mixtures aged at 135°C. Therefore, the authors recommended an optimal loose mix aging temperature of 95°C.

Although the above-mentioned aging protocols seemed promising for use in mix design and performance testing, their correlations with field aging have not been identified. In addition, the traditional expectation of LTOA to simulate 7–10 years of field aging may not be appropriate to evaluate TDC, since most of these cracks were reported to develop within a significantly shorter period (i.e., 3–5 years) (*1*, *2*, *14*). Therefore, the objectives of this study were to: (1) identify a critical field aging condition under which TDC starts to develop; and (2) select a representative laboratory aging protocol to age asphalt mixtures for the NCAT TDC experiment.

Identification of Critical Field Aging for TDC

Pavement in-service time at the time of coring was a commonly used parameter to quantify field aging of asphalt mixtures. This parameter, however, failed to differentiate the aging of pavements with different climates and construction dates. Generally, field aging is more severe in warmer climate than in colder climate given the same aging time. In NCHRP project 09-52, the cumulative degree-days (CDD) parameter was proposed to overcome this shortcoming (*7*). As expressed in Equation 1, CDD was defined as the accumulation of the daily high temperature above freezing for all the days being considered from the time of construction to the time of coring. Compared to the parameter of pavement in-service time, CDD has the advantage of considering both temperature and time when characterizing the field aging of asphalt pavements:

$$\text{CDD} = \left\{ \begin{array}{l} \sum (T_{\text{dmax}} - 32), \quad \text{if } T_{\text{dmax}} \geq 32 \\ 0, \quad \text{if } T_{\text{dmax}} < 32 \end{array} \right\} \quad (1)$$

where T_{dmax} is daily maximum temperature, °F.

In the recently completed NCHRP project 09-49A, the long-term field performance of 53 warm-mix asphalt (WMA) and 30 HMA pavement sections were monitored (*15*). Two rounds of pavement surveys were conducted in 2012/2013 and 2014/2015, respectively, to collect information on field distresses, including rutting, transverse cracking, and wheel-path longitudinal cracking. Figure 1 presents the wheel-path longitudinal crack length versus the corresponding pavement CDD at the time of pavement surveys.

For the first survey results, shown in Figure 1*a*, pavements with fewer than 70,000 CDD showed no cracking, while those with more than 70,000 CDD had measurable cracking (crack length over 20 feet). A similar trend was also observed for the second survey results, as shown in Figure 1*b*. Examination of field cores sampled from pavements with wheel-path longitudinal cracks showed that these cracks initiated from the pavement surface and stopped within the asphalt layer, which was indicative of top-down fatigue cracking (*15*). Based on the survey results in Figure 1, TDC seemed to develop in asphalt

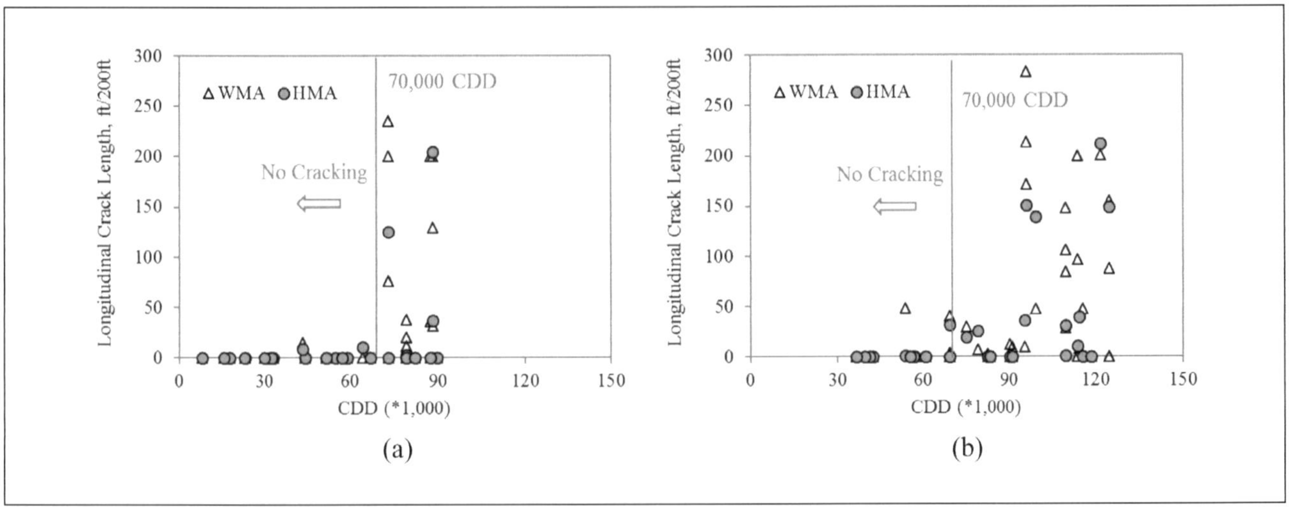

Figure 1. Wheel-path longitudinal crack survey results: (*a*) first survey in 2012/2013; (*b*) second survey in 2014/2015 (*15*).

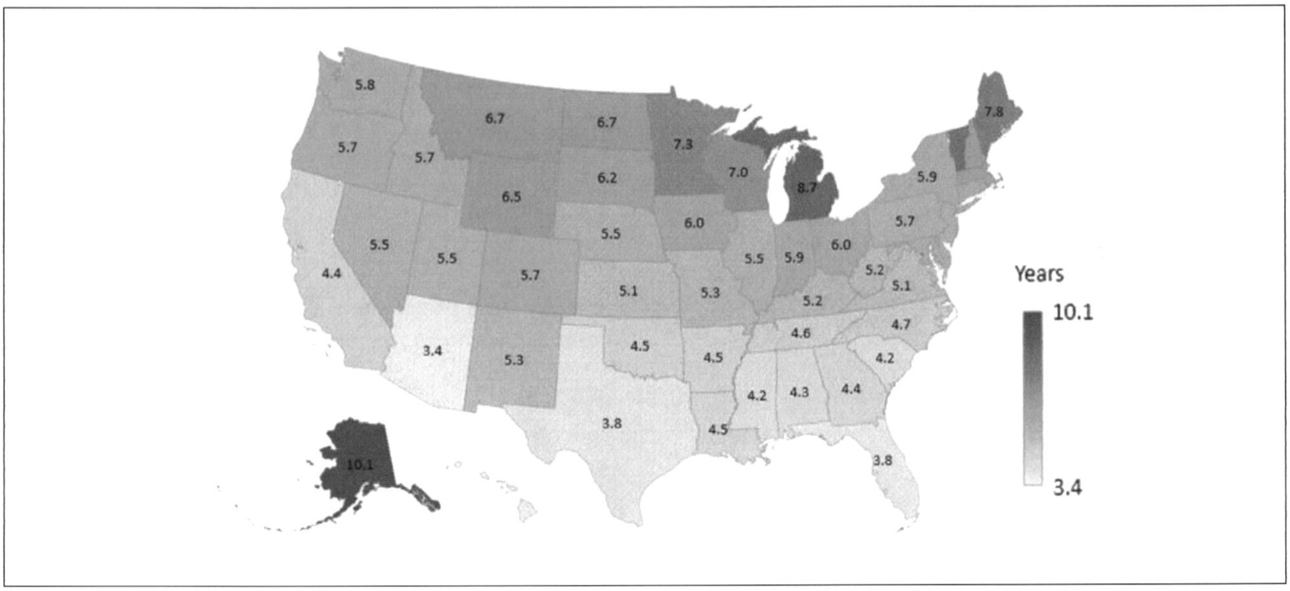

Figure 2. A map showing the number of years to reach 70,000 CDD.

pavements with more than approximately 70,000 CDD of field aging. A similar finding was also reported by the 2009 NCAT Test Track research group experiment, in which pavement cracks were found to start developing approximately four years after construction in Alabama weather (65,000 CDD) (*16*). Thus, 70,000 CDD was proposed as a preliminary critical field aging condition to evaluate TDC. Figure 2 presents a map showing the number of years required in different states to reach this preliminary CDD value. In general, warm-climate states (e.g., Alabama, Florida, Texas) need approximately 4 years to achieve 70,000 CDD, while a significantly longer time of up to 8 years is required for states with cold climates (e.g., Michigan, Minnesota, Wisconsin).

Experimental Design to Select Representative Aging Protocols

Figure 3 presents the research methodology used to select a laboratory aging protocol to simulate 70,000 CDD of field aging. Materials used in the experiment were from five field projects in Washington, Michigan, and Alabama. For each field project, plant production mix and post-construction field cores were sampled and tested to characterize field aging. Four loose mix aging protocols of 5 days at 95°C, 6 h at 135°C, 12 h at 135°C, and 24 h at 135°C were evaluated. In this study, only binder testing was conducted because of the limited number of field cores available. Asphalt binders extracted

Table 1. Summary of Filed Projects

	Mix ID				
	Mix 1	Mix 2	Mix 3	Mix 4	Mix 5
Location	Rapid River, Michigan	Walla Walla, Washington	Auburn, Alabama	Auburn, Alabama	Auburn, Alabama
Mixture type	HMA	HMA	HMA	HMA	HMA
Virgin binder	PG 52-34	PG 64-28	PG 76-22 (SBS)	PG 67-22	PG 67-22
NMAS	12.5 mm	12.5 mm	9.5 mm	12.5 mm	12.5 mm
Gradation	Dense-graded	Dense-graded	Dense-graded	Dense-graded	Dense-graded
RAP	17%	20%	0%	20%	0%
Construction date	July 2010	April 2010	July 2009	September 2006	July 2000
Post-construction coring date	September 2016	October 2016	October 2016	October 2016	January 2015
In-service time	6 years, 2 months	6 years, 6 months	7 years, 3 months	10 years, 1 month	14 years, 6 months
CDD	48,000	80,000	114,000	157,000	235,000

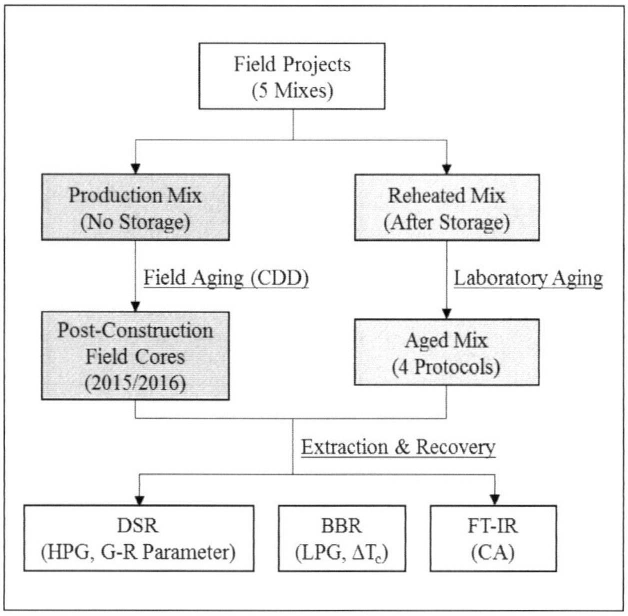

Figure 3. Research methodology.

and recovered from plant mix and field cores were tested in the dynamic shear rheometer (DSR) and bending beam rheometer (BBR), and using Fourier transform infrared spectroscopy (FT-IR), to determine their rheological and oxidation properties. Test results were analyzed to compare the different aging protocols and determine their correlations with field aging.

Materials

Materials used in the experiment were from five field projects that covered a wide range of pavement ages, climatic conditions, and mixture components. These projects were selected based on the availability of plant loose mix that were sampled during production. Two projects were from the NCHRP project 09-47A; one was on US-12 in Walla Walla, Washington, and the other was on County Road 513 near Rapid River, Michigan. The other three projects were from test sections on the NCAT Test Track, but they were constructed in different research cycles. Table 1 provides a brief summary of mixture components and construction and coring information for the five field projects.

Sample Preparation

For each field project, sufficient amounts of plant mix were sampled during production and were stored in the NCAT laboratory. For specimen preparation, buckets with plant mix were first placed in an oven at 150°C for approximately 3 h. Afterwards, the reheated mix was batched into individual samples of approximately 2,500 g, which were then spread out in shallow pans and placed back in the oven for further aging at the four conditions noted above. Mix samples aged for 5 days at 95°C were stirred once every 24 h to ensure aging uniformity. As mentioned previously, post-construction field cores were included in the experiment to represent field aging. Considering the non-uniform aging of asphalt pavements with depth (*10*, *17*, *18*), only the top 1 inch of the field cores were tested. In this study, asphalt binders were extracted (using trichloroethylene) and recovered from reheated plant mixes, laboratory aged mixes, and post-construction field cores in accordance with AASTHO T164 and ASTM D 5404, respectively. The extracted and recovered binders were then subjected to DSR, BBR, and FT-IR testing without additional aging using the rolling thin-film oven (RTFO) and pressure aging vessel (PAV). It should be noted that the impact of asphalt extraction and recovery on the rheological and chemical properties of asphalt binders was not considered in the study.

Laboratory Binder Tests

Dynamic Shear Rheometer. DSR was used to characterize the rheological properties of asphalt binders by measuring the complex shear modulus (G^*) and phase angle (δ) at a specific temperature and frequency. Extracted asphalt binders were first tested to determine the continuous high-temperature performance grade (HPG) per AASHTO R29 and M 320. In addition, the limited DSR frequency sweep test was conducted at three temperatures of 20, 30, and 40°C, and with an angular frequency range of 0.1 to 10 rad/s. The maximum oscillation strain was controlled at 1% to ensure the asphalt binder was in the linear viscoelastic range. For data analysis, G^* and δ master curves were constructed by fitting the individual G^* and δ results at each temperature and frequency to the Christensen–Anderson–Marasteanu (CAM) model (*19*), as expressed in Equation 2. The master curves were then utilized to calculate the Glover–Rowe (G-R) parameter using Equation 3 (*20*, *21*). The G-R parameter considers both binder stiffness and embrittlement and offers an indication of the cracking potential at intermediate temperature. Asphalt binders with higher G-R parameters are expected to have experienced a greater level of aging than those with lower G-R parameters:

$$G^* = G_g\left[1 + \left(\frac{\omega_c}{\omega}\right)^v\right]^{-\frac{w}{v}}$$
$$\delta = \frac{90w}{\left[1 + \left(\frac{\omega}{\omega_c}\right)^v\right]} \tag{2}$$

where:

G_g = glass modulus, assumed equal to 1 GPa
ω_c = crossover frequency
ω = reduced frequency
v and w = model coefficients

$$\text{G-R parameter} = \left\{\frac{G^*\left[\cos(\delta)^2\right]}{\sin(\delta)}\right\}_{T = 15^\circ C, f = 0.005\ \text{rad/s}} \tag{3}$$

Bending Beam Rheometer. A BBR was used to characterize the resistance of asphalt binders to thermal cracking at low temperatures. The test was conducted to determine the low-temperature performance grade (LPG) based on the creep stiffness (S) and m-value results per AASHTO T313. The ΔT_c parameter, which is defined as the numerical difference between the continuous LPG determined from the S criteria and the grade determined from the m-value criteria (*22*), was also determined. The ΔT_c parameter is indicative of the cracking potential of asphalt binders; specifically, binders with a more negative ΔT_c are more likely to be susceptible to block cracking than those with a less negative ΔT_c.

Fourier Transform Infrared Spectroscopy. FT-IR was used to measure the infrared spectrum of asphalt binders and determine their compositional changes with aging. The test was conducted using a Nicolet 6700 FT-IR spectrometer with attenuated total reflectance (ATR). The carbonyl area (CA), defined as the integrated peak area for the wavelength range from 1,820 to 1,650 cm^{-1} (*23*), was used to evaluate the oxidation level of asphalt binders extracted from mixes with various aging protocols. Asphalt binders with higher CA are expected to have experienced a greater level of oxidative aging than those with lower CA.

Tests Results and Data Analysis

This section presents the DSR, BBR, and FT-IR results of asphalt binders extracted from laboratory aged mixes and post-construction field cores. Test results were analyzed to evaluate the effects of laboratory loose mix aging on binder rheological and oxidation properties. Additionally, the correlation of field aging with these aging protocols was explored. Finally, protocols that were representative of 70,000 CDD of field aging were selected for the NCAT TDC experiment.

Comparisons of Laboratory Aging Protocols

Continuous Performance Grade Results. Figure 4 presents the continuous HPG and LPG results of asphalt binders extracted from reheated and laboratory aged mixes. As shown in Figure 4*a*, for all five mixes, asphalt binders from laboratory aged mixes showed consistently higher HPG than those from the corresponding reheated mixes, which indicated that the aging protocols evaluated in this study yielded significant levels of asphalt aging. The average increase in the continuous HPG for different aging protocols varied from 8.4 to 24.1. As expected, longer aging times at 135°C showed greater increases in the continuous HPG. For all mixes except Mix 4, the 5-day, 95°C protocol yielded a slightly higher increase in the HPG than the 6-h, 135°C protocol. The average difference in the continuous HPG between these two aging protocols was approximately 1.1, which was not considered practically significant. Similar trends were also observed for the continuous LPG results in Figure 4*b*. In general, the 24-h, 135°C protocol yielded the greatest increase in both HPG and LPG, followed by the 12-h, 135°C protocol, 5-day, 95°C protocol, and 6-h, 135°C protocol, respectively.

Cracking Parameter Results. Figure 5 presents the G-R parameter results in Black space, where the G^* at 15°C and 0.005 rad/s of extracted asphalt binders were plotted against the corresponding δ values. A consistent trend

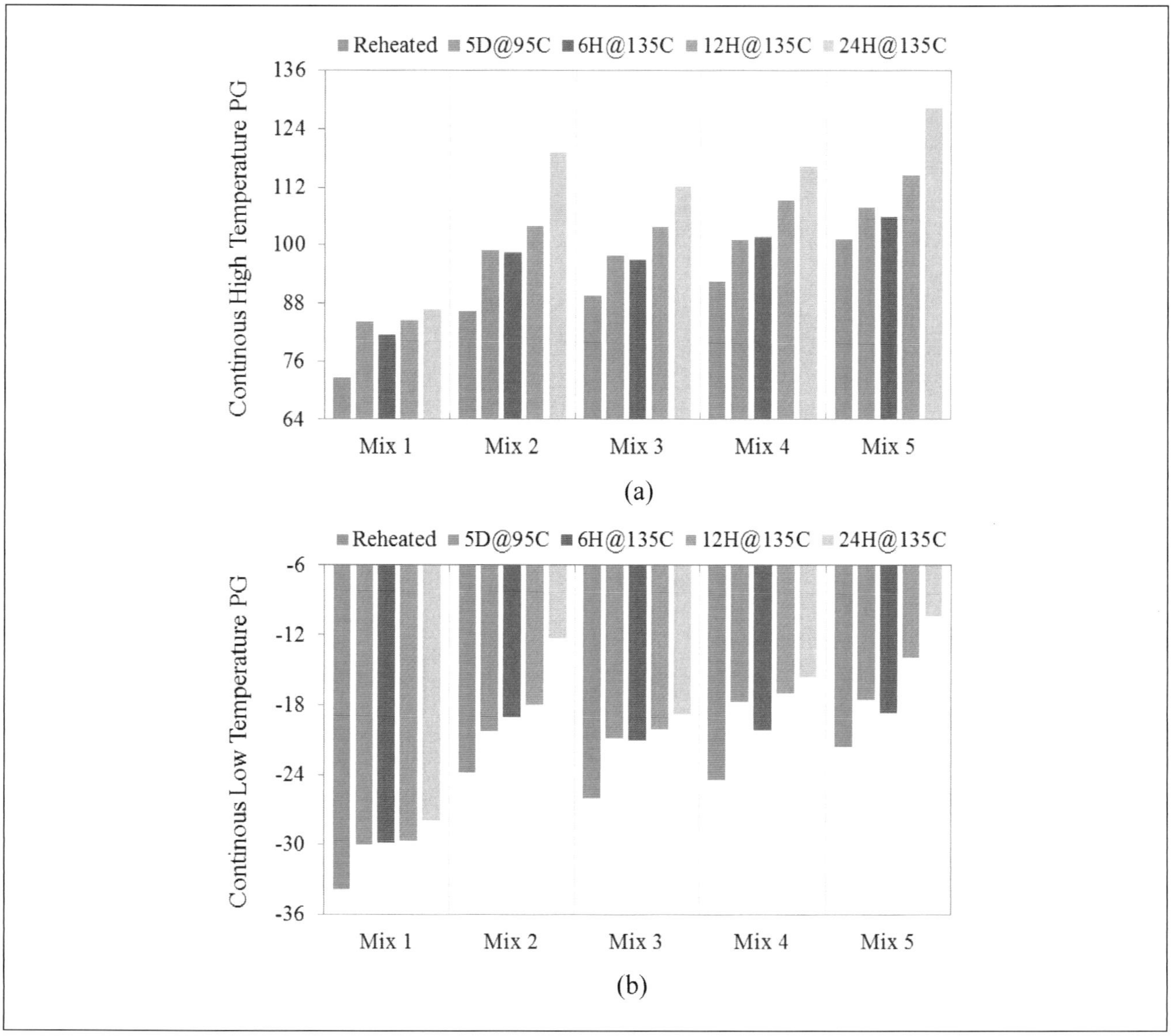

Figure 4. Continuous PG results of extracted asphalt binders with various loose mix aging protocols: (*a*) high-temperature; (*b*) low temperature.

was observed for all mixes: Asphalt binders extracted from laboratory aged mixes were located closer to the upper left corner on the Black space diagram than were those from the reheated mixes, which indicated that asphalt binders after aging had increased stiffness and loss of relaxation properties. Based on the "traveled" distances on the Black space diagram, the 24-h, 135°C protocol produced the most significant level of asphalt aging, followed by the 12-h, 135°C protocol, 5-day, 95°C protocol, and 6-h, 135°C protocol, respectively. These results were in agreement with the continuous PG results in Figure 4.

Figure 6 presents the BBR ΔT_c results of asphalt binders extracted from reheated and laboratory aged mixes. As mentioned previously, asphalt binders with a more negative ΔT_c are more susceptible to cracking owing to reduced relaxation properties. As can be seen, the ΔT_c of asphalt binders from laboratory aged mixes was consistently lower (i.e., more negative) than those of the corresponding reheated mixes. The 12-h and 24-h, 135°C protocols exhibited the two lowest ΔT_c values for all mixes. No consistent trend was found for the ΔT_c comparisons between the 5-day, 95°C and 6-h, 135°C protocols; but in most cases, the difference in the ΔT_c between these two aging protocols was less than 1.0.

FT-IR CA Results. Figure 7 presents the FT-IR CA results of asphalt binders extracted from reheated and laboratory aged mixes. In general, the results fell in line with

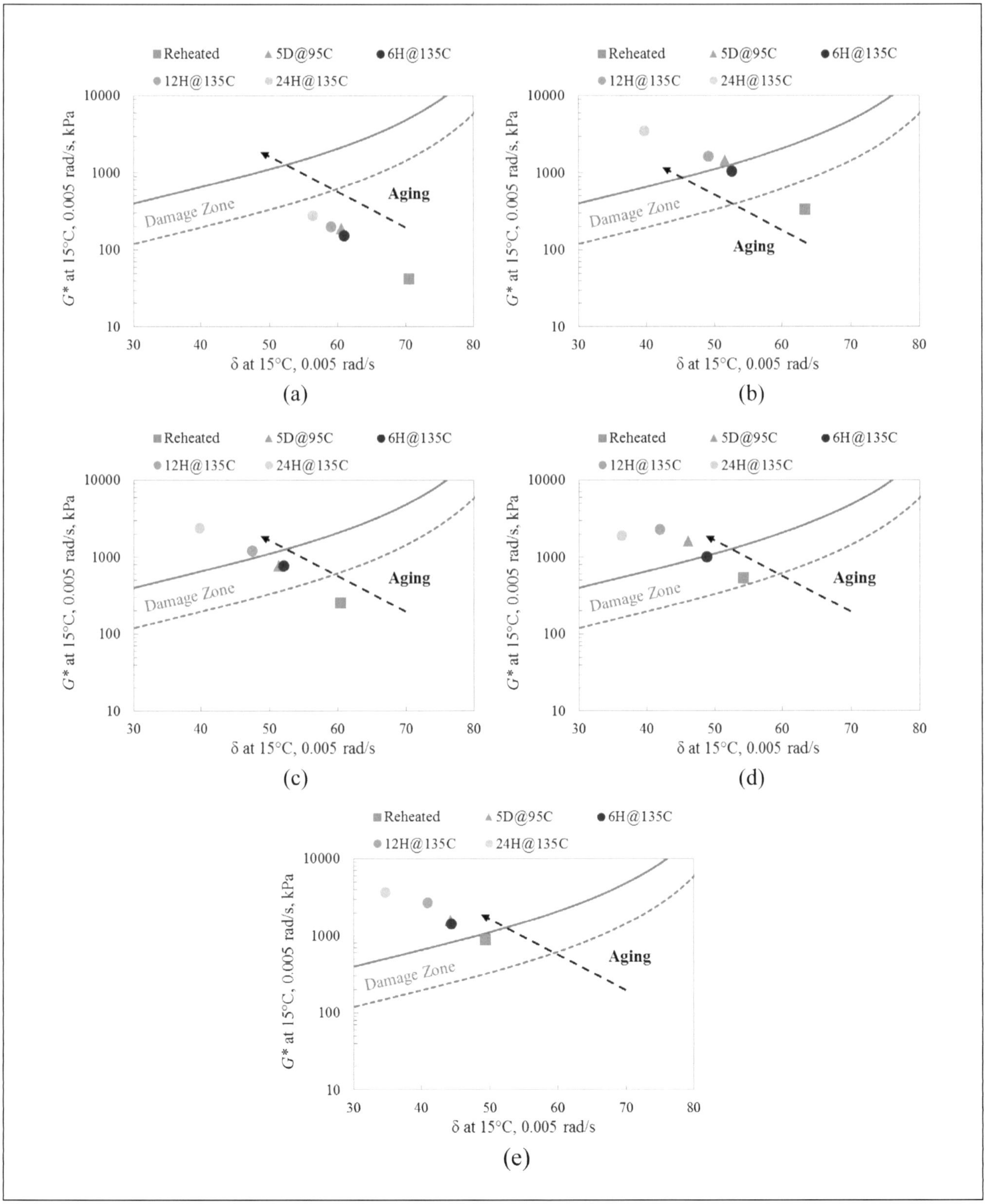

Figure 5. G-R parameter results in Black space diagram: (*a*) Mix 1; (*b*) Mix 2; (*c*) Mix 3; (*d*) Mix 4; (*e*) Mix 5.

the continuous PG and G-R parameter results shown in Figures 4 and 5, respectively. Asphalt binders extracted from 135°C mixes showed a consistent increase in CA with extensions in aging time, which indicated that more polar oxygen-containing functional groups were formed during the process. For all mixes except Mix 4, extracted

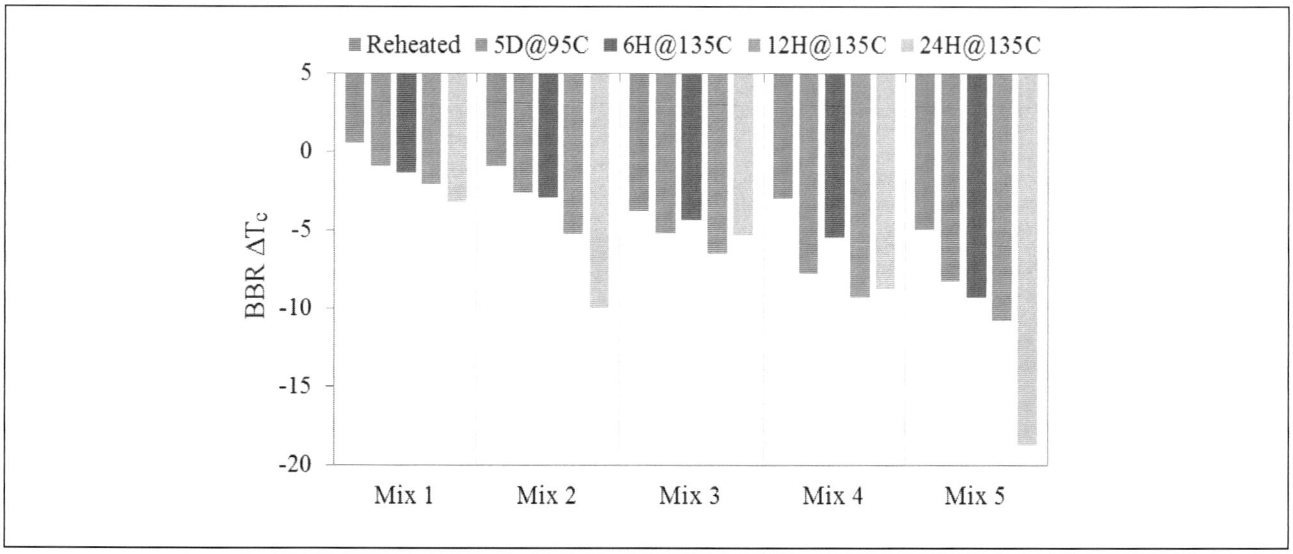

Figure 6. BBR ΔT_c results of extracted asphalt binders with various loose mix aging protocols.

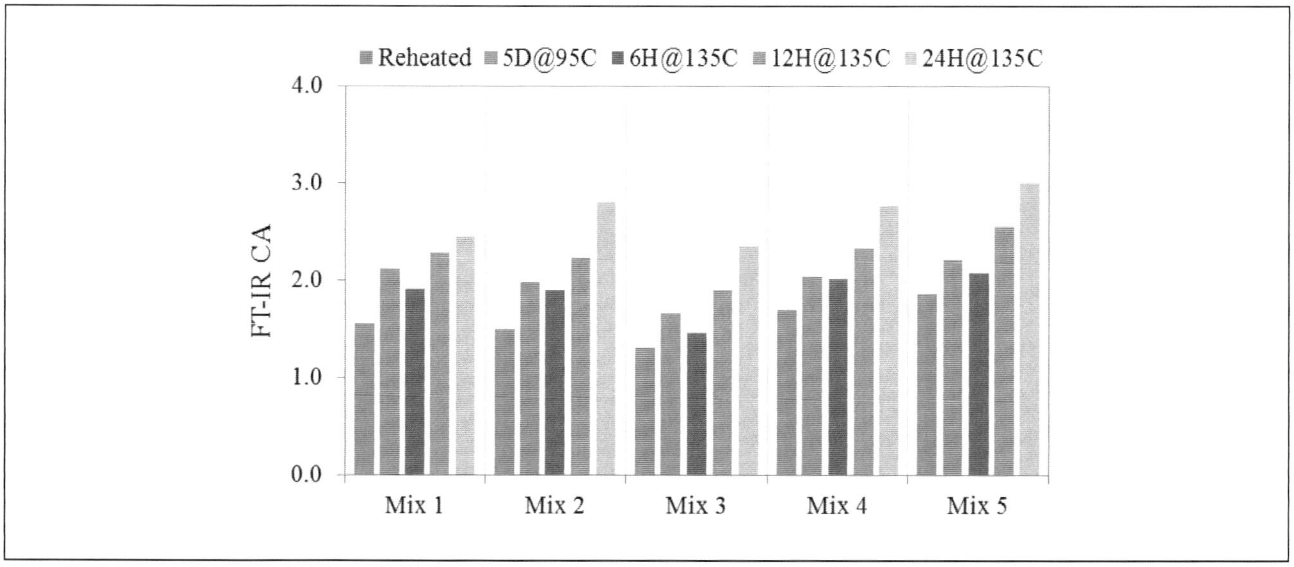

Figure 7. FT-IR CA results of extracted asphalt binders with various loose mix aging protocols.

binders from the 5-day, 95°C mix had a CA that was between those of binders from the 6-h and 12-h, 135°C mixes. Based on the FT-IR CA results, the four loose mix aging protocols were ranked in the following order based on the resultant aging level: 6-h, 135°C < 5-day, 95°C < 12-h, 135°C < 24-h, 135°C.

Hardening Susceptibility Results. Previous literature reported that aging at 135°C could change the relationship between chemical oxidation and physical hardening of asphalt binders owing to the physicochemical disruption of its microstructure at such an elevated temperature (*24*). This oxidation–hardening relationship was typically referred to as hardening susceptibility (*20*, *25*). In this study, the Glover–Rowe hardening susceptibility (G-R HS) parameter was used to explore the impact of loose mix aging on the oxidation–hardening relationship of asphalt binders. As expressed in Equation 4, the G-R HS was defined as the ratio of change in the logarithm of the G-R parameter over the change in the FT-IR CA with aging (*26*):

$$G\text{-R HS} = \frac{d[\log(\mathrm{G} - \mathrm{R})]}{d(\mathrm{CA})} \tag{4}$$

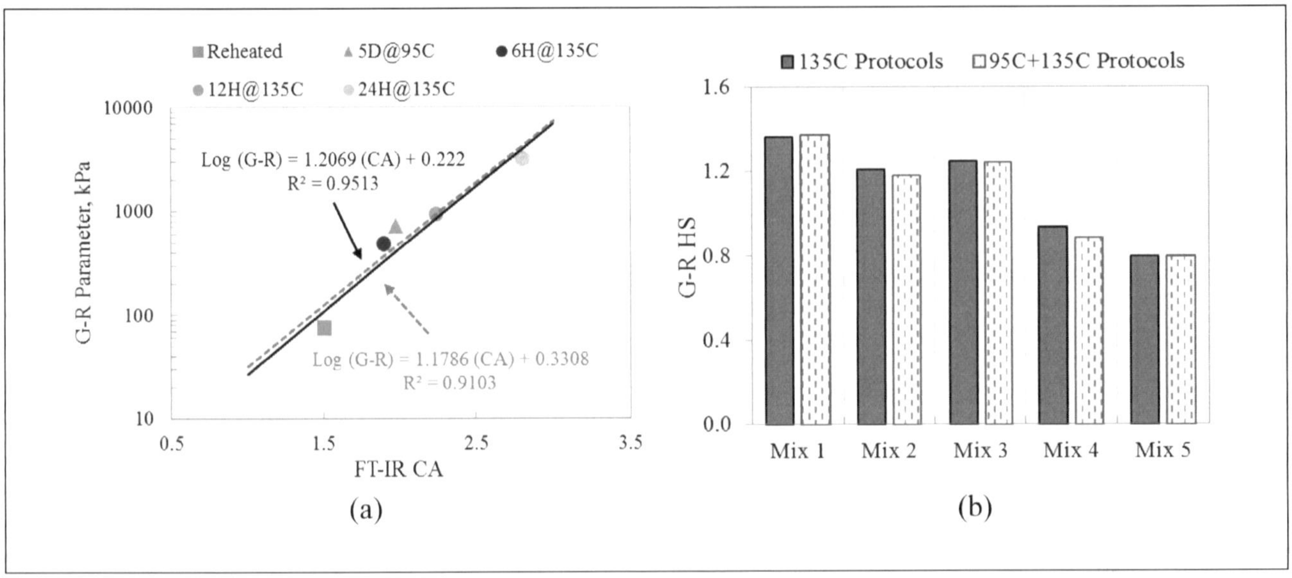

Figure 8. G-R HS results: (*a*) example of Mix 2 data; (*b*) comparison of loose mix aging protocols at 95°C and 135°C.

Figure 8*a* presents an example to illustrate the determination of G-R HS. In the figure, the G-R parameter and CA results were fitted using a semi-log linear relationship and the slope of the trend line represents the G-R HS. The dashed line represents the trend line determined based on all aging protocols evaluated in this study, while the solid trend line is only for protocols at 135°C. The coefficient of determination (i.e., R^2) of both trend lines is higher than 0.9, indicating the goodness of fit. In addition, the two trend lines almost overlap. Similar trends were also shown for the other four mixes, but the results are not presented because of space limitations. Figure 8*b* summarizes the comparison of G-R HS results for all aging protocols versus those only at 135°C. As shown, for the five mixes evaluated in this study, the two groups of aging protocols exhibited similar G-R HS results, indicating no significant difference in the oxidation–hardening relationship of asphalt binders aged at 95°C versus 135°C.

Correlation of Field Aging with Laboratory Aging Protocols

Figure 9 presents the continuous HPG results to determine the correlation of field aging with laboratory loose mix aging protocols. The pattern-filled bars represent the difference in the continuous HPG of asphalt binders extracted from post-construction field cores versus the corresponding plant mixes that were tested immediately after sampling, and the solid-filled bars represent the difference between the laboratory aged mixes and reheated mixes. Mix 3 was excluded from the analysis as it uses styrene–butadiene–styrene (SBS) polymer modified binder, which has been extensively reported as having significantly different aging characteristics compared to neat binders (*27–29*).

As shown in Figure 9*a*, the 5-day, 95°C protocol was more severe than field aging of approximately 48,000 CDD (Mix 1), as indicated by a greater increase in the continuous HPG. The opposite trend, however, was shown for the other mixes, with field aging of approximately 80,000, 157,000, and 235,000 CDD, respectively. These results indicate that the 5-day, 95°C protocol yielded asphalt aging that was representative of field aging between 48,000 and 80,000 CDD. The same approach was applied for the other three loose mix aging protocols (Figure 9*b–d*), and their representative CDD ranges are summarized as follows:

- 5-day, 95°C protocol: 48,000 to 80,000 CDD
- 6-h, 135°C protocol: approximately 48,000 CDD
- 12-h, 135°C protocol: 80,000 to 157,000 CDD
- 24-h, 135°C protocol: greater than 235,000 CDD

As discussed previously, 70,000 CDD was identified as a preliminary critical field aging condition for evaluating TDC. Therefore, based on the data shown in Figure 9, loose mix aging of 5 days at 95°C was the most representative aging protocol.

Selection of Alternative Aging Protocol at 135°C

Although loose mix aging of 5 days at 95°C was identified as the most appropriate protocol to simulate 70,000 CDD of field aging, this protocol might not be suitable for practical implementation owing to long time span.

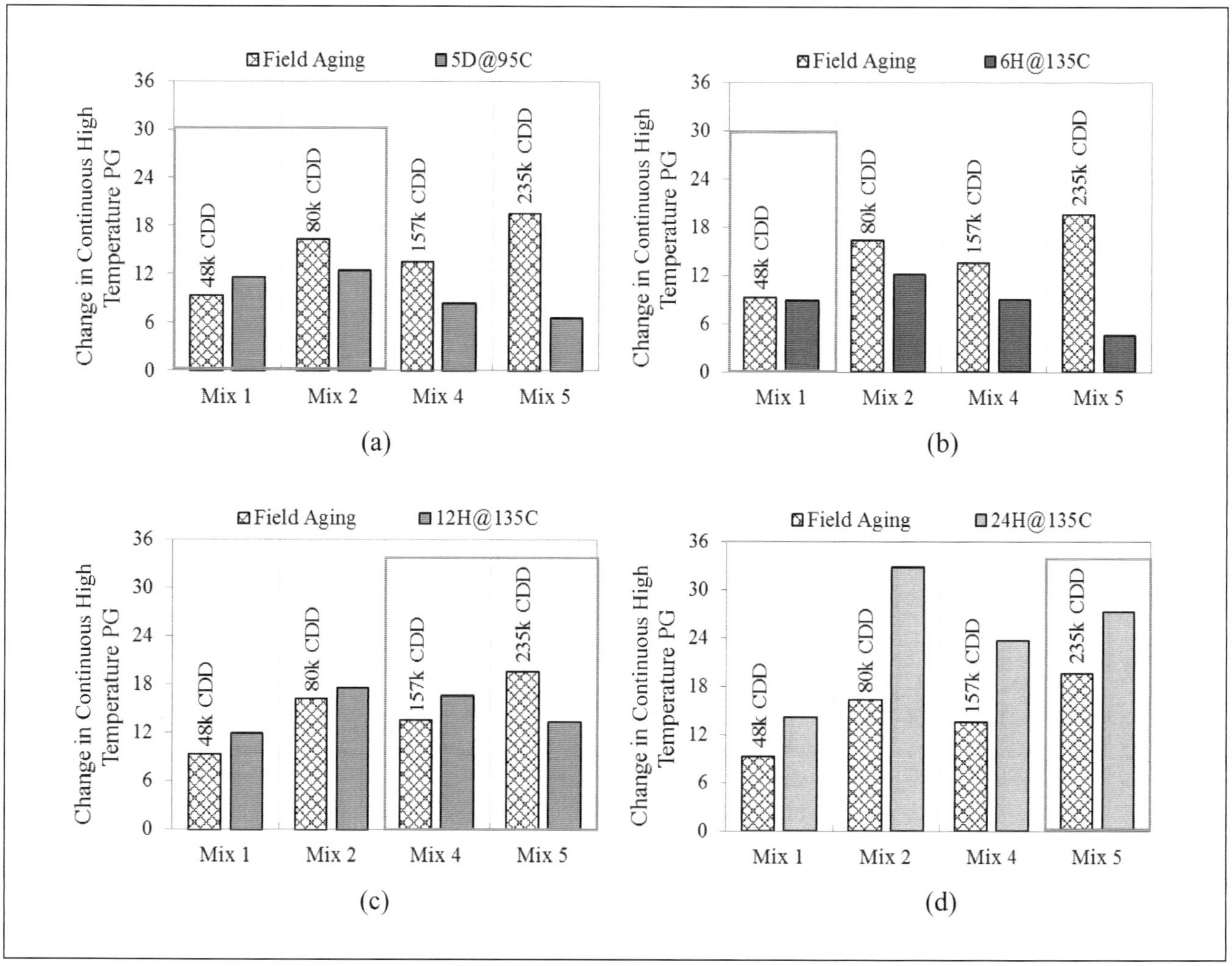

Figure 9. Comparison of field aging and laboratory loose mix aging protocols: (*a*) 5-day, 95°C; (*b*) 6-h, 135°C; (*c*) 12-h, 135°C; (*d*) 24-h, 135°C.

Therefore, test results were further analyzed to determine an alternative aging protocol at 135°C that yielded an equivalent level of aging as the 5-day, 95°C protocol. Figure 10*a* presents an example of the FT-IR CA results of Mix 2. To determine the alternative aging protocol, the fast-rate constant-rate oxidation kinetic model (*30*), described in Equation 5, was first used to fit the CA results of laboratory aged mixes at 135°C. Once the model coefficients were determined, the equivalent aging time at 135°C was then calculated based on the measured CA of the 5-day, 95°C protocol (Figure 10*a*). For the continuous HPG and G-R results, linear and semi-log linear models were used to determine the equivalent aging times, respectively. As shown in Figure 10*b*, for both oxidation and rheological parameters, similar level of asphalt aging was achieved by loose mix aging protocols of 5 days at 95°C and approximately 8 h at 135°C. Thus, the 8-h, 135°C protocol was recommended as an alternative aging protocol to simulate 70,000 CDD of field aging:

$$CA = CA_0 + k_c t + M(1 - e^{-k_f t}) \quad (5)$$

where:

CA_0 = CA of asphalt binder extracted from the reheated mix

t = aging time at 135°C

k_c, k_f, and M = model coefficients

Conclusions and Recommendations

The objective of this study was to select a laboratory mix aging protocol for the NCAT TDC experiment. Four different loose mix aging protocols were evaluated based on their effects on rheological and oxidation properties of asphalt binders. In addition, their correlations with field

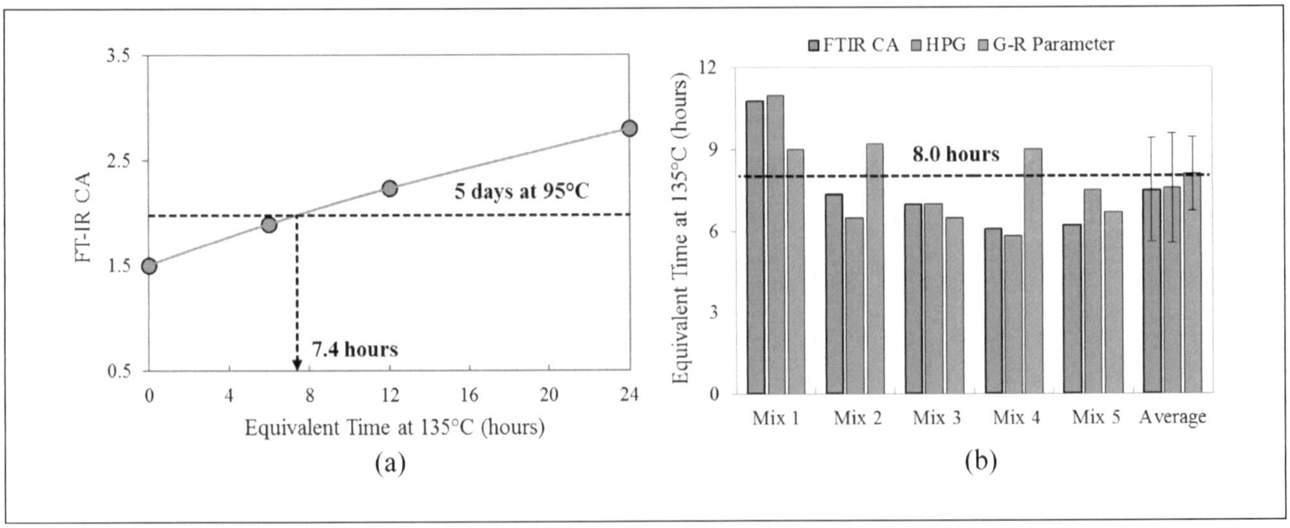

Figure 10. Determination of equivalent aging time at 135°C: (*a*) example of Mix 2 FI-IR CA results; (*b*) summary of all results.

aging were investigated based on the concept of CDD. Based on the results of this study, the following conclusions were obtained:

- TDC was found to develop in asphalt pavements with over approximately 70,000 CDD; thus, this CDD value was identified as the critical field aging condition for evaluating TDC.
- Based on the rheological and oxidation results, the 24-h, 135°C protocol yielded the most significant level of asphalt aging, followed by the 12-h, 135°C protocol, 5-day, 95°C protocol, and 6-h, 135°C protocol, respectively.
- The G-R HS parameter was used to explore the effect of loose mix aging on the oxidation–hardening relationship of asphalt binders. For the five mixes evaluated in this study, no significant difference in the G-R HS results was shown for mixes aged at 95°C versus 135°C.
- Among the four loose mix aging protocols evaluated in this study, the 5-day, 95°C protocol was most representative of 70,000 CDD of field aging.
- Based on the HPG, G-R parameter, and CA results, the loose mix aging protocol of 8 h at 135°C was anticipated to yield an equivalent level of aging as that of 5 days at 95°C. Therefore, the 8-h, 135°C protocol was recommended as an alternative, but more practical, protocol to simulate 70,000 CDD of field aging.

sIn this study, the representative loose mix aging protocols were determined based on a limited number of asphalt mixtures and were primarily based on the HPG results. Thus, future research is needed to verify the selected protocols using additional mixtures and binder properties. Additionally, there is a need for future research to evaluate the aging characteristics of polymer modified asphalt binders. Furthermore, mixture performance testing is recommended to validate the equivalency between the 5-day, 95°C and 8-h, 135°C loose mix aging protocols. Finally, it is recommended to continue monitoring the performance of pavements with TDC to validate the preliminary critical field aging of 70,000 CDD.

Acknowledgments

The authors would like to acknowledge the following sponsors for their support of the NCAT/MnROAD Cracking Group experiment: Alabama Department of Environmental Management, Alabama Department of Transpiration (DOT), Collaborative Aggregates LLC, Colorado DOT, Federal Highway Administration, FP2 Inc., Florida DOT, Georgia DOT, Illinois DOT, Kentucky Transportation Cabinet, Michigan DOT, Mississippi DOT, Minnesota DOT, Missouri DOT, New York DOT, North Carolina DOT, Oklahoma DOT, South Carolina DOT, Tennessee DOT, Virginia DOT, and Wisconsin DOT. Acknowledgment also goes to Carolina Rodezno, Adam Taylor, Jason Nelson, and Nathan Moore at NCAT for their contributions to the study.

Author Contributions

The authors confirm contribution to the paper as follows: study conception and design: FY, RCW; data collection: CC, PT; analysis and interpretation of results: CC, FY, NT; draft manuscript preparation: CC, FY. All authors reviewed the results and approved the final version of the manuscript.

References

1. Myers, L. A., R. Roque, and B. E. Ruth. Mechanisms of Surface-Initiated Longitudinal Wheel Path Cracks in

High-Type Bituminous Pavements. *Journal of the Association of Asphalt Paving Technologists*, Vol. 67, 1998.
2. Uhlmeyer, J., K. Willoughby, L. Pierce, and J. Mahoney. Top-down Cracking in Washington State Asphalt Concrete Wearing Courses. *Transportation Research Record: Journal of the Transportation Research Board*, 2000. 1730: 110–116.
3. Svasdisant, T., M. Schorsch, G. Baladi, and S. Pinyosunun. Mechanistic Analysis of Top-down Cracks in Asphalt Pavements. *Transportation Research Record: Journal of the Transportation Research Board*, 2002. 1809: 126–136.
4. Han, G., C. S. Clopotel, and H. U. Bahia. Effects of High Modulus Asphalt Binders on Performance of Typical Asphalt Pavement Structures. *Construction and Building Materials*, Vol. 44, 2013, pp. 207–213.
5. Moore, N. *Evaluation of Laboratory Cracking Tests Related to Top-Down Cracking in Asphalt Pavements*. Auburn University, Auburn, AL, 2016.
6. Bell, C. A., Y. AbWahab, M. E. Cristi, and D. Sosnovske. *Selection of Laboratory Aging Procedures for Asphalt-aggregate Mixtures*. No. SHRP-A-383. Strategic Highway Research Program, 1994.
7. Newcomb, D., A. E. Martin, F. Yin, E. Arambula, E. S. Park, A. Chowdhury, R. Brown, C. Rodezno, N. Tran, E. Coleri, and D. Jones. *Short-term Laboratory Conditioning of Asphalt Mixtures*. No. 09-52. National Cooperative Highway Research Program, 2015.
8. Islam, M. R., M. I. Hossain, and R. A. Tarefder. A Study of Asphalt Aging Using Indirect Tensile Strength Test. *Construction and Building Materials*, Vol. 95, 2015, pp. 218–223.
9. Howard, I. L., and J. D. Doyle. Durability Indexes via Cantabro Testing for Unaged, Laboratory-Conditioned, and One-Year Outdoor-Aged Asphalt Concrete. Presented at 94th Annual Meeting of the Transportation Research Board, Washington, D.C., 2015.
10. Elwardany, M. D., F. Yousefi Rad, C. Castorena, and Y. R. Kim. Evaluation of Asphalt Mixture Laboratory Long-term Ageing Methods for Performance Testing and Prediction. *Road Materials and Pavement Design*, Vol. 18, 2017, pp. 28–61.
11. Braham, A. F., W. G. Buttlar, T. R. Clyne, M. O. Marasteanu, and M. I. Turos. The Effect of Long-term Laboratory Aging on Asphalt Concrete Fracture Energy. *Journal of the Association of Asphalt Paving Technologists*, Vol. 78, 2009.
12. Reinke, G., A. Hanz, D. Herlitzka, S. Engber, and M. Ryan. Further Investigations into the Impact of REOB & Paraffinic Oils on the Performance of Bituminous Mixtures. In *Federal Highway Administration Expert Task Group Meeting*, Fall River, MA, 2015, April.
13. Kim, Y. R., C. Castorena, M. D. Elwardany, F. Y. Rad, S. Underwood, A. Gundla, P. Gudipudi, M. J. Farrar, and R. R. Glaser. *Long-Term Aging of Asphalt Mixtures for Performance Testing and Prediction*. No. 9-54. National Cooperative Highway Research Program, 2015.
14. Dauzats, M., and A. Rampal. Mechanism of Surface Cracking in Wearing Courses. In *Sixth International Conference, Structural Design of Asphalt Pavements*, Ann Arbor, MI, July 1987.
15. Shen, S. H., H. F. Wen, L. Mohammad, N. Lund, A. Faheem, W. G. Zhang, and S. H. Wu. *Performance of WMA Technologies: Stage II – Long-Term Field Performance*. Transportation Research Board of the National Academies, Washington, D.C., 2017.
16. Tran, N. *Evaluation of Warm Mix Asphalt with High RAP on NCAT Pavement Test Track*. NCAT, Auburn, AL, 2010.
17. Turner, F. *WRI Research Related to the Optimal Timing of Preventive Maintenance for Addressing Environmental Aging*. Western Research Institute, Laramie, WY, 2008.
18. Yin, F., A. Epps Martin, E. Arambula-Mercado, and D. Newcomb. Characterization of Non-Uniform Field Aging in Asphalt Pavements. *Construction and Building Materials*. Vol. 153, 2017, pp. 607–615.
19. Santagata, E., D. A. Anderson, and M. Marasteanu. A Rheological Investigation on the Physical Hardening of Modified Asphalt Binders. *Eurasphalt & Eurobitume Congress*, Strasbourg, May 7–10, 1996, Vol. 3, Paper E&E. 5.151.
20. Glover, C. J., R. R. Davison, C. H. Domke, Y. Ruan, P. Juristyarini, D. B. Knorr, and S. H. Jung. Development of a New Method for Assessing Asphalt Binder Durability with Field Validation. Report FHWA/TX-05/1872-2. Texas Department of Transport, Austin, TX, 2005.
21. Rowe, G. M. Prepared Discussion for the AAPT Paper by Anderson et al.: Evaluation of the Relationship Between Asphalt Binder Properties and Non-Load Related Cracking. *Journal of the Association of Asphalt Paving Technologists*, Vol. 80, 2011, pp. 649–662.
22. Anderson, R. M., G. N. King, D. I. Hanson, and P. B. Blankenship. Evaluation of the Relationship Between Asphalt Binder Properties and Non-load Related Cracking. *Journal of the Association of Asphalt Paving Technologists*, Vol. 80, 2011.
23. Liu, M., M. A. Ferry, R. R. Davison, C. J. Glover, and J. A. Bullin. Oxygen Uptake as Correlated to Carbonyl Growth in Aged Asphalts and Asphalt Corbett Fractions. *Industrial & Engineering Chemistry Research*, Vol. 37, No. 12, 1998, pp. 4669–4674.
24. Petersen, J. C. *Transportation Research E-Circular (E-C140): A Review of the Fundamentals of Asphalt Oxidation: Chemical, Physicochemical, Physical Property, and Durability Relationships*. Transportation Research Board of the National Academies, Washington, D.C., 2009.
25. Domke, C., R. Davison, and C. Glover. Effect of Oxidation Pressure on Asphalt Hardening Susceptibility. *Transportation Research Record: Journal of the Transportation Research Board*, 1999. 1661: 114–121.
26. Yin, F., F. Kaseer, E. Arámbula-Mercado, and A. E. Martin. Characterizing the Long-Term Rejuvenating Effectiveness of Recycling Agents on Asphalt Blends and Mixtures with High RAP and RAS Contents. *Road Materials and Pavement Design*, Vol. 18, 2017, pp. 273–292.
27. Sun, L., Y. Wang, and Y. Zhang. Aging Mechanism and Effective Recycling Ratio of SBS Modified Asphalt. *Construction and Building Materials*, Vol. 70, 2014, pp. 26–35.

28. Ruan, Y., R. R. Davison, and C. J. Glover. The Effect of Long-term Oxidation on the Rheological Properties of Polymer Modified Asphalts. *Fuel*, Vol. 82, No. 14, 2003, pp. 1763–1773.
29. Ouyang, C., S. Wang, Y. Zhang, and Y. Zhang. Improving the Aging Resistance of Styrene–Butadiene–Styrene Tri-Block Copolymer Modified Asphalt by Addition of Antioxidants. *Polymer Degradation and Stability*, Vol. 91, No. 4, 2006, pp. 795–804.
30. Jin, X., R. Han, Y. Cui, and C. J. Glover. Fast-Rate–Constant-Rate Oxidation Kinetics Model for Asphalt Binders. *Industrial & Engineering Chemistry Research*, Vol. 50, No. 23, 2011, pp. 13373–13379.

The Standing Committee on Asphalt Binders (AFK20) peer-reviewed this paper (18-04203).

Article

TRR
JOURNAL OF THE TRANSPORTATION RESEARCH BOARD

Transportation Research Record
2018, Vol. 2672(28) 372–382

DOI: 10.1177/0361198118782795
journals.sagepub.com/home/trr
SAGE

Sine versus Haversine Displacement Waveform Comparison for Hot Mix Asphalt Four-Point Bending Fatigue Testing

Angel Mateos[1], Rongzong Wu[2], Erik Denneman[3], and John Harvey[2]

Abstract

An experimental study was conducted to determine the effect of haversine and sine displacement waveforms on four-point flexural beam fatigue test results for hot mix asphalt. Seven asphalt mixtures with different gradations, binder types, and binder contents were tested for this study. Four of the mixes were tested in California and three in Australia. The mixes were tested at different strain levels under both waveforms in displacement-controlled mode without rest periods. The comparison showed no differences between haversine and sine testing modes for six of the seven mixtures. Fatigue life, the shape of the stiffness reduction curves, initial dynamic moduli and phase angles, and Black diagrams were essentially the same for both testing modes. This similarity was attributed to the viscoelastic nature of the asphalt mix. Because of asphalt's viscoelasticity, the beam at-rest position in haversine testing will rapidly move to halfway between zero and maximum displacement, and so the same stress is produced by haversine and sine displacement waveforms as soon as the peak-to-peak amplitudes are equal. For one of the seven mixes, sinusoidal testing produced considerably longer fatigue lives, although it is believed this mix is an outlier and can be ignored. Based on these results, it was concluded that there is no compelling reason to recommend sinusoidal over haversine testing mode, or vice versa. A practical reason to recommend the sine testing mode is that the stress in the haversine alternative becomes sinusoidal after a small number of load repetitions.

The four-point flexural beam fatigue (4PFBF) test has long been used to evaluate fatigue performance of asphalt concrete at intermediate pavement temperatures (*1*). This test measures the evolution of the flexural stiffness of an asphalt concrete beam subjected to repeated flexural bending. Current U.S. practice for 4PFBF follows one of two standard methods: ASTM D7460-10 or AASHTO T 321-14. Both of these were developed from research findings made by the SHRP A-003A project (*1*), completed in 1994, which selected 4PFBF, after rigorous evaluation, as the best method for evaluating asphalt concrete fatigue performance.

Over the years the ASTM and AASHTO test methods have undergone several rounds of changes, rendering the current versions very similar to each other. However, the way that each specifies how load is to be applied to the beam remains a major difference between them: the ASTM method states that the loading device shall be capable of providing "cyclic haversine loading," while the AASHTO method requires the loading device to be capable of providing "repeated sinusoidal loading." Meanwhile, the European standard for 4PFBF, EN 12697-24:2012 Annex D, specifies that "the applied load shall vary sinusoidally."

Since both ASTM and AASHTO methods are run in displacement-controlled mode, the words "haversine" and "sinusoidal" in the loading descriptions refer to the waveforms of the displacements imparted to the beam by the loading device. The "cyclic haversine loading" of the ASTM standard is understood to be the displacement waveform in the shape of a displaced sine wave with full amplitude on one side of zero. The "repeated sinusoidal loading" of the AASHTO standard is understood as a sine wave displacement oscillating around zero. The same is applicable to the EN standard when tests are conducted in displacement-control mode. Consistent definition of strain amplitude is critical in comparing the fatigue results from the different test meth-

[1]University of California Pavement Research Center, Institute of Transportation Studies, University of California, Berkeley, Richmond, CA
[2]University of California Pavement Research Center, Department of Civil and Environmental Engineering, University of California, Davis, CA
[3]Australian Asphalt Pavement Association (formerly with ARRB Group Ltd), Queensland, Australia

Corresponding Author:
Address correspondence to Angel Mateos: angel-mateos@berkeley.edu

ods. All strain levels in this paper are expressed in peak-to-peak values.

The 4PFBF tests were conducted in the SHRP A-003A project using a haversine displacement waveform (*1*). Conceptually, a haversine waveform simulates field conditions better than a sinusoidal waveform because the passing of a truck will mainly push a pavement downward. Using a haversine displacement waveform has been the practice for the University of California Pavement Research Center (UCPRC) since 1991, and also for most researchers around the world (*2*).

Some concerns have been raised about the appropriateness of using the haversine waveform in 4PFBF testing (*3*). Mamlouk et al. concluded (*2*) that 4PFBF tests run with a haversine displacement waveform "produce erroneous fatigue results since the test assumptions do not match the actual test conditions," although they did not show any data supporting this statement. Their point is that both the haversine and the sinusoidal displacement waveforms produce a sinusoidal load waveform for nearly all of a given test, with the haversine load wave changing to a sinusoidal shape soon after initiation of the test. As a matter of fact, there is no contradiction between that phenomenon and the hypotheses supporting the haversine testing mode. Mateos et al. (*4*) found the same fatigue life versus peak-to-peak strain resulted from haversine 4PFBF testing conducted according to the ASTM standard and sinusoidal 4PFBF testing conducted according to the EN standard. Criqui and Wendell (*5*) caution against replacing the haversine waveform with the sinusoidal waveform altogether before sufficient substantiating data is obtained.

The concerns raised about use of the haversine waveform are significant because they affect how the large amount of historical data accumulated by many researchers around the world should be used. Furthermore, the UCPRC has used the stiffness degradation curves recorded during 4PFBF testing to determine the parameters of the incremental-recursive fatigue model implemented in CalME (*6*). CalME is the California Department of Transportation's (Caltrans) mechanistic–empirical design method for asphalt pavements. If there is a difference between results obtained using haversine and sinusoidal waveforms, then a change by Caltrans and the UCPRC to use of the sinusoidal waveform would require a recalibration of the models.

The issues raised by the differences in displacement waveforms specified by the ASTM and AASHTO test methods are the reason why, in April 2015, the Federal Highway Administration asphalt mixture expert task group charged an ad hoc committee to standardize the 4PFBF test. After its work the committee did not reach a consensus on a specific displacement waveform. Instead it recommended that both sinusoidal and haversine waveforms could be used, but that the selected waveform must be reported. The same issues caused Caltrans to task the UCPRC to conduct an experimental study to investigate the effect of the two displacement waveforms on the results of 4PFBF testing. The results of that study are summarized in this paper.

Objective

The main objective of the study presented in this paper is to investigate the effect of two displacement waveforms, namely haversine and sinusoidal waveforms, on the results of 4PFBF tests that do not include rest periods between load pulses. The ultimate goal is to determine which of the two displacement waveforms, haversine, or sine, should be used in beam fatigue tests.

Experiment Design

Seven asphalt mixes were selected and tested in 4PFBF following both haversine and sinusoidal displacement-controlled modes. The results from the two types of loading were then compared from different perspectives, following these steps:

1. Analysis of the classic fatigue laws. A fatigue law (Wöhler curve) was fitted to the fatigue lives determined for each mix. Then, the sine-versus-haversine comparison was conducted in terms of different parameters of the fatigue law.
2. Analysis of variance (ANOVA) for each individual mix. The fatigue lives obtained with sine and haversine testing modes were compared with each other. The comparison was conducted for each mix independently of the other mixes.
3. Analysis of the shape of the stiffness degradation curves. The shapes of the stiffness degradation curves obtained in sine and haversine testing modes were compared with each other. The reason for including this comparison is that the stiffness degradation curves contain valuable information that cannot be captured by just using the failure point.
4. Analysis of test results in the Black space. In this step, the evolution of dynamic modulus versus phase angle during fatigue testing in haversine and sine testing modes was examined. The reason for this comparison is that damage accumulation in fatigue tests is strongly related to dissipated energy, which at the same time depends on dynamic modulus and phase angle.
5. ANOVA for all mixes. The fatigue lives obtained with sine and haversine testing modes were compared with each other, as in step 3. In step 5, the comparison was conducted for all mixes simultaneously, by including "mix type" as a factor in the ANOVA. The goal of step 5 comparison was to capture information that could have been missed in the step 3 piecewise comparison.

Table 1. Summary of Mixes and Testing Conditions

Mix ID	Mix source	NMAS[a]	Binder	Binder content (twm)	Test device	Temp /freq	Strain		Replicates	
							Strain level[b]	Peak-to-peak strain	Haversine	Sine
RHMA-G	Calif.	19 mm (¾ in.)	ARB[c]	8.8%	Cox & Sons	20°C /10 Hz	Low	470 με	3	3
							Med.	580 με	3	3
HMA	Calif.	19 mm (¾ in.)	PG 70-10	5.9%	Cox & Sons	20°C /10 Hz	Low	270 με	3	3
							Med.	315 με	3	3
HMA-PM	Calif.	19 mm (¾ in.)	PG 64-28	5.0%	Cox & Sons	20°C /10 Hz	Low	505 με	3	3
							Med.	655 με	3	3
HMA-RB	Calif.	19 mm (¾ in.)	PG 64-10	5.9%	Cox & Sons	20°C /10 Hz	Low	350 με	3	3
							Med.	430 με	3	3
DG10-C320	Australia	10 mm (0.4 in.)	Class 320[d]	5.6%	IPC	20°C /10 Hz	Low	200 με	3	3
							Med.	400 με	3	3
							High	800 με	3	3
DG20-C600	Australia	20 mm (¾ in.)	Class 600[d]	4.7%	IPC	30°C[e] /10 Hz	Low	480 με	4	3
							Med.	800 με	3	3
							High[f]	950 με	3	3
EME	Australia	14 mm (0.55 in.)	15/25[g]	5.6%	IPC	30°C[e] /10 Hz	Low	460 με	3	8
							Med.	600 με	4	6
							High[f]	760 με	3	6

[a]Nominal maximum aggregate size.
[b]The designation of "strain level" is specific to each mix.
[c]Asphalt rubber binder, high viscosity, with PG 64-16 base binder and 18% minimum crumb rubber.
[d]Australian binder specification: viscosity in Pa. s at 60°C.
[e]The test temperature was 30°C (instead of 20°C) because the mixes DG20-C600 and EME were tested within a larger study investigating the influence of temperature on fatigue performance.
[f]Additional tests were conducted with the DG20-C600 and EME mixtures at strain levels other than those reflected in the table. Results from these other tests were used in the study to estimate fatigue law parameters.
[g]Low penetration binder following European standard EN 13924. 15/25 is penetration specification interval.

Four of the asphalt mixtures selected for this study are representative of the variety used by Caltrans, while the other three are representative of Australian materials. Californian mixes included a gap-graded mix with asphalt rubber binder (wet process-high viscosity), referred to as RHMA-G in this paper, and three dense-graded mixes, referred to as HMA, HMA-PM, and HMA-RB, respectively. HMA used PG 70-10 neat binder, HMA-PM used polymer-modified binder meeting PG 64-28 requirements, and HMA-RB is a rich bottom mix meeting Caltrans performance-related specifications for a long-life project on Interstate 5. Australian materials included a high-modulus asphalt concrete (referred to as EME in this paper) and two other dense-graded mixes: DG10-C320 and DG20-C600. The DG10-C320 binder met Australian requirements for Class 320 (viscosity of approximately 320 Pa.s at 60°C), while the DG20-C600 binder met the requirements for Class 600 (viscosity of approximately 600 Pa.s at 60°C). Details of these mixes can be found in Table 1 and Mateos et al. (*7*).

Caltrans mixes were tested at the UCPRC, while Australian mixes were tested at ARRB. Each of the Caltrans mixes was tested at two different strain levels, targeted to fail at 0.1 to 0.5 million and at 0.5 to 1 million repetitions. Each of the Australian mixes was tested at three strain levels, targeting 10^4, 10^5, and 10^6 repetitions to failure. Table 1 shows a summary of testing conditions. Beam displacement measurement locations and all other features of the testing conducted on the two manufacturers' devices were standardized to the recommendations of the ad hoc FHWA Mix expert task group. Note that all strain levels are reported as peak-to-peak values.

Fatigue tests conducted at the UCPRC were continued until beam stiffness (quantified as dynamic modulus) diminished to 20% of the initial value. For those tests, two different failure criteria were used: 50% stiffness reduction (classic approach) and maximum of stiffness multiplied by the number of cycles. The later criterion is referred to as $(SRxN)_{Max}$ in this paper. $(SRxN)_{Max}$ is a simplified energy ratio criterion proposed by Rowe and Bouldin (*8*) that has been adopted by both the ASTM D7460-10 and AASHTO T 321-14 standards. Fatigue tests conducted at ARRB were continued until 50% stiffness reduction, so only the first of the two failure criteria could be directly compared. For one of the Australian mixes, DG10-C320, the $(SRxN)_{Max}$ failure point could be extrapolated in most tests, since this failure point was only slightly beyond the 50% stiffness reduction cycle.

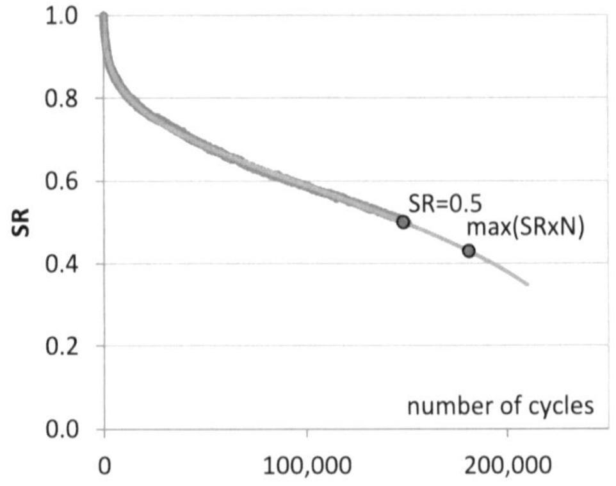

(a)

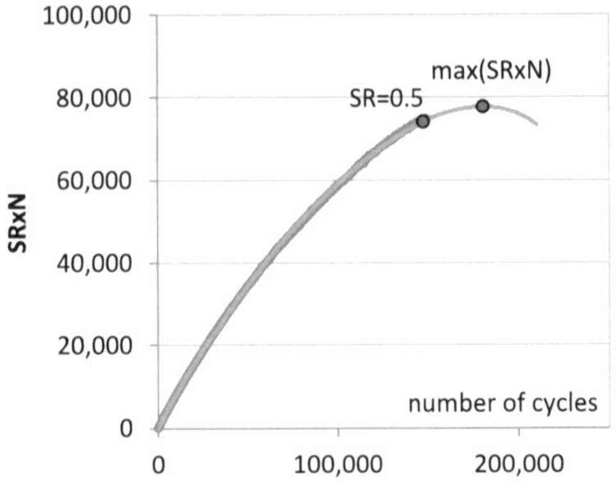

(b)

• measured SR —— logit model ● Nf

Logit model:

$$SR = SR_0 - \beta \cdot \text{logit}(p^{\gamma}) = SR_0 - \beta \cdot \ln\left(\frac{p^{\gamma}}{1 - p^{\gamma}}\right)$$

$$p = \frac{n - N_0}{N_L - N_0}$$

SR_0, N_L, β, and γ are independent (fitting) parameters

N_0 is a dependent parameter

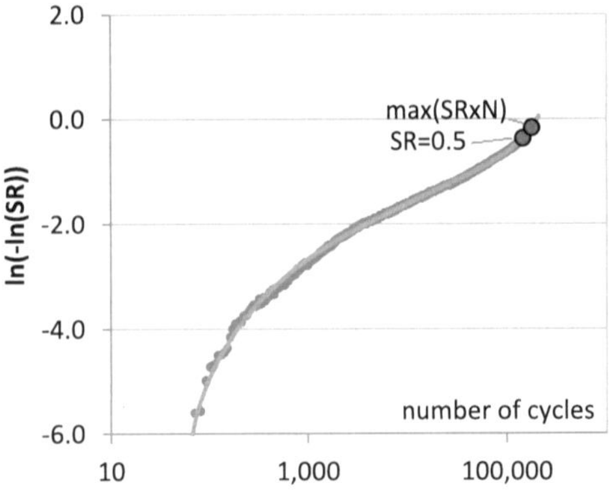

(c)

Figure 1. Curve fitting to extrapolate fatigue life (DG10-C320, 400 με peak-to-peak): (*a*) SR evolution in the linear scale; (*b*) SRxN evolution in the linear scale; (*c*) SR evolution in the Weibull space.
Note: Experimental data (SR versus N) were fit with the logit model, whose equation is shown in this figure. By doing so, the parameters of the logit model (SR_0, N_L, β, and γ) were determined for each particular test. Then, the fitted model was used to extrapolate SR beyond the 50% stiffness reduction cycle, as shown in plot (*a*). The predicted SR was multiplied by N, as shown in plot (*b*). Data in plot (*b*) were used to determine the number of cycles to failure based on $(SRxN)_{Max}$ criterion.

Peak-to-peak strains were calculated by using the same formulations for both haversine and sinusoidal tests. In both cases, peak-to-peak strain was determined using the peak-to-peak displacement, and peak-to-peak stress was determined using the peak-to-peak load.

Data Analysis and Results

Complex modulus (E*) degradation curves were analyzed in order to determine several characteristic parameters. First, initial dynamic modulus (|E*|) and phase angle (φ) were determined at cycle 50, as indicated by ASTM D7460-10 and AASHTO T 321-14. The initial dynamic modulus, $|E^*|_{INI}$, was used to calculate the stiffness ratio (SR) for each test cycle ($SR=|E^*|/|E^*|_{INI}$). Then, the number of cycles to failure was determined for the two different failure criteria. In most tests, the number of cycles to failure could be easily determined from the stiffness degradation curves by using simple Excel formulas. In other cases, curve fitting was required to determine the fatigue life according to one or two failure criteria. The reasons for conducting curve fitting included the need to interpolate between two loading cycles, to determine the maximum of the curve SRxN versus number of cycles when it was not clearly defined, to screen out noisy data, and to extrapolate SR. It should be noted that only short extrapolations were conducted, like the example shown in Figure 1, where the evolution of stiffness versus number of cycles up to the failure point could be clearly inferred from the stiffness measured in previous cycles.

The curve fitting was based on the logit model (*9*). This model has been shown to reproduce the complete stiffness reduction curve up to the transition from micro- to macrocracking, where a sharp decrease in beam stiffness takes place (*10*). This approach was preferred to using splines or polynomials, as recommended by the ASTM and AASHTO 4PFBF standards, because it is easy and repeatable.

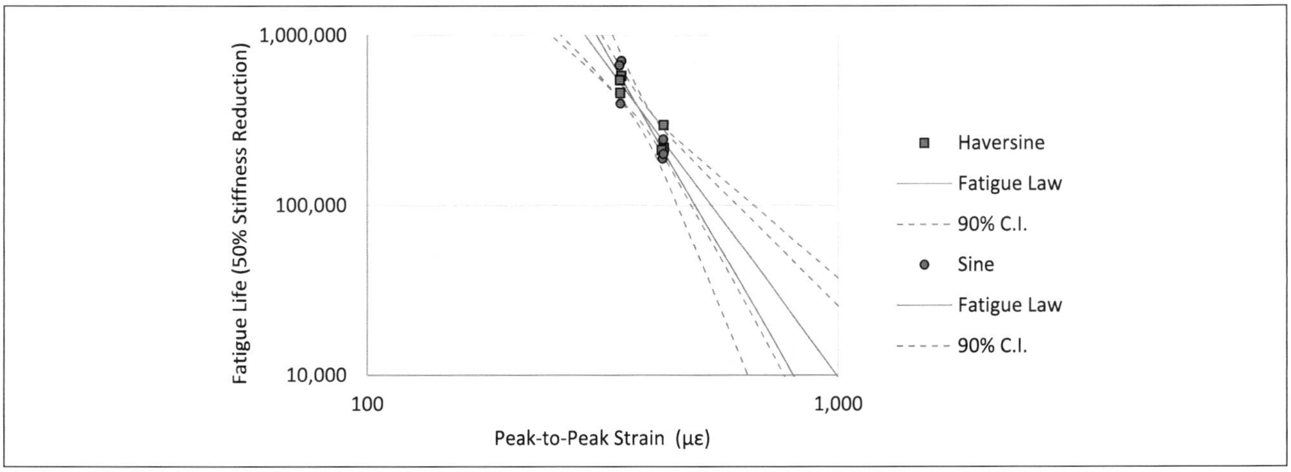

Figure 2. Fatigue laws for mix HMA-RB.

Analysis of the Classic Fatigue Laws

A classic fatigue law (Wöhler curve) was fitted to the fatigue lives determined for each mix using the 50% stiffness reduction failure criterion. This failure criterion could be quantified for all the mixes. For six of the seven mixes, the haversine and sine fatigue lives seemed to follow the same pattern; an example is shown in Figure 2. Clear differences were observed for the DG20-C600 mix, however, as Figure 3 shows. For this mix sine mode testing resulted in a longer fatigue life than haversine mode testing, approximately one to two times longer in the linear scale.

A comparative analysis was also conducted in terms of two main parameters defining the fatigue law of each mix: ε_6, strain corresponding to a fatigue life of 10^6 cycles, and the slope of the classic fatigue relation. The analysis used the 50% stiffness reduction fatigue lives. The results of this analysis are shown in Figure 4. Confidence intervals determined for ε_6 the haversine and sine test modes overlap for six of the seven mixes. Again, the DG20-C600 mix was the exception. For the slope, similar values were obtained for haversine and sine in all the mixes, including DG20-C600. This indicates that for this specific mix the testing mode effect may manifest as a shift in the fatigue law in the double logarithmic scale (log Nf versus log strain). This outcome agrees with results shown in Figure 3. The standard error of the model predictions varied considerably from one mix to another. Nevertheless, similar errors were obtained for the haversine and sine data in six of the seven mixes. Again, mix DG20-C600 differed from the others, with the errors for haversine and sine testing being considerably different from each other.

Analysis of Variance for Each Individual Mix

ANOVA was conducted for each mix, independent of the others. The study variable was the number of cycles to 50%

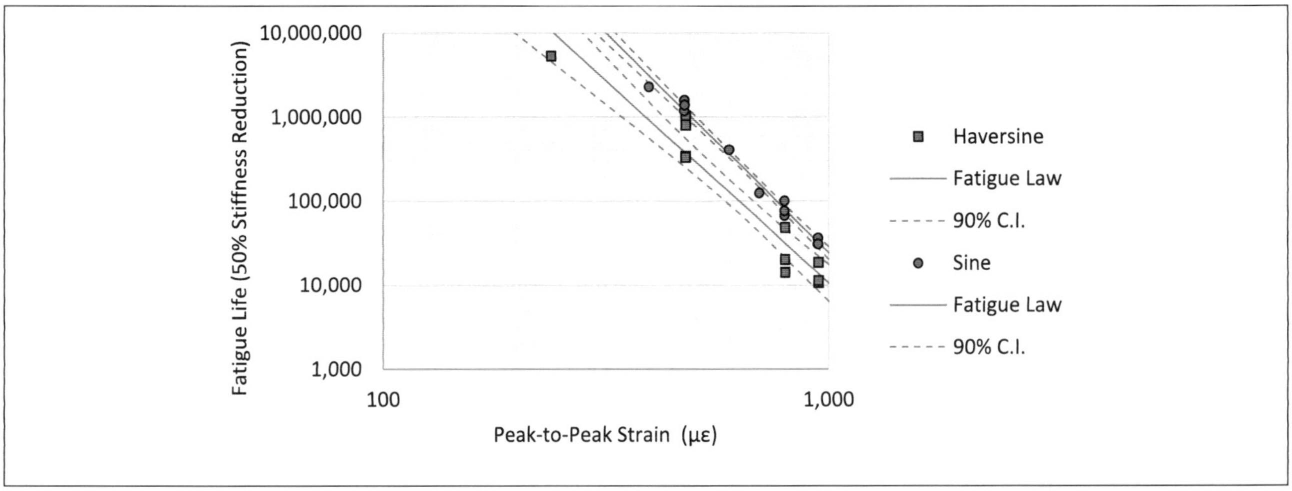

Figure 3. Fatigue laws for mix DG20-C600.

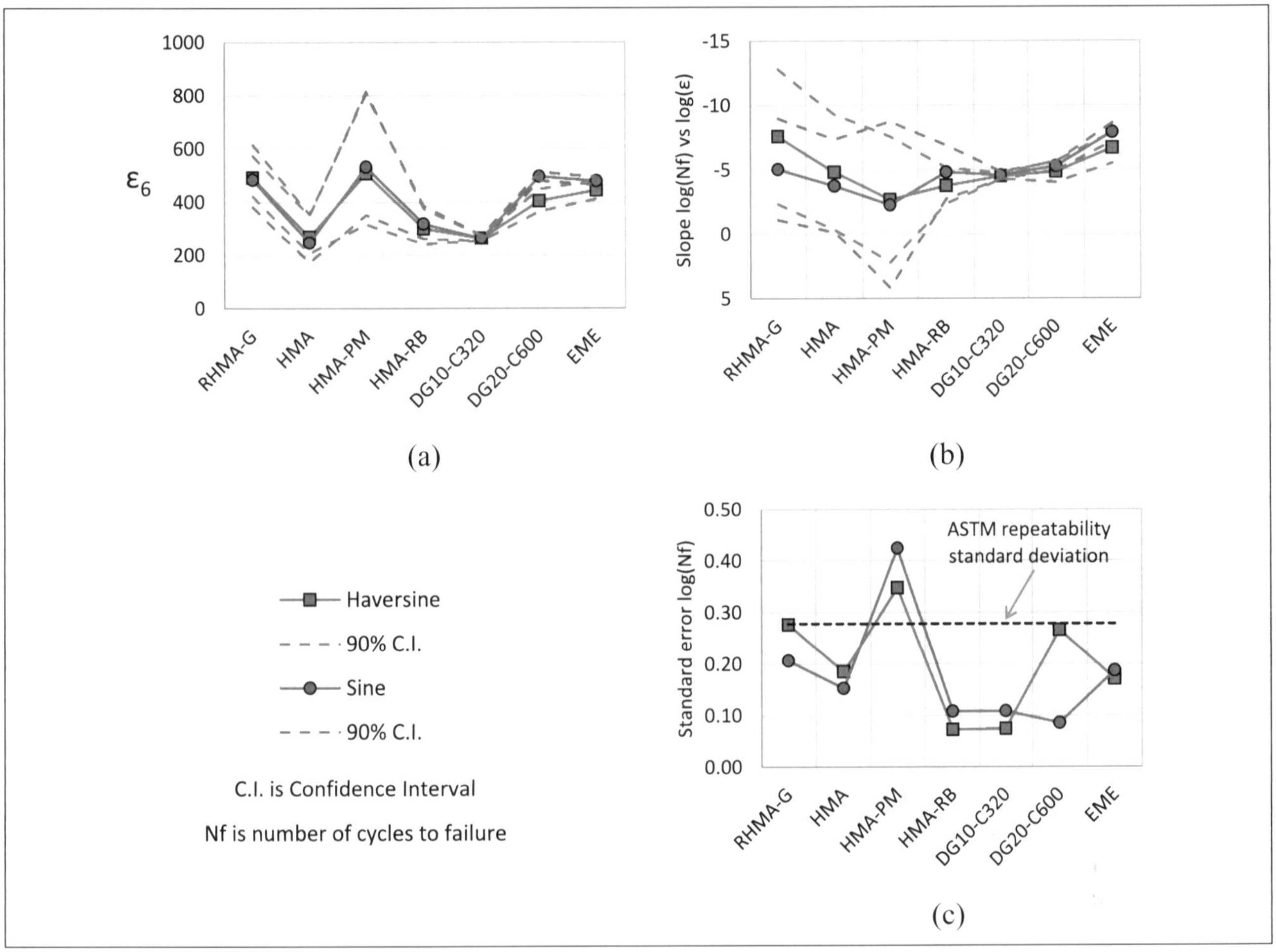

Figure 4. Comparison of sine versus haversine fatigue life results (50% stiffness reduction): (*a*) sine versus haversine comparison (ε_6); (*b*) sine versus haversine comparison (slope); (*c*) sine versus haversine comparison (fitting error).
Note: The relatively wide confidence intervals obtained for the slope of the Californian mixes are due to the fact that only two strain levels were tested and the two were relatively close.

stiffness reduction, which was the only fatigue life that could be determined for all the mixes. Two factors were considered in this ANOVA: strain level and test mode. The test mode effect was only significant (p-value =.0001) for the DG20-C600 mix. The interaction between test mode and strain was not significant in any of the cases. This means that the test mode effect (haversine versus sine) in DG20-C600 mix did not differ from one strain level to another. This result is in line with the earlier observation that haversine and sine fatigue laws for this mix seem to be parallel to each other.

Analysis of the Shape of the Stiffness Degradation Curves

The stiffness degradation curves, as used in the CalME program, were fitted for all the fatigue tests conducted at medium strain level (Table 1). The fitting was conducted using either the logit model, shown in Figure 1, or the logit-logistic model. The former is a modified version of the logit model which can be used to fit stiffness reduction data beyond the failure phase (*9*). In order to define a unique stiffness reduction curve representative of each mix and testing mode, the fitted functions were averaged (average of all replicates for medium strain level). Then the average haversine and sine stiffness degradation curves were scaled (in the cycles axis) so that both resulted in 50% stiffness reduction for the same number of cycles. This latter step was taken to make it easier to observe differences in damage (and reversible phenomena) accumulation during the tests. The resulting functions are presented in three different formats in Figure 5: (*a*) SR versus number of cycles (N) in the linear scale, (*b*) SR versus N in the logarithmic scale, which shows the stiffness reduction at the beginning of the tests better than the linear scale does, and (*c*) SR evolution based on the Weibull survivor function, which is recommended by the ASTM 4PFBF standard in order to extrapolate SR data (*11*).

Figure 5 shows that for six of the seven mixes the haversine and sine stiffness degradation curves are almost identical. Some differences for the HMA-RB and HMA-PM mixes appear beyond the transition from micro- to macro-cracking,

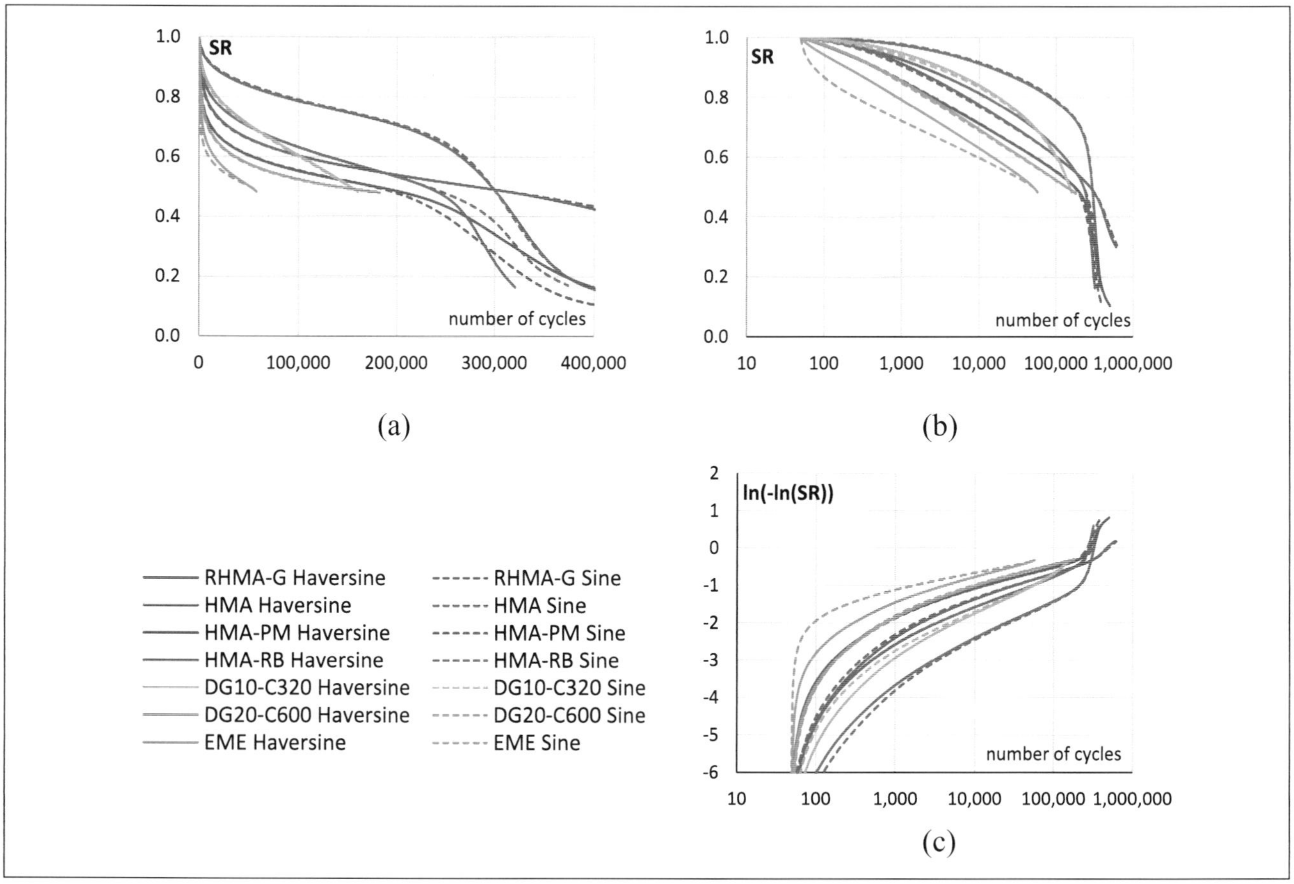

Figure 5. Shape of the stiffness degradation curves (SR versus number of cycles): (*a*) SR evolution in the linear scale; (*b*) SR evolution in the logarithmic scale; (c) SR evolution in the Weibull space.
Note: Haversine and sine stiffness degradation curves have been scaled (constant factor in the cycles axle) so that both coincide at the 50% stiffness reduction point.

and it is suspected that the differences are related both to the random nature of the macro-cracking process and the fact that only three replicates were tested for each mix and testing mode. As before, a different pattern was observed for mix DG20-C600, where the haversine and sine stiffness degradation curves clearly differed from each other. Stiffness reduction at the beginning of the sine mode testing exceeded that at the beginning of haversine mode testing for mix DG20-C600. That difference is opposite to what would have been expected, as explained below. The haversine mode testing results in higher tensile stresses than the sinusoidal mode testing, for the same peak-to-peak strain, during the first cycles of fatigue tests (*3*). For that reason, initial stiffness reduction could be expected to be greater in haversine tests than in sine tests, which is opposite to what happened for mix DG20-C600.

Analysis of Test Results in the Black Space

Damage accumulation in fatigue tests has long been known to be strongly related to dissipated energy (*12*). Therefore, the joint evolution of dynamic modulus and phase angle, upon which dissipated energy depends, was plotted in the Black space (dynamic modulus versus phase angle evolution during fatigue testing), providing another way to compare the results of haversine versus sine testing. For six of the seven mixes, the differences between average haversine and sine curves were considerably smaller than the variability between the tests, as shown in Figure 6, for example. However, as in the other tests where contradictory results were obtained, for mix DG20-C600 the differences between the average haversine and sine curves were greater than those for the test-to-test variability. For this particular mix, complex modulus evolution followed a different path depending on whether the test was conducted in haversine or sine mode. The main difference came from the phase angle, which was consistently higher in the sine testing than in the haversine testing. This higher phase angle appeared in the earliest cycles. This can also be seen in Figure 7, where initial dynamic modulus and phase angle are plotted for all mixes. In this figure, each point corresponds to a fatigue test. And again, haversine and sine seem to follow the same pattern for all the mixes except DG20-C600.

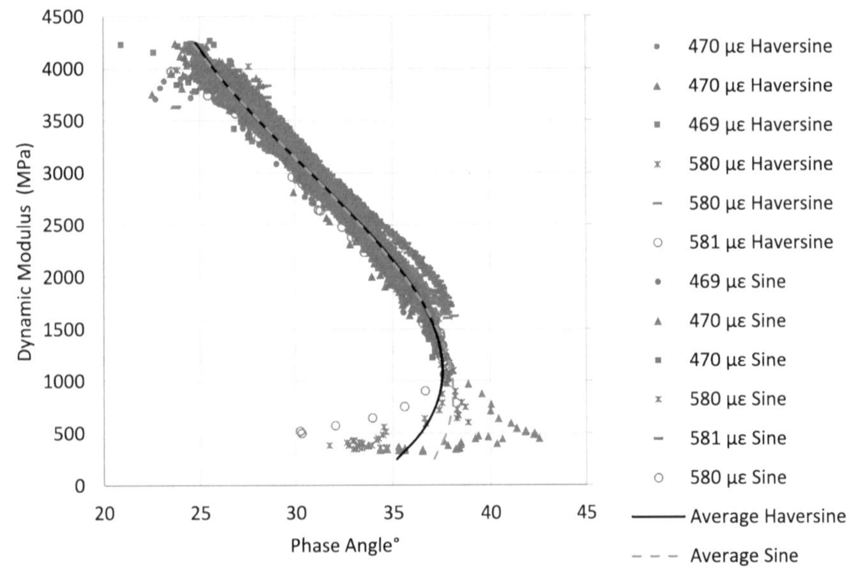

Figure 6. Haversine versus sine comparison in the Black space (RHMA-G).

Analysis of Variance for All Mixes

A series of ANOVAs was conducted for the seven mixes, in an attempt to capture information that could have been missed in the piecewise comparison described above (ANOVA of each individual mix). The overall analysis includes mix type, test mode (haversine or sine), and strain level, as fixed factors, and log fatigue life as the dependent variable. Mix type and testing mode were regarded as crossed factors, while strain level was regarded as a factor nested in mix type. This is because the meaning of strain level differs from one mix to another.

The ANOVA was first conducted using the 50% stiffness reduction fatigue lives. The ANOVA indicated that the test mode effect was not significant (p-value =.105), although the first order interaction "Mix*Test Mode" was significant (p-value =.003). This indicates that the test mode effect may vary among the different mixtures, which is in line with the significance levels obtained for the individual mixes.

A second ANOVA that did not include mix DG20-C600 was also conducted. In this second analysis, neither test mode (p-value =.955) nor its interaction with mix type (p-value =.438) were significant. This indicates that haversine and sine

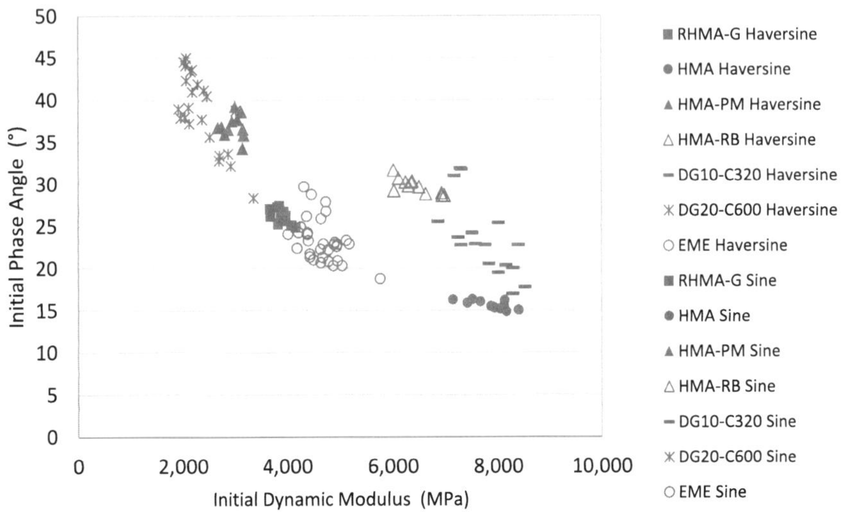

Figure 7. Haversine versus sine comparison in terms of initial complex modulus (at cycle 50).

did not result in statistically different 50% stiffness reduction fatigue lives.

The fact that the ANOVA was conducted using the 50% stiffness reduction fatigue lives was considered a limitation, since this failure criterion was regarded as somewhat arbitrary and not tied to a change of the damage phenomenon. Compared with the 50% stiffness reduction criterion, fatigue lives determined using the $(SRxN)_{MAX}$ criterion are much better correlated to the transition from micro- to macrocracking (*7*). For this reason, a third ANOVA was conducted using fatigue lives determined according to $(SRxN)_{MAX}$ failure criterion. Only one of the Australian mixes could be included in this ANOVA: DG-C320. As explained above, the $(SRxN)_{MAX}$ failure point could be extrapolated in most tests conducted on DG10-C320 mix, since this failure point was only slightly beyond the 50% stiffness reduction cycle. For the other two Australian mixes, the extrapolation was not regarded as reliable. Results of this third ANOVA indicated that neither test mode (p-value =.965) nor its interaction with mix type (p-value =.472) were significant.

The mean square error of the ANOVA analysis was 0.04 (0.20 standard deviation) when the failure criterion was 50% stiffness reduction, and it was 0.068 (0.26 standard deviation) when $(SRxN)_{MAX}$ criterion was used. These values are somewhat smaller than the repeatability standard deviation of 0.278 reported by ASTM D7460-10.

Discussion of Results

The analyses presented above showed no differences between haversine and sine test results for six of the seven asphalt mixtures evaluated in this study. For these six mixes, the haversine and sine fatigue lives did not seem to differ from each other when plotted versus strain in a double logarithmic graph (i.e., log [fatigue life] versus log[strain]). In particular, 90% confidence intervals determined for ε_6 for the haversine and sine testing modes overlapped. The slope of the curves and the fitting errors were similar between the haversine and sine testing modes even though both parameters changed considerably from one mix to another. The ANOVA analyses did not show statistically significant differences between the two testing modes, in terms of the fatigue lives determined for two different failure criteria, for these six mixes. In addition, the shapes of the stiffness degradation curves (residual SR versus number of cycles) were almost identical when haversine and sine were compared with each other. A similar conclusion was reached when the joint evolution of dynamic modulus and phase angle during fatigue testing (Black diagrams) was considered for these six mixes.

This similar performance is believed to be strongly related to the viscoelastic nature of the asphalt mix. Because of this viscoelastic behavior, a flexural bending test that is initiated as haversine rapidly changes into a sinusoidal test conducted on a bended beam. This rapid change is illustrated in an example in Figure 8, and it has been shown by other authors before (*2*, *3*). As Figure 8 shows, the initial peak downward load is higher in the haversine case, since initial displacement (versus zero position) is twice the value of the sine test. Nonetheless, the asphalt mix flows due to its viscoelastic nature so the beam at-rest position will progressively deflect too. Consequently, the load required to bend the beam to the maximum displacement will lessen while a negative force will be required to bring the beam back to the original at-rest position. After a number of cycles, around 50 in the example shown in Figure 8, the beam at-rest position will be half way between zero and maximum displacement,

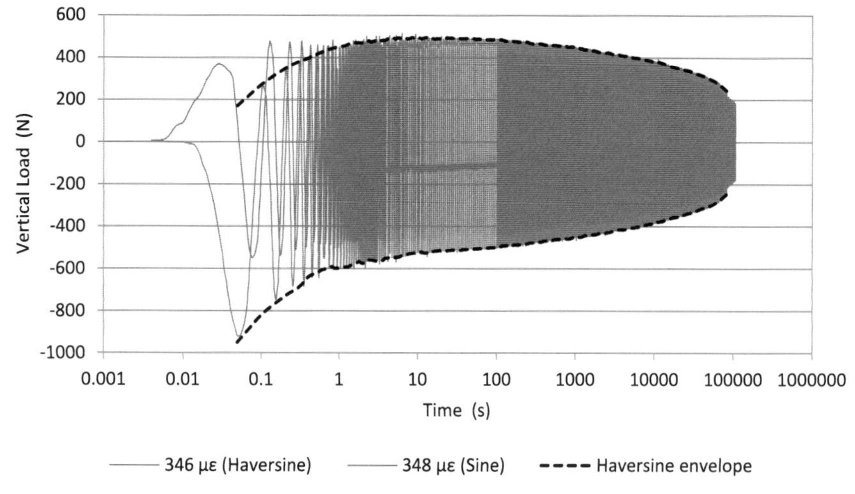

Figure 8. Evolution from haversine to "sine on deflected beam" (HMA-RB, 350 με peak-to-peak).
Note: Haversine mode loading seems to turn into sine in 50 cycles, approximately. After this transition, the peak-to-peak load differences are attributable to the fact that two different specimens are compared in this figure.

so the force required to reach both extremes will be the same magnitude but with the opposite sign. This means that, stress wise, there is no difference between haversine and sine fatigue testing after a number of cycles have been applied. The fact that initial stiffness and phase angle (determined at cycle 50) were similar between haversine and sine testing for these six mixes (Figure 7) supports the observation that the transition from haversine to "sine on deflected beam" takes place very rapidly, i.e., the beam rapidly "forgets" what its original position was.

As has been shown in all the results, one of the seven mixtures tested, DG20-C600, yielded a different set of conclusions in the sine-versus-haversine comparison. DG20-C600 is a dense-graded Australian mix typically used as a base layer, and it was tested at 30°C. For this particular mix, sinusoidal testing resulted in considerably longer fatigue lives than haversine testing (one to two times longer). These differences were shown to be statistically significant. The shape of the stiffness degradation curves showed clear differences between the haversine and sine testing modes for this particular mix. Clear differences were also observed in the Black space, since phase angle was systematically higher in sinusoidal testing compared with the haversine. Haversine testing mode resulted in much larger variability than the sine mode in this particular mix, and it is believed the relatively high variability might explain the different performance of this mix compared with the other six mixes. After a rigorous evaluation, no definitive explanation was found for the performance of this mix. In particular, the transition from haversine to "sine on deflected beam" was verified for this mix, and it took place even faster than in most of the other mixes (*7*). For these reasons, it is believed this mix is an outlier and can be ignored.

It is important to bear in mind that the key to the similarity between haversine and sine testing is the viscoelastic nature of the asphalt mix. Beam flexural testing conducted at low temperatures or on mixes whose binder has undergone considerable aging may be sensitive to whether haversine or sinusoidal displacement is used. Something similar might happen for very hard binders tested at intermediate and low temperatures. However, none of these possible explanations apply to mix DG20-C600.

Summary, Conclusions, and Recommendations

Summary

An experimental study was conducted to determine the effect of displacement waveform on fatigue test results for hot mix asphalt. Seven asphalt mixtures, comprising a wide variety of gradations, binder types, and binder contents were selected for this study. Four of the mixes were tested at the UCPRC, and three were tested by the Australian ARRB Group. The mixes were tested at different strain levels under both haversine and sinusoidal displacement-controlled modes, without introducing rest periods between load cycles, and results from the two testing modes were compared with each other from different perspectives. This comparison allowed a detailed evaluation of the effects each testing mode has on fatigue testing results.

- For six of the seven mixes, no evidence was found to indicate that the haversine and sine testing modes resulted in different beam stiffness degradation. Fatigue life, the shape of the stiffness reduction curves, initial dynamic moduli and phase angles, and Black diagrams were essentially the same for haversine and sine testing.
- The reason for the similarity of the results from haversine and sine testing is attributable to the viscoelastic nature of the asphalt mix. Experimental data showed that haversine tests rapidly change into sine tests conducted on a bended beam, so the same stress is produced by haversine and sine displacement waveforms as soon as the peak-to-peak strain amplitudes are equal.
- The findings indicate that a displacement-controlled haversine test produces the same effects on the tested beam as a displacement-controlled sinusoidal test where the zero-to-peak displacement is half of the of peak-to-peak displacement applied in the haversine test.
- For one of the seven mixes, sinusoidal testing produced considerably longer fatigue lives than haversine testing (one to two times longer). No definitive explanation was found for the outcome of the testing on this mix and it is believed this mix is an outlier and can be ignored.

Conclusion

Based on these results, there is no compelling reason to recommend sinusoidal over haversine testing mode, or vice versa. This is because the same fatigue life and stiffness degradation curves were found from the two testing modes when rest periods are not included between loading cycles. A practical reason to use the sine testing mode is that the stress in the haversine alternative becomes sinusoidal after a small number of load repetitions.

It was also concluded that none of the historical fatigue model parameters calibrated by UCPRC (using haversine mode) would need to be adjusted if sinusoidal mode of testing was adopted in place of the haversine mode that has been used for nearly 25 years.

It must be noted that the results obtained in this experimental study should not be extrapolated to conditions where the nature of the asphalt mix is more elastic than viscoelastic; examples of these conditions include testing the mix at low temperatures or conducting tests on mixes with very hard binders or whose binder has undergone considerable aging. Under these conditions, 4PFBF test results may indeed be sensitive to whether haversine or sinusoidal displacement is used.

Recommendations

It is recommended that the ASTM and AASHTO test methods be harmonized to use either sinusoidal or haversine testing mode. Although sinusoidal mode is preferred for reasons of simplicity, the UCPRC should follow the test mode selected by the harmonized ASTM and AASHTO test methods.

To avoid confusion, it is recommended that whether the strain in 4PFBF testing refers to peak-to-peak or zero-to-peak displacement amplitude should be reported systematically.

Acknowledgments

This paper describes research activities that were requested and sponsored by the California Department of Transportation (Caltrans) and the Queensland Department of Transport and Main Roads. The sponsorship of both is gratefully acknowledged.

Author Contributions

The authors confirm contribution to the paper as follows: study conception and design: Angel Mateos, Rongzong Wu, Erik Denneman, and John Harvey; data collection: Angel Mateos, Rongzong Wu, Erik Denneman; analysis and interpretation of results: Angel Mateos, Rongzong Wu, Erik Denneman, and John Harvey; draft manuscript preparation: Angel Mateos. All authors reviewed the results and approved the final version of the manuscript.

References

1. Deacon, J., A. Tayebali, J. Coplantz, F. Finn, and C. Monismith. *Report SHRP-A-404: Fatigue Response of Asphalt-Aggregate Mixes*. Institute of Transportation Studies, University of California, Berkeley, CA, 1994.
2. Mamlouk, M. S., M. I. Souliman, W. A. Zeiada, and K. E. Kaloush. Refining Conditions of Fatigue Testing of Hot Mix Asphalt. *Advances in Civil Engineering Materials*, Vol. 1, No. 1, 2012, pp. 1–13.
3. Pronk, A. C., M. R. Poot, M. M. J. Jacobs, and R. F. Gelpke. Haversine Fatigue Testing in Controlled Deflection Mode: Is It Possible? Presented at 89th Annual Meeting of the Transportation Research Board, Washington, D.C., 2010.
4. Mateos, A., J. Ayuso, and B. Cadavid. Shift Factors for Asphalt Fatigue from Full-Scale Testing. *Transportation Research Record: Journal of the Transportation Research Board*, 2011. 2225: 128–136.
5. Criqui, W. B., and C. D. Wendell. Flexural Beam Fatigue Equipment Nomenclature Issues. *Proc., 3rd Conference on Four-Point Bending*, Davis, CA, 2012.
6. Wu, R., and J. Harvey. Calibration of Asphalt Concrete Cracking Models for California Mechanistic-Empirical Design (CalME). *Proc., 7th RILEM International Conference on Cracking in Pavements: Mechanisms, Modeling, Testing, Detection and Prevention Case Histories*, Delft, the Netherlands, Springer Netherlands, 2012, pp. 537–547.
7. Mateos, A., R. Wu, E. Denneman, A. Liu, and J. Harvey. *Evaluation of the Effect of Deflection Waveform on Fatigue Test Results for Hot Mix Asphalt*. University of California Pavement Research Center Technical Memorandum: UCPRC-TM-2015-03, 2017.
8. Rowe, G. M., and M. G. Bouldin. Improved Techniques to Evaluate the Fatigue Resistance of Asphaltic Mixtures. *Proc. 2nd Eurasphalt and Eurobitume Congress*, Barcelona, Spain, Vol. 1, Breukelen, 2000, pp. 754–763.
9. Mateos, A., R. Wu, J. Harvey, E. Denneman, and A. Fan. The Logit Model and the Need to Reproduce the Stiffness Degradation Curve of Asphalt Specimens During Fatigue Testing. *Transportation Research Record: Journal of the Transportation Research Board*, 2017. 2631: 105–113.
10. Di Benedetto, H., C. De La Roche, H. Baaj, A. Pronk, and R. Lundström. Fatigue of Bituminous Mixtures. *Materials and Structures*, Vol. 37, No. 3, 2004. pp. 202–216.
11. Tsai, B. W., J. Harvey, and C. L. Monismith. Application of Weibull Theory in Prediction of Asphalt Concrete Fatigue Performance. *Transportation Research Record: Journal of the Transportation Research Board*, 2003. 1832: 121–130.
12. Van Dijk, W., and W. Visser. Energy Approach to Fatigue for Pavement Design. *Association of Asphalt Paving Technologists Proceedings*, Vol. 46, 1977, pp. 1–40.

The Standing Committee on Structural Requirements of Asphalt Mixtures (AFK50) peer-reviewed this paper (18-05276).

Article

TRR
JOURNAL OF THE TRANSPORTATION RESEARCH BOARD

Transportation Research Record
2018, Vol. 2672(28) 383–393

DOI: 10.1177/0361198118792114
journals.sagepub.com/home/trr
SAGE

Impact of Specimen Configuration and Characteristics on Illinois Flexibility Index

Jose Rivera-Perez[1], **Hasan Ozer**[1], **and Imad L. Al-Qadi**[1]

Abstract

The Illinois Department of Transportation adopted the Illinois Flexibility Index Test (I-FIT) to evaluate the cracking vulnerability of asphalt concrete (AC) mixtures that was often shown to increase with the addition of recycled materials such as reclaimed asphalt pavement and recycled asphalt shingles. The test consists of a semi-circular AC sample that has a notch, loaded along the symmetric axis. Fracture energy (FE), post-peak slope, and the flexibility index (FI) are computed from the load displacement curve. These results can be influenced by specimen geometry and test parameters such as loading rate, AC voids content, and so forth. Therefore, this study investigated the effect of notch length, specimen thickness, loading rate, and AC air void content on the I-FIT results. It was found that an increase in the specimen thickness or loading rate resulted in a steeper post-peak slope without affecting the FE. As a result, the FI decreased. An increase in the notch length or AC air void content resulted in a flatter post-peak slope, thus, increasing the FI. From the results, it was concluded that existing correction factors to address the variations caused by specimen thickness and air void content are appropriate. A correction factor to address notch length variations is proposed. A unique correction factor for loading rate could not be developed because of the varying rate dependency of each AC mixture.

Asphalt pavements are among the top recycled products in the United States. According to the National Asphalt Pavement Association, over 90% of asphalt pavements materials are recycled annually (*1*). The use of recycled materials, such as reclaimed asphalt pavements (RAP) and recycled asphalt shingles (RAS), helps achieve sustainable pavements goals. However, the use of recycled products can compromise the performance of asphalt pavements. Use of RAP leads to a decrease in rutting at intermediate and high temperatures, but it increases low-temperature cracking potential (*2*, *3*) as a result of the aged and brittle binder that stiffens the asphalt concrete (AC) mix. In addition, RAP stockpiles do not have consistent aggregate gradations, and the selection of a representative bulk specific gravity can be difficult. Thus, the use of RAP makes it difficult to achieve the desired volumetric for AC mixtures (*4*). As a result of these challenges, Superpave volumetric requirements may not be sufficient to achieve well-performing AC, especially when using higher percentages of RAP. The Hamburg Wheel Tracking test is commonly used by agencies to ensure rutting resistance of AC mixes (*5*). However, there is a need for a cracking prediction test, along with the Superpave Volumetric and the Hamburg Wheel Tracking test, to design balanced AC mixtures that can withstand both cracking and rutting during pavements design life.

In 2015, the Illinois Center for Transportation at the University of Illinois at Urbana-Champaign (UIUC) published the report "Testing Protocols to Ensure Performance of High Asphalt Binder Replacement Mixes Using RAP and RAS," proposing a test methodology to evaluate the cracking prediction of AC mixtures (*4*, *6*, *7*). In the study, multiple fracture tests were evaluated to identify a method that can detect changes in cracking performance among AC mixes with variable recycled contents. One of the primary goals of the study was to develop a practical and low-cost test method that can be easily implemented by the industry. The Illinois Flexibility Index Testing (I-FIT) protocol was developed as a result of this study and adopted by the American Association of State Highway and Transportation Officials Provisional Standard TP 124-16 (AASHTO TP

[1]Department of Civil and Environmental Engineering, University of Illinois at Urbana-Champaign, Urbana, IL

Corresponding Author:
Address correspondence to Jose Rivera-Perez: riverap2@illinois.edu

124-16). The I-FIT protocol uses the semi-circle bending test geometry with a vertical notch at the center bottom of the beam to create mode-I fracture. The test is run at a temperature of 25°C and a loading rate of 50 mm/min, thus allowing specimen testing without the need for an environmental chamber.

The study found that fracture energy (FE) was insufficient for discriminating cracking performance within the AC mixtures (*4*). The FE can be similar between AC mixtures showing different crack propagation (*4*). To address this issue, the flexibility index (FI) was developed based on the total fracture resistance and post-failure crack propagation. The calculation of FI includes both FE and the slope at the inflection point of the post-peak curve (*6*). This parameter was shown to discriminate cracking performance within AC mixtures of variable recycled material contents (*7*).

In general, most of the fracture and strength tests are size and geometry dependent and sometimes rate dependent for materials like AC mixtures. Small deviations from thickness, notch depth, density, and loading rate can lead to erroneous reporting of results. Therefore, it is very important to follow the standard test geometry and specifications. Although, laboratory-compacted specimens can be prepared to meet the I-FIT test requirements, field cores may not meet such requirements because of insufficient lift thicknesses. Also, contractors might lack proper equipment needed for preparation of the notch, thus failing to meet the standard notch length requirement. In addition to the geometric challenges, the field cores have variations in the air void content that can affect FI results (*8*). Previous studies proposed correction factors for the thickness and air void content variation (*4, 8*). These studies helped address the test dependence of the FI, but other factors are yet to be considered. There is a need for a study to evaluate sensitivity of various factors affecting I-FIT results.

Objectives

This study aims at evaluating the effect of various geometric and test parameters on I-FIT results. Specifically, the following three goals were identified:

- Evaluate the effect of AC specimen thickness, notch length, and air void content, and the loading rate on the FI, post-peak slope, and FE.
- Conduct a statistical analysis to determine the sensitivity of I-FIT results to specimen geometric and loading parameters.
- Develop a correction factor to be used for specimens that deviate from the thickness, notch length, and air void content specified by AASHTO TP 124-16.

To meet these objectives, I-FIT specimens were produced using a laboratory-designed AC mixture with controlled volumetric properties and no recycled content. The specimens were then tested at multiple thicknesses, notch length, loading rates, and air void contents separately, and results were statistically analyzed.

Materials and Experimental Methodology

The Illinois Flexibility Index Test for cracking resistance characterization (AASHTO TP 124-16) was used to evaluate the impact of geometric and loading characteristics on the tests results of AC mixes. Details of the testing method and materials are described here.

Illinois Flexibility Index Test

The Illinois Flexibility Index Test (AASHTO TP 124-16) uses a semi-circular asphalt specimen subjected to three-point bending, as shown in Figure 1. The specimen has 50.0 ± 1.0 mm thickness and 150.0 ± 1.0 diameter with a 15.0 ± 1.0 mm notch made at the center of the specimen (*9, 10*).

During the test, the specimen is placed over two rollers, with a monotonic load applied along the vertical diameter of the specimen at a displacement rate of 50.0 ± 1.0 mm/min controlled by linear variable differential transformer (LVDT). From the load vs. displacement curve, the work of fracture and FE can be computed as follows (*9*):

Work of Fracture:

$$W_f = \int_0^{u_f} P\,du\,(\text{Joules}) \tag{1}$$

Fracture Energy:

$$G_f = \frac{W_f}{Crack\ ligament\ Area} X10^6 \left(\frac{Joules}{m^2}\right) \tag{2}$$

where

P = the load;
u = LVDT displacement;
W_f = fracture energy;
G_f = work of fracture;
u_f = LVDT displacement at the end of the test.

The FI is computed based on the fracture energy and slope of the post-peak load versus displacement curve as (*9*):

Post-Peak Slope:

$$m = \frac{dP}{du}\,(at\ Post\ Peak\ Inflection\ Point) \tag{3}$$

FI:

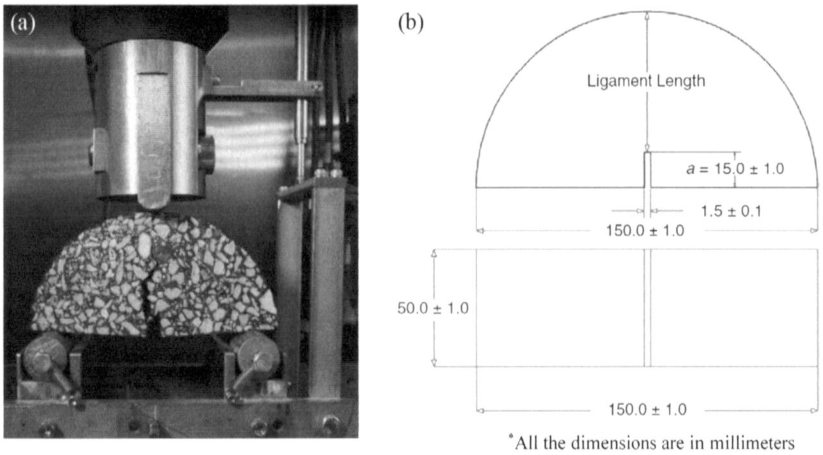

Figure 1. Illinois Flexibility Index Test: (*a*) fixture and (*b*) specimen geometry.

$$FI = \frac{G_f}{m} XA \tag{4}$$

where m = slope (kN/mm);

A = unit conversion factor and scaling coefficient (taken as 0.01).

The FE indicates the energy required to create a unit surface area of a crack, and the slope is an indication of the crack propagation speed; a higher post-peak slope indicates a faster (brittle) crack propagation and vice versa. As the FI increases, flexibility of the AC mixture increases and cracking vulnerability decreases. Currently, the Illinois Department of Transportation considers a FI of 8 as a threshold for AC mix acceptance.

Table 1. Control Mixture Volumetric Properties

ID	Mix 1
Binder	PG 64-22
Binder content	6.0%
Bulk specific gravity at 90 gyrations	2.394
Voids in total mixture	3.9%
Voids in mineral aggregate	15.2%
Dust proportion	0.85
Voids filled with asphalt	74.40%
Effective specific gravity	2.739
Absorbed asphalt	1.20%
Effective asphalt	4.9%
Mixing temperature	155°C
Compacting temperature	147°C

Materials

One control AC mixture was designed and produced at the Advanced Transportation Research and Engineering Laboratory of the UIUC to be used in this study. In contrast to plant-produced AC mixtures, laboratory-designed AC allows strict control of the volumetric properties. The mix design target was 4% air void content at 90 gyrations. The mixture was designed using local aggregates: dolomite, crushed dolomite, natural sand, and mineral filler. The aggregate gradation resulted in the volumetric properties shown in Table 1.

I-FIT samples were prepared from the cylindrical pills of the laboratory-designed AC mixtures. The pills were compacted using a Superpave gyratory compactor at 7.0 ± 0.5% air void content. Then, the cylindrical samples of 150 mm diameter and 180 mm height were cut into semi-circular samples. When the pills were cut, the top and bottom surfaces of the pill were removed, and two circular slices (50.0 ± 1.0 mm thickness) were cut from the middle portion of the pill. Finally, the circular slides were divided into two semi-circular specimens and a 15.0 ± 1.0 mm notch (15 mm length 1.5 ± 1.0 mm width) was carved in the middle. As a result, the standard I-FIT samples had a 50.0 ± 1.0 mm thickness, 15.0 ± 1.0 mm notch length, and 7.0 ± 0.5% air void content. Then, depending on the parameter to be studied, properties of the specimen and the test were modified according to Table 2. For each test, replicate specimens were tested.

Statistical Analysis

An analysis of variance test (ANOVA) was used to determine the statistical significance of the results found in this study. Typically, the results showed that the dependent variable remained constant or varied linearly with respect to the independent variable. Therefore, two

Table 2. Summary of Conducted Tests

Tested parameter	Test No.	Thickness (mm)	Air void (%)	Loading rate (mm/min)	Notch length (mm)	Binder grade
Thickness	1	25	7	50	15	PG 64-22
	2	30				
	3	40				
	4	45				
	5	50				
	6	60				
Air void (VTM)	7	50	2	50	15	PG 64-22
	8		4			
	9		6			
	10		8			
	11		10			
Loading rate	12	50	7	0.7	15	PG 64-22
	13			6.25		
	14			25		
	15			50		
Notch length	16	50	7	50	10	PG 64-22
	17				15	
	18				20	
	19				35	

types of analysis were conducted to determine if the trends were statistically significant. For the results that showed no variation, a one-way ANOVA analysis was conducted using IBM SPSS Statistic 22 to compute the p-value. The results are not statistically different if the p-value is larger than 0.05. For results that showed a linear variation, a regression was fit using Excel 2016 to perform an F-Test. Then, the coefficient of determination (r-square) and the significance F (probability that the regression does not fit the data) was computed. If the significant F value is lower than 0.1, the results fits to a linear regression trend.

Results and Discussion

Thickness Impact

I-FIT specimens were prepared with target thicknesses of 25, 30, 40, 45, 50, and 60 mm to evaluate the thickness impact on the FI. Figure 2*a* presents the load versus displacement curves of representative specimens from each thickness. The peak load, post-peak slope, and the area under the curve (work of fracture) increased as the thickness increased. As a result, additional work of fracture was needed to break the fracture area where the crack propagated when the thickness increased. However, as the thickness increased, the specimen appeared to be less compliant in the pre-peak portion of the load versus displacement curve.

The average I-FIT results are shown in Figure 2, *b–d*. It was expected that the FE increases with larger thicknesses as a result of the higher confining stresses at the crack fronts (*8*). However, the results did not indicate a significant change in FE. Using the average values shown in Figure 2*b*, a one-way ANOVA analysis resulted in a p-value of 0.54, indicating that there is no significant variation in FE. Combining results from the six thicknesses, an average FE of 2301 J/m^2 with a standard deviation of 196 J/m^2 and a coefficient of variation (COV) of 8.5% were computed. The COV value is within the acceptable range. The post-peak slope increased linearly with the increased thickness, however, resulting in a decrease in the FI, as shown in Figure 2*c*. Statistical analysis suggested that the slope and the FI fitted a linear regression relationship with an r-square of 0.901 and 0.800, respectively. The F values are 3.1×10^{-7} and 1.8×10^{-7} for the slope and FI, respectively. Therefore, a correction factor for the slope is needed when the specimen thickness varies from that of the specification.

The correction factor for the slope as presented by Barry (*8*) is shown in Equation 5:

$$m_{50} = \frac{50}{t} X\ m \tag{5}$$

Corrected results for slope and FI are presented in Figure 3. The one-way ANOVA analysis of the corrected slopes resulted in a p-value of 0.57. Therefore, the variation in slope was corrected. The average corrected post-peak slope was 1.32 (considering the results from the various thicknesses) with a standard deviation of 0.16 and a COV of 12.2%, respectively. The calculated corrected FI = 17.1 is shown in Figure 3*b* (standard deviation: 2.9 and COV of 16.5%). The p-value for the corrected FI was 0.47 and, therefore, the correction factor from the literature can effectively address the variation in thickness.

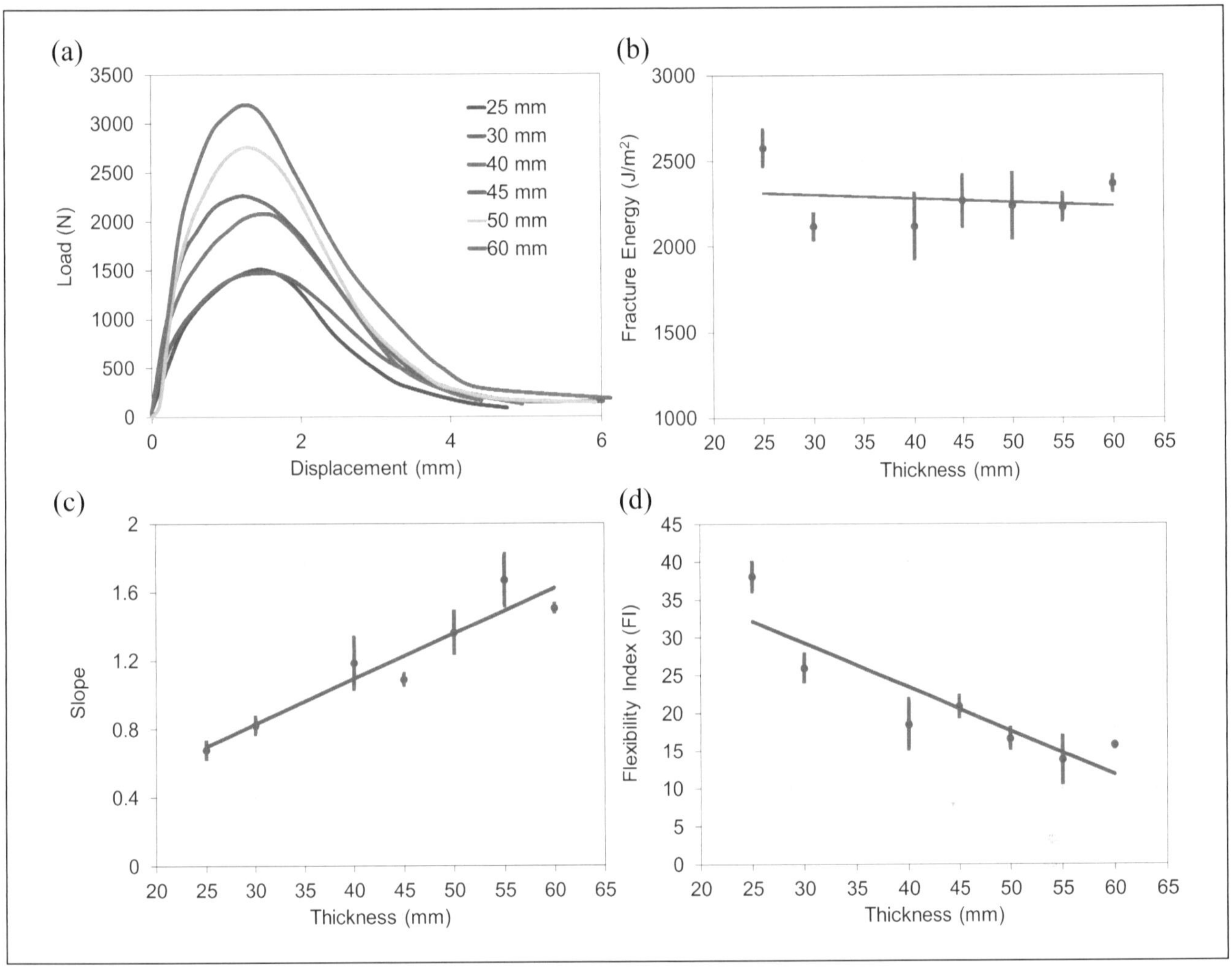

Figure 2. I-FIT results for the thickness impact: (*a*) load vs displacement curve, (*b*) FE, (*c*) slope, and (*d*) FI.

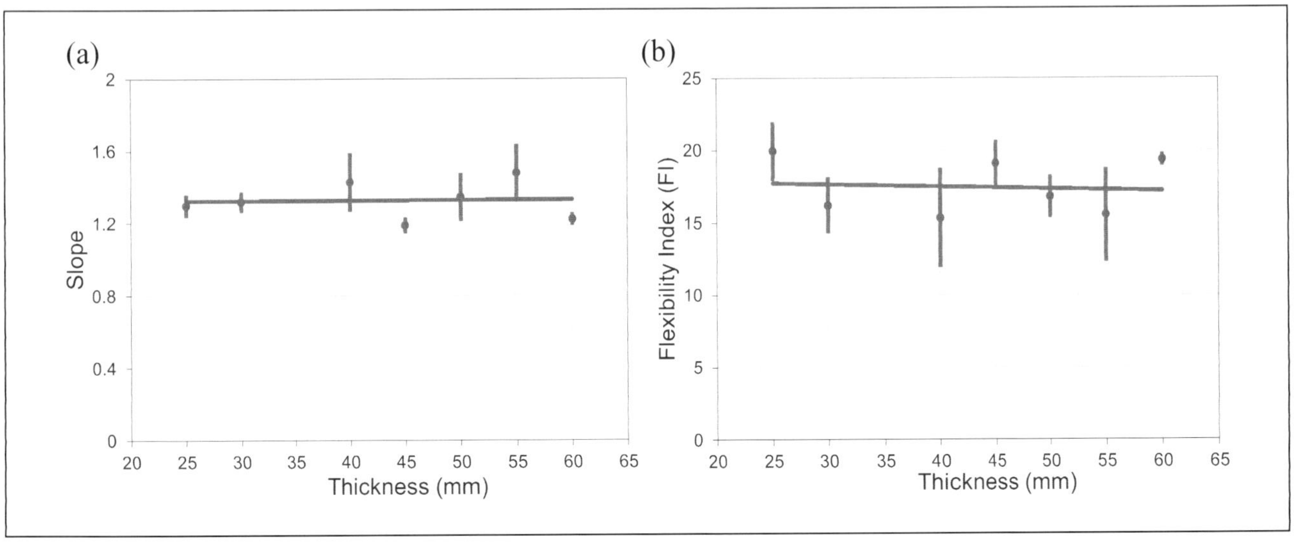

Figure 3. (*a*) Thickness corrected slope and (*b*) Thickness corrected FI.

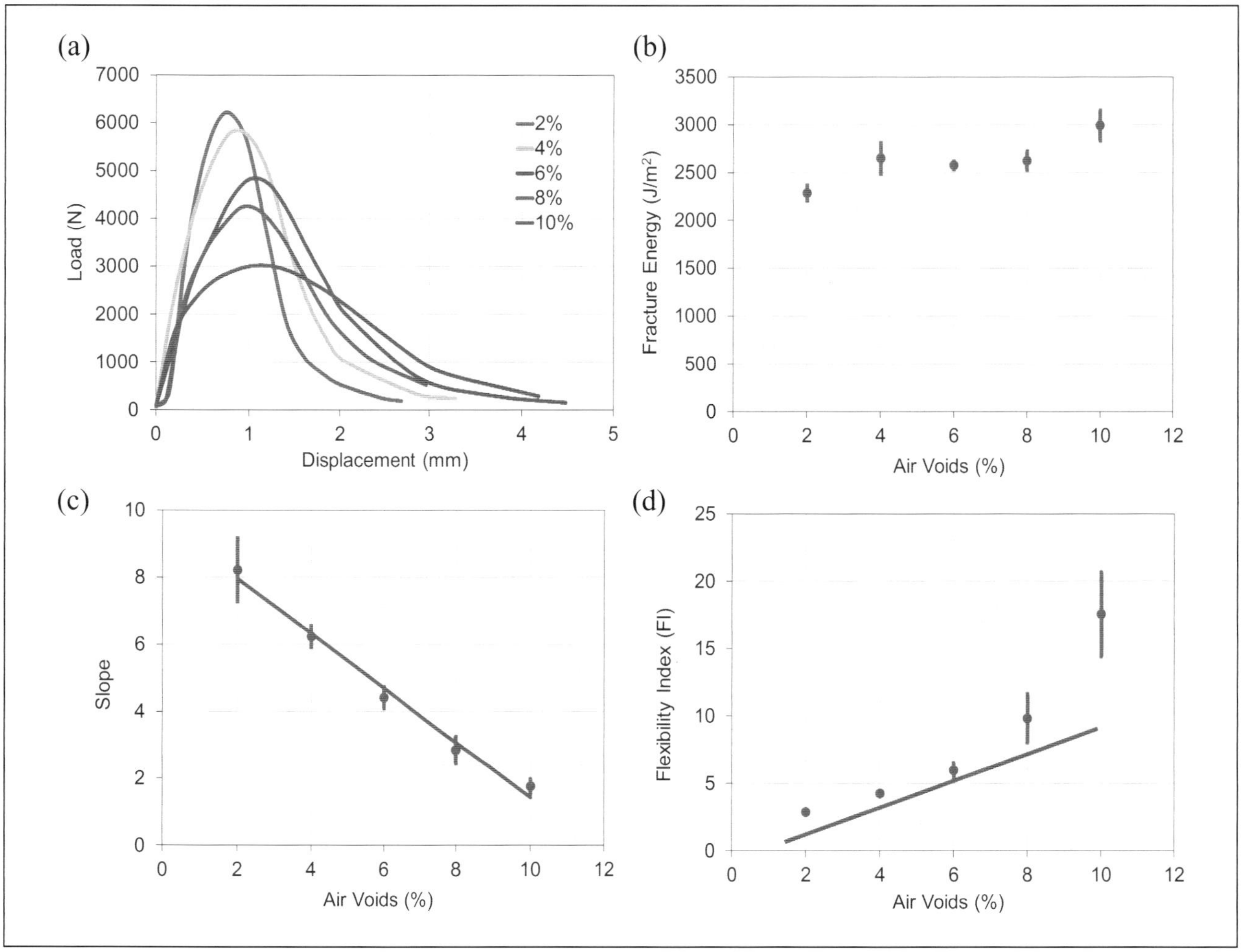

Figure 4. I-FIT results for the air void content impact: (*a*) load vs. displacement curve, (*b*) FE, (*c*) slope, and (*d*) FI.

Air Void Content Impact

I-FIT specimens were prepared with target air void content of 2, 4, 6, 8, and 10% using the Mix 1. Volumetrics were measured for each specimen to determine the air void content. The specimens were prepared and tested under the same conditions to allow for a comparison of the air void content. Figure 4*a* shows the load versus displacement curve for several representative specimens. As the air void content increases, both peak load and post-peak slope decrease. This was because of the increased rate of damage propagation induced by the specimen's air void content.

The I-FIT analysis results are shown in Figure 4*b*. Initially, it seemed that the FE slightly increased with the increase in the air void content. However, the one-way ANOVA statistical analysis resulted in a *p*-value of 0.486, suggesting that FE did not statistically vary with respect to air void content. The average FE was 2258 J/m^2 and the standard deviation was 218 J/m^2. This results in a COV of 9.7%, which is within the acceptable ranges for FE. The post-peak slope linearly decreased as the air void content increased (Figure 4*c*). A linear regression was fit to slope values versus specimen's air void content, resulting in a significant *F* lower than 0.1 (resulting in an accepted fit). Therefore, the slope decreased, resulting in an increase in the FI. As shown in Figure 4*d*, the FI linearly increased until 8% air void content then experienced an exponential increase at 10% air void. The effect of air void on fracture resistance is complex. As air void increases, the mixture tends to deteriorate faster because of the reduction in the load-carrying capacity as well as the increased effects of moisture and oxidation penetration. However, it can also be expected that the increase in air void would make the mixture more ductile (damage can be diffused at the crack tip, thus the delay in the initiation of a major crack) to a certain extent. That is why the slope values, as an indicator of crack initiation and propagation speed, decrease with the increase in air void. A global and microstructural evaluation of such energy dissipation mechanisms was conducted using the

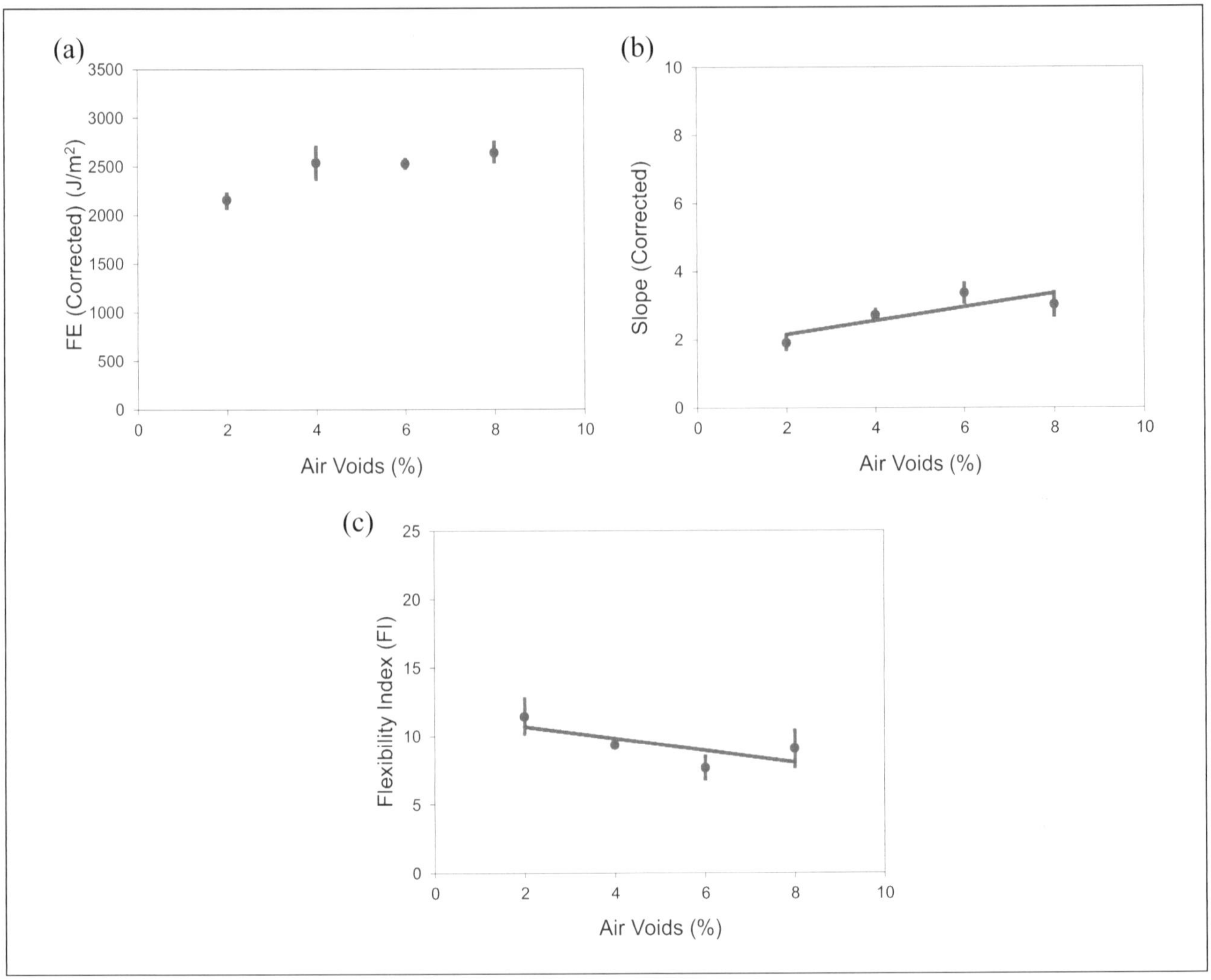

Figure 5. Air void content corrected: (*a*) FE, (*b*) slope, and (*c*) FI.

digital image correlation technique with high-resolution cameras elsewhere (*11*, *12*). Therefore, one must be cautious when interpreting the FI values with changing air voids. FI is an indicator of ductility of the mixture but should be balanced with the stiffness and strength of the mixture.

In addition, the increase in FI is attributed to additional inelastic mechanisms of energy dissipation away from the crack front that may also delay fracture initiation and propagation. Such inelastic mechanisms can include plasticity and viscoelasticity, damage occurring in the bulk of the specimen outside the crack tip. The exponential increase in the FI as a result of notable changes in the load displacement curve at 10% air void indicates that this test can no longer be accepted as a fracture test when specimens are experiencing such high levels of non-fracture-related energy dissipation. As a result, all values up to 8% air void content were included within the correction factor for air voids. Specimens with air void content above 8% may not provide proper results at this point.

The correction factors proposed by Barry (*8*) were used to correct for the air void content variation and results are shown in Figure 5. Results indicate that the correction factor for the FE did not have a significant impact; the corrected and uncorrected FE values are similar, Figure 5*a*. On the other hand, the slope was corrected using the Barry (*8*) approach (Figure 5*b*). The average corrected slope was 2.82 with a standard deviation of 0.59 and 21% COV, respectively. Consequently, the average corrected FI, using the results of Figure 5*c*, becomes 9.52 with a standard deviation of 2.25 and 23.6% COV, respectively. The ANOVA analysis indicates a *p*-value of 0.588 and 0.081 for the slope and the FI, respectively (p-value > 0.05 results in successful correction). It was concluded that the air void content

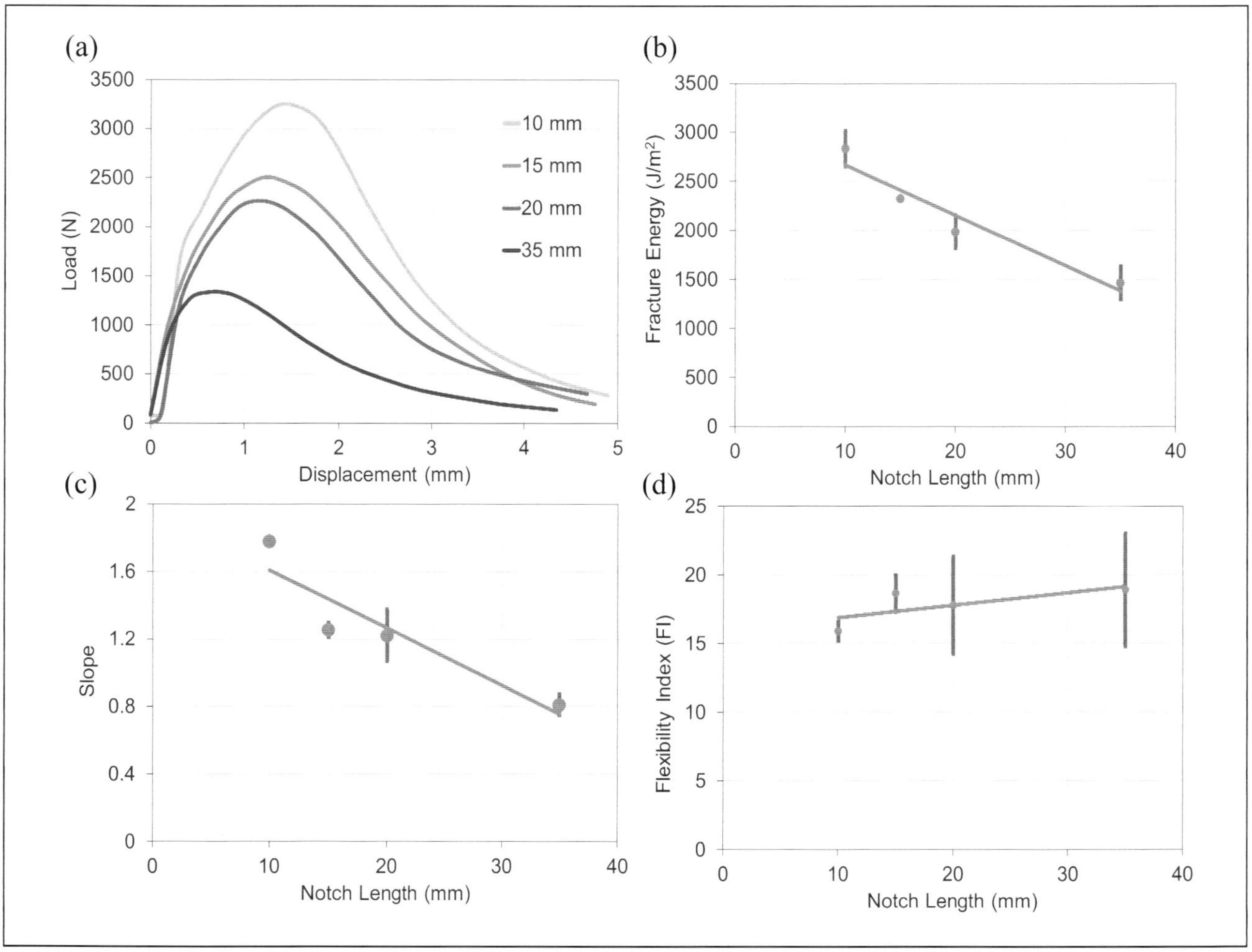

Figure 6. I-FIT results for the notch length: (*a*) load vs. displacement curve, (*b*) FE, (*c*) slope, and (*d*) FI.

correction factor for the slope as proposed by Barry (*8*) could address the variation in the FI without the need of correcting the FE.

Notch Length Impact

Sixteen I-FIT specimens with notch lengths of 10, 15, 20, and 35 mm were prepared from Mix 1 to evaluate the impact of the notch length. Figure 6*a* shows the load versus displacement curves of representative specimens for each notch length. A decrease in the peak load, post-peak slope, and the area under the load versus displacement curve (work of fracture) was observed when the notch length increased. The peak load and the work of fracture decreased because of the reduction in the ligament area (less work of fracture was needed to fracture a smaller area). Figure 6, *b–d* shows the results for FE, slope, and FI. Contrary to the expectation, FE did not remain constant as a result of normalization by the ligament area; the values decreased linearly with notch length increase. This variation was studied by evaluating the fracture process zone using digital image correlation. It was found that crack tip fields can be influenced by the compression arch under the loading head when notch length is too close to the boundary affecting the fracture response. The results from the ANOVA analysis indicate a significant linear trend for both the FE and slope with an r-square and F value of 0.767 and 0.0029, respectively, for the FE. For the slope, the r-square and F value resulted in 0.613 and 0.0074, respectively. On the other hand, the one-way ANOVA analysis for FI resulted in a p-value of 0.03. Thus, the FI did not demonstrate any significant variation with respect to the change of notch length because the reduction in the FE and slope compensated for such variation. A correction for the FE and slope was recommended.

The FE and slope were linearly normalized to the 15 mm notch length. The correction factors are presented in Equations 6 and 7:

Correction factor for fracture energy (G_f):

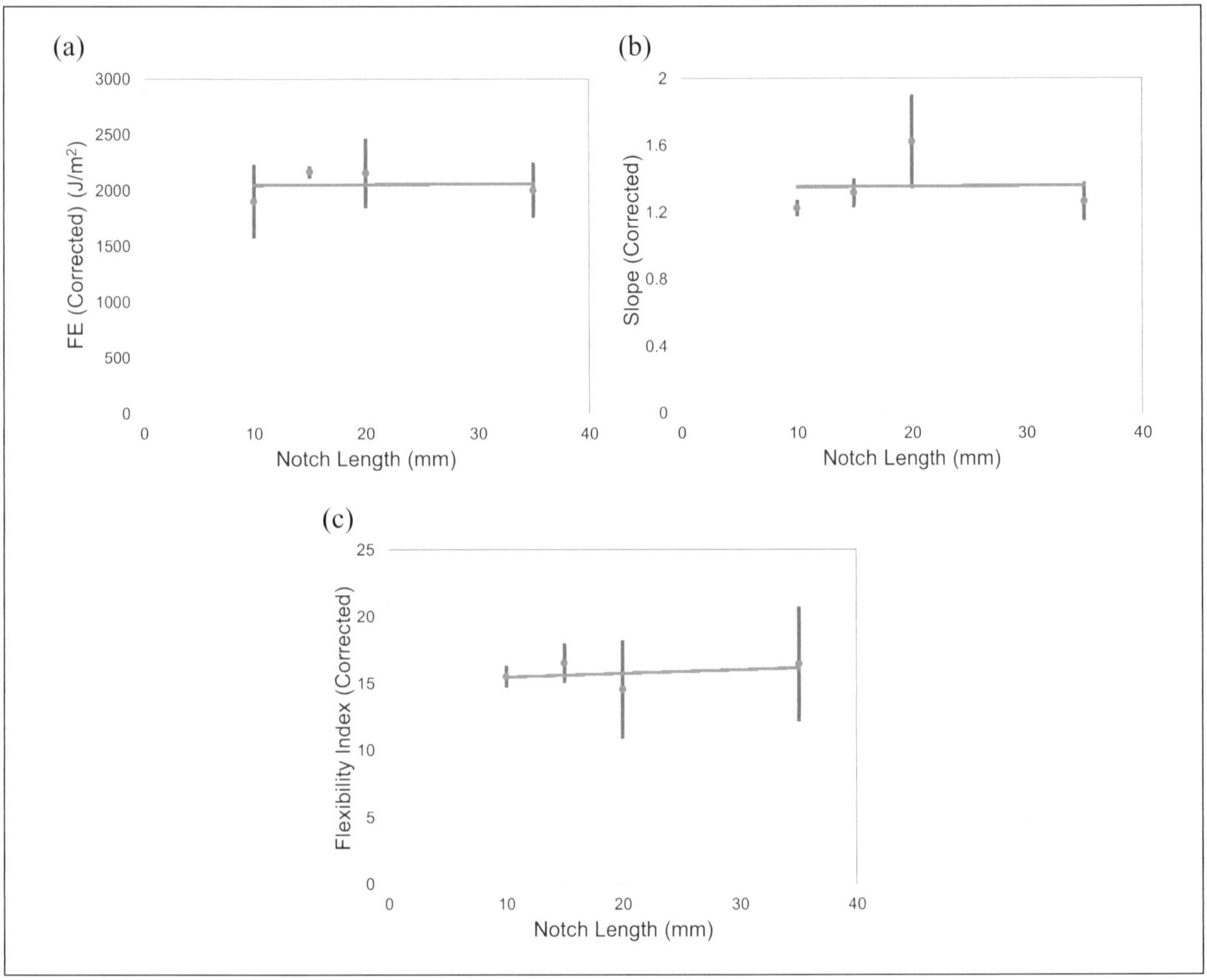

Figure 7. I-FIT results corrected for notch length: (*a*) FE, (*b*) slope, and (*c*) FI.

$$G_{f\mathrm{NL}_{\mathrm{corrected}}} = G_f * \frac{Notch\ Length(mm)}{15} \tag{6}$$

Correction factor for post-peak slope (m):

$$m_{\mathrm{NL}_{\mathrm{corrected}}} = m * \frac{Notch\ Length(mm)}{15} \tag{7}$$

Equations 5 and 6 were applied to the results as shown in Figure 7, *a* and *b*. The average FE was 2062 J/m^2 with a standard deviation of 255 and 12% COV, respectively. The average slope was 1.36 with a standard deviation of 0.37 and 28% COV. Consequently, the average FI led to 15.6, a standard deviation of 2.86 and 18% COV. A one-way ANOVA analysis was run to the corrected FE, slope, and FI indicating non-statistical variation and successful correction with p-values of 0.151, 0.157, and 0.088, respectively.

Loading Rate Impact

The loading rate impact was studied using 20 samples tested at 0.7, 6.25, 25, and 50 mm/min. Figure 8*a* shows the variation of the load versus displacement curves from representative specimens. As expected, the peak load, work of fracture, and post-peak slope increased at faster loading rates. Because of its viscoelastic behavior, AC became stiffer and more brittle at higher loading rates, thus resulting in an increase in the peak load and faster crack propagation (represented by the high post-peak slope). In addition, as the loading rate increased, the material became less compliant in the pre-peak portion of the load versus displacement curve. The I-FIT results for the loading rate variation are shown in Figure 8*b*. As the rate increased, the FE and post-peak slope increased. The substantial increment of the slope resulted in a lower

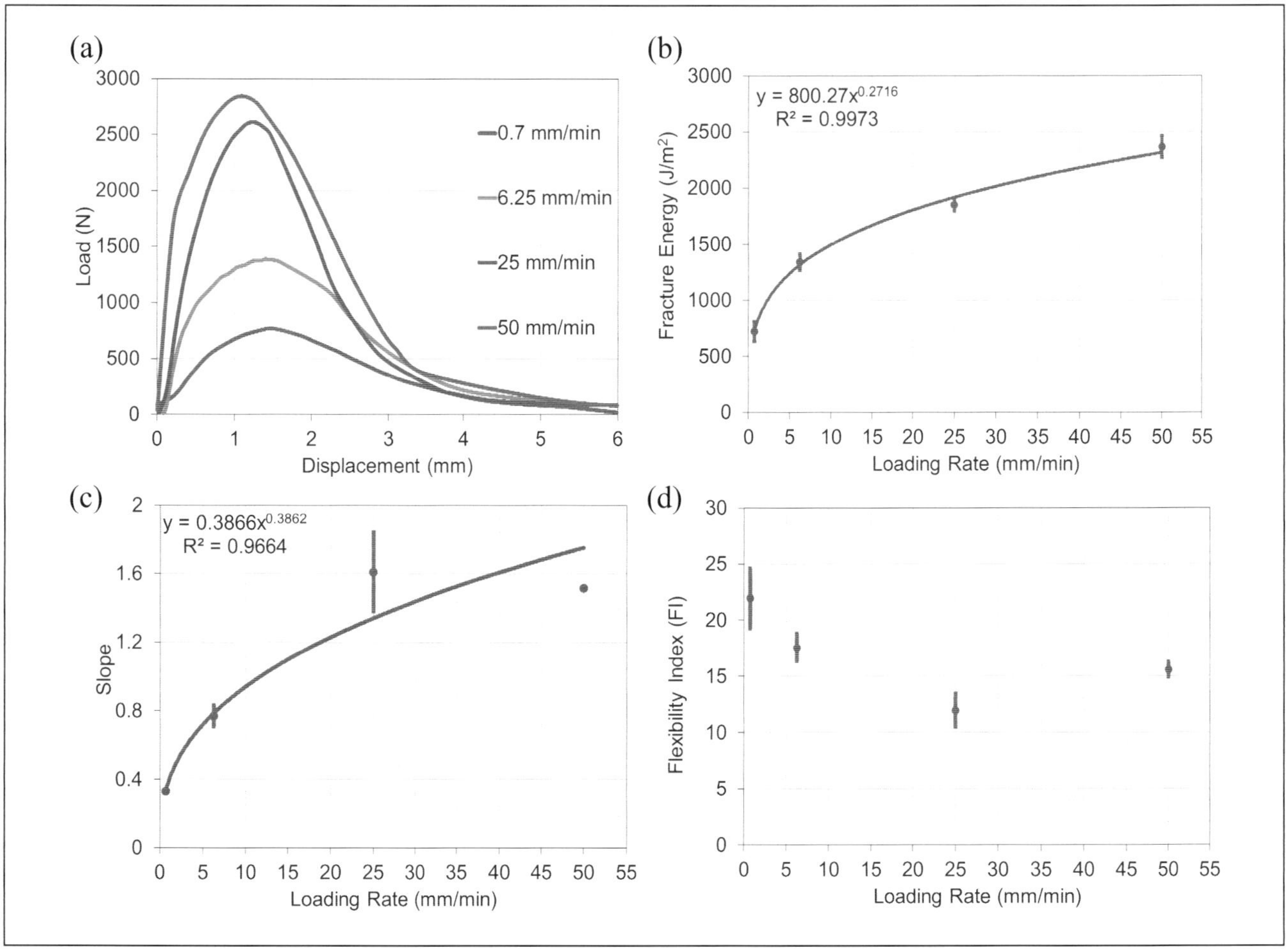

Figure 8. I-FIT results for the loading rate evaluation: (*a*) load vs. displacement curve, (*b*) FE, (*c*) slope, and (*d*) FI.

FI, indicating a stiffer response with higher loading rates. As AC mixtures are highly susceptible to the loading rate with varying degrees of sensitivity depending on the mixtures' viscoelastic characteristics, a generalized correction factor cannot be determined until a comprehensive list of mixtures with varying viscoelastic characteristics is tested. As a result, the correction factor may not be applicable to all the spectrum of AC mixes. Therefore, it was recommended to avoid any modification to the standard loading rate (50.0 ± 1.0 mm/min).

Conclusion and Recommendations

The effects of various geometric factors and test parameters affecting the I-FIT results were evaluated using a comprehensive experimental program. The dependency of the I-FIT response including the FE, post-peak slope (m-value), and FI were characterized with variations in thickness, air void content, notch length, and loading rate. The correction factors available for addressing the variation caused by specimen thickness and air void content were found to be appropriate. A new correction factor was proposed to account for the notch length. Following is a summary of the experimental findings:

- The increase in AC specimen thickness caused an increase in the post-peak slope, resulting in a decrease of the FI. The correction factor for the post-peak slope presented by Al-Qadi et al. (*4*) was found to be sufficient for addressing this variation.
- The post-peak slope decreased as the air void content increased. The correction factor for the post-peak slope proposed by Barry (*8*) addressed the variability in the FI. The effect of air void on the fracture resistance and FI was discussed along with the energy dissipation mechanisms at the crack front and away from the crack front that can delay initiation and propagation of major crack when air void increases. When there is excessive energy dissipation observed with specimen at air voids higher than 8%, the test can no longer be

considered as a fracture test. Therefore, the suggested correction should be limited to air void content less than 8%.

- Increasing the notch length from 10 mm to 35 mm resulted in a decrease in the FE and post-peak slope. Therefore, a correction factor was proposed to correct the FE and post-peak slope results.
- At faster loading rates, the FE and post-peak slope increased. The data were fit into a power function. The power was used to correct the FE and slope resulting in a corrected FI. Because of the varying viscoelastic nature of each AC mixture, a generalized correction factor solely based on rate could not be developed. Therefore, it is recommended to maintain the testing loading rate constant as specified in the test.

Acknowledgments

The support by the Illinois Center for Transportation and the Dwight David Eisenhower Transportation Fellowship Program to the graduate studies of Jose Rivera is acknowledged. The input of Jim Meister, Greg Renshaw, and Marc Killion is greatly appreciated. The authors would also like to acknowledge the help and support of Open Roads Paving that supplied the materials used in the laboratory study. The authors confirm contribution to the paper as follows: study conception and design: Jose Rivera-Perez, Hasan Ozer, Imad L. Al-Qadi; data collection: Jose Rivera-Perez; analysis and interpretation of results: Jose Rivera-Perez, Hasan Ozer, Imad L. Al-Qadi; draft manuscript preparation: Jose Rivera-Perez, Hasan Ozer. All authors reviewed the results and approved the final version of the manuscript.

Author Contributions

The authors confirm contribution to the paper as follows: study conception and design: J. Rivera-Perez, H. Ozer, I.L. Al-Qadi ; data collection: J. Rivera-Perez; analysis and interpretation of results: J. Rivera-Perez, H. Ozer, I.L. Al-Qadi; draft manuscript preparation: J. Rivera-Perez, H. Ozer, I.L. Al-Qadi. All authors reviewed the results and approved the final version of the manuscript.

References

1. Hansen, K. R. *Asphalt Pavement Industry Survey on Recycled Materials and Warm-Mix Asphalt Usage 2014*. Information Series 138, National Asphalt Pavement Association, 2015.
2. Al-Qadi, I. L., S. Carpenter, G. Roberts, H. Ozer, Q. Aurangzeb, M. Elseifi, and J. Trepanier. *Determination of Usable Residual Asphalt Binder in RAP*. Report FHWA-ICT-09-031. Illinois Center for Transportation, 2009.
3. Al-Qadi, I. L., Q. Aurangzeb, and S. Carpenter. *Impact of High RAP Contents on Structural and Performance Properties of Asphalt Mixtures*. Report FHWA-ICT-12-002. Illinois Center for Transportation, 2012.
4. Al-Qadi, I. L., H. Ozer, J. Lambros, A. EL Khatib, P. Singhvi, T. Khan, J. Rivera-Perez, and B. Doll. *Testing Protocols to Ensure Performance of High Asphalt Binder Replacement Mixes Using RAP and RAS*. Report FHWA-ICT-15-017. Illinois Center for Transportation, 2015.
5. AASHTO T 324. *Hamburg Wheel-Track Testing of Compacted Hot Mix Asphalt (HMA)*. Washington, D.C.: American Association of State Highway and Transportation Officials, 2016.
6. Ozer, H., I. L. Al-Qadi, J. Lambros, A. El-Khatib, P. Singhvi, and B. Doll. Development of the Fracture-Based Flexibility Index for Asphalt Concrete Cracking Potential Using Modified Semi-Circle Bending Test Parameters. *Construction and Building Materials*, Vol. 115, 2016, pp. 390–401. https://doi.org/10.1016/j.conbuildmat.2016.03.144
7. Ozer, H., I. Al-Qadi, P. Singhvi, T. Khan, J. Rivera-Perez, and A. El-Khatib. Fracture Characterization of Asphalt Mixtures with High Recycled Content Using Illinois Semicircular Bending Test Method and Flexibility Index. *Transportation Research Record: Journal of the Transportation Research Board*, 2016. 2575: 130–137.
8. Barry, M. *An Analysis of Impact Factors on the Illinois Flexibility Index Test*. MS thesis. University of Illinois at Urbana-Champaign, Urbana, Ill., 2016.
9. AASHTO TP 124-16. *Determining the Fracture Potential of Asphalt Mixtures Using Semicircular Bend Geometry (SCB) at Intermediate Temperature*. Washington, D.C.: American Association of State Highway and Transportation Officials, 2016.
10. AASHTO 105-13. *Determining the Fracture Energy of Asphalt Mixtures Using the Semicircular Bend Geometry*. Washington, D.C.: American Association of State Highway and Transportation Officials, 2013.
11. Doll, B., H. Ozer, J. Rivera-Perez, I. L. Al-Qadi, and J. Lambros. Investigation of Viscoelastic Fracture Fields in Asphalt Mixtures Using Digital Image Correlation. *International Journal of Fracture*, Vol. 205, 2017, pp. 37–56. https://doi.org/10.1007/s10704-017-0180-8
12. Doll, B., H. Ozer, J. Rivera-Perez, I. L. Al-Qadi, and J. Lambros. Damage Zone Development in Heterogeneous Asphalt Concrete. *Engineering Fracture Mechanics*, Vol. 182, 2017, pp. 356–371. https://doi.org/10.1016/j.engfracmech.2017.06.002

The Standing Committee on Structural Requirements of Asphalt Mixtures (AFK50) peer-reviewed this paper (18-03186).

Article

TRR
JOURNAL OF THE TRANSPORTATION RESEARCH BOARD

Transportation Research Record
2018, Vol. 2672(28) 394–403

DOI: 10.1177/0361198118793276
journals.sagepub.com/home/trr
SAGE

Indirect Tensile Test (IDT) to Determine Asphalt Mixture Performance Indicators during Quality Control Testing in New Jersey

Thomas Bennert[1], Edwin Haas[1], and Edward Wass[1]

Abstract

In recent years, there has been a growing interest in incorporating performance testing during the mixture design and quality control testing of asphalt mixtures. The move toward utilizing performance related specifications (PRS) and balanced mixture design concepts have pushed the need for asphalt mixture performance testing to the forefront. Numerous researchers have proposed a variety of laboratory tests that have showed promise at predicting asphalt mixture performance, yet most of these test methods are never adopted due to a number of issues often cited by the asphalt industry: (1) equipment cost; (2) equipment/test method complexity; and (3) time/labor effort required. The research presented here summarizes the effort to evaluate the indirect tensile test (IDT) as a potential performance indicator for hot-mix asphalt that can be easily utilized during quality control testing at an asphalt plant. Utilizing the same test equipment and basic procedure, both high-temperature rutting and intermediate-temperature fatigue cracking can be evaluated in a timely manner. Comparison testing to more standardized and accepted rutting and fatigue cracking test methods have shown excellent agreement, indicating that the suite of IDT tests have potential for adoption within a quality control testing program. Examples of criteria are given utilizing the New Jersey Department of Transportation's (NJDOT) PRS.

Since the development of a formal asphalt mixture design system in the United States, performance testing was intended to be utilized to evaluate the final mixture performance. One of the original performance tests utilized during asphalt mixture design and production came in the 1920s with the Hubbard stability test. In the 1930s, Francis Hveem developed the Hveem stabilometer and the cohensionmeter to measure the mechanical strength of the asphalt mixtures once the optimum asphalt content had been determined. The primitive triaxial strength test was thought to be capable of determining stable versus unstable asphalt mixtures, and threshold values were eventually developed. During the 1940s, Bruce Marshall was finalizing the Marshall method. The Marshall method, widely used until the development of Superpave, utilized an indirect tensile approach to measure the strength of asphalt mixtures. Known as stability and flow, the strength parameters provided an idea of the strength and relative brittleness during the strength testing (i.e., too low of a flow value indicated it was too brittle; too high of a flow value indicated it was too soft).

The Strategic Highway Research Program (SHRP) of the late 1980s to the early 1990s was intended to develop a performance-based asphalt binder and mixture design system. Both asphalt binders and mixtures were to be evaluated for their respective performance for high-temperature permanent deformation, intermediate-temperature fatigue cracking, and low-temperature thermal cracking. Although the asphalt binder procedures eventually became the performance grading system known under AASHTO M323, because of the complexity and cost of the asphalt mixture test methods SHRP proposed, Superpave was left without a formal system of performance tests to balance the permanent deformation and cracking performance of the designed asphalt mixture.

[1]Center for Advanced Infrastructure and Transportation (CAIT), Rutgers University, Piscataway, NJ

Corresponding Author:
Address correspondence to Thomas Bennert: bennert@soe.rutgers.edu

Indirect Tensile Test (IDT): History and Procedures

A significant amount of work has been conducted by researchers over the last 20 years on the development of new and promising asphalt mixture test procedures. Extensive literature reviews (*1–4*) and heavily funded research studies under the National Cooperative Highway Research Program (*5–7*) projects have provided tremendous insight into the advantages and disadvantages of the multitude of performance tests available. One test method that continued to be explored and considered was the IDT, for assessing both permanent deformation and fatigue cracking.

The IDT test was originally developed by Carneiro when he proposed that the test method could be used to determine the tensile strength of concrete (*8*). Similar research was also being conducted in Japan by Akazawa (*9*), but because of World War II published information related to the work in Japan became unavailable. Carneiro determined that by testing the concrete specimen in a diametral position, a relatively uniform tensile stress develops perpendicular to and along the diametral (*8*). Before 1962, the IDT test was primarily used for concrete. It was not until after 1962 that other researchers began utilizing the IDT test method for other materials. Livneh and Shklarsky reported the use of the IDT test to evaluate the anisotropic cohesion of asphalt mixtures (*10*). Thompson began utilizing the test method to evaluate lime-stabilized soils and asphalt mixtures (*11*). Soon after in 1966, Messina conducted research at the University of Texas utilizing the IDT procedure to again evaluate asphalt mixtures (*12*). In the same year, Breen and Stephens began utilizing the IDT procedure to evaluate the low-temperature tensile properties of asphalt mixtures (*13*). A series of reports conducted by Kennedy and his associates at the University of Texas critically evaluated the IDT test procedure for asphalt materials under both static and dynamic loading conditions and made the IDT an accepted test method to evaluate asphalt mixtures (*14*).

More recently, under the movement to develop a "simple" performance test for asphalt mixtures, researchers re-evaluated the IDT test and its relationship to various asphalt mixture properties. Gokhale evaluated the use of the IDT test method to better understand the shear strength properties of asphalt mixtures and concluded that the IDT strength was highly correlated to asphalt mixture cohesion under the Mohr–Coloumb shear strength failure envelope theory (*15*). Christensen et al. found that the IDT test, conducted at a test temperature representative of the critical pavement temperature, was found to be a simple, inexpensive, and effective test for evaluating the rutting resistance of asphalt mixtures (*16*). The authors also provided preliminary threshold values that were developed based on correlations from testing FHWA Accelerated Loading Facility (ALF) mixtures. Unfortunately, the final test method required a loading rate of 3.75 mm/min, requiring a more specialized loading system. However, utilizing updated models and test data from NCHRP projects 9-25, 9-31, and 9-33, Christensen and Bonaquist refined the test method to be conducted at 50 mm/min and a test temperature 10°C lower than the yearly, 7-day average, maximum pavement temperature 20 mm below the pavement surface as determined by Long-Term Pavement Performance Bind (LTPPBind) (*17*, *18*). These changes greatly improved the IDT test, now called high-temperature IDT (HT-IDT), by providing standardized guidance on test temperature, as well as allowing more readily available test equipment to conduct the test. Proposed HT-IDT guidelines to minimize pavement rutting are shown in Table 1 (*17*).

Table 1. Proposed Guideline for High Temperature IDT (HT-IDT) to Minimize Pavement Rutting (*17*)

Design traffic level* (ESALs)	Rut resistance category	IDT strength range (kPa)
–	Very poor	<50
<0.3	Poor	50 to <110
0.3 to <3	Minimal	110 to <170
3 to <10	Fair	170 to <2.70
10 to <30	Good	270 to <430
30 to <100	Very good	430 to <660
100 to <300	Excellent	660

*At 70 km/h (44 mph). To adjust estimated traffic level to 70 km/h, multiply by (70/*v*), where *v* is the average traffic speed in km/h.

As previously described, the IDT, when conducted at the appropriate test temperature and loading rate, shows great promise as a means for assessing rutting potential. However, its use as a fatigue cracking indicator is generally what most asphalt technologists first think about when considering using the IDT test. At intermediate temperature, the IDT test method has been traditionally used to measure maximum tensile strength and fracture energy (i.e., area under the load–displacement curve). Witczak et al. compared the IDT parameters at intermediate temperature to measured field performance at MnRoad, FHWA ALF, and WesTrack, and found a poor and site-dependent relationship between the IDT parameters the researchers identified and the field performance (*5*). Meanwhile, work conducted by Wen and Kim (*19*) found that fracture energy obtained from viscoelastic analysis of the IDT strength correlated very well with the field performance of WesTrack mixtures. Other researchers have also found that the IDT fracture energy correlated well with fracture performance of recycled asphalt mixtures (*20*), while others have utilized the IDT fracture energy concept with additional IDT configuration testing

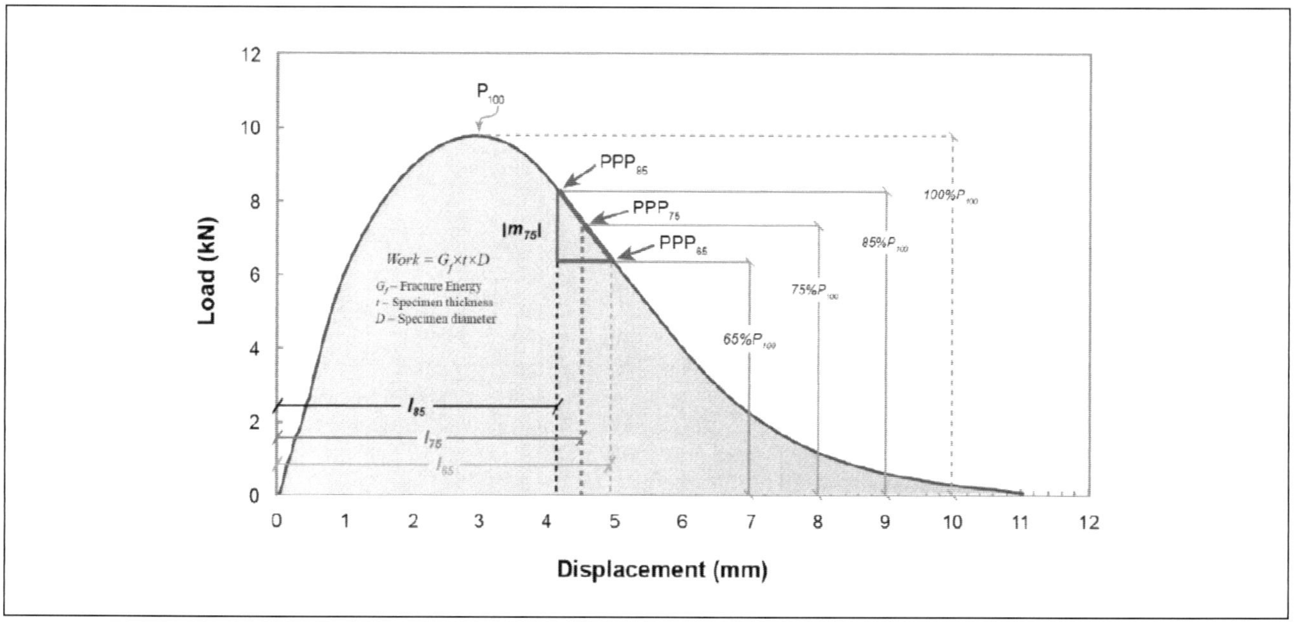

Figure 1. Illustration of IDEAL-CT data analysis of the load–displacement curve.

to develop the new asphalt mixture parameters for top-down cracking analysis (*21–23*).

Under NCHRP Project 9-57, *Experimental Design for Field Validation of Laboratory Tests to Assess Cracking Resistance of Asphalt Mixtures*, Zhou et al. identified different fatigue cracking tests based on their "desirable" features (*7*):

1. Simplicity: lack of instrumentation, cutting, gluing, drilling, and notching
2. Practicality: minimal training needed
3. Efficiency: test completion time within 1 min
4. Test equipment cost: minimal financial investment by laboratory
5. Repeatability: coefficient of variation (COV) less than 25%
6. Sensitivity: sensitive to asphalt volumetrics and composition
7. Correlation to field performance: good relationship to field cracking

Although many of the tests Zhou et al. evaluated met a majority of these characteristics (*7*), the researchers did not find a fatigue cracking test meeting all seven. Clearly identifying a need in the asphalt industry, Zhou et al. developed a test method called IDEAL-CT (*24*). The IDEAL-CT is similar to the traditional IDT test, conducted at 25°C on cylindrical test specimens at a loading rate of 50 mm/min. Although the researchers recommend using a 150 mm diameter and 62 mm thick specimen, they have shown that any size cylindrical specimens with various diameters and thicknesses can be used. The test analysis of the resultant IDT load–displacement curve was inspired by the laws of crack propagation (*25, 26*). The final test value, called CT_{Index}, is defined by Equation 1 and schematically shown as Figure 1:

$$CT_{Index} = \frac{G_f}{|m_{75}|} \times \left(\frac{l_{75}}{D}\right) \quad (1)$$

where, G_f is the work of fracture (total area under the load–displacement curve); m_{75} is the slope calculated as $\left|\frac{P_{85}-P_{65}}{l_{85}-l_{65}}\right|$ (see Figure 1); P_{85} is 85% of the peak load at the post-peak stage; P_{75} is 75% of the peak load at the post-peak stage; P_{65} is 65% of the peak load at the post-peak stage; l_{85} is the displacement corresponding to 85% of the peak load at the post-peak stage; l_{75} is the displacement corresponding to 75% of the peak load at the post-peak stage; l_{65} is displacement corresponding to 65% of the peak load at the post-peak stage; and D = specimen diameter (mm).

When test specimens are not 62 mm thick, Equation 2 should be used:

$$CT_{Index} = \frac{G_f}{|m_{75}|} \times \left(\frac{l_{75}}{D}\right) \times \left(\frac{t}{62}\right) \quad (2)$$

where t is specimen thickness (mm).

Zhou et al. showed that the IDEAL-CT test was sensitive to asphalt binder type, volumetrics, recycled asphalt content, and aging conditions, and also correlated well to cracking observed in the field (*24*). With the IDEAL-CT capable of being conducted on readily available testing equipment generally found at most asphalt plants, the

Table 2. NJDOT PRS Asphalt Mixtures and Criteria

NJDOT PRS asphalt mixture	Asphalt Pavement Analyzer rutting requirement	Overlay Tester fatigue cracking requirement
High-performance thin overlay	<4 mm	>700 cycles
Bituminous-rich intermediate course	<6 mm	>700 cycles
High RAP—surface course	<4 mm	>275 cycles
High RAP—inter/base course	<7 mm	>150 cycles

researchers believed they had satisfied the seven "desirable" characteristics of a fatigue cracking test.

With the strong evidence showing that the IDT test method can be utilized for both a rutting and fatigue cracking test, while being conducted on conventional laboratory equipment commonly found at asphalt plant quality control laboratories, the IDT test method has the potential to be used as a performance-related quality control tool during asphalt mixture design and production.

Objectives

The New Jersey Department of Transportation (NJDOT) has been utilizing performance-related specifications (PRS) for almost 10 years. Unfortunately, most of the asphalt industry in New Jersey has been reluctant to embrace the performance testing on account of the cost, complexity, and time required to conduct many of the rutting and fatigue cracking tests currently available. To aid the asphalt industry in the adoption of an inexpensive and quick performance test that can be conducted during routine quality control testing, a research study was conducted with the following objectives:

- Identify an inexpensive suite of test methods for rutting and fatigue cracking that can be conducted quickly under a quality control testing environment.
- Verify the test method(s) that correlate to standardized test methods used in New Jersey, either currently adopted or that show a strong relationship to field performance.
- Propose preliminary criteria that can be used during mixture design and quality control testing for the NJDOT's PRS.

New Jersey PRS

The NJDOT has been utilizing PRS since 2006 for their specialty asphalt mixtures. Within the NJDOT PRS, the asphalt supplier must submit asphalt specimens for both rutting and fatigue cracking tests during mixture design and plant production. Details on the NJDOT PRS can be found elsewhere (*27*). Table 2 shows the different NJDOT PRS asphalt mixtures and associated performance criteria. As a rutting test procedure, the NJDOT utilizes the Asphalt Pavement Analyzer (AASHTO T340) at a test temperature of 64°C. Meanwhile, for a fatigue cracking test, the NJDOT utilizes the Overlay Tester (NJDOT B-10) at a test temperature of 25°C.

The New Jersey asphalt industry has been somewhat apprehensive in fully adopting the PRS for several reasons, including:

- Lack of availability of equipment
- Cost of purchasing testing equipment utilized by the NJDOT
- Complexity and time associated with specimen preparation, testing time, and data analysis

The Asphalt Pavement Analyzer can generally be conducted within a 24 h period. However, most fatigue cracking tests, including the Overlay Tester, require the cutting and trimming of test specimens, and may even require gluing to test platens to achieve the tensile strains required for fatigue cracking analysis. Depending on the performance of some mixtures, actual testing time can be extended, further delaying the results and confirmation of whether or not the asphalt mixture has met the performance requirements. Although these tests have proven field performance relationships and have been reported to be an excellent means for quality acceptance, some fatigue cracking tests are difficult to implement during the rapid and chaotic pace of asphalt plant quality control testing.

Laboratory Testing Program

The HT-IDT and IDEAL-CT test methods were evaluated and compared to the Asphalt Pavement Analyzer and the Overlay Tester, respectively, using a variety of plant-produced and laboratory-produced asphalt mixtures. In addition to the IDEAL-CT test, the Semi-Circular Bend Flexibility Index (SCB FI) test (AASHTO TP124) was compared to the Overlay Tester. The SCB FI test was also evaluated as it is an improvement in testing speed when compared to the Overlay Tester and can still be conducted on conventional test equipment usually found at an asphalt plant quality control laboratory. Although

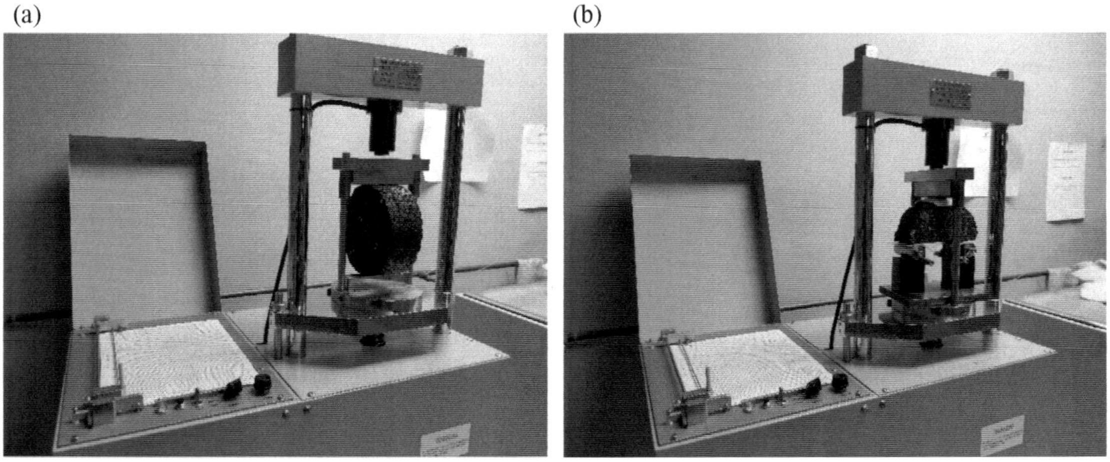

Figure 2. Marshall compression machine test set-up: (*a*) HT-IDT and IDEAL-CT test; (*b*) SCB Flexibility Index test.

some specimen cutting is required for the SCB FI, the test method is envisioned as a potential quality control test.

Since relationships between different test methods was the main goal in the laboratory study, a wide variety of asphalt mixtures were utilized from both plant and laboratory production. This resulted in a wide range of asphalt binder grades, volumetrics, aggregate gradations and types, and recycled asphalt binder percentages. The test specimens prepared, tested, and compared to each of the performance tests were compacted to within ± 0.5% of the specified mix type. For example, all dense-graded and high RAP asphalt mixtures were compacted to a target of 6.5% air voids; high-performance thin overlay were compacted to a target of 5% air voids; and bituminous-rich intermediate course were compacted to a target of 3.5% air voids.

Rutting Evaluation

As previously mentioned, the NJDOT utilizes the Asphalt Pavement Analyzer as a means of determining rutting potential of asphalt mixtures. The specifications developed consider not only the traffic conditions, but also the location in the pavement (i.e., surface, intermediate, base) where the asphalt mixture is to be placed. Therefore, if the HT-IDT test is to be utilized as a quality control test, a relationship over a wide range of rutting potential must be conducted. NJDOT specifications require the Asphalt Pavement Analyzer to be conducted at a test temperature of 64°C, hose pressure of 100 psi, and a vertical wheel load of 100 lb. For the HT-IDT, average New Jersey conditions indicate that the yearly 7-day average, maximum pavement temperature 20 mm below the pavement surface as determined by LTPPBind 3.1 is 54°C. This results in a test temperature of 44°C to be utilized during HT-IDT testing.

To simulate the performance testing at an asphalt plant's quality control laboratory, HT-IDT test specimens were tested in a Pine Instruments compression machine traditionally used for testing Marshall test specimens (Figure 2*a*). The test data were captured on the standard pen-and-paper system typically employed with most Marshall compression machines and simply transposed into an Excel spreadsheet for data analysis. A standard Lottman test head, meeting the specifications of AASHTO T283, was used during testing. Test specimens were compacted to a height of 95 mm in the gyratory compactor, again to maintain consistency with current test practices for AASHTO T283. Compacted air voids of the test specimens were dependent on the NJDOT mixture type evaluated, as the air void requirements differ slightly depending on the mixture type specified. However, the respective mix type test specimens to compare the HT-IDT and Asphalt Pavement Analyzer rutting were within an air void range of ± 0.5%. Before testing, HT-IDT test specimens were conditioned in a forced draft oven at 44°C for 2 h. The resultant comparison between the Asphalt Pavement Analyzer and the HT-IDT tests are shown in Figure 3. The test results shown in the figure clearly indicate a strong relationship between the test results (black solid line). Superimposed in the figure are recommended criteria from NCHRP Project 9-33 (open symbols), which match well with the test data developed in this study.

The data shown in Figure 3 include error bars that represent a single operator COV. The HT-IDT test results generated in the study showed that the average COV was 6.0%. The average COV for the Asphalt Pavement Analyzer (AASHTO T340) was determined in a previous study to be 9.6% (*28*).

With the ultimate goal of using the HT-IDT as a quality control test in lieu of the Asphalt Pavement Analyzer,

Table 3. NJDOT Proposed Equivalent Rutting Requirement for HT-IDT Testing

NJDOT PRS asphalt mixture	Asphalt Pavement Analyzer rutting requirement	HT-IDT strength requirement
High-performance thin overlay	<4 mm	>47 psi
Bituminous-rich intermediate course	<6 mm	>30 psi
High RAP—surface course	<4 mm	>47 psi
High RAP—inter/base course	<7 mm	>25 psi

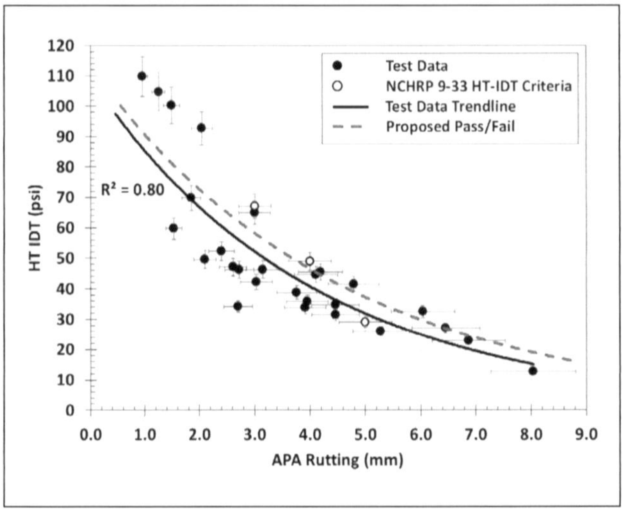

Figure 3. High-temperature IDT compared to Asphalt Pavement Analyzer rutting.

criteria are proposed using the test data trendline established in Figure 3. In an effort to be conservative when establishing a proposed pass/fail criteria with the HT-IDT, the COV determined when using the HT-IDT test was incorporated in the test data trendline by simply adding the COV to the HT-IDT portion of the trendline. The result is a slightly more conservative pass/fail criteria that the NJDOT can modify in the future when more experience with the test data has developed (Table 3).

Fatigue Cracking Evaluation

Similar to the high-temperature permanent-deformation evaluation, a comparison of fatigue cracking tests was also conducted. The different fatigue test specimens were once again compacted to a target air void level, ± 0.5%, based on the mixture type tested.

The primary laboratory performance test used by the NJDOT to evaluate the fatigue cracking potential of asphalt mixtures is the Overlay Tester (NJDOT B-10). The NJDOT B-10 test specification mirrors that of the Texas Department of Transportation (TxDOT) Tex 248F test specification. Five test specimens are tested at 25°C with the high and low values eliminated, and the final result is an average of the middle three results. The test procedure uses a displacement-controlled, triangular waveform with the maximum opening displacement of 0.025 inches. The test specimen is "pulled" to maximum displacement in 5 s and then "pushed" back to zero placement in 5 s.

The IDEAL-CT test method is also conducted at 25°C and a displacement-controlled test at a rate of 50 mm/min. The loading rate allows for testing to be conducted on traditional Marshall compression machines (Figure 2*a*). Before testing, test specimens were conditioned for 2 h in an environmental chamber at 25°C. However, it is also envisioned that conditioning could occur in the AASHTO T166 water bath after the test specimen has been sealed in a zip-lock or CoreLok bag to maintain a dry test specimen.

The resultant relationship developed between the Overlay Tester and IDEAL-CT test is shown in Figure 4. The figure shows a very strong relationship between the test results (black solid line). Similar to the HT-IDT data shown earlier, the average single operator COV in the test data is shown as error bars on the data points. The COV for the IDEAL-CT and Overlay Tester were 16.5% and 24.5%, respectively.

With no previous history using the IDEAL-CT test method, proposed criteria were established using the trendline relationship generated in Figure 4, as well as adding the average COV from the IDEAL-CT test. This results in a slightly conservative approach to establishing criteria until further experience and field performance relationships are developed. The proposed equivalent fatigue cracking requirement for the IDEAL-CT test for the NJDOT is shown in Table 4.

Because of the testing conditions established with the SCB FI (25°C, 50 mm/min displacement rate), the SCB FI test method was also included in the study. The test method allows for the use of conventional asphalt plant laboratory testing (Figure 2*b*), but does include the added complexity and time of cutting and notching test specimens. However, with recent literature showing a good relationship to field performance (*29, 30*), the SCB FI could not be ignored in the evaluation as the total testing time is still faster than the Overlay Tester.

SCB FI test specimens were cut from gyratory specimens compacted to 110 mm tall. A 50 mm specimen was cut from the center of the gyratory sample in an attempt

Table 4. NJDOT Proposed Equivalent Fatigue Cracking Requirement for IDEAL-CT Testing

NJDOT PRS asphalt mixture	Overlay Tester fatigue cracking requirement	IDEAL-CT fatigue cracking requirement
High-performance thin overlay	>700 cycles	>245
Bituminous-rich intermediate course	>700 cycles	>245
High RAP—surface course	>275 cycles	>180
High RAP—inter/base course	>150 cycles	>140

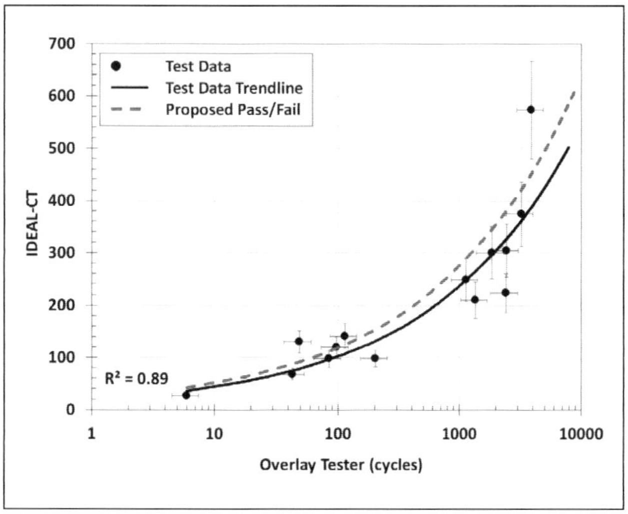

Figure 4. Overlay Tester compared to the IDEAL-CT test.

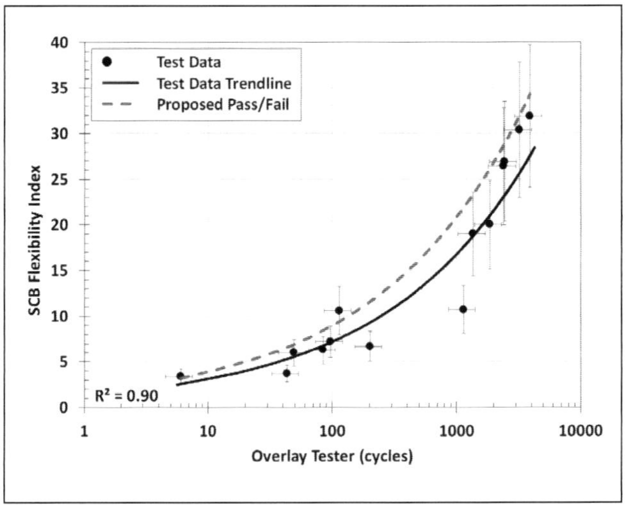

Figure 5. Overlay Tester compared to SCB Flexibility Index.

to eliminate any air void gradients at the top and bottom of the gyratory specimen. The 50 mm slice was then cut in half to create two semi-circular specimens from one gyratory cylinder. Finally, a notch was cut with a wet tile saw in accordance with AASHTO TP124. To expedite test specimen drying, a CoreDry unit was employed to quickly dry the wet-sawed specimens for conditioning and testing. Similar to the IDEAL-CT test specimens, an environmental chamber set at 25°C was used to condition the test specimens for 2 h before testing.

The comparison between the Overlay Tester and SCB FI test methods is shown in Figure 5. The figure shows an excellent relationship between the two test methods (black solid line). The single operator average COV for the SCB FI and Overlay Tester were 24.4% and 24.5%, respectively. Again, using a conservative approach to develop an initial set of performance criteria, the average COV was added to the trendline in Figure 5 (red dashed line) to establish the proposed SCB FI fatigue cracking requirement shown in Table 5.

Practical Considerations

The environment of an asphalt plant quality control laboratory during asphalt mixture production can be quite chaotic, especially if the asphalt plant is running multiple jobs in the same day/evening. Therefore, consistency with respect to sample preparation and testing equipment is extremely important during quality control testing. Additionally, since most asphalt plant quality control labs are not sized and housed like research laboratories, the ability to utilize a simple and rather small testing apparatus for multiple testing purposes is also important. Lastly, because the asphalt industry is in desperate need of a suite of performance tests that can quickly provide a "pass/fail" answer to asphalt mixture performance, whether this is during mixture design or quality control, extravagate data analysis and modeling is not preferred.

The suite of IDT test methods illustrated in the study fits the criteria established under NCHRP Project 9-57 (*7*). The IDT test methods evaluated in the study were shown to be:

1. Simple and practical with respect to equipment, sample preparation, and analysis
2. Efficient with time and cost investment on equipment
3. Repeatability, as both IDT test methods resulted in lower COV when compared to currently NJDOT-specified test methods

Table 5. NJDOT Proposed Equivalent Fatigue Cracking Requirement for SCB FI Testing

NJDOT PRS asphalt mixture	Overlay Tester fatigue cracking requirement	SCB FI
High-performance thin overlay	>700 cycles	>18
Bituminous-rich intermediate course	>700 cycles	>18
High RAP—surface course	>275 cycles	>13
High RAP—inter/base course	>150 cycles	>11

4. Sensitive to asphalt mixture volumetrics, composition, binder grade, and aging
5. Strongly correlated to field performance—in the case of this study, a strong correlation to test methods that are known to have a strong correlation to field performance

Although the SCB FI showed a strong correlation to the Overlay Tester and, when using the test on the Marshall compression machine was found to be simple and practical, it was not as efficient with respect to the time required for sample preparation. To implement both the rutting and fatigue cracking IDT-based performance tests, some modifications would need to be made to the current quality control testing practice in New Jersey. First, an additional six test specimens (three for HT-IDT and three for IDEAL-CT) would need to be compacted to target air voids and appropriate specimen heights for the respective test. Second, after the specimens have cooled and tested for air voids, the specimens must undergo conditioning at the respective test temperature. In this study, it was found that 2 h was sufficient time to achieve the desired test temperatures in a standard environmental chamber. However, this should be verified at each respective laboratory. There is the potential to use water as the conditioning medium, but this was not attempted in this study. Testing time in the Marshall compression machine is fairly quick, and although cumbersome, manually entering test data into an Excel spreadsheet is also quick. Incorporating a data acquisition system with a computer to the compression machine could speed up the analysis process. And although the addition of this extra work may seem to put an additional burden on the laboratory technicians, the work effort is very similar to when most state agencies were still utilizing Marshall mixture design and required stability and flow testing.

Conclusions

As the asphalt industry moves into PRS and balanced mixture design, the demands of performance testing become more of a burden on the asphalt plant quality control laboratory. Testing time, equipment cost, and test complexity have often stalled the implementation of several quality asphalt mixture test methods that have been found to correlate well with field performance. The research study presented evaluated the use of a suite of rutting and fatigue cracking tests, both of which can be conducted using conventional asphalt plant quality control equipment utilizing the IDT. The results of the study showed:

- The NJDOT has utilized the Asphalt Pavement Analyzer for PRS with confidence for almost 10 years. However, because of testing time and equipment expense, the HT-IDT test was evaluated as a potential surrogate test for the Asphalt Pavement Analyzer during quality control testing. The HT-IDT showed a strong correlation to Asphalt Pavement Analyzer rutting when evaluating asphalt mixtures of varying volumetrics, asphalt binder grades, and gradations. The study showed that the HT-IDT test was also more repeatable than the Asphalt Pavement Analyzer, as their respective single operator COVs were 6.0% and 9.6%. It should be noted that the HT-IDT test was conducted in accordance with the recommendations in NCHRP Project 9-33 (*18*).
- The NJDOT has been utilizing the Overlay Tester for almost seven years and has found it to strongly correlate to asphalt mixture fatigue performance in New Jersey. Both the IDEAL-CT and the SCB FI test methods were compared to the Overlay Tester using asphalt mixtures of varying volumetrics, asphalt binder grades, and gradations. Both the IDEAL-CT and SCB FI test methods were found to strongly correlate to the Overlay Tester. The IDEAL-CT was found to be more repeatable than the Overlay Tester, with a single operator COV of 16.5%, while the SCB FI was found to have a similar single operator COV of 24.4%. The Overlay Tester resulted in a COV of 24.5%.
- Although the SCB FI test method can be conducted on conventional asphalt plant equipment, similar to the IDEAL-CT test, the additional time and equipment required to cut and notch the SCB FI test specimens may add work to the already hectic environment in an asphalt plant quality

control laboratory during production. Therefore, for fatigue cracking evaluation, it is recommended that the IDEAL-CT be utilized over the SCB FI.

- The IDT approach of using the HT-IDT for rutting evaluation and IDEAL-CT for fatigue cracking evaluation appears to provide a quick and inexpensive means to conduct asphalt mixture performance testing during mixture design and quality control testing. For agencies already having established testing protocols, such as New Jersey, the IDT approach was found to have the potential for use as a surrogate test that would allow an asphalt supplier to verify the asphalt mixture is being produced in accordance with performance requirements during typical quality control testing. For state agencies that are looking at implementing performance testing, the IDT approach can provide a quick and repeatable suite of tests that relate to both rutting and fatigue cracking behavior of asphalt mixtures.

Author Contributions

The authors confirm contribution to the paper as follows: study conception and design: TB; data collection: TB, EH, EW; analysis and interpretation of results: TB, EH; draft manuscript preparation: TB. All authors reviewed the results and approved the final version of the manuscript.

References

1. Brown, E. R., P. Kandhal, and J. Zhang. *Performance Testing for Hot Mix Asphalt*. NCAT Report No. 01-05. National Center for Asphalt Technology, Auburn, AL, 2001.
2. *Transportation Research Circular E-C068: New Simple Performance Tests for Asphalt Mixes*. Transportation Research Board of the National Academies, Washington, D.C., 2004. http://onlinepubs.trb.org/Onlinepubs/circulars/ec068.pdf.
3. Mohammad, L. (ed.) *Performance Tests for Hot Mix Asphalt (HMA) Including Fundamental and Empirical Procedures*. ASTM Special Technical Publication (STP) 1469. West Conshohocken, PA, 2006.
4. McCarthy, L., J. Callans, R., R. Quigley, and S. Scott. *NCHRP Synthesis 492: Performance Specifications for Asphalt Mixtures: A Synthesis of Highway Practice*. Transportation Research Board of the National Academies, Washington, D.C., 2016.
5. Witczak, M., K. Kaloush, T. Pellinen, M. El-Basyouny, and H. Von Quintus. *NCHRP Report 465: Simple Performance Test for Superpave Mix Design*. HRB, National Research Council, Washington, D.C., 2002.
6. Bonaquist, R., D. Christensen, and W. Stump. *NCHRP Report 513: Simple Performance Tester for Superpave Mix Design: First-Article Development*. HRB, National Research Council, Washington, D.C., 2003.
7. Zhou, F., D. Newcomb, C. Gurganus, S. Banihashermrad, M. Sakhaefar, E. S. Park, and R. Lytton. *NCHRP Research Results Digest 399: Field Validation of Laboratory Tests to Assess Cracking Resistance of Asphalt Mixtures: An Experimental Design*. Transportation Research Board of the National Academies, Washington, D.C., 2016.
8. Carneiro, F. A New Method to Determine the Tensile Strength of Concrete. *Proc., First Congress International Society of Rock Mechanics*, Lisbon, Portugal, 1943.
9. Akazawa, R. New Test Method for Evaluating Internal Stress Due to Compression of Concrete: The Splitting Tension Test. *Journal of Japanese Society of Civil Engineering*, Vol. 29, 1943, pp. 777–787.
10. Livneh, M., and E. Shklarsky. The Splitting Test for Determination of Bituminous Concrete Strength. *Proceedings of the Meeting of the Association of Asphalt Paving Technologists (AAPT)*, Vol. 31, 1962, pp. 457–476.
11. Thompson, M. The Split Cylinder Test for Evaluation of the Tensile Strength of Asphalt Concrete Mixtures. *Highway Research Record*, No. 92, 1965, pp. 69–79.
12. Messina, R. *Split Cylinder Test for Evaluation of the Tensile Strength of Asphalt Concrete Mixtures*. Master of Science thesis. Department of Civil Engineering, University of Texas at Austin, Austin, TX, 1966.
13. Breen, J., and J. E. Stephens. Split Cylinder Test Applied to Bituminous Mixtures at Low Temperatures. *Journal of Materials*, Vol. 1, No. 1, 1966, pp 66–76.
14. Kennedy, T., and J. Anagnos. *Procedures for the Static and Repeated Load Indirect Tensile Test*. Research Report Number 183-14, Research Project 3-9-72-183. Tensile Characterization of Highway Pavement Materials, Texas State Department of Highways and Public Transportation, Austin, TX, 1983.
15. Gokhale, S. *Simplified Protocol for Triaxial Testing of Hot Mix Asphalt Concrete*. Master of Science thesis, Pennsylvania State University, Old Main, State College, PA, 2001.
16. Christensen, D., R. Bonaquist, D. Anderson, and S. Gokhale. *Transportation Research Circular EC068: Indirect Tension Strength as a Simple Performance Test. New Simple Performance Tests for Asphalt Mixtures*. Transportation Research Board of the National Academies, Washington, D.C., 2004, pp. 44–57.
17. Christensen, D., and R. Bonaquist. *Transportation Research Circular E-C124: Using the Indirect Tensile Test to Evaluate Rut Resistance in Developing Hot-Mix Asphalt Designs. Practical Approaches to Hot-Mix Asphalt Mix Design and Production Quality Control Testing*. Transportation Research Board, of the National Academies, Washington, D.C., 2007, pp. 62–77.
18. Advanced Asphalt Technologies, LLC. *NCHRP Report 673: A Manual for Design of Hot Mix Asphalt with Commentary*. HRB, National Research Council, Washington, D.C., 2011.
19. Wen, H., and >Y. Kim. Simple Performance Test for Fatigue Cracking and Validation with WesTrack Mixtures. *Transportation Research Record: Journal of the Transportation Research Board*, 2002. 1789: 66–72.
20. West, R., J. Willis, and M. Marasteanu. *NCHRP Report 752: Improved Mix Design, Evaluation, and Materials Man-*

agement Practices for Hot Mix Asphalt with High Reclaimed Asphalt Pavement Content. HRB, National Research Council, Washington, D.C., 2013.

21. Roque, R., B. Birgisson, Z. Zhang, B. Sandpetngam, and T. Grant. *Implementation of SHRPP Indirect Tension Tester to Mitigate Cracking in Asphalt Pavements and Overlays*. Florida Department of Transportation (FDOT), Tallahassee, FL, 2002.
22. Roque, R., B. Birigisson, C. Drakos, and B. Dietrich. Development and Field Evaluation of Energy-Based Criteria for Top-down Cracking Performance of Hot Mix Asphalt. *Journal of the Association of Asphalt Paving Technologists (AAPT)*, Vol. 73, 2004, pp. 229–260.
23. Zou, J., and R. Roque. Top-Down Cracking: Enhanced Performance Model and Improved Understanding of Mechanisms. *Journal of the Association of Asphalt Paving Technologists (AAPT)*, Vol. 80, 2011, pp. 255–287.
24. Zhou, F., S. Im, L. Sun, and T. Scullion. Development of an IDEAL Cracking Test for Asphalt Mix Design and QC/QA. *Journal of the Association of Asphalt Paving Technologists (AAPT)*, Vol. 86, 2017.
25. Paris, P., and F. Erdogan. A Critical Analysis of Crack Propagation Laws. *Journal of Basic Engineering, Transactions of the American Society of Mechanical Engineers*, 1963, pp. 528–534.
26. Bazant, Z., and P. Prat. Effect of Temperature and Humidity on Fracture Energy of Concrete. *ACI Materials Journal*, Vol. 85, No. 4, 1998, pp. 262–271.
27. Bennert, T., E. Sheehy, R. Blight, S. Gresavage, and F. Fee. *Transportation Research Circular E-C189: Implementation of Performance-Based Specifications for Asphalt Mix Design and Production Quality Control for New Jersey*. Transportation Research Board of the National Academies, Washington, D.C., 2014, pp. 13–31.
28. Bennert, T., E. Wass, Jr., and B. DiFrancesco. *Subtask 1.3: Round Robin Testing Program for the Asphalt Pavement Analyzer*. AASHTO T340. Pavement Support Program 2016, New Jersey Department of Transportation, Federal Highway Administration (FHWA), Trenton, NJ, 2016.
29. Al-Qadi, I., H. Ozer, J. Lambros, A. El Khatib, P. Singhvi, T. Khan, J. Rivera-Perez, and B. Doll. *Testing Protocols to Ensure Performance of High Asphalt Binder Replacement Mixes Using RAP and RAS*. Research Report No. FHWA-ICT-15-017. Illinois Center for Transportation (ICT), Rantoul, IL, 2015.
30. Bennert, T., C. Ericson, D. Pezeshki, R. Shamborovskyy, and C. Bognacki. Moving Towards Asphalt Binder and Mixture Protocols to Minimize Fatigue Cracking on Asphalt Airfields. *Transportation Research Record: Journal of the Transportation Research Board*, 2017. 2633: 117–126.

The Standing Committee on Structural Requirements of Asphalt Mixtures (AFK50) peer-reviewed this paper (18-05836).

Article

TRR
JOURNAL OF THE TRANSPORTATION RESEARCH BOARD

Transportation Research Record
2018, Vol. 2672(28) 404–415

Article reuse guidelines:
sagepub.com/journals-permissions
DOI: 10.1177/0361198118794713
journals.sagepub.com/home/trr
SAGE

Comparison of Different Micromechanical Models for Predicting the Effective Properties of Open Graded Mixes

Hong Zhang[1], Kumar Anupam[1], Athanasios Scarpas[1,2], and Cor Kasbergen[1]

Abstract

ZOAB (Zeer Open Asphalt Beton) is the most widely used asphalt mixture in the Netherlands. As a type of open asphalt mixture, it is known to suffer from raveling distress. In order to analyze the propensity of raveling, micromechanical models are considered effective. However, most of the research work about micromechanical models has focused on dense asphalt mixture and the application of these models on ZOAB mixes has not been paid adequate attention. Therefore, in this research study, the performance of various micromechanical models for predicting mechanical properties of ZOAB was evaluated. The predicted results were compared with the measured values from a dynamic uniaxial compression test. The analysis results showed that none of the applied micromechanical models could obtain acceptable predicted results of the dynamic Young's modulus and phase angle of ZOAB. On one hand, the Dilute model, the Mori-Tanaka model, the generalized self-consistent model and the Lielens' model provided lower values of dynamic Young's modulus and higher values of phase angle, whereas, for the self-consistent model, the predicted results of dynamic Young's modulus were higher, and the values of phase angle were lower. On the other hand, the shapes of the predicted master curves of both dynamic modulus and phase angle of ZOAB could not match well with the experimental results. The further research on the differential scheme method showed that at lower frequencies the predicted mechanical properties of ZOAB mixes by the applied micromechanical models could not be improved even by following this scheme.

In the Netherlands, ZOAB (Zeer Open Asphalt Beton) is the most commonly used road surfacing material (*1*). It is a kind of porous asphalt mixture, which provides a significant noise reduction and a good skid resistance. Due to these advantages, more than 70% of the major roads in the Netherlands have been surfaced with ZOAB and this percentage is still increasing (*1*).

However, due to the high air voids content characteristic, ZOAB always suffers from a particular distress known as raveling (*2*), which occurs as individual aggregate particles separated from the pavement surface. The propensity of raveling for a given ZOAB mix can be analyzed on the basis of the stresses and strains information of its individual phases (*3*).

In the recent past, computational models based on finite element methods (FEM) and/or discrete element methods (DEM) have been proposed as a means to obtain the stresses and/or strains at mix component level of ZOAB mixes (*4*). Although FEM/DEM based models are able to handle complex compositions and almost realistic mix component material properties, they require the development of very large FEM meshes (usually by means of post-processing of the results of CT scans) and very large-scale computational facilities (*4*).

Homogenization methods offer an attractive alternative to the above issues. By means of a homogenization technique, the effective micromechanical properties of a mix are determined from the properties of its different phases, which further allows the calculation of the stress/strain field of the different phases based on the applied

[1]Section of Road Engineering, Faculty of Civil Engineering & Geosciences, Delft University of Technology, Delft, the Netherlands
[2]Khalifa University of Science and Technology, Abu Dhabi, UAE

Corresponding Author:
Address correspondence to Hong Zhang: h.zhang-4@tudelft.nl

loading condition (*5*). This information, in combination with appropriate laboratory tests, can be utilized to evaluate the propensity of a given ZOAB mix for raveling.

On the basis of the homogenization theory, micromechanical models, i.e., the Dilute model (*6*), the Mori-Tanaka (MT) model (*7*), the self-consistent (SC) model (*8*), the generalized self-consistent (GSC) model (*9*), the Lielens' model (*10*), etc., have been developed by different researchers. Micromechanical models do not only consider the mechanical properties of the constituents but also take the geometrical characteristics into consideration. The determination of the effective properties of a mix by these models does not require the calibration factor from laboratory tests. This is a significant advantage over general homogenization methods. Since micromechanical models were initially developed for elastic materials, these models are not directly applicable to asphaltic materials. Several researchers (*11–18*) have tried to extend these models for asphalt pavements by considering viscoelasticity with limited success.

Researchers (*11*) presented the predicted shear modulus of asphalt mastic by the Composite Spheres model (*12*) and the GSC model. They pointed out that these models generally underestimate the measured results and it was explained that these models were not developed to account for stiffening effects beyond volume filling. Other researchers (*13*) have also investigated the capabilities of the Composite Spheres model, the GSC model, and the generalized Nielsen's model (*14*) for estimating the properties of mastic. Their study showed that the predicted results agree with the experimental data only for mixes with low volume concentrations of filler. They also pointed out that the particle interactions, potential physicochemical reinforcement between bitumen and filler, and irregular particle shape and texture roughness may account for the inaccurate prediction for a mix with a high volume concentration of inclusions. In the later studies, researchers (*15*) examined various mastic types with different volume concentrations of filler at sub-zero temperatures. It was concluded that the SC model performs better than the Dilute model, the MT model and the GSC model. The limitation of their work is that it was performed at low temperatures, so the conclusions are only valid in sub-zero temperature conditions. On the other hand, other researchers (*16*, *17*) who conducted their experiments on densely filled mixes concluded that the GSC model is more applicable than the MT model. They also proposed that the MT model can predict better results if the "volume fractions" of inclusions are replaced by the "effective volume fractions." However, they observed that the acceptable results are only valid for tensile creep tests, and not for the dynamic tests, especially those which are performed at high temperatures. Recently, researchers (*18*) evaluated the currently available micromechanical models for the prediction of asphalt mastic shear modulus and it was found that none of them can accurately capture the temperature- and frequency- related stiffening of mastic.

As described in the above paragraphs, it can be concluded that with the increased use of ZOAB, it is of vital importance for pavement engineers to investigate the stresses/strains of individual phases in ZOAB mixes. For this purpose, the micromechanical models are more effective and accessible in comparison with the FEM/DEM. However, although research studies have been conducted on predicting the mechanical properties of asphalt materials by various micromechanical models, there are no commonly accepted models within the pavement engineering community. Moreover, most of the research activities related to the micromechanical modeling focus on the dense asphalt materials, and not much attention has been paid to evaluating the performance of ZOAB.

Objective and Scope

Therefore, the aim of this paper is to compare various micromechanical models in predicting the mechanical properties of ZOAB-type mixes on the basis of laboratory tests.

The scope of the proposed study includes:

- Determination of viscoelastic behavior of ZOAB over a wide range of frequency from a dynamic uniaxial compression test.
- Prediction of viscoelastic properties of ZOAB using five commonly used micromechanical models, i.e., the Dilute model, the MT model, the SC model, the GSC model, and the Lielens' model.
- Frequency/temperature sensitivity analysis to evaluate the performance of various micromechanical models in predicting the mechanical properties of ZOAB.
- Study of the performance of these micromechanical models by following the differential scheme method in which the constituents are included in steps rather than concurrently.

Background Knowledge

Homogenization Theory

Homogenization theory was developed to evaluate the effective properties of a mix given the properties of its constituents. In this theory, a representative volume element (RVE) which represents the overall mechanical response of the mix is chosen. In an RVE, the volume fraction of each phase (i.e., mastic, aggregate, and air void), obviously sums up to unity, as

$$f_{\text{mas}} + f_{\text{agg}} + f_{\text{void}} = 1 \tag{1}$$

where f_{mas} is the volume fraction of the mastic phase; f_{agg} and f_{void} are the volume fractions of the aggregate phase and the air void phase, respectively.

In an RVE, the constitutive relationship between the average stress and average strain of each phase is shown in Equation 2, where the average stress and strain of each phase can be further related to the average stress and strain of the mix by Equations 3 and 4.

$$<\boldsymbol{\sigma}>_{mas} = \mathbf{C}_{mas}^{*} : <\boldsymbol{\varepsilon}>_{mas}; <\boldsymbol{\sigma}>_{agg} = \mathbf{C}_{agg} : <\boldsymbol{\varepsilon}>_{agg} \quad (2)$$

$$<\boldsymbol{\sigma}>_{mix} = f_{mas} <\boldsymbol{\sigma}>_{mas} + f_{agg} <\boldsymbol{\sigma}>_{agg} + f_{void} <\boldsymbol{\sigma}>_{void} \quad (3)$$

$$<\boldsymbol{\varepsilon}>_{mix} = f_{mas} <\boldsymbol{\varepsilon}>_{mas} + f_{agg} <\boldsymbol{\varepsilon}>_{agg} + f_{void} <\boldsymbol{\varepsilon}>_{void} \quad (4)$$

where

$<\boldsymbol{\sigma}>_{mas}$ and $<\boldsymbol{\varepsilon}>_{mas}$ = average stress and average strain of the mastic phase, respectively;

$\mathbf{C}_{mas}^{*}$ = stiffness tensor of the mastic phase;

$<\boldsymbol{\sigma}>_{agg}$ and $<\boldsymbol{\varepsilon}>_{agg}$ = average stress and average strain of the aggregate phase, respectively;

$\mathbf{C}_{agg}$ = stiffness tensor of the aggregate phase;

$<\boldsymbol{\sigma}>_{void}$ and $<\boldsymbol{\varepsilon}>_{void}$ = average stress and average strain of the air void phase, respectively; and

$<\boldsymbol{\sigma}>_{mix}$ and $<\boldsymbol{\varepsilon}>_{mix}$ = average stress and average strain of the mix, respectively.

On the basis of the relationship between $<\boldsymbol{\sigma}>_{mix}$ and $<\boldsymbol{\varepsilon}>_{mix}$, the effective stiffness tensor of the mix $\mathbf{C}_{mix}^{*}$ is defined by

$$<\boldsymbol{\sigma}>_{mix} = \mathbf{C}_{mix}^{*} : <\boldsymbol{\varepsilon}>_{mix} \quad (5)$$

In order to enable the calculation of $\mathbf{C}_{mix}^{*}$, various researchers (*6–10*) have proposed different relationships between the average strain of each phase and $<\boldsymbol{\varepsilon}>_{mix}$, as

$$<\boldsymbol{\varepsilon}>_{mas} = \mathbf{A}_{mas} : <\boldsymbol{\varepsilon}>_{mix}; <\boldsymbol{\varepsilon}>_{agg} = \mathbf{A}_{agg} : <\boldsymbol{\varepsilon}>_{mix}; <\boldsymbol{\varepsilon}>_{void} = \mathbf{A}_{void} : <\boldsymbol{\varepsilon}>_{mix} \quad (6)$$

where $\mathbf{A}_{mas}$ is the strain localization tensor of the mastic phase; and $\mathbf{A}_{agg}$ and $\mathbf{A}_{void}$ denote the strain localization tensors of the aggregate phase and the void phase, respectively. If Equation 6 is substituted into Equation 4, the relationship between the strain localization tensor of each phase can be expressed by

$$f_{mas}\mathbf{A}_{mas} + f_{agg}\mathbf{A}_{agg} + f_{void}\mathbf{A}_{void} = \mathbf{I} \quad (7)$$

where $\mathbf{I}$ is the unit fourth-order tensor.

Once the strain localization tensor of each phase is known, the value of $\mathbf{C}_{mix}^{*}$ can be obtained by combing Equations 2–7, as

$$\mathbf{C}_{mix}^{*} = \mathbf{C}_{mas}^{*} + f_{agg}(\mathbf{C}_{agg} - \mathbf{C}_{mas}^{*}) : \mathbf{A}_{agg} + f_{void}(-\mathbf{C}_{mas}^{*}) : \mathbf{A}_{void} \quad (8)$$

Micromechanical Models

As described in the earlier section, the value of $\mathbf{C}_{mix}^{*}$ can be calculated by using the values of $\mathbf{A}_{agg}$ and $\mathbf{A}_{void}$. Micromechanical models are utilized to calculate the values of $\mathbf{A}_{agg}$ and $\mathbf{A}_{void}$ on the basis of the continuum mechanics theory. For these models, a mix can be considered to consist of a matrix and various types of inclusions embedding into this matrix. The fundamental inhomogeneity problem in which a spherical or elliptical inclusion is embedded into an infinite matrix was solved by Eshelby (*6*). Based on the Eshelby's solution, various micromechanical models have been developed. In the following section, some of these micromechanical models will be introduced. It is noted that although all the micromechanical models were initially developed for elastic materials, their applications have been extended to viscoelastic mixes according to the elastic–viscoelastic correspondence principle (*19*).

Dilute Model. In the Dilute model, the interactions between inclusions are supposed to be negligible and thus ignored, also, inclusions are assumed to be embedded in an infinite matrix. Based on these assumptions, the strain localization tensors of the aggregate phase and the air void phase for the Dilute model, $\mathbf{A}_{agg}^{DM}$ and $\mathbf{A}_{void}^{DM}$, can be directly obtained from the Eshelby's solution, as

$$\mathbf{A}_{agg}^{DM} = [\mathbf{I} + \mathbf{S}_{mas} : (\mathbf{C}_{mas}^{*})^{-1} : (\mathbf{C}_{agg} - \mathbf{C}_{mas}^{*})]^{-1}; \quad \mathbf{A}_{void}^{DM} = [\mathbf{I} + \mathbf{S}_{mas} : (\mathbf{C}_{mas}^{*})^{-1} : (-\mathbf{C}_{mas}^{*})]^{-1} \quad (9)$$

where $\mathbf{S}_{mas}$ is the Eshelby's tensor as obtained from the properties of the mastic phase (*6*). As the Dilute model neglects the interaction between inclusions, the major limitation of the model is that it is only applicable for mixes which have low contents of inclusions.

Mori-Tanaka Model. The MT model assumes that each inclusion is surrounded by the matrix with a uniform averaged strain. As a comparison with the Dilute model, this model takes the interactions between inclusions into consideration. Therefore, it is applicable for mixes with a higher volume fraction of inclusions. In the MT model, the strain localization tensors of the aggregate phase and the air void phase, $\mathbf{A}_{agg}^{MT}$ and $\mathbf{A}_{void}^{MT}$, are given by

$$\mathbf{A}_{agg}{}^{MT} = \mathbf{A}_{agg}{}^{DM} : \left(f_{mas}\mathbf{I} + f_{agg}\mathbf{A}_{agg}{}^{DM} + f_{void}\mathbf{A}_{void}{}^{DM}\right)^{-1}$$
$$\mathbf{A}_{void}{}^{MT} = \mathbf{A}_{void}{}^{DM} : \left(f_{mas}\mathbf{I} + f_{agg}\mathbf{A}_{agg}{}^{DM} + f_{void}\mathbf{A}_{void}{}^{DM}\right)^{-1} \quad (10)$$

Self-Consistent Model. In the SC model, each inclusion is assumed to be embedded into an infinite media which has the same properties with the mix itself and the strain localization tensors of the aggregate phase and the air void phase, $\mathbf{A}_{agg}{}^{SC}$ and $\mathbf{A}_{void}{}^{SC}$, are expressed as

$$\mathbf{A}_{agg}{}^{SC} = [\mathbf{I} + \widetilde{\mathbf{S}}_{mix} : (\widetilde{\mathbf{C}}_{mix}{}^{*})^{-1} : (\mathbf{C}_{agg} - \widetilde{\mathbf{C}}_{mix}{}^{*})]^{-1};$$
$$\mathbf{A}_{void}{}^{SC} = [\mathbf{I} + \widetilde{\mathbf{S}}_{mix} : (\widetilde{\mathbf{C}}_{mix}{}^{*})^{-1} : (-\widetilde{\mathbf{C}}_{mix}{}^{*})]^{-1} \quad (11)$$

where $\widetilde{\mathbf{S}}_{mix}$ is the Eshelby's tensor as obtained from the properties of the mix, and $\widetilde{\mathbf{C}}_{mix}{}^{*}$ is the effective stiffness tensor of the mix in each iteration.

Since inclusions are assumed to be embedded into the mix itself in the SC model, the effective stiffness tensor is also required. Thus, the explicit expression of the mix stiffness cannot be obtained directly, and Equation 11 needs to be solved by an iterative procedure.

Generalized Self-Consistent Model. The GSC model is similar to the SC model; however, it further adds a spherical annulus of matrix material out of inclusions. For a two-phase mix, Christensen and Lo (*9*) proposed the following solution for the effective shear modulus of the mix $\mu_{mix}{}^{*}$

$$A\left(\frac{\mu_{mix}{}^{*}}{\mu_{mas}{}^{*}}\right)^2 + 2B\left(\frac{\mu_{mix}{}^{*}}{\mu_{mas}{}^{*}}\right) + C = 0 \quad (12)$$

where $\mu_{mas}{}^{*}$ is the shear modulus of the mastic phase; and the values of A, B and C are related to the mechanical properties and volume fraction of each phase. The detailed formulations of these parameters can be found elsewhere (*9*). The solution for the effective bulk modulus of the GSC model corresponds to that of the Composite Sphere model (*20*).

For a three-phase mix, the solution to the GSC model is too cumbersome to be solved analytically (*21*). In order to employ the classical two-phase GSC model for a three-phase mix, researchers (*17*) normally divided the prediction procedure into two steps: i) either the aggregate phase or the air void phase is added into the mastic phase; ii) the other phase is added into the composite solution of step 1 by using the same expressions. It is highlighted here that in this research study all the orders of phase additions were considered which will be described in the later section titled "Upscaling by Various Micromechanical Models".

Lielens' Model. The Lielens' model considers interpolation in the lower bound and the upper bound of the strain localization tensor. The lower bound is calculated in the case where the stiff phase is embedded into the soft phase. For this case, the strain localization tensor of the aggregate phase, $\mathbf{A}_{agg}{}^{lower}$, is calculated by

$$\mathbf{A}_{agg}{}^{lower} = [\mathbf{I} + \mathbf{S}_{mas} : (\mathbf{C}_{mas}{}^{*})^{-1} : (\mathbf{C}_{agg} - \mathbf{C}_{mas}{}^{*})]^{-1} \quad (13)$$

The other case in which the soft phase is embedded into the stiff phase yields the upper bound and the strain localization tensor of the aggregate phase, $\mathbf{A}_{agg}{}^{upper}$, is

$$\mathbf{A}_{agg}{}^{upper} = [\mathbf{I} + \mathbf{S}_{agg} : \mathbf{C}_{agg}{}^{-1} : (\mathbf{C}_{mas}{}^{*} - \mathbf{C}_{agg})]^{-1} \quad (14)$$

where $\mathbf{S}_{agg}$ is the Eshelby's tensor as obtained based on the properties of the aggregate phase.

With interpolating $\mathbf{A}_{agg}{}^{lower}$ and $A_{agg}{}^{upper}$, the strain localization of the aggregate phase for the Lielens' model, $A_{agg}{}^{LM}$, is

$$\mathbf{A}_{agg}{}^{LM} = \hat{\mathbf{A}}_{agg}{}^{LM} : \left(f_{mas}\mathbf{I} + f_{agg}\hat{\mathbf{A}}_{agg}{}^{LM} + f_{void}\mathbf{A}_{void}{}^{LM}\right)^{-1} \quad (15)$$

where

$$\hat{\mathbf{A}}_{agg}{}^{LM} = \left\{(1 - a_{agg})\left[\mathbf{A}_{agg}{}^{lower}\right]^{-1} + a_{agg}\left[\mathbf{A}_{agg}{}^{upper}\right]^{-1}\right\}^{-1} \quad (16)$$

The interpolation factor a_{agg} is related to the volume fraction of the aggregate phase and is calculated by

$$a_{agg} = \frac{1}{2}f_{agg}(1 + f_{agg}) \quad (17)$$

It is noted here that Equations 13–17 were used to obtain $\mathbf{A}_{agg}{}^{LM}$. $\mathbf{A}_{void}{}^{LM}$ was assumed to be the same as $\mathbf{A}_{void}{}^{MT}$ (see Equation 10) since the value of $\mathbf{C}_{void}{}^{-1}$ cannot be calculated.

Differential Scheme. Many research studies (*13, 15, 18*) have pointed out that the above micromechanical models are not suitable for a mix with higher volume faction of inclusions. In order to improve the accuracy of the predicted results for dense inclusions, the Differential Scheme (DS) approach (*22, 23*) was proposed. In this approach, inclusions are added in steps and for each step the value of $\mathbf{C}_{mix}{}^{*}$ can be calculated by

$$\mathbf{C}_{mix}{}^{*}(i+1) = \mathbf{C}_{mix}{}^{*}(i) + \frac{\Delta f_{agg}(i+1)}{f_{mas}(i)}(\mathbf{C}_{agg} - \mathbf{C}_{mix}{}^{*}(i)) : \mathbf{A}_{agg}(i) + \frac{\Delta f_{void}(i+1)}{f_{mas}(i)}(-\mathbf{C}_{mix}{}^{*}(i)) : \mathbf{A}_{void}(i) \quad (18)$$

Table 1. Properties of Different Materials

Type of material	Grade/size	Bulk density (kg/m^3)	Young's modulus (MPa)	Poisson's ratio
Asphalt binder	70/100 pen	1030	–	0.50
Corse aggregate	2 mm–16 mm	2678	53000	0.27
Sand	0.063 mm–2 mm	2658	53000	0.27
Filler (Wigro 60K)	<0.063 mm	2638	53000	0.27

Table 2. Gradation of Aggregates

Size (mm)	16	11.2	8	5.6	2	0.063	Filler
Gradation (% Passing)	98	77	44	22	15	4	0

Table 3. Volume Fraction of Each Phase

Phase	Mastic	Sand	Corse aggregate	Air voids
Volume fraction	0.116	0.080	0.624	0.179

where

$\mathbf{C}_{mix}^*(i + 1)$ and $\mathbf{C}_{mix}^*(i)$ are the values of $\mathbf{C}_{mix}^*$ in step $i + 1$ and i, respectively;

$\mathbf{A}_{agg}(i)$ and $\mathbf{A}_{void}(i)$ are the values of A_{agg} and $\mathbf{A}_{void}$ in step i, respectively, which can be obtained through any of the above models (*24*);

$f_{mas}(i)$ is the value of f_{mas} in step i; and

$\Delta f_{agg}(i + 1)$ and $\Delta f_{void}(i + 1)$ are the increments of f_{agg} and f_{void} in step $i + 1$, respectively.

In this research work, after the sensitivity analysis about the effect of different calculation steps on the predicted results, a total of 50 steps were finally conducted to calculate the value of $\mathbf{C}_{mix}^*$. The initial condition for $\mathbf{C}_{mix}^*$ is that when $f_{agg} = f_{void} = 0$, $\mathbf{C}_{mix}^* = \mathbf{C}_{mas}^*$. For each step, the values of Δf_{agg} and Δf_{void} were identical to $f_{agg}/50$ and $f_{void}/50$, respectively.

Material and Testing Procedure

Materials and Specimen Preparation

Material Properties. The material properties for making ZOAB specimens are shown in Table 1, and the corresponding aggregate gradation is shown in Table 2. Since mastic is broadly accepted as the basic functional matrix within asphalt mixture (*25*), in this research ZOAB was considered as a three-phase mix of the aggregate phase (excluding filler) and the air void phase embedding into the mastic phase. On the basis of the properties of each phase and the composition of ZOAB, the volume fraction of each phase, as shown in Table 3, was calculated.

Specimen Preparation of Mortar. The mortar column specimens were prepared according to the past research work carried out at TU Delft (*26*). The dimension of the specimens can be found in Figure *a*. The maximum particle size of mortar was chosen to be 2 mm as suggested by authors (*25*).

Specimen Preparation of ZOAB. According to the AASHTO T 342-11 standard method (*27*), specimens for the measurement of dynamic modulus of asphalt mixtures were prepared. The initial specimen, the size of which was 170 mm in height and 150 mm in diameter, was compacted using a gyratory compactor. This specimen was further cored and cut to the test specimen with a height of 150 mm and diameter of 100 mm.

Laboratory Tests

DSR Test. Frequency sweep tests were conducted by using dynamic shear rheology (DSR) to measure the shear modulus of mastic and mortar. The parallel-plate configuration was applied to test mastic specimens, while mortar specimens were tested by using the so-called "Bitumen Column Configuration" (*28*), Figure 1*b* and *c*. The cylindrical mortar specimens were fixed at the bottom and torsional forces were applied at the upper fixture. All the tests were performed in a frequency range of 50 Hz to 0.1 Hz, at four different temperatures, –10°C, 4°C, 21°C and 37°C with 20 uε–5000 uε of strain amplitudes (depending upon the test temperature).

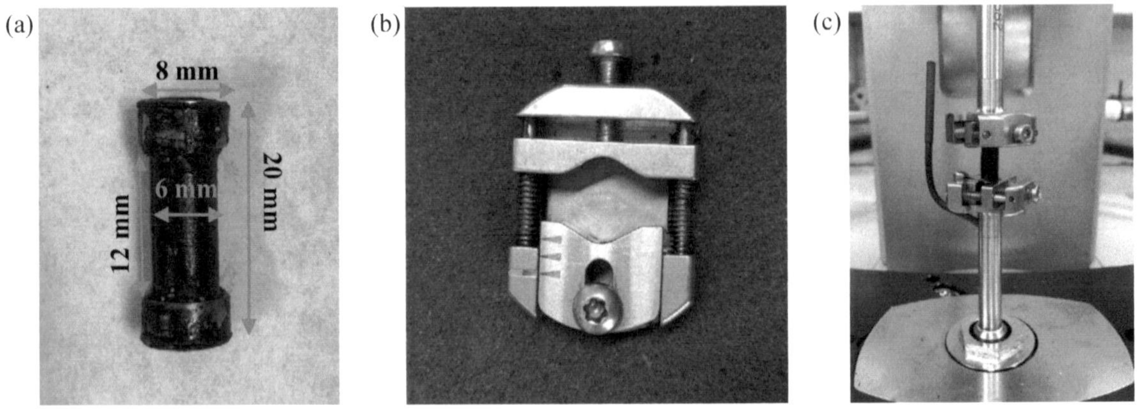

Figure 1. DSR setup for mortar tests: (*a*) specimen, (*b*) fixture, and (*c*) configuration.

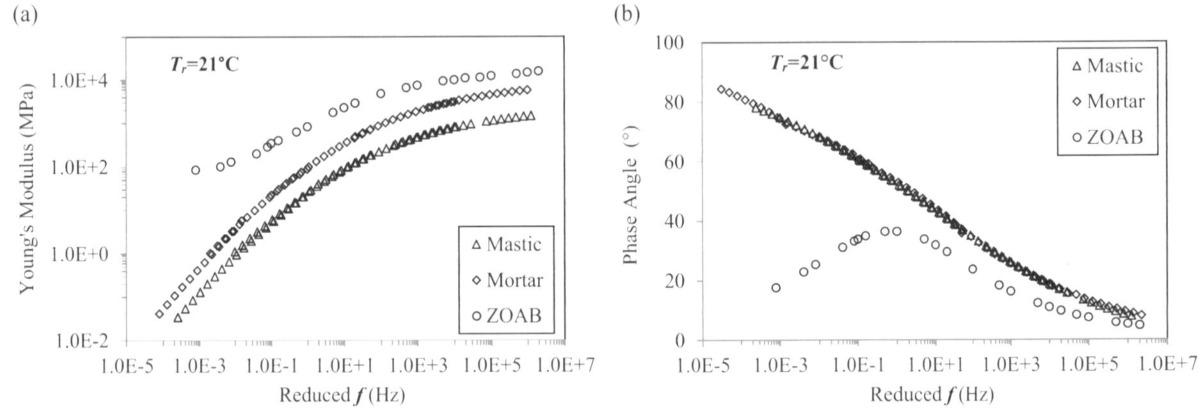

Figure 2. Experimental results of mastic, mortar, and ZOAB for (*a*) Young's modulus, and (*b*) phase angle.

Uniaxial Compression Test. The Young's modulus of ZOAB was measured under dynamic uniaxial compression load by universal testing machine (UTM) at different frequencies and temperatures. In accordance with the AASHTO T 342-11 standard method (*27*), four temperatures, −10°C, 4°C, 21°C and 37°C, were applied during the test and for each temperature, six different frequencies, which were 20 Hz, 10 Hz, 5 Hz, 1 Hz, 0.5 Hz and 0.1 Hz, were performed. The load applied on the specimen was under stress-controlled mode and three linear variable differential transformers (LVDT) were installed to measure the displacements of the specimen. The obtained results and analyses will be discussed in the following section.

Results and Discussion

The tests performed provided us with a background for investigating different aspects of the performances of various micromechanical models. In general, to represent the data in a frequency spectrum, dynamic modulus/phase angle master curves were obtained according to the Time-Temperature Superposition Principle (TTSP) at a reference temperature (T_r) of 21°C. These curves were prepared on the basis of three samples per test that was conducted.

Comparison of Viscoelastic Behavior of Mastic, Mortar, and ZOAB

It is expected that the major part of the viscoelastic characteristic of mortar and ZOAB comes from the mastic phase. That is why it is logical to study the behavior at the mastic scale and compare it with the similar plots at the mortar- and mix- scale.

The master curves of mastic, mortar, and ZOAB are plotted in Figure 2. The dynamic Young's moduli of mastic ($|E^*|_{mas}$), mortar ($|E^*|_{mor}$), and ZOAB ($|E^*|_{mix}$) are shown in Figure 2*a*. The values of $|E^*|_{mas}$ and $|E^*|_{mor}$

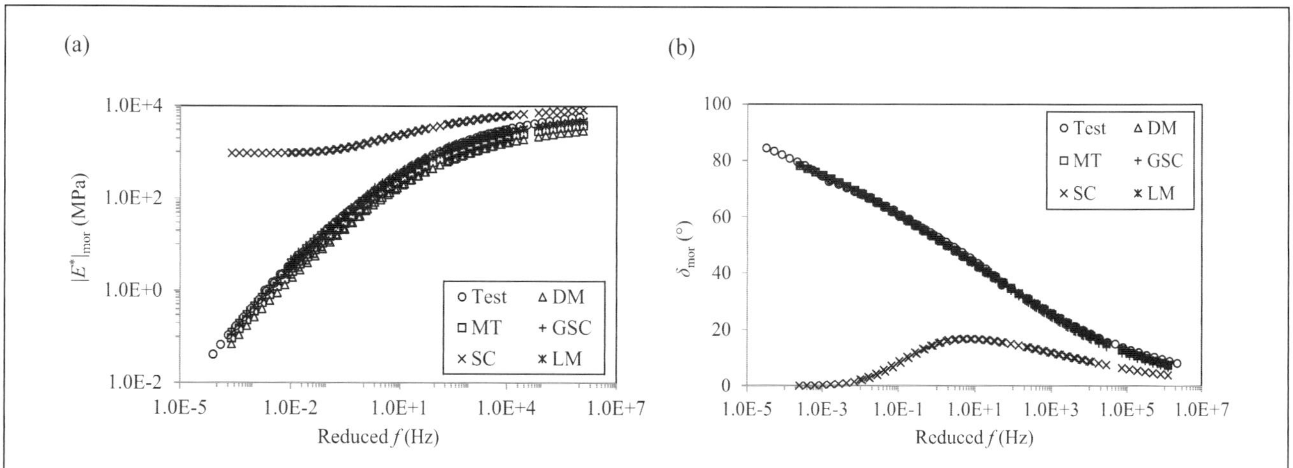

Figure 3. Predicted results of mortar properties: (*a*) dynamic Young's modulus, and (*b*) phase angle.

were calculated in accordance with the relationship between the Young's modulus and the shear modulus (in elasticity). It is noted here that the Poisson's ratios of mastic and mortar are assumed to be 0.495 and 0.35, respectively (*29–31*).

Figure 2*b* shows the master curves of phase angles of mastic (δ_{mas}), mortar (δ_{mor}), and ZOAB (δ_{mix}). The values of δ_{mas} and δ_{mor} were obtained directly from DSR tests and the calculation of δ_{mix} was carried out in accordance with AASHTO T 342-11 standard (*27*).

As shown in Figure 2*a* and *b*, it can be observed that the shapes of mastic and mortar curves are similar for both $|E^*|$ and δ. This can be interpreted by the fact that there is no solid body (aggregates, sand, etc.) packing in mortar, thus the behavior of mortar is mainly dominated by the mastic phase.

From these figures, it can also be seen that ZOAB shows a similar asymptotic behavior as that of mastic at higher frequencies ($>10^5$ Hz) both for $|E^*|_{mix}$ and δ_{mix}. However, with the decrease in frequencies, both the curves start to differ significantly. While $|E^*|_{mas}$ and δ_{mas} keep following the same trend with decreasing frequencies, $|E^*|_{mix}$ reaches a lower bound of the asymptotic value at around 1 Hz and δ_{mix} reaches a peak value before declining again.

Above observations can be hypothesized to the fact that at higher frequencies the effect of aggregate interactions is less pronounced and the behavior of ZOAB is governed by the behavior of mastic. This results in the similar nature for both mix- and mastic- curves. At lower frequencies, mastic keeps on deforming under the applied load, while the interlocked aggregate network of ZOAB limits the deformation beyond a critical frequency (1 Hz in this case). These noticeable differences between the mastic scale and the mix scale are important in assessing the performance of micromechanical models which will be discussed later in the paper.

Upscaling by Various Micromechanical Models

Upscaling from the mastic scale to the mortar scale to the mix scale was implemented in this research work. As an intermediate check to judge the performance of micromechanical models, the obtained results of mortar (by upscaling mastic to mortar) were compared against the experimental data, which will be discussed in the following subsection.

Upscaling Results of the Mortar Scale. On the basis of the volume fractions and mechanical properties of mastic and sand (see Tables 1 and 3), different micromechanical models were utilized to predict $|E^*|_{mor}$ and δ_{mor}. Figure 3*a* and *b* show the comparison of the predicted $|E^*|_{mor}$ and δ_{mor} against the experimental data (labeled as "Test") respectively. With the exception of the SC model, in general, the predicted results are in good agreement with the experimental results.

Upscaling Results of the Mix Scale. Comparisons between the predicted results of $|E^*|_{mix}$ and δ_{mix} with the experimental data (labeled as "Test") are shown in Figure 4. In these figures, "GSC 1" represents the case in which aggregates are added in the first step and "GSC 2" represents the case in which air voids are added in the first step, as described in earlier section.

As shown in Figure 4*a* and *b*, with the exception of the SC model and the "GSC 1" model, all models underestimate $|E^*|_{mix}$ and overestimate δ_{mix}. Corresponding figures also show that the Dilute model ("DM") results in the lowest values of $|E^*|_{mix}$ while the SC model

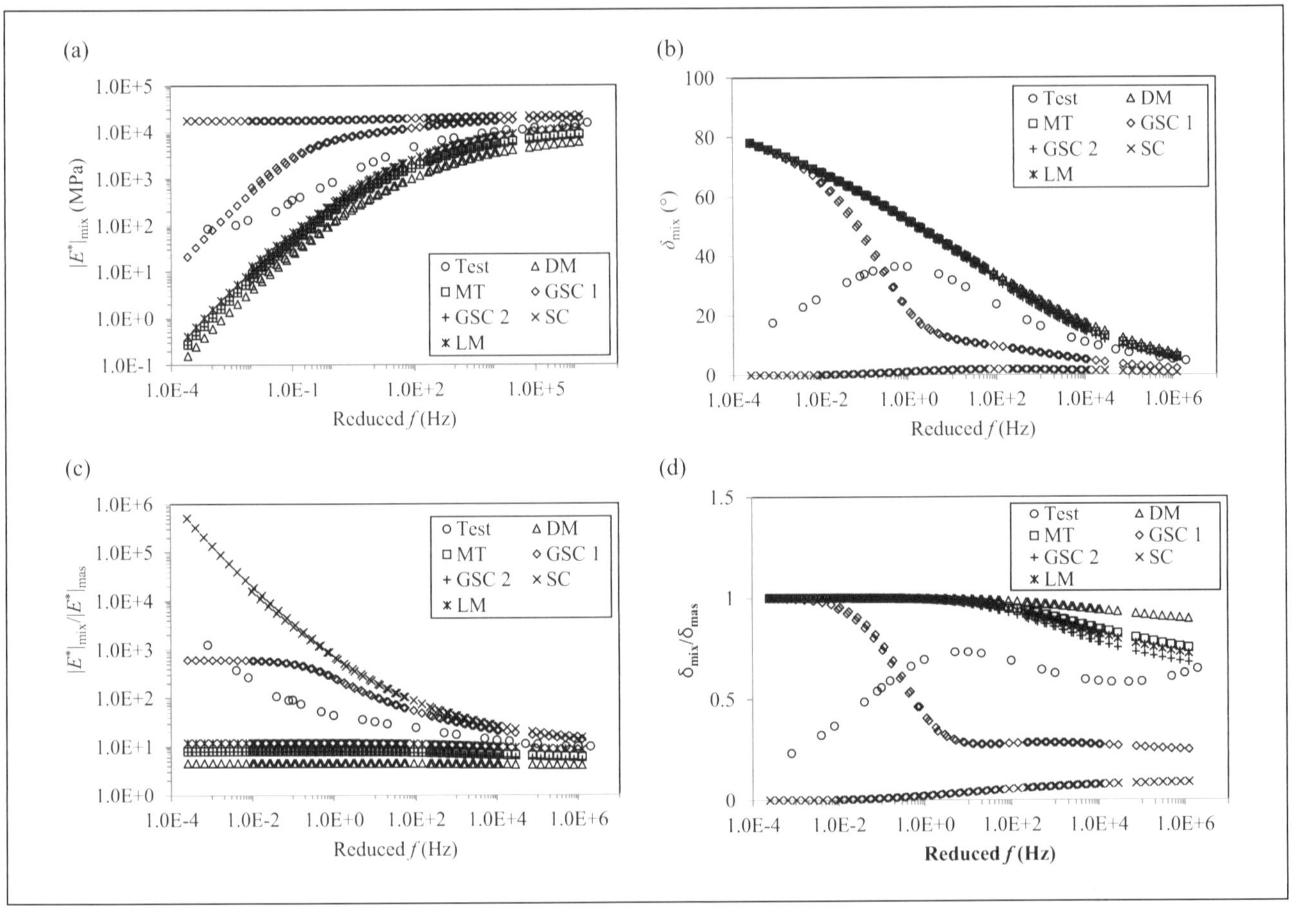

Figure 4. Predicted results of ZOAB properties: (*a*) dynamic Young's modulus, (*b*) phase angle, (*c*) ratio of the dynamic modulus of ZOAB to mastic, and (*d*) ratio of the phase angle of ZOAB to mastic.

predicts the highest. It can also be observed that the Lielens' model ("LM") performs better particularly at high frequencies and produces $|E^*|_{mix}$ and δ_{mix} values closer to the experimental data. Figure 4*c* and *d* show a comparison of the predicted $|E^*|_{mix}/|E^*|_{mas}$ and $\delta_{mix}/\delta_{mas}$ against the experimental data. These plots show that generally the values of predicted $|E^*|_{mix}/|E^*|_{mas}$ and $\delta_{mix}/\delta_{mas}$ almost remain constant in the whole frequency range, whereas the experimental results of $|E^*|_{mix}/|E^*|_{mas}$ increase and the values of $\delta_{mix}/\delta_{mas}$ decrease with the decrease of frequency. Overall, it can be concluded that none of the models can predict acceptable results of ZOAB properties at lower frequencies.

The above observations indicate that better performances of micromechanical models can be obtained in mortar- against mix- scale. It can be associated with the fact that these models were primarily developed to account for the stiffening effect caused by the embedded inclusions in a mix with minimal particle interactions. In other words, these models tend to be applicable for dispersed suspensions, the behavior of which is dominated by the matrix phase. Since mortar is expected to be a dispersed suspension while for ZOAB, particularly at lower frequencies, a dispersed suspension is not expected because of the packing aggregates, the predicted results by these models are acceptable for mortar but for ZOAB they are inaccurate.

In order to address the above issue, researchers (*24, 32*) have proposed that micromechanical models based upon the DS can provide better predictions of mechanical properties of asphalt mixes, which is expected to avoid the interactions between aggregate particles. In the following section, the applicability of this scheme will be judged in the context of ZOAB mixes.

Predicted Results based on the Differential Scheme. The predicted results of $|E^*|_{mix}$ and δ_{mix} on the basis of the DS approach are shown in Figure 5*a–f*. In order to understand the difference between the DS approach and the non-DS approach, the plots from the non-DS approach (Figure 4) are plotted against predictions by the DS approach in the same figures. It can be observed that with the exception of the "GSC 2" model, the performance of all the models improves significantly especially

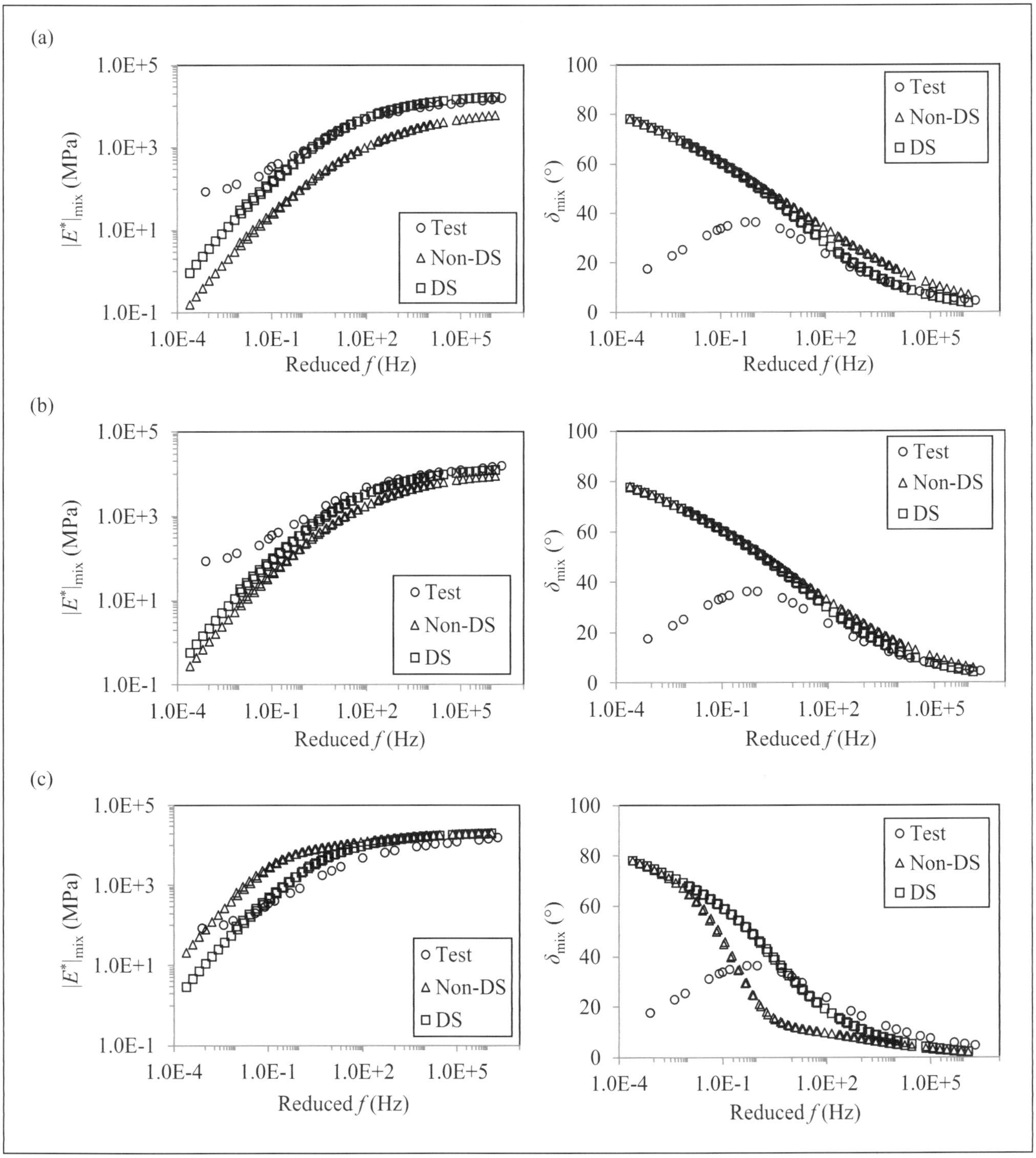

Figure 5. Effect of the differential scheme on the performance of micromechanical models: (*a*) Dilute, (*b*) MT, (*c*) GSC 1, (*d*) GSC 2, (*e*) SC, and (*f*) Lielens'.

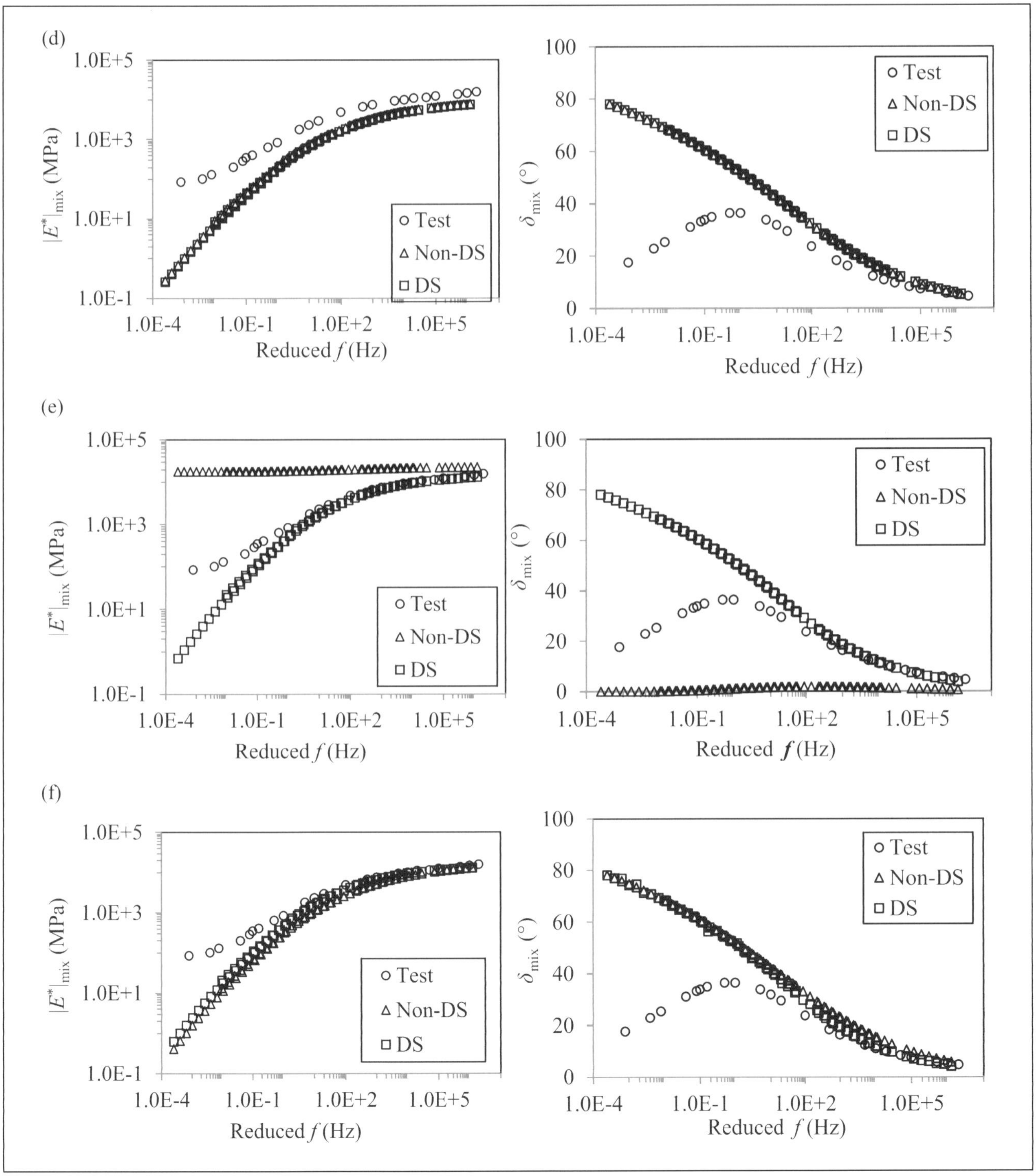

Figure 5. (continued)

at higher frequencies. However, it is also observed that the DS approach does not change the shape of the curve but more or less has a scaling effect. At lower frequencies, even the DS approach fails to predict $|E^*|_{mix}$ and δ_{mix} accurately. The failure of the DS can be attributed to the fact that it does not change the fundamental theory of the micromechanical models.

In general, it can be concluded that there is a need for more robust models/methodologies which take into account the aggregate interactions in a better way.

Conclusion

ZOAB is a commonly used mix on the major road network of the Netherlands. Despite several advantages of ZOAB over dense asphalt mixtures, the mix is known to suffer from raveling distress during its design life. One way to study the raveling propensity of a given mix is by utilizing micromechanics-based models. Although many research attempts have been made to study such models for various mixes, not much attention has been paid to study the suitability of such models for ZOAB-type mixes. Therefore, this paper examined and presented the performance of various commonly utilized micromechanical models in estimating the mechanical properties of ZOAB. The following general conclusions can be made:

- None of the applied micromechanical models could adequately predict the mechanical properties of ZOAB, particularly at low frequencies. This can be related to the fact that these models are applicable for a dispersed suspension with minimal particle interactions.
- At lower frequencies, the performance of the models did not improve after the introduction of the differential scheme, whereas, the predicted results were better at higher frequencies. This can be attributed to the fact that this scheme does not change the fundamental theory of the micromechanical models.
- In order to improve the performance of micromechanical models for ZOAB mixes, factors which reflect the effect of the packing aggregates more adequately should be introduced.

Recommendation for Future Research

Since all the micromechanical models that were considered in this study failed to predict the mechanical properties of ZOAB mixes, there is a need for more robust models/methodology which take into account the interactions between the packing aggregates in a better way.

Author Contributions

The authors confirm contribution to the paper as follows: study conception and design: Athanasios Scarpas, Kumar Anupam, Hong Zhang; data collection: Hong Zhang; analysis and interpretation of results: Hong Zhang, Kumar Anupam, Athanasios Scarpas, Cor Kasbergen; draft manuscript preparation: Kumar Anupam, Hong Zhang. All authors reviewed and the results and approved the final version of the manuscript.

References

1. Hagos, E. T. *The Effect of Aging on Binder Properties of Porous Asphalt Concrete*. Delft University of Technology, 2008.
2. Mo, L., M. Huurman, S. Wu, and A. A. A. Molenaar. Ravelling Investigation of Porous Asphalt Concrete Based on Fatigue Characteristics of Bitumen–Stone Adhesion and Mortar. *Materials & Design*, Vol. 30, No. 1, 2009, pp. 170–179.
3. Kringos, N., and A. Scarpas. Raveling of Asphaltic Mixes Due to Water Damage: Computational Identification of Controlling Parameters. *Transportation Research Record: Journal of the Transportation Research Board*, 2005. 1929: 79–87.
4. Anupam, K., S. K. Srirangam, A. Varveri, C. Kasbergen, and A. Scarpas. Microstructural Analysis of Porous Asphalt Concrete Mix Subjected to Rolling Truck Tire Loads. *Transportation Research Record: Journal of the Transportation Research Board*, 2016. 2575: 113–122.
5. Ghossein, E. *Numerical Validation of Analytical Homogenization Models for the Case of Randomly Distributed and Oriented Ellipsoidal Fibers Reinforced Composites*. Université de Montréal, 2014.
6. Eshelby, J. D. The Determination of the Elastic Field of an Ellipsoidal Inclusion and Related Problems. *Proceedings of the Royal Society London A*, Vol. 241, No. 1226, 1957, pp. 376–396.
7. Mori, T., and K. Tanaka. Average Stress in Matrix and Average Elastic Energy of Materials with Misfitting Inclusions. *Acta Metallurgica*, Vol. 21, No. 5, 1973, pp. 571–574.
8. Hill, R. A Self-consistent Mechanics of Composite Materials. *Journal of the Mechanics and Physics of Solids*, Vol. 13, No. 4, 1965, pp. 213–222.
9. Christensen, R. M., and K. H. Lo. Solutions for Effective Shear Properties in Three Phase Sphere and Cylinder Models. *Journal of Mechanics and Physics of Solids*, Vol. 27, No. 4, 1979, pp. 315–330.
10. Lielens, G., P. Pirotte, A. Couniot, F. Dupret, and R. Keunings. Prediction of Thermo-mechanical Properties for Compression Moulded Composites. *Composites Part A: Applied Science and Manufacturing*, Vol. 29, No. 1–2, 1998, pp. 63–70.
11. Buttlar, W. G., D. Bozkurt, G. G. Al-khateeb, and A. S. Waldhoff. Understanding Asphalt Mastic Behavior through Micromechanics. *Transportation Research Record: Journal of the Transportation Research Board*, 1999. 1681: 157–169.

12. Hashin, Z. The Elastic Moduli of Heterogeneous Materials. *Journal of Applied Mechanics*, Vol. 29, No. 1, 1962, pp. 143–150.
13. Kim, Y. R., and D. N. Little. Linear Viscoelastic Analysis of Asphalt Mastics. *Journal of Materials in Civil Engineering*, Vol. 16, No. 2, 2004, pp. 122–132.
14. Nielsen, L. E. Generalized Equation for the Elastic Moduli of Composite Materials. *Journal of Applied Physics*, Vol. 41, No. 11, 1970, pp. 4626–4627.
15. Yin, H. M., W. G. Buttlar, G. H. Paulino, and H. Di Benedetto. Assessment of Existing Micro-mechanical Models for Asphalt Mastics Considering Viscoelastic Effects. *Road Materials and Pavement Design*, Vol. 9, No. 1, 2008, pp. 31–57.
16. Aigner, E., R. Lackner, and C. Pichler. Multiscale Prediction of Viscoelastic Properties of Asphalt Concrete. *Journal of Materials in Civil Engineering*, Vol. 21, No. 12, 2009, pp. 771–780.
17. Pichler, C., R. Lackner, and E. Aigner. Generalized Self-consistent Scheme for Upscaling of Viscoelastic Properties of Highly-filled Matrix-inclusion Composites-Application in the Context of Multiscale Modelling of Bituminous Mixtures. *Composites Part B: Engineering*, Vol. 43, No. 2, 2012, pp. 457–464.
18. Underwood, B. S., and Y. R. Kim. A Four Phase Micromechanical Model for Asphalt Mastic Modulus. *Mechanics of Materials*, Vol. 75, 2014, pp. 13–33.
19. Hashin, Z. Complex Moduli of Viscoelastic Composites—I. General Theory and Application to Particulate Composites. *International Journal of Solids and Structures*, Vol. 6, No. 5, 1970, pp. 539–552.
20. Christensen, R. M. A Critical Evaluation for a Class of Micro-Mechanics Models. *Journal of Mechanics and Physics of Solids*, Vol. 38, No. 3, 1990, pp. 379–404.
21. Benveniste, Y. Revisiting the Generalized Self-consistent Scheme in Composites: Clarification of Some Aspects and a New Formulation. *Journal of the Mechanics and Physics of Solids*, Vol. 56, No. 10, 2008, pp. 2984–3002.
22. McLaughlin, R. A Study of the Differential Scheme for Composite Materials. *International Journal of Engineering Science*, Vol. 15, No. 4, 1977, pp. 237–244.
23. Norris, A. N. A Differential Scheme for the Effective Moduli of Composites. *Mechanics of Materials*, Vol. 4, No. 1, 1985, pp. 1–16.
24. Yin, H., and Y. Zhao. *Introduction to the Micromechanics of Composite Materials*. CRC Press, Boca Raton, 2016.
25. Underwood, B. S. *Multiscale Constitutive Modelling of Asphalt Concrete*. North Carolina State University, Raleigh, 2011.
26. Huurman, M. *Lifetime Optimisation Tool, LOT*. Delft University of Technology, 2007.
27. American Association of State Highway Transportation Officials (AASHTO). *T 342-11: Standard Method of Test for Determining Dynamic Modulus of Hot Mix Asphalt (HMA)*. Washington, D.C., 2015.
28. Mukandila, E. M., W. J. Steyn, and T. I. Milne. Modelling of Cohesion and Adhesion Damage of Seal Based on Dynamic Shear Rheometer Testing. *International Journal of Pavement Engineering*, Vol. 19, No. 9, 2018, pp. 786–797.
29. Zhang, J., Z. Fan, J. Pei, R. Li, and M. Chang. Multiscale Validation of the Applicability of Micromechanical Models for Asphalt Mixture. *Advances in Materials Science and Engineering*, Vol. 2015, 2015, pp. 1–8.
30. Ma, T., H. Wang, D. Zhang, and Y. Zhang. Heterogeneity Effect of Mechanical Property on Creep Behavior of Asphalt Mixture Based on Micromechanical Modeling and Virtual Creep Test. *Mechanics of Materials*, Vol. 104, 2017, pp. 49–59.
31. Chen, J., L. Wang, and X. Huang. Micromechanical Modeling of Asphalt Concrete Fracture Using a User-defined Three-dimensional Discrete Element Method. *Journal of Central South University*, Vol. 19, No. 12, 2012, pp. 3595–3602.
32. Kim, M., and W. G. Buttlar. Differential Scheme Effective Medium Theory for Hot-Mix Asphalt $|E^*|$ Prediction. *Journal of Materials in Civil Engineering*, Vol. 23, No. 1, 2011, pp. 69–78.

The Standing Committee on Structural Requirements of Asphalt Mixtures (AFK50) peer-reviewed this paper (18-03275).

Article

TRR
JOURNAL OF THE TRANSPORTATION RESEARCH BOARD

Transportation Research Record
2018, Vol. 2672(28) 416–425

DOI: 10.1177/0361198118787973
journals.sagepub.com/home/trr
SAGE

Comparison of Asphalt Mixtures Crack Resistance at Intermediate Temperatures using Advanced Test Methods and Theories

Wei Cao[1], Louay N. Mohammad[1,2], Peyman Barghabany[1,2], Samuel B. Cooper, III[1], and Saman Salari[1]

Abstract

Due to the increased use of recycled materials in pavement, the crack resistance of asphalt mixtures has become a major concern. This study evaluated the crack resistance of asphalt mixtures containing up to 40% reclaimed asphalt pavement (RAP) at intermediate temperatures. Three groups of materials consisting of seven plant-produced hot-mix asphalt mixtures were evaluated. A suite of test methods with various mechanistic backgrounds were employed, namely: semi-circular bend, simplified viscoelastic continuum damage, indirect tension, Texas overlay, and Illinois flexibility index tests. The corresponding parameters obtained for assessing crack resistance were: critical strain energy release rate J_c, material fatigue sensitivity, dissipated creep strain energy, number of cycles to failure $N_{f,OT}$, the interaction plot of critical fracture energy versus crack progression rate, and the flexibility index, respectively. Test results indicated that mixtures with higher RAP content exhibited higher stiffness and indirect tensile strength with lower compliance than the counterpart materials. In addition, all evaluation parameters, except for the use of the interaction plot from the Texas overlay test, provided consistent observations that incorporation of more RAP deteriorated the materials' crack resistance. The mixture-discriminating capabilities of all test methodologies were also investigated using the paired *t*-test on the experimental data of each mixture group. The results presented herein were part of Pooled Fund TPF-5(294) study titled *Develop Mix Design and Analysis Procedures for Asphalt Mixtures Containing High-RAP and/or RAS Contents.*

Traffic-induced cracking is a major concern for asphalt pavements, which may appear on the pavement surface in the forms of longitudinal and alligator cracking. The presence of cracks affects the roughness and thus driving quality of pavement. In addition, cracks allow for moisture infiltration, which may weaken the materials inside the structure and reduce the pavement's bearing capacity. Satisfactory pavement performance in resisting cracking requires adequate structural design and construction. In addition, material selection and mixture design are of equal importance. In recent years, an increasing amount and variety of materials are being used in pavement mixtures (*1*). Examples include reclaimed asphalt pavement (RAP), recycled asphalt shingles (RAS), and various additives and modifiers. The incorporation of recycled asphalt binder from RAP and/or RAS in particular reduces the stress relaxation capability of the mixture, thereby making it more prone to cracking (*2*). In order to alleviate such adverse effects introduced by the aged binder, a soft base binder may be employed. The use of warm-mix technologies with RAP and/or RAS has also been shown to provide comparable performance with conventional hot-mix asphalt mixtures (*3*, *4*).

For mixtures containing recycled materials, the traditional volumetrics-based Superpave mixture design method should be complemented with mechanistic tests for crack resistance (*5*). To date, a number of mechanistic test methods and modeling schemes have been developed, such as semi-circular bend (SCB) test, simplified viscoelastic continuum damage (S-VECD) characterization, indirect tension (IDT) test, Texas overlay test, and Illinois flexibility index test (I-FIT). The mechanistic analyses are based on fracture mechanics, continuum damage mechanics, and the concept of dissipated energy, to list a few. Specimens used in these tests can be easily obtained using the Superpave gyratory compactor (SGC) or from field cores. This paper presents an experimental effort using the above-mentioned five test methods on the evaluation of crack resistance of asphalt

[1]Louisiana Transportation Research Center, Louisiana State University, Baton Rouge, LA
[2]Department of Civil and Environmental Engineering, Louisiana State University, Baton Rouge, LA

Corresponding Author:
Address correspondence to Louay N. Mohammad: louaym@lsu.edu

Table 1. Asphalt Mixture Composition

Mixture designation	Mixture source	RAP RBR, %	Design asphalt binder content, %	PG of base binder
CO-V	Colorado DOT	0	5.3	64-22
CO-18RAP		18	5.3	58-28
FL-17RAP	Florida DOT	17	5.2	58-22
FL-36RAP		36	5.0	52-28
ALF-L1-V	FHWA ALF	0	5.0	64-22
ALF-L6-20RAP		20	5.0	64-22
ALF-L5-40RAP		40	5.0	64-22

Note: RAP = reclaimed asphalt pavement; RBR = recycled binder ratio; PG = performance grade.

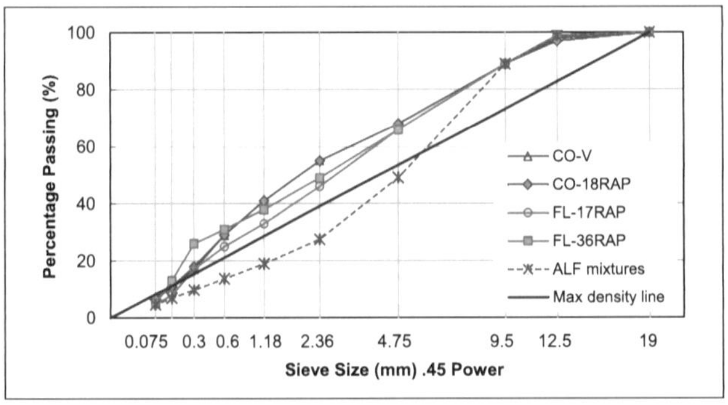

Figure 1. Aggregate gradation.

mixtures containing up to 40% RAP. Since these test methods and data interpretations are established upon different mechanics theories and concepts, it is also of substantial importance and interest to examine their sensitivity to material compositions and to find out if consistent results would be yielded.

Objectives

The objectives of this study are to:

- Investigate the effect of incorporation of RAP on the crack resistance of asphalt mixtures using advanced test methods and theories
- Evaluate and compare the sensitivity of each test method in capturing the effect of material compositions on cracking
- Ascertain if these various tests and corresponding evaluation parameters would yield consistent observations regarding crack resistance

Materials

To achieve the objectives of this study, a total of seven plant-produced hot-mix asphalt mixtures with 12.5-mm nominal maximum aggregate size were employed. The RAP content was expressed in terms of recycled binder ratio (RBR), which is defined as the percentage of binder from recycled materials in the total binder. As shown in Table 1, the mixtures provided by the Colorado Department of Transportation (DOT) contained 0% and 18% RAP, the mixtures supplied by the Florida DOT utilized 17% and 36% RAP, and those sampled during the construction of the FHWA Accelerated Loading Facility (ALF) test lanes (i.e., Lanes 1, 6, 5) incorporated 0%, 20%, and 40% RAP. Note that for the Colorado and Florida mixtures, the increase in the RAP content was addressed with a lower grade base asphalt. The aggregate gradation is shown in Figure 1. Within each group of mixtures, the aggregate type and RAP source were kept constant.

Test Methods

This section briefly describes the five tests used in this study. Note that different intermediate test temperatures and various specimen geometries were used by following the corresponding protocols; see Table 2. In addition, it is noted that different tests account for material behaviors during either one or both phases of crack initiation and propagation. The plant-produced loose mixtures were reheated and compacted in the laboratory using an SGC and then trimmed to obtain the required geometries. Specimens were then dried using the CoreDry method in accordance with ASTM D7227, *Standard Practice for Rapid Drying of Compacted Asphalt Specimens Using Vacuum Drying Apparatus*. Specimens with air voids within the target range, 7.0% ± 0.5%, were then long-term oven aged within 24 hours after cutting at 85°C for 120 hours in accordance with AASHTO R30, *Standard Practice for Mixture Conditioning of Hot Mix Asphalt*. All specimens were tested within 24 hours after completion of the aging process. The IDT test was conducted using a closed-loop servohydraulic materials test system, while all other tests were performed using the Asphalt Mixture Performance Tester (AMPT).

Semi-Circular Bend Test

The SCB test was introduced in asphalt concrete (*9*) to obtain the critical strain energy release rate, J_c, which is based on nonlinear fracture mechanics principles. J_c measures the strain energy required to form a unit area of new surface that is fractured in a medium. A larger J_c value represents higher

Table 2. Mixture Performance Tests

Test	Temp.	Protocol	Test mode	Specimen geometry	Cracking phases	Engineering properties	Remarks
SCB	25°C	ASTM D8044	Monotonic, 0.5 mm/min.	Semi-circular, ϕ150 mm × 57 mm	Crack propagation	• Critical strain energy release rate	• Additional replicates may be needed due to untrimmed surface
S-VECD	18°C	• AASHTO TP79 • AASHTO TP107	Cyclic, 10 Hz	• ϕ100 mm × 150 mm for modulus • ϕ100 mm × 130 mm for fatigue	Crack initiation	• Dynamic modulus • Damage characteristic curve • Material fatigue sensitivity	• Steep learning curve for the theory • Extra sample preparation time due to gluing of studs & end plates
IDT	10°C	Roque and Buttlar (*6*), Buttlar and Roque (*7*), Roque et al. (*8*)	• Cyclic for modulus, 10 Hz • Monotonic for strength test, 50 mm/min.	ϕ150 mm × 38 mm	Crack initiation	• Dynamic modulus • Creep compliance • Indirect tensile strength • Dissipated creep strain energy	• A criterion is needed for data quality (variability) of creep compliance • Special sample conditioning system is required for energy ratio
Texas overlay	25°C	Tex-248-F	Cyclic, 0.1 Hz, 0.6-mm maximum opening displacement	Oval brick, ϕ150 mm × 76 mm × 38 mm	Crack initiation & propagation	• Number of cycles to failure • Critical fracture energy • Crack progression rate	• Difficult cutting without a double-blade saw • Proper gluing of plates is challenging • Use of interaction plot needs further validation
I-FIT	25°C	AASHTO TP124	Monotonic, 50 mm/min.	Semi-circular, ϕ150 mm × 50 mm	Crack propagation	• Flexibility index	• High data sampling rate required • Use of flexibility index needs further validation

toughness, and thus is desired for crack-resistant mixtures. The SCB test and the validity of J_c has been widely investigated and verified in numerical analysis (*10*), laboratory experiments (*11*), and field evaluation (*12*). This test has been favored by many researchers and practitioners due to the ease of sample preparation and simple testing procedure.

The SCB test was conducted at 25°C in accordance with ASTM D8044, *Standard Test Method for Evaluation of Asphalt Mixture Cracking Resistance Using the Semi-Circular Bend (SCB) Test at Intermediate Temperatures*. Three notch depths were used: 25.4 mm, 31.8 mm, and 38.1 mm, and for each four semi-circular specimens were prepared. During testing, the specimen was loaded monotonically with a displacement rate of 0.5 mm/min. until fracture in a three-point bending configuration.

S-VECD Characterization

The viscoelastic continuum damage (VECD) theory has been applied in modeling the constitutive and fatigue behaviors of materials under monotonic and cyclic loading, respectively. In this framework, the effect of microcrack damage was delineated from the material's intrinsic nonlinearity through the use of pseudo-strain based elastic-viscoelastic correspondence principle (*13*). The capability and efficacy of the model have been demonstrated in laboratory material evaluation as well as fatigue performance prediction of pavements (*14*, *15*). Note that the S-VECD approach can also be applied to address existing pavements via the use of small-geometry specimens obtained from field cores (*15*).

The experimental effort of S-VECD characterization consists of dynamic modulus and direct tension cyclic fatigue tests. Three replicates were prepared for each test. The dynamic modulus test was performed in accordance with AASHTO TP79, *Standard Method of Test for Determining the Dynamic Modulus and Flow Number for Hot Mix Asphalt (HMA) Using the Asphalt Mixture Performance Tester (AMPT)*. The fatigue test was conducted according to AASHTO TP107, *Standard Method of Test for Determining the Damage Characteristic Curve of Asphalt Mixtures from Direct Tension Cyclic Fatigue Test*.

Indirect Tension Test

The IDT test was performed by subjecting the specimen to a compression load diametrically, thereby creating a tensile zone perpendicular to the loading direction around the specimen center. For this test, the dissipated creep strain energy (DCSE) as a material parameter was proposed as the threshold defining the development of (non-healable) macrocracks. Using this threshold concept, the fracture mechanics model for hot-mix asphalt was developed (*16*), which accounted for the effect of pavement structure and provided the basis for the energy ratio (ER) parameter (*17*). Based on a comprehensive evaluation of 22 field pavements that were in service for 10 years throughout the state of Florida, Roque et al. (*17*) recommended two criteria based on DCSE and ER to control top-down cracking.

Given the unavailability of a well-established national or state standard for the IDT test, this test was conducted following the procedures outlined in existing literature (*6–8*). The IDT test includes three consecutive steps: dynamic modulus, creep, and strength tests. In this study, all test steps were performed at 10°C on the same specimen with a minimum interval of 5 min. The dynamic modulus test was conducted at a frequency of 10 Hz; the load level was adjusted to yield a nominal strain (defined as the ratio of displacement between gauge points to the original gauge length) in the horizontal direction within 50 ± 5 microstrain. The creep load was carefully selected such that at the end of loading (i.e., 1000 s) the horizontal nominal strain was between 150 and 300 microstrain. The strength test was controlled in the monotonic mode with a displacement rate of 50 mm/min.

Texas Overlay Test

The Texas overlay test was originally developed to ascertain a material's resistance to reflective cracking in asphalt overlays (*18*). This test has also been employed to evaluate the use of geosynthetics in asphalt layers in mitigating reflective cracking (*19*). Advanced analysis has also been attempted to obtain the fracture and healing properties of asphalt mixture based on fracture mechanics theories and the concept of dissipated pseudo-strain energy (*18–20*).

The test jig employed in this study used oval brick specimens that were trimmed from SGC samples. Sample fabrication and test procedures were performed in accordance with the Texas DOT Tex-248-F, *Test Procedure for Overlay Test*. Five oval brick specimens were tested for each mixture. The test was conducted at 25°C under the displacement controlled mode using a triangular waveform with 0.6 mm amplitude at 0.1 Hz.

Illinois Flexibility Index Test

The Illinois flexibility index test (I-FIT) was recently developed by Al-Qadi et al. (*21*) using the SCB geometry via an investigation involving various temperatures and loading rates. The resulting flexibility index (FI) parameter is used to represent the mixture's crack resistance. Materials with higher FI values are expected to exhibit higher resistance to cracking. To date, limited studies have been conducted with respect to this test. In some studies, the FI obtained was found to yield reasonable correlations with field performance of pavements (*21*). Others provided both positive and negative conclusions regarding the capability of this test in discriminating mixtures (*22, 23*). In a more recent study, FI was found to be highly variable and was not sensitive to the presence and type of polymer modifiers (*24*).

The I-FIT test was conducted in accordance with AASHTO TP 124, *Determining the Fracture Potential of Asphalt Mixtures Using Semicircular Bend Geometry (SCB) at Intermediate Temperature*. For each mixture, six semicircular specimens were prepared with notches that were 15 mm deep and 1.5 mm wide. Loading was performed monotonically with a displacement rate of 50 mm/min. until fracture at 25°C.

Results and Discussion

This section presents the test results for the evaluation of crack resistance as well as stiffness properties of the mixtures considered in this study.

Semi-Circular Bend Test

The critical strain energy release rate J_c was computed using the SCB load and displacement data via:

$$J_c = -\frac{1}{w}\frac{dU}{dh} \quad (1)$$

where J_c is the critical strain energy release rate (kJ/m^2), w is specimen thickness (m), h is notch depth (m), and U is strain energy up to failure (kJ). Mixtures with higher J_c values are expected to exhibit higher crack resistance.

Table 3 presents the J_c results along with the averaged coefficient of variation (CoV) of corresponding fracture energy per unit thickness of specimen. The CoVs for all mixtures were below 15% which indicates a good test repeatability. It is noted that within each group of the CO, FL, and ALF mixtures, the increase in the RAP content decreased J_c and thus the crack resistance.

S-VECD Characterization

The dynamic modulus master curves were constructed using the time–temperature superposition principle at a reference temperature of 21.1°C, and the results are illustrated in Figure 2*a*. The stiffening effect of including more RAP is demonstrated by the Colorado DOT (CO) and ALF mixtures,

Table 3. SCB Test Results

Mix designation	J_c (kJ/m^2)	CoV
CO-V	0.44	10%
CO-18RAP	0.37	12%
FL-17RAP	0.39	14%
FL-36RAP	0.35	12%
ALF-L1-V	0.55	10%
ALF-L6-20RAP	0.41	12%
ALF-L5-40RAP	0.34	10%

whereas the Florida DOT (FL) mixtures containing 17% and 36% RAP exhibited very similar stiffness properties.

The damage characteristic curve represents a unique material relationship describing the loss in material integrity (C) as a result of damage accumulation (S). Figure 2*b* presents the fitted $C(S)$ curves using the exponential form:

$$C = e^{aS^b} \tag{2}$$

where a and b are regression coefficients. Typically, stiff mixtures would have $C(S)$ curves lying above those for soft materials, as evidenced by the CO and ALF mixtures. Meanwhile, the two FL mixtures having similar stiffness properties yielded comparable $C(S)$ curves.

It has been recognized that the $C(S)$ relationship cannot be directly used to ascertain a material's crack resistance, as it is complicated by the role of stiffness and does not contain information on the material's failure (*25*). As such, the recently developed parameter, material fatigue sensitivity (MFS), was used to address the evaluation and comparison of mixtures' fatigue. Details regarding the development and computation of this parameter is given elsewhere (*25*). The obtained MFS results are presented in Figure 2*c*. Note that lower MFS represents higher crack resistance. For all three groups of mixtures, increase in the RAP content yielded higher MFS values, indicating deterioration in the mixtures' crack resistance.

Indirect Tension Test

The experimental data from the dynamic modulus test step were analyzed using the viscoelastic solutions derived by Kim et al. (*26*). Details regarding data reduction for creep compliance and DCSE can be found elsewhere (*6*, *16*, *17*). The obtained creep compliance is fitted using a power-law form as:

$$D(t) = D_0 + D_1 t^m \tag{3}$$

where $D(t)$ is creep compliance (GPa^{-1}), t is time (s), and D_0, D_1, and m are regression coefficients. The DCSE parameter can be determined from the IDT strength test data via:

$$DCSE = FE - \frac{1}{2}\frac{S_T^{\,2}}{E} \tag{4}$$

where FE is fracture energy (kJ/m^3), S_T is tensile strength (kPa), and E denotes dynamic modulus (kPa).

Table 4 summarizes the IDT test results for the seven mixtures. Note that evaluation using the ER parameter was not pursued in this study because this parameter accounts for pavement structural effects, which is beyond the scope of this study. As shown in Table 4, within each mixture group, increase in the RAP content improved the materials' stiffness and tensile strength while reducing the compliance. As indicated by DCSE, increased RBR saw a reduction in mixtures' crack resistance.

Texas Overlay Test

In the Texas overlay test, fatigue life as denoted by $N_{f,OT}$ is defined as the number of cycles to reach 93% reduction in the maximum tensile load (*27*). As suggested by Walubita et al. (*28*), $N_{f,OT}$ can be determined from three replicates which were selected out of the five to yield the minimum CoV. Garcia et al. (*29*) have recently developed a new analysis approach involving two parameters: critical fracture energy and crack progression rate. Critical fracture energy measures a material's resistance to crack initiation, while crack progression rate is the rate of crack propagation under cyclic loading after the crack is initiated.

Table 5 summarizes the parameters obtained from the Texas overlay test. Note that $N_{f,OT}$ was obtained using the three replicates that yielded the minimum CoV, while for comparison it was also determined using all five replicates. The critical fracture energy and crack progression rate were computed using all replicates given their low variability. It is observed that the two sets of $N_{f,OT}$ provided consistent ranking for each group of mixtures; increase in RAP content reduced the materials' fatigue life. By comparing the critical fracture energy values it was observed that within each group incorporation of more RAP enhanced the materials' resistance to crack initiation. Additionally, higher RBR resulted in higher crack progression rate in all mixture groups.

The critical fracture energy and crack progression rate were then cross plotted to obtain the interaction plot in Figure 3. Based on the plot, the CO-18RAP mixture is considered to have crack resistance similar to CO-V due to its higher critical fracture energy as well as the higher crack progression rate. Increase of RBR in FL mixtures substantially decreased crack resistance, as evidenced by the considerably higher crack progression rate of FL-36RAP. Within the ALF mixtures, use of 20% RAP appeared to improve the crack resistance by comparing the Lane 1 and Lane 6 mixtures. When the RBR was increased to 40%, as in Lane 5, however, the crack resistance was significantly reduced, as evidenced by the dramatically elevated crack progression rate.

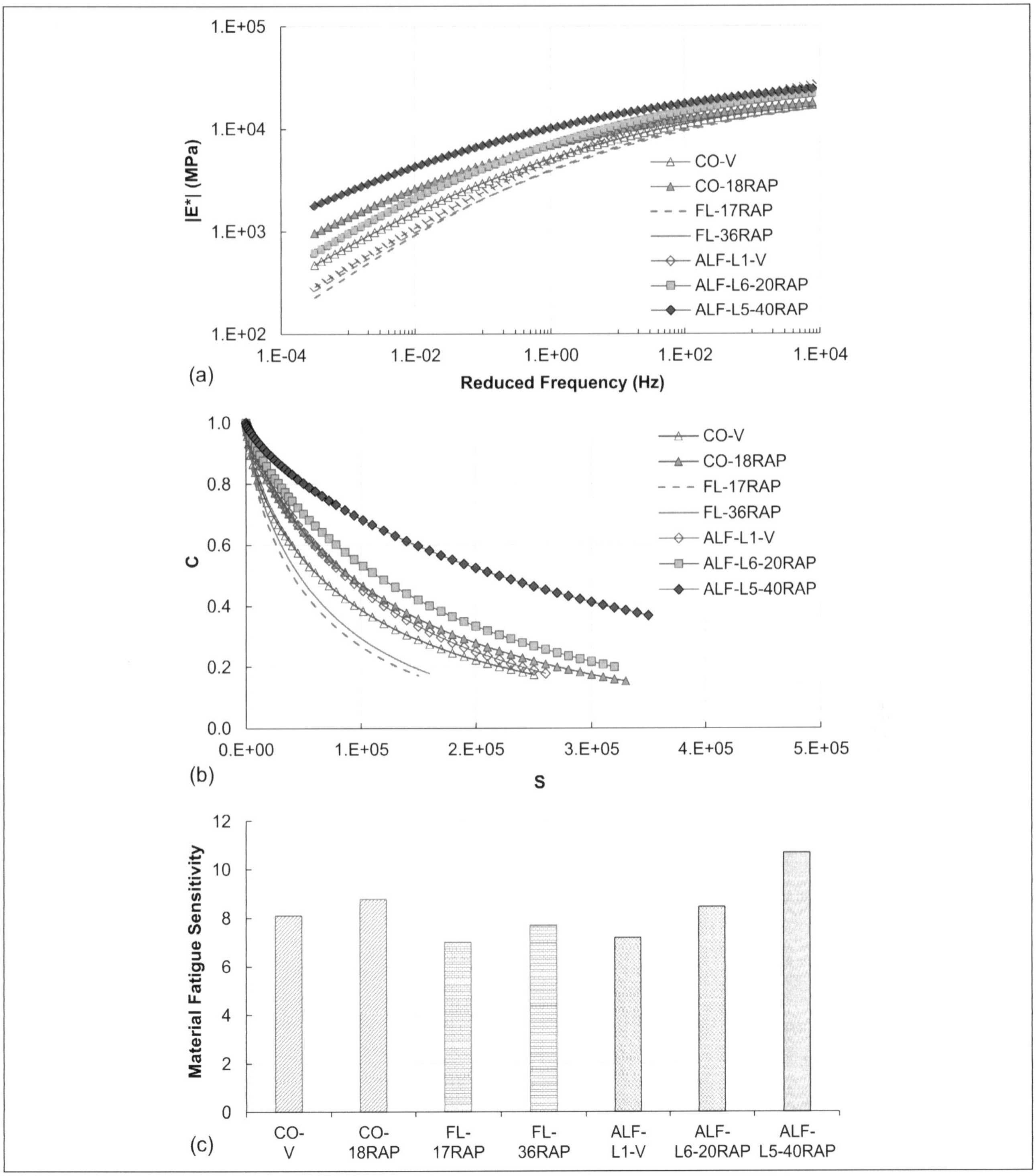

Figure 2. S-VECD fatigue characterization results: (*a*) dynamic modulus master curves; (*b*) damage characteristic curves; (c) material fatigue sensitivity.

Illinois Flexibility Index Test

The FI was calculated using the collected load and displacement data as:

$$FI = \frac{G_f}{n} \times \phi \quad \text{with} \quad G_f = \frac{W_f}{A_{lig}} \tag{5}$$

Table 4. IDT Test Results

Mix designation	Statistics	|E*| (MPa)	S_T (kPa)	m	D(t) at 200 s (GPa^{-1})	DCSE (kJ/m^3)
CO-V	Average	11,909	2,173	0.533	1.95	4.84
	CoV	3%	3%	–	–	11%
CO-18RAP	Average	12,527	2,392	0.514	0.78	3.46
	CoV	7%	6%	–	–	24%
FL-17RAP	Average	11,761	2,221	0.477	0.90	4.33
	CoV	4%	3%	–	–	8%
FL-36RAP	Average	12,276	2,600	0.385	0.56	3.79
	CoV	7%	10%	–	–	14%
ALF-L1-V	Average	14,884	2,614	0.542	0.87	3.86
	CoV	11%	2%	–	–	8%
ALF-L6-20RAP	Average	14,786	2,690	0.500	0.57	2.48
	CoV	12%	9%	–	–	30%
ALF-L5-40RAP	Average	19,151	3,302	0.366	0.19	1.83
	CoV	4%	5%	–	–	15%

Note: – = not applicable. Statistical analysis of D(t) and m is not applicable because in computing the creep compliance the horizontal and vertical displacements at both faces of all specimens were trimmed and averaged.

Table 5. Texas Overlay Test Results

Mix designation	Statistics	$N_{f,OT}$ (All five)	$N_{f,OT}$ (Best three)	Critical fracture energy (kJ/m^2) (All five)	Crack progression rate (All five)
CO-V	Average	204	212	0.40	0.323
	CoV	28%	8%	5%	9%
CO-18RAP	Average	86	78	0.50	0.385
	CoV	40%	21%	5%	11%
FL-17RAP	Average	254	218	0.52	0.321
	CoV	41%	8%	4%	5%
FL-36RAP	Average	11	7	0.59	1.375
	CoV	48%	7%	3%	16%
ALF-L1-V	Average	208	189	0.55	0.372
	CoV	58%	37%	8%	8%
ALF-L6-20RAP	Average	139	93	0.62	0.375
	CoV	55%	13%	8%	8%
ALF-L5-40RAP	Average	50	62	0.72	0.881
	CoV	63%	39%	3%	15%

where G_f is fracture energy (J/m^2); W_f (in J) is work of fracture obtained as the area under the load versus displacement curve up to the post-peak cut-off load (0.1 kN); A_{lig} is the ligament area (m^2); n (in kN/mm) is the absolute value of slope at the post-peak inflection point; and ϕ is a scaling factor set at 0.01. In order to obtain the slope, the post-peak load versus displacement curve was fitted using the generalized logistic function:

$$F(\delta) = \kappa + \frac{\mu - \kappa}{\left[\chi + \gamma e^{-\varphi(\delta - \tau)}\right]^{1/\lambda}} \tag{6}$$

where F is load (kN), δ is displacement (mm), and κ, μ, χ, γ, φ, τ, and λ are fitting coefficients. The data used in fitting were truncated by setting the cut-off load at 0.5 kN. For all tests, the resulting R^2-values of fitting were above 99.9%. The slope n

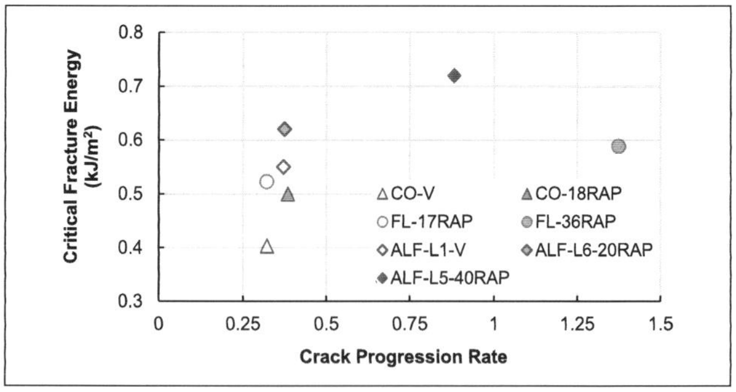

Figure 3. The Texas overlay interaction plot of crack resistance.

Table 6. Illinois Flexibility Index Test Results

Mix designation	Statistics	Fracture energy (J/m^2)	Slope n (kN/mm)	FI
CO-V	Average	1618	4.4	3.7
	CoV	4%	9%	10%
CO-18RAP	Average	1276	11.9	1.1
	CoV	8%	16%	17%
FL-17RAP	Average	1720	8.2	2.1
	CoV	8%	14%	11%
FL-36RAP	Average	1238	12.0	1.0
	CoV	8%	12%	13%

Table 7. Statistical Comparison

		CO mixtures (% RBR)		FL mixtures (% RBR)		ALF mixtures (% RBR)		
Test	Evaluation parameter	0	18	17	36	0	20	40
SCB	J_c	A	B	A	B	A	B	C
S-VECD	MFS	–	–	–	–	–	–	–
IDT	DCSE	A	B	A	A	A	B	B
Texas overlay	$N_{f,OT}$ (best three)	A	B	A	B	A	A/B	B
I-FIT	FI	A	B	A	B	–	–	–

Note: RBR = recycled binder ratio; – = not applicable. Statistical analysis is not applicable to MFS as it was obtained from simulation. Statistical grouping results are indicated by letters A, B, and C, representing significantly distinct crack resistance from best to worst.

was then numerically determined as the absolute value of the minimum of the first derivative of $F(\delta)$ in Equation 6.

Table 6 presents the I-FIT results. Note that due to the availability of materials, only the CO and FL mixtures were characterized using this test method. For each mixture the four tests yielding the minimum CoV of FI were used in the final analysis. As shown in Table 6, for both the CO and FL mixtures, increase in the RAP content decreased fracture energy and FI while increasing the post-peak slope, which indicates reduced crack resistance and more brittle material response.

Statistical Comparison of Different Tests

It is observed that, except for the use of the Texas overlay interaction plot, all the other parameters yielded the same observation that within each group of the CO, FL, and ALF mixtures increasing the RAP content deteriorated the materials' crack resistance. As shown in Table 2, different test methodologies account for either one or both phases of crack initiation and propagation. In order to identify the mixture-discriminating capability of each test, the paired *t*-test was performed and the results are summarized in Table 7. For each test, mixtures within each group were compared with respect to the effect of material composition. The letters A, B, and C were used to represent statistically distinct crack resistance from best to worst. Note that statistical analysis is not applicable to the MFS parameter because it was obtained from S-VECD fatigue simulation. It is seen that the parameters from all other tests were able to discriminate the two CO mixtures. The J_c, $N_{f,OT}$, and FI parameters from SCB, Texas overlay, and I-FIT tests, respectively, differentiated the two FL mixtures, while only J_c successfully distinguished the three ALF mixtures.

Summary and Conclusion

This study assessed the crack resistance of asphalt mixtures containing up to 40% RAP. A suite of advanced test methods with different mechanistic backgrounds were performed, namely: semi-circular bend, simplified viscoelastic continuum damage, indirect tension, Texas overlay, and Illinois flexibility index tests. Various intermediate temperatures and specimen geometries were employed. The following conclusions were drawn based on the findings in this study:

- Based on the parameters of J_c, MFS, DCSE, $N_{f,OT}$, and FI obtained from the SCB, S-VECD, IDT, Texas overlay, and I-FIT tests, respectively, higher RBR yielded mixtures with higher crack susceptibility within all the CO, FL, and ALF mixture groups.
- According to the dynamic modulus test as part of the S-VECD characterization, the mixtures with higher RBR within each group exhibited higher stiffness.
- According to the IDT test, increasing the RAP content improved materials' stiffness and indirect tensile

strength while reducing the compliance within each group of mixtures.

- For the Texas overlay test, the two parameters, critical fracture energy and crack progression rate, in general exhibited lower variability than $N_{f,OT}$. Within each mixture group, increase in the RAP content improved mixtures' resistance to crack initiation but also yielded faster crack propagation once the crack is initiated.
- According to the I-FIT, within each mixture group higher RBR resulted in a more brittle material response.
- Except the S-VECD MFS parameter to which the statistical analysis is not applicable, all single parameters from other tests correctly distinguished the two CO mixtures. The SCB, Texas overlay, and I-FIT tests were able to discriminate the two FL mixtures, while only the SCB J_c parameter successfully differentiated the three ALF mixtures. Overall, within the tests considered in this study, the SCB J_c was able to differentiate consistently the effect of RAP content for the three groups of mixtures.

Acknowledgments

The research presented herein is part of Pooled Fund TPF-5(294) "Develop Mix Design and Analysis Procedures for Asphalt Mixtures Containing High-RAP and/or RAS Contents." The authors acknowledge the support of the Colorado Department of Transportation, Florida Department of Transportation, and Louisiana Transportation Research Center. The assistance of Paragon Technical Services, Inc., in notching I-FIT specimens is appreciated.

Author Contributions

The authors confirm contribution to the paper as follows: study conception and design: Mohammad, Cao, Cooper; data collection: Barghabany, Cao; analysis and interpretation of results: Cao, Mohammad, Salari; draft manuscript preparation: Cao, Mohammad. All authors reviewed the results and approved the final version of the manuscript.

References

1. Hansen, K. R., and A. Copeland. *Information Series 138: 5th Annual Asphalt Pavement Industry Survey on Recycled Materials and Warm-Mix Asphalt Usage: 2014*. National Asphalt Pavement Association, Lanham, Maryland, 2015.
2. Daniel, J. D., N. Gibson, S. Tarbox, A. Copeland, and A. Andriescu. Effect of Long Term Aging on RAP Mixtures: Laboratory Evaluation of Plant Produced Mixtures. *Journal of the Association of Asphalt Paving Technologists*, Vol. 82, 2013, pp. 327–365.
3. Hong, F., D. H. Chen, and M. M. Mikhail. Long-Term Performance Evaluation of Recycled Asphalt Pavement Results from Texas: Pavement Studies Category 5 Sections from the Long-Term Pavement Performance Program. *Transportation Research Record: Journal of the Transportation Research Board*, 2010. 2180: 58–66.
4. Wu, S., K. Zhang, H. Wen, J. DeVol, and K. Kelsey. Performance Evaluation of Hot Mix Asphalt Containing Recycled Asphalt Shingles in Washington State. *Journal of Materials in Civil Engineering*, Vol. 28, No. 1, 2016, p. 04015088.
5. Mogawer, W. S., A. Austerman, R. Roque, B. S. Underwood, L. Mohammad, and J. Zou. Ageing and Rejuvenators: Evaluating Their Impact on High RAP Mixtures Fatigue Cracking Characteristics Using Advanced Mechanistic Models and Testing Methods. *Road Materials and Pavement Design*, Vol. 16, No. Suppl. 2, 2015, pp. 1–28.
6. Roque, R., and W. G. Buttlar. The Development of a Measurement and Analysis System to Accurately Determine Asphalt Concrete Properties Using the Indirect Tensile Mode. *Journal of Association of Asphalt Paving Technologists*, Vol. 61, 1992, pp. 304–332.
7. Buttlar, W. G., and R. Roque. Development and Evaluation of the Strategic Highway Research Program Measurement and Analysis System for Indirect Tensile Testing of Asphalt Mixtures at Low Temperatures. *Transportation Research Record: Journal of the Transportation Research Board*, 1994. 1454: 163–171.
8. Roque, R., B. Birgisson, B. Sangpetngam, and Z. Zhang. Hot Mix Asphalt Fracture Mechanics: a Fundamental Crack Growth Law for Asphalt Mixtures. *Journal of the Association of Asphalt Paving Technologists*, Vol. 71, 2002, pp. 816–827.
9. Mull, M. A., K. Stuart, and A. Yehia. Fracture Resistance Characterization of Chemically Modified Crumb Rubber Asphalt Pavement. *Journal of Materials Science*, Vol. 37, No. 3, 2002, pp. 557–566.
10. Mahmoud, E., S. Saadeh, H. Hakimelahi, and J. Harvey. Extended Finite-Element Modelling of Asphalt Mixtures Fracture Properties Using the Semi-Circular Bending Test. *Road Materials and Pavement Design*, Vol. 15, No. 1, 2014, pp. 153–166.
11. Cooper, S. B., Jr., , L. N. Mohammad, and M. A. Elseifi. Laboratory Performance of Asphalt Mixtures Containing Recycled Asphalt Shingles. *Transportation Research Record: Journal of the Transportation Research Board*, 2014. 2445: 94–102.
12. Kim, M., L. N. Mohammad, and M. A. Elseifi. Characterization of Fracture Properties of Asphalt Mixtures as Measured by Semicircular Bend Test and Indirect Tension Test. *Transportation Research Record: Journal of the Transportation Research Board*, 2012. 2296: 115–124.
13. Schapery, R. A. Correspondence Principles and a Generalized J-Integral for Large Deformation and Fracture Analysis of Viscoelastic Media. *International Journal of Fracture*, Vol. 25, No. 3, 1984, pp. 195–223.
14. Cao, W., A. Norouzi, and Y. R. Kim. Application of Viscoelastic Continuum Damage Approach to Predict Fatigue Performance of Binzhou Perpetual Pavements. *Journal of Traffic and Transportation Engineering*, Vol. 3, No. 2, 2016, pp. 104–115.
15. Park, H. J., M. Eslaminia, and Y. R. Kim. Mechanistic Evaluation of Cracking in In-Service Asphalt Pavements. *Materials and Structures*, Vol. 47, No. 8, 2014, pp. 1339–1358.
16. Zhang, Z., R. Roque, B. Birgisson, and B. Sangpetgnam. Identification and Verification of a Suitable Crack Growth Law for Asphalt Mixtures. *Journal of the Association of Asphalt Paving Technologists*, Vol. 70, 2001, pp. 206–241.
17. Roque, R., B. Birgisson, C. Drakos, and B. Dietrich. Development and Field Evaluation of Energy-Based Criteria

for Top-down Cracking Performance of Hot Mix Asphalt. *Journal of the Association of Asphalt Paving Technologists*, Vol. 73, 2004, pp. 229–260.

18. Germann, F. P., and R. L. Lytton. *Methodology for Predicting the Reflection Cracking Life of Asphalt Concrete Overlays*. Report No. FHWA-TX-79-09+207-5. Texas State Department of Highways and Public Transportation, Austin, 1979.
19. Cleveland, G., R. Lytton, and J. Button. Using Pseudostrain Damage Theory to Characterize Reinforcing Benefits of Geosynthetic Materials in Asphalt Concrete Overlays. *Transportation Research Record: Journal of the Transportation Research Board*, 2003. 1849: 202–211.
20. Zhou, F., S. Hu, T. Scullion, D. H. Chen, X. Qi, and G. Claros. Development and Verification of the Overlay Tester Based Fatigue Cracking Prediction Approach. *Journal of the Association of Asphalt Paving Technologists*, Vol. 76, 2007, pp. 627–662.
21. Al-Qadi, I. L., H. Ozer, J. Lambros, A. E. Khatib, P. Singhvi, T. Khan, J. Rivera-Perez, and B. Doll. *Testing Protocols to Ensure Performance of High Asphalt Binder Replacement Mixes Using RAP and RAS*. Report No. FHWA-ICT-15-017. Illinois Center for Transportation, Urbana, 2015.
22. Ling, C., D. Swiertz, T. Mandal, P. Teymourpour, and H. Bahia. Sensitivity of the Illinois Flexibility Index Test to Mixture Design Factors. *Transportation Research Record: Journal of the Transportation Research Board*, 2017. 2631: 153–159.
23. West, R. NCHRP 9-55: Recycled Asphalt Shingles in Asphalt Mixtures with Warm Mix Asphalt Technologies. Presented at 2017 Spring Meeting of the Asphalt Mix Expert Task Groups, Ames, Iowa, 2017.
24. Hanz, A. Effect of Polymer Modification on I-FIT Parameters. Presented at 2017 Fall Meeting of the Asphalt Binder Expert Task Groups, Bozeman, Montana, 2017.
25. Cao, W., L. N. Mohammad, and M. Elseifi. Assessing the Effects of RAP, RAS, and Warm-Mix Technologies on Fatigue Performance of Asphalt Mixtures and Pavements Using Viscoelastic Continuum Damage Approach. *Road Materials and Pavement Design*, Vol. 18, No. Suppl. 4, 2017, pp. 353–371.
26. Kim, Y. R., Y. Seo, M. King, and M. Momen. Dynamic Modulus Testing of Asphalt Concrete in Indirect Tension Mode. *Transportation Research Record: Journal of the Transportation Research Board*, 2004. 1891: 163–173.
27. Zhou, F., S. Hu, D. H. Chen, and T. Scullion. Overlay Tester: Simple Performance Test for Fatigue Cracking. *Transportation Research Record: Journal of the Transportation Research Board*, 2007. 2001: 1–8.
28. Walubita, L. F., A. N. Faruk, G. Das, H. A. Tanvir, J. Zhang, and T. Scullion. *The Overlay Tester: a Sensitivity Study to Improve Repeatability and Minimize Variability in the Test Results*. Report No. FHWA/TX-12/0-6607-1. Texas Transportation Institute, College Station, 2012.
29. Garcia, V., A. Miramontes, J. Garibay, I. Abdallah, and S. Nazarian. *Improved Overlay Tester for Fatigue Cracking Resistance of Asphalt Mixtures*. Report No. TxDOT 0-6815-1. Center for Transportation Infrastructure Systems, The University of Texas at El Paso, 2017.

The Standing Committee on Structural Requirements of Asphalt Mixtures (AFK50) peer-reviewed this paper (18-04854).

Article

TRR
JOURNAL OF THE TRANSPORTATION RESEARCH BOARD

Influence of Production Considerations on Balanced Mixture Designs

Transportation Research Record
2018, Vol. 2672(28) 426–437

DOI: 10.1177/0361198118786826
journals.sagepub.com/home/trr
SAGE

Alexander J. Austerman[1], Walaa S. Mogawer[1], and Kevin D. Stuart[2]

Abstract

Measuring or predicting mixture performance prior to placement has become essential for many state transportation agencies due to incorporation of more recycled materials into mixtures, utilization of binders formulated with various modifiers, and utilization of new and unproven technologies. Mixtures designed incorporating these elements in accordance with the current Superpave mixture design method have unknown performance as Superpave was never designed to compensate for them. Thus, agencies have interest in implementing the balanced mixture concept that requires testing for acceptable mixture performance during the design phase. The effect of production considerations on a balanced mixture design is not known. It is unknown if a mixture remains performance balanced during production due to factors like asphalt content, gradation and binder source variations. This study was undertaken to understand the influence of these production considerations on a balanced mixture design. Two Superpave mixtures, a 9.5 mm and 12.5 mm, were developed using the balanced mixture design approach using volumetric properties, one rutting test and one cracking test. These mixtures were then modified to represent production variations in asphalt binder content, gradation passing the No. 200 sieve, and asphalt binder source. All mixtures remained performance balanced when examining the production considerations in terms of rutting and cracking using the selected tests. Examination of other performance results, like moisture susceptibility, suggest it is possible to effectively unbalance the mixture design during production. This underscores the importance of proper selection of performance tests to address the performance needs of the agency.

Most state transportation agencies currently require the use of the Superpave methodology to design their asphalt mixtures, specifically dense graded mixtures. The Superpave mixture design method was developed under the first SHRP which ran from 1987 to 1993. It was originally intended that a performance-based specification for both asphalt binder and mixture would be developed and implemented as part of this initiative (*1*). The performance-based asphalt binder specification is readily utilized today as outlined in the performance-graded (PG) asphalt binder specification AASHTO M320 "Standard Specification for Performance-Graded Asphalt Binder." The Superpave mixture design system was developed with three levels of mixture design, known as Level 1, Level 2 and Level 3 (*1*). Level 2 and Level 3 mixture design included performance-based mixture tests, however, these design levels were never implemented because the testing and analysis were considered too complex (*1*). Although the performance-based mixture design Level 2 and Level 3 were never implemented, the Superpave mixture design method was established based on Level 1 and is readily utilized today as outlined in AASHTO M323 "Standard Specification for Superpave Volumetric Mix Design."

Significance of Problem

Although this Level 1 version of the Superpave mixture design method increased the likelihood of good performing mixtures, it was unable to quantifiably measure or predict the expected mixture performance. Currently, being able to measure or predict mixture performance prior to placement has become essential for many state transportation agencies due to many recent changes in the asphalt paving industry. These changes include: the incorporation of more recycled materials (reclaimed asphalt pavement, recycled asphalt shingles, ground tire rubber, etc.) into mixtures, utilization of binders formulated with various modifiers (re-refined engine oil bottoms, air blown asphalt, rubber, polymers, polyphosphoric acid, etc.) versus conventionally neat asphalt binders, and utilization of innovative technologies (warm mix asphalt,

[1]Highway Sustainability Research Center (HSRC), Civil and Environmental Engineering Department, University of Massachusetts, Fall River, MA
[2]Consultant, Phoenixville, PA

Corresponding Author:
Address correspondence to Walaa S. Mogawer: wmogawer@umassd.edu

asphalt rejuvenators, bio-binders, etc.). Mixtures designed with respect to these changes in the paving industry, and in accordance with the current Superpave mixture design method, have unknown performance as Superpave was never designed to compensate for these factors. Thus, state transportation agencies may be approving mixture designs that meet the Superpave design method but may ultimately exhibit subpar performance. This fact, combined with the availability of laboratory-based mixture performance testing equipment that was not available during the SHRP, has renewed interest in developing a performance-related specification using a balanced mixture concept as opposed to solely relying on the volumetric properties derived from the Superpave mixture design method .

Currently, the balanced mixture design concept is still in its infancy with limited utilization by a few state transportation agencies. It has been broken into three approaches for simplicity by the FHWA Balanced Mix Design Task Force and outlined as follows (*2*):

Approach 1: Volumetric Design with Performance Verification

This approach starts with the conventional Superpave mixture design methodology followed by performance testing. If the performance is not acceptable, the mixture is redesigned, and the process starts over again.

Approach 2: Performance-Modified Volumetric Design

In the second approach, the initial design binder content is selected using the Superpave mixture design methodology and the mixtures are tested for performance. Based on performance results, the mixture may be adjusted (gradation and/or binder content) to meet performance criteria. In this approach, the volumetric properties may or may not meet Superpave requirements.

Approach 3: Performance Design

In the third approach, performance tests are conducted at varying binder contents with the design content being selected so all the target performance criteria are met. Volumetric properties may or may not be tested for and would not be required to meet Superpave requirements.

It is important to note that state transportation agencies are using all three approaches and the specific performance testing requirements vary by individual state (*2*). What is similar in each approach is the fact that some measure of performance is required to have an approved balanced mixture design. Typically, state transportation agencies are utilizing a rutting and cracking test for whichever approach they utilize. Other tests for moisture damage or specific types of cracking (fatigue, low temperature, reflective, top-down, bottom-up, etc.) may also be utilized.

Perhaps an overlooked component of a balanced mixture design is the influence of production considerations. Simply speaking, the result of a balanced mixture design procedure is an approved mixture design gradation with an associated optimum binder content that provides acceptable performance in the selected laboratory mixture performance tests. The question that remains is what happens to a balanced mixture design during production. Binder content and mixture gradations are dynamic during production and allowed to fall within specific ranges or tolerance. The source of the required PG asphalt binders for a specific mixture are changed during the production season as refinery supplies change resulting in a switch in binder source. It is currently unknown what happens to the balanced mixture design with respect to these types of production considerations. Ultimately it should be known whether the balanced mixture remains balanced during production, otherwise, revisions to the design approach are required. The study presented herein represents a first step at understanding the influence of these production considerations on a balanced mixture design.

Objective

The objective of this study was to determine the influence of production considerations on mixtures developed using a balanced mixture design approach. Specifically of interest was whether the balanced mixture design remains performance balanced under production considerations like asphalt binder content production variation, production gradation variation of the percent passing the No. 200 sieve, and changes in PG asphalt binder source.

Research Methodology

To meet the objective of this study, an experimental plan was developed as outlined in Figure 1. Using Approach 1 outlined above, balanced mixture designs were developed for two different nominal maximum aggregate size (NMAS) surface mixtures, a 9.5 mm and 12.5 mm. Two different mixtures were utilized to account for difference in gradation, binder content, and volumetric properties based on mixture NMAS.

Each mixture was designed to meet the requirement of the Superpave mixture design methodology. Initial binder contents were determined, and volumetric analysis was performed to establish the optimum binder content. Mixture performance testing in terms of rutting and cracking was performed at the optimum binder content as part of the balanced mixture design approach. Mixtures failing the performance tests were redesigned, while those passing were further evaluated with respect to production considerations. Production considerations that were evaluated were asphalt binder content production variation, production gradation variation of the percent passing the No. 200 sieve, and changes in PG asphalt binder source. Each mixture was tested with respect to these production considerations using the previous rutting

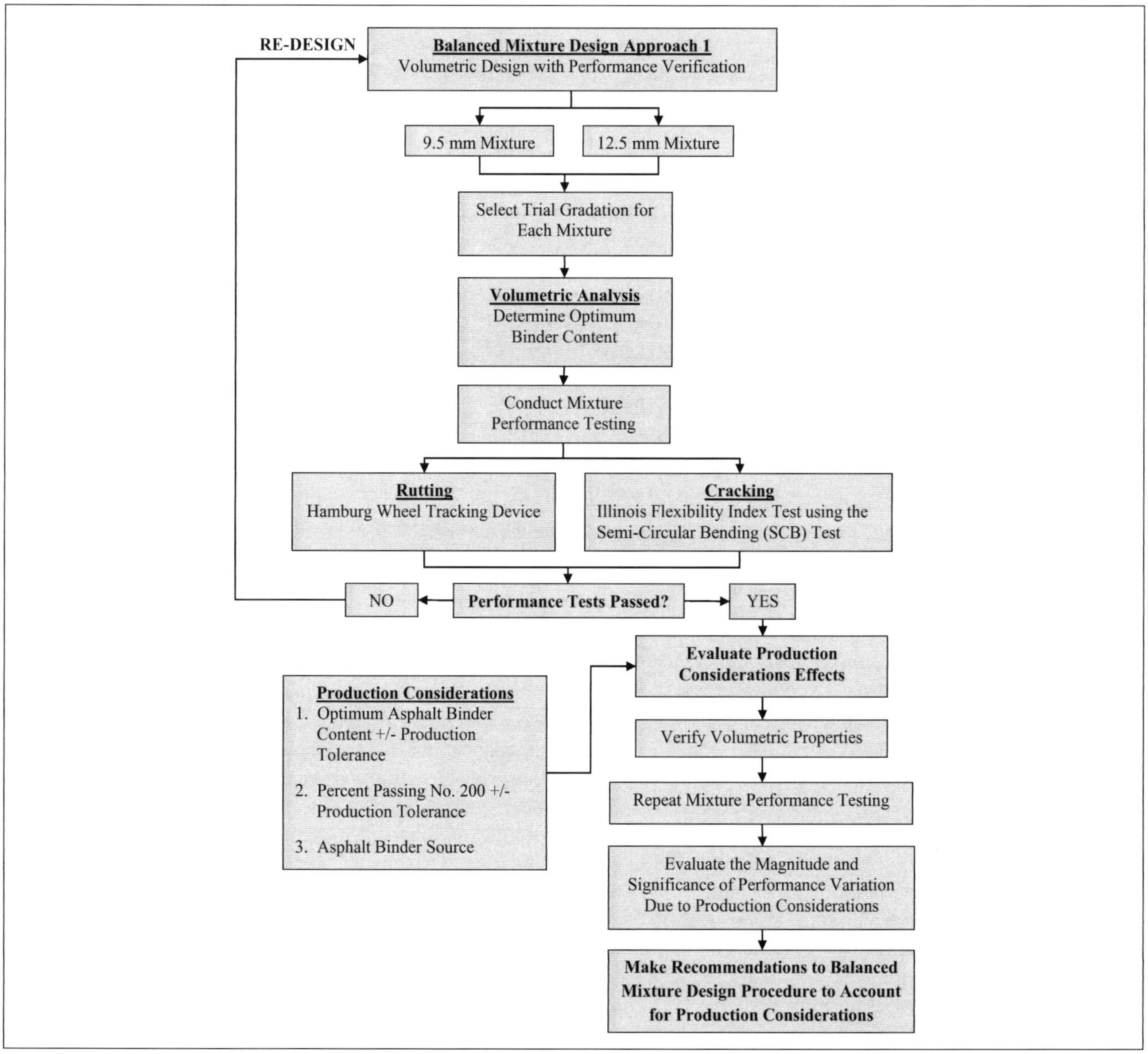

Figure 1. Experimental plan.

and cracking performance tests. Additionally, the volumetric properties with respect to each production consideration were also determined. Based on the performance results, recommendations for the balanced mixture design approach were made.

Production Considerations

As previously noted, three distinct production considerations were evaluated for each balanced mixture design. These three production considerations were selected as they were considered to have the largest potential impact on mixture performance and ultimately the balanced mixture design. They are described in greater detail in the following sections. The current Massachusetts Department of Transportation (MassDOT) Quality Assurance Specification for Hot Mix Asphalts Section 450 (*3*) was utilized to determine the acceptable tolerances and criteria for each production consideration. Additionally, the same specification was utilized to determine the acceptable variation in mixture volumetric properties for this study.

Asphalt Binder Content during Production

Binder content can have a significant impact on the performance of a mixture. Inadequate amounts of binder will make the mixture more prone to cracking, but excessive amounts of binder will make the mixture more prone to rutting. With respect to balanced mixture design, it was important to investigate the effects of asphalt binder content production tolerance on the performances of the mixtures. In accordance with MassDOT quality assurance specification, the binder content tolerance during production should be within ±0.3% of the design optimum. This tolerance was used for this study.

Variation of Gradation Passing No. 200 Sieve

The amount of aggregate passing the No. 200 sieve can have a significant impact on the performance of a mixture because it can affect both the stiffness and the quantity of the resultant mastic in a mixture. Higher amounts might also reduce stone-on-stone contact. Therefore, understanding the impact of gradation variation during production for the percent passing the No. 200 sieve was important to include in this study. In accordance with MassDOT quality assurance specification, the combined aggregate gradation should be within ±1.0% of the design job mix formula (JMF) for the percent passing the No. 200 sieve. This tolerance was used for this study.

Asphalt Binder Source

As noted previously, asphalt binders are currently being formulated with various modifiers (re-refined engine oil bottoms, air blown asphalt, rubber, polymers, polyphosphoric acid, etc.) versus conventionally neat asphalt binders. Because of formulation, it is possible to have two different asphalt binders that have the same PG in accordance with AASHTO M320 that were developed entirely differently. These differences in formulation can have a significant impact on the resultant mixture performance properties (*4*). To understand the potential performance differences between two similarly graded PG binders, two different PG asphalt binders from different sources were utilized for the performance evaluations in this study, designated as A and B.

Materials

Binder

The typical asphalt grade specified in Massachusetts is a PG64-28. Two different PG64-28 binders were obtained from separate sources for this study to evaluate the effect of asphalt binder source on balanced mixture performance as discussed previously. Since large discrepancies between the continuous grades might lead to variations in mixture performance, each binder was graded in accordance with AASHTO M320 to ensure the continuous grades were similar and the performance grades were the same. The binder from Source A had a continuous grade of 66.2-28.4 which was a PG64-28. The binder from Source B had a continuous grade of 65.0-28.9 which was a PG64-28. The discrepancies between continuous grades were small with no practical significance, thus, the binders were essentially graded to be equal. Finally, the viscosity of each binder was determined to establish mixture mixing and compaction temperatures. The mixing temperature was 158°C and the compaction temperature was 146°C for both binder sources.

Aggregates

Aggregates and reclaimed asphalt pavement (RAP) were obtained from a local contractor for use in the mixture design and subsequent performance testing. The aggregate stockpiles obtained were 1/2-in. crushed stone, 3/8-in. crushed stone, stone sand, and stone dust. These stockpiles were sieved into individual size fractions to accurately maintain the mixture gradations for this study.

Performance Testing for Balanced Mixture Design

As noted earlier, state transportation agencies typically utilize a rutting and cracking test in the balanced mixture design approach. A test for each of these was utilized in this study as outlined in the following.

Rutting—Hamburg Wheel Tracking Device Test

The Hamburg Wheel Tracking Device (HWTD) test can be considered both a rutting and moisture susceptibility performance test. This test is conducted in accordance with AASHTO T324 "Standard Method of Test for Hamburg Wheel-Track Testing of Compacted Hot Mix Asphalt (HMA)" (*5*) and is utilized to determine the failure susceptibility of a mixture due to weakness in the aggregate structure, excessive asphalt binder content, inadequate binder stiffness and moisture damage. Compacted mixture specimens are submerged in heated water, typically at 40–50°C, and subjected to repeated loading provided by a 705N steel wheel. As the steel wheel loads the specimen, the corresponding rut depth of the specimen is recorded. From this rut depth data, the stripping inflection point (SIP) can be determined. The SIP gives an indication of the onset of moisture damage in the mixture. Prior to the SIP, or in the absence of a SIP, the rut depth can give an indication of the rutting performance of the mixture.

The HWTD test was included in this study as it is already included in MassDOT specifications as a requirement for mixture approval. All new mixture designs must meet the MassDOT specification criteria after acceptable volumetric criteria properties are obtained for a mixture. The MassDOT HWTD criteria is a maximum rut depth of 12.5 mm after 20,000 passes combined with no SIP before 15,000 passes. Currently, HWTD testing of mixtures is only required every three years for existing mixture designs unless significant mixture design changes are made.

The specimens needed for this test were fabricated using the Superpave gyratory compactor (SGC) to an air void level of 7.0 ± 1.0% as required by AASHTO T324. Testing was conducted at a test temperature of 45°C as required by the MassDOT specification. Specimens were tested at a rate of 52 passes/min after a soak time of 30 min at 45°C.

Cracking at Intermediate Temperature—Illinois Flexibility Index Test

The cracking test selected for use in this study was the Illinois Flexibility Index Test (IFIT). This test is conducted in accordance with AASHTO TP124 "Standard Method of Test for Determining the Fracture Potential of Asphalt Mixtures Using Semicircular Bend Geometry (SCB) at Intermediate Temperature" (*5*) to determine different properties related to the intermediate cracking potential of the mixture. Typically, this test is performed at 25°C. In this test, a monotonic load is applied along the vertical diameter of a notched semicircular mixture specimen at a displacement rate of 50 mm/min (*6*) and a load versus displacement curve is obtained. From this curve two cracking properties are obtained, fracture energy (FE) and flexibility index (FI). FE is calculated using the work of fracture method by finding the area under the load-displacement curve and dividing it by the crack propagation area. The FI accounts for both fracture energy and the post-peak behavior of a mixture. Methods for determining FE and FI are detailed in other studies (*6*). Both FE and FI give an indication of the mixture cracking performance at intermediate temperature with higher value being more desirable. In the absence of other data or guidance, a previous study has shown that mixtures with FI value less than 2 correlate to poor-performing mixtures and mixtures with FI value around 10 have correlated to good-performing field mixtures (*6*). The developers of the test have suggested that a FI of greater than 8.0 be used as a pass/fail criterion to distinguish between mixtures (*7*).

The IFIT test was included in this study as MassDOT has expressed interest in using this device to evaluate the cracking performance of mixtures. Additionally, MassDOT is interested in implementing the IFIT test because it is relatively simple, and the results are obtained quickly unlike other fatigue tests such as the flexural beam fatigue.

The specimens needed for this test were compacted in the SGC to a height of 180 mm. Two 50 mm slices were cut from the middle of each specimen. Each slice was cut in half across the diameter to yield four specimens. Subsequently, each specimen had a 15 mm vertical notch cut up from midpoint of the base of each specimen. The air void levels in the test specimens were 7.0 ± 1.0%. A minimum of eight specimens were tested for each mixture in this study.

Balanced Mixture Designs

Two mixture designs, a 9.5 mm and 12.5 mm NMAS, were developed using the first balanced mixture design approach discussed at the start of this paper, which was volumetric design with performance verification. Trial aggregate gradations were developed using existing state approved mixture designs obtained from the same contractor where the aggregates were obtained. These designs included the use of 15% RAP.

The first step in the balanced mixture design approach was to develop each mixture in accordance with the Superpave design method. The target gradations, shown in Table 1, were developed to meet Superpave requirements as outlined in AASHTO M323 "Superpave Volumetric Mix Design" and AASHTO R35 "Superpave Volumetric Design for Hot Mix Asphalt." (*5*) The design equivalent single axle loads (ESALs) for each mixture was 0.3 to <3 million. The design Superpave gyratory compactive effort for this ESALs was N_{design} = 75 gyrations. Each mixture was developed with the PG64-28 binder from Source A. As shown in Table 1, the volumetric properties of the mixture at the optimum binder content were within Superpave and MassDOT specification limits. This completed the first step of the balanced mixture design approach.

The second step in the balanced mixture design approach was to test each mixture for rutting and cracking using the previously described performance tests. Specimens were fabricated using the PG64-28 binder from Source A at the optimum binder content. As outlined in Table 1, each mixture provided acceptable rutting and cracking performance in references to the MassDOT criteria for the HWTD tests and the preliminary IFIT FI criteria. Thus, these two mixture designs were considered balanced with respect to volumetric properties and the performance tests conducted.

Evaluation of Performance Considerations with Respect to Performance

The developed mixtures were further evaluated to determine the effects of the performance considerations on the balanced mixture design. It should be noted that the volumetric properties of each mixture were determined with respect to each production consideration to determine any changes that may be outside acceptable production targets. The MassDOT specification (*3*) only requires monitoring of air voids for acceptance testing. The mixture air voids must be within ±1.3% of the target of 4%. The other Superpave volumetric properties of VMA, VFA, and dust-to-binder ratio are only required to be verified during the mixture design process and not during production. Thus, during production, only the air void percentage outside of the tolerance would trigger a corrective action by the contractor. The average air voids of three replicate specimens for each mixture with respect to each production consideration are shown in Table 2. Excluding one mixture, the average air voids of the mixtures were within the specification limits and therefore would be an accepted mixture during production. The lone mixture failing this

Table 1. Balanced Mixture Design Information

Sieve size (mm)	Percent passing by weight			
	9.5 mm mixture	9.5 mm Superpave specification	12.5 mm mixture	12.5 mm Superpave specification
19.0	100	–	100	100 min
12.5	100	100 min	94.0	90–100
9.5	94.0	90–100	86.0	90 max
4.75 (No. 4)	64.0	90 max	61.0	–
2.36 (No. 8)	42.0	32–67	42.0	28–58
1.18 (No. 16)	30.0	–	29.0	–
0.60 (No. 30)	20.0	–	19.0	–
0.30 (No. 50)	13.0	–	13.0	–
0.15 (No. 100)	8.0	–	7.0	–
0.075 (No. 200)	4.2	2–10	4.0	2–10
Optimum binder content, %	5.9%	–	5.5%	–
Volumetric properties				
Air voids, %	4.1%	4%	4.3%	4%
Voids in mineral aggregate (VMA), %	16.3%	16% min.*	15.5%	15% min.*
Voids filled with asphalt (VFA), %	74.8%	73–76%	72.1%	65–78%
Dust to binder ratio	0.78	0.6–1.2	0.82	0.6–1.2
Performance testing results for rutting				
HWTD rutting at 10,000 passes, mm	1.8	–	1.1	–
HWTD rutting at 20,000 passes, mm	2.9	<12.5 mm**	1.6	<12.5 mm**
HWTD SIP	NONE	–	NONE	–
Performance testing results for cracking				
IFIT FI	13.1	>8.0	9.0	>8.0
IFIT FE, J/m^2	1,892	–	1,622	–

Note: *MassDOT specifications require a 1% increase in VMA which has been presented.
**MassDOT specification criteria.

Table 2. Average Mixture Air Voids

Mixture	Production consideration	Average air voids, %	
		Binder source A	Binder source B
9.5 mm	Asphalt binder content during production		
	Binder lower limit (−0.3%)	5.0	4.0
	Optimum binder content	4.1	3.2
	Binder upper limit (+0.3%)	2.9	2.4*
	Variation of gradation passing No. 200 sieve		
	−1% No. 200	3.6	4.2
	Design No. 200	4.1	3.2
	+1% No. 200	3.1	3.4
12.5 mm	Asphalt binder content during production		
	Binder lower limit (−0.3%)	5.1	4.6
	Optimum binder content	4.3	4.1
	Binder upper limit (+0.3%)	3.6	3.0
	Variation of gradation passing No. 200 sieve		
	−1% No. 200	3.7	3.9
	Design No. 200	4.2	4.1
	+1% No. 200	3.1	3.6

Note: *Average mixture air voids outside of 4 ± 1.3% production tolerance.

Table 3. Rutting Performance Test Results for Asphalt Binder Content Production Consideration

Mixture	Rutting at 10,000 passes, mm	Rutting at 20,000 passes, mm	HWTD SIP
9.5 mm binder lower limit (−0.3%)	2.8	3.5	NONE
9.5 mm optimum binder content	1.8	2.9	NONE
9.5 mm binder upper limit (+0.3%)	4.1	8.4	10,400
12.5 mm binder lower limit (−0.3%)	2.3	2.7	NONE
12.5 mm optimum binder content	1.1	1.6	NONE
12.5 mm binder upper limit (+0.3%)	3.3	3.7	NONE

requirement was included in the remainder of this study to determine if it would pass subsequent performance testing.

Asphalt Binder Content during Production

In accordance with MassDOT specification, the binder content tolerance during production should be within ±0.3% optimum determined during the mixture design. Specimens of each mixture were fabricated at the lower limit (−0.3%) and upper limit (+0.3%) binder contents and tested for their performance using the same rutting and cracking tests used for the balanced mixture design development.

The rutting performance test results are shown in Table 3 while the fatigue cracking performance test results are shown in Figure 2. The error bars in the figure indicate the standard deviation of the measurements. Asphalt binder Source A was used in these tests.

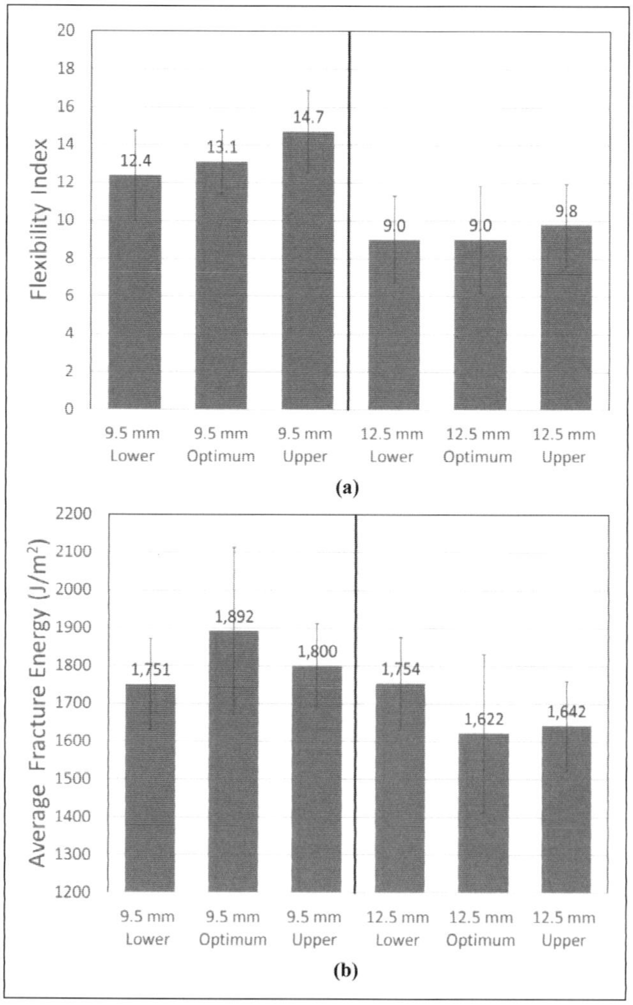

Figure 2. Cracking performance test result for asphalt binder content production consideration: (*a*) FI, and (*b*) FE.

Discussion of Binder Content Evaluation. Generally, all mixtures provided acceptable rutting and cracking performance in reference to the MassDOT criteria for the HWTD tests and the preliminary IFIT FI criteria. This indicated that the mixtures remained balanced with respect to asphalt binder content during production. Generally, mixtures designed at optimum binder content exhibited less rutting than at the production tolerance limits. In terms of cracking, FI values increased with increasing binder content for the 9.5 mm mixture but remained nearly constant for the 12.5 mm mixture. FE values did not follow this trend and varied based on mixture type.

For the 9.5 mm mixture at the upper limit binder content (+0.3%), it is noteworthy that the mixture was no longer balanced to the MassDOT specification requirements as a SIP was determined for the mixture. The MassDOT specification requires that there be no SIP before 15,000 passes. This requirement was not met even though the mixture rutting was acceptable. This underscores the importance of performance-related specification test selection as this degraded mixture performance may not have been detected if only a rutting test had been used.

Variation of Gradation Passing No. 200 Sieve

In accordance with MassDOT specification, the variation of the gradation passing the No. 200 sieve during production should be within ±1.0% of the design percentage determined during the mixture design. Specimens of each mixture were fabricated at the lower limit (−1.0%) and upper limit (+1.0%) percent passing the No. 200 and tested for their performance using the same rutting and cracking tests used for the balanced mixture design development.

The rutting performance test results are shown in Table 4 while the fatigue cracking performance test results are shown

Table 4. Rutting Performance Test Results for Variation of Gradation Passing No. 200 Sieve Production Consideration

Mixture	Rutting at 10,000 passes, mm	Rutting at 20,000 passes, mm	HWTD SIP
9.5 mm −1% No. 200	1.2	1.7	NONE
9.5 mm Design No. 200	1.8	2.9	NONE
9.5 mm +1% No. 200	3.4	4.0	NONE
12.5 mm −1% No. 200	1.6	2.0	NONE
12.5 mm Design No. 200	1.1	1.6	NONE
12.5 mm +1% No. 200	3.1	4.0	NONE

in Figure 3. The error bars in the figure indicate the standard deviation of the measurements. Asphalt binder Source A was used in these tests.

Discussion of Variation of Gradation Evaluation. Generally, all mixtures provided acceptable rutting and cracking performance in reference to the MassDOT criteria for the HWTD tests and the preliminary IFIT FI criteria. This indicated that the mixtures remained balanced with respect to variation of the gradation passing the No. 200 sieve. Generally, mixtures designed with +1% passing the No. 200 sieve exhibited more rutting. In terms of cracking, the FI values were generally the same based on mixture type. FE values did not follow this trend and varied based on mixture type.

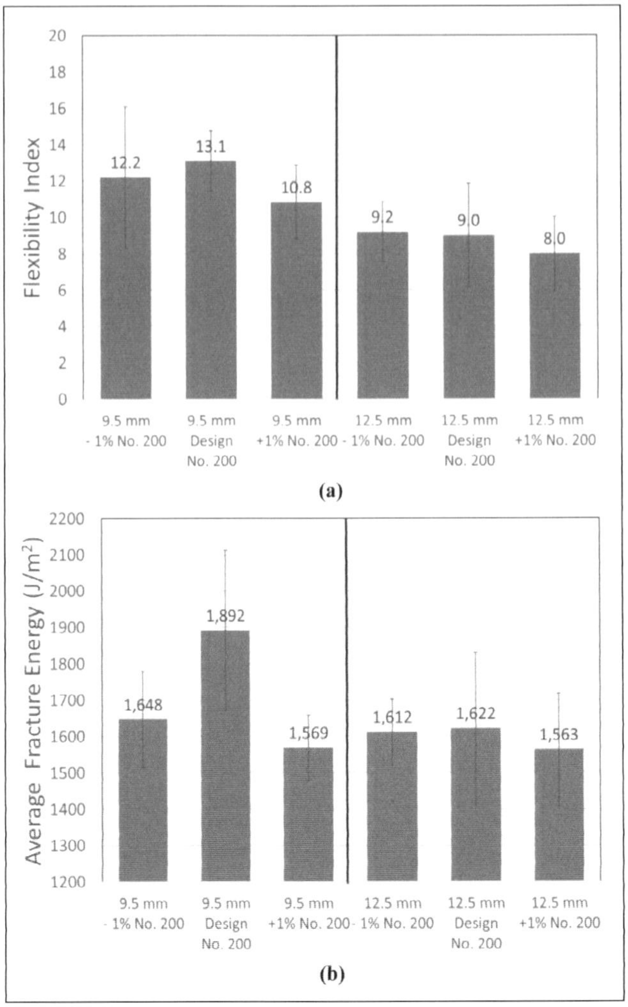

Figure 3. Cracking performance test result for variation of gradation passing No. 200 sieve production consideration: (*a*) FI, and (*b*) FE.

Asphalt Binder Source

In this section, the potential performance differences in the balance mixture designs were evaluated using two similarly graded PG binders from different sources.

The rutting performance test results are shown in Table 5. The fatigue cracking performance test results with respect to asphalt binder content is shown in Figure 4 while the test results with respect to the gradation passing the No. 200 sieve is shown in Figure 5. The error bars in the figures indicate the standard deviation of the measurements.

Discussion of Binder Source Evaluation. Generally, all mixtures provided acceptable rutting and cracking performance in reference to the MassDOT criteria for the HWTD tests and the preliminary IFIT FI criteria. No criterion is currently available for the FE values obtained from IFIT test. This indicated that mixtures remained balanced with respect to asphalt binder source.

Rutting data did not give an indication that asphalt binder source had a major impact on performance. In terms of cracking, asphalt binder source had mixed effects on FI values. Generally, FI values were larger for 9.5 mm mixture with asphalt binder Source A while the opposite trend was noted for the 12.5 mm mixture with asphalt binder Source B providing larger FI values. A two-way analysis of variance (ANOVA) at a level of significance $\alpha = 0.05$ was performed. A null hypothesis was that changing binder content and the minus 200 within the allowable tolerance and changing the binder source will have no effect on the FI values. This hypothesis was rejected as for certain mixtures it was found that changing the binder content or the binder source can have a significant effect on the FI values, however, they all passed.

Although no pass/fail criteria exist for the FE values, the trend was clear. Regardless of mixture type or production

Table 5. Rutting Performance Test Results for Asphalt Source Production Consideration

Mixture	Binder source	Rutting at 10,000 passes, mm	Rutting at 20,000 passes, mm	HWTD SIP
9.5 mm binder lower limit (−0.3%)	Source A	2.8	3.5	NONE
	Source B	1.5	1.9	NONE
9.5 mm optimum binder content	Source A	1.8	2.9	NONE
	Source B	2.4	3.1	NONE
9.5 mm binder upper limit (+0.3%)	Source A	4.1	8.4	10,400
	Source B*	3.3	4.5	NONE
12.5 mm binder lower limit (−0.3%)	Source A	2.3	2.7	NONE
	Source B	2.8	3.7	NONE
12.5 mm optimum binder content	Source A	1.1	1.6	NONE
	Source B	1.5	2.0	NONE
12.5 mm binder upper limit (+0.3%)	Source A	3.3	3.7	NONE
	Source B	3.1	5.2	NONE
9.5 mm −1% No. 200	Source A	1.2	1.7	NONE
	Source B	2.3	3.0	NONE
9.5 mm Design No. 200	Source A	1.8	2.9	NONE
	Source B	2.4	3.1	NONE
9.5 mm +1% No. 200	Source A	3.4	4.0	NONE
	Source B	1.4	1.9	NONE
12.5 mm −1% No. 200	Source A	1.6	2.0	NONE
	Source B	1.7	2.7	NONE
12.5 mm Design No. 200	Source A	1.1	1.6	NONE
	Source B	1.5	2.0	NONE
12.5 mm +1% No. 200	Source A	3.1	4.0	NONE
	Source B	2.5	3.6	NONE

Note: *Mixture failed volumetric requirement for air voids.

consideration, mixtures fabricated with asphalt binder Source B consistently exhibited lower FE values as compared to those fabricated with asphalt binder Source A. Since the PG of the asphalt binders was essentially the same on a continuous scale, the FE results show that two mixtures that should exhibit similar performance did not. In terms of a balanced mixture design, if FE was utilized as the pass/fail criteria of the cracking test, it would be possible that many of the mixtures fabricated with asphalt binder Source B may fail depending on what threshold value was considered acceptable. This shows the importance of selecting appropriate performance tests and criteria for balanced mixture designs that correlate to field performance. The FE results suggest that it is conceptually possible to effectively unbalance the mixture design by simply changing the source of asphalt binder.

Also, noteworthy was that the one mixture that did not meet the required air voids passed the rutting and cracking performance tests.

Conclusions and Recommendations

Based on the testing and analysis conducted in the study, the following conclusions and recommendations were made:

- The two different mixtures remained performance balanced when evaluating the production considerations. Acceptable performance was referenced to the MassDOT criteria for the HWTD test and the preliminary IFIT FI criteria, respectively. Production considerations examined were asphalt binder content production variation, production gradation variation of the percent passing the No. 200 sieve, and changes in PG asphalt binder source.
- Except for one mixture, all mixtures met the required production volumetric criteria for air voids with respect to the production considerations. The one mixture that did not meet the required air voids passed the rutting and cracking performance tests. This mixture would be acceptable under a balanced mixture design approach that did not require the mixture to meet volumetric properties.
- HWTD results identified one mixture failing to meet the MassDOT criteria for SIP when the asphalt binder content was increased to the upper limit allowed during production. If moisture susceptibility had been selected as a performance criterion, the results suggest it is possible to effectively unbalance the mixture design by simply changing the amount of asphalt binder during production. This underscores the importance of proper selection of performance tests to address performance needs or distresses encountered by a specific agency.
- Although no pass/fail criteria exist for the FE values, the results showed that mixtures fabricated with one

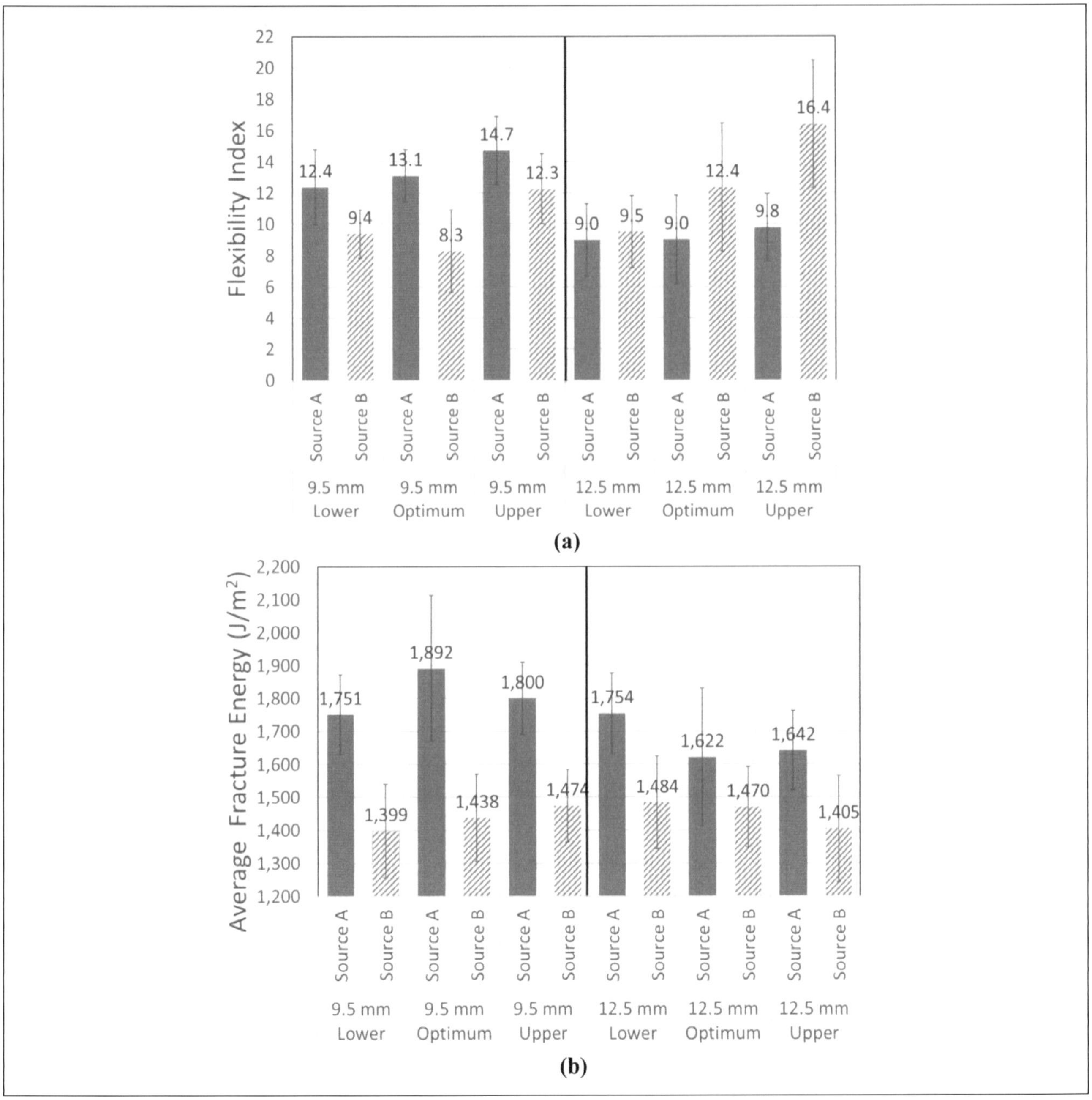

Figure 4. Cracking performance test result for asphalt binder source production consideration with respect to asphalt binder content: (*a*) FI, and (*b*) FE.

asphalt binder source had consistently lower FE values. Since the PG of the asphalt binders was essentially the same on a continuous scale, the FE results indicate that the two mixtures may not have equivalent performance. If FE was utilized as the pass/fail criteria for the cracking test, it would be possible that the mixtures may fail the cracking requirement depending on what threshold value would be considered acceptable. This suggests it is conceptually possible to effectively unbalance the mixture design by simply changing the source of asphalt binder.

- Overall, the balanced mixture design concept is still in its infancy. This study showed that certain production considerations can affect volumetric properties and the performance of a mixture developed using the balanced mixture design concept. It underscores the importance of understanding the effect of these considerations and compensating for them when developing mixtures using the balanced mixture design concept.
- This research is part of an on-going larger study which includes evaluating the effects of the interactions

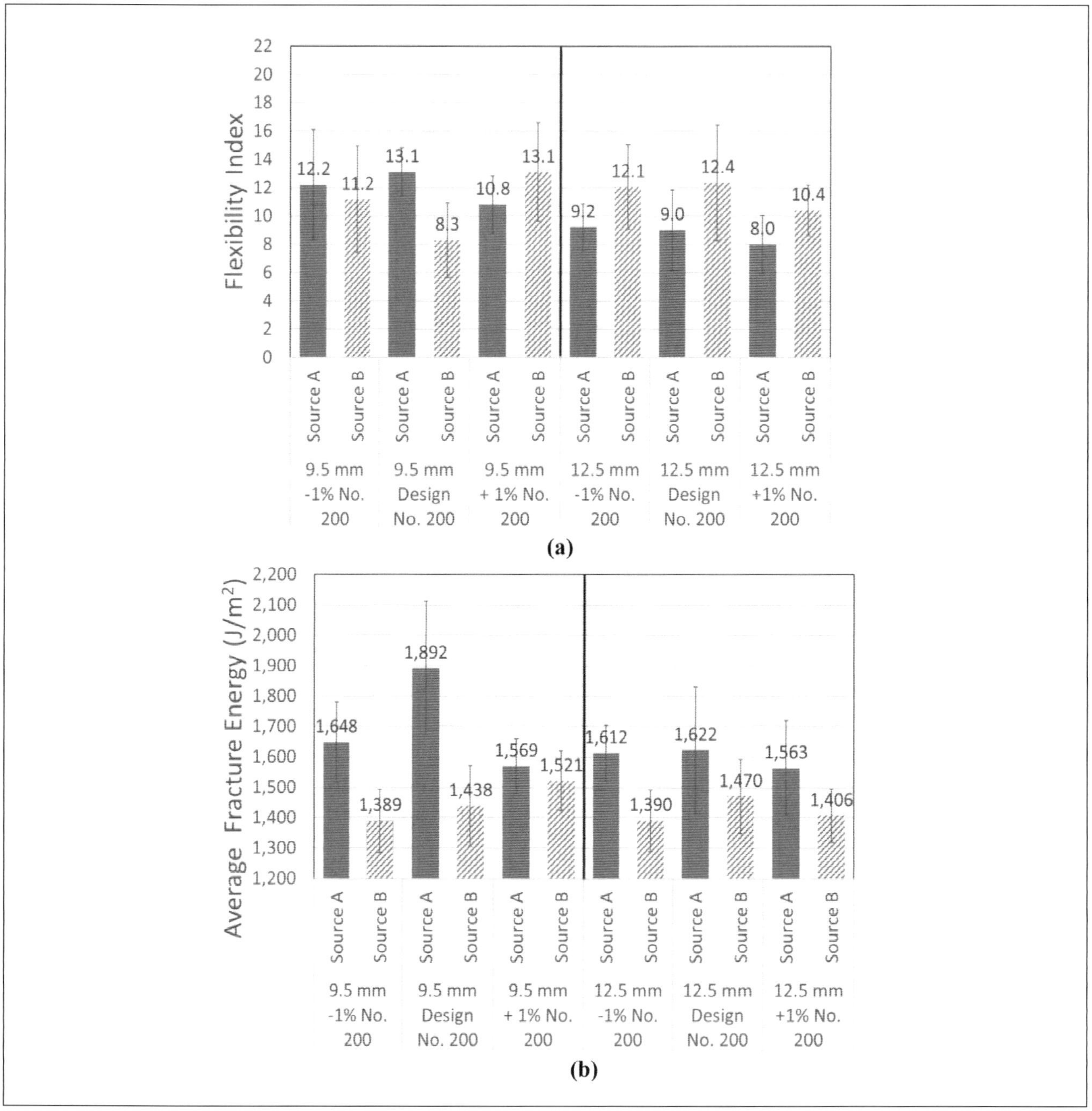

Figure 5. Cracking performance test result for asphalt binder source production consideration with respect to percent passing No. 200 sieve: (*a*) FI, and (*b*) FE.

between different production considerations simultaneously on a balanced mixture design.

Author Contributions

The authors confirm contribution to the paper as follows: study conception and design: Austerman, Mogawer, Stuart; data collection: Austerman, Mogawer, Stuart; analysis and interpretation of results: Austerman, Mogawer, Stuart; draft manuscript preparation: Austerman, Mogawer, Stuart. All authors reviewed the results and approved the final version of the manuscript.

References

1. Huber G. *History of Asphalt Mix Design in North America, Part 2 Superpave*. http://asphaltmagazine.com/history-of-asphalt-mix-design-in-north-america-part-2/. Accessed July 6, 2017.
2. Buchanan, S. *Balanced Mix Design (BMD) for Asphalt Mixtures*. Presentation for SHRP2 Peer to Peer Exchange 2016, 2016. http://shrp2.transportation.org/documents/presentations/10_Buchanan_SHRP2BalancedMixDesignforAsphaltMixtures.pdf. Accessed July 7, 2017.

3. Massachusetts Department of Transportation Highway Division. Quality Assurance Specifications for Hot Mix Asphalt Pavement, Section 450. Massachusetts Department of Transportation Highway Division, 2010.
4. Mogawer, W. S., A. J. Austerman, K. D. Stuart, M. Marasteanu, and M. Nazzal. Influence of Asphalt Binder Formulation and Source on the Performances of Binders with the Same Continuous PG. *Journal of Association of Asphalt Paving Technologist*, 2017. Accepted for publication.
5. American Association of State Highway and Transportation Officials. *Standard Specifications for Transportation Materials and Methods of Sampling and Testing*. American Association of State Highway and Transportation Officials, Washington, D.C., 2015.
6. Ozer, H., I. L. Al-Qadi, J. Lambros, A. Khatib, P. Singhvi, and T. Khan. Development of a Fracture-Based Flexibility Index for Asphalt Concrete Cracking Potential Using Modified Semi-Circle Bending Test Parameters. *Construction and Building Materials*, Vol. 115, 2015, pp. 390–401. https://doi.org/10.1016/j.conbuildmat.2016.03.144.
7. Al-Qadi, I., H. Ozer, J. Lambros, A. El Khatib, P. Singhvi, T. Khan, J. Rivera, and B. Doll. *Testing Protocols to Ensure Performance of High Asphalt Binder Replacement Mixes Using RAP and RAS*. Publication FHWA-ICT-15-017, Illinois Department of Transportation, 2015.

The Standing Committee on Structural Requirements of Asphalt Mixtures (AFK50) peer-reviewed this paper (18-01234).

Article

TRR
JOURNAL OF THE TRANSPORTATION RESEARCH BOARD

Transportation Research Record
2018, Vol. 2672(28) 438–450

DOI: 10.1177/0361198118790845
journals.sagepub.com/home/trr
SAGE

Optimization of the Laboratory Fabrication of Small Specimens for Asphalt Mixture Performance Testing

Sonja Pape[1], Kangjin Lee[1], Cassie Castorena[1], and Y. Richard Kim[1]

Abstract

The use of 38-mm-diameter small specimens for uniaxial dynamic modulus and cyclic fatigue asphalt mixture performance testing offers a significant opportunity to improve the efficiency of laboratory-fabricated specimen testing because multiple test specimens can be extracted per Superpave gyratory-compacted (SGC) sample. This study seeks to optimize the procedure used for the extraction of small specimens from SGC samples for dynamic modulus and cyclic fatigue tests. To this end, small cylindrical specimens were cored horizontally and vertically from SGC samples and subjected to performance testing. The dynamic modulus and fatigue test results indicate that the effects of anisotropy are minimal. However, all of the horizontally extracted small specimens exhibited fatigue failure at the specimen ends, outside the range of the gauges; the failure was likely due to the peripheral air void gradients in the SGC samples. Therefore, the authors concluded that small specimens should be vertically cored from SGC samples for the laboratory fabrication of small specimens. Specifically, four small specimens were cored vertically from the inner 100 mm of SGC samples where the air void content is relatively uniform. Four mixtures with different nominal maximum aggregate sizes (NMASs) were used to prepare small specimens using the proposed extraction procedure. These specimens were subjected to dynamic modulus and cyclic fatigue testing. The results demonstrate an increase in specimen-to-specimen variability with an increase in NMAS, which also is expected in large specimen testing.

Current standards for uniaxial dynamic modulus testing (*1*) and cyclic fatigue testing (*2*) use 100-mm-diameter test specimens with heights of 150 mm and 130 mm, respectively. These large specimen geometries allow for the preparation of one test specimen per Superpave gyratory-compacted (SGC) sample (*3*). However, the 100-mm diameter of the test specimens prevents forensic testing of many pavement layers, and the 150-mm height precludes the extraction of test specimens from standard 6-inch (150-mm) diameter field cores. Kim et al. (*4*) developed an indirect tension (IDT) test protocol that utilizes 150-mm-diameter by 38-mm-thick samples to enable the testing of individual pavement layers within field cores. The thin disk-shaped specimens used in the IDT test also allow multiple test specimens to be obtained from a single SGC sample. The more recently proposed semi-circular bend (SCB) testing geometry for evaluating the cracking resistance of asphalt mixtures also uses thin disk specimens that allow multiple test specimens to be extracted per SGC sample (*5*, *6*). Thus, the IDT and SCB test geometries offer an attractive means to improve the efficiency of laboratory-fabricated specimen testing compared with the standard uniaxial test geometry. However, the IDT and SCB test geometries induce a complicated stress state in specimens and can induce confounding effects from permanent deformation at the site of the load application. In contrast, uniaxial testing imposes a uniform stress state in samples. Consequently, uniaxial test results can be used to obtain fundamental properties for use in models. Due to the advantages of uniaxial testing, Kutay et al. (*7*) proposed the use of 38-mm-diameter cylindrical specimens (termed "small specimens" here) to enable the uniaxial testing of field cores.

By comparing the performance test results of large and small specimens, several researchers (*7–12*) have demonstrated that the small specimen geometry satisfies the representative volume element criterion for

[1]Department of Civil, Construction, and Environmental Engineering, North Carolina State University, Raleigh, NC

Corresponding Author:
Address correspondence to Cassie Castorena: cahintz@ncsu.edu

performance testing in many circumstances. Li and Gibson (*8*) evaluated small specimen geometry for dynamic modulus and cyclic fatigue tests using 10 mixtures with the nominal maximum aggregate size (NMAS) ranging from 4.75 mm to 19.0 mm. Bowers et al. (*9*) evaluated dynamic modulus test results of 16 mixtures with NMAS values ranging from 9.5 mm to 25.0 mm. Lee et al. (*10*) evaluated dynamic modulus and cyclic fatigue test results for mixtures with the NMAS ranging from 9.5 mm to 25.0 mm (*10*). All of these researchers found that the small specimen dynamic modulus and cyclic fatigue test results agreed with those of large specimens at low and intermediate temperatures. However, the results for small specimens generally deviated from those for large specimens at high temperatures. Consequently, researchers recommend that dynamic modulus tests of small specimens using mixtures from Virginia and North Carolina should be limited to a maximum test temperature of 40°C (*9*, *10*); this is consistent with the maximum testing temperature specified in AASHTO PP 61 for preparing the dynamic modulus mastercurve. Thus, past efforts suggest that small specimen testing can successfully be used to develop dynamic modulus mastercurves.

Although small specimens were used initially to enable the testing of field cores, they also offer a significant opportunity to improve the efficiency of laboratory-fabricated uniaxial specimen tests. Multiple small specimens can be extracted from a single SGC sample (*7–12*). The small sample size also reduces the time required for thermal equilibration, which in turn reduces the overall testing time and thus improves efficiency. In cyclic fatigue tests, the small specimen diameter allows end platens to be attached with quick-setting epoxy, which allows the specimens to be tested one hour after gluing. By contrast, large specimens generally require a slower setting epoxy due to the larger gluing surface areas (*2*).

Although past studies have demonstrated that small specimen testing can capture bulk asphalt mixture dynamic modulus and cyclic fatigue behavior, their small specimen fabrication procedures have varied, thus indicating the need to determine the optimal procedure that can be standardized for efficiency. Some researchers have extracted specimens vertically (*7–8*, *10–12*) from SGC samples whereas others have extracted specimens horizontally (*9*, *11*). Coring small specimens vertically follows the current practice for laboratory-fabricated large specimen uniaxial testing. However, pavements experience tension perpendicular to the direction of compaction, and therefore, the horizontal extraction of small specimens better reflects field conditions. For example, Bowers et al. (*9*) found that small specimens should be horizontally extracted from SGC samples to best mimic specimens obtained from field cores. Park (*11*) evaluated the effect of anisotropy on small specimens by comparing 100-mm-tall small specimens that had been extracted both horizontally and vertically from SGC samples. Park (*11*) concluded that the anisotropic effects were insignificant, but his study was limited to a single mixture.

The process used to extract small specimens vertically from SGC samples also has varied among researchers. SGC samples inherently contain higher air void contents at their peripheries than at the specimen center (*13*). Consequently, several past efforts have vertically extracted three small specimens from the inner 100-mm diameter of SGC samples (where large specimens also are obtained) in an effort to avoid the peripheral air void gradient (*7*, *10–12*). Li and Gibson (*8*) vertically extracted six small specimens per SGC sample—a center specimen surrounded by a concentric ring of five specimens. Coring six specimens per SGC sample led to the extraction of small specimens from outside the inner 100-mm diameter where air voids are expected to be relatively uniform. Li and Gibson (*8*) found that the five outer ring specimens had similar bulk air void contents but the center specimen had a comparably lower air void content. This finding led them to adjust the coring pattern to a ring of five specimens closer to the center of the gyratory-compacted sample. Li and Gibson (*8*) and Bowers et al. (*9*) also varied the height of small specimens and found the height of 110 mm to be optimal because standard asphalt mixture performance tester (AMPT) gauges cannot be used with shorter specimens and taller specimens proved to be problematic.

This study investigated both the effect of the coring direction on small specimen performance and the air void distribution in SGC samples to identify the optimal extraction pattern/process for small specimens obtained from SGC samples. The specimen-to-specimen variability found in the small specimen performance test results also was evaluated after optimizing the small specimen laboratory fabrication procedure.

Objectives

The objectives of this study are to:

1. Optimize the laboratory fabrication of small specimens extracted from SGC samples.
2. Evaluate the specimen-to-specimen variability found in performance test results obtained using the optimized small specimen laboratory fabrication procedure.

Experimental Plan

Asphalt Mixtures

Table 1 presents the mixtures evaluated in this study and the experimental plan. The mixtures evaluated are typical

Table 1. Asphalt Mixture Properties and Experimental Plan

	Mixture name	RB25.0B	RI19.0B-1	RI19.0B-2	SM12.5A	RS9.5D
Material properties	NMAS (mm)	25	19	19	12.5	9.5
	Binder grade	PG 64-22	PG 64-22	PG 64-22	PG 64-22	PG 76-22
	Binder content (%)	4.2	4.4	4.5	5.3	5.4
	RAP content (%)	30	20	40	30	20
	Recycled binder ratio	0.28	0.23	0.38	*	0.17
	VMA (%)	13.6	14.5	14	*	15.8
	VFA (%)	71.5	71.5	73	*	75
Evaluation of coring direction	Vertical core dynamic modulus		X			X
	Vertical core fatigue		X			
	Horizontal core dynamic modulus		X			X
	Horizontal core fatigue		X			
Analysis of specimen-to-specimen variability	Air void analysis				X	
	Dynamic modulus + fatigue	X		X	X	X
	Air void content (%)	4.7 ± 0.7	N/A	4.3 ± 0.7	4.6 ± 0.6	3.8 ± 0.3

Note: RAP = reclaimed asphalt pavement; VMA = voids in mineral aggregate; VFA = voids filled with asphalt.
*Limited information was available for the SM12.5A mixture.

materials sourced from Raleigh, North Carolina and the surrounding area, except for the SM12.5A mixture which is from Virginia. The mixtures were selected to encompass a wide range of NMAS. The experimental work was comprised of two phases: (1) evaluation of the effect of the coring direction to select the optimal pattern/process for extracting small specimens from SGC samples and (2) analysis of specimen-to-specimen variability in the optimized procedure. Note that there are two RI19.0B mixtures: the RI19.0B-1 mixture supply was exhausted during the coring direction evaluation, which necessitated the use of the new RI19.0B-2 mixture for the specimen-to-specimen variability study.

Test Methods

Dynamic modulus tests were conducted in an AMPT at 4°C, 20°C, and 40°C, with frequencies of 25, 10, 5, 1, 0.5, and 0.1 Hz applied at each temperature. The target average on-specimen peak-to-peak strain was 63 $\mu\varepsilon$, with an allowable range of 50 to 75 $\mu\varepsilon$ to limit nonlinear effects and minimize the potential for aggregate mobilization at high temperatures (*14*); otherwise, testing was conducted in accordance with AASHTO TP 79 (*1*). Three replicates were tested for each mixture. Dynamic modulus and phase angle mastercurves were produced by means of the time–temperature superposition principle.

Cyclic direct tension tests were conducted in accordance with AASHTO TP 107, which specifies constant crosshead displacement amplitude loading at a frequency of 10 Hz and a test temperature of 18°C for the mixtures evaluated (*2*). Simplified viscoelastic continuum damage (S-VECD) analysis was used to develop damage characteristic curves (i.e., pseudo stiffness [*C*] versus damage [*S*] curves) and failure criterion. In this work, a new fatigue failure criterion, referred to as D^R, was used to evaluate the mixtures. The D^R criterion is the average reduction in pseudo stiffness up to failure. Wang and Kim demonstrated that D^R is a material constant that is independent of mode of loading, temperature, and stress/strain history (*15*).

Effect of Coring Direction on Small Specimens Extracted from SGC Samples

Performance Testing. Performance tests were conducted using horizontally extracted small specimens made from the RI19.0B-1 and RS9.5D mixtures to study the effect of coring direction, as detailed in Table 1. The RI19.0B-1 mixture was selected because it had the largest NMAS and offered complete fatigue characterization from prior work (*10*). Dynamic modulus tests of the RS9.5D mixture were conducted in anticipation of further cyclic fatigue testing, but this idea was abandoned due to the

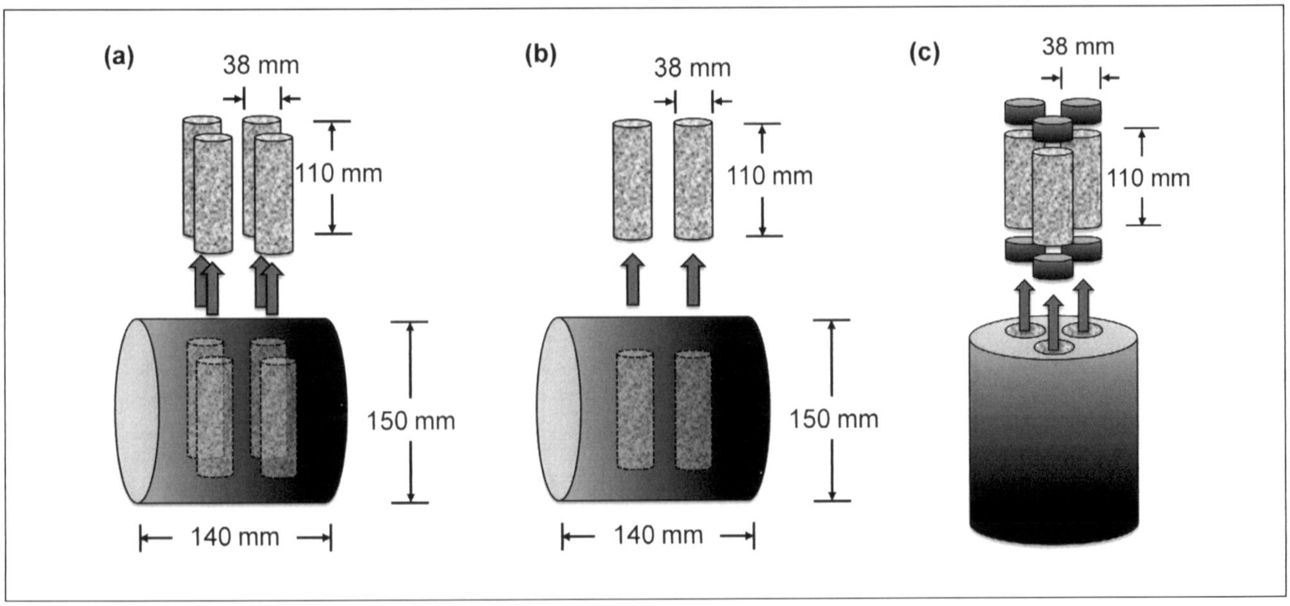

Figure 1. Depictions of core extractions from gyratory-compacted samples for the evaluation of coring direction: (*a*) four-core horizontal pattern; (*b*) two-core horizontal pattern; (*c*) three-core vertical pattern.

difficulties encountered in the cyclic fatigue testing of the RI19.0B-1 mixture, which are detailed in the results. The results for horizontally extracted specimens were compared with those for vertically extracted specimens documented in previous research (*10*).

The SGC samples from which the small samples were horizontally extracted were 140 mm tall. Shorter heights were tried, but a significant difference was observed in the air void contents of specimens of different heights that were extracted from different locations. Initially, four cores were extracted from each 140-mm-tall SGC sample, as illustrated in Figure 1*a*, to try to maximize the efficiency of specimen fabrication. Extracting two cores at the same height from the gyratory-compacted sample led to regions of high air void contents at the specimen ends, which led to end failure in fatigue tests of the RI19.0B-1 mixture. Therefore, the extraction process was adjusted to a two-core extraction process, with one sample taken from each lift, in order to extract specimens further away from the edges of the gyratory-compacted sample, as shown in Figure 1*b*. Fatigue tests of the RI19.0B-1 mixture also were conducted using the two-core extraction process. All of the evaluated small specimens had total air void contents within the range of 4.0 ± 0.5%. The extraction pattern for vertically extracted specimens is shown in Figure 1*c* for comparison.

Air Void Analysis. Air void analysis was conducted to determine any differences in the general trends in air void distribution between the horizontally extracted specimens and the vertically extracted specimens. The SM12.5A mixture was analyzed due to the unavailability of the RI19.0B-1 mixture. Three SGC samples were fabricated for each coring direction. Small specimens were extracted using the two-core horizontal pattern shown in Figure 1*b* and the three-core vertical pattern shown in Figure 1*c*. Each small specimen was sawn into four sections of equal height along the length of the sample. The bulk specific gravity of each sample section was measured to infer the horizontal and vertical air void gradients in the small specimens.

Specimen-to-Specimen Variability Investigation

To develop a reliable small specimen fabrication procedure for performance testing, the vertical extraction of small specimens from gyratory-compacted samples is required (and discussed in the results section of this paper). Large specimens are extracted from the inner 100 mm of gyratory-compacted specimens. The extraction of four small specimens from the inner 100 mm of 178-mm-tall gyratory-compacted samples was selected as the optimal small specimen laboratory fabrication procedure, as shown in Figure 2*a*. The specimen-to-specimen variability in the dynamic modulus and cyclic fatigue test results of specimens fabricated using the optimized procedure was evaluated using all of the mixtures listed in Table 1, except the RI19.0B-1 mixture.

Some variation in the air void contents among small specimens extracted from a given SGC sample is inevitable. Generally, asphalt mixture performance testing specifications (*1–2, 5–6*) stipulate that test specimens must

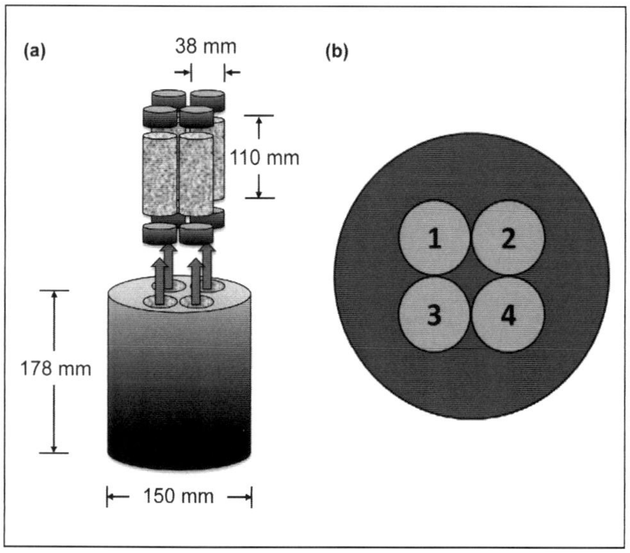

Figure 2. (*a*) Depiction of vertical small specimen extraction procedure and (*b*) sample extraction pattern for four vertically cored specimens.

have bulk air void contents within ±0.5% of the target. In this study, all small specimens produced from SGC samples were tested, regardless of whether the air void content was within ±0.5% of the target, to evaluate whether specimens outside of this tolerance range warrant rejection.

SGC Sample Fabrication. It is important to minimize the variability in the air void contents of small specimens extracted from a given SGC sample. In previous work, 36.5% (19 out of 52) of the initial samples produced fell outside of the widely accepted tolerance of ±0.5% variability in terms of air void content (*10*). During the initial effort, specimens were fabricated with no attention given to the way the loose mixture was poured into the gyratory mold. When loose mixture is not evenly distributed within the mold, the mixture will densify non-uniformly under compaction. Thus, small specimens extracted from different sides of an SGC sample can have dramatically different air void contents.

This study found that pouring the loose mixture directly into the center of the gyratory mold greatly reduced small specimen air void variability within the SGC samples. Therefore, in the specimen-to-specimen variability analysis, the loose mixture was poured into the center of the gyratory mold. When the loose mixture was poured into the center of the mold, only 18.8% (9 out of 48) of the small specimens fell outside of the widely accepted air void tolerance of ±0.5% from the target.

Test Specimen Fabrication. Each mixture was used to prepare three SGC samples. The sample labeling system has two numbers: the first number indicates the specific gyratory-compacted sample, and the second number indicates the location of the core, as shown in Figure 2*b*, to track if any bias may be associated with the location from which the small specimens were extracted. All gyratory-compacted samples were prepared with a diameter of 150 mm and height of 178 mm.

One of the three SGC samples was used to obtain dynamic modulus test specimens. Three replicate dynamic modulus tests were conducted using specimens with air void contents that were closest to the average air void content. The remaining two SGC samples were used to obtain fatigue test specimens. All eight small specimens from the two SGC samples were subjected to fatigue testing. Note that the data acquisition software experienced two failures when storing the electronic data; however, this problem did not prohibit visual observations of the ultimate failure.

Results

Effect of Small Specimen Extraction Direction

Performance Test Results. Figure 3 presents comparisons of the dynamic modulus test results obtained from horizontally and vertically extracted specimens made from the RS9.5D and RI19.0B mixtures. These small specimens are designated with either an *H* for horizontally extracted or a *V* for vertically extracted. Good agreement can be observed for the dynamic modulus values of the horizontally and vertically extracted small specimens made with the RS9.5D mixture. In the RI19.0B mixture results, visual comparisons between the vertically and horizontally extracted small specimens show that the horizontally extracted small specimens exhibit slightly higher average dynamic modulus values at high reduced frequencies and slightly higher phase angle values at a low reduced frequencies. Note that the strain used in dynamic modulus testing in this study is lower than that specified in AASHTO TP 79 (*1*). This may have helped negate the effects of anisotropy by minimizing aggregate mobilization.

Table 2 presents the analysis of variance (ANOVA) results used to determine the statistical significance of observed differences in the dynamic modulus test results for horizontally and vertically extracted specimens. To evaluate whether the variance in dynamic modulus between geometries was equal, F-Tests were conducted; the F-Tests determined that the variance was equal with 98% confidence. Therefore, an equal variance Student's *t*-test was conducted to evaluate the significance of differences in the average dynamic modulus test results between geometries. The results that indicate significant differences between the geometries at a 98% confidence level are highlighted. For the RS9.5D mixture, the

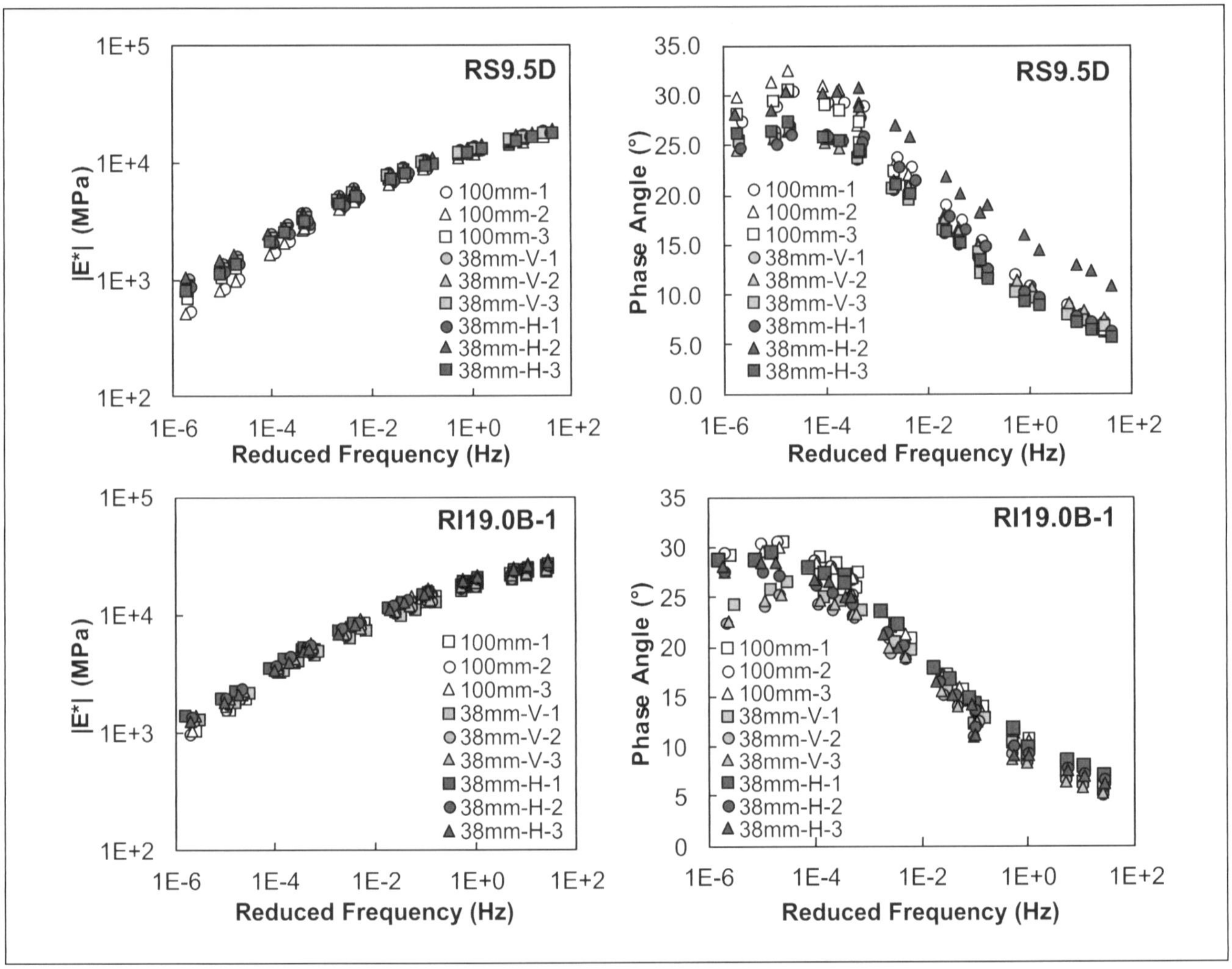

Figure 3. Dynamic modulus and phase angle mastercurves for small specimens extracted vertically and horizontally.

dynamic modulus test results for the vertically and horizontally extracted small specimens are statistically equivalent. In contrast, for the RI19.0B mix, a statistically significant difference is evident between the dynamic modulus test results for the small specimens that were vertically and horizontally extracted at the low and intermediate temperatures. However, the dynamic modulus test results for the horizontally extracted small specimens are statistically equivalent to the dynamic modulus test results for vertically extracted large specimens, indicating that the anisotropic effects are negligible.

All seven horizontally extracted RI19.0B-1 small specimens that were subjected to cyclic fatigue tests failed at the specimen end, outside the range of the gauge points. Fatigue failure outside the range of the gauge points prohibits failure detection by the linear variable differential transducers (LVDTs), which in turn prohibits the development of a failure criterion. Figure 4*a* presents photographs of four of the horizontally extracted RI19.0B-1 small specimens after fatigue testing. The occurrence of end failure was speculated to be caused by a relatively high air void content at the edge of the horizontally extracted specimens, because regions with higher air voids are more susceptible to fatigue than areas with low air void contents. However, the length of the AMPT LVDTs prohibits the use of shorter specimens to avoid the peripheral region of the SGC sample.

Although ultimate fatigue failure cannot be captured when end failure occurs, data before failure can still be analyzed to generate damage characteristic curves. Thus, in this study, damage characteristic curves were developed for both sets of test results that correspond to the horizontally and vertically extracted test specimens. However, two of the horizontally extracted specimens failed in less than 500 cycles, which did not provide sufficient data for analysis according to AASHTO TP 107. Figure 4*b* presents the damage characteristic curves for the horizontally and vertically extracted small specimen

Table 2. ANOVA p-value Results for Evaluation of Coring Direction

Temp (°C)	Frequency (Hz)	RS9.5D Small/Horiz. f-Test	RS9.5D Small/Horiz. t-Test	RS9.5D Large/Horiz. f-Test	RS9.5D Large/Horiz. t-Test	RI19B Small/Horiz. f-Test	RI19B Small/Horiz. t-Test	RI19B Large/Horiz. f-Test	RI19B Large/Horiz. t-Test
4.4	25	0.18	0.527	0.26	0.022	0.77	0.004	0.14	0.437
	10	0.25	0.743	0.27	0.027	0.63	0.004	0.11	0.445
	5	0.32	0.946	0.33	0.034	0.56	0.004	0.11	0.453
	1	0.50	0.598	0.58	0.075	0.46	0.004	0.10	0.476
	0.5	0.58	0.461	0.71	0.110	0.40	0.004	0.09	0.486
	0.1	0.70	0.289	0.95	0.230	0.21	0.004	0.05	0.498
21.1	25	0.84	0.235	0.73	0.225	0.99	0.009	0.17	0.668
	10	0.79	0.202	0.65	0.288	0.92	0.013	0.21	0.564
	5	0.79	0.191	0.62	0.312	0.83	0.018	0.25	0.505
	1	0.85	0.191	0.61	0.276	0.53	0.063	0.46	0.471
	0.5	0.91	0.199	0.63	0.235	0.43	0.117	0.60	0.517
	0.1	0.92	0.230	0.69	0.136	0.31	0.379	0.97	0.854
37.8	25	0.83	0.226	0.65	0.129	0.50	0.168	0.79	0.933
	10	0.95	0.243	0.70	0.094	0.48	0.312	0.98	0.636
	5	0.95	0.258	0.74	0.073	0.51	0.459	0.88	0.396
	1	0.70	0.296	0.82	0.042	0.73	0.860	0.68	0.069
	0.5	0.60	0.312	0.86	0.034	0.91	0.973	0.66	0.026
	0.1	0.42	0.343	0.93	0.024	0.56	0.676	0.82	0.003
54.4	25	0.64	0.321	0.99	0.030	0.18	0.089	0.60	0.001
	10	0.52	0.329	1.00	0.024	0.12	0.158	0.63	0.000
	5	0.44	0.334	1.00	0.021	0.09	0.237	0.64	0.000
	1	0.31	0.341	0.98	0.017	0.19	0.520	0.82	0.000
	0.5	0.28	0.341	0.97	0.016	0.31	0.657	0.58	0.000
	0.1	0.28	0.337	0.96	0.015	0.64	0.904	0.32	0.002

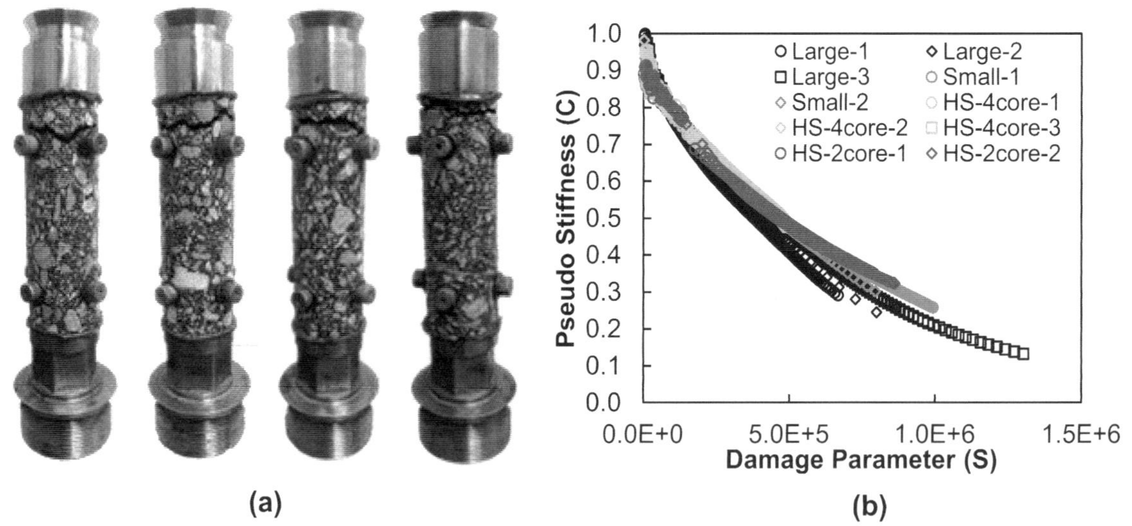

Figure 4. (*a*) Photographs of horizontally cored RI19.0B-1 small specimens obtained from gyratory-compacted samples after fatigue testing and (*b*) damage characteristic curves of RI19.0B-1 small specimens extracted vertically and horizontally.

test results. Damage characteristic curves represent the relationship between material integrity (*C*) and damage (*S*). A damage characteristic curve should be independent of loading and temperature history according to AASHTO TP 107. The results demonstrate good agreement between the results of the horizontally and

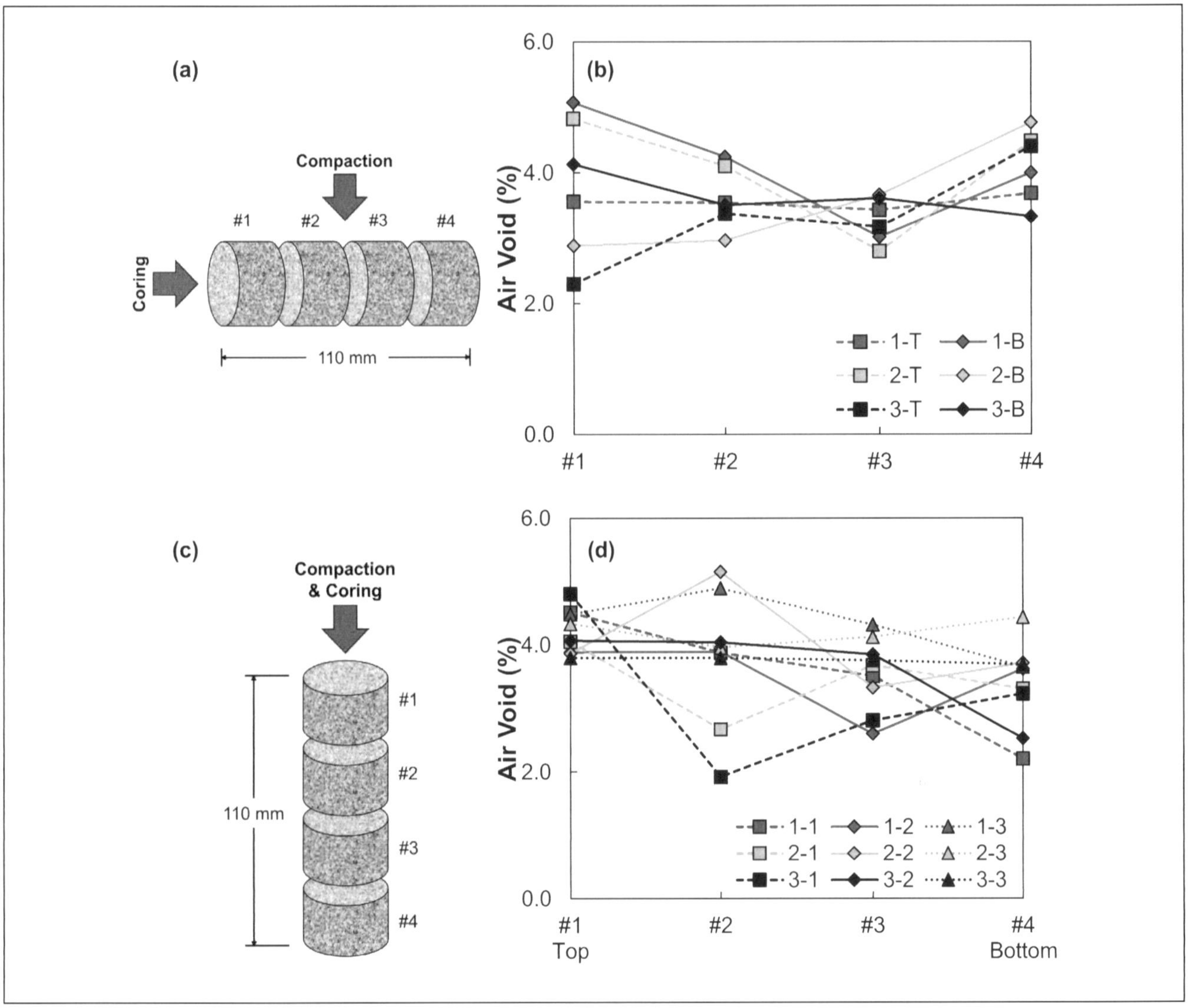

Figure 5. Air void analysis results and explanatory schematics for the (*a*, *b*) horizontally extracted specimens and (*c*, *d*) vertically extracted specimens.

vertically extracted specimens, suggesting that anisotropic effects are negligible.

Air Void Analysis. The air void distribution within the horizontally and vertically extracted small specimens was analyzed to better understand the observed occurrence of end failure in the horizontally extracted specimen fatigue test results. Figure 5 presents the air void analysis results of the SM12.5A mixture. Each test specimen was given a unique identifier based on the gyratory-compacted sample and sliced into four parts from which the bulk air void contents were measured. The fractional specimens are likely too small to be representative; however, differences in bulk specific gravity indicate differences in material composition which may not be entirely due to air voids. Figure 5, *a* and *c* show the orientation of the specimen extraction with respect to cutting and coring and show the section numbering used on the horizontal axis of Figure 5, *b* and *d*.

Figure 5*b* presents the results of the air void analysis that was conducted using horizontally extracted small specimens obtained from the SM12.5A mixture. The number in the title of each data series (e.g., the *1* in *1-T*) corresponds to the gyratory-compacted sample. The *T* or *B* in the title of the data series indicates that the small specimen was extracted from the top or bottom portion of the gyratory-compacted sample, respectively. All test specimens evaluated exhibited the highest air void content at their end, which explains the end failure results for the fatigue tests. Except for one specimen, the decrease in air void content found from the section with the highest air void content compared with the neighboring section is greater than 0.5%, indicating that this difference in air void content would make a difference in the failure location based on the typical acceptance tolerance for the variability in the bulk air void content of test specimens used for performance testing (*3, 4, 6*).

Figure 5*d* presents the results of the air void analysis that was conducted using vertically extracted small specimens obtained from the SM12.5A mixture. The first number in the title of each data series (e.g., the *1* in *1-2*) corresponds to the gyratory-compacted sample. The second number in the title of each data series (e.g., the *2* in *1-2*) corresponds to the specimen extracted from a specific gyratory-compacted specimen. Three small specimens were extracted from the inner 100 mm diameter of the gyratory-compacted samples. Three out of the nine specimens evaluated exhibited the highest air void contents within the middle section of the sample. In addition, three out of the six specimens with the highest air void contents at the end exhibited less than 0.5% air void content difference between the section with the highest air void content and the neighboring section. Based on current standards for bulk air void tolerance variability for performance test specimens, the authors do not anticipate that ± 0.5% air voids would result in significant differences in performance. Thus, the air void content results suggest a considerably higher probability of failure within the middle portion of a specimen in fatigue tests when small specimens are vertically cored, which corroborates the fatigue test findings. In addition, the average difference between the air void content of the two middle sections, where deformation is measured by the LVDTs, are comparable in vertically extracted specimens (0.7%) and horizontally extracted specimens (0.6%).

Summary. Based on the results presented, anisotropy does not significantly affect dynamic modulus test results. Given the problems associated with the air void gradient and cyclic fatigue testing in horizontally extracted specimens, this study recommends that small specimens should be vertically extracted from SGC samples for both dynamic modulus and cyclic fatigue testing, for consistency. Large specimens are extracted from the inner 100 mm of SGC samples. Due to concerns about peripheral air void gradients, which are supported by horizontally extracted specimen results, it is recommended to core all test specimens within the inner 100 mm of the gyratory sample. A maximum of four small specimens can be vertically extracted from the inner 100 mm of a SGC sample, which therefore constitutes the recommended procedure. The findings suggest a significant opportunity to improve the efficiency of dynamic modulus and axial fatigue test specimen fabrication compared with the existing large specimen fabrication method. In the next section, the specimen-to-specimen variability found in small specimens that were fabricated using the optimized approach is discussed.

Specimen-to-Specimen Variability Analysis

Dynamic Modulus. Figure 6 presents the dynamic modulus test results for each mixture that was evaluated in the specimen-to-specimen variability study. The legends within each figure indicate the location from which the core was extracted and the air void content of the specimen. The coefficient of variation of each mixture was calculated and found to be satisfactory in accordance with AASHTO TP 79, except for the RB25.0B mixture at 4°C. The variability observed in the RB25.0B mixture dynamic modulus test results could have been caused by its relatively large NMAS or its relatively large specimen-to-specimen variability in terms of air void content. The air void contents of the test specimens for the RS9.5D, SM12.5A, and RI19.0B-2 mixtures had a maximum range of 0.4%, whereas the air void contents of the RB25.0B specimens had a considerably higher span of 1.2%.

Cyclic Fatigue. Figure 7 presents the fatigue damage characteristic curves and D^R failure criterion results for each mixture evaluated within the specimen-to-specimen variability study. The D^R variability in the test replicates can be observed in the plot of the summation of (*1-C*) up to the point of failure versus the number of cycles to failure. If the D^R values of the test replicates show low variability, the data points align to form a line that passes through the origin. The legends in Figure 7 indicate the location at which the core was extracted and the air void content of the specimen, as shown in Figure 2.

Generally, the results demonstrate good agreement among the damage characteristic curves, with an increase in variability corresponding to an increase in the mixture NMAS. Similarly, the failure criterion plots indicate generally low scatter, with a small increase in scatter corresponding to an increase in the mixture NMAS. A low incidence of end failure also is evident, with two end failures occurring in the RI19.0B-2 mixture, one in the SM12.5A mixture, and zero in the other two mixtures. These findings demonstrate no clear relationship between end failure and mixture NMAS. Note that the faded damage characteristic curves and corresponding failure criterion values shown in Figure 7 correspond to irregular failures, including end failure, and the presence of large voids or broken coarse aggregate on the failure surface. Generally, such irregular failures correspond to outlier behavior. All data are shown in the damage characteristic curve plots; however, end failures are not

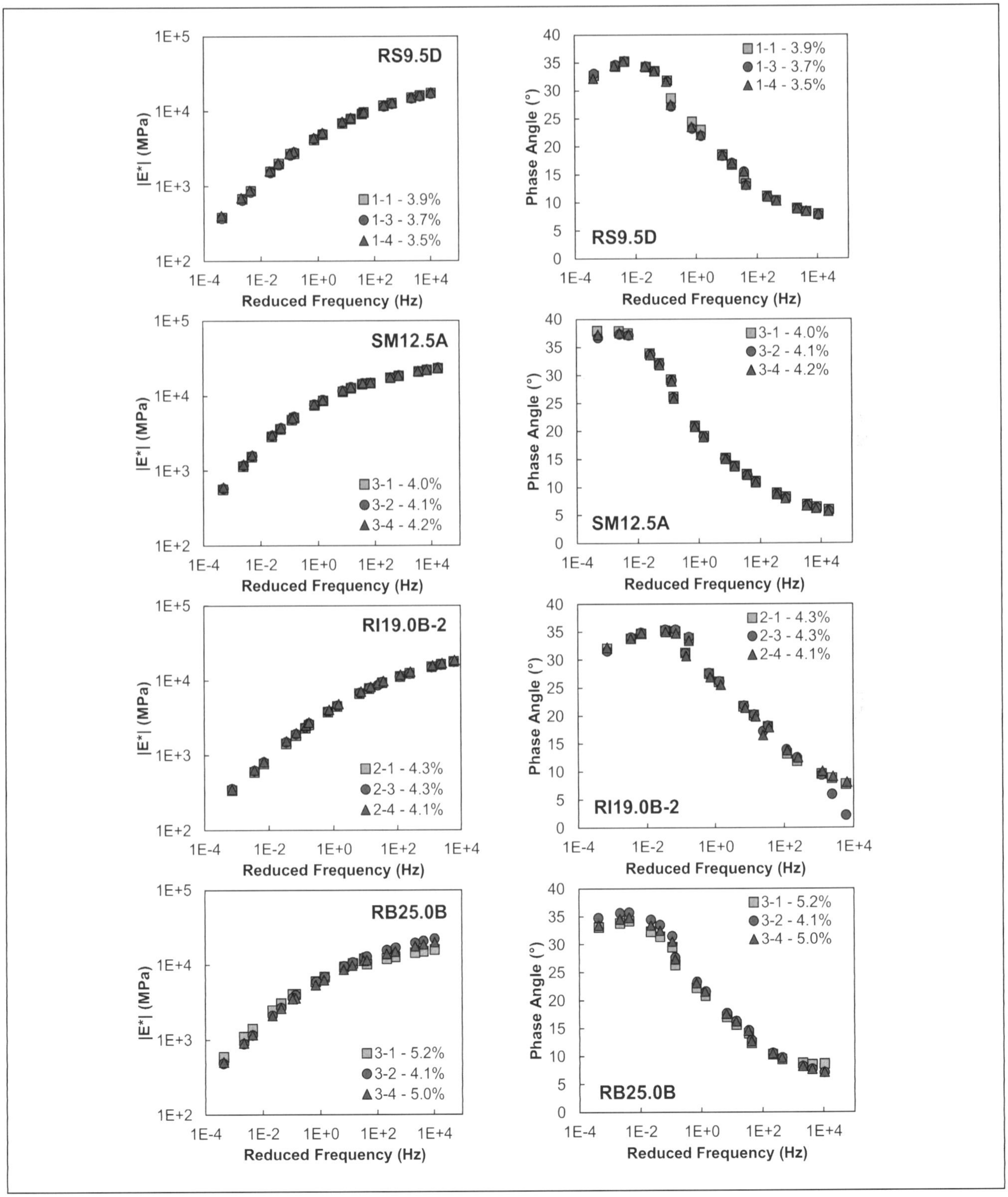

Figure 6. Dynamic modulus and phase angle mastercurves for sample-to-sample variability analysis.

shown in the failure criterion plots. While other irregular failure data are shown in the failure criterion plots, they are not used in fitting the linear regression. No clear relationship can be observed between the fatigue testing variability and the small specimen air void contents within the observed specimen air void content range of $\pm 0.7\%$. It should be noted that the plant operator who provided the RI19.0B-2 mixture indicated that the

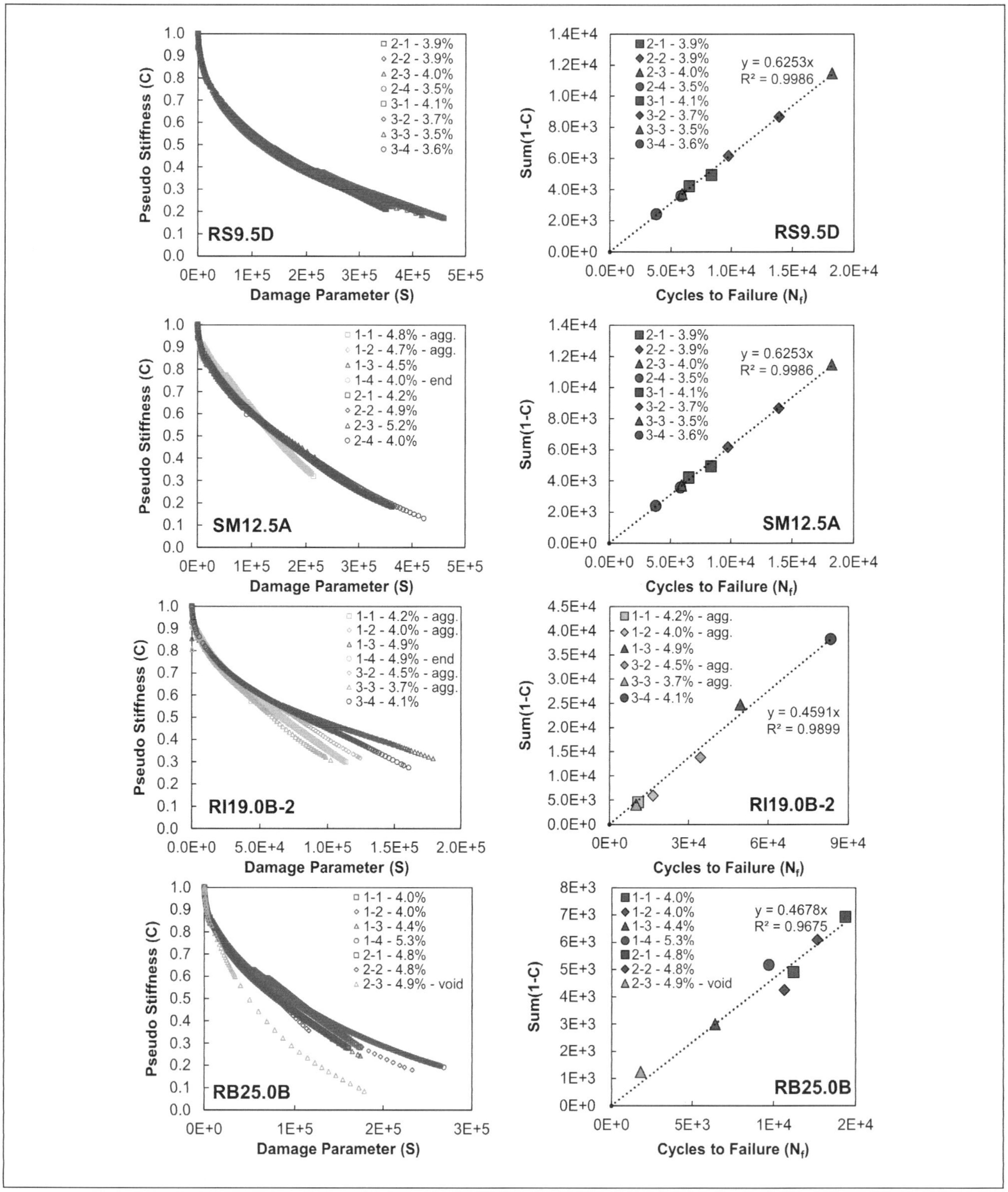

Figure 7. Damage characteristic curves and D^R failure criterion results for sample-to-sample variability analysis.

mixture was of poor quality, which likely contributed to the observed scatter in the results.

Given that no clear relationship is apparent between the small specimen air void contents and the performance test results in this study, the authors recommend that initial small specimen provisional standards should include an air void content tolerance of ±0.7%, which then should be refined on ruggedness testing.

Conclusions and Recommendations

The following conclusions and recommendations are drawn from this study:

- Small specimens should be extracted vertically rather than horizontally from SGC samples. Anisotropy does not affect dynamic modulus test results or damage characteristic curves that have been developed from cyclic fatigue test results. Horizontal coring infringes on the peripheral region of gyratory-compacted specimens that have relatively high air void contents, which leads to end failure and prevents the development of a fatigue failure criterion. This problem is alleviated by vertical coring.
- Four 38-mm-diameter small specimens can be vertically extracted from the inner 100 mm of gyratory-compacted samples to produce test specimens with generally acceptable air void uniformity and to produce test results consistent with 100-mm-diameter specimens.
- No clear relationship is evident between small specimen air void content and performance test results within the observed specimen air void content range of ±0.7%. Therefore, this study recommends that initial small specimen provisional standards should include an air void content tolerance of ±0.7%, which should then be refined on ruggedness testing.

Acknowledgments

This paper is based on work supported by the National Cooperative Highway Research Program (NCHRP) under project NCHRP IDEA N181 and the National Science Foundation (NSF) through the NSF Graduate Research Fellowship.

Author Contributions

The authors confirm contributing to the paper as follows: study conception and design: Cassie Castorena, Richard Kim; data collection: Sonja Pape, Kangjin Lee; analysis and interpretation of results: Sonja Pape, Kangjin Lee, Cassie Castorena; draft manuscript preparation: Sonja Pape, Cassie Castorena, Richard Kim. All authors reviewed the results and approved the final version of the manuscript.

References

1. American Association of State Highway and Transportation Officials. *Determining the Dynamic Modulus and Flow Number of Asphalt Mixtures Using the Asphalt Mixture Performance Tester (AMPT)*. AASHTO TP 79. American Association of State Highway and Transportation Officials, Washington, D.C., 2015.
2. American Association of State Highway and Transportation Officials. *Determining the Damage Characteristic Curve of Asphalt Concrete from Direct Tension Cyclic Fatigue Tests*, AASHTO TP 107. American Association of State Highway and Transportation Officials, Washington, D.C., 2015.
3. Bonaquist, R. *NCHRP Report 614: Refining the Simple Performance Tester for Use in Routine Practice*. Transportation Research Board of the National Academies, Washington, D.C., 2008.
4. Kim, Y. R., Y. Seo, M. King, and M. Momen. Dynamic Modulus Testing of Asphalt Concrete in Indirect Tension Mode. *Transportation Research Record: Journal of the Transportation Research Board*, 2004. 1891: 163–173.
5. American Association of State Highway and Transportation Officials. *Standard Method of Test for Determining the Fracture Energy of Asphalt Mixtures Using the Semicircular Bend Geometry (SCB)*. AASHTO TP 105. American Association of State Highway and Transportation Officials, Washington, D.C., 2013.
6. ASTM International. *Evaluation of Asphalt Mixture Cracking Resistance Using the Semi-Circular Bend Test (SCB) at Intermediate Temperatures*. ASTM D8044. ASTM International, West Conshohocken, PA, 2016.
7. Kutay, M. E., N. Gibson, J. Youtcheff, and R. Dongre. Use of Small Samples to Predict Fatigue Lives of Field Cores: Newly Developed Formulation Based on Viscoelastic Continuum Damage Theory. *Transportation Research Record: Journal of the Transportation Research Board*, 2009. 2127: 90–97.
8. Li, X., and N. H. Gibson. Using Small Scale Specimens for AMPT Dynamic Modulus and Fatigue Tests. *Journal of the Association of Asphalt Paving Technologists*, Vol. 82, 2013, pp. 579–616.
9. Bowers, B. F., B. K. Diefenderfer, and S. D. Diefenderfer. Evaluation of Dynamic Modulus in Asphalt Paving Mixtures Utilizing Small-Scale Specimen Geometries. *Journal of the Association of Asphalt Paving Technologists*, Vol. 84, 2015, pp. 497–526.
10. Lee, K., S. Pape, C. Castorena, and Y. R. Kim. Evaluation of Small Specimen Geometries for Asphalt Mixture Performance Testing and Pavement Performance Prediction. *Transportation Research Record: Journal of the Transportation Research Board*, 2017. 2631: 74–82.
11. Park, H., and Y. R. Kim. Investigation into Top-Down Cracking of Asphalt Pavements in North Carolina. *Transportation Research Record: Journal of the Transportation Research Board*, 2013. 2368: 45–55.
12. Park, H. *Investigation of Primary Causes of Cracking in Asphalt Concrete*. PhD dissertation. North Carolina State University, Raleigh, NC, 2013.
13. Chehab, G. R., E. O'Quinn, and Y. R. Kim. Specimen Geometry Study for Direct Tension Test Based on Mechanical Tests and Air Void Variation in Asphalt Concrete Specimens Compacted by Superpave Gyratory Compactor. *Transportation Research Record: Journal of the Transportation Research Board*, 2000. 1723: 125–132.
14. American Association of State Highway and Transportation Officials. *Determining the Fracture Potential of Asphalt Mixtures Using Semicircular Bend Geometry (SCB) at*

Intermediate Temperatures. AASHTO TP 124. American Association of State Highway and Transportation Officials, Washington, D.C., 2016.

15. Wang, Y., and Y. R. Kim. Development of a pseudo strain energy-based fatigue failure criterion for asphalt mixtures. *International Journal of Pavement Engineering*, 2017. https://doi.org/10.1080/10298436.2017.1394100.

The Standing Committee on Structural Requirements of Asphalt Mixtures (AFK50) peer-reviewed this paper (18-04736).

Any opinion, findings, conclusions, or recommendations expressed in this material are those of the authors and do not necessarily reflect the views of the NSF.

Article

TRR
JOURNAL OF THE TRANSPORTATION RESEARCH BOARD

Transportation Research Record
2018, Vol. 2672(28) 451–461

DOI: 10.1177/0361198118773892
journals.sagepub.com/home/trr
SAGE

Effect of Loading Waveform Pattern and Rest Period on Fatigue Life of Asphalt Concrete Using Viscoelastic Continuum Damage Model

Waleed Abdelaziz Zeiada[1,2], Padmini P. Gudipudi[3], B. Shane Underwood[4], and Mena I. Souliman[5]

Abstract
Fatigue cracking is one of the most critical types of distress in asphalt pavements and is due to actions of repetitive traffic loading over time. The fatigue life of asphalt concrete is often estimated from laboratory experiments where the performance depends directly on the test method, loading conditions, temperature, rest period, and aging in addition to the composition and properties of the mixture itself. The uniaxial fatigue test has become a popular method for developing constitutive models that describe the fatigue behavior of asphalt concrete mixture owing to the uniform states of stress across the specimen section. This study investigates the effect of the loading waveform (sinusoidal versus haversine) and rest period (continuous versus intermittent) on the laboratory fatigue life of asphalt concrete mixtures. The fatigue analysis was performed using the simplified viscoelastic continuum damage (S-VECD) approach where the damage characteristic (C-S) curves were established for all the cases, and then used to estimate the fatigue laws through simulated predictions. The proposed uniaxial fatigue test and analysis method were able to determine the fatigue life relationships of asphalt concrete mixture at different waveform and rest period conditions with a reduced testing time compared to other traditional testing and analysis methods. Overall, both rest period and waveform pattern were found to affect the laboratory fatigue life of asphalt concrete mixture. Model predictions show that pulse-rest loading yields an equivalent fatigue life to continuous loading at strain values that are approximately four times greater.

Load-associated fatigue cracking is one of the most critical types of distress in flexible pavements. The action of repeated traffic loading induces tensile and shear stresses in the asphalt concrete (AC) layers and ultimately leads to loss of their structural integrity. The continued loss of integrity leads to visual surface cracking, which can allow water to enter into the supporting layers of the pavement and further accelerate structural failure. Owing to the importance of this distress mechanism, researchers have sought methods to investigate the inherent resistance of AC materials to damage growth and failure. In the laboratory, these tests are generally performed using a repeating loading pattern of sinusoidal or haversine controlled displacement or controlled force (*1–4*). The decision on whether to perform sinusoidal or haversine loading ultimately affects the observed fatigue performance and the reporting of results. For example, in the case of haversine controlled displacement loading the strain remains constant and in tension, but the viscoelastic nature of the material means that the bending stress becomes fully reversed and sinusoidal, which may complicate how results are reported and interpreted relative to controlled displacement sinusoidal loading (*5, 6*). This decision if not properly considered will affect the formation of fatigue life performance models (*7*).

Another factor that complicates laboratory fatigue studies is that under real traffic conditions the pavement

[1]Department of Civil and Environmental Engineering, University of Sharjah, Sharjah, United Arab Emirates
[2]Department of Public Works Engineering, College of Engineering, Mansoura University, Mansoura City, Egypt
[3]Flatiron Constructors, Inc., Technical Services Group, Large Projects Division, Broomfield, CO
[4]Civil, Environmental and Sustainable Engineering Program, Arizona State University, Tempe, AZ
[5]Departments of Civil Engineering and Construction Management, The University of Texas at Tyler, TX

Corresponding Author:
Address correspondence to Waleed Abdelaziz Zeiada: wzeiada@sharjah.ac.ae

structure is subjected to a succession of load pulses as the traffic flows past a particular point on the road. The duration and spacing of the load pulses depends on several factors, including vehicle speed, axle configuration, and headway between following vehicles (*8*). Although the earliest studies on asphalt fatigue considered this effect in the experiment, most modern studies use continuous loading to save time (*9*). When tested for rest period effects, two different ways of introducing the rest periods have emerged. In the first method, referred to as block loading-rest, rest intervals are introduced after a certain pre-defined or random number of continuous loading cycles (*10–12*). The second method, referred to as pulse-rest, follows each loading cycle with a rest period of pre-defined duration (*2*, *8*, *13*). The second method with intermittent loading resembles more closely the sequence of traffic pulses in the field than the first method.

The literature shows inconsistent records with respect to the specific effect of rest periods (*9*, *13–16*). What is certain is that two mechanisms can occur when rest periods are applied: linear viscoelastic relaxation/recovery and healing. There is debate on the relative influence of these two mechanisms with support for one or the other related to the specific factors incumbent to the study in question, including mode of loading (stress versus strain), loading waveform pattern (sinusoidal versus haversine), and possibly material. For example, Monismith et al. (*9*) assessed the effect of rest period by conducting repeated flexure tests on elastically supported beams with rest periods from 1 to 19 s. The results indicated that increasing the rest period from 1 to 19 s had no effect on the fatigue performance. Khaki et al. (*15*) studied the effects of haversine and triangular loading waveforms on the fatigue life of AC using different load durations and rest periods and concluded that the fatigue life of triangular waveform was higher than that of haversine, that the load duration was inversely proportional to the fatigue life, and that the fatigue life increased as the rest period increased (from 4 to 9 s). Other researchers have conducted similar studies to identify the effect of rest period on damage, healing, and the endurance limit (*3*, *13–16*).

The fatigue behavior of AC is a complex result of both the linear viscoelastic response of the material and its inherent resistance to mechanisms that yield a reduction in modulus (commonly referred to as damage resistance, but also described using heating mechanisms, thixotropy, and others). It is affected by several factors such as mixture properties (gradation, binder content, binder grade, mixture volumetric, etc.), test type (simple flexure, uniaxial, triaxial, etc.), and test conditions (mode of loading, waveform pattern, temperature, frequency etc.). Although a remarkable amount of research has been conducted to investigate the effect of those factors on laboratory fatigue behaviors, only a limited number of studies have investigated the effect of waveform patterns on AC fatigue life especially if it is coupled with the effect of rest period. This study uses the advantages of recently developed modeling techniques using the simplified viscoelastic continuum damage (S-VECD) model to incorporate the healing effect by introducing intermittent rest periods which helps in reducing the required testing time to investigate the effect of waveform pattern and intermittent rest periods on fatigue life performance models of AC mixture.

Objective

This study investigates the combined effect of waveform patterns, such as haversine and sinusoidal, and rest periods on the fatigue life of AC mixture by employing the S-VECD model using uniaxial fatigue test results. Toward this goal, an improved S-VECD model was utilized to incorporate healing effect for better fatigue prediction of asphalt pavements.

Materials and Specimen Manufacturing

The AC gradation used in this study is a Superpave 19 mm nominal maximum size of aggregate mixture with a dense gradation from Phoenix, Arizona, and was in accord with specifications from the Arizona Department of Transportation (ADOT) (*17*). This mixture was designed using PG 64-22 asphalt binder and 6.0% target asphalt content. Details of this mixture, its composition, constituent material properties, and fabrication protocols are available elsewhere (*18*). All test specimens used in this study were tested at an air void content in the range of 4.0 ± 0.5%.

Test Methods and Results

In this study, AC mixtures were subjected to two different experimental protocols: dynamic modulus ($|E^*|$) and axial fatigue. For both tests, brass buttons were glued onto the sample surface 120° apart. Brackets were attached to these buttons and used to secure three sets of linear variable displacement transducers (LVDTs). $|E^*|$ and axial fatigue testing was conducted using the IPC Global UTM 25 servo-hydraulic load frame with environmental chamber. Fatigue test samples were glued to end plates using a custom designed gluing jigs that ensured proper vertical alignment.

Dynamic Modulus ($|E^*|$) Test

The $|E^*|$ tests were conducted per AASHTO T 342-11 at five temperatures: 10, 4.4, 21.1, 37.8, 54.4°C and six load frequencies: 25, 10, 5, 1, 0.5 and 0.1 Hz. Tests were conducted in an increasing order of temperature and in a decreasing order of loading frequency to minimize any potential damage to the specimen before the next sequential test. The stress levels were varied with the frequency and temperature to keep the measured on-specimen strain amplitude below 70 $\mu\varepsilon$. At each temperature and frequency, the $|E^*|$ and phase angle, δ, were measured and used to compute the storage modulus, E′. This modulus was then fit to a sigmoidal function,

$$\log E' = \delta + \frac{\alpha}{1 + \frac{1}{e^{\beta + \gamma(\log \omega_r)}}} \quad (1)$$

which was then used with the collocation method to determine the coefficients of the Prony form of relaxation modulus (*19*), $E(\xi)$,

$$E(\xi) = E_\infty + \sum_{m=1}^{N} E_m e^{-\xi/\rho_m} \quad (2)$$

where

ω_r = reduced frequency of loading,

δ = logarithm of the minimum value of E',

$\delta + \alpha$ = logarithm of the maximum value of E',

β, γ = parameters describing the shape of the sigmoidal function

E_m = modulus of prony term m,

ρ_m = relaxation time of Prony term m, and

ξ = reduced time, s.

The $E(\xi)$ was later used in analyzing the fatigue tests (continuous and rest period) and characterizing the functions for the VECD model.

Axial Repeated Loading Test

Axial repeated loading tests were conducted to evaluate the effect of different loading configurations on the fatigue damage of AC specimens. The loading configurations included haversine (tension), sinusoidal (tension-compression), and with or without rest periods. Feedback control for the tests was based on the average on-specimen displacement measurements. To achieve this consistent control, the test software dynamically adjusts the actuator movement based on the feedback from the three LVDTs. Prior to performing the fatigue loading, a low strain magnitude $|E^*|$ test was conducted at the fatigue test temperature (21.1°C) and at 10 Hz only. The $|E^*|$ calculated from this loading is termed the fingerprint modulus, $|E^*|_{FP}$. The following four cases were considered in the axial fatigue testing experiment:

1. Continuous haversine fatigue test without rest period (designated as: H-0 RP).
2. Continuous sinusoidal fatigue test without rest period (designated as: S-0 RP).
3. Intermittent haversine fatigue test with a 3 s rest period (designated as: H-3 RP).
4. Intermittent sinusoidal fatigue test with a 3 s rest period (designated as: S-3 RP).

The 3 s rest period was determined based upon previous experience, which showed that rest periods longer than 3 s did not result in any additional increase of fatigue life or healing (*20*). This conclusion was reached by comparing the stiffness recovery and healing at different rest periods of 1, 5, and 10 s. Figures 1 and 2 show the typical displacement waveform input and corresponding load waveform response respectively for the haversine and sinusoidal conditions without and with rest period. Care was taken for each test to ensure that the on-specimen displacement/strain magnitudes were in fact consistent for each of the loading patterns and that their shape maintained the target sinusoidal or haversine pattern.

Simplified VECD Model Analysis

Damage Characteristic Function

The viscoelastic continuum damage (VECD) model approach attempts to identify a unique relationship between the amount of damage, S, and material integrity, quantified as pseudo-stiffness, C. This relationship is named the damage characteristic curve. The VECD modeling approach was developed based on three important principles (*21–23*); the elastic-viscoelastic correspondence principle; the continuum damage mechanics-based work potential theory; and the temperature-time superposition principle. The VECD modeling approach has shown an accurate prediction of AC fatigue when the material is subjected to simple loading such as constant rate tests. For cyclic loading fatigue test, the regular VECD formulations and calculations are more challenging and prone to additional errors in damage calculation. To address this issue, Underwood et al. (*19*), developed a simplified viscoelastic continuum damage model (S-VECD). The S-VECD model comprises a damage calculation in two stages: stage one accounts for damage accumulation during the first loading path and stage two accounts for the damage under additional repetitive loading. This approach is adopted in the current research effort to predict fatigue behavior of AC mixture using different conditions of continuous waveform patterns. The S-VECD model is described in detail elsewhere (*19*), but in summary it consists of the following essential components:

- Pseudostrain (ε^R) function in time and steady-state frequency domains;

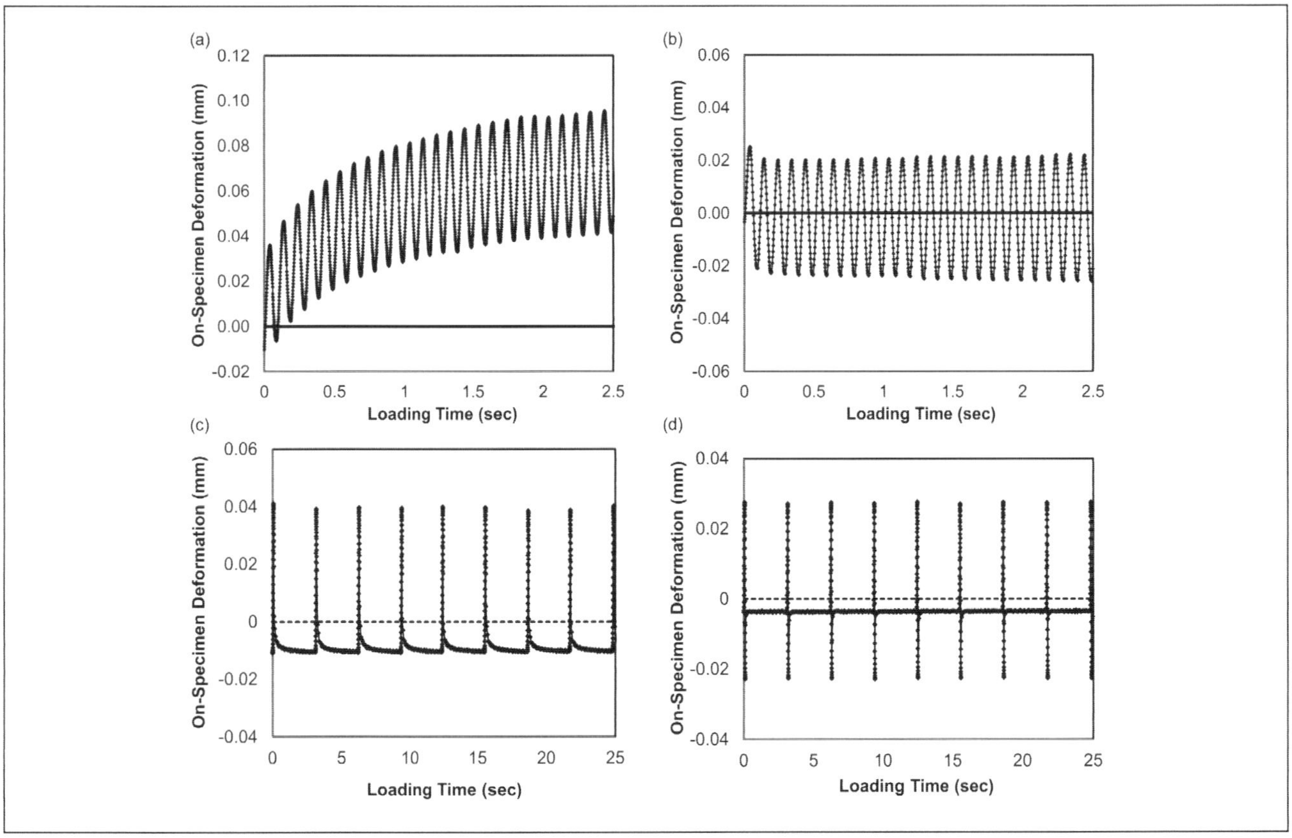

Figure 1. Typical displacement waveform input over time: (*a*) continuous haversine; (*b*) continuous sinusoidal; (*c*) intermittent haversine; and (*d*) intermittent sinusoidal.

$$\varepsilon^R = \begin{cases} \varepsilon^R = \dfrac{1}{E_R}\displaystyle\int_0^{\xi} E(\xi-\tau)\frac{d\varepsilon}{d\tau}d\tau & \xi \leq \xi_p \\ \left(\varepsilon^R_{0,ta}\right)_{cycle\ i} = \dfrac{1}{E_R}\cdot\dfrac{\beta+1}{2}\left(\left(\varepsilon_{0,pp}\right)_i\cdot|S^*|_{LVE}\right) & \xi > \xi_p \end{cases} \tag{3}$$

- Stress (σ) – (ε^R) relationship in time and steady-state frequency domains; and

$$C = \begin{cases} \dfrac{\sigma}{\varepsilon^R\cdot DMR} & \xi \leq \xi_p \\ \dfrac{\sigma_{0,ta}}{\varepsilon^R_{0,ta}\cdot DMR} & \xi > \xi_p \end{cases} \tag{4}$$

- Damage evolution law in time and steady-state frequency domains.

$$dS = \begin{cases} (dS)_{time\ Step\ j} = \left(-\dfrac{1}{2}\left(\varepsilon^R\right)_j^2\dfrac{\partial C}{\partial S}\right)^{\alpha}\cdot(d\xi)_j & \xi \leq \xi_p \\ (dS)_{cycle\ i} = \left(-\dfrac{1}{2}\left(\varepsilon^R_{0,ta}\right)_j^2\dfrac{\partial C}{\partial S}\right)^{\alpha}\cdot(d\xi_p)\cdot(K_1) & \xi > \xi_p \end{cases} \tag{5}$$

where

E_R = reference modulus (taken as 1),
$E(\xi)$ = material's linear viscoelastic relaxation modulus, Equation 2,
$\varepsilon^R_{0,ta}$ = pseudo strain tension amplitude,
τ = integration term,
ξ = reduced time,
ξ_p = pulse reduced time,
$\varepsilon_{0,pp}$ = peak-to-peak measured strain amplitude,
$|S^*|_{LVE}$ = material's effective linear viscoelastic dynamic modulus,
β = wave shape factor (1 for tension, 0 for tension-compression, and −1 for compression),
DMR = dynamic modulus ratio,
$\sigma_{0,ta}$ = stress tension amplitude,
C = pseudo stiffness (material integrity),
S = damage,
α = damage evolution rate, and
K_1 = loading shape factor.

In the case of continuous loading the effective linear viscoelastic modulus, $|S^*|_{LVE}$ is the same as the $|E^*|$ value at the test temperature and frequency. However, under intermittent loading, the steady-state condition cannot be assumed and the effective modulus must be calculated according to (*24*)

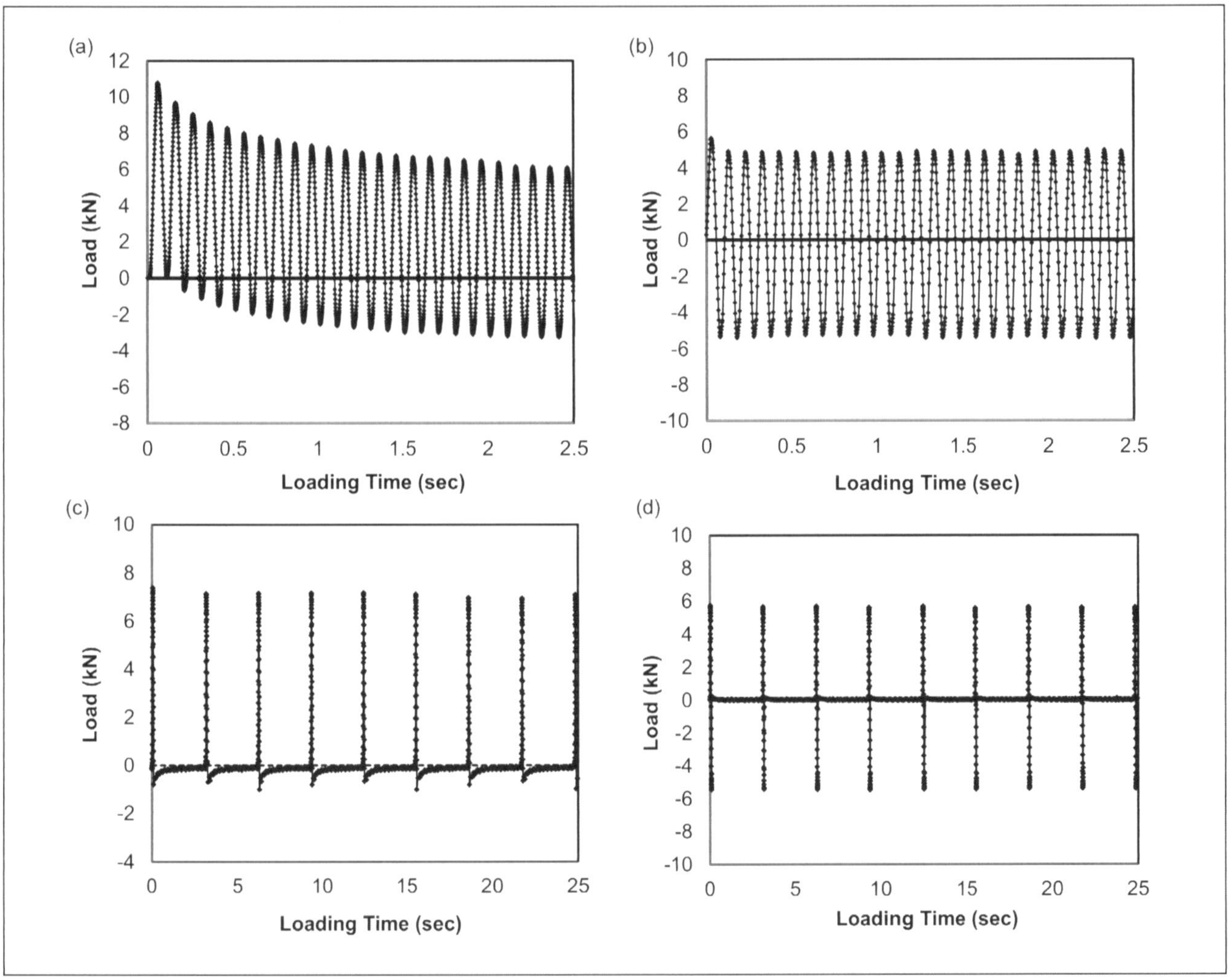

Figure 2. Typical load waveform response over time: (*a*) continuous haversine; (*b*) continuous sinusoidal; (*c*) intermittent haversine; and (*d*) intermittent sinusoidal.

$$|S*| = \frac{\max\left(\int_0^{\xi} E(\xi-\tau)\frac{d(\varepsilon_{o,pp}f(\xi))}{d\tau}d\tau\right) - \min\left(\int_0^{\xi} E(\xi-\tau)\frac{d(\varepsilon_{o,pp}f(\xi))}{d\tau}d\tau\right)}{\varepsilon_{o,pp}} \tag{6}$$

where the function $f(\xi)$ expresses the time varying pattern of either the haversine or sinusoidal loading pulse.

This equation is an approximation, but was found to produce pseudostrain values that are accurate and comparable to those values calculated using a full-history loading data with only 1% difference (*24*). Damage and the material integrity are computed using Equations 3 to 5 for the tests in this study. Once the values of S and C are known, they are plotted with respect to one another and are described analytically using the power form function

$$C = 1 - C_1 S^{C_2} \tag{7}$$

where C_1 and C_2 are regression constants.

Fatigue Life Prediction

The S-VECD model equations can be rearranged to yield predictive equations that estimate the number of cycles to failure under a prescribed repetitive loading pattern (*25*). In this study attention is given to simulating controlled sinusoidal or haversine inputs from the damage functions. In these cases, the following equation is used to predict the number of cycles to failure for a prescribed strain amplitude, and for the case of haversine or sinusoidal loading:

$$N_f = \frac{(f_{red})(2^{\alpha})\left(\frac{(1-C_f)}{C_1}\right)^{\frac{(\alpha-\alpha C_2+1)}{C_2}}}{(\alpha-\alpha C_2+1)(C_1C_2)^{\alpha}\left[\left(\varepsilon_{0,pp}\right)\left(|S*|_{LVE}\right)\frac{\beta+1}{2}\right]^{2\alpha}K_1} \tag{8}$$

where

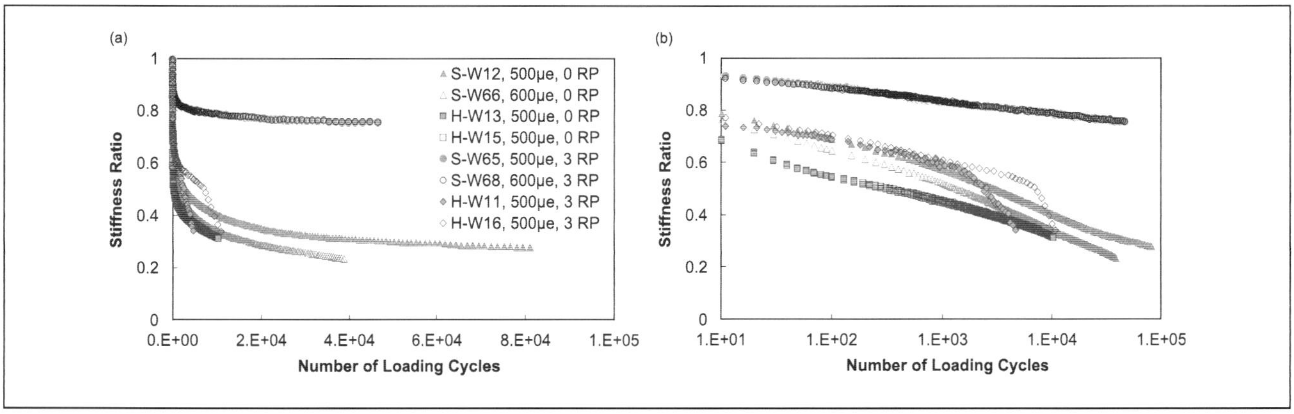

Figure 3. Axial fatigue test results: (*a*) cycle number in arithmetic scale; and (*b*) cycle number in logarithmic scale.

Table 1. Summary of Finger Print Modulus and Axial Fatigue Test Results

Test configuration	Specimen ID	Peak-to-peak strain, $\mu\varepsilon$	$\lvert E^*\rvert_{FP}$, MPa	Machine compliance factor, *MCF*	Cycles to failure, N_f	C at failure
Continuous haversine	H-W13	500	6,458	4.04	10,292	0.31
	H-W15	500	6,525	3.95	7,625	0.34
Continuous sinusoidal	S-W12	500	6,607	4.20	139,051	0.25
	S-W66	600	6,168	3.84	38,784	0.23
Intermittent	H-W11	500	5,995	3.74	4,591	0.34
Haversine	H-W16	500	5,652	3.47	10,483	0.34
Intermittent	S-W65	500	6,919	4.23	48,047*	0.75[a]
Sinusoidal	S-W68	500	6,452	4.04	46,773*	0.75[a]

[a]Specimens tested under intermittent sinusoidal condition did not reach failure and the tests were stopped at reported *N*.

N_f = predicted cycle number of cycles to failure,
f_{red} = reduced frequency for the condition being simulated, and
C_f = stiffness ratio at failure.

These simulations require estimates for the fatigue failure criteria. Here, the representative *C* at failure approach is adopted because of the limited number of replicate tests available.

Effect of Waveform Pattern and Rest Period on Fatigue Life

Stiffness Deterioration under Different Loading Modes

To compare the stiffness degradation between tests with different waveform patterns and rest period conditions, the stiffness ratio parameter was used instead of the absolute stiffness values. In this study stiffness ratio was calculated as the stiffness value at any loading cycle, $(\sigma_0/\varepsilon_0)_j$, divided by the linear viscoelastic dynamic modulus, $|E^*|_{LVE}$,

$$Stiffness\ Ratio = \frac{\left(\frac{\sigma_0}{\varepsilon_0}\right)_j}{|E*|_{LVE}} \quad (9)$$

Mathematically, the value of the stiffness ratio is identical to the definition of *C* for the repetitive loading cases.

Figure 3 shows the stiffness ratio versus number of loading cycles for different waveform patterns and rest period combinations, while Table 1 includes a summary of the test results of those test combinations. In Figure 3, continuous fatigue tests without rest periods are labeled "0 RP", while intermitted fatigue tests with 3 s rest periods are labeled "3 RP". It is worth mentioning that the target strains were chosen in a way to target cycles to failure within 50,000 cycles due to prolonged test time required for test cases with rest periods. It can be observed from Figure 3 and Table 1 that successful failures were reached for all the cases except for the sinusoidal waveform with rest period. In this case it was decided to stop the test at around 50,000 cycles. Moreover, for the sinusoidal continuous test case, the target strain was increased from 500 $\mu\varepsilon$ for the first replicate to 600 $\mu\varepsilon$ and, as a consequence, the number of loading cycles at failure was reduced from 139,051 to 38,784.

The failure of the sample in the test was assessed using the drop in *Stiffness Ratio* x *Cycle Number* approach (*26*). As can be observed from Table 1, it is quite interesting that haversine fatigue tests (continuous and intermittent) reached a failure condition at an average stiffness ratio of 0.34 which was higher compared with the continuous sinusoidal fatigue tests that failed at stiffness ratio close to 0.24. A strain level (or more precisely energy level) dependence on the stiffness ratio at failure, is consistent with the findings of others, which suggest that a more accurate assessment of failure can be obtained by examining the rate of energy dissipation in stress-pseudostrain space (*27*). This effect could not be examined in detail here since only two replicate tests were carried out at each test condition. Since the intermittent sinusoidal was not carried out until failure, it was necessary to assume some value for the stiffness ratio at failure in order to use S-VECD simulations. It is assumed, based on the similarity in stiffness ratio at failure from the intermittent and continuous haversine loading tests, that the stiffness ratio at failure in the intermittent sinusoidal test is the same as that from the continuous sinusoidal testing.

From Figure 3 it can be observed for the sinusoidal fatigue tests that introducing intermittent rest periods between the loading appears to lessen the modulus degradation during the test. It is not immediately apparent from the data in the figure whether this effect is due to stress relaxation during the rest period or actual healing in the mixture. Only after accounting for linear viscoelastic effects is it possible to differentiate these two mechanisms. It is necessary to examine the smeared damage characteristic function (shown in the next section) to make this determination.

It is also observed from Figure 3 that in the case of the haversine loading test although the pattern of modulus reduction is different when rest periods are introduced, the actual number of cycles to failure is approximately the same as the continuous loading case. Until approximately cycle 4,000 to 10,000 the intermittent rest period tests have stiffness ratios greater than the continuous loading, afterwards though the stiffness ratio in the intermittent rest period tests reduce quickly (a clear change in slope is observed) and failure occurs at approximately the same cycle number. This effect is easier to see in Figure 3*b*. This pattern can be interpreted from the perspective afforded by the S-VECD theory to be a combination of damage, the pseudostrain energy that causes that damage, viscoelastic relaxation, and (though not readily apparent from this figure), what is believed to be healing. With respect to damage and relaxation, under continuous haversine tensile strain the stress response is the superposition of the individual responses from the sinusoidal variations and the mean strain offset. The response due to the strain level offset (i.e., the mean strain of the haversine input) is a relaxation of stress and the response from the sinusoidal component of the haversine strain input is also sinusoidal. When superimposed, these responses result in a total stress that begins as haversine but transitions to sinusoidal as the relaxation of the mean strain component occurs.

The significance of this relaxation mechanism during the loading is important because of the contrasting behavior that occurs when an intermittent rest period is introduced. In this case, the mean strain relaxation has only a small effect on any given cycle and so the magnitude of the stress response for each cycle of the intermittent rest period loading will be greater than for the same strain level and cycle number in the continuous case. This effect is also apparent when observing Figures 1 and 2. Within the damage model formulation, this effect is represented by a β value that is approximately equal to one for the pulse-rest loading and is closer to zero for the continuous loading. In this case it becomes apparent that the damage growth rate will be higher for a given cycle of haversine pulse-rest loading than it will be for haversine continuous loading. Despite this effect the stiffness ratio in pulse-rest loading remains higher than the continuous loading until between 4,000 and 10,000 cycles. It is acknowledged that this effect could be the result of thixotropic or heating mechanisms, but here it is interpreted to be the result of healing that takes place during the rest period.

Damage Characteristic Curves

All samples are analyzed using Equations 3 to 5 to develop damage characteristic curves (C-S) for each loading configuration. In the case of the continuous loading the resulting curves represent the correct relationship between C and S, while the curves from the intermittent tests represent a combination of damage and (as it is interpreted in this paper) healing. The simple effects of linear viscoelastic relaxation do not affect either curve since the elastic-viscoelastic correspondence principle is applied to remove the time effects from the data. The damage characteristic curves for each replicate from each test condition are plotted until the failure cycle in Figure 4. Part (*a*) shows the replicate and best fit data for the continuous loading cases, part (*b*) shows the curves for the intermittent rest period cases, and part (*c*) shows the best fit functions for all cases together. Note that a major portion of the curve for the sinusoidal rest period case is extrapolated.

First, the damage curves for continuous loading cases were the same regardless of whether the pattern was sinusoidal or haversine. Of course, as mentioned earlier, the haversine and sinusoidal loading patterns result in different C values at failure and so the end point in the damage characterization plots for the tests is slightly different. The intermittent rest period testing exhibits damage characteristic curves positioned above those for the continuous loading case. This positioning reflects the fact

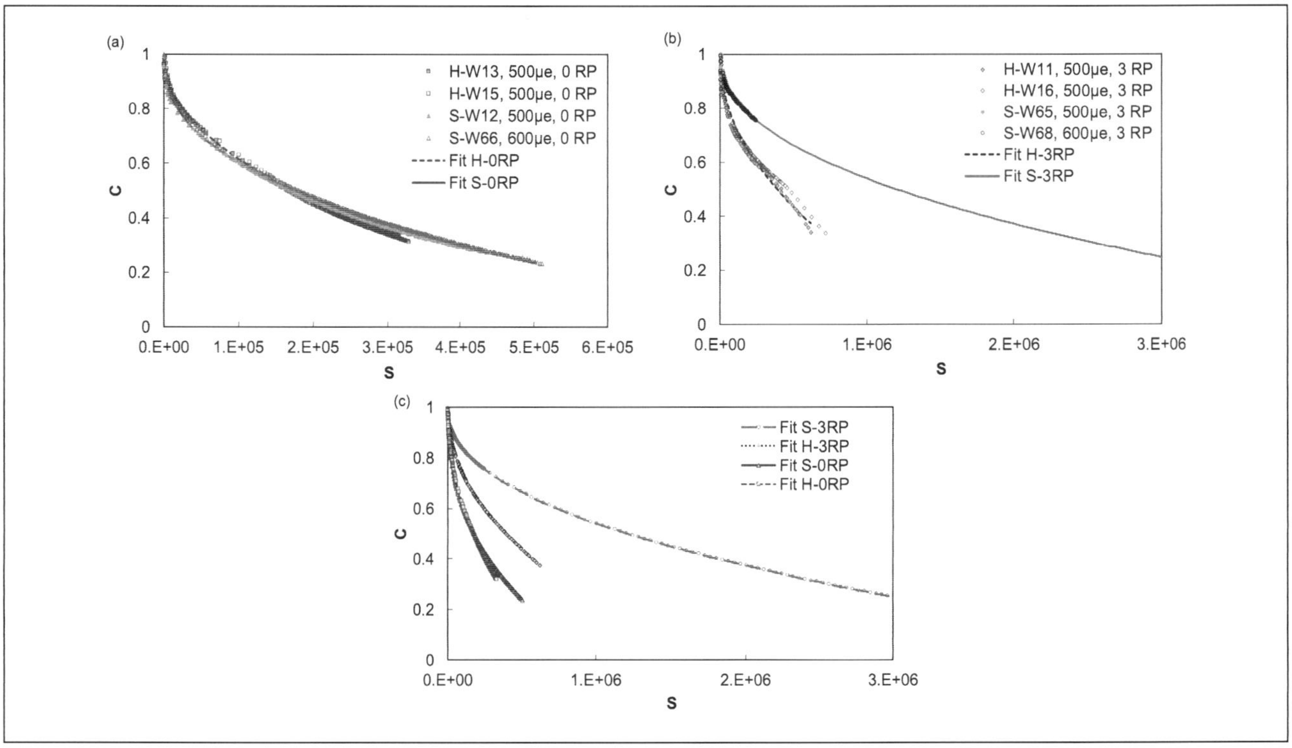

Figure 4. Damage characteristic curves: (*a*) continuous loading cases; (*b*) intermittent loading cases only; and (*c*) best fit for all cases.

that the effect of other, non-damage mechanisms (taken to be healing in the authors' interpretation) occur during these tests. In the case of the haversine rest period tests this effect is not apparent in observing the evolution of stiffness ratio with cycles (Figure 3) because under a phenomenological perspective of the test data, both rest and the continuous loading cases have the same strain magnitude. However, in developing the damage function, the fact that stress is greater in the intermittent haversine loading test than it is in the continuous loading haversine test is explicitly accounted for and in this case the presence of some mechanism that slows down the overall rate of damage can be detected. The use of this apparent damage function provides some additional insights into the behaviors of the intermittent haversine test as well. From Figure 4*b* the slope of the damage function (dC/dS) in the intermittent test case becomes slightly steeper as the tests approach failure. This pattern is different than what is observed from the continuous loading tests where the slope of the damage function decreases as the tests approach failure. Based on this increased slope, and the increased rate of stiffness ratio reduction in Figure 3, it is inferred that at some point the damage volume becomes too high for healing mechanisms to counterbalance and at this point damage accumulates relatively quickly.

Since the constitutive model (as presented here) does not explicitly account for these mechanisms (and the ones occurring in the intermittent sinusoidal testing) they become smeared into the resultant damage curve. The manifestation of these healing behaviors differs for the sinusoidal and haversine cases, most likely because the pattern of loading is different and more severe in the haversine case as was already noted, and so the resultant benefits from healing differ. The practical significance of this observation is that one should carefully apply apparent damage functions to only loading conditions similar to the ones used when developing the apparent damage function. This means that the apparent damage function resulting from the intermittent rest period test should ideally only be used to predict behaviors from other intermittent rest period, haversine loading conditions.

To demonstrate this fact more explicitly, fatigue simulations are carried out for sinusoidal and haversine strain histories using Equation 8. To simulate a sinusoidal loading condition (with or without rest periods) the value of β is set equal to zero and strain values are adjusted until the number of predicted cycles to failure is approximately 1,000, 10,000, and 100,000 cycles. The simulation results are shown in Figure 5.The results are compiled based on both the average values and probabilistically taking the variation in $|E^*|$, *C-S* curves, and failure conditions in the characterization process. The probabilistic assessment is done following the method given elsewhere (*18*), and involves using 100,000 standard Monte Carlo simulations with normal and uniform distributions for the

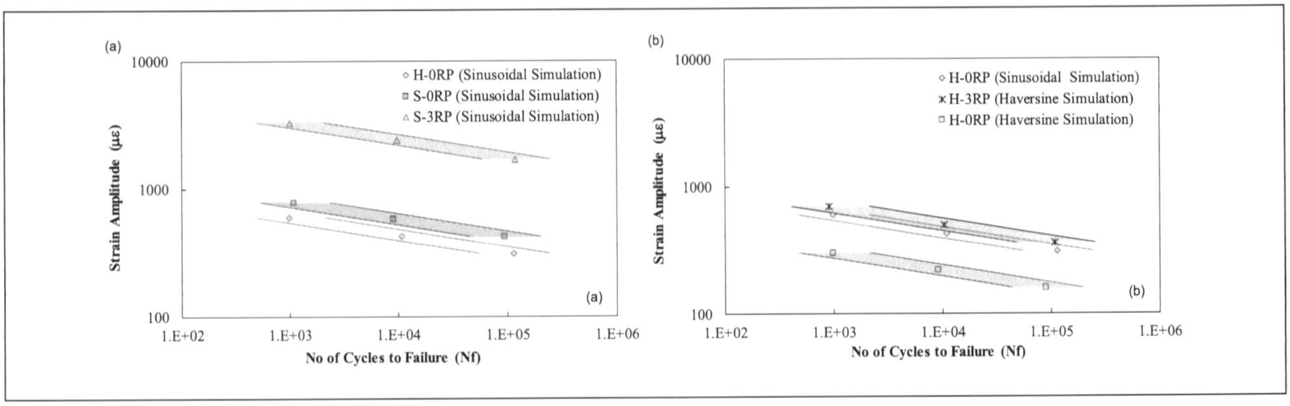

Figure 5. Fatigue life plots based on damage parameters from continuous and intermittent testing: (*a*) sinusoidal strain simulation; and (*b*) haversine strain simulation using only haversine-based characterization tests.

model variables. For this paper, the mean predictions are shown along with the 95% confidence intervals from the simulations (shaded areas in Figure 5) to provide a perspective and indication of the statistical significance of the comparisons.

Figure 5*a* shows the sinusoidal input fatigue predictions. There are two important things to observe from this figure as it relates to interpretation of fatigue test data under different loading histories. First, the models from H-0RP and S-0RP yield different relationships between strain and failure cycles. This difference is outside the 95% confidence interval based on the model uncertainty and suggests a systematic bias from the H-0RP based model. This difference is due to the differences in the failure limit identified from the two loading modes. As was mentioned earlier, the failure criterion adopted for the analysis here was based on the *C* value when laboratory experiments failed and that H-0RP experiments showed a C at failure that was higher than the S-0RP tests. Since the S-0RP experiments are closer to the simulation conditions (i.e., sinusoidal) it is likely that the predictions from that model are the most accurate. Application of a failure criterion that is independent of the mode of loading would likely solve this inconsistency and result in nearly identical predictions from both H-0RP and S-0RP models (*27*). Such a failure criterion can be more cumbersome to use, and in lieu of accepting this complication one would need to carefully consider the types of loading that are to be simulated and choose a test method similar to this condition. The second observation from this figure is that using a damage model that includes intermittent rest periods results in predicted failure cycles that are substantially greater than those predicted when continuous loading based functions are used. Only the model from the S-3RP tests is used to simulate the effect of intermittent rest periods for sinusoidal variations since the loading pattern in the experiment is similar to the one being simulated. Recall that the experiments could not complete in a reasonable amount of time. The simulations on the other hand suggest that the effect of rest period in sinusoidal loading is such that tests with rest periods can be carried out at strain levels approximately four times higher than tests without rest periods and still achieve similar cycles to failure.

Figure 5*b* shows simulation results for the haversine cases and the sinusoidal simulations based on the H-0RP model (labeled as H-0RP [Sinusoidal Simulation]). This data series is identical to the same named series in Figure 5*a* and is shown to explain the effect of the mean strain relaxation effect described earlier. Simulations for haversine input are carried out by setting β in Equation 8 equal to one. First, however, compare the data series showing haversine simulations. The difference between the H-3RP and H-0RP series reflects the actual beneficial role of healing in the case of intermittent rest periods under haversine loading. The effect is substantially less than the sinusoidal case, which is a result of the fact that repeated haversine at a given strain magnitude loading subjects the sample to overall greater stresses than repeated sinusoidal loading. Now, compare the H-3RP haversine simulation with the H-0RP sinusoidal simulation. This comparison is very similar to the conditions that existed in the experimental study. Recall that, in the H-0RP test, while the strain was always tensile the stress (and pseudostrain) is tension-compression. Thus, the pseudostrain energy that drives damage in these tests is more similar to the case where β is zero. The prediction comparison confirms what was observed in the experiment (similar cycles to failure for H-0RP and H-3RP experiments), but provides more direct insight into the causes. In this case the model suggests that the healing that takes place in the intermittent test is counteracted by a more severe stress condition. The model based interpretation also suggests that the benefits of the healing that takes place in intermittent haversine loading is less than in sinusoidal loading. Knowing that these two counterbalancing mechanisms exist provides a more complete perspective on the fatigue tests that may be necessary since neither simple haversine nor sinusoidal loading represent the conditions under which AC is subjected in real pavements.

Conclusions

This paper presented research to evaluate the effect of waveform pattern (sinusoidal versus haversine) and rest period (continuous versus intermittent) on the laboratory fatigue performance of an AC mixture. Dynamic modulus tests were conducted to determine the viscoelastic properties of the tested AC mixture. Controlled on-specimen displacement uniaxial fatigue tests were conducted with different waveforms and with and without rest periods. The uniaxial fatigue test results were then further processed and used to characterize the damage behavior in accordance with the S-VECD model. The following conclusions are made based on the outcomes from this study:

- For continuous uniaxial fatigue tests, a haversine strain input results in a stress response that begins as haversine but transitions to a sinusoidal waveform. This effect is due to the material's viscoelastic behavior. On the other hand, the sinusoidal stress waveform remained the same throughout the test.
- For intermittent uniaxial fatigue tests with a 3 s rest period, the stress waveforms maintained the same overall shape throughout testing, suggesting that the applied rest periods were sufficiently long that viscoelastic memory effects were nearly completely faded between load cycles.
- The rest period seems to increase the fatigue life remarkably when the continuous and intermittent sinusoidal fatigue tests are compared. However, for haversine fatigue tests, the rest period seems to have less effect.
- Fatigue life simulations confirmed the above conclusions and showed that the intermittent sinusoidal fatigue tests have a superior fatigue life followed by the intermittent haversine fatigue tests. For continuous loading conditions, the haversine fatigue tests showed the lowest fatigue life, which is related to how failure is defined in the simulations.

In addition to trying to elucidate (or speculate on) the mechanisms occurring in haversine and sinusoidal loading histories with or without rest periods, a key outcome of the testing and analysis here is to demonstrate that empirical fatigue laws are strongly dependent upon the exact nature of loading. As such, one must exercise caution when simply adopting and applying these empirical laws to perform pavement structural fatigue analysis since embedded into the function will be assumptions of loading type (sinusoidal or haversine) and rest period (continuous or intermittent).

Author Contributions

The authors confirm contribution to the paper as follows: study conception and design: Waleed Zeiada, Shane Underwood, and Mena Souliman; data collection: Waleed Zeiada, and Padmini Gudipudi; analysis and interpretation of results: Waleed Zeiada, Shane Underwood, Padmini Gudipudi, and Mena Souliman; draft manuscript preparation: Waleed Zeiada, Shane Underwood, Padmini Gudipudi, and Mena Souliman. All authors reviewed the results and approved the final version of the manuscript.

References

1. Tangella, S. R., J. Craus, J. A. Deacon, and C. L. Monismith. *Summary Report on Fatigue Response of Asphalt Mixtures*. Report No. SHRP-A-312, Institute of Transportation Studies, University of California, Berkley, 1990.
2. Castro, M., and J. A. Sanchez. Fatigue and Healing of Asphalt Mixtures: Discriminate Analysis of Fatigue Curves. *Journal of Transportation Engineering, ASCE*, Vol. 132, No. 2, 2006, pp. 168–174.
3. Bonnaure, F., A. Huibers, and A. A. Boonders. Laboratory Investigation of the Influence of Rest Periods on the Fatigue Characteristics of Bituminous Mixes. *Journal of the Association of Asphalt Paving Technologists*, Vol. 51, 1982, pp.104–128.
4. Raithby, K. D., and A. B. Sterling. The Effect of Rest Periods on the Fatigue Performance of Hot-Rolled Asphalt under Reversed Axial Loading. *Proceedings of the Association of Asphalt Paving Technologists (AAPT)*, Vol. 39, 1970, pp. 134–147.
5. Pronk, A. C., M. R. Poot, M. M. J. Jacobs, and R. F. Gelpke. Haversine Fatigue Testing in Controlled Deflection Mode: Is It Possible? Presented at 89th Annual Meeting of the Transportation Research Board, Washington, D.C., 2010.
6. Mamlouk, M. S., M. I. Souliman, W. A. Zeiada, and K. E. Kaloush. Refining HMA Beam Fatigue Testing Conditions. *ASTM Journal of Advances in Civil Engineering Materials*, Vol. 1, No. 1, 2012, pp. 1–13.
7. Mamlouk, M., and M. I. Souliman. Reducing Inconsistency of HMA Flexure Fatigue Testing. *Journal of Materials in Civil Engineering ASCE*, Vol. 28, No 2, 2016, p.04015131.
8. Witczak, M., M. Mamlouk, M. Souliman, and W. A. Zeiada. *NCHRP Report 762: Laboratory Validation of an Endurance Limit for HMA Pavements*. Transportation Research Board of the National Academies, Washington, D.C., 2013.
9. Monismith, C. L., K. E. Secor, and E. W. Blackner. Asphalt Mixture Behavior in Repeated Flexure.

Proceedings of the Association of Asphalt Paving Technologists (AAPT), Vol. 30, 1961, pp. 188–222.
10. Daniel, J. S., and Y. R. Kim. Laboratory Evaluation of Fatigue Damage and Healing of Asphalt Mixtures. *Journal of Materials in Civil Engineering*, Vol. 13, No. 6, 2001, pp. 434–440.
11. Kim, B., and R. Roque. Evaluation of Healing Property of Asphalt Mixtures. *Transportation Research Record: Journal of the Transportation Research Board*, 2006. 1970: 84–91.
12. Al-Rub, R. K. A., M. K. Darabi, D. N. Little, and E. A. Masad. A Micro-damage Healing Model that Improves Prediction of Fatigue Life in Asphalt Mixes. *International Journal of Engineering Science*, Vol. 48, No. 11, 2010, pp. 966–990.
13. Carpenter, S., and S. Shen. Dissipated Energy Approach to Study Hot-mix Asphalt Healing in Fatigue. *Transportation Research Record: Journal of the Transportation Research Board*, 2006. 1970: 178–185.
14. Roque, R., R. Simms, Y. Chen, C. Koh, and G. Lopp. *Development of Test Method That Will Allow Evaluation and Quantification of the Effects of Healing on Asphalt Mixture*. University of Florida, FL, 2012.
15. Khaki, A. M., A. Samdzadeh, and M. Jebalbarezi. Study of Loading Waveform, Loading Duration, Rest Period and Stress Level on Fatigue Life of Asphalt Mixtures. *Engineering Solid Mechanics*, Vol. 3, 2015, pp. 93–102.
16. Raithby, K. D., and A. B. Sterling. The Effect of Rest Periods on the Fatigue Performance of Hot-Rolled Asphalt under Reversed Axial Loading. *Proceedings of the Association of Asphalt Paving Technologists*, Vol. 39, 1970, pp. 134–147.
17. ADOT Standard. *Standard Specifications for Road and Bridge Construction*, Arizona Department of Transportation, Arizona, 2008.
18. Gudipudi, P. P., and B. S. Underwood. Reliability Analysis of Fatigue Life Prediction from the Viscoelastic Continuum Damage Model. *Transportation Research Record: Journal of the Transportation Research Board*, 2016. 2576: 91–99.
19. Underwood, B. S., Y. R. Kim, and M. N. Guddati. Characterization and Performance Prediction of ALF Mixtures Using a Viscoelastoplastic Continuum Damage Model. *Journal of the Association of Asphalt Paving Technologists*, Vol. 75, 2006, pp. 577–636.
20. Zeiada, W. A., M. I. Souliman, K. E. Kaloush, and M. Mamlouk. Endurance Limit for HMA based on Healing Concept Using Uniaxial Tension-Compression Fatigue Test. *Journal of Materials in Civil Engineering*, Vol. 26, No. 8, 2013, p.04014036.
21. Schapery, R. A. Correspondence Principles and a Generalized J Integral for Large Deformation and Fracture Analysis of Viscoelastic Media. *International Journal of Fracture*, Vol. 25, 1984, pp. 195–223.
22. Schapery, R. A. A Theory of Mechanical Behavior of Elastic Media with Growing Damage and other Changes in Structure. *Journal of Mechanics and Physics of Solids*, Vol. 38, 1990, pp. 215–253.
23. Chehab, G. R., Y. R. Kim, R. A. Schapery, M. W. Witczak, and R. Bonaquist. Time-Temperature Superposition Principle for Asphalt Concrete Mixtures with Growing Damage in Tension State. *Journal of the Association of Asphalt Paving Technologists*, Vol. 72, 2002, pp. 559–593.
24. Underwood, B., and W. A. Zeiada. Characterization of Microdamage Healing in Asphalt Concrete with a Smeared Continuum Damage Approach. *Transportation Research Record: Journal of the Transportation Research Board*, 2014. 2447: 126–135.
25. Underwood, B., C. Baek, and Y. Kim. Simplified Viscoelastic Continuum Damage Model as Platform for Asphalt Concrete Fatigue Analysis. *Transportation Research Record: Journal of the Transportation Research Board*, 2012. 2296: 36–45.
26. Safaei, F., J. Lee, L.A. H. Nascimento, C. Hintz, and Y.R. Kim. Implications of Warm-Mix Asphalt on Long Term Oxidative Aging and Fatigue Performance of Asphalt Binders and Mixtures. *Road Materials and Pavement Design*, Vol. 15, 2014, pp. 45–61.
27. Sabouri, M., and Y. Kim. Development of a Failure Criterion for Asphalt Mixtures under Different Modes of Fatigue Loading. *Transportation Research Record: Journal of the Transportation Research Board*, 2014. 2447: 117–125.

The Standing Committee on Structural Requirements of Asphalt Mixtures (AFK50) peer-reviewed this paper (18-03274).

Article

TRR
JOURNAL OF THE TRANSPORTATION RESEARCH BOARD

Transportation Research Record
2018, Vol. 2672(28) 462–470

DOI: 10.1177/0361198118787389
journals.sagepub.com/home/trr
SAGE

Practical Method to Determine the Effect of Air Voids on the Dynamic Modulus of Asphalt Mixture

Xinjun Li[1] and Jack Youtcheff[2]

Abstract

This study presents a practical method for estimating the effect of air voids on the dynamic modulus of asphalt mixture. Dynamic modulus was predicted for mixes with a large range of air void contents using the construction mix volumetric and binder rheological data from 10 accelerated loading facility (ALF) lanes, following the Witczak and Hirsch methods. A large variety of plant-produced and laboratory-prepared mixtures, including hot- and warm-mix asphalt (HMA and WMA), reclaimed asphalt pavement, and recycled asphalt shingles, was tested for dynamic modulus at different air void contents. The experimentally measured and normalized correction factors were found to be more dependent on test temperature than the frequency. The predicted correction factors were found to match with the experimental data at lower temperature but to be clearly lower at high temperature. A set of correction factors for each test temperature is recommended to practitioners correcting dynamic modulus with variation in air voids in asphalt pavement.

NCHRP Project 9-19 Superpave Support and Performance Models Management recommended dynamic modulus as one of the simple performance tests to be incorporated in the Superpave volumetric mixture design process (*1*). Mechanistic-empirical pavement design uses dynamic modulus as the key parameter in flexible pavement design that controls permanent deformation and fatigue cracking (*2*). It is fair to conclude that dynamic modulus is a key engineering property of asphalt mixture for use in routine practices, including mixture design and pavement structural design.

Numerous research efforts based on empirical or theoretical methods were made in the past decades to experimentally measure or numerically predict the dynamic modulus of asphalt mixtures. Various testing methods to directly or indirectly measure the dynamic modulus were developed from these efforts. One of the most important outcomes of these research activities is the development of a standard dynamic modulus test protocol, as is now widely adopted by state highway agencies and industry (*3*). This test protocol calls for the use of axial compression on 100-mm diameter by 150-mm tall specimens that are cored from gyratory compacted mixtures (*4*).

One of the more critical issues related to the standardized dynamic modulus test is the specimen size when it comes to the characterization of as-built pavements. As most construction lifts are only a few inches thick, preparation of the standard dynamic modulus test specimens (100 mm by 150 mm) is not possible. To address this problem, Kim et al developed an analytical solution for the dynamic modulus of asphalt mixtures tested in the indirect tension (IDT) mode using the theory of linear viscoelasticity (*5*). The IDT mode dynamic modulus test method employs specimens as thin as 38 mm, making it a promising alternative test for measuring the dynamic modulus of field-compacted mixtures. This modification necessitates test specimens with smooth and parallel surfaces to ensure that a quality measurement can be achieved with this IDT test setup. The IDT test method requires an extra test fixture, and the loading mode is not compatible with the existing asphalt mixture performance tester (AMPT), as is now widely adopted in the asphalt pavement community for its value performing various performance tests and its role in mechanistic-empirical pavement distress prediction (*6*). As a solution, a recent research effort proposed a dynamic modulus test method using small-scale cylindrical specimens that can be manufactured from both laboratory or field-compacted asphalt mixture (*7*). One important advantageous feature of this method is that the small-scale specimen (38-mm diameter by 110-mm height) can be easily horizontally drilled from overlay pavement that is as thin as 2 in. More critically, this proposed testing method, including sample preparation and loading mode, is satisfactorily

[1]SES Group and Associates, Inc., Turner-Fairbank Highway Research Center, McLean, VA
[2]Federal Highway Administration, Turner-Fairbank Highway Research Center, McLean, VA

Corresponding Author:
Address correspondence to Xinjun Li: xinjun.li.ctr@dot.gov

compatible with the AMPT. Consequently, the small-scale dynamic modulus test (and cyclic fatigue test) method was adopted by other researchers and is currently under development to form an AASHTO testing standard (*8*, *9*).

It should be noted that significant difference may exist when performing dynamic modulus testing with the laboratory-compacted and field cored specimens. As required in AASHTO TP79, only laboratory manufactured specimens meeting the target air voids will be tested for dynamic modulus. However, for the specimens sampled from field-compacted pavement, it would be very difficult, if not impossible, to require the uniformity of the volumetric properties within the specimen replicates. This can be easily attributed to the construction variation within the pavement, especially when it comes to the relatively thin pavement overlay. In other words, compared with the laboratory generated specimens, the dynamic modulus testing specimens from field-compacted mixtures have a much bigger range of volumetric properties, including the air void content. The variation of the volumetric properties should be taken into consideration when utilizing the dynamic modus data tested with field cores to accurately evaluate the existing mixture and pavement structure. Two dynamic modulus predictive models, the Witczak model (*10*) and the Hirsch model (*11*), are two useful tools for calculating the dynamic modulus of mix when the mix volumetrics and the binder shear modulus are known. The Witczak method is a nonlinear regression model with more than 10 variables based on the 7,400 data points from 346 mixtures, whereas the Hirsch model is more of an analytical method incorporating the binder modulus, voids in the mineral aggregate (VMA), and voids filled with asphalt (VFA). Although providing satisfactory prediction of dynamic modulus of asphalt mixture, both models ask for a fair amount of volumetric and binder data, which can be very time consuming or costly. This is even more obvious for the field-compact mix in which the volumetric parameter most concerned is the air void content, although variations in gradation, fines content, mineral filler, and binder content have commonly been noted during plant material production. Therefore, it is of great interest to develop a quick and practical method to determine the effect of the air void on the dynamic modulus when performing the dynamic modulus testing on the AMPT with field-compacted mix.

Objectives

The objectives of this paper are to present a practical procedure for estimating the impact of air void content on the dynamic modulus. Based on a large dynamic modulus testing database of various asphalt mixtures tested at the FHWA, this approach is intended to provide the practitioners with guidance for selecting a factor for correcting the dynamic modulus tested from field specimens with different air voids.

Materials Tested

FHWA's Turner-Fairbank Highway Research Center (TFHRC) has been working with prototype methodologies of dynamic modulus testing for over 15 years. Since then, more than 100 asphalt mixtures have been tested. TFHRC has been involved in the verification, improvement and development, and implementation of dynamic modulus testing methods. The work was initially conducted on a universal servo-hydraulic loading system and later transferred to the more unified AMPT equipment.

In generating a robust database for this study, a large variety of materials was considered, which included both plant-produced mixture and laboratory-prepared mixtures, produced as hot-mix asphalt (HMA) and warm-mix asphalt (WMA), and various percentages of reclaimed asphalt pavement (RAP), and recycled asphalt shingles (RAS). Nominal maximum aggregate sizes (NMAS) and different air void contents were the other variables tested. Both the field-compacted and laboratory-compacted mixtures were included in this study. It should be noted that the database would be more robust if asphalt mix with polymer modified binder, such as the widely-used Styrene-Butadiene-Styrene (SBS), modified asphalt. Unfortunately, the polymer modified binder and mix was not included in this study because of the limitations of material availability.

Field Cores

The field-compacted samples were cored from the 10 test lanes built in 2013 at the FHWA accelerated loading facility (ALF) in McLean, Virginia. The objective of the 2013 ALF experiment is to establish realistic boundaries for high-RAP and RAS mixtures, employing WMA technologies based on percent binder replacement and binder grade changes when combined together. The mixtures in the 2013 ALF study were Superpave 12.5 mm NMAS designed with 65 gyrations. Table 1 presents the volumetrics, binder properties, and structural properties of the 10 lanes. Additional details of the 2013 ALF lanes can be found elsewhere (*12*).

It should be noted that all 10 ALF lanes were designed to have 4-in. asphalt mix and the asphalt mix was paved and compacted with two 2-in. lifts. Production mix was sampled behind the paver. Field cores (6-in. diameter) were sampled a few days after construction. Six field cores were randomly chosen for performance testing. Two small-scale specimens (38 mm in diameter and 110 mm long) were generated from the top and bottom lift separately by horizontally drilling, generating four small-scale specimens from each field core, two from top lift and two from bottom lift. Figure 1 illustrates how the small-scale specimens were cored and drilled from the ALF field cores.

The small-scale specimens from all 10 lanes were tested for bulk specific gravity following saturated surface dry (SSD) method (AASHTO T166). Air void content was

Table 1. Properties for Each ALF Lane

Lane	L1	L2	L3	L4	L5	L6	L7	L8	L9	L11
G_{mm}	2.750	2.723	2.744	2.740	2.743	2.741	2.740	2.729	2.736	2.722
Air voids	4.3	4.3	3.3	4.4	5.2	3.6	4.1	4.9	3.7	4.9
G_{mb}	2.633	2.606	2.653	2.620	2.601	2.642	2.630	2.597	2.601	2.588
P_b	5.08	5.07	4.98	4.95	4.60	4.91	4.91	4.95	4.98	4.89
VMA	16.1	16.1	14.6	15.6	15.9	14.9	15.3	16.4	15.1	16.5
VFA	73.4	73.6	77.3	72.0	67.3	75.7	73.6	70.3	75.5	70.5
DB ratio	1.04	1.15	1.28	1.12	1.38	1.17	1.25	1.14	1.15	1.18
Core AV	7.2	6.0	8.5	7.5	7.4	7.2	7.9	7.2	5.2	8.0
Specified PG	64–22	58–28	64–22	64–22	64–22	64–22	58–28	58–28	64–22	58–28
Virgin PG (tank)	68–27	–	68–27	66–26	66–26	66–26	–	62–28	65–24	64–30
Recovered PG	72–23	–	91–11	71–22	86–12	71–22	–	75–18	77–18	86–12
Avg. asphalt thick. (in.)	5.0	4.8	4.5	4.9	3.9	5.0	4.3	4.6	4.0	3.9

Note: Avg. = average; Pb = asphalt content; AV = air voids; PG = performance grade.

Figure 1. Small-scale field specimens cored from ALF.

calculated and presented in Table 2 and plotted in Figures 2 and 3 for the top and bottom lifts, respectively. Also presented in these figures are the air voids of the full-size (150-mm diameter) cores sampled for the acceptance after the construction but separately tested for bulk specific gravity by lift. It may be beneficial to note that the error bar shown in the plot represents one standard deviation of tested replicates.

The air voids data from both the small-scale and full-size specimens on the top lift show that all lanes meet the target air void range (7% ± 1%) when taking the standard deviation into consideration, indicating satisfactory compaction was applied to the top lift during construction. No significant air void difference was found between the small-scale and full-size specimens. However, the bottom lift specimens tell a slightly different story. Although in average no significant difference was found between the small-scale and full-size specimens, the air void data do indicate that three lanes could be over compacted on the bottom asphalt lift. It was also found that the bottom lift has an obviously higher variation with the air void data. For example, specimens from Lane 9 were found to have the highest, 9.21%, and the lowest, 3.01%, air voids content, making a wide range of more than 6% air void difference. This significant variation could be attributed to the fact that compaction temperature was purposely set very low to evaluate the WMA technology employed in Lane 9. However, Lane 1, a typical HMA with 0% recycled material, was found to have an air void range as wide as 4% (5–9%) in the field bottom lift specimens. It is beneficial to note that each ALF lane was paved only after the volumetric criteria of the mix was approved from a prepaved test strip. The wide range of the air voids from the field samples in a well-constructed pavement illustrates the significance of this study.

Laboratory-Tested Materials

Table 3 lists the information of mixtures that were tested for the dynamic modulus in the laboratory. A total of 16 different mixtures, including plant-produced and laboratory-prepared mixes, were tested in this study. Each mix was gyratory compacted with two or three target air void contents for the testing specimens, as shown in the last column in Table 3. For example, the mix "ALF 2013 L1" was first gyratory

Table 2. Air Void Content of the ALF Field Specimens

	Top lift				Bottom lift			
	Small-size specimen		Full-size specimen		Small-size specimen		Full-size specimen	
Lane	Avg. VA (%)	SD (%)	Avg. VA (%)	SD (%)	Avg. VA (%)	SD (%)	Avg. VA (%)	SD (%)
L1	7.36	0.92	7.23	0.58	6.26	0.80	5.76	0.20
L2	7.19	1.20	7.58	1.82	4.76	1.25	5.48	1.63
L3	6.95	1.01	7.93	0.93	7.15	1.22	7.70	1.42
L4	7.68	1.19	8.74	0.77	5.75	1.19	5.99	1.35
L5	6.60	0.73	7.55	0.30	6.81	1.30	6.88	0.68
L6	6.47	0.87	7.17	0.89	4.91	0.30	5.41	0.29
L7	7.61	1.08	8.25	0.86	6.55	1.35	7.63	1.48
L8	7.53	0.87	8.46	0.57	5.81	0.77	5.99	0.64
L9	6.97	1.52	7.19	1.39	5.48	2.73	5.29	2.50
L11	7.21	1.21	7.78	1.21	6.77	0.63	7.38	0.41

Note: Avg. = average; SD = standard deviation; VA = air voids.

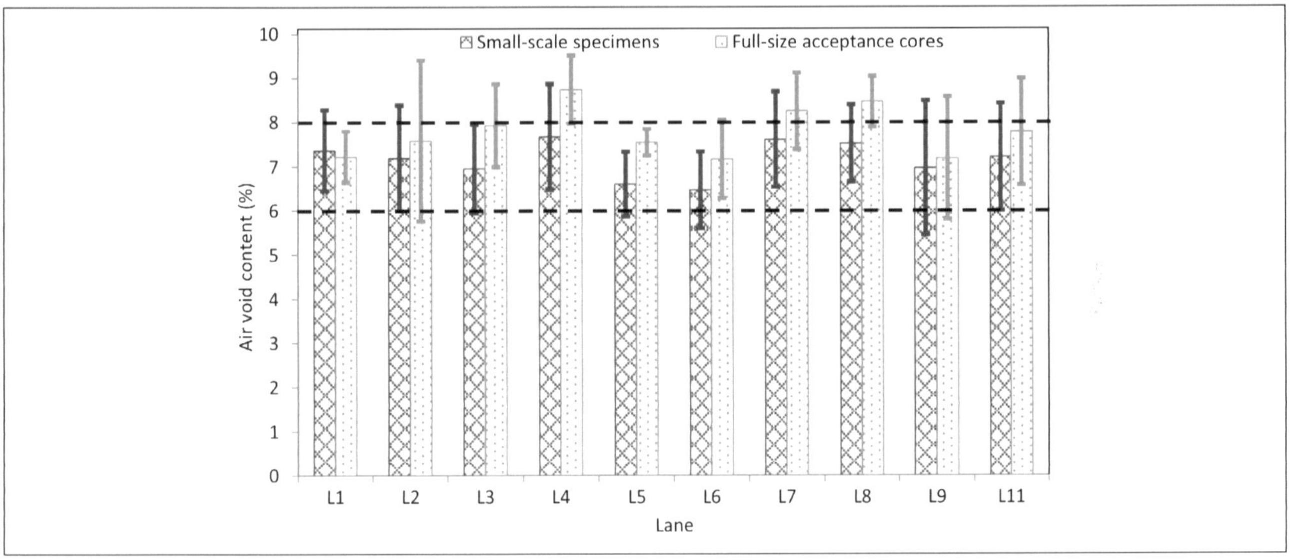

Figure 2. Air void content of field specimens from ALF top lift.

compacted to produce 150-mm by 185-mm cylinders with three different compaction efforts (gyrations). The 150-mm × 185-mm cylinders were then cored and cut to form 100-mm by 150-mm testing specimens. The 100-mm × 150-mm testing specimens were checked for the air void contents following the SSD method. Only the specimens that meet the target air void (4% ± 0.5%, 7% ± 0.5%, 10% ± 0.5%) were tested for the dynamic modulus. In this study, dynamic modulus was measured at 4.4°C, 21.1°C, 37.8°C, and 54.4°C, and frequencies of 20 Hz, 10 Hz, 5 Hz, 1 Hz, 0.5 Hz, and 0.1 Hz on three replicates for each mix at each air void content. The test was performed on an AMPT following the AASHTO TP79 test procedure.

Experimental Results And Data Analysis

Field Cores

The mix volumetric data, including the asphalt content, percentage of passing (remaining) on each sieve size, VMA, and VFA, were obtained for each ALF lane from the construction acceptance testing. The asphalt binder of each lane sampled from the mix production was characterized with rheological tests including the dynamic shear testing at different temperatures and frequencies. The dynamic modulus of the small-scale field sample was then predicted with the mix volumetric data and the binder shear modulus data at various

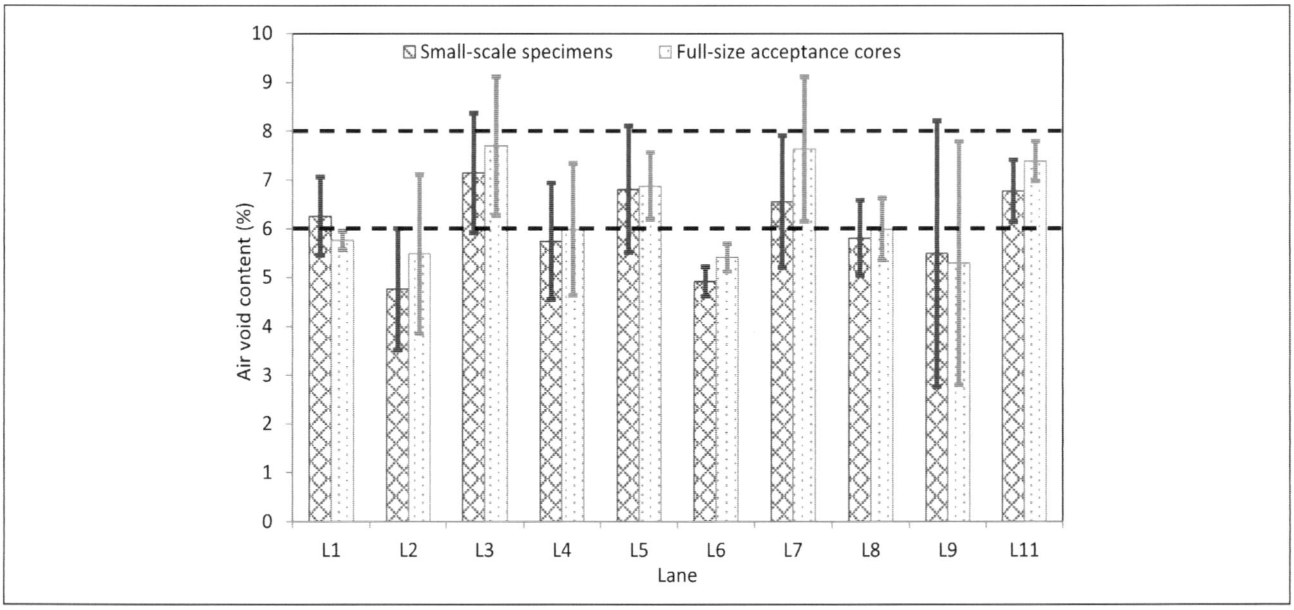

Figure 3. Air void content of field specimens from ALF bottom lift.

Table 3. Information of Laboratory-Tested Mixtures

Mixture ID	Source	Material type	NMAS (mm)	Binder PG	Binder content (%)	VMA (%)	RAP/RAS	HMA/ WMA	Target air void content (%)
ALF 2013 L1	FHWA	Plant mix	12.5	64–22	5.08	16.1	0% RAP	HMA	4, 7, 10
ALF 2013 L3	FHWA	Plant mix	12.5	64–22	4.98	14.6	6% RAS	HMA	4, 7, 10
ALF 2013 L4	FHWA	Plant mix	12.5	64–22	4.95	15.6	22% RAP	WMA	4, 7, 10
ALF 2013 L5	FHWA	Plant mix	12.5	64–22	4.60	15.9	44% RAP	HMA	4, 7, 10
WFL loose	WFLHD	Plant mix	12.5	64–28	5.60	15.9	20% RAP	HMA	4, 7, 10
WFL batch	WFLHD	Lab prepared	12.5	64–28	5.53	15.9	20% RAP	HMA	4, 7, 10
ALF 2016	FHWA	Plant mix	19	64–22	4.20	14.4	25% RAP	HMA	7, 9, 11
PBMD AB	FHWA	Lab prepared	12.5	64–22	4.20	15.0	22% RAP	HMA	7, 9
PBMD CDE	FHWA	Lab prepared	12.5	64–22	4.50	14.5	22% RAP	HMA	5, 7, 9
PBMD FG	FHWA	Lab prepared	12.5	64–22	4.90	14.7	22% RAP	HMA	5, 7
PBMD HI	FHWA	Lab prepared	12.5	64–22	3.80	14.1	22% RAP	HMA	7, 9
PBMD JKL	FHWA	Lab prepared	12.5	64–22	4.10	13.5	22% RAP	HMA	5, 7, 9
PBMD MN	FHWA	Lab prepared	12.5	64–22	4.40	13.7	22% RAP	HMA	5, 7
PBMD OP	FHWA	Lab prepared	12.5	64–22	3.20	12.9	22% RAP	HMA	7, 9
PBMD QRS	FHWA	Lab prepared	12.5	64–22	3.60	12.5	22% RAP	HMA	5, 7, 9
PBMD TU	FHWA	Lab prepared	12.5	64–22	3.90	12.8	22% RAP	HMA	5, 7

Note: WFL = West Federal Land Highway Division; PBMD = performance based mix design.

temperatures and frequencies following the Witczak and Hirsch methods.

For simplification purposes, the dynamic shear modulus data at three testing conditions, 7°C and 100 rad/sec (15.9 Hz) (designated as high stiffness), 19°C and 6.3 rad/sec (1.0 Hz) (designated as intermediate stiffness), and 58°C and 0.63 rad/sec (0.1 Hz) (designated as low stiffness), were used to predict the mix dynamic modulus at low temperature and high frequency, intermediate temperature and frequency, and high temperature and low frequency conditions, respectively. The air void data measured from all small-scale field samples are used as the input of air void content in the prediction. The mix dynamic modulus at 7% air void content, as widely accepted as the target compaction air voids for asphalt pavement construction, is predicted and used as a reference value to normalize the dynamic modulus predicted with other air void contents. A correction factor, as is the ratio of the dynamic modulus at one air void content to that at 7% air voids, is then calculated.

Results from the Witczak Model. Figure 4 plots the correction factors predicted with the Witczak method and normalized to

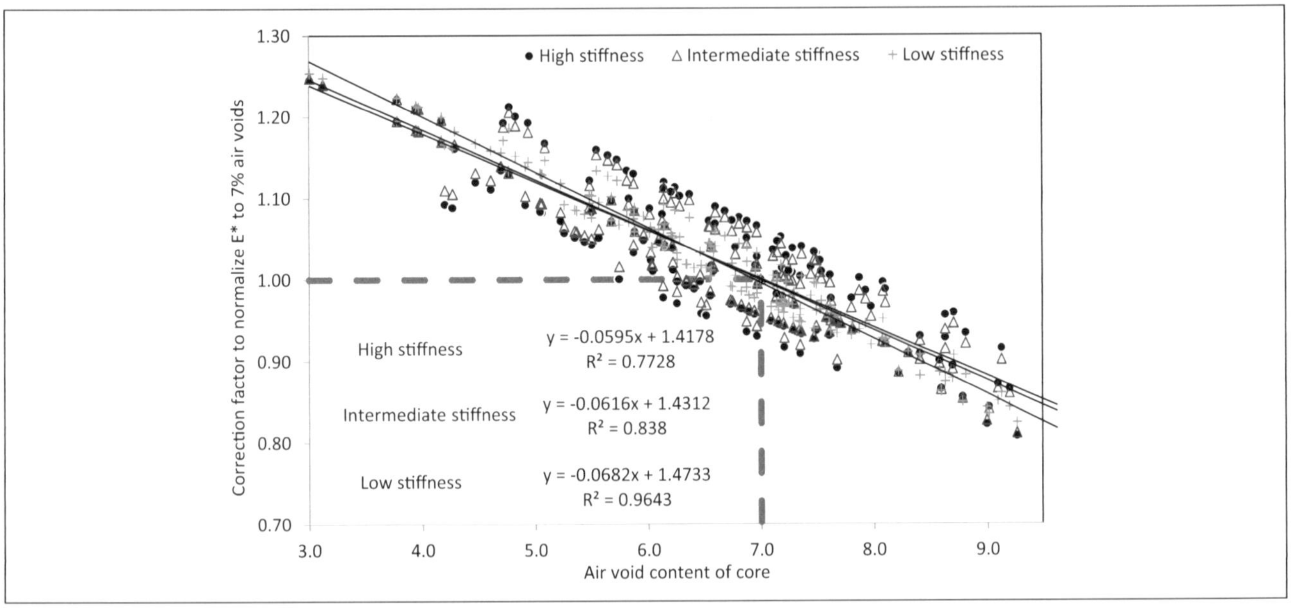

Figure 4. Normalized correction factor predicted with Witczak model.

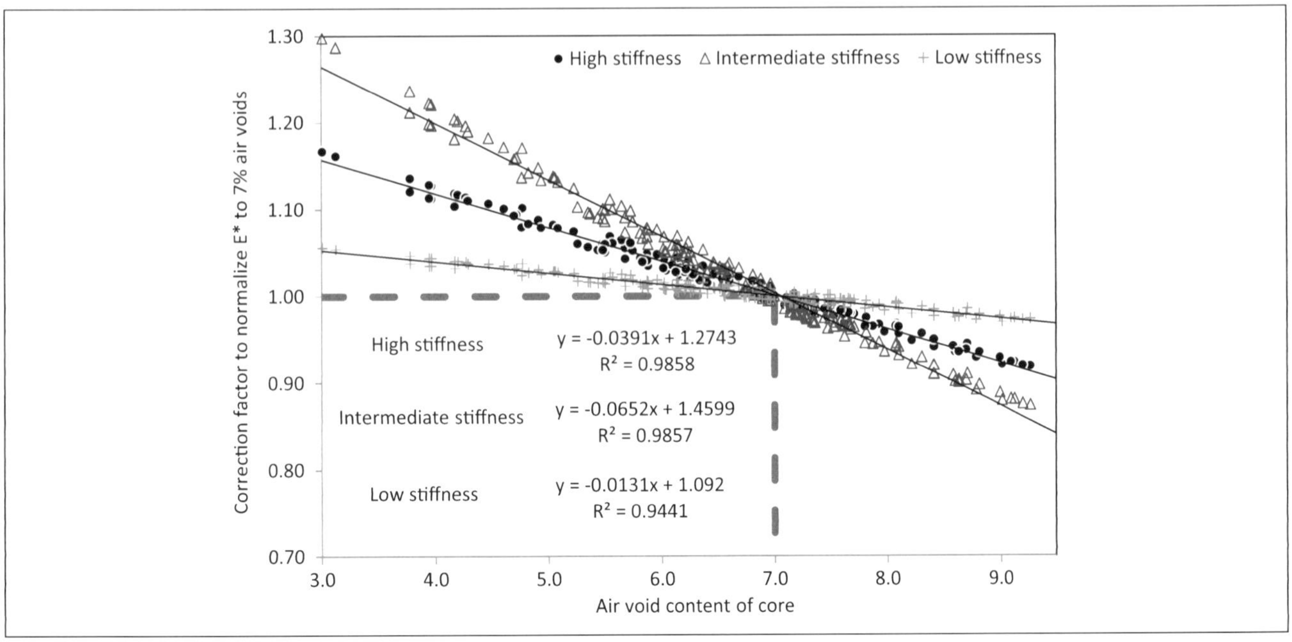

Figure 5. Normalized correction factor predicted with Hirsch model.

the 7% air void content as a function of the air void content. As aforementioned, binder shear modulus data at three different conditions (temperature and frequency), designed as high stiffness, intermediate stiffness, and low stiffness, represent three different conditions of the mix. As a result, the predicted mix dynamic modulus data are divided into three groups correspondingly. The normalized correction factors were found not only to decrease with the increase of the air void content, but to have very good linear regression with the air void content for all three groups. The slope of the linear regression indicates the change of dynamic modulus caused by 1% air void change. The predicted results illustrate that the mix dynamic modulus is going to change by 5.95%, 6.16%, and 6.82% when the asphalt binder is of high, intermediate, and low stiffness, respectively. In other words, 1% air void content change is going to bring 6%, 6.2%, and 6.8% change in dynamic modulus when the mix is at low (7°C), intermediate (21°C), and high (58°C) temperatures.

Table 4. Typical Correction Factor Normalized to 7% Air Voids

Mix	Avg. VA (%)	Temp (°C)	Frequency (Hz) 25	10	5	1	0.5	0.1	Avg. correction factor	COV (%)
ALF 2013 L1	4.1	4.4	1.13	1.14	1.14	1.15	1.16	1.20	1.15	2.0
	6.8		1.00	1.00	1.00	1.00	1.00	1.00	1.00	0.0
	10.2		0.82	0.81	0.81	0.80	0.80	0.79	0.80	1.4
	4.1	21	1.23	1.24	1.26	1.30	1.33	1.39	1.29	4.7
	6.8		1.00	1.00	1.00	1.00	1.00	1.00	1.00	0.0
	10.2		0.81	0.80	0.80	0.79	0.78	0.78	0.79	1.4
	4.1	37.8	1.33	1.38	1.42	1.49	1.52	1.57	1.45	6.3
	6.8		1.00	1.00	1.00	1.00	1.00	1.00	1.00	0.0
	10.2		0.78	0.78	0.78	0.78	0.78	0.78	0.78	0.3
	4.1	54.4	1.38	1.45	1.47	1.48	1.45	1.37	1.44	3.3
	6.8		1.00	1.00	1.00	1.00	1.00	1.00	1.00	0.0
	10.2		0.74	0.75	0.74	0.76	0.76	0.77	0.75	1.4
ALF 2013 L3	4.1	4.4	1.15	1.16	1.17	1.18	1.19	1.21	1.18	1.8
	7.1		1.00	1.00	1.00	1.00	1.00	1.00	1.00	0.0
	10.0		0.84	0.85	0.85	0.86	0.86	0.87	0.86	1.2
	4.1	21	1.24	1.26	1.28	1.31	1.33	1.37	1.30	3.7
	7.1		1.00	1.00	1.00	1.00	1.00	1.00	1.00	0.0
	10.0		0.89	0.90	0.91	0.92	0.93	0.94	0.91	1.7
	4.1	37.8	1.35	1.38	1.39	1.43	1.43	1.45	1.40	2.7
	7.1		1.00	1.00	1.00	1.00	1.00	1.00	1.00	0.0
	10.0		0.95	0.95	0.96	0.97	0.97	0.97	0.96	0.9
	4.1	54.4	1.40	1.44	1.45	1.46	1.46	1.48	1.45	1.8
	7.1		1.00	1.00	1.00	1.00	1.00	1.00	1.00	0.0
	10.0		0.96	0.98	0.99	0.99	0.99	0.96	0.98	1.5

Results from the Hirsch Model. Similarly to the Witczak model, dynamic modulus was predicted for all small-scale field specimens with the Hirsch model. The predicted correction factors that are normalized to the 7% air void content are plotted in Figure 5 as a function of air void content. Again, the normalized correction factors were found to decrease with the increase of the air voids, and very satisfactory linear trend lines were added to all three groups. The prediction results illustrate that 1% air void change alters dynamic modulus by 3.91%, 6.51%, and 1.31% at low, intermediate, and high temperatures.

Laboratory-Tested Materials

As mentioned above, full-size specimens (100 mm × 150 mm) from 16 different mixtures were laboratory manufactured and tested for dynamic modulus. As shown in Table 3, all tested mixtures include a target air void content of 7%. Similar to the predicted results, the experimentally measured dynamic modulus data are also normalized to the 7% air void content, and correction factors are calculated for each temperature and frequency. Table 4 presents the results for two mixtures, as is typical for all tested materials. Also listed in Table 4 is the averaged correction factor at one same temperature but at six different frequencies. It is of significance to have found that the normalized correction factors at the same temperature but different frequencies are very similar. Coefficient of variance (COV) is calculated for all mixtures at each testing temperature and a maximum COV value of 8.4% was found (mixture "PBMD CDE" 5% air voids and 54°C). This clearly indicates that the normalized correction factor is mainly determined by testing temperature, not the frequency when the material is tested for dynamic modulus.

Figure 6 plots the normalized correction factors for the mix ALF 2013 L1, as listed in Table 4, as a function of the air voids. In this plot, the correction factors are grouped by temperature. It can be found that very good linear regression may exist for the correction factor as a function of air voids for each temperature. Consequently, all normalized correction factors from all 16 mixtures were plotted together and presented in Figures 7 and 8 by temperatures (4.4°C and 21.1°C in one plot; 37.8°C and 54.4°C in the other plot). Both plots present very satisfactory linear correlation between the normalized correction factor and air void content. The regression results indicate that 6.1%, 7.1%, 8.4%, and 9.6% correction should be added to the dynamic

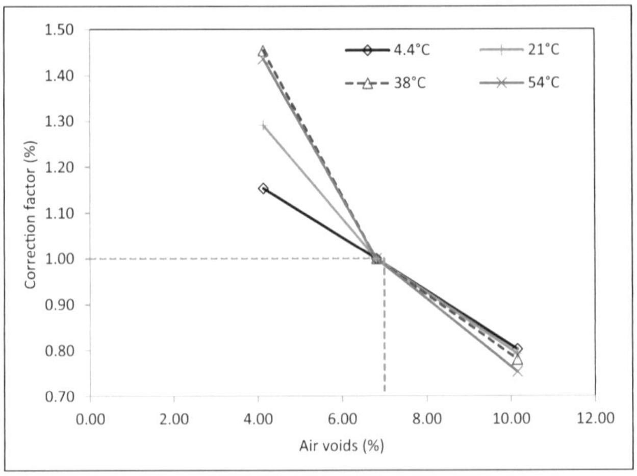

Figure 6. Typical measured correction factor versus air voids.

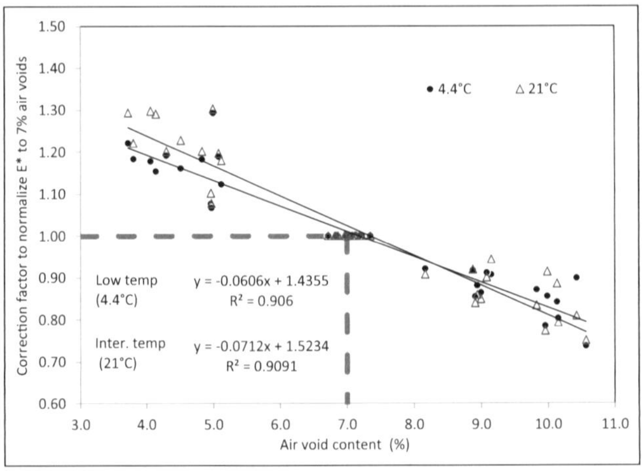

Figure 7. Measured correction factor at two lower temperatures versus air voids.

modulus with 1% air void content change when the mix is at 4.4°C, 21.1°C, 37.8°C, and 54.4°C, respectively.

Table 5 summarizes the normalized correction factors that are obtained from both prediction and laboratory testing in this study. It was found that at the lowest temperature (4.4°C /7.0°C) and the intermediate temperature (19°C /21°C), the correction factors predicted from both the Witczak and Hirsch models satisfactorily matched with that from the laboratory testing data. However, the predicted factors are significantly lower than the experimentally measured values when the temperature is very high (54°C /58°C). The experimentally obtained factors of 6.1%, 7.1%, 8.4%, and 9.6% are recommended to correct the dynamic modulus for 1% air void content change at the four test temperatures from low to high.

Summary And Conclusion

The following can be summarized and concluded:

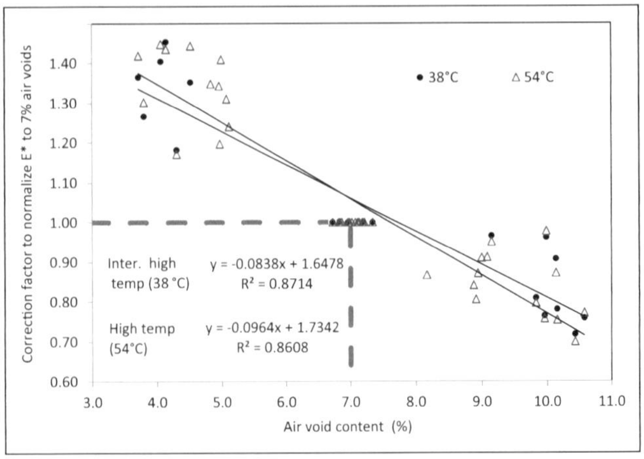

Figure 8. Measured correction factor at two higher temperatures versus air voids.

Table 5. Summarized Correction Factors from Different Methods

Temperature (°C)	Correction factor from each method: Witczak model (%)	Hirsch model (%)	Laboratory data (%)
4.4/7.0	6.0	3.9	6.1
19/21.1	6.2	6.5	7.1
37.8	NA	NA	8.4
54.4/58	6.8	1.3	9.6

1. FHWA ALF construction data show that a wide variation range of air void content exists in well-controlled asphalt pavement. This indicates the necessity and significance of a practical method to normalize the effect of the air void on the dynamic modulus to more accurately evaluate the existing pavement material.
2. Dynamic modulus with a large range of air void content was predicted for 10 ALF lanes using the construction mix volumetric data and binder rheological data, following the Witczak and Hirsch methods.
3. A large variety of materials, including both plant-produced mixture and laboratory-prepared mixture, both HMA and WMA, RAP (various percentage) and RAS, were tested for dynamic modulus at different air voids.
4. The experimentally measured correction factors were normalized to 7% air voids and found to be more dependent on the testing temperature than the frequency.
5. It was found that the correction factors from both prediction and experiment match very well at the two lower temperatures; the predicted factors are significantly lower than the experimentally measured values at high temperature.

6. The experimentally obtained factors of 6%, 7%, 8%, and 9.6% that are normalized to 7% air voids are recommended to correct the dynamic modulus for every 1% air void content change at the four test temperatures, from low to high, if users are expected to perform dynamic modulus testing and solely correct for air void content.

Acknowledgments

The authors want to thank Dr Nelson Gibson from the Transportation Research Board, formerly of FHWA, for his critical contribution to this study. Scott Parobeck and Frank Davis at TFHRC are also thanked for providing their expert attention in the performance data collection, specimen preparation, and laboratory testing.

Author Contributions

The authors confirm contribution to the paper as follows: study conception and design: Sean (Xinjun) Li, Jack Youtcheff; data collection: Sean (Xinjun) Li; analysis and interpretation of results: Sean (Xinjun) Li, Jack Youtcheff; draft manuscript preparation: Sean (Xinjun) Li. All authors reviewed the results and approved the final version of the manuscript.

References

1. Witczak, M. W., K. Kaloush, T. Pellinen, M. El-Basyouny, and H. Von Quintus. *NCHRP Report 465: Simple Performance Test for Superpave Mix Design*. Transportation Research Board of the National Academies, Washington, D.C., 2002.
2. Advanced Research Associates. *NCHRP 1-37A Project: 2002 Design Guide: Design of New and Rehabilitated Pavement Structures*. Transportation Research Board of the National Academies, Washington, D.C., 2004.
3. American Association of State Highway and Transportation Officials. *Determining the Dynamic Modulus and Flow Number for Asphalt Mixtures Using the Asphalt Mixture Performance Tester*. AASHTO TP79-15. American Association of State Highway and Transportation Officials, Washington, D.C., 2016.
4. American Association of State Highway and Transportation Officials. *Preparation of Cylindrical Performance Test Specimens Using the Superpave Gyratory Compactor*. AASHTO PP60-14. American Association of State Highway and Transportation Officials, Washington, D.C., 2016.
5. Kim, R., Y. Seo, M. King, and M. Momen. Dynamic Modulus Testing of Asphalt Concrete in Indirect Tension Mode. *Transportation Research Record: Journal of the Transportation Research Board*, 2004. 1891: 163–173.
6. FHWA and Transportation Pooled Fund Study TPF–5(178). *Implementation of the AMPT for Superpave Validation*, Washington, D.C., 2010.
7. Li, X., and N. Gibson. Using Small Scale Specimens for AMPT Dynamic Modulus and Fatigue Tests. *Journal of Association of Asphalt Paving Technologists*, Vol. 82, 2013, pp. 579–615.
8. Bowers, B., B. K. Difenderfer, and S. D. Difenderfer. Evaluation of Dynamic Modulus in Asphalt Paving Mixtures Utilizing Small-Scale Specimen Geometries. *Journal of the Association of Asphalt Paving Technologists*, Vol. 84, 2015, pp. 497–526.
9. Hintz, C. *NCHRP Idea 181: Development of Small Specimen Geometry for Asphalt Mixture Performance Testing*. Transportation Research Board of the National Academies, Washington, D.C., 2018.
10. Bari, J., and M. W. Witczak. Development of a New Revised Version of the Witczak E* Predictive Models for Hot Mix Asphalt Mixtures. *Journal of the Association of Asphalt Paving Technologists*, Vol. 75, 2006, pp. 381–423.
11. Christensen, D. W., T. Pellinen, and R. F. Bonaquist. Hirsch Model for Estimating Modulus of Asphalt Concrete. *Journal of the Association of Asphalt Paving Technologists*, Vol. 72, 2003, pp. 97–121.
12. Li, X., and N. Gibson. Comparison of Asphalt Mixture Performance Tester Fatigue Characteristics with Full Scale Pavement Cracking for Recycled and Warm Mix Asphalts. *Transportation Research Record: Journal of the Transportation Research Board*, 2016. 2576: 100–108.

The Standing Committee on Structural Requirements of Asphalt Mixtures (AFK50) peer-reviewed this paper (18-06331).

Article

TRR
JOURNAL OF THE TRANSPORTATION RESEARCH BOARD

Transportation Research Record
2018, Vol. 2672(28) 471–480

DOI: 10.1177/0361198118797826
journals.sagepub.com/home/trr
SAGE

Effect of Mix Design Variables on Thermal Cracking Performance Parameters of Asphalt Mixtures

Mirkat Oshone[1], Debaroti Ghosh[2], Eshan V. Dave[1], Jo Sias Daniel[1], Joseph M. Voels[2], and Shongtao Dai[2]

Abstract

To address asphalt pavement thermal cracking, researchers have developed performance-based evaluation tools for asphalt mixtures. A minimum fracture energy obtained from a disc-shaped compact tension test and Black space parameters determined by the stiffness and relaxation properties of asphalt mixtures are two such methods to ensure good thermal cracking resistance. Mix specifiers and producers strive to meet the requirements set by these performance-based criteria by adjusting their mix designs. However, there is a lack of information and consensus on the effect of mix design variables (such as binder grade and mix volumetrics) on thermal cracking performance of mixtures as it relates to fracture energy and Black space location. This study strives to fill this gap by quantifying the effect of: (1) recycled asphalt content, (2) effective binder content, (3) air voids, (4) asphalt film thickness, (5) voids in mineral aggregates, and (6) PG low and high temperature grades on thermal cracking resistance. A large dataset, 90 mixtures from the Minnesota Department of Transportation and 81 mixtures from University of New Hampshire database, was used for the study. The results indicate a strong correlation between binder related properties (binder content, asphalt film thickness, PG spread) and fracture energy. The correlation coefficients obtained from this study for PG spread, effective binder content, and air void can be confidently employed to achieve targeted fracture energy thresholds. The same can be achieved for the Glower-Rowe parameter at 15°C by employing the correlation coefficients obtained for PG low temperature, virgin asphalt content, and voids in the mineral aggregate.

Thermal cracking is one of the most prevalent form of distress in asphalt pavements in climates with cooler temperatures and high cooling rates. Tensile stresses induced by significant changes in temperature is the main cause of thermal cracking. As the temperature drops, the pavement wants to contract but is restrained by friction from underlying layers and lack of contraction joints, subsequently causing tensile thermal stress to build up. When this stress exceeds the tensile strength, microcracks are initiated and these can coalesce into macro-cracks leading to the formation of a thermal crack that typically forms in a transverse direction on the pavement surface. An asphalt mixture's fracture and viscoelastic properties play a significant role in controlling the ability of the mixture to limit thermal stresses and maintain material integrity as stresses approach material stress capacity. These cracks allow water to infiltrate the pavement structure; this subsequently compromises its performance and structural integrity.

Efforts have been made to develop performance-based binder and mixture specifications to lower the propensity of thermal cracking distress. As part of this endeavor, several researchers have developed mixture based thermal cracking performance evaluation tools (*1–4*). This study focuses on two recently developed lab measured mechanistic thermal cracking evaluation parameters. The first method uses the fracture energy determined from disc compact tension (DCT) testing to limit thermal cracking by ensuring a minimum fracture toughness of the material, and the second method combines the stiffness and relaxation properties of asphalt mixtures in Black space to limit thermal stresses.

Recent studies have shown the correlation between different fracture parameters and thermal cracking resistance of asphalt mixtures. One such parameter is the fracture energy of asphalt mixtures that could be

[1]University of New Hampshire, Durham, NH
[2]Office of Materials and Road Research, Maplewood, MN

Corresponding Author:
Address correspondence to Eshan V. Dave: eshan.dave@unh.edu

determined from DCT or other geometries, such as the semi-circular bend test (*5–7*). Some agencies have implemented DCT testing as a requirement and established different limits for fracture energy in their specifications to ensure the thermal cracking resistance of asphalt mixtures at design and production stage. For example, the Minnesota Department of Transportation (MnDOT) limits minimum 450 J/m^2 for traffic levels 1, 2, and 3 or 500 J/m^2 for traffic levels 4 and 5 during mix design phase, and 400 J/m^2 for traffic levels 1, 2, and 3 or 450 J/m^2 for traffic levels 4 and 5 for quality assurance.

Black space is defined as the cross-plot of dynamic modulus ($|E^*|$ or $|G^*|$) and phase angle (ϕ) of material at a given temperature or frequency. Recently the Glover-Rowe (G-R) parameter (*8*) was identified as a tool to describe the thermal cracking resistance of asphalt binders in a Black space. The parameter is determined from the $|G^*|$ and phase angle of binders at 15ºC and a frequency of 0.005 rad/s. Values of 180 kPa and 450 kPa were set as limits to indicate the onset of cracking and significant cracking, respectively. Since the binder evaluation in Black space neglects the mix properties, Mensching et al. (*9*) expanded the study to mixtures and developed a single parameter (mix G-R) in Black space as an indicator for thermal cracking performance of mixtures. Their work utilized thermal cracking performance from a thermal stress restrained specimen test.

A study by Gaw (*10*) identified climate, subgrade type, mix design properties, pavement age, and traffic as the main factors that influence the thermal cracking resistance of asphalt pavements. Kallas (*11*) showed that aggregate type affects fracture strength. A study employed by Haas et al. (*12*) using multiple regression models concluded that minimum pavement temperature, coefficient of thermal contraction, and pavement layer thickness have the highest correlation to fracture energy. Abu Abdo et al. (*13*) investigated the effect of mix variables on fracture toughness and concluded that a decrease in asphalt content is associated with lower fracture toughness values. A study by Braham et al. (*14*) performed a statistical analysis and concluded the significance of binder content, aggregate type, and temperature on fracture energy of mixtures. As part of an effort to develop a statistical model to predict fracture energy from the DCT test, Marasteanu et al. (*15*) identified PG high temperature (PGHT), PG low temperature (PGLT), and aggregate type as significant variables to fit the model. A study by Li et al. (*16*) showed the dependency of fracture energy on temperature, type of aggregate, and binder modifier. In general, the literature review carried out as part of this study indicated a lack of consensus in the conclusions drawn from different studies and limited information on the effect of mix design variables (all previous studies were limited to less than 15 asphalt mixtures) on thermal properties as it relates to fracture energy and Black space location (mix G-R value).

Mix specifiers and producers strive to design and produce mixtures that meet established threshold values for fracture energy and mix G-R value by adjusting their mix design variables. Current adjustments are mainly employed on mix design variables that are included in the specification and controlled through current quality control and acceptance procedures. Due to a lack of reliable guidance, the adjustment of mix design variables is usually a trial and error process. Therefore, this study was designed to obtain a better understanding of the topic to provide guidance to mix specifiers and producers on changes they should consider making to the composition of asphalt mixture to achieve specification requirements as it relates to fracture energy and mix G-R value.

The objective of the study is to identify mix design variables that potentially affect the thermal cracking performance properties of asphalt mixtures. Databases developed by MnDOT with data from 90 mixtures and University of New Hampshire (UNH) containing data for 81 mixtures were used to determine the statistical significance and correlation between common mix design variables, including recycled asphalt amount, mix volumetric properties, and binder grade, to the fracture energy and mix G-R values.

Thermal Cracking Performance Evaluation Parameters

Fracture Energy from a DCT Test

The DCT test was developed at the University of Illinois to measure the fracture energy of asphalt mixtures (*6*). The test is performed on notched 150 mm diameter specimens by applying a constant crack mouth opening displacement at a rate of 1 mm/min. The fracture energy is determined from the test by normalizing the area under the load displacement curve by the fractured face area. In general, a higher fracture energy value is desirable and is expected to indicate a better thermal cracking resistance in the field. A 400 J/m^2 threshold is commonly employed to ensure the thermal cracking resistance of asphalt mixtures.

The frequency distribution plot presented in Figure 1 depicts ranges of fracture energy values used in the study. This shows the wide range of fracture energy measurements used in the analysis and subsequently the increased confidence in the validity of the study conclusions.

Mix G-R Parameter

Mixture Black space diagrams assess the stiffness and relaxation capability of mixtures from a plot of dynamic

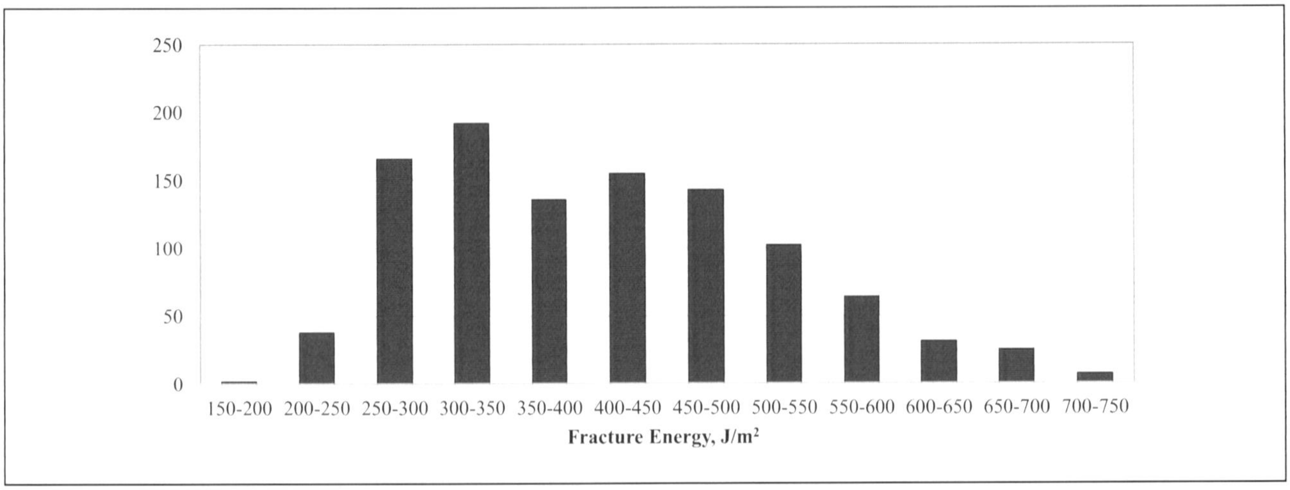

Figure 1. Frequency distribution plot for fracture energy.

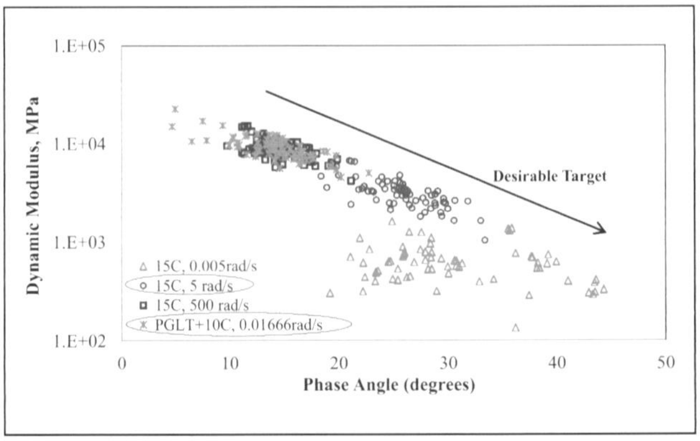

Figure 2. Black space diagram for study mixtures.

modulus ($|E^*|$) versus phase angle (ϕ). Mixtures that plot further to the right have more viscous behavior, while the lower phase angle values indicate more elastic behavior. The combination of lower phase angle (less relaxation capability) and higher dynamic modulus (more stiffness) may indicate that the mixture is more susceptible to cracking. A recent study by Mensching et al. (*9*) evaluated the correlation between mix G-R values in Black space (calculated using Equation 1) to thermal cracking performance. Later a preliminary threshold of 3.684E04 MPa is proposed for mix G-R values corresponding to temperature and frequency combination of PGLT + 10°C and 0.01666 rad/s (*9*). A lower G-R value is desirable as it indicates better thermal cracking resistance:

$$\frac{|E^*|(cos\varnothing)^2}{sin\varnothing} \quad (1)$$

Figure 2 shows the Black space plot for the 81 mixes considered in this study at four temperature and frequency combinations. The plot shows that more separation between mixes is observed at 15°C and 5 rad/s, and PGLT + 10°C and 0.01666 rad/s. Thus, these two frequency and temperature combinations were used for further analysis. It should be noted that at 15ºC, mixture evaluation is done at a constant temperature, whereas normalization is done with respect to the low temperature grade of the mixtures for evaluation done at PGLT + 10°C. The mix G-R values were computed at these frequency and temperature combinations and their relationship with various mix design variables is studied.

Research Approach and Materials

Materials

To determine the effect of mix design variables on fracture energy, a database developed by MnDOT's Office of Materials and Road Research containing various information for 90 mixtures and their corresponding fracture energy measurement determined from approximately 1,170 DCT tests is utilized. The data include virgin asphalt content, total binder content, effective binder content, air void (AV), recycled asphalt pavement (RAP) amount, asphalt film thickness (AFT), PGLT, PGHT, and PG spread (PGHT-PGLT). The DCT data are from testing conducted on production material that correspond to the data collected at construction stage. DCT test specimen production and testing is carried out following the test procedure of ASTM D7313-13/MnDOT modified specification (*17*). The number of replicates tested for a specific mix range from 12 to 16. DCT test information available in the database includes number of replicates, test temperature, replicate's fracture energy, replicate peak load, average fracture energy, and average peak load.

Table 1. Overview of Mix Design Variables Used for the Study

Performance criteria	Mix design variable	Acronym	Range of mix variable
Fracture energy	Virgin asphalt content	$P_{b,V}$	3–5%
	Total binder content	P_b	4–6%
	Effective binder content	P_{be}	4.0–5.5%
	Air voids	AV	3–5%
	Recycled asphalt pavement content	% RAP	10–30%
	Voids in the mineral aggregate	VMA	13–16
	Asphalt film thickness	AFT	7–10
	PG low temperature	PGLT	–22 to −34
	PG high temperature	PGHT	52–64
	PG spread (PGHT-PGLT)	ΔPG	86–98
	Nominal maximum aggregate size	NMAS	9.5, 12.5 mm
Mix G-R Parameter	Virgin asphalt content	$P_{b,V}$	4–7%
	Total binder content	P_b	4.5–7%
	Effective binder content	P_{be}	4.0–6.5%
	Air voids	AV	3–10%
	Recycled asphalt pavement content	% RAP	0–40%
	Voids in the mineral aggregate	VMA	13–18
	Asphalt film thickness	AFT	5.5–9
	PG low temperature	PGLT	–22 to −34
	PG high temperature	PGHT	52–64
	PG spread (PGHT-PGLT)	ΔPG	86–92
	Nominal maximum aggregate size	NMAS	9.5–19 mm

To understand the relationship between different mix design variables and mix G-R value, a database developed at UNH is utilized. The database contains diverse volumetric and test information (dynamic modulus and phase angle) for 81 mixtures representing material from the New England region in the United States. Test specimen production is done following AASHTO PP 60 (*18*) and complex modulus testing is performed according to AASHTO T 342 (*19*) on three replicates to determine the dynamic modulus and phase angle master curves of the mixtures and subsequently the mix G-R values.

Table 1 is presented to show the breadth of information available as well as the mix design variables considered for the study.

Data Analysis Methodology

This section describes the data collection approach and statistical analysis performed to determine the effect of different mix variables on thermal cracking performance parameters. The mix description in the database is used to identify the different mixtures and map mix variable and test information to a specific mix. For each mixture, the average values of the parameters were calculated and used for analysis. The statistical analysis is done in three phases described below using JMP® statistical software package.

Explore and Remove Outliers. Exploring and removing outliers is an important part of statistical analysis, particularly due to anticipated errors during data measurement and collection. The step is vital as inclusion of outliers in a statistical analysis could cause bias in the conclusions drawn from the analysis. In this study, JMP® is used to locate the outliers by employing the Mahalanobis distance approach. A recent study by Nemati and Dave (*20*) used a similar approach for removing outliers in complex modulus datasets. The outliers were consequently removed from the input file. It should be noted that the fracture energy measurements obtained from the MnDOT database contained only three outliers, whereas four outliers were present on the G-R value calculated based on data from UNH. The presence of the small number of outliers confirms the lower variability encountered during DCT and complex modulus testing.

Determine Significance of Mix Design Variables. A stepwise regression analysis was used to evaluate the significance of different mix design variables on the thermal cracking properties of asphalt mixture. The analysis makes an inference about a larger population to recognize mix design variables with a statistically significant effect on fracture energy and mix G-R value. This is accomplished by performing a stepwise regression analysis and assessing p-values. The p-value provides information on the probability of the existence of a relationship between different mix design variables and thermal cracking properties as it relates to fracture energy and mix G-R value. The conclusions drawn from this analysis will inform mix specifiers and producers about the most important mix design variables related to thermal cracking properties.

Throughout the analysis, the null hypothesis assumes that there is no statistically significant relationship between the variables and the thermal cracking evaluation parameter. For this study, the common practice of utilizing p-value <0.05 is adopted and the null hypothesis is rejected for a p-value <0.05 indicating the parameter has contributed significantly to the thermal cracking performance of a mixture. In other words, the relatively low p-value indicates the presence of a mathematical relationship between the mix parameter and fracture property such that a linear function of this parameter can predict the fracture property of the mixture within a 95% confidence level of the parameter data.

Determine Pearson Correlation Coefficient. In this study, the Pearson correlation is used to understand how the thermal cracking properties of mixtures, specifically fracture energy and mix G-R value, are affected by changes in mix design variables. The Pearson correlation factor is the most widely known type of correlation and is used to measure the degree of relationship between linearly related variables and the direction of the relationship based on the data provided. Based on the strength of the relationship, the value of the correlation coefficient varies between $+1$ and -1. A correlation of $+1$ indicates a linear positive relationship whereas -1 indicates a linear negative (inverse) relationship between the variables. As the correlation factor moves toward zero from both directions, the relationship becomes weaker. The analysis is done in JMP® by pairing the results of a mix variable to the corresponding fracture energy/mix G-R values. Based on the correlation factor obtained, the relationship is defined as weak/strong and the direction of the impact is identified.

Results and Discussion

Statistical Significance between Mix Variables and Thermal Cracking Performance Parameters

The p-values from stepwise regression analysis corresponding to different mix variables and fracture energy are presented in Table 2*a*. The *p*-values indicate whether a statistically significant relationship between mix design

Table 2. Statistical Significance (*p*-Values) between Mix Design Variables and Thermal Cracking Performance Parameters

(*a*) Fracture energy

Mix design variable	Prob > F	Significance (Yes/No)
PG low temperature	<0.0001	Yes
PG high temperature	<0.0001	Yes
PG spread	<0.0001	Yes
Voids in the mineral aggregate	<0.0001	Yes
Asphalt film thickness	0.0006	Yes
Air voids	0.001	Yes
Virgin asphalt content	0.002	Yes
Effective binder content	0.0124	Yes
Total binder content	0.024	Yes
Recycled asphalt pavement content	0.093	No
Nominal maximum aggregate size	0.830	No

(*b*) Mix Glover-Rowe parameter

Mix design variable	Prob > F		Significance (Yes/No)	
	15°C, 5 rad/s	PGLT + 10°C, 0.01666 rad/s	15°C, 5 rad/s	PGLT + 10°C, 0.01666 rad/s
PG low temperature	<0.0001	0.738	Yes	No
Virgin asphalt content	0.001	0.017	Yes	Yes
Voids in the mineral aggregate	0.001	0.133	Yes	No
Total binder content	0.001	0.001	Yes	Yes
Nominal maximum aggregate size	0.027	0.0002	Yes	Yes
PG spread	0.052	0.008	No	Yes
Recycled asphalt pavement	0.152	0.568	No	No
Effective binder content	0.205	0.034	No	Yes
PG high temperature	0.217	0.066	No	No
Asphalt film thickness	0.360	0.160	No	No
Air voids	0.376	0.070	No	No

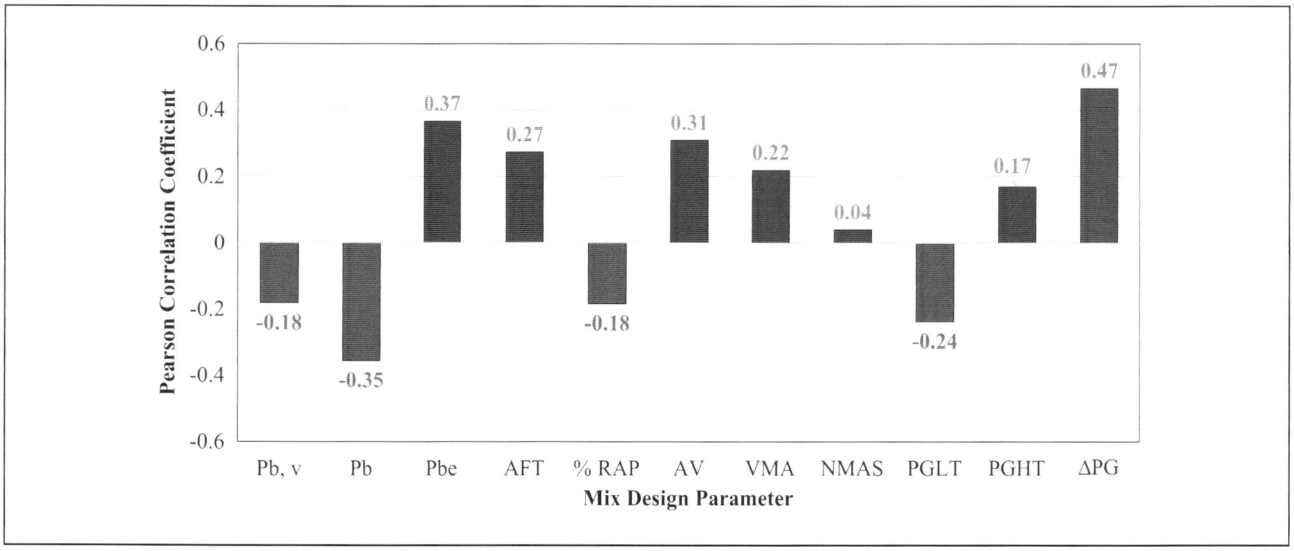

Figure 3. Pearson correlation coefficient between mix variables and fracture energy.

variables and fracture energy exists (designated as "Yes") or not (designated as "No"). Overall, the p-values corresponding to mix design variables were low indicating a statistically significant relationship between the mix design variables and fracture energy. The exceptions to this are recycled asphalt content ($p=0.093$) and nominal maximum aggregate size ($p=0.830$) which have a higher p-value than the significance threshold. Therefore, it is concluded that mix specifiers and producers can consider changing PGLT and PGHT grades, PG spread, voids in the mineral aggregate, asphalt film thickness, air void, virgin asphalt content, effective binder content, and total binder content to adjust mixes to achieve set threshold values effectively.

Table 2*b* displays the p-values between the different mix design variables and mix G-R values corresponding to the two cases selected for this analysis (15°C and PGLT + 10°C). Virgin asphalt content, total binder content, and nominal maximum aggregate size (NMAS) demonstrated a significant effect on mix G-R value at both temperatures. In addition, PGLT and voids in the mineral aggregate were significant at 15°C, whereas PG spread and effective binder content have a significant relationship to mix G-R at PGLT + 10°C. The significant difference observed for the two cases implies a dependence of the effect of mix variables on the temperature and frequency combination considered as it relates to mix G-R. This shows that mix variables adjustment should be considered differently and should be made based on results obtained for that specific temperature and frequency combination. Notably, the PGLT is significant at 15°C but not significant at PGLT + 10°C. The non-significance observed at PGLT + 10°C could be a result of the normalization done with respect to the low temperature contrary to the 15°C which does the evaluation at a constant temperature.

Pearson Correlation Coefficients of Mix Design Variables and Fracture Energy

Figure 3 displays the Pearson correlation coefficients between mix design variables and fracture energy of mixtures. The coefficient value represents the mean change in fracture energy for a one-unit increase in the mix design variable. Based on the results, total binder content, effective binder content, asphalt film thickness, air void, and PG spread showed a stronger correlation to fracture energy compared with the other parameters included in the study. Out of the five parameters, which indicated a strong correlation, four of them are related to binder. This indicates that binder properties have a major effect on the thermal cracking properties of asphalt mixtures.

The positive correlation of effective binder content, asphalt film thickness, air void, voids in the mineral aggregate, PGHT, and PG spread indicates that an increase in these parameters improves the fracture properties of asphalt mixtures. The stronger positive correlation of effective binder content and asphalt film thickness leads to the conclusion that the availability of more asphalt to coat aggregate particles in the mix helps with the relaxation capacity that the pavement requires during temperature fluctuation. Moreover, considering the significance of the correlation between effective binder content and fracture energy (Table 3), the researchers recommend the inclusion of effective binder content in the specification control since it represents actual binder content available to the mixture. The same conclusion drawn with respect to voids in the mineral aggregate indicates that the more space available to form asphalt film benefits thermal cracking resistance. The positive correlation observed between air void and fracture energy infers an increase in air void and improves the ability of asphalt mixture to contract with less thermal stress build-up. The range of air void evaluated in this study (3% to 5%) can limit the observed effect and the relationship might differ as the air void increases beyond 5%. It is also essential to give attention to PG spread of mixtures as it relates to thermal cracking which has the strongest correlation to fracture energy compared with other mix variables. NMAS displayed a weak positive correlation. This could be due to the use of fracture energy only to evaluate wear courses that resulted in only two NMAS levels (9.5 and 12.5 mm) being included in the analysis.

The statistical analysis performed indicated that total virgin binder, total binder content, RAP content, and PGLT grade have a negative impact on fracture energy, implying an increase in these variables results in a potential for thermal cracking related problems. While the total binder contents (P_b and $P_{b,\ v}$) have a negative effect, the effective binder content (P_{be}) has a positive effect on fracture energy. The authors believe that the presence of binder from RAP in the total binder content could be the cause of the observed negative impact of total binder content on fracture energy. The total virgin binder accounts for absorbed asphalt in addition to the effective asphalt content. In most cases, an increment in absorbed asphalt content is related to finer aggregate, which is hypothesized as the reason for the impact of virgin binder on fracture energy in a negative manner. The negative effect of RAP amount agrees with other studies showing aged binder from RAP having negative effects on thermal cracking performance. The negative effect of increasing PGLT is expected: as PGLT lowers (a better low temperature grade) the fracture energy improves.

The correlation probability of mix variables to fracture energy is displayed in Table 3. It is expressed in terms of probability level and indicates how unlikely it will be for a given correlation coefficient to occur if there is no relation in the population. Therefore, a smaller p-value indicates the likelihood of an observed correlation, by chance, to be minimal (to be at less than 5% confidence level), and it can be assumed that the observed correlation applies to not only the study data but to the general population. Overall, the observed correlation probability of all mix variables to fracture is very low, demonstrating the very low probability of observing the correlations by chance. The higher reliability obtained for PG spread, total binder content, effective binder content, and air void implies that by extending the correlation obtained for these variables to other sets of mixtures one can achieve the required threshold values of fracture energy. While the correlation probability values of other variables are higher than the significance threshold, adjusted asphalt film thickness and PGLT show *p*-values very close to the significance level.

Table 3. Significance of Pearson Correlation Coefficient between Mix Design Variables and Fracture Energy

Mix variables	Acronym	Correlation probability
PG spread	ΔPG	0.0003
Total binder content	P_b	0.017
Effective binder content	P_{be}	0.018
Air void	AV	0.038
Asphalt film thickness	AFT	0.072
PG low temperature	PGLT	0.070
Voids in the mineral aggregate	VMA	0.151
PG high temperature	PGHT	0.199
Recycled asphalt pavement content	% RAP	0.233
Virgin asphalt content	$P_{b,\ v}$	0.239
Nominal maximum aggregate size	NMAS	0.790

Pearson Correlation Coefficient of Mix Design Variables and Mix G-R Value

Figure 4 shows the Pearson correlation coefficient between various mix design variables and mix G-R value. The relationship will be discussed based on the results obtained for the two cases (15°C and PGLT + 10°C). Total binder content, void in the mineral aggregate, and PGLT were found to have a stronger correlation to mix G-R compared with the other mix variables considered in this study.

Recycled asphalt content and NMAS showed a similar positive correlation to mix G-R parameters in both cases, inferring their potential negative effect on thermal cracking resistance. It is also interesting to see NMAS having a big influence on thermal cracking performance. This impact could be due to the loss of asphalt mixture flexibility with NMAS increase, typically because of lower asphalt binder content in the mixture.

The negative correlation of total binder content, virgin binder content, effective binder content, air void, voids in mineral aggregate, and PG spread in both cases on mix G-R values indicates that an increase in these parameters helps with the thermal cracking property. Both an increase in binder content and air void are expected to decrease stiffness and increase phase angle of a mixture, moving mix G-R parameter to more desirable values.

The direction of impact on thermal performance due to PGLT and PGHT were opposite for 15°C and PGLT + 10°C. However, results from the correlation probability show the lower reliability of the correlation values observed for PGLT and PGHT as it relates to

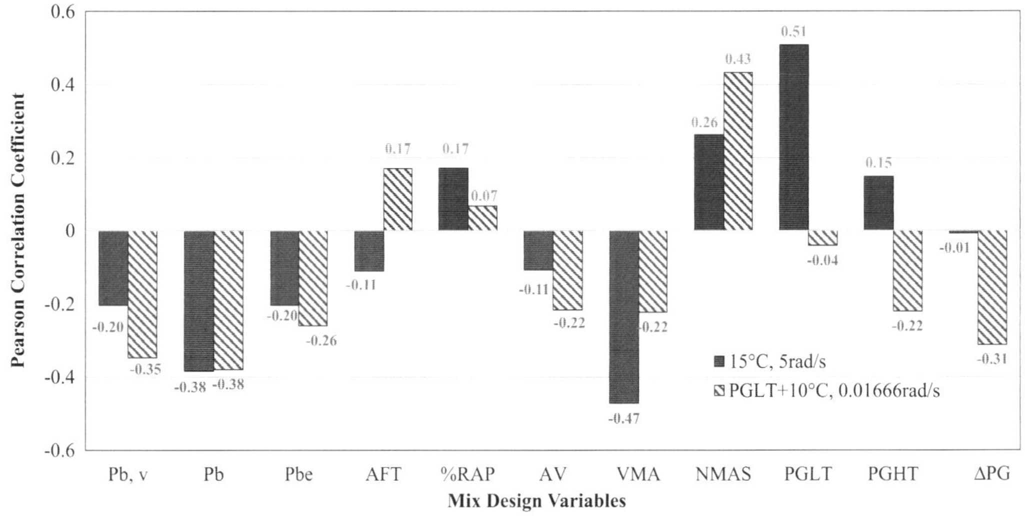

Figure 4. Pearson correlation coefficient between mix variables and mix G-R value.

PGLT + 10°C. The very low correlation probability observed for PGLT for 15°C demonstrates the very low chance of observing correlations related to these variables by chance. Thus, a conclusion can be drawn that the PGLT and mix G-R parameter have a strong positive correlation. This means that as the PGLT grade decreases the mix G-R parameter reduces, improving the thermal cracking property of asphalt mixtures. For asphalt film thickness, the correlation values were opposite for the two cases.

In general, the correlation probability displayed in Table 4 shows that the higher reliability of correlation coefficients is obtained at 15°C as opposed to PGLT + 10°C. Overall, values related to 15°C were lower than the significance threshold for PGLT, virgin asphalt content, and voids in mineral aggregate indicating that the correlation coefficient obtained from this study applies to a different population as it relates to these variables. However, for PGLT + 10°C, the values were above the significance threshold, with the exception of NMAS, implying higher probability of the observed correlation due to chance.

Table 4. Significance of Pearson Correlation Coefficient between Mix Design Variables and Mix G-R Value

Mix design variables	Correlation probability: 15°C, 5 rad/s	Correlation probability: PGLT + 10°C, 0.01666 rad/s
PG low temperature	0.013	0.855
Virgin asphalt content	0.019	0.105
Voids in the mineral aggregate	0.023	0.308
Total binder content	0.071	0.321
Voids filled with asphalt	0.105	0.299
Effective binder content	0.149	0.064
Nominal maximum aggregate size	0.225	0.039
PG spread	0.288	0.148
Asphalt film thickness	0.350	0.160
Recycled asphalt pavement	0.434	0.755
PG high temperature	0.499	0.314
Air voids	0.628	0.321

Summary and Conclusion

This study investigated the effects of different mix design variables on thermal cracking performance properties of asphalt mixtures in terms of Black space location and fracture energy. The primary objective of this work is to provide insight and tools for mix designers and specifiers in terms of the effects of mix properties on performance properties. Black space location controls thermal stress build-up due to consideration of asphalt mixture stiffness and relaxation capabilities. Fracture energy provides a measure of crack resistance when thermal stresses approach and exceed material strength. The study utilized 90 mixtures from a MnDOT database and 81 mixtures from a UNH database. A stepwise regression analysis, which accounts for a broader population, is used to determine the statistical significance of mix variables. The Pearson correlation coefficient is determined to quantify and gain insight to the direction and extent of the effect that a mix variable would have on performance property. Based on the results of this study, the following conclusions can be drawn:

- The p-values from the stepwise regression analysis indicated a significant relationship between fracture energy and PG spread, voids in the mineral aggregate, asphalt film thickness, air void, virgin asphalt content, effective binder content, total binder content PGLT, and PGHT grade. This indicates that manipulation of any of these parameters could have a potential effect on thermal cracking performance. The extent of the impact is variable and is assessed through correlation analysis.
- The p-values between mix variables and G-R parameter indicated a significant relationship between virgin asphalt content, total binder content, NMAS, and mix G-R value at both 15°C and PGLT + 10°C. For the other mix variables, the significance of the effect is dependent on the temperature and frequency combination used to determine the mix G-R value. Due to the normalization done with respect to the PGLT + 10°C, PGLT was found to be insignificant on the mix G-R value for this case.
- The result from the Pearson correlation coefficient indicated a stronger correlation of binder related mix design variables (total binder content [negative], effective binder content [positive], asphalt film thickness [positive], PG spread [positive]) to fracture energy compared with the other mix design variables. This verifies the vital role binder plays in thermal cracking performance.
- Effective binder content, asphalt film thickness, air void, voids in the mineral aggregate, PGHT, and PG spread showed a positive correlation to fracture energy implying that an increase in one or more of these variables is expected to result in improved thermal cracking performance. However, a negative correlation is observed between total virgin binder, total binder content, recycled asphalt pavement content, and PGLT with fracture energy. The correlation probability corresponding to PG spread, total binder content, effective binder content, and air void were found to be lower than the threshold, implying that these variables can be employed confidently to obtain the required fracture energy level. The results support the consideration for using effective binder content to improve thermal cracking performance as opposed to total binder content.
- PGLT, recycled asphalt content, and NMAS displayed a positive correlation to mix G-R value. A negative correlation was found between total binder content, virgin binder content, effective binder content, air void, voids in the mineral aggregate, and mix G-R parameter, implying that an increase in these parameters could potentially improve thermal performance of asphalt mixtures. The correlation probability indicated the better reliability of correlation coefficients obtained at 15°C for PGLT, virgin asphalt content, and voids in the mineral aggregate.

The findings from this study give additional insight into the influence of different mix variables on fracture energy and mix G-R value. This provides information to mix specifiers and producers in determining how to adjust mix composition effectively and efficiently to meet minimum threshold values and subsequently result in more crack resistant pavements.

It is possible for two variables to have a zero linear relationship and a strong curvilinear relationship at the same time, so future research should evaluate the existence of a nonlinear relationship between the mix variables and thermal parameters. Future research efforts are also needed to adapt the findings from the study in actual projects to validate and make any necessary adjustments to the conclusions drawn. Finally, work presented here serves as a foundation for developing predictive models for thermal cracking performance properties in the future.

Author Contributions

The authors confirm contributing to the paper as follows: study conception and design: All authors; data collection: MO, DG, JMV, SD; analysis and interpretation of results: MO, EVD, JSD; draft manuscript preparation: MO, EVD, JSD. All authors reviewed the results and approved the final version of the manuscript.

References

1. Dave, E. V., S. Leon, and K. Park. Thermal Cracking Prediction Model and Software for Asphalt Pavements. In *Transportation and Development Institute Congress 2011: Integrated Transportation and Development for a Better Tomorrow*, Chicago, IL., 2011, pp. 667–676.
2. Anderson, D., L. Lapalu, M. Marasteanu, Y. Hir, J. P. Planche, and D. Martin. Low-Temperature Thermal Cracking of Asphalt Binders as Ranked by Strength and Fracture Properties. *Transportation Research Record: Journal of the Transportation Research Board*, 2001. 1766: 1–6.
3. Dave, E. V., B. Behnia, S. Ahmed, W. G. Buttlar, and H. Reis. Low Temperature Fracture Evaluation of Asphalt Mixtures using Mechanical Testing and Acoustic Emissions Techniques. *Journal of the Association of Asphalt Paving Technologists*, Vol. 80, 2011, pp. 193–220.
4. Olard, F., H. Di Benedetto, B. Eckmann, and J. Vaniscote. Failure Behavior of Bituminous Binders and Mixes at Low Temperatures. *Proc., 3rd Euroasphalt and Eurobitumen Congress Held at Vienna*, Vol. 2. Foundation Eurasphalt, Breukelen, the Netherlands, 2004.

5. Marasteanu, M., S. Dai, J. Labuz, and X. Li. Determining the Low-Temperature Fracture Toughness of Asphalt Mixtures. *Transportation Research Record: Journal of the Transportation Research Board*, 2002. 1789: 191–199.
6. Wagoner, M., W. Buttlar, G. Paulino, and P. Blankenship. Investigation of the Fracture Resistance of Hot-Mix Asphalt Concrete using a Disk-Shaped Compact Tension Test. *Transportation Research Record: Journal of the Transportation Research Board*, 2005. 1929: 183–192.
7. Wagoner, M. P., W. G. Buttlar, and G. H. Paulino. Development of a Single-Edge Notched Beam Test for Asphalt Concrete Mixtures. *Journal of Testing and Evaluation*, Vol. 33, No. 6, 2005, pp. 452–460.
8. Anderson, R. M., G. N. King, D. I. Hanson, and P. B. Blankenship. Evaluation of the Relationship between Asphalt Binder Properties and Non-Load Related Cracking. *Journal of the Association of Asphalt Paving Technologists*, Vol. 80, 2011, pp. 615–664.
9. Mensching, D. J., G. M. Rowe, and J. Sias Daniel. A Mixture-Based Black Space Parameter for Low-Temperature Performance of Hot Mix Asphalt. *Road Materials and Pavement Design*, Vol. 18, 2017, pp. 404–425.
10. Gaw, W. J. *Design Techniques to Minimize Low-Temperature Asphalt Pavement Transverse Cracking*. Research Report No. 81-1. Asphalt Institute, Lexington, KY, 1981.
11. Kallas, B. F. *Low-Temperature Mechanical Properties of Asphalt Concrete*. Research Report No. 82-3. Asphalt Institute, Lexington, KY, 1982.
12. Haas, R., F. Meyer, G. Assaf, and H. Lee. A Comprehensive Study of Cold Climate Airport Pavement Cracking. *Journal of the Association of Asphalt Paving Technologists*, Vol. 56, 1987, pp. 198–245.
13. Abu Abdo, A. M., F. Eckwright, S. J. Jung, F. Bayomy, and R. Nielsen. Semi-Circular Notched Beam Testing Procedure for Hot Mixture Asphalt. *Proc., Institution of Civil Engineers-Transport*, Vol. 167, No. 1, Thomas Telford Ltd, 2014, pp. 48–58.
14. Braham, A., W. Buttlar, and M. Marasteanu. Effect of Binder Type, Aggregate, and Mixture Composition on Fracture Energy of Hot-Mix Asphalt in Cold Climates. *Transportation Research Record: Journal of the Transportation Research Board*, 2007. 2001: 102–109.
15. Marasteanu, M., A. Zofka, M. Turos, X. Li, R. Velasquez, X. Li, W. Buttlar, G. Paulino, A. Braham, E. Dave, and J. Ojo. *Investigation of Low Temperature Cracking in Asphalt Pavements National Pooled Fund Study*. Report No. 776. Minnesota Department of Transportation, Saint Paul, MN, 2007.
16. Li, X., A. F. Braham, M. O. Marasteanu, W. G. Buttlar, and R. C. Williams. Effect of Factors Affecting Fracture Energy of Asphalt Concrete at Low Temperature. *Road Materials and Pavement Design*, Vol. 9, 2008, pp. 397–416.
17. ASTM International. *Standard Test Method for Determining Fracture Energy of Asphalt-Aggregate Mixtures using the Disk-Shaped Compact Tension Geometry*. ASTM D7313-13. ASTM International, Washington, D.C., 2013.
18. American Association of State Highway Transportation Officials. *Standard Practice for Preparation of Cylindrical Performance Test Specimens Using the Superpave Gyratory Compactor (SGC)*. AASHTO PP 60. American Association of State Highway Transportation, Officials, Washington, D.C., 2011.
19. American Association of State Highway Transportation Officials. *Standard Method of Test for Determining Dynamic Modulus of Hot-Mix Asphalt Concrete Mixtures*. AASHTO T 342. American Association of State Highway Transportation Officials, Washington, D.C., 2015.
20. Nemati, R., and E. Dave. Nominal Property Based Predictive Models for Asphalt Mixture Complex Modulus (Dynamic Modulus and Phase Angle). *Construction and Building Materials*, Vol. 158, 2018, pp. 308–319.

The Standing Committee on Structural Requirements of Asphalt Mixtures (AFK50) peer-reviewed this paper (18-05397).

TRR
JOURNAL OF THE TRANSPORTATION RESEARCH BOARD

Article

Transportation Research Record
2018, Vol. 2672(28) 481–492

DOI: 10.1177/0361198118782033
journals.sagepub.com/home/trr
SAGE

A Straightforward Procedure to Characterize Nonlinear Viscoelastic Response of Asphalt Concrete at High Temperatures

Mohammad Bazzaz[1], Masoud K. Darabi[1], Dallas N. Little[2], and Navneet Garg[3]

Abstract

This paper proposes a straightforward procedure to characterize the nonlinear viscoelastic response of asphalt concrete materials. Furthermore, a model is proposed to estimate the nonlinear viscoelastic parameters as a function of the triaxiality ratio, which accounts for both confinement and deviatoric stress levels. The simplified procedure allows for easy characterization of linear viscoelastic (LVE) and nonlinear viscoelastic (NVE) responses. First, Schapery's nonlinear viscoelastic model is used to represent the viscoelastic behavior. Dynamic modulus tests are performed to calibrate LVE properties. Repeated creep-recovery tests at variable deviatoric stress levels (RCRT-VS) were designed and conducted to calibrate the nonlinear viscoelastic properties of four types of mixtures used in the Federal Aviation Administration's National Airport Pavement and Materials Research Center test sections. The RCRT-VS were conducted at 55°C, 140 kPa initial confinement pressure, and wide range of deviatoric stress levels; mimicking the stress levels induced in a pavement structure under traffic. Once calibrated, the model was validated by comparing the model predictions and experimental measurements at different deviatoric stress levels. The predictions indicate that the proposed method is capable of characterizing NVE response of asphalt concrete materials.

Rutting and fatigue damage are two major forms of distress that cause the most concern in asphalt concrete pavements, as they affect both serviceability and safety over the performance life of asphalt pavements. Rutting—permanent deformation under the wheels' path—is more dominant at high temperatures, and fatigue damage—evolution of micro-cracks and micro-voids—generally is more problematic at low and intermediate temperatures (*1–9*). Several prominent studies have been conducted regarding constitutive modeling of rutting and fatigue damage mechanisms resulting in proposed (visco-)plastic and (visco-)damage constitutive relationships (*10–19*). These mechanisms result because of the temperature dependency of asphalt binders, the complex composite microstructure of asphalt concrete, and localization of strain in the binder phase as a result of significant differences between the moduli of the asphalt binder and the aggregate phases (*20–22*). However, the nonlinear viscoelastic (NVE) response of asphalt concrete (i.e., the mechanism responsible for recoverable strain response) plays a pivotal role in the evolution rate and magnitude of distress induced in asphalt pavements. Therefore, prediction of these fatigue damage and rutting events require accurate characterization of NVE responses. Schapery's (*23*) nonlinear viscoelastic model has been widely used to represent the viscoelastic response of asphalt concrete materials and structures (*24–37*).

Although significant progress has been made during the past several years in mechanistic-based constitutive modeling of different mechanisms of asphalt concrete, the pavement community has been reluctant to fully implement and use these sophisticated constitutive relationships in design and refined analysis of asphalt concrete pavements and structures. One of the main

[1]Department of Civil, Environmental, and Architectural Engineering, University of Kansas, Lawrence, KS
[2]Zachry Department of Civil Engineering, Texas A&M University, College Station, TX
[3]National Airport Pavement & Materials Research Center (NAPMRC), FAA William J. Hughes Technical Center, Atlantic City, NJ

Corresponding Author:
Address correspondence to Masoud K. Darabi: m-darabi@ku.edu

obstacles that prevents pavement engineers from using these models is the lack of straightforward and simple procedures to extract the material properties associated with these models. Available characterization methods either are over-simplified or require extensive laboratory tests and complex analysis procedures to extract the material properties associated with viscoelastic as well as distress mechanisms observed in pavements. This paper aims to provide a straightforward procedure to extract NVE parameters of asphalt concrete materials while maintaining the required level of accuracy by incorporating the important factors affecting this mechanism.

Accurate estimation of NVE responses of asphalt concrete materials requires precise characterization of complex mechanical responses. Deviatoric, confinement, and shear stress levels are factors affecting the NVE response and significantly contribute to the level of induced distress. Furthermore, a location within the pavement structure experiences a multi-axial state of stress under traffic loading. To mimic realistic stress states in pavement structures under traffic, a repeated creep-recovery test at variable deviatoric stress levels (RCRT-VS) was designed and used in this study to identify NVE response of asphalt mixtures. The proposed characterization method was then applied to four different asphalt mixtures used in the Federal Aviation Administration (FAA)'s National Airport Pavement and Materials Research Center (NAPMRC) test sections to evaluate the efficacy of the method. The first stress invariant and von Mises equivalent stress were considered as surrogates to confinement level and level of deviatoric stress, respectively. It is shown that the NVE parameters strongly correlate with the triaxiality ratio that accounts for both confinement level, deviatoric stress level, and multi-axial states of stress. Finally, a model is proposed by which to relate NVE parameters to the level of triaxiality.

Objectives

The primary objectives of the study presented in this paper are to:

- Develop a straightforward procedure that allows for identification of NVE response of asphalt concrete;
- Incorporate the effect of confinement and deviatoric stress levels into the proposed procedure;
- Introduce a lab test that can be used systematically to extract NVE parameters while mimicking the realistic stress states observed in a pavement structure;
- Apply the proposed method to extract NVE parameters associated with asphalt mixtures used in the FAA's NAPMRC test sections;
- Propose a mechanistic-based model for NVE parameters of asphalt concrete when subjected to general multi-axial state of stresses.

Materials and Testing Procedures

Materials

Four different asphalt mixtures used in the FAA's NAPMRC test sections were studied. The aggregate blend was the same for all four mixtures according to the job mix formula (JMF). This aggregate blend included 30% #57 stone, 26% #8 stone, 36% #10 stone, and 8% natural sand. The testing was conducted on two hot mix asphalts (HMA) containing 6% PG 76-22 (polymer modified) and PG 64-22 (unmodified) binders, and two warm mix asphalts (WMA) containing 6% PG 76-22 and PG 64-22 binders with Evotherm as the additive. Table 1 lists the properties of the mixtures used in this study. These mixtures were used as a surface course with a nominal maximum aggregate size of 19 mm.

Specimen Preparation

Cylindrical specimens were prepared in the laboratory using a Superpave Gyratory Compactor (SGC) (*38*, *39*). The SGC was used to compact 15 cm diameter by 17.8 cm-high specimens. These specimens were cored and cut to a 10.2 cm diameter and height of 15.2 cm to ensure uniform distribution of air voids within the specimen. Figure 1 illustrates specimen preparation steps.

Table 1. Properties of Asphalt Mixtures used in NAPMRC Test Sections Based on JMF

Material	Source	Percent used in final aggregate blend	Bulk specific gravity (G_{sb})	Maximum specific gravity of the mix (G_{mm})
CA #57	Springhouse, Springhouse, PA	30	2.697	2.657
CA #10	Springhouse, Springhouse, PA	36	2.619	
CA #8	Springhouse, Springhouse, PA	26	2.673	
FA Nat. Sand	Hanson, Berlin, NJ	8	2.628	
Binder	PG 64-22	6	1.039	2.467
	PG 76-22		1.033	2.477

Figure 1. Specimen preparation steps: (*a*) SGC compacted the mixtures, (*b*) 15 cm diameter by 17.8 cm-high specimens, (*c*) Coring the specimens to 10.2 cm diameter, (*d*) Cutting the specimens to 15.2 cm high, (e) Jig saw; cored and cut specimens, and (*f*) Final specimen.

Test Procedure

Dynamic Modulus Test. Dynamic modulus test data were analyzed to identify linear viscoelastic (LVE) material properties, including Prony series coefficients and time–temperature shift factors. This test was conducted in accordance with AASHTO T342 (*40*, *41*), using an asphalt mixture performance tester. Three axial linear variable differential transformers (LVDTs) were mounted radially on the specimen at 120° spacing to capture axial strain responses during testing. This test was conducted in an unconfined compression mode at five different temperatures (−10, 4.4, 21.1, 37.8, and 54.4°C) and six loading frequencies (25, 10, 5, 1, 0.5, and 0.1 Hz) at each temperature. The effect of confinement was captured through analysis of the RCRT-VS test results. The applied strain level was between 50 and 75 micro strains to avoid damage and to ensure the responses remained within the LVE region. Two replicates were tested for each test and the averaged values were reported.

Repeated Creep-Recovery Test at Variable Stress Levels (RCRT-VS). A repeated creep-recovery test protocol over a range of stress levels was developed to identify the NVE and viscoplastic responses of the asphalt concrete materials. A material testing system with a triaxial cell and environmental chamber was used to conduct this test in compression at 55°C and 140 kPa initial confinement pressure. A test temperature of 55°C was selected to minimize the contribution of fatigue cracking in the test results. Each loading block consisted of eight creep-recovery cycles with increasing applied deviatoric stress level. The initial confinement level of 140 kPa was determined to mimic the confinement level that a specific point may experience in a pavement structure under the realistic loads (i.e., 137–275 kPa). The loading time of 0.4 s with an unloading time of 30 s remained constant throughout the entire test for each loading cycle. The loading time of 0.4 s represents slow speed of traffic. The unloading time of 30 s ensures full recovery of the viscoelastic strain during the rest time.

Specimens were pre-conditioned for 2 h under the confining pressure until the strain response reached a constant value. After 2 h, the cyclic creep and recovery loading history was applied. Figure 2 shows the test

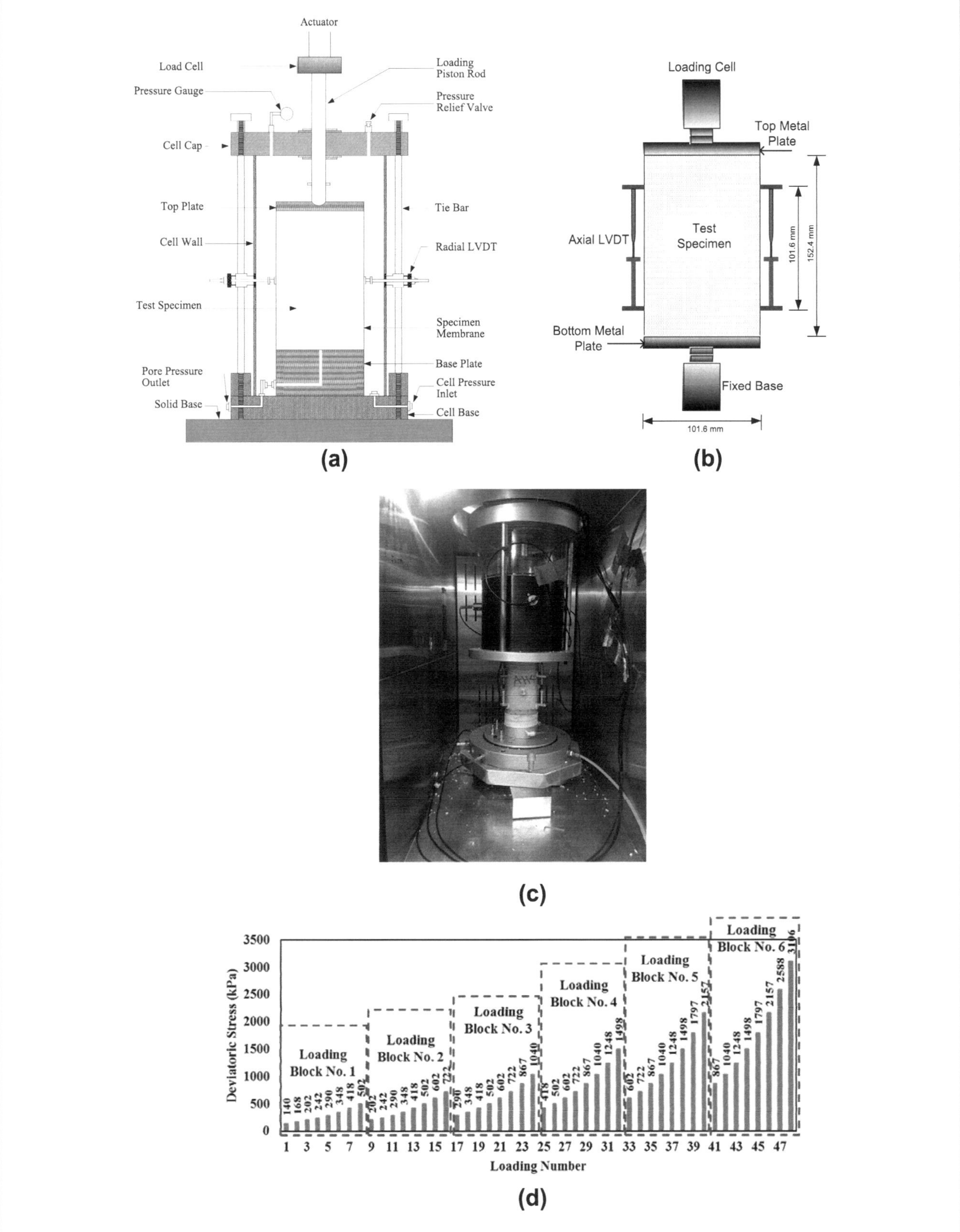

Figure 2. Experimental test setup for RCRT-VS: (*a*) Schematic view of typical triaxial cell with through-the-wall radial LVDTs, (*b*) Schematic view of test specimen with mounted axial LVDTs, (*c*) Triaxial cell equipped with radial LVDTs inside environmental chamber, and (*d*) Applied deviatoric stress in the loading blocks of the RCRT-VS.

setup as well as the applied deviatoric stress levels during RCRT-VS test. The deviatoric stress started from 140 kPa and increased by a factor of $1.2^{(n-1)}$ for eight loading cycle (i.e., first loading block) to gradually increase the deviatoric stress level to provide data on nonlinear response of the material over a wide range of stress levels. The first deviatoric stress of the subsequent loading blocks was chosen to be equal to the third stress level in the preceding loading block, and increased by the same factor of $1.2^{(n-1)}$ for the loading cycles within that block. The test was conducted on two replicates and the averaged values were reported.

Analysis Approaches

The total strain is decomposed into viscoelastic and viscoplastic components, such that:

$$\boldsymbol{\varepsilon}^{\text{total}} = \boldsymbol{\varepsilon}^{\text{rec(ve)}} + \boldsymbol{\varepsilon}^{\text{irrec(vp)}} \quad (1)$$

where $\boldsymbol{\varepsilon}^{\text{total}}$, $\boldsymbol{\varepsilon}^{\text{rec(ve)}}$, and $\boldsymbol{\varepsilon}^{\text{irrec(vp)}}$ are total, viscoelastic, and viscoplastic strain tensors, respectively. Schapery's (*23*) nonlinear viscoelastic constitutive relationship was used to represent the nonlinear recoverable strain, such that:

$$\varepsilon^{\text{rec(ve)}}(t) = g_0^{\sigma_t} D_0 \sigma^t + g_1^{\sigma_t} \int_0^t \Delta D(\psi^t - \psi^\tau) \frac{d(g_2^{\sigma_\tau} \sigma^\tau)}{d\tau} d\tau, \quad (2)$$

where superscripts t and τ designate responses at a specific time over the history of loading; D_0 and ΔD are the instantaneous and transient creep compliances; g_0, g_1, and g_2 are nonlinear viscoelastic parameters, which are a function of the stress level. The parameters g_0 and g_1 capture the effect of stress level on instantaneous and transient compliances, respectively, while g_2 incorporates the nonlinearity caused by the rate of loading. The term $\psi^t = \int_0^t \frac{d\xi}{a_T}$ is the reduced time, where a_T is the time–temperature shift factor. The transient compliance is expressed in terms of the Prony series, such that:

$$\Delta D(t) = \sum_{n=1}^{N} D_n [1 - \exp(-\lambda_n \psi^t)] \quad (3)$$

where D_n and λ_n are the nth compliance and retardation time, respectively, and N is the number of Prony series components.

Identification of Linear and Nonlinear Viscoelastic Properties

Identification of Linear Viscoelastic Properties. The dynamic modulus master curve at the reference temperature of 20°C was produced by analyzing complex dynamic

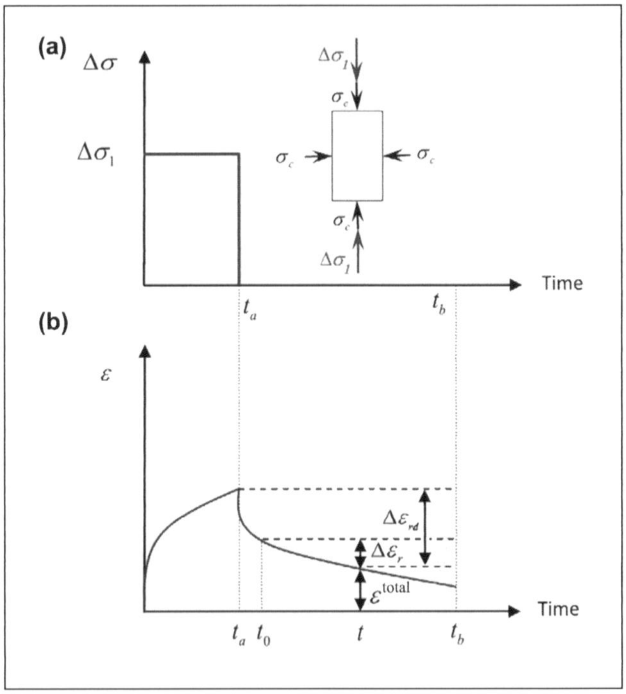

Figure 3. Schematic representation of the strain response during a cycle of RCRT-VS test.

modulus (i.e., $E^*(\omega)$) and phase angle (i.e., δ) at different frequencies and temperatures. In this study, a sigmoidal-type function proposed by Pellinen (*42*) was used to construct the dynamic modulus master curve. The nonlinear generalized reduced gradient method (*43*) was used to minimize the error between the experimental measurements and model predictions of the storage and loss moduli (i.e, $E'_{Exp}(\omega) = |E^*(\omega)| \cos\delta$ and $E''_{Exp}(\omega) = |E^*(\omega)| \sin\delta$, respectively) from which the Prony series coefficients were obtained.

Identification of Nonlinear Viscoelastic Properties. To provide data over a wide range of deviatoric stresses, RCRT-VS was conducted and analyzed to extract NVE properties. Figure 3 schematically represents one cycle of the RCRT-VS. The components of the stress tensor during the RCRT-VS test are expressed as:

$$\sigma_{11} = \sigma_c + \Delta\sigma_1; \; \sigma_{22} = \sigma_{33} = \sigma_c; \; \sigma_{12} = \sigma_{23} = \sigma_{13} = 0 \quad (4)$$

where σ_c is the initial confinement pressure and $\Delta\sigma_1$ is the additional applied axial stress. Although initial confinement pressure is constant (i.e., σ_c), the mean pressure (i.e., $\frac{I_1}{3} = \sigma_c + \frac{\Delta\sigma_1}{3}$) during each loading cycle changes, which means that the mean pressure varies as the additional applied axial stress changes. The deviatoric stress tensor and the mean pressure (i.e., averaged volumetric stress during RCRT-VS can be calculated as:

$$\frac{1}{3}\sigma_{kk}\delta_{ij} = \left(\sigma_c + \frac{\Delta\sigma_1}{3}\right)\delta_{ij};$$
$$S_{ij} = \begin{bmatrix} \frac{2}{3}\Delta\sigma_1 & 0 & 0 \\ 0 & -\frac{1}{3}\Delta\sigma_1 & 0 \\ 0 & 0 & -\frac{1}{3}\Delta\sigma_1 \end{bmatrix} \quad (5)$$

The experimental measurements for the viscoelastic recovery strain response $\Delta\varepsilon_r^{Exp}$ between times t_0 and an arbitrary time t during the rest period can be calculated, such that:

$$\Delta\varepsilon_r^{Exp} = \varepsilon_{11}^{total}(t_0) - \varepsilon_{11}^{total}(t);\ t_o \leq t \leq t_b \quad (6)$$

It should be noted that the viscoelastic recovery strain $\Delta\varepsilon_r$ which is the difference between total strains at two points during the rest period (e.g., between times t_0 and t as shown in Figure 3), is analyzed to characterize the nonlinear viscoelastic properties. The viscoelastic recovery strain, Equation 6, can be simplified as:

$$\Delta\varepsilon_r = \left[\varepsilon_{11}^{NVE}(t_0) + \varepsilon_{11}^{vp}(t_0)\right] - \left[\varepsilon_{11}^{NVE}(t) + \varepsilon_{11}^{vp}(t)\right] \quad (7)$$

It is argued that the permanent deformation and viscoplastic strain do not evolve during the rest period and therefore $\varepsilon_{11}^{vp}(t_0) = \varepsilon_{11}^{vp}(t)$, such that $\Delta\varepsilon_r$ can be expressed as follows using Schapery's nonlinear viscoelasticity yields:

$$\begin{aligned}\Delta\varepsilon_r &= \varepsilon_{11}^{NVE}(t_0) - \varepsilon_{11}^{NVE}(t) \\ &= g_2[\Delta D(\psi^{t_0}) - \Delta D(\psi^{t_0} - \psi^{t_a})] \\ &\Delta\sigma_1 - g_2[\Delta D(\psi^{t}) - \Delta D(\psi^{t} - \psi^{t_a})]\Delta\sigma_1;\ t_o \leq t \leq t_b\end{aligned} \quad (8)$$

where g_2 is the nonlinear parameter which is a function of the stress level at the end of the loading time. The parameter g_2 is the only unknown in Equation 8 and can be identified by minimizing the error between experimental measurements and model predictions for viscoelastic recovery strain (i.e., $\Delta\varepsilon_r^{Exp}$ and $\Delta\varepsilon_r$, respectively). Once g_2 is identified, the total recovered viscoelastic strain $\Delta\varepsilon_{rd}^{Exp}$ at any time t during the rest period can be calculated by subtracting the total strain at time t from the total strain at the end of the creep (i.e., $\varepsilon_{11}^{total}(t_a^-)$), such that:

$$\Delta\varepsilon_{rd}^{Exp} = \varepsilon_{11}^{total}(t_a^-) - \varepsilon_{11}^{total}(t);\ t_a \leq t \leq t_b \quad (9)$$

Similarly, the viscoplastic strain during the rest time remains equal to the viscoplastic strain at the end of the creep loading, such that one can use nonlinear Schapery's viscoelastic model to calculate Equation 9, such that:

$$\begin{aligned}\Delta\varepsilon_{rd} &= \varepsilon_{11}^{NVE}(t_a^-) - \varepsilon_{11}^{NVE}(t) \\ &= [(D_0 + g_1 g_2 \Delta D(\psi^{t_a}))\Delta\sigma_1] \\ &- [g_2(\Delta D(\psi^{t}) - \Delta D(\psi^{t} - \psi^{t_a}))\Delta\sigma_1];\ t_a \leq t \leq t_b\end{aligned} \quad (10)$$

Having the parameter g_2 in hand, the parameter g_1 can be identified by minimizing the error between experimental measurements for total recovered strain $\Delta\varepsilon_{rd}^{Exp}$ and model representation of the same measure $\Delta\varepsilon_{rd}$. This analysis procedure was repeated for each loading cycle of RCRT-VS test. The nonlinear parameter g_0 was assumed to be one from previous studies (*33*) based on the fact that: (1) it is difficult and in some cases impractical to accurately measure instantaneous response during recovery; and (2) instantaneous strain response at high temperatures is negligible as compared with the total viscoelastic strain. Therefore, viscoelastic nonlinearity was captured through the parameters g_1 and g_2 (*35*). The nonlinear parameter $g_1$controls the nonlinearity in the transient compliance term, which affects the viscoelastic nonlinearity mostly during the loading stages. The nonlinear parameter g_2 affects both loading and recovery strain responses.

The confinement level and deviatoric stress levels change during each loading cycle of the RCRT-VS test as the additional axial stress increases. To incorporate the effect of confinement level and deviatoric stress level, the identified nonlinear parameters were plotted as a function of the triaxiality ratio (*35*), η. The parameter η is the ratio of the first stress invariant to the second deviatoric stress invariant (the ratio of hydrostatic pressure over the von Mises equivalent stress), such that:

$$\eta = \frac{I_1/3}{\sqrt{3J_2}} \quad (11)$$

where I_1 is the first stress invariant and J_2 is the second deviatoric stress invariant. The sign of the triaxiality ratio is the same as the sign of the first stress invariant, such that negative triaxiality ratio indicates compressive mean pressure and positive triaxiality ratio indicates tensile mean pressure. For ductile materials, the triaxiality ratio strongly affects the level of permanent strain and this ratio increases as the confinement level dominants the Mises equivalent stress. Therefore, high confinement levels or low values of the Misses equivalent stress result in higher values of η. In case of RCRT-VS test, triaxiality ratio η is the mean pressure over the deviatoric applied stress, such that:

$$\eta = \frac{\sigma_c + \Delta\sigma/3}{\Delta\sigma} = \frac{\sigma_c}{\Delta\sigma} + \frac{1}{3} \quad (12)$$

As shown in the next section, the analysis results indicate that the nonlinear parameters are strong functions of the triaxiality ratio.

Results and Discussions

Linear Viscoelastic Properties

Figure 4 shows the constructed master curve at the reference temperature of 20°C for the four types of asphalt mixtures used in the NAPMRC test sections with 4.5% air void content for PG 76-22 and 3.5% air void content for PG 64-22 cored specimens. The Prony series coefficients, shown in Equation 3, were determined by interconversion relationships (*44*) between the Prony series coefficients and the loss and storage dynamic compliances. Table 2 lists the identified Prony series coefficients and time–temperature shift factors.

As shown in Figure 4, Evotherm causes a decrease in moduli at different frequencies except for PG 76-22 at high frequencies. Furthermore, PG 76-22 resulted in higher moduli as compared with moduli for PG 64-22, as expected. As it will be shown, the impact of Evotherm on NVE parameters is much more pronounced than its impact on LVE parameters.

Nonlinear Viscoelastic Properties

Figures 5 and 6 present the extracted NVE parameters. According to these figures, g_1 increases as the deviatoric stress level increases while g_2 decreases as the deviatoric stress level increases. The product of g_1 by g_2 (i.e., $g_1 \times g_2$) controls the nonlinearity of the strain response during the loading stages whereas g_2 dominates the nonlinearity of the strain response during the recovery. As shown in Figures 5 and 6, both $g_1 \times g_2$ and g_2 values decrease as the deviatoric stress level increases. This observation is physically sound since higher deviatoric stress levels cause permanent rotation and sliding of the aggregates to form a stiffer microstructure, which builds up more resistant against the applied load. According to Equation 2, reduction in the value of $g_1 \times g_2$ leads to a drop in the effective transient compliance (i.e., $g_1 \times g_2 \Delta D$). Therefore, as the deviatoric stress level increases, the effective transient compliance decreases.

Figure 7 compares the experimental measurements and the calibrated NVE model predictions for recovered strain. As shown in Figure 7, the predictions of the calibrated NVE model agree well with experimental measurements. It is observed that HMA PG76-22 yields the lowest level of recovered strain. This could be owing to high viscosity of modified PG 76-22 which results in higher values of the dynamic modulus E^*. For the same stress level, as the dynamic modulus increases the induced NVE strain and therefore the recovered strain decreases.

Figures 5 and 6 show the variation of the NVE parameters with respect to the deviatoric stress. However, as mentioned, both confinement and deviatoric stress levels affect the NVE response of asphalt concrete. Figures 8 and 9 illustrate the variation of NVE parameters with respect to the triaxiality ratio. The triaxiality ratio, η, accounts for the combined effects of confinement pressure, deviatoric stress level, and multi-axial state of

Table 2. Identified Prony Series Coefficients and Time–Temperature Shift Factors

	HMA PG 76-22		WMA PG 76-22		HMA PG 64-22		WMA PG 64-22	
n	D_n (MPa^{-1})	λ_n (1/s)	D_n (MPa^{-1})	λ_n(1/s)	D_n (MPa^{-1})	λ_n(1/s)	D_n (MPa^{-1})	λ_n(1/s)
0	5.58×10^{-5}	–	5.63×10^{-5}	–	6.20×10^{-5}	–	3.85×10^{-5}	–
1	2.54×10^{-5}	4.35×10^{3}	2.37×10^{-5}	4.21×10^{3}	3.70×10^{-5}	1.65×10^{3}	3.54×10^{-5}	7.47×10^{4}
2	2.73×10^{-5}	4.01×10^{2}	2.40×10^{-5}	3.93×10^{2}	3.89×10^{-5}	1.78×10^{2}	4.33×10^{-5}	5.95×10^{3}
3	7.23×10^{-5}	3.69×10	6.23×10^{-5}	3.67×10	1.00×10^{-4}	1.92×10	6.91×10^{-5}	4.73×10^{2}
4	1.45×10^{-4}	3.40	1.23×10^{-4}	3.42	2.08×10^{-4}	2.07	1.52×10^{-4}	3.76×10
5	4.14×10^{-4}	3.13×10^{-1}	3.43×10^{-4}	3.19×10^{-1}	6.39×10^{-4}	2.24×10^{-1}	3.82×10^{-4}	2.99
6	9.94×10^{-4}	2.88×10^{-2}	8.12×10^{-4}	2.98×10^{-2}	1.26×10^{-3}	2.41×10^{-2}	1.13×10^{-3}	2.38×10^{-1}
7	2.94×10^{-3}	2.65×10^{-3}	2.41×10^{-3}	2.78×10^{-3}	4.97×10^{-3}	2.60×10^{-3}	3.61×10^{-3}	1.89×10^{-2}
8	5.96×10^{-3}	2.44×10^{-4}	4.23×10^{-3}	2.59×10^{-4}	8.35×10^{-3}	2.81×10^{-4}	1.77×10^{-2}	1.51×10^{-3}
9	2.07×10^{-2}	2.25×10^{-5}	1.50×10^{-2}	2.42×10^{-5}	2.44×10^{-2}	3.03×10^{-5}	2.39×10^{-2}	1.20×10^{-4}
				Shift parameters				
Temp	a_T	Log (a_T)	a_T	Log (a_T)	a_T	Log (a_T)	a_T	Log (a_T)
−10.0	13762.86	4.1387	11578.01	4.0636	5813.25	3.7644	4023.92	3.6046
4.4	112.96	2.0529	106.61	2.0278	76.11	1.8815	65.54	1.8165
21.1	0.7301	−0.1366	0.7311	−0.1360	0.7474	−0.1265	0.7527	−0.1234
37.8	0.0083	−2.0799	0.0082	−2.0880	0.0113	−1.9456	0.0120	−1.9194
54.4	0.0002	−3.7674	0.0002	−3.8184	0.0003	−3.5667	0.0003	−3.5622

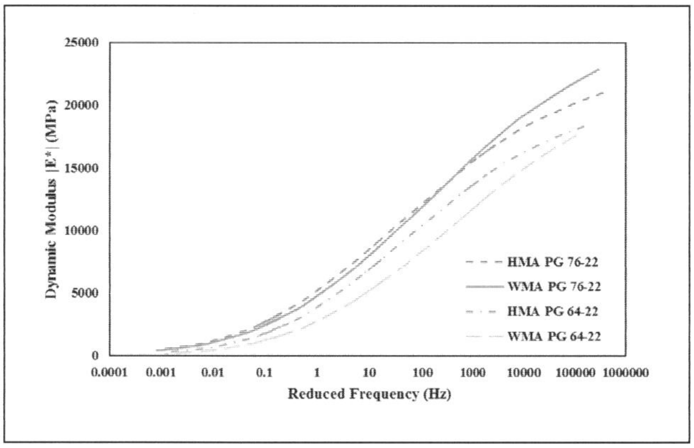

Figure 4. Master curve according to AASHTO T342 (*41*) and ASTM D3497-79 (*40*) for the dynamic modulus,E^*, using the sigmoidal fitting function in log–log scale.

stresses within a single scalar parameter. Figures 8 and 9 clearly show that the nonlinear parameters are highly correlated to the level of the triaxiality ratio and can be represented as a power law function of this parameter. Therefore, the following model is proposed for the NVE properties, such that:

$$g_i(\eta) = a_i\eta^{b_i},\ i = 1, 2;\ g_1g_2 = \alpha\eta^{\beta} \quad (13)$$

where a_i, b_i, α, and β are model parameters. Figures 8 and 9 clearly show that as the triaxiality ratio increases, the variation of the nonlinear parameters decreases. A high triaxiality ratio indicates a high level of confining pressure or low level of deviatoric stresses. Obviously, asphalt concrete demonstrates less variation in deformation at high confining pressures and low deviatoric stress levels. Moreover, as the confinement pressure increases, the same level of the deviatoric stress has a lesser effect on the nonlinear response of asphalt concrete materials. Therefore, asphalt concrete materials show less time-dependent response at high confinement levels.

The results presented in this study agree with the results reported by previous researchers (*35*, *45*). In addition, expressing the NVE responses using g_1 and g_2 through the expression of triaxiality ratio will considerably simplify the implementation of the NVE response of asphalt concrete materials. Eventually, application of the triaxiality ratio enhances the Schapery's (*23*) NVE model by considering the NVE response of asphalt concrete materials and structures subjected to the multi-axial state of stresses. Utilization of the NVE response as a function of the effective von Mises stress is a common practice for implementing Schapery's nonlinear model (*23*). In addition, the effect of mode of loading under multi-axial stress states can be included in Schapery's (*23*) model. This improvement is imperative for pressure-sensitive materials such as asphalt concrete, geomaterials, and polymer composites.

Conclusion

This study presents a straightforward procedure to characterize NVE response of asphalt concrete materials.

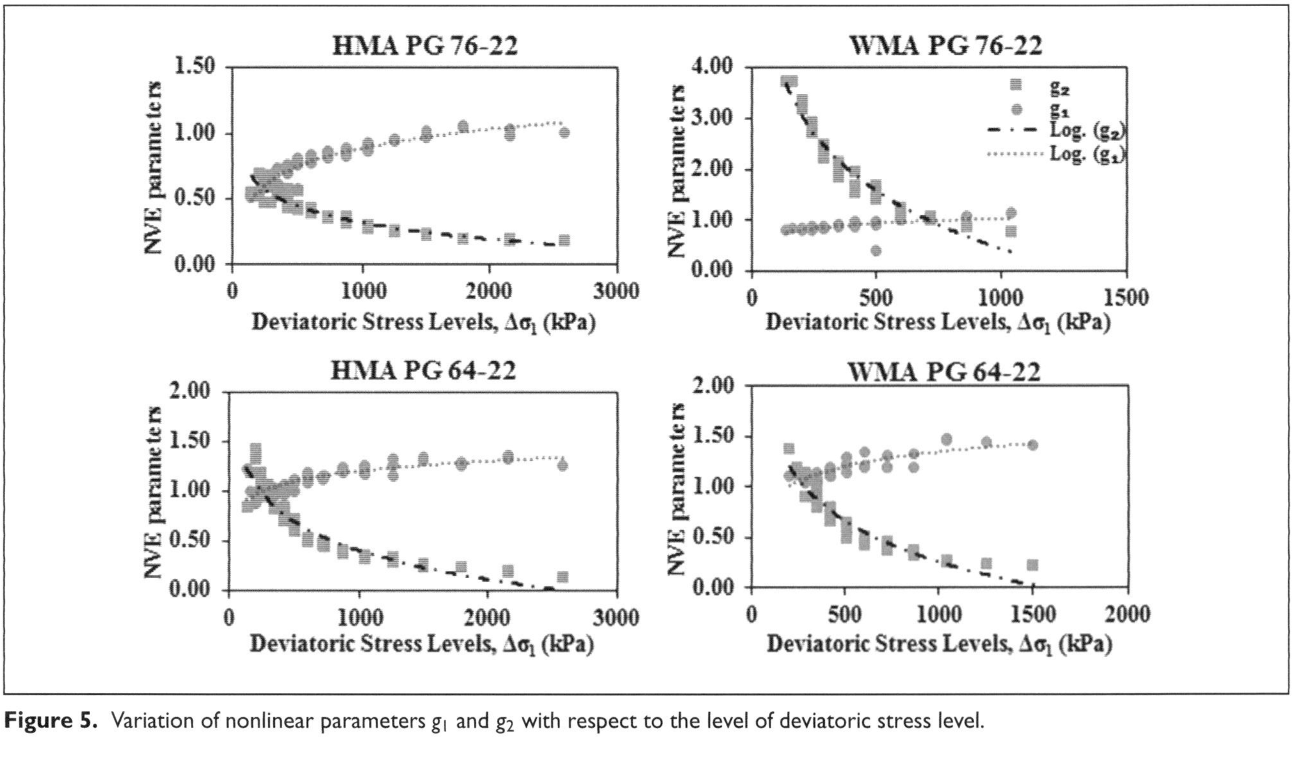

Figure 5. Variation of nonlinear parameters g_1 and g_2 with respect to the level of deviatoric stress level.

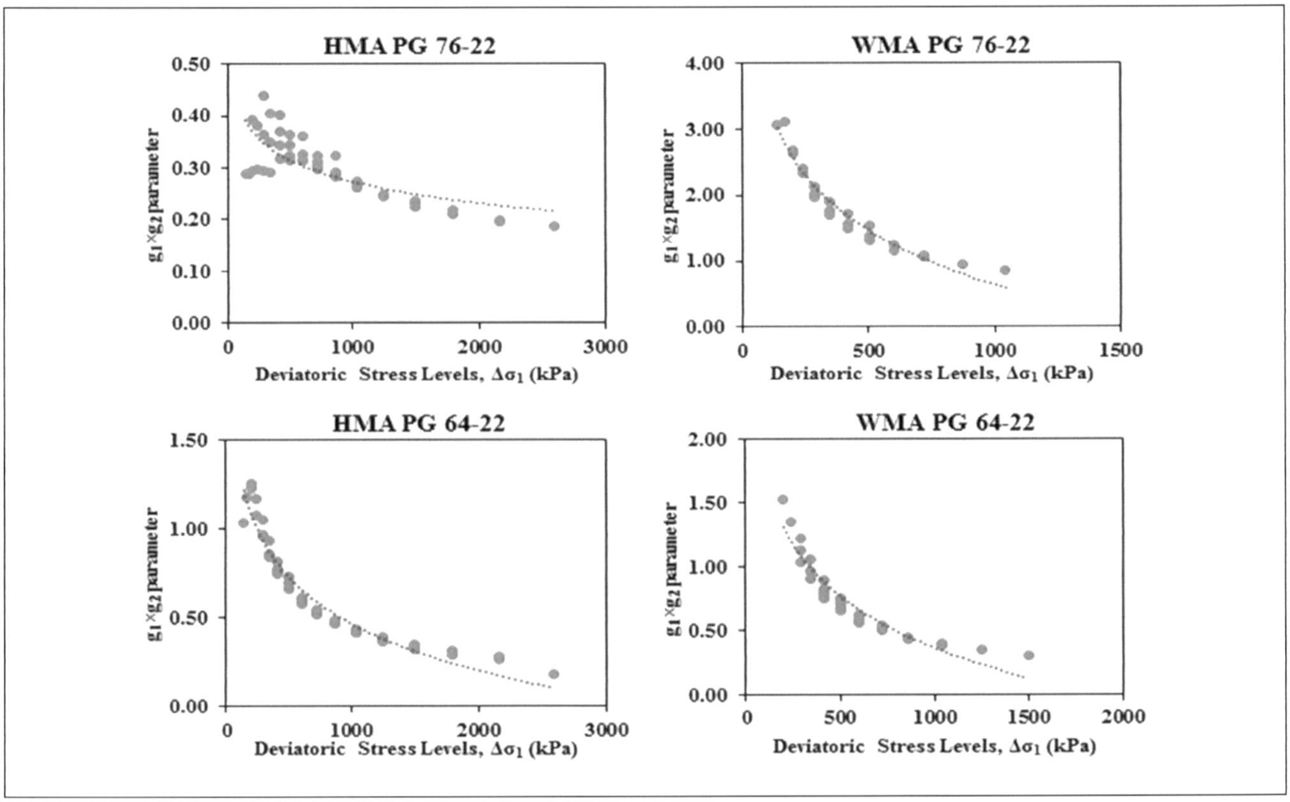

Figure 6. Variation of the nonlinear parameter $g_1 \times g_2$ with respect to the level of deviatoric stress level.

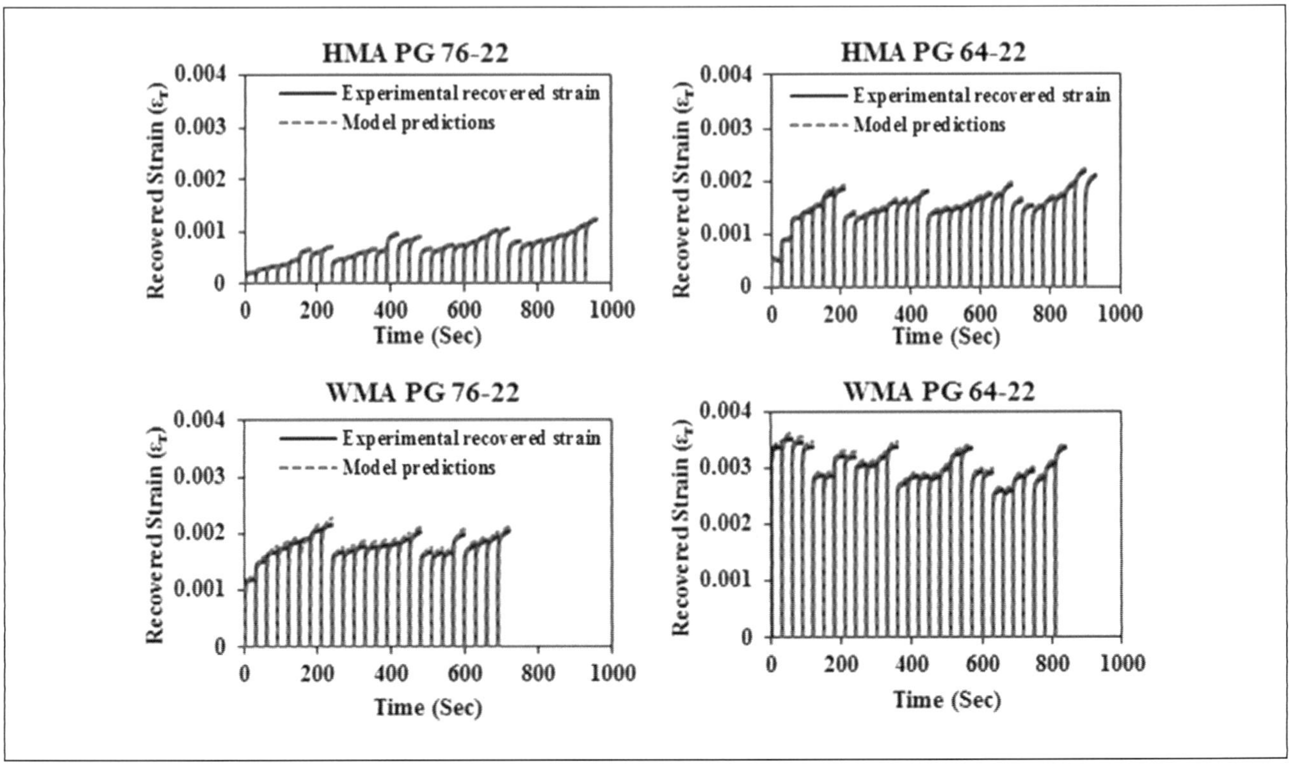

Figure 7. Comparison of the experimental measurements and the model predictions for the recovered strain response.

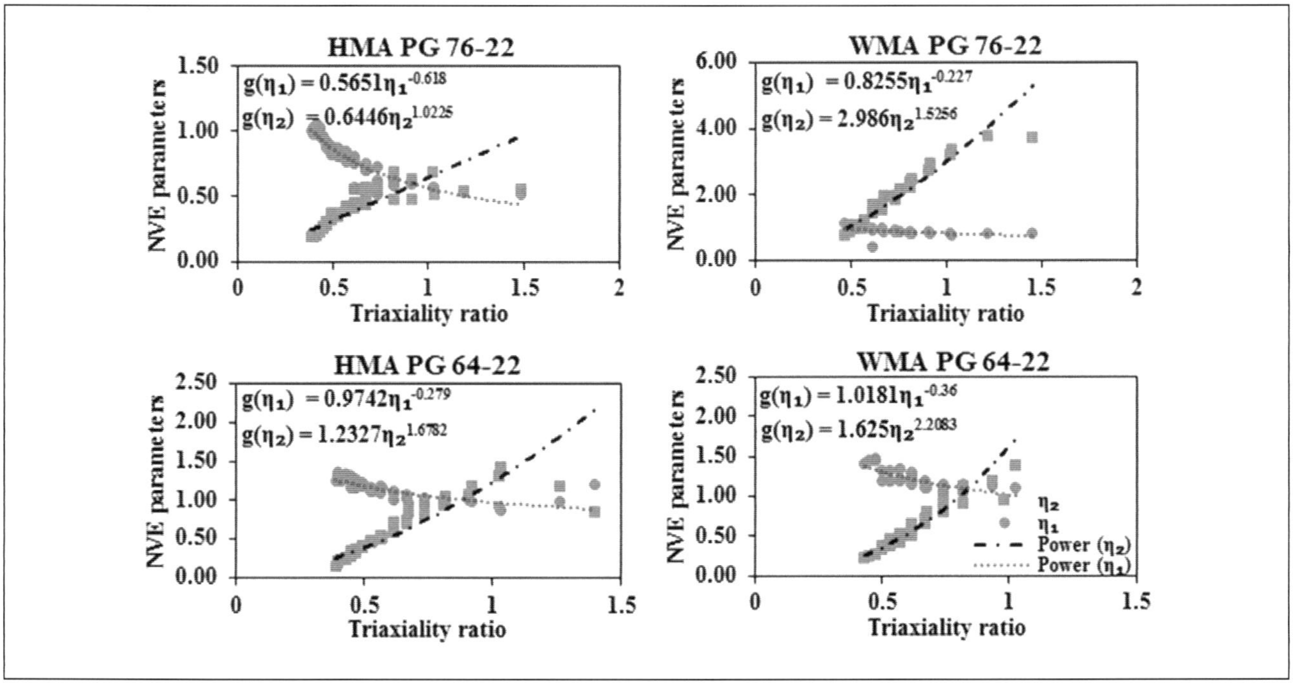

Figure 8. Variation of the nonlinear viscoelastic parameters g_1 and g_2 with respect to the triaxiality ratio.

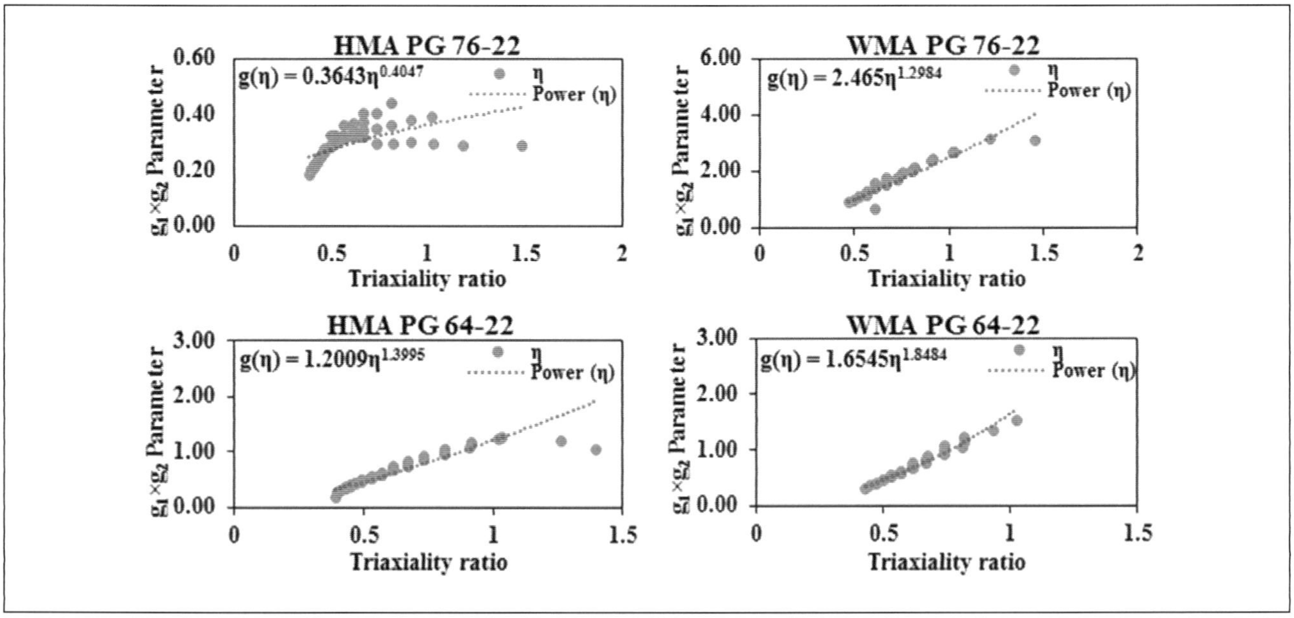

Figure 9. Variation of the nonlinear viscoelastic parameter $g_1 \times g_2$ with respect to the triaxiality ratio.

The proposed procedure was applied to extract LVE and NVE properties of asphalt mixtures used in the FAA's NAPMRC test sections. The nonlinear parameters were expressed in terms of the triaxiality ratio based on which a model was proposed. Finally, model predictions were compared with experimental measurements for validations purposes. The experimental studies, theoretical developments, and data analysis presented in this paper lead to the following conclusions:

- The proposed procedure along with the dynamic modulus and RCRT-VS tests can be used effectively to characterize NVE responses of asphalt concrete materials;

- The NVE response of asphalt concrete strongly depends on confining pressure, deviatoric stress level, and multi-axial state of stresses. Therefore, uniaxial tests such as uniaxial creep-recovery tests cannot be used accurately to characterize NVE properties of asphalt concrete.
- The NVE properties are highly correlated to the triaxiality ratio that accounts for confinement pressure, deviatoric stress level, and multi-axial stress states;
- The proposed model for NVE properties as a function of triaxiality ratio can be used effectively to simulate the nonlinear response of pavements under traffic when specific locations in pavement structures are subjected to general multi-axial state of stresses.
- The NVE response of asphalt concrete materials can be effectively expressed as a function of the triaxiality ratio.
- The nonlinear parameter g_1 decreases with the increase in triaxiality ratio leading to less creep strain either as the confinement level increases or the deviatoric stress level decreases.
- The nonlinear parameter g_2 increases with the increase in triaxiality ratio leading to higher recovery of viscoelastic strain during rest period as the confinement level increases and the deviatoric stress level decreases.
- Comparisons between the characterized NVE model and experimental measurements illustrate the capabilities of the proposed model and characterization procedure in capturing NVE response of asphalt concrete materials.

Acknowledgments

The work described in this paper was supported by the FAA's Airport Technology Research and Development Branch through Grant 15-G-011.

References

1. Morrison, G. R., J. K. Lee, and S. A. M. Hesp. Chlorinated Polyolefins for Asphalt Binder Modification. *Journal of Applied Polymer Science*, Vol. 54, No. 2, 1994, pp. 231–240.
2. Tashman, L., E. Masad, D. Little, and H. Zbib. A Microstructure-Based Viscoplastic Model for Asphalt Concrete. *International Journal of Plasticity*, Vol. 21, No. 9, 2005, pp. 1659–1685.
3. Uzan, J. Asphalt Concrete Characterization for Pavement Performance Prediction. *Asphalt Paving Technology*, Vol. 65, 1996, pp. 573–607.
4. Arabali, P., M. Sakhaeifar, T. Freeman, B. Wilson, and J. Borowiec. Decision-Making Tool for the Selection of Pavement Preservation Treatments in General Aviation Airport Pavements. In *International Conference on Transportation and Development 2016*, ASCE, Houston, Texas, 2016, pp. 30–41.
5. Arabali, P., M. S. Sakhaeifar, T. J. Freeman, B. T. Wilson, and J. D. Borowiec. Decision-Making Guideline for Preservation of Flexible Pavements in General Aviation Airport Management. *Journal of Transportation Engineering, Part B: Pavements*, Vol. 143, No. 2, 2017, p. 04017006.
6. Anand, P., A. Nahvi, H. Ceylan, V. D. Pyrialakou, K. Gkritza, K. Gopalakrishnan, S. Kim, and P. C. Taylor. Energy and Financial Viability of Hydronic Heated Pavement Systems. In *Automated Vehicles Symposium*, DOT/FAA/TC-17/47, San Francisco, CA, 2017, 95p.
7. Nahvi, A., S. S. Sadati, K. Cetin, H. Ceylan, A. Sassani, and S. Kim. Towards Resilient Infrastructure Systems for Winter Weather Events: Integrated Stochastic Economic Evaluation of Electrically Conductive Heated Airfield Pavements. *Sustainable Cities and Society*, Vol. 41, 2018, pp. 195-204.
8. Shen, W., H. Ceylan, K. Gopalakrishnan, S. Kim, and A. Nahvi. Sustainability Assessment of Alternative Snow-Removal Methods for Airport Apron Paved Surfaces. *Automated Vehicles Symposium*, DOT/FAA/TC-17/34, San Francisco, CA, 2017, 67p.
9. Nobakht, M., M. S. Sakhaeifar, and D. E. Newcomb. Selection of Structural Overlays Using Asphalt Mixture Performance. *Journal of Materials in Civil Engineering*, Vol. 29, No. 11, 2017, p. 04017209.
10. Perzyna, P. Thermodynamic Theory of Viscoplasticity. *Advances in Applied Mechanics*, Vol. 11, 1971, pp. 313–354.
11. Masad, E., L. Tashman, D. Little, and H. Zbib. Viscoplastic Modeling of Asphalt Mixes with the Effects of Anisotropy, Damage and Aggregate Characteristics. *Mechanics of Materials*, Vol. 37, No. 12, 2005, pp. 1242–1256.
12. Saadeh, S., E. Masad, and D. Little. Characterization of Asphalt Mix Response under Repeated Loading Using Anisotropic Nonlinear Viscoelastic-Viscoplastic Model. *Journal of Materials in Civil Engineering*, Vol. 19, No. 10, 2007, pp. 912–924.
13. Darabi, M. K., R. K. A. Al-Rub, E. A. Masad, C.-W. Huang, and D. N. Little. A Thermo-Viscoelastic–Viscoplastic–Viscodamage Constitutive Model for Asphaltic Materials. *International Journal of Solids and Structures*, Vol. 48, No. 1, 2011, pp. 191–207.
14. Kim, Y. R., and D. N. Little. One-Dimensional Constitutive Modeling of Asphalt Concrete. *Journal of Engineering Mechanics*, Vol. 116, No. 4, 1990, pp. 751–772.
15. Seibi, A. C., M. G. Sharma, G. A. Ali, and W. J. Kenis. Constitutive Relations for Asphalt Concrete under High Rates of Loading. *Transportation Research Record: Journal of the Transportation Research Board*, 2001. 1767: 111–119.
16. Park, S. W., Y. R. Kim, and R. A. Schapery. A Viscoelastic Continuum Damage Model and Its Application to Uniaxial Behavior of Asphalt Concrete. *Mechanics of Materials*, Vol. 24, No. 4, 1996, pp. 241–255.
17. Lee, H. J., J. S. Daniel, and Y. R. Kim. Continuum Damage Mechanics-Based Fatigue Model of Asphalt Concrete. *Journal of Materials in Civil Engineering*, Vol. 12, No. 2, 2000, pp. 105–112.

18. Drescher, A., N. Kringos, and T. Scarpas. On the Behavior of a Parallel Elasto-Visco-Plastic Model for Asphaltic Materials. *Mechanics of Materials*, Vol. 42, No. 2, 2010, pp. 109–117.
19. Villani, M. M., C. Kasbergen, A. Scarpas, and D. Lo Presti. Mechanistic Procedure for Parameter Determination of Multiplicative Decomposition Based Constitutive Models. *International Journal of Pavement Engineering*, Vol. 18, No. 5, 2017, pp. 391–403.
20. Kose, S., M. Guler, H. Bahia, and E. Masad. Distribution of Strains Within Hot-Mix Asphalt Binders: Applying Imaging and Finite-Element Techniques. *Transportation Research Record: Journal of the Transportation Research Board*, 2000. 1728: 21–27.
21. Deshpande, V., and D. Cebon. Steady-State Constitutive Relationship for Idealised Asphalt Mixes. *Mechanics of Materials*, Vol. 31, No. 4, 1999, pp. 271–287.
22. Masad, E., and N. Somadevan. Microstructural Finite-Element Analysis of Influence of Localized Strain Distribution on Asphalt Mix Properties. *Journal of Engineering Mechanics*, Vol. 128, No. 10, 2002, pp. 1105–1114.
23. Schapery, R. A. On the Characterization of Nonlinear Viscoelastic Materials. *Polymer Engineering & Science*, Vol. 9, No. 4, 1969, pp. 295–310.
24. Huang, C.-W., E. Masad, A. H. Muliana, and H. Bahia. Nonlinearly Viscoelastic Analysis of Asphalt Mixes Subjected to Shear Loading. *Mechanics of Time-Dependent Materials*, Vol. 11, No. 2, 2007, pp. 91–110.
25. Huang, C.-W., R. K. Abu Al-Rub, E. A. Masad, D. N. Little, and G. D. Airey. Numerical Implementation and Validation of a Nonlinear Viscoelastic and Viscoplastic Model for Asphalt Mixes. *International Journal of Pavement Engineering*, Vol. 12, No. 4, 2011, pp. 433–447.
26. Haj-Ali, R. M., and A. H. Muliana. Numerical Finite Element Formulation of the Schapery Non-Linear Viscoelastic Material Model. *International Journal for Numerical Methods in Engineering*, Vol. 59, No. 1, 2004, pp. 25–45.
27. Karimi, M. M., N. Tabatabaee, H. Jahanbakhsh, and B. Jahangiri. Development of a Stress-Mode Sensitive Viscoelastic Constitutive Relationship for Asphalt Concrete: Experimental and Numerical Modeling. *Mechanics of Time-Dependent Materials*, Vol. 21, No. 3, 2017, pp. 383–417.
28. Karimi, M. M., N. Tabatabaee, B. Jahangiri, and M. K. Darabi. Constitutive Modeling of Hardening-Relaxation Response of Asphalt Concrete in Cyclic Compressive Loading. *Construction and Building Materials*, Vol. 137, 2017, pp. 169–184.
29. Schapery, R. Nonlinear Viscoelastic Solids. *International Journal of Solids and Structures*, Vol. 37, No. 1, 2000, pp. 359–366.
30. Schapery, R. A. *Further Development of a Thermodynamic Constitutive Theory: Stress Formulation*. Purdue University School of Aeronautics, Astronautics and Engineering Sciences, 1969.
31. Christensen, R. M. On Obtaining Solutions in Nonlinear Viscoelasticity. *Journal of Applied Mechanics*, Vol. 35, No. 1, 1968, pp. 129–133.
32. Kim, Y.-R., D. Allen, and D. Little. Computational Constitutive Model for Predicting Nonlinear Viscoelastic Damage and Fracture Failure of Asphalt Concrete Mixtures. *International Journal of Geomechanics*, Vol. 7, No. 2, 2007, pp. 102–110.
33. Masad, E. A., C.-W. Huang, J. DAngelo, and D. N. Little. Characterization of Asphalt Binder Resistance to Permanent Deformation Based on Nonlinear Viscoelastic Analysis of Multiple Stress Creep Recovery (MSCR) Test. *Journal of the Association of Asphalt Paving Technologists*, Vol. 78, 2009.
34. Ye, Y., X. Yang, and C. Chen. Modified Schapery's Model for Asphalt Sand. *Journal of Engineering Mechanics*, Vol. 136, No. 4, 2010, pp. 448–454.
35. Rahmani, E., M. K. Darabi, R. K. A. Al-Rub, E. Kassem, E. A. Masad, and D. N. Little. Effect of Confinement Pressure on the Nonlinear-Viscoelastic Response of Asphalt Concrete at High Temperatures. *Construction and Building Materials*, Vol. 47, 2013, pp. 779–788.
36. Motamed, A., A. Bhasin, and K. M. Liechti. Interaction Nonlinearity in Asphalt Binders. *Mechanics of Time-Dependent Materials*, Vol. 16, No. 2, 2012, pp. 145–167.
37. Motamed, A., A. Bhasin, and K. M. Liechti. Constitutive Modeling of the Nonlinearly Viscoelastic Response of Asphalt Binders; Incorporating Three-Dimensional Effects. *Mechanics of Time-Dependent Materials*, Vol. 17, No. 1, 2013, pp. 83–109.
38. ASTM C136/C136M-14. *Standard Test Method for Sieve Analysis of Fine and Coarse Aggregates*. ASTM International, West Conshohocken, Pa., 2014.
39. AASHTO T27-14. Standard Method of Test for Sieve Analysis of Fine and Coarse Aggregates. *Standard Specifications for Transportation Materials and Methods of Sampling and Testing*, Washington, D.C., 2016.
40. ASTM D3497-79. *Standard Test Method for Dynamic Modulus of Asphalt Mixtures*. ASTM International, West Conshohocken, Pa., 2003.
41. AASHTO T342-11. *Standard Method of Test for Determining Dynamic Modulus of Hot-Mix Asphalt Concrete Mixtures*. American Association of State Highway and Transportation Officials, Washington, D.C., 2011.
42. Pellinen, T. K., M. W. Witczak, and R. F. Bonaquist. Asphalt Mix Master Curve Construction using Sigmoidal Fitting Function with Non-Linear Least Squares Optimization. In *Recent Advances in Materials Characterization and Modeling of Pavement Systems*, 15th Engineering Mechanics Division Conference, 2004. pp. 83–101.
43. Glover, C. J., R. R. Davison, and N. Vassiliev. *A New Method for Simulating Hot-Mix Plant Asphalt Aging*. Texas Transportation Institute, the Texas A&M University System, 2001.
44. Park, S., and R. Schapery. Methods of Interconversion between Linear Viscoelastic Material Functions. Part I—A Numerical Method Based on Prony Series. *International Journal of Solids and Structures*, Vol. 36, No. 11, 1999, pp. 1653–1675.
45. Pellinen, T., and M. Witczak. Stress Dependent Master Curve Construction for Dynamic (Complex) Modulus (With Discussion). *Journal of the Association of Asphalt Paving Technologists*, Vol. 71, 2002, pp. 281–309.

The Standing Committee on Structural Requirements of Asphalt Mixtures (AFK50) peer-reviewed this paper (18-03704).

Article

TRR
JOURNAL OF THE TRANSPORTATION RESEARCH BOARD

Transportation Research Record
2018, Vol. 2672(28) 493–502

DOI: 10.1177/0361198118797808
journals.sagepub.com/home/trr
SAGE

Improving the Reliability of Damage Characteristic Curves in the Simplified Viscoelastic Continuum Damage Model

Kangjin Lee[1], Cassie Castorena[1], and Y. Richard Kim[1]

Abstract

One of the major advantages of the cyclic fatigue test (AASHTO TP 107) is that the results can be used to calibrate the Simplified Viscoelastic Continuum Damage (S-VECD) model, which is used for mechanistic pavement performance predictions. The crux of the S-VECD model is the damage characteristic curve, which has been shown to be independent of mode of loading, loading history, and temperature. Consequently, a model can be fitted to the damage characteristic curve and used to predict the damage response for any given loading history of interest using limited test results. AASHTO TP 107 currently lacks a specific procedure for fitting a model to the damage characteristic curve and evaluating the repeatability of test replicates. In this study, a robust and practical method is proposed for fitting a power law model to the damage characteristic curve. The proposed fitting method was verified using cyclic fatigue test results of 19 mixtures sourced from the United States, Canada, and South Korea. In addition, a means to evaluate the specimen-to-specimen variability of damage characteristic curves using a shape factor is proposed. Thresholds for acceptable variability in the shape factor were derived using confidence interval analysis and verified through FlexPAVE™ pavement performance predictions. The findings of this study can be used to improve the reliability of the damage characteristic curves derived from cyclic fatigue tests for pavement performance predictions.

The accurate prediction of pavement performance is crucial to the design and evaluation of pavement materials and structures. Fatigue cracking is one of the major distress types in asphalt pavements, and thus, a significant amount of effort has been devoted to developing a test protocol and associated model to characterize the fatigue cracking resistance of asphalt mixtures. Traditionally, asphalt concrete fatigue performance has been modeled using empirical approaches that require extensive testing at multiple loading amplitudes (*1*). The empiricism and extensive testing requirements of traditional approaches for fatigue analysis have prompted researchers to apply mechanistic models to characterize the fatigue performance of asphalt mixtures.

Cyclic fatigue testing based on AASHTO TP 107 enables the practical, mechanistic performance characterization of asphalt concrete using the Asphalt Mixture Performance Tester (AMPT). The AMPT was developed under NCHRP Project 9-29 to enable the practical performance characterization of asphalt concrete (*2*). One of the major advantages of cyclic fatigue testing is that the results can be used to calibrate the Simplified Viscoelastic Continuum Damage (S-VECD) model, which is used for mechanistic asphalt mixture and pavement performance predictions. The crux of the S-VECD model is the so-called damage characteristic curve that is independent of mode of loading, loading history, and temperature. Consequently, the prediction of the damage response to any given loading history of interest is possible using limited test results, thus making the fatigue cracking characterization of asphalt mixtures much more efficient than empirical methods do.

AASHTO TP 107 requires the testing of at least three specimens. With the use of the newly developed 38-mm diameter small specimen geometry, the required number of test replicates can be completed within one working day, making the test even more efficient for comprehensive fatigue characterization (*3*). Previously, 100-mm diameter specimens were used for cyclic fatigue testing, which yielded a single test specimen per gyratory sample.

[1]Department of Civil, Construction, and Environmental Engineering, North Carolina State University, Raleigh, NC

Corresponding Author:
Address correspondence to Y. Richard Kim: kim@ncsu.edu

In contrast, using the 38-mm diameter small specimen geometry, four small specimens can be extracted from a single gyratory sample. In addition, the small sample size reduces the time required for thermal equilibration. The small sample size also allows the use of quick-setting epoxy to affix end platens to small specimens that require only 1 h of curing prior to testing. Large specimens, by contrast, require the use of a slower setting epoxy to allow sufficient working time that takes 24 h for the specimen to cure (*4*).

The S-VECD damage characteristic curve represents the relationship between the pseudo secant modulus (C), which is commonly referred to simply as pseudo stiffness, and the internal state variable that represents microstructural damage (S). The pseudo stiffness initiates from a value of 1.0 when the material is undamaged; this value decreases as a result of damage growth under repeated cyclic loading (*5*). The accurate fitting of the damage characteristic curve using reliable data is a critical component of the use of cyclic fatigue test results for performance predictions. However, AASHTO TP 107 currently lacks a standard procedure for fitting a model to damage characteristic curves.

AASHTO TP 107 also lacks provisions for the assessment of the repeatability of test replicates. Consequently, a method is needed that can evaluate the repeatability of test replicates and then identify outliers that should be neglected in the analysis.

In this study, a robust and practical method is proposed for fitting a power law model to damage characteristic curves. In addition, a means to evaluate the specimen-to-specimen variability in damage characteristic curves using a new shape factor is proposed.

Objectives

The objectives of this study are to:

1. Optimize the fitting of a model to the S-VECD damage characteristic curves obtained from AMPT cyclic fatigue testing.
2. Develop a means to evaluate the specimen-to-specimen variability in damage characteristic curves and develop corresponding thresholds for acceptable variability for reliable performance predictions.

Materials

The North Carolina State University (NCSU) research group investigated 19 mixtures with various nominal maximum aggregate sizes (NMASs) and asphalt binder types for this study. The study was performed under the National Cooperative Highway Research Program Innovations Deserving Exploratory Analysis (NCHRP IDEA) N181 project, the Federal Highway Administration (FHWA) -sponsored Hot-Mix Asphalt Performance-Related Specifications project, and the Transportation Pooled Fund Project TPF-5(230). The NCHRP IDEA N181 mixtures, RS9.5D, SM12.5A, RI19.0B, and RB25.0B, were used to optimize the damage characteristic curve model-fitting procedure. The other mixtures were selected to evaluate specimen-to-specimen variability. These mixtures were selected because the results of test replicates demonstrated relatively high variability relative to the other mixtures in the NCSU database. Table 1 provides details about the mixtures evaluated in this study. The test results for these mixtures have been presented previously in research papers (*6–10*).

Pavement Performance Predictions

Proposed thresholds for acceptable specimen-to-specimen variability were derived using confidence interval analysis. The practical significance of the thresholds was evaluated by analyzing the sensitivity of the FlexPAVE™ pavement performance predictions to the variability in the damage characteristic curves.

The FlexPAVE™ program was recently developed under the FHWA-sponsored Hot-Mix Asphalt Performance-Related Specifications project (*8*). FlexPAVE™ performs three-dimensional analysis of pavement structures in a computationally efficient manner to determine the stress and strain in a pavement system and to predict pavement performance using mechanics-based models. To this end, FlexPAVE™ utilizes layered viscoelastic analysis and a time-scale separation scheme to compute stress and strain under moving traffic simulations. It can also capture the effects of viscoelasticity, thermal stress, and traffic loading. Furthermore, the program combines the concepts of Fourier transform and finite element discretization in order to provide simulation times that are orders of magnitude shorter than those required in conventional three-dimensional finite element methods (*11*). The entire procedure, from entering inputs to viewing outputs, is operated via a user-friendly graphic interface that is similar to that found in Pavement ME Design software. The FlexPAVE™ program has been designed to use pavement temperature data provided by the Enhanced Integrated Climate Model (EICM). FlexPAVE™ has been used in various studies to investigate the effects of different material and pavement structures on pavement performance (*7, 8, 12, 13*). Several of these studies have demonstrated that FlexPAVE™ can capture the effect of different pavement design parameters on pavement distress development by comparing the simulation results to

Table 1. Asphalt Mixture Information

Mixture type	NMAS (mm)	Asphalt binder	Source
NCHRP IDEA-RS9.5D	9.5	PG 76-22	North Carolina
NCHRP IDEA-SM12.5A	12.5	PG 64-22	Virginia
NCHRP IDEA-RI19.0B	19	PG 64-22	North Carolina
NCHRP IDEA-RB25.0B	25	PG 64-22	North Carolina
NCAT-OGFC	12.5	PG 76-22	Alabama
KEC-BB3	25	PG 64-22	South Korea
MIT-C1	16	PG 58-28	Manitoba, Canada
MIT-E1	16	PG 58-28	Manitoba, Canada
MIT-E2	16	PG 58-28	Manitoba, Canada
MIT-R50R	16	PG 58-28	Manitoba, Canada
NHe20LC	12.5	PG 64-28	New Hampshire
VTe00LC	9.5	PG 64-28	Vermont
VTe20LC	9.5	PG 64-28	Vermont
VTe30LC	9.5	PG 64-28	Vermont
VTe40LC	9.5	PG 64-28	Vermont
NH-5820	12.5	PG 58-28	New Hampshire
NH-6420	12.5	PG 64-28	New Hampshire
NH-6440	12.5	PG 64-28	New Hampshire
RAPCAP	12.5	PG 64-28	South Korea

field observations of in-service pavements. Detailed information about the FlexPAVE™ program can be found in the aforementioned research papers (*7*, *8*, *12*, *13*).

Fitting of Damage Characteristic Curve

According to AASHTO TP 107, the damage characteristic curve can be fitted using the exponential and power law models presented in equations 1 and 2, respectively:

$$C = e^{aS^b} \quad (1)$$

$$C = 1 - yS^z \quad (2)$$

where

C is pseudo stiffness;

S is the damage parameter;

a and b are the fitting coefficients for the exponential model; and

y and z are the fitting coefficients for the power law model (*2*).

A logarithmic transformation can be used to linearize the power law model, as shown in:

$$\log(1 - C) = \log(y) + z\log(S) \quad (3)$$

Equation 3 indicates that log(y) is the intercept of a line formed as log($1-C$) versus log(S), and z is the slope of this line. Thus, the values of the power law model-fitting coefficients can be calculated easily.

In contrast, error minimization optimization must be performed to determine the values of the exponential model-fitting coefficients. Consequently, a risk of bias can emerge in the results of the error minimization as to the initial guess of the model coefficient values. Figure 1 presents comparisons of the power law and exponential damage characteristic curves for the NCHRP IDEA N-181 project mixtures. These graphs show that the power law model better represents the shape of the damage characteristic curves than the exponential model. In addition, it should be noted that the power law model can be easily differentiated and integrated, which allows for easier derivation of models for performance predictions compared with the exponential model. Based on these advantageous characteristics, the power law model is proposed for use in practice.

AASHTO TP 107 requires a minimum of three tests with different strain levels to achieve a range of fatigue lives (N_f) for the failure criterion characterization. However, data sampling frequency remains the same for the fatigue tests with different fatigue lives. Consequently, without appropriate adjustments, longer tests produce more data points and the fitting of the damage characteristic curve may be biased toward longer tests. Damage (S) is calculated using numerical integration, and AASHTO TP 107 specifies the calculation of damage using smaller time increments at the start of the fatigue test when damage growth is fast and using larger time increments after significant fatigue damage has accumulated. Consequently, without appropriate filtering of the data, the model fitting of the damage characteristic curve will be biased toward the portion of the data that corresponds to low damage levels. Figure 2*a* demonstrates this phenomenon by the damage characteristic curve data and corresponding power law model fits for two of the NCHRP IDEA N181 mixtures. The figure

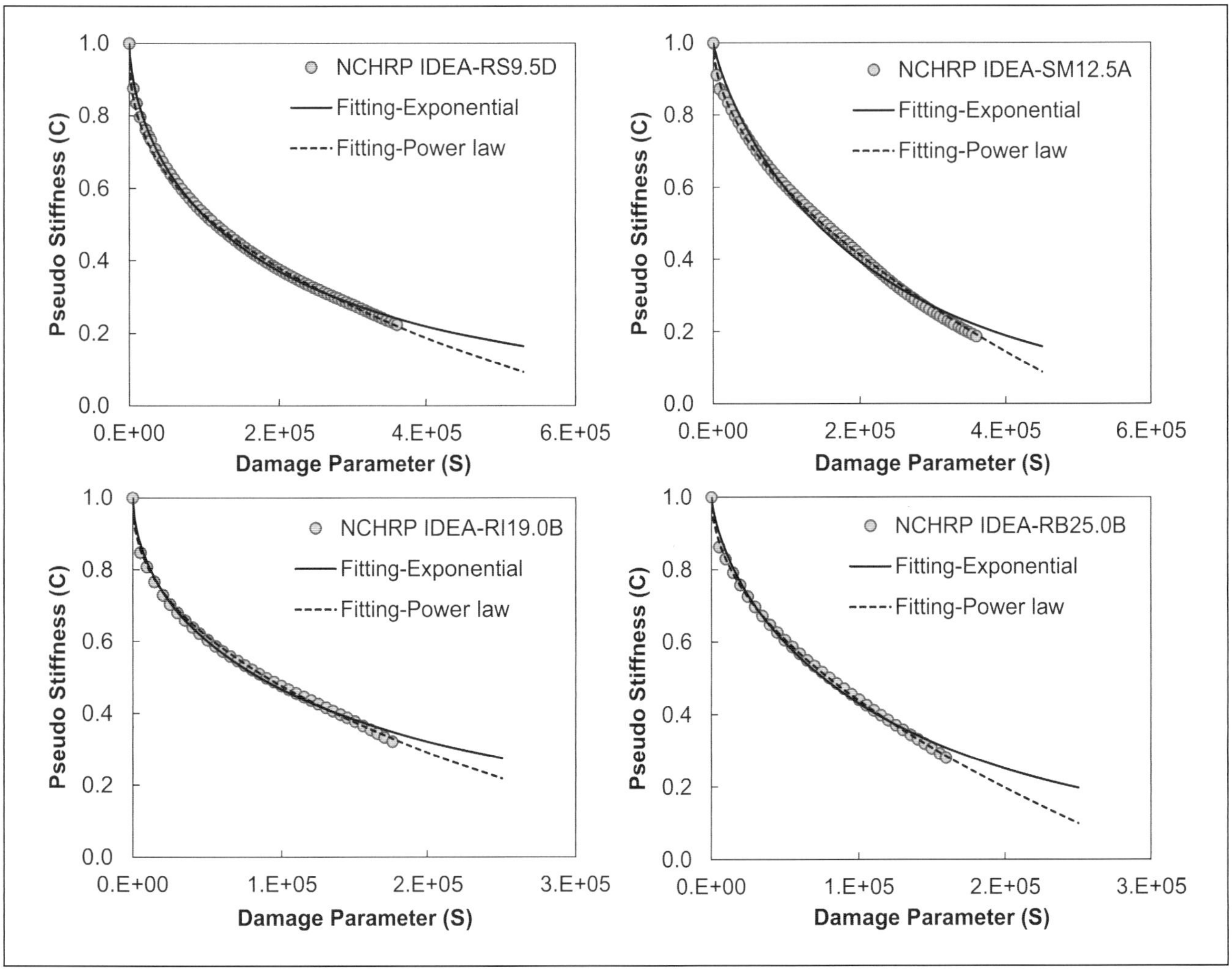

Figure 1. Comparisons of exponential and power law models for fitting damage characteristic curve.

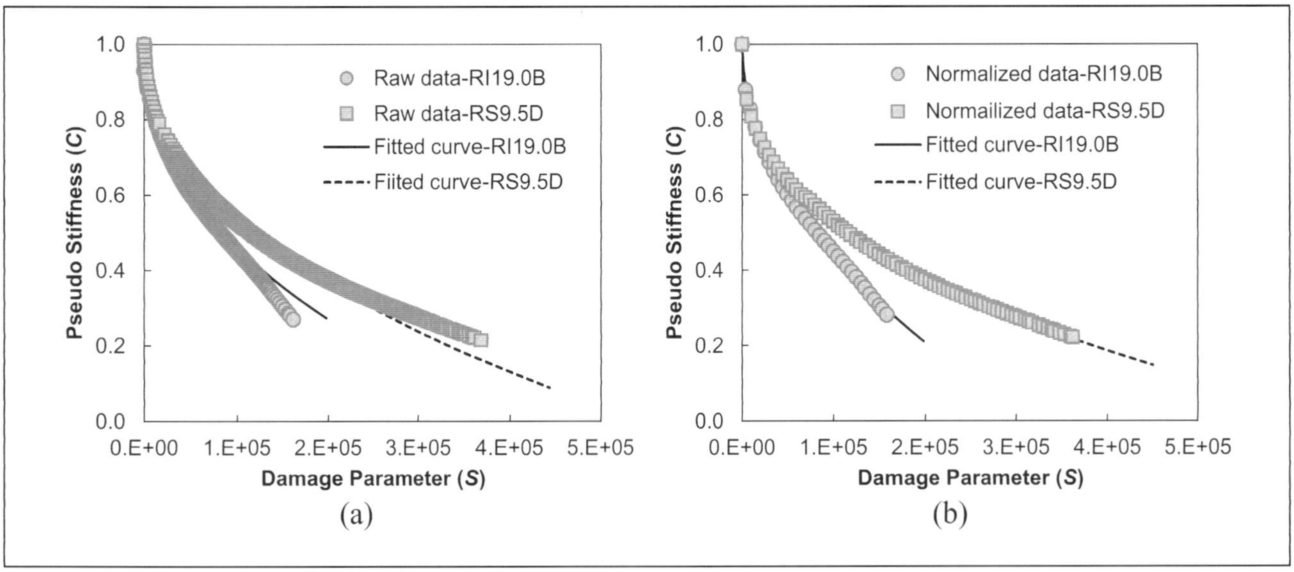

Figure 2. RS9.5D and RI19.0B mixture damage characteristic curves: model fitting using (*a*) raw data and (*b*) normalized data.

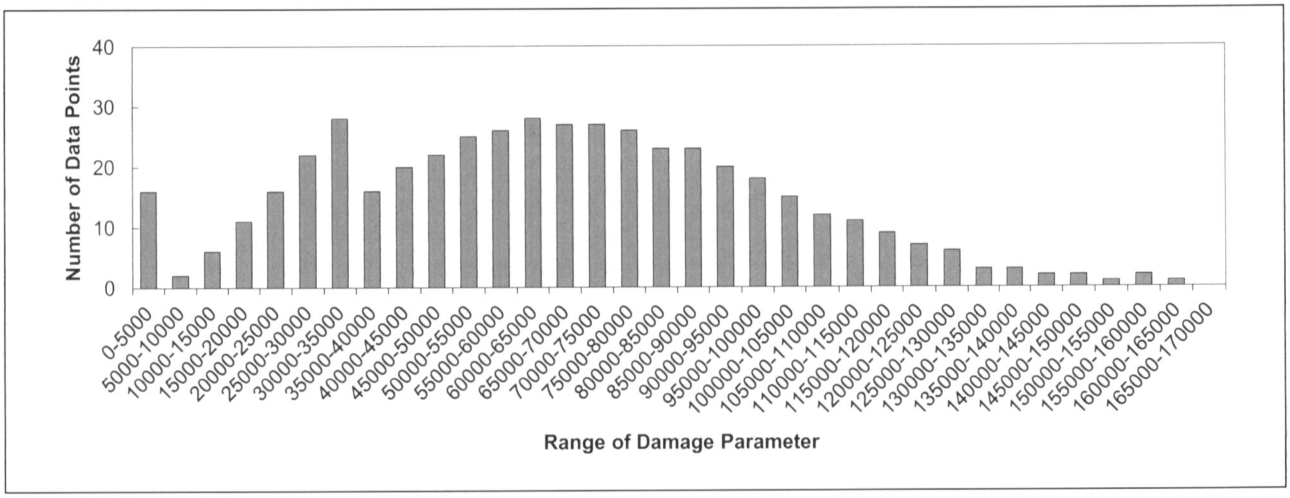

Figure 3. Data point distribution on damage characteristic curves for NCHRP IDEA-RI19.0B mixture.

shows that the power law model fits the data well at low damage (S) values but not at high damage (S) values. Figure 3 presents a histogram of the number of data points as a function of damage. The figure shows non-uniform distribution in the number of data points that correspond to different ranges of S values, which explains the bias in the fitting.

To remove the bias of the model fit toward the low S values, the data were filtered prior to model fitting using equal S increments of 5,000. The S increments of 5,000 were determined based on NCHRP IDEA N181 mixtures and verified using other mixtures evaluated in this study. Although the shorter S spacing is better for model fitting, the maximum S spacing between two adjacent data points in the damage characteristic curves was approximately 5,000 in this study. In order to fit the damage characteristic curve properly, at least one data point should be present in the S intervals. Therefore, S increments of 5,000 were used to normalize the data points for all the mixtures evaluated in this study. Figure 2*b* presents the normalized data points and the corresponding fitted curves. This figure shows that normalization using equal S increments greatly improves the quality of the model fit. Therefore, it is recommended that data should be filtered to produce S increments of 5,000 prior to fitting the power law model to the damage characteristic curves. This recommended damage characteristic curve model-fitting procedure was applied to all of the mixtures evaluated in this study.

Quantifying Specimen-to-Specimen Variability

Shape Factor

According to AASHTO TP 107, a minimum of three cyclic fatigue tests are required to calibrate the damage characteristic curve model. Currently, AASHTO TP 107 does not include guidance for assessing the repeatability of the test results and identifying outliers. Therefore, developing an effective analysis protocol to identify and exclude outliers could improve the reliability of fatigue performance predictions. In this study, when the damage characteristic curve model-fitting procedure was applied to the results of several test replicates, all of the data were aggregated to determine the fitting coefficients using the linearization of the power law model given in Equation 3. A shape factor is proposed in this study to evaluate the specimen-to-specimen variability of the damage characteristic curves based on the difference between the damage characteristic curves of individual test replicates and the power law model fit using all the test replicates. The shape factor is calculated as the ratio of the area above the damage characteristic curve for an individual test replicate (i.e., $A_{measured}$) to the area above the fitted model damage characteristic curve (i.e., $A_{predicted}$), as shown in Figure 4. Equations 4 and 5 can be used to calculate the area above the damage characteristic curve and the corresponding shape factor, respectively:

$$A = \int_0^{S_f} (1 - C)dS \tag{4}$$

$$Shape\ Factor = \frac{A_{measured}}{A_{predicted}} \tag{5}$$

where A is the area above the damage characteristic curve, and S_f is the damage at failure, defined by the peak in phase angle in cyclic fatigue tests.

The shape factor, therefore, is an indicator of the difference between the fitted curve and the results of an individual test replicate. If the shape factor is close to 1.0, the replicate has good agreement with the fitted

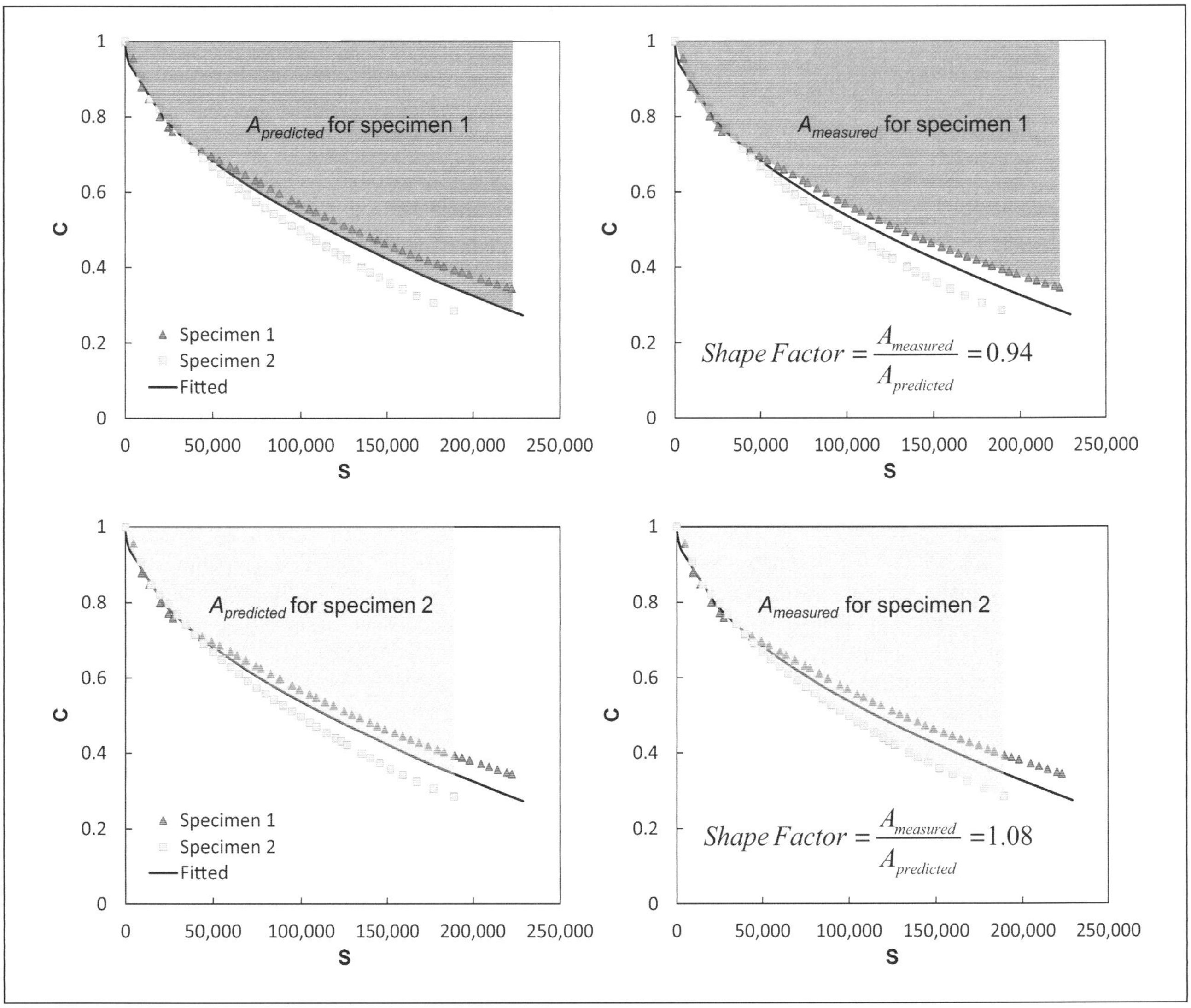

Figure 4. Illustrations of shape factors.

curve and, thus, indicates a valid specimen. However, if one of the replicates has a shape factor that is significantly higher or lower than 1.0, then the results for that replicate are significantly different from the fitted curve and, hence, other test replicates. Figure 5 presents the damage characteristic curves and shape factors (values shown in the same color as the corresponding damage characteristic curve) for individual specimen test results for each mixture evaluated. Two to four specimens were investigated for each mixture, and the shape factors were determined based on the fitted curves. In general, the outlier test results fall under the fitted curve, which means that the shape factor is higher than 1.0. The damage characteristic curves with shape factors that are significantly greater than 1.0 generally correspond to very short fatigue tests. Short fatigue tests can be caused by a pre-existing flaw in a specimen (e.g., a large internal void) that serves as a site for crack initiation and, thus, often corresponds to outlier behavior. It should be noted that the mixtures considered in this study were selected from the NCSU database based on their relatively high specimen-to-specimen variability compared with normal cases. Thus, the specimen-to-specimen variability observed is generally lower than that shown in Figure 5.

Confidence Interval Analysis to Establish Thresholds for Acceptable Variability

Confidence interval analysis was conducted to establish thresholds for shape factor values that correspond to acceptable variability. In this study, 16 mixtures were used to obtain shape factors, but the KEC-BB3 mixture had only two replicates, which means that the number of samples for the KEC-BB3 mixture was not sufficient for confidence interval analysis. Consequently, 15 mixtures were used in the analysis. Table 2 presents the confidence

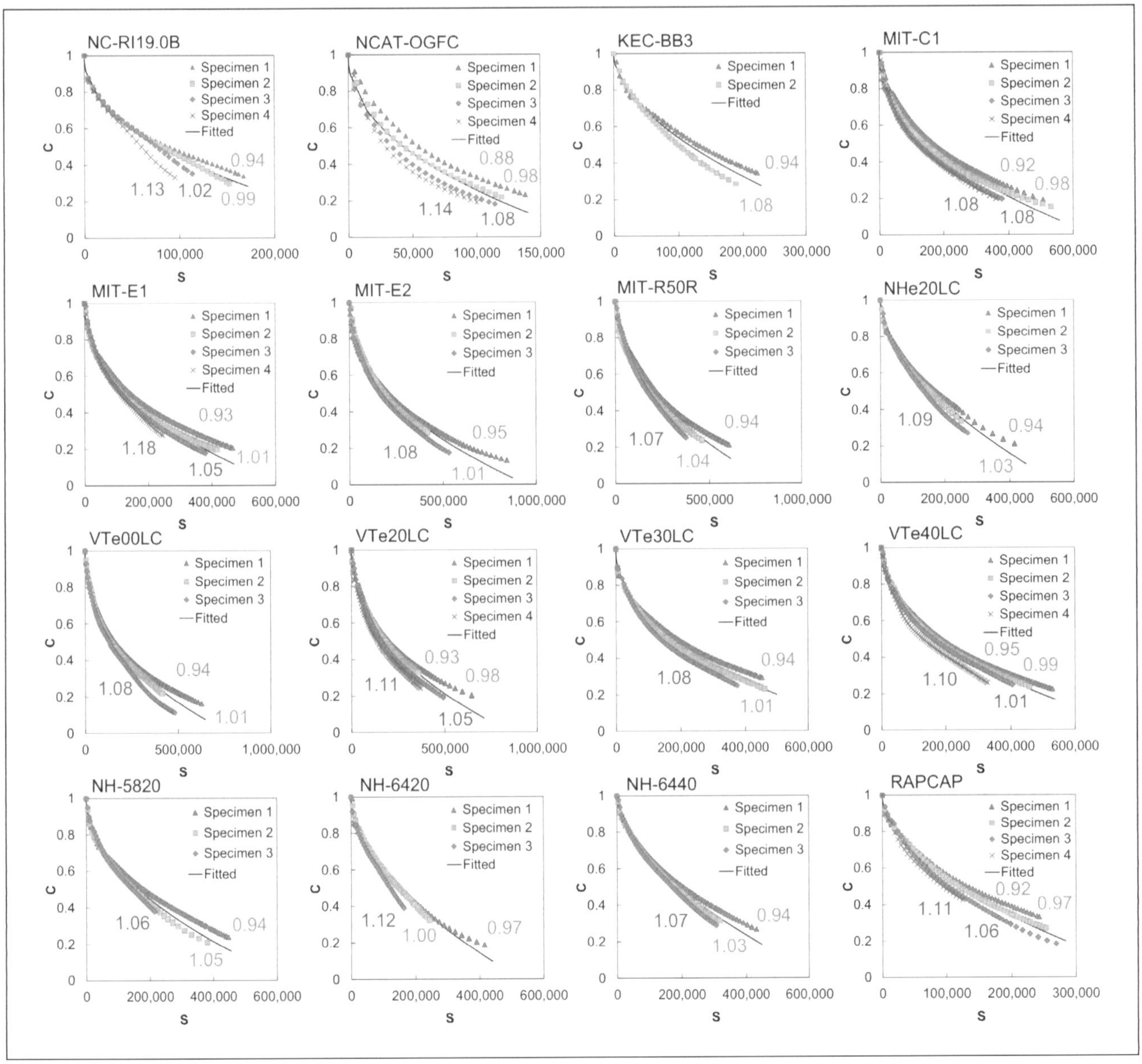

Figure 5. Damage characteristic curves and shape factors (values shown in each graph) for each mixture.

interval analysis results for the 15 mixtures. The average confidence interval value for the shape factor that corresponds to the 98% confidence level is 1.0 ± 0.1. Therefore, the replicates with shape factors greater than 1.1 or less than 0.9 are considered outliers and should be neglected from analyses.

Pavement Performance Predictions to Verify Acceptance Threshold for Shape Factors

FlexPAVE™ pavement performance simulations were conducted to evaluate the practical significance of changes in shape factor values. Dynamic modulus and cyclic fatigue performance test results of the 16 mixtures evaluated were used within the FlexPAVE™ program to predict fatigue damage in pavements. In order to evaluate the effect of the damage characteristic curves' specimen-to-specimen variability on the pavement performance predictions, FlexPAVE™ simulations were conducted using individual test specimen damage characteristic curves and the mixture representative damage characteristic curve that was obtained by fitting the combined results of all the test replicates. In order to isolate the effect of damage characteristic curve variability on the performance prediction results, the averaged dynamic modulus values and failure criteria properties were used for each mixture.

Table 2. Confidence Interval Analysis Results

Mixture type	No. of samples	Standard deviation	98% confidence interval	Average 98% confidence interval
NCHRP IDEA-RI19.0B	4	0.084	± 0.10	± 0.10
NCAT-OGFC	4	0.117	± 0.14	
MIT-C1	4	0.081	± 0.09	
MIT-E1	4	0.115	± 0.13	
MIT-E2	3	0.067	± 0.09	
MIT-R50R	3	0.071	± 0.10	
NHe20LC	3	0.079	± 0.11	
VTe00LC	3	0.071	± 0.10	
VTe20LC	4	0.081	± 0.90	
VTe30LC	3	0.071	± 0.10	
VTe40LC	4	0.065	± 0.08	
NH-5820	3	0.070	± 0.09	
NH-6420	3	0.087	± 0.12	
NH-6440	3	0.069	± 0.09	
RAPCAP	4	0.088	± 0.10	

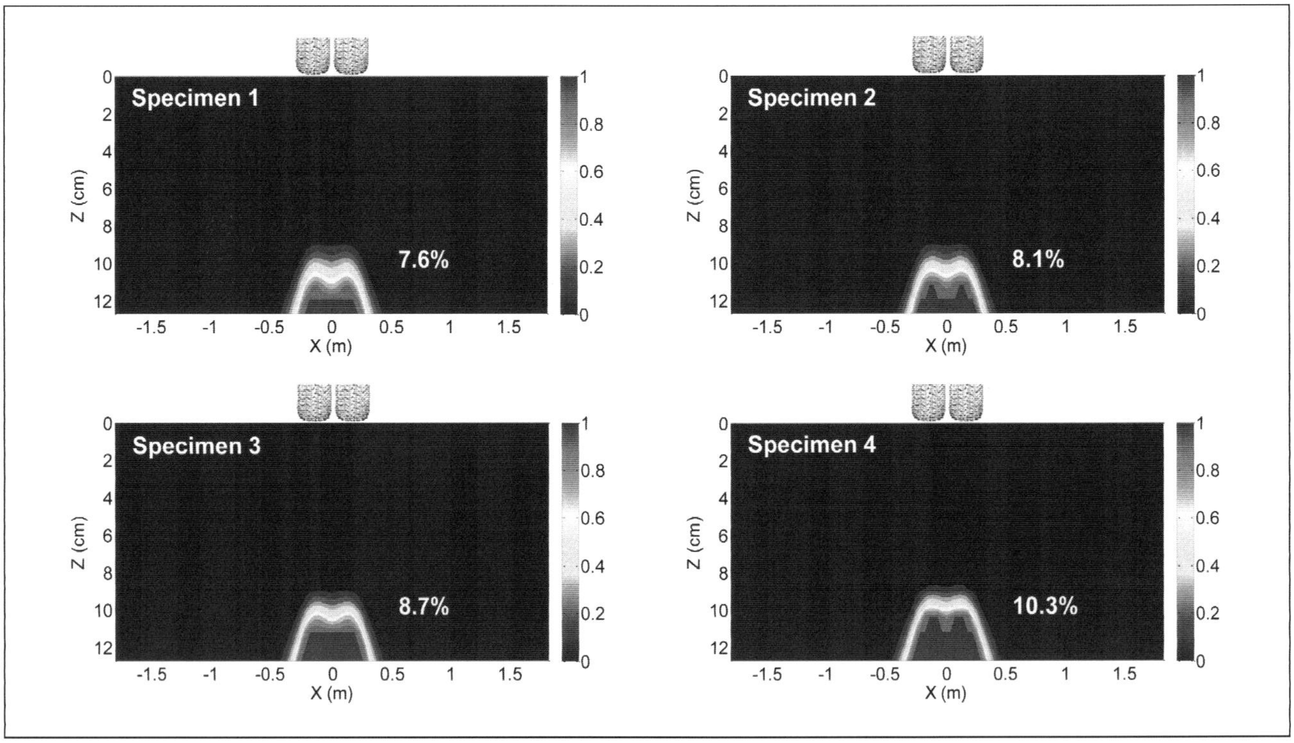

Figure 6. Damage contours for each specimen of NCHRP IDEA-RI19.0B mixture at pavement service life of 20 years.

In this study, a simple pavement structure was used for all the analyses: a 12.7-cm (5-in.) asphalt concrete layer over a 20-cm (8-in.) aggregate base and 380-cm (15-in.) subgrade. Pavement temperature data were obtained from the EICM database based on the location from which the mixture was sourced. The EICM database provides hourly pavement temperature as a function of depth. The traffic input for this structure was 350 daily equivalent single-axle loads with no growth. The traffic volume was chosen such that the fatigue damage would be apparent in the analysis within a performance prediction period of 20 years.

Figure 6 shows the FlexPAVE™ predicted distribution of fatigue damage within a cross-section of the asphalt concrete pavement layers after 20 years of simulated traffic loading for the NCHRP IDEA-RI19.0B mixture. A color scale that represents values from zero to one was used to present the damage contours. Blue (zero)

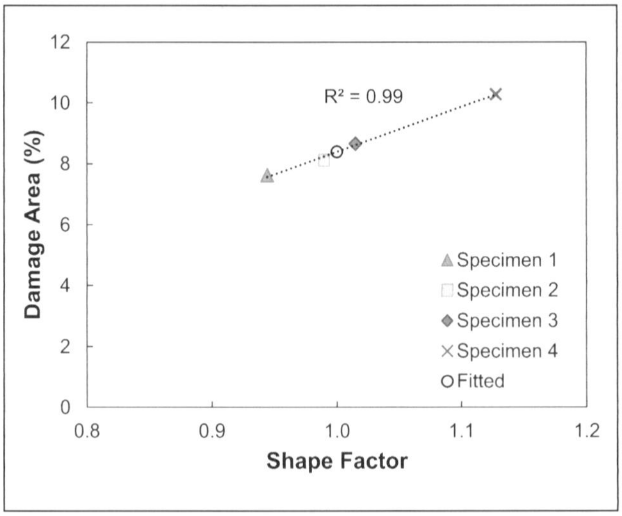

Figure 7. Fatigue damage area versus shape factor for NCHRP IDEA-RI19.0B mixture.

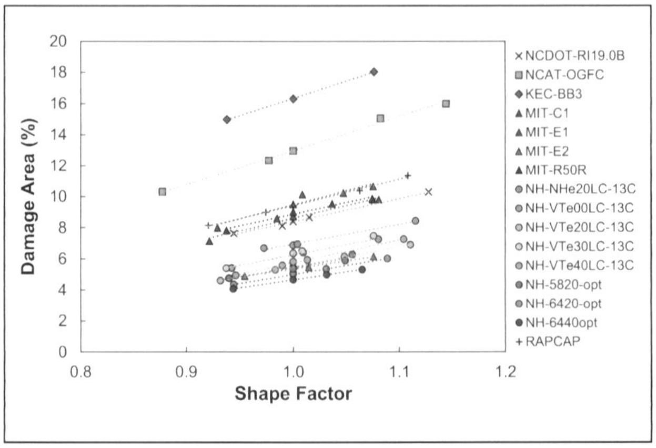

Figure 8. Fatigue damage area versus shape factor for 16 mixtures.

represents the intact pavement condition and red (one) represents the fatigue failure within the damage contours. The wheel path is directly above the region of damage localization. The shape factors for the NCHRP IDEA-RI19.0B mixture, shown at the left-most of the top row in Figure 5, are 0.94, 0.99, 1.02, and 1.13, and the corresponding predicted damage areas are 7.6%, 8.1%, 8.7%, and 10.3%, respectively. The "damage area" is defined as the ratio of the damaged area to the area of a trapezoid of which the upper base is between the outer edges of the loading wheels and the length of the bottom base is half of the width of the load axle. This study found that the amount of fatigue damage predicted increases as the shape factor increases for all cases evaluated. It is noted that a higher shape factor value represents a lower damage characteristic curve, i.e., a lower C value for a given S value.

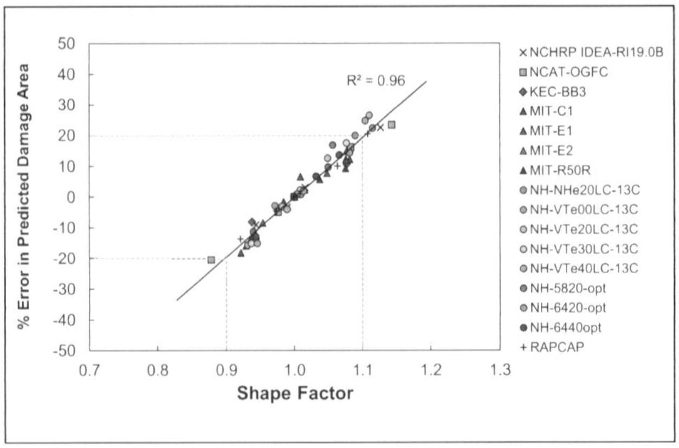

Figure 9. "Percent error" of damage area versus shape factor for 16 mixtures.

Figure 7 presents the relationship between the fatigue damage area and shape factor for the NCHRP IDEA N-181 RI19.0B mixture. A strong, positive linear relationship is evident. Figure 8 presents the relationship between the fatigue damage area and shape factor for each mixture evaluated. Figure 8 shows that a positive, linear relationship exists between the fatigue damage area and shape factor for each mixture evaluated; however, because of the differences in the amount of fatigue damage predicted for each mixture, these relationships are mixture-dependent.

To produce a mixture-independent relationship, the percentage of error ("percent error") in the fatigue damage area predicted by FlexPAVE™ that is due to the variation in the shape factor was calculated for each test replicate using the fitted curve as the reference. The results are presented in Figure 9, which shows a unique linear relationship between the percent error in the predicted fatigue damage area and the shape factor. It should be noted that the linear function can be changed by FlexPAVE™ inputs such as pavement thickness or traffic volume. However, it should also be noted that the climate history differed among the mixtures evaluated, and thus, temperature did not affect the relationship between the percent error in the predicted damage area and the shape factor. Based on a 98% confidence level, the acceptance range for the shape factor values is 0.9 to 1.1. Figure 9 shows that the corresponding error in the damage area predicted by FlexPAVE™ is approximately $\pm 20\%$ for the pavement structures and traffic conditions evaluated in this study.

Conclusions and Recommendations

This study presents a robust and practical method for fitting a model to the S-VECD damage characteristic curves obtained from cyclic fatigue testing. In addition, this study proposes a framework to evaluate the repeatability of the damage characteristic curve results. Based on the results, the following conclusions and recommendations are drawn from this study:

- A power law model can be used to represent damage characteristic curves obtained from AMPT cyclic fatigue tests. It is recommended that the power law model should be fitted to the test results after filtering the data to produce damage (*S*) increments of 5,000. The filtering of the data removes bias toward low *S* values in the model fit.
- A new index called the shape factor, can be used effectively to evaluate the specimen-to-specimen variability in damage characteristic curves. The shape factor is linearly related to the error in the fatigue damage predicted by FlexPAVE™ as a result of the variability in the damage characteristic curve.
- Based on confidence interval analysis using 98% reliability, the shape factors within the range of 0.9 to 1.1 represent acceptable specimen-to-specimen variability. Shape factors of 0.9 and 1.1 lead to approximately ±20% error in the FlexPAVE™ predictions of fatigue damage. The linear relationship between the shape factors and the "percent error" in the predicted damage area is independent of the asphalt mixture properties. Therefore, the relationship can be used to determine specimen-to-specimen variability tolerance or data quality control.

Acknowledgments

The authors acknowledge the financial support from the Federal Highway Administration under the DTFH61-08-H-00005 Project.

Author Contributions

The authors confirm contribution to the paper as follows: study conception and design: KL, CC, RK; data collection: KL; analysis and interpretation of results: KL, CC, RK; draft manuscript preparation: KL, CC, RK. All authors reviewed the results and approved the final version of the manuscript.

References

1. Monismith, C. L., J. A. Epps, D. A. Kasianchuk, and D. B. McLean. *Asphalt Mixture Behavior in Repeated Flexure.* Report No. TE 70-5. Institute of Transportation and Traffic Engineering, University of California, Berkeley, CA, 1970.
2. Bonaquist, R. *NCHRP Report 614: Refining the Simple Performance Tester for Use in Routine Practice.* Transportation Research Board of the National Academies, Washington, D.C., 2008.
3. American Association of State Highway and Transportation Officials. *Determining the Damage Characteristic Curve of Asphalt Concrete from Direct Tension Cyclic Fatigue Tests.* AASHTO TP 107. American Association of State Highway and Transportation Officials, Washington, D.C., 2015.
4. Pape, S., K. Lee, C. Castorena, and Y. R. Kim. Optimization of the Laboratory Fabrication of Small Specimens for Asphalt Mixture Performance Testing. *Transportation Research Record: Journal of the Transportation Research Board*, 2018. https://doi.org/10.1177/0361198118790845.
5. Underwood, B. S., Y. R. Kim, and M. N. Guddati. Improved Calculation Method of Damage Parameter in Viscoelastic Continuum Damage Model. *International Journal of Pavement Engineering*, Vol. 11, No. 6, 2010, pp. 459–476.
6. Sabouri, M., and Y. R. Kim. Development of a Failure Criterion for Asphalt Mixtures under Different Modes of Fatigue Loading. *Transportation Research Record: Journal of the Transportation Research Board*, 2014. 2447: 117–125.
7. Norouzi, A., D. Kim, and Y. R. Kim. Numerical Evaluation of Pavement Design Parameters for the Fatigue Cracking and Rutting Performance of Asphalt Pavements. *Materials and Structures*, Vol. 49, No. 9, 2015, pp. 3619–3634.
8. Wang, Y., A. H. Norouzi, and Y. R. Kim. Comparison of Fatigue Cracking Performance of Asphalt Pavements Predicted by Pavement ME and LVECD Programs. *Transportation Research Record: Journal of the Transportation Research Board*, 2016. 2590: 44–55.
9. Sabouri, M., T. Bennert, J. S. Daniel, and Y. R. Kim. A Comprehensive Evaluation of the Fatigue Behaviour of Plant-Produced RAP Mixtures. *Journal of Road Materials and Pavement Design*, Vol. 16, Suppl. 2, 2015, pp. 29–54.
10. Sabouri, M., Y. Choi, Y. Wang, S. Hwang, C. Baek, and Y. R. Kim. Effect of Rejuvenator on Performance Properties of WMA Mixtures with High RAP Content. *Proc., 8th RILEM International Symposium on Testing and Characterization of Sustainable and Innovative Bituminous Materials*, Springer, Dordrecht, the Netherlands, 2016, pp. 473–484.
11. Eslaminia, M., S. Thirunavukkarasu, M. N. Guddati, and Y. R. Kim. Accelerated Pavement Performance Modeling using Layered Viscoelastic Analysis. *Proc., 7th RILEM International Conference on Cracking in Pavements*, Springer, Dordrecht, the Netherlands, 2012, pp. 497–506.
12. Park, H. J., M. Eslaminia, and Y. R. Kim. Mechanistic Evaluation of Cracking in In-Service Asphalt Pavements. *Materials and Structures*, Vol. 47, No. 8, 2014, pp. 1339–1358.
13. Lee, K., S. Pape, C. Castorena, and Y. R. Kim. Evaluation of Small Specimen Geometries for Asphalt Mixture Performance Testing and Pavement Performance Prediction. *Transportation Research Record: Journal of the Transportation Research Board*, 2017. 2631: 74–82.

The Standing Committee on Structural Requirements of Asphalt Mixtures (AFK50) peer-reviewed this paper (18-05389).

Article

TRR
JOURNAL OF THE TRANSPORTATION RESEARCH BOARD

Transportation Research Record
2018, Vol. 2672(28) 503–513

DOI: 10.1177/0361198118795012
journals.sagepub.com/home/trr
SAGE

Evaluation of Viscoelastic and Fracture Properties of Asphalt Mixtures with Long-Term Laboratory Conditioning

Reyhaneh Rahbar-Rastegar[1], Jo Sias Daniel[2], and Eshan V. Dave[2]

Abstract

Aging affects the properties of asphalt mixtures in different ways; increase of stiffness, decrease of relaxation capability, and the increase of brittleness, resulting in changes in cracking behavior of asphalt mixtures. In this study, ten plant-produced, lab-compacted mixtures with various compositions (recycled materials, binder grades, binder source, and nominal maximum aggregate size) are evaluated at different long-term aging levels (24 hours at 135°C, 5 days at 95°C, and 12 days at 95°C on loose mix and 5 days at 85°C on compacted specimens). The asphalt mixture linear viscoelastic properties (|E*| and δ) and master curve shape parameters measured from complex modulus testing and fracture properties (measured from disc-shaped compact tension and semi-circular bending fracture testing) are compared at different levels of aging. The results indicate that the mixture exposure time to aging is proportional to the dynamic modulus and phase angle changes. Generally, the fracture parameters of mixtures become worse when aging level changes from 5 to 12 days aging. In spite of the similar viscoelastic properties, the mixtures with 24 hours at 135°C and 12 days at 95°C aging do not show similar fracture parameters.

Cracking has always been a challenging issue for asphalt pavements that negatively impacts the ride quality and pavement service life. Typically, cracking susceptibility of asphalt mixtures changes over time as asphalt materials age. Asphalt materials undergo aging during production, construction, and over the service life of the pavement. The aging process is the change of binder chemistry due to two primary processes: volatilization and oxidation.

Volatilization is the evaporation of lighter fractions (hydrocarbons) resulting in the increase of asphalt specific gravity. Volatilization occurs primarily during the production and construction stages where the binder temperature is very high (about 150°C). The volatilization rate increases dramatically with the increase of temperature (*1*).

Oxidation occurs due to the chemical reaction of asphalt hydrocarbons with oxygen that occurs over the pavement's service life. The interaction of hydrocarbons with hydroatoms like oxygen causes an imbalance in electrochemical forces and the polarity increases in the binder molecules. More polarity results in stronger intermolecular forces, and accordingly, the elastic modulus and viscosity of the asphalt increase. It is well known that the ambient temperature has a significant effect on aging rate. Other environmental conditions (e.g., pressure and moisture), traffic loading, and mix volumetrics also affect the aging process.

Aging causes physical property changes to asphalt mixtures by increasing stiffness and brittleness and decreasing relaxation capability. Consequently, the cracking resistance of aged mixtures is expected to be lower than that of unaged mixtures. Considering the importance of performance-based design methodologies, the evaluation of fatigue and thermal cracking properties of aged asphalt mixtures is desired during the mix design stage. To this aim, the aging of asphalt materials must be simulated in the laboratory. Limited laboratory conditioning methods directly applicable to the methods used in this study are presented here, but more comprehensive literature review is presented elsewhere (*2*):

1. The current standard to simulate short- and long-term aging of asphalt mixtures is AASHTO R30. In this standard practice, the loose-mix asphalt is

[1]Purdue University, West Lafayette, IN
[2]University of New Hampshire, Durham, NH

Corresponding Author:
Address correspondence to Jo Sias Daniel: jo.daniel@unh.edu

placed in a forced-draft oven for 4 hours at a temperature of 275°F (135°C) to simulate short-term aging (*3*). For long-term aging, short-term aged mixtures are compacted (following AASHTO T 312) into a specimen that is then conditioned in a forced-draft oven for 5 days at 85°C. Studies have shown that this laboratory aging method only simulates 5–10 years of field aging (*4*). Only one conditioning time and temperature is considered for all locations and climate conditions (*4*).
2. The Asphalt Institute procedure proposed by Blankenship et al. recommends loose-mix asphalt conditioning in an oven for 24 hours at 135°C, (*5*). This level of conditioning is expected to simulate 7–10 years of aging in the field. Braham et al. suggested that this level of aging might be slightly conservative for fracture evaluation of asphalt mixtures (*6*).
3. The recent findings of the NCHRP 09-54 project on long-term aging of asphalt mixtures for performance evaluation suggests the aging of loose-mix asphalt at 95°C for various times depending upon the climatic location of the pavement to be simulated (*7*). These findings are based on temperature conditioning of asphalt for both compacted and loose mix with and without pressure. Volumetric, stiffness, and fatigue properties of the mixtures were compared. Oven aging on loose-mix asphalt was recommended because of the uniformity of aging gradient in the final test specimen. Various research indicates that a conditioning temperature above 100°C impacts binder chemistry and differences in the response of the mixtures to damage (*8*, *9*). Yousefi Rad et al. recommended the conditioning temperature of 95°C as an optimal temperature for aging of loose-mix asphalt (*10*). The conditioning time should be adjusted based on climate and depth in the pavement; for example, their results show 8.2 days aging of loose mix at 95°C can match 17 years of aging for the top 6 mm of a pavement in Marathon County, WI in terms of binder rheology.

The objective of this study is to evaluate how the viscoelastic and fracture properties of asphalt mixtures change as a function of conditioning level.

Mixtures and Materials

This study includes testing on nine different recycled mixtures and one virgin mixture. Table 1 shows the mixtures evaluated and the combinations of aging levels and testing conducted for each mixture. Insufficient materials were available for a full factorial experiment. Letters in the cells indicate the testing conducted at each of the mixture-aging combinations: All tests, only Complex modulus tests, and No tests conducted. In the results and discussion section, 18.9% recycled asphalt pavement (RAP), and 28.3% and 31.3% RAP mixtures are labelled as 20% RAP and 30% RAP for simplicity. The mixtures containing both RAP and recycled asphalt shingles (RAS) are labelled as RAP/RAS mixtures.

Methodology

Aging

The asphalt materials (both loose mix and compacted specimens) were conditioned in ovens for long-term aging. The aging of compacted specimens was performed

Table 1. Summary of Mixtures and Testing Conducted at Different Aging Levels

			LTOA			
Binder PG grade	NMAS (mm)	%Total binder replacement (% RAP/ % RAS)	5 days 85°C compacted	24 hr 135°C loose	12 days 95°C loose	5 days 95°C loose
76-22	9.5	0	N	A	C	C
58-28	12.5	18.9 (18.9/0)	C	A	A	N
	12.5	18.5 (7.4/11.1)	N	A	A	A
	12.5	28.3 (28.3/0)	C	A	A	N
	19	20.4 (8.2/12.2)	N	A	A	A
52-34	12.5	18.9 (18.9/0)	C	A	A	A
	12.5	18.5 (7.4/11.1)	N	A	A	A
	12.5	28.3 (28.3/0)	C	A	A	A
	19	20.4 (8.2/12.2)	N	A	A	A
	19	31.3 (31.3/0)	N	A	A	A

Note: A = complex modulus, DCT, and SCB testing; C = only complex modulus; N = no test; NMAS = nominal maximum aggregate size; LTOA = long-term oven aged.

following AASHTO R-30. Compacted specimens were cored and trimmed to final test-specimen dimensions and wrapped in wire mesh with clamps to prevent any changes in the shape of the specimens during aging.

Three different conditioning protocols for loose-mix asphalt were evaluated in this study:

- 24 hours at 135°C
- 12 days at 95°C
- 5 days at 95°C

The 24 hours at 135°C follows the Asphalt Institute procedure. The selection of two other aging levels is based on preliminary recommendations from the NCHRP 09-54 project (*10*). Based on the mixtures used in this study, 12 days at 95°C is selected to match with 24 hours at 135°C, and 5 days at 95°C is considered as intermediate aging level evaluated to determine evolution of properties with aging duration.

The loose-mix asphalt was spread in steel pans at an approximate depth of 1 inch. The materials were stirred every other day and the pans were rotated around the oven to obtain a consistent aging condition in all materials. The materials were reheated at 135°C for 2 hours before compaction to achieve final test specimens with air void contents of 6 ± 0.5%.

Testing Methods

The complex modulus testing was conducted following AASHTO T 342 using an asphalt mixture performance tester (AMPT). The raw data were analyzed using Abatech RHEA® software. Dynamic modulus and phase angle master curves were constructed based on the time–temperature superposition principle.

Two common testing methods to characterize the fracture behavior of asphalt materials in laboratory were used in this study: disc-shaped compact tension (DCT) and Illinois Flexibility Index using semi-circular bending (SCB) testing. The asphalt mixture materials follow a quasi-brittle behavior under the fracture process. The typical load displacement curve obtained from both tests is shown in Figure 1.

The DCT test (ASTM D 7313) was conducted to compare the thermal cracking behavior of the various mixtures and aging levels. The appropriate low temperature PG grade for the recycled mixtures was determined to be −28°C using LTPPBind software based on typical construction location for these mixtures. DCT testing was performed at −18°C (10°C warmer than low temperature PG grade requirement for the pavement location) for these mixtures and −12°C for the virgin mixture. This test was developed to measure the fracture energy of circular notched specimens under a tension load, which provides an oriented crack propagation along the notch. The fracture work is defined as the area under the load versus crack mouth opening displacement (CMOD) curve. Fracture energy is determined by normalizing the fracture work by specimen thickness and ligament length. The fracture energy is the amount of energy required to develop a unit surface fracture of the asphalt mixture. The fracture strain tolerance (FST), a new parameter suggested by Zhu et al. (2017), is calculated by normalizing the fracture energy of mixture with the fracture strength (G_f/S_f), (*11*).

The SCB fracture test (AASHTO TP 124) is performed at an intermediate temperature (25°C) and evaluates the resistance of asphalt mixtures to fatigue cracking. The load is applied to a notched semi-circular specimen at a displacement rate of 50 mm/min. The crack propagates along the notch in the middle of the specimen (*12*). The measured data are analyzed using the IFIT software developed by Illinois Center of Transportation (ICT), to calculate the fracture energy and flexibility index (*FI*) parameters defined by Equations 1 and 2:

$$G_f = \frac{W_f}{t \times a} \quad (1)$$

$$\mathrm{FI} = \frac{G_f}{m_{\text{Inflection Point}}} \quad (2)$$

where W_f is fracture work, t is the thickness of specimen, and a is ligament length. $m_{\text{Inflection Point}}$ is the slope of the post-peak softening curve at an inflection point near the middle of the post-peak region.

Results and Discussion

Linear Viscoelastic Parameters

Example results of dynamic modulus and phase angle master curves for different aging levels are presented as the average of three replicates for one mixture (PG 52-34, 12.5 mm, 28.3% RAP) in Figure 2. The overall trend is similar for all mixtures evaluated in this study: as the asphalt materials age the stiffness (|E*|) increases while the relaxation capability of mixtures, as represented by phase angle (δ), decreases. Statistical testing (t-test) was conducted for dynamic modulus and phase angle results at each temperature and frequency combination using the measured data obtained from three replicates of each mixture. With a confidence level of 95%, there is a significant difference between |E*| and δ of short-term oven aged (STOA) mixtures with all levels of long-term aging. Statistical significance testing also indicates neither dynamic modulus nor phase angle show a statistical difference between 24 hour and 12 days aged mixtures. Also, two shorter levels of aging (5 days on loose mix and 5 days on compacted samples) are not statistically

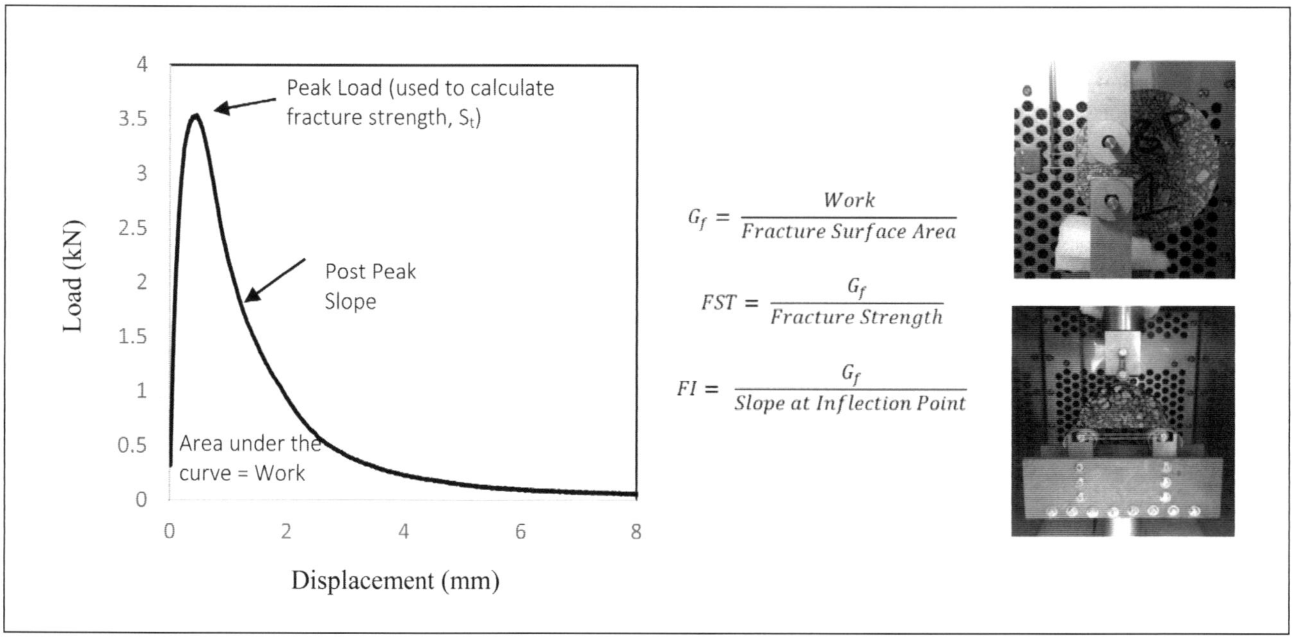

Figure 1. Typical load-displacement curve of fracture tests.

different. It should be mentioned that this comparison was conducted for only four available mixtures. Comparing 12 days and 5 days aging levels, there is a significant difference for all mixtures, except the two 19 mm, PG 52-34 mixtures.

One interesting observation is that the peak phase angle value moves down and left (lower frequencies) as materials age, so that for the two high levels of aging (24 hour and 12 days) the peak phase angle was not measured within the standard testing temperatures (4.4°C, 21.1°C, and 37.8°C) and frequencies (25, 10, 5, 1, 0.5, and 0.1 Hz). To capture the peak point for these two levels of aging, the complex modulus testing also included 0.01 Hz at 37.8°C test in addition to standard frequencies and temperatures.

Figure 3 compares the average dynamic modulus and average phase angle master curves (in the frequency range of 10^{-5} to 10^{5} Hz) at different long-term aging levels versus short-term aging values for all of the mixtures evaluated in this study. As expected, all LTOA mixtures have higher dynamic modulus than STOA mixtures. This shows the clear difference between the two intermediate aging levels and the two longer-term aging levels, and the similarities of the two long-term aging levels at the intermediate frequencies. At the very low and high frequencies, the $|E^*|$ of aged mixtures becomes closer to the line of equality, while the difference is greater at intermediate frequencies. Lines representing double and triple the $|E^*|_{(STOA)}$ values are drawn in Figure 4*a* for added perspective. At frequencies higher than 10 Hz, the dynamic modulus of long-term aged mixtures are less than twice the $|E^*|_{(STOA)}$, while at frequencies around 0.01 Hz, the dynamic modulus of long-term aged mixtures increase to as high as six times the dynamic modulus of short-term aged condition. The reason is that the response of asphalt mixtures at very high frequencies is dominated by elastic behavior which appears to have not been as impacted by long-term aging.

The phase angle values of all LTOA mixtures are lower than those of STOA mixture at low and intermediate temperature, which is expected as the relaxation capabilities of mixtures reduce with long-term aging. As shown in Figure 2*b*, a horizontal shift is observed in phase angle master curves as the aging level increases. At the lower frequencies, the phase angle of STOA mixtures begins to decrease after the inflection point, while the phase angle values of LTOA mixtures are still increasing. At the frequencies lower than the intersection point of STOA and LTOA master curves, the phase angle of STOA mixtures are lower than those of LTOA mixtures (as shown in Figure 3*b*). As the aging level increases, two curves intersect at a lower frequency. Although it changes from one mixture to another, the longer-aged mixtures intersect with STOA mixtures' master curves between 0.001 to 0.01 Hz. The intersection of intermediate aged and short-term aged mixtures is between 0.01 to 0.1 Hz. At the frequencies lower than these values, the phase angle of LTOA mixtures are higher than phase angle of STOA mixtures.

To capture the combination of stiffness and relaxation capability of mixtures in a single plot, Black space diagrams are shown in Figure 4. The figure shows how

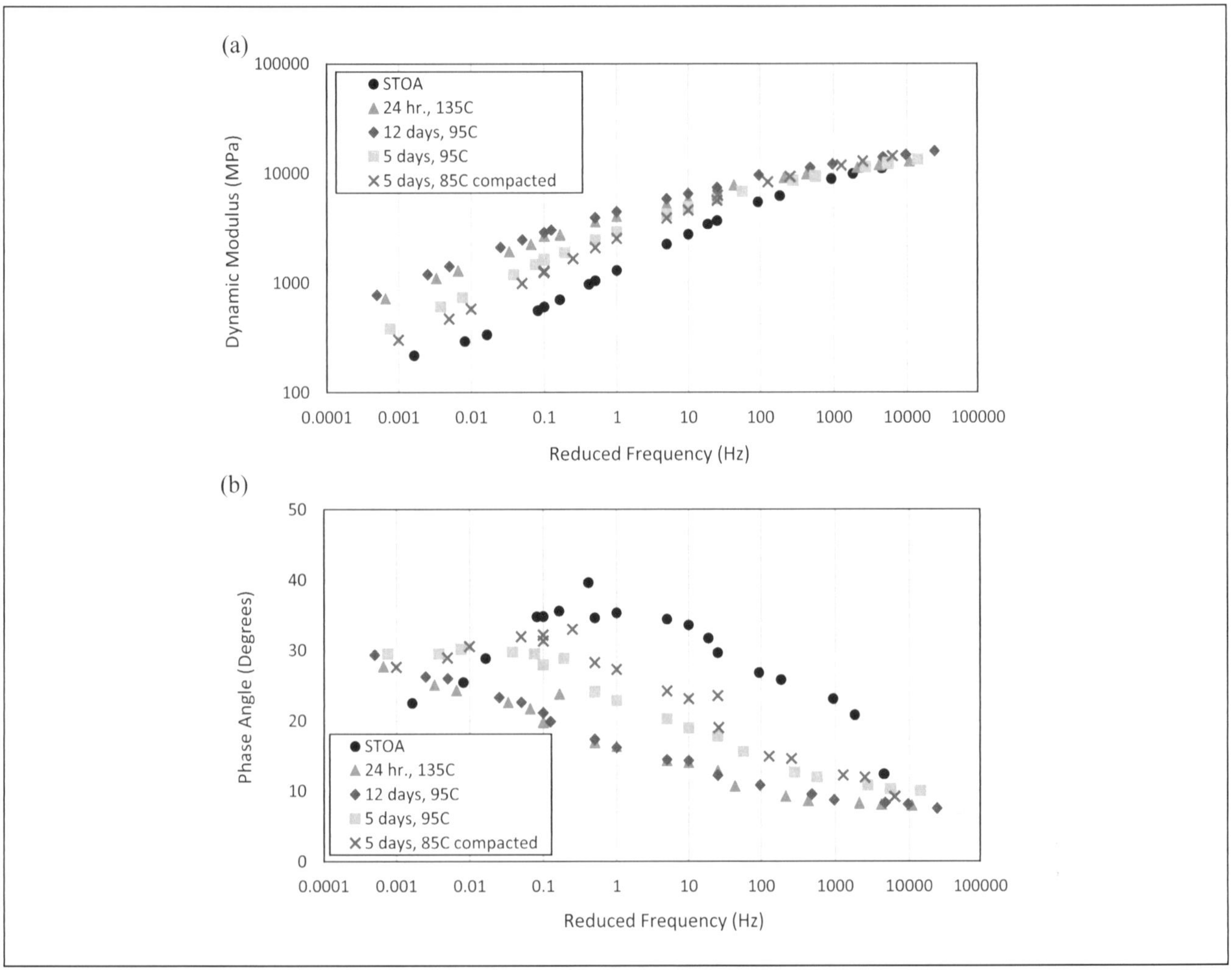

Figure 2. Example of (*a*) dynamic modulus and (*b*) phase angle master curves for sample mixture (PG 52-34, 12.5 mm, 28.3% RAP) at different aging levels (reference temperature 21.1°C).

Black space curves move with additional amount of aging. The inflection point moves to the bottom left side as more aging occurs. The observations in Black space diagram can be used to estimate thermal cracking susceptibility of asphalt mixtures. Generally, a mixture with higher stiffness at a constant phase angle is expected to incur greater thermal stress values. If the relaxation capability (phase angle) of this mixture is lower, the mixture relieves the thermal stress at a lower rate, resulting in higher thermal cracking potential. In Figure 4, higher phase angle for STOA with decreasing phase angle values are seen for STOA condition as compared with long-term aged condition at constant value of stiffness (|E*|). This indicates that even for same level of thermal stress, relaxation capabilities of asphalt mixtures would diminish with increasing aging levels. Thus, aged mixtures would be more prone to cracking at a lower cooling rate than short-term aged mixtures.

Generally, a standard or generalized sigmoidal model is used to fit the dynamic modulus master curve. In this study, the standard sigmoidal model is employed:

$$\log|E^*| = \delta + \frac{\alpha}{1 + e^{\beta + \gamma \log(\omega)}} \tag{3}$$

where $|E^*|$ is dynamic modulus, ω is frequency, and δ, α, β, and γ are the fit coefficients that describe the shape of dynamic modulus master curve. As the asphalt materials age, the shape of master curve changes, resulting in a variation in fit coefficients. Accordingly, these coefficients can be the indicators of aging level. The α and δ parameters are related to the equilibrium modulus (lower asymptote) and glassy modulus (upper asymptote) of master curve, respectively. The γ value controls the width of relaxation spectra, and the frequency of the inflection point can be calculated from $10^{-\beta/\gamma}$. As the asphalt material ages, the $|E^*|$master curve tends to flatten and the inflection point is shifts to lower frequencies (*13*).

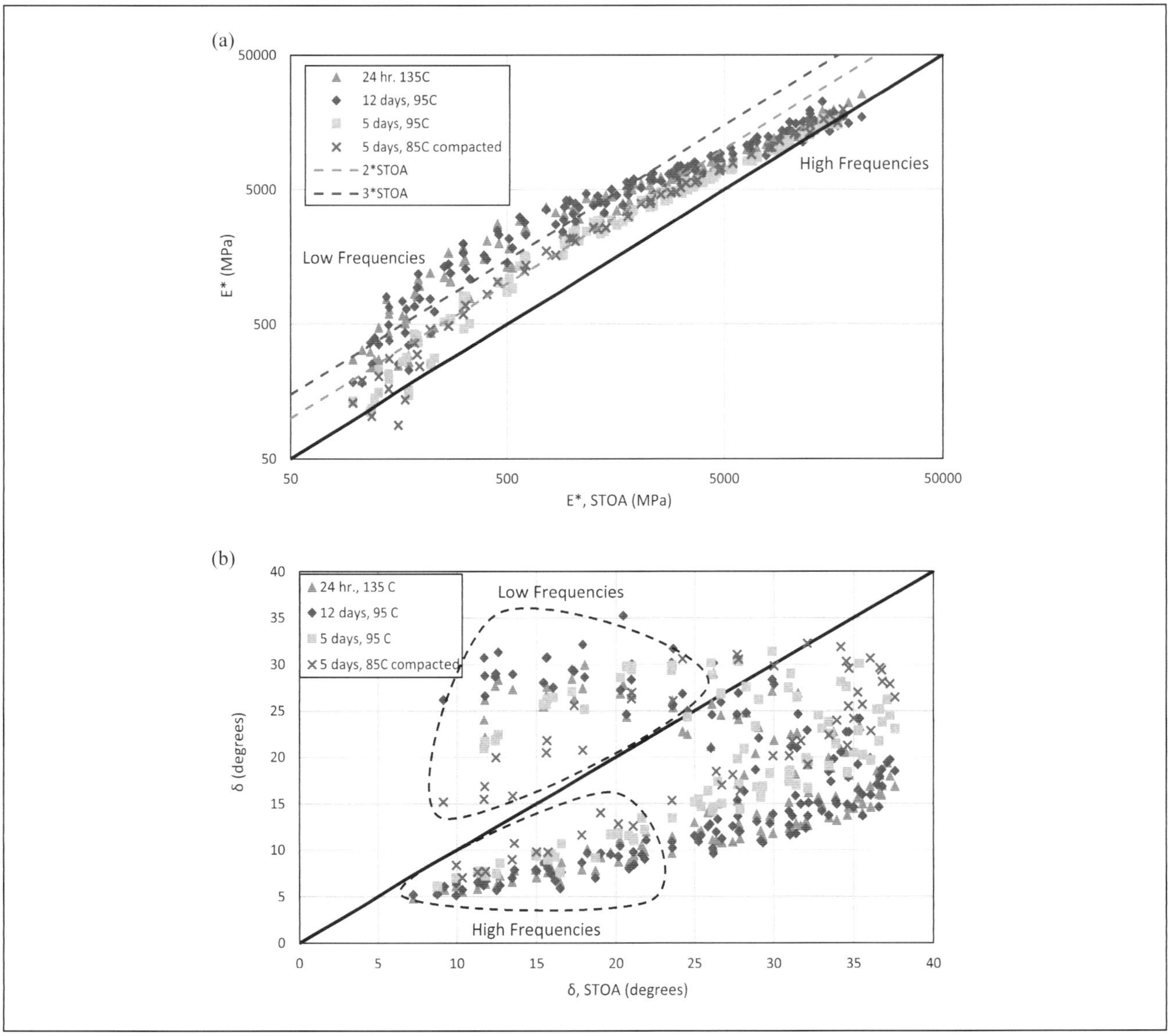

Figure 3. Linear viscoelastic (LVE) properties: (*a*) dynamic modulus; (*b*) phase angle of LTOA mixtures versus STOA mixtures for different mixtures.

The inflection point parameter ($-\beta/\gamma$) versus relaxation spectra width parameter (γ) plot for mixtures (Figure 5*a*) is similar in concept to a crossover frequency versus rheological index (R value) plot for binders. The $^{-\beta}/_{\gamma}$ parameter decreases and γ increases, moving points further towards the lower right as more aging occurs. The dashed lines connecting the points at different aging levels are drawn for three mixtures to show the trend from STOA to higher levels of aging. The parameter $^{-\beta}/_{\gamma}$ for all the short-term aged mixtures (except virgin mix) is about zero. This parameter for 5 days compacted and 5 days loose aged mixtures varies between -1.1 to -1.5, and -2 to -2.9, respectively, while the variation of $^{-\beta}/_{\gamma}$ for two more severely aged mixtures (24 hours and 12 days) is greater. There is a gap between $^{-\beta}/_{\gamma}$ values of 24 hour aged mixtures which splits the mixtures into two groups. All the PG 52-34, 24 hour aged mixtures have higher $^{-\beta}/_{\gamma}$ than PG 58-28 mixtures, indicating less impact from aging on the viscoelastic properties for these mixtures. As a hypothesis, the severe conditioning of 135°C in a short duration (24 hours) might have different effects on mixtures with different binder grades than 12 days aging at the lower temperature. It should be noted that the results of binder testing on extracted and recovered binders from short-term aged mixtures showed elevated zinc levels in two 19 mm mixtures with PG 52-34 binder, indicating that re-refined engine oil bottoms (REOB) may have been used in the production of the

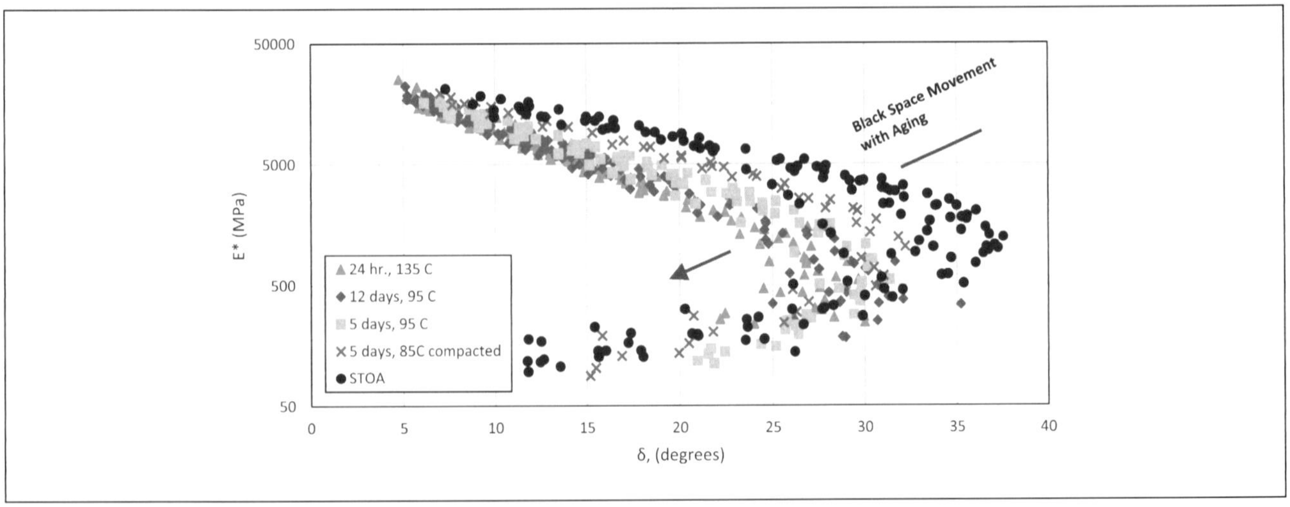

Figure 4. Black space diagrams of different aging levels.

virgin binder. One of the concerns about using REOB in asphalt mixtures is that it might increase the aging of binder (*14*).

A Lorentzian equation (Equation 4) has been shown to accurately model the phase angle master curve (*15*) and is used in this study:

$$\delta = \frac{a.b^2}{\left[(\log(\omega) - c)^2 + b^2\right]} \tag{4}$$

where δ is phase angle (degree), ω is frequency (Hz), and a, b, and c are the fit coefficients as follows: "a" shows the peak value, "b" controls the width of transition, and "c" is related to the horizontal position of the peak point. As the testing results show (Figure 2*b*), the phase angle master curves shift vertically and horizontally with different aging conditions. Therefore, the variation of vertical position of peak (a) and the parameter related to horizontal position of peak (c) were selected for aging evaluation and are designated as "vertical peak" and "horizontal peak", respectively. Figure 5*b* shows how both vertical and horizontal peak values decrease with increased aging level, moving the points towards the bottom left of the plot. The plot can be an indicator of the relaxation capability of asphalt mixtures. The mixtures with higher horizontal and vertical peak values are expected to have higher relaxation capability and better fatigue and fracture behavior. Similar to what was observed with the dynamic modulus coefficients, for 24 hour aging, PG 52-34 mixtures (except PG 52-34, 12.5 mm, RAP/RAS) are separate from PG 58-28 mixtures with a higher horizontal peak value, shown with two circles in Figure 5*a*. The mixtures containing REOB (two PG 52-34, 19 mm mixtures) show lower vertical peak (a) values in all levels of aging. However, the decrease of horizontal peak (c) for these mixtures was less than the other mixtures.

Mensching et al. (*13*) developed a parameter to evaluate the cracking performance of asphalt mixture in the format of the binder Glover-Rowe (G-R) parameter ($\frac{|E^*|\cos\delta^2}{\sin\delta}$). In this study, the parameter is calculated at the temperature–frequency combination of 15°C-0.005 rad/s to be consistent with the binder G-R parameter. The ratio of mixture G-R parameter in LTOA condition to the STOA condition is presented in Figure 6. As expected, the mixture G-R parameter increases as the level of aging changes from short term to intermediate and then to high aging levels, indicating higher susceptibility to cracking. There is a substantial increase in mixture G-R ratio when aging level increases from 5 days to 12 days at the same temperature. The intermediate aging levels increase the mixture G-R parameter from 1 to 3 times, but this ratio is from 3 to more than 7 for two high aging levels. This ratio is smaller for 19 mm, PG 52-34 mixtures (mixtures with REOB) and agrees with the variation of horizontal peak in Figure 6*b*.

Fracture Parameters

The results of DCT and SCB fracture testing is presented and discussed in this section. Figure 7 shows the average fracture energy and fracture strain tolerance for the various mixtures at different aging levels. The error bars show the standard deviation of three replicates tested for each mixture. A threshold value of 400 J/m^2 for fracture energy of DCT has been proposed by previous researchers (*16*, *17*) for short-term aged mixtures and is shown for visual comparison. Most of the high-aged mixtures have the fracture energies less than this limit. There is not a significant difference between the fracture

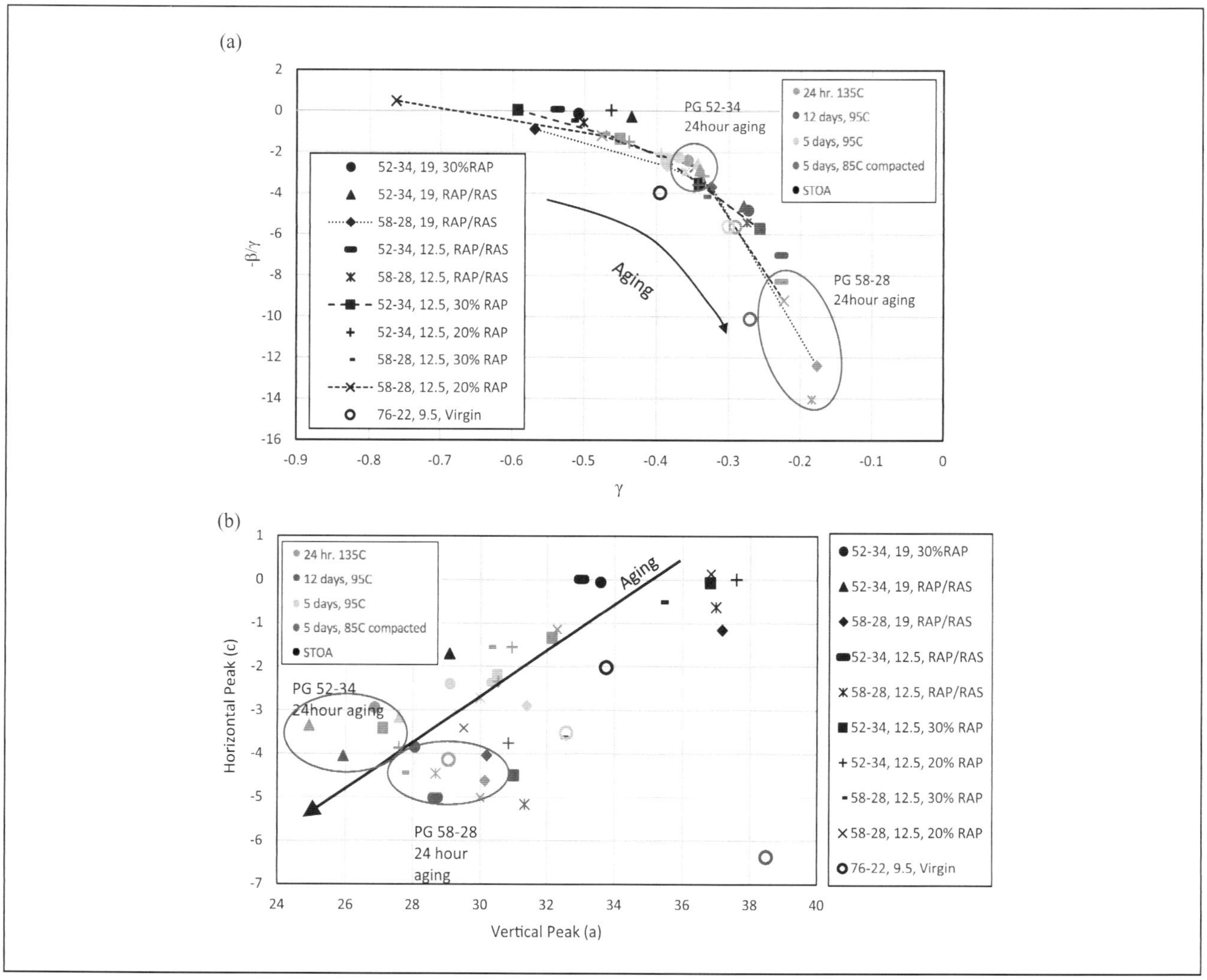

Figure 5. Shape parameters: (*a*) variation of dynamic modulus master curve parameter; (*b*) variation of phase angle master curve parameters with aging.

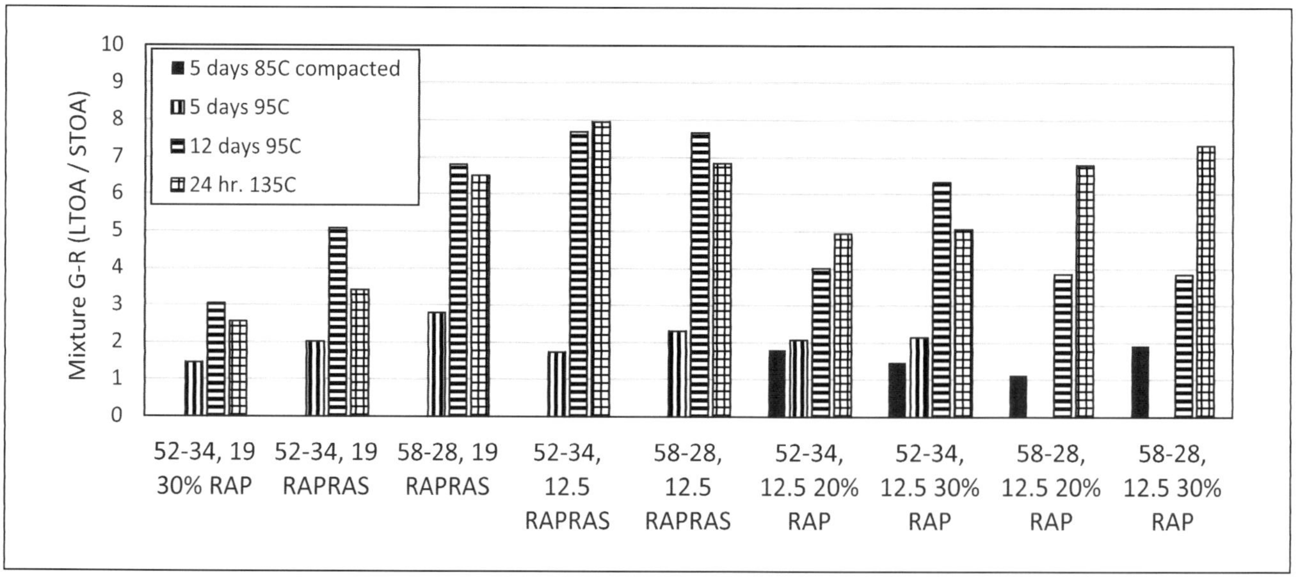

Figure 6. Ratio of mixture G-R_{LTOA} /mixture G-R_{STOA} (15°C and 0.005 rad/sec).

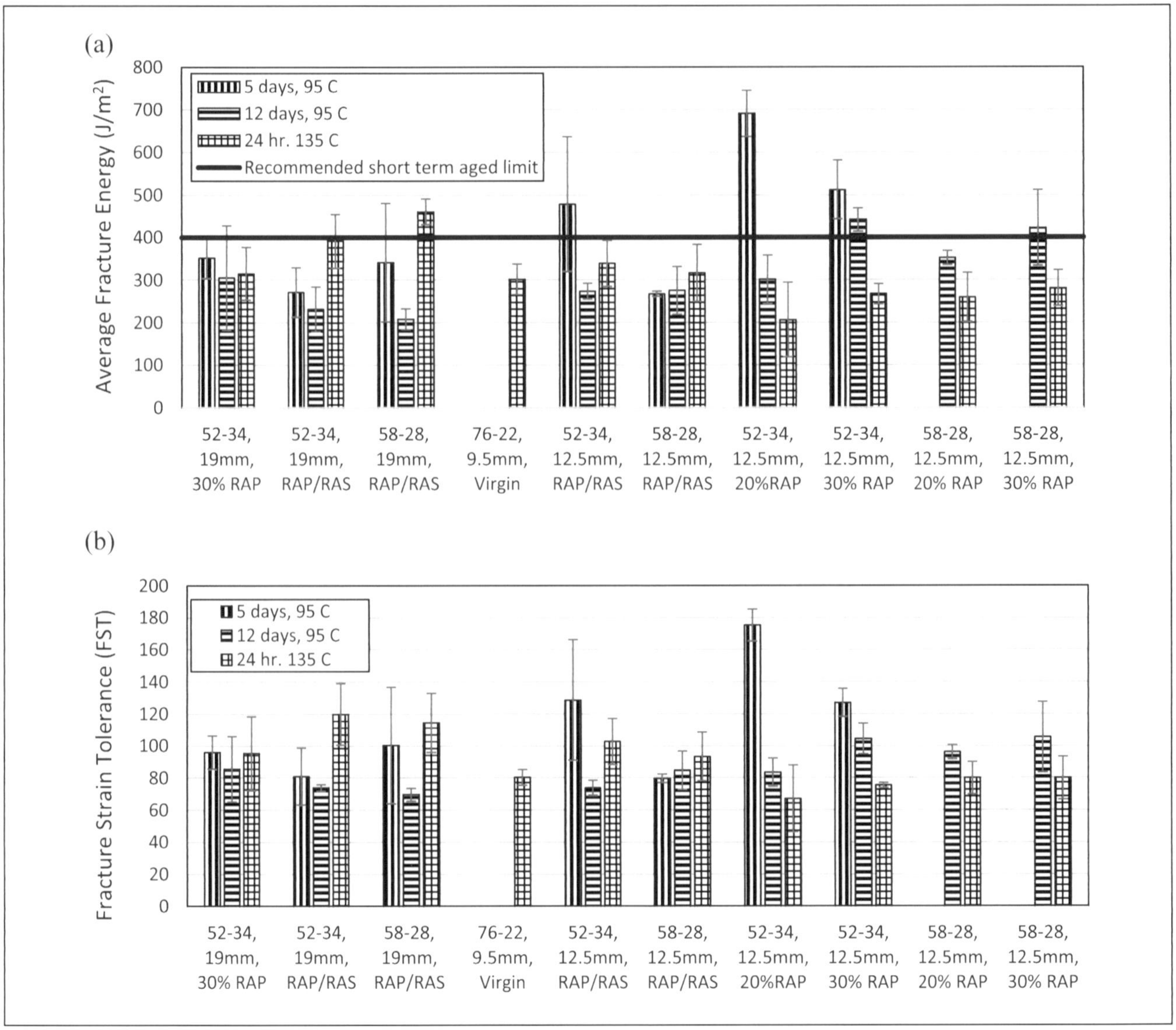

Figure 7. (*a*) fracture energy; (*b*) FST values (DCT testing).

parameters of intermediate and high aging levels for two PG 52-34, 19 mm mixtures (with REOB). This agrees with the mixture G-R and phase angle shape parameters, indicating that the LVE and fracture properties of these mixtures do not increase much with aging. The trend of fracture strain tolerance (FST) is similar to fracture energy for these mixtures. For all the 12.5 mm only RAP mixtures, the trend is that both G_f and FST decrease when aging level changes from 5 days to 12 days, and 24 hours, while for the RAP/RAS mixtures, 24 hour mixtures show better fracture parameters than 12 days aged mixtures. The 24 hour aging level seems to be less detrimental to fracture energy than 12 days aging for all RAP/RAS mixtures. A potential reason for this behavior of RAP/RAS mixtures might be the greater amount of already aged and oxidized asphalt binder present in these mixtures, which is not as prone to a more severe aging temperature as other mixtures. There is a significant difference between the fracture properties of 5 days and 24 hour aging for PG 52-34, 12.5 mm mixtures, while fracture energy and FST of 5 days aged, PG 58-28, 12.5 mm, RAP/RAS mixtures are very close to those of high-aged mixtures.

Figure 8 shows the flexibility index (FI) parameter which is the average of 3 to 4 replicates for each mixture, with the standard deviation error bars. The FI values of 5 days aged mixtures are higher than 24 hour and 12 days aged values for all mixtures, with higher differences observed for RAP/RAS mixtures. The FI of PG 52-34 mixtures is generally higher than the similar PG 58-28 mixtures, especially for 5 days aging level. The fracture properties obtained from SCB testing do not show a similar trend with the results of DCT testing. It is not surprising since the loading mode and testing temperature

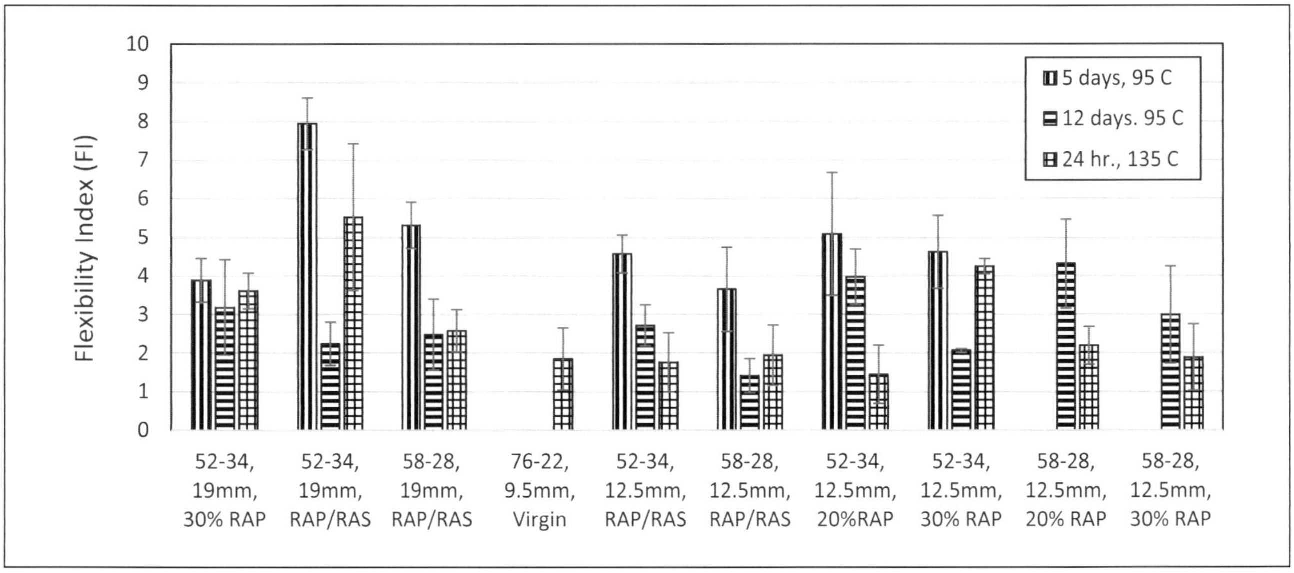

Figure 8. Average FI values (SCB testing).

are different in these two fracture tests. Results shown here agree with recent work by Haslett et al. (*18*) that showed that a single 25°C test temperature for SCB testing may not as clearly distinguish between mixtures with different low temperature binder grades.

Summary and Conclusion

The main objective of this research was to investigate how mixtures' properties change with different long-term aging levels (5 days at 95°C, 12 days at 95°C, and 24 hours at 135°C on loose mix and 5 days at 85°C on compacted samples). This study includes nine recycled mixtures and one virgin mixture evaluated by complex modulus, DCT, and SCB fracture testing. The following conclusions can be drawn from the results of the testing and analysis:

- All levels of long-term aging have made a significant difference on LVE properties ($|E^*|$ and δ) as compared with the properties measured at the STOA level. There was a similar trend in the variation of dynamic modulus and phase angle at different aging levels for the various mixtures. Based on the Black space diagram, the combination of higher dynamic modulus (at constant phase angle) and lower phase angle (at constant dynamic modulus) can be translated to higher thermal stress and higher relaxation capability, respectively.
- For the mixtures available in this study, 24 hour aged mixtures show very similar dynamic modulus and phase angle values with 12 days aged mixtures. Although 24 hour and 12 days aging create the similar effects on LVE properties, the fracture properties of asphalt mixtures obtained from SCB and DCT testing are different for these two aging levels.
- The shape parameters from dynamic modulus and phase angle master curves can indicate the relative aging levels and cracking behavior in mixtures. The evolution of characteristic shape parameters can be utilized in the future to develop aging models.
- Generally, the fracture properties of asphalt mixtures (fracture energy, FI, and FST) decrease as the aging level changes from 5 days to higher levels of aging, but there is not an evident trend between the fracture properties of 24 hours and 12 days. For the RAP/RAS mixtures, the 24 hour aged mixtures show better fracture properties than 12 days aged mixtures, while there is an inverse trend for most of the RAP only mixtures.

Future Work

Additional mixture testing (i.e., uniaxial fatigue testing) and analysis are underway to compare the fatigue properties of long-term aged and short-term aged mixtures. Most of the mixtures were placed in the field during the 2013 construction season and are being monitored. The cracking performance of field aged asphalt mixtures will be evaluated by laboratory testing on field core samples. Also, additional mixtures from different areas and with a wider range of binder grades and recycled materials are being evaluated.

Further analysis is planned to investigate the correlation between the viscoelastic characteristics, damage coefficients, and different cracking mechanisms including fatigue and reflective cracking and their relationship with field performance. Work is being conducted on the development of an aging prediction model for LVE and fracture properties of asphalt mixtures and comparison with other existing models such as global aging model (Mirza and Witczak, 1995) that is used in Pavement ME.

Acknowledgments

The authors would like to acknowledge New Hampshire Department of Transportation (NHDOT) for sponsoring this study, and Pike Industries Inc. for their cooperation in producing and sampling materials. Acknowledgment is also extended to Katie Haslett for assistance with testing and analysis.

Author Contributions

The authors confirm contribution to the paper as follows: study conception and design: Reyhaneh Rahbar-Rastegar, Jo Sias Daniel, Eshan V. Dave; data collection: Reyhaneh Rahbar-Rastegar; analysis and interpretation of results: Reyhaneh Rahbar-Rastegar, Jo Sias Daniel, Eshan V. Dave; draft manuscript preparation: Reyhaneh Rahbar-Rastegar. All authors reviewed the results and approved the final version of the manuscript.

References

1. Lavin, P. G. *Asphalt Pavement: A Practical Guide to Design, Production, and Maintenance for Engineers and Architects*. CRC Press, Taylor and Francis Group, New York, NY, 2003.
2. Rahbar-Rastegar, R. *Cracking in Asphalt Pavements: Impact of Component Properties and Aging on Fatigue and Thermal Cracking*. PhD dissertation. University of New Hampshire, Durham, NH, 2017.
3. Bell, C. A., Y. AbWahab, M. E. Cristi, and D. Sognovske. *Selection of Laboratory Aging Procedures for Asphalt-Aggregate Mixtures*. Strategic Highway Research Program, National Research Council, Washington, D.C., 1994.
4. Kim, R. Y., C. Hintz, F. Yousefi Rad, M. Elwardany, S. Underwood, M. J. Farrer, and R. R. Glaser. *NCHRP Report 871: Long-Term Aging of Asphalt Mixtures for Performance Testing and Prediction*. Transportation Research Board of the National Academies, Washington, D.C., 2018.
5. Blankenship, P. B., M. A. Anderson, G. N. King, and D. I. Hanson. *A Laboratory and Field Investigation to Develop Test Procedures for Predicting Non-Load Associated Cracking of Airfield HMA Pavements*. AAPTP Project 06-01 Report. Airfield Asphalt Pavement Technology Program, Auburn, AL, 2010.
6. Braham, A. F., W. G. Buttlar, T. R. Clyne, M. O. Marasteanu, and M. I. Turos. The Effect of Long-Term Laboratory Aging on Asphalt Concrete Fracture Energy. *Journal of the Association of Asphalt Paving Technologists*, Vol. 78, 2009, pp. 417–454.
7. Elwardany, M. D., F. Yousefi Rad, C. Castorena, and R. Y. Kim. Evaluation of Asphalt Mixture Laboratory Long-Term Aging Methods for Performance Testing and Prediction. *Road Materials and Pavement Design*, Vol. 18, 2017, pp. 28–61.
8. Petersen, J. C., and P. Harnsberger. Asphalt Aging: Dual Oxidation Mechanism and Its Interrelationships with Asphalt Composition and Oxidative Age Hardening. *Transportation Research Record: Journal of the Transportation Research Board*, 1998. 1638: 47–55.
9. Glaser, R., J. Schabron, T. Turner, J. P. Planche, S. Salmans, and J. Loveridge. Low Temperature Oxidation Kinetics of Asphalt Binders. *Transportation Research Record: Journal of the Transportation Research Board*, 2013. 2370: 63–68.
10. Rad, F. Y., M. D. Elwardany, C. Castorena, and R. Y. Kim. Investigation of Proper Long-term Laboratory Aging Temperature for Performance Testing of Asphalt Concrete. *Construction and Building Materials*, Vol. 147, 2017, pp. 616–629.
11. Zhu, Y., E. V. Dave, R. Rahbar-Rastegar, J. S. Daniel, and A. Zofka. Comprehensive Evaluation of Low Temperature Cracking Fracture Indices for Asphalt Mixtures. *Road Materials and Pavement Design*, Vol. 18, 2017, pp. 467–490.
12. Al-Qadi, I., H. Ozer, J. Lambros, A. El Khatib, P. Singhvi, T. Khan, J. Rivera-Perez, and B. Doll. *Testing Protocols to Ensure Performance of High Asphalt Binder Replacement Mixes Using RAP and RAS*. Report No. FHWA-ICT-15-017. Illinois Center of Transportation, Rantoul, IL, 2015.
13. Mensching, D. J., G. M. Rowe, and J. S. Daniel. A Mixture-Based Black Space Parameter for Low Temperature Performance of Hot Mix Asphalt. *Road Materials and Pavement Design*, Vol. 18, 2017, pp. 404–425.
14. Mogawer, W. S., A. Austerman, I. L. Al-Qadi, W. Buttlar, H. Ozer, and B. Hill. Using Binder and Mixture Space Diagrams to Evaluate the Effect of Re-refined Engine Oil Bottoms on Binders and Mixtures after Ageing. *Road Materials and Pavement Design*, Vol. 18, 2017, pp. 154–182.
15. Nemati, R., and E. V. Dave. Nominal Property Based Predictive Models for Asphalt Mixture Complex Modulus (Dynamic Modulus and Phase Angle). *Construction and Building Materials*, Vol. 158, 2017, pp. 308–319.
16. Dave, E. V., and C. Hoplin. Flexible Pavement Thermal Cracking Performance Sensitivity to Fracture Energy Variation of Asphalt Mixtures. *Road Materials and Pavement Design*, Vol. 16, 2015, pp. 423–441.
17. Van Deusen, D., L. Johanneck, J. Geib, J. Garrity, C. Hanson, and E. V. Dave. *DCT Low Temperature Fracture Testing Pilot Project*. Final Report 2015-20. Minnesota Department of Transportation, Saint Paul, MN, 2015.
18. Haslett, K. E., E. V. Dave, and J. S. Daniel. Exploration of Temperature and Loading Rate Interdependency for Fracture Properties of Asphalt Mixtures. *Proc., International Conference on Advances in Construction Materials and Systems, ICACMS, RILEM: International Union of Laboratories and Experts in Construction Materials, Systems, and Structures, Chennai*, India, 2017.

The Standing Committee on Structural Requirements of Asphalt Mixtures (AFK50) peer-reviewed this paper (18-04108).

Sage Publications USA
sagepub.com

ISBN 978-1-5108-7832-7

9 781510 878327

CURRAN ASSOCIATES INC.
proceedings.com

16th Annual Meeting of the Special Interest Group on Discourse and Dialogue (SIGDIAL 2015)

Prague, Czech Republic
2 – 4 September 2015